Third Edition

**ROBERT A. ADAMS**
**Department of Mathematics**
**University of British Columbia**

**Addison-Wesley Publishers Limited**
Don Mills, Ontario • Reading, Massachusetts • Menlo Park, California •
New York • Wokingham, England • Amsterdam • Bonn •
Sydney • Singapore • Tokyo • Madrid • San Juan

SPONSORING EDITOR: Ron Doleman, Publisher
MANAGING EDITOR: Linda Scott
COPY EDITORS: Valerie Jones, First Folio Resource Group Incorporated
David Hamilton and Robert Templeton, Suzanne Schaan
REVIEWS COORDINATOR: Abdulqafar Abdullahi
DESIGN: Pronk & Associates
PRODUCTION COORDINATOR: Melanie van Rensburg
MANUFACTURING COORDINATOR: Sharon Latta Paterson

**Canadian Cataloguing in Publication Data**

Adams, Robert A. (Robert Alexander), 1940-
Calculus of several variables

3rd ed.
Includes index.
ISBN 0-201-88195-0

1. Functions of several real variables. 2. Vector analysis.
I. Title.

QA303.A32 1995 515'.84 C95-932101-2



ISBN 0-201-88195-0

Printed and bound in Canada

B C D E - FP - 99 98 97

to Anne

# Preface

This third edition of *Calculus of Several Variables* constitutes the second half of the author's *Calculus: A Complete Course* and represents a significant expansion and revision of the previous edition. The principal changes are as follows:

- The inclusion of a 25-page introductory review of the main ideas of single-variable calculus (Chapter R). The presentation of this material is very brief; there are no exercises and few examples in it. It is intended to help students who have forgotten some of what they learned in a previous course, and would like a handy reference to that material.
- Chapters 1 and 2 on numerical sequences and series and on power series have been included to meet the needs of those courses that cover this material in the third semester instead of the second semester at the end of the study of single-variable calculus. A brief introduction to Fourier series has been added to the end of Chapter 2 on power series. With very minor exceptions, the material in these two chapters is not required for the chapters that follow, and they can be omitted or delayed. The main thrust of this book, the development of multivariable and vector calculus, starts in Chapter 3.
- Some changes have been made to ensure that the first examples of new concepts are as simple as possible.
- The number and variety of applied examples and exercises has been increased to make the book more interesting and more useful for students in various disciplines. Applications to physics and mechanics abound; the book was designed to be especially appropriate for students in engineering and the sciences.
- Several *EXPLORE* modules, spread throughout the text, encourage the student to make use of technology to discover mathematical relationships and solve problems that would be too difficult to do by hand. Some of these modules require the use of scientific calculators or programmable calculators. Some require graphing utilities (either graphing calculators or mathematical graphing software on personal computers). Some require the use of computer algebra software such as Derive, Maple, or Mathematica. Some require the use of spreadsheet programs like Lotus 1-2-3, Microsoft Excel, or Borland Quattro Pro. With a few exceptions, these *EXPLORE* modules assume that the student knows how to use the hardware or software involved, and does not try to teach how to use it. There are numerous computer and calculator supplements for calculus on the market.

- Each chapter, except the preliminary chapter (R) which is itself a review, has a Chapter Review at the end. Typically, these consist of four parts:
  - *Key Ideas*, short questions that review the main ideas of the chapter.
  - *Review Exercises*, a collection of exercises based on the main ideas of the chapter.
  - *Express Yourself*, consisting of a few questions or topics requiring longer answers or discussions in the form of sentences and paragraphs, avoiding the use of symbols wherever possible. They encourage the student to think through complex ideas and express them in his/her own words.
  - *Challenging Problems*, a collection of longer and more difficult problems for the more mathematically gifted students.
- The exercise sets have been expanded and diversified. There are still "drill" exercises to help develop calculation and graphing skills, but there are now more exercises aimed at testing and improving the understanding of concepts, developing reasoning skills, modelling applications, developing planning skills for attacking multistep problems, and developing skills in the use of mathematical computer software and advanced calculators.
- A separate nine-section chapter (Chapter 10) has been added to provide a complete, self-contained introduction to first- and second-order ordinarily differential equations. This chapter includes sections on nonhomogeneous, linear equations and on the use of numerical methods.
- A brief introduction to complex numbers is included in two Appendices. Appendix I deals with complex arithmetic and Appendix II with complex functions of a complex variable. The latter treats mapping properties, differentiability, the complex exponential function, and the Fundamental Theorem of Algebra.

Most features of previous editions have been carried over to this one, and in some cases expanded.

- There is an emphasis on geometry and the use of geometric reasoning in solving problems.
- There is an increased emphasis on numerical applications and approximations using calculators and computers.
- There is an increased emphasis on linear algebra and vector techniques. Although it remains an optional topic, we now define the derivative of a transformation from $m$-space to $n$-space as the linear transformation given by the Jacobian matrix of the first partial derivatives of the components of the transformation.

This edition maintains the clear but concise writing style of the previous editions.

# To the Student

When you began studying calculus you were dealing with functions that depended on only one variable and had numerical values. You had to assimilate many new concepts: limits, derivatives, and integrals to name a few, and learn to calculate and apply them efficiently. Calculus of several variables does not involve as many *new* ideas, but rather extends the existing ones to functions that depend on more than one variable and have values that are vectors with more than one component rather than just single numbers. While the ideas (limits, differentiation, and integration) are not new, the new contexts make for more *complication* than you experienced with single-variable calculus, and you must learn to deal with this complication, frequently without the aid of simple $x$-$y$ graphs to convey the properties of the functions you are using.

You might find multivariable and vector calculus harder than single-variable calculus at first, but once you understand the similarities between these subjects and what you already know, the difficulties will diminish. The skills you acquire here will help you to build more realistic models of real-world phenomena, in which behaviour of systems depends on many factors rather than just one. It is worth your while to acquire those skills, but, like any other worthwhile task, this one requires much effort on your part. No book or instructor can remove that requirement.

In writing this book I have tried to organize material in such a way as to make it as easy as possible, but not at the expense of "sweeping the real difficulties under the rug." You may find some of the concepts difficult to understand when they are first introduced. If so, *reread* the material slowly, if necessary several times; *think about it*; formulate questions to ask fellow students, your TA, or your instructor. Don't delay. It is important to resolve your problems as soon as possible. If you don't understand today's topic, you may not understand how it applies to tomorrow's either. Mathematics is a "linear discipline"; it builds from one idea to the next.

Doing exercises is the best way to deepen your understanding of calculus and to convince yourself that you understand it. There are numerous exercises in this text — too many for you to try them all. Some are "drill" exercises to help you develop your skills in calculation. More important, however, are the problems that develop reasoning skills and your ability to apply the techniques you have learned to concrete situations. In some cases you will have to plan your way through a problem that requires several diffferent "steps" before you can get to the answer. Other exercises are designed to extend the theory developed in the text and therefore enhance your understanding of the concepts of calculus.

The exercises vary in difficulty. Usually, the more difficult ones occur towards the end of exercise sets, but these sets are not strictly graded in this way because exercises on a specific topic tend to be grouped together. Some exercises are marked with an asterisk "*." This symbol indicates that the exercise is *either* somewhat more theoretical, *or* somewhat more difficult than most. The theoretical ones need not be difficult; sometimes they are quite easy. Most of the problems in the *Challenging Problems* section forming part of the *Chapter Review* at the end of most chapters are also on the difficult side.

Do not be discouraged if you can't do *all* the exercises. Some are very difficult indeed, and only a few very gifted students will be able to do them. However, you should be able to do the vast majority of the exercises. Some will take much more effort than others. When you encounter difficulties with problems proceed as follows:

1. Read and reread the problem until you understand exactly what information it gives you and what you are expected to find or do.
2. If appropriate, draw a diagram illustrating the relationships between the quantities involved in the problem.
3. If necessary, introduce symbols to represent the quantities in the problem. Use appropriate letters (e.g., $V$ for volume, $t$ for time, etc.) Don't feel you have to use $x$ or $y$ for everything.
4. Develop a plan of attack. This is usually the hardest part. Look for known relationships; try to recognize patterns; look back at the worked examples in the current or relevant previous sections; try to think of possible connections between the problem at hand and others you have seen before. Can the problem be simplified by making some extra assumptions? If you can solve a simplified version it may help you decide what to do in the given problem. Can the problem be broken down into several cases, each of which is a simpler problem? When reading the examples in the text, be alert for methods that may turn out to be useful in other contexts later.
5. Try to carry out the steps in your plan. If you have trouble with one or more of them you may have to alter the plan.
6. When you arrive at an answer for a problem, always ask yourself whether it is *reasonable*. If it isn't, look back to determine places where you may have made an error.

Answers for most of the odd-numbered exercises are provided at the back of the book. Exceptions are exercises that don't have short answers, for example "Prove that ..." or "Show that ..." problems where the answer is the whole solution. A Student Solutions Manual that contains detailed solutions to even-numbered exercises is available.

Besides * used to mark more difficult or theoretical problems, the following symbols are used to mark exercises of special types:

❖ exercises pertaining to differential equations or initial-value problems. (It is not used in Chapter 10, which is wholely concerned with DEs.)

problems requiring the use of a calculator. Often a scientific calculator is needed. Some such problems may require a programmable calculator.

problems requiring the use of either a graphing calculator or mathematical graphing software on a personal computer.

problems requiring the use of a computer. Typically, these will require either computer algebra software (e.g., Maple, Mathematica, Derive), or a spreadsheet program (e.g., Lotus 1-2-3, Microsoft Excel, Borland Quattro Pro).

# To the Instructor

Real-variable calculus is usually covered in three or four semesters of study, the first two of which are typically devoted to the differentiation and integration of functions of a single real variable, and the remainder to functions of several real variables and to vector-valued functions. This book is intended to cover the second half of this program, and contains, with very minor modifications, the material of Chapters 10–19 of the author's *Calculus: A Complete Course*. There is a wealth of material here, perhaps too much to cover in your courses. Some selection will have to be made on what to include and what to omit. This selection should take into account the background and needs of your students.

Any division of material into "core" and "optional" is necessarily somewhat arbitrary. The background discussion of vectors and geometry in Sections 3.1–3.4, the basic development of partial differentiation in Sections 4.1–4.8, the application to extreme value problems in Sections 5.1–5.3, and most of Chapter 6 on multiple integration should be considered as core.

For vector calculus, Sections 7.1–7.3, all of Chapter 8, and Sections 9.1–9.4 might be considered core.

The remaining sections and even some of the material and examples in core sections is optional in the sense that (with minor exceptions) its prior coverage is not necessary for any of what follows. It is up to individual instructors to decide what is most appropriate for their classes.

The text is designed for courses for science and engineering students, and for mathematics majors and honours students. The material covered presupposes that the students have a good background in single-variable calculus. The preliminary chapter (Chapter R) reviews this material very briefly. Chapters 1 and 2 are also "single-variable" chapters dealing with numerical series and power-series representations of functions. These chapters are included here because they are sometimes not covered until the third semester. Some optional material is more subtle and/or theoretical, and is intended mainly for stronger students.

Several supplements are available for use with *Calculus of Several Variables*. Among these are

- an **Instructor's Solutions Manual** with detailed solutions to all the exercises, prepared by the author.
- a **Student Solutions Manual** with detailed solutions to all the even-numbered exercises, prepared by the author.
- a **Maple Lab Manual** by Robert Israel (University of British Columbia) with instruction, examples, and problems dealing with the effective use of the computer algebra system Maple as a tool for calculus.

# Acknowledgments

The first two editions of this material have been used, since publication, for classes of general science, engineering, and mathematics majors and honours students at the University of British Columbia. I am grateful to colleagues at UBC, in particular Jim Carrell, John Fournier, Ed Granirer, Robert Israel, Peter Kiernan, and John Walsh, and at many other institutions where these books have been used, for their encouragement and useful comments and criticisms. I am also grateful to many of my students who have made helpful suggestions.

In preparing the revision of this text I have had guidance and advice from many users who responded to our surveys. As well, dedicated reviewers provided new insight and direction to my writing. I am especially grateful to the following people:

| | | | |
|---|---|---|---|
| F. A. Baragar | University of Alberta | Richard Lee | University of New Brunswick |
| Robert Bryan | University of Western Ontario | Peter McLure | University of Manitoba |
| Masao Chen | Vanier College | Niels W. Pedersen | AalBorg University Center |
| Karim Dahao | Royal Institute of Technology | John Rhodes | Bates College |
| Bertil Ericsson | Royal Institute of Technology | R. A. Ross | University of Toronto |
| Thomas Gustafsson | University of Lulea | Viena Stastna | University of Calgary |
| Ferdinand Haring | North Dakota State University | Lars Svensson | Royal Institute of Technology |
| Olof Hédan | Royal Institute of Technology | Björn Textorius | University of Linkoping |
| Matthew Ikle | University of Nevada | Hans Tranberg | Royal Institute of Technology |

Thanks also go to Brian Forrest (University of Waterloo) and to Robert Israel (University of British Columbia) who contributed new problems for this edition, to Ken MacKenzie (McGill University) who checked the answers for all the new exercises, and to Valerie Jones and David Hamilton for their careful copy editing.

I typeset this volume using TeX and PostScript on a NeXT computer and also generated all of the figures in PostScript using the mathematical graphics software package **MG** developed by my colleague Robert Israel and myself.

Finally, I wish thank several people at Addison-Wesley for their assistance and encouragement. These include College Publisher Ron Doleman who has supervised the publication of all of my Addison-Wesley books for the past twelve years, and who first introduced me to TeX and PostScript, Linda Scott, Managing Editor, and Abdulqafar Abdullahi and Suzanne Schaan. I would also like to thank the sales and marketing staff of all Addison-Wesley divisions around the world for making the previous editions so successful.

Despite all the excellent help I have received, I am not so naïve as to believe that the text is free of errors and obscurities, and I accept full responsibility for any that remain. Any comments, corrections, and suggestions for future revisions received from readers will be much appreciated.

*R.A.A.*
*Vancouver, Canada*
*July, 1995*

# Contents

CHAPTER R

# A Brief Review of Single-Variable Calculus

**INTRODUCTION** Single-variable calculus in concerned with studying the behaviour of functions of a single real variable. Such a function $f$ is a rule or *machine* that generates a real number output $f(x)$ from each real number $x$ in its *domain* of possible input values. Calculus builds upon algebra and geometry, which also study functions. What distinguishes calculus is the introduction of a new concept called a *limit* that enables calculus to study the way function values change in a dynamic way. Calculus is divided into two basic branches, differential calculus and integral calculus, both of which have limits at their cores.

## Limits and Continuity

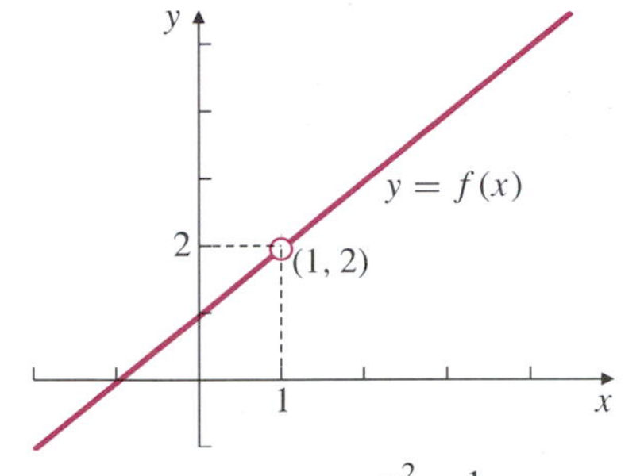


**Figure R.1** If $f(x) = \dfrac{x^2-1}{x-1}$, then $\lim_{x\to 1} f(x) = 2$ even though $f(1)$ is not defined

We say that the function $f(x)$ has **limit** $L$ as $x$ approaches $x_0$, and we write

$$\lim_{x\to x_0} f(x) = L,$$

provided that we can ensure that the number $f(x)$ is *as close as we want* to the number $L$ by taking $x$ *close enough* (but not equal to) $x_0$. Expressed in terms of symbols, this condition says that for every positive number $\epsilon$ (no matter how small), there exists a positive number $\delta$ (depending on $\epsilon$) such that

$$0 < |x - x_0| < \delta \quad \Rightarrow \quad |f(x) - L| < \epsilon.$$

We can also define the **left limit** and **right limit** of $f$ at $x_0$ by requiring $x < x_0$ or $x > x_0$ in the above definition. The limit exists if and only if both the left and right limits exist and the two are equal.

Limits of combinations of functions can be calculated by using the following **limit rules**: if $\lim_{x\to a} f(x) = L$ and $\lim_{x\to a} g(x) = M$, then

$$\begin{aligned} &\lim_{x\to a}[f(x)+g(x)] = L + M && \lim_{x\to a} kf(x) = kL \\ &\lim_{x\to a}[f(x)-g(x)] = L - M && \lim_{x\to a}\frac{f(x)}{g(x)} = \frac{L}{M}, \quad \text{if } M \neq 0 \\ &\lim_{x\to a} f(x)g(x) = LM && \lim_{x\to a}\big[f(x)\big]^r = L^r, \quad \text{if } L > 0. \end{aligned}$$

If $b < a < c$, $f(x) \le g(x) \le h(x)$ on $[b, c]$, and $\lim_{x\to a} f(x) = \lim_{x\to a} h(x) = L$, then $\lim_{x\to a} g(x) = L$ also.

The limit concept can be extended to allow for **limits at infinity** and **infinite limits**. For example,

$$\lim_{x\to 1}\frac{1}{(x-1)^2} = \infty, \qquad \lim_{x\to\infty}\frac{3x+2}{x+5} = \lim_{x\to\infty}\frac{3+(2/x)}{1+(5/x)} = \frac{3}{1} = 3.$$

A function $f$ is **continuous** at a point $c$ in its domain if

$$\lim_{x\to c} f(x) = f(c).$$

(If $f(x)$ is only defined on one side of $c$, the limit need only be one-sided.) Functions that are continuous wherever they are defined are called continuous functions.

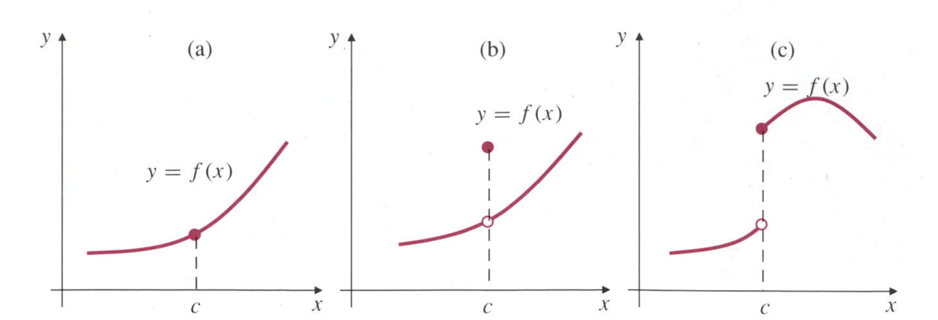


**Figure R.2**

(a) $f$ is continuous at $c$

(b) $f$ is not continuous at $c$ because $\lim_{x \to c} f(x) \neq f(c)$

(c) $f$ is not continuous at $c$ because $\lim_{x \to c} f(x)$ does not exist

Most functions encountered previously are continuous. For example the following functions are continuous wherever they are defined:

(a) all polynomials

(b) all rational functions

(c) all rational powers $x^{m/n} = \sqrt[n]{x^m}$

(d) the trigonometric functions sine, cosine, tangent, secant, cosecant, and cotangent

(e) the absolute value function $|x|$.

The Heaviside function $H(x) = \begin{cases} 1 & \text{if } x \geq 0 \\ 0 & \text{if } x < 0 \end{cases}$ is defined for all real $x$, but is discontinuous at $x = 0$.

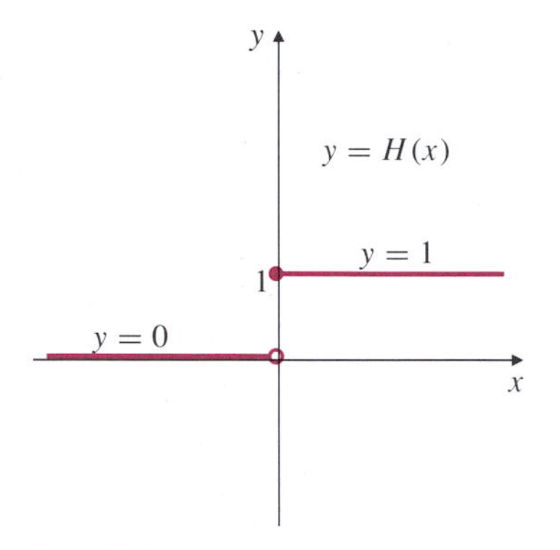


**Figure R.3** $H(x)$ is discontinuous at $x = 0$ because $\lim_{x \to 0-} H(x) = 0$ and $\lim_{x \to 0+} H(x) = 1$

Sums, products, constant multiples, and compositions of continuous functions are continuous. So are quotients, except, of course, at points where the denominator is zero. The following theorems are very important ones in calculus.

**THEOREM 1**

**The Max-Min Theorem**

If $f(x)$ is continuous on the closed, finite interval $[a, b]$, then there exist numbers $x_1$ and $x_2$ in $[a, b]$ such that for all $x$ in $[a, b]$,

$$f(x_1) \leq f(x) \leq f(x_2).$$

Thus $f$ has a unique absolute minimum value $m = f(x_1)$ and a unique absolute maximum value $M = f(x_2)$. The points $x_1$ and $x_2$ are not necessarily unique.

**THEOREM 2**

**The Intermediate-Value Theorem**

If $f(x)$ is continuous on the interval $[a, b]$ and if $s$ is a number between $f(a)$ and $f(b)$, then there exists a number $c$ in $[a, b]$ such that $f(c) = s$.

## Tangents and Derivatives

The line joining the two points $P = (x_0, f(x_0))$ and $Q = (x_0 + h, f(x_0 + h))$ on the graph $y = f(x)$ of the function $f$ has slope

$$\frac{f(x_0 + h) - f(x_0)}{h}.$$

If this slope has a limit $m$ as $h$ approaches 0 (i.e., as $Q$ approaches $P$ along the graph), then the line through $P$ with slope $m$ is **tangent** to the curve $y = f(x)$ at $P$. For example, $y = 2x - 1$ is tangent to $y = x^2$ at $(1, 1)$ because it passes through that point and has slope

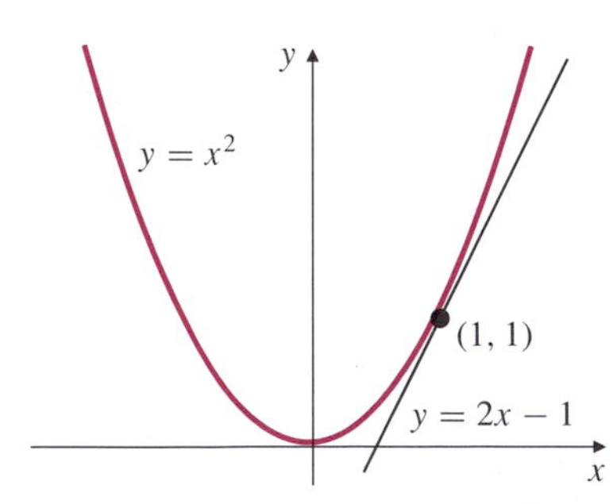


**Figure R.4** The tangent to $y = x^2$ at $(1, 1)$

$$\lim_{h \to 0} \frac{(1 + h)^2 - 1^2}{h} = \lim_{h \to 0} \frac{2h + h^2}{h} = \lim_{h \to 0} (2 + h) = 2.$$

If the (two-sided) limit of the slope of $PQ$ does not exist, but is either $\infty$ or $-\infty$, then the vertical line $x = x_0$ through $P$ is tangent to the graph $y = f(x)$ at $P$. Curves have tangents only at points where they are "smooth."

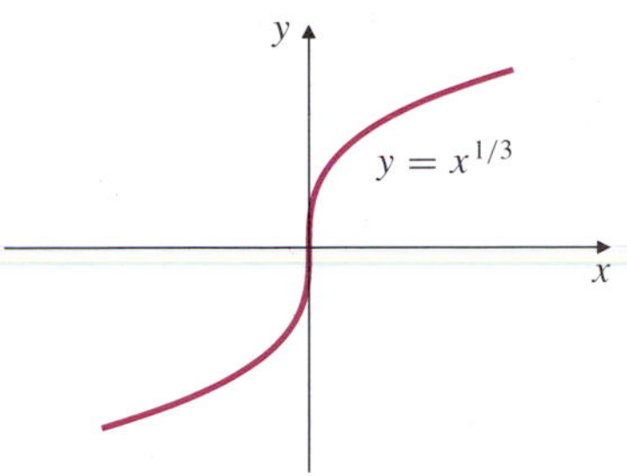


**Figure R.5** The $y$-axis is tangent to $y = x^{1/3}$ at the origin

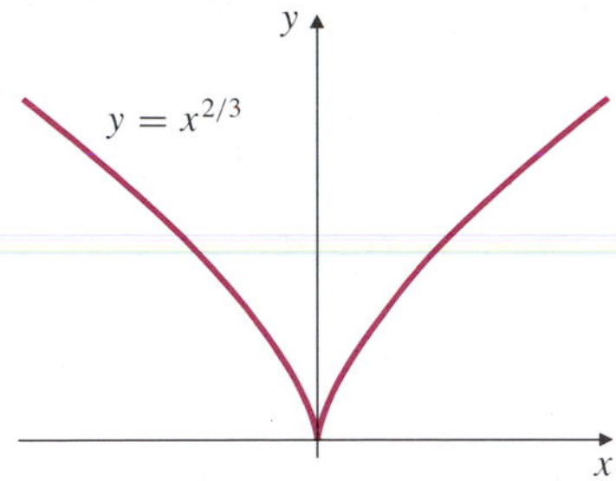


**Figure R.6** $y = x^{2/3}$ has no tangent at the origin

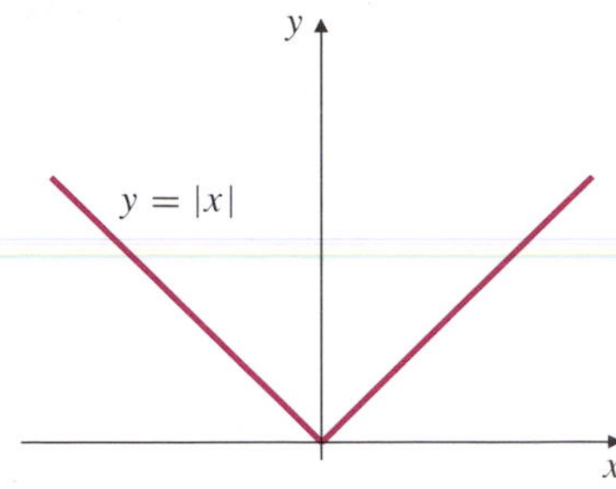


**Figure R.7** $y = |x|$ has no tangent at the origin

If a curve has a nonvertical tangent at a point $P$, then the slope of the curve at $P$ is the slope of that tangent. In general the slope of the curve $y = f(x)$ varies from point to point on the curve, so is itself a function of $x$. We call this function **the derivative of** $f$, and denote it by $f'$.

The **derivative** of a function $f$ is the function $f'$ defined by

$$f'(x) = \lim_{h\to 0} \frac{f(x+h) - f(x)}{h}$$

at all points $x$ for which the limit exists (i.e., is a finite real number). If $f'(x)$ exists, we say that $f$ is **differentiable** at $x$.

A function is continuous at any point where it is differentiable. Its graph is smooth at such a point, and has a nonvertical tangent there.

**Table 1.** Some elementary functions and their derivatives

| $f(x)$ | $f'(x)$ |
|---|---|
| $c$ (constant) | $0$ |
| $x$ | $1$ |
| $x^2$ | $2x$ |
| $\dfrac{1}{x}$ | $-\dfrac{1}{x^2}$ $(x \neq 0)$ |
| $\sqrt{x}$ | $\dfrac{1}{2\sqrt{x}}$ $(x > 0)$ |
| $x^r$ | $r\,x^{r-1}$ $(x^{r-1}$ real$)$ |
| $\lvert x\rvert$ | $\dfrac{x}{\lvert x\rvert} = \operatorname{sgn} x$ |

If $y = f(x)$ we can use the dependent variable $y$ to represent the function, and we can denote the derivative of the function in any of the following ways:

$$D_x y = y' = \frac{dy}{dx} = \frac{d}{dx} f(x) = f'(x) = Df(x).$$

The value of the derivative of a function at a particular number $x_0$ in its domain can also be expressed in several ways:

$$D_x y\Big|_{x=x_0} = y'\Big|_{x=x_0} = \frac{dy}{dx}\Big|_{x=x_0} = \frac{d}{dx}f(x)\Big|_{x=x_0} = f'(x_0) = Df(x_0).$$

The calculation of derivatives of combinations of functions is carried out with the aid of the following differentiation rules:

**THEOREM 3**

**Differentiation Rules**

If $f$ and $g$ are differentiable at $x$ then so are $f+g$, $f-g$, $Cf$ (where $C$ is a constant), $fg$, and $1/g$ and $f/g$ (provided $g(x) \neq 0$). Moreover,

| | |
|---|---|
| $(f+g)'(x) = f'(x) + g'(x)$ | **The Sum Rule** |
| $(f-g)'(x) = f'(x) - g'(x)$ | **The Difference Rule** |
| $(Cf)'(x) = Cf'(x)$ | **The Constant Multiple Rule** |
| $(fg)'(x) = f'(x)g(x) + f(x)g'(x)$ | **The Product Rule** |
| $\left(\frac{1}{g}\right)'(x) = \frac{-g'(x)}{(g(x))^2}$ | **The Reciprocal Rule** |
| $\left(\frac{f}{g}\right)'(x) = \frac{g(x)f'(x) - f(x)g'(x)}{(g(x))^2}$ | **The Quotient Rule** |

If $f(u)$ is differentiable at $u = g(x)$, and $g(x)$ is differentiable at $x$, then the composite function $f \circ g(x) = f(g(x))$ is differentiable at $x$, and

$$(f \circ g)'(x) = f'(g(x))g'(x) \qquad \textbf{The Chain Rule}$$

In terms of Leibniz notation, if $y = f(u)$ where $u = g(x)$, then

$$\frac{dy}{dx} = \frac{dy}{du}\frac{du}{dx}, \qquad \text{where } \frac{dy}{du} \text{ is evaluated at } u = g(x).$$

An equation involving two variables $x$ and $y$ can be regarded as defining one of the variables, say $y$, implicitly as a function of the other, and hence differentiated by the Chain Rule. For instance

$$y^2 + x^3 = 5xy \qquad \Rightarrow \qquad 2y\frac{dy}{dx} + 3x^2 = 5x\frac{dy}{dx} + 5y.$$

## The Trigonometric Functions

The functions $\cos t$ and $\sin t$ are defined to be, respectively, the $x$- and $y$-coordinates of the point $P_t$ on the circle $x^2 + y^2 = 1$ obtained by starting at the point $A = (1, 0)$ on the circle and proceeding a distance $|t|$ measured along the circle in a counterclockwise direction if $t > 0$, or in a clockwise direction if $t < 0$. The number $t$ is the radian measure of the angle $AOP_t$. Since the circumference of the circle is $2\pi$, we have

$$\pi \text{ radians} = 180°.$$

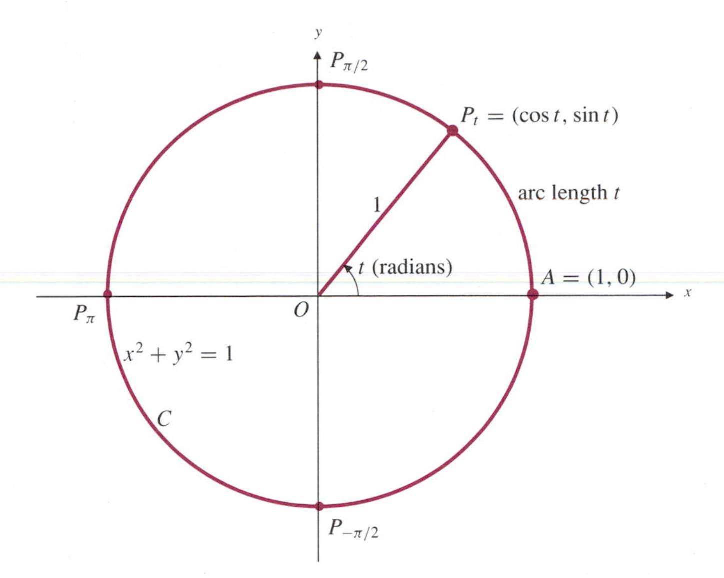


**Figure R.8** Angle $AOP_t = t$ radians, where $t$ is the length of arc $AP_t$

The graphs of cosine and sine indicate that these functions are defined and continuous at all real numbers $x$, and are periodic with period $2\pi$.

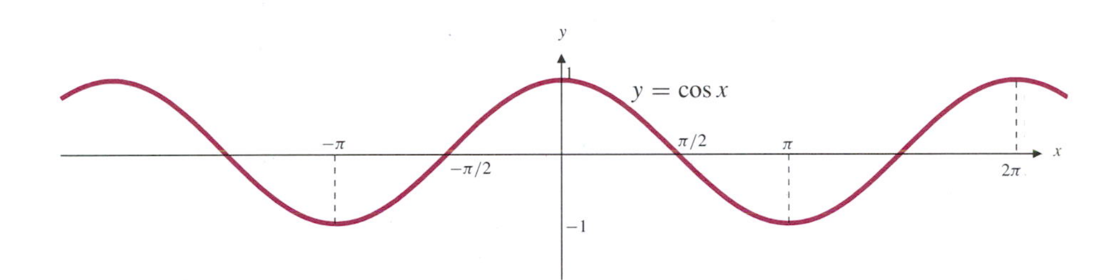


**Figure R.9** The graph of $\cos x$

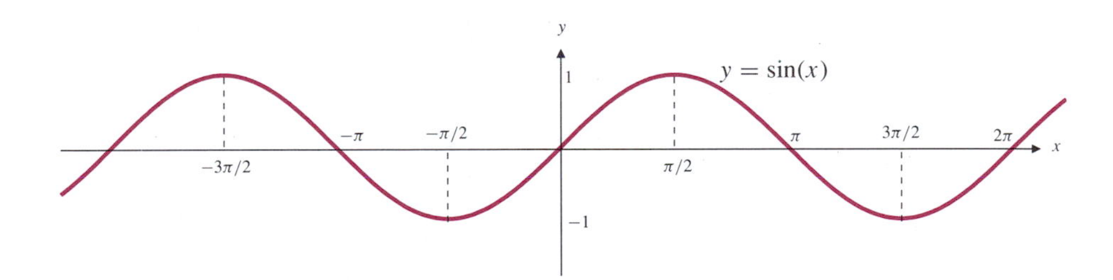


**Figure R.10** The graph of $\sin x$

Cosine and sine satisfy numerous identities, among them $\cos^2 t + \sin^2 t = 1$, and

**Addition formulas for cosine and sine**

$$\cos(s + t) = \cos s \cos t - \sin s \sin t$$
$$\sin(s + t) = \sin s \cos t + \cos s \sin t$$
$$\cos(s - t) = \cos s \cos t + \sin s \sin t$$
$$\sin(s - t) = \sin s \cos t - \cos s \sin t.$$

**Double- and half-angle formulas**

$$\begin{aligned} \sin 2t &= 2\sin t \cos t \qquad \text{and} \\ \cos 2t &= \cos^2 t - \sin^2 t \\ &= 2\cos^2 t - 1 \qquad (\text{using } \sin^2 t + \cos^2 t = 1) \\ &= 1 - 2\sin^2 t \\ \cos^2 t &= \frac{1 + \cos 2t}{2}, \qquad \sin^2 t = \frac{1 - \cos 2t}{2}. \end{aligned}$$

Four other trigonometric functions are defined in terms of cosine and sine:

$$\tan t = \frac{\sin t}{\cos t} \qquad \sec t = \frac{1}{\cos t}$$
$$\cot t = \frac{\cos t}{\sin t} = \frac{1}{\tan t} \qquad \csc t = \frac{1}{\sin t}.$$

The derivatives of the trigonometric functions follow from the addition formulas for sine and cosine, and the important limit

$$\lim_{\theta \to 0} \frac{\sin \theta}{\theta} = 1 \qquad (\text{where } \theta \text{ is in radians}).$$

They are as follows:

$$\frac{d}{dx}\sin x = \cos x \qquad \frac{d}{dx}\cos x = -\sin x$$
$$\frac{d}{dx}\tan x = \sec^2 x \qquad \frac{d}{dx}\sec x = \sec x \tan x$$
$$\frac{d}{dx}\cot x = -\csc^2 x \qquad \frac{d}{dx}\csc x = -\csc x \cot x.$$

## Higher-Order Derivatives

If the derivative $y' = f'(x)$ of a function $y = f(x)$ is itself differentiable at $x$, we can calculate *its* derivative, which we call the **second derivative** of $f$ and denote by $y'' = f''(x)$. Some of the more common notations for second derivatives are

$$y'' = f''(x) = \frac{d^2y}{dx^2} = \frac{d}{dx}\frac{d}{dx}f(x) = \frac{d^2}{dx^2}f(x) = D_x^2 y = D_x^2 f(x).$$

Similarly, third-, fourth-, and in general $n$th-order derivatives can be calculated. The prime notation is inconvenient for derivatives of high order, so we denote the order by a superscript in parentheses (to distinguish it from an exponent): the $n$th derivative of $y = f(x)$ is

$$y^{(n)} = f^{(n)}(x) = \frac{d^n y}{dx^n} = \frac{d^n}{dx^n}f(x) = D_x^n y = D_x^n f(x),$$

and it is defined to be the derivative of the $(n-1)$st derivative. For $n = 1, 2$, and 3, primes are still normally used: $f^{(2)}(x) = f''(x)$, $f^{(3)}(x) = f'''(x)$. It is sometimes convenient to denote $f^{(0)}(x) = f(x)$, that is, to regard a function as its own zeroth-order derivative.

## Antiderivatives and Indefinite Integrals

An **antiderivative** of a function $f$ on an interval $I$ is another function $F$ satisfying

$$F'(x) = f(x) \quad \text{for } x \text{ in } I.$$

Antiderivatives are not unique but any two antiderivatives of the same function on the same interval differ by a constant; if $f'(x) = 0$ on an interval $I$, then $f(x) = C$ (a constant function) on $I$. The **indefinite integral** of $f(x)$ on interval $I$ is the *general* antiderivative of $f$ on $I$, that is,

$$\int f(x)\,dx = F(x) + C \qquad \text{on } I,$$

where $F'(x) = f(x)$ for all $x$ in $I$. Here is a list of some elementary indefinite integrals:

(a) $\displaystyle\int dx = \int 1\,dx = x + C$ (b) $\displaystyle\int x\,dx = \frac{x^2}{2} + C$

(c) $\displaystyle\int x^2\,dx = \frac{x^3}{3} + C$ (d) $\displaystyle\int \frac{1}{x^2}\,dx = \int \frac{dx}{x^2} = -\frac{1}{x} + C$

(e) $\displaystyle\int \frac{1}{\sqrt{x}}\,dx = 2\sqrt{x} + C$ (f) $\displaystyle\int x^r\,dx = \frac{x^{r+1}}{r+1} + C \quad (r \neq -1)$

(g) $\displaystyle\int \sin x\,dx = -\cos x + C$ (h) $\displaystyle\int \cos x\,dx = \sin x + C$

(i) $\displaystyle\int \sec^2 x\,dx = \tan x + C$ (j) $\displaystyle\int \csc^2 x\,dx = -\cot x + C$

(k) $\displaystyle\int \sec x \tan x\,dx = \sec x + c$ (l) $\displaystyle\int \csc x \cot x\,dx = -\csc x + C.$

A **differential equation** (abbreviated DE) is an equation involving one or more derivatives of an unknown function. Any function whose derivatives satisfy the differential equation *identically on an interval* is called a **solution** of the equation on that interval. For instance, the function $y = x^3 - x$ is a solution of the differential equation

$$\frac{dy}{dx} = 3x^2 - 1$$

on the whole real line. This differential equation has more than one solution; in fact, $y = x^3 - x + C$ is a solution for any value of the constant $C$; this is called a **general solution** of the equation.

The **order** of a differential equation is the order of the highest-order derivative appearing in the equation. The general solution of a DE of order $n$ typically involves $n$ arbitrary constants.

An **initial-value problem** (abbreviated IVP) is a problem that consists of

(i) a differential equation (to be solved for an unknown function), and

(ii) prescribed values for the solution and enough of its derivatives at a particular point (the initial point) to determine values for all the arbitrary constants in the general solution of the DE and so yield a unique particular solution.

## Maximum and Minimum Values

The function $f$ has an **absolute maximum value** $f(x_0)$ at the point $x_0$ in its domain if $f(x) \leq f(x_0)$ holds for every $x$ in the domain of $f$. Similarly, $f$ has an **absolute minimum value** $f(x_1)$ at the point $x_1$ in its domain if $f(x) \geq f(x_1)$ holds for every $x$ in the domain of $f$. Theorem 1 implies that a continuous function defined on a finite, closed interval has absolute maximum and minimum values.

The function $f$ has a local maximum (or minimum) value at $x_0$ if it has an absolute maximum (or minimum) value at $x_0$ when its domain is restricted to points sufficiently near $x_0$. Geometrically, the graph of $f$ is at least as high (or low) at $x_0$ as it is at nearby points.

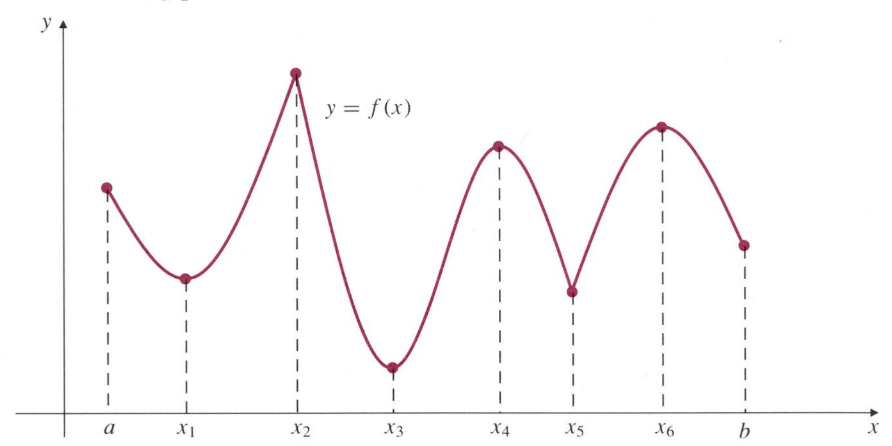


**Figure R.11** $f$ has local maximum values at $a$, $x_2$, $x_4$, and $x_6$, and local minimum values at $x_1$, $x_3$, $x_5$, and $b$. Its absolute maximum value is $f(x_2)$ and its absolute minimum value is $f(x_3)$

If $f$ has a local maximum or minimum value at $x$ in its domain, then $x$ must be either

(i) a **critical point** of $f$ (a point where $f'(x) = 0$), or

(ii) a **singular point** of $f$ (a point where $f'(x)$ is not defined), or

(iii) an **endpoint** of the domain of $f$ (a point in $D(f)$ that does not belong to any open interval contained in $D(f)$).

## Linear Approximations and Newton's Method

The **linearization**, or **best linear approximation**, of the function $f$ near $x = a$ is the function $L(x)$ defined by

$$L(x) = f(a) + f'(a)(x - a).$$

If $|f''(t)| \leq K$ for all $t$ in an interval containing $a$ and $x$, then the error $E(x) = f(x) - L(x)$ in the linear approximation $f(x) \approx L(x)$ satisfies

$$|E(x)| \leq \frac{K}{2}(x - a)^2.$$

Newton's method is a procedure for approximating a **root** of the equation $f(x) = 0$, that is, a number $r$ such that $f(r) = 0$, where $f$ is differentiable near the root. Start by guessing an initial approximation $x_0$ for the root, and then calculating successive approximations $x_1, x_2, x_3, \ldots$ according to the formula

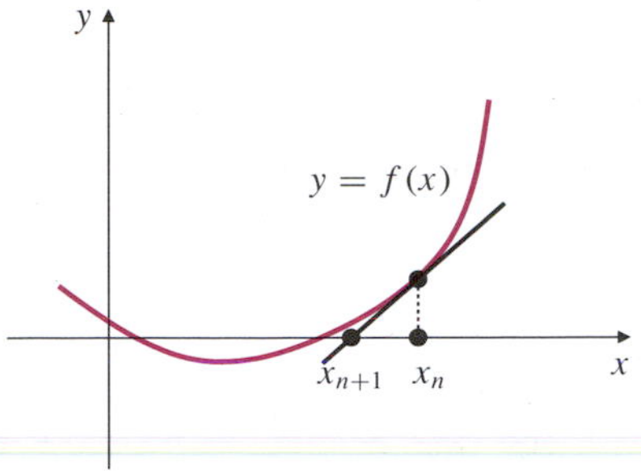

**Figure R.12** The geometric construction of the Newton's method formula

$$x_{n+1} = x_n - \frac{f(x_n)}{f'(x_n)}.$$

Note that $x_{n+1}$ is the $x$-intercept of the line tangent to $y = f(x)$ at $x = x_n$.

## Inverse Functions

A function $f$ is **one-to-one** if $f(x_1) \neq f(x_2)$ whenever $x_1$ and $x_2$ belong to the domain of $f$ and $x_1 \neq x_2$. Such a function $f$ has an **inverse function**, denoted $f^{-1}$. The value of $f^{-1}(x)$ is the unique number $y$ in the domain of $f$ for which $f(y) = x$. Thus

$$y = f^{-1}(x) \quad \Longleftrightarrow \quad x = f(y).$$

Inverse functions have the following properties:

1. $y = f^{-1}(x) \quad \Longleftrightarrow \quad x = f(y).$
2. The domain of $f^{-1}$ is the range of $f$.
3. The range of $f^{-1}$ is the domain of $f$.
4. $f^{-1}\big(f(x)\big) = x$ for all $x$ in the domain of $f$.
5. $f\big(f^{-1}(x)\big) = x$ for all $x$ in the domain of $f^{-1}$.
6. $(f^{-1})^{-1}(x) = f(x)$ for all $x$ in the domain of $f$.
7. The graph of $f^{-1}$ is the reflection of the graph of $f$ in the line $x = y$.
8. $\dfrac{d}{dx} f^{-1}(x) = \dfrac{1}{f'\big(f^{-1}(x)\big)}.$

## The Natural Logarithm and Exponential

One important pair of inverse functions consists of the **natural logarithm** function $\ln x$, defined for $x > 0$, and the **exponential** function, $\exp x = e^x$, where $e = exp(1) = 2.7\,1828\,1828\,45\,90\,45\ldots$.

The natural logarithm satisfies

$$\ln 1 = 0, \qquad \text{and} \qquad \frac{d}{dx} \ln x = \frac{1}{x},$$

from which the following properties of logarithms can be deduced:

$$\text{(i)}\quad \ln(xy) = \ln x + \ln y \qquad\qquad \text{(ii)}\quad \ln\left(\frac{1}{x}\right) = -\ln x$$

$$\text{(iii)}\quad \ln\left(\frac{x}{y}\right) = \ln x - \ln y \qquad\qquad \text{(iv)}\quad \ln\left(x^r\right) = r\,\ln x$$

$$\text{(v)}\quad \lim_{x\to\infty} \ln x = \infty \qquad\qquad \text{(vi)}\quad \lim_{x\to 0+} \ln x = -\infty.$$

On any interval not containing 0, we have $\displaystyle\int \frac{1}{x}\,dx = \ln|x| + C.$

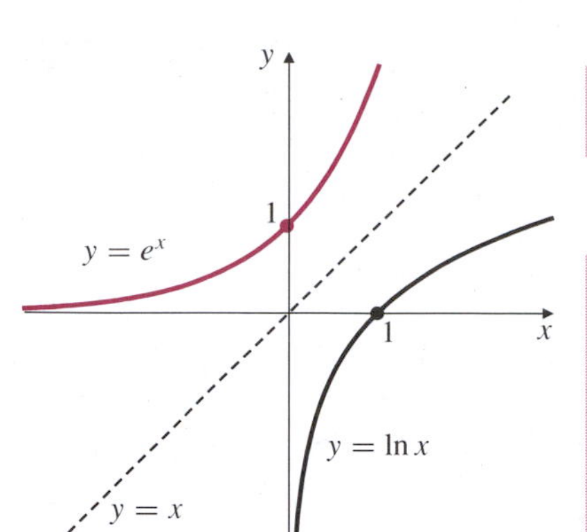


**Figure R.13** The graphs of $y = e^x$ and $y = \ln x$

Being the inverse of the natural logarithm, the exponential function satisfies

$$y = \exp x = e^x \iff x = \ln y \quad (y > 0).$$

Therefore, we have $e^0 = 1$ and

$$\text{(i)}\quad e^{x+y} = e^x e^y \qquad \text{(ii)}\quad (e^x)^r = e^{rx}$$

$$\text{(iii)}\quad e^{-x} = \frac{1}{e^x} \qquad \text{(iv)}\quad e^{x-y} = \frac{e^x}{e^y}$$

$$\text{(v)}\quad \lim_{x\to-\infty} e^x = 0 \qquad \text{(vi)}\quad \lim_{x\to\infty} e^x = \infty.$$

Also

$$\frac{d}{dx}e^x = e^x, \quad \text{and} \quad \int e^x\,dx = e^x + C \quad \text{on any interval.}$$

For every real number $x$, $e^x = \lim_{n\to\infty}\left(1 + \frac{x}{n}\right)^n$.

The initial-value problem $\begin{cases} \dfrac{dy}{dt} = ky \\ y(0) = y_0 \end{cases}$ has solution $y = y_0 e^{kt}$.

For $a > 0$, the **general exponential function** $a^x$ is defined by $a^x = e^{x \ln a}$. Its inverse is the **base $a$ logarithm**, $\log_a x = \dfrac{\ln x}{\ln a}$. We have

$$\frac{d}{dx}a^x = a^x \ln a, \quad \text{and} \quad \frac{d}{dx}\log_a x = \frac{1}{x \ln a}.$$

## The Inverse Trigonometric Functions

The **inverse sine** function, $\sin^{-1}x$ or arcsin $x$, and the **inverse tangent** function, $\tan^{-1}x$ or arctan $x$, are defined by

$$y = \sin^{-1}x \iff x = \sin y \quad \text{and} \quad -\frac{\pi}{2} \le y \le \frac{\pi}{2}$$

$$y = \tan^{-1}x \iff x = \tan y \quad \text{and} \quad -\frac{\pi}{2} < y < \frac{\pi}{2}.$$

We have

$$\frac{d}{dx}\sin^{-1}x = \frac{1}{\sqrt{1-x^2}} \qquad \int \frac{1}{\sqrt{a^2-x^2}}\,dx = \sin^{-1}\frac{x}{a} + C \quad (a > 0)$$

$$\frac{d}{dx}\tan^{-1}x = \frac{1}{1+x^2} \qquad \int \frac{dx}{a^2+x^2} = \frac{1}{a}\tan^{-1}\left(\frac{x}{a}\right) + C.$$

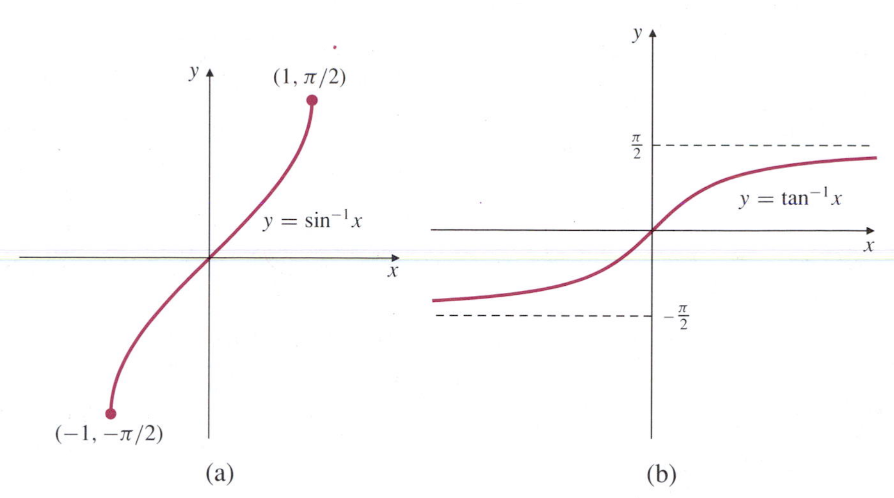


**Figure R.14**
(a) The graph of $\sin^{-1}x$
(b) The graph of $\tan^{-1}x$

The other inverse trigonometric functions can be expressed in terms of $\sin^{-1}$ and $\tan^{-1}$. For example,

$$\cos^{-1}x = \frac{\pi}{2} - \sin^{-1}x \qquad \text{for} \quad -1 \leq x \leq 1$$

$$\sec^{-1}x = \cos^{-1}\left(\frac{1}{x}\right) = \frac{\pi}{2} - \sin^{-1}\left(\frac{1}{x}\right) \qquad \text{for} \quad |x| \geq 1.$$

The derivatives of these functions are

$$\frac{d}{dx}\cos^{-1}x = -\frac{1}{\sqrt{1-x^2}}, \qquad \frac{d}{dx}\sec^{-1}x = \frac{1}{|x|\sqrt{x^2-1}}.$$

## Hyperbolic Functions

The hyperbolic cosine, sine, and tangent are defined in terms of the exponential functions:

$$\cosh x = \frac{e^x + e^{-x}}{2}, \qquad \sinh x = \frac{e^x - e^{-x}}{2}, \qquad \tanh x = \frac{e^x - e^{-x}}{e^x + e^{-x}}.$$

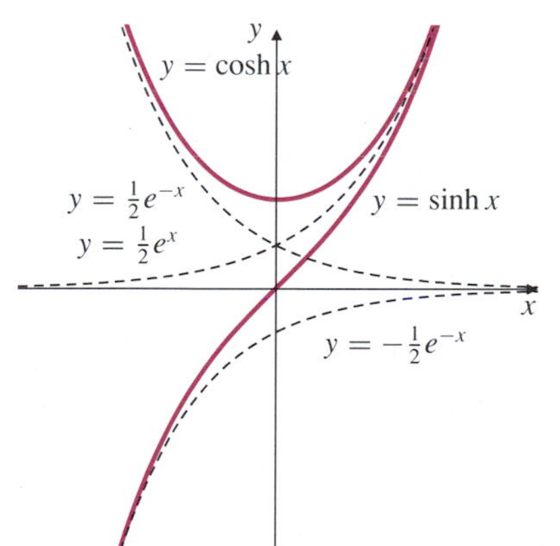


**Figure R.15** The graphs $\cosh x$ and $\sinh x$ and some exponential graphs to which they are asymptotic

It follows that $\cosh^2 x - \sinh^2 x = 1$, $(d/dx)\cosh x = \sinh x$, and $(d/dx)\sinh x = \cosh x$. The three other hyperbolic functions sech, csch, and coth are the reciprocals of cosh, sinh, and tanh, respectively.

The inverses of the hyperbolic functions can be expressed in terms of natural logarithms:

$$\cosh^{-1}x = \ln\left(x + \sqrt{x^2-1}\right), \qquad \text{for } x \geq 1,$$

$$\sinh^{-1}x = \ln\left(x + \sqrt{x^2+1}\right) \qquad \text{for all } x,$$

$$\tanh^{-1}x = \frac{1}{2}\ln\left(\frac{1+x}{1-x}\right) \qquad \text{for } -1 < x < 1.$$

## The Shape of a Graph

Examining the sign of the derivative of a function gives information about the behaviour of that function. For example, if $f$ is continuous on an interval $I$ and differentiable on the open interval $J$ consisting of the points of $I$ that are not endpoints of $I$, then the following assertions are valid:

(a) If $f'(x) = 0$ for all $x$ in $J$, then $f$ is constant on $I$.

(b) If $f'(x) > 0$ for all $x$ in $J$, then $f$ is increasing on $I$.

(c) If $f'(x) < 0$ for all $x$ in $J$, then $f$ is decreasing on $I$.

(d) If $f'(x) \geq 0$ for all $x$ in $J$, then $f$ is nondecreasing on $I$.

(e) If $f'(x) \leq 0$ for all $x$ in $J$, then $f$ is nonincreasing on $I$.

All of these follow from the following theorem.

**THEOREM 4**

**The Mean-Value Theorem**

Suppose that the function $f$ is continuous on the closed, finite interval $[a, b]$ and that it is differentiable on the open interval $(a, b)$. Then there exists a point $c$ in the open interval $(a, b)$ such that

$$\frac{f(b) - f(a)}{b - a} = f'(c).$$

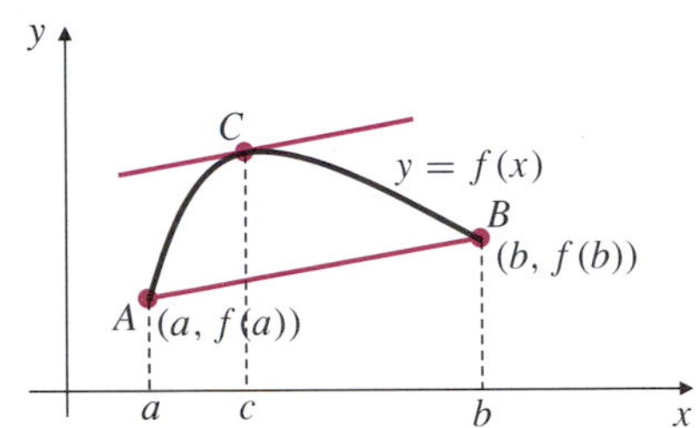


**Figure R.16** There is a point $C$ on the curve where the tangent is parallel to the chord $AB$

This says that the slope of the chord line joining the points $(a, f(a))$ and $(b, f(b))$ is equal to the slope of the tangent line to the curve $y = f(x)$ at the point $(c, f(c))$, so the two lines are parallel.

Knowledge of where a function is increasing and decreasing can tell us whether the function has a local maximum or local minimum value (or neither) at a critical point, a singular point, or an endpoint of its domain. For example, if $f$ is continuous at $x_0$, where $a < x_0 < b$, and $f'(x) > 0$ on $(a, x_0)$ and $f'(x) < 0$ on $(x_0, b)$, then $f$ must have a local maximum value at $x_0$. If $f'(x) < 0$ on $(a, x_0)$ and $f'(x) > 0$ on $(x_0, b)$, then $f$ must have a local minimum value at $x_0$. This way of classifying critical or singular points of $f$ is called the **first derivative test.**

We say that the function $f$ is **concave up** on an open interval $I$ if it is differentiable there and the derivative $f'$ is an increasing function on $I$. Similarly, $f$ is **concave down** on $I$ if $f'$ exists and is decreasing on $I$.

We say that the point $\big(x_0, f(x_0)\big)$ is an **inflection point** of the curve $y = f(x)$ (or that the function $f(x)$ has an inflection point at $x = x_0$) if the following two conditions are satisfied:

(a) the graph $y = f(x)$ has a tangent line at $x_0$, and

(b) the concavity of $f$ is opposite on opposite sides of $x_0$.

Concavity is determined by the sign of the second derivative of a function:

(a) If $f''(x) > 0$ on interval $I$, then $f$ is concave up on $I$.

(b) If $f''(x) < 0$ on interval $I$, then $f$ is concave down on $I$.

If $f$ has an inflection point at $x_0$, and if $f''(x_0)$ exists, then $f''(x_0) = 0$.

The **second derivative test** can be used to classify a critical point of a function:

(a) If $f'(x_0) = 0$ and $f''(x_0) < 0$, then $f$ has a local maximum value at $x_0$.

(b) If $f'(x_0) = 0$ and $f''(x_0) > 0$, then $f$ has a local minimum value at $x_0$.

(c) If $f'(x_0) = 0$ and $f''(x_0) = 0$, no conclusion can be drawn; $f$ may have a local maximum at $x_0$, or a local minimum, or it may have an inflection point instead.

## Indeterminate Forms and l'Hôpital's Rule

The laws for calculating limits are indecisive when they lead to values like "0/0" or "$\infty - \infty$" that are undefined in the real number system. The following table gives examples of various types of such **indeterminate forms**:

**Table 2.** Types of indeterminate forms

| Type | Example |
| --- | --- |
| $[0/0]$ | $\lim_{x\to 1} \frac{\ln x}{x-1}$ |
| $[\infty/\infty]$ | $\lim_{x\to 0} \frac{\ln(1/x^2)}{\cot(x^2)}$ |
| $[0\cdot\infty]$ | $\lim_{x\to 0+} x \ln \frac{1}{x}$ |
| $[\infty-\infty]$ | $\lim_{x\to(\pi/2)-} \left(\tan x - \frac{1}{\pi - 2x}\right)$ |
| $[0^0]$ | $\lim_{x\to 0+} x^x$ |
| $[\infty^0]$ | $\lim_{x\to(\pi/2)-} (\tan x)^{\cos x}$ |
| $[1^\infty]$ | $\lim_{x\to\infty} \left(1+\frac{1}{x}\right)^x$ |

Limits of type $[0/0]$ or $[\infty/\infty]$ can sometimes be evaluated using the following theorem:

**THEOREM 5**

**l'Hôpital's Rule**

Suppose the functions $f$ and $g$ are differentiable on the interval $(a, b)$, and $g'(x) \neq 0$ there. Suppose also that the following two conditions hold:

(i) either $\lim_{x\to a+} f(x) = \lim_{x\to a+} g(x) = 0$, or $\lim_{x\to a+} f(x) = \pm\infty$ and $\lim_{x\to a+} g(x) = \pm\infty$

(ii) $\lim_{x\to a+} \frac{f'(x)}{g'(x)} = L$ (where $L$ is finite or $\infty$ or $-\infty$).

Then $\lim_{x\to a+} \frac{f(x)}{g(x)} = L$.

Similar results hold for $\lim_{x\to b-}$ and for $\lim_{x\to c}$, and the cases $a = -\infty$ and $b = \infty$ are allowed.

The other types of indeterminate forms in the table can usually be put in the form $[0/0]$ or $[\infty/\infty]$ by algebraic manipulation and the taking of logarithms.

## Higher-Order Approximations

The polynomial of degree $n$ that best approximates $f(x)$ near $x = a$ is

$$P_n(x) = f(a) + f'(a)(x-a) + \frac{f''(a)}{2!}(x-a)^2 + \cdots + \frac{f^{(n)}(a)}{n!}(x-a)^n,$$

since, at $x = a$, the values of this polynomial and all its derivatives of order up to $n$ are equal to the corresponding values of $f$ and its derivatives. The following theorem provides a means for estimating the error $E_n(x) = f(x) - P_n(x)$ in the approximation $f(x) \approx P_n(x)$.

**THEOREM 6** **Taylor's Theorem with Lagrange Remainder**

If the $(n+1)$st-order derivative, $f^{(n+1)}(t)$, exists for all $t$ in an interval containing $a$ and $x$, then

$$f(x) = f(a) + f'(a)(x-a) + \frac{f''(a)}{2!}(x-a)^2 + \cdots + \frac{f^{(n)}(a)}{n!}(x-a)^n + E_n(x), \quad \textbf{Taylor's Formula},$$

where the error term $E_n(x)$ is given by

$$E_n(x) = \frac{f^{(n+1)}(X)}{(n+1)!}(x-a)^{n+1}, \qquad \textbf{Lagrange Remainder},$$

for some number $X$ between $a$ and $x$.

## Riemann Sums and the Definite Integral

If $m$ and $n$ are integers with $m \le n$, and if $f$ is a function defined at the integers $m$, $m+1, m+2, \ldots, n$, the symbol $\sum_{i=m}^{n} f(i)$ represents the sum of the values of $f$ at those integers:

$$\sum_{i=m}^{n} f(i) = f(m) + f(m+1) + f(m+2) + \cdots + f(n).$$

The explicit sum appearing on the right side of this equation is the **expansion** of the sum represented in sigma notation on the left side. The values of some particular sums can be expressed as simple formulas:

$$\text{(a)} \quad \sum_{i=1}^{n} 1 = \underbrace{1+1+1+\cdots+1}_{n \text{ terms}} = n$$

$$\text{(b)} \quad \sum_{i=1}^{n} i = 1+2+3+\cdots+n = \frac{n(n+1)}{2}$$

$$\text{(c)} \quad \sum_{i=1}^{n} i^2 = 1^2+2^2+3^2+\cdots+n^2 = \frac{n(n+1)(2n+1)}{6}$$

$$\text{(d)} \quad \sum_{i=1}^{n} r^{i-1} = 1+r+r^2+r^3+\cdots+r^{n-1} = \frac{r^n-1}{r-1} \quad \text{if } r \ne 1.$$

A **partition** of the interval $[a, b]$ is a finite set of points arranged in order between $a$ and $b$ on the real line, say

$$P = \{x_0, x_1, x_2, x_3, \ldots, x_{n-1}, x_n\},$$

where $a = x_0 < x_1 < x_2 < x_3 < \cdots < x_{n-1} < x_n = b$. It divides $[a, b]$ into $n$ subintervals of which the $i$th, $[x_{i-1}, x_i]$, has length $\Delta x_i = x_i - x_{i-1}$. Let $f$ be continuous on $[a, b]$, and let $c_i$ be a point in the $i$th subinterval of $P$. The sum

$$\begin{aligned} R(f, P) &= \sum_{i=1}^{n} f(c_i)\,\Delta x_i \\ &= f(c_1)\,\Delta x_1 + f(c_2)\,\Delta x_2 + f(c_3)\,\Delta x_3 + \cdots + f(c_n)\,\Delta x_n \end{aligned}$$

is called a **Riemann sum** of $f$ on $[a, b]$.

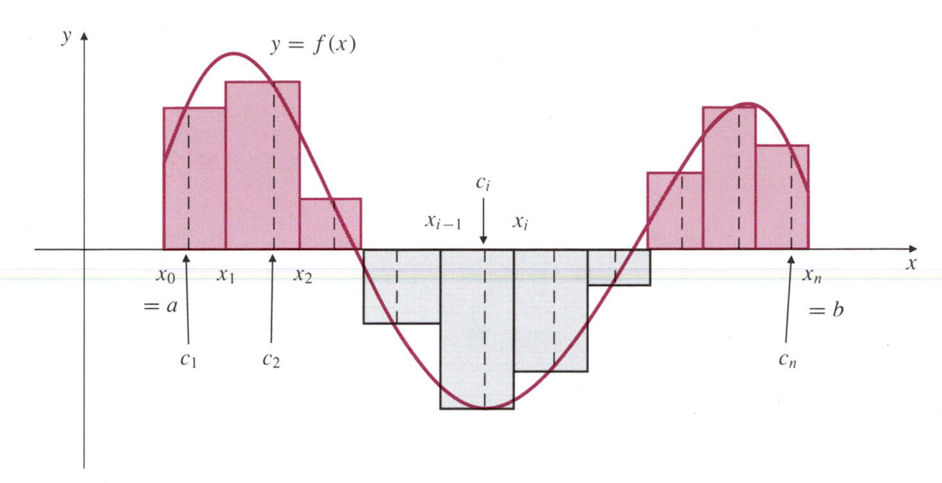


**Figure R.17** The Riemann sum $R(f, P)$ is the sum of areas of the rectangles shaded in colour minus the sum of the areas of the rectangles shaded in grey

Because $f$ is continuous on $[a, b]$, there is a single number $I$ such that we can ensure that $R(f, P)$ is as close as we want to $I$, independently of how the numbers $c_i$ are chosen, by choosing $P$ so that all its subintervals are sufficiently short. We call $I$ the definite integral of $f$ on $[a, b]$, and write it

$$I = \int_a^b f(x)\,dx.$$

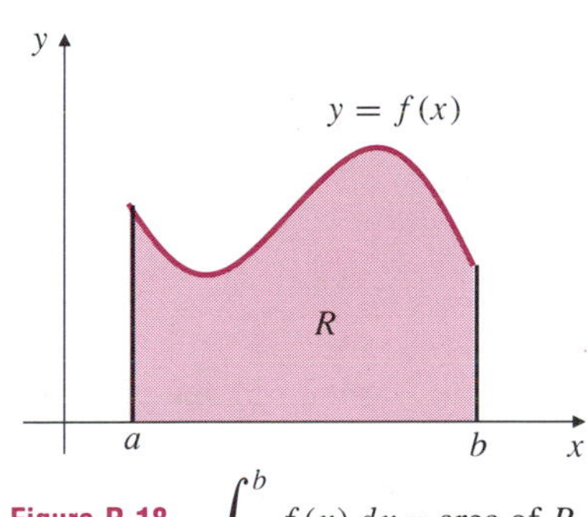


**Figure R.18** $\int_a^b f(x)\,dx$ = area of $R$

The definite integral has the following properties:

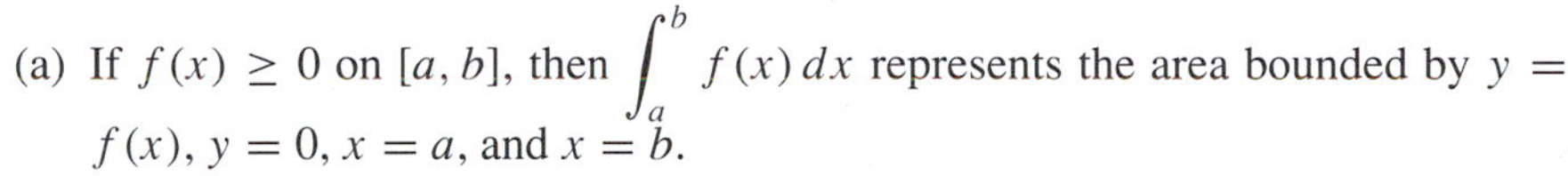

(a) If $f(x) \geq 0$ on $[a, b]$, then $\int_a^b f(x)\,dx$ represents the area bounded by $y = f(x)$, $y = 0$, $x = a$, and $x = b$.

(b) An integral over an interval of zero length is zero. $\int_a^a f(x)\,dx = 0$.

(c) Reversing the limits of integration changes the sign of the integral.

$$\int_b^a f(x)\,dx = -\int_a^b f(x)\,dx.$$

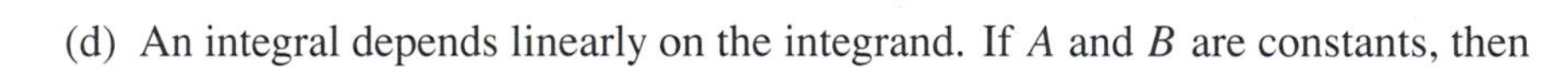

(d) An integral depends linearly on the integrand. If $A$ and $B$ are constants, then

$$\int_a^b \big(Af(x) + Bg(x)\big)\,dx = A\int_a^b f(x)\,dx + B\int_a^b g(x)\,dx.$$

(e) An integral depends additively on the interval of integration.

$$\int_a^b f(x)\,dx + \int_b^c f(x)\,dx = \int_a^c f(x)\,dx.$$

(f) If $f(x) \leq g(x)$ and $a \leq b$, then $\int_a^b f(x)\,dx \leq \int_a^b g(x)\,dx$.

(g) The **triangle inequality** for definite integrals: if $a \leq b$, then

$$\left|\int_a^b f(x)\,dx\right| \leq \int_a^b |f(x)|\,dx.$$

(h) If $f$ is an odd function ($f(-x) = -f(x)$), then $\int_{-a}^{a} f(x)\,dx = 0$.

(i) If $f$ is an even function ($f(-x) = f(x)$), then $\int_{-a}^{a} f(x)\,dx = 2\int_{0}^{a} f(x)\,dx$.

If $f$ is continuous on $[a, b]$, then there exists a point $c$ in $[a, b]$ such that

$$\int_a^b f(x)\,dx = (b-a)f(c).$$

We call $f(c)$ the **average value** or **mean value** of $f$ on $[a, b]$, and denote it by $\overline{f}$.

$$\overline{f} = \frac{1}{b-a}\int_a^b f(x)\,dx.$$

A function $f$ is **piecewise continuous** on $[a, b]$ if there exists a partition $a = c_0 < c_1 < c_2 < \cdots < c_n = b$ of $[a, b]$ and functions $F_i(x)$ continuous on $[c_{i-1}, c_i]$ such that $f(x) = F_i(x)$ for $c_{i-1} < x < c_i$. The definite integral can be extended to such functions $f$ by defining

$$\int_a^b f(x)\,dx = \sum_{i=1}^{n}\int_{c_{i-1}}^{c_i} F_i(x)\,dx.$$

**THEOREM 7**

**The Fundamental Theorem of Calculus**

Suppose that the function $f$ is continuous on an interval $I$ containing the point $a$.

PART I. Let the function $F$ be defined on $I$ by

$$F(x) = \int_a^x f(t)\,dt.$$

Then $F$ is differentiable on $I$ and $F'(x) = f(x)$ there; that is, $F$ is an antiderivative of $f$ on $I$. Expressed in terms of symbols,

$$\frac{d}{dx}\int_a^x f(t)\,dt = f(x).$$

PART II. If $G(x)$ is *any* antiderivative of $f(x)$ on $I$, so that $G'(x) = f(x)$ on $I$, then for any $b$ in $I$ we have

$$\int_a^b f(x)\,dx = G(b) - G(a).$$

# Integration Formulas and Techniques

**Some elementary integrals**

1. $\int 1\,dx = x + C$
2. $\int x\,dx = \frac{1}{2}x^2 + C$
3. $\int x^2\,dx = \frac{1}{3}x^3 + C$
4. $\int \frac{1}{x^2}\,dx = -\frac{1}{x} + C$
5. $\int \sqrt{x}\,dx = \frac{2}{3}x^{3/2} + C$
6. $\int \frac{1}{\sqrt{x}}\,dx = 2\sqrt{x} + C$
7. $\int x^r\,dx = \frac{1}{r+1}x^{r+1} + C \quad (r \neq -1)$
8. $\int \frac{1}{x}\,dx = \ln|x| + C$
9. $\int \sin ax\,dx = -\frac{1}{a}\cos ax + C$
10. $\int \cos ax\,dx = \frac{1}{a}\sin ax + C$
11. $\int \sec^2 ax\,dx = \frac{1}{a}\tan ax + C$
12. $\int \csc^2 ax\,dx = -\frac{1}{a}\cot ax + C$
13. $\int \sec ax\,\tan ax\,dx = \frac{1}{a}\sec ax + C$
14. $\int \csc ax\,\cot ax\,dx = -\frac{1}{a}\csc ax + C$
15. $\int \frac{1}{\sqrt{a^2 - x^2}}\,dx = \sin^{-1}\frac{x}{a} + C \quad (a > 0)$
16. $\int \frac{1}{a^2 + x^2}\,dx = \frac{1}{a}\tan^{-1}\frac{x}{a} + C$
17. $\int e^{ax}\,dx = \frac{1}{a}e^{ax} + C$
18. $\int b^{ax}\,dx = \frac{1}{a\ln b}b^{ax} + C$
19. $\int \cosh ax\,dx = \frac{1}{a}\sinh ax + C$
20. $\int \sinh ax\,dx = \frac{1}{a}\cosh ax + C$

The following techniques can be used to evaluate more complicated integrals.

**The Method of Substitution**. Let $u = g(x)$. Then $du = g'(x)\,dx$, and

$$\int f\big(g(x)\big)\,g'(x)\,dx = \int f(u)\,du.$$

Suppose that $g$ is a differentiable function on $[a, b]$, and satisfies $g(a) = A$ and $g(b) = B$. Also suppose that $f$ is continuous on the range of $g$. Then

$$\int_a^b f\big(g(x)\big)\,g'(x)\,dx = \int_A^B f(u)\,du.$$

Substitution and trigonometric identities provide values for the following integrals:

$$\int \tan x\,dx = \ln|\sec x| + C,$$

$$\int \cot x\,dx = \ln|\sin x| + C = -\ln|\csc x| + C,$$

$$\int \sec x\,dx = \ln|\sec x + \tan x| + C,$$

$$\int \csc x\,dx = -\ln|\csc x + \cot x| + C = \ln|\csc x - \cot x| + C,$$

$$\int \cos^2 x\,dx = \frac{1}{2}(x + \sin x\cos x) + C,$$

$$\int \sin^2 x\,dx = \frac{1}{2}(x - \sin x\cos x) + C.$$

**Integration by Parts.** If $U = U(x)$, $dV = V'(x)\,dx$, $dU = U'(x)\,dx$, and $V = V(x)$, then $d(UV) = [U'(x)V(x) + U(x)V'(x)]\,dx$, and so

$$\int U(x)V'(x)\,dx = \int U\,dV = UV - \int V\,dU = U(x)V(x) - \int V(x)U'(x)\,dx.$$

**Inverse Trigonometric Substitutions.**

1. Integrals involving $\sqrt{a^2 - x^2}$ (where $a > 0$) can frequently be reduced to a simpler form by means of the substitution

   $$x = a\sin\theta \quad \text{or, equivalently,} \quad \theta = \sin^{-1}\frac{x}{a}.$$

   In this case, $\sqrt{a^2 - x^2} = a\cos\theta$ and $dx = a\cos\theta\,d\theta$.

2. Integrals involving $\sqrt{a^2 + x^2}$ or $\dfrac{1}{x^2 + a^2}$ (where $a > 0$) are often simplified by the substitution

   $$x = a\tan\theta \quad \text{or, equivalently,} \quad \theta = \tan^{-1}\frac{x}{a}.$$

   In this case, $\sqrt{a^2 + x^2} = a\sec\theta$ and $dx = a\sec^2\theta\,d\theta$.

3. Integrals involving $\sqrt{x^2 - a^2}$ (where $a > 0$) can frequently be simplified by using the substitution

   $$x = a\sec\theta \quad \text{or, equivalently,} \quad \theta = \sec^{-1}\frac{x}{a}.$$

   If $x \geq a$, then $\sqrt{x^2 - a^2} = a\tan\theta$; if $x \leq -a$, then $\sqrt{x^2 - a^2} = -a\tan\theta$. In either case $dx = a\sec\theta\tan\theta\,d\theta$.

**Integrals of Rational Functions.** If $P(x)$ and $Q(x)$ are polynomials such that the degree of $P$ is less than the degree of $Q$ and $Q$ can be factored into a product of linear and quadratic factors, then the fraction $P/Q$ can be written as a sum of other such fractions whose denominators are the factors of $Q$. For example,

$$\frac{1}{x^2 - a^2} = \frac{1}{2a}\,\frac{1}{x - a} - \frac{1}{2a}\,\frac{1}{x + a}.$$

The integral of $P(x)/Q(x)$ can thereby be expressed as a sum of simpler integrals. The following integrals are useful in this context:

$$\int \frac{1}{ax + b}\,dx = \frac{1}{a}\ln|ax + b| + C,$$

$$\int \frac{x\,dx}{x^2 + a^2} = \frac{1}{2}\ln(x^2 + a^2) + C,$$

$$\int \frac{x\,dx}{x^2 - a^2} = \frac{1}{2}\ln|x^2 - a^2| + C,$$

$$\int \frac{dx}{x^2 + a^2} = \frac{1}{a}\tan^{-1}\frac{x}{a} + C,$$

$$\int \frac{dx}{x^2 - a^2} = \frac{1}{2a}\ln\left|\frac{x - a}{x + a}\right| + C.$$

## Improper Integrals

Improper integrals are ones that correspond to areas of unbounded regions in the $xy$-plane. Such integrals occur if either (or both) of the limits of integration are infinite, or if $f$ is unbounded near a point of $[a, b]$ (and hence is not continuous on $[a, b]$). In this context, the ordinary definite integral of a continuous function over a closed, finite interval is called a **proper integral**.

**Improper Integrals of Type I.** If $f$ is continuous on $[a, \infty)$, we define the improper integral of $f$ over $[a, \infty)$ as a limit of proper integrals:

$$\int_a^\infty f(x)\,dx = \lim_{R\to\infty} \int_a^R f(x)\,dx.$$

Similarly, if $f$ is continuous on $(-\infty, b]$ then we define

$$\int_{-\infty}^b f(x)\,dx = \lim_{R\to-\infty} \int_R^b f(x)\,dx.$$

In either case, if the limit exists (is a finite number), we say that the improper integral **converges**; if the limit does not exist, we say that the improper integral **diverges**. If the limit is $\infty$ (or $-\infty$), we say the improper integral **diverges to infinity** (or **diverges to negative infinity**).

**Improper Integrals of Type II.** If $f$ is continuous on the interval $(a, b]$, and is possibly unbounded near $a$, we define the improper integral

$$\int_a^b f(x)\,dx = \lim_{c\to a+} \int_c^b f(x)\,dx.$$

Similarly, if $f$ is continuous on $[a, b)$, and is possibly unbounded near $b$, we define

$$\int_a^b f(x)\,dx = \lim_{c\to b-} \int_a^c f(x)\,dx.$$

These improper integrals may converge, diverge, diverge to infinity, or diverge to negative infinity.

For example, if $0 < a < \infty$, then

$$\int_a^\infty x^{-p}\,dx \begin{cases} \text{converges to } \dfrac{a^{1-p}}{p-1} & \text{if } p > 1 \\ \text{diverges to } \infty & \text{if } p \le 1 \end{cases}$$

$$\int_0^a x^{-p}\,dx \begin{cases} \text{converges to } \dfrac{a^{1-p}}{1-p} & \text{if } p < 1 \\ \text{diverges to } \infty & \text{if } p \ge 1. \end{cases}$$

To determine convergence or divergence and estimate the size of improper integrals for which no antiderivative can be found you can compare the integral with a simpler one that can be evaluated:

Let $-\infty \le a < b \le \infty$ and suppose that functions $f$ and $g$ are continuous on the interval $(a, b)$ and satisfy $0 \le f(x) \le g(x)$. If $\int_a^b g(x)\,dx$ converges, then so does $\int_a^b f(x)\,dx$, and

$$\int_a^b f(x)\,dx \le \int_a^b g(x)\,dx.$$

Equivalently, if $\int_a^b f(x)\,dx$ diverges to $\infty$, then so does $\int_a^b g(x)\,dx$.

## Numerical Integration

We assume that $f(x)$ is continuous on $[a, b]$ and subdivide $[a, b]$ into $n$ subintervals of equal length $h = (b - a)/n$ using the $n + 1$ points

$$x_0 = a, \quad x_1 = a + h, \quad x_2 = a + 2h, \quad \ldots, \quad x_n = a + nh = b.$$

We also assume that the value of $f(x)$ at each of these points is known:

$$y_0 = f(x_0), \quad y_1 = f(x_1), \quad y_2 = f(x_2), \quad \ldots, \quad y_n = f(x_n).$$

The $n$-subinterval **Trapezoid Rule** approximation to $\displaystyle\int_a^b f(x)\,dx$, denoted $T_n$, is given by

$$\begin{aligned} T_n &= h\left(\frac{1}{2}y_0 + y_1 + y_2 + y_3 + \cdots + y_{n-1} + \frac{1}{2}y_n\right) \\ &= \frac{h}{2}\left(y_0 + 2\sum_{j=1}^{n-1} y_j + y_n\right). \end{aligned}$$

Let $m_j = a + \left(j - \frac{1}{2}\right)h$ for $1 \le j \le n$. The **Midpoint Rule** approximation to $\int_a^b f(x)\,dx$, denoted $M_n$, is given by

$$M_n = h\big(f(m_1) + f(m_2) + \cdots + f(m_n)\big) = h\sum_{j=1}^{n} f(m_j).$$

Observe that $T_{2n} = \frac{1}{2}(T_n + M_n)$.

The **Simpson's Rule** approximation to $\int_a^b f(x)\,dx$ based on a subdivision of $[a, b]$ into an even number $n$ of subintervals of equal length $h = (b - a)/n$ is denoted $S_n$, and is given by:

$$\begin{aligned} &\int_a^b f(x)\,dx \approx S_n \\ &= \frac{h}{3}\big(y_0 + 4y_1 + 2y_2 + 4y_3 + 2y_4 + \cdots + 2y_{n-2} + 4y_{n-1} + y_n\big) \\ &= \frac{h}{3}\big(y_{\text{"ends"}} + 4y_{\text{"odds"}} + 2y_{\text{"evens"}}\big) \end{aligned}$$

The numbers $y_i$ are as specified for the Trapezoid Rule.

Observe that $S_{2n} = \dfrac{T_n + 2M_n}{3}$.

The following error estimates hold for these approximations:

If $f$ has a continuous second derivative on $[a, b]$ satisfying $|f''(x)| \le K$ there, then

$$\left| \int_a^b f(x)\,dx - T_n \right| \le \frac{K(b-a)}{12} h^2 = \frac{K(b-a)^3}{12n^2},$$

$$\left| \int_a^b f(x)\,dx - M_n \right| \le \frac{K(b-a)}{24} h^2 = \frac{K(b-a)^3}{24n^2}.$$

If $f$ has a continuous fourth derivative on the interval $[a, b]$, satisfying $|f^{(4)}(x)| \le K$ there, then

$$\left| \int_a^b f(x)\,dx - S_n \right| \le \frac{K(b-a)}{180} h^4 = \frac{K(b-a)^5}{180n^4}.$$

## Areas and Volumes

If $a < b$ and $f(x) \le g(x)$ on $[a, b]$, then the area $A$ of the region $R$ bounded by the curves $y = f(x)$ and $y = g(x)$ between the vertical lines $x = a$ and $x = b$ is given by

$$A = \int_a^b \big(g(x) - f(x)\big)\,dx.$$

It is useful to regard this formula as expressing $A$ as the "sum" (that is, the integral) of *infinitely many* **area elements**

$$dA = (g(x) - f(x))\,dx,$$

corresponding to values of $x$ between $a$ and $b$.

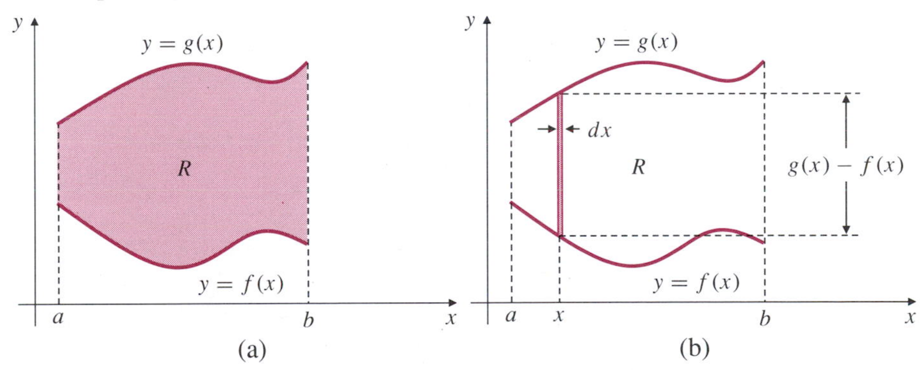

**Figure R.19**
(a) The region $R$ lying between two graphs
(b) An area element of the region $R$

The volume $V$ of a solid between $x = a$ and $x = b$ having cross-sectional area $A(x)$ at position $x$ is

$$V = \int_a^b A(x)\,dx.$$

Here the integral "sums" volume elements $dV = A(x)\,dx$, which are the volumes of "infinitely thin slices" perpendicular to the $x$-axis.

For **solids of revolution** the cross-sections are circular disks. If a vertical strip $0 < y < f(x)$ between $x$ and $x + dx$ is rotated about the $x$-axis, it generates a circular disk of radius $f(x)$ and thickness $dx$. If the strip is rotated about the $y$-axis, it generates a cylindrical shell of height $f(x)$, radius $x$, and thickness $dx$.

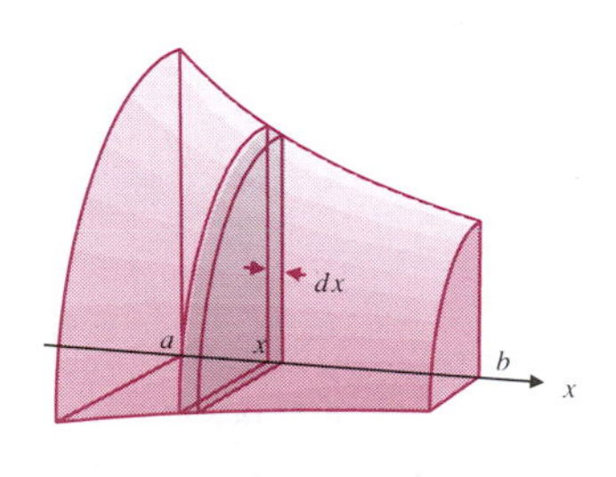


**Figure R.20** The volume element

The volume of the solid obtained by rotating the plane region $0 \le y \le f(x)$, $a < x < b$ about the $x$-axis is

$$V = \pi \int_a^b (f(x))^2 \, dx.$$

The volume of the solid obtained by rotating the plane region $0 \le y \le f(x)$, $0 \le a < x < b$ about the $y$-axis is

$$V = 2\pi \int_a^b x \, f(x) \, dx.$$

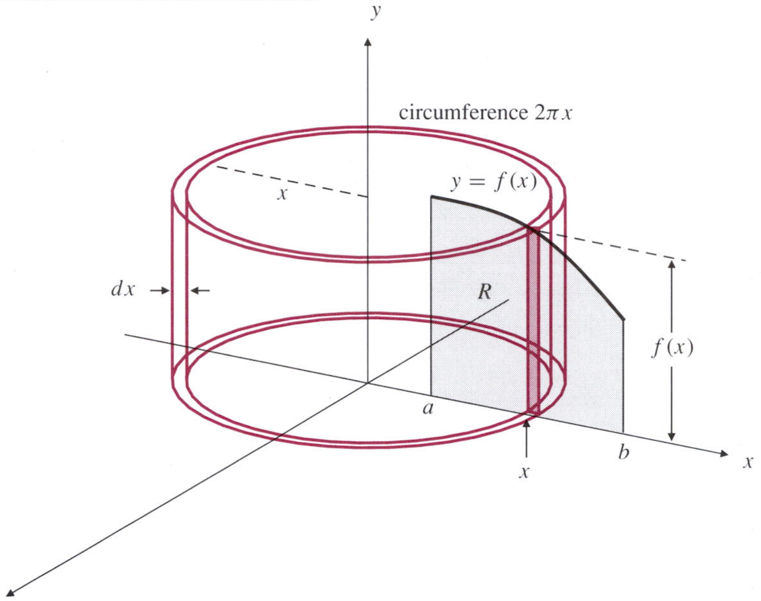

**Figure R.21** When rotated around the $y$-axis, the area element of width $dx$ under $y = f(x)$ at $x$ generates a cylindrical shell of height $f(x)$, circumference $2\pi x$, and hence volume $dV = 2\pi x \, f(x) \, dx$

## Arc Length and Surface Area

The arc length $s$ of the curve $y = f(x)$ from $x = a$ to $x = b$ is given by

$$s = \int_a^b \sqrt{1 + (f'(x))^2} \, dx = \int_a^b \sqrt{1 + \left(\frac{dy}{dx}\right)^2} \, dx.$$

You can regard this integral formula as giving the arc length $s$ of the curve as a "sum" of **arc length elements**

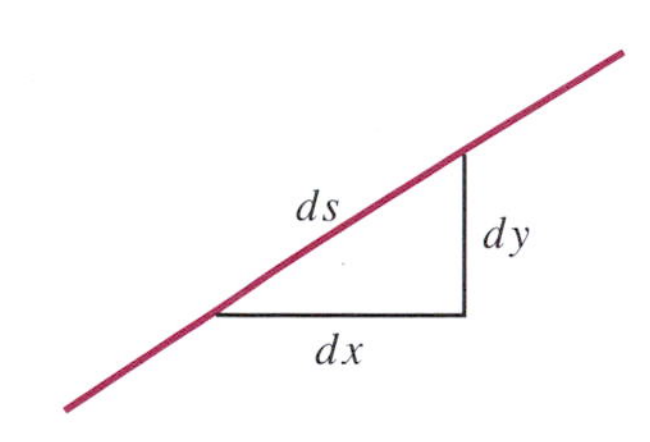


**Figure R.22** A differential triangle

$$s = \int_{x=a}^{x=b} ds \qquad \text{where} \qquad ds = \sqrt{(dx)^2 + (dy)^2} = \sqrt{1 + \left(\frac{dy}{dx}\right)^2} \, dx.$$

When a plane curve is rotated (in three dimensions) about a line in the plane of the curve, it sweeps out a **surface of revolution**. If the radius of rotation of an arc length element $ds$ on the curve is $r$, then it generates, on rotation, a circular band of width $ds$ and length (circumference) $2\pi r$. The area of this band is, therefore,

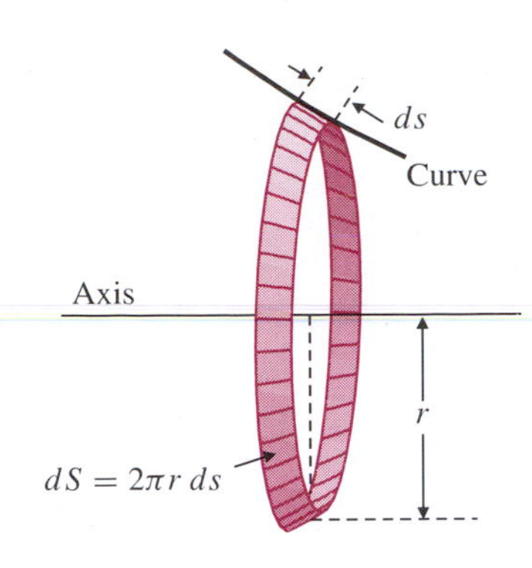


**Figure R.23** The circular band generated by rotating arc length element $ds$ about the axis

$$dS = 2\pi r\, ds.$$

For example, if $f'(x)$ is continuous on $[a, b]$ and the curve $y = f(x)$ is rotated about the $x$-axis, then the area of the surface of revolution so generated is

$$S = 2\pi \int_{x=a}^{x=b} |y|\, ds = 2\pi \int_a^b |f(x)|\sqrt{1 + (f'(x))^2}\, dx.$$

If the rotation is about the $y$-axis, the surface area is

$$S = 2\pi \int_{x=a}^{x=b} |x|\, ds = 2\pi \int_a^b |x|\sqrt{1 + (f'(x))^2}\, dx.$$

## Centres of Mass and Centroids

Suppose that matter is distributed on the interval $[a, b]$ with *line density* (mass per unit length) $\delta(x)$, which depends on position $x$ in the interval. The mass, $m$, and moment about $x = 0$, $M_{x=0}$, of the matter in the interval, are given by:

$$m = \int_a^b \delta(x)\, dx, \qquad M_{x=0} = \int_a^b x\delta(x)\, dx.$$

The centre of mass of the distribution is the point $\bar{x} = M_{x=0}/m$ at which the whole mass $m$ could be put to have the same moment about $x = 0$. If the density is constant, the centre of mass is the midpoint of the interval. We call it the **centroid** of the interval.

The centroid of any region is the centre of mass of an object of constant density $\delta = 1$ occupying that region. The centroid of the plane region $a \le x \le b$, $0 \le y \le f(x)$, is $(\bar{x}, \bar{y})$, where $\bar{x} = \dfrac{M_{x=0}}{A}$, $\bar{y} = \dfrac{M_{y=0}}{A}$, and

$$A = \int_a^b f(x)\, dx, \quad M_{x=0} = \int_a^b x f(x)\, dx, \quad M_{y=0} = \frac{1}{2}\int_a^b (f(x))^2\, dx.$$

For example, the coordinates of the centroid of a triangle are the averages of the corresponding coordinates of the three vertices of the triangle. The triangle with vertices $(x_1, y_1)$, $(x_2, y_2)$, and $(x_3, y_3)$ has centroid

$$(\bar{x}, \bar{y}) = \left(\frac{x_1 + x_2 + x_3}{3}, \frac{y_1 + y_2 + y_3}{3}\right).$$

**THEOREM 8**

**Pappus's Theorem**

(a) If a plane region $R$ lies on one side of a line $L$ in that plane, and if $R$ is rotated about $L$ to generate a solid of revolution, the volume $V$ of that solid is the product of the area of $R$ and the distance travelled by the centroid of $R$ under the rotation; that is,

$$V = 2\pi \bar{r} A,$$

where $A$ is the area of $R$ and $\bar{r}$ is the perpendicular distance from the centroid of $R$ to $L$.

(b) If a plane curve $C$, lying on one side of a line $L$ in the plane, is rotated about that line to generate a surface of revolution, then the area $S$ of that surface is the length of $C$ times the distance travelled by the centroid of $C$;

$$S = 2\pi \bar{r} s,$$

where $s$ is the length of the curve $C$ and $\bar{r}$ is the perpendicular distance from the centroid of $C$ to the line $L$.

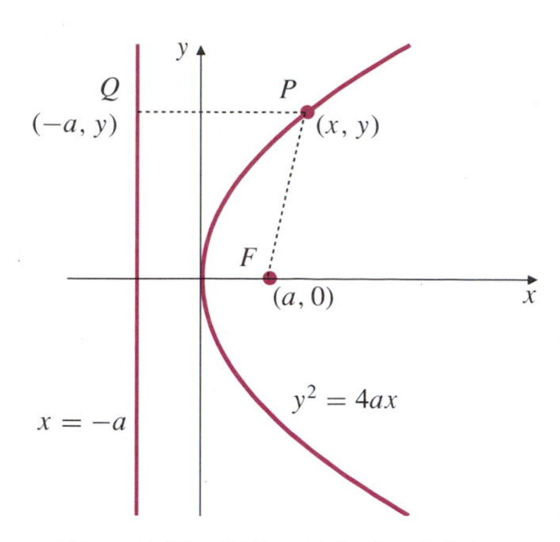


**Figure R.24** $PF = PQ$: the defining property of a parabola

## Conics

A **parabola** consists of points in the plane that are equidistant from a given point (the **focus** of the parabola) and a given straight line (the **directrix** of the parabola). The line through the focus perpendicular to the directrix is called the **principal axis** (or simply the **axis**) of the parabola. The **vertex** of the parabola is the point where the parabola crosses its principal axis. It is on the axis halfway between the focus and the directrix.

**Table 3.** Some parabolas with vertex at the origin

| Focus | Directrix | Equation |
|---|---|---|
| $(a, 0)$ | $x = -a$ | $y^2 = 4ax$ |
| $(-a, 0)$ | $x = a$ | $y^2 = -4ax$ |
| $(0, a)$ | $y = -a$ | $x^2 = 4ay$ |
| $(0, -a)$ | $y = a$ | $x^2 = -4ay$ |

An **ellipse** consists of all points in the plane, the sum of whose distances from two fixed points (the **foci**) is constant.

If $a > b > 0$, then the equation $\dfrac{x^2}{a^2} + \dfrac{y^2}{b^2} = 1$ represents an ellipse with centre at the origin and foci at $(\pm c, 0)$, where $c^2 = a^2 - b^2$. The sum of the distances from $P = (x, y)$ on the ellipse to the two foci is $2a$.

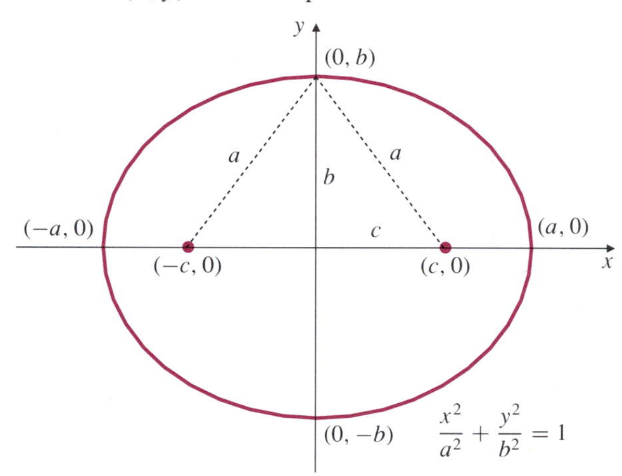


**Figure R.25** An ellipse

$a$ is the **semi-major axis**

$b$ is the **semi-minor axis**

$c = \sqrt{a^2 - b^2}$ is the **semi-focal separation**

$\epsilon = \dfrac{c}{a} < 1$ is the **eccentricity**.

A **hyperbola** consists of all points in the plane, the difference of whose distances from two fixed points (the **foci**) is constant.

The equation $\dfrac{x^2}{a^2} - \dfrac{y^2}{b^2} = 1$ represents a hyperbola with centre at the origin and foci at $(\pm c, 0)$, where $c^2 = a^2 + b^2$. The difference of the distances from $P = (x, y)$ on the hyperbola to the two foci is $2a$. The points $(a, 0)$ and $(-a, 0)$ are called the **vertices** of the hyperbola. The line through the centre, the vertices, and the foci is the **transverse axis**. The line through the centre perpendicular to the transverse axis is the **conjugate axis**. The hyperbola has two asymptotes with equations $(x/a) \pm (y/b) = 0$. A **rectangular** hyperbola is one whose asymptotes are perpendicular to each other.

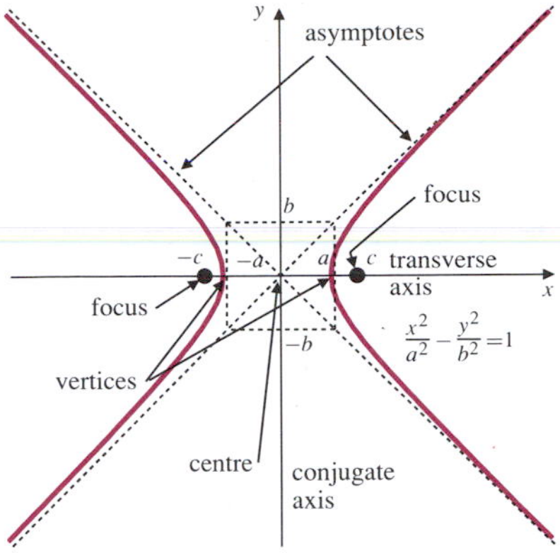

**Figure R.26** Terms associated with a hyperbola

$a$ is the **semi-transverse axis**

$b$ is the **semi-conjugate axis**

$c = \sqrt{a^2 + b^2}$ is the **semi-focal separation**

$\epsilon = \dfrac{c}{a} > 1$ is the **eccentricity**.

## Parametric Curves

A **parametric curve** $C$ in the plane consists of an ordered pair $(f, g)$ of continuous functions each defined on the same interval $I$. The equations

$$x = f(t), \qquad y = g(t), \qquad \text{for } t \text{ in } I$$

are called **the parametric equations** of the curve $C$. The independent variable $t$ is called the **parameter**. We often regard the parametric curve as being *the set of points* $\big(f(t), g(t)\big)$ corresponding to values of $t$ in $I$ rather than as the pair of functions itself.

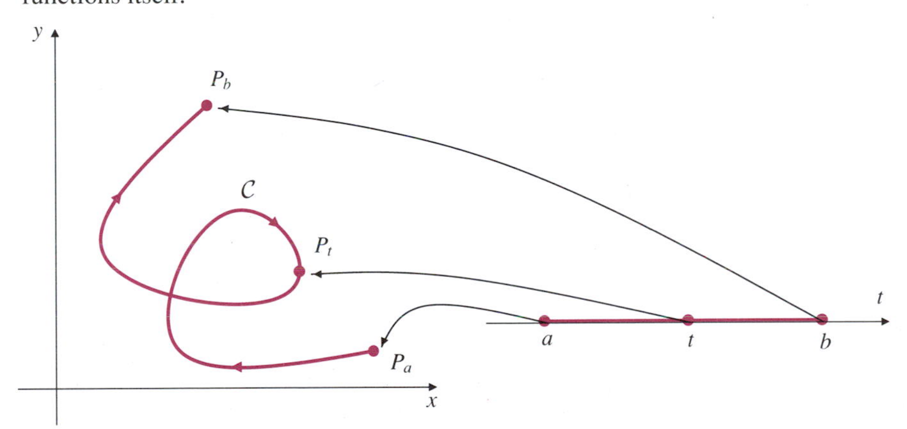


**Figure R.27** As $t$ increases from $a$ to $b$, the point $(x, y) = \big(f(t), g(t)\big)$ traces the curve $C$ in the $xy$-plane in the direction indicated

For example, the path followed by a particle moving in the $xy$-plane can be regarded as a parametric curve with parametric equations giving the position of the particle at time $t$.

The arc length element for a parametric curve is given by

$$ds = \sqrt{\left(\frac{dx}{dt}\right)^2 + \left(\frac{dy}{dt}\right)^2}\, dt,$$

so that the length of the curve $C$: $x = f(t)$, $y = g(t)$, $(a \le t \le b)$ is

$$s = \int_{t=a}^{t=b} ds = \int_a^b \sqrt{\big(f'(t)\big)^2 + \big(g'(t)\big)^2}\, dt.$$

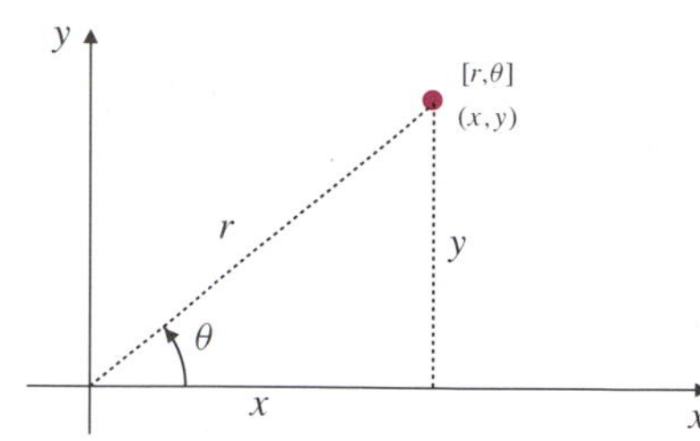


**Figure R.28** Relating Cartesian and polar coordinates of a point

## Polar Coordinates

Instead of using Cartesian coordinates $(x, y)$, we can locate a point in the plane by using **polar coordinates** $[r, \theta]$, where

(i) $r$ is the distance from $O$ to $P$, and

(ii) $\theta$ is the angle that the ray $OP$ makes with the positive $x$-axis (counterclockwise angles being considered positive).

The polar and Cartesian coordinates of a point are linked by the following equations:

$$x = r\cos\theta \qquad r^2 = x^2 + y^2$$
$$y = r\sin\theta \qquad \tan\theta = \frac{y}{x}.$$

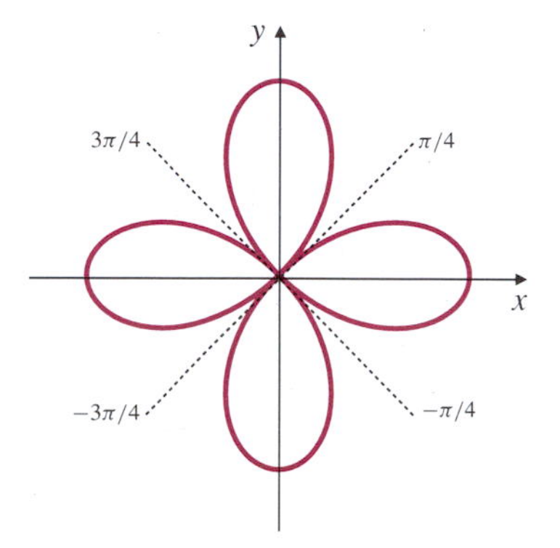


**Figure R.29** The polar curve $r = \cos(2\theta)$

The equations for $r^2$ and $\tan\theta$ cannot be solved uniquely for $r$ and $\theta$; the Cartesian coordinates of a point are unique, but its polar coordinates are not.

Polar graphs typically have equations of the form $r = f(\theta)$. Such a graph can approach the origin only from those directions $\theta$ for which $f(\theta) = 0$.

At any point $P$ other than the origin on the polar curve $r = f(\theta)$, the angle $\psi$ between the radial line from the origin to $P$ and the tangent to the curve satisfies

$$\tan\psi = \frac{f(\theta)}{f'(\theta)}.$$

In particular, $\psi = \pi/2$ if $f'(\theta) = 0$.

If $f(\theta_0) = 0$ and the curve has a tangent line at $\theta_0$ then that tangent line has polar equation $\theta = \theta_0$.

The region bounded by $r = f(\theta)$ and the rays $\theta = \alpha$ and $\theta = \beta$, $(\alpha < \beta)$, has area

$$A = \frac{1}{2}\int_{\alpha}^{\beta} \big(f(\theta)\big)^2\, d\theta.$$

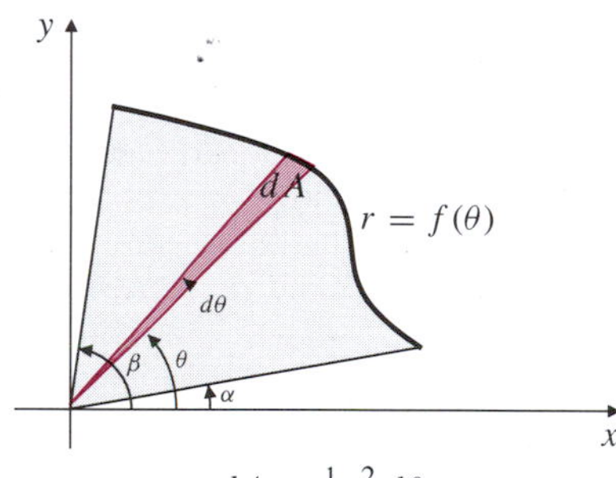


**Figure R.30** $dA = \frac{1}{2}r^2\, d\theta = \frac{1}{2}\big(f(\theta)\big)^2\, d\theta$

The arc length element for the polar curve $r = f(\theta)$ is

$$ds = \sqrt{\left(\frac{dr}{d\theta}\right)^2 + r^2}\, d\theta = \sqrt{\big(f'(\theta)\big)^2 + \big(f(\theta)\big)^2}\, d\theta.$$

CHAPTER 1

# Sequences and Series

**INTRODUCTION** An infinite series is a sum that involves infinitely many terms. Since addition is carried out on two numbers at a time, the evaluation of the sum of an infinite series necessarily involves finding a limit. Complicated functions $f(x)$ can frequently be expressed as series of simpler functions. For example, many of the transcendental functions we have encountered can be expressed as series of powers of $x$ so that they resemble polynomials of infinite degree. Such series can be differentiated and integrated term by term, and they play a very important role in the study of calculus. This role is developed in Chapter 2.

Some of the material in this chapter can be omitted without hindering the understanding of power series in Chapter 2. However, you should understand the concepts of sequence, series, and convergence as presented in Sections 1.1 and 1.2, and be familiar with some elementary series such as geometric series and harmonic series (Section 1.2) and $p$-series (Section 1.3). The ratio test (Section 1.3) and the concept of absolute convergence and the alternating series test (Section 1.4) will also be used.

The discussion of infinite sequences and series of real numbers presented in this chapter involves notions of limit and convergence that are similar to, but somewhat simpler than, those for functions and integrals. Accordingly, we proceed with slightly more rigour in developing them.

## 1.1 SEQUENCES AND CONVERGENCE

By a **sequence** (or an **infinite sequence**) we mean an ordered list having a first element but no last element. For our purposes the elements (called **terms**) of a sequence will always be real numbers, though much of our discussion could be applied to complex numbers as well. Examples of sequences are:

$\{1, 2, 3, 4, 5, \ldots\}$ the sequence of positive integers,

$\left\{-\frac{1}{2}, \frac{1}{4}, -\frac{1}{8}, \frac{1}{16}, \ldots\right\}$ the sequence of positive integer powers of $-\frac{1}{2}$.

The terms of a sequence are usually listed in braces { } as shown. The ellipsis (...) should be read "and so on."

An infinite sequence is a special kind of function, one whose domain is a set of integers extending from some starting integer to infinity. The starting integer is usually 1, so the domain is the set of positive integers. The sequence $\{a_1, a_2, a_3, a_4, \ldots\}$ is the function $f$ that takes the value $f(n) = a_n$ at each positive integer $n$. A sequence can be specified in three ways:

(i) We can list the first few terms followed by ... *if the pattern is obvious.*

(ii) We can provide a formula for the **general term** $a_n$ as a function of $n$.

(iii) We can provide a formula for calculating the term $a_n$ as a function of earlier terms $a_1, a_2, \ldots, a_{n-1}$ and specify enough of the beginning terms so the process of computing higher terms can begin.

In each case it must be possible to determine any term of the sequence, though it may be necessary to calculate all the preceding terms first.

**EXAMPLE 1** **Some examples of sequences**

(a) $\{n\} = \{1, 2, 3, 4, 5, \ldots\}$

(b) $\left\{\left(-\frac{1}{2}\right)^n\right\} = \left\{-\frac{1}{2}, \frac{1}{4}, -\frac{1}{8}, \frac{1}{16}, \ldots\right\}$

(c) $\left\{\frac{n-1}{n}\right\} = \left\{0, \frac{1}{2}, \frac{2}{3}, \frac{3}{4}, \frac{4}{5}, \ldots\right\}$

(d) $\{(-1)^{n-1}\} = \{\cos((n-1)\pi)\} = \{1, -1, 1, -1, 1, \ldots\}$

(e) $\left\{\frac{n^2}{2^n}\right\} = \left\{\frac{1}{2}, 1, \frac{9}{8}, 1, \frac{25}{32}, \frac{36}{64}, \frac{49}{128}, \ldots\right\}$

(f) $\left\{\left(1+\frac{1}{n}\right)^n\right\} = \left\{2, \left(\frac{3}{2}\right)^2, \left(\frac{4}{3}\right)^3, \left(\frac{5}{4}\right)^4, \ldots\right\}$

(g) $\left\{\frac{\cos(n\pi/2)}{n}\right\} = \left\{0, -\frac{1}{2}, 0, \frac{1}{4}, 0, -\frac{1}{6}, 0, \frac{1}{8}, 0, \ldots\right\}$

(h) $a_1 = 1, \quad a_{n+1} = \sqrt{6 + a_n}, \quad (n = 1, 2, 3, \ldots)$.
In this case $\{a_n\} = \{1, \sqrt{7}, \sqrt{6+\sqrt{7}}, \ldots\}$.
Note that there is no *obvious* formula for $a_n$ as an explicit function of $n$ here, but we can still calculate $a_n$ for any desired value of $n$ provided we first calculate all the earlier values $a_2, a_3, \ldots, a_{n-1}$.

(i) $a_1 = 1, a_2 = 1, a_{n+2} = a_n + a_{n+1}, \quad (n = 1, 2, 3, \ldots)$
Here $\{a_n\} = \{1, 1, 2, 3, 5, 8, 13, 21, \ldots\}$.
This is called the **Fibonacci sequence**. Each term after the second is the sum of the previous two terms. ■

In parts (a)–(g) of Example 1, the formulas on the left sides define the general term of each sequence $\{a_n\}$ as an explicit function of $n$. In parts (h) and (i) we say the sequence $\{a_n\}$ is defined **recursively** or **inductively**; each term must be calculated from previous ones rather than directly as a function of $n$.

The following definitions introduce terminology used to describe various properties of sequences.

**DEFINITION 1**

**Terms for describing sequences**

(a) The sequence $\{a_n\}$ is **bounded below** by $L$, and $L$ is a **lower bound** for $\{a_n\}$, if $a_n \ge L$ for every $n = 1, 2, 3, \ldots$. The sequence is **bounded above** by $M$, and $M$ is an **upper bound**, if $a_n \le M$ for every such $n$.

The sequence $\{a_n\}$ is **bounded** if it is both bounded above and bounded below. In this case there is a constant $K$ such that $|a_n| \le K$ for every $n = 1, 2, 3, \ldots$. (We can take $K$ to be the larger of $-L$ and $M$.)

(b) The sequence $\{a_n\}$ is **positive** if it is bounded below by zero, that is, if $a_n \ge 0$ for every $n = 1, 2, 3, \ldots$; it is **negative** if $a_n \le 0$ for every $n$.

(c) The sequence $\{a_n\}$ is **increasing** if $a_{n+1} \ge a_n$ for every $n = 1, 2, 3, \ldots$; it is **decreasing** if $a_{n+1} \le a_n$ for every such $n$. The sequence is said to be **monotonic** if it is either increasing or decreasing. (Note that the terminology here is looser than that we have used for functions, where we would have used *nondecreasing* and *nonincreasing* to describe this behaviour.)

(d) The sequence $\{a_n\}$ is **alternating** if $a_n a_{n+1} < 0$ for every $n = 1, 2, \ldots$, that is, if any two consecutive terms have opposite sign.

**EXAMPLE 2** **Describing some sequences**

(a) The sequence $\{n\} = \{1, 2, 3, \ldots\}$ is positive, increasing, and bounded below. A lower bound for the sequence is 1, or any smaller number. The sequence is not bounded above.

(b) $\left\{\dfrac{n-1}{n}\right\} = \left\{0, \dfrac{1}{2}, \dfrac{2}{3}, \dfrac{3}{4}, \ldots\right\}$ is positive, bounded, and increasing. Here 0 is a lower bound and 1 is an upper bound.

(c) $\left\{\left(-\dfrac{1}{2}\right)^n\right\} = \left\{-\dfrac{1}{2}, \dfrac{1}{4}, -\dfrac{1}{8}, \dfrac{1}{16}, \ldots\right\}$ is bounded and alternating. Here $-1/2$ is a lower bound and $1/4$ is an upper bound.

(d) $\{(-1)^n n\} = \{-1, 2, -3, 4, -5, \ldots\}$ is alternating but not bounded either above or below. ■

When you want to show that a sequence is increasing you can try to show that the inequality $a_{n+1} - a_n \ge 0$ holds for $n \ge 1$. Alternatively, if $a_n = f(n)$ for a differentiable function $f(x)$, you can show that $f$ is an increasing function on $[1, \infty)$ by showing that $f'(x) \ge 0$ there. Similar approaches are useful for showing that a sequence is decreasing.

**EXAMPLE 3** If $a_n = \dfrac{n}{n^2+1}$, show that the sequence $\{a_n\}$ is decreasing.

***SOLUTION*** Since $a_n = f(n)$, where $f(x) = \dfrac{x}{x^2+1}$ and

$$f'(x) = \frac{(x^2+1)(1) - x(2x)}{(x^2+1)^2} = \frac{1-x^2}{(x^2+1)^2} \le 0 \quad \text{for } x \ge 1,$$

the function $f(x)$ is decreasing on $[1, \infty)$ and therefore $\{a_n\}$ is a decreasing sequence. ■

The sequence $\left\{\frac{n^2}{2^n}\right\} = \left\{\frac{1}{2}, 1, \frac{9}{8}, 1, \frac{25}{32}, \frac{36}{64}, \frac{49}{128}, \ldots\right\}$ is obviously positive therefore bounded below. It seems clear that from the fourth term on, all the terms are getting smaller. However, $a_2 > a_1$ and $a_3 > a_2$. Since $a_{n+1} \leq a_n$ only if $n \geq 3$, we say that this sequence is **ultimately decreasing**. The adverb *ultimately* is used to describe any termwise property of a sequence that the terms have from some point on, but not necessarily at the beginning of the sequence. Thus the sequence

$$\{n - 100\} = \{-99, -98, \ldots, -2, -1, 0, 1, 2, 3, \ldots\}$$

is *ultimately positive* even though the first 99 terms are negative, and the sequence

$$\left\{(-1)^n + \frac{4}{n}\right\} = \left\{3, 3, \frac{1}{3}, 2, -\frac{1}{5}, \frac{5}{3}, -\frac{3}{7}, \frac{3}{2}, \ldots\right\}$$

is *ultimately alternating* even though the first few terms do not alternate.

## Convergence of Sequences

Central to the study of sequences is the notion of convergence. The concept of the limit of a sequence is a special case of the concept of the limit of a function $f(x)$ as $x \to \infty$. We say that the sequence $\{a_n\}$ **converges to the limit** $L$, and we write $\lim_{n\to\infty} a_n = L$, provided the distance from $a_n$ to $L$ on the real line approaches 0 as $n$ increases toward $\infty$. We state this definition more formally as follows:

**DEFINITION 2**

**Limit of a sequence**

We say that sequence $\{a_n\}$ converges to the limit $L$, and we write $\lim_{n\to\infty} a_n = L$, if for every positive real number $\epsilon$ there exists an integer $N$ (which may depend on $\epsilon$) such that if $n > N$, then $|a_n - L| < \epsilon$.

This definition is illustrated in Figure 1.1.

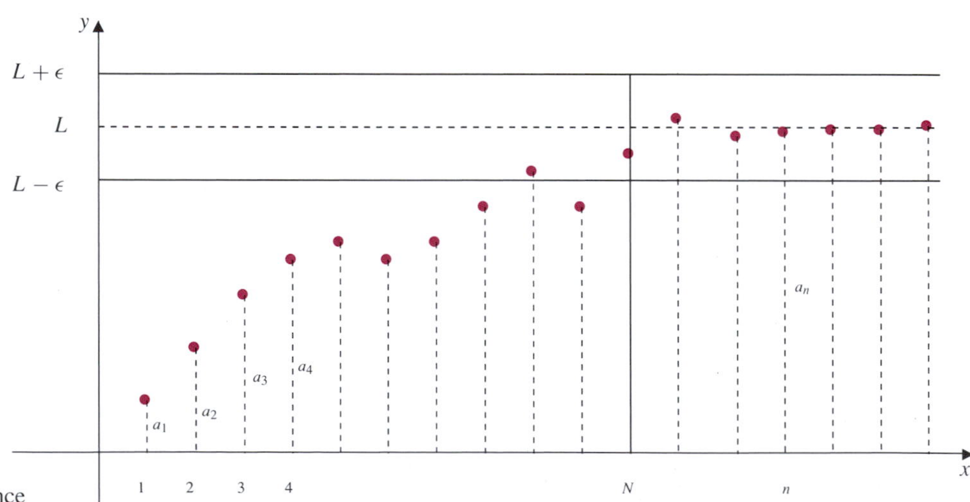


**Figure 1.1** A convergent sequence

**EXAMPLE 4** Show that $\lim_{n\to\infty} \frac{c}{n^p} = 0$ for any real number $c$ and any $p > 0$.

**SOLUTION** Let $\epsilon > 0$ be given. Then

$$\left|\frac{c}{n^p}\right| < \epsilon \quad \text{if} \quad n^p > \frac{|c|}{\epsilon},$$

that is, if $n > N = \left(\frac{|c|}{\epsilon}\right)^{1/p}$. By Definition 2, $\lim_{n\to\infty} \frac{c}{n^p} = 0$. ■

Every sequence $\{a_n\}$ must either **converge** to a finite limit $L$ or **diverge**. That is, either $\lim_{n\to\infty} a_n = L$ exists (is a real number), or $\lim_{n\to\infty} a_n$ does not exist. If $\lim_{n\to\infty} a_n = \infty$, we can say that the sequence diverges to $\infty$; if $\lim_{n\to\infty} a_n = -\infty$, we can say that it diverges to $-\infty$. If $\lim_{n\to\infty} a_n$ simply does not exist (but is not $\infty$ or $-\infty$), we can only say that the sequence diverges.

### ■ EXAMPLE 5 Examples of convergent and divergent sequences

(a) $\{(n-1)/n\}$ converges to 1; $\lim_{n\to\infty}(n-1)/n = \lim_{n\to\infty}\big(1-(1/n)\big) = 1$.

(b) $\{n\} = \{1, 2, 3, 4, \ldots\}$ diverges to $\infty$.

(c) $\{-n\} = \{-1, -2, -3, -4, \ldots\}$ diverges to $-\infty$.

(d) $\{(-1)^n\} = \{-1, 1, -1, 1, -1, \ldots\}$ simply diverges.

(e) $\{(-1)^n n\} = \{-1, 2, -3, 4, -5, \ldots\}$ diverges (but not to $\infty$ or $-\infty$ even though $\lim_{n\to\infty} |a_n| = \infty$). ■

The limit of a sequence is equivalent to the limit of a function as its argument approaches infinity:

$$\text{If } \lim_{x\to\infty} f(x) = L \text{ and } a_n = f(n), \text{ then } \lim_{n\to\infty} a_n = L.$$

Because of this, the standard rules for limits of functions also hold for limits of sequences, with the appropriate changes of notation. Thus, if $\{a_n\}$ and $\{b_n\}$ converge, then

$$\lim_{n\to\infty}(a_n \pm b_n) = \lim_{n\to\infty} a_n \pm \lim_{n\to\infty} b_n,$$

$$\lim_{n\to\infty} ca_n = c \lim_{n\to\infty} a_n,$$

$$\lim_{n\to\infty} a_n b_n = (\lim_{n\to\infty} a_n)(\lim_{n\to\infty} b_n),$$

$$\lim_{n\to\infty} \frac{a_n}{b_n} = \frac{\lim_{n\to\infty} a_n}{\lim_{n\to\infty} b_n} \qquad \text{assuming } \lim_{n\to\infty} b_n \neq 0.$$

If $a_n \le b_n$ ultimately, then $\lim_{n\to\infty} a_n \le \lim_{n\to\infty} b_n$.

If $a_n \le b_n \le c_n$ ultimately, and $\lim_{n\to\infty} a_n = L = \lim_{n\to\infty} c_n$, then $\lim_{n\to\infty} b_n = L$.

The limits of many explicitly defined sequences can be evaluated using these properties in a manner similar to the methods used for limits of the form $\lim_{x\to\infty} f(x)$.

### ■ EXAMPLE 6 Calculate the limits of the sequences

$$\text{(a) } \left\{\frac{2n^2-n-1}{5n^2+n-3}\right\}, \quad \text{(b) } \left\{\frac{\cos n}{n}\right\}, \quad \text{(c) } \{\sqrt{n^2+2n}-n\}.$$

***SOLUTION***

(a) We divide the numerator and denominator of the expression for $a_n$ by the highest power of $n$ in the denominator, that is, by $n^2$:

$$\lim_{n\to\infty}\frac{2n^2-n-1}{5n^2+n-3}=\lim_{n\to\infty}\frac{2-(1/n)-(1/n^2)}{5+(1/n)-(3/n^2)}=\frac{2-0-0}{5+0-0}=\frac{2}{5},$$

since $\lim_{n\to\infty} 1/n = 0$ and $\lim_{n\to\infty} 1/n^2 = 0$. The sequence converges and its limit is 2/5.

(b) Since $|\cos n| \le 1$ for every $n$, we have

$$-\frac{1}{n}\le\frac{\cos n}{n}\le\frac{1}{n}\quad\text{for}\quad n\ge 1$$

Now $\lim_{n\to\infty} -1/n = 0$ and $\lim_{n\to\infty} 1/n = 0$, so by the sequence version of the Squeeze Theorem, $\lim_{n\to\infty}(\cos n)/n = 0$ also. The sequence converges to 0.

(c) For this sequence we multiply the numerator and the denominator (which is 1) by the conjugate of the expression in the numerator:

$$\lim_{n\to\infty}(\sqrt{n^2+2n}-n)=\lim_{n\to\infty}\frac{(\sqrt{n^2+2n}-n)(\sqrt{n^2+2n}+n)}{\sqrt{n^2+2n}+n}$$

$$=\lim_{n\to\infty}\frac{2n}{\sqrt{n^2+2n}+n}=\lim_{n\to\infty}\frac{2}{\sqrt{1+(2/n)}+1}=1.$$

The sequence converges to 1. ■

■ **EXAMPLE 7** Evaluate $\lim_{n\to\infty} n\tan^{-1}\left(\frac{1}{n}\right)$.

***SOLUTION*** For this example it is best to replace the $n$th term of the sequence by the corresponding function of a real variable $x$ and take the limit as $x\to\infty$. We use l'Hôpital's rule:

$$\lim_{n\to\infty} n\tan^{-1}\left(\frac{1}{n}\right)=\lim_{x\to\infty} x\tan^{-1}\left(\frac{1}{x}\right)$$

$$=\lim_{x\to\infty}\frac{\tan^{-1}\left(\frac{1}{x}\right)}{\frac{1}{x}}\qquad\left[\frac{0}{0}\right]$$

$$=\lim_{x\to\infty}\frac{\frac{1}{1+(1/x^2)}\left(-\frac{1}{x^2}\right)}{-\left(\frac{1}{x^2}\right)}=\lim_{x\to\infty}\frac{1}{1+\frac{1}{x^2}}=1.$$ ■

**THEOREM 1** If $\{a_n\}$ converges, then $\{a_n\}$ is bounded.

***PROOF*** Suppose $\lim_{n\to\infty} a_n = L$. According to Definition 2, for $\epsilon = 1$ there exists a number $N$ such that if $n \geq N$ then $|a_n - L| < 1$; therefore $|a_n| < 1 + |L|$ for such $n$. (Why is this true?) If $K$ denotes the largest of the numbers $|a_1|$, $|a_2|$, ..., $|a_{N-1}|$, and $1 + |L|$, then $|a_n| \leq K$ for every $n = 1, 2, 3, \ldots$. Hence $\{a_n\}$ is bounded.

The converse of Theorem 1 is false; the sequence $\{(-1)^n\}$ is bounded but does not converge.

The *completeness property* of the real number system can be reformulated in terms of sequences to read as follows:

**Bounded monotonic sequences converge**

If the sequence $\{a_n\}$ is bounded above and is (ultimately) increasing, then it converges. The same conclusion holds if $\{a_n\}$ is bounded below and is (ultimately) decreasing.

Thus, a bounded, ultimately monotonic sequence is convergent. (See Figure 1.2.)

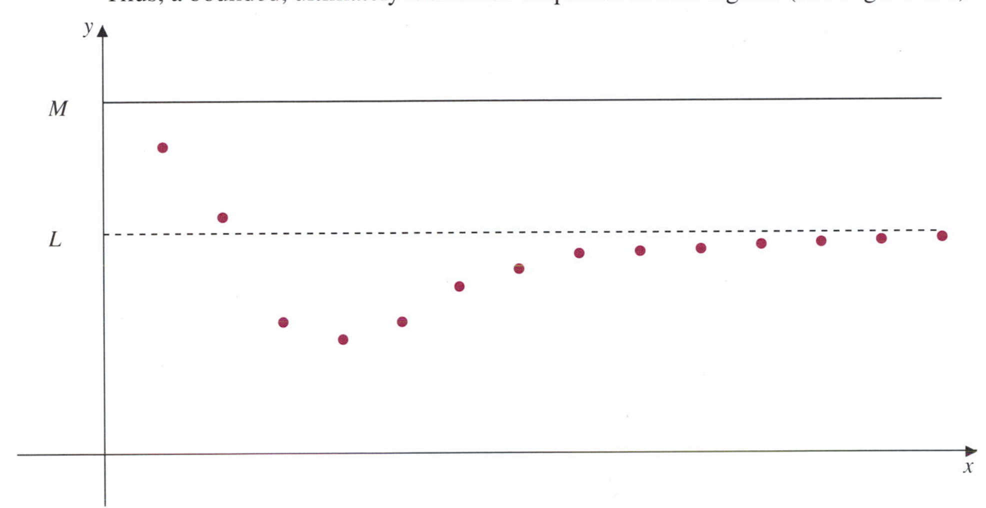


**Figure 1.2** An ultimately increasing sequence that is bounded above

**EXAMPLE 8** Let $a_n$ be defined recursively by

$$a_1 = 1, \qquad a_{n+1} = \sqrt{6 + a_n} \qquad (n = 1, 2, 3, \ldots).$$

Show that $\lim_{n\to\infty} a_n$ exists and find its value.

***SOLUTION*** Observe that $a_2 = \sqrt{6+1} = \sqrt{7} > a_1$. If $a_{k+1} > a_k$, then $a_{k+2} = \sqrt{6 + a_{k+1}} > \sqrt{6 + a_k} = a_{k+1}$, so $\{a_n\}$ is increasing, by induction. Now observe that $a_1 = 1 < 3$. If $a_k < 3$, then $a_{k+1} = \sqrt{6 + a_k} < \sqrt{6+3} = 3$, so $a_n < 3$ for every $n$ by induction. Since $\{a_n\}$ is increasing and bounded above, $\lim_{n\to\infty} a_n = a$ exists, by completeness. Since $\sqrt{6+x}$ is a continuous function of $x$, we have

$$a = \lim_{n\to\infty} a_{n+1} = \lim_{n\to\infty} \sqrt{6 + a_n} = \sqrt{6 + \lim_{n\to\infty} a_n} = \sqrt{6 + a}.$$

Thus $a^2 = 6 + a$, or $a^2 - a - 6 = 0$, or $(a-3)(a+2) = 0$. This quadratic has roots $a = 3$ and $a = -2$. Since $a_n \geq 1$ for every $n$, we must have $a \geq 1$. Therefore, $a = 3$ and $\lim_{n\to\infty} a_n = 3$. ■

**EXAMPLE 9** Does $\left\{\left(1+\frac{1}{n}\right)^n\right\}$ converge or not?

***SOLUTION*** We could make an effort to show that the given sequence is, in fact, increasing and bounded above. (See Exercise 32 at the end of this section.) However, we already know the answer.

$$\lim_{n\to\infty}\left(1+\frac{1}{n}\right)^n = e^1 = e.$$

**THEOREM 2** If $\{a_n\}$ is (ultimately) increasing, then either it is bounded above, and therefore convergent, or else it is not bounded above and diverges to infinity.

The proof of this theorem is left as an exercise. A corresponding result holds for (ultimately) decreasing sequences.

The following theorem evaluates two important limits that find frequent application in the study of series.

**THEOREM 3**

(a) If $|x| < 1$, then $\lim_{n\to\infty} x^n = 0$.

(b) If $x$ is any real number, then $\lim_{n\to\infty} \frac{x^n}{n!} = 0$.

***PROOF*** For part (a) observe that

$$\lim_{n\to\infty} \ln |x|^n = \lim_{n\to\infty} n \ln |x| = -\infty,$$

since $\ln |x| < 0$ when $|x| < 1$. Accordingly, since $e^x$ is continuous,

$$\lim_{n\to\infty} |x|^n = \lim_{n\to\infty} e^{\ln |x|^n} = e^{\lim_{n\to\infty} \ln |x|^n} = 0.$$

Since $-|x|^n \le x^n \le |x|^n$, we have $\lim_{n\to\infty} x^n = 0$ by the Squeeze Theorem.

For part (b), pick any $x$ and let $N$ be an integer such that $N > |x|$. If $n > N$ we have

$$\begin{aligned}\left|\frac{x^n}{n!}\right| &= \frac{|x|}{1}\,\frac{|x|}{2}\,\frac{|x|}{3}\cdots\frac{|x|}{N-1}\,\frac{|x|}{N}\,\frac{|x|}{N+1}\cdots\frac{|x|}{n}\\ &< \frac{|x|^{N-1}}{(N-1)!}\,\frac{|x|}{N}\,\frac{|x|}{N}\,\frac{|x|}{N}\cdots\frac{|x|}{N}\\ &= \frac{|x|^{N-1}}{(N-1)!}\left(\frac{|x|}{N}\right)^{n-N+1} = K\left(\frac{|x|}{N}\right)^n,\end{aligned}$$

where $K = \frac{|x|^{N-1}}{(N-1)!}\left(\frac{|x|}{N}\right)^{1-N}$ is a constant that is independent of $n$. Since $|x|/N < 1$, we have $\lim_{n\to\infty}(|x|/N)^n = 0$ by part (a). Thus $\lim_{n\to\infty} |x^n/n!| = 0$, and so $\lim_{n\to\infty} x^n/n! = 0$.

**EXAMPLE 10** Find $\lim_{n\to\infty} \frac{3^n + 4^n + 5^n}{5^n}$.

***SOLUTION*** $\lim_{n\to\infty} \frac{3^n + 4^n + 5^n}{5^n} = \lim_{n\to\infty}\left[\left(\frac{3}{5}\right)^n + \left(\frac{4}{5}\right)^n + 1\right] = 0 + 0 + 1 = 1,$ by Theorem 3(a).

## EXERCISES 1.1

In Exercises 1–13, determine whether the given sequence is (a) bounded (above or below), (b) positive or negative (ultimately), (c) increasing, decreasing, or alternating, and (d) convergent, divergent, divergent to $\infty$ or $-\infty$.

**1.** $\left\{\dfrac{2n^2}{n^2+1}\right\}$

**2.** $\left\{\dfrac{2n}{n^2+1}\right\}$

**3.** $\left\{4-\dfrac{(-1)^n}{n}\right\}$

**4.** $\left\{\sin\dfrac{1}{n}\right\}$

**5.** $\left\{\dfrac{n^2-1}{n}\right\}$

**6.** $\left\{\dfrac{e^n}{\pi^n}\right\}$

**7.** $\left\{\dfrac{e^n}{\pi^{n/2}}\right\}$

**8.** $\left\{\dfrac{(-1)^n n}{e^n}\right\}$

**9.** $\left\{\dfrac{2^n}{n^n}\right\}$

**10.** $\left\{\dfrac{(n!)^2}{(2n)!}\right\}$

**11.** $\left\{n\cos\left(\dfrac{n\pi}{2}\right)\right\}$

**12.** $\left\{\dfrac{\sin n}{n}\right\}$

**13.** $\{1, 1, -2, 3, 3, -4, 5, 5, -6, \ldots\}$

In Exercises 14–29, evaluate, wherever possible, the limit of the sequence $\{a_n\}$.

**14.** $a_n = \dfrac{5-2n}{3n-7}$

**15.** $a_n = \dfrac{n^2-4}{n+5}$

**16.** $a_n = \dfrac{n^2}{n^3+1}$

**17.** $a_n = (-1)^n\dfrac{n}{n^3+1}$

**18.** $a_n = \dfrac{n^2-2\sqrt{n}+1}{1-n-3n^2}$

**19.** $a_n = \dfrac{e^n-e^{-n}}{e^n+e^{-n}}$

**20.** $a_n = n\sin\dfrac{1}{n}$

**21.** $a_n = \left(\dfrac{n-3}{n}\right)^n$

**22.** $a_n = \dfrac{n}{\ln(n+1)}$

**23.** $a_n = \sqrt{n+1}-\sqrt{n}$

**24.** $a_n = n-\sqrt{n^2-4n}$

**25.** $a_n = \sqrt{n^2+n}-\sqrt{n^2-1}$

**26.** $a_n = \left(\dfrac{n-1}{n+1}\right)^n$

**27.** $a_n = \dfrac{(n!)^2}{(2n)!}$

**28.** $a_n = \dfrac{n^2 2^n}{n!}$

**29.** $a_n = \dfrac{\pi^n}{1+2^{2n}}$

**30.** Let $a_1 = 1$ and $a_{n+1} = \sqrt{1+2a_n}$ $(n = 1, 2, 3, \ldots)$. Show that $\{a_n\}$ is increasing and bounded above. (*Hint:* show that 3 is an upper bound.) Hence conclude that the sequence converges, and find its limit.

* **31.** Repeat Exercise 30 for the sequence defined by $a_1 = 3$, $a_{n+1} = \sqrt{15+2a_n}$, $n = 1, 2, 3, \ldots$. This time you will have to guess an upper bound.

* **32.** Let $a_n = \left(1+\dfrac{1}{n}\right)^n$ so that $\ln a_n = n\ln\left(1+\dfrac{1}{n}\right)$. Use properties of the logarithm function to show that (a) $\{a_n\}$ is increasing and (b) $e$ is an upper bound for $\{a_n\}$.

* **33.** Prove Theorem 2. Also, state an analogous theorem pertaining to ultimately decreasing sequences.

**34.** If $\{|a_n|\}$ is bounded, prove that $\{a_n\}$ is bounded.

**35.** If $\lim_{n\to\infty}|a_n| = 0$, prove that $\lim_{n\to\infty}a_n = 0$.

* **36.** Which of the following statements are TRUE and which are FALSE? Justify your answers.

(a) If $\lim_{n\to\infty}a_n = \infty$ and $\lim_{n\to\infty}b_n = L > 0$, then $\lim_{n\to\infty}a_n b_n = \infty$.

(b) If $\lim_{n\to\infty}a_n = \infty$ and $\lim_{n\to\infty}b_n = -\infty$, then $\lim_{n\to\infty}(a_n+b_n) = 0$.

(c) If $\lim_{n\to\infty}a_n = \infty$ and $\lim_{n\to\infty}b_n = -\infty$, then $\lim_{n\to\infty}a_n b_n = -\infty$.

(d) If neither $\{a_n\}$ nor $\{b_n\}$ converges, then $\{a_n b_n\}$ does not converge.

(e) If $\{|a_n|\}$ converges, then $\{a_n\}$ converges.

## 1.2 INFINITE SERIES

An **infinite series**, usually just called a **series**, is a formal sum of infinitely many terms; for instance,

$$a_1+a_2+a_3+a_4+\cdots$$

is a series formed by adding the terms of the sequence $\{a_n\}$. This series is also denoted $\sum_{n=1}^{\infty}a_n$:

$$\sum_{n=1}^{\infty}a_n = a_1+a_2+a_3+a_4+\cdots$$

For example,

$$\sum_{n=1}^{\infty} \frac{1}{n} = 1 + \frac{1}{2} + \frac{1}{3} + \frac{1}{4} + \cdots$$
$$\sum_{n=1}^{\infty} \frac{(-1)^{n-1}}{2^{n-1}} = 1 - \frac{1}{2} + \frac{1}{4} - \frac{1}{8} + \frac{1}{16} - \cdots .$$

It is sometimes necessary or useful to start the sum from some index other than 1:

$$\sum_{n=0}^{\infty} a^n = 1 + a + a^2 + a^3 + \cdots$$
$$\sum_{n=2}^{\infty} \frac{1}{\ln n} = \frac{1}{\ln 2} + \frac{1}{\ln 3} + \frac{1}{\ln 4} + \cdots .$$

Note that the latter series would make no sense if we had started the sum from $n = 1$; the first term would have been undefined.

When necessary, we can change the index of summation to start at a different value. For instance, using the substitution $n = m - 2$, we can rewrite $\sum_{n=1}^{\infty} a_n$ in the form $\sum_{m=3}^{\infty} a_{m-2}$. Both sums give rise to the same expansion

$$\sum_{n=1}^{\infty} a_n = a_1 + a_2 + a_3 + \cdots = \sum_{m=3}^{\infty} a_{m-2}.$$

Addition is an operation that is carried out on two numbers at a time. If we want to calculate the finite sum

$$a_1 + a_2 + a_3,$$

we could proceed by adding $a_1 + a_2$ and then adding $a_3$ to this sum, or else we might first add $a_2 + a_3$ and then add $a_1$ to the sum. Of course the associative law for addition assures us we will get the same answer both ways. This is the reason the symbol $a_1 + a_2 + a_3$ makes sense; we would otherwise have to write $(a_1 + a_2) + a_3$ or $a_1 + (a_2 + a_3)$. This reasoning extends to any sum $a_1 + a_2 + \cdots + a_n$ of finitely many terms, but it is not obvious what should be meant by a sum with infinitely many terms:

$$a_1 + a_2 + a_3 + a_4 + \cdots .$$

We no longer have any assurance that the terms can be added up in any order to yield the same sum. In fact, we will see in Section 1.4 that in certain circumstances, changing the order of terms in a series can actually change the sum of the series. The interpretation we place on the infinite sum is that of adding from left to right, as suggested by the grouping

$$\cdots ((((a_1 + a_2) + a_3) + a_4) + a_5) + \cdots .$$

We accomplish this by defining a new sequence $\{s_n\}$, called the **sequence of partial sums** of the series $\sum_{n=1}^{\infty} a_n$ so that $s_n$ is the sum of the first $n$ terms of the series:

$$\begin{aligned}
s_1 &= a_1 \\
s_2 &= s_1 + a_2 = a_1 + a_2 \\
s_3 &= s_2 + a_3 = a_1 + a_2 + a_3 \\
&\vdots \\
s_n &= s_{n-1} + a_n = a_1 + a_2 + a_3 + \cdots + a_n = \sum_{j=1}^{n} a_j. \\
&\vdots
\end{aligned}$$

We then define the sum of the infinite series to be the limit of this sequence of partial sums.

**DEFINITION 3**

**Convergence of a series**

We say that the series $\sum_{n=1}^{\infty} a_n$ **converges to the sum** $s$, and we write

$$\sum_{n=1}^{\infty} a_n = s,$$

if $\lim_{n\to\infty} s_n = s$, where $s_n$ is the $n$th partial sum of $\sum_{n=1}^{\infty} a_n$:

$$s_n = a_1 + a_2 + a_3 + \cdots + a_n = \sum_{j=1}^{n} a_j.$$

Thus, the *series* $\sum_{n=1}^{\infty} a_n$ converges if and only if the *sequence* $\{s_n\}$ of its partial sums converges.

Similarly, a series is said to diverge to infinity, diverge to negative infinity, or simply diverge, if its sequence of partial sums does so. It must be stressed that the convergence of the series $\sum_{n=1}^{\infty} a_n$ depends on the convergence of the sequence $\{s_n\} = \{\sum_{j=1}^{n} a_j\}$, *not* the sequence $\{a_n\}$.

## Geometric Series

**DEFINITION 4**

**Geometric series**

A series of the form

$$\sum_{n=1}^{\infty} a\,r^{n-1} = a + ar + ar^2 + ar^3 + \cdots,$$

whose $n$th term is $a_n = a\,r^{n-1}$ is called a **geometric series**. The number $a$ is the first term. The number $r$ is called the **common ratio of** the series since it is the value of the ratio of the $(n+1)$st term to the $n$th term for any $n \geq 1$:

$$\frac{a_{n+1}}{a_n} = \frac{ar^n}{ar^{n-1}} = r \qquad n = 1, 2, 3, \ldots.$$

The $n$th partial sum $s_n$ of a geometric series is calculated as follows:

$$\begin{aligned} s_n &= a + ar + ar^2 + ar^3 + \cdots + ar^{n-1} \\ rs_n &= \phantom{a + {}} ar + ar^2 + ar^3 + \cdots + ar^{n-1} + ar^n. \end{aligned}$$

The second equation is obtained by multiplying the first by $r$. Subtracting these two equations (note the cancellations), we get $(1 - r)s_n = a - ar^n$. If $r \neq 1$, we can divide by $1 - r$ and get a formula for $s_n$.

**Partial sums of geometric series**

If $r \neq 1$, then the $n$th partial sum of a geometric series $\sum_{n=1}^{\infty} ar^{n-1}$ is

$$s_n = a + ar + ar^2 + \cdots + ar^{n-1} = \frac{a(1 - r^n)}{1 - r}.$$

If $r = 1$, we have, instead, $s_n = na$.

If $a = 0$ then $s_n = 0$ for every $n$, and $\lim_{n\to\infty} s_n = 0$. Now suppose $a \neq 0$. If $|r| < 1$, then $\lim_{n\to\infty} r^n = 0$, so $\lim_{n\to\infty} s_n = a/(1 - r)$. If $r > 1$, then $\lim_{n\to\infty} r^n = \infty$, and $\lim_{n\to\infty} s_n = \infty$ if $a > 0$, or $\lim_{n\to\infty} s_n = -\infty$ if $a < 0$. The same conclusion holds if $r = 1$, since $s_n = na$ in this case. If $r \leq -1$, $\lim_{n\to\infty} r^n$ does not exist and neither does $\lim_{n\to\infty} s_n$. Hence we conclude that

$$\sum_{n=1}^{\infty} ar^{n-1} \begin{cases} \text{converges to } 0 & \text{if } a = 0 \\ \text{converges to } \dfrac{a}{1-r} & \text{if } |r| < 1 \\ \text{diverges to } \infty & \text{if } r \geq 1 \text{ and } a > 0 \\ \text{diverges to } -\infty & \text{if } r \geq 1 \text{ and } a < 0 \\ \text{diverges} & \text{if } r \leq -1. \end{cases}$$

The representation of the function $1/(1 - x)$ as the sum of a geometric series,

$$\frac{1}{1-x} = \sum_{n=0}^{\infty} x^n = 1 + x + x^2 + x^3 + \cdots \quad \text{for } -1 < x < 1,$$

will play an important role in our discussion of power series in Chapter 2.

**EXAMPLE 1 Examples of geometric series and their sums**

(a) $1 + \dfrac{1}{2} + \dfrac{1}{4} + \dfrac{1}{8} + \cdots = \displaystyle\sum_{n=1}^{\infty} \left(\frac{1}{2}\right)^{n-1} = \frac{1}{1 - \frac{1}{2}} = 2$. Here $a = 1$ and $r = \dfrac{1}{2}$.

Since $|r| < 1$, the series converges.

(b) $\pi - e + \dfrac{e^2}{\pi} - \dfrac{e^3}{\pi^2} + \cdots = \displaystyle\sum_{n=1}^{\infty} \pi \left(-\frac{e}{\pi}\right)^{n-1}$ Here $a = \pi$ and $r = -\dfrac{e}{\pi}$.

$$= \frac{\pi}{1 - \left(-\frac{e}{\pi}\right)} = \frac{\pi^2}{\pi + e}.$$

The series converges since $\left|-\dfrac{e}{\pi}\right| < 1$.

(c) $1 + 2^{1/2} + 2 + 2^{3/2} + \cdots = \sum_{n=1}^{\infty} (\sqrt{2})^{n-1}$. This series diverges to $\infty$ since $a = 1 > 0$ and $r = \sqrt{2} > 1$.

(d) $1 - 1 + 1 - 1 + 1 - \cdots = \sum_{n=1}^{\infty} (-1)^{n-1}$. This series diverges since $r = -1$.

(e) $\displaystyle\sum_{n=1}^{\infty} \frac{3^{n+1}}{4^n} = \sum_{n=1}^{\infty} \frac{9}{4} \left(\frac{3}{4}\right)^{n-1} = \frac{9}{4} \frac{1}{1 - \frac{3}{4}} = 9.$

This series converges since $r = 3/4$ satisfies $|r| < 1$.

(f)
$$\sum_{n=5}^{\infty} 3^{n+1} 2^{-(n-2)} = \frac{3^6}{2^3} + \frac{3^7}{2^4} + \frac{3^8}{2^5} + \cdots$$
$$= \sum_{n=1}^{\infty} \frac{3^6}{8} \left(\frac{3}{2}\right)^{n-1}.$$

This series diverges to $\infty$ since $a = 3^6/8 > 0$ and $r = 3/2 > 1$.

(g) Let $x = 0.323232 \cdots = 0.\overline{32}$; then

$$x = \frac{32}{100} + \frac{32}{100^2} + \frac{32}{100^3} + \cdots = \sum_{n=1}^{\infty} \frac{32}{100} \left(\frac{1}{100}\right)^{n-1} = \frac{32}{100} \frac{1}{1 - \frac{1}{100}} = \frac{32}{99}.$$

■

■ **EXAMPLE 2** If money earns interest at a constant effective rate of 8% per year, how much should you pay today for an annuity that will pay you (a) \$1,000 at the end of each of the next 10 years and (b) \$1,000 at the end of every year forever?

***SOLUTION*** A payment of \$1,000 that is due to be received $n$ years from now has present value $\$1{,}000 \times \left(\frac{1}{1.08}\right)^n$ (since $\$A$ would grow to $\$A(1.08)^n$ in $n$ years). Thus \$1,000 payments at the end of each of the next $n$ years are worth $\$s_n$ at the present time, where

$$s_n = 1{,}000 \left[\frac{1}{1.08} + \left(\frac{1}{1.08}\right)^2 + \cdots + \left(\frac{1}{1.08}\right)^n\right]$$
$$= \frac{1{,}000}{1.08} \left[1 + \frac{1}{1.08} + \left(\frac{1}{1.08}\right)^2 + \cdots + \left(\frac{1}{1.08}\right)^{n-1}\right]$$
$$= \frac{1{,}000}{1.08} \frac{1 - \left(\frac{1}{1.08}\right)^n}{1 - \frac{1}{1.08}} = \frac{1{,}000}{0.08} \left[1 - \left(\frac{1}{1.08}\right)^n\right]$$

(a) The present value of 10 future payments is $\$s_{10} = \$6{,}710.08$.

(b) The present value of future payments continuing forever is

$$\$ \lim_{n \to \infty} s_n = \frac{\$1{,}000}{0.08} = \$12{,}500.$$

■

## Telescoping Series and Harmonic Series

**EXAMPLE 3** Show that the series

$$\sum_{n=1}^{\infty} \frac{1}{n(n+1)} = \frac{1}{1\times 2} + \frac{1}{2\times 3} + \frac{1}{3\times 4} + \frac{1}{4\times 5} + \cdots$$

converges and find its sum.

***SOLUTION*** Since $\dfrac{1}{n(n+1)} = \dfrac{1}{n} - \dfrac{1}{n+1}$ we can write the partial sum $s_n$ in the form

$$\begin{aligned} s_n &= \frac{1}{1\times 2} + \frac{1}{2\times 3} + \frac{1}{3\times 4} + \cdots + \frac{1}{(n-1)n} + \frac{1}{n(n+1)} \\ &= \left(1 - \frac{1}{2}\right) + \left(\frac{1}{2} - \frac{1}{3}\right) + \left(\frac{1}{3} - \frac{1}{4}\right) + \cdots \\ &\quad \cdots + \left(\frac{1}{n-1} - \frac{1}{n}\right) + \left(\frac{1}{n} - \frac{1}{n+1}\right) \\ &= 1 - \frac{1}{2} + \frac{1}{2} - \frac{1}{3} + \frac{1}{3} - \cdots - \frac{1}{n} + \frac{1}{n} - \frac{1}{n+1} \\ &= 1 - \frac{1}{n+1}. \end{aligned}$$

Therefore $\lim_{n\to\infty} s_n = 1$ and the series converges to 1:

$$\sum_{n=1}^{\infty} \frac{1}{n(n+1)} = 1.$$

This is an example of a **telescoping series**, so called because the partial sums *fold up* into a simple form when the terms are expanded in partial fractions. Other examples can be found in the exercises at the end of this section. As these examples show, the method of partial fractions can be a useful tool for series as well as for integrals. ■

**EXAMPLE 4** Show that the **harmonic series**

$$\sum_{n=1}^{\infty} \frac{1}{n} = 1 + \frac{1}{2} + \frac{1}{3} + \frac{1}{4} + \cdots$$

diverges to infinity.

***SOLUTION*** If $s_n$ is its $n$th partial sum of the harmonic series, then

$$\begin{aligned} s_n &= 1 + \frac{1}{2} + \frac{1}{3} + \cdots + \frac{1}{n} \\ &= \text{sum of areas of rectangles shaded in Figure 1.3} \\ &> \text{area under } y = \frac{1}{x} \text{ from 1 to } n+1 \\ &= \int_1^{n+1} \frac{dx}{x} = \ln(n+1). \end{aligned}$$

Now $\lim_{n\to\infty} \ln(n+1) = \infty$. Therefore $\lim_{n\to\infty} s_n = \infty$ and

$$\sum_{n=1}^{\infty} \frac{1}{n} = 1 + \frac{1}{2} + \frac{1}{3} + \cdots \qquad \text{diverges to infinity.}$$

■

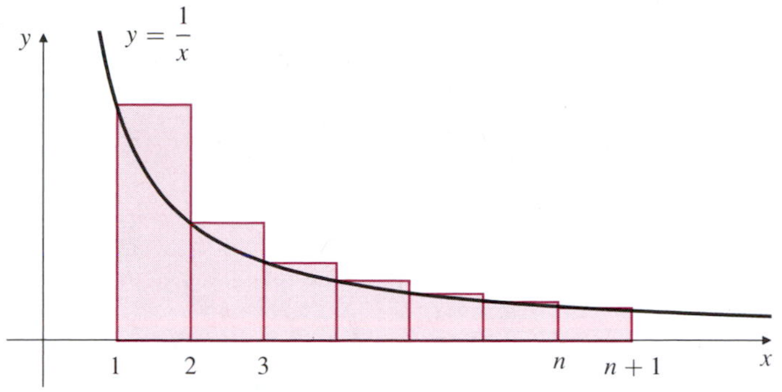

Figure 1.3 A partial sum of the harmonic series

Like geometric series, the harmonic series will often be encountered in subsequent sections.

## Some Theorems About Series

**THEOREM 4**

If $\sum_{n=1}^{\infty} a_n$ converges, then $\lim_{n\to\infty} a_n = 0$.

***PROOF*** If $s_n = a_1 + a_2 + \cdots + a_n$, then $s_n - s_{n-1} = a_n$. If $\sum_{n=1}^{\infty} a_n$ converges, then $\lim_{n\to\infty} s_n = s$ exists, and $\lim_{n\to\infty} s_{n-1} = s$. Hence $\lim_{n\to\infty} a_n = s - s = 0$.

***REMARK*** Theorem 4 is *very important* for the understanding of infinite series. Students often err either in forgetting that *a series cannot converge if its terms do not approach zero* or in confusing this result with its *converse*, which is false. The converse would say that if $\lim_{n\to\infty} a_n = 0$, then $\sum_{n=1}^{\infty} a_n$ must converge. The harmonic series is a counterexample showing the falsehood of this assertion:

$$\lim_{n\to\infty} \frac{1}{n} = 0 \qquad \text{but} \qquad \sum_{n=1}^{\infty} \frac{1}{n} \text{ diverges to infinity.}$$

When considering whether a given series converges, the first question you should ask yourself is "Does the $n$th term approach 0 as $n$ approaches $\infty$?" If the answer is *no*, then the series does *not* converge. If the answer is *yes*, then the series *may or may not* converge. If the sequence of terms $\{a_n\}$ tends to a nonzero limit $L$, then $\sum_{n=1}^{\infty} a_n$ diverges to infinity if $L > 0$ and diverges to negative infinity if $L < 0$.

**EXAMPLE 5**

(a) $\displaystyle\sum_{n=1}^{\infty} \frac{n}{2n-1}$ diverges to infinity since $\lim_{n\to\infty} \dfrac{n}{2n+1} = 1/2 > 0$.

(b) $\sum_{n=1}^{\infty} (-1)^n n \sin(1/n)$ diverges since

$$\lim_{n\to\infty} \left| (-1)^n n \sin \frac{1}{n} \right| = \lim_{n\to\infty} \frac{\sin(1/n)}{1/n} = \lim_{x\to 0+} \frac{\sin x}{x} = 1 \neq 0.$$

The following theorem asserts that it is only the *ultimate* behaviour of $\{a_n\}$ that determines whether $\sum_{n=1}^{\infty} a_n$ converges. Any finite number of terms can be dropped from the beginning of a series without affecting the convergence; the convergence depends only on the *tail* of the series. Of course, the actual sum of the series depends on *all* the terms.

**THEOREM 5**

$\sum_{n=1}^{\infty} a_n$ converges if and only if $\sum_{n=N}^{\infty} a_n$ converges for any integer $N \geq 1$.

**THEOREM 6**

If $\{a_n\}$ is ultimately positive, then the series $\sum_{n=1}^{\infty} a_n$ must either converge (if its partial sums are bounded above) or diverge to infinity (if its partial sums are not bounded above).

The proofs of these two theorems are posed as exercises at the end of this section. The following theorem is just a reformulation of standard laws of limits.

**THEOREM 7**

If $\sum_{n=1}^{\infty} a_n$ and $\sum_{n=1}^{\infty} b_n$ converge to $A$ and $B$, respectively, then

(a) $\sum_{n=1}^{\infty} ca_n$ converges to $cA$ (where $c$ is any constant);

(b) $\sum_{n=1}^{\infty}(a_n \pm b_n)$ converges to $A \pm B$;

(c) If $a_n \le b_n$ for all $n = 1, 2, 3, \ldots$, then $A \le B$.

**EXAMPLE 6** Find the sum of the series $\displaystyle\sum_{n=1}^{\infty} \frac{1+2^{n+1}}{3^n}$.

***SOLUTION*** The given series is the sum of two geometric series,

$$\sum_{n=1}^{\infty} \frac{1}{3^n} = \sum_{n=1}^{\infty} \frac{1}{3}\left(\frac{1}{3}\right)^{n-1} = \frac{1/3}{1-(1/3)} = \frac{1}{2} \quad \text{and}$$

$$\sum_{n=1}^{\infty} \frac{2^{n+1}}{3^n} = \sum_{n=1}^{\infty} \frac{4}{3}\left(\frac{2}{3}\right)^{n-1} = \frac{4/3}{1-(2/3)} = 4.$$

Thus its sum is $\dfrac{1}{2} + 4 = \dfrac{9}{2}$ by Theorem 7(b). ■

## EXERCISES 1.2

In Exercises 1–18, find the sum of the given series, or show that the series diverges (possibly to infinity or negative infinity). Exercises 11–14 are telescoping series and should be done by partial fractions as suggested in Example 3 in this section.

**1.** $\displaystyle \frac{1}{3} + \frac{1}{9} + \frac{1}{27} + \cdots = \sum_{n=1}^{\infty} \frac{1}{3^n}$

**2.** $\displaystyle 3 - \frac{3}{4} + \frac{3}{16} - \frac{3}{64} + \cdots = \sum_{n=1}^{\infty} 3\left(-\frac{1}{4}\right)^{n-1}$

**3.** $\displaystyle \sum_{n=5}^{\infty} \frac{1}{(2+\pi)^{2n}}$

**4.** $\displaystyle \sum_{n=0}^{\infty} \frac{5}{10^{3n}}$

**5.** $\displaystyle \sum_{n=2}^{\infty} \frac{(-5)^n}{8^{2n}}$

**6.** $\displaystyle \sum_{n=0}^{\infty} \frac{1}{e^n}$

**7.** $\displaystyle \sum_{k=0}^{\infty} \frac{2^{k+3}}{e^{k-3}}$

**8.** $\displaystyle \sum_{j=1}^{\infty} \pi^{j/2}\cos(j\pi)$

**9.** $\displaystyle \sum_{n=1}^{\infty} \frac{3+2^n}{2^{n+2}}$

**10.** $\displaystyle \sum_{n=0}^{\infty} \frac{3+2^n}{3^{n+2}}$

**11.** $\displaystyle \sum_{n=1}^{\infty} \frac{1}{n(n+2)} = \frac{1}{1\times 3} + \frac{1}{2\times 4} + \frac{1}{3\times 5} + \cdots$

**12.** $\displaystyle \sum_{n=1}^{\infty} \frac{1}{(2n-1)(2n+1)} = \frac{1}{1\times 3} + \frac{1}{3\times 5} + \frac{1}{5\times 7} + \cdots$

**13.** $\displaystyle \sum_{n=1}^{\infty} \frac{1}{(3n-2)(3n+1)} = \frac{1}{1\times 4} + \frac{1}{4\times 7} + \frac{1}{7\times 10} + \cdots$

* **14.** $\displaystyle \sum_{n=1}^{\infty} \frac{1}{n(n+1)(n+2)} = \frac{1}{1\times 2\times 3} + \frac{1}{2\times 3\times 4} + \frac{1}{3\times 4\times 5} + \cdots$

**15.** $\displaystyle \sum_{n=1}^{\infty} \frac{1}{2n-1}$

**16.** $\displaystyle \sum_{n=1}^{\infty} \frac{n}{n+2}$

**17.** $\displaystyle \sum_{n=1}^{\infty} n^{-1/2}$

**18.** $\displaystyle \sum_{n=1}^{\infty} \frac{2}{n+1}$

**19.** Obtain a simple expression for the partial sum $s_n$ of the series $\sum_{n=1}^{\infty}(-1)^n$ and use it to show that this series diverges.

* **20.** Find a simple expression for the partial sum $s_n$ of the series $\displaystyle\sum_{n=2}^{\infty}\left(3e^{-n} - \frac{2}{n^2-1}\right)$. Also find the sum of the series.

**21.** When dropped, an elastic ball bounces back up to a height three-quarters of that from which it fell. If the ball is dropped from a height of 2 m and allowed to bounce up and down indefinitely, what is the total distance it travels before coming to rest?

**22.** If a bank account pays 10% simple interest into an account once a year, what is the balance in the account at the end of 8 years if \$1,000 is deposited into the account at the beginning of each of the eight years? (Assume there was no balance in the account initially.)

* **23.** Prove Theorem 5.

* **24.** Prove Theorem 6.

* **25.** State a theorem analogous to Theorem 6 but for a negative sequence.

In Exercises 26–31, decide whether the given statement is TRUE or FALSE. If it is true, prove it. If it is false, give a counterexample showing the falsehood.

* **26.** If $a_n = 0$ for every $n$, then $\sum a_n$ converges.

* **27.** If $\sum a_n$ converges, then $\sum(1/a_n)$ diverges to infinity.

* **28.** If $\sum a_n$ and $\sum b_n$ both diverge, then so does $\sum(a_n + b_n)$.

* **29.** If $a_n \ge c > 0$ for every $n$, then $\sum a_n$ diverges to infinity.

* **30.** If $\sum a_n$ diverges and $\{b_n\}$ is bounded, then $\sum a_n b_n$ diverges.

* **31.** If $a_n > 0$ and $\sum a_n$ converges, then $\sum(a_n)^2$ converges.

## 1.3 CONVERGENCE TESTS FOR POSITIVE SERIES

In the previous section we saw a few examples of convergent series (geometric and telescoping series) whose sums could be determined exactly because the partial sums $s_n$ could be expressed in closed form as explicit functions of $n$ whose limits as $n \to \infty$ could be evaluated. It is not usually possible to do this with a given series, and therefore it is not usually possible to determine the sum of the series exactly. However, there are many techniques for determining whether a given series converges and, if it does, for approximating the sum to any desired degree of accuracy. We shall investigate some of these techniques in the remaining sections of this chapter.

In this section we deal exclusively with *positive series*, that is, series of the form

$$\sum_{n=1}^{\infty} a_n = a_1 + a_2 + a_3 + \cdots$$

where $a_n \ge 0$ for all $n \ge 1$. As noted in Theorem 6, such a series will converge if its partial sums are bounded above and will diverge to infinity otherwise. All our results apply equally well to *ultimately* positive series since convergence or divergence depends only on the *tail* of a series.

### The Integral Test

The integral test provides a means for determining whether an ultimately positive series converges or diverges by comparing it with an improper integral that behaves similarly. Example 4 in Section 1.2 is an example of the use of this technique. We formalize the method in the following theorem.

**THEOREM 8**

**The integral test**

Suppose that $a_n = f(n)$, where $f(x)$ is positive, continuous, and nonincreasing on an interval $[N, \infty)$ for some positive integer $N$. Then

$$\sum_{n=1}^{\infty} a_n \qquad \text{and} \qquad \int_N^{\infty} f(t)\,dt$$

either both converge or both diverge to infinity.

**PROOF** Let $s_n = a_1 + a_2 + \cdots + a_n$. If $n > N$, we have

$$\begin{aligned} s_n &= s_N + a_{N+1} + a_{N+2} + \cdots + a_n \\ &= s_N + f(N+1) + f(N+2) + \cdots + f(n) \\ &= s_N + \text{ sum of areas of rectangles shaded in Figure 1.4(a)} \\ &\le s_N + \int_N^\infty f(t)\,dt. \end{aligned}$$

If the improper integral $\int_N^\infty f(t)\,dt$ converges, then the sequence $\{s_n\}$ is bounded above and $\sum_{n=1}^\infty a_n$ converges.

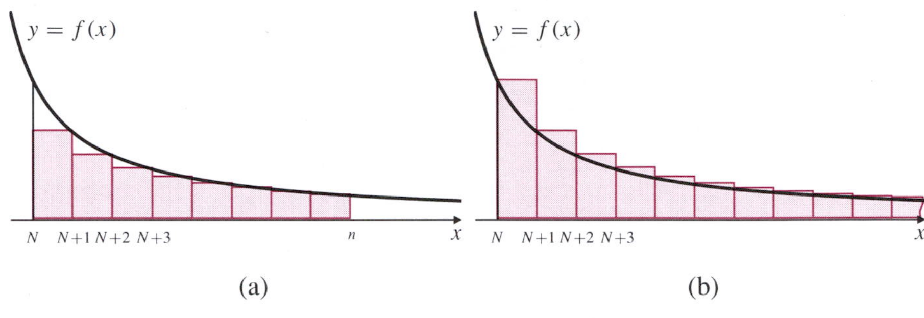


Figure 1.4

Conversely, suppose that $\sum_{n=1}^\infty a_n$ converges to the sum $s$. Then

$$\begin{aligned} \int_N^\infty f(t)\,dt = &\text{ area under } y = f(t) \text{ above } y = 0 \text{ from } t = N \text{ to } t = \infty \\ &\le \text{ sum of areas of shaded rectangles in Figure 1.4(b)} \\ &= a_N + a_{N+1} + a_{N+2} + \cdots \\ &= s - s_{N-1} < \infty, \end{aligned}$$

so the improper integral represents a finite area and is thus convergent. (We omit the remaining details showing that $\lim_{R\to\infty} \int_N^R f(t)\,dt$ exists; like the series case, the argument depends on the completeness of the real numbers.)

**REMARK** If $a_n = f(n)$, where $f$ is positive, continuous, and nonincreasing on $[1, \infty)$, then Theorem 8 assures us that $\sum_{n=1}^\infty a_n$ and $\int_1^\infty f(x)\,dx$ both converge or both diverge to infinity. It does *not* tell us that the sum of the series is equal to the value of the integral. The two are not likely to be equal in the case of convergence. However, see Section 1.5 for examples where the integrals are used to approximate the sums of series.

The principal use of the integral test is to establish the result of the following example concerning the series $\sum_{n=0}^\infty n^{-p}$, which is called a ***p*-series**. This result should be memorized; we will frequently compare the behaviour of other series with $p$-series later in this and subsequent sections.

### EXAMPLE 1 *p*-series

Show that

$$\sum_{n=1}^\infty n^{-p} = \sum_{n=1}^\infty \frac{1}{n^p} \begin{cases} \text{converges if } p > 1 \\ \text{diverges to infinity if } p \le 1. \end{cases}$$

***SOLUTION*** Observe that if $p > 0$, then $f(x) = x^{-p}$ is positive, continuous, and decreasing on $[1, \infty)$. By the integral test, the $p$-series converges for $p > 1$ and diverges for $0 < p \leq 1$ by comparison with $\int_1^\infty x^{-p}\,dx$. If $p \leq 0$, then $\lim_{n\to\infty}(1/n^p) \neq 0$, so the series cannot converge in this case. Being a positive series, it must diverge to infinity. ■

***REMARK*** The harmonic series $\sum_{n=1}^\infty n^{-1}$ (the case $p = 1$ of the $p$-series) is on the borderline between convergence and divergence, although it diverges. While its terms decrease towards 0 as $n$ increases, they do not decrease *fast enough* to allow the sum of the series to be finite. If $p > 1$, the terms of $\sum_{n=1}^\infty n^{-p}$ decrease towards zero fast enough that their sum is finite. We can refine the distinction between convergence and divergence at $p = 1$ by using terms that decrease faster than $1/n$, but not as fast as $1/n^q$ for any $q > 1$. Recall that $\ln n$ grows more slowly than any positive power of $n$ as $n$ increases. Accordingly, if $p > 0$ and $q > 1$, then

$$\frac{1}{n^q} < \frac{1}{n(\ln n)^p} < \frac{1}{n}.$$

The question now arises whether $\sum_{n=2}^\infty 1/(n(\ln n)^p)$ converges or not.

■ **EXAMPLE 2** For what values of $p$ does the series $\displaystyle\sum_{n=2}^\infty \frac{1}{n(\ln n)^p}$ converge?

***SOLUTION*** We had to start the series at $n = 2$ because $\ln 1 = 0$. Let $p > 0$. Since $f(x) = 1/(x(\ln x)^p))$ is positive, continuous, and decreasing on $[2, \infty)$, the given series should be compared with

$$\int_2^\infty \frac{dx}{x(\ln x)^p}.$$

The substitution $u = \ln x$ transforms this integral to the equivalent one

$$\int_{\ln 2}^\infty \frac{du}{u^p},$$

which converges if $p > 1$ and diverges if $0 < p < 1$.Therefore the given series converges if $p > 1$ and diverges if $0 < p < 1$. For $p = 0$ the given series is the harmonic series, so it diverges to infinity. For $p < 0$ the terms are ultimately greater than those of the harmonic series, so the partial sums of the series are not bounded above, and the series diverges to infinity in this case also. ■

This process of fine-tuning Example 1 exhibited in Example 2 can be extended even further. (See Exercise 30 below.)

## Comparison Tests

The next test we consider for positive series is analogous to the comparison theorem for improper integrals. It enables us to determine the convergence or divergence of one series by comparing it with another series that is known to converge or diverge.

**THEOREM 9**

**A comparison test**

Let $\{a_n\}$ and $\{b_n\}$ be sequences for which there exists a positive constant $K$ such that, ultimately, $0 \leq a_n \leq K b_n$.

(a) If the series $\sum_{n=1}^\infty b_n$ converges, then so does the series $\sum_{n=1}^\infty a_n$.

(b) If the series $\sum_{n=1}^\infty a_n$ diverges to infinity, then so does the series $\sum_{n=1}^\infty b_n$.

DANGER

Theorem 9 does *not* say that if $\sum a_n$ converges, then $\sum b_n$ converges. It is possible that the *smaller* sum may be finite while the *larger* one is infinite. (Do not confuse a theorem with its converse.)

DANGER

***PROOF*** Since a series converges if and only if its tail converges (Theorem 5), we can assume, without loss of generality, that the condition $0 \le a_n \le K b_n$ holds for all $n \ge 1$. Let $s_n = a_1 + a_2 + \cdots + a_n$ and $S_n = b_1 + b_2 + \cdots + b_n$. Then $s_n \le K S_n$. If $\sum b_n$ converges, then $\{S_n\}$ is convergent and hence is bounded by Theorem 1. Hence $\{s_n\}$ is bounded above. By Theorem 6, $\sum a_n$ converges. Since the convergence of $\sum b_n$ guarantees that of $\sum a_n$, if the latter series diverges to infinity, then the former cannot converge either, so it must diverge to infinity too.

**EXAMPLE 3** Do the following series converge or not? Give reasons for your answer.

(a) $\displaystyle\sum_{n=1}^{\infty} \frac{1}{2^n+1}$, (b) $\displaystyle\sum_{n=1}^{\infty} \frac{3n+1}{n^3+1}$, (c) $\displaystyle\sum_{n=2}^{\infty} \frac{1}{\ln n}$.

***SOLUTION*** In each case we must find a suitable comparison series that we already know converges or diverges.

(a) Since $0 < \dfrac{1}{2^n+1} < \dfrac{1}{2^n}$ for $n = 1, 2, 3, \ldots$, and since $\sum_{n=1}^{\infty}(1/2^n)$ is a convergent geometric series, the series $\sum_{n=1}^{\infty}(1/(2^n+1))$ also converges by comparison.

(b) Observe that $\dfrac{3n+1}{n^3+1}$ behaves like $\dfrac{3}{n^2}$ for large $n$, so we would expect to compare the series with the convergent $p$-series $\sum_{n=1}^{\infty} n^{-2}$. We have, for $n \ge 1$,

$$\frac{3n+1}{n^3+1} = \frac{3n}{n^3+1} + \frac{1}{n^3+1} < \frac{3n}{n^3} + \frac{1}{n^3} < \frac{3}{n^2} + \frac{1}{n^2} = \frac{4}{n^2}.$$

Thus the given series converges by Theorem 9.

(c) For $n = 2, 3, 4, \ldots$, we have $0 < \ln n < n$. Thus $\dfrac{1}{\ln n} > \dfrac{1}{n}$. Since $\sum_{n=2}^{\infty} \dfrac{1}{n}$ diverges to infinity (it is a harmonic series), so does $\sum_{n=2}^{\infty} \dfrac{1}{\ln n}$ by comparison. ■

If a given positive series is compared with one whose sum we know, we can often use the comparison to get an upper bound for the sum of the given series.

**EXAMPLE 4** Find an upper bound for the sum of the series $\displaystyle\sum_{n=1}^{\infty} \frac{1}{n^2}$.

***SOLUTION*** We know from the integral test that the $p$-series $\sum_{n=1}^{\infty} 1/n^2$ converges, but we don't know what its sum is. However, we do know from Example 3 of Section 1.2 that

$$\sum_{n=2}^{\infty} \frac{1}{(n-1)n} = \sum_{m=1}^{\infty} \frac{1}{m(m+1)} = 1.$$

Since $n^2 > n^2 - n = (n-1)n > 0$ for $n \ge 2$, we have

$$0 < \frac{1}{n^2} < \frac{1}{n^2-n} = \frac{1}{(n-1)n} \qquad \text{for all } n \ge 2.$$

Therefore,

$$\sum_{n=1}^{\infty} \frac{1}{n^2} = 1 + \sum_{n=2}^{\infty} \frac{1}{n^2} < 1 + \sum_{n=2}^{\infty} \frac{1}{(n-1)n} = 1 + 1 = 2.$$

(It can be shown by methods not yet available to us that $\sum_{n=1}^{\infty} 1/n^2 = \pi^2/6$.) ■

**EXPLORE!**

## Using a Calculator to Find the Sum of a Series

Suppose you want to find the value of $S = \sum_{n=1}^{\infty} 1/n^2$ correct to three decimal places with the aid of a calculator. How many terms should you add?

Since $1/n^2 \le 1/10{,}000 = 0.0001$ if $n \ge 100$, you might be *tempted* to use the first 100 terms. Use a programmable calculator or one with a built-in function for summing the terms of a sequence to calculate $\sum_{n=1}^{100} 1/n^2$. Round the answer to three decimal places.

Now, as a check, calculate $\sum_{n=1}^{500} 1/n^2$ and round the answer to three decimal places. Was your first answer correct to three decimal places? If not, why not?

The number of terms of a convergent series needed to estimate the sum to within a given error tolerance depends on the behaviour of the terms. It is usually *not* sufficient to add terms until the terms are smaller than the tolerance. See Section 1.5 for a discussion of how to determine the number of terms needed.

The following theorem provides a version of the comparison test that is not quite as general as Theorem 9, but is often easier to apply in specific cases.

**THEOREM 10**

**A limit comparison test**

Suppose that $\{a_n\}$ and $\{b_n\}$ are positive sequences and that

$$\lim_{n\to\infty} \frac{a_n}{b_n} = L,$$

where $L$ is either a nonnegative finite number or $+\infty$.

(a) If $L < \infty$ and $\sum_{n=1}^{\infty} b_n$ converges, then $\sum_{n=1}^{\infty} a_n$ also converges.

(b) If $L > 0$ and $\sum_{n=1}^{\infty} b_n$ diverges to infinity, then so does $\sum_{n=1}^{\infty} a_n$.

***PROOF*** If $L < \infty$, then for $n$ sufficiently large, we have $b_n > 0$ and

$$0 \le \frac{a_n}{b_n} \le L + 1,$$

so $0 \le a_n \le (L+1)b_n$. Hence $\sum_{n=1}^{\infty} a_n$ converges if $\sum_{n=1}^{\infty} b_n$ converges, by Theorem 9(a).

If $L > 0$, then for $n$ sufficiently large

$$\frac{a_n}{b_n} \ge \frac{L}{2}.$$

Therefore $0 < b_n \le (2/L)a_n$, and $\sum_{n=1}^{\infty} a_n$ diverges to infinity if $\sum_{n=1}^{\infty} b_n$ does, by Theorem 9(b).

**EXAMPLE 5** Do the following series converge or not? Give reasons for your answers.

(a) $\displaystyle\sum_{n=1}^{\infty} \frac{1}{1+\sqrt{n}}$, (b) $\displaystyle\sum_{n=1}^{\infty} \frac{n+5}{n^3-2n+3}$, (c) $\displaystyle\sum_{n=1}^{\infty} \frac{2^n+1}{3^n-1}$.

**SOLUTION** Again we must make appropriate choices for comparison series.

(a) The terms of this series decrease like $1/\sqrt{n}$. Observe that

$$L = \lim_{n\to\infty} \frac{\dfrac{1}{1+\sqrt{n}}}{\dfrac{1}{\sqrt{n}}} = \lim_{n\to\infty} \frac{\sqrt{n}}{1+\sqrt{n}} = \lim_{n\to\infty} \frac{1}{(1/\sqrt{n})+1} = 1.$$

Since the $p$-series $\sum_{n=1}^{\infty} 1/\sqrt{n}$ diverges to infinity ($p = 1/2$), so does the series $\sum_{n=1}^{\infty} 1/(1+\sqrt{n})$, by the limit comparison test.

(b) For large $n$, the terms behave like $n/n^3$ so let us compare the series with the $p$-series $\sum_{n=1}^{\infty} 1/n^2$, which we know converges.

$$L = \lim_{n\to\infty} \frac{\dfrac{n+5}{n^3-2n+3}}{\dfrac{1}{n^2}} = \lim_{n\to\infty} \frac{n^3+5n^2}{n^3-2n+3} = 1.$$

Since $L < \infty$, the series $\displaystyle\sum_{n=1}^{\infty} \frac{n+5}{n^3-2n+3}$ also converges by the limit comparison test.

(c) In this case the terms behave like $(2/3)^n$ for large $n$. We have

$$L = \lim_{n\to\infty} \frac{\dfrac{2^n+1}{3^n-1}}{\left(\dfrac{2}{3}\right)^n} = \lim_{n\to\infty} \frac{2^n+1}{2^n}\,\frac{3^n}{3^n-1} = \lim_{n\to\infty} \frac{1+\dfrac{1}{2^n}}{1-\dfrac{1}{3^n}} = 1.$$

Again $L < \infty$ and the series $\displaystyle\sum_{n=1}^{\infty} \frac{2^n+1}{3^n-1}$ converges by comparison with the convergent geometric series $\displaystyle\sum_{n=1}^{\infty} \left(\frac{2}{3}\right)^n$. ■

In order to apply the original version of the comparison test (Theorem 9) successfully, it is important to have an intuitive feeling for whether the given series converges or diverges. The form of the comparison will depend on whether you are trying to prove convergence or divergence. For instance, if you did not know intuitively that

$$\sum_{n=1}^{\infty} \frac{1}{100n + 20{,}000}$$

would have to diverge to infinity, you might try to argue that

$$\frac{1}{100n+20{,}000} < \frac{1}{n} \qquad \text{for } n = 1, 2, 3, \ldots.$$

While true, this doesn't help at all. $\sum_{n=1}^{\infty} 1/n$ diverges to infinity, and therefore Theorem 9 yields no information from this comparison. We could, of course, argue instead that

$$\frac{1}{100n+20{,}000} \ge \frac{1}{20{,}100n} \qquad \text{if } n \ge 1,$$

and conclude by Theorem 9 that $\sum_{n=1}^{\infty}(1/(100n+20{,}000))$ diverges to infinity by comparison with the divergent series $\sum_{n=1}^{\infty} 1/n$. An easier way is to use Theorem 10 and the fact that

$$L = \lim_{n\to\infty} \frac{\dfrac{1}{100n+20{,}000}}{\dfrac{1}{n}} = \lim_{n\to\infty} \frac{n}{100n+20{,}000} = \frac{1}{100} > 0.$$

However, the limit comparison test Theorem 10 has a disadvantage when compared to the ordinary comparison test Theorem 9. It can fail in certain cases because the limit $L$ does not exist. In such cases it is possible that the ordinary comparison test may still work.

**EXAMPLE 6** Test the series $\displaystyle\sum_{n=1}^{\infty} \frac{1+\sin n}{n^2}$ for convergence.

***SOLUTION*** Since

$$\lim_{n\to\infty} \frac{\dfrac{1+\sin n}{n^2}}{\dfrac{1}{n^2}} = \lim_{n\to\infty}(1+\sin n)$$

does not exist, the limit comparison test gives us no information. However, since $\sin n \le 1$, we have

$$0 \le \frac{1+\sin n}{n^2} \le \frac{2}{n^2} \qquad \text{for } n = 1, 2, 3, \ldots.$$

The given series does, in fact, converge by comparison with $\sum_{n=1}^{\infty} 1/n^2$, using the ordinary comparison test. ■

## The Ratio and Root Tests

**THEOREM 11**

**The ratio test**

Suppose that $a_n > 0$ (ultimately) and that $\rho = \lim_{n\to\infty} \dfrac{a_{n+1}}{a_n}$ exists or is $+\infty$.

(a) If $0 \le \rho < 1$, then $\sum_{n=1}^{\infty} a_n$ converges.

(b) If $1 < \rho \le \infty$, then $\lim_{n\to\infty} a_n = \infty$ and $\sum_{n=1}^{\infty} a_n$ diverges to infinity.

(c) If $\rho = 1$, this test gives no information; the series may either converge or diverge to infinity.

***PROOF*** Here $\rho$ is the lower case Greek letter *rho*, (pronounced *row*).

(a) Suppose $\rho < 1$. Pick a number $r$ such that $\rho < r < 1$. Since we are given that $\lim_{n\to\infty} a_{n+1}/a_n = \rho$, therefore $a_{n+1}/a_n \le r$ for $n$ sufficiently large; that is, $a_{n+1} \le r a_n$ for $n \ge N$, say. In particular,

$$\begin{aligned}
a_{N+1} &\le r a_N \\
a_{N+2} &\le r a_{N+1} \le r^2 a_N \\
a_{N+3} &\le r a_{N+2} \le r^3 a_N \\
&\vdots \\
a_{N+k} &\le r^k a_N \qquad (k = 0, 1, 2, 3, \ldots)
\end{aligned}$$

Hence $\sum_{n=N}^{\infty} a_n$ converges by comparison with the convergent geometric series $\sum_{k=0}^{\infty} r^k$. It follows that $\sum_{n=1}^{\infty} a_n = \sum_{n=1}^{N-1} a_n + \sum_{n=N}^{\infty} a_n$ must also converge.

(b) Now suppose that $\rho > 1$. Pick a number $r$ such that $1 < r < \rho$. Since $\lim_{n\to\infty} a_{n+1}/a_n = \rho$, therefore $a_{n+1}/a_n \ge r$ for $n$ sufficiently large, say for $n \ge N$. We assume $N$ is chosen large enough that $a_N > 0$. It follows by an argument similar to that used in part (a) that $a_{N+k} \ge r^k a_N$ for $k = 0, 1, 2, \ldots$, and since $r > 1$, $\lim_{n\to\infty} a_n = \infty$. Therefore $\sum_{n=1}^{\infty} a_n$ diverges to infinity.

(c) If $\rho$ is computed for the series $\sum_{n=1}^{\infty} 1/n$ and $\sum_{n=1}^{\infty} 1/n^2$, we get $\rho = 1$ in each case. Since the first series diverges to infinity and the second converges, the ratio test cannot distinguish between convergence and divergence if $\rho = 1$.

All $p$-series fall into the indecisive category where $\rho = 1$, as does $\sum_{n=1}^{\infty} a_n$, where $a_n$ is any rational function of $n$. The ratio test is most useful for series whose terms decrease at least exponentially fast. The presence of factorials in a term also suggest that the ratio test might be useful.

**EXAMPLE 7** Test for convergence:

$$\text{(a)}\ \sum_{n=1}^{\infty} \frac{99^n}{n!}, \qquad \text{(b)}\ \sum_{n=1}^{\infty} \frac{n^5}{2^n}, \qquad \text{(c)}\ \sum_{n=1}^{\infty} \frac{n!}{n^n}, \qquad \text{(d)}\ \sum_{n=1}^{\infty} \frac{(2n)!}{(n!)^2}$$

***SOLUTION*** We use the ratio test for each of these series.

$$\text{(a)}\ \rho = \lim_{n\to\infty} \frac{99^{n+1}}{(n+1)!} \Bigg/ \frac{99^n}{n!} = \lim_{n\to\infty} \frac{99}{n+1} = 0 < 1.$$

Thus $\sum_{n=1}^{\infty} (99^n/n!)$ converges.

$$\text{(b)}\ \rho = \lim_{n\to\infty} \frac{(n+1)^5}{2^{n+1}} \Bigg/ \frac{n^5}{2^n} = \lim_{n\to\infty} \frac{1}{2}\left(\frac{n+1}{n}\right)^5 = \frac{1}{2} < 1.$$

Hence $\sum_{n=1}^{\infty} (n^5/2^n)$ converges.

$$\text{(c)}\ \rho = \lim_{n\to\infty} \frac{(n+1)!}{(n+1)^{n+1}} \Bigg/ \frac{n!}{n^n} = \lim_{n\to\infty} \frac{(n+1)!\,n^n}{(n+1)^{n+1} n!} = \lim_{n\to\infty} \left(\frac{n}{n+1}\right)^n$$

$$= \lim_{n\to\infty} \frac{1}{\left(1 + \dfrac{1}{n}\right)^n} = \frac{1}{e} < 1.$$

Thus $\sum_{n=1}^{\infty} (n!/n^n)$ converges.

$$\text{(d)}\ \rho = \lim_{n\to\infty} \frac{(2(n+1))!}{((n+1)!)^2} \Bigg/ \frac{(2n)!}{(n!)^2} = \lim_{n\to\infty} \frac{(2n+2)(2n+1)}{(n+1)^2} = 4 > 1.$$

Thus $\sum_{n=1}^{\infty} (2n)!/(n!)^2$ diverges to infinity. ■

The following theorem is very similar to the ratio test, but is less frequently used. Its proof is left as an exercise. (See Exercise 31.) For examples of series to which it can be applied, see Exercises 32 and 33.

**THEOREM 12**

**The root test**

Suppose that $a_n > 0$ (ultimately) and that $\sigma = \lim_{n\to\infty} (a_n)^{1/n}$ exists or is $+\infty$.

(a) If $0 \le \sigma < 1$, then $\sum_{n=1}^{\infty} a_n$ converges.

(b) If $1 < \sigma \le \infty$, then $\lim_{n\to\infty} a_n = \infty$ and $\sum_{n=1}^{\infty} a_n$ diverges to infinity.

(c) If $\sigma = 1$, this test gives no information; the series may either converge or diverge to infinity.

## EXERCISES 1.3

In Exercises 1–28, determine whether the given series converges or diverges by using any appropriate test. The $p$-series can be used for comparison, as can geometric series. Be alert for series whose terms do not approach 0.

**1.** $\displaystyle\sum_{n=1}^{\infty} \frac{1}{n^2+1}$

**2.** $\displaystyle\sum_{n=1}^{\infty} \frac{n}{n^4-2}$

**3.** $\displaystyle\sum_{n=1}^{\infty} \frac{n^2+1}{n^3+1}$

**4.** $\displaystyle\sum_{n=1}^{\infty} \frac{\sqrt{n}}{n^2+n+1}$

**5.** $\displaystyle\sum_{n=1}^{\infty} \left|\sin \frac{1}{n^2}\right|$

**6.** $\displaystyle\sum_{n=8}^{\infty} \frac{1}{\pi^n+5}$

**7.** $\displaystyle\sum_{n=2}^{\infty} \frac{1}{(\ln n)^3}$

**8.** $\displaystyle\sum_{n=1}^{\infty} \frac{1}{\ln(3n)}$

**9.** $\displaystyle\sum_{n=1}^{\infty} \frac{1}{\pi^n - n^\pi}$

**10.** $\displaystyle\sum_{n=0}^{\infty} \frac{1+n}{2+n}$

**11.** $\displaystyle\sum_{n=1}^{\infty} \frac{1+n^{4/3}}{2+n^{5/3}}$

**12.** $\displaystyle\sum_{n=1}^{\infty} \frac{n^2}{1+n\sqrt{n}}$

**13.** $\displaystyle\sum_{n=3}^{\infty} \frac{1}{n\ln n\sqrt{\ln\ln n}}$

**14.** $\displaystyle\sum_{n=2}^{\infty} \frac{1}{n\ln n(\ln\ln n)^2}$

**15.** $\displaystyle\sum_{n=1}^{\infty} \frac{1-(-1)^n}{n^4}$

**16.** $\displaystyle\sum_{n=1}^{\infty} \frac{1+(-1)^n}{\sqrt{n}}$

**17.** $\displaystyle\sum_{n=1}^{\infty} \frac{2+\cos n}{n+\ln n}$

**18.** $\displaystyle\sum_{n=1}^{\infty} \frac{e^n\cos^2 n}{1+\pi^n}$

**19.** $\displaystyle\sum_{n=1}^{\infty} \frac{1}{2^n(n+1)}$

**20.** $\displaystyle\sum_{n=1}^{\infty} \frac{n^4}{n!}$

**21.** $\displaystyle\sum_{n=1}^{\infty} \frac{n!}{n^2 e^n}$

**22.** $\displaystyle\sum_{n=1}^{\infty} \frac{(2n)!6^n}{(3n)!}$

**23.** $\displaystyle\sum_{n=2}^{\infty} \frac{\sqrt{n}}{3^n\ln n}$

**24.** $\displaystyle\sum_{n=0}^{\infty} \frac{n^{100}2^n}{\sqrt{n!}}$

**25.** $\displaystyle\sum_{n=1}^{\infty} \frac{(2n)!}{(n!)^3}$

**26.** $\displaystyle\sum_{n=1}^{\infty} \frac{1+n!}{(1+n)!}$

**27.** $\displaystyle\sum_{n=4}^{\infty} \frac{2^n}{3^n-n^3}$

**28.** $\displaystyle\sum_{n=1}^{\infty} \frac{n^n}{\pi^n n!}$

**29.** Use the integral test to show that $\displaystyle\sum_{n=1}^{\infty} \frac{1}{1+n^2}$ converges. Show that the sum $s$ of the series is less than $\pi/2$.

* **30.** Show that $\sum_{n=3}^{\infty}(1/(n\ln n(\ln\ln n)^p))$ converges if and only if $p > 1$. Generalize this result to series of the form

$$\sum_{n=N}^{\infty} \frac{1}{n(\ln n)(\ln\ln n)\cdots(\ln_j n)(\ln_{j+1} n)^p}$$

where $\ln_j n = \underbrace{\ln\ln\ln\ln\cdots\ln}_{j\ \ln' s} n$.

* **31.** Prove the root test. *Hint:* mimic the proof of the ratio test.

**32.** Use the root test to show that $\displaystyle\sum_{n=1}^{\infty} \frac{2^{n+1}}{n^n}$ converges.

* **33.** Use the root test from Exercise 32 to test for convergence

$$\sum_{n=1}^{\infty}\left(\frac{n}{n+1}\right)^{n^2}.$$

**34.** Repeat Exercise 32, but use the ratio test instead of the root test.

* **35.** Try to use the ratio test to determine whether $\displaystyle\sum_{n=1}^{\infty} \frac{2^{2n}(n!)^2}{(2n)!}$ converges. What happens? Now observe that

$$\frac{2^{2n}(n!)^2}{(2n)!} = \frac{[2n(2n-2)(2n-4)\cdots 6\times 4\times 2]^2}{2n(2n-1)(2n-2)\cdots 4\times 3\times 2\times 1}$$
$$= \frac{2n}{2n-1}\times\frac{2n-2}{2n-3}\times\cdots\times\frac{4}{3}\times\frac{2}{1}.$$

Does the given series converge or not? Why?

* **36.** Determine whether the series $\displaystyle\sum_{n=1}^{\infty} \frac{(2n)!}{2^{2n}(n!)^2}$ converges. (*Hint:* proceed as in Exercise 35. Show that $a_n \geq 1/(2n)$.)

## 1.4 ABSOLUTE AND CONDITIONAL CONVERGENCE

All of the series $\sum_{n=1}^{\infty} a_n$ considered in the previous section were ultimately positive; that is, $a_n \geq 0$ for $n$ sufficiently large. We now drop this restriction and allow arbitrary real terms $a_n$. We can, however, always obtain a positive series from any given series by replacing all the terms with their absolute values.

**DEFINITION 5**

**Absolute convergence**

The series $\sum_{n=1}^{\infty} a_n$ is said to be **absolutely convergent** if $\sum_{n=1}^{\infty} |a_n|$ converges.

The series

$$s = \sum_{n=1}^{\infty} \frac{(-1)^n}{n^2} = -1 + \frac{1}{4} - \frac{1}{9} + \frac{1}{16} - \cdots$$

converges absolutely since

$$S = \sum_{n=1}^{\infty} \left| \frac{(-1)^n}{n^2} \right| = \sum_{n=1}^{\infty} \frac{1}{n^2} = 1 + \frac{1}{4} + \frac{1}{9} + \frac{1}{16} + \cdots$$

converges. It seems reasonable that the first series must converge, and its sum $s$ should satisfy $-S \leq s \leq S$. In general, the cancellation that occurs because some terms are negative and others positive makes it *easier* for a series to converge than if all the terms are of one sign. We verify this insight in the following theorem.

**THEOREM 13**

If a series converges absolutely, then it converges.

***PROOF*** Let $\sum_{n=1}^{\infty} a_n$ be absolutely convergent, and let $b_n = a_n + |a_n|$ for each $n$. Since $-|a_n| \leq a_n \leq |a_n|$, we have $0 \leq b_n \leq 2|a_n|$ for each $n$. Thus $\sum_{n=1}^{\infty} b_n$ converges by the comparison test. Therefore $\sum_{n=1}^{\infty} a_n = \sum_{n=1}^{\infty} b_n - \sum_{n=1}^{\infty} |a_n|$ also converges.

Again you are cautioned not to confuse the statement of Theorem 13 with the converse statement, which is false. We will show later in this section that the **alternating harmonic series**

**DANGER**

Although absolute convergence implies convergence, convergence does *not* imply absolute convergence.

**DANGER**

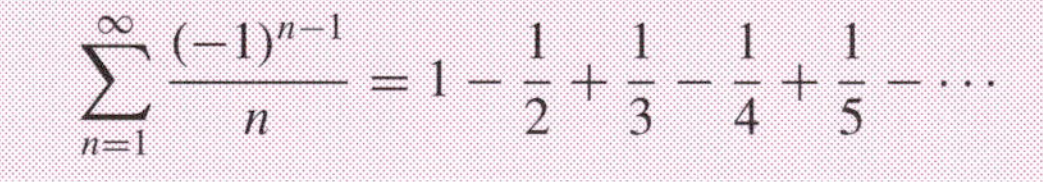

$$\sum_{n=1}^{\infty} \frac{(-1)^{n-1}}{n} = 1 - \frac{1}{2} + \frac{1}{3} - \frac{1}{4} + \frac{1}{5} - \cdots$$

converges, although it does not converge absolutely. If we replace all the terms by their absolute values we get the divergent harmonic series:

$$\sum_{n=1}^{\infty} \frac{1}{n} = 1 + \frac{1}{2} + \frac{1}{3} + \frac{1}{4} + \cdots = \infty.$$

**DEFINITION 6**

**Conditional convergence**

If $\sum_{n=1}^{\infty} a_n$ is convergent, but not absolutely convergent, then we say that it is **conditionally convergent** or that it **converges conditionally**.

The alternating harmonic series is an example of a conditionally convergent series.

The comparison tests, the integral test and the ratio test, can each be used to test for absolute convergence. They should be applied to the series $\sum_{n=1}^{\infty} |a_n|$. For the ratio test we calculate $\rho = \lim_{n\to\infty} |a_{n+1}/a_n|$. If $\rho < 1$, then $\sum_{n=1}^{\infty} a_n$ converges absolutely. If $\rho > 1$, then $\lim_{n\to\infty} |a_n| = \infty$, so both $\sum_{n=1}^{\infty} |a_n|$ and $\sum_{n=1}^{\infty} a_n$ must diverge. If $\rho = 1$ we get no information; the series $\sum_{n=1}^{\infty} a_n$ may converge absolutely, it may converge conditionally, or it may diverge.

**EXAMPLE 1** Test for absolute convergence:

$$\text{(a)}\quad \sum_{n=1}^{\infty} \frac{(-1)^{n-1}}{2n-1}, \qquad \text{(b)}\quad \sum_{n=1}^{\infty} \frac{n\cos(n\pi)}{2^n}$$

***SOLUTION***

(a) $\displaystyle \lim_{n\to\infty} \left|\frac{(-1)^{n-1}}{2n-1}\right| \Big/ \frac{1}{n} = \lim_{n\to\infty} \frac{n}{2n-1} = \frac{1}{2} > 0.$ Since the harmonic series $\sum_{n=1}^{\infty}(1/n)$ diverges to infinity, therefore $\sum_{n=1}^{\infty}((-1)^{n-1}/(2n-1))$ does not converge absolutely (comparison test).

(b) $\displaystyle \rho = \lim_{n\to\infty} \left| \frac{(n+1)\cos((n+1)\pi)}{2^{n+1}} \Big/ \frac{n\cos(n\pi)}{2^n} \right| = \lim_{n\to\infty} \frac{n+1}{2n} = \frac{1}{2} < 1.$

(Note that $\cos(n\pi)$ is just a fancy way of writing $(-1)^n$.) Therefore (ratio test) $\sum_{n=1}^{\infty}((n\cos(n\pi))/2^n)$ converges absolutely.

## The Alternating Series Test

We cannot use any of the previously developed tests to show that the alternating harmonic series converges; all of those tests apply only to (ultimately) positive series, and so they can test only for absolute convergence. Demonstrating convergence that is not absolute is generally harder to do. We present only one test that can establish such convergence; this test can only be used on a very special kind of series.

**THEOREM 14**

**The alternating series test**

Suppose that the sequence $\{a_n\}$ is positive, decreasing, and converges to 0, that is, suppose that

(i) $a_n \ge 0$ for $n = 1, 2, 3, \ldots$;

(ii) $a_{n+1} \le a_n$ for $n = 1, 2, 3, \ldots$;

(iii) $\lim_{n\to\infty} a_n = 0$.

Then the alternating series

$$\sum_{n=1}^{\infty} (-1)^{n-1} a_n = a_1 - a_2 + a_3 - a_4 + a_5 - \cdots$$

converges.

**DANGER**
Read this proof slowly and think about why each statement is true.
**DANGER**

***PROOF*** Since the sequence $\{a_n\}$ is decreasing, we have $a_{2n+1} \ge a_{2n+2}$. Therefore $s_{2n+2} = s_{2n} + a_{2n+1} - a_{2n+2} \ge s_{2n}$ for $n = 1, 2, 3, \ldots$; the even partial sums $\{s_n\}$ form an increasing sequence. Similarly $s_{2n+1} = s_{2n-1} - a_{2n} + a_{2n+1} \le s_{2n-1}$, so the odd partial sums $\{s_{2n-1}\}$ form a decreasing sequence. Since $s_{2n} = s_{2n-1} - a_{2n} \le s_{2n-1}$, we can say, for any $n \ge 1$, that

$$s_2 \le s_4 \le s_6 \le \cdots \le s_{2n} \le s_{2n-1} \le s_{2n-3} \le \cdots \le s_5 \le s_3 \le s_1.$$

Hence $s_2$ is a lower bound for the decreasing sequence $\{s_{2n-1}\}$ and $s_1$ is an upper bound for the increasing sequence $\{s_{2n}\}$. Both of these sequences therefore converge by the completeness of the real numbers:

$$\lim_{n\to\infty} s_{2n-1} = s_{\text{odd}}, \qquad \lim_{n\to\infty} s_{2n} = s_{\text{even}}.$$

Now $a_{2n} = s_{2n-1} - s_{2n}$, so $0 = \lim_{n\to\infty} a_{2n} = \lim_{n\to\infty}(s_{2n-1} - s_{2n}) = s_{\text{odd}} - s_{\text{even}}$. Therefore $s_{\text{odd}} = s_{\text{even}} = s$, say. Every partial sum $s_n$ is either of the form $s_{2n-1}$ or of the form $s_{2n}$. Thus $\lim_{n\to\infty} s_n = s$ exists and the series $\sum(-1)^{n-1}a_n$ converges to this sum $s$.

***REMARK*** Note that the series $\sum_{n=1}^{\infty}(-1)^{n-1}a_n$ begins with a positive term, $a_1$. The conclusion of Theorem 14 also holds for the series $\sum_{n=1}^{\infty}(-1)^n a_n$, which starts with a negative term, $-a_1$. (It is just the negative of the first series.) The theorem also remains valid if the conditions that $\{a_n\}$ is positive and decreasing are replaced by the corresponding *ultimate* versions:

$$\text{(i)} \quad a_n \geq 0 \quad \text{and} \quad \text{(ii)}\; a_{n+1} \leq a_n \quad \text{for } n = N, N+1, N+2, \ldots.$$

***REMARK*** The proof of Theorem 14 shows that every even partial sum is less than or equal to $s$ and every odd partial sum is greater than or equal to $s$. (The reverse is true if $(-1)^n$ is used instead of $(-1)^{n-1}$.) That is, $s$ lies between $s_{2n}$ and either $s_{2n-1}$ or $s_{2n+1}$. It follows that, for odd or even $n$,

$$|s - s_n| \leq |s_{n+1} - s_n| = a_{n+1}.$$

This proves the following theorem.

**THEOREM 15**

**Error estimate for alternating series**

If $\sum_{n=1}^{\infty} a_n$ is an alternating series, (that is, $\{a_n\}$ is an alternating sequence), and if the sequence $\{|a_n|\}$ is decreasing and converges to 0 so that $\sum_{n=1}^{\infty} a_n$ converges to sum $s$, then, for any $n \geq 1$, the $n$th partial sum $s_n$ of the series satisfies

$$|s - s_n| \leq |a_{n+1}|.$$

That is, the size of the error involved in using $s_n$ as an approximation to $s$ is less than the size of the first omitted term. If $|a_{n+1}| \leq |a_n|$ only holds ultimately, say for $n \geq N$, then the inequality $|s - s_n| \leq |a_{n+1}|$ also holds only for $n \geq N$.

**EXAMPLE 2** How many terms of the series $\displaystyle\sum_{n=1}^{\infty} \frac{(-1)^n}{1+2^n}$ are needed to compute the sum of the series with error less that 0.001?

***SOLUTION*** This series satisfies the hypotheses for Theorem 15 apart from an extra minus sign in each term. If we use the partial sum of the first $n$ terms of the series to approximate the sum of the series, the error will satisfy

$$|\text{error}| \leq |\text{first omitted term}| = \frac{1}{1+2^{n+1}}.$$

This error is less than 0.001 if $1 + 2^{n+1} > 1,000$. Since $2^{10} = 1,024$, $n + 1 = 10$ will do; we need 9 terms of the series to compute the sum to within 0.001 of its actual value. ■

When determining the convergence of a given series, it is best to consider first whether the series converges absolutely. If it does not, then there remains the possibility of conditional convergence.

**EXAMPLE 3** Test the following series for absolute and conditional convergence:

$$\text{(a)} \quad \sum_{n=1}^{\infty} \frac{(-1)^{n-1}}{n}, \qquad \text{(b)} \quad \sum_{n=2}^{\infty} \frac{\cos(n\pi)}{\ln n}, \qquad \text{(c)} \quad \sum_{n=1}^{\infty} \frac{(-1)^{n-1}}{n^4}$$

***SOLUTION*** The absolute values of the terms in series (a) and (b) are $1/n$ and $1/(\ln n)$, respectively. Since $1/(\ln n) > 1/n$, and $\sum_{n=k}^{\infty} 1/n$ diverges to infinity, neither series (a) nor (b) converges absolutely. However, both $\{1/n\}$ and $\{1/(\ln n)\}$ are positive, decreasing sequences that converge to 0. Therefore both (a) and (b) converge by Theorem 14. Each of these series is conditionally convergent.

Series (c) is absolutely convergent because $|(-1)^{n-1}/n^4| = 1/n^4$, and $\sum_{n=1}^{\infty} 1/n^4$ is a convergent $p$-series ($p = 4 > 1$). We could establish its convergence using Theorem 14, but there is no need to do that since every absolutely convergent series is convergent (Theorem 13). ■

**EXAMPLE 4** For what values of $x$ does the series $\displaystyle\sum_{n=1}^{\infty} \frac{(x-5)^n}{n\,2^n}$ converge absolutely? converge conditionally? diverge?

***SOLUTION*** For such series whose terms involve functions of a variable $x$, it is usually wisest to begin testing for absolute convergence with the ratio test. We have

$$\rho = \lim_{n\to\infty} \left| \frac{(x-5)^{n+1}}{(n+1)2^{n+1}} \Big/ \frac{(x-5)^n}{n\,2^n} \right| = \lim_{n\to\infty} \frac{n}{n+1} \left| \frac{x-5}{2} \right| = \left| \frac{x-5}{2} \right|.$$

The series converges absolutely if $|(x-5)/2| < 1$. This inequality is equivalent to $|x-5| < 2$, (the distance from $x$ to 5 is less than 2), that is, $3 < x < 7$. If $x < 3$ or $x > 7$, then $|(x-5)/2| > 1$. The series diverges; its terms do not approach zero.

If $x = 3$, the series is $\sum_{n=1}^{\infty}((-1)^n/n)$, which converges conditionally (it is an alternating harmonic series); if $x = 7$, the series is the harmonic series $\sum_{n=1}^{\infty} 1/n$, which diverges to infinity. Hence the given series converges absolutely on the open interval $(3, 7)$, converges conditionally at $x = 3$, and diverges everywhere else. ■

**EXAMPLE 5** For what values of $x$ does the series $\displaystyle\sum_{n=0}^{\infty}(n+1)^2\left(\frac{x}{x+2}\right)^n$ converge absolutely? converge conditionally? diverge?

***SOLUTION*** Again we begin with the ratio test.

$$\begin{aligned}\rho &= \lim_{n\to\infty} \left| (n+2)^2 \left(\frac{x}{x+2}\right)^{n+1} \Big/ (n+1)^2 \left(\frac{x}{x+2}\right)^n \right| \\ &= \lim_{n\to\infty} \left(\frac{n+2}{n+1}\right)^2 \left|\frac{x}{x+2}\right| = \left|\frac{x}{x+2}\right| = \frac{|x|}{|x+2|}\end{aligned}$$

The series converges absolutely if $|x|/|x+2| < 1$. This condition says that the distance from $x$ to 0 is less than the distance from $x$ to $-2$. Hence $x > -1$. The series diverges if $|x|/|x+2| > 1$, that is, if $x < -1$. If $x = -1$, the series is $\sum_{n=0}^{\infty}(-1)^n(n+1)^2$, which diverges. We conclude that the series converges absolutely for $x > -1$, converges conditionally nowhere, and diverges for $x \le -1$. ■

When using the alternating series test, it is important to verify (at least mentally) that *all three conditions* (i)–(iii) are satisfied. (As mentioned above, conditions (i) and (ii) need only be satisfied ultimately.)

**EXAMPLE 6** Test for convergence:

(a) $$\sum_{n=1}^{\infty}(-1)^{n-1}\frac{n+1}{n},$$

(b) $$1-\frac{1}{4}+\frac{1}{3}-\frac{1}{16}+\frac{1}{5}-\cdots=\sum_{n=1}^{\infty}(-1)^{n-1}a_n, \quad \text{where}$$

$$a_n=\begin{cases}1/n & \text{if } n \text{ is odd,}\\ 1/n^2 & \text{if } n \text{ is even.}\end{cases}$$

***SOLUTION***

(a) Here $a_n = (n+1)/n$ is positive and decreases as $n$ increases. However, $\lim_{n\to\infty} a_n = 1 \neq 0$. The alternating series test does not apply. In fact, the given series diverges, because its terms do not approach 0.

(b) This series alternates, $a_n$ is positive, and $\lim_{n\to\infty} a_n = 0$. However, $\{a_n\}$ is not decreasing (even ultimately). Once again the alternating series test cannot be applied. In fact, since

$$-\frac{1}{4}-\frac{1}{16}-\cdots-\frac{1}{(2n)^2}-\cdots \qquad \text{converges, and}$$

$$1+\frac{1}{3}+\frac{1}{5}+\cdots+\frac{1}{2n-1}+\cdots \qquad \text{diverges to infinity,}$$

it is readily seen that the given series diverges to infinity. ■

## Rearranging the Terms in a Series

The basic difference between absolute and conditional convergence is that when a series $\sum_{n=1}^{\infty} a_n$ converges absolutely it does so because its terms $\{a_n\}$ decrease in size fast enough that their sum can be finite even if no cancellation occurs due to terms of opposite sign. If cancellation is required to make the series converge (because the terms decrease slowly), then the series can only converge conditionally.

Consider the alternating harmonic series

$$1-\frac{1}{2}+\frac{1}{3}-\frac{1}{4}+\frac{1}{5}-\frac{1}{6}+\cdots.$$

This series converges, but only conditionally. If we take the subseries containing only the positive terms, we get the series

$$1+\frac{1}{3}+\frac{1}{5}+\frac{1}{7}+\cdots,$$

which diverges to infinity. Similarly, the subseries of negative terms

$$-\frac{1}{2}-\frac{1}{4}-\frac{1}{6}-\frac{1}{8}-\cdots$$

diverges to negative infinity.

If a series converges absolutely, the subseries consisting of positive terms and the subseries consisting of negative terms must each converge to a finite sum. If a series converges conditionally, the positive and negative subseries will both diverge, to $\infty$ and $-\infty$, respectively.

Using these facts we can answer a question raised at the beginning of Section 1.2. If we rearrange the terms of a convergent series so that they are added in a different order, must the rearranged series converge or not, and if it does will it converge to the same sum? The answer depends on whether the original series was absolutely convergent or merely conditionally convergent.

**THEOREM 16** **Convergence of rearrangements of a series**

(a) If the terms of an absolutely convergent series are rearranged so that addition occurs in a different order, the rearranged series still converges to the same sum as the original series.

(b) If a series is conditionally convergent, and $L$ is any real number, then the terms of the series can be rearranged so as to make the series converge (conditionally) to the sum $L$. It can also be rearranged so as to diverge to $\infty$ or to $-\infty$, or just to diverge.

Part (b) shows that conditional convergence is a rather suspect kind of convergence, being dependent on the order in which the terms are added. We will not present a formal proof of the theorem, but will give an example suggesting what is involved. (See also Exercise 34 below.)

**EXAMPLE 7** In Chapter 2 we will show that the alternating harmonic series

$$\sum_{n=1}^{\infty} \frac{(-1)^{n-1}}{n} = 1 - \frac{1}{2} + \frac{1}{3} - \frac{1}{4} + \frac{1}{5} - \frac{1}{6} + \frac{1}{7} - \cdots$$

converges (conditionally) to the sum $\ln 2$. Describe how to rearrange its terms so that it converges to 8 instead.

***SOLUTION*** Start adding terms of the positive subseries

$$1 + \frac{1}{3} + \frac{1}{5} + \cdots$$

and keep going until the partial sum exceeds 8. (It will, eventually, because the positive subseries diverges to infinity.) Then add the first term $-1/2$ of the negative subseries

$$-\frac{1}{2} - \frac{1}{4} - \frac{1}{6} - \cdots.$$

This will reduce the partial sum below 8 again. Now resume adding terms of the positive subseries until the partial sum climbs above 8 once more. Then add the second term of the negative subseries and the partial sum will drop below 8. Keep repeating this procedure, alternately adding terms of the positive subseries to force the sum above 8 and then terms of the negative subseries to force it below 8. Since both subseries have infinitely many terms and diverge to $\infty$ and $-\infty$, respectively, eventually every term of the original series will be included, and the partial sums of the new series will oscillate back and forth around 8, converging to that number. Of course, any number other than 8 could also be used in place of 8. ■

## EXERCISES 1.4

Determine whether the series in Exercises 1–12 converge absolutely, converge conditionally, or diverge.

**1.** $\displaystyle\sum_{n=1}^{\infty} \frac{(-1)^{n-1}}{\sqrt{n}}$

**2.** $\displaystyle\sum_{n=1}^{\infty} \frac{(-1)^{n}}{n^2 + \ln n}$

**3.** $\displaystyle\sum_{n=1}^{\infty} \frac{\cos(n\pi)}{(n+1)\ln(n+1)}$

**4.** $\displaystyle\sum_{n=1}^{\infty} \frac{(-1)^{2n}}{2^n}$

**5.** $\displaystyle\sum_{n=0}^{\infty} \frac{(-1)^{n}(n^2-1)}{n^2+1}$

**6.** $\displaystyle\sum_{n=1}^{\infty} \frac{(-2)^{n}}{n!}$

**7.** $\displaystyle\sum_{n=1}^{\infty} \frac{(-1)^n}{n\pi^n}$ **8.** $\displaystyle\sum_{n=0}^{\infty} \frac{-n}{n^2+1}$

**9.** $\displaystyle\sum_{n=1}^{\infty} (-1)^n \frac{20n^2-n-1}{n^3+n^2+33}$ **10.** $\displaystyle\sum_{n=1}^{\infty} \frac{100\cos(n\pi)}{2n+3}$

**11.** $\displaystyle\sum_{n=1}^{\infty} \frac{n!}{(-100)^n}$ **12.** $\displaystyle\sum_{n=10}^{\infty} \frac{\sin(n+1/2)\pi}{\ln\ln n}$

For the series in Exercises 13–16, find the smallest integer $n$ that ensures that the partial sum $s_n$ approximates the sum $s$ of the series with error less than 0.001 in absolute value.

**13.** $\displaystyle\sum_{n=1}^{\infty} (-1)^{n-1} \frac{n}{n^2+1}$ **14.** $\displaystyle\sum_{n=0}^{\infty} \frac{(-1)^n}{(2n)!}$

**15.** $\displaystyle\sum_{n=1}^{\infty} (-1)^{n-1} \frac{n}{2^n}$ **16.** $\displaystyle\sum_{n=0}^{\infty} (-1)^n \frac{3^n}{n!}$

Determine the values of $x$ for which the series in Exercises 17–28 converge absolutely, converge conditionally, or diverge.

**17.** $\displaystyle\sum_{n=0}^{\infty} \frac{x^n}{\sqrt{n+1}}$ **18.** $\displaystyle\sum_{n=1}^{\infty} \frac{(x-2)^n}{n^2 2^{2n}}$

**19.** $\displaystyle\sum_{n=0}^{\infty} (-1)^n \frac{(x-1)^n}{2n+3}$ **20.** $\displaystyle\sum_{n=1}^{\infty} \frac{1}{2n-1}\left(\frac{3x+2}{-5}\right)^n$

**21.** $\displaystyle\sum_{n=2}^{\infty} \frac{x^n}{2^n \ln n}$ **22.** $\displaystyle\sum_{n=1}^{\infty} \frac{(4x+1)^n}{n^3}$

**23.** $\displaystyle\sum_{n=1}^{\infty} \frac{(2x+3)^n}{n^{1/3}4^n}$ **24.** $\displaystyle\sum_{n=1}^{\infty} \frac{1}{n}\left(1+\frac{1}{x}\right)^n$

**25.** $\displaystyle\sum_{n=1}^{\infty} \frac{1}{n^2}\left(1-\frac{1}{x}\right)^n$ **26.** $\displaystyle\sum_{n=1}^{\infty} \frac{(x^2-1)^n}{\sqrt{n}}$

**27.** $\displaystyle\sum_{n=0}^{\infty} n(x^2-x-1)^n$ **28.** $\displaystyle\sum_{n=1}^{\infty} \frac{\sin^n x}{n}$

* **29.** Does the alternating series test apply directly to the series $\sum_{n=1}^{\infty}(1/n)\sin(n\pi/2)$? Determine whether the series converges.

* **30.** Show that the series $\sum_{n=1}^{\infty} a_n$ converges absolutely if $a_n = 10/n^2$ for even $n$ and $a_n = -1/10n^3$ for odd $n$.

* **31.** Which of the following statements are TRUE and which are FALSE? Justify your assertion of truth, or give a counterexample to show falsehood.

(a) If $\sum_{n=1}^{\infty} a_n$ converges, then $\sum_{n=1}^{\infty}(-1)^n a_n$ converges.

(b) If $\sum_{n=1}^{\infty} a_n$ converges and $\sum_{n=1}^{\infty}(-1)^n a_n$ converges, then $\sum_{n=1}^{\infty} a_n$ converges absolutely.

(c) If $\sum_{n=1}^{\infty} a_n$ converges absolutely, then

$\sum_{n=1}^{\infty}(-1)^n a_n$ converges absolutely.

* **32.** (a) Use a Riemann sum argument to show that

$$\ln n! \geq \int_1^n \ln t\, dt = n\ln n - n + 1.$$

(b) For what values of $x$ does the series $\sum_{n=1}^{\infty} \dfrac{n!x^n}{n^n}$ converge absolutely? converge conditionally? diverge? (*Hint:* first use the ratio test. To test the cases where $\rho = 1$, you may find the inequality in part (a) useful.)

* **33.** For what values of $x$ does the series $\sum_{n=1}^{\infty} \dfrac{(2n)!x^n}{2^{2n}(n!)^2}$ converge absolutely? converge conditionally? diverge? (*Hint:* see Exercise 36 of Section 1.3.)

* **34.** Devise procedures for rearranging the terms of the alternating harmonic series so that the rearranged series (a) diverges to $\infty$, (b) converges to $-2$.

## 1.5 ESTIMATING THE SUM OF A SERIES

So far we have been concerned mainly with determining whether a given series converges or diverges, that is, whether or not it has a finite sum. Except for very special series (geometric series, telescoping series) we have not actually calculated the sums of the series we have shown to converge. Indeed, there are no elementary techniques for calculating the sums of many series, though we will learn how to find the sums of some other series in the next chapter.

When it is not practical to find the exact sum of a convergent series, we can still approximate that sum by using a partial sum $s_n$ of the series:

$$s = \sum_{k=1}^{\infty} a_k = a_1 + a_2 + a_3 + \cdots$$

$$s \approx s_n = \sum_{k=1}^{n} a_k = a_1 + a_2 + a_3 + \cdots + a_n.$$

Of course, such an approximation is of little use unless we know how good an approximation it is, that is, unless we have a bound for the size of the error, $|s - s_n|$. Theorem 15 in the previous section provided such an error estimate for series that satisfied the conditions of the alternating series test. In this section we will obtain some error estimates for positive series.

Observe that the error $s - s_n$ in the approximation $s \approx s_n$ is just the *tail* of the given series; it is the sum of the rest of the terms of the series beyond those included in $s_n$:

$$s - s_n = \sum_{k=n+1}^{\infty} a_k = a_{n+1} + a_{n+2} + a_{n+3} + \cdots$$

If the given series converges, then the error $s - s_n$ approaches 0 as $n$ increases towards infinity (since $\lim s_n = s$), but we would still like to know how fast that approach is and how large $n$ need be to ensure that the error is within tolerable bounds. Various convergence tests involve, implicitly or explicitly, techniques for finding bounds for the error. We shall investigate error-bounding techniques arising from the integral test and the ratio test.

## Integral Bounds

Suppose that $a_k = f(k)$ for $k = n+1,\ n+2,\ n+3, \ldots$, where $f$ is a positive, continuous function, decreasing at least on the interval $[n, \infty)$. We have:

$$\begin{aligned} s - s_n &= \sum_{k=n+1}^{\infty} f(k) \\ &= \text{sum of areas of rectangles shaded in Figure 1.5(a)} \\ &\le \int_n^{\infty} f(x)\,dx. \end{aligned}$$

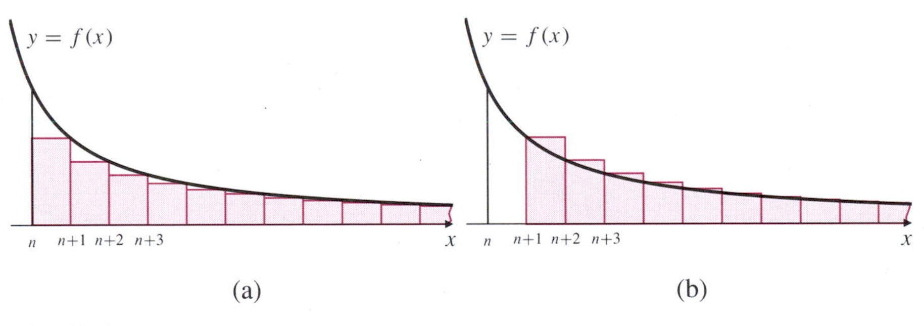


Figure 1.5

Similarly,

$$\begin{aligned} s - s_n &= \text{sum of areas of rectangles in Figure 1.5(b)} \\ &\ge \int_{n+1}^{\infty} f(x)\,dx. \end{aligned}$$

If we define

$$A_n = \int_n^{\infty} f(x)\,dx$$

then we can combine the above inequalities to obtain

$$A_{n+1} \le s - s_n \le A_n,$$

or, equivalently:

$$s_n + A_{n+1} \le s \le s_n + A_n.$$

The error in the approximation $s \approx s_n$ satisfies $0 \le s - s_n \le A_n$. However, since $s$ must lie in the interval $[s_n + A_{n+1}, s_n + A_n]$, we can do better by using the midpoint $s_n^*$ of this interval as an approximation for $s$. The error is then less than half the length $A_n - A_{n+1}$ of the interval:

**A better integral approximation**

The error $|s - s_n^*|$ in the approximation

$$s \approx s_n^* = s_n + \frac{A_{n+1} + A_n}{2}, \qquad \text{where} \quad A_n = \int_n^\infty f(x)\,dx,$$

satisfies $\quad |s - s_n^*| \le \dfrac{A_n - A_{n+1}}{2}.$

(Whenever a quantity is known to lie in a certain interval, the midpoint of that interval can be used to approximate the quantity, and the absolute value of the error in that approximation does not exceed half the length of the interval.)

**EXAMPLE 1** Find the best approximation $s_n^*$ to the sum $s$ of the series $\sum_{n=1}^{\infty} 1/n^2$, making use of the partial sum $s_n$ of the first $n$ terms. How large would $n$ have to be to ensure that the approximation $s \approx s_n^*$ has error less than 0.001 in absolute value? How large would $n$ have to be to ensure that the approximation $s \approx s_n$ has error less than 0.001 in absolute value?

***SOLUTION*** Since $f(x) = 1/x^2$ is positive, continuous, and decreasing on $[1, \infty)$ for any $n = 1, 2, 3, \dots$, we have

$$s_n + A_{n+1} \le s \le s_n + A_n$$

where

$$A_n = \int_n^\infty \frac{dx}{x^2} = \lim_{R\to\infty} \left(-\frac{1}{x}\right)\Bigg|_n^\infty = \frac{1}{n}.$$

The best approximation to $s$ using $s_n$ is

$$\begin{aligned} s_n^* &= s_n + \frac{1}{2}\left(\frac{1}{n+1} + \frac{1}{n}\right) = s_n + \frac{2n+1}{2n(n+1)} \\ &= 1 + \frac{1}{4} + \frac{1}{9} + \cdots + \frac{1}{n^2} + \frac{2n+1}{2n(n+1)}. \end{aligned}$$

The error in this approximation satisfies

$$|s - s_n^*| \le \frac{1}{2}\left(\frac{1}{n} - \frac{1}{n+1}\right) = \frac{1}{2n(n+1)} \le 0.001$$

provided $2n(n+1) \ge 1/0.001 = 1,000$. It is easily checked that this condition is satisfied if $n \ge 22$; the approximation

$$s \approx s_{22}^* = 1 + \frac{1}{4} + \frac{1}{9} + \cdots + \frac{1}{22^2} + \frac{45}{44 \times 23}$$

will have error with absolute value not exceeding 0.001.

Had we used the approximation $s \approx s_n$ we could only have concluded that

$$0 \le s - s_n \le A_n = \frac{1}{n} < 0.001,$$

provided $n > 1000$; we would need 1000 terms of the series to get the desired accuracy. ■

**EXAMPLE 2** Find the sum of the series $\sum_{n=1}^{\infty} 1/n^4$ with error less than 0.001.

***SOLUTION*** Let $f(x) = 1/x^4$, which is positive, continuous, and decreasing on $[1, \infty)$. Let

$$A_n = \int_n^{\infty} \frac{dx}{x^4} = \lim_{R\to\infty} \left(-\frac{1}{3x^3}\right)\bigg|_n^R = \frac{1}{3n^3}.$$

We use the approximation

$$s \approx s_n^* = s_n + \frac{1}{2}\left(\frac{1}{3(n+1)^3} + \frac{1}{3n^3}\right).$$

The error satisfies

$$|s - s_n^*| \le \frac{1}{2}\left(\frac{1}{3n^3} - \frac{1}{3(n+1)^3}\right) = \frac{1}{6}\,\frac{(n+1)^3 - n^3}{n^3(n+1)^3} = \frac{1}{6}\,\frac{3n^2+3n+1}{n^3(n+1)^3} < \frac{7}{6n^4}.$$

We have used $3n^2 + 3n + 1 \le 7n^2$ and $n^3(n+1)^3 > n^6$ to obtain the last inequality. We will have $|s - s_n^*| < 0.001$ provided

$$\frac{7}{6n^4} < 0.001,$$

that is, if $n^4 > 7000/6$. Since $6^4 = 1296 > 7000/6$, $n = 6$ will do. Thus

$$\begin{aligned}\sum_{n=1}^{\infty} \frac{1}{n^4} &\approx s_6^* = 1 + \frac{1}{2^4} + \frac{1}{3^4} + \frac{1}{4^4} + \frac{1}{5^4} + \frac{1}{6^4} + \frac{1}{6}\left(\frac{1}{7^3} + \frac{1}{6^3}\right)\\ &\approx 1.082 \qquad \text{with error less than 0.001 in absolute value.}\end{aligned}$$

Observe that only six terms are required to enable us to estimate the sum of the series $\sum_{n=1}^{\infty} 1/n^4$ in Example 2 within the same tolerance for which we needed 22 terms of the series $\sum_{n=1}^{\infty} 1/n^2$ in Example 1. Evidently the series $\sum_{n=1}^{\infty} 1/n^2$ *converges more slowly* than does $\sum_{n=1}^{\infty} 1/n^4$. The use of integrals to bound the tails of series is appropriate for those series for which the integral test would be appropriate to demonstrate convergence. In particular, this class includes series whose terms are rational functions of $n$.

## Geometric Bounds

Suppose that an inequality of the form

$$0 \le a_k \le Kr^k$$

holds for $k = n+1, n+2, n+3, \ldots$, where $K$ and $r$ are constants and $r < 1$. We can then use a geometric series to bound the tail of $\sum_{n=1}^{\infty} a_n$.

$$\begin{aligned}0 \le s - s_n &= \sum_{k=n+1}^{\infty} a_k \le \sum_{k=n+1}^{\infty} Kr^k\\ &= Kr^{n+1}(1 + r + r^2 + \cdots)\\ &= \frac{Kr^{n+1}}{1-r}.\end{aligned}$$

Since $r < 1$ the series converges and the error approaches 0 at an exponential rate as $n$ increases.

**EXAMPLE 3** In Chapter 2 we will show that

$$e = \frac{1}{0!} + \frac{1}{1!} + \frac{1}{2!} + \frac{1}{3!} + \cdots = \sum_{n=0}^{\infty} \frac{1}{n!}$$

(Recall that $0! = 1$.) Estimate the error if the sum $s_n$ of the first $n$ terms of the series is used to approximate $e$. Find $e$ to three-decimal-place accuracy using the series.

***SOLUTION*** We have

$$\begin{aligned} s_n &= \frac{1}{0!} + \frac{1}{1!} + \frac{1}{2!} + \frac{1}{3!} + \cdots + \frac{1}{(n-1)!} \\ &= 1 + 1 + \frac{1}{2} + \frac{1}{6} + \frac{1}{24} + \cdots + \frac{1}{(n-1)!}. \end{aligned}$$

(Since the series starts with the term for $n = 0$, the $n$th term is $1/(n-1)!$.) We can estimate the error in the approximation $s \approx s_n$ as follows:

$$\begin{aligned} 0 < s - s_n &= \frac{1}{n!} + \frac{1}{(n+1)!} + \frac{1}{(n+2)!} + \frac{1}{(n+3)!} + \cdots \\ &= \frac{1}{n!}\left(1 + \frac{1}{n+1} + \frac{1}{(n+1)(n+2)} + \frac{1}{(n+1)(n+2)(n+3)} + \cdots\right) \\ &< \frac{1}{n!}\left(1 + \frac{1}{n+1} + \frac{1}{(n+1)^2} + \frac{1}{(n+1)^3} + \cdots\right) \end{aligned}$$

since $n + 2 > n + 1$, $n + 3 > n + 1$, and so on. The latter series is geometric, so

$$0 < s - s_n < \frac{1}{n!}\,\frac{1}{1 - \dfrac{1}{n+1}} = \frac{n+1}{n!n}.$$

If we want to evaluate $e$ accurately to three decimal places, then we must ensure that the error is less than 5 in the fourth decimal place, that is, that the error is less than 0.0005. Hence we want

$$\frac{n+1}{n}\,\frac{1}{n!} < 0.0005 = \frac{1}{2000}.$$

Since $7! = 5040$, but $6! = 720$, we can use $n = 7$ but no smaller. We have

$$\begin{aligned} e &\approx 1 + 1 + \frac{1}{2!} + \frac{1}{3!} + \frac{1}{4!} + \frac{1}{5!} + \frac{1}{6!} \\ &= 2 + \frac{1}{2} + \frac{1}{6} + \frac{1}{24} + \frac{1}{120} + \frac{1}{720} \approx 2.718 \quad \text{to three decimal places.} \end{aligned}$$ ■

It is appropriate to use geometric series to bound the tails of positive series whose convergence would be demonstrated by the ratio test. Such series converge ultimately faster than any $p$-series $\sum_{n=1}^{\infty} n^{-p}$, for which the limit ratio is $\rho = 1$.

## EXERCISES 1.5

In Exercises 1–6, let $s_n$ denote the sum of the first $n$ terms of the given series. Using $s_n$, find the smallest interval that you can be sure contains the sum $s$ of the series. If the midpoint $s_n^*$ of this interval is used to approximate $s$, how large should $n$ be chosen to ensure that the error is less than 0.001?

**1.** $\displaystyle\sum_{k=1}^{\infty} \frac{1}{k^{10}}$ **2.** $\displaystyle\sum_{k=1}^{\infty} \frac{1}{k^3}$

**3.** $\displaystyle\sum_{k=1}^{\infty} \frac{1}{k^{3/2}}$ **4.** $\displaystyle\sum_{k=2}^{\infty} \frac{1}{k(\ln k)^2}$

**5.** $\displaystyle\sum_{k=1}^{\infty} \frac{1}{e^k}$ **6.** $\displaystyle\sum_{k=1}^{\infty} \frac{1}{k^2+4}$

For each positive series in Exercises 7–10, find the best upper bound you can for the error $s - s_n$ encountered if the partial sum $s_n$ is used to approximate the sum $s$ of the series. How many terms of each series do you need to be sure that the approximation has error less than 0.001?

**7.** $\displaystyle\sum_{k=1}^{\infty} \frac{1}{2^k k!}$ **8.** $\displaystyle\sum_{n=1}^{\infty} \frac{1}{(2n-1)!}$

**9.** $\displaystyle\sum_{n=0}^{\infty} \frac{2^n}{(2n)!}$ **10.** $\displaystyle\sum_{n=1}^{\infty} \frac{1}{n^n}$

* **11.** (a) Show that if $k > 0$ and $n$ is a positive integer, then $n < \frac{1}{k}(1+k)^n$.

(b) Use the estimate in (a) with $0 < k < 1$ to obtain an upper bound for the sum of the series $\sum_{n=0}^{\infty} n/2^n$. For what value of $k$ is this bound least?

(c) If we use the sum $s_n$ of the first $n$ terms to approximate the sum $s$ of the series in (b), obtain an upper bound for the error $s - s_n$ using the inequality from (a). For given $n$, find $k$ to minimize this upper bound.

* **12. (Improving the convergence of a series)** We know that $\sum_{n=1}^{\infty} 1/(n(n+1)) = 1$. (See Example 3 of Section 1.2.) Since

$$\frac{1}{n^2} = \frac{1}{n(n+1)} + c_n, \quad \text{where} \quad c_n = \frac{1}{n^2(n+1)},$$

we have $\displaystyle\sum_{n=1}^{\infty} \frac{1}{n^2} = 1 + \sum_{n=1}^{\infty} c_n.$

The series, $\sum_{n=1}^{\infty} c_n$ converges more rapidly than does $\sum_{n=1}^{\infty} 1/n^2$ because its terms decrease like $1/n^3$. Hence fewer terms of that series will be needed to compute $\sum_{n=1}^{\infty} 1/n^2$ to any desired degree of accuracy than would be needed if we calculated with $\sum_{n=1}^{\infty} 1/n^2$ directly. Using integral upper and lower bounds, determine a value of $n$ for which the modified partial sum $s_n^*$ for the series $\sum_{n=1}^{\infty} c_n$ approximates the sum of that series with error less than 0.001 in absolute value. Hence determine $\sum_{n=1}^{\infty} 1/n^2$ to within 0.001 of its true value.
(The technique exibited in this exercise is known as *improving the convergence* of a series. It can be applied to estimating the sum $\sum a_n$ if we know the sum $\sum b_n$ and if $a_n - b_n = c_n$, where $|c_n|$ decreases faster than $|a_n|$ as $n$ tends to infinity.)

**13.** Consider the series $s = \sum_{n=1}^{\infty} 1/(2^n + 1)$, and the partial sum $s_n$ of its first $n$ terms.

(a) How large need $n$ be taken to ensure that the error in the approximation $s \approx s_n$ is less than 0.001 in absolute value?

(b) The geometric series $\sum_{n=1}^{\infty} 1/2^n$ converges to 1. If

$$b_n = \frac{1}{2^n+1} - \frac{1}{2^n}$$

for $n = 1, 2, 3, \ldots$, how many terms of the series $\sum_{n=1}^{\infty} b_n$ are needed to calculate its sum to within 0.001?

(c) Use the result of part (b) to calculate the $\sum_{n=1}^{\infty} 1/(2^n+1)$ to within 0.001.

# CHAPTER REVIEW

## Key Ideas

- **What does it mean to say that the sequence $\{a_n\}$**
  - ◇ is bounded above?
  - ◇ is ultimately positive?
  - ◇ is alternating?
  - ◇ is increasing?
  - ◇ converges?
  - ◇ diverges to infinity?
  - ◇ is the sequence of partial sums of $\sum_{n=1}^{\infty} b_n$?
- **What does it mean to say that the series $\sum_{n=1}^{\infty} a_n$**
  - ◇ converges?
  - ◇ diverges?
  - ◇ is geometric?
  - ◇ is harmonic?
  - ◇ is telescoping?
  - ◇ is a $p$-series?
  - ◇ is positive?
  - ◇ converges absolutely?
  - ◇ converges conditionally?
- **State the following convergence tests for series.**
  - ◇ the integral test
  - ◇ the comparsion test
  - ◇ the limit comparison test
  - ◇ the ratio test
  - ◇ the alternating series test

- **How can you find bounds for the tail of a series?**
- **What is a bound for the tail of an alternating series?**

## Review Exercises

In Exercises 1–6, determine whether the given sequence converges or not, and find its limit if it does converge.

**1.** $\left\{\dfrac{n+\sqrt{n}}{\sqrt{n}-3n}\right\}$  **2.** $\left\{\dfrac{n^{100}+2^n\pi}{2^n}\right\}$

**3.** $\left\{\dfrac{(-1)^n e^n}{n!}\right\}$  **4.** $\left\{\dfrac{(2n)!}{(n!)^2}\right\}$

**5.** $\left\{\dfrac{\ln n}{\tan^{-1} n}\right\}$  **6.** $\left\{\dfrac{(-1)^n n^2}{\pi n(n-\pi)}\right\}$

**7.** Let $a_1 > \sqrt{2}$, and let

$$a_{n+1} = \frac{a_n}{2} + \frac{1}{a_n} \quad \text{for} \quad n = 1.2, 3, \ldots$$

Show that $\{a_n\}$ is decreasing and that $a_n > \sqrt{2}$ for $n \geq 1$. Why must $\{a_n\}$ converge? Find $\lim_{n\to\infty} a_n$.

**8.** Find the limit of the sequence $\{\ln\ln(n+1) - \ln\ln n\}$.

Evaluate the sums of the series in Exercises 9–12.

**9.** $\displaystyle\sum_{n=1}^{\infty} 2^{-(n-5)/2}$  **10.** $\displaystyle\sum_{n=0}^{\infty} \frac{4^{n-1}}{(\pi-1)^{2n}}$

**11.** $\displaystyle\sum_{n=1}^{\infty} \frac{1}{n^2-\frac{1}{4}}$  **12.** $\displaystyle\sum_{n=1}^{\infty} \frac{1}{n^2-\frac{9}{4}}$

Determine whether the series in Exercises 13–22 converge or not. Give reasons for your answers.

**13.** $\displaystyle\sum_{n=1}^{\infty} \frac{n-1}{n^3}$  **14.** $\displaystyle\sum_{n=1}^{\infty} \cos\frac{1}{n}$

**15.** $\displaystyle\sum_{n=1}^{\infty} \frac{n+2^n}{n^3+2^n}$  **16.** $\displaystyle\sum_{n=1}^{\infty} \frac{n+2^n}{1+3^n}$

**17.** $\displaystyle\sum_{n=1}^{\infty} \frac{n}{(1+n)(1+n\sqrt{n})}$  **18.** $\displaystyle\sum_{n=1}^{\infty} \frac{n^2}{(1+n)(1+n\sqrt{n})}$

**19.** $\displaystyle\sum_{n=1}^{\infty} \frac{n\sqrt{n}}{(1+n)(1+n\sqrt{n})}$  **20.** $\displaystyle\sum_{n=1}^{\infty} \frac{n^2}{(1+2^n)(1+n\sqrt{n})}$

**21.** $\displaystyle\sum_{n=1}^{\infty} \frac{3^{2n+1}}{n!}$  **22.** $\displaystyle\sum_{n=1}^{\infty} \frac{n!}{(n+2)!+1}$

Do the series in Exercises 23–26 converge absolutely, converge conditionally, or diverge?

**23.** $\displaystyle\sum_{n=1}^{\infty} \frac{(-1)^{n-1}}{1+n^3}$  **24.** $\displaystyle\sum_{n=1}^{\infty} \frac{(-1)^n}{2^n-n}$

**25.** $\displaystyle\sum_{n=10}^{\infty} \frac{(-1)^{n-1}}{\ln\ln n}$  **26.** $\displaystyle\sum_{n=1}^{\infty} \frac{n^2\cos(n\pi)}{1+n^3}$

For what values of $x$ do the series in Exercises 27–30 converge absolutely? converge conditionally? diverge?

**27.** $\displaystyle\sum_{n=1}^{\infty} \frac{(x-2)^n}{3^n\sqrt{n}}$  **28.** $\displaystyle\sum_{n=1}^{\infty} \frac{(5-2x)^n}{n}$

**29.** $\displaystyle\sum_{n=1}^{\infty} \frac{(\sin x)^n}{\ln(n+1)}$  **30.** $\displaystyle\sum_{n=0}^{\infty} \frac{1}{\sqrt{n+1}}\left(1-\frac{1}{x}\right)^n$

Determine the sums of the series in Exercises 31–34 to within 0.001.

**31.** $\displaystyle\sum_{n=1}^{\infty} \frac{1}{n^3}$  **32.** $\displaystyle\sum_{n=1}^{\infty} \frac{(-1)^{n-1}}{n^2}$

**33.** $\displaystyle\sum_{n=0}^{\infty} \frac{(-1)^n}{1+e^n}$  **34.** $\displaystyle\sum_{n=1}^{\infty} \frac{1}{4+n^2}$

## Express Yourself

Answer these questions in words, using complete sentences and, if necessary, paragraphs. Use mathematical symbols only where necessary.

1. Discuss the difference between the concepts of convergence of a sequence and convergence of the series obtained by adding the terms of the sequence. What does each type of convergence imply about the other?

2. What kinds of series are best tested for convergence with the ratio test? For what kinds of series will the ratio test not provide useful information?

3. What does it mean to say that one series converges more slowly than another series? Why might it be important to know that a series converges quickly?

4. Why is the sum of a conditionally convergent series less to be trusted than the sum of an absolutely convergent series?

## Challenging Problems

1. **(A refinement of the ratio test)** Suppose $a_n > 0$ and $a_{n+1}/a_n \geq n/n+1$ for all $n$. Show that $\sum_{n=1}^{\infty} a_n$ diverges. (*Hint:* $a_n \geq K/n$ for some constant $K$.)

* 2. **(Summation by parts)** Let $\{u_n\}$ and $\{v_n\}$ be two sequences and let $s_n = \sum_{k=1}^{n} v_k$.

   (a) Show that $\sum_{k=1}^{n} u_k v_k = u_{n+1}s_n + \sum_{k=1}^{n}(u_k - u_{k+1})s_n$. (*Hint:* write $v_n = s_n - s_{n-1}$, with $s_0 = 0$, and rearrange the sum.)

   (b) If $\{u_n\}$ is positive, decreasing, and convergent to 0, and if $\{v_n\}$ has bounded partial sums, $|s_n| \leq K$ for all $n$, where $K$ is a constant, show that $\sum_{n=1}^{\infty} u_n v_n$ converges. (*Hint:* show that the series $\sum_{n=1}^{\infty}(u_n - u_{n+1})s_n$ converges by comparing it to the telescoping series $\sum_{n=1}^{\infty}(u_n - u_{n+1})$.)

* **3.** Show that $\sum_{n=1}^{\infty}(1/n)\sin(nx)$ converges for every $x$. *Hint:* if $x$ is an integer multiple of $\pi$ all the terms in the series are 0 so there is nothing to prove. Otherwise, $\sin(x/2)\neq 0$. In this case show that

$$\sum_{n=1}^{N}\sin(nx)=\frac{\cos(x/2)-\cos((N+1/2)x)}{2\sin(x/2)}$$

using the identity

$$\sin a\sin b=\frac{\cos(a-b)-\cos(a+b)}{2}$$

to make the sum telescope. Then apply the result of Exercise 2(b) with $u_n=1/n$ and $v_n=\sin(nx)$.

**4.** Let $a_1, a_2, a_3, \ldots$ be those positive integers that do not contain the digit 0 in their decimal representations. Thus $a_1=1$, $a_2=2, \ldots a_9=9$, $a_{10}=11, \ldots a_{18}=19$, $a_{19}=21, \ldots$ $a_{90}=99$, $a_{91}=111$, etc. Show that the series $\sum_{n=1}^{\infty}1/a_n$ converges, and that the sum is less than 90. (*Hint:* how many of these integers have $m$ digits? Each term $1/a_n$, where $a_n$ has $m$ digits, is less than $10^{-m+1}$.)

* **5. (Using an integral to improve convergence)** Recall the error formula for the Midpoint Rule, according to which

$$\int_{k-1/2}^{k+1/2}f(x)\,dx-f(k)=\frac{f''(c)}{24},$$

where $k-(1/2)\le c\le k+(1/2)$.

(a) If $f''(x)$ is a decreasing function of $x$, show that

$$f'(k+\tfrac{3}{2})-f'(k+\tfrac{1}{2})\le f''(c)\le f'(k-\tfrac{1}{2})-f'(k-\tfrac{3}{2}).$$

(b) If (i) $f''(x)$ is a decreasing function of $x$, (ii) $\int_{N+1/2}^{\infty}f(x)\,dx$ converges, and (iii) $f'(x)\to 0$ as $x\to\infty$, show that

$$\frac{f'(N-\frac{1}{2})}{24}\le\sum_{n=N+1}^{\infty}f(n)-\int_{N+1/2}^{\infty}f(x)\,dx\le\frac{f'(N+\frac{3}{2})}{24}.$$

(c) Use the result of part (b) to approximate $\sum_{n=1}^{\infty}1/n^2$ to within 0.001.

**6.** Use the Wallis product

$$\lim_{n\to\infty}\frac{2}{1}\cdot\frac{2}{3}\cdot\frac{4}{3}\cdot\frac{4}{5}\cdot\frac{6}{5}\cdot\frac{6}{7}\cdots\frac{2n}{2n-1}\cdot\frac{2n}{2n+1}=\frac{\pi}{2}$$

to help you evaluate $\lim_{n\to\infty}2^{2n}(n!)^2/((2n)!\sqrt{n})$.

* **7. (A series for $e$)** Given that $e=\lim_{n\to\infty}\left(1+\frac{1}{n}\right)^n$, prove that

$$e=\sum_{j=0}^{\infty}\frac{1}{j!}=\lim_{n\to\infty}\sum_{j=0}^{n}\frac{1}{j!}$$
$$=\lim_{n\to\infty}\left(1+\frac{1}{1!}+\frac{1}{2!}+\cdots+\frac{1}{n!}\right)$$

by verifying the squeezing inequalities

$$\left(1+\frac{1}{n}\right)^n\le\sum_{j=0}^{n}\frac{1}{j!}<e$$

for all positive integers $n$. *Hint:* use the Binomial Theorem to expand

$$\left(1+\frac{1}{n}\right)^n$$
$$=1+\frac{n}{1!}\frac{1}{n}+\frac{n(n-1)}{2!}\frac{1}{n^2}+\cdots+\frac{n!}{n!}\frac{1}{n^n}$$
$$=1+\frac{1}{1!}+\frac{1}{2!}\left(1-\frac{1}{n}\right)+\frac{1}{3!}\left(1-\frac{1}{n}\right)\left(1-\frac{2}{n}\right)$$
$$+\cdots+\frac{1}{n!}\left(1-\frac{1}{n}\right)\left(1-\frac{2}{n}\right)\cdots\left(1-\frac{n-1}{n}\right)$$

The left inequality follows because the factors $(1-j/n)$ do not exceed 1. For the right inequality hold $n$ fixed and write a similar expansion for $(1+1/m)^m$, where $m>n$. Then take the limit as $m\to\infty$.

* **8. (The number $e$ is irrational.)** Assume that $e=\sum_{n=0}^{\infty}1/n!$, as proved in Exercise 7, and again in Chapter 2.

(a) Use the technique of Example 3 in Section 1.5 to show that for any $n>0$,

$$0<e-\sum_{j=0}^{n}\frac{1}{j!}<\frac{1}{n!n}.$$

(Note that the sum here has $n+1$ terms rather than $n$ terms.)

(b) Suppose that $e$ is a rational number, say $e=M/N$ for certain positive integers $M$ and $N$. Show that $N!\left(e-\sum_{j=0}^{N}(1/j!)\right)$ is an integer.

(c) Combine parts (a) and (b) to show that there is an integer between 0 and $1/N$. Why is this not possible? Conclude that $e$ cannot be a rational number.

CHAPTER 2

# Power Series

**INTRODUCTION** The main reason for studying infinite series in an elementary calculus course is that many of the most important elementary functions $f(x)$ have useful representations as series whose terms are constant multiples of nonnegative integer powers of $x$. Such series are called power series or, when associated with a particular function, Taylor series or Maclaurin series. They generalize polynomials that are sums of finitely many such terms. Among the functions studied in calculus, polynomials are the easiest to manipulate, to calculate with, to differentiate, and to integrate. Power series, which behave somewhat like *polynomials of infinite degree,* inherit many of these benign qualities. Therefore, the representation of such transcendental functions as $e^x$, $\sin x$, $\tan^{-1} x$, and others as power series renders these functions even more useful in the mathematical modelling of concrete problems. Power series can be used to solve many problems that are intractable by other methods.

In this chapter we develop the machinery of power series and establish several power series representations of elementary functions. We begin by introducing the concept of a power series in Section 2.1. This section requires some familiarity with $p$-series and the ratio test from from Section 1.3 and absolute convergence from Section 1.4. Most of the material in this chapter assumes a basic understanding of the convergence of sequences and series from Sections 1.1 and 1.2.

An alternative approach is provided in Section 2.4. There we define Taylor polynomial approximations to functions, and use the very important Taylor's Theorem (see Section 5.6 in *Single-Variable Calculus*) to obtain Taylor series representations of the functions as the limits of such polynomials as the degree of the polynomial approaches infinity. It is possible to start with Section 2.4 and then proceed to applications in Section 2.3 without covering the material on power series in Sections 2.1 and 2.2. In this case, most of Chapter 1 is not needed either.

In addition to power series, there are many other useful series representations of functions, in particular trigonometric series (Fourier series), that are briefly introduced in Section 2.6. They are encountered in higher level courses in mathematical analysis and differential equations and, like power series, constitute a basic tool of mathematicians, engineers, and other mathematically oriented scientists.

## 2.1 POWER SERIES

This section is concerned with a special kind of infinite series called a *power series* that may be thought of as a polynomial of infinite degree.

**DEFINITION 1**

**Power series**

A series of the form

$$\sum_{n=0}^{\infty} a_n(x-c)^n = a_0 + a_1(x-c) + a_2(x-c)^2 + a_3(x-c)^3 + \cdots$$

is called a **power series in powers of** $x-c$ or a **power series about the point** $x=c$. The constants $a_0, a_1, a_2, \ldots$ are called the **coefficients** of the power series.

Since the terms of a power series are functions of a variable $x$, the series may or may not converge for each value of $x$. For those values of $x$ for which the series does converge, the sum defines a function of $x$. For example, if $-1 < x < 1$, then

$$1 + x + x^2 + x^3 + \cdots = \frac{1}{1-x}.$$

The geometric series on the left side is a power series *representation* of the function $1/(1-x)$ in powers of $x$ (or about the point $x=0$). Note that the representation is valid only in the open interval $(-1, 1)$ even though $1/(1-x)$ is defined for all real $x$ except $x=1$. For $x=-1$ and for $|x|>1$ the series does not converge, so it cannot represent $1/(1-x)$ at these points.

The point $c$ is the **centre of convergence** of the power series $\sum_{n=0}^{\infty} a_n(x-c)^n$. The series certainly converges (to $a_0$) at $x=c$. (All the terms except possibly the first are 0.) Theorem 1 below shows that if the series converges anywhere else, then it converges on an interval (possibly infinite) centred at $x=c$, and it converges absolutely everywhere on that interval except possibly at one or both of the endpoints if the interval is finite. The geometric series

$$1 + x + x^2 + x^3 + \cdots$$

is an example of this behaviour. It has centre of convergence $c=0$, and converges only on the interval $(-1, 1)$, centred at 0. The convergence is absolute at every point of the interval. Another example is the series

$$\sum_{n=1}^{\infty} \frac{1}{n\,2^n}(x-5)^n = \frac{x-5}{2} + \frac{(x-5)^2}{2\times 2^2} + \frac{(x-5)^3}{3\times 2^3} + \cdots,$$

which we discussed in Example 4 of Section 1.4. We showed that this series converges on the interval $[3, 7)$, an interval with centre $x=5$, and that the convergence is absolute on the open interval $(3, 7)$, but is only conditional at the endpoint $x=3$.

**THEOREM 1**

For any power series $\sum_{n=0}^{\infty} a_n\,(x-c)^n$ one of the following alternatives must hold:

(i) the series may converge only at $x=c$, or

(ii) the series may converge at every real number $x$, or

(iii) there may exist a positive real number $R$ such that the series converges at every $x$ satisfying $|x-c|<R$ and diverges at every $x$ satisfying $|x-c|>R$. In this case the series may or may not converge at either of the two *endpoints* $x=c-R$ and $x=c+R$.

In each of these cases the convergence is absolute except possibly at the endpoints $x=c-R$ and $x=c+R$ in case (iii).

***PROOF*** We observed above that every power series converges at its centre of convergence; only the first term can be nonzero so the convergence is absolute. To prove the rest of this theorem, it suffices to show that if the series converges at any number $x_0 \neq c$, then it converges absolutely at every number $x$ closer to $c$ than $x_0$ is, that is, at every $x$ satisfying $|x - c| < |x_0 - c|$. This means that convergence at any $x_0 \neq c$ implies absolute convergence on $(c - x_0, c + x_0)$, so the set of points $x$ where the series converges must be an interval centred at $c$.

Suppose, therefore, that $\sum_{n=0}^{\infty} a_n(x_0 - c)^n$ converges. Then $\lim a_n(x_0 - c)^n = 0$, and so $|a_n(x_0 - c)^n| \le K$ for all $n$, where $K$ is some constant (Theorem 1 of Section 1.1). If $r = |x - c|/|x_0 - c| < 1$, then

$$\sum_{n=0}^{\infty} |a_n(x - c)^n| = \sum_{n=0}^{\infty} |a_n(x_0 - c)^n| \left| \frac{x - c}{x_0 - c} \right|^n \le K \sum_{n=0}^{\infty} r^n = \frac{K}{1 - r} < \infty.$$

Thus $\sum_{n=0}^{\infty} a_n(x - c)^n$ converges absolutely. ◆

By Theorem 1, the set of values $x$ for which the power series $\sum_{n=0}^{\infty} a_n(x - c)^n$ converges is an interval centred at $x = c$. We call this interval the **interval of convergence** of the power series. It must be of one of the following forms:

(i) the isolated point $x = c$ (a degenerate closed interval $[c, c]$),

(ii) the entire line $(-\infty, \infty)$,

(iii) a finite interval centred at $c$:
$[c - R, c + R]$, or $[c - R, c + R)$, or $(c - R, c + R]$, or $(c - R, c + R)$.

The number $R$ in (iii) is called the **radius of convergence** of the power series. In case (i) we say the radius of convergence is $R = 0$; in case (ii) it is $R = \infty$.

The radius of convergence, $R$, can often be found by using the ratio test on the power series: if

$$\rho = \lim_{n\to\infty} \left| \frac{a_{n+1}(x - c)^{n+1}}{a_n(x - c)^n} \right| = \left( \lim_{n\to\infty} \left| \frac{a_{n+1}}{a_n} \right| \right) |x - c|$$

exists, then the series $\sum_{n=0}^{\infty} a_n(x - c)^n$ converges absolutely where $\rho < 1$, that is, where

$$|x - c| < R = \frac{1}{L} = 1 \Big/ \lim_{n\to\infty} \left| \frac{a_{n+1}}{a_n} \right|.$$

The series diverges if $|x - c| > R$.

**Radius of convergence**

Suppose that $L = \lim_{n\to\infty} \left| \dfrac{a_{n+1}}{a_n} \right|$ exists or is $\infty$. Then the power series $\sum_{n=0}^{\infty} a_n(x - c)^n$ has radius of convergence $R = 1/L$ (If $L = 0$ then $R = \infty$; if $L = \infty$ then $R = 0$.)

**EXAMPLE 1** Determine the centre, radius, and interval of convergence of

$$\sum_{n=0}^{\infty} \frac{(2x + 5)^n}{(n^2 + 1)3^n}.$$

***SOLUTION*** The series can be rewritten

$$\sum_{n=0}^{\infty}\left(\frac{2}{3}\right)^n \frac{1}{n^2+1}\left(x+\frac{5}{2}\right)^n.$$

The centre of convergence is $x = -5/2$. The radius of convergence, $R$, is given by

$$\frac{1}{R} = L = \lim\left|\frac{\left(\frac{2}{3}\right)^{n+1}\frac{1}{(n+1)^2+1}}{\left(\frac{2}{3}\right)^n\frac{1}{n^2+1}}\right| = \lim\frac{2}{3}\,\frac{n^2+1}{(n+1)^2+1} = \frac{2}{3}.$$

Thus $R = 3/2$. The series converges absolutely on $(-5/2-3/2, -5/2+3/2) = (-4, -1)$, and it diverges on $(-\infty, -4)$ and on $(-1, \infty)$. At $x = -1$ the series is $\sum_{n=0}^{\infty} 1/(n^2+1)$; at $x = -4$ it is $\sum_{n=0}^{\infty}(-1)^n/(n^2+1)$. Both series converge (absolutely). The interval of convergence of the given power series is therefore $[-4, -1]$. ■

■ **EXAMPLE 2** Determine the radii of convergence of the series

(a) $\displaystyle\sum_{n=0}^{\infty}\frac{x^n}{n!}$ and (b) $\displaystyle\sum_{n=0}^{\infty} n!x^n$.

***SOLUTION***

(a) $L = \left|\lim \dfrac{1}{(n+1)!}\Big/\dfrac{1}{n!}\right| = \lim\dfrac{n!}{(n+1)!} = \lim\dfrac{1}{n+1} = 0.$ Thus $R = \infty$.

This series converges (absolutely) for all $x$. The sum is $e^x$ as will be shown in Example 1 in Section 2.2.

(b) $L = \left|\lim\dfrac{(n+1)!}{n!}\right| = \lim(n+1) = \infty.$ Thus $R = 0$

This series converges only at its centre of convergence, $x = 0$. ■

## Algebraic Operations on Power Series

To simplify the following discussion, we will consider only power series with $x = 0$ as centre of convergence, that is, series of the form

$$\sum_{n=0}^{\infty} a_n x^n = a_0 + a_1 x + a_2 x^2 + a_3 x^3 + \cdots.$$

Any properties we demonstrate for such series extend automatically to power series of the form $\sum_{n=0}^{\infty} a_n(y-c)^n$ via the change of variable $x = y - c$.

First we observe that series having the same centre of convergence can be added or subtracted on whatever interval is common to their intervals of convergence. The following theorem is a simple consequence of Theorem 7 of Section 1.2 and does not require a proof.

**THEOREM 2**

Let $\sum_{n=0}^{\infty} a_n\, x^n$ and $\sum_{n=0}^{\infty} b_n\, x^n$ be two power series with radii of convergence $R_a$ and $R_b$ respectively, and let $c$ be a constant. Then

(i) $\sum_{n=0}^{\infty}(ca_n)\, x^n$ has radius of convergence $R_a$, and

$$\sum_{n=0}^{\infty}(ca_n)\, x^n = c\sum_{n=0}^{\infty} a_n\, x^n$$

wherever the series on the right converges.

(ii) $\sum_{n=0}^{\infty}(a_n + b_n)\,x^n$ has radius of convergence $R$ at least as large as the smaller of $R_a$ and $R_b$ ($R \geq \min\{R_a, R_b\}$), and

$$\sum_{n=0}^{\infty}(a_n + b_n)\,x^n = \sum_{n=0}^{\infty} a_n\,x^n + \sum_{n=0}^{\infty} b_n\,x^n$$

wherever both series on the right converge.

The situation regarding multiplication and division of power series is more complicated. We will mention only the results and will not attempt any proofs of our assertions. A textbook in mathematical analysis will provide more details.

Long multiplication of the form

$$\begin{aligned}(a_0 + a_1x + a_2x^2 + \cdots)(b_0 + b_1x + b_2x^2 + \cdots)\\ = a_0b_0 + (a_0b_1 + a_1b_0)x + (a_0b_2 + a_1b_1 + a_2b_0)x^2 + \cdots\end{aligned}$$

leads us to conjecture the formula

$$\left(\sum_{n=0}^{\infty} a_nx^n\right)\left(\sum_{n=0}^{\infty} b_nx^n\right) = \sum_{n=0}^{\infty} c_nx^n,$$

where

$$c_n = a_0b_n + a_1b_{n-1} + \cdots + a_nb_0 = \sum_{j=0}^{n} a_jb_{n-j}.$$

The series $\sum_{n=0}^{\infty} c_nx^n$ is called the **Cauchy product** of the series $\sum_{n=0}^{\infty} a_nx^n$ and $\sum_{n=0}^{\infty} b_nx^n$. Like the sum, the Cauchy product also has radius of convergence at least equal to the lesser of those of the factor series.

**EXAMPLE 3** Since

$$\frac{1}{1-x} = 1 + x + x^2 + x^3 + \cdots = \sum_{n=0}^{\infty} x^n$$

holds for $-1 < x < 1$, we can determine a power series representation for $1/(1-x)^2$ by taking the Cauchy product of this series with itself. Since $a_n = b_n = 1$ for $n = 0, 1, 2, \ldots$ we have

$$c_n = \sum_{j=0}^{n} 1 = n + 1$$

and

$$\frac{1}{(1-x)^2} = 1 + 2x + 3x^2 + 4x^3 + \cdots \sum_{n=0}^{\infty}(n+1)x^n,$$

which must also hold for $-1 < x < 1$. The same series can be obtained by direct long multiplication of the series:

$$\begin{array}{rcccccccc} & 1 & + & x & + & x^2 & + & x^3 & + \cdots \\ \times & 1 & + & x & + & x^2 & + & x^3 & + \cdots \\ \hline & 1 & + & x & + & x^2 & + & x^3 & + \cdots \\ & & & x & + & x^2 & + & x^3 & + \cdots \\ & & & & & x^2 & + & x^3 & + \cdots \\ & & & & & & & x^3 & + \cdots \\ & & & & & & & & \cdots \\ \hline & 1 & + & 2x & + & 3x^2 & + & 4x^3 & + \cdots \end{array}$$

■

Long division can also be performed on power series, but there is no simple rule for determining the coefficients of the quotient series. The radius of convergence of the quotient series is not less than the least of the three numbers $R_1$, $R_2$, and $R_3$, where $R_1$ and $R_2$ are the radii of convergence of the divisor and dividend series and $R_3$ is the distance from the centre of convergence to the nearest complex number where the divisor series has sum equal to 0.

**EXAMPLE 4** To illustrate this point, observe that 1 and $1 - x$ are both power series with infinite radii of convergence:

$$\begin{aligned} 1 &= 1 + 0x + 0x^2 + 0x^3 + \cdots && \text{for all } x, \\ 1 - x &= 1 - \ x + 0x^2 + 0x^3 + \cdots && \text{for all } x. \end{aligned}$$

Their quotient, $1/(1 - x)$, however, only has radius of convergence 1, the distance from the centre of convergence $x = 0$ to the point $x = 1$ where the denominator vanishes:

$$\frac{1}{1-x} = 1 + x + x^2 + x^3 + \cdots \qquad \text{for} \quad |x| < 1.$$

## Differentiation and Integration of Power Series

If a power series has a positive radius of convergence, it can be differentiated or integrated term by term. The resulting series will converge to the appropriate derivative or integral of the sum of the original series everywhere except possibly at the endpoints of the interval of convergence of the original series. This very important fact ensures that, for purposes of calculation, power series behave just like polynomials, the easiest functions to differentiate and integrate. We formalize the differentiation and integration properties of power series in the following theorem.

**THEOREM 3**

**Term by term differentiation and integration of power series**

If the series $\sum_{n=0}^{\infty} a_n x^n$ converges to the sum $f(x)$ on an interval $(-R, R)$, where $R > 0$; that is,

$$f(x) = \sum_{n=0}^{\infty} a_n x^n = a_0 + a_1 x + a_2 x^2 + a_3 x^3 + \cdots, \qquad (-R < x < R),$$

then $f$ is differentiable on $(-R, R)$ and

$$f'(x) = \sum_{n=1}^{\infty} n a_n x^{n-1} = a_1 + 2a_2 x + 3a_3 x^2 + \cdots, \qquad (-R < x < R).$$

Also, $f$ is integrable over any closed subinterval of $(-R, R)$, and if $|x| \le R$, then

$$\int_0^x f(t)\,dt = \sum_{n=0}^{\infty} \frac{a_n}{n+1} x^{n+1} = a_0 x + \frac{a_1}{2} x^2 + \frac{a_2}{3} x^3 + \cdots.$$

***PROOF*** Let $x$ satisfy $-R < x < R$ and choose $H > 0$ such that $|x| + H < R$. By Theorem 1, we then have[1]

$$\sum_{n=1}^{\infty} |a_n|(|x| + H)^n = K < \infty.$$

The Binomial Theorem (see Section 2.5) shows that if $n \geq 1$, then

$$(x+h)^n = x^n + nx^{n-1}h + \sum_{k=2}^{n} \binom{n}{k} x^{n-k}h^k.$$

Therefore, if $|h| \leq H$ we have

$$\begin{aligned}
|(x+h)^n - x^n - nx^{n-1}h| &= \left| \sum_{k=2}^{n} \binom{n}{k} x^{n-k}h^k \right| \\
&\leq \sum_{k=2}^{n} \binom{n}{k} |x|^{n-k} \frac{|h|^k}{H^k} H^k \\
&\leq \frac{|h|^2}{H^2} \sum_{k=0}^{n} \binom{n}{k} |x|^{n-k} H^k \\
&= \frac{|h|^2}{H^2} (|x| + H)^n.
\end{aligned}$$

Also

$$|nx^{n-1}| = \frac{n|x|^{n-1}H}{H} \leq \frac{1}{H}(|x| + H)^n.$$

Thus

$$\sum_{n=1}^{\infty} |na_n x^{n-1}| \leq \frac{1}{H} \sum_{n=1}^{\infty} |a_n|(|x| + H)^n = \frac{K}{H} < \infty$$

so the series $\sum_{n=1}^{\infty} na_n x^{n-1}$ converges (absolutely) to $g(x)$, say. Now

$$\begin{aligned}
\left| \frac{f(x+h) - f(x)}{h} - g(x) \right| &= \left| \sum_{n=1}^{\infty} \frac{a_n(x+h)^n - a_n x^n - na_n x^{n-1}h}{h} \right| \\
&\leq \frac{1}{|h|} \sum_{n=1}^{\infty} |a_n||(x+h)^n - x^n - nx^{n-1}h| \\
&\leq \frac{|h|}{H^2} \sum_{n=1}^{\infty} |a_n|(|x| + H)^n \leq \frac{K|h|}{H^2}.
\end{aligned}$$

Letting $h$ approach zero, we obtain $|f'(x) - g(x)| \leq 0$, so $f'(x) = g(x)$, as required.

Now observe that since $|a_n/(n+1)| \leq |a_n|$, the series

$$h(x) = \sum_{n=0}^{\infty} \frac{a_n}{n+1} x^{n+1}$$

---

[1] This proof is due to R. Výborný, *American Mathematical Monthly*, April 1987.

converges (absolutely) at least on the interval $(-R, R)$. Using the differentiation result proved above, we obtain

$$h'(x) = \sum_{n=0}^{\infty} a_n x^n = f(x).$$

Since $h(0) = 0$, we have

$$\int_0^x f(t)\,dt = \int_0^x h'(t)\,dt = h(t)\Big|_0^x = h(x),$$

as required.

Together, these results imply that the termwise differentiated or integrated series have the same radius of convergence as the given series. In fact, as the following examples illustrate, the interval of convergence of the differentiated series is the same as that of the original series except for the *possible* loss of an endpoint if the original series converges at an endpoint of its interval of convergence. Similarly, the integrated series will converge everywhere on the interval of convergence of the original series and possibly at one or both endpoints of that interval, even if the original series does not converge at the endpoints.

Being differentiable on $(-R, R)$, where $R$ is the radius of convergence, the sum $f(x)$ of a power series is necessarily continuous on that open interval. If the series happens to converge at either or both of the endpoints $-R$ and $R$, then $f$ is also continuous (on one side) up to these endpoints. This result is stated formally in the following theorem. We will not prove it here; the interested reader is referred to textbooks on mathematical analysis for a proof.

**THEOREM 4**

**Abel's Theorem**

The sum of a power series is a continuous function everywhere on the interval of convergence of the series. In particular, if $\sum_{n=0}^{\infty} a_n R^n$ converges for some $R > 0$, then

$$\lim_{x\to R-} \sum_{n=0}^{\infty} a_n x^n = \sum_{n=0}^{\infty} a_n R^n,$$

and if $\sum_{n=0}^{\infty} a_n (-R)^n$ converges, then

$$\lim_{x\to -R+} \sum_{n=0}^{\infty} a_n x^n = \sum_{n=0}^{\infty} a_n (-R)^n.$$

The following examples show how the above theorems are applied to obtain power series representations for functions.

**EXAMPLE 5** Find power series representations for the functions

(a) $\dfrac{1}{(1-x)^2}$, (b) $\dfrac{1}{(1-x)^3}$, (c) $\ln(1+x)$

by starting with the geometric series

$$\frac{1}{1-x} = \sum_{n=0}^{\infty} x^n = 1 + x + x^2 + x^3 + \cdots \qquad (-1 < x < 1)$$

and using differentiation, integration, and substitution. Where is each series valid?

***SOLUTION***

(a) Differentiate the geometric series term by term according to Theorem 3 to obtain

$$\frac{1}{(1-x)^2} = \sum_{n=1}^{\infty} n x^{n-1} = 1 + 2x + 3x^2 + 4x^3 + \cdots \qquad (-1 < x < 1).$$

This is the same result obtained by multiplication of series in Example 3 above.

(b) Differentiate again to get, for $-1 < x < 1$,

$$\frac{2}{(1-x)^3} = \sum_{n=2}^{\infty} n(n-1)\,x^{n-2} = (1 \times 2) + (2 \times 3)x + (3 \times 4)x^2 + \cdots$$

Now divide by 2:

$$\frac{1}{(1-x)^3} = \sum_{n=2}^{\infty} \frac{n(n-1)}{2} x^{n-2} = 1 + 3x + 6x^2 + 10x^3 + \cdots \quad (-1 < x < 1).$$

(c) Substitute $-t$ in place of $x$ in the original geometric series:

$$\frac{1}{1+t} = \sum_{n=0}^{\infty} (-1)^n t^n = 1 - t + t^2 - t^3 + t^4 - \cdots \qquad (-1 < t < 1).$$

Integrate from 0 to $x$, where $|x| < 1$, to get

$$\ln(1+x) = \int_0^x \frac{dt}{1+t} = \sum_{n=0}^{\infty} (-1)^n \int_0^x t^n \, dt$$
$$= \sum_{n=0}^{\infty} (-1)^n \frac{x^{n+1}}{n+1} = x - \frac{x^2}{2} + \frac{x^3}{3} - \frac{x^4}{4} + \cdots \quad (-1 < x \le 1).$$

Note that the latter series converges (conditionally) at the endpoint $x = 1$ as well as on the interval $-1 < x < 1$. Since $\ln(1+x)$ is continuous at $x = 1$, Theorem 4 assures us the series must converge to that function at $x = 1$ also. In particular, therefore, the alternating harmonic series converges to $\ln 2$:

$$\ln 2 = 1 - \frac{1}{2} + \frac{1}{3} - \frac{1}{4} + \frac{1}{5} - \cdots = \sum_{n=0}^{\infty} \frac{(-1)^n}{n+1}.$$

This would not, however, be a very useful formula for calculating the value of $\ln 2$. (Why not?) ■

■ **EXAMPLE 6** Use the geometric series of the previous example to find a power series representation for $\tan^{-1} x$.

***SOLUTION*** Substitute $-t^2$ for $x$ in the geometric series. Since $0 \le t^2 < 1$ whenever $-1 < t < 1$, we obtain

$$\frac{1}{1+t^2} = 1 - t^2 + t^4 - t^6 + t^8 - \cdots \qquad (-1 < t < 1).$$

Now integrate from 0 to $x$, where $|x| < 1$:

$$\begin{aligned}\tan^{-1} x = \int_0^x \frac{dt}{1+t^2} &= \int_0^x (1 - t^2 + t^4 - t^6 + t^8 - \cdots)\, dt \\ &= x - \frac{x^3}{3} + \frac{x^5}{5} - \frac{x^7}{7} + \frac{x^9}{9} - \cdots \\ &= \sum_{n=0}^{\infty} (-1)^n \frac{x^{2n+1}}{2n+1} \qquad (-1 < x < 1).\end{aligned}$$

However, note that the series also converges (conditionally) at $x = -1$ and 1. Since $\tan^{-1}$ is continuous at $\pm 1$, the above series representation for $\tan^{-1} x$ also holds for these values, by Theorem 4. Letting $x = 1$ we get another interesting result:

$$\frac{\pi}{4} = 1 - \frac{1}{3} + \frac{1}{5} - \frac{1}{7} + \frac{1}{9} - \cdots.$$

Again, however, this would not be a good formula with which to calculate a numerical value of $\pi$. (Why not?) ■

**EXAMPLE 7** Find the sum of the series $\sum_{n=1}^{\infty} \frac{n^2}{2^n}$ by first finding the sum of the power series

$$\sum_{n=1}^{\infty} n^2 x^n = x + 4x^2 + 9x^3 + 16x^4 + \cdots.$$

***SOLUTION*** Observe (in Example 5(a)) how the process of differentiating the geometric series produces a series with coefficients 1, 2, 3, .... Start with the series obtained for $1/(1-x)^2$ and multiply it by $x$ to obtain

$$\sum_{n=1}^{\infty} n x^n = x + 2x^2 + 3x^3 + 4x^4 + \cdots = \frac{x}{(1-x)^2}.$$

Now differentiate again to get a series with coefficients $1^2$, $2^2$, $3^2$, ...

$$\sum_{n=1}^{\infty} n^2 x^{n-1} = 1 + 4x + 9x^2 + 16x^3 + \cdots = \frac{d}{dx} \frac{x}{(x-1)^2} = \frac{1+x}{(1-x)^3}.$$

Multiplication by $x$ again gives the desired power series

$$\sum_{n=1}^{\infty} n^2 x^n = x + 4x^2 + 9x^3 + 16x^4 + \cdots = \frac{x(1+x)}{(1-x)^3}.$$

Differentiation and multiplication by $x$ do not change the radius of convergence, so this series converges to the indicated function for $-1 < x < 1$. Putting $x = 1/2$, we get

$$\sum_{n=1}^{\infty} \frac{n^2}{2^n} = \frac{\frac{1}{2} \times \frac{3}{2}}{\frac{1}{8}} = 6.$$ ■

The following example illustrates how substitution can be used to obtain power series representations of functions with centres of convergence different from 0.

**EXAMPLE 8** Find a series representation of $f(x) = 1/(2+x)$ in powers of $x-1$. What is the interval of convergence of this series?

***SOLUTION*** Let $t = x - 1$ so that $x = t + 1$. We have

$$\begin{aligned}
\frac{1}{2+x} &= \frac{1}{3+t} = \frac{1}{3}\,\frac{1}{1+\dfrac{t}{3}} \\
&= \frac{1}{3}\left(1 - \frac{t}{3} + \frac{t^2}{3^2} - \frac{t^3}{3^3} + \cdots\right) && (-1 < t/3 < 1) \\
&= \sum_{n=0}^{\infty} (-1)^n \frac{t^n}{3^{n+1}} && (-3 < t < 3) \\
&= \sum_{n=0}^{\infty} (-1)^n \frac{(x-1)^n}{3^{n+1}} && (-2 < x < 4).
\end{aligned}$$

Note that the radius of convergence of this series is 3, the distance from the centre of convergence, 1, to the point $-2$ where the denominator is 0. We could have predicted this in advance. ■

## EXERCISES 2.1

Determine the centre, radius, and interval of convergence of each of the power series in Exercises 1–8.

**1.** $\displaystyle\sum_{n=0}^{\infty} \frac{x^{2n}}{\sqrt{n+1}}$

**2.** $\displaystyle\sum_{n=0}^{\infty} 3n\,(x+1)^n$

**3.** $\displaystyle\sum_{n=1}^{\infty} \frac{1}{n}\left(\frac{x+2}{2}\right)^n$

**4.** $\displaystyle\sum_{n=1}^{\infty} \frac{(-1)^n}{n^4 2^{2n}} x^n$

**5.** $\displaystyle\sum_{n=0}^{\infty} n^3 (2x-3)^n$

**6.** $\displaystyle\sum_{n=1}^{\infty} \frac{e^n}{n^3} (4-x)^n$

**7.** $\displaystyle\sum_{n=0}^{\infty} \frac{(1+5^n)}{n!} x^n$

**8.** $\displaystyle\sum_{n=1}^{\infty} \frac{(4x-1)^n}{n^n}$

**9.** Use multiplication of series to find a power series representation of $1/(1-x)^3$ valid in the interval $(-1, 1)$.

**10.** Determine the Cauchy product of the series $1 + x + x^2 + x^3 + \cdots$ and $1 - x + x^2 - x^3 + \cdots$. On what interval and to what function does the product series converge?

**11.** Determine the power series expansion of $1/(1-x)^2$ by formally dividing $1 - 2x + x^2$ into 1.

Starting with the power series representation

$$\frac{1}{1-x} = 1 + x + x^2 + x^3 + \cdots \qquad (-1 < x < 1)$$

determine power series representations for the functions indicated in Exercises 12–20. On what interval is each representation valid?

**12.** $\dfrac{1}{2-x}$ in powers of $x$

**13.** $\dfrac{1}{(2-x)^2}$ in powers of $x$

**14.** $\dfrac{1}{1+2x}$ in powers of $x$

**15.** $\ln(2-x)$ in powers of $x$

**16.** $\dfrac{1}{x}$ in powers of $x-1$

**17.** $\dfrac{1}{x^2}$ in powers of $x+2$

**18.** $\dfrac{1-x}{1+x}$ in powers of $x$

**19.** $\dfrac{x^3}{1-2x^2}$ in powers of $x$

**20.** $\ln x$ in powers of $x-4$

Determine the interval of convergence and the sum of each of the series in Exercises 21–26.

**21.** $1 - 4x + 16x^2 - 64x^3 + \cdots = \displaystyle\sum_{n=0}^{\infty} (-1)^n (4x)^n$

*** 22.** $3 + 4x + 5x^2 + 6x^3 + \cdots = \displaystyle\sum_{n=0}^{\infty} (n+3)x^n$

*** 23.** $\dfrac{1}{3} + \dfrac{x}{4} + \dfrac{x^2}{5} + \dfrac{x^3}{6} + \cdots = \displaystyle\sum_{n=0}^{\infty} \frac{x^n}{n+3}$

*** 24.** $1 \times 3 - 2 \times 4x + 3 \times 5x^2 - 4 \times 6x^3 + \cdots$
$\qquad = \displaystyle\sum_{n=0}^{\infty} (-1)^n (n+1)(n+3)\,x^n$

*** 25.** $2 + 4x^2 + 6x^4 + 8x^6 + 10x^8 + \cdots = \displaystyle\sum_{n=0}^{\infty} 2(n+1)\,x^{2n}$

*** 26.** $1 - \dfrac{x^2}{2} + \dfrac{x^4}{3} - \dfrac{x^6}{4} + \dfrac{x^8}{5} - \cdots = \displaystyle\sum_{n=0}^{\infty} \frac{(-1)^n x^{2n}}{n+1}$

Use the technique (or the result) of Example 7 to find the sums of the numerical series in Exercises 27–32.

**27.** $\sum_{n=1}^{\infty} \frac{n}{3^n}$

**28.** $\sum_{n=0}^{\infty} \frac{n+1}{2^n}$

* **29.** $\sum_{n=0}^{\infty} \frac{(n+1)^2}{\pi^n}$

* **30.** $\sum_{n=1}^{\infty} \frac{(-1)^n n(n+1)}{2^n}$

**31.** $\sum_{n=1}^{\infty} \frac{(-1)^{n-1}}{n2^n}$

**32.** $\sum_{n=3}^{\infty} \frac{1}{n2^n}$

## 2.2 TAYLOR AND MACLAURIN SERIES

If a power series $\sum_{n=0}^{\infty} a_n(x-c)^n$ has a positive radius of convergence $R$, then the sum of the series defines a function $f(x)$ on the interval $(c-R, c+R)$. We say that the power series is a **representation** of $f(x)$ on that interval. What relationship exists between the function $f(x)$ and the coefficients $a_0$, $a_1$, $a_2$, ... of the power series? The following theorem answers this question.

**THEOREM 5**

Suppose the series

$$f(x) = \sum_{n=0}^{\infty} a_n(x-c)^n = a_0 + a_1(x-c) + a_2(x-c)^2 + a_3(x-c)^3 + \cdots$$

converges to $f(x)$ for $c - R < x < c + R$, where $R > 0$. Then

$$a_k = \frac{f^{(k)}(c)}{k!} \qquad \text{for } k = 0, 1, 2, 3, \ldots.$$

***PROOF*** This proof requires that we differentiate the series for $f(x)$ term by term several times, a process justified by Theorem 3 (suitably reformulated for powers of $x - c$):

$$f'(x) = \sum_{n=1}^{\infty} na_n(x-c)^{n-1} = a_1 + 2a_2(x-c) + 3a_3(x-c)^2 + \cdots$$

$$f''(x) = \sum_{n=2}^{\infty} n(n-1)a_n(x-c)^{n-2} = 2a_2 + 6a_3(x-c) + 12a_4(x-c)^2 + \cdots$$

$$\vdots$$

$$\begin{aligned} f^{(k)}(x) &= \sum_{n=k}^{\infty} n(n-1)(n-2)\cdots(n-k+1)a_n(x-c)^{n-k} \\ &= k!a_k + \frac{(k+1)!}{1!}a_{k+1}(x-c) + \frac{(k+2)!}{2!}a_{k+2}(x-c)^2 + \cdots. \end{aligned}$$

Each series converges for $c - R < x < c + R$. Setting $x = c$, we obtain $f^{(k)}(c) = k!a_k$, which proves the theorem.

Theorem 5 shows that a function $f(x)$ that has a power series representation with centre at $c$ and positive radius of convergence must have derivatives of all orders in an interval around $x = c$, and it can have only one representation as a power series in powers of $x - c$, namely

$$f(x) = \sum_{n=0}^{\infty} \frac{f^{(n)}(c)}{n!}(x-c)^n = f(c) + f'(c)(x-c) + \frac{f''(c)}{2!}(x-c)^2 + \cdots.$$

Such a series is called a Taylor series or, if $c = 0$, a Maclaurin series.

**DEFINITION 2**

**Taylor and Maclaurin series**

If $f(x)$ has derivatives of all orders at $x = c$ (that is, if $f^{(k)}(c)$ exists for $k = 0, 1, 2, 3, \ldots$), then the series

$$\sum_{k=0}^{\infty} \frac{f^{(k)}(c)}{k!}(x-c)^k$$
$$= f(c) + f'(c)(x-c) + \frac{f''(c)}{2!}(x-c)^2 + \frac{f^{(3)}(c)}{3!}(x-c)^3 + \cdots$$

is called the **Taylor series of $f$ about $x = c$** (or the **Taylor series of $f$ in powers of $x - c$**). If $c = 0$, the term **Maclaurin series** is usually used in place of Taylor series.

The Taylor series is a power series as defined in Section 2.1. Theorem 1 implies that $c$ must be the centre of any interval on which such a series converges, but the definition of Taylor series makes no requirement that the series should converge anywhere except at the point $x = c$ where the series is just $f(0) + 0 + 0 + \cdots$. The series exists provided all the derivatives of $f$ exist at $x = c$; in practice this means that each derivative must exist in an open interval containing $x = c$. (Why?) However, the series may converge nowhere except at $x = c$, and if it does converge elsewhere, it may converge to something other than $f(x)$. (See Exercise 40 at the end of this section for an example where this happens.) If the Taylor series does converge to $f(x)$ in an open interval containing $c$, then we will say that $f$ is analytic at $x = c$.

**DEFINITION 3**

**Analytic functions**

A function $f(x)$ is **analytic** at $x = c$ if $f(x)$ is the sum of a power series in powers of $x - c$ having positive radius of convergence. (The series is its Taylor series.) If $f$ is analytic at each point of an open interval $I$, then we say it is analytic on the interval $I$.

Most, but not all, of the elementary functions encountered in calculus are analytic wherever they have derivatives of all orders. On the other hand, whenever a power series converges on an open interval containing $c$, then its sum $f(x)$ is analytic at $c$, and the given series is the Taylor series of $f(x)$ about $x = c$.

## Maclaurin Series for Some Elementary Functions

Calculating Taylor and Maclaurin series for a function $f$ directly from Definition 2 is practical only when we can find a formula for the $n$th derivative of $f$. Examples of such functions include $(ax+b)^r$, $e^{ax+b}$, $\ln(ax+b)$, $\sin(ax+b)$, $\cos(ax+b)$, and sums of such functions.

**EXAMPLE 1** Find the Taylor series for $e^x$ about $x = c$. Where does the series converge to $e^x$? Where is $e^x$ analytic? What is the Maclaurin series for $e^x$?

**SOLUTION** Since all the derivatives of $f(x) = e^x$ are $e^x$, we have $f^{(n)}(c) = e^c$ for every integer $n \geq 0$. Thus the Taylor series for $e^x$ about $x = c$ is

$$\sum_{n=0}^{\infty} \frac{e^c}{n!}(x-c)^n = e^c + e^c(x-c) + \frac{e^c}{2!}(x-c)^2 + \frac{e^c}{3!}(x-c)^3 + \cdots.$$

The radius of convergence $R$ of this series is given by

$$\frac{1}{R} = \lim_{n\to\infty} \left| \frac{e^c/(n+1)!}{e^c/n!} \right| = \lim_{n\to\infty} \frac{n!}{(n+1)!} = \lim_{n\to\infty} \frac{1}{n+1} = 0.$$

Thus the radius of convergence is $R = \infty$ and the series converges for all $x$. Suppose the sum is $g(x)$:

$$g(x) = e^c + e^c(x-c) + \frac{e^c}{2!}(x-c)^2 + \frac{e^c}{3!}(x-c)^3 + \cdots.$$

By Theorem 3, we have

$$\begin{aligned} g'(x) &= 0 + e^c + \frac{e^c}{2!}2(x-c) + \frac{e^c}{3!}3(x-c)^2 + \cdots \\ &= e^c + e^c(x-c) + \frac{e^c}{2!}(x-c)^2 + \cdots = g(x) \end{aligned}$$

Also, $g(c) = e^c + 0 + 0 + \cdots = e^c$. Since $g(x)$ satisfies the differential equation $g'(x) = g(x)$ of exponential growth, we have $g(x) = Ce^x$. Substituting $x = c$ gives $e^c = g(c) = Ce^c$, so $C = 1$. Thus the Taylor series for $e^x$ in powers of $x - c$ converges to $e^x$ for every real number $x$:

$$\begin{aligned} e^x &= \sum_{n=0}^{\infty} \frac{e^c}{n!}(x-c)^n \\ &= e^c + e^c(x-c) + \frac{e^c}{2!}(x-c)^2 + \frac{e^c}{3!}(x-c)^3 + \cdots \qquad \text{(for all } x\text{)}. \end{aligned}$$

In particular, setting $c = 0$ we obtain the Maclaurin series for $e^x$:

$$e^x = \sum_{n=0}^{\infty} \frac{x^n}{n!} = 1 + x + \frac{x^2}{2!} + \frac{x^3}{3!} + \cdots \qquad \text{(for all } x\text{)}.$$

■

■ **EXAMPLE 2** Find the Maclaurin series for (a) $\sin x$, and (b) $\cos x$. Where does each series converge?

**SOLUTION** Let $f(x) = \sin x$. Then we have $f(0) = 0$ and:

$$\begin{aligned} f'(x) &= \cos x & f'(0) &= 1 \\ f''(x) &= -\sin x & f''(0) &= 0 \\ f^{(3)}(x) &= -\cos x & f^{(3)}(0) &= -1 \\ f^{(4)}(x) &= \sin x & f^{(4)}(0) &= 0 \\ f^{(5)}(x) &= \cos x & f^{(5)}(0) &= 1 \\ &\vdots & &\vdots \end{aligned}$$

Thus, the Maclaurin series for $\sin x$ is

$$\begin{aligned} g(x) &= 0 + x + 0 - \frac{x^3}{3!} + 0 + \frac{x^5}{5!} + 0 - \cdots \\ &= x - \frac{x^3}{3!} + \frac{x^5}{5!} - \frac{x^7}{7!} + \cdots = \sum_{n=0}^{\infty} \frac{(-1)^n}{(2n+1)!} x^{2n+1}. \end{aligned}$$

We have denoted the sum by $g(x)$ since we don't yet know whether the series converges to $\sin x$. The series does converge for all $x$ by the ratio test:

$$\begin{aligned} \lim_{n\to\infty} \left| \frac{\dfrac{(-1)^{n+1}}{(2(n+1)+1)!} x^{2(n+1)+1}}{\dfrac{(-1)^n}{(2n+1)!} x^{2n+1}} \right| &= \lim_{n\to\infty} \frac{(2n+1)!}{(2n+3)!} |x|^2 \\ &= \lim_{n\to\infty} \frac{|x|^2}{(2n+3)(2n+2)} = 0. \end{aligned}$$

Now we can differentiate the function $g(x)$ twice to get

$$\begin{aligned} g'(x) &= 1 - \frac{x^2}{2!} + \frac{x^4}{4!} - \frac{x^6}{6!} + \cdots \\ g''(x) &= -x + \frac{x^3}{3!} - \frac{x^5}{5!} + \frac{x^7}{7!} - \cdots = -g(x). \end{aligned}$$

Thus $g(x)$ satisfies the differential equation $g''(x) + g(x) = 0$ of simple harmonic motion. The general solution of this equation is

$$g(x) = A \cos x + B \sin x.$$

Observe, from the series, that $g(0) = 0$ and $g'(0) = 1$. These values determine that $A = 0$ and $B = 1$. Thus $g(x) = \sin x$ and $g'(x) = \cos x$ for all $x$.

We have therefore demonstrated that

$$\sin x = \sum_{n=0}^{\infty} \frac{(-1)^n}{(2n+1)!} x^{2n+1} = x - \frac{x^3}{3!} + \frac{x^5}{5!} - \frac{x^7}{7!} + \cdots \qquad \text{(for all } x\text{)}$$

$$\cos x = \sum_{n=0}^{\infty} \frac{(-1)^n}{(2n)!} x^{2n} = 1 - \frac{x^2}{2!} + \frac{x^4}{4!} - \frac{x^6}{6!} + \cdots \qquad \text{(for all } x\text{)}$$

■

Theorem 5 shows that we can use any available means to find a power series converging to a given function on an interval, and the series obtained will turn out to be the Taylor series. In Section 2.1 several series were constructed by manipulating a geometric series. These include:

**Some Maclaurin series**

$$\frac{1}{1-x} = \sum_{n=0}^{\infty} x^n = 1 + x + x^2 + x^3 + \cdots \qquad (-1 < x < 1)$$

$$\frac{1}{(1-x)^2} = \sum_{n=1}^{\infty} n x^{n-1} = 1 + 2x + 3x^2 + 4x^3 + \cdots \qquad (-1 < x < 1)$$

$$\ln(1+x) = \sum_{n=1}^{\infty} \frac{(-1)^{n-1}}{n} x^n = x - \frac{x^2}{2} + \frac{x^3}{3} - \frac{x^4}{4} + \cdots \qquad (-1 < x \le 1)$$

$$\tan^{-1} x = \sum_{n=0}^{\infty} \frac{(-1)^n}{2n+1} x^{2n+1} = x - \frac{x^3}{3} + \frac{x^5}{5} - \frac{x^7}{7} + \cdots \qquad (-1 \le x \le 1)$$

These series, together with the intervals on which they converge, are frequently used hereafter and should be memorized.

## Other Maclaurin and Taylor Series

Series can be combined in various ways to generate new series. For example, we can find the Maclaurin series for $e^{-x}$ by replacing $x$ with $-x$ in the Maclaurin series for $x$:

$$e^{-x} = \sum_{n=0}^{\infty} \frac{(-1)^n}{n!} x^n = 1 - x + \frac{x^2}{2!} - \frac{x^3}{3!} + \cdots \qquad \text{(for all } x.)$$

The series for $e^x$ and $e^{-x}$ can then be subtracted or added and the results divided by 2 to obtain Maclaurin series for the hyperbolic functions $\sinh x$ and $\cosh x$:

$$\sinh x = \frac{e^x - e^{-x}}{2} = \sum_{n=0}^{\infty} \frac{x^{2n+1}}{(2n+1)!} = x + \frac{x^3}{3!} + \frac{x^5}{5!} + \cdots \qquad \text{(for all real } x)$$

$$\cosh x = \frac{e^x + e^{-x}}{2} = \sum_{n=0}^{\infty} \frac{x^{2n}}{(2n)!} = 1 + \frac{x^2}{2!} + \frac{x^4}{4!} + \cdots \qquad \text{(for all real } x).$$

**REMARK** Observe the similarity between the series for $\sin x$ and $\sinh x$ and between those for $\cos x$ and $\cosh x$. If we were to allow complex numbers (numbers of the form $z = x + iy$, where $i^2 = -1$ and $x$ and $y$ are real — see Appendix I) as arguments for our functions, and if we were to demonstrate that our operations on series could be extended to series of complex numbers, we would see that $\cos x = \cosh(ix)$ and $\sin x = -i \sinh(ix)$. In fact,

$$e^{ix} = \cos x + i \sin x \qquad \text{and} \qquad e^{-ix} = \cos x - i \sin x,$$

so

$$\cos x = \frac{e^{ix} + e^{-ix}}{2} \qquad \text{and} \qquad \sin x = \frac{e^{ix} - e^{-ix}}{2i}.$$

Such formulas are encountered in the study of functions of a complex variable; from the complex point of view the trigonometric and exponential functions are just different manifestations of the same basic function, a complex exponential $e^z = e^{x+iy}$. We content ourselves here with having mentioned the interesting relationships above and invite the reader to verify them formally by calculating with series. (Such formal calculations do not, of course, constitute a proof since we have not established the various rules covering series of complex numbers.)

**EXAMPLE 3** Obtain Maclaurin series for the following functions:

(a) $e^{-x^2/3}$, (b) $\dfrac{\sin(x^2)}{x}$, (c) $\sin^2 x$.

***SOLUTION***

(a) We substitute $-x^2/3$ for $x$ in the Maclaurin series for $e^x$:

$$\begin{aligned}e^{-x^2/3} &= 1 - \frac{x^2}{3} + \frac{1}{2!}\left(\frac{x^2}{3}\right)^2 - \frac{1}{3!}\left(\frac{x^2}{3}\right)^3 + \cdots \\ &= \sum_{n=0}^{\infty}(-1)^n \frac{1}{3^n n!} x^{2n} \qquad \text{(for all real } x\text{)}.\end{aligned}$$

(b) For all $x \neq 0$ we have

$$\begin{aligned}\frac{\sin x^2}{x} &= \frac{1}{x}\left(x^2 - \frac{(x^2)^3}{3!} + \frac{(x^2)^5}{5!} - \cdots\right) \\ &= x - \frac{x^5}{3!} + \frac{x^9}{5!} - \cdots = \sum_{n=0}^{\infty}(-1)^n \frac{x^{4n+1}}{(2n+1)!}.\end{aligned}$$

Note that $f(x) = (\sin(x^2))/x$ is not defined at $x = 0$ but does have a limit (namely 0) as $x$ approaches 0. If we define $f(0) = 0$ (the continuous extension of $f(x)$ to $x = 0$), then the series converges to $f(x)$ for all $x$.

(c) We use a trigonometric identity to express $\sin^2 x$ in terms of $\cos 2x$ and then use the Maclaurin series for $\cos x$ with $x$ replaced by $2x$.

$$\begin{aligned}\sin^2 x &= \frac{1 - \cos 2x}{2} = \frac{1}{2} - \frac{1}{2}\left(1 - \frac{(2x)^2}{2!} + \frac{(2x)^4}{4!} - \cdots\right) \\ &= \frac{1}{2}\left(\frac{(2x)^2}{2!} - \frac{(2x)^4}{4!} + \frac{(2x)^6}{6!} - \cdots\right) \\ &= \sum_{n=0}^{\infty}(-1)^n \frac{2^{2n+1}}{(2n+2)!} x^{2n+2} \qquad \text{(for all real } x\text{)}.\end{aligned}$$

Taylor series about points other than 0 can often be obtained from known Maclaurin series by a change of variable.

**EXAMPLE 4** Find the Taylor series for $\ln x$ in powers of $x - 2$. Where does the series converge to $\ln x$?

***SOLUTION*** Note that

$$\ln x = \ln\big(2 + (x - 2)\big) = \ln 2\left(1 + \frac{x-2}{2}\right) = \ln 2 + \ln(1 + t),$$

where $t = (x - 2)/2$. We use the known Maclaurin series for $\ln(1 + t)$:

$$\begin{aligned}\ln x &= \ln 2 + \ln(1 + t) \\ &= \ln 2 + t - \frac{t^2}{2} + \frac{t^3}{3} - \frac{t^4}{4} - \cdots \\ &= \ln 2 + \frac{x-2}{2} - \frac{(x-2)^2}{2 \times 2^2} + \frac{(x-2)^3}{3 \times 2^3} - \frac{(x-2)^4}{4 \times 2^4} + \cdots \\ &= \ln 2 + \sum_{n=1}^{\infty} \frac{(-1)^{n-1}}{n\, 2^n}(x - 2)^n.\end{aligned}$$

Since the series for $\ln(1 + t)$ is valid for $-1 < t \le 1$, this series for $\ln x$ is valid for $-1 < (x - 2)/2 \le 1$, that is, for $0 < x \le 4$.

**EXAMPLE 5** Find the Taylor series for $\cos x$ about the point $x = \pi/3$. Where is the series valid?

***SOLUTION*** We use the addition formula for cosine:

$$\begin{aligned}\cos x &= \cos\left(x - \frac{\pi}{3} + \frac{\pi}{3}\right) = \cos\left(x - \frac{\pi}{3}\right)\cos\frac{\pi}{3} - \sin\left(x - \frac{\pi}{3}\right)\sin\frac{\pi}{3} \\ &= \frac{1}{2}\left[1 - \frac{1}{2!}\left(x - \frac{\pi}{3}\right)^2 + \frac{1}{4!}\left(x - \frac{\pi}{3}\right)^4 - \cdots\right] \\ &\quad - \frac{\sqrt{3}}{2}\left[\left(x - \frac{\pi}{3}\right) - \frac{1}{3!}\left(x - \frac{\pi}{3}\right)^3 + \cdots\right] \\ &= \frac{1}{2} - \frac{\sqrt{3}}{2}\left(x - \frac{\pi}{3}\right) - \frac{1}{2}\frac{1}{2!}\left(x - \frac{\pi}{3}\right)^2 + \frac{\sqrt{3}}{2}\frac{1}{3!}\left(x - \frac{\pi}{3}\right)^3 \\ &\quad + \frac{1}{2}\frac{1}{4!}\left(x - \frac{\pi}{3}\right)^4 - \cdots .\end{aligned}$$

This series representation is valid for all $x$. A similar calculation would enable us to expand $\cos x$ or $\sin x$ in powers of $x - c$ for any real $c$; both functions are analytic at every point of the real line. ■

Sometimes it is quite difficult, if not impossible, to find a formula for the general term of a Maclaurin or Taylor series. In such cases it is usually possible to obtain the first few terms before the calculations get too cumbersome. Had we attempted to solve Example 3(c) by multiplying the series for $\sin x$ by itself we might have found ourselves in this bind. Other examples occur when it is necessary to substitute one series into another or to divide one by another.

**EXAMPLE 6** Obtain the first three nonzero terms of the Maclaurin series for (a) $\ln\cos x$, and (b) $\tan x$.

***SOLUTION***

$$\begin{aligned}\text{(a)}\quad \ln\cos x &= \ln\left(1 + \left(-\frac{x^2}{2!} + \frac{x^4}{4!} - \frac{x^6}{6!} + \cdots\right)\right) \\ &= \left(-\frac{x^2}{2!} + \frac{x^4}{4!} - \frac{x^6}{6!} + \cdots\right) - \frac{1}{2}\left(-\frac{x^2}{2!} + \frac{x^4}{4!} - \frac{x^6}{6!} + \cdots\right)^2 \\ &\quad + \frac{1}{3}\left(-\frac{x^2}{2!} + \frac{x^4}{4!} - \frac{x^6}{6!} + \cdots\right)^3 - \cdots \\ &= -\frac{x^2}{2} + \frac{x^4}{24} - \frac{x^6}{720} + \cdots - \frac{1}{2}\left(\frac{x^4}{4} - \frac{x^6}{24} + \cdots\right) \\ &\quad + \frac{1}{3}\left(-\frac{x^6}{8} + \cdots\right) - \cdots \\ &= -\frac{x^2}{2} - \frac{x^4}{12} - \frac{x^6}{45} - \cdots\end{aligned}$$

Note that at each stage of the calculation we kept only enough terms to ensure that we could get all the terms with powers up to $x^6$. Being an even function, $\ln\cos x$ has only even powers in its Maclaurin series. We cannot find the general term of this series, and only with considerable computational effort can we find many more terms than we have already found. We could also try to calculate terms by using the formula $a_k = f^{(k)}(0)/k!$ but even this becomes difficult after the first few values of $k$.

(b) $\tan x = (\sin x)/(\cos x)$. We can obtain the first three terms of the Maclaurin series for $\tan x$ by long division of the series for $\cos x$ into that for $\sin x$:

$$\begin{array}{r|l}
 & x + \dfrac{x^3}{3} + \dfrac{2}{15}x^5 + \cdots \\
\hline
1 - \dfrac{x^2}{2} + \dfrac{x^4}{24} & x - \dfrac{x^3}{6} + \dfrac{x^5}{120} - \cdots \\
 & x - \dfrac{x^3}{2} + \dfrac{x^5}{24} - \cdots \\
\cline{2-2}
 & \qquad \dfrac{x^3}{3} - \dfrac{x^5}{30} + \cdots \\
 & \qquad \dfrac{x^3}{3} - \dfrac{x^5}{6} + \cdots \\
\cline{2-2}
 & \qquad\qquad \dfrac{2x^5}{15} - \cdots \\
 & \qquad\qquad \dfrac{2x^5}{15} - \cdots \\
\cline{2-2}
\end{array}$$

Thus $\tan x = x + \dfrac{1}{3}x^3 + \dfrac{2}{15}x^5 + \cdots$.

Again we cannot easily find all the terms of the series. This Maclaurin series for $\tan x$ converges for $|x| < \pi/2$, but we cannot demonstrate this fact by the techniques we have at our disposal now. Note that the series for $\tan x$ could also have been derived from that of $\ln \cos x$ obtained in part (a) because we have $\tan x = -\dfrac{d}{dx} \ln \cos x$. ■

## EXERCISES 2.2

Find Maclaurin series representations for the functions in Exercises 1–14. For what values of $x$ is each representation valid?

**1.** $e^{3x+1}$

**2.** $\cos(2x^3)$

**3.** $\sin(x - \pi/4)$

**4.** $\cos(2x - \pi)$

**5.** $x^2 \sin(x/3)$

**6.** $\cos^2(x/2)$

**7.** $\sin x \cos x$

**8.** $\tan^{-1}(5x^2)$

**9.** $\dfrac{1 + x^3}{1 + x^2}$

**10.** $\ln(2 + x^2)$

**11.** $\ln \dfrac{1 - x}{1 + x}$

**12.** $(e^{2x^2} - 1)/x^2$

**13.** $\cosh x - \cos x$

**14.** $\sinh x - \sin x$

Find the required Taylor series representations of the functions in Exercises 15–26. Where is each series representation valid?

**15.** $f(x) = e^{-2x}$ about the point $x = -1$

**16.** $f(x) = \sin x$ about the point $x = \pi/2$

**17.** $f(x) = \cos x$ in powers of $x - \pi$

**18.** $f(x) = \ln x$ in powers of $x - 3$

**19.** $f(x) = \ln(2 + x)$ in powers of $x - 2$

**20.** $f(x) = e^{2x+3}$ in powers of $x + 1$

**21.** $f(x) = \sin x - \cos x$ about $x = \dfrac{\pi}{4}$

**22.** $f(x) = \cos^2 x$ about $x = \dfrac{\pi}{8}$

**23.** $f(x) = 1/x^2$ in powers of $x + 2$

**24.** $f(x) = \dfrac{x}{1 + x}$ in powers of $x - 1$

**25.** $f(x) = x \ln x$ in powers of $x - 1$

**26.** $f(x) = xe^x$ in powers of $x + 2$

Find the first three nonzero terms in the Maclaurin series for the functions in Exercises 27–30.

**27.** $\sec x$

**28.** $\sec x \tan x$

**29.** $\tan^{-1}(e^x - 1)$

**30.** $e^{\tan^{-1} x} - 1$

* **31.** Use the fact that $(\sqrt{1 + x})^2 = 1 + x$ to find the first three nonzero terms of the Maclaurin series for $\sqrt{1 + x}$.

**32.** Does $\csc x$ have a Maclaurin series? Why? Find the first three nonzero terms of the Taylor series for $\csc x$ about the point $x = \pi/2$.

Find the sums of the series in Exercises 33–36.

**33.** $1 + x^2 + \dfrac{x^4}{2!} + \dfrac{x^6}{3!} + \dfrac{x^8}{4!} + \cdots$

* **34.** $x^3 - \dfrac{x^9}{3! \times 4} + \dfrac{x^{15}}{5! \times 16} - \dfrac{x^{21}}{7! \times 64} + \dfrac{x^{27}}{9! \times 256} - \cdots$

**35.** $1 + \dfrac{x^2}{3!} + \dfrac{x^4}{5!} + \dfrac{x^6}{7!} + \dfrac{x^8}{9!} + \cdots$

* **36.** $1 + \dfrac{1}{2 \times 2!} + \dfrac{1}{4 \times 3!} + \dfrac{1}{8 \times 4!} + \cdots$

**37.** Let $P(x) = 1 + x + x^2$. Find (a) the Maclaurin series for $P(x)$ and (b) the Taylor series for $P(x)$ about $x = 1$.

* **38.** Verify by direct calculation that $f(x) = 1/x$ is analytic at $x = a$ for every $a \neq 0$.

* **39.** Verify by direct calculation that $\ln x$ is analytic at $x = a$ for every $a > 0$.

* **40.** The function

$$f(x) = \begin{cases} e^{-1/x^2} & \text{if } x \neq 0 \\ 0 & \text{if } x = 0 \end{cases}$$

has derivatives of all orders at every point of the real line, and $f^{(k)}(0) = 0$ for every positive integer $k$. What is the Maclaurin series for $f(x)$? What is the interval of convergence of this Maclaurin series? On what interval does the series converge to $f(x)$? Is $f$ analytic at $x = 0$?

* **41.** By direct multiplication of the Maclaurin series for $e^x$ and $e^y$ show that $e^x e^y = e^{x+y}$.

# 2.3 APPLICATIONS OF TAYLOR AND MACLAURIN SERIES

## Approximating the Values of Functions

Partial sums of Taylor and Maclaurin series can be used as polynomial approximations to more complicated functions. Example 4 of Section 5.6 in *Single-Variable Calculus* is an example of this. In that example the Lagrange remainder in Taylor's Formula is used to determine how many terms of the Maclaurin series for $e^x$ are needed to calculate $e^1 = e$ correct to three decimal places. We will reconsider Taylor's Formula in the next section. In the following examples we use other methods (previously described in Sections 1.4 and 1.5) to estimate the error when a partial sum is used to approximate an infinite series.

**EXAMPLE 1** Use Maclaurin series to find $\sqrt{e}$ correct to four decimal places, that is, with error less than 0.00005 in absolute value.

***SOLUTION*** For the approximation we use the sum of the first $n$ terms of the Maclaurin series:

$$\begin{aligned}\sqrt{e} = e^{1/2} &= 1 + \frac{1}{2} + \frac{1}{2!}\frac{1}{2^2} + \frac{1}{3!}\frac{1}{2^3} + \cdots \\ &\approx 1 + \frac{1}{2} + \frac{1}{2!}\frac{1}{2^2} + \frac{1}{3!}\frac{1}{2^3} + \cdots + \frac{1}{(n-1)!}\frac{1}{2^{n-1}} = s_n\end{aligned}$$

The error in this approximation can be bounded by a geometric series if we use the inequalities $n + 1 < n + 2 < n + 3 < \cdots$.

$$\begin{aligned}0 < \sqrt{e} - s_n &= \frac{1}{n!}\frac{1}{2^n} + \frac{1}{(n+1)!}\frac{1}{2^{n+1}} + \frac{1}{(n+2)!}\frac{1}{2^{n+2}} + \cdots \\ &= \frac{1}{n!}\frac{1}{2^n}\left[1 + \frac{1}{2(n+1)} + \frac{1}{2^2(n+1)(n+2)} + \frac{1}{2^3(n+1)(n+2)(n+3)} + \cdots\right] \\ &< \frac{1}{n!}\frac{1}{2^n}\left[1 + \frac{1}{2(n+1)} + \left(\frac{1}{2(n+1)}\right)^2 + \left(\frac{1}{2(n+1)}\right)^3 + \cdots\right] \\ &= \frac{1}{n!}\frac{1}{2^n}\frac{1}{1 - \dfrac{1}{2(n+1)}} = \frac{2(n+1)}{(2n+1)n!\,2^n}.\end{aligned}$$

We want this error to be less than $0.00005 = 1/20,000$. Thus, we want

$$\frac{2^n n!(2n+1)}{2n+2} > 20,000.$$

This is true if $n = 6$; $2^6 6!(13)/14 \approx 42,789 > 20,000$. Thus

$$\sqrt{e} \approx 1 + \frac{1}{2} + \frac{1}{8} + \frac{1}{48} + \frac{1}{384} + \frac{1}{3840} \approx 1.6487,$$

rounded to four decimal places. ■

When the terms $a_n$ of a series (i) alternate in sign, (ii) decrease steadily in size, and (iii) approach zero as $n \to \infty$, then the error involved in using a partial sum of the series as an approximation to the sum of the series has the same sign as, and is smaller in absolute value than the first omitted term. This was proved in Section 1.4 and provides an easy estimate for the error when the three conditions above are satisfied.

**■ EXAMPLE 2** Find $\cos 43^\circ$ with error less than $1/10,000$.

***SOLUTION*** We give two alternative solutions:

METHOD I. We can use the Maclaurin series:

$$\cos 43^\circ = \cos \frac{43\pi}{180} = 1 - \frac{1}{2!}\left(\frac{43\pi}{180}\right)^2 + \frac{1}{4!}\left(\frac{43\pi}{180}\right)^4 - \cdots.$$

Now $43\pi/180 \approx 0.75049\cdots < 1$, so the series above must satisfy the conditions (i)–(iii) mentioned above. If we truncate the series after the $n$th term

$$(-1)^{n-1}\frac{1}{(2n-2)!}\left(\frac{43\pi}{180}\right)^{2n-2},$$

then the error $E$ will satisfy

$$|E| \le \frac{1}{(2n)!}\left(\frac{43\pi}{180}\right)^{2n} < \frac{1}{(2n)!}.$$

The error will not exceed $1/10{,}000$ if $(2n)! > 10{,}000$, so $n = 4$ will do; $(8! = 40{,}320)$.

$$\cos 43^\circ \approx 1 - \frac{1}{2!}\left(\frac{43\pi}{180}\right)^2 + \frac{1}{4!}\left(\frac{43\pi}{180}\right)^4 - \frac{1}{6!}\left(\frac{43\pi}{180}\right)^6 \approx 0.73135\cdots$$

METHOD II. Since $43^\circ$ is close to $45^\circ$ we can do a bit better by using the Taylor series about $x = \pi/4$ instead of the Maclaurin series.

$$\begin{aligned}
\cos 43^\circ &= \cos\left(\frac{\pi}{4} - \frac{\pi}{90}\right) \\
&= \cos\frac{\pi}{4}\cos\frac{\pi}{90} + \sin\frac{\pi}{4}\sin\frac{\pi}{90} \\
&= \frac{1}{\sqrt{2}}\left[\left(1 - \frac{1}{2!}\left(\frac{\pi}{90}\right)^2 + \frac{1}{4!}\left(\frac{\pi}{90}\right)^4 - \cdots\right)\right. \\
&\qquad \left. + \left(\frac{\pi}{90} - \frac{1}{3!}\left(\frac{\pi}{90}\right)^3 + \cdots\right)\right].
\end{aligned}$$

Since

$$\frac{1}{4!}\left(\frac{\pi}{90}\right)^4 < \frac{1}{3!}\left(\frac{\pi}{90}\right)^3 < \frac{1}{20{,}000},$$

we need only the first two terms of the first series and the first term of the second series:

$$\cos 43^\circ \approx \frac{1}{\sqrt{2}}\left(1 + \frac{\pi}{90} - \frac{1}{2}\left(\frac{\pi}{90}\right)^2\right) \approx 0.731358\cdots.$$

(In fact, $\cos 43^\circ = 0.7313537\cdots$.) ■

When finding approximate values of functions, it is best, whenever possible, to use a power series about a point as close as possible to the point where the approximation is desired.

## Functions Defined by Integrals

Many functions that can be expressed as simple combinations of elementary functions cannot be antidifferentiated by elementary techniques; their antiderivatives are not simple combinations of elementary functions. We can, however, often find the Taylor series for the antiderivatives of such functions and hence approximate their definite integrals.

■ **EXAMPLE 3** Find the Maclaurin series for

$$E(x) = \int_0^x e^{-t^2}\,dt$$

and use it to evaluate $E(1)$ correct to three decimal places.

***SOLUTION*** The Maclaurin series for $E(x)$ is given by

$$\begin{aligned} E(x) &= \int_0^x \left(1 - t^2 + \frac{t^4}{2!} - \frac{t^6}{3!} + \frac{t^8}{4!} - \cdots\right) dt \\ &= \left(t - \frac{t^3}{3} + \frac{t^5}{5\times 2!} - \frac{t^7}{7\times 3!} + \frac{t^9}{9\times 4!} - \cdots\right)\Bigg|_0^x \\ &= x - \frac{x^3}{3} + \frac{x^5}{5\times 2!} - \frac{x^7}{7\times 3!} + \frac{x^9}{9\times 4!} - \cdots = \sum_{n=0}^{n}(-1)^n\frac{x^{2n+1}}{(2n+1)n!}, \end{aligned}$$

and is valid for all real $x$ because the series for $e^{-t^2}$ is valid for all real $t$. Therefore

$$\begin{aligned} E(1) &= 1 - \frac{1}{3} + \frac{1}{5\times 2!} - \frac{1}{7\times 3!} + \cdots \\ &\approx 1 - \frac{1}{3} + \frac{1}{5\times 2!} - \frac{1}{7\times 3!} + \cdots + \frac{(-1)^{n-1}}{(2n-1)(n-1)!}. \end{aligned}$$

We stopped with the $n$th term. The error in this approximation does not exceed the first omitted term, so it will be less than 0.0005, provided $(2n+1)n! > 2,000$. Since $13 \times 6! = 9360$, $n = 6$ will do. Thus

$$E(1) \approx 1 - \frac{1}{3} + \frac{1}{10} - \frac{1}{42} + \frac{1}{216} - \frac{1}{1320} \approx 0.747,$$

rounded to three decimal places. ■

### Indeterminate Forms

Maclaurin and Taylor series can provide a useful alternative to l'Hôpital's Rule for evaluating the limits of indeterminate forms.

**EXAMPLE 4** Evaluate (a) $\lim_{x\to 0} \dfrac{x-\sin x}{x^3}$ and (b) $\lim_{x\to 0} \dfrac{(e^{2x}-1)\ln(1+x^3)}{(1-\cos 3x)^2}$

***SOLUTION***

(a)
$$\lim_{x\to 0} \frac{x-\sin x}{x^3} \quad \left[\frac{0}{0}\right]$$
$$= \lim_{x\to 0} \frac{x-\left(x-\dfrac{x^3}{3!}+\dfrac{x^5}{5!}-\cdots\right)}{x^3}$$
$$= \lim_{x\to 0} \frac{\dfrac{x^3}{3!}-\dfrac{x^5}{5!}+\cdots}{x^3}$$
$$= \lim_{x\to 0}\left(\frac{1}{3!}-\frac{x^2}{5!}+\cdots\right) = \frac{1}{3!} = \frac{1}{6}.$$

(b)
$$\lim_{x\to 0} \frac{(e^{2x}-1)\ln(1+x^3)}{(1-\cos 3x)^2} \quad \left[\frac{0}{0}\right]$$
$$= \lim_{x\to 0} \frac{\left(1+(2x)+\dfrac{(2x)^2}{2!}+\dfrac{(2x)^3}{3!}+\cdots-1\right)\left(x^3-\dfrac{x^6}{2}+\cdots\right)}{\left(1-\left(1-\dfrac{(3x)^2}{2!}+\dfrac{(3x)^4}{4!}-\cdots\right)\right)^2}$$
$$= \lim_{x\to 0} \frac{2x^4+2x^5+\cdots}{\left(\dfrac{9}{2}x^2-\dfrac{3^4}{4!}x^4+\cdots\right)^2}$$
$$= \lim_{x\to 0} \frac{2+2x+\cdots}{\left(\dfrac{9}{2}-\dfrac{3^4}{4!}x^2+\cdots\right)^2} = \frac{2}{\left(\dfrac{9}{2}\right)^2} = \frac{8}{81}.$$

You can check that the second of these examples is much more difficult if attempted using l'Hôpital's Rule. ■

## EXERCISES 2.3

Use Maclaurin or Taylor series to calculate the function values indicated in Exercises 1–12, with error less than $5\times 10^{-5}$ in absolute value.

**1.** $e^{0.2}$

**2.** $1/e$

**3.** $e^{1.2}$

**4.** $\sin(0.1)$

**5.** $\cos 5^\circ$

**6.** $\ln(6/5)$

**7.** $\ln(0.9)$

**8.** $\sin 80^\circ$

**9.** $\cos 65^\circ$

**10.** $\tan^{-1} 0.2$

**11.** $\cosh(1)$

**12.** $\ln(3/2)$

Find Maclaurin series for the functions in Exercises 13–17.

**13.** $I(x) = \displaystyle\int_0^x \frac{\sin t}{t}\,dt$

**14.** $J(x) = \displaystyle\int_0^x \frac{e^t-1}{t}\,dt$

**15.** $K(x) = \displaystyle\int_1^{1+x} \frac{\ln t}{t-1}\,dt$

**16.** $L(x) = \displaystyle\int_0^x \cos(t^2)\,dt$

**17.** $M(x) = \displaystyle\int_0^x \frac{\tan^{-1} t^2}{t^2}\,dt$

**18.** Find $L(0.5)$ correct to three decimal places, with $L$ defined as in Exercise 16.

**19.** Find $I(1)$ correct to three decimal places, with $I$ defined as in Exercise 13.

Evaluate the limits in Exercises 20–25.

**20.** $\lim_{x\to 0} \dfrac{\sin(x^2)}{\sinh x}$

**21.** $\lim_{x\to 0} \dfrac{1-\cos(x^2)}{(1-\cos x)^2}$

**22.** $\lim_{x\to 0} \dfrac{(e^x-1-x)^2}{x^2-\ln(1+x^2)}$

**23.** $\lim_{x\to 0} \dfrac{2\sin 3x - 3\sin 2x}{5x-\tan^{-1} 5x}$

**24.** $\lim_{x\to 0} \dfrac{\sin(\sin x)-x}{x(\cos(\sin x)-1)}$

**25.** $\lim_{x\to 0} \dfrac{\sinh x - \sin x}{\cosh x - \cos x}$

❖ **26.** Series can provide a useful tool for solving differential equations. The method is developed in Section 10.9, but for now you can experiment as follows: calculate the first two derivatives of the series $y = \sum_{n=0}^{\infty} a_n x^n$ and substitute the series for $y$, $y'$ and $y''$ into the differential equation

$$y'' + xy' + y = 0.$$

Thereby determine a relationship between $a_{n+2}$ and $a_n$ that enables all the coefficients in the series to be determined in terms of $a_0$ and $a_1$. Solve the initial-value problem

$$\begin{cases} y'' + xy' + y = 0 \\ y(0) = 1 \\ y'(0) = 0. \end{cases}$$

❖ **27.** Solve the initial-value problem

$$\begin{cases} y'' + xy' + 2y = 0 \\ y(0) = 1 \\ y'(0) = 2 \end{cases}$$

by the method suggested in Exercise 26.

## 2.4 TAYLOR POLYNOMIALS AND TAYLOR'S FORMULA

For most applications that involve calculations, such as those in Section 2.3, it is not necessary to use Taylor series, but only finite sums of terms called Taylor polynomials that we define as follows.

**DEFINITION 4**

**Taylor polynomials**

If the function $f(x)$ has derivatives up to and including order $n$ at the point $x = c$, we can form the polynomial

$$\begin{aligned} P_n(x) &= f(c) + f'(c)(x-c) + \frac{f''(c)}{2!}(x-c)^2 + \cdots + \frac{f^{(n)}(c)}{n!}(x-c)^n \\ &= \sum_{k=0}^{n} \frac{f^{(k)}(c)}{k!}(x-c)^k, \end{aligned}$$

called the **Taylor polynomial of degree** $n$ for $f(x)$ *about* $x = c$, or *in powers of* $(x-c)$. (If $c = 0$, $P_n$ is called a **Maclaurin polynomial.**)

Note that $P_n(x)$ is a polynomial of degree *at most* $n$. If it happens that $f^{(n)}(c) = 0$ then $P_n(x)$ will have degree less than $n$, though we will still call it the $n$th degree Taylor polynomial. If $f(x)$ has a Taylor series about $x = c$, then $P_n(x)$ is just a partial sum of that series; it is the partial sum $s_{n+1}$ of the first $n+1$ terms. Among all polynomials of degree at most $n$, the Taylor polynomial $P_n(x)$ best describes the behaviour of $f(x)$ near $x = c$ in the sense that

$$P_n(c) = f(c), \quad P_n'(c) = f'(c), \quad P_n''(c) = f''(c), \quad \ldots, \quad P_n^{(n)}(c) = f^{(n)}(c).$$

You can verify that $P^{(k)}(c) = f^{(k)}(c)$ for $0 \le k \le n$ by differentiating the formula for $P_n(x)$ in Definition 4 $k$ times and then substituting $x = c$.

This section can be used as a self-contained alternative introduction to Taylor and Maclaurin approximations for students who do not want to undertake a more thorough study of numerical and power series (Chapter 1 and Sections 2.1 and 2.2). The polynomials can then be used in calculations like those in Section 2.3. To achieve this, in the following examples we calculate Taylor and Maclaurin polynomials directly from the definition. However, if you know the corresponding Taylor or Maclaurin series, you can write the polynomials immediately as the appropriate partial sums of the series.

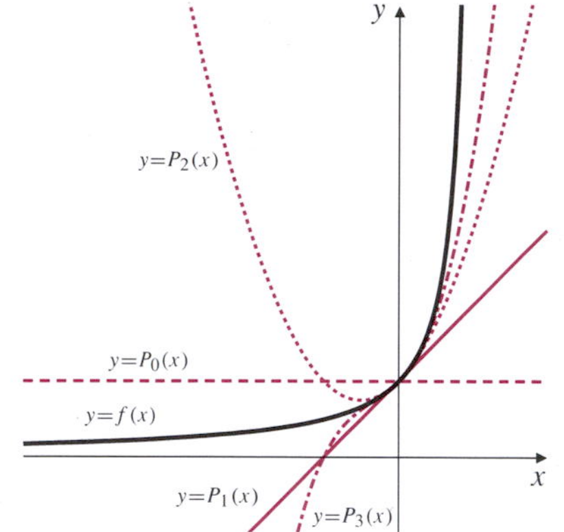


**Figure 2.1** Graphs of $f(x) = 1/(1-x)$, $P_0(x) = 1$, $P_1(x) = 1 + x$, $P_2(x) = 1 + x + x^2$, $P_3(x) = 1 + x + x^2 + x^3$

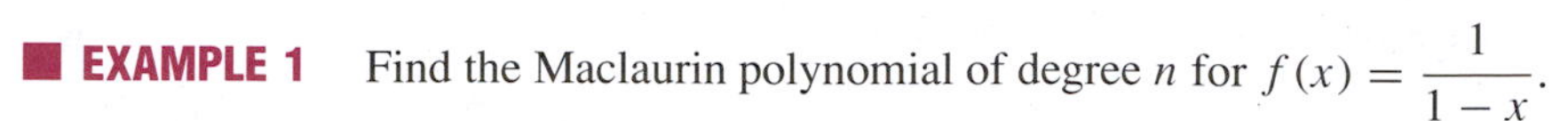

**EXAMPLE 1** Find the Maclaurin polynomial of degree $n$ for $f(x) = \dfrac{1}{1-x}$.

***SOLUTION*** We have

$$f'(x) = \frac{1}{(1-x)^2}, \quad f''(x) = \frac{2!}{(1-x)^3}, \quad f'''(x) = \frac{3!}{(1-x)^4}, \quad \ldots$$

$$f^{(k)}(x) = \frac{k!}{(1-x)^{k+1}}.$$

Therefore $f^{(k)}(0) = k!$ for $k = 0, 1, 2, \ldots$. The Maclaurin polynomial of degree $n$ for $f$ is

$$P_n(x) = \sum_{k=0}^{n} \frac{k!}{k!} x^k = \sum_{k=0}^{n} x^k = 1 + x + x^2 + x^3 + \cdots + x^n.$$

The graphs of $f(x)$ and of the Maclaurin polynomials $P_0(x)$, $P_1(x)$, $P_2(x)$, and $P_3(x)$ are shown in Figure 2.1. Observe how the graph of $P_n(x)$ gets closer to the graph of $f$ near $x = 0$ as $n$ increases. ■

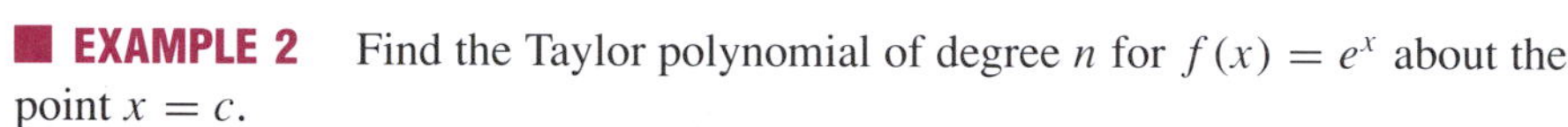

**EXAMPLE 2** Find the Taylor polynomial of degree $n$ for $f(x) = e^x$ about the point $x = c$.

***SOLUTION*** Since $f^{(k)}(x) = e^x$ for $k = 0, 1, 2, \ldots$, therefore $f^{(k)}(c) = e^c$ and the required Taylor polynomial is

$$\begin{aligned} P_n(x) &= \sum_{k=0}^{n} \frac{e^c}{k!}(x-c)^k \\ &= e^c + e^c(x-c) + \frac{e^c}{2!}(x-c)^2 + \cdots + \frac{e^c}{n!}(x-c)^n. \end{aligned}$$ ■

**EXAMPLE 3** Find the Maclaurin polynomials of degree 5 and 6 for $f(x) = \sin x$.

***SOLUTION*** We have

$$f'(x) = \cos x, \qquad f''(x) = -\sin x, \qquad f'''(x) = -\cos x,$$
$$f^{(4)}(x) = \sin x, \qquad f^{(5)}(x) = \cos x, \qquad f^{(6)}(x) = -\sin x.$$

Hence $f(0) = f''(0) = f^{(4)}(0) = f^{(6)}(0) = 0$; and $f'(0) = 1$, $f'''(0) = -1$, $f^{(5)}(0) = 1$. Both the Maclaurin polynomials $P_5$ and $P_6$ therefore have nonzero terms up to degree 5 only:

$$P_5(x) = P_6(x) = x - \frac{x^3}{3!} + \frac{x^5}{5!} = x - \frac{x^3}{6} + \frac{x^5}{120}.$$

The Maclaurin polynomials for the *odd function* $\sin x$ involve only *odd* powers of $x$. Graphs of the Maclaurin polynomials $P_1(x)$, $P_3(x)$, $P_5(x)$, and $P_{13}(x)$ for $\sin x$ are shown in Figure 2.2. ■

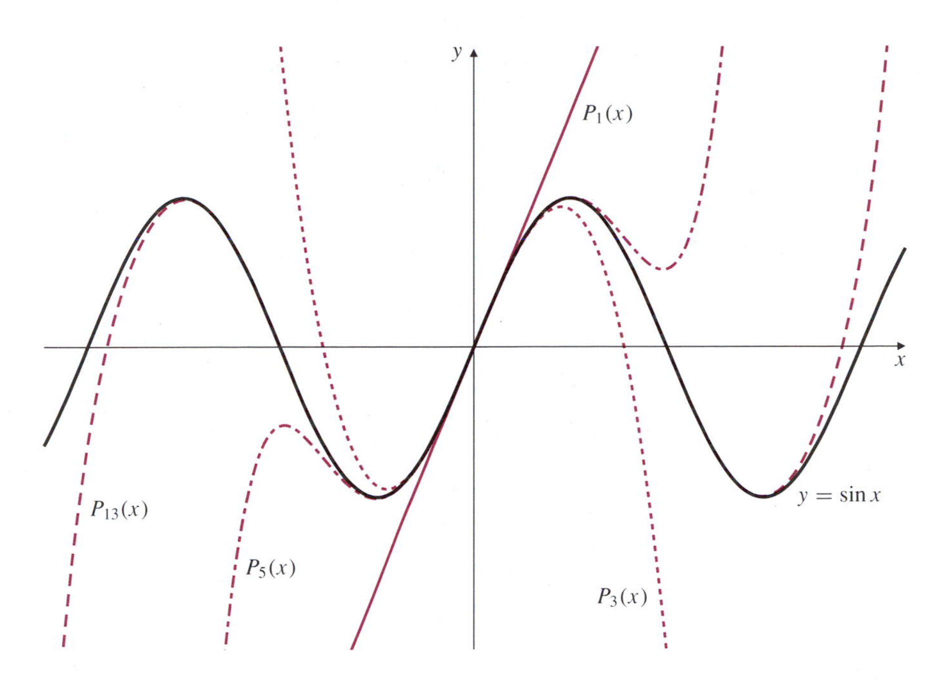


**Figure 2.2** Some Maclaurin polynomials for $f(x) = \sin x$

Taylor and Maclaurin polynomials are not always as easy to calculate as those in the above examples. Calculating higher derivatives can become very difficult.

**EXAMPLE 4** Find the Maclaurin polynomial of degree 5 for $f(x) = \tan x$.

***SOLUTION*** Since $\tan x$ is an odd function, its Maclaurin polynomials involve only odd powers of $x$. If $f(x) = \tan x$, we have $f(0) = 0$ and

$$\begin{aligned}
f'(x) &= \sec^2 x & f'(0) &= 1 \\
f''(x) &= 2\sec^2 x \tan x & f''(0) &= 0 \\
f^{(3)}(x) &= 2\sec^4 x + 4\sec^2 x \tan^2 x & f^{(3)}(0) &= 2 \\
f^{(4)}(x) &= 16\sec^4 x \tan x + 8\sec^2 x \tan^3 x & f^{(4)}(0) &= 0 \\
f^{(5)}(x) &= 88\sec^4 x \tan^2 x + 16\sec^6 x + 16\sec^2 x \tan^4 x. & f^{(5)}(0) &= 16
\end{aligned}$$

Thus the Maclaurin polynomial of degree 5 is

$$P_5(x) = x + \frac{2}{3!}x^3 + \frac{16}{5!}x^5 = x + \frac{x^3}{3} + \frac{2x^5}{15}.$$

## EXPLORE! Maclaurin Polynomials

(a) Use a graphing utility to plot simultaneously the graphs

$$y = \ln(1+x) \quad \text{and} \quad y = P_{10}(x) = \sum_{n=1}^{10} \frac{(-1)^{n-1}x^n}{n},$$

the 10th degree Maclaurin polynomial for $\ln(1+x)$. Plot in the window $-2 \le x \le 2$, $-4 \le y \le 2$. On what interval does the graph suggest that $P_{10}(x)$ is a reasonably good approximation to $\ln(1+x)$?

(b) Repeat (a) for the function $y = e^x$ and its Maclaurin polynomials $P_3(x)$, $P_6(x)$, and $P_9(x)$. (Here $P_k(x) = \sum_{n=0}^{k} x^n/n!$.) Experiment with different window settings to best determine how far on each side of $x = 0$ each polynomial well approximates $e^x$.

Because the derivatives of the Taylor polynomial $P_n(x)$ up to order $n$ match those of $f(x)$ at $x = c$, $P_n(x)$ is the polynomial of degree $n$ which *best approximates* $f(x)$ near $x = c$. Let

$$R_n(x) = f(x) - P_n(x)$$

denote the *error* in this approximation. The formula

$$f(x) = P_n(x) + R_n(x) = \sum_{k=0}^{n} \frac{f^{(k)}(c)}{k!}(x-c)^k + R_n(x)$$

is known as **Taylor's Formula with Remainder**, or simply just **Taylor's Formula**. Various versions of **Taylor's Theorem** provide specific expressions for the remainder term, $R_n(x)$, in Taylor's Formula. One version of Taylor's Theorem is Theorem 13 in Section 5.6 of *Single-Variable Calculus* (where $R_n$ was denoted $E_n$), which we restate in our current context as follows.

**THEOREM 6**

**Taylor's Theorem (with Lagrange remainder)**

If the $(n+1)$st derivative of $f$ exists on an interval containing $c$ and $x$, and if $P_n(x)$ is the Taylor polynomial of degree $n$ for $f$ about the point $x = c$, then the remainder $R_n(x) = f(x) - P_n(x)$ in Taylor's Formula is given by

$$R_n(x) = \frac{f^{(n+1)}(X)}{(n+1)!}(x-c)^{n+1}$$

(called the **Lagrange remainder**) for some $X$ between $c$ and $x$.

This theorem can beproved by induction on $n$.

Observe that the Lagrange form of the remainder, $R_n(x)$, looks just like the $(n+1)$st degree term in $P_{n+1}(x)$, except that $c$ in $f^{(n+1)}(c)$ has been replaced by an unknown number $X$ between $c$ and $x$. The cases $n = 0$ and $n = 1$ of Taylor's Formula with Lagrange remainder are just the Mean-Value Theorem and the error formula for linear approximation, respectively. An example using the Lagrange remainder for Taylor's Theorem to calculate $e$ to three-decimal-place accuracy is given in Section 5.6 of *Single-Variable Calculus*. Here are two more examples.

**EXAMPLE 5** Use Taylor's Theorem to determine how many terms of the Maclaurin series for $\cos x$ are needed to calculate $\cos 10°$ correctly to five decimal places.

***SOLUTION*** Being an even function, $f(x) = \cos x$ has only even degree terms in its Maclaurin series. Calculating the values of the derivatives of $f(x)$ at $x = 0$ as was done for $\sin x$ in Example 3 we obtain the Maclaurin polynomials

$$P_{2n}(x) = P_{2n+1}(x) = \sum_{j=0}^{n} \frac{(-1)^j}{(2j)!} x^{2j}$$
$$1 - \frac{x^2}{2!} + \frac{x^4}{4!} - \cdots + \frac{(-1)^n}{(2n)!} x^{2n}.$$

It makes good sense to use the remainder $R_{2n+1}$ rather than the remainder $R_{2n}$; it is likely to be smaller, and therefore assure us of more accuracy for any given value of $n$. Since $f^{(2n+2)}(x) = (-1)^n \cos x$, we have, for some $X$ between 0 and $x$,

$$|R_{2n+1}(x)| = \left| \frac{(-1)^n \cos X}{(2n+2)!} x^{2n+2} \right| \le \frac{|x|^{2n+2}}{(2n+2)!}.$$

For $x = 10^\circ = \pi/18 \approx 0.174533 < 0.2$ radians, we will have five decimal place accuracy if

$$\frac{0.2^{2n+2}}{(2n+2)!} < 0.000005.$$

This is satisfied if $n = 2$ $(0.2^6/6! < 9 \times 10^{-8})$, but not $n = 1$. Thus

$$\cos 10^\circ = \cos \frac{\pi}{18} \approx 1 - \frac{1}{2}\left(\frac{\pi}{18}\right)^2 + \frac{1}{24}\left(\frac{\pi}{18}\right)^4 \approx 0.98481$$

to five decimal places. ∎

**EXAMPLE 6** Estimate the error in the approximation $\tan(0.5) \approx 0.5 + \dfrac{(0.5)^3}{3}$.

***SOLUTION*** As observed in Example 4 above, the Maclaurin series for $\tan x$ involves only odd powers of $x$ and the Maclaurin polynomials $P_3(x)$ and $P_4(x)$ coincide:

$$P_3(x) = P_4(x) = x + \frac{x^3}{3}.$$

If we use the approximation $\tan(0.5) \approx P_4(0.5) = 0.5 + (0.5)^3/3$, then the error is

$$R_4(0.5) = \frac{f^{(5)}(X)}{5!}(0.5)^5 \qquad \text{for some } X \text{ between 0 and 0.5.}$$

Since $0.5 < \pi/6$, if $0 \le X \le 0.5$ we have

$$0 \le \sec X \le \sec(0.5) < \sec\frac{\pi}{6} = \frac{2}{\sqrt{3}},$$
$$0 \le \tan X \le \tan(0.5) < \tan\frac{\pi}{6} = \frac{1}{\sqrt{3}}.$$

Using the formula for $f^{(5)}(x)$ obtained in Example 4 we estimate

$$\begin{aligned} |f^{(5)}(X)| &= |88\sec^4 X \tan^2 X + 16\sec^6 X + 16\sec^2 X \tan^4 X| \\ &\le 88\left(\frac{2}{\sqrt{3}}\right)^4\left(\frac{1}{\sqrt{3}}\right)^2 + 16\left(\frac{2}{\sqrt{3}}\right)^6 + 16\left(\frac{2}{\sqrt{3}}\right)^2\left(\frac{1}{\sqrt{3}}\right)^4 \\ &= \frac{1}{27}(88 \times 16 + 16 \times 64 + 16 \times 4) \approx 92.44 < 93. \end{aligned}$$

Thus

$$|R_4(0.5)| < \frac{93}{5!}(0.5)^5 \approx 0.024$$

and the error in the approximation $\tan(0.5) \approx 0.5 + (0.5)^3/3 \approx 0.542$ is less than 0.025 in absolute value. ∎

## Using Taylor's Theorem to Find Taylor and Maclaurin Series

If a function $f$ has derivatives of all orders, we can write Taylor's Formula for any $n$:

$$f(x) = P_n(x) + R_n(x).$$

If we can show that $\lim_{n\to\infty} R_n(x) = 0$ for all $x$ in some interval $I$, then we are entitled to conclude, for $x$ in $I$, that

$$f(x) = \lim_{n\to\infty} P_n(x) = \sum_{k=0}^{\infty} \frac{f^{(k)}(c)}{k!}(x-c)^k,$$

that is, we will have expressed $f(x)$ as the sum of an infinite series of terms which are multiples of positive integer powers of $x - c$, and the series converges for all $x$ in $I$. This series is the **Taylor series** representation of $f$ in powers of $x - c$ (or the **Maclaurin series** if $c = 0$).

**EXAMPLE 7** Use Taylor's Theorem to find the Maclaurin series for $f(x) = e^x$. Where does the series converge to $f(x)$?

***SOLUTION*** Since $e^x$ is positive and increasing, $e^X \le e^{|x|}$ for any $X \le |x|$. Since $f^{(k)}(x) = e^x$ for any $k$ we have, taking $c = 0$ in the Lagrange remainder in Taylor's Formula,

$$\begin{aligned} |R_n(x)| &= \left| \frac{f^{(n+1)}(X)}{(n+1)!} x^{n+1} \right| \\ &\le \frac{e^X}{(n+1)!}|x|^{n+1} \le e^{|x|}\frac{|x|^{n+1}}{(n+1)!} \to 0 \text{ as } n \to \infty \end{aligned}$$

for any real $x$, as shown in Theorem 3(b) of Section 1.1. Therefore

$$e^x = \sum_{k=0}^{\infty} \frac{x^k}{k!} = 1 + x + \frac{x^2}{2!} + \frac{x^3}{3!} + \cdots,$$

and the series converges to $e^x$ for all real numbers $x$. ■

## Taylor's Theorem with Integral Remainder

The following theorem is another version of Taylor's Theorem, where the remainder in Taylor's Formula is expressed as an integral.

**THEOREM 7**

**Taylor's Theorem (with integral remainder)**

If the $n + 1$st derivative of $f$ is continuous on an interval containing $c$ and $x$, and if $P_n(x)$ is the Taylor polynomial of degree $n$ for $f$ about the point $x = c$, then the remainder $R_n(x) = f(x) - P_n(x)$ in Taylor's Formula is given by

$$R_n(x) = \frac{1}{n!}\int_c^x (x-t)^n f^{(n+1)}(t)\, dt.$$

**PROOF** We start with the Fundamental Theorem of Calculus written in the form

$$f(x) = f(c) + \int_c^x f'(t)\,dt = P_0(x) + R_0(x).$$

Note that the Fundamental Theorem is just the special case $n = 0$ of Taylor's Formula with integral remainder. We now apply integration by parts to the integral, setting

$$\begin{aligned} U &= f'(t), & dV &= dt \\ dU &= f''(t)\,dt, & V &= -(x-t). \end{aligned}$$

(We have broken our usual rule about not including a constant of integration with $V$. In this case we have included the constant $-x$ in $V$ in order to have $V$ vanish when $t = x$.) We have

$$\begin{aligned} f(x) &= f(c) - f'(t)(x-t)\Big|_{t=c}^{t=x} + \int_c^x (x-t)f''(t)\,dt \\ &= f(c) + f'(c)(x-c) + \int_c^x (x-t)f''(t)\,dt \\ &= P_1(x) + R_1(x). \end{aligned}$$

We have thus proved the case $n = 1$ of Taylor's Formula with integral remainder.

Let us complete the proof for general $n$ by mathematical induction. Suppose that Taylor's Formula holds with Integral Remainder for some $n = k$:

$$f(x) = P_k(x) + R_k(x) = P_k(x) + \frac{1}{k!}\int_c^x (x-t)^k f^{(k+1)}(t)\,dt.$$

Again we integrate by parts. Let

$$\begin{aligned} U &= f^{(k+1)}(t), & dV &= (x-t)^k\,dt, \\ dU &= f^{(k+2)}(t)\,dt, & V &= \frac{-1}{k+1}(x-t)^{k+1}. \end{aligned}$$

We have

$$\begin{aligned} f(x) &= P_k(x) + \frac{1}{k!}\left(-\frac{f^{(k+1)}(t)(x-t)^{k+1}}{k+1}\Bigg|_{t=c}^{t=x} + \int_c^x \frac{(x-t)^{k+1}f^{(k+2)}(t)}{k+1}\,dt\right) \\ &= P_k(x) + \frac{f^{(k+1)}(c)}{(k+1)!}(x-c)^{k+1} + \frac{1}{(k+1)!}\int_c^x (x-t)^{k+1}f^{(k+2)}(t)\,dt \\ &= P_{k+1}(x) + R_{k+1}(x). \end{aligned}$$

Thus Taylor's Formula with integral remainder is valid for $n = k + 1$ if it is valid for $n = k$. Having been shown to be valid for $n = 0$ (and $n = 1$), it must therefore be valid for every positive integer $n$.

◆

**REMARK** Using one or the other of the versions of Taylor's Theorem given in this section, all the basic Maclaurin and Taylor series given in Section 2.2 can be verified without having to use the theory of power series.

## EXERCISES 2.4

Find the Taylor or Maclaurin polynomials of the specified degrees for the functions in Exercises 1–10.

1. $f(x) = 1/x$ in powers of $x - 2$, degree 4
2. $f(x) = \sqrt{x}$ about $x = 1$, degree 4
3. $f(x) = \cos x$ in powers of $x$, degree 4
4. $f(x) = \sin x$ in powers of $x$, degree 5
5. $f(x) = \sin x$ in powers of $x - (\pi/4)$, degree 6
6. $f(x) = e^{-x}$ in powers of $x$, degree $n$
7. $f(x) = e^{-x}$ in powers of $x + 2$, degree $n$
8. $f(x) = \ln x$ about $x = 2$, degree $n$
9. $f(x) = \sec x$ about $x = 0$, degree 4
10. $f(x) = \tan^{-1} x$ in powers of $x - 1$, degree 3
11. Let $f(x) = x^2 + 2x + 3$. Find
    (a) the Maclaurin polynomial of degree 2 for $f(x)$,
    (b) the Maclaurin polynomial of degree 99 for $f(x)$,
    (c) the Taylor polynomial of degree 2 for $f(x)$ in powers of $x + 2$,
    (d) the Taylor polynomial of degree 99 for $f(x)$ in powers of $x - 3$.
12. Find the Taylor polynomial of degree 6 for $f(x) = x^6$ about the point of $x = 1$.
13. For what functions $f(x)$ is it true that $f(x)$ is identically equal to its Maclaurin polynomial of degree $n$?
14. For what functions $f(x)$ is it true that the Taylor polynomial of degree $n$ in powers of $x - a$ vanishes identically?
15. Use a graphing utility to plot $y = \cos x$ and its Maclaurin polynomials $P_4(x)$, $P_8(x)$, and $P_{12}(x)$. Using the graph, estimate how far away from $x = 0$ each polynomial approximates $\cos x$ to within 0.05.
16. Use a graphing utility to plot $y = \tan^{-1} x$ and its Maclaurin polynomials $P_5(x)$, $P_{10}(x)$, and $P_{15}(x)$. Using the graph, estimate how far away from $x = 0$ each polynomial approximates $\tan^{-1} x$ to within 0.05.
17. Estimate the error if the Maclaurin polynomial of degree 5 for $\sin x$ is used to approximate $\sin(0.2)$.
18. Estimate the error if the Maclaurin polynomial of degree 6 for $\cos x$ is used to approximate $\cos(1)$.
19. Estimate the error if the Maclaurin polynomial of degree 4 for $e^{-x}$ is used to approximate $e^{-0.5}$.
20. Estimate the error if the Maclaurin polynomial of degree 2 for $\sec x$ is used to approximate $\sec(0.2)$.
21. Estimate the error if the Maclaurin polynomial of degree 3 for $\ln(\cos x)$ is used to approximate $\ln(\cos 0.1)$.
22. Estimate the error if the Taylor polynomial of degree 3 for $\tan^{-1} x$ in powers of $x - 1$ is used to approximate $\tan^{-1} 0.99$.
23. Estimate the error if the Taylor polynomial of degree 4 for $\ln x$ in powers of $x - 2$ is used to approximate $\ln(1.95)$.

Use Taylor's Formula to establish the Maclaurin series for the functions in Exercises 24–31.

24. $e^{-x}$
25. $2^x$
26. $\cos x$
27. $\sin x$
28. $\sin^2 x$
29. $\dfrac{1}{1-x}$

* 30. $\ln(1 + x)$ (Use the integral remainder.)

* 31. $\dfrac{x}{2 + 3x}$ (Use Exercise 29.)

Use Taylor's Formula to obtain the Taylor series indicated in Exercises 32–37.

32. for $e^x$ in powers of $x - a$
33. for $\sin x$ in powers of $x - (\pi/6)$
34. for $\cos x$ in powers of $x - (\pi/4)$

* 35. for $\ln x$ in powers of $x - 1$ (Use the integral remainder.)

* 36. for $\ln x$ in powers of $x - 2$

37. for $1/x$ in powers of $x + 2$ (Use Exercise 29.)

## 2.5 THE BINOMIAL THEOREM AND BINOMIAL SERIES

**EXAMPLE 1** Use Taylor's Formula to prove the Binomial Theorem: if $n$ is a positive integer then

$$(a+x)^n = a^n + n a^{n-1} x + \frac{n(n-1)}{2!} a^{n-2} x^2 + \cdots + n a x^{n-1} + x^n$$
$$= \sum_{k=0}^{n} \binom{n}{k} a^{n-k} x^k,$$

where $\displaystyle \binom{n}{k} = \frac{n!}{(n-k)!k!}$.

**SOLUTION** Let $f(x) = (a+x)^n$. Then

$$f'(x) = n(a+x)^{n-1} = \frac{n!}{(n-1)!}(a+x)^{n-1}$$
$$f''(x) = \frac{n!}{(n-1)!}(n-1)(a+x)^{n-2} = \frac{n!}{(n-2)!}(a+x)^{n-2}$$
$$\vdots$$
$$f^{(k)}(x) = \frac{n!}{(n-k)!}(a+x)^{n-k} \qquad (0 \le k \le n).$$

In particular, $f^{(n)}(x) = \dfrac{n!}{0!}(a+x)^{n-n} = n!$, a constant, and

$$f^{(k)}(x) = 0 \qquad \text{for all } x, \text{ if } k > n.$$

For $0 \le k \le n$ we have $f^{(k)}(0) = \dfrac{n!}{(n-k)!}a^{n-k}$. Thus, by Taylor's Theorem with Lagrange remainder,

$$\begin{aligned}(a+x)^n = f(x) &= \sum_{k=0}^{n} \frac{f^{(k)}(0)}{k!}x^k + \frac{f^{(n+1)}(X)}{(n+1)!}x^{n+1} \\ &= \sum_{k=0}^{n} \frac{n!}{(n-k)!k!}a^{n-k}x^k + 0 = \sum_{k=0}^{n} \binom{n}{k} a^{n-k}x^k.\end{aligned}$$

This is, in fact, the Maclaurin *series* for $(a+x)^n$, not just the Maclaurin polynomial of degree $n$. Since all higher-degree terms are zero, the series has only finitely many nonzero terms and so converges for all $x$. ■

**REMARK** If $f(x) = (a+x)^r$ where $a > 0$ and $r$ is any real number, then calculations similar to those above show that the Maclaurin polynomial of degree $n$ for $f$ is

$$P_n(x) = a^r + \sum_{k=1}^{n} \frac{r(r-1)(r-2)\cdots(r-k+1)}{k!}a^{r-k}x^k.$$

However, if $r$ is not a positive integer, then there will be no positive integer $n$ for which the remainder $R_n(x) = f(x) - P_n(x)$ vanishes identically, and the corresponding Maclaurin series will not be a polynomial.

## The Binomial Series

To simplify the discussion of the function $(a+x)^r$ when $r$ is not a positive integer, we take $a = 1$ and consider the function $(1+x)^r$. Results for the general case follow via the identity

$$(a+x)^r = a^r\left(1+\frac{x}{a}\right)^r,$$

valid for any $a > 0$.

If $r$ is any real number and $x > -1$, then the $k$th derivative of $(1+x)^r$ is

$$r(r-1)(r-2)\cdots(r-k+1)(1+x)^{r-k}, \qquad (k = 1, 2, \ldots).$$

Thus the Maclaurin series for $(1+x)^r$ is

$$1+\sum_{k=1}^{n}\frac{r(r-1)(r-2)\cdots(r-k+1)}{k!}x^k,$$

which is called the **binomial series**. The following theorem shows that the binomial series does, in fact, converge to $(1+x)^r$ if $|x|<1$. We could accomplish this by writing Taylor's Formula for $(1+x)^r$ with $c=0$ and showing that the remainder $R_n(x)\to 0$ as $n\to\infty$. (We would need to use the integral form of the remainder to prove this for all $|x|<1$.) However, we will use an easier method, similar to the one used for the exponential and trigonometric functions in Section 2.2.

**THEOREM 8**

**The binomial series**

If $|x|<1$, then

$$\begin{aligned}(1+x)^r &= 1+rx+\frac{r(r-1)}{2!}x^2+\frac{r(r-1)(r-2)}{3!}x^3+\cdots\\ &= 1+\sum_{n=1}^{\infty}\frac{r(r-1)(r-2)\cdots(r-n+1)}{n!}x^n \quad (-1<x<1).\end{aligned}$$

***PROOF*** If $|x|<1$, then the series

$$f(x)=1+\sum_{n=1}^{\infty}\frac{r(r-1)(r-2)\cdots(r-n+1)}{n!}x^n$$

converges by the ratio test, since

$$\begin{aligned}\rho &= \lim_{n\to\infty}\left|\frac{\dfrac{r(r-1)(r-2)\cdots(r-n+1)(r-n)}{(n+1)!}x^{n+1}}{\dfrac{r(r-1)(r-2)\cdots(r-n+1)}{n!}x^n}\right|\\ &= \lim_{n\to\infty}\left|\frac{r-n}{n+1}\right||x| = |x|<1.\end{aligned}$$

Note that $f(0)=1$. We need to show that $f(x)=(1+x)^r$ for $|x|<1$.

By Theorem 3, we can differentiate the series for $f(x)$ termwise on $|x|<1$ to obtain

$$\begin{aligned}f'(x) &= \sum_{n=1}^{\infty}\frac{r(r-1)(r-2)\cdots(r-n+1)}{(n-1)!}x^{n-1}\\ &= \sum_{n=0}^{\infty}\frac{r(r-1)(r-2)\cdots(r-n)}{n!}x^n.\end{aligned}$$

We have replaced $n$ with $n+1$ to get the second version of the sum from the first version. Adding the second version to $x$ times the first version, we get

$$\begin{aligned}(1+x)f'(x) &= \sum_{n=0}^{\infty}\frac{r(r-1)(r-2)\cdots(r-n)}{n!}x^n\\ &\quad +\sum_{n=1}^{\infty}\frac{r(r-1)(r-2)\cdots(r-n+1)}{(n-1)!}x^n\\ &= r+\sum_{n=1}^{\infty}\frac{r(r-1)(r-2)\cdots(r-n+1)}{n!}x^n\left[(r-n)+n\right]\\ &= r\,f(x).\end{aligned}$$

The differential equation $(1+x)f'(x) = rf(x)$ implies that

$$\frac{d}{dx}\frac{f(x)}{(1+x)^r} = \frac{(1+x)^r f'(x) - r(1+x)^{r-1}f(x)}{(1+x)^{2r}} = 0$$

for all $x$ satisfying $|x| < 1$. Thus $f(x)/(1+x)^r$ is constant on that interval, and since $f(0) = 1$, the constant must be 1. Thus $f(x) = (1+x)^r$.

***REMARK*** For some values of $r$ the binomial series may converge at the endpoints $x = 1$ or $x = -1$. As observed above, if $r$ is a positive integer, the series has only finitely many nonzero terms, and so converges for all $x$.

**EXAMPLE 2** Find the Maclaurin series for $\dfrac{1}{\sqrt{1+x}}$.

***SOLUTION*** Here $r = -(1/2)$:

$$\begin{aligned}
\frac{1}{\sqrt{1+x}} &= (1+x)^{-1/2} \\
&= 1 - \frac{1}{2}x + \frac{1}{2!}\left(-\frac{1}{2}\right)\left(-\frac{3}{2}\right)x^2 + \frac{1}{3!}\left(-\frac{1}{2}\right)\left(-\frac{3}{2}\right)\left(-\frac{5}{2}\right)x^3 + \cdots \\
&= 1 - \frac{1}{2}x + \frac{1\times 3}{2^2 2!}x^2 - \frac{1\times 3\times 5}{2^3 3!}x^3 + \cdots \\
&= 1 + \sum_{n=1}^{\infty}(-1)^n \frac{1\times 3\times 5\times\cdots\times(2n-1)}{2^n n!}x^n.
\end{aligned}$$

This series converges for $-1 < x \le 1$. (Use the alternating series test to get the endpoint $x = 1$.)

**EXAMPLE 3** Find the Maclaurin series for $\sin^{-1} x$.

***SOLUTION*** Replace $x$ with $-t^2$ in the series obtained in the previous example to get

$$\frac{1}{\sqrt{1-t^2}} = 1 + \sum_{n=1}^{\infty}\frac{1\times 3\times 5\times\cdots\times(2n-1)}{2^n n!}t^{2n} \qquad (-1 < x < 1).$$

Now integrate $t$ from 0 to $x$:

$$\begin{aligned}
\sin^{-1} x &= \int_0^x \frac{dt}{\sqrt{1-t^2}} = \int_0^x \left(1 + \sum_{n=1}^{\infty}\frac{1\times 3\times 5\times\cdots\times(2n-1)}{2^n n!}t^{2n}\right)dt \\
&= x + \sum_{n=1}^{\infty}\frac{1\times 3\times 5\times\cdots\times(2n-1)}{2^n n!(2n+1)}x^{2n+1} \\
&= x + \frac{x^3}{6} + \frac{3}{40}x^5 + \cdots \qquad (-1 < x < 1).
\end{aligned}$$

## EXERCISES 2.5

Find Maclaurin series representations for the functions in Exercises 1–6. Use the binomial series to calculate the answers.

**1.** $\sqrt{1+x}$

**2.** $x\sqrt{1-x}$

**3.** $\sqrt{4+x}$

**4.** $\dfrac{1}{\sqrt{4+x^2}}$

**5.** $(1-x)^{-2}$

**6.** $(1+x)^{-3}$

* **7. (Binomial coefficients)** Show that the binomial coefficients

$$\binom{n}{k} = \frac{n!}{k!\,(n-k)!}$$

satisfy

i) $\binom{n}{0} = \binom{n}{n} = 1$ for every $n$, and

ii) if $0 \le k \le n$, then $\binom{n}{k-1} + \binom{n}{k} = \binom{n+1}{k}$.

It follows that, for fixed $n \ge 1$, the binomial coefficients $\binom{n}{0}, \binom{n}{1}, \binom{n}{2}, \ldots, \binom{n}{n}$ are the elements of the $n$th row of **Pascal's triangle**

```
        1     1
     1     2     1
   1    3     3    1
 1   4     6     4   1
1  5   10    10   5   1
   .                 .
 .                     .
```

where each element with value $> 1$ is the sum of the two diagonally above it.

* **8. (An inductive proof of the Binomial Theorem)** Use mathematical induction and the results of Exercise 7 to prove the Binomial Theorem:

$$\begin{aligned}(a+b)^n &= \sum_{k=0}^{n} \binom{n}{k} a^{n-k} b^k \\ &= a^n + na^{n-1}b + \binom{n}{2}a^{n-2}b^2 + \binom{n}{3}a^{n-3}b^3 + \cdots + b^n.\end{aligned}$$

* **9. (The Leibniz Rule)** Use mathematical induction, the Product Rule, and Exercise 7 to verify the Leibniz Rule for the $n$th derivative of a product of two functions:

$$\begin{aligned}(fg)^{(n)} &= \sum_{k=0}^{n} \binom{n}{k} f^{(n-k)} g^{(k)} \\ &= f^{(n)}g + nf^{(n-1)}g' + \binom{n}{2} f^{(n-2)}g'' \\ &\quad + \binom{n}{3} f^{(n-3)}g^{(3)} + \cdots + fg^{(n)}.\end{aligned}$$

# 2.6 FOURIER SERIES

As we have seen, power series representations of functions make it possible to approximate those functions as closely as we want in intervals near a particular point of interest by using partial sums of the series, that is, polynomials. However, in many important applications of mathematics, the functions involved are required to be periodic. For example, much of electronic engineering is concerned with the analysis and manipulation of *waveforms*, which are periodic functions of time. Polynomials are not periodic functions, and for this reason power series are not well suited to representing such functions.

Much more appropriate for the representations of periodic functions over extended intervals are certain infinite series of periodic functions called Fourier series.

## Periodic Functions

Recall that a function $f$ defined on the real line is **periodic** with period $T$ if

$$f(t+T) = f(t) \quad \text{for all real } t. \qquad (*)$$

This implies that $f(t + mT) = f(t)$ for any integer $m$, so that if $T$ is a period of $f$, then so is any multiple $mT$ of $T$. The smallest positive number $T$ for which $(*)$ holds is called **the fundamental period**, or simply **the period** of $f$.

The entire graph of a function with period $T$ can be obtained by shifting the part of the graph in any half-open interval of length $T$ (for example, the interval $[0, T)$) to the left or right by integer multiples of the period $T$. Figure 2.3 shows the graph of a function of period 2.

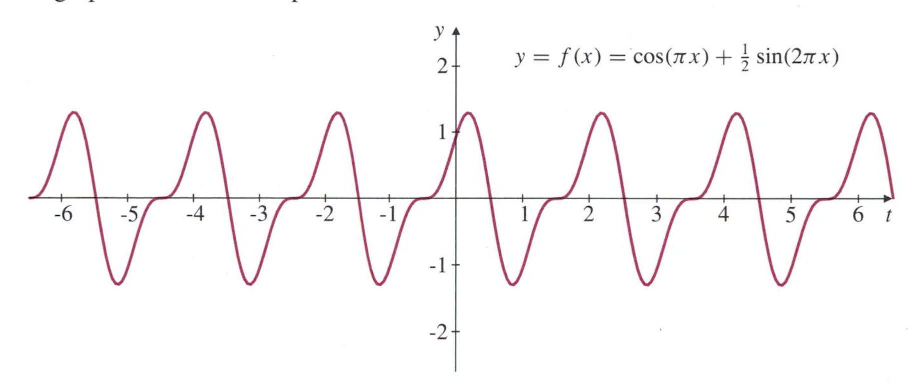


**Figure 2.3** This function has period 2. Observe how the graph repeats the part in the interval [0, 2) over and over to the left and right

**EXAMPLE 1** The functions $g(t) = \cos(\pi t)$ and $h(t) = \sin(\pi t)$ are both periodic with period 2:

$$g(t + 2) = \cos(\pi t + 2\pi) = \cos(\pi t) = g(t).$$

The function $k(t) = \sin(2\pi t)$ also has period 2, but this is not its fundamental period. The fundamental period is 1:

$$k(t + 1) = \sin(2\pi t + 2\pi) = \sin(2\pi t) = k(t).$$

The sum $f(t) = g(t) + \frac{1}{2}k(t) = \cos(\pi t) + \frac{1}{2}\sin(2\pi t)$, graphed in Figure 2.3, has period 2, the least common multiple of the periods of its two terms. ■

**EXAMPLE 2** For any positive integer $n$, the functions

$$f_n(t) = \cos(n\omega t) \qquad \text{and} \qquad g_n(t) = \sin(n\omega t)$$

all have period $T = 2\pi/\omega$; the fundamental periods of $f_n$ and $g_n$ are $T/n$. The subject of Fourier series is concerned with expressing general functions with period $T$ as series whose terms are real multiples of these functions. ■

## Fourier Series

It can be shown (but we won't do it here) that if $f(t)$ is periodic with fundamental period $T$, is continuous, and has a piecewise continuous derivative on the real line, then $f(t)$ is everywhere the sum of a series of the form

$$f(t) = \frac{a_0}{2} + \sum_{n=1}^{\infty}\big(a_n \cos(n\omega t) + b_n \sin(n\omega t)\big), \qquad (**)$$

called the **Fourier series** of $f$, where $\omega = 2\pi/T$ and the sequences $\{a_n\}_{n=0}^{\infty}$ and $\{b_n\}_{n=1}^{\infty}$ are the **Fourier coefficients** of $f$. Determining the values of these coefficients for a given such function $f$ is made possible by the following identities, valid for integers $m$ and $n$, which are easily proved by using the addition formulas for sine and cosine. (See Exercises 57–59 in Section 6.6 of *Single-Variable Calculus.*)

$$\int_0^T \cos(n\omega t)\,dt = \begin{cases} 0 & \text{if } n \neq 0 \\ T & \text{if } n = 0 \end{cases}$$

$$\int_0^T \sin(n\omega t)\,dt = 0$$

$$\int_0^T \cos(m\omega t)\cos(n\omega t)\,dt = \begin{cases} 0 & \text{if } m \neq n \\ T/2 & \text{if } m = n \end{cases}$$

$$\int_0^T \sin(m\omega t)\sin(n\omega t)\,dt = \begin{cases} 0 & \text{if } m \neq n \\ T/2 & \text{if } m = n \end{cases}$$

$$\int_0^T \cos(m\omega t)\sin(n\omega t)\,dt = 0.$$

If we multiply equation $(**)$ by $\cos(m\omega t)$ (or by $\sin(m\omega t)$) and integrate the resulting equation over $[0, T]$ term by term, all the terms on the right except the one involving $a_m$ (or $b_m$) will be 0. (The term-by-term integration requires justification, but we won't try to do that here either.) The integration results in

$$\int_0^T f(t)\cos(m\omega t)\,dt = \frac{1}{2}T a_m$$

$$\int_0^T f(t)\sin(m\omega t)\,dt = \frac{1}{2}T b_m.$$

(Note that the first of these formulas is even valid for $m = 0$ because we chose to call the constant term in the Fourier series $a_0/2$ instead of $a_0$.) Since the integrands are all periodic with period $T$, the integrals can be taken over any interval of length $T$; it is often convenient to use $[-T/2, T/2]$ instead of $[0, T]$. The Fourier coefficients of $f$ are therefore given by

$$a_n = \frac{2}{T}\int_{-T/2}^{T/2} f(t)\cos(n\omega t)\,dt \quad (n = 0, 1, 2, \ldots)$$

$$b_n = \frac{2}{T}\int_{-T/2}^{T/2} f(t)\sin(n\omega t)\,dt \quad (n = 1, 2, 3, \ldots),$$

where $\omega = 2\pi/T$.

**EXAMPLE 3** Find the Fourier series of the sawtooth function $f(t)$ of period $2\pi$ whose values in the interval $[-\pi, \pi]$ are given by $f(t) = \pi - |t|$. (See Figure 2.4.)

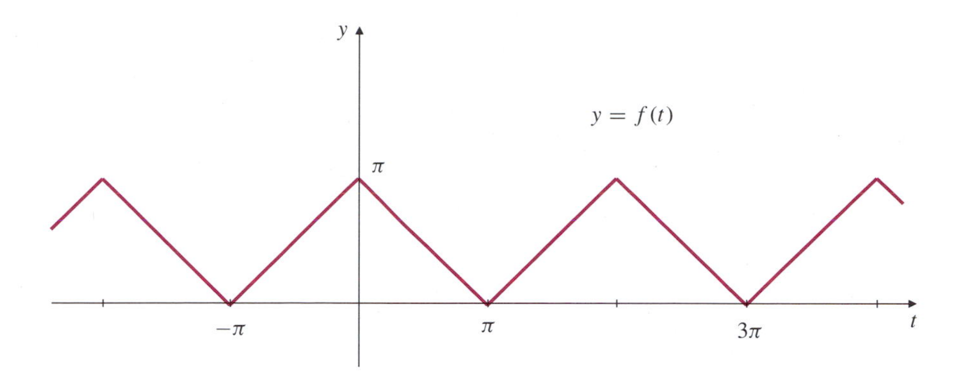


**Figure 2.4** A sawtooth function of period $2\pi$

***SOLUTION*** Here $T = 2\pi$ and $\omega = 2\pi/(2\pi) = 1$. Since $f(t)$ is an even function, therefore $f(t)\sin(nt)$ is odd, and so all the Fourier sine coefficients $b_n$ are zero:

$$b_n = \frac{2}{2\pi}\int_{-\pi}^{\pi} f(t)\sin(nt)\,dt = 0.$$

Also, $f(t)\cos(nt)$ is an even function, so

$$\begin{aligned} a_n &= \frac{2}{2\pi}\int_{-\pi}^{\pi} f(t)\cos(nt)\,dt \\ &= \frac{2}{\pi}\int_{0}^{\pi} (\pi - t)\cos(nt)\,dt \\ &= \begin{cases} \pi & \text{if } n = 0 \\ 0 & \text{if } n \neq 0 \text{ and } n \text{ is even} \\ 4/(\pi n^2) & \text{if } n \text{ is odd} \end{cases} \end{aligned}$$

Since odd positive integers $n$ are of the form $n = 2k - 1$, where $k$ is a positive integer, the Fourier series of $f$ is given by

$$f(t) = \frac{\pi}{2} + \sum_{k=1}^{\infty} \frac{4}{\pi(2k-1)^2}\cos\big((2k-1)t\big).$$ ■

## Convergence of Fourier Series

The partial sums of a Fourier series are called Fourier polynomials because they can be expressed as polynomials in $\sin(\omega t)$ and $\cos(\omega t)$, although we will not actually try to write them that way. The Fourier polynomial of order $m$ of the periodic function $f$ having period $T$ is

$$f_m(t) = \frac{a_0}{2} + \sum_{n=1}^{m}\big(a_n\cos(n\omega t) + b_n\sin(n\omega t)\big),$$

where $\omega = 2\pi/T$.

■ **EXAMPLE 4** The Fourier polynomial of order 3 of the sawtooth function of Example 3 is

$$f_3(t) = \frac{\pi}{2} + \frac{4}{\pi}\cos t + \frac{4}{9\pi}\cos(3t).$$

The graph of this function is shown in Figure 2.5. Observe that it appears to be a reasonable approximation to the graph of $f$ in Figure 2.4, but, being a finite sum of differentiable functions, $f_3(t)$ is itself differentiable everywhere, even at the integer multiples of $\pi$ where $f$ is not differentiable. ■

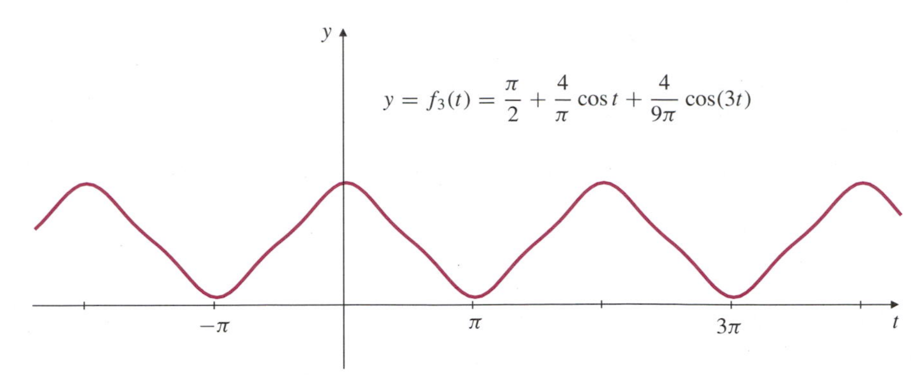


**Figure 2.5** The Fourier polynomial approximation $f_3(t)$ to the sawtooth function of Example 3

As noted earlier, the Fourier series of a function $f(t)$ that is periodic, continuous, and has a piecewise continuous derivative on the real line converges to $f(t)$ at each real number $t$. However, the Fourier coefficients (and hence the Fourier series) can be calculated (by the formulas given above) for periodic functions with piecewise continuous derivative even if the functions are not themselves continuous, but only piecewise continuous.

Recall that $f(t)$ is piecewise continuous on the interval $[a, b]$ if there exists a partition $\{a = x_0 < x_1 < x_2 < \cdots < x_k = b\}$ of $[a, b]$ and functions $F_1, F_2, \ldots, F_k$, such that

(i) $F_i$ is continuous on $[x_{i-1}, x_i]$, and

(ii) $f(t) = F_i(t)$ on $(x_{i-1}, x_i)$.

The integral of such a function $f$ is the sum of integrals of the functions $F_i$:

$$\int_a^b f(t)\,dt = \sum_{i=1}^{k} \int_{x_{i-1}}^{x_i} F_i(t)\,dt.$$

Since $f(t)\cos(n\omega t)$ and $f(t)\sin(n\omega t)$ are piecewise continuous if $f$ is, the Fourier coefficients of a piecewise continuous, periodic function can be calculated by the same formulas given for a continuous periodic function. The question of where and to what the Fourier series converges in this case is answered by the following theorem, proved in textbooks on Fourier analysis.

**THEOREM 9** The Fourier series of a piecewise continuous, periodic function $f$ with piecewise continuous derivative converges to that function at every point $t$ where $f$ is continuous. Moreover, if $f$ is discontinuous at $t = c$, then $f$ has different, but finite, left and right limits at $c$:

$$\lim_{t \to c-} f(t) = f(c-), \qquad \text{and} \qquad \lim_{t \to c+} f(t) = f(c+).$$

The Fourier series of $f$ converges at $t = c$ to the average of these left and right limits:

$$\frac{a_0}{2} + \sum_{n=1}^{\infty} \big(a_n \cos(n\omega c) + b_n \sin(n\omega c)\big) = \frac{f(c-) + f(c+)}{2},$$

where $\omega = 2\pi/T$.

**EXAMPLE 5** Calculate the Fourier series for the periodic function $f$ with period 2 satisfying

$$f(t) = \begin{cases} -1 & \text{if } -1 < x < 0 \\ 1 & \text{if } 0 < x < 1. \end{cases}$$

Where does $f$ fail to be continuous? To what does the Fourier series of $f$ converge at these points?

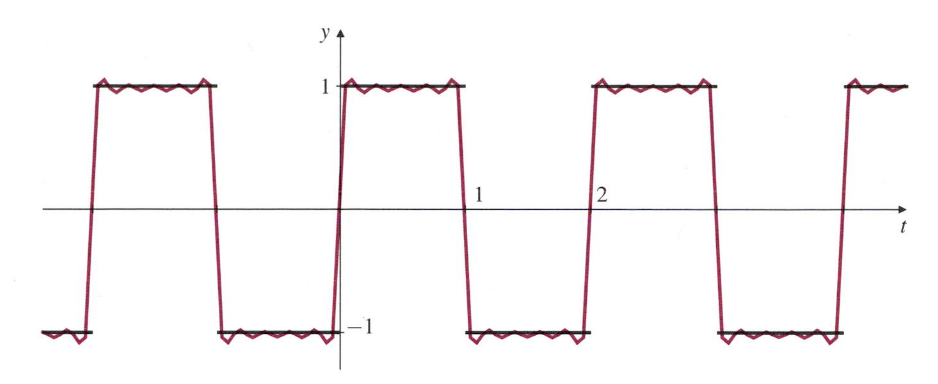


**Figure 2.6** The piecewise continuous function $f$ (black) of Example 5 and its Fourier polynomial $f_{15}$ (colour)

$$f_{15}(t) = \sum_{k=1}^{8} \frac{4\sin\big((2k-1)\pi t\big)}{(2k-1)\pi}$$

***SOLUTION*** Here $T = 2$ and $\omega = 2\pi/2 = \pi$. Since $f$ is an odd function, its cosine coefficients are all zero:

$$\begin{aligned} a_n &= \int_{-1}^{1} f(t)\cos(n\pi t)\,dt \\ &= -\int_{-1}^{0} \cos(n\pi t)\,dt + \int_{0}^{1} \cos(n\pi t)\,dt = 0. \end{aligned}$$

The same symmetry implies that

$$\begin{aligned} b_n &= \int_{-1}^{1} f(t)\sin(n\pi t)\,dt \\ &= 2\int_{0}^{1} \sin(n\pi t)\,dt = -\frac{2\cos(n\pi t)}{n\pi}\bigg|_0^1 \\ &= -\frac{2}{n\pi}\big((-1)^n - 1\big) = \begin{cases} 4/(n\pi) & \text{if } n \text{ is odd} \\ 0 & \text{if } n \text{ is even.} \end{cases} \end{aligned}$$

Odd integers $n$ are of the form $n = 2k - 1$ for $k = 1, 2, 3, \ldots$. Therefore, the Fourier series of $f$ is

$$\begin{aligned} &\frac{4}{\pi}\sum_{k=1}^{\infty} \frac{1}{2k-1}\sin\big((2k-1)\pi t\big) \\ &= \frac{4}{\pi}\left(\sin(\pi t) + \frac{1}{3}\sin(3\pi t) + \frac{1}{5}\sin(5\pi t) + \cdots\right). \end{aligned}$$

Note that $f$ is continuous except at the points where $t$ is an integer. At each of these points $f$ jumps from $-1$ to 1 or from 1 to $-1$, so the average of the left and right limits of $f$ at these points is 0. Observe that the sum of the Fourier series is 0 at integer values of $t$, in accordance with Theorem 9.

The graphs of $f$ (black) and its Fourier polynomial $f_{15}$ (colour) are shown in Figure 2.6. ■

## Fourier Cosine and Sine Series

As observed in Example 3 and Example 5, even functions have no sine terms in their Fourier series, and odd functions have no cosine terms (including the constant term $a_0/2$). It is often necessary in applications to find a Fourier series representation of a given function defined on a finite interval $[0, a]$ having either no sine terms (a **Fourier cosine series**) or no cosine terms (a **Fourier sine series**). This is accomplished by extending the domain of $f$ to $[-a, 0)$ so as to make $f$ either even or odd on $[-a, a]$,

$$\begin{aligned} f(-t) &= f(t) \text{ if } -a \le t < 0 \text{ for the even extension} \\ f(-t) &= -f(t) \text{ if } -a \le t < 0 \text{ for the odd extension,} \end{aligned}$$

and then calculating its Fourier series considering the extended $f$ to have period $2a$. (If we want the odd extension, we may have to redefine $f(0)$ to be 0.)

**EXAMPLE 6** Find the Fourier cosine series of $g(t) = \pi - t$ defined on $[0, \pi]$.

***SOLUTION*** The even extension of $g(t)$ to $[-\pi, \pi]$ is the function $f$ of Example 3. Thus the Fourier cosine series of $g$ is

$$\frac{\pi}{2} + \sum_{k=1}^{\infty} \frac{4}{\pi(2k-1)^2} \cos\big((2k-1)t\big).$$

■

**EXAMPLE 7** Find the Fourier sine series of $h(t) = 1$ defined on $[0, 1]$.

***SOLUTION*** If we redefine $h(0) = 0$, then the odd extension of $h$ to $[-1, 1]$ coincides with the function $f(t)$ of Example 5 except that the latter function is undefined at $t = 0$. The Fourier sine series of $h$ is the series obtained in Example 5, namely

$$\frac{4}{\pi} \sum_{k=1}^{\infty} \frac{1}{2k-1} \sin\big((2k-1)\pi t\big).$$

■

***REMARK*** Fourier cosine and sine series are treated from a different perspective in Section 5.4.

## EXERCISES 2.6

In Exercises 1–4, what is the fundamental period of the given function?

**1.** $f(t) = \sin(3t)$

**2.** $g(t) = \cos(3 + \pi t)$

**3.** $h(t) = \cos^2 t$

**4.** $k(t) = \sin(2t) + \cos(3t)$

In Exercises 5–8, find the Fourier series of the given function.

**5.** $f(t) = t$, $-\pi < t \le \pi$, $f$ has period $2\pi$.

**6.** $f(t) = \begin{cases} 0 & \text{if } 0 \le t < 1 \\ 1 & \text{if } 1 \le t < 2, \end{cases}$ $f$ has period 2.

**7.** $f(t) = \begin{cases} 0 & \text{if } -1 \le t < 0 \\ t & \text{if } 0 \le t < 1, \end{cases}$ $f$ has period 2.

**8.** $f(t) = \begin{cases} t & \text{if } 0 \le t < 1 \\ 1 & \text{if } 1 \le t < 2 \\ 3 - t & \text{if } 2 \le t < 3, \end{cases}$ $f$ has period 3.

**9.** What is the Fourier cosine series of the function $h(t)$ of Example 7?

**10.** Calculate the Fourier sine series of the function $g(t)$ of Example 6.

**11.** Find the Fourier sine series of $f(t) = t$ on $[0, 1]$.

**12.** Find the Fourier cosine series of $f(t) = t$ on $[0, 1]$.

**13.** Use the result of Example 3 to evaluate

$$\sum_{n=1}^{\infty} \frac{1}{(2n-1)^2} = 1 + \frac{1}{3^2} + \frac{1}{5^2} + \cdots.$$

**14.** Verify that if $f$ is an even function of period $T$, then the Fourier sine coefficients $b_n$ of $f$ are all zero and the Fourier cosine coefficients $a_n$ of $f$ are given by

$$a_n = \frac{4}{T} \int_0^{T/2} f(t) \cos(n\omega t)\, dt, \qquad n = 0, 1, 2, \ldots,$$

where $\omega = 2\pi/T$. State and verify the corresponding result for odd functions $f$.

# CHAPTER REVIEW

## Key Ideas

- **What do the following phrases mean?**
  - ◇ a power series
  - ◇ interval of convergence
  - ◇ radius of convergence
  - ◇ centre of convergence
  - ◇ a Taylor series
  - ◇ a Maclaurin series
  - ◇ a Taylor polynomial
  - ◇ a binomial series
  - ◇ an analytic function
- **Where is the sum of a power series differentiable?**
- **Where does the integral of a power series converge?**
- **Where is the sum of a power series continuous?**

- **State Taylor's Theorem with Lagrange remainder**
- **State Taylor's Theorem with integral remainder**
- **What is the binomial theorem?**
- **What is a Fourier series?**

## Review Exercises

In Exercises 1–10, find Maclaurin series for the given functions. State where each series converges to the function.

**1.** $\dfrac{1}{3-x}$

**2.** $\dfrac{x}{3-x^2}$

**3.** $\ln(e+x^2)$

**4.** $\ln\dfrac{1}{e^2-x}$

**5.** $\dfrac{\sin(2x^2)}{x}$

**6.** $\dfrac{1-e^{-2x}}{x}$

**7.** $x\cos^2 x$

**8.** $\sin(x+(\pi/3))$

**9.** $(8+x)^{-1/3}$

**10.** $(1+x)^{1/3}$

Find Taylor series for the functions in Exercises 11–14 about the indicated points $x=c$.

**11.** $1/x, \quad c=\pi$

**12.** $\dfrac{x}{(x-4)^2}, \quad c=2$

**13.** $xe^x, \quad c=-1$

**14.** $\sin x+\cos x, \quad c=\pi/4$

Find the Maclaurin polynomial of the indicated degree for the functions in Exercises 15–20.

**15.** $e^{x^2+2x}$, degree 3

**16.** $\sin(1+x)$, degree 3

**17.** $\cos(\sin x)$, degree 4

**18.** $\ln(1+xe^x)$, degree 4

**19.** $\dfrac{x}{\cos x}$, degree 5

**20.** $\sqrt{1+\sin x}$, degree 4

**21.** What function has Maclaurin series

$$1-\frac{x}{2!}+\frac{x^2}{4!}-\cdots=\sum_{n=0}^{\infty}\frac{(-1)^n x^n}{(2n)!}?$$

**22.** A function $f(x)$ has Maclaurin series

$$1+x^2+\frac{x^4}{2^2}+\frac{x^6}{3^2}+\cdots=1+\sum_{n=1}^{\infty}\frac{x^{2n}}{n^2}.$$

Find $f^{(k)}(0)$ for all positive integers $k$.

Find the sums of the series in 23–26.

**23.** $\displaystyle\sum_{n=0}^{\infty}\frac{n+1}{\pi^n}$

* **24.** $\displaystyle\sum_{n=0}^{\infty}\frac{n^2}{\pi^n}$

**25.** $\displaystyle\sum_{n=1}^{\infty}\frac{1}{ne^n}$

* **26.** $\displaystyle\sum_{n=2}^{\infty}\frac{(-1)^n\pi^{2n-4}}{(2n-1)!}$

**27.** If $S(x)=\displaystyle\int_0^x \sin(t^2)\,dt$, find $\displaystyle\lim_{x\to 0}\frac{x^3-3S(x)}{x^7}$.

**28.** Use series to evaluate $\displaystyle\lim_{x\to 0}\frac{(x-\tan^{-1}x)(e^{2x}-1)}{2x^2-1+\cos(2x)}$.

**29.** How many nonzero terms in the Maclaurin series for $e^{-x^4}$ are needed to evaluate $\int_0^{1/2} e^{-x^4}\,dx$ correct to five decimal places? Evaluate the integral to that accuracy.

**30.** Estimate the size of the error if the Taylor polynomial of degree 4 about $x=\pi/2$ for $f(x)=\ln\sin x$ is used to approximate $\ln\sin(1.5)$.

**31.** Find the Fourier sine series of $f(t)=\pi-t$ on $[0,\pi]$.

**32.** Find the Fourier series of $f(t)=\begin{cases}1 & \text{if } \pi<t\le 0\\ t & \text{if } 0<t\le\pi.\end{cases}$

## Express Yourself

Answer these questions in words, using complete sentences and, if necessary, paragraphs. Use mathematical symbols only where necessary.

**1.** Why are power series representations of functions useful?

**2.** If a function has a Taylor series about a point, is the function analytic at that point? Explain.

**3.** Give some properties of a function that are easy to calculate if you know the Taylor series of the function about a particular point. Also give some properties that can't easily be calculated from a Taylor series.

**4.** What happens to the graphs of Maclaurin polynomials of an analytic function as the degrees of the polynomials increase?

## Challenging Problems

**1.** Let

$$f(x)=\sum_{k=0}^{\infty}\frac{2^{2k}k!}{(2k+1)!}x^{2k+1}$$
$$=x+\frac{2}{3}x^3+\frac{4}{3\times 5}x^5+\frac{8}{3\times 5\times 7}x^7+\cdots.$$

(a) Find the radius of convergence of this power series.

(b) Show that $f'(x)=1+2xf(x)$.

(c) What is $\dfrac{d}{dx}\left(e^{-x^2}f(x)\right)$?

(d) Express $f(x)$ in terms of an integral.

* **2.** **(The number $\pi$ is irrational)** The last problem in the Chapter Review at the end of Chapter 1 shows how to prove that $e$ is irrational by assuming the contrary and deducing a contradiction. In this problem you will show that $\pi$ is also irrational. The proof for $\pi$ is also by contradiction, but is rather more complicated and will be broken down into several parts.

(a) Let $f(x)$ be a polynomial, and let

$$g(x)=f(x)-f''(x)+f^{(4)}(x)-f^{(6)}(x)+\cdots$$
$$=\sum_{j=0}^{\infty}(-1)^j f^{(2j)}(x).$$

(Since $f$ is a polynomial all but a finite number of terms in the above sum are identically zero, so there are no convergence problems.) Verify that

$$\frac{d}{dx}\left(g'(x)\sin x - g(x)\cos x\right) = f(x)\sin x,$$

and hence that

$$\int_0^{\pi} f(x)\sin x\,dx = g(\pi) + g(0).$$

(b) Suppose that $\pi$ is rational, say $\pi = m/n$, where $m$ and $n$ are positive integers. You will show that this leads to a contradiction, and thus cannot be true. Choose a positive integer $k$ such that $(\pi m)^k/k! < 1/2$. (Why is this possible?) Consider the polynomial

$$f(x) = \frac{x^k(m-nx)^k}{k!} = \frac{1}{k!}\sum_{j=0}^{k}\binom{k}{j} m^{k-j}(-n)^j x^{j+k}.$$

Show that $0 < f(x) < 1/2$ for $0 < x < \pi$, and hence that

$$0 < \int_0^{\pi} f(x)\sin x\,dx < 1.$$

Thus $0 < g(\pi) + g(0) < 1$, where $g(x)$ is defined as in part (a).

(c) Show that the $i$th derivative of $f(x)$ is given by

$$f^{(i)}(x) = \frac{1}{k!}\sum_{j=0}^{k}\binom{k}{j} m^{k-j}(-n)^j \frac{(j+k)!}{(j+k-i)!}x^{j+k-i}.$$

(d) Show that $f^{(i)}(0)$ is an integer for $i = 0, 1, 2, \ldots$. (*Hint:* observe for $i < k$ that $f^{(i)}(0) = 0$, and for $i > 2k$ that $f^{(i)}(x) = 0$ for all $x$. For $k \le i \le 2k$, show that only one term in the sum for $f^{(i)}(0)$ is not 0, and that this term is an integer. You will need the fact that the binomial coefficients $\binom{k}{j}$ are integers.)

(e) Show that $f(\pi - x) = f(x)$ for all $x$, and hence that $f^{(i)}(\pi)$ is also an integer for each $i = 0, 1, 2, \ldots$. Therefore, if $g(x)$ is defined as in (a), then $g(\pi) + g(0)$ is an integer. This contradicts the conclusion of part (b), and so shows that $\pi$ cannot be rational.

* **3. (An asymptotic series)** Use integration by parts to show that

$$\int_0^x e^{-1/t}\,dt = e^{-1/x}\sum_{n=2}^{N}(-1)^n(n-1)!x^n + (-1)^{N+1}N!\int_0^x t^{N-1}e^{-1/t}\,dt.$$

Why can't you just use a Maclaurin series to approximate this integral? Using $N = 5$, find an approximate value for $\int_0^{0.1} e^{-1/t}\,dt$, and estimate the error. Estimate the error for $N = 10$ and $N = 20$.

Note that the series $\sum_{n=2}^{\infty}(-1)^n(n-1)!x^n$ **diverges** for any $x \neq 0$. This is an example of what is called an **asymptotic series**. Even though it diverges, a properly chosen partial sum gives a good approximation to our function when $x$ is small.

CHAPTER 3

# Coordinate Geometry and Vectors in 3-Space

**INTRODUCTION** A complete real-variable calculus program involves the study of

(i) real-valued functions of a single real variable,

(ii) real-valued functions of a real vector variable,

(iii) vector-valued functions of a single real variable,

(iv) vector-valued functions of a real vector variable.

Chapters 1–2 and all of *Single-Variable Calculus* were concerned with item (i). The remaining chapters deal with items (ii), (iii), and (iv). Specifically, Chapters 4–6 are concerned with the differentiation and integration of real-valued functions of several real variables, that is, of a real vector variable. *Single-Variable Calculus* dealt briefly with vector-valued functions of a real variable. This treatment is greatly amplified and extended in Chapter 7 and part of Chapter 8. Most of Chapters 8 and 9 present aspects of the calculus of functions whose domains and ranges both have dimension greater than one, that is, vector-valued functions of a vector variable. Most of the time we will limit our attention to vector functions with domains and ranges in the plane, or in 3-dimensional space.

In this chapter we will lay the foundation for multi-variable calculus by extending the concepts of analytic geometry and vectors to three and more dimensions. We also introduce matrices, as these will prove useful for formulating some of the concepts of calculus. This chapter is not intended to be a course in linear algebra. We develop only those aspects that we will use in later chapters, and omit most proofs.

## 3.1 ANALYTIC GEOMETRY IN THREE AND MORE DIMENSIONS

We say that the physical world in which we live is three dimensional because through any point there can pass three, and no more, straight lines that are **mutually perpendicular**, that is to say, each of them is perpendicular to the other two. This is equivalent to the fact that we require three numbers to locate a point in space with respect to some reference point (the **origin**). One way to use three numbers to locate a point is by having them represent (signed) distances from the origin, measured in the directions of three mutually perpendicular lines passing through the origin. We call such a set of lines a Cartesian coordinate system, and each of the lines is called a coordinate axis. We usually call these axes the $x$-axis, the $y$-axis, and the $z$-axis, regarding the $x$- and $y$-axes as lying in a horizontal plane and the $z$-axis as vertical. Moreover, the coordinate system should have a **right-handed orientation**. This means that the thumb, forefinger, and middle finger of the right hand can be extended so as to point respectively in the directions of the positive $x$-axis, the positive $y$-axis, and the positive $z$-axis. For the more mechanically minded, a right-handed screw will advance in the positive $z$ direction if twisted in the direction of rotation from the positive $x$-axis towards the positive $y$-axis. (See Figure 3.1(a).)

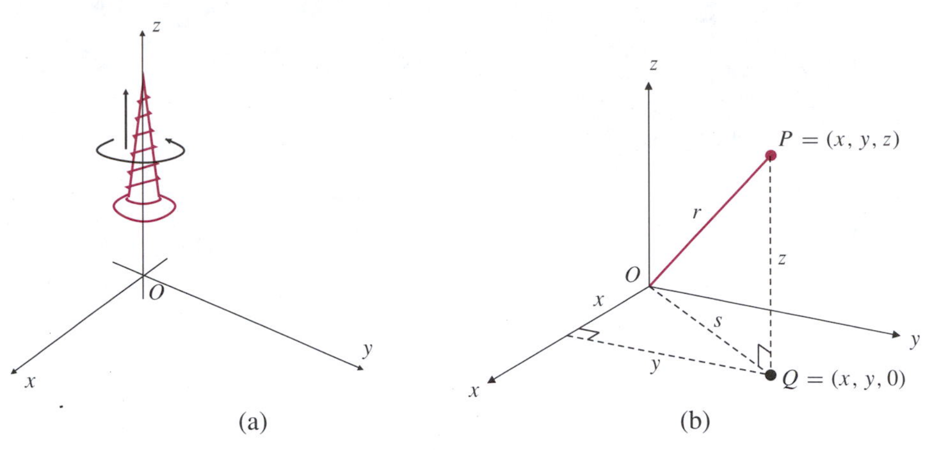


**Figure 3.1**

(a) The screw moves upward when twisted counterclockwise as seen from above

(b) The three coordinates of a point in 3-space

With respect to such a Cartesian coordinate system, the **coordinates** of a point $P$ in 3-space constitute an ordered triple of real numbers, $(x, y, z)$. The numbers $x$, $y$, and $z$ are, respectively, the signed distances of $P$ from the origin, measured in the directions of the $x$-axis, the $y$-axis, and the $z$-axis. (See Figure 3.1(b).)

Let $Q$ be the point with coordinates $(x, y, 0)$. Then $Q$ lies in the $xy$-plane (the plane containing the $x$- and $y$-axes) directly under (or over) $P$. We say that $Q$ is the vertical projection of $P$ onto the $xy$-plane. If $r$ is the distance from the origin $O$ to $P$ and $s$ is the distance from $O$ to $Q$, then, using two right-angled triangles, we have

$$s^2 = x^2 + y^2 \qquad \text{and} \qquad r^2 = s^2 + z^2 = x^2 + y^2 + z^2.$$

Thus the distance from $P$ to the origin is given by

$$r = \sqrt{x^2 + y^2 + z^2}.$$

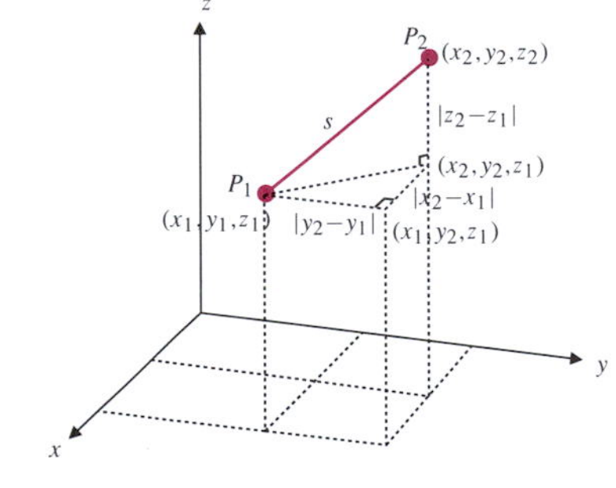


**Figure 3.2**

Similarly, the distance $s$ between points $P_1 = (x_1, y_1, z_1)$ and $P_2 = (x_2, y_2, z_2)$ (see Figure 3.2) is

$$s = \sqrt{(x_2 - x_1)^2 + (y_2 - y_1)^2 + (z_2 - z_1)^2}.$$

**EXAMPLE 1** Show that the triangle with vertices $A = (1, -1, 2)$, $B = (3, 3, 8)$, and $C = (2, 0, 1)$ has a right angle.

***SOLUTION*** We calculate the lengths of the three sides of the triangle:

$$\begin{aligned} a &= |BC| = \sqrt{(2-3)^2 + (0-3)^2 + (1-8)^2} = \sqrt{59} \\ b &= |AC| = \sqrt{(2-1)^2 + (0+1)^2 + (1-2)^2} = \sqrt{3} \\ c &= |AB| = \sqrt{(3-1)^2 + (3+1)^2 + (8-2)^2} = \sqrt{56} \end{aligned}$$

By the cosine law, $a^2 = b^2 + c^2 - 2bc\cos A$. In this case $a^2 = 59 = 3 + 56 = b^2 + c^2$, so that $2bc\cos A$ must be 0. Therefore $\cos A = 0$ and $A = 90^\circ$.

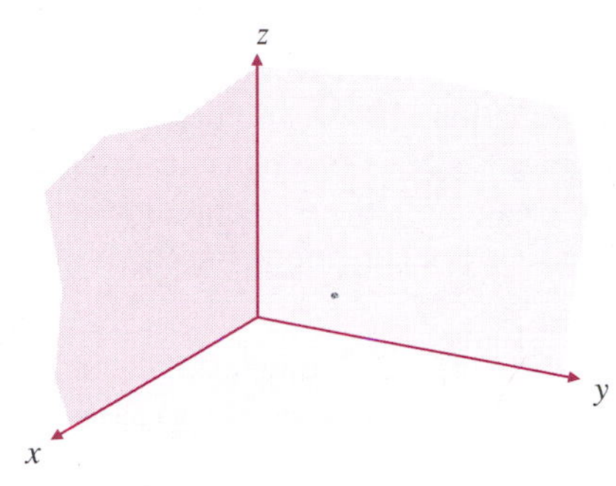


Figure 3.3 The first octant

Just as the $x$- and $y$-axes divide the $xy$-plane into four quadrants, so also the three **coordinate planes** in 3-space (the $xy$-plane, the $xz$-plane, and the $yz$-plane) divide 3-space into eight **octants**. We call the octant in which $x \geq 0$, $y \geq 0$, and $z \geq 0$ the **first octant**. When drawing graphs in 3-space it is sometimes easier to draw only the part lying in the first octant (Figure 3.3).

An equation or inequality involving the three variables $x$, $y$, and $z$ defines a subset of points in 3-space whose coordinates satisfy the equation or inequality. A single equation usually represents a surface (a two-dimensional object) in 3-space.

## EXAMPLE 2 Some equations and the surfaces they represent

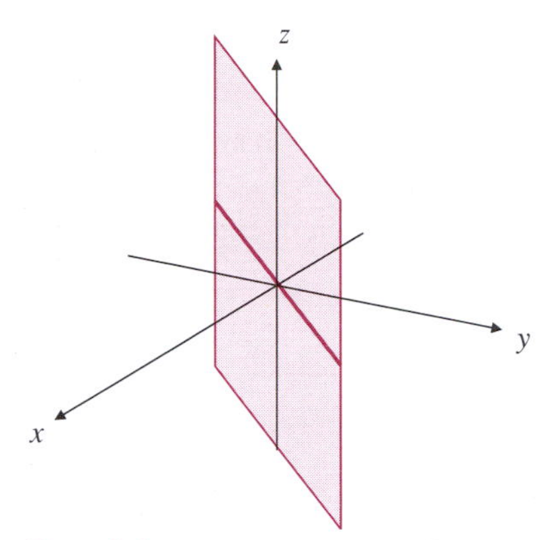


Figure 3.4 Equation $x = y$ defines a vertical plane

(a) The equation $z = 0$ represents all points with coordinates $(x, y, 0)$, that is, the $xy$-plane. The equation $z = -2$ represents all points with coordinates $(x, y, -2)$, that is, the horizontal plane passing through the point $(0, 0, -2)$ on the $z$-axis.

(b) The equation $x = y$ represents all points with coordinates $(x, x, z)$. This is a vertical plane containing the straight line with equation $x = y$ in the $xy$-plane. The plane contains the $z$-axis. (See Figure 3.4.)

(c) The equation $x + y + z = 1$ represents all points the sum of whose coordinates is 1. This set is a plane that passes through the three points $(1, 0, 0)$, $(0, 1, 0)$, and $(0, 0, 1)$. These points are not collinear (they do not lie on a straight line) so there is only one plane passing through all three. (See Figure 3.5.) The equation $x + y + z = 0$ represents a plane parallel to the one with equation $x + y + z = 1$, but passing through the origin.

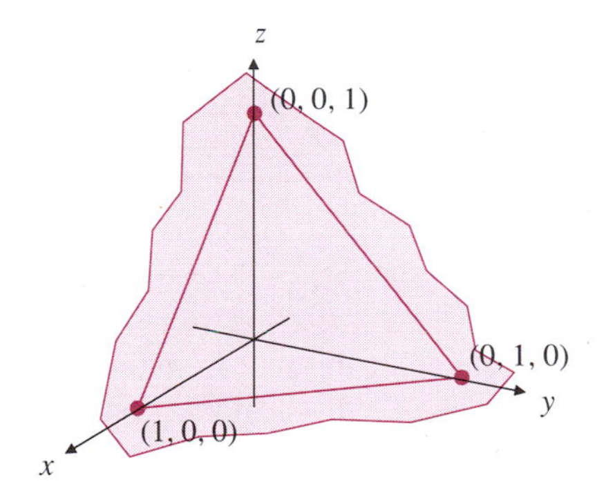


Figure 3.5 The plane with equation $x + y + z = 1$

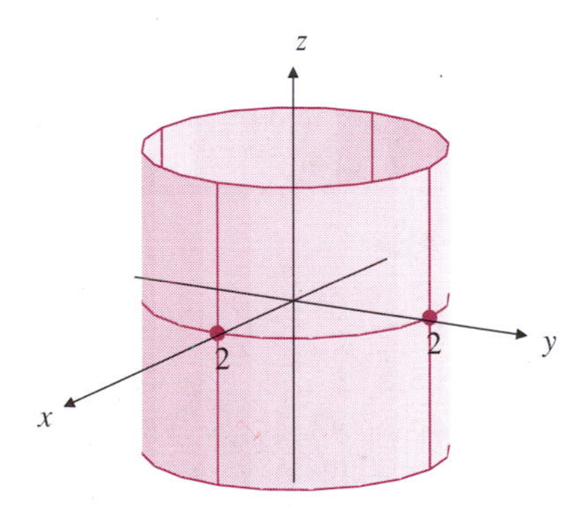


Figure 3.6 The circular cylinder with equation $x^2 + y^2 = 4$

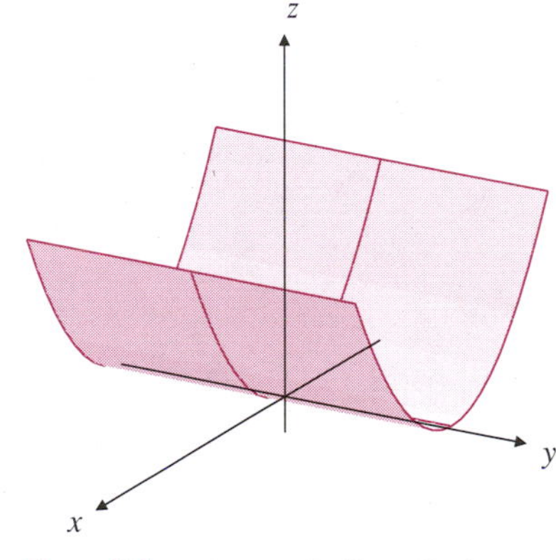


Figure 3.7 The parabolic cylinder with equation $z = x^2$

(d) The equation $x^2 + y^2 = 4$ represents all points on the vertical circular cylinder containing the circle with equation $x^2 + y^2 = 4$ in the $xy$-plane. This cylinder has radius 2 and axis along the $z$-axis. (See Figure 3.6.)

(e) The equation $z = x^2$ represents all points with coordinates $(x, y, x^2)$. This surface is a parabolic cylinder tangent to the $xy$-plane along the $y$-axis. (See Figure 3.7.)

(f) The equation $x^2 + y^2 + z^2 = 25$ represents all points $(x, y, z)$ at distance 5 from the origin. This set of points is a *sphere* of radius 5 centred at the origin. ■

Observe that equations in $x$, $y$, and $z$ need not involve each variable explicitly. When one of the variables is missing from the equation, the equation represents a surface *parallel to* the axis of the missing variable. Such a surface may be a plane or a cylinder. For example, if $z$ is absent from the equation, the equation represents in 3-space a vertical (that is, parallel to the $z$-axis) surface containing the curve with the same equation in the $xy$-plane.

Occasionally a single equation may not represent a two-dimensional object (a surface). It can represent a one-dimensional object (a line or curve), a zero-dimensional object (one or more points), or even nothing at all.

**EXAMPLE 3** Identify the graphs of the equations: (a) $y^2 + (z-1)^2 = 4$, (b) $y^2 + (z-1)^2 = 0$, (c) $x^2 + y^2 + z^2 = 0$, and (d) $x^2 + y^2 + z^2 = -1$.

**SOLUTION**

(a) Since $x$ is absent, the equation $y^2 + (z-1)^2 = 4$ represents an object parallel to the $x$-axis. In the $yz$-plane the equation represents a circle of radius 2 centred at $(y, z) = (0, 1)$. In 3-space it represents a horizontal circular cylinder, parallel to the $x$-axis, with axis one unit above the $x$-axis.

(b) Since squares cannot be negative, the equation $y^2 + (z-1)^2 = 0$ implies that $y = 0$ and $z = 1$, so it represents points $(x, 0, 1)$. All these points lie on the line parallel to the $x$-axis and one unit above it.

(c) As in part (b), $x^2 + y^2 + z^2 = 0$ implies that $x = 0$, $y = 0$, and $z = 0$. The equation represents only one point, the origin.

(d) The equation $x^2 + y^2 + z^2 = -1$ is not satisfied by any real numbers $x$, $y$, and $z$, so represents no points at all. ■

A single inequality in $x$, $y$, and $z$ typically represents points lying on one side of the surface represented by the corresponding equation (together with points on the surface if the inequality is not strict).

**EXAMPLE 4**

(a) The inequality $z > 0$ represents all points above the $xy$-plane.

(b) The inequality $x^2 + y^2 \geq 4$ says that the square of the distance from $(x, y, z)$ to the nearest point $(x, y, 0)$ on the $z$-axis is at least 4. This inequality represents all points lying on or outside the cylinder of Example 2(d).

(c) The inequality $x^2 + y^2 + z^2 \leq 25$ says that the square of the distance from $(x, y, z)$ to the origin is no greater than 25. It represents the solid ball of radius 5 centred at the origin, which consists of all points lying inside or on the sphere of Example 2(f). ■

Two equations in $x$, $y$, and $z$ normally represent a one-dimensional object, the line or curve along which the two surfaces represented by the two equations intersect. Any point whose coordinates satisfy both equations must lie on both the surfaces, so must lie on their intersection.

**EXAMPLE 5** What sets of points in 3-space are represented by the pairs of equations?

(a) $\begin{cases} x + y + z = 1 \\ y - 2x = 0 \end{cases}$ (b) $\begin{cases} x^2 + y^2 + z^2 = 1 \\ x + y = 1 \end{cases}$

**SOLUTION**

(a) The equation $x + y + z = 1$ represents the oblique plane of Example 2(c), and the equation $y - 2x = 0$ represents a vertical plane through the origin and the point $(1, 2, 0)$. Together these two equations represent the line of intersection of the two planes. This line passes through, for example, the points $(0, 0, 1)$ and $(\frac{1}{3}, \frac{2}{3}, 0)$. (See Figure 3.8(a).)

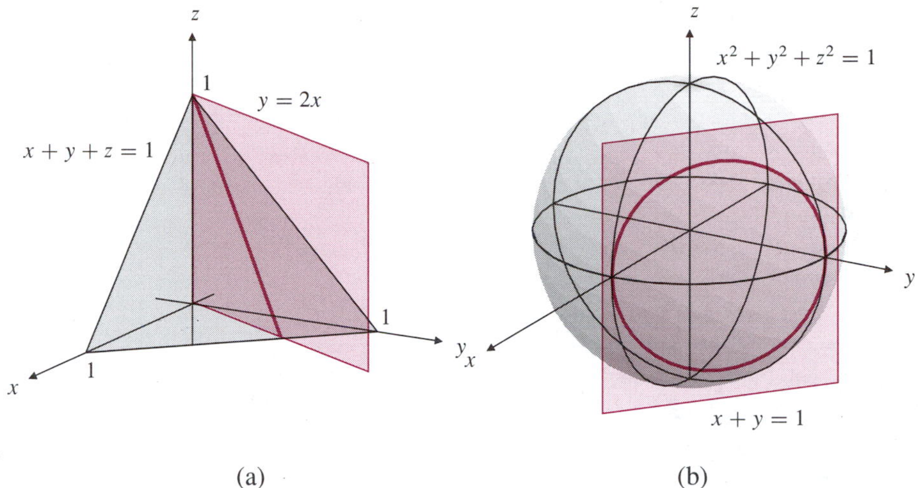


**Figure 3.8**

(a) The two planes intersect in a straight line

(b) The plane intersects the sphere in a circle

(b) The equation $x^2 + y^2 + z^2 = 1$ represents a sphere of radius 1 with centre at the origin, and $x + y = 1$ represents a vertical plane through the points $(1, 0, 0)$ and $(0, 1, 0)$. The two surfaces intersect in a circle, as shown in Figure 3.8(b). The line from $(1, 0, 0)$ to $(0, 1, 0)$ is a diameter of the circle, so the centre of the circle is $(\frac{1}{2}, \frac{1}{2}, 0)$, and its radius is $\sqrt{2}/2$. ■

In Sections 3.4 and 3.5 we will see many more examples of geometric objects in 3-space represented by simple equations.

## Euclidean *n*-Space

Mathematicians and users of mathematics frequently need to consider ***n*-dimensional space** where $n$ is greater than three, and may even be infinite. Students sometimes have difficulty in visualizing a space of dimension four or higher. The secret to dealing with these spaces is to regard the points in $n$-space as *being* ordered $n$-tuples of real numbers; that is, $(x_1, x_2, \dots, x_n)$ is a point in $n$-space instead of just being the coordinates of such a point. We stop thinking of points as existing in physical space and start thinking of them as algebraic objects. We usually denote $n$-space by the symbol $\mathbb{R}^n$ to show that its points are $n$-tuples of *real* numbers. Thus $\mathbb{R}^2$ and $\mathbb{R}^3$ denote the plane and 3-space, respectively. Note that in passing from $\mathbb{R}^3$ to $\mathbb{R}^n$ we have altered the notation a bit — in $\mathbb{R}^3$ we called the coordinates $x$, $y$, and $z$ while in $\mathbb{R}^n$ we called them $x_1, x_2, \dots$ and $x_n$ so as not to run out of letters. We could, of course, talk about coordinates $(x_1, x_2, x_3)$ in $\mathbb{R}^3$, and $(x_1, x_2)$ in the plane $\mathbb{R}^2$, but $(x, y, z)$ and $(x, y)$ are traditionally used there.

Although we think of points in $\mathbb{R}^n$ as $n$-tuples rather than geometric objects, we do not want to lose all sight of the underlying geometry. By analogy with the two- and three-dimensional cases, we still consider the quantity

$$\sqrt{(y_1 - x_1)^2 + (y_2 - x_2)^2 + \cdots + (y_n - x_n)^2}$$

as representing the *distance* between the points with coordinates $(x_1, x_2, \ldots, x_n)$ and $(y_1, y_2, \ldots, y_n)$. Also, we call the $(n-1)$-dimensional set of points in $\mathbb{R}^n$ which satisfy the equation $x_n = 0$ a **hyperplane**, by analogy with the plane $z = 0$ in $\mathbb{R}^3$.

## EXERCISES 3.1

Find the distance between the pairs of points in Exercises 1–4.

**1.** $(0, 0, 0)$ and $(2, -1, -2)$

**2.** $(-1, -1, -1)$ and $(1, 1, 1)$

**3.** $(1, 1, 0)$ and $(0, 2, -2)$

**4.** $(3, 8, -1)$ and $(-2, 3, -6)$

**5.** What is the shortest distance from the point $(x, y, z)$ to (a) the $xy$-plane? (b) the $x$-axis?

**6.** Show that the triangle with vertices $(1, 2, 3)$, $(4, 0, 5)$, and $(3, 6, 4)$ has a right angle.

**7.** Find the angle $A$ in the triangle with vertices $A = (2, -1, -1)$, $B = (0, 1, -2)$, and $C = (1, -3, 1)$.

**8.** Show that the triangle with vertices $(1, 2, 3)$, $(1, 3, 4)$, and $(0, 3, 3)$ is equilateral.

**9.** Find the area of the triangle with vertices $(1, 1, 0)$, $(1, 0, 1)$, and $(0, 1, 1)$.

**10.** What is the distance from the origin to the point $(1, 1, \ldots, 1)$ in $\mathbb{R}^n$?

**11.** What is the distance from the point $(1, 1, \ldots, 1)$ in $n$-space to the closest point on the $x_1$-axis?

In Exercises 12–23, describe (and sketch if possible) the set of points in $\mathbb{R}^3$ which satisfy the given equation or inequality.

**12.** $z = 2$

**13.** $y \geq -1$

**14.** $z = x$

**15.** $x + y = 1$

**16.** $x^2 + y^2 + z^2 = 4$

**17.** $(x-1)^2 + (y+2)^2 + (z-3)^2 = 4$

**18.** $x^2 + y^2 + z^2 = 2z$

**19.** $y^2 + z^2 \leq 4$

**20.** $x^2 + z^2 = 4$

**21.** $z = y^2$

**22.** $z \geq \sqrt{x^2 + y^2}$

**23.** $x + 2y + 3z = 6$

In Exercises 24–33, describe (and sketch if possible) the set of points in $\mathbb{R}^3$ which satisfy the given pair of equations or inequalities.

**24.** $\begin{cases} x = 1 \\ y = 2 \end{cases}$

**25.** $\begin{cases} x = 1 \\ y = z \end{cases}$

**26.** $\begin{cases} x^2 + y^2 + z^2 = 4 \\ z = 1 \end{cases}$

**27.** $\begin{cases} x^2 + y^2 + z^2 = 4 \\ x^2 + y^2 + z^2 = 4x \end{cases}$

**28.** $\begin{cases} x^2 + y^2 + z^2 = 4 \\ x^2 + z^2 = 1 \end{cases}$

**29.** $\begin{cases} x^2 + y^2 = 1 \\ z = x \end{cases}$

**30.** $\begin{cases} y \geq x \\ z \leq y \end{cases}$

**31.** $\begin{cases} x^2 + y^2 \leq 1 \\ z \geq y \end{cases}$

**32.** $\begin{cases} x^2 + y^2 + z^2 \leq 1 \\ \sqrt{x^2 + y^2} \leq z \end{cases}$

**33.** $\begin{cases} z \geq x^2 \\ x^2 + y^2 + z^2 \leq 1 \end{cases}$

## 3.2 VECTORS IN 3-SPACE

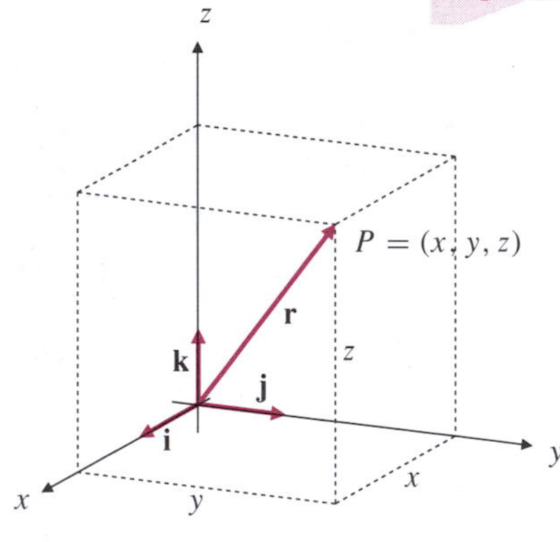


**Figure 3.9** The standard basis vectors **i**, **j**, and **k**

The concept of a vector as a quantity possessing magnitude and direction is introduced in Section 9.5 of *Single-Variable Calculus*. There we were concerned only with vectors lying in a plane, typically the $xy$-plane. However, the algebra and geometry of vectors described there extends to spaces of any number of dimensions; we can still think of vectors as represented by arrows, and sums and scalar multiples are formed just as for plane vectors.

In this section we state the properties of vectors for vectors in 3-space. It will be evident how extensions can be made to any number of dimensions.

Given a Cartesian coordinate system in 3-space, we define three **standard basis vectors**, **i**, **j**, and **k**, represented by arrows from the origin to the points $(1, 0, 0)$, $(0, 1, 0)$, and $(0, 0, 1)$, respectively. (See Figure 3.9.) Any vector in 3-space can be written as a *linear combination* of these basis vectors; for instance, the vector **r** from the origin to the point $(x, y, z)$ is given by

$$\mathbf{r} = x\mathbf{i} + y\mathbf{j} + z\mathbf{k}.$$

We say that $\mathbf{r}$ has **components** $x$, $y$, and $z$. The length of $\mathbf{r}$ is

$$|\mathbf{r}| = \sqrt{x^2 + y^2 + z^2}.$$

Such a vector from the origin to a point is called the **position vector** of that point. Generally, however, vectors do not have any specific location; two vectors are considered to be equal if they have the same length and the same direction, that is, if they have the same components.

If $P_1 = (x_1, y_1, z_1)$ and $P_2 = (x_2, y_2, z_2)$ are two points in 3-space, then the vector $\mathbf{v} = \overrightarrow{P_1P_2}$ from $P_1$ to $P_2$ has components $x_2 - x_1$, $y_2 - y_1$, and $z_2 - z_1$, and is therefore represented in terms of the standard basis vectors by

$$\mathbf{v} = \overrightarrow{P_1P_2} = (x_2 - x_1)\mathbf{i} + (y_2 - y_1)\mathbf{j} + (z_2 - z_1)\mathbf{k}.$$

Sums and scalar multiples of vectors are easily expressed in terms of components. If $\mathbf{u} = u_1\mathbf{i} + u_2\mathbf{j} + u_3\mathbf{k}$ and $\mathbf{v} = v_1\mathbf{i} + v_2\mathbf{j} + v_3\mathbf{k}$, and if $t$ is a scalar (i.e. a real number) then

$$\begin{aligned} \mathbf{u} + \mathbf{v} &= (u_1 + v_1)\mathbf{i} + (u_2 + v_2)\mathbf{j} + (u_3 + v_3)\mathbf{k}, \\ t\mathbf{u} &= (tu_1)\mathbf{i} + (tu_2)\mathbf{j} + (tu_3)\mathbf{k}. \end{aligned}$$

The zero vector is $\mathbf{0} = 0\mathbf{i} + 0\mathbf{j} + 0\mathbf{k}$. It has length zero and no specific direction. For any vector $\mathbf{u}$ we have $0\mathbf{u} = \mathbf{0}$. A **unit vector** is a vector of length 1. The standard basis vectors $\mathbf{i}$, $\mathbf{j}$, and $\mathbf{k}$ are unit vectors. Given any nonzero vector $\mathbf{v}$, we can form a unit vector $\hat{\mathbf{v}}$ in the same direction as $\mathbf{v}$ by multiplying $\mathbf{v}$ by the reciprocal of its length (a scalar):

$$\hat{\mathbf{v}} = \left(\frac{1}{|\mathbf{v}|}\right)\mathbf{v}.$$

**EXAMPLE 1** If $\mathbf{u} = 2\mathbf{i} + \mathbf{j} - 2\mathbf{k}$ and $\mathbf{v} = 3\mathbf{i} - 2\mathbf{j} - \mathbf{k}$, find $\mathbf{u} + \mathbf{v}$, $\mathbf{u} - \mathbf{v}$, $3\mathbf{u} - 2\mathbf{v}$, $|\mathbf{u}|$, $|\mathbf{v}|$, and a unit vector $\hat{\mathbf{u}}$ in the direction of $\mathbf{u}$.

***SOLUTION*** We have

$$\begin{aligned} \mathbf{u} + \mathbf{v} &= (2 + 3)\mathbf{i} + (1 - 2)\mathbf{j} + (-2 - 1)\mathbf{k} = 5\mathbf{i} - \mathbf{j} - 3\mathbf{k} \\ \mathbf{u} - \mathbf{v} &= (2 - 3)\mathbf{i} + (1 + 2)\mathbf{j} + (-2 + 1)\mathbf{k} = -\mathbf{i} + 3\mathbf{j} - \mathbf{k} \\ 3\mathbf{u} - 2\mathbf{v} &= (6 - 6)\mathbf{i} + (3 + 4)\mathbf{j} + (-6 + 2)\mathbf{k} = 7\mathbf{j} - 4\mathbf{k} \\ |\mathbf{u}| &= \sqrt{4 + 1 + 4} = 3, \qquad |\mathbf{v}| = \sqrt{9 + 4 + 1} = \sqrt{14} \\ \hat{\mathbf{u}} &= \left(\frac{1}{|\mathbf{u}|}\right)\mathbf{u} = \frac{2}{3}\mathbf{i} + \frac{1}{3}\mathbf{j} - \frac{2}{3}\mathbf{k}. \end{aligned}$$

■

Although the following example is only two-dimensional, it illustrates the way vectors can be used to solve problems involving relative velocities. If $A$ moves with velocity $\mathbf{v}_{A \text{ rel } B}$ relative to $B$, and $B$ moves with velocity $\mathbf{v}_{B \text{ rel } C}$ relative to $C$, then $A$ moves with velocity $\mathbf{v}_{A \text{ rel } C}$ relative to $C$, where

$$\mathbf{v}_{A \text{ rel } C} = \mathbf{v}_{A \text{ rel } B} + \mathbf{v}_{B \text{ rel } C}.$$

**EXAMPLE 2** An aircraft cruises at a speed of 300 km/h in still air. If the wind is blowing from the east at 100 km/h, in what direction should the aircraft head in order to fly in a straight line from city $P$ to city $Q$, 400 km north northeast of $P$? How long will the trip take?

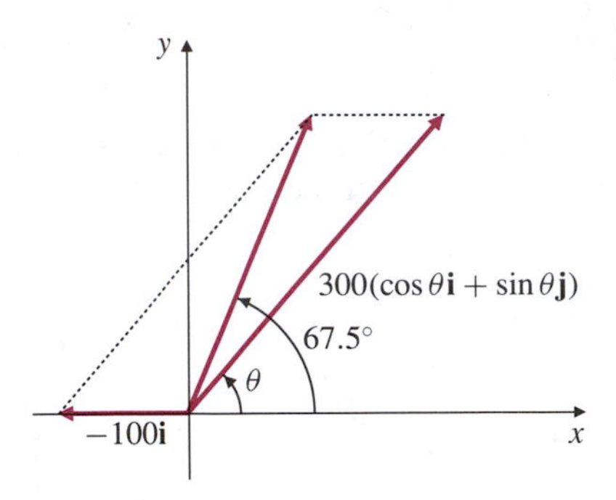


**Figure 3.10** Velocity diagram for the aircraft in Example 2

***SOLUTION*** The problem is two-dimensional, so we use plane vectors. Let us choose our coordinate system so that the $x$- and $y$-axes point east and north, respectively. Figure 3.10 illustrates the three velocities that must be considered. The velocity of the air relative to the ground is

$$\mathbf{v}_{\text{air rel ground}} = -100\,\mathbf{i}.$$

If the aircraft heads in a direction making angle $\theta$ with the positive direction of the $x$-axis then the velocity of the aircraft relative to the air is

$$\mathbf{v}_{\text{aircraft rel air}} = 300\cos\theta\,\mathbf{i} + 300\sin\theta\,\mathbf{j}.$$

Thus the velocity of the aircraft relative to the ground is

$$\begin{aligned}\mathbf{v}_{\text{aircraft rel ground}} &= \mathbf{v}_{\text{aircraft rel air}} + \mathbf{v}_{\text{air rel ground}}\\ &= (300\cos\theta - 100)\,\mathbf{i} + 300\sin\theta\,\mathbf{j}.\end{aligned}$$

We want this latter velocity to be in a north-northeasterly direction, that is, in the direction making angle $3\pi/8 = 67.5^\circ$ with the positive direction of the $x$-axis. Thus we will have

$$\mathbf{v}_{\text{aircraft rel ground}} = v\,[(\cos 67.5^\circ)\ \mathbf{i} + (\sin 67.5^\circ)\ \mathbf{j}],$$

where $v$ is the actual groundspeed of the aircraft. Comparing the two expressions for $\mathbf{v}_{\text{aircraft rel ground}}$ we obtain

$$\begin{aligned}300\cos\theta - 100 &= v\ \cos 67.5^\circ\\ 300\sin\theta &= v\ \sin 67.5^\circ.\end{aligned}$$

Eliminating $v$ between these two equations we get

$$300\ \cos\theta\ \sin 67.5^\circ - 300\ \sin\theta\ \cos 67.5^\circ = 100\ \sin 67.5^\circ,$$

or,

$$3\ \sin(67.5^\circ - \theta) = \sin 67.5^\circ.$$

Therefore the aircraft should head in direction $\theta$ given by

$$\theta = 67.5^\circ - \text{Arcsin}\left(\frac{1}{3}\ \sin 67.5^\circ\right) \approx 49.56^\circ,$$

that is, $49.56^\circ$ north of east. The groundspeed is now seen to be

$$v = 300\sin\theta/\sin 67.5^\circ \approx 247.15\text{ km/h}.$$

Thus the 400 km trip will take about $400/247.15 \approx 1.616$ hours, or, about 1 hour and 37 minutes. ■

## The Dot Product and Projections

The dot product of two plane vectors, $\mathbf{u} = u_1\mathbf{i} + u_2\mathbf{j}$ and $\mathbf{v} = v_1\mathbf{i} + v_2\mathbf{j}$, is defined to be the *number*

$$\mathbf{u} \bullet \mathbf{v} = u_1 v_1 + u_2 v_2.$$

A similar formula with one more term serves to define the dot product of two vectors in 3-space:

**DEFINITION 1**

**The dot product**

If $\mathbf{u} = u_1\mathbf{i} + u_2\mathbf{j} + u_3\mathbf{k}$ and $\mathbf{v} = v_1\mathbf{i} + v_2\mathbf{j} + v_3\mathbf{k}$, then the **dot product $\mathbf{u} \bullet \mathbf{v}$** is the number

$$\mathbf{u} \bullet \mathbf{v} = u_1 v_1 + u_2 v_2 + u_3 v_3.$$

The dot product has the following algebraic and geometric properties:

$$\begin{aligned}
&\mathbf{u} \bullet \mathbf{v} = \mathbf{v} \bullet \mathbf{u} && \text{(commutative law)}\\
&\mathbf{u} \bullet (\mathbf{v} + \mathbf{w}) = \mathbf{u} \bullet \mathbf{v} + \mathbf{u} \bullet \mathbf{w} && \text{(distributive law)}\\
&(t\mathbf{u}) \bullet \mathbf{v} = \mathbf{u} \bullet (t\mathbf{v}) = t(\mathbf{u} \bullet \mathbf{v}) && \text{(for real } t\text{)}\\
&\mathbf{u} \bullet \mathbf{u} = |\mathbf{u}|^2 &&\\
&\mathbf{u} \bullet \mathbf{v} = |\mathbf{u}||\mathbf{v}| \cos\theta, &&
\end{aligned}$$

where $\theta$ is the angle between $\mathbf{u}$ and $\mathbf{v}$. Again we note that two vectors are perpendicular if and only if their dot product is zero. The angle between the vectors is acute if the dot product is positive, and obtuse if the dot product is negative.

**EXAMPLE 3** Find the angle $\theta$ between the vectors $\mathbf{u} = 2\mathbf{i} + \mathbf{j} - 2\mathbf{k}$ and $\mathbf{v} = 3\mathbf{i} - 2\mathbf{j} - \mathbf{k}$.

***SOLUTION*** Solving the formula $\mathbf{u} \bullet \mathbf{v} = |\mathbf{u}||\mathbf{v}| \cos\theta$ for $\theta$, we obtain

$$\begin{aligned}
\theta = \cos^{-1} \frac{\mathbf{u} \bullet \mathbf{v}}{|\mathbf{u}||\mathbf{v}|} &= \cos^{-1}\left(\frac{(2)(3) + (1)(-2) + (-2)(-1)}{3\sqrt{14}}\right)\\
&= \cos^{-1} \frac{2}{\sqrt{14}} \approx 57.69^\circ.
\end{aligned}$$

It is sometimes useful to project one vector along another. We define both scalar and vector projections of $\mathbf{u}$ in the direction of $\mathbf{v}$:

**DEFINITION 2**

**Scalar projection**

The **scalar projection** $s$ of any vector $\mathbf{u}$ in the direction of a nonzero vector $\mathbf{v}$ is the dot product of $\mathbf{u}$ with a unit vector in the direction of $\mathbf{v}$. Thus it is the *number*

$$s = \frac{\mathbf{u} \bullet \mathbf{v}}{|\mathbf{v}|} = |\mathbf{u}| \cos\theta$$

where $\theta$ is the angle between $\mathbf{u}$ and $\mathbf{v}$.

Note that $|s|$ is the length of the line segment along the line of $\mathbf{v}$ obtained by dropping perpendiculars to that line from the tail and head of $\mathbf{u}$. (See Figure 3.11.) Also, $s$ is negative if $\theta > 90°$.

**DEFINITION 3**

**Vector projection**

The **vector projection**, $\mathbf{u_v}$, of $\mathbf{u}$ in the direction of $\mathbf{v}$ (see Figure 3.11) is the scalar multiple of a unit vector $\hat{\mathbf{v}}$ in the direction of $\mathbf{v}$, by the scalar projection of $\mathbf{u}$ in the direction of $\mathbf{v}$, that is,

$$\text{vector projection of } \mathbf{u} \text{ along } \mathbf{v} = \mathbf{u_v} = \frac{\mathbf{u} \bullet \mathbf{v}}{|\mathbf{v}|}\hat{\mathbf{v}} = \frac{\mathbf{u} \bullet \mathbf{v}}{|\mathbf{v}|^2}\mathbf{v}.$$

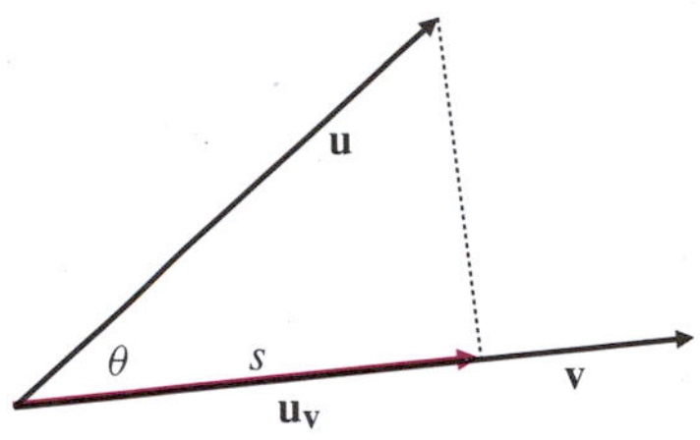

**Figure 3.11** The scalar projection $s$ and the vector projection $\mathbf{u_v}$ of $\mathbf{u}$ along $\mathbf{v}$

It is often necessary to express a vector as a sum of two other vectors parallel and perpendicular to a given direction.

**EXAMPLE 4** Express the vector $3\mathbf{i} + \mathbf{j}$ as a sum of vectors $\mathbf{u} + \mathbf{v}$, where $\mathbf{u}$ is parallel to the vector $\mathbf{i} + \mathbf{j}$ and $\mathbf{v}$ is perpendicular to $\mathbf{u}$.

***SOLUTION***

METHOD I (Using vector projection): Note that $\mathbf{u}$ must be the vector projection of $3\mathbf{i} + \mathbf{j}$ in the direction of $\mathbf{i} + \mathbf{j}$. Thus

$$\mathbf{u} = \frac{(3\mathbf{i} + \mathbf{j}) \bullet (\mathbf{i} + \mathbf{j})}{|\mathbf{i} + \mathbf{j}|^2}(\mathbf{i} + \mathbf{j}) = \frac{4}{2}(\mathbf{i} + \mathbf{j}) = 2\mathbf{i} + 2\mathbf{j}$$
$$\mathbf{v} = 3\mathbf{i} + \mathbf{j} - \mathbf{u} = \mathbf{i} - \mathbf{j}.$$

METHOD II (From basic principles): Since $\mathbf{u}$ is parallel to $\mathbf{i} + \mathbf{j}$ and $\mathbf{v}$ is perpendicular to $\mathbf{u}$, we have

$$\mathbf{u} = t(\mathbf{i} + \mathbf{j}) \qquad \text{and} \qquad \mathbf{v} \bullet (\mathbf{i} + \mathbf{j}) = 0,$$

for some scalar $t$. We want $\mathbf{u} + \mathbf{v} = 3\mathbf{i} + \mathbf{j}$. Take the dot product of this equation with $\mathbf{i} + \mathbf{j}$:

$$\mathbf{u} \bullet (\mathbf{i} + \mathbf{j}) + \mathbf{v} \bullet (\mathbf{i} + \mathbf{j}) = (3\mathbf{i} + \mathbf{j}) \bullet (\mathbf{i} + \mathbf{j})$$
$$t(\mathbf{i} + \mathbf{j}) \bullet (\mathbf{i} + \mathbf{j}) + 0 = 4.$$

Thus $2t = 4$, and so $t = 2$. Therefore,

$$\mathbf{u} = 2\mathbf{i} + 2\mathbf{j} \qquad \text{and} \qquad \mathbf{v} = 3\mathbf{i} + \mathbf{j} - \mathbf{u} = \mathbf{i} - \mathbf{j}.$$

All of the above ideas make sense for vectors in spaces of any dimension. Vectors in $\mathbb{R}^n$ can expressed as linear combinations of the $n$ unit vectors

| | | |
|---|---|---|
| $\mathbf{e_1}$ | from the origin to the point | $(1, 0, 0, \ldots, 0)$ |
| $\mathbf{e_2}$ | from the origin to the point | $(0, 1, 0, \ldots, 0)$ |
| | $\vdots$ | |
| $\mathbf{e_n}$ | from the origin to the point | $(0, 0, 0, \ldots, 1)$. |

These vectors constitute a *standard basis* in $\mathbb{R}^n$. The $n$-vector $\mathbf{x}$ with components $x_1, x_2, \ldots, x_n$ is expressed in the form

$$\mathbf{x} = x_1\mathbf{e_1} + x_2\mathbf{e_2} + \cdots + x_n\mathbf{e_n}.$$

The length of $\mathbf{x}$ is $|\mathbf{x}| = \sqrt{x_1{}^2 + x_2{}^2 + \cdots + x_n{}^2}$. The angle between two vectors $\mathbf{x}$ and $\mathbf{y}$ is defined to be

$$\theta = \cos^{-1}\frac{\mathbf{x} \bullet \mathbf{y}}{|\mathbf{x}||\mathbf{y}|},$$

where

$$\mathbf{x} \bullet \mathbf{y} = x_1 y_1 + x_2 y_2 + \cdots + x_n y_n.$$

We will not make much use of $n$-vectors for $n > 3$ but you should be aware that everything said up until now for 2-vectors or 3-vectors extends to $n$-vectors.

## EXERCISES 3.2

In Exercises 1–3, calculate the following for the given vectors $\mathbf{u}$ and $\mathbf{v}$:

(a) $\mathbf{u} + \mathbf{v}$, $\mathbf{u} - \mathbf{v}$, $2\mathbf{u} - 3\mathbf{v}$,

(b) the lengths $|\mathbf{u}|$ and $|\mathbf{v}|$,

(c) unit vectors $\hat{\mathbf{u}}$ and $\hat{\mathbf{v}}$ in the directions of $\mathbf{u}$ and $\mathbf{v}$, respectively,

(d) the dot product $\mathbf{u} \bullet \mathbf{v}$,

(e) the angle between $\mathbf{u}$ and $\mathbf{v}$,

(f) the scalar projection of $\mathbf{u}$ in the direction of $\mathbf{v}$,

g) the vector projection of $\mathbf{v}$ along $\mathbf{u}$.

**1.** $\mathbf{u} = 3\mathbf{i} + 4\mathbf{j} - 5\mathbf{k}$ and $\mathbf{v} = 3\mathbf{i} - 4\mathbf{j} - 5\mathbf{k}$

**2.** $\mathbf{u} = \mathbf{i} - \mathbf{j}$ and $\mathbf{v} = \mathbf{j} + 2\mathbf{k}$

**3.** $\mathbf{u} = \mathbf{i} - 2\mathbf{j}$ and $\mathbf{v} = 2\mathbf{i} + 2\mathbf{j} - 3\mathbf{k}$

**4.** Solve the equation $2\mathbf{x} + 3\mathbf{i} - 5\mathbf{j} = 5\mathbf{x} + \mathbf{j} - 9\mathbf{k}$ for $\mathbf{x}$.

**5.** A weather vane mounted on the top of a car moving due north at 50 km/h indicates that the wind is coming from the west. When the car doubles its speed, the weather vane indicates that the wind is coming from the northwest. From what direction is the wind coming, and what is its speed?

**6.** A straight river 500 m wide flows due east at a constant speed of 3 km/h. If you can row your boat at a speed of 5 km/h in still water, in what direction should you head if you wish to row from point $A$ on the south shore to point $B$ on the north shore directly north of $A$? How long will the trip take?

* **7.** In what direction should you head to cross the river in Exercise 6 if you can only row at 2 km/h, and you wish to row from $A$ to point $C$ on the north shore, $k$ km downstream from $B$. For what values of $k$ is the trip not possible?

**8.** A certain aircraft flies with an airspeed of 750 km/h. In what direction should it head in order to make progress in a true easterly direction if the wind is from the northeast at 100 km/h? How long will it take to complete a trip to a city 1500 km from its starting point?

**9.** For what value of $t$ is the vector $2t\mathbf{i} + 4\mathbf{j} - (10 + t)\mathbf{k}$ perpendicular to the vector $\mathbf{i} + t\mathbf{j} + \mathbf{k}$?

**10.** Find the angle between a diagonal of a cube and one of the edges of the cube.

**11.** Find the angle between a diagonal of a cube and a diagonal of one of the faces of the cube. Give all possible answers.

**12.** If a vector $\mathbf{u}$ in $\mathbb{R}^3$ makes angles $\alpha$, $\beta$, and $\gamma$ with the coordinate axes, show that

$$\hat{\mathbf{u}} = \cos\alpha\mathbf{i} + \cos\beta\mathbf{j} + \cos\gamma\mathbf{k}$$

is a unit vector in the direction of $\mathbf{u}$. Hence show that

$$\cos^2\alpha + \cos^2\beta + \cos^2\gamma = 1.$$

**13.** Find a unit vector that makes equal angles with the three coordinate axes.

**14.** Find the three angles of the triangle with vertices $(1, 0, 0)$, $(0, 2, 0)$, and $(0, 0, 3)$.

**15.** If $\mathbf{r}_1$ and $\mathbf{r}_2$ are the position vectors of two points, $P_1$ and $P_2$, and $\lambda$ is a real number, show that

$$\mathbf{r} = (1 - \lambda)\mathbf{r}_1 + \lambda\mathbf{r}_2$$

is the position vector of a point $P$ on the straight line joining $P_1$ and $P_2$. Where is $P$ if $\lambda = 1/2$? if $\lambda = 2/3$? if $\lambda = -1$? if $\lambda = 2$?

**16.** Let $\mathbf{a}$ be a nonzero vector. Descibe the set of all points in 3-space whose position vectors $\mathbf{r}$ satisfy $\mathbf{a} \bullet \mathbf{r} = 0$.

**17.** Let $\mathbf{a}$ be a nonzero vector and let $b$ be any real number. Descibe the set of all points in 3-space whose position vectors $\mathbf{r}$ satisfy $\mathbf{a} \bullet \mathbf{r} = b$.

In Exercises 18–20, $\mathbf{u} = 2\mathbf{i} + \mathbf{j} - 2\mathbf{k}$, $\mathbf{v} = \mathbf{i} + 2\mathbf{j} - 2\mathbf{k}$, and $\mathbf{w} = 2\mathbf{i} - 2\mathbf{j} + \mathbf{k}$.

**18.** Find two unit vectors each of which is perpendicular to $\mathbf{u}$ and $\mathbf{v}$.

**19.** Find a vector $\mathbf{x}$ satisfying the system of equations $\mathbf{x} \bullet \mathbf{u} = 9$, $\mathbf{x} \bullet \mathbf{v} = 4$, $\mathbf{x} \bullet \mathbf{w} = 6$.

**20.** Find two unit vectors each of which makes equal angles with $\mathbf{u}$, $\mathbf{v}$, and $\mathbf{w}$.

**21.** Find a unit vector that bisects the angle between any two nonzero vectors $\mathbf{u}$ and $\mathbf{v}$.

**22.** Given two nonparallel vectors $\mathbf{u}$ and $\mathbf{v}$, describe the set of all points whose position vectors $\mathbf{r}$ are of the form $\mathbf{r} = \lambda\mathbf{u} + \mu\mathbf{v}$, where $\lambda$ and $\mu$ are arbitrary real numbers.

**23.** **(The triangle inequality)** Let $\mathbf{u}$ and $\mathbf{v}$ be two vectors.

(a) Show that $|\mathbf{u} + \mathbf{v}|^2 = |\mathbf{u}|^2 + 2\mathbf{u} \bullet \mathbf{v} + |\mathbf{v}|^2$.

(b) Show that $\mathbf{u} \bullet \mathbf{v} \le |\mathbf{u}||\mathbf{v}|$.

(c) Deduce from (a) and (b) that $|\mathbf{u} + \mathbf{v}| \le |\mathbf{u}| + |\mathbf{v}|$.

**24.** (a) Why is the inequality in Exercise 23(c) called a triangle inequality?

(b) What conditions on $\mathbf{u}$ and $\mathbf{v}$ imply that $|\mathbf{u} + \mathbf{v}| = |\mathbf{u}| + |\mathbf{v}|$?

**25.** **(Orthonormal bases)** Let $\mathbf{u} = \frac{3}{5}\mathbf{i} + \frac{4}{5}\mathbf{j}$, $\mathbf{v} = \frac{4}{5}\mathbf{i} - \frac{3}{5}\mathbf{j}$, and $\mathbf{w} = \mathbf{k}$.

(a) Show that $|\mathbf{u}| = |\mathbf{v}| = |\mathbf{w}| = 1$ and $\mathbf{u} \bullet \mathbf{v} = \mathbf{u} \bullet \mathbf{w} = \mathbf{v} \bullet \mathbf{w} = 0$. The vectors $\mathbf{u}$, $\mathbf{v}$, and $\mathbf{w}$ are mutually perpendicular unit vectors, and as such are said to constitute an **orthonormal basis** for $\mathbb{R}^3$.

(b) If $\mathbf{r} = x\mathbf{i} + y\mathbf{j} + z\mathbf{k}$, show by direct calculation that

$$\mathbf{r} = (\mathbf{r} \bullet \mathbf{u})\mathbf{u} + (\mathbf{r} \bullet \mathbf{v})\mathbf{v} + (\mathbf{r} \bullet \mathbf{w})\mathbf{w}.$$

**26.** Show that if $\mathbf{u}$, $\mathbf{v}$, and $\mathbf{w}$ are any three mutually perpendicular unit vectors in $\mathbb{R}^3$ and $\mathbf{r} = a\mathbf{u} + b\mathbf{v} + c\mathbf{w}$, then $a = \mathbf{r} \bullet \mathbf{u}$, $b = \mathbf{r} \bullet \mathbf{v}$, and $c = \mathbf{r} \bullet \mathbf{w}$.

**27.** **(Resolving a vector in perpendicular directions)** If $\mathbf{a}$ is a nonzero vector and $\mathbf{w}$ is any vector, find vectors $\mathbf{u}$ and $\mathbf{v}$ such that $\mathbf{w} = \mathbf{u} + \mathbf{v}$, $\mathbf{u}$ is parallel to $\mathbf{a}$, and $\mathbf{v}$ is perpendicular to $\mathbf{a}$.

**28.** **(Expressing a vector as a linear combination of two other vectors with which it is coplanar)** Suppose that $\mathbf{u}$, $\mathbf{v}$, and $\mathbf{r}$ are position vectors of points $U$, $V$, and $P$ respectively, that $\mathbf{u}$ is not parallel to $\mathbf{v}$, and that $P$ lies in the plane containing the origin, $U$ and $V$. Show that there exist numbers $\lambda$ and $\mu$ such that $\mathbf{r} = \lambda\mathbf{u} + \mu\mathbf{v}$. (*Hint:* resolve both $\mathbf{v}$ and $\mathbf{r}$ as sums of vectors parallel and perpendicular to $\mathbf{u}$ as suggested in Exercise 27.)

* **29.** Given constants $r$, $s$ and $t$, with $r \ne 0$ and $s \ne 0$, and given a vector $\mathbf{a}$ satisfying $|\mathbf{a}|^2 > 4rst$, solve the system of equations

$$\begin{cases} r\mathbf{x} + s\mathbf{y} = \mathbf{a} \\ \mathbf{x} \bullet \mathbf{y} = t \end{cases}$$

for the unknown vectors $\mathbf{x}$ and $\mathbf{y}$.

## 3.3 THE CROSS PRODUCT IN 3-SPACE

There is defined, *in 3-space only*, another kind of product of vectors called a cross product or vector product, and denoted $\mathbf{u}\times\mathbf{v}$.

> For any vectors $\mathbf{u}$ and $\mathbf{v}$ in $\mathbb{R}^3$, the **cross product** $\mathbf{u}\times\mathbf{v}$ is the unique vector satisfying the three conditions:
>
> (i) $(\mathbf{u}\times\mathbf{v}) \bullet \mathbf{u} = 0$ and $(\mathbf{u}\times\mathbf{v}) \bullet \mathbf{v} = 0$,
>
> (ii) $|\mathbf{u}\times\mathbf{v}| = |\mathbf{u}||\mathbf{v}| \sin\theta$, where $\theta$ is the angle between $\mathbf{u}$ and $\mathbf{v}$, and
>
> (iii) $\mathbf{u}$, $\mathbf{v}$, and $\mathbf{u}\times\mathbf{v}$ form a right-handed triad.

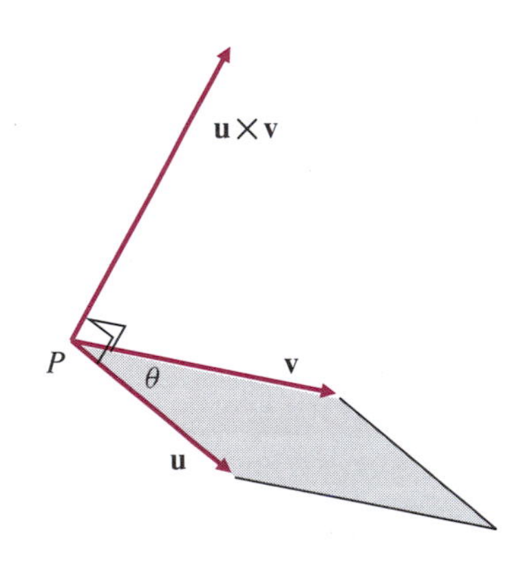


**Figure 3.12** $\mathbf{u}\times\mathbf{v}$ is perpendicular to both $\mathbf{u}$ and $\mathbf{v}$ and has length equal to the area of the shaded parallelogram

If $\mathbf{u}$ and $\mathbf{v}$ are parallel, condition (ii) says that $\mathbf{u}\times\mathbf{v} = \mathbf{0}$, the zero vector. Otherwise, through any point in $\mathbb{R}^3$ there is a unique straight line that is perpendicular to both $\mathbf{u}$ and $\mathbf{v}$. Condition (i) says that $\mathbf{u}\times\mathbf{v}$ is parallel to this line. Condition (iii) determines which of the two directions along this line is the direction of $\mathbf{u}\times\mathbf{v}$; a right-handed screw advances in the direction of $\mathbf{u}\times\mathbf{v}$ if rotated in the direction from $\mathbf{u}$ towards $\mathbf{v}$. (This is equivalent to saying that the thumb, forefinger, and middle finger of the right hand can be made to point in the directions of $\mathbf{u}$, $\mathbf{v}$, and $\mathbf{u}\times\mathbf{v}$, respectively.)

If $\mathbf{u}$ and $\mathbf{v}$ have their tails at the point $P$, then $\mathbf{u}\times\mathbf{v}$ is normal (i.e., perpendicular) to the plane through $P$ in which $\mathbf{u}$ and $\mathbf{v}$ lie, and, by condition (ii), $\mathbf{u}\times\mathbf{v}$ has length equal to the area of the parallelogram spanned by $\mathbf{u}$ and $\mathbf{v}$. (See Figure 3.12.) These properties make the cross product very useful for the description of tangent planes and normal lines to surfaces in $\mathbb{R}^3$.

The definition of cross product given above does not involve any coordinate system and therefore does not directly show the components of the cross product with respect to the standard basis. These components are provided by the following theorem:

**THEOREM 1**

**Components of the cross product**

If $\mathbf{u} = u_1\mathbf{i} + u_2\mathbf{j} + u_3\mathbf{k}$ and $\mathbf{v} = v_1\mathbf{i} + v_2\mathbf{j} + v_3\mathbf{k}$ then

$$\mathbf{u}\times\mathbf{v} = (u_2v_3 - u_3v_2)\mathbf{i} + (u_3v_1 - u_1v_3)\mathbf{j} + (u_1v_2 - u_2v_1)\mathbf{k}.$$

***PROOF*** First we observe that the vector

$$\mathbf{w} = (u_2v_3 - u_3v_2)\mathbf{i} + (u_3v_1 - u_1v_3)\mathbf{j} + (u_1v_2 - u_2v_1)\mathbf{k}$$

is perpendicular to both $\mathbf{u}$ and $\mathbf{v}$ since

$$\mathbf{u} \bullet \mathbf{w} = u_1(u_2v_3 - u_3v_2) + u_2(u_3v_1 - u_1v_3) + u_3(u_1v_2 - u_2v_1) = 0,$$

and similarly $\mathbf{v} \bullet \mathbf{w} = 0$. Thus $\mathbf{u}\times\mathbf{v}$ is parallel to $\mathbf{w}$. Next we show that $\mathbf{w}$ and $\mathbf{u}\times\mathbf{v}$ have the same length. In fact

$$\begin{aligned} |\mathbf{w}|^2 =& (u_2v_3 - u_3v_2)^2 + (u_3v_1 - u_1v_3)^2 + (u_1v_2 - u_2v_1)^2 \\ =& u_2^2v_3^2 + u_3^2v_2^2 - 2u_2v_3u_3v_2 + u_3^2v_1^2 + u_1^2v_3^2 \\ & - 2u_3v_1u_1v_3 + u_1^2v_2^2 + u_2^2v_1^2 - 2u_1v_2u_2v_1, \end{aligned}$$

while

$$\begin{aligned} |\mathbf{u}\times\mathbf{v}|^2 =& |\mathbf{u}|^2|\mathbf{v}|^2 \sin^2\theta \\ =& |\mathbf{u}|^2|\mathbf{v}|^2 (1 - \cos^2\theta) \\ =& |\mathbf{u}|^2|\mathbf{v}|^2 - (\mathbf{u} \bullet \mathbf{v})^2 \\ =& (u_1^2 + u_2^2 + u_3^2)(v_1^2 + v_2^2 + v_3^2) - (u_1v_1 + u_2v_2 + u_3v_3)^2 \\ =& u_1^2v_1^2 + u_1^2v_2^2 + u_1^2v_3^2 + u_2^2v_1^2 + u_2^2v_2^2 + u_2^2v_3^2 + u_3^2v_1^2 + u_3^2v_2^2 + u_3^2v_3^2 \\ & - u_1^2v_1^2 - u_2^2v_2^2 - u_3^2v_3^2 - 2u_1v_1u_2v_2 - 2u_1v_1u_3v_3 - 2u_2v_2u_3v_3 \\ =& |\mathbf{w}|^2. \end{aligned}$$

Since $\mathbf{w}$ is parallel to, and has the same length as $\mathbf{u}\times\mathbf{v}$, we must have either $\mathbf{u}\times\mathbf{v} = \mathbf{w}$ or $\mathbf{u}\times\mathbf{v} = -\mathbf{w}$. It remains to be shown that the first of these is the correct choice. To see this, suppose that the triad of vectors $\mathbf{u}$, $\mathbf{v}$, and $\mathbf{w}$ are rigidly rotated in 3-space so that $\mathbf{u}$ points in the direction of the positive $x$-axis and $\mathbf{v}$ lies in the upper half of the $xy$-plane. Then $\mathbf{u} = u_1\mathbf{i}$, and $\mathbf{v} = v_1\mathbf{i} + v_2\mathbf{j}$, where $u_1 > 0$ and $v_2 > 0$. By the "right-hand rule" $\mathbf{u}\times\mathbf{v}$ must point in the direction of the positive $z$-axis. But $\mathbf{w} = u_1v_2\mathbf{k}$ does point in that direction, so $\mathbf{u}\times\mathbf{v} = \mathbf{w}$, as asserted.

The formula for the cross product in terms of components may seem awkward and asymmetric. As we shall see, however, it can be written more easily in terms of a determinant. We introduce determinants later in this section.

**EXAMPLE 1** **Calculating cross products**

(a)
$$\mathbf{i}\times\mathbf{i} = \mathbf{0}, \qquad \mathbf{i}\times\mathbf{j} = \mathbf{k}, \qquad \mathbf{j}\times\mathbf{i} = -\mathbf{k},$$
$$\mathbf{j}\times\mathbf{j} = \mathbf{0}, \qquad \mathbf{j}\times\mathbf{k} = \mathbf{i}, \qquad \mathbf{k}\times\mathbf{j} = -\mathbf{i},$$
$$\mathbf{k}\times\mathbf{k} = \mathbf{0}, \qquad \mathbf{k}\times\mathbf{i} = \mathbf{j}, \qquad \mathbf{i}\times\mathbf{k} = -\mathbf{j}.$$

(b)
$$\begin{aligned}(2\mathbf{i} + \mathbf{j} - 3\mathbf{k})\times(-2\mathbf{j} + 5\mathbf{k}) &= \big((1)(5) - (-2)(-3)\big)\mathbf{i} + \big((-3)(0) - (2)(5)\big)\mathbf{j} + \big((2)(-2) - (1)(0)\big)\mathbf{k} \\ &= -\mathbf{i} - 10\mathbf{j} - 4\mathbf{k}.\end{aligned}$$

The cross product has some, but not all of the properties we usually ascribe to products. We summarize its algebraic properties as follows:

**Properties of the cross product**

If $\mathbf{u}$, $\mathbf{v}$, and $\mathbf{w}$ are any vectors in $\mathbb{R}^3$, and $t$ is a real number (a scalar), then

(i) $\mathbf{u}\times\mathbf{u} = \mathbf{0}$,

(ii) $\mathbf{u}\times\mathbf{v} = -\mathbf{v}\times\mathbf{u}$, (The cross product is **anticommutative**.)

(iii) $(\mathbf{u} + \mathbf{v})\times\mathbf{w} = \mathbf{u}\times\mathbf{w} + \mathbf{v}\times\mathbf{w}$,

(iv) $\mathbf{u}\times(\mathbf{v} + \mathbf{w}) = \mathbf{u}\times\mathbf{v} + \mathbf{u}\times\mathbf{w}$,

(v) $(t\mathbf{u})\times\mathbf{v} = \mathbf{u}\times(t\mathbf{v}) = t(\mathbf{u}\times\mathbf{v})$,

(vi) $\mathbf{u} \bullet (\mathbf{u}\times\mathbf{v}) = \mathbf{v} \bullet (\mathbf{u}\times\mathbf{v}) = 0$.

These identities are all easily verified using the components or the definition of the cross product or by using properties of determinants discussed below. They are left as exercises for the reader. Note the absence of an associative law. The cross product is not associative. (See Exercise 21 at the end of this section.) In general,

$$\mathbf{u}\times(\mathbf{v}\times\mathbf{w}) \neq (\mathbf{u}\times\mathbf{v})\times\mathbf{w}.$$

## Determinants

In order to simplify certain formulas such as the component representation of the cross product we introduce $2 \times 2$ and $3 \times 3$ **determinants**. General $n \times n$ determinants are normally studied in courses on linear algebra; we will encounter them in Section 3.6. In this section we will outline enough of the properties of determinants to enable us to use them as shorthand in some otherwise complicated formulas.

A determinant is an expression that involves the elements of a square array (matrix) of numbers. The determinant of the $2\times2$ array of numbers

$$\begin{matrix} a & b \\ c & d \end{matrix}$$

is denoted by enclosing the array between vertical bars, and its value is the number $ad - bc$:

$$\begin{vmatrix} a & b \\ c & d \end{vmatrix} = ad - bc.$$

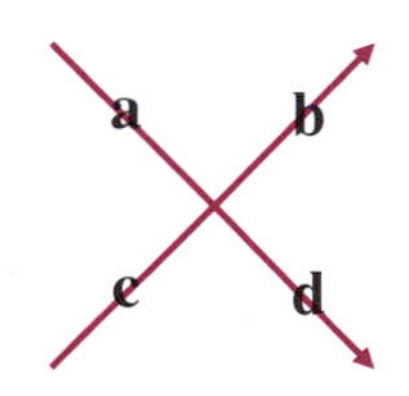

Figure 3.13

This is the product of elements in the *downward diagonal* of the array, minus the product of elements in the *upward diagonal* as shown in Figure 3.13. For example,

$$\begin{vmatrix} 1 & 2 \\ 3 & 4 \end{vmatrix} = (1)(4) - (2)(3) = -2.$$

Similarly, the determinant of a $3\times3$ array of numbers is defined by

$$\begin{vmatrix} a & b & c \\ d & e & f \\ g & h & i \end{vmatrix} = aei + bfg + cdh - gec - hfa - idb.$$

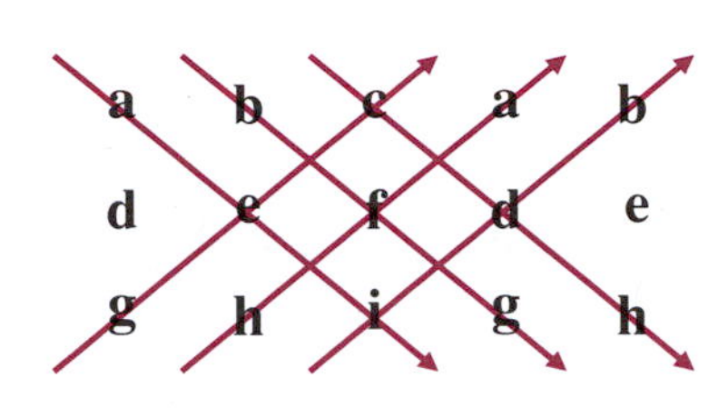


Figure 3.14

Observe that each of the six products in the value of the determinant involves exactly one element from each row and exactly one from each column of the array. As such, each term is the product of elements in a *diagonal* of an *extended* array obtained by repeating the first two columns of the array to the right of the third column, as shown in Figure 3.14. The value of the determinant is the sum of products corresponding to the three complete *downward* diagonals minus the sum corresponding of the three *upward* diagonals. With practice you will be able to form these diagonal products without having to write the extended array.

If we group the terms in the expansion of the determinant to factor out the elements of the first row we obtain

$$\begin{aligned}\begin{vmatrix} a & b & c \\ d & e & f \\ g & h & i \end{vmatrix} &= a(ei - fh) - b(di - fg) + c(dh - eg) \\ &= a\begin{vmatrix} e & f \\ h & i \end{vmatrix} - b\begin{vmatrix} d & f \\ g & i \end{vmatrix} + c\begin{vmatrix} d & e \\ g & h \end{vmatrix}.\end{aligned}$$

The $2\times2$ determinants appearing here (called *minors* of the given $3\times3$ determinant) are obtained by deleting the row and column containing the corresponding element from the original $3\times3$ determinant. This process is called *expanding* the $3\times3$ determinant *in minors* about the first row.

Such expansions in minors can be carried out about any row or column. Note that a minus sign appears in any term whose minor is obtained by deleting the $i$th row and $j$th column, where $i + j$ is an *odd* number. For example, we can expand the above determinant in minors about the second column as follows:

$$\begin{aligned}\begin{vmatrix} a & b & c \\ d & e & f \\ g & h & i \end{vmatrix} &= -b\begin{vmatrix} d & f \\ g & i \end{vmatrix} + e\begin{vmatrix} a & c \\ g & i \end{vmatrix} - h\begin{vmatrix} a & c \\ d & f \end{vmatrix} \\ &= -bdi + bfg + eai - ecg - haf + hcd.\end{aligned}$$

(Of course this is the same value as the one obtained previously.)

**EXAMPLE 2**

$$\begin{aligned}\begin{vmatrix} 1 & 4 & -2 \\ -3 & 1 & 0 \\ 2 & 2 & -3 \end{vmatrix} &= 3\begin{vmatrix} 4 & -2 \\ 2 & -3 \end{vmatrix} + 1\begin{vmatrix} 1 & -2 \\ 2 & -3 \end{vmatrix} \\ &= 3(-8) + 1 = -23.\end{aligned}$$

We expanded about the second row; the third column would also have been a good choice. (Why?) ■

Any row (or column) of a determinant may be regarded as the components of a vector. Then the determinant is a *linear function* of that vector. For example

$$\begin{vmatrix} a & b & c \\ d & e & f \\ sx + tl & sy + tm & sz + tn \end{vmatrix} = s\begin{vmatrix} a & b & c \\ d & e & f \\ x & y & z \end{vmatrix} + t\begin{vmatrix} a & b & c \\ d & e & f \\ l & m & n \end{vmatrix}$$

because the determinant is a linear function of its third row. This and other properties of determinants follow directly from the definition. Some other properties are summarized below. These are stated for rows and for $3 \times 3$ determinants but similar statements can be made for columns and for determinants of any order.

**Properties of determinants**

(i) If two rows of a determinant are interchanged, then the determinant changes sign:

$$\begin{vmatrix} d & e & f \\ a & b & c \\ g & h & i \end{vmatrix} = -\begin{vmatrix} a & b & c \\ d & e & f \\ g & h & i \end{vmatrix}.$$

(ii) If two rows of a determinant are equal, the determinant has value 0:

$$\begin{vmatrix} a & b & c \\ a & b & c \\ g & h & i \end{vmatrix} = 0.$$

(iii) If a multiple of one row of a determinant is added to another row, the value of the determinant remains unchanged:

$$\begin{vmatrix} a & b & c \\ d+ta & e+tb & f+tc \\ g & h & i \end{vmatrix} = \begin{vmatrix} a & b & c \\ d & e & f \\ g & h & i \end{vmatrix}.$$

## The Cross Product as a Determinant

The elements of a determinant are usually numbers, because they have to be multiplied to get the value of the determinant. However, it is possible to use vectors as the elements of *one row* (or column) of a determinant. When expanding in minors about that row (or column) the minor for each vector element is a number that determines the scalar multiple of the vector. The formula for the cross product of

$$\mathbf{u} = u_1\mathbf{i} + u_2\mathbf{j} + u_3\mathbf{k} \quad \text{and} \quad \mathbf{v} = v_1\mathbf{i} + v_2\mathbf{j} + v_3\mathbf{k}$$

presented in Theorem 1 can be expressed symbolically as a determinant with the standard basis vectors as the elements of the first row:

$$\mathbf{u}\times\mathbf{v} = \begin{vmatrix} \mathbf{i} & \mathbf{j} & \mathbf{k} \\ u_1 & u_2 & u_3 \\ v_1 & v_2 & v_3 \end{vmatrix} = \begin{vmatrix} u_2 & u_3 \\ v_2 & v_3 \end{vmatrix}\mathbf{i} - \begin{vmatrix} u_1 & u_3 \\ v_1 & v_3 \end{vmatrix}\mathbf{j} + \begin{vmatrix} u_1 & u_2 \\ v_1 & v_2 \end{vmatrix}\mathbf{k}.$$

The formula for the cross product given in that theorem is just the expansion of this determinant in minors about the first row.

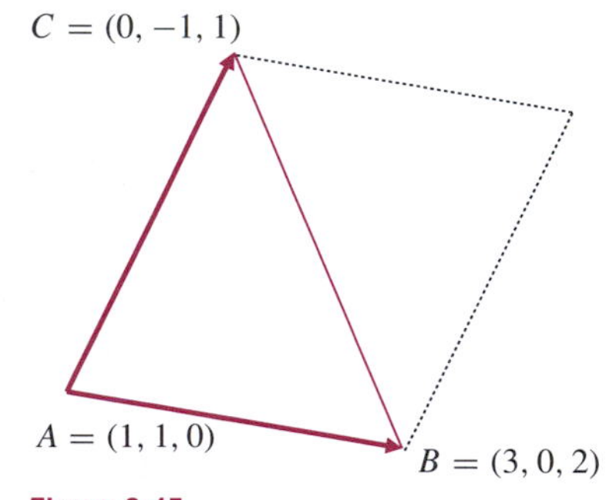

Figure 3.15

**EXAMPLE 3** Find the area of the triangle with vertices at the three points $A = (1, 1, 0)$, $B = (3, 0, 2)$, and $C = (0, -1, 1)$.

***SOLUTION*** Two sides of the triangle (Figure 3.15) are given by the vectors:

$$\overrightarrow{AB} = 2\mathbf{i} - \mathbf{j} + 2\mathbf{k} \qquad \text{and} \qquad \overrightarrow{AC} = -\mathbf{i} - 2\mathbf{j} + \mathbf{k}.$$

The area of the triangle is half the area of the parallelogram spanned by $\overrightarrow{AB}$ and $\overrightarrow{AC}$. By the definition of cross product, the area of the triangle must therefore be

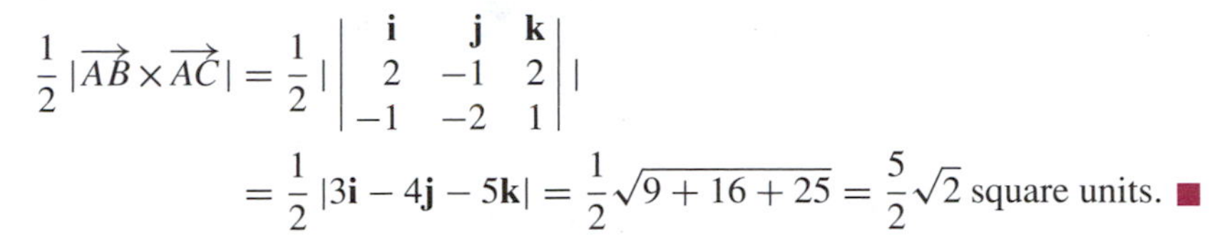

$$\frac{1}{2}|\overrightarrow{AB}\times\overrightarrow{AC}| = \frac{1}{2}\left|\begin{vmatrix} \mathbf{i} & \mathbf{j} & \mathbf{k} \\ 2 & -1 & 2 \\ -1 & -2 & 1 \end{vmatrix}\right|$$

$$= \frac{1}{2}|3\mathbf{i} - 4\mathbf{j} - 5\mathbf{k}| = \frac{1}{2}\sqrt{9+16+25} = \frac{5}{2}\sqrt{2} \text{ square units. } \blacksquare$$

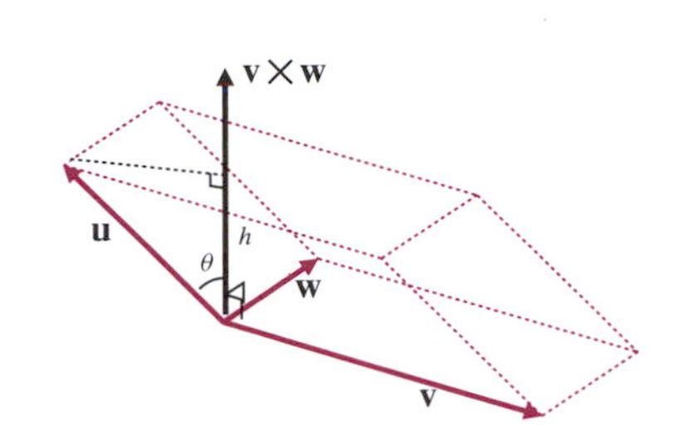


Figure 3.16

A **parallelepiped** is the three-dimensional analogue of a parallelogram. It is a solid with three pairs of parallel planar faces. Each face is in the shape of a parallelogram. A rectangular brick is a special case of a parallelepiped in which nonparallel faces intersect at right angles. We say that a parallelepiped is **spanned** by three vectors coinciding with three of its edges that meet at one vertex. (See Figure 3.16.)

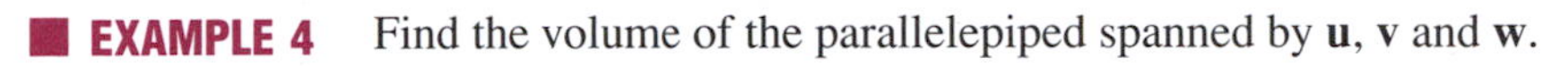

**EXAMPLE 4** Find the volume of the parallelepiped spanned by **u**, **v** and **w**.

***SOLUTION*** The volume of the parallelepiped is equal to the area of one of its faces, say the face spanned by **v** and **w**, multiplied by the height of the parallelepiped measured in a direction perpendicular to that face. The area of the face is $|\mathbf{v}\times\mathbf{w}|$. Since $\mathbf{v}\times\mathbf{w}$ is perpendicular to the face, the height $h$ of the parallelepiped will be the absolute value of the scalar projection of **u** along $\mathbf{v}\times\mathbf{w}$. If $\theta$ is the angle between **u** and $\mathbf{v}\times\mathbf{w}$, then the volume of the parallelepiped is given by

$$\text{Volume} = |\mathbf{u}||\mathbf{v}\times\mathbf{w}||\cos\theta| = |\mathbf{u}\bullet(\mathbf{v}\times\mathbf{w})| \text{ cubic units.} \quad \blacksquare$$

**DEFINITION 5**

The quantity $\mathbf{u}\bullet(\mathbf{v}\times\mathbf{w})$ is called the **scalar triple product** of the vectors **u**, **v**, and **w**.

The scalar triple product is easily expressed in terms of a determinant. If $\mathbf{u} = u_1\mathbf{i} + u_2\mathbf{j} + u_3\mathbf{k}$, and similar representations hold for **v** and **w**, then

$$\mathbf{u}\bullet(\mathbf{v}\times\mathbf{w}) = u_1\begin{vmatrix} v_2 & v_3 \\ w_2 & w_3 \end{vmatrix} - u_2\begin{vmatrix} v_1 & v_3 \\ w_1 & w_3 \end{vmatrix} + u_3\begin{vmatrix} v_1 & v_2 \\ w_1 & w_2 \end{vmatrix}$$

$$= \begin{vmatrix} u_1 & u_2 & u_3 \\ v_1 & v_2 & v_3 \\ w_1 & w_2 & w_3 \end{vmatrix}.$$

The volume of the parallelepiped spanned by **u**, **v**, and **w** is the absolute value of this determinant.

Using the properties of the determinant, it is easily verified that

$$\mathbf{u}\bullet(\mathbf{v}\times\mathbf{w}) = \mathbf{v}\bullet(\mathbf{w}\times\mathbf{u}) = \mathbf{w}\bullet(\mathbf{u}\times\mathbf{v}).$$

(See Exercise 18 below.) Note that **u**, **v**, and **w** remain in the same *cyclic order* in these three expressions. Reversing the order would introduce a factor $-1$:

$$\mathbf{u}\bullet(\mathbf{v}\times\mathbf{w}) = -\mathbf{u}\bullet(\mathbf{w}\times\mathbf{v}).$$

Three vectors in 3-space are said to be **coplanar** if the parallelepiped they span has zero volume; if their tails coincide, three such vectors must line in the same plane.

$$\mathbf{u},\ \mathbf{v},\ \text{and}\ \mathbf{w}\ \text{are coplanar} \iff \mathbf{u} \bullet (\mathbf{v}\times\mathbf{w}) = 0$$
$$\iff \begin{vmatrix} u_1 & u_2 & u_3 \\ v_1 & v_2 & v_3 \\ w_1 & w_2 & w_3 \end{vmatrix} = 0.$$

Three vectors are certainly coplanar if any of them is $\mathbf{0}$, or if any pair of them is parallel. If neither of these degenerate conditions apply, they are only coplanar if any one of them can be expressed as a linear combination of the other two. (See Exercise 20 below.)

## Applications of Cross Products

Cross products are of considerable importance in mechanics and electromagnetic theory, as well as in the study of motion in general. For example:

(a) The linear velocity $\mathbf{v}$ of a particle located at position $\mathbf{r}$ in a body rotating with angular velocity $\boldsymbol{\Omega}$ about the origin is given by $\mathbf{v} = \boldsymbol{\Omega}\times\mathbf{r}$. (See Section 7.1 for more details.)

(b) The angular momentum of a planet of mass $m$ moving with velocity $\mathbf{v}$ in its orbit around the sun is given by $\mathbf{h} = \mathbf{r}\times m\mathbf{v}$, where $\mathbf{r}$ is the position vector of the planet relative to the sun as origin. (See Section 7.4.)

(c) If a particle of electric charge $q$ is travelling with velocity $\mathbf{v}$ through a magnetic field whose strength and direction are given by vector $\mathbf{B}$, then the force that the field exerts on the particle is given by $\mathbf{F} = q\mathbf{v}\times\mathbf{B}$. The electron beam in a television tube is controlled by magnetic fields using this principle.

(d) The torque $\mathbf{T}$ of a force $\mathbf{F}$ applied at the point $P$ with position vector $\mathbf{r}$ about another point $P_0$ with position vector $\mathbf{r}_0$ is defined to be

$$\mathbf{T} = \overrightarrow{P_0P}\times\mathbf{F} = (\mathbf{r} - \mathbf{r}_0)\times\mathbf{F}.$$

This torque measures the effectiveness of the force $\mathbf{F}$ in causing rotation about $P_0$. The direction of $\mathbf{T}$ is along the axis through $P_0$ about which $\mathbf{F}$ acts to rotate $P$.

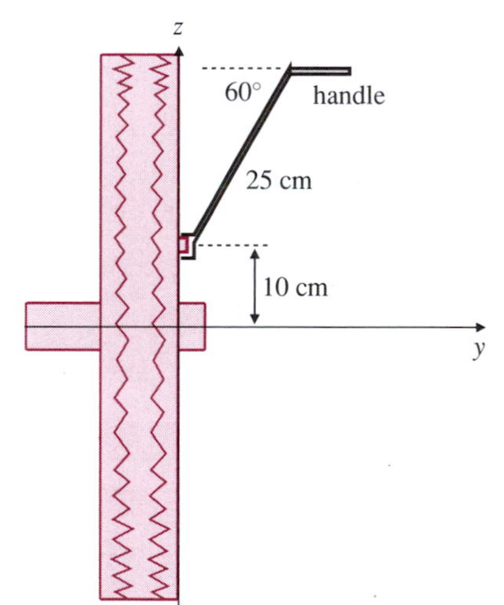

**Figure 3.17** The force on the handle is 500 N in a direction directly towards you

**EXAMPLE 5** An automobile wheel has centre at the origin and axle along the $y$-axis. One of the retaining nuts holding the wheel is at position $P_0 = (0, 0, 10)$. (Distances are measured in centimetres.) A bent tire wrench with arm 25 cm long and inclined at an angle of 60° to the direction of its handle is fitted to the nut in an upright direction, as shown in Figure 3.17. If a horizontal force $\mathbf{F} = 500\mathbf{i}$ newtons is applied to the handle of the wrench, what is its torque on the nut? What part (component) of this torque is effective in trying to rotate the nut about its horizontal axis? What is the effective torque trying to rotate the wheel?

***SOLUTION*** The nut is at position $\mathbf{r}_0 = 10\mathbf{k}$, and the handle of the wrench is at position

$$\mathbf{r} = 25\cos 60^\circ\mathbf{j} + (10 + 25\sin 60^\circ)\mathbf{k} \approx 12.5\mathbf{j} + 31.65\mathbf{k}.$$

The torque of the force $\mathbf{F}$ on the nut is

$$\begin{aligned}\mathbf{T} &= (\mathbf{r} - \mathbf{r}_0)\times\mathbf{F} \\ &\approx (12.5\mathbf{j} + 21.65\mathbf{k})\times 500\mathbf{i} \approx 10{,}825\mathbf{j} - 6{,}250\mathbf{k},\end{aligned}$$

which is at right angles to **F** and to the arm of the wrench. Only the horizontal component of this torque is effective in turning the nut. This component is 10,825 N·cm or 108.24 N·m in magnitude. For the effective torque on the wheel itself, we have to replace $\mathbf{r}_0$ by $\mathbf{0}$, the position of the centre of the wheel. In this case the horizontal torque is

$$31.65\mathbf{k}\times 500\mathbf{i} \approx 15{,}825\mathbf{j},$$

that is, about 158.25 N·m.

## EXERCISES 3.3

1. Calculate $\mathbf{u}\times\mathbf{v}$ if $\mathbf{u} = \mathbf{i} - 2\mathbf{j} + 3\mathbf{k}$ and $\mathbf{v} = 3\mathbf{i} + \mathbf{j} - 4\mathbf{k}$.
2. Calculate $\mathbf{u}\times\mathbf{v}$ if $\mathbf{u} = \mathbf{j} + 2\mathbf{k}$ and $\mathbf{v} = -\mathbf{i} - \mathbf{j} + \mathbf{k}$.
3. Find the area of the triangle with vertices $(1, 2, 0)$, $(1, 0, 2)$, and $(0, 3, 1)$.
4. Find a unit vector perpendicular to the plane containing the points $(a, 0, 0)$, $(0, b, 0)$, and $(0, 0, c)$. What is the area of the triangle with these vertices?
5. Find a unit vector perpendicular to the vectors $\mathbf{i} + \mathbf{j}$ and $\mathbf{j} + 2\mathbf{k}$.
6. Find a unit vector with positive **k** component that is perpendicular to both $2\mathbf{i} - \mathbf{j} - 2\mathbf{k}$ and $2\mathbf{i} - 3\mathbf{j} + \mathbf{k}$.

Verify the identities in Exercises 7–11, either by using the definition of cross product or by using properties of determinants.

7. $\mathbf{u}\times\mathbf{u} = \mathbf{0}$
8. $\mathbf{u}\times\mathbf{v} = -\mathbf{v}\times\mathbf{u}$
9. $(\mathbf{u} + \mathbf{v})\times\mathbf{w} = \mathbf{u}\times\mathbf{w} + \mathbf{v}\times\mathbf{w}$
10. $(t\mathbf{u})\times\mathbf{v} = \mathbf{u}\times(t\mathbf{v}) = t(\mathbf{u}\times\mathbf{v})$
11. $\mathbf{u}\bullet(\mathbf{u}\times\mathbf{v}) = \mathbf{v}\bullet(\mathbf{u}\times\mathbf{v}) = 0$
12. Obtain the addition formula

$$\sin(\alpha - \beta) = \sin\alpha\cos\beta - \cos\alpha\sin\beta$$

by examining the cross product of the two unit vectors $\mathbf{u} = \cos\beta\mathbf{i} + \sin\beta\mathbf{j}$ and $\mathbf{v} = \cos\alpha\mathbf{i} + \sin\alpha\mathbf{j}$. Assume $0 \le \alpha - \beta \le \pi$. (*Hint:* regard **u** and **v** as position vectors. What is the area of the parallelogram they span?)

13. If $\mathbf{u} + \mathbf{v} + \mathbf{w} = \mathbf{0}$ show that $\mathbf{u}\times\mathbf{v} = \mathbf{v}\times\mathbf{w} = \mathbf{w}\times\mathbf{u}$.
14. **(Volume of a tetrahedron)** A **tetrahedron** is a pyramid with a triangular base and three other triangular faces. It has four vertices and six edges. Like any pyramid or cone, its volume is equal to $\frac{1}{3}Ah$, where $A$ is the area of the base and $h$ is the height measured perpendicular to the base. If **u**, **v**, and **w** are vectors coinciding with the three edges of a tetrahedron that meet at one vertex, show that the tetrahedron has volume given by

$$\text{Volume} = \frac{1}{6}|\mathbf{u}\bullet(\mathbf{v}\times\mathbf{w})| = \frac{1}{6}\left|\begin{vmatrix} u_1 & u_2 & u_3 \\ v_1 & v_2 & v_3 \\ w_1 & w_2 & w_3 \end{vmatrix}\right|.$$

Thus the volume of a tetrahedron spanned by three vectors is one-sixth of the volume of the parallelepiped spanned by the same vectors.

15. Find the volume of the tetrahedron with vertices $(1, 0, 0)$, $(1, 2, 0)$, $(2, 2, 2)$, and $(0, 3, 2)$.
16. Find the volume of the parallelepiped spanned by the diagonals of the three faces of a cube of side $a$ that meet at one vertex of the cube.
17. For what value of $k$ do the four points $(1, 1, -1)$, $(0, 3, -2)$, $(-2, 1, 0)$, and $(k, 0, 2)$ all lie on a plane?
18. Verify the identities

$$\mathbf{u}\bullet(\mathbf{v}\times\mathbf{w}) = \mathbf{v}\bullet(\mathbf{w}\times\mathbf{u}) = \mathbf{w}\bullet(\mathbf{u}\times\mathbf{v}).$$

19. If $\mathbf{u}\bullet(\mathbf{v}\times\mathbf{w}) \ne 0$ and **x** is an arbitrary 3-vector, find the numbers $\lambda$, $\mu$, and $\nu$ such that

$$\mathbf{x} = \lambda\mathbf{u} + \mu\mathbf{v} + \nu\mathbf{w}.$$

20. If $\mathbf{u}\bullet(\mathbf{v}\times\mathbf{w}) = 0$ but $\mathbf{v}\times\mathbf{w} \ne \mathbf{0}$, show that there are constants $\lambda$ and $\mu$ such that

$$\mathbf{u} = \lambda\mathbf{v} + \mu\mathbf{w}.$$

(*Hint:* use the result of Exercise 19 with **u** in place of **x** and $\mathbf{v}\times\mathbf{w}$ in place of **u**.)

21. Calculate $\mathbf{u}\times(\mathbf{v}\times\mathbf{w})$ and $(\mathbf{u}\times\mathbf{v})\times\mathbf{w}$, given that $\mathbf{u} = \mathbf{i} + 2\mathbf{j} + 3\mathbf{k}$, $\mathbf{v} = 2\mathbf{i} - 3\mathbf{j}$, and $\mathbf{w} = \mathbf{j} - \mathbf{k}$. Why would you not expect these to be equal?
22. Does the notation $\mathbf{u}\bullet\mathbf{v}\times\mathbf{w}$ make sense? Why? How about the notation $\mathbf{u}\times\mathbf{v}\times\mathbf{w}$?
23. **(The vector triple product)** The product $\mathbf{u}\times(\mathbf{v}\times\mathbf{w})$ is called a **vector triple product**. Since it is perpendicular to $\mathbf{v}\times\mathbf{w}$ it must lie in the plane of **v** and **w**. Show that

$$\mathbf{u}\times(\mathbf{v}\times\mathbf{w}) = (\mathbf{u}\bullet\mathbf{w})\mathbf{v} - (\mathbf{u}\bullet\mathbf{v})\mathbf{w}.$$

(*Hint:* this can be done by direct calculation of the components of both sides of the equation, but the job is much easier if you choose coordinate axes so that **v** lies along the $x$-axis and **w** lies in the $xy$-plane.)

**24.** If $\mathbf{u}$, $\mathbf{v}$, and $\mathbf{w}$ are mutually perpendicular vectors, show that $\mathbf{u}\times(\mathbf{v}\times\mathbf{w}) = \mathbf{0}$. What is $\mathbf{u}\bullet(\mathbf{v}\times\mathbf{w})$ in this case?

**25.** Show that $\mathbf{u}\times(\mathbf{v}\times\mathbf{w}) + \mathbf{v}\times(\mathbf{w}\times\mathbf{u}) + \mathbf{w}\times(\mathbf{u}\times\mathbf{v}) = \mathbf{0}$.

**26.** Find all vectors $\mathbf{x}$ that satisfy the equation

$$(-\mathbf{i}+2\mathbf{j}+3\mathbf{k})\times\mathbf{x} = \mathbf{i}+5\mathbf{j}-3\mathbf{k}.$$

**27.** Show that the equation

$$(-\mathbf{i}+2\mathbf{j}+3\mathbf{k})\times\mathbf{x} = \mathbf{i}+5\mathbf{j}$$

has no solutions for the unknown vector $\mathbf{x}$.

**28.** What condition must be satisfied by the nonzero vectors $\mathbf{a}$ and $\mathbf{b}$ to guarantee that the equation $\mathbf{a}\times\mathbf{x} = \mathbf{b}$ has a solution for $\mathbf{x}$? Is the solution unique?

# 3.4 PLANES AND LINES

A single equation in the three variables, $x$, $y$, and $z$, constitutes a single constraint on the freedom of the point $P = (x, y, z)$ to lie anywhere in 3-space. Such a constraint usually results in the loss of exactly one *degree of freedom* and so forces $P$ to lie on a two-dimensional surface. For example, the equation

$$x^2 + y^2 + z^2 = 4$$

states that the point $(x, y, z)$ is at distance 2 from the origin. All points satisfying this condition lie on a **sphere** (that is, the surface of a ball) of radius 2 centred at the origin. The equation above therefore represents that sphere, and the sphere is the graph of the equation. In this section we will investigate the graphs of linear equations in three variables.

## Planes in 3-Space

Let $P_0 = (x_0, y_0, z_0)$ be a point in $\mathbb{R}^3$ with position vector

$$\mathbf{r}_0 = x_0\mathbf{i} + y_0\mathbf{j} + z_0\mathbf{k}.$$

If $\mathbf{n} = A\mathbf{i} + B\mathbf{j} + C\mathbf{k}$ is any given *nonzero* vector, then there exists exactly one **plane** (flat surface) passing through $P_0$ and perpendicular to $\mathbf{n}$. We say that $\mathbf{n}$ is a **normal vector** to the plane. The plane is the set of all points $P$ for which $\overrightarrow{P_0P}$ is perpendicular to $\mathbf{n}$. (See Figure 3.18.)

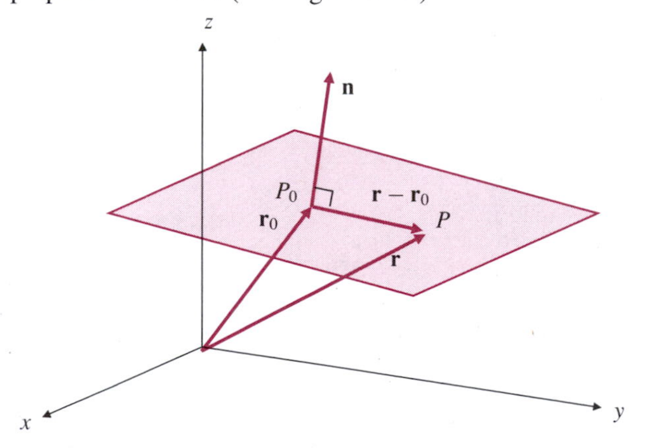


**Figure 3.18** The plane through $P_0$ with normal $\mathbf{n}$ contains all points $P$ for which $\overrightarrow{P_0P}$ is perpendicular to $\mathbf{n}$

If $P = (x, y, z)$ has position vector $\mathbf{r}$, then $\overrightarrow{P_0P} = \mathbf{r} - \mathbf{r}_0$. This vector is perpendicular to $\mathbf{n}$ if and only if $\mathbf{n}\bullet(\mathbf{r}-\mathbf{r}_0) = 0$. This is the equation of the plane in vector form. We can rewrite it in terms of coordinates to obtain the corresponding scalar equation.

**The equations of a plane**

The plane having nonzero normal vector $\mathbf{n} = A\mathbf{i} + B\mathbf{j} + C\mathbf{k}$, and passing through the point $P_0 = (x_0, y_0, z_0)$ with position vector $\mathbf{r}_0$ has equation

$$\mathbf{n} \bullet (\mathbf{r} - \mathbf{r}_0) = 0$$

in vector form, or, equivalently,

$$A(x - x_0) + B(y - y_0) + C(z - z_0) = 0$$

in scalar form

The scalar form can be written more simply as $Ax + By + Cz = D$, where $D = Ax_0 + By_0 + Cz_0$.

If at least one of the constants $A$, $B$, and $C$ is not zero, then the *linear equation* $Ax + By + Cz = D$ always represents a plane in $\mathbb{R}^3$. For example, if $A \neq 0$, it represents the plane through $(D/A, 0, 0)$ with normal vector $\mathbf{n} = A\mathbf{i} + B\mathbf{j} + C\mathbf{k}$. A vector normal to a plane can always be determined from the coefficients of $x$, $y$ and $z$. If the constant term $D = 0$, then the plane must pass through the origin.

**EXAMPLE 1** **Recognizing and writing the equations of planes**

(a) The equation $2x - 3y - 4z = 0$ represents a plane that passes through the origin and is normal (perpendicular) to the vector $\mathbf{n} = 2\mathbf{i} - 3\mathbf{j} - 4\mathbf{k}$.

(b) The plane that passes through the point $(2, 0, 1)$ and is perpendicular to the straight line passing through the points $(1, 1, 0)$ and $(4, -1, -2)$ has normal vector $\mathbf{n} = (4 - 1)\mathbf{i} + (-1 - 1)\mathbf{j} + (-2 - 0)\mathbf{k} = 3\mathbf{i} - 2\mathbf{j} - 2\mathbf{k}$. Therefore its equation is $3(x-2) - 2(y-0) - 2(z-1) = 0$, or, more simply, $3x - 2y - 2z = 4$.

(c) The plane with equation $2x - y = 1$ has a normal $2\mathbf{i} - \mathbf{j}$ that is perpendicular to the $z$-axis. The plane is therefore parallel to the $z$-axis. Note that the equation is independent of $z$. In the $xy$-plane the equation $2x - y = 1$ represents a straight line; in 3-space it represents a plane containing that line and parallel to the $z$-axis. What does the equation $y = z$ represent in $\mathbb{R}^3$? The equation $y = -2$?

(d) The equation $2x + y + 3z = 6$ represents a plane with normal $\mathbf{n} = 2\mathbf{i} + \mathbf{j} + 3\mathbf{k}$. In this case we cannot directly read from the equation the coordinates of a particular point on the plane, but it is not difficult to discover some points. For instance, if we put $y = z = 0$ in the equation we get $x = 3$, and so $(3, 0, 0)$ is a point on the plane. We say that the **$x$-intercept** of the plane is 3 since $(3, 0, 0)$ is the point where the plane intersects the $x$-axis. Similarly, the $y$-intercept is 6 and the $z$-intercept is 2 because the plane intersects the $y$- and $z$-axes at $(0, 6, 0)$ and $(0, 0, 2)$, respectively.

(e) In general, if $a$, $b$, and $c$ are all nonzero, the plane with equation

$$\frac{x}{a} + \frac{y}{b} + \frac{z}{c} = 1$$

has intercepts $a$, $b$, and $c$ on the coordinate axes. (See Figure 3.19.) ■

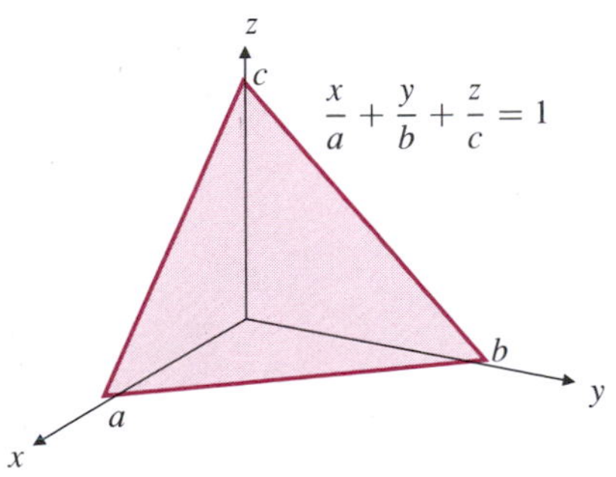

**Figure 3.19** The plane with intercepts $a$, $b$, and $c$ on the coordinate axes

**EXAMPLE 2** Find an equation of the plane that passes through the three points $P = (1, 1, 0)$, $Q = (0, 2, 1)$, and $R = (3, 2, -1)$.

**SOLUTION** We need to find a vector, $\mathbf{n}$, normal to the plane. Such a vector will be perpendicular to the vectors $\overrightarrow{PQ} = -\mathbf{i} + \mathbf{j} + \mathbf{k}$ and $\overrightarrow{PR} = 2\mathbf{i} + \mathbf{j} - \mathbf{k}$. Therefore we can use

$$\mathbf{n} = \overrightarrow{PQ} \times \overrightarrow{PR} = \begin{vmatrix} \mathbf{i} & \mathbf{j} & \mathbf{k} \\ -1 & 1 & 1 \\ 2 & 1 & -1 \end{vmatrix} = -2\mathbf{i} + \mathbf{j} - 3\mathbf{k}.$$

We can use this normal vector together with the coordinates of any one of the three given points to write the equation of the plane. Using point $P$ leads to the equation $-2(x-1) + 1(y-1) - 3(z-0) = 0$, or

$$2x - y + 3z = 1.$$

You can check that using either $Q$ or $R$ leads to the same equation. (If the cross product $\overrightarrow{PQ} \times \overrightarrow{PR}$ had been the zero vector what would have been true about the three points $P$, $Q$, and $R$? Would they have determined a unique plane?) ■

■ **EXAMPLE 3** Show that the two planes

$$x - y = 3 \qquad \text{and} \qquad x + y + z = 0$$

intersect and find a vector, $\mathbf{v}$, parallel to their line of intersection.

**SOLUTION** The two planes have normal vectors

$$\mathbf{n}_1 = \mathbf{i} - \mathbf{j} \qquad \text{and} \qquad \mathbf{n}_2 = \mathbf{i} + \mathbf{j} + \mathbf{k},$$

respectively. Since these vectors are not parallel, the planes are not parallel, and so they intersect in a straight line perpendicular to both $\mathbf{n}_1$ and $\mathbf{n}_2$. This line must therefore be parallel to

$$\mathbf{v} = \mathbf{n}_1 \times \mathbf{n}_2 = \begin{vmatrix} \mathbf{i} & \mathbf{j} & \mathbf{k} \\ 1 & -1 & 0 \\ 1 & 1 & 1 \end{vmatrix} = -\mathbf{i} - \mathbf{j} + 2\mathbf{k}.$$ ■

**Figure 3.20** A pencil of planes

A family of planes intersecting in a straight line is called a **pencil of planes**. (See Figure 3.20.) Such a pencil of planes is determined by any two nonparallel planes in it, since these have a unique line of intersection. If the two nonparallel planes have equations

$$A_1x + B_1y + C_1z = D_1 \qquad \text{and} \qquad A_2x + B_2y + C_2z = D_2$$

then, for any value of the real number $\lambda$, the equation

$$A_1x + B_1y + C_1z - D_1 + \lambda(A_2x + B_2y + C_2z - D_2) = 0$$

represents a plane in the pencil. To see this, observe that the equation is linear, and so represents a plane, and that any point $(x, y, z)$ satisfying the equations of both given planes also satisfies this equation for any value of $\lambda$. Any plane in the pencil except the second defining plane, $A_2x + B_2y + C_2z = D_2$, can be obtained by suitably choosing $\lambda$.

**EXAMPLE 4** Find an equation of the plane passing through the line of intersection of the two planes

$$x + y - 2z = 6 \qquad \text{and} \qquad 2x - y + z = 2$$

and also passing through the point $(-2, 0, 1)$.

***SOLUTION*** For any constant $\lambda$, the equation

$$x + y - 2z - 6 + \lambda(2x - y + z - 2) = 0$$

represents a plane, and is satisfied by the coordinates of all points on the line of intersection of the given planes. This plane passes through the point $(-2, 0, 1)$ if $-2 - 2 - 6 + \lambda(-4 + 1 - 2) = 0$, that is, if $\lambda = -2$. The equation of the required plane therefore simplifies to $3x - 3y + 4z + 2 = 0$. (This solution would not have worked if the given point had been on the second plane, $2x - y + z = 2$. Why?)

## Lines in 3-Space

As we observed above, any two nonparallel planes in $\mathbb{R}^3$ determine a unique (straight) line of intersection, and a vector parallel to this line can be obtained by taking the cross product of normal vectors to the two planes.

Suppose that $\mathbf{r}_0 = x_0\mathbf{i} + y_0\mathbf{j} + z_0\mathbf{k}$ is the position vector of point $P_0$ and $\mathbf{v} = a\mathbf{i} + b\mathbf{j} + c\mathbf{k}$ is a nonzero vector. There is a unique line passing through $P_0$ parallel to $\mathbf{v}$. If $\mathbf{r} = x\mathbf{i} + y\mathbf{j} + z\mathbf{k}$ is the position vector of any other point, $P$, on the line then $\mathbf{r} - \mathbf{r}_0$ lies along the line and so is parallel to $\mathbf{v}$. (See Figure 3.21.) Thus $\mathbf{r} - \mathbf{r}_0 = t\mathbf{v}$ for some real number $t$. This equation, usually rewritten in the form

$$\mathbf{r} = \mathbf{r}_0 + t\mathbf{v},$$

is called the **vector parametric equation of the straight line**. All points on the line can be obtained as $t$ ranges from $-\infty$ to $\infty$. The vector $\mathbf{v}$ is called a **direction vector** of the line.

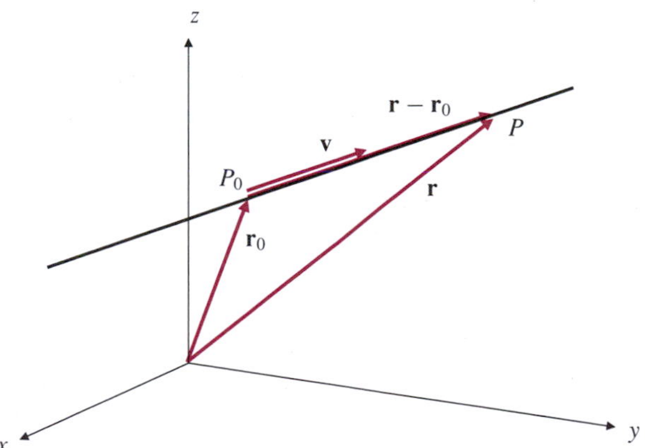


**Figure 3.21** The line through $P_0$ parallel to $\mathbf{v}$

Breaking the vector parametric equation down into its components yields the **scalar parametric** equations of the line:

$$\begin{cases} x = x_0 + at \\ y = y_0 + bt \\ z = z_0 + ct. \end{cases} \qquad (-\infty < t < \infty)$$

These appear to be *three* linear equations, but the parameter $t$ can be eliminated to give *two* linear equations in $x$, $y$, and $z$. If $a \neq 0$, $b \neq 0$, and $c \neq 0$, then we can solve each of the scalar equations for $t$, and so obtain

$$\frac{x - x_0}{a} = \frac{y - y_0}{b} = \frac{z - z_0}{c},$$

which is called the **standard form** for the equations of the straight line through $(x_0, y_0, z_0)$ parallel to $\mathbf{v}$. The standard form must be modified if any component of $\mathbf{v}$ vanishes. For example, if $c = 0$ the equations are

$$\frac{x - x_0}{a} = \frac{y - y_0}{b}, \qquad z = z_0.$$

Note that none of above equations for straight lines is unique; each depends on the particular choice of the point $(x_0, y_0, z_0)$ on the line. In general, you can always use the equations of two nonparallel planes to represent their line of intersection.

**EXAMPLE 5** **Equations of straight lines**

(a) The equations

$$\begin{cases} x = 2 + t \\ y = 3 \\ z = -4t \end{cases}$$

represent the straight line through $(2, 3, 0)$ parallel to the vector $\mathbf{i} - 4\mathbf{k}$.

(b) The straight line through $(1, -2, 3)$ perpendicular to the plane $x - 2y + 4z = 5$ is parallel to the normal vector $\mathbf{i} - 2\mathbf{j} + 4\mathbf{k}$ of the plane. Therefore the line has vector parametric equation

$$\mathbf{r} = \mathbf{i} - 2\mathbf{j} + 3\mathbf{k} + t(\mathbf{i} - 2\mathbf{j} + 4\mathbf{k}),$$

or scalar parametric equations

$$\begin{cases} x = 1 + t \\ y = -2 - 2t \\ z = 3 + 4t \end{cases}$$

Its standard form equations are

$$\frac{x - 1}{1} = \frac{y + 2}{-2} = \frac{z - 3}{4}.$$

■

**EXAMPLE 6** Find a direction vector for the line of intersection of the two planes

$$x + y - z = 0 \qquad \text{and} \qquad y + 2z = 6,$$

and find a set of equations for the line in standard form.

***SOLUTION*** The two planes have respective normals $\mathbf{n}_1 = \mathbf{i}+\mathbf{j}-\mathbf{k}$ and $\mathbf{n}_2 = \mathbf{j}+2\mathbf{k}$. Thus a direction vector of their line of intersection is

$$\mathbf{v} = \mathbf{n}_1 \times \mathbf{n}_2 = 3\mathbf{i} - 2\mathbf{j} + \mathbf{k}.$$

We need to know one point on the line in order to write equations in standard form. We can find a point by assigning a value to one coordinate and calculating the other two from the given equations. For instance, taking $z = 0$ in the two equations we are led to $y = 6$ and $x = -6$, so $(-6, 6, 0)$ is one point on the line. Thus the line has standard form equations

$$\frac{x+6}{3} = \frac{y-6}{-2} = z.$$

This answer is not unique; the coordinates of any other point on the line could be used in place of $(-6, 6, 0)$. You could even find a direction vector $\mathbf{v}$ by subtracting the position vectors of two different points on the line. ■

## Distances

The **distance** between two geometric objects always means the minimum distance between two points, one in each object. In the case of *flat* objects like lines or planes defined by linear equations, such minimum distances can usually be determined by geometric arguments without having to use calculus.

**■ EXAMPLE 7 Distance from a point to a plane**

(a) Find the distance from the point $P_0 = (x_0, y_0, z_0)$ to the plane $\mathcal{P}$ having equation $Ax + By + Cz = D$.

(b) What is the distance from $(2, -1, 3)$ to the plane $2x - 2y - z = 9$?

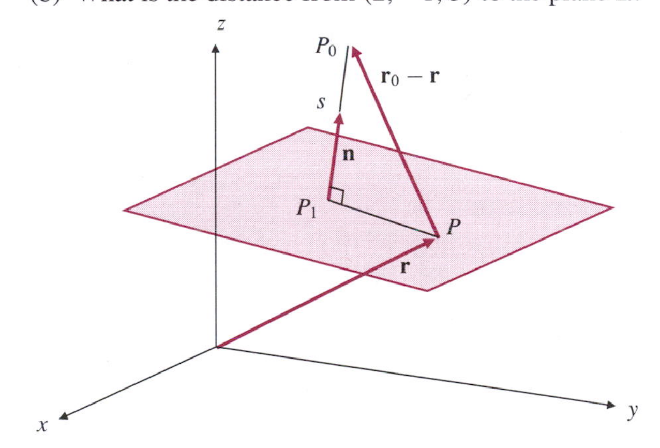


**Figure 3.22** The distance from $P_0$ to the plane $\mathcal{P}$ is the length of the vector projection of $PP_0$ along the normal $\mathbf{n}$ to $\mathcal{P}$, where $P$ is any point on $\mathcal{P}$

***SOLUTION***

(a) Let $\mathbf{r}_0$ be the position vector of $P_0$ and let $\mathbf{n} = A\mathbf{i} + B\mathbf{j} + C\mathbf{k}$, be the normal to $\mathcal{P}$. Let $P_1$ be the point on $\mathcal{P}$ that is closest to $P_0$. Then $PP_0$ is perpendicular to $\mathcal{P}$, and so is parallel to $\mathbf{n}$. The distance from $P_0$ to $\mathcal{P}$ is $s = |\overrightarrow{P_1P_0}|$. If $P$, having position vector $\mathbf{r}$, is any point on $\mathcal{P}$ then $s$ is the length of the projection of $\overrightarrow{PP_0} = \mathbf{r}_0 - \mathbf{r}$ in the direction of $\mathbf{n}$. (See Figure 3.22.) Thus

$$s = \left| \frac{\overrightarrow{PP_0} \bullet \mathbf{n}}{|\mathbf{n}|} \right| = \frac{|(\mathbf{r}_0 - \mathbf{r}) \bullet \mathbf{n}|}{|\mathbf{n}|} = \frac{|\mathbf{r}_0 \bullet \mathbf{n} - \mathbf{r} \bullet \mathbf{n}|}{|\mathbf{n}|}.$$

Since $P = (x, y, z)$ lies on $\mathcal{P}$ we have $\mathbf{r} \bullet \mathbf{n} = Ax + By + Cz = D$. In terms of the coordinates $(x_0, y_0, z_0)$ of $P_0$, we can therefore represent the distance from $P_0$ to $\mathcal{P}$ as

$$s = \frac{|Ax_0 + By_0 + Cz_0 - D|}{\sqrt{A^2 + B^2 + C^2}}.$$

(b) The distance from $(2, -1, 3)$ to the plane $2x - 2y - z = 9$ is

$$s = \frac{|2(2) - 2(-1) - 1(3) - 9|}{\sqrt{2^2 + (-2)^2 + (-1)^2}} = \frac{|-6|}{3} = 2 \text{ units.}$$

**EXAMPLE 8** **Distance from a point to a line**

(a) Find the distance from the point $P_0$ to the straight line $\mathcal{L}$ through $P_1$ parallel to the nonzero vector $\mathbf{v}$.

(b) What is the distance from $(2, 0, -3)$ to the line $\mathbf{r} = \mathbf{i} + (1 + 3t)\mathbf{j} - (3 - 4t)\mathbf{k}$?

**SOLUTION**

(a) Let $\mathbf{r}_0$ and $\mathbf{r}_1$ be the position vectors of $P_0$ and $P_1$ respectively. The point $P_2$ on $\mathcal{L}$ that is closest to $P_0$ is such that $P_2P_0$ is perpendicular to $\mathcal{L}$. The distance from $P_0$ to $\mathcal{L}$ is

$$s = |P_2P_0| = |P_1P_0| \sin\theta = |\mathbf{r}_0 - \mathbf{r}_1| \sin\theta,$$

where $\theta$ is the angle between $\mathbf{r}_0 - \mathbf{r}_1$ and $\mathbf{v}$. (See Figure 3.23(a).) Since

$$|(\mathbf{r}_0 - \mathbf{r}_1) \times \mathbf{v}| = |\mathbf{r}_0 - \mathbf{r}_1|\,|\mathbf{v}|\,\sin\theta,$$

we have

$$s = \frac{|(\mathbf{r}_0 - \mathbf{r}_1) \times \mathbf{v}|}{|\mathbf{v}|}.$$

(b) The line $\mathbf{r} = \mathbf{i} + (1 + 3t)\mathbf{j} - (3 - 4t)\mathbf{k}$ passes through $P_1 = (1, 1, -3)$ and is parallel to $\mathbf{v} = 3\mathbf{j} + 4\mathbf{k}$. The distance from $P_0 = (2, 0, -3)$ to this line is

$$\begin{aligned} s &= \frac{\big|\big((2-1)\mathbf{i} + (0-1)\mathbf{j} + (-3+3)\mathbf{k}\big) \times (3\mathbf{j} + 4\mathbf{k})\big|}{\sqrt{3^2 + 4^2}} \\ &= \frac{|(\mathbf{i} - \mathbf{j}) \times (3\mathbf{j} + 4\mathbf{k})|}{5} = \frac{|-4\mathbf{i} - 4\mathbf{j} + 3\mathbf{k}|}{5} = \frac{\sqrt{41}}{5} \text{ units.} \end{aligned}$$

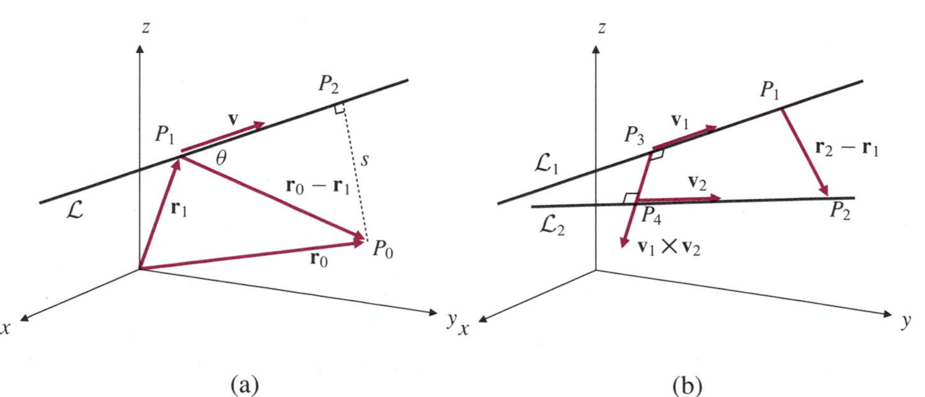


**Figure 3.23**

(a) The distance from $P_0$ to the line $\mathcal{L}$ is $s = |P_0P_1| \sin\theta$

(b) The distance between lines is the length of the projection of $P_1P_2$ along the vector $\mathbf{v}_1 \times \mathbf{v}_2$

**EXAMPLE 9** **The distance between two lines**

Find the distance between the two lines $\mathcal{L}_1$ through point $P_1$ parallel to vector $\mathbf{v}_1$ and $\mathcal{L}_2$ through point $P_2$ parallel to vector $\mathbf{v}_2$.

***SOLUTION*** Let $\mathbf{r}_1$ and $\mathbf{r}_2$ be the position vectors of points $P_1$ and $P_2$, respectively. If $P_3$ and $P_4$ (with position vectors $\mathbf{r}_3$ and $\mathbf{r}_4$) are the points on $\mathcal{L}_1$ and $\mathcal{L}_2$, respectively, that are closest to one another, then $\overrightarrow{P_3P_4}$ is perpendicular to both lines, and is therefore parallel to $\mathbf{v}_1 \times \mathbf{v}_2$. (See Figure 3.23(b).) $\overrightarrow{P_3P_4}$ is the vector projection of $\overrightarrow{P_1P_2} = \mathbf{r}_2 - \mathbf{r}_1$ along $\mathbf{v}_1 \times \mathbf{v}_2$. Therefore, the distance $s = |\overrightarrow{P_3P_4}|$ between the lines is given by

$$s = |\mathbf{r}_4 - \mathbf{r}_3| = \frac{|(\mathbf{r}_2 - \mathbf{r}_1) \bullet (\mathbf{v}_1 \times \mathbf{v}_2)|}{|\mathbf{v}_1 \times \mathbf{v}_2|}.$$

## EXERCISES 3.4

**1.** A single equation involving the coordinates $(x, y, z)$ need not always represent a two-dimensional "surface" in $\mathbb{R}^3$. For example, $x^2 + y^2 + z^2 = 0$ represents the single point $(0, 0, 0)$, which has dimension zero. Give examples of single equations in $x$, $y$, and $z$ that represent

(a) a (one-dimensional) straight line,

(b) the whole of $\mathbb{R}^3$,

(c) no points at all (that is, the empty set).

In Exercises 2–9, find equations of the planes satisfying the given conditions.

**2.** Passing through $(0, 2, -3)$ and normal to the vector $4\mathbf{i} - \mathbf{j} - 2\mathbf{k}$

**3.** Passing through the origin and having normal $\mathbf{i} - \mathbf{j} + 2\mathbf{k}$

**4.** Passing through $(1, 2, 3)$ and parallel to the plane $3x + y - 2z = 15$

**5.** Passing through the three points $(1, 1, 0)$, $(2, 0, 2)$, and $(0, 3, 3)$

**6.** Passing through the three points $(-2, 0, 0)$, $(0, 3, 0)$, and $(0, 0, 4)$

**7.** Passing through $(1, 1, 1)$ and $(2, 0, 3)$ and perpendicular to the plane $x + 2y - 3z = 0$

**8.** Passing through the line of intersection of the planes $2x + 3y - z = 0$ and $x - 4y + 2z = -5$, and passing through the point $(-2, 0, -1)$

**9.** Passing through the line $x + y = 2$, $y - z = 3$, and perpendicular to the plane $2x + 3y + 4z = 5$

**10.** Under what geometric condition will three distinct points in $\mathbb{R}^3$ not determine a unique plane passing through them? How can this condition be expressed algebraically in terms of the position vectors, $\mathbf{r}_1$, $\mathbf{r}_2$, and $\mathbf{r}_3$, of the three points?

**11.** Give a condition on the position vectors of four points that guarantees that the four points are *coplanar*, that is, all lie on one plane.

Describe geometrically the one-parameter families of planes in Exercises 12–14. ($\lambda$ is a real parameter.)

**12.** $x + y + z = \lambda$.

* **13.** $x + \lambda y + \lambda z = \lambda$.

* **14.** $\lambda x + \sqrt{1 - \lambda^2}\, y = 1$.

In Exercises 15–19, find equations of the line specified in vector and scalar parametric forms and in standard form.

**15.** Through the point $(1, 2, 3)$ and parallel to $2\mathbf{i} - 3\mathbf{j} - 4\mathbf{k}$

**16.** Through $(-1, 0, 1)$ and perpendicular to the plane $2x - y + 7z = 12$

**17.** Through the origin and parallel to the line

$$x + 2y - z = 2, \qquad 2x - y + 4z = 5$$

**18.** Through $(2, -1, -1)$ and parallel to each of the two planes $x + y = 0$ and $x - y + 2z = 0$

**19.** Through $(1, 2, -1)$ and making equal angles with the positive directions of the coordinate axes

In Exercises 20–22, find the equations of the given line in standard form.

**20.** $\mathbf{r} = (1 - 2t)\mathbf{i} + (4 + 3t)\mathbf{j} + (9 - 4t)\mathbf{k}$.

**21.** $\begin{cases} x = 4 - 5t \\ y = 3t \\ z = 7 \end{cases}$

**22.** $\begin{cases} x - 2y + 3z = 0 \\ 2x + 3y - 4z = 4 \end{cases}$

**23.** Under what conditions on the position vectors of four distinct points $P_1$, $P_2$, $P_3$, and $P_4$ will the straight line through $P_1$ and $P_2$ intersect the straight line through $P_3$ and $P_4$ at a unique point?

Find the required distances in Exercises 24–27.

**24.** From the origin to the plane $x + 2y + 3z = 4$

**25.** From $(1, 2, 0)$ to the plane $3x - 4y - 5z = 2$

**26.** From the origin to the line $x + y + z = 0$, $2x - y - 5z = 1$

**27.** Between the lines

$$\begin{cases} x + 2y = 3 \\ y + 2z = 3 \end{cases} \quad \text{and} \quad \begin{cases} x + y + z = 6 \\ x - 2z = -5 \end{cases}$$

**28.** Show that the line

$$x - 2 = \frac{y+3}{2} = \frac{z-1}{4}$$

is parallel to the plane $2y - z = 1$. What is the distance between the line and the plane?

In Exercises 29–30, describe the one-parameter families of straight lines represented by the given equations. ($\lambda$ is a real parameter.)

* **29.** $(1 - \lambda)(x - x_0) = \lambda(y - y_0)$, $z = z_0$.

* **30.** $\dfrac{x - x_0}{\sqrt{1 - \lambda^2}} = \dfrac{y - y_0}{\lambda} = z - z_0$.

**31.** Why does the factored second-degree equation

$$(A_1x + B_1y + C_1z - D_1)(A_2x + B_2y + C_2z - D_2) = 0$$

represent a pair of planes rather than a single straight line?

## 3.5 QUADRIC SURFACES

The most general second-degree equation in three variables is

$$Ax^2 + By^2 + Cz^2 + Dxy + Exz + Fyz + Gx + Hy + Iz = J.$$

We will not attempt the (rather difficult) task of classifying all the surfaces that can be represented by such an equation, but will examine some interesting special cases. Let us observe at the outset that if the above equation can be factored in the form

$$(A_1x + B_1y + C_1z - D_1)(A_2x + B_2y + C_2z - D_2) = 0$$

then the graph is, in fact, a pair of planes,

$$A_1x + B_1y + C_1z = D_1 \qquad \text{and} \qquad A_2x + B_2y + C_2z = D_2,$$

or one plane if the two linear equations represent the same plane. This is considered a degenerate case. Where such factorization is not possible, the surface (called a **quadric surface**) will not be flat, although there may still be straight lines that lie on the surface. Nondegenerate quadric surfaces fall into the following six categories.

**Spheres.** The equation $x^2 + y^2 + z^2 = a^2$ represents a sphere of radius $a$ centred at the origin. More generally,

$$(x - x_0)^2 + (y - y_0)^2 + (z - z_0)^2 = a^2$$

represents a sphere of radius $a$ centred at the point $(x_0, y_0, z_0)$. If a quadratic equation in $x$, $y$, and $z$ has equal coefficients for the $x^2$, $y^2$, and $z^2$ terms and has no other second-degree terms, then it will represent, if any surface at all, a sphere. The centre can be found by completing the squares as for circles in the plane.

**Cylinders.** The equation $x^2 + y^2 = a^2$, being independent of $z$, represents a **right-circular cylinder** of radius $a$ and axis along the $z$-axis. (See Figure 3.24(a).) The intersection of the cylinder with the horizontal plane $z = k$ is the circle with equations

$$\begin{cases} x^2 + y^2 = a^2 \\ z = k. \end{cases}$$

Quadric cylinders also come in other shapes — elliptic, parabolic and hyperbolic. For instance, $z = x^2$ represents a parabolic cylinder with vertex line along the $y$-axis. (See Figure 3.24(b).) In general, an equation in two variables only will represent a cylinder in 3-space.

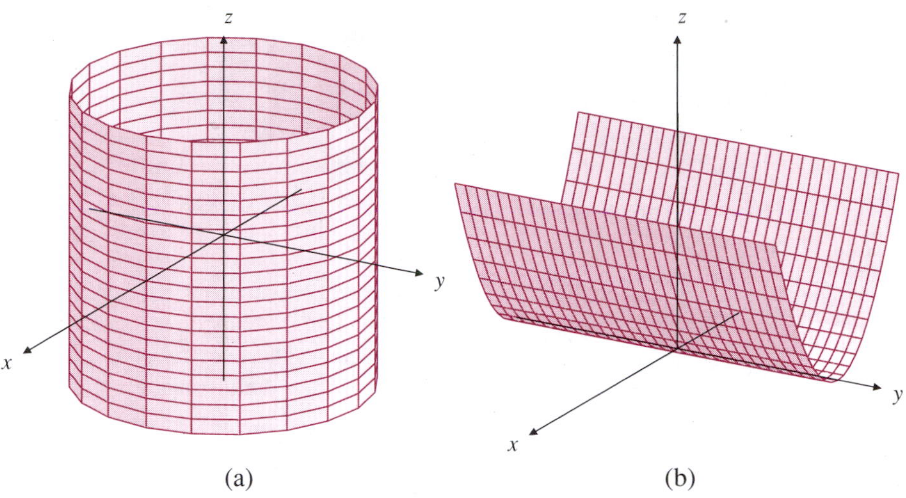


**Figure 3.24**
(a) The circular cylinder $x^2 + y^2 = a^2$
(b) The parabolic cylinder $z = x^2$

**Cones.** The equation $z^2 = x^2 + y^2$ represents a **right-circular cone** with axis along the $z$-axis. The surface is generated by rotating about the $z$-axis the line $z = y$ in the $yz$-plane. This *generator* makes an angle of 45° with the axis of the cone. Cross-sections of the cone in planes parallel to the $xy$-plane are circles. (See Figure 3.25(a).) The equation $x^2 + y^2 = a^2z^2$ also represents a right circular cone with vertex at the origin and axis along the $z$-axis, but having semi-vertical angle $\alpha = \tan^{-1} a$. A circular cone has plane cross-sections that are elliptical, parabolic, and hyperbolic. Conversely, any nondegenerate quadric cone has a direction perpendicular to which the cross-sections of the cone are circular. In that sense, every quadric cone is a circular cone, though it may be *oblique* rather than right-circular in that the line joining the centres of the circular cross-sections need not be perpendicular to those cross-sections. (See Exercise 24 at the end of this section.)

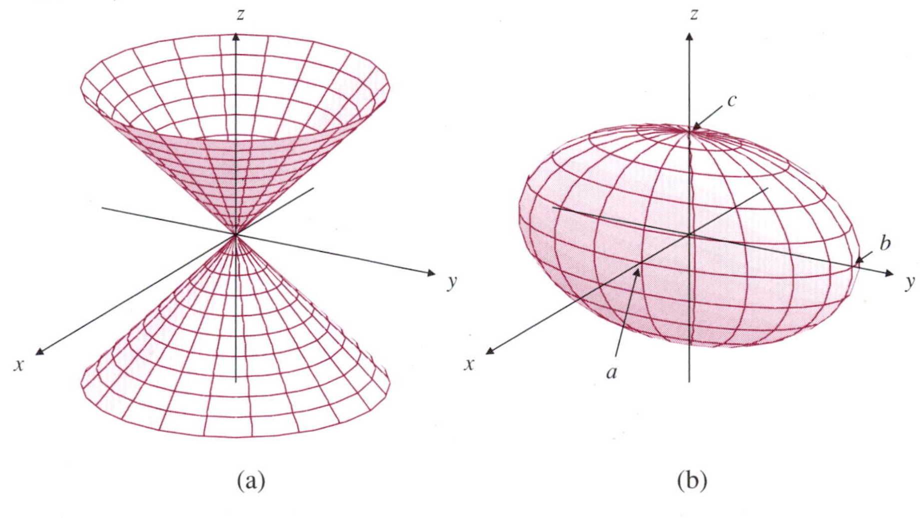


**Figure 3.25**
(a) The circular cone $a^2z^2 = x^2 + y^2$
(b) The ellipsoid $\dfrac{x^2}{a^2} + \dfrac{y^2}{b^2} + \dfrac{z^2}{c^2} = 1$

**Ellipsoids.** The equation

$$\frac{x^2}{a^2} + \frac{y^2}{b^2} + \frac{z^2}{c^2} = 1$$

represents an **ellipsoid** with *semi-axes* $a$, $b$, and $c$. (See Figure 3.25(b).) The surface is oval, and it is enclosed inside the rectangular parallelepiped $-a \le x \le a$, $-b \le y \le b$, $-c \le z \le c$. If $a = b = c$ the ellipsoid is a sphere. In general, all plane cross-sections of ellipsoids are ellipses. This is easy to see for cross-sections parallel to coordinate planes, but somewhat harder to see for other planes.

**Paraboloids.** The equations

$$z = \frac{x^2}{a^2} + \frac{y^2}{b^2} \qquad \text{and} \qquad z = \frac{x^2}{a^2} - \frac{y^2}{b^2}$$

represent, respectively, an **elliptic paraboloid** and a **hyperbolic paraboloid**. (See Figures 3.26(a) and (b).) Cross-sections in planes $z = k$ ($k$ a positive constant) are ellipses (circles if $a = b$) and hyperbolas, respectively. Parabolic reflective mirrors have the shape of circular paraboloids. The hyperbolic paraboloid is a **ruled surface**. (A ruled surface is one through every point of which there passes a straight line lying wholly on the surface. Cones and cylinders are also examples of ruled surfaces.) There are two one-parameter families of straight lines that lie on the hyperbolic paraboloid, namely

$$\begin{cases} \lambda z = \dfrac{x}{a} - \dfrac{y}{b} \\ \dfrac{1}{\lambda} = \dfrac{x}{a} + \dfrac{y}{b} \end{cases} \qquad \text{and} \qquad \begin{cases} \mu z = \dfrac{x}{a} + \dfrac{y}{b} \\ \dfrac{1}{\mu} = \dfrac{x}{a} - \dfrac{y}{b} \end{cases}$$

where $\lambda$ and $\mu$ are real parameters. Every point on the hyperbolic paraboloid lies on one line of each family.

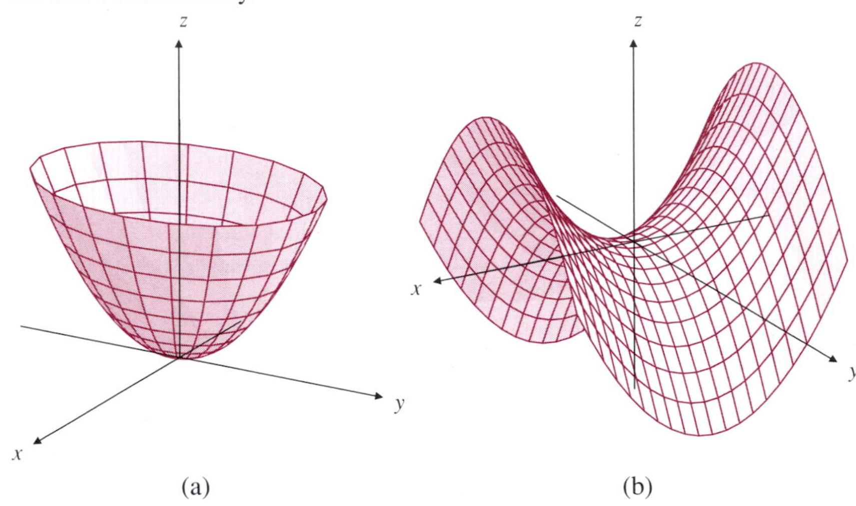


**Figure 3.26**
(a) The elliptic paraboloid $z = \dfrac{x^2}{a^2} + \dfrac{y^2}{b^2}$
(b) The hyperbolic paraboloid $z = \dfrac{x^2}{a^2} - \dfrac{y^2}{b^2}$

**Hyperboloids.** The equation

$$\frac{x^2}{a^2} + \frac{y^2}{b^2} - \frac{z^2}{c^2} = 1$$

represents a surface called a **hyperboloid of one sheet**. (See Figure 3.27(a).) The equation

$$\frac{x^2}{a^2} + \frac{y^2}{b^2} - \frac{z^2}{c^2} = -1$$

represents a **hyperboloid of two sheets**. (See Figure 3.27(b).) Both surfaces have elliptical cross-sections in horizontal planes and hyperbolic cross-sections in vertical planes. Both are *asymptotic* to the elliptic cone with equation

$$\frac{x^2}{a^2} + \frac{y^2}{b^2} = \frac{z^2}{c^2};$$

they approach arbitrarily close to the cone as they recede arbitrarily far away from the origin. Like the hyperbolic paraboloid, the hyperboloid of one sheet is a ruled surface.

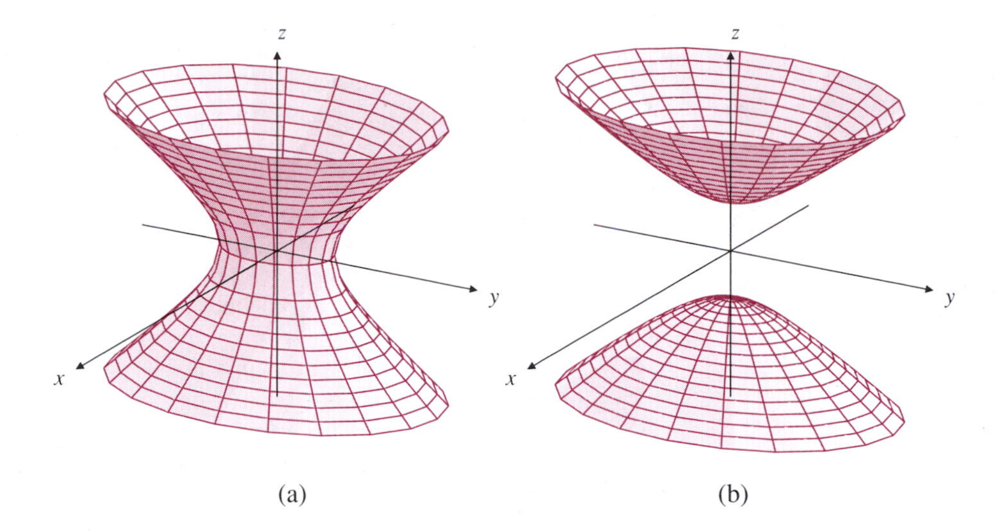


**Figure 3.27**

(a) The hyperboloid of one sheet
$$\frac{x^2}{a^2} + \frac{y^2}{b^2} - \frac{z^2}{c^2} = 1$$
(b) The hyperboloid of two sheets
$$\frac{x^2}{a^2} + \frac{y^2}{b^2} - \frac{z^2}{c^2} = -1$$

## EXERCISES 3.5

Identify the surfaces represented by the equations in Exercises 1–15 and sketch their graphs.

**1.** $x^2 + 4y^2 + 9z^2 = 36$

**2.** $x^2 + y^2 + 4z^2 = 4$

**3.** $2x^2 + 2y^2 + 2z^2 - 4x + 8y - 12z + 27 = 0$

**4.** $x^2 + 4y^2 + 9z^2 + 4x - 8y = 8$

**5.** $z = x^2 + 2y^2$

**6.** $z = x^2 - 2y^2$

**7.** $x^2 - y^2 - z^2 = 4$

**8.** $-x^2 + y^2 + z^2 = 4$

**9.** $z = xy$

**10.** $x^2 + 4z^2 = 4$

**11.** $x^2 - 4z^2 = 4$

**12.** $y = z^2$

**13.** $x = z^2 + z$

**14.** $x^2 = y^2 + 2z^2$

**15.** $(z - 1)^2 = (x - 2)^2 + (y - 3)^2$

**16.** $(z - 1)^2 = (x - 2)^2 + (y - 3)^2 + 4$

Describe and sketch the geometric objects represented by the systems of equations in Exercises 17–20.

**17.** $\begin{cases} x^2 + y^2 + z^2 = 4 \\ x + y + z = 1 \end{cases}$

**18.** $\begin{cases} x^2 + y^2 = 1 \\ z = x + y \end{cases}$

**19.** $\begin{cases} z^2 = x^2 + y^2 \\ z = 1 + x \end{cases}$

**20.** $\begin{cases} x^2 + 2y^2 + 3z^2 = 6 \\ y = 1 \end{cases}$

**21.** Find two one-parameter families of straight lines that lie on the hyperboloid of one sheet

$$\frac{x^2}{a^2} + \frac{y^2}{b^2} - \frac{z^2}{c^2} = 1.$$

**22.** Find two one-parameter families of straight lines that lie on the hyperbolic paraboloid $z = xy$.

**23.** The equation $2x^2 + y^2 = 1$ represents a cylinder with elliptical cross-sections in planes perpendicular to the $z$-axis. Find a vector **a** perpendicular to which the cylinder has circular cross-sections.

* **24.** The equation $z^2 = 2x^2 + y^2$ represents a cone with elliptical cross-sections in planes perpendicular to the $z$-axis. Find a vector **a** perpendicular to which the cone has circular cross-sections. (*Hint:* do Exercise 23 first, and use its result.)

## 3.6 A LITTLE MATRIX ALGEBRA

Differential calculus is essentially the study of linear approximations to functions. The tangent line to the graph $y = f(x)$ at $x = x_0$ provides the "best linear approximation" to $f(x)$ near $x_0$. Differentiation of functions of several variables can also be viewed as a process of finding *best linear approximations*. Therefore the language of linear algebra can be very useful for expressing certain concepts in the calculus of several variables.

Linear algebra is a vast subject, and is usually studied independently of calculus. This is unfortunate, because understanding the relationship between the two subjects can greatly enhance your understanding and appreciation of each of them. Knowledge of linear algebra, and therefore familiarity with the material covered in this section, is *not essential* for fruitful study of the rest of this book. However, we shall from time to time comment on the significance of the subject at hand from the point of view of linear algebra. To this end we need only a little of the terminology and content of linear algebra, especially that part pertaining to matrix manipulation and systems of linear equations. In the rest of this section we present an outline of this material. Some students will already be familiar with it; others will encounter it later. We make no attempt at completeness here, and refer interested students to standard linear algebra texts for proofs of some assertions. Students proceeding beyond this book to further study of advanced calculus and differential equations will certainly need a much more extensive background in linear algebra.

## Matrices

An $m \times n$ **matrix** $\mathcal{A}$ is a rectangular array of $mn$ numbers arranged in $m$ rows and $n$ columns. If $a_{ij}$ is the element in the $i$th row and the $j$th column, then

$$\mathcal{A} = \begin{pmatrix} a_{11} & a_{12} & \cdots & a_{1n} \\ a_{21} & a_{22} & \cdots & a_{2n} \\ \vdots & \vdots & & \vdots \\ a_{m1} & a_{m2} & \cdots & a_{mn} \end{pmatrix}.$$

Sometimes, as a shorthand notation, we write $\mathcal{A} = (a_{ij})$. In this case $i$ is assumed to range from 1 to $m$ and $j$ from 1 to $n$. If $m = n$ we say that $\mathcal{A}$ is a square matrix. The elements $a_{ij}$ of the matrices we use in this book will always be real numbers.

The **transpose** of an $m \times n$ matrix $\mathcal{A}$ is the $n \times m$ matrix $\mathcal{A}^T$ whose rows are the columns of $\mathcal{A}$:

$$\mathcal{A}^T = \begin{pmatrix} a_{11} & a_{21} & \cdots & a_{m1} \\ a_{12} & a_{22} & \cdots & a_{m2} \\ \vdots & \vdots & & \vdots \\ a_{1n} & a_{2n} & \cdots & a_{mn} \end{pmatrix}.$$

Matrix $\mathcal{A}$ is called **symmetric** if $\mathcal{A}^T = \mathcal{A}$. Symmetric matrices are necessarily square. Observe that $(\mathcal{A}^T)^T = \mathcal{A}$ for *every* matrix $\mathcal{A}$. Frequently we want to consider an $n$-vector $\mathbf{x}$ as an $n \times 1$ matrix having $n$ rows and one column:

$$\mathbf{x} = \begin{pmatrix} x_1 \\ x_2 \\ \vdots \\ x_n \end{pmatrix}.$$

As such, $\mathbf{x}$ is called a **column vector**. $\mathbf{x}^T$ has one row and $n$ columns, and so is called a **row vector**:

$$\mathbf{x}^T = (x_1\ x_2\ \cdots\ x_n).$$

Most of the usefulness of matrices depends on the following definition of matrix multiplication, which enables two arrays to be combined into a single one in a manner that preserves linear relationships.

**DEFINITION 6**

**Multiplying matrices**

If $\mathcal{A} = (a_{ij})$ is an $m \times n$ matrix and $\mathcal{B} = (b_{ij})$ is an $n \times p$ matrix, then the product $\mathcal{AB}$ is the $m \times p$ matrix $\mathcal{C} = (c_{ij})$ with elements given by

$$c_{ij} = \sum_{k=1}^{n} a_{ik} b_{kj}, \qquad i = 1, \ldots, m, \qquad j = 1, \ldots, p.$$

That is, $c_{ij}$ is the *dot product* of the $i$th row of $\mathcal{A}$ and the $j$th column of $\mathcal{B}$ (both of which are $n$-vectors).

Note that only *some* pairs of matrices can be multiplied. The product $\mathcal{AB}$ is only defined if the number of columns of $\mathcal{A}$ is equal to the number of rows of $\mathcal{B}$.

## EXAMPLE 1

$$\begin{pmatrix} 1 & 0 & 3 \\ 2 & 1 & -1 \end{pmatrix} \begin{pmatrix} 2 & 1 & 1 & 0 \\ 0 & -1 & 3 & 1 \\ 1 & 0 & 4 & 5 \end{pmatrix} = \begin{pmatrix} 5 & 1 & 13 & 15 \\ 3 & 1 & 1 & -4 \end{pmatrix}$$

The left factor has 2 rows and 3 columns, and the right factor has 3 rows and 4 columns. Therefore the product has 2 rows and 4 columns. The element in the first row and third column of the product, 13, is the dot product of the first row, (1,0,3), of the left factor and the third column, (1,3,4), of the second factor:

$$1 \times 1 + 0 \times 3 + 3 \times 4 = 13.$$

With a little practice you can easily calculate the elements of a matrix product by simultaneously running your left index finger across rows of the left factor and your right index finger down columns of the right factor while taking the dot products.

## EXAMPLE 2

$$\begin{pmatrix} 1 & 2 & 3 \\ 0 & 1 & -1 \\ -2 & 3 & 0 \end{pmatrix} \begin{pmatrix} x \\ y \\ z \end{pmatrix} = \begin{pmatrix} x + 2y + 3z \\ y - z \\ -2x + 3y \end{pmatrix}$$

The product of a $3 \times 3$ matrix with a column 3-vector is a column 3-vector.

Matrix multiplication is *associative*. This means that

$$\mathcal{A}(\mathcal{BC}) = (\mathcal{AB})\mathcal{C}$$

(provided $\mathcal{A}$, $\mathcal{B}$, and $\mathcal{C}$ have dimensions compatible with the formation of the various products) and therefore it makes sense to write $\mathcal{ABC}$. However, matrix multiplication is *not commutative*. Indeed, if $\mathcal{A}$ is an $m \times n$ matrix and $\mathcal{B}$ is an $n \times p$ matrix then the product $\mathcal{AB}$ is defined, but the product $\mathcal{BA}$ is not defined unless $m = p$. Even if $\mathcal{A}$ and $\mathcal{B}$ are square matrices of the same size it is not necessarily true that $\mathcal{AB} = \mathcal{BA}$.

### EXAMPLE 3

$$\begin{pmatrix} 1 & 2 \\ 3 & 0 \end{pmatrix}\begin{pmatrix} 1 & -1 \\ 1 & 1 \end{pmatrix} = \begin{pmatrix} 3 & 1 \\ 3 & -3 \end{pmatrix} \quad \text{but} \quad \begin{pmatrix} 1 & -1 \\ 1 & 1 \end{pmatrix}\begin{pmatrix} 1 & 2 \\ 3 & 0 \end{pmatrix} = \begin{pmatrix} -2 & 2 \\ 4 & 2 \end{pmatrix}$$

The reader should verify that if the product $\mathcal{AB}$ is defined then the transpose of the product is the product of the transposes *in the reverse order*:

$$(\mathcal{AB})^T = \mathcal{B}^T \mathcal{A}^T.$$

## Determinants and Matrix Inverses

In Section 3.3 we introduced $2 \times 2$ and $3 \times 3$ determinants as certain algebraic expressions associated with $2 \times 2$ and $3 \times 3$ square arrays of numbers. In general, it is possible to define the determinant $\det(\mathcal{A})$ for any square matrix. For an $n \times n$ matrix $\mathcal{A}$ we continue to denote

$$\det(\mathcal{A}) = \begin{vmatrix} a_{11} & a_{12} & \cdots & a_{1n} \\ a_{21} & a_{22} & \cdots & a_{2n} \\ \vdots & \vdots & \ddots & \vdots \\ a_{n1} & a_{n2} & \cdots & a_{nn} \end{vmatrix}.$$

We will not attempt to give a formal definition of the determinant here, but will note that the properties of determinants stated for the $3 \times 3$ case in Section 3.3 continue to be true. In particular, an $n \times n$ determinant can be expanded in minors about any row or column and so expressed as a sum of multiples of $(n-1) \times (n-1)$ determinants. Continuing this process, we can eventually reduce the evaluation of any $n \times n$ determinant to the evaluation of (perhaps many) $2 \times 2$ or $3 \times 3$ determinants. It is important to realize that the "diagonal" method for evaluating $2 \times 2$ or $3 \times 3$ determinants does not extend to $4 \times 4$ or higher-order determinants.

### EXAMPLE 4

$$\begin{aligned}\begin{vmatrix} 2 & 1 & 0 & 1 \\ 1 & 0 & 1 & 1 \\ 3 & 0 & 0 & 2 \\ -1 & 1 & 1 & 0 \end{vmatrix} &= -\begin{vmatrix} 2 & 1 & 1 \\ 3 & 0 & 2 \\ -1 & 1 & 0 \end{vmatrix} - \begin{vmatrix} 2 & 1 & 1 \\ 1 & 0 & 1 \\ 3 & 0 & 2 \end{vmatrix} \\ &= -\left(-3\begin{vmatrix} 1 & 1 \\ 1 & 0 \end{vmatrix} - 2\begin{vmatrix} 2 & 1 \\ -1 & 1 \end{vmatrix}\right) - \left(-1\begin{vmatrix} 1 & 1 \\ 3 & 2 \end{vmatrix}\right) \\ &= 3(0-1) + 2(2+1) + 1(2-3) = 2.\end{aligned}$$

We expanded the $4 \times 4$ determinant in minors about the third column to obtain the two $3 \times 3$ determinants. The first of these was expanded about the second row, the other about the second column.

In addition to the properties stated in Section 3.3, determinants have two other very important properties, which are stated in the following theorem.

**THEOREM 2**

If $\mathcal{A}$ and $\mathcal{B}$ are $n \times n$ matrices then

(a) $\det(\mathcal{A}^T) = \det(\mathcal{A})$.

(b) $\det(\mathcal{AB}) = \det(\mathcal{A})\det(\mathcal{B})$.

We will not attempt any proof of this or other theorems in this section. The reader is referred to texts on linear algebra. Part (a) is not very difficult to prove, even in the case of general $n$. Part (b) cannot really be proved in general without a formal definition of determinant. However, the reader should verify (b) for $2 \times 2$ matrices by direct calculation.

We say that the square matrix $\mathcal{A}$ is **singular** if $\det(\mathcal{A}) = 0$. If $\det(\mathcal{A}) \neq 0$ we say that $\mathcal{A}$ is **nonsingular** or **invertible**.

***REMARK*** If $\mathcal{A}$ is a $3 \times 3$ matrix then $\det(\mathcal{A})$ is the scalar triple product of the rows of $\mathcal{A}$, and its absolute value is the volume of the parallelepiped spanned by those rows. Therefore $\mathcal{A}$ is nonsingular if and only if its rows span a parallelepiped of positive volume; the row vectors cannot all lie in the same plane. The same may be said of the columns of $\mathcal{A}$.

In general, an $n \times n$ matrix is singular if its rows (or columns), considered as vectors, satisfy one or more linear equations of the form

$$c_1\mathbf{x}_1 + c_2\mathbf{x}_2 + \cdots + c_n\mathbf{x}_n = \mathbf{0},$$

with at least one nonzero coefficient $c_i$. A set of vectors satisfying such a linear equations is called **linearly dependent** because one of the vectors can always be expressed as a linear combination of the others; if $c_1 \neq 0$ then

$$\mathbf{x}_1 = -\frac{c_2}{c_1}\mathbf{x}_2 - \frac{c_3}{c_1}\mathbf{x}_3 - \cdots - \frac{c_n}{c_1}\mathbf{x}_n.$$

All linear combinations of the vectors in a linearly dependent set of $n$ vectors in $\mathbb{R}^n$ must lie in a **subspace** of dimension lower than $n$.

The $n \times n$ **identity matrix** is the matrix

$$\mathcal{I} = \begin{pmatrix} 1 & 0 & \cdots & 0 \\ 0 & 1 & \cdots & 0 \\ \vdots & \vdots & \ddots & \vdots \\ 0 & 0 & \cdots & 1 \end{pmatrix}$$

with "1" in every position on the **main diagonal** and "0" in every other position. Evidently $\mathcal{I}$ commutes with every $n \times n$ matrix: $\mathcal{I}\mathcal{A} = \mathcal{A}\mathcal{I} = \mathcal{A}$. Also $\det(\mathcal{I}) = 1$. The identity matrix plays the same role in matrix algebra that the number 1 plays in arithmetic.

Any nonzero number $x$ has a reciprocal $x^{-1}$ such that $xx^{-1} = x^{-1}x = 1$. A similar situation holds for square matrices. The **inverse** of a *nonsingular* square matrix $\mathcal{A}$ is a nonsingular square matrix $\mathcal{A}^{-1}$ satisfying

$$\mathcal{A}\mathcal{A}^{-1} = \mathcal{A}^{-1}\mathcal{A} = \mathcal{I}.$$

**THEOREM 3** Every nonsingular square matrix $\mathcal{A}$ has a *unique* inverse $\mathcal{A}^{-1}$. Moreover, the inverse satisfies

(a) $\det(\mathcal{A}^{-1}) = \frac{1}{\det(\mathcal{A})}$,

(b) $(\mathcal{A}^{-1})^T = (\mathcal{A}^T)^{-1}$.

We will not have much cause to calculate inverses, but we note that it can be done by solving systems of linear equations, as the following simple example illustrates.

**EXAMPLE 5** Show that the matrix $\mathcal{A} = \begin{pmatrix} 1 & -1 \\ 1 & 1 \end{pmatrix}$ is nonsingular and find its inverse.

**SOLUTION** $\det(\mathcal{A}) = \begin{vmatrix} 1 & -1 \\ 1 & 1 \end{vmatrix} = 1 + 1 = 2$. Therefore $\mathcal{A}$ is nonsingular and invertible. Let $\mathcal{A}^{-1} = \begin{pmatrix} a & b \\ c & d \end{pmatrix}$. Then

$$\begin{pmatrix} 1 & 0 \\ 0 & 1 \end{pmatrix} = \begin{pmatrix} 1 & -1 \\ 1 & 1 \end{pmatrix}\begin{pmatrix} a & b \\ c & d \end{pmatrix} = \begin{pmatrix} a-c & b-d \\ a+c & b+d \end{pmatrix},$$

so $a$, $b$, $c$, and $d$ must satisfy the systems of equations

$$\begin{cases} a - c = 1 \\ a + c = 0 \end{cases} \qquad \begin{cases} b - d = 0 \\ b + d = 1. \end{cases}$$

Evidently $a = b = d = 1/2$, $c = -1/2$, and

$$\mathcal{A}^{-1} = \begin{pmatrix} \frac{1}{2} & \frac{1}{2} \\ -\frac{1}{2} & \frac{1}{2} \end{pmatrix}.$$

Generally, matrix inversion is not carried out by the method of the above example, but rather by an orderly process of performing operations on the rows of the matrix to transform it into the identity. When the same operations are performed on the rows of the identity matrix, the inverse of the original matrix results. See a text on linear algebra for a description of the method.

## Linear Transformations

A function $F$ whose domain is the $m$-dimensional space $\mathbb{R}^m$ and whose range is contained in the $n$-dimensional space $\mathbb{R}^n$ is called a **linear transformation from $\mathbb{R}^m$ to $\mathbb{R}^n$** if it satisfies

$$F(\lambda \mathbf{x} + \mu \mathbf{y}) = \lambda F(\mathbf{x}) + \mu F(\mathbf{y})$$

for all points $\mathbf{x}$ and $\mathbf{y}$ in $\mathbb{R}^m$ and all real numbers $\lambda$ and $\mu$. To such a linear transformation $F$ there corresponds an $n \times m$ matrix $\mathcal{F}$ such that for all $\mathbf{x}$ in $\mathbb{R}^m$,

$$F(\mathbf{x}) = \mathcal{F}\mathbf{x},$$

or expressed in terms of the components of $\mathbf{x}$,

$$F(x_1, x_2, \cdots, x_m) = \mathcal{F}\begin{pmatrix} x_1 \\ x_2 \\ \vdots \\ x_m \end{pmatrix}.$$

We say that $\mathcal{F}$ is a **matrix representation** of the linear transformation $F$. If $m = n$ so that $F$ maps $\mathbb{R}^m$ into itself, then $\mathcal{F}$ is a square matrix. In this case $\mathcal{F}$ is nonsingular if and only if $F$ is one-to-one and has the whole of $\mathbb{R}^m$ as range.

A composition of linear transformations is still a linear transformation and so will have a matrix representation. The real motivation lying behind the definition of matrix multiplication is that the matrix representation of a *composition* of linear transformations is the *product* of the individual matrix representations of the transformations being composed.

**THEOREM 4**

If $F$ is a linear transformation from $\mathbb{R}^m$ to $\mathbb{R}^n$ represented by the $n \times m$ matrix $\mathcal{F}$, and if $G$ is a linear transformation from $\mathbb{R}^n$ to $\mathbb{R}^p$ represented by the $p \times n$ matrix $\mathcal{G}$, then the composition $G \circ F$ defined by

$$G \circ F(x_1, x_2, \ldots, x_m) = G\Big(F(x_1, x_2, \ldots, x_m)\Big)$$

is itself a linear transformation from $\mathbb{R}^m$ to $\mathbb{R}^p$ represented by the $p \times m$ matrix $\mathcal{G}\mathcal{F}$. That is,

$$G\Big(F(\mathbf{x})\Big) = \mathcal{G}\mathcal{F}\mathbf{x}.$$

## Linear Equations

A system of $n$ linear equations in $n$ unknowns:

$$\begin{aligned} a_{11}x_1 + a_{12}x_2 + \cdots + a_{1n}x_n &= b_1 \\ a_{21}x_1 + a_{22}x_2 + \cdots + a_{2n}x_n &= b_2 \\ &\vdots \\ a_{n1}x_1 + a_{n2}x_2 + \cdots + a_{nn}x_n &= b_n \end{aligned}$$

can be written compactly as a single matrix equation,

$$\mathcal{A}\mathbf{x} = \mathbf{b},$$

where

$$\mathcal{A} = \begin{pmatrix} a_{11} & a_{12} & \cdots & a_{1n} \\ a_{21} & a_{22} & \cdots & a_{2n} \\ \vdots & \vdots & \ddots & \vdots \\ a_{n1} & a_{n2} & \cdots & a_{nn} \end{pmatrix}, \qquad \mathbf{x} = \begin{pmatrix} x_1 \\ x_2 \\ \vdots \\ x_n \end{pmatrix}, \quad \text{and} \quad \mathbf{b} = \begin{pmatrix} b_1 \\ b_2 \\ \vdots \\ b_n \end{pmatrix}.$$

Compare the equation $\mathcal{A}\mathbf{x} = \mathbf{b}$ with the equation $ax = b$ for a single unknown $x$. The equation $ax = b$ has the unique solution $x = a^{-1}b$ provided $a \neq 0$. By analogy, the linear system $\mathcal{A}\mathbf{x} = \mathbf{b}$ has a unique solution given by

$$\mathbf{x} = \mathcal{A}^{-1}\mathbf{b},$$

provided $\mathcal{A}$ is nonsingular. To see this, just multiply both sides of the equation $\mathcal{A}\mathbf{x} = \mathbf{b}$ on the left by $\mathcal{A}^{-1}$.

If $\mathcal{A}$ is singular, then the system $\mathcal{A}\mathbf{x} = \mathbf{b}$ may or may not have a solution, and if a solution exists it will not be unique. Consider the case $\mathbf{b} = \mathbf{0}$ (the zero vector). Then any vector $\mathbf{x}$ perpendicular to all the rows of $\mathcal{A}$ will satisfy the system. Since the rows of $\mathcal{A}$ lie in a space of dimension less than $n$ (because $\det(\mathcal{A}) = 0$) there will be at least a line of such vectors $\mathbf{x}$. Thus solutions of $\mathcal{A}\mathbf{x} = \mathbf{0}$ are not unique if $\mathcal{A}$ is singular. The same must be true of the system $\mathcal{A}^T\mathbf{y} = \mathbf{0}$; there will be nonzero vectors $\mathbf{y}$ satisfying it if $\mathcal{A}$ is singular. But then, if the system $\mathcal{A}\mathbf{x} = \mathbf{b}$ has any solution $\mathbf{x}$ we must have

$$(\mathbf{y} \bullet \mathbf{b}) = \mathbf{y}^T\mathbf{b} = \mathbf{y}^T\mathcal{A}\mathbf{x} = (\mathbf{x}^T\mathcal{A}^T\mathbf{y})^T = (\mathbf{x}^T\mathbf{0})^T = (0).$$

Hence $\mathcal{A}\mathbf{x} = \mathbf{b}$ can only have solutions for those vectors $\mathbf{b}$ that are perpendicular to *every* solution $\mathbf{y}$ of $\mathcal{A}^T\mathbf{y} = \mathbf{0}$.

A system of $m$ linear equations in $n$ unknowns may or may not have any solutions if $n < m$. It will have solutions if some $m - n$ of the equations are *linear combinations* (sums of multiples) of the other $n$ equations. If $n > m$, then we can try to solve the $m$ equations for $m$ of the variables, allowing the solutions to depend on the other $n - m$ variables. Such a solution exists if the determinant of the coefficients of the $m$ variables for which we want to solve is not zero. This is a special case of the **implicit function theorem** that we will examine in Section 4.8.

**EXAMPLE 6** Solve $\begin{cases} 2x + y - 3z = 4 \\ x + 2y + 6z = 5 \end{cases}$ for $x$ and $y$ in terms of $z$.

***SOLUTION*** The system can be expressed in the form

$$\mathcal{A}\begin{pmatrix} x \\ y \end{pmatrix} = \begin{pmatrix} 4 + 3z \\ 5 - 6z \end{pmatrix} \qquad \text{where} \quad \mathcal{A} = \begin{pmatrix} 2 & 1 \\ 1 & 2 \end{pmatrix}$$

$\mathcal{A}$ has determinant 3 and inverse $\mathcal{A}^{-1} = \begin{pmatrix} 2/3 & -1/3 \\ -1/3 & 2/3 \end{pmatrix}$. Thus

$$\begin{pmatrix} x \\ y \end{pmatrix} = \mathcal{A}^{-1}\begin{pmatrix} 4 + 3z \\ 5 - 6z \end{pmatrix} = \begin{pmatrix} 2/3 & -1/3 \\ -1/3 & 2/3 \end{pmatrix}\begin{pmatrix} 4 + 3z \\ 5 - 6z \end{pmatrix} = \begin{pmatrix} 1 + 4z \\ 2 - 5z \end{pmatrix}$$

The solution is $x = 1 + 4z$, $y = 2 - 5z$. (Of course, this solution could have been found by elimination of $x$ or $y$ from the given equations without using matrix methods.) ■

## EXPLORE! Linear Algebra Using Calculators or Computers

Computer algebra programs and some advanced calculators have built-in routines for handling matrices and solving linear systems. Among other features, such utilities can

(i) set up, transpose, and multiply rectangular matrices,

(ii) calculate determinants and inverses of square matrices, and

(iii) solve systems of linear equations.

Use such a utility to find $\det(\mathcal{A})$ and $\mathcal{A}^{-1}$, and to solve the system $\mathcal{A}\mathbf{x} = \mathbf{b}$, where

$$\mathcal{A} = \begin{pmatrix} 3 & 2 & 1 & -2 \\ 2 & 0 & 1 & 1 \\ 2 & 1 & 1 & 0 \\ -1 & -2 & 1 & 1 \end{pmatrix} \qquad \text{and} \qquad \mathbf{b} = \begin{pmatrix} 5 \\ 2 \\ 10 \\ -5 \end{pmatrix}.$$

We conclude this section by stating a result of some theoretical importance expressing the solution of the system $\mathcal{A}\mathbf{x} = \mathbf{b}$ for nonsingular $\mathcal{A}$ in terms of determinants.

**THEOREM 5** **Cramer's Rule**

Let $\mathcal{A}$ be a nonsingular $n \times n$ matrix. Then the solution $\mathbf{x}$ of the system

$$\mathcal{A}\mathbf{x} = \mathbf{b}$$

has components given by

$$x_1 = \frac{\det(\mathcal{A}_1)}{\det(\mathcal{A})}, \qquad x_2 = \frac{\det(\mathcal{A}_2)}{\det(\mathcal{A})}, \qquad \ldots, \qquad x_n = \frac{\det(\mathcal{A}_n)}{\det(\mathcal{A})},$$

where $\mathcal{A}_j$ is the matrix $\mathcal{A}$ with its $j$th column replaced by the column vector $\mathbf{b}$. That is

$$\det(\mathcal{A}_j) = \begin{vmatrix} a_{11} & \cdots & a_{1(j-1)} & b_1 & a_{1(j+1)} & \cdots & a_{1n} \\ a_{21} & \cdots & a_{2(j-1)} & b_2 & a_{2(j+1)} & \cdots & a_{2n} \\ \vdots & & \vdots & \vdots & \vdots & & \vdots \\ a_{n1} & \cdots & a_{n(j-1)} & b_n & a_{n(j+1)} & \cdots & a_{nn} \end{vmatrix}.$$

The following example provides a concrete illustration of use of Cramer's Rule to solve a specific linear system. However, Cramer's Rule is primarily used in a more general (theoretical) context; it is not efficient to use determinants to calculate solutions of linear systems.

**EXAMPLE 7** Find the point of intersection of the three planes

$$\begin{aligned} x + y + 2z &= 1 \\ 3x + 6y - z &= 0 \\ x - y - 4z &= 3. \end{aligned}$$

***SOLUTION*** The solution of the linear system above provides the coordinates of the intersection point. The determinant of the coefficient matrix of this system is

$$\det(\mathcal{A}) = \begin{vmatrix} 1 & 1 & 2 \\ 3 & 6 & -1 \\ 1 & -1 & -4 \end{vmatrix} = -32,$$

so the system does have a unique solution. We have

$$x = \frac{1}{-32}\begin{vmatrix} 1 & 1 & 2 \\ 0 & 6 & -1 \\ 3 & -1 & -4 \end{vmatrix} = \frac{-64}{-32} = 2,$$

$$y = \frac{1}{-32}\begin{vmatrix} 1 & 1 & 2 \\ 3 & 0 & -1 \\ 1 & 3 & -4 \end{vmatrix} = \frac{32}{-32} = -1,$$

$$z = \frac{1}{-32}\begin{vmatrix} 1 & 1 & 1 \\ 3 & 6 & 0 \\ 1 & -1 & 3 \end{vmatrix} = \frac{0}{-32} = 0.$$

The intersection point is $(2, -1, 0)$. ■

## EXERCISES 3.6

Evaluate the matrix products in Exercises 1–4.

**1.** $\begin{pmatrix} 3 & 0 & -2 \\ 1 & 1 & 2 \\ -1 & 1 & -1 \end{pmatrix}\begin{pmatrix} 2 & 1 \\ 3 & 0 \\ 0 & -2 \end{pmatrix}$

**2.** $\begin{pmatrix} 1 & 1 & 1 \\ 0 & 1 & 1 \\ 0 & 0 & 1 \end{pmatrix}\begin{pmatrix} 1 & 1 & 1 \\ 0 & 1 & 1 \\ 0 & 0 & 1 \end{pmatrix}$

**3.** $\begin{pmatrix} a & b \\ c & d \end{pmatrix}\begin{pmatrix} w & x \\ y & z \end{pmatrix}$

**4.** $\begin{pmatrix} w & x \\ y & z \end{pmatrix}\begin{pmatrix} a & b \\ c & d \end{pmatrix}$

**5.** Evaluate $\mathcal{A}\mathcal{A}^T$ and $\mathcal{A}^2 = \mathcal{A}\mathcal{A}$ where

$$\mathcal{A} = \begin{pmatrix} 1 & 1 & 1 & 1 \\ 0 & 1 & 1 & 1 \\ 0 & 0 & 1 & 1 \\ 0 & 0 & 0 & 1 \end{pmatrix}$$

**6.** Evaluate $\mathbf{x}\mathbf{x}^T$, $\mathbf{x}^T\mathbf{x}$, and $\mathbf{x}^T\mathcal{A}\mathbf{x}$ where

$$\mathbf{x} = \begin{pmatrix} x \\ y \\ z \end{pmatrix} \quad \text{and} \quad \mathcal{A} = \begin{pmatrix} a & p & q \\ p & b & r \\ q & r & c \end{pmatrix}.$$

Evaluate the determinants in Exercises 7–8.

**7.** $\begin{vmatrix} 2 & 3 & -1 & 0 \\ 4 & 0 & 2 & 1 \\ 1 & 0 & -1 & 1 \\ -2 & 0 & 0 & 1 \end{vmatrix}$

**8.** $\begin{vmatrix} 1 & 1 & 1 & 1 \\ 1 & 2 & 3 & 4 \\ -2 & 0 & 2 & 4 \\ 3 & -3 & 2 & -2 \end{vmatrix}$

**9.** Show that if $\mathcal{A} = (a_{ij})$ is an $n \times n$ matrix for which $a_{ij} = 0$ whenever $i > j$, then $\det(\mathcal{A}) = \prod_{k=1}^{n} a_{kk}$, the product of the elements on the main diagonal of $\mathcal{A}$.

**10.** Show that $\begin{vmatrix} 1 & 1 \\ x & y \end{vmatrix} = y - x$, and

$$\begin{vmatrix} 1 & 1 & 1 \\ x & y & z \\ x^2 & y^2 & z^2 \end{vmatrix} = (y - x)(z - x)(z - y).$$

Try to generalize this result to the $n \times n$ case.

**11.** Verify the associative law $(\mathcal{A}\mathcal{B})\mathcal{C} = \mathcal{A}(\mathcal{B}\mathcal{C})$ by direct calculation for three arbitrary $2 \times 2$ matrices.

**12.** Show that $\det(\mathcal{A}^T) = \det(\mathcal{A})$ for $n \times n$ matrices by induction on $n$. Start with the $2 \times 2$ case.

**13.** Verify by direct calculation that $\det(\mathcal{A}\mathcal{B}) = \det(\mathcal{A})\det(\mathcal{B})$ holds for two arbitrary $2 \times 2$ matrices.

**14.** Let $\mathcal{A}_\theta = \begin{pmatrix} \cos\theta & \sin\theta \\ -\sin\theta & \cos\theta \end{pmatrix}$. Show that $(\mathcal{A}_\theta)^T = (\mathcal{A}_\theta)^{-1} = \mathcal{A}_{-\theta}$.

Find the inverses of the matrices in Exercises 15–16.

**15.** $\begin{pmatrix} 1 & 1 & 1 \\ 0 & 1 & 1 \\ 0 & 0 & 1 \end{pmatrix}$ **16.** $\begin{pmatrix} 1 & 0 & -1 \\ -1 & 1 & 0 \\ 2 & 1 & 3 \end{pmatrix}$

**17.** Use your result from Exercise16 to solve the linear system

$$\begin{cases} x - z = -2 \\ -x + y = 1 \\ 2x + y + 3z = 13. \end{cases}$$

**18.** Solve the system of Exercise 17 by using Cramer's Rule.

**19.** Solve the system $\begin{cases} x_1 + x_2 + x_3 + x_4 = 0 \\ x_1 + x_2 + x_3 - x_4 = 4 \\ x_1 + x_2 - x_3 - x_4 = 6 \\ x_1 - x_2 - x_3 - x_4 = 2. \end{cases}$

**20.** Verify Theorem 4 for the special case where $F$ and $G$ are linear transformations from $\mathbb{R}^2$ to $\mathbb{R}^2$.

# CHAPTER REVIEW

## Key Ideas

- **What is each of the following?**
  - ◇ a vector in 3-space
  - ◇ the dot product of two 3-vectors
  - ◇ a scalar projection ◇ a vector projection
  - ◇ the cross product of two 3-vectors
  - ◇ a scalar triple product ◇ a vector triple product
  - ◇ a matrix ◇ a determinant
  - ◇ a plane ◇ a straight line
  - ◇ a cone ◇ a cylinder
  - ◇ an ellipsoid ◇ a sphere
  - ◇ a paraboloid ◇ a hyperboloid
  - ◇ the transpose of a matrix ◇ the inverse of a matrix
  - ◇ a linear transformation from $\mathbb{R}^3$ to $\mathbb{R}^3$
- **What is the angle between the vectors u and v?**
- **How do you calculate u×v, given the components of u and v?**
- **What is an equation of the plane through $P_0$ having normal vector N?**
- **What is an equation of the straight line through $P_0$ parallel to a?**
- **Given two $3 \times 3$ matrices $A$ and $B$, how do you calculate $AB$?**
- **What is the distance from $P_0$ to the plane $Ax + By + Cz + D = 0$?**
- **What is Cramer's Rule and how is it used?**

## Review Exercises

Describe the sets of points in 3-space that satisfy the given equations or inequalities in Exercises 1–18.

**1.** $x + 3z = 3$ **2.** $y - z \geq 1$

**3.** $x + y + z \geq 0$ **4.** $x - 2y - 4z = 8$

**5.** $y = 1 + x^2 + z^2$ **6.** $y = z^2$

**7.** $x = y^2 - z^2$ **8.** $z = xy$

**9.** $x^2 + y^2 + 4z^2 < 4$ **10.** $x^2 + y^2 - 4z^2 = 4$

**11.** $x^2 - y^2 - 4z^2 = 0$ **12.** $x^2 - y^2 - 4z^2 = 4$

* **13.** $(x - z)^2 + y^2 = 1$ * **14.** $(x - z)^2 + y^2 = z^2$

**15.** $\begin{cases} x + 2y = 0 \\ z = 3 \end{cases}$ **16.** $\begin{cases} x + y + 2z = 1 \\ x + y + z = 0 \end{cases}$

**17.** $\begin{cases} x^2 + y^2 + z^2 = 4 \\ x + y + z = 3 \end{cases}$ **18.** $\begin{cases} x^2 + z^2 \leq 1 \\ x - y \geq 0 \end{cases}$

Find equations of the planes and lines specified in Exercises 19–28.

**19.** The plane through the origin perpendicular to the line

$$\frac{x-1}{2} = \frac{y+3}{-1} = \frac{z+2}{3}$$

**20.** The plane through $(2, -1, 1)$ and $(1, 0, -1)$ parallel to the line in Exercise 19

**21.** The plane through $(2, -1, 1)$ perpendicular to the planes $x - y + z = 0$ and $2x + y - 3z = 2$

**22.** The plane through $(-1, 1, 0)$, $(0, 4, -1)$, and $(2, 0, 0)$

**23.** The plane containing the line of intersection of the planes $x + y + z = 0$ and $2x + y - 3z = 2$, and passing through the point $(2, 0, 1)$

**24.** The plane containing the line of intersection of the planes $x + y + z = 0$ and $2x + y - 3z = 2$, and perpendicular to the plane $x - 2y - 5z = 17$

**25.** The vector parametric equation of the line through $(2, 1, -1)$ and $(-1, 0, 1)$

**26.** Standard form equations of the line through $(1, 0, -1)$ parallel to each of the planes $x - y = 3$ and $x + 2y + z = 1$

**27.** Scalar parametric equations of the line through the origin perpendicular to the plane $3x - 2y + 4z = 5$

**28.** The vector parametric equation of the line that joins points on the two lines

$$\mathbf{r} = (1 + t)\mathbf{i} - t\mathbf{j} - (2 + 2t)\mathbf{k}$$
$$\mathbf{r} = 2t\mathbf{i} + (t - 2)\mathbf{j} - (1 + 3t)\mathbf{k}$$

and is perpendicular to both those lines

Express the given conditions or quantities in Exercises 29–30 in terms of dot and cross products.

**29.** The three points with position vectors $\mathbf{r}_1$, $\mathbf{r}_2$, and $\mathbf{r}_3$ all lie on a straight line

**30.** The four points with position vectors $\mathbf{r}_1$, $\mathbf{r}_2$, $\mathbf{r}_3$, and $\mathbf{r}_4$ do not all lie on a plane

**31.** Find the area of the triangle with vertices $(1, 2, 1)$, $(4, -1, 1)$, and $(3, 4, -2)$.

**32.** Find the volume of the tetrahedron with vertices $(1, 2, 1)$, $(4, -1, 1)$, $(3, 4, -2)$, and $(2, 2, 2)$.

**33.** Show that the matrix

$$\mathcal{A} = \begin{pmatrix} 1 & 0 & 0 & 0 \\ 2 & 1 & 0 & 0 \\ 3 & 2 & 1 & 0 \\ 4 & 3 & 2 & 1 \end{pmatrix}$$

has an inverse, and find the inverse $\mathcal{A}^{-1}$.

**34.** Let $\mathcal{A} = \begin{pmatrix} 1 & 1 & 1 \\ 2 & 1 & 0 \\ 1 & 0 & -1 \end{pmatrix}$. What condition must the vector $\mathbf{b}$ satsify in order that the equation $\mathcal{A}\mathbf{x} = \mathbf{b}$ has solutions $\mathbf{x}$? What are the solutions $\mathbf{x}$ if $\mathbf{b}$ satisfies the condition?

## Express Yourself

Answer these questions in words, using complete sentences and, if necessary, paragraphs. Use mathematical symbols only where necessary.

1. What do we mean when we say that space is three-dimensional?
2. Why does a single equation in three variables typically represent a surface rather than a curve?
3. Why is the cross product of two vectors defined only in three-dimensional space and not, say, in the plane or in four-dimensional space?
4. What is meant by the distance between two straight lines in 3-space? Between a point and a plane? Between a point and a line? When does it make sense to say that the distance from a line to a plane is 3 units?
5. What kinds of geometric objects can be represented by a single quadratic equation in three variables?
6. What is the difference between a square matrix and a determinant?

## Challenging Problems

**1.** Show that the distance $d$ from point $P$ to the line $AB$ can be expressed in terms of the position vectors of $P$, $A$, and $B$ by

$$d = \frac{|(\mathbf{r}_A - \mathbf{r}_P) \times (\mathbf{r}_B - \mathbf{r}_P)|}{|\mathbf{r}_A - \mathbf{r}_B|}$$

**2.** For any vectors $\mathbf{u}$, $\mathbf{v}$, $\mathbf{w}$, and $\mathbf{x}$, show that

$$\begin{aligned}(\mathbf{u} \times \mathbf{v}) \times (\mathbf{w} \times \mathbf{x}) &= \big((\mathbf{u} \times \mathbf{v}) \bullet \mathbf{x}\big)\mathbf{w} - \big((\mathbf{u} \times \mathbf{v}) \bullet \mathbf{w}\big)\mathbf{x} \\ &= \big((\mathbf{w} \times \mathbf{x}) \bullet \mathbf{u}\big)\mathbf{v} - \big((\mathbf{w} \times \mathbf{x}) \bullet \mathbf{v}\big)\mathbf{u}.\end{aligned}$$

In particular, show that

$$(\mathbf{u} \times \mathbf{v}) \times (\mathbf{u} \times \mathbf{w}) = \big((\mathbf{u} \times \mathbf{v}) \bullet \mathbf{w}\big)\mathbf{u}.$$

**3.** Show that the area $A$ of a triangle with vertices $(x_1, y_1, 0)$, $(x_2, y_2, 0)$, and $(x_3, y_3, 0)$, in the $xy$-plane is given by

$$A = \frac{1}{2}\left| \begin{vmatrix} x_1 & y_1 & 1 \\ x_2 & y_2 & 1 \\ x_3 & y_3 & 1 \end{vmatrix} \right|.$$

**4.** (a) If $L_1$ and $L_2$ are two **skew** (that is, non-parallel and non-intersecting) lines, show that there is a pair of parallel planes $P_1$ and $P_2$ such that $L_1$ lies in $P_1$ and $L_2$ lies in $P_2$.

(b) Find parallel planes containing the lines $L_1$ through $(1, 1, 0)$ and $(2, 0, 1)$ and $L_2$ through $(0, 1, 1)$ and $(1, 2, 2)$.

**5.** What condition must the vectors $\mathbf{a}$ and $\mathbf{b}$ satisfy to ensure that the equation $\mathbf{a} \times \mathbf{x} = \mathbf{b}$ has solutions? If this condition is satisfied, find all solutions of the equation. Describe the set of solutions.

CHAPTER 4

# Partial Differentiation

**INTRODUCTION** This chapter is concerned with extending the idea of the derivative to real functions of a vector variable, that is, to functions depending on several real variables.

## 4.1 FUNCTIONS OF SEVERAL VARIABLES

The notation $y = f(x)$ is used to indicate that the variable $y$ depends on the single real variable $x$, that is, that $y$ is a function of $x$. The domain of such a function $f$ is a set of real numbers. Many quantities can be regarded as depending on more than one real variable, and thus to be functions of more than one variable. For example, the volume of a circular cylinder of radius $r$ and height $h$ is given by $V = \pi r^2 h$; we say that $V$ is a function of the two variables $r$ and $h$. If we choose to denote this function by $f$, then we would write $V = f(r, h)$ where

$$f(r, h) = \pi r^2 h, \qquad (r > 0, \quad h > 0).$$

Thus $f$ is a function of two variables having as *domain* the set of points in the $rh$-plane with coordinates $(r, h)$ satisfying $r > 0$ and $h > 0$. Similarly, the relationship $w = f(x, y, z) = x + 2y - 3z$ defines $w$ as a function of the three variables $x$, $y$, and $z$, with domain the whole of $\mathbb{R}^3$, or, if we state explicitly, some particular subset of $\mathbb{R}^3$.

By analogy with the corresponding definition for functions of one variable, we define a function of $n$ variables as follows:

**DEFINITION 1**

A **function** $f$ of $n$ real variables is a rule that assigns a *unique* real number $f(x_1, x_2, \ldots, x_n)$ to each point $(x_1, x_2, \ldots, x_n)$ in some subset $\mathcal{D}(f)$ of $\mathbb{R}^n$. $\mathcal{D}(f)$ is called the **domain** of $f$. The set of real numbers $f(x_1, x_2, \ldots, x_n)$ obtained from points in the domain is called the **range** of $f$.

As for functions of one variable, it is conventional to take the domain of such a function of $n$ variables to be the largest set of points $(x_1, x_2, \ldots, x_n)$ for which $f(x_1, x_2, \ldots, x_n)$ makes sense as a real number, unless the domain is explicitly stated to be a smaller set.

Most of the examples we consider hereafter will be functions of two or three independent variables. When a function $f$ depends on two variables we will usually call these independent variables $x$ and $y$, and use $z$ to denote the dependent variable which represents the value of the function; that is, $z = f(x, y)$. We will normally use $x$, $y$, and $z$ as the independent variables of a function of three variables, and $w$ as the value of the function; $w = f(x, y, z)$. Some definitions will be given, and some theorems will be stated (and proved) only for the two-variable case, but extensions to three or more variables will usually be obvious.

## Graphical Representations

The graph of a function $f$ of one variable (that is, the graph of the equation $y = f(x)$) is the set of points in the $xy$-plane having coordinates $(x, f(x))$, where $x$ is in the domain of $f$. Similarly, the graph of a function $f$ of 2 variables, (the graph of the equation $z = f(x, y)$), is the set of points in 3-space having coordinates $(x, y, f(x, y))$, where $(x, y)$ belongs to the domain of $f$. This graph is a surface in $\mathbb{R}^3$ lying above (if $f(x, y) > 0$) or below (if $f(x, y) < 0$) the domain of $f$ in the $xy$-plane. (See Figure 4.1.) The graph of a function of three variables is a three-dimensional *hypersurface* in 4-space, $\mathbb{R}^4$. In general, the graph of a function of $n$ variables is an $n$-dimensional *surface* in $\mathbb{R}^{n+1}$. We will not attempt to *draw* graphs of functions of more than two variables!

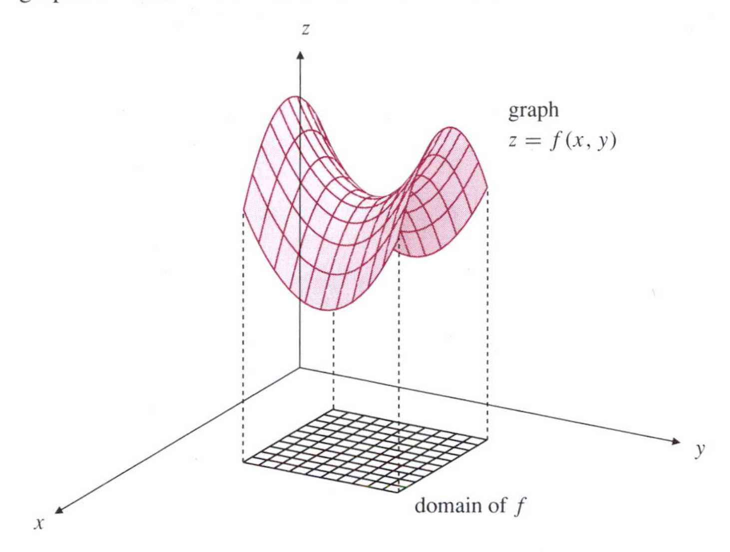


**Figure 4.1** The graph of $f(x, y)$ is the surface with equation $z = f(x, y)$ defined for points $(x, y)$ in the domain of $f$

**EXAMPLE 1** Consider the function

$$f(x, y) = 3\left(1 - \frac{x}{2} - \frac{y}{4}\right), \qquad (0 \le x \le 2, \quad 0 \le y \le 4 - 2x).$$

The graph of $f$ is the plane triangular surface with vertices at $(2, 0, 0)$, $(0, 4, 0)$, and $(0, 0, 3)$. (See Figure 4.2.) If the domain of $f$ had not been explicitly stated to be a particular set in the $xy$-plane, the graph would have been the whole plane through these three points. ■

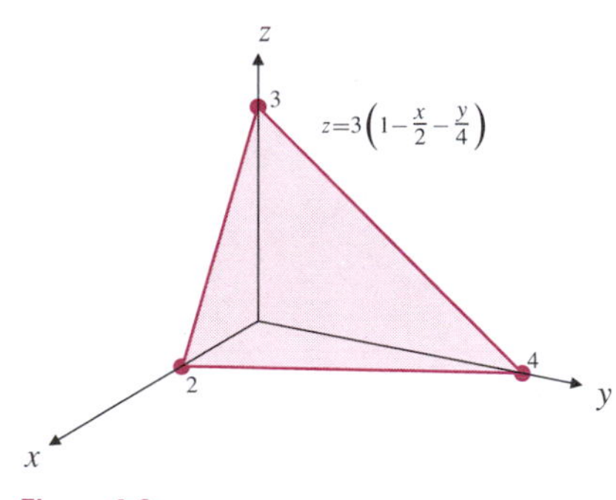


**Figure 4.2**

**EXAMPLE 2** Consider $f(x, y) = \sqrt{9 - x^2 - y^2}$. The expression under the square root can not be negative, so the domain is the disk $x^2 + y^2 \le 9$ in the $xy$-plane. If we square the equation $z = \sqrt{9 - x^2 - y^2}$ we can rewrite the result in the form $x^2 + y^2 + z^2 = 9$. This is a sphere of radius 3 centred at the origin. However, the graph of $f$ is only the upper hemisphere where $z \ge 0$. (See Figure 4.3.) ■

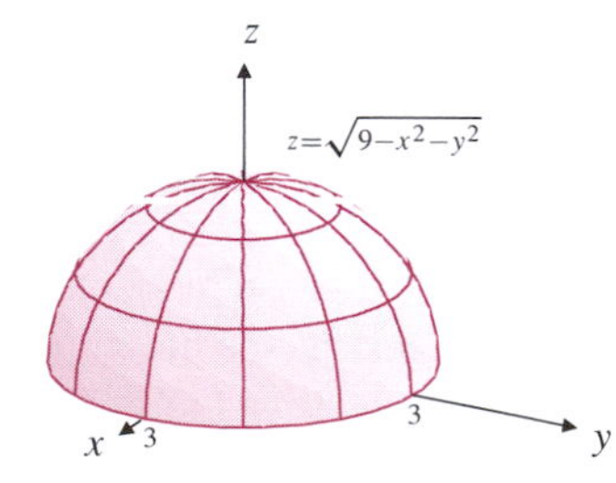


**Figure 4.3**

Since it is necessary to project the surface $z = f(x, y)$ onto a two-dimensional page, most such graphs are difficult to sketch without considerable artistic talent and training. Nevertheless, you should always try to visualize such a graph, and sketch it as best you can. Sometimes it is convenient to sketch only part of a graph, for instance, the part lying in the first octant. It is also helpful to determine (and sketch) the intersections of the graph with various planes, especially the coordinate planes, and planes parallel to the coordinate planes. (See Figure 4.1.)

## EXPLORE! Graphs of Functions of Two Variables

Some mathematical software packages will produce plots of three-dimensional graphs to help you get a feeling for how the corresponding functions behave. Figure 4.1 is an example of such a computer-drawn graph. Along with all the other mathematical graphics in this book, it was produced using the mathematical graphics software package **MG**. Two more examples are given in Figures 4.4–4.5.

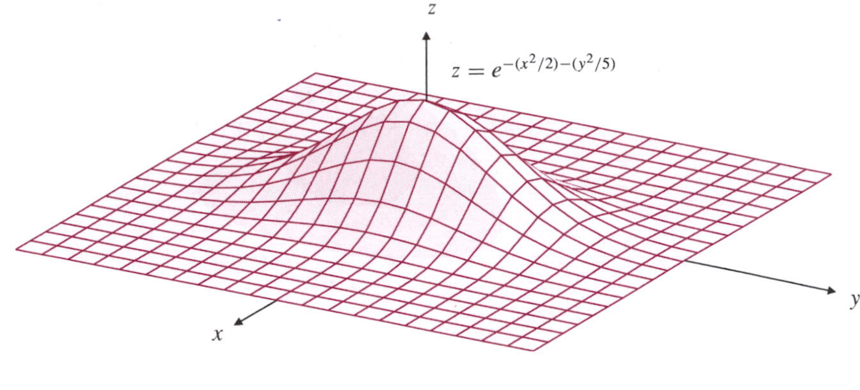


**Figure 4.4** The graph of $e^{-(x^2/2)-(y^2/5)}$

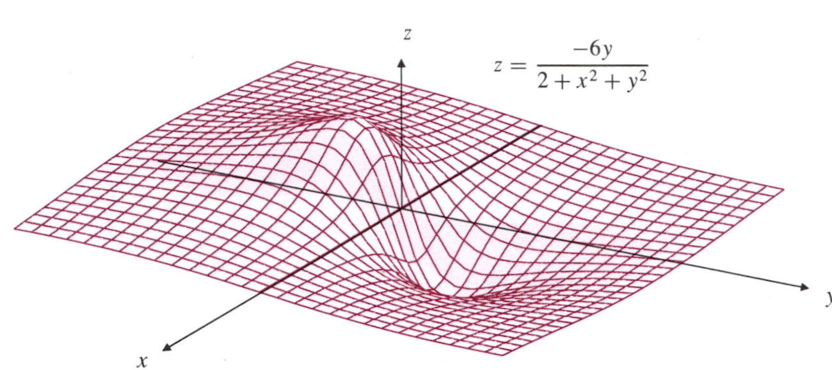


**Figure 4.5** The graph of $\dfrac{-6y}{2+x^2+y^2}$

Use a graphics program with such capabilities to explore the behaviour of the following functions. In each case you will have to make suitable choices for intervals of the variables and for other graphing parameters. Some experimentation will be necessary; don't expect to get a good graph showing all the important behaviour on your first try.

**1.** $y^3 - 3x^2y$

**2.** $(x^2 - y^2)^2$

**3.** $\sin(x - y)$

**4.** $\sin x \sin y$

**5.** $0.1/\sqrt{x^2 + y^2} - 0.2x$

**6.** $0.1y/\sqrt{(x^2 + (y^2 - 1)^2}$

Another way to represent the function $f(x, y)$ graphically is to produce a two-dimensional *topographic map* of the surface $z = f(x, y)$. In the $xy$-plane we sketch the curves $f(x, y) = C$ for various values of the constant $C$. These curves are called **level curves** of $f$ because they are the vertical projections onto the $xy$-plane of the curves in which the graph $z = f(x, y)$ intersects the horizontal (level) planes $z = C$. The graph and some level curves of the function $f(x, y) = x^2 + y^2$ are shown in Figure 4.6. The graph is a circular paraboloid; the level curves are circles centred at the origin.

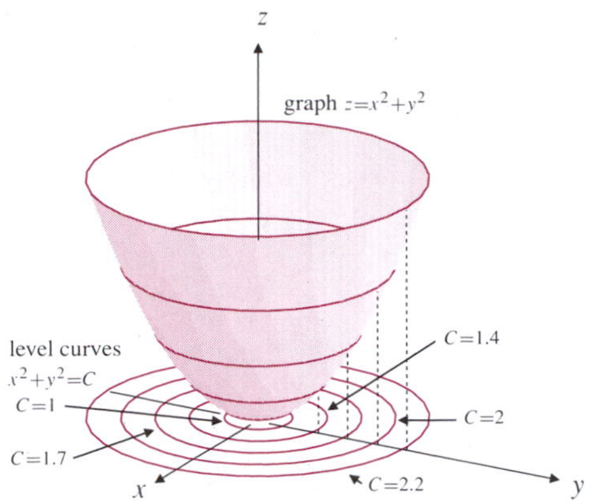

**Figure 4.6** The graph of $f(x, y) = x^2 + y^2$ and some of level curves of $f$

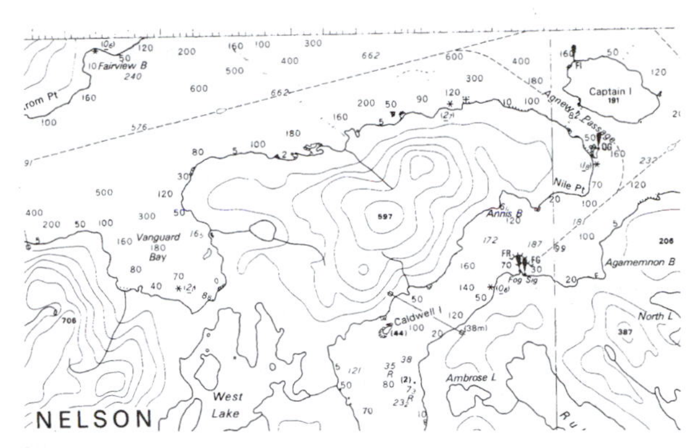


**Figure 4.7** Level curves (contours) representing elevation in a topographic map

The contour curves in the topographic map in Figure 4.7 show the elevations, in 100 m increments above sea level, on part of Nelson Island on the British Columbia coast. Since these contours are drawn for equally spaced values of $C$, the spacing of the contours themselves conveys information about the relative steepness at various places on the mountains; the land is steepest where the contour lines are closest together. Observe also that the streams shown cross the contours at right angles. They take the route of steepest descent. Isotherms (curves of constant temperature) and isobars (curves of constant pressure) on weather maps are also examples of level curves.

**EXAMPLE 3** The level curves of the function $f(x, y) = 3\left(1 - \frac{x}{2} - \frac{y}{4}\right)$ of Example 1 are the segments of the straight lines

$$3\left(1 - \frac{x}{2} - \frac{y}{4}\right) = C \qquad \text{or} \qquad \frac{x}{2} + \frac{y}{4} = 1 - \frac{C}{3}, \qquad (0 \le C \le 3)$$

that lie in the first quadrant. Several such level curves are shown in Figure 4.8(a). They correspond to equally spaced values of $C$, and their equal spacing indicates the uniform steepness of the graph of $f$. ■

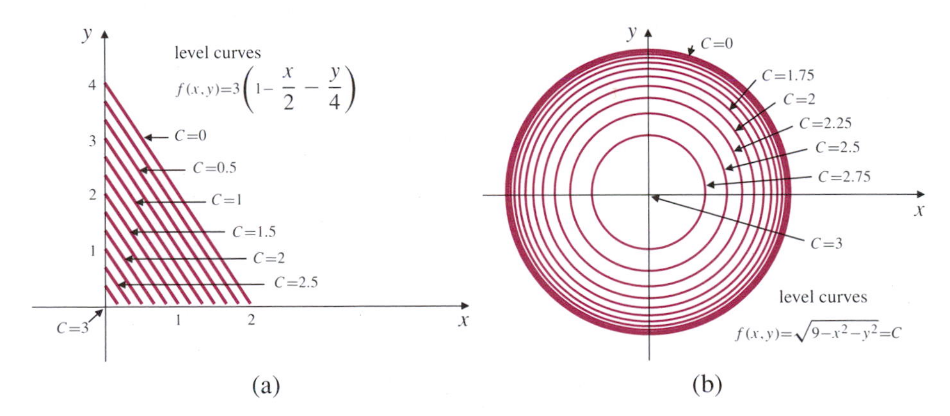


**Figure 4.8**
(a) Level curves of $3\left(1 - \frac{x}{2} - \frac{y}{4}\right)$
(b) Level curves of $\sqrt{9 - x^2 - y^2}$

**EXAMPLE 4** The level curves of the function $f(x, y) = \sqrt{9 - x^2 - y^2}$ of Example 2 are the concentric circles

$$\sqrt{9 - x^2 - y^2} = C \qquad \text{or} \qquad x^2 + y^2 = 9 - C^2, \qquad (0 \le C \le 3).$$

Observe the spacing of these circles in Figure 4.8(b); they are plotted for several equally spaced values of $C$. The bunching of the circles as $C \to 0+$ indicates the steepness of the hemispherical surface that is the graph of $f$. ■

A function determines its level curves with any given spacing between consecutive values of $C$. However, level curves only determine the function if *all of them* are known.

**EXAMPLE 5** The level curves of the function $f(x, y) = x^2 - y^2$ are the curves $x^2 - y^2 = C$. For $C = 0$ the level "curve" is the pair of straight lines $x = y$ and $x = -y$. For other values of $C$ the level curves are rectangular hyperbolas with these lines as asymptotes. (See Figure 4.9(a).) The graph of $f$ is the saddle-like hyperbolic paraboloid in Figure 4.9(b). ■

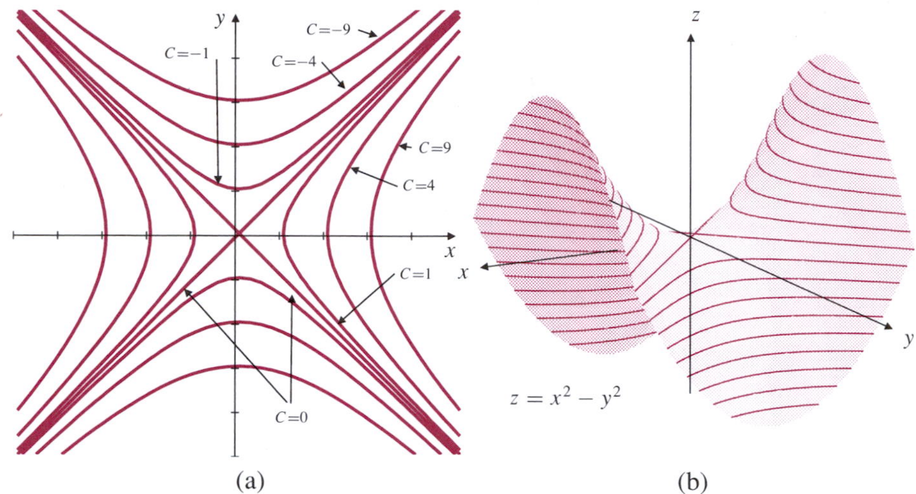


**Figure 4.9**
(a) Level curves of $x^2 - y^2$
(b) The graph of $x^2 - y^2$

**EXAMPLE 6** Describe and sketch some level curves of the function $z = g(x, y)$ defined by $z \geq 0$, $x^2 + (y - z)^2 = 2z^2$. Also sketch the graph of $g$.

***SOLUTION*** The level curve $z = g(x, y) = C$ (where $C$ is a positive constant) has equation $x^2 + (y - C)^2 = 2C^2$, and is, therefore, a circle of radius $\sqrt{2}C$ centred at $(0, C)$. Level curves for $C$ in increments of 0.1 from 0 to 1 are shown in Figure 4.10(a). These level curves intersect rays from the origin at equal spacing (the spacing is different for different rays) indicating that the surface $z = g(x, y)$ is an oblique circular cone. See Figure 4.10(b). ■

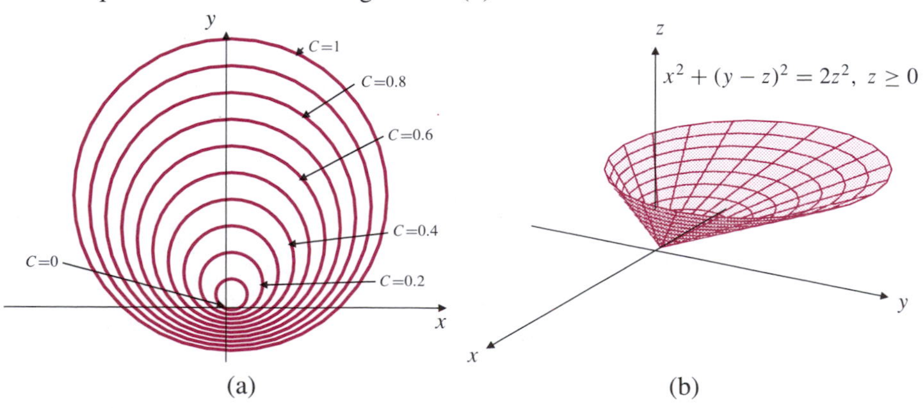


**Figure 4.10**
(a) Level curves of $z = g(x, y)$ for Example 6
(b) The graph of $z = g(x, y)$

**EXPLORE!** Level Curves Using Graphics Software

Computer programs that are able to plot graphs of functions of two variables can usually also plot level curves of such functions. Use such software to plot many level curves of the functions whose graphs appear in Figures 4.4–4.5, and the six functions given in the **EXPLORE** on page 152.

Although the *graph* of a function $f(x, y, z)$ of three variables cannot easily be drawn (it is a three-dimensional *hypersurface* in 4-space), such a function has **level surfaces** in 3-space that can, perhaps be drawn. These level surfaces have equations $f(x, y, z) = C$ for various choices of the constant $C$. For instance, the level surfaces of the function $f(x, y, z) = x^2+y^2+z^2$ are concentric spheres centred at the origin. Figure 4.11 shows a few level surfaces of the function $f(x, y, z) = x^2 - z$. They are parabolic cylinders.

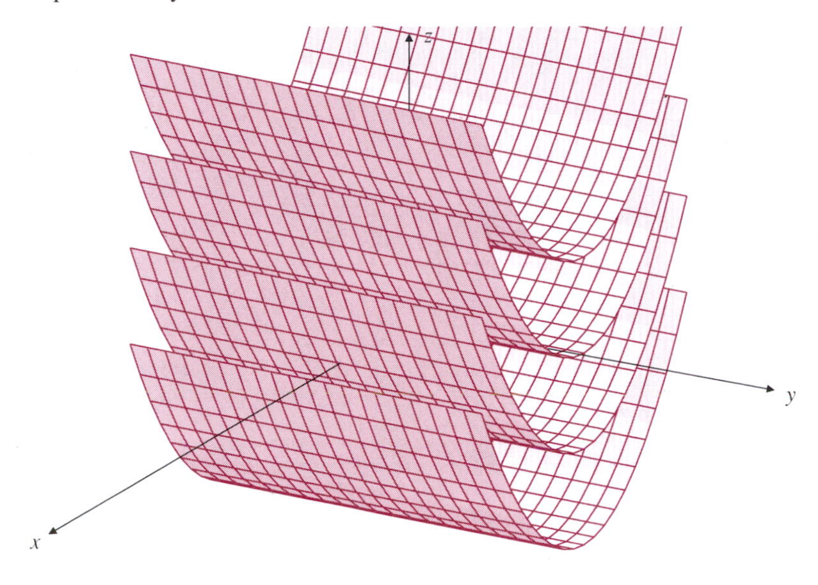


**Figure 4.11** Level surfaces of $f(x, y, z) = x^2 - z$

## EXERCISES 4.1

Specify the domains of the functions in Exercises 1–10.

**1.** $f(x, y) = \dfrac{x+y}{x-y}$

**2.** $f(x, y) = \sqrt{xy}$

**3.** $f(x, y) = \dfrac{x}{x^2+y^2}$

**4.** $f(x, y) = \dfrac{xy}{x^2-y^2}$

**5.** $f(x, y) = \sqrt{4x^2+9y^2-36}$

**6.** $f(x, y) = \dfrac{1}{\sqrt{x^2-y^2}}$

**7.** $f(x, y) = \ln(1+xy)$

**8.** $f(x, y) = \sin^{-1}(x+y)$

**9.** $f(x, y, z) = \dfrac{xyz}{x^2+y^2+z^2}$

**10.** $f(x, y, z) = \dfrac{e^{xyz}}{\sqrt{xyz}}$

Sketch the graphs of the functions in Exercises 11–18.

**11.** $f(x, y) = x, \quad (0 \le x \le 2, \quad 0 \le y \le 3)$

**12.** $f(x, y) = \sin x, \quad (0 \le x \le 2\pi, \quad 0 \le y \le 1)$

**13.** $f(x, y) = y^2, \quad (-1 \le x \le 1, \quad -1 \le y \le 1)$

**14.** $f(x, y) = 4 - x^2 - y^2, \quad (x^2+y^2 \le 4,\ x \ge 0,\ y \ge 0)$

**15.** $f(x, y) = \sqrt{x^2+y^2}$

**16.** $f(x, y) = 4 - x^2$

**17.** $f(x, y) = |x| + |y|$

**18.** $f(x, y) = 6 - x - 2y$

Sketch some of the level curves of the functions in Exercises 19–26.

**19.** $f(x, y) = x - y$

**20.** $f(x, y) = x^2 + 2y^2$

**21.** $f(x, y) = xy$

**22.** $f(x, y) = \dfrac{x^2}{y}$

**23.** $f(x, y) = \dfrac{x-y}{x+y}$

**24.** $f(x, y) = \dfrac{y}{x^2+y^2}$

**25.** $f(x, y) = xe^{-y}$

**26.** $f(x, y) = \sqrt{\dfrac{1}{y} - x^2}$

Exercises 27–28 refer to Figure 4.12, which shows contours of a hilly region with heights given in metres.

**27.** At which of the points $A$ and $B$ is the landscape steepest? How do you know?

**28.** Describe the topography of the region near point $C$.

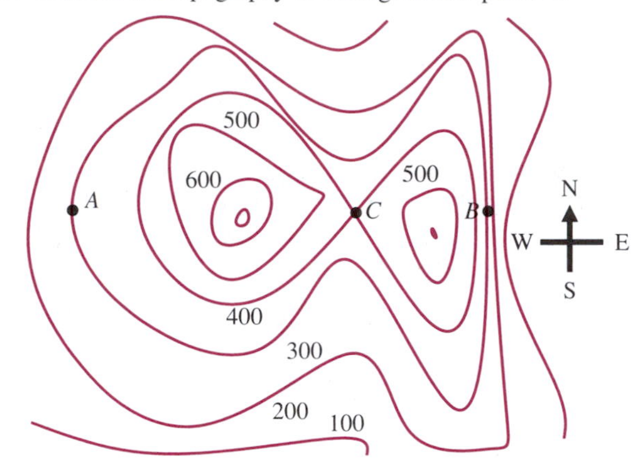


Figure 4.12

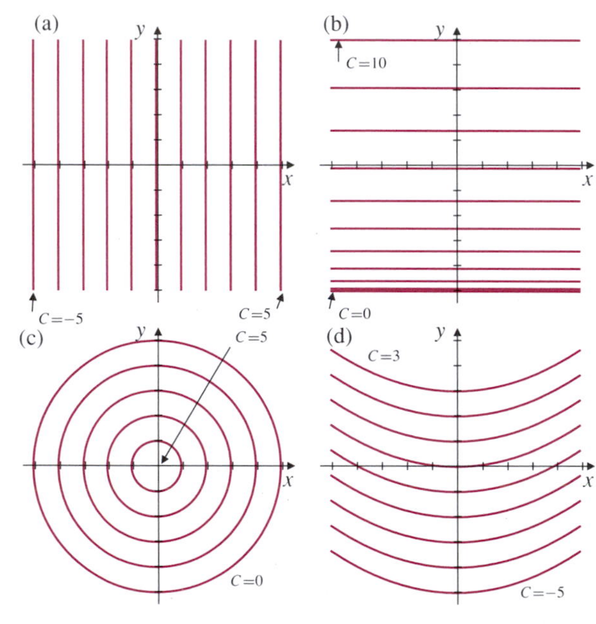


Figure 4.13

Describe the graphs of the functions $f(x, y)$ for which families of level curves $f(x, y) = C$ are shown in the figures referred to in Exercises 29–32. Assume that each family corresponds to equally spaced values of $C$, and that the behaviour of the family is representative of all such families for the function.

**29.** See Figure 4.13(a). **30.** See Figure 4.13(b).

**31.** See Figure 4.13(c). **32.** See Figure 4.13(d).

**33.** Are the curves $y = (x - C)^2$ level curves of a function $f(x, y)$? What property must a family of curves in a region of the $xy$-plane have to be the family of level curves of a function defined in the region?

**34.** If we assume $z \geq 0$, the equation $4z^2 = (x - z)^2 + (y - z)^2$ defines $z$ as a function of $x$ and $y$. Sketch some level curves of this function. Describe its graph.

**35.** Find $f(x, y)$ if each level curve $f(x, y) = C$ is a circle centred at the origin and having radius
(a) $C$, (b) $C^2$, (c) $\sqrt{C}$, (d) $\ln C$.

**36.** Find $f(x, y, z)$ if for each constant $C$ the level surface $f(x, y, z) = C$ is a plane having intercepts $C^3$, $2C^3$, and $3C^3$ on the $x$-axis, the $y$-axis and the $z$-axis respectively.

Describe the level surfaces of the functions specified in Exercises 37–41.

**37.** $f(x, y, z) = x^2 + y^2 + z^2$

**38.** $f(x, y, z) = x + 2y + 3z$

**39.** $f(x, y, z) = x^2 + y^2$ **40.** $f(x, y, z) = \dfrac{x^2 + y^2}{z^2}$

**41.** $f(x, y, z) = |x| + |y| + |z|$

**42.** Describe the "level hypersurfaces" of the function

$$f(x, y, z, t) = x^2 + y^2 + z^2 + t^2.$$

Use 3-dimensional computer graphing software to examine the graphs and the level curves of the functions in Exercises 43–50.

**43.** $\dfrac{1}{1 + x^2 + y^2}$ **44.** $\dfrac{\cos x}{1 + y^2}$

**45.** $\dfrac{y}{1 + x^2 + y^2}$ **46.** $\sin(xy)$

**47.** $\dfrac{\sin\sqrt{x^2 + y^2}}{1 + x^2 + y^2}$ **48.** $\dfrac{x}{(x^2 - 1)^2 + y^2}$

**49.** $xy$ **50.** $\dfrac{1}{xy}$

## 4.2 LIMITS AND CONTINUITY

Before defining the limit of a function of several variables, we introduce some terminology for describing sets of points. For simplicity, we phrase these definitions for sets in the plane $\mathbb{R}^2$, but only minor modifications are necessary for sets in 3-space or spaces of higher dimension.

**DEFINITION 2**

> If $P_0 = (x_0, y_0)$ is a point in the plane and if $r > 0$, then the set of points $P = (x, y)$ in the plane whose distances from $P_0$ are less than $r$ is called an **open disk** of radius $r$ about $P_0$, or, more simply, a **neighbourhood** of $P_0$. If we denote this neighbourhood by $N_r(P_0)$, then
>
> $$N_r(P_0) = \{(x, y) \,:\, (x - x_0^2 + (y - y_0)^2 < r^2\}.$$

**DEFINITION 3**

> Let $S$ be a set of points in the plane.
>
> (a) We say that a point $P_0$ is a **boundary point** of $S$ if every neighbourhood of $P_0$ contains at least one point in $S$ and at least one point not in $S$. The set of all boundary points of $S$ is called **the boundary** of $S$.
>
> (b) We say that $S$ is **closed** (or is a **closed set**) if every boundary point of $S$ belongs to $S$.
>
> (c) We say that $S$ is **open** (or is an **open set**) if no boundary point of $S$ belongs to $S$.
>
> (d) The **interior** of $S$ is the set of all points in $S$ that are not boundary points of $S$. The **exterior** of $S$ is the set of all points that are not in $S$ and not boundary points of $S$.

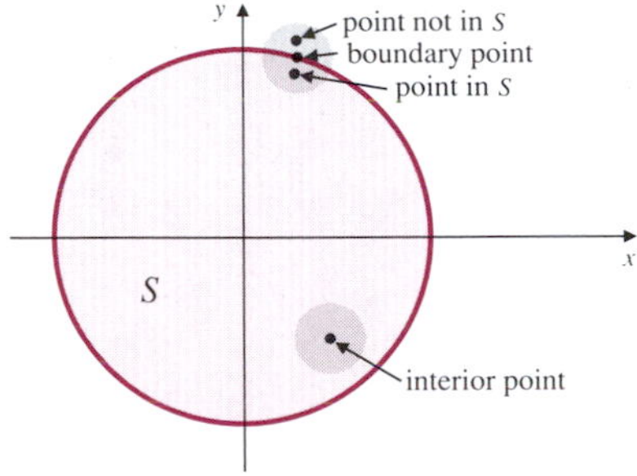


**Figure 4.14** The closed disk $x^2 + y^2 \le 1$, its boundary circle, and interior open disk. Observe the neighbourhoods of the boundary point and the interior point

**EXAMPLE 1** Consider the disk $S$ whose points $(x, y)$ satisfy the inequality $x^2 + y^2 \le 1$. The boundary of $S$ is the circle $x^2 + y^2 = 1$; any neighbourhood of a point on this circle contains points in $S$ and points not in $S$, as shown in Figure 4.14. Since all the boundary points belong to $S$, $S$ is a closed set. The interior of $S$ consists of all points $(x, y)$ that satisfy $x^2 + y^2 < 1$; it is an open disk of radius 1 centred at the origin. The exterior consists of points satisfying $x^2 + y^2 > 1$, which is also an open set. ■

***REMARK*** You should compare the notions of open and closed sets and boundary points with the similar one-dimensional notions of open and closed intervals and endpoints. The definitions above can all be extended to sets in 3-space or spaces of higher dimension modifying the definition of *neighbourhood* to be a *ball* of the appropriate dimension. The neighbourhood $N_r(P_0)$ of a point $P_0 = (x_0, y_0, z_0)$ in 3-space is the open ball

$$N_r(P_0) = \{(x, y, z) \,:\, (x - x_0^2 + (y - y_0)^2 + (z - z_0)^2 < r^2\}.$$

The concept of the limit of a function of several variables is similar to that for functions of one variable. For clarity we present the definition for functions of two variables only; the general case is similar.

We say that $f(x, y)$ approaches the limit $L$ as the point $(x, y)$ approaches the point $(a, b)$, and we write

$$\lim_{(x,y)\to(a,b)} f(x, y) = L,$$

if all points of a neighbourhood of $(a, b)$, except possibly the point $(a, b)$ itself, belong to the domain of $f$, and if $f(x, y)$ approaches $L$ as $(x, y)$ approaches $(a, b)$. We make this more formal in the following definition.

**DEFINITION 4**

We say that $\lim_{(x,y)\to(a,b)} f(x, y) = L$ if and only if for every positive number $\epsilon$ there exists a positive number $\delta = \delta(\epsilon)$, depending on $\epsilon$, such that

$$|f(x, y) - L| < \epsilon \quad \text{whenever} \quad 0 < \sqrt{(x-a)^2 + (y-b)^2} < \delta.$$

If a limit exists it is unique. For a single-variable function $f$, the existence of $\lim_{x\to a} f(x)$ implies that $f(x)$ approaches the same finite number as $x$ approaches $a$ from either the right or the left. Similarly, for a function of two variables, $\lim_{(x,y)\to(a,b)} f(x, y) = L$ exists only if $f(x, y)$ approaches the same number $L$ no matter how $(x, y)$ approaches $(a, b)$ in the $xy$-plane. In particular, $(x, y)$ can approach $(a, b)$ along any curve through $(a, b)$. It is not necessary that $L = f(a, b)$ even if $f(a, b)$ is defined. The examples below illustrate these assertions.

All the usual laws of limits extend to functions of several variables in the obvious way. For example, if $\lim_{(x,y)\to(a,b)} f(x, y) = L$ and $\lim_{(x,y)\to(a,b)} g(x, y) = M$, then

$$\lim_{(x,y)\to(a,b)} \big(f(x, y) \pm g(x, y)\big) = L \pm M,$$

$$\lim_{(x,y)\to(a,b)} f(x, y)\, g(x, y) = LM,$$

$$\lim_{(x,y)\to(a,b)} \frac{f(x, y)}{g(x, y)} = \frac{L}{M}, \qquad \text{provided } M \neq 0.$$

Also, if $F(t)$ is continuous at $t = L$, then

$$\lim_{(x,y)\to(a,b)} F\big(f(x, y)\big) = F(L).$$

**EXAMPLE 2**

(a) $\displaystyle\lim_{(x,y)\to(2,3)} 2x - y^2 = 4 - 9 = -5,$

(b) $\displaystyle\lim_{(x,y)\to(a,b)} x^2 y = a^2 b,$

(c) $\displaystyle\lim_{(x,y)\to(\pi/3,2)} y \sin(x/y) = 2\sin(\pi/6) = 1.$ ■

The following examples show that the requirement that $f(x, y)$ approach the same limit *no matter how* $(x, y)$ approaches $(a, b)$ can be very restrictive.

**EXAMPLE 3** Investigate the limiting behaviour of $f(x, y) = \dfrac{2xy}{x^2 + y^2}$ as $(x, y)$ approaches $(0, 0)$.

***SOLUTION*** Note that $f(x, y)$ is defined at all points of the $xy$-plane except the origin $(0, 0)$. We can still ask whether $\lim_{(x,y)\to(0,0)} f(x, y)$ exists or not. If we let $(x, y)$ approach $(0, 0)$ along the $x$-axis ($y = 0$), then $f(x, y) = f(x, 0) \to 0$ (because $f(x, 0) = 0$ identically). Thus $\lim_{(x,y)\to(0,0)} f(x, y)$ must be 0 if it exists at all. Similarly, at all points of the $y$-axis we have $f(x, y) = f(0, y) = 0$. However, at points of the line $x = y$, $f$ has a different constant value; $f(x, x) = 1$. Since the limit of $f(x, y)$ is 1 as $(x, y)$ approaches $(0, 0)$ along this line, it follows that $f(x, y)$ cannot have a unique limit at the origin. That is,

$$\lim_{(x,y)\to(0,0)} \frac{2xy}{x^2+y^2} \quad \text{does not exist.}$$

Observe that $f(x, y)$ has a constant value on any ray from the origin (on the ray $y = kx$ the value is $2k/(1 + k^2)$), but these values differ on different rays. The level curves of $f$ are the rays from the origin (with the origin itself removed). It is difficult to sketch the graph of $f$ near the origin. The first-octant part of the graph is the "hood-shaped" surface in Figure 4.15(a). ■

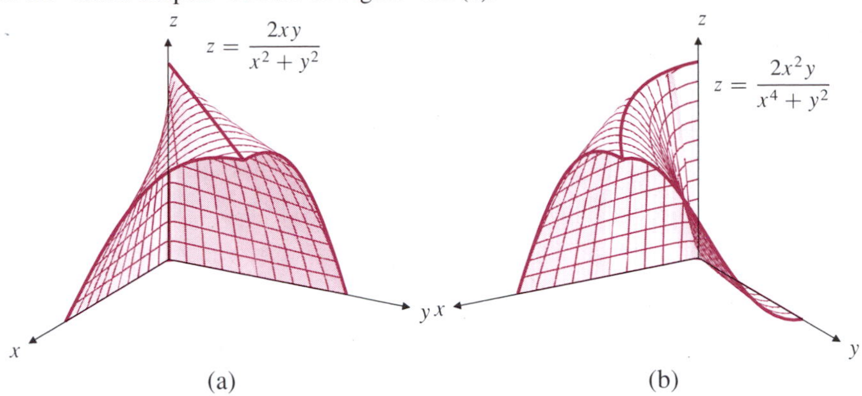

**Figure 4.15**

(a) $f(x, y)$ has different limits as $(x, y) \to (0, 0)$ along different straight lines

(b) $f(x, y)$ has the same limit 0 as $(x, y) \to (0, 0)$ along any straight line, but has limit 1 as $(x, y) \to (0, 0)$ along $y = x^2$

**■ EXAMPLE 4** Investigate the limiting behaviour of $f(x, y) = \dfrac{2x^2y}{x^4+y^2}$ as $(x, y)$ approaches $(0, 0)$.

***SOLUTION*** As in Example 3, $f(x, y)$ vanishes identically on the coordinate axes so $\lim_{(x,y)\to(0,0)} f(x, y)$ must be 0 if it exists at all. If we examine $f(x, y)$ at points of the ray $y = kx$, we obtain

$$f(x, kx) = \frac{2kx^3}{x^4+k^2x^2} = \frac{2kx}{x^2+k^2} \to 0, \qquad \text{as} \quad x \to 0 \quad (k \neq 0).$$

Thus $f(x, y) \to 0$ as $(x, y) \to (0, 0)$ along *any* straight line through the origin. We might be tempted to conclude, therefore, that $\lim_{(x,y)\to(0,0)} f(x, y) = 0$, but this is incorrect. Observe the behaviour of $f(x, y)$ along the curve $y = x^2$:

$$f(x, x^2) = \frac{2x^4}{x^4+x^4} = 1.$$

Thus $f(x, y)$ does not approach 0 as $(x, y)$ approaches the origin along this curve, and so $\lim_{(x,y)\to(0,0)} f(x, y)$ does not exist. The level curves of $f$ are pairs of parabolas of the form $y = kx^2$, $y = x^2/k$ with the origin removed. See Figure 4.15(b) for the first octant part of the graph of $f$. ■

**■ EXAMPLE 5** Show that the function $f(x, y) = \dfrac{x^2y}{x^2+y^2}$ does have a limit at the origin; specifically,

$$\lim_{(x,y)\to(0,0)} \frac{x^2y}{x^2+y^2} = 0.$$

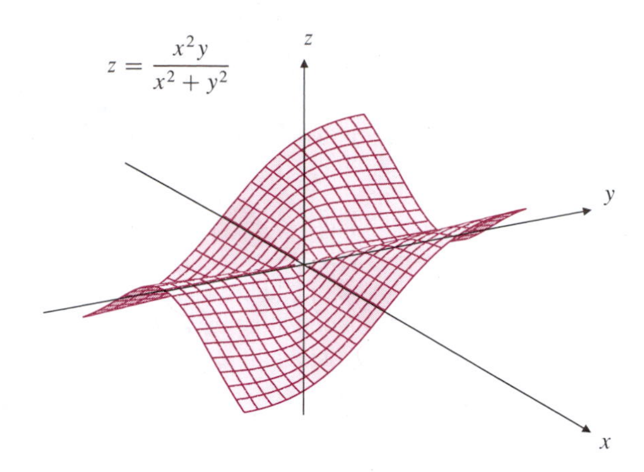


**Figure 4.16** $\lim_{(x,y)\to(0,0)} \frac{x^2 y}{x^2 + y^2} = 0$

***SOLUTION*** This function is also defined everywhere except at the origin. Observe that since $x^2 \le x^2 + y^2$, we have

$$|f(x, y) - 0| = \left| \frac{x^2 y}{x^2 + y^2} \right| \le |y| \le \sqrt{x^2 + y^2},$$

which approaches zero as $(x, y) \to (0, 0)$. (See Figure 4.16.) Formally, if $\epsilon > 0$ is given and we take $\delta = \epsilon$, then $|f(x, y) - 0| < \epsilon$ whenever $0 < \sqrt{x^2 + y^2} < \delta$, so $f(x, y)$ has limit 0 as $(x, y) \to (0, 0)$ by Definition 4. ■

As for functions of one variable, continuity at an interior point of the domain is defined directly in terms of the limit.

**DEFINITION 5**

The function $f(x, y)$ is continuous at the point $(a, b)$ if

$$\lim_{(x,y)\to(a,b)} f(x, y) = f(a, b).$$

It remains true that sums, differences, products, quotients, and compositions of continuous functions are continuous. The functions of Examples 3 and 4 above are continuous wherever they are defined, that is, at all points except the origin. There is no way to define $f(0, 0)$ so that these functions become continuous at the origin. They show that the continuity of the single-variable functions $f(x, b)$ at $x = a$ and $f(a, y)$ at $y = b$ does *not* imply that $f(x, y)$ is continuous at $(a, b)$. In fact, even if $f(x, y)$ is continuous along every straight line through $(a, b)$ it still need not be continuous at $(a, b)$. (See Exercises 24–25 below.) Note, however, that the function $f(x, y)$ of Example 5, though not defined at the origin, has a continuous extension to that point. If we extend the domain of $f$ by defining $f(0, 0) = \lim_{(x,y)\to(0,0)} f(x, y) = 0$, then $f$ is continuous on the whole $xy$-plane.

As for functions of one variable, the existence of a limit of a function at a point does not imply that the function is continuous at that point. The function

$$f(x, y) = \begin{cases} 0 & \text{if } (x, y) \ne (0, 0) \\ 1 & \text{if } (x, y) = (0, 0) \end{cases}$$

satisfies $\lim_{(x,y)\to(0,0)} f(x, y) = 0$, which is not equal to $f(0, 0)$, so $f$ is not continuous at $(0, 0)$. Of course we can *make* $f$ continuous at $(0, 0)$ by redefining its value at that point to be 0.

Definition 4 requires that $f(x, y)$ be defined at all points of a neighbourhood of $(a, b)$ (except the point $(a, b)$ itself) in order that $\lim_{(x,y)\to(a,b)} f(x, y)$ can exist. This means that continuity is defined by Definition 5 only for *interior* points of the domain of $f$. We can modify the definition of limit to allow for limits at points on the *boundary* of the domain of $f$ in a manner somewhat similar to that used for *one-sided limits* of functions of one variable. Definition 6 below is thus an extension of Definition 4. With this extended definition of limit, Definition 5 can be used to determine the continuity of a function at a point on the boundary of its domain.

**DEFINITION 6**

**Extended definition of limit**

We say that $\lim_{(x,y)\to(a,b)} f(x, y) = L$, provided that

(i) every neighbourhood of $(a, b)$ contains points of the domain of $f$ different from $(a, b)$, and

(ii) for every positive number $\epsilon$ there exists a positive number $\delta = \delta(\epsilon)$ such that $|f(x, y) - L| < \epsilon$ holds whenever $(x, y)$ is in the domain of $f$ and satisfies $0 < \sqrt{(x-a)^2 + (y-b)^2} < \delta$.

Condition (i) is included in Definition 6 because it is not appropriate to consider limits at *isolated* points of the domain of $f$, that is, points with neighbourhoods that contain no other points of the domain.

**EXAMPLE 6** The function $f(x, y) = \sqrt{1 - x^2 - y^2}$ is continuous at all points of its domain, the *closed* disk $x^2 + y^2 \le 1$. Of course $(x, y)$ can approach points of the bounding circle $x^2 + y^2 = 1$ only from within the disk. ■

## EXERCISES 4.2

In Exercises 1–4, specify the boundary and the interior of the plane sets $S$ whose points $(x, y)$ satisfy the given conditions. Is $S$ open, closed, or neither?

**1.** $0 < x^2 + y^2 < 1$

**2.** $x \ge 0, \quad y < 0$

**3.** $x + y = 1$

**4.** $|x| + |y| \le 1$

In Exercises 5–8, specify the boundary and the interior of the sets $S$ in 3-space whose points $(x, y, z)$ satisfy the given conditions. Is $S$ open, closed, or neither?

**5.** $1 \le x^2 + y^2 + z^2 \le 4$

**6.** $x \ge 0, \quad y > 1, \quad z < 2$

**7.** $(x - z)^2 + (y - z)^2 = 0$

**8.** $x^2 + y^2 < 1, \ y + z > 2$

In Exercises 9–20, evaluate the indicated limit or explain why it does not exist.

**9.** $\displaystyle\lim_{(x,y)\to(2,-1)} xy + x^2$

**10.** $\displaystyle\lim_{(x,y)\to(0,0)} \sqrt{x^2 + y^2}$

**11.** $\displaystyle\lim_{(x,y)\to(0,0)} \frac{x^2 + y^2}{y}$

**12.** $\displaystyle\lim_{(x,y)\to(0,0)} \frac{x}{x^2 + y^2}$

**13.** $\displaystyle\lim_{(x,y)\to(1,\pi)} \frac{\cos(xy)}{1 - x - \cos y}$

**14.** $\displaystyle\lim_{(x,y)\to(0,1)} \frac{x^2(y-1)^2}{x^2 + (y-1)^2}$

**15.** $\displaystyle\lim_{(x,y)\to(0,0)} \frac{y^3}{x^2 + y^2}$

**16.** $\displaystyle\lim_{(x,y)\to(0,0)} \frac{\sin(x - y)}{\cos(x + y)}$

**17.** $\displaystyle\lim_{(x,y)\to(0,0)} \frac{\sin(xy)}{x^2 + y^2}$

**18.** $\displaystyle\lim_{(x,y)\to(1,2)} \frac{2x^2 - xy}{4x^2 - y^2}$

**19.** $\displaystyle\lim_{(x,y)\to(0,0)} \frac{x^2y^2}{x^2 + y^4}$

**20.** $\displaystyle\lim_{(x,y)\to(0,0)} \frac{x^2y^2}{2x^4 + y^4}$

**21.** How can the function

$$f(x, y) = \frac{x^2 + y^2 - x^3y^3}{x^2 + y^2}, \qquad (x, y) \ne (0, 0),$$

be defined at the origin so that it becomes continuous at all points of the $xy$-plane?

**22.** How can the function

$$f(x, y) = \frac{x^3 - y^3}{x - y}, \qquad (x \ne y),$$

be defined along the line $x = y$ so that the resulting function is continuous on the whole $xy$-plane?

**23.** What is the domain of

$$f(x, y) = \frac{x - y}{x^2 - y^2}?$$

Does $f(x, y)$ have a limit by Definition 4 as $(x, y) \to (1, 1)$? By Definition 6? Can the domain of $f$ be extended so that the resulting function is continuous at $(1, 1)$? Can the domain be extended so that the resulting function is continuous everywhere in the $xy$-plane?

* **24.** Given a function $f(x, y)$ and a point $(a, b)$ in its domain, define single-variable functions $g$ and $h$ as follows:

$$g(x) = f(x, b), \qquad h(y) = f(a, y).$$

If $g$ is continuous at $x = a$ and $h$ is continuous at $y = b$ does it follow that $f$ is continuous at $(a, b)$? Conversely, does the continuity of $f$ at $(a, b)$ guarantee the continuity of $g$ at $a$ and the continuity of $h$ at $b$? Justify your answers.

* **25.** Let $\mathbf{u} = u\mathbf{i} + v\mathbf{j}$ be a unit vector, and let

$$f_{\mathbf{u}}(t) = f(a + tu, b + tv)$$

be the single-variable function obtained by restricting the domain of $f(x, y)$ to points of the straight line through $(a, b)$ parallel to $\mathbf{u}$. If $f_{\mathbf{u}}(t)$ is continuous at $t = 0$ for every unit vector $\mathbf{u}$, does it follow that $f$ is continuous at $(a, b)$? Conversely, does the continuity of $f$ at $(a, b)$ guarantee the continuity of $f_{\mathbf{u}}(t)$ at $t = 0$? Justify your answers.

* **26.** What condition must the nonnegative integers $m$, $n$, and $p$ satisfy to guarantee that $\lim_{(x,y)\to(0,0)} x^m y^n/(x^2 + y^2)^p$ exists? Prove your answer.

* **27.** What condition must the constants $a$, $b$, and $c$ satisfy to guarantee that $\lim_{(x,y)\to(0,0)} xy/(ax^2 + bxy + cy^2)$ exists in the sense of Definition 6? Prove your answer.

* **28.** Can the function $f(x, y) = \dfrac{\sin x \sin^3 y}{1 - \cos(x^2 + y^2)}$ be defined at $(0, 0)$ in such a way that it becomes continuous there? If so, how?

**29.** Use 2- and 3-dimensional mathematical graphing software to examine the graph and level curves of the function $f(x, y)$ of Example 3 on the region $-1 \le x \le 1$, $-1 \le y \le 1$, $(x, y) \ne (0, 0)$ How would you describe the behaviour of the graph near $(x, y) = (0, 0)$?

**30.** Use 2- and 3-dimensional mathematical graphing software to examine the graph and level curves of the function $f(x, y)$ of Example 4 on the region $-1 \le x \le 1$, $-1 \le y \le 1$, $(x, y) \ne (0, 0)$ How would you describe the behaviour of the graph near $(x, y) = (0, 0)$?

**31.** The graph of a single-variable function $f(x)$ that is continuous on an interval is a curve that has no *breaks* in it there and that intersects any vertical line through a point in the interval exactly once. What analogous statement can you make about the graph of a bivariate function $f(x, y)$ that is continuous on a region of the $xy$-plane?

## 4.3 PARTIAL DERIVATIVES

In this section we begin the process of extending the concepts and techniques of single-variable calculus to functions of more than one variable. It is convenient to begin by considering the rate of change of such functions with respect to one variable at a time. Thus a function of $n$ variables has $n$ *first-order partial derivatives,* one with respect to each of its independent variables. For a function of two variables we make this precise in the following definition.

**DEFINITION 7**

The **first partial derivatives** of the function $f(x, y)$ **with respect to the variables** $x$ and $y$ are the functions $f_1(x, y)$ and $f_2(x, y)$ given by

$$f_1(x, y) = \lim_{h\to 0} \frac{f(x + h, y) - f(x, y)}{h}$$

$$f_2(x, y) = \lim_{k\to 0} \frac{f(x, y + k) - f(x, y)}{k}$$

provided these limits exist.

Each of the two partial derivatives is the limit of a Newton quotient in one of the variables. Observe that $f_1(x, y)$ is just the ordinary first derivative of $f(x, y)$ considered as a function of $x$ only, regarding $y$ as a constant parameter. Similarly, $f_2(x, y)$ is the first derivative of $f(x, y)$ considered as a function of $y$ alone, with $x$ held fixed.

**EXAMPLE 1** If $f(x, y) = x^2 \sin y$, then

$$f_1(x, y) = 2x \sin y \qquad \text{and} \qquad f_2(x, y) = x^2 \cos y.$$

The subscripts "1" and "2" in the notations for the partial derivatives refer to the "first" and "second" variables of $f$. For functions of one variable we use the notation $f'$ for the derivative; the *prime* ($'$) denotes differentiation with respect to the only variable on which $f$ depends. For functions $f$ of two variables we use $f_1$ or $f_2$ to show the variable of differentiation. Do not confuse these subscripts with subscripts used for other purposes, for example to denote the components of vectors.

The partial derivative $f_1(a, b)$ measures the rate of change of $f(x, y)$ with respect to $x$ at $x = a$ while $y$ is held fixed at $b$. In graphical terms, the surface $z = f(x, y)$ intersects the vertical plane $y = b$ in a curve. If we take horizontal and vertical lines through the point $(0, b, 0)$ as coordinate axes in the plane $y = b$, then the curve has equation $z = f(x, b)$, and its slope at $x = a$ is $f_1(a, b)$. (See Figure 4.17.) Similarly, $f_2(a, b)$ represents the rate of change of $f$ with respect to $y$ at $y = b$ with $x$ held fixed at $a$. The surface $z = f(x, y)$ intersects the vertical plane $x = a$ in a curve $z = f(a, y)$ whose slope at $y = b$ is $f_2(a, b)$. (See Figure 4.18.)

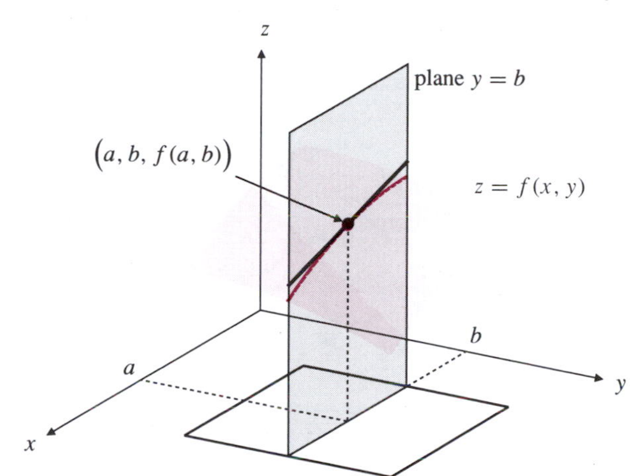


**Figure 4.17** $f_1(a, b)$ is the slope of the curve of intersection of $z = f(x, y)$ and the vertical plane $y = b$ at $x = a$

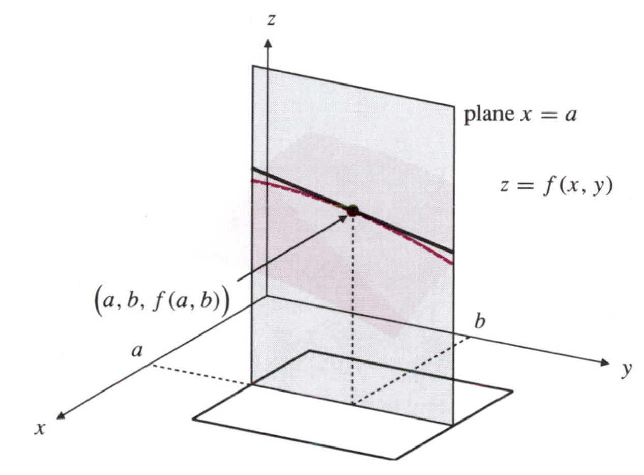


**Figure 4.18** $f_2(a, b)$ is the slope of the curve of intersection of $z = f(x, y)$ and the vertical plane $x = a$ at $y = b$

Various notations can be used to denote the partial derivatives of $z = f(x, y)$ considered as functions of $x$ and $y$:

**Notations for first partial derivatives**

$$\frac{\partial z}{\partial x} = \frac{\partial}{\partial x} f(x, y) = f_1(x, y) = D_1 f(x, y)$$

$$\frac{\partial z}{\partial y} = \frac{\partial}{\partial y} f(x, y) = f_2(x, y) = D_2 f(x, y)$$

The symbol "$\partial$" should be read as "partial" so "$\partial z/\partial x$" is "partial $z$ with respect to $x$." The reason for distinguishing "$\partial$" from the "$d$" of ordinary derivatives of single-variable functions will be made clear later.

Values of partial derivatives at a particular point $(a, b)$ are denoted similarly:

**Values of partial derivatives**

$$\left.\frac{\partial z}{\partial x}\right|_{(a,b)} = \left.\left(\frac{\partial}{\partial x} f(x,y)\right)\right|_{(a,b)} = f_1(a,b) = D_1 f(a,b)$$

$$\left.\frac{\partial z}{\partial y}\right|_{(a,b)} = \left.\left(\frac{\partial}{\partial y} f(x,y)\right)\right|_{(a,b)} = f_2(a,b) = D_2 f(a,b)$$

Some authors prefer to use $f_x$ or $\partial f/\partial x$ and $f_y$ or $\partial f/\partial y$ instead of $f_1$ and $f_2$. However, this can lead to problems of ambiguity when compositions of functions arise. For instance, suppose $f(x,y) = x^2 y$. By $f_1(x^2, xy)$ we clearly mean

$$\left.\left(\frac{\partial}{\partial x} f(x,y)\right)\right|_{(x^2,xy)} = \left. 2xy \right|_{(x^2,xy)} = (2)(x^2)(xy) = 2x^3 y.$$

But does $f_x(x^2, xy)$ mean the same thing? One could argue that $f_x(x^2, xy)$ should mean

$$\frac{\partial}{\partial x}\Big(f(x^2, xy)\Big) = \frac{\partial}{\partial x}\Big((x^2)^2(xy)\Big) = \frac{\partial}{\partial x}(x^5 y) = 5x^4 y.$$

In order to avoid such ambiguities we usually prefer to use $f_1$ and $f_2$ instead of $f_x$ and $f_y$. (However, in some situations where no confusion is likely to occur we may still use the notations $f_x$ and $f_y$, and also $\partial f/\partial x$ and $\partial f/\partial y$.)

All the standard differentiation rules for sums, products, reciprocals, and quotients continue to apply to partial derivatives.

**EXAMPLE 2** Find $\partial z/\partial x$ and $\partial z/\partial y$ if $z = x^3 y^2 + x^4 y + y^4$.

***SOLUTION*** $\partial z/\partial x = 3x^2 y^2 + 4x^3 y$ and $\partial z/\partial y = 2x^3 y + x^4 + 4y^3$. ■

**EXAMPLE 3** Find $f_1(0, \pi)$ if $f(x,y) = e^{xy}\cos(x+y)$.

***SOLUTION***

$$f_1(x,y) = y\,e^{xy}\cos(x+y) - e^{xy}\sin(x+y),$$
$$f_1(0,\pi) = \pi\, e^0 \cos(\pi) - e^0 \sin(\pi) = -\pi.$$ ■

The single-variable version of the Chain Rule also continues to apply to, say, $f\big(g(x,y)\big)$:

$$\frac{\partial}{\partial x} f\big(g(x,y)\big) = f'\big(g(x,y)\big)\, g_1(x,y), \quad \frac{\partial}{\partial y} f\big(g(x,y)\big) = f'\big(g(x,y)\big)\, g_2(x,y).$$

We will develop versions of the Chain Rule for more complicated compositions of multivariate functions in Section 4.5.

**EXAMPLE 4** If $f$ is an everywhere differentiable function of one variable show that $z = f(x/y)$ satisfies the *partial differential equation*

$$x\frac{\partial z}{\partial x} + y\frac{\partial z}{\partial y} = 0.$$

***SOLUTION*** By the (single-variable) Chain Rule,

$$\frac{\partial z}{\partial x} = f'\left(\frac{x}{y}\right)\left(\frac{1}{y}\right) \quad \text{and} \quad \frac{\partial z}{\partial y} = f'\left(\frac{x}{y}\right)\left(\frac{-x}{y^2}\right).$$

Hence

$$x\frac{\partial z}{\partial x} + y\frac{\partial z}{\partial y} = f'\left(\frac{x}{y}\right)\left(x \times \frac{1}{y} + y \times \frac{-x}{y^2}\right) = 0.$$

■

Definition 7 can be extended in the obvious way to cover functions of more than two variables. If $f$ is a function of $n$ variables $x_1, x_2, \ldots, x_n$ then $f$ has $n$ first partial derivatives, $f_1(x_1, x_2, \ldots, x_n)$, $f_2(x_1, x_2, \ldots, x_n)$, ..., $f_n(x_1, x_2, \ldots, x_n)$, one with respect to each variable.

**EXAMPLE 5**

$$\frac{\partial}{\partial z}\left(\frac{2xy}{1 + xz + yz}\right) = -\frac{2xy}{(1 + xz + yz)^2}(x + y).$$

Again all the standard differentiation rules are applied to calculate partial derivatives.

■

***REMARK*** If a single-variable function $f(x)$ has a derivative $f'(a)$ at $x = a$, then $f$ is necessarily continuous at $x = a$. This property does *not* extend to partial derivatives. Even if all the first partial derivatives of a function of several variables exist at a point, the function may still fail to be continuous at that point. See Exercise 36 below.

## Tangent Planes and Normal Lines

If the graph $z = f(x, y)$ is a "smooth" surface near the point $P$ with coordinates $(a, b, f(a, b))$, then that graph will have a **tangent plane** and a **normal line** at $P$. The normal line is the line through $P$ that is perpendicular to the surface; for instance, a line joining a point on a sphere to the centre of the sphere is normal to the sphere. Any nonzero vector that is parallel to the normal line at $P$ is called a normal vector to the surface at $P$. The tangent plane to the surface $z = f(x, y)$ at $P$ is the plane through $P$ that is perpendicular to the normal line at $P$.

Let us assume that the surface $z = f(x, y)$ has a *nonvertical* tangent plane (and therefore a *nonhorizontal* normal line) at point $P$. (Later in this chapter we will state precise conditions that guarantee that the graph of a function has a nonvertical tangent plane at a point.) The tangent plane intersects the vertical plane $y = b$ in a straight line that is tangent at $P$ to the curve of intersection of the surface $z = f(x, y)$ and the plane $y = b$. (See Figures 4.17 and 4.19.) This line has slope $f_1(a, b)$, and so is parallel to the vector $\mathbf{T}_1 = \mathbf{i} + f_1(a, b)\mathbf{k}$. Similarly, the tangent plane intersects the vertical plane $x = a$ in a straight line having slope $f_2(a, b)$. This line is therefore parallel to the vector $\mathbf{T}_2 = \mathbf{j} + f_2(a, b)\mathbf{k}$. It follows that the tangent plane, and therefore the surface $z = f(x, y)$ itself, has normal vector

$$\mathbf{n} = \mathbf{T}_2 \times \mathbf{T}_1 = \begin{vmatrix} \mathbf{i} & \mathbf{j} & \mathbf{k} \\ 0 & 1 & f_2(a, b) \\ 1 & 0 & f_1(a, b) \end{vmatrix} = f_1(a, b)\mathbf{i} + f_2(a, b)\mathbf{j} - \mathbf{k}.$$

**Normal vector to $z = f(x, y)$ at $(a, b, f(a, b))$**

$$\mathbf{n} = f_1(a, b)\mathbf{i} + f_2(a, b)\mathbf{j} - \mathbf{k}$$

Since the tangent plane passes through $P = (a, b, f(a, b))$, it has equation

$$f_1(a, b)(x - a) + f_2(a, b)(y - b) - (z - f(a, b)) = 0,$$

or, equivalently,

**Equation of the tangent plane to $z = f(x, y)$ at $\big(a, b, f(a, b)\big)$**

$$z = f(a, b) + f_1(a, b)(x - a) + f_2(a, b)(y - b).$$

We shall obtain this result by a different method in Section 4.7.

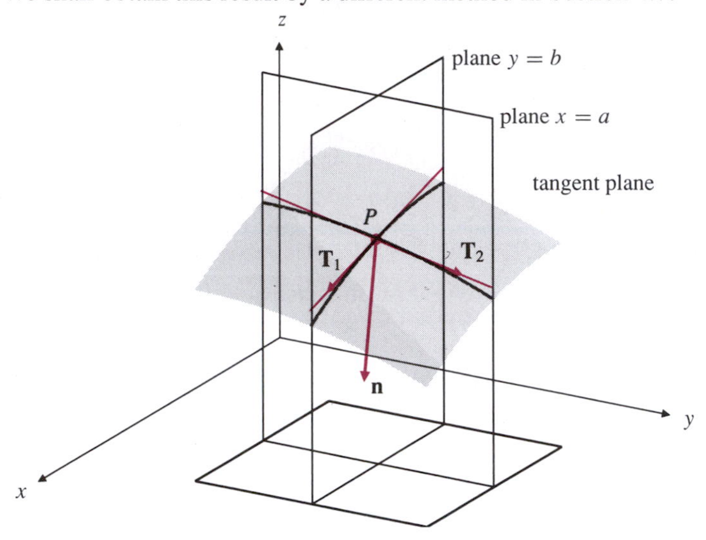


**Figure 4.19** The tangent plane and a normal vector to $z = f(x, y)$ at $P = \big(a, b, f(a, b)\big)$

We can write equations for the normal line to the graph of $f$ at $P$:

The normal line to $z = f(x, y)$ at $\big(a, b, f(a, b)\big)$ has direction vector $f_1(a, b)\mathbf{i} + f_2(a, b)\mathbf{j} - \mathbf{k}$, and so has equations

$$\frac{x - a}{f_1(a, b)} = \frac{y - b}{f_2(a, b)} = \frac{z - f(a, b)}{-1}$$

with suitable modifications if either $f_1(a, b) = 0$ or $f_2(a, b) = 0$.

**EXAMPLE 6** Find a normal vector and equations of the tangent plane and normal line to the graph $z = \sin(xy)$ at the point where $x = \pi/3$ and $y = -1$.

***SOLUTION*** The point on the graph has coordinates $(\pi/3, -1, -\sqrt{3}/2)$ Now

$$\frac{\partial z}{\partial x} = y\cos(xy) \qquad \text{and} \qquad \frac{\partial z}{\partial y} = x\cos(xy).$$

At $(\pi/3, -1)$ we have $\partial z/\partial x = -1/2$ and $\partial z/\partial y = \pi/6$. Therefore the surface has normal vector $\mathbf{n} = -(1/2)\mathbf{i} + (\pi/6)\mathbf{j} - \mathbf{k}$, and tangent plane

$$z = \frac{-\sqrt{3}}{2} - \frac{1}{2}\left(x - \frac{\pi}{3}\right) + \frac{\pi}{6}(y + 1),$$

or, more simply, $3x - \pi y + 6z = 2\pi - 3\sqrt{3}$. The normal line has equation

$$\frac{x - \frac{\pi}{3}}{\frac{-1}{2}} = \frac{y + 1}{\frac{\pi}{6}} = \frac{z + \frac{\sqrt{3}}{2}}{-1} \quad \text{or} \quad \frac{6x - 2\pi}{-3} = \frac{6y + 6}{\pi} = \frac{6z + 3\sqrt{3}}{-6}.$$

**EXAMPLE 7** What horizontal plane is tangent to the surface

$$z = x^2 - 4xy - 2y^2 + 12x - 12y - 1$$

and what is the point of tangency?

***SOLUTION*** A plane is horizontal only if its equation is of the form $z = k$, that is, it is independent of $x$ and $y$. Therefore we must have $\partial z/\partial x = \partial z/\partial y = 0$ at the point of tangency. The equations

$$\frac{\partial z}{\partial x} = 2x - 4y + 12 = 0$$
$$\frac{\partial z}{\partial y} = -4x - 4y - 12 = 0$$

have solution $x = -4$, $y = 1$. For these values we have $z = -31$ so the required tangent plane has equation $z = -31$ and the point of tangency is $(-4, 1, -31)$. ■

## Distance from a Point to a Surface: A Geometric Example

**EXAMPLE 8** Find the distance from the point $(3, 0, 0)$ to the hyperbolic paraboloid with equation $z = x^2 - y^2$.

***SOLUTION*** This is an optimization problem of a sort we will deal with in a more systematic way in the next chapter. However, such problems involving minimizing distances from points to surfaces can frequently be solved using geometric methods. If $Q = (X, Y, Z)$ is the point on the surface $z = x^2 - y^2$ that is closest to $P = (3, 0, 0)$ then the vector $\overrightarrow{PQ} = (X - 3)\mathbf{i} + Y\mathbf{j} + Z\mathbf{k}$ must be normal to the surface at $Q$. (See Figure 4.20(a).) Using the partial derivatives of $z = x^2 - y^2$ we know that the vector $\mathbf{n} = 2X\mathbf{i} - 2Y\mathbf{j} - \mathbf{k}$ is normal to the surface at $Q$. Thus $\overrightarrow{PQ}$ must be parallel to $\mathbf{n}$, and so $\overrightarrow{PQ} = t\mathbf{n}$ for some scalar $t$. Separated into components, this vector equation states that

$$X - 3 = 2Xt, \qquad Y = -2Yt, \qquad \text{and} \qquad Z = -t.$$

The middle equation implies that either $Y = 0$ or $t = -\frac{1}{2}$. We must consider both of these possibilities.

CASE I: If $Y = 0$, then

$$X = \frac{3}{1 - 2t} \qquad \text{and} \qquad Z = -t.$$

But $Z = X^2 - Y^2$, so we must have

$$-t = \frac{9}{(1 - 2t)^2}.$$

This is a cubic equation in $t$, so we might expect to have to solve it numerically, for instance by using Newton's method. However, if we try small integer values of $t$, we will quickly discover that $t = -1$ is a solution. The graphs of both sides of the equation are shown in Figure 4.20(b). They show that $t = -1$ is the only real solution. Calculating the corresponding values of $X$ and $Z$, we obtain $(1, 0, 1)$ as a candidate for $Q$. The distance from this point to $P$ is $\sqrt{5}$.

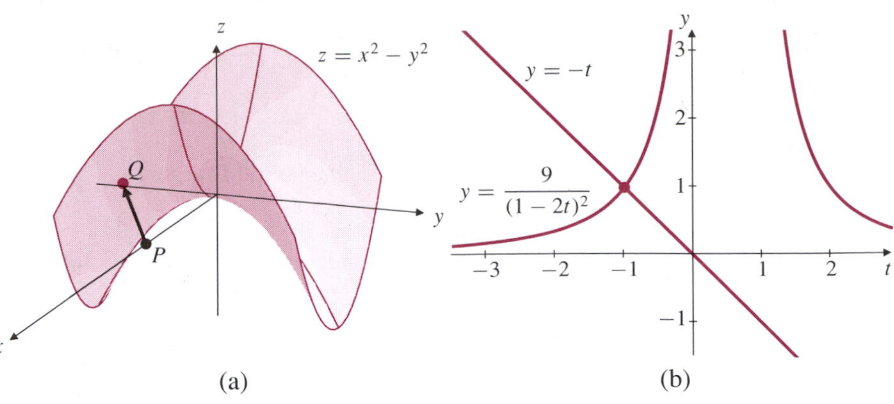


**Figure 4.20**

(a) If $Q$ is the point on $z = x^2 - y^2$ closest to $P$ then $\overrightarrow{PQ}$ is normal to the surface

(b) Equation $-t = \dfrac{9}{(1-2t)^2}$ has only one real root, $t = -1$

CASE II: If $t = -1/2$, then $X = 3/2$, $Z = 1/2$, and $Y = \pm\sqrt{X^2 - Z} = \pm\sqrt{7}/2$, and the distance from these points to $P$ is $\sqrt{17}/2$.

Since $\frac{17}{4} < 5$, the points $(3/2, \pm\sqrt{7}/2, 1/2)$ are the points on $z = x^2 - y^2$ closest to $(3, 0, 0)$, and the distance from (3,0,0) to the surface is $\sqrt{17}/2$ units. ■

## EXERCISES 4.3

In Exercises 1–10, find all the first partial derivatives of the functions specified and evaluate them at the given point.

**1.** $f(x, y) = x - y + 2, \quad (3, 2)$

**2.** $f(x, y) = xy + x^2, \quad (2, 0)$

**3.** $f(x, y, z) = x^3y^4z^5, \quad (0, -1, -1)$

**4.** $g(x, y, z) = \dfrac{xz}{y+z}, \quad (1, 1, 1)$

**5.** $z = \tan^{-1}\left(\dfrac{y}{x}\right), \quad (-1, 1)$

**6.** $w = \ln(1 + e^{xyz}), \quad (2, 0, -1)$

**7.** $f(x, y) = \sin(x\sqrt{y}), \quad \left(\dfrac{\pi}{3}, 4\right)$

**8.** $f(x, y) = \dfrac{1}{\sqrt{x^2+y^2}}, \quad (-3, 4)$

**9.** $w = x^{(y \ln z)}, \quad (e, 2, e)$

**10.** $g(x_1, x_2, x_3, x_4) = \dfrac{x_1 - x_2^2}{x_3 + x_4^2}, \quad (3, 1, -1, -2)$

In Exercises 11–12, calculate the first partial derivatives of the given functions at (0, 0). You will have to use Definition 7.

**11.** $f(x, y) = \begin{cases} \dfrac{2x^3 - y^3}{x^2 + 3y^2} & \text{if } (x, y) \neq (0, 0) \\ 0 & \text{if } (x, y) = (0, 0). \end{cases}$

**12.** $f(x, y) = \begin{cases} \dfrac{x^2 - 2y^2}{x - y} & \text{if } x \neq y \\ 0 & \text{if } x = y. \end{cases}$

In Exercises 13–22, find equations of the tangent plane and normal line to the graph of the given function at the point with specified values of $x$ and $y$.

**13.** $f(x, y) = x^2 - y^2$ at $(-2, 1)$

**14.** $f(x, y) = \dfrac{x-y}{x+y}$ at $(1, 1)$

**15.** $f(x, y) = \cos(x/y)$ at $(\pi, 4)$

**16.** $f(x, y) = e^{xy}$ at $(2, 0)$

**17.** $f(x, y) = \dfrac{x}{x^2+y^2}$ at $(1, 2)$

**18.** $f(x, y) = y\,e^{-x^2}$ at $(0, 1)$

**19.** $f(x, y) = \ln(x^2 + y^2)$ at $(1, -2)$

**20.** $f(x, y) = \dfrac{2xy}{x^2+y^2}$ at $(0, 2)$

**21.** $f(x, y) = \tan^{-1}(y/x)$ at $(1, -1)$

**22.** $f(x, y) = \sqrt{1 + x^3y^2}$ at $(2, 1)$

**23.** Find the coordinates of all points on the surface with equation $z = x^4 - 4xy^3 + 6y^2 - 2$ where the surface has a horizontal tangent plane.

**24.** Find all horizontal planes that are tangent to the surface with equation $z = xye^{-(x^2+y^2)/2}$. At what points are they tangent?

In Exercises 25–31, show that the given function satisfies the given partial differential equation.

❖ **25.** $z = x\,e^y, \quad x\dfrac{\partial z}{\partial x} = \dfrac{\partial z}{\partial y}$

❖ **26.** $z = \dfrac{x+y}{x-y}, \quad x\dfrac{\partial z}{\partial x} + y\dfrac{\partial z}{\partial y} = 0$

❖ **27.** $z = \sqrt{x^2+y^2}, \quad x\dfrac{\partial z}{\partial x} + y\dfrac{\partial z}{\partial y} = z$

❖ **28.** $w = x^2 + yz, \quad x\dfrac{\partial w}{\partial x} + y\dfrac{\partial w}{\partial y} + z\dfrac{\partial w}{\partial z} = 2w$

❖ **29.** $w = \dfrac{1}{x^2+y^2+z^2}, \quad x\dfrac{\partial w}{\partial x} + y\dfrac{\partial w}{\partial y} + z\dfrac{\partial w}{\partial z} = -2w$

❖ **30.** $z = f(x^2 + y^2)$, where $f$ is any differentiable function of one variable,

$$y\frac{\partial z}{\partial x} - x\frac{\partial z}{\partial y} = 0.$$

◆ **31.** $z = f(x^2 - y^2)$, where $f$ is any differentiable function of one variable,

$$y\frac{\partial z}{\partial x} + x\frac{\partial z}{\partial y} = 0.$$

**32.** Give a formal definition of the three first partial derivatives of the function $f(x, y, z)$.

**33.** What is an equation of the "tangent hyperplane" to the graph $w = f(x, y, z)$ at $\big(a, b, c, f(a, b, c)\big)$?

* **34.** Find the distance from the point $(1, 1, 0)$ to the circular paraboloid with equation $z = x^2 + y^2$.

* **35.** Find the distance from the point $(0, 0, 1)$ to the elliptic paraboloid having equation $z = x^2 + 2y^2$.

* **36.** Let $f(x, y) = \begin{cases} \dfrac{2xy}{x^2 + y^2}, & \text{if } (x, y) \neq (0, 0) \\ 0, & \text{if } (x, y) = (0, 0). \end{cases}$

Note that $f$ is not continuous at (0,0). (See Example 3 of Section 4.2.) Therefore its graph is not smooth there. Show, however, that $f_1(0, 0)$ and $f_2(0, 0)$ both exist. Hence the existence of partial derivatives does not imply that a function of several variables is continuous. This is in contrast to the single-variable case.

**37.** Determine $f_1(0, 0)$ and $f_2(0, 0)$ if they exist, where

$$f(x, y) = \begin{cases} (x^3 + y)\sin\dfrac{1}{x^2 + y^2} & \text{if } (x, y) \neq (0, 0) \\ 0 & \text{if } (x, y) = (0, 0). \end{cases}$$

**38.** Calculate $f_1(x, y)$ for the function in Exercise 37. Is $f_1(x, y)$ continuous at $(0, 0)$?

* **39.** Let $f(x, y) = \begin{cases} \dfrac{x^3 - y^3}{x^2 + y^2}, & \text{if } (x, y) \neq (0, 0) \\ 0, & \text{if } (x, y) = (0, 0). \end{cases}$

Calculate $f_1(x, y)$ and $f_2(x, y)$ at all points $(x, y)$ in the plane. Is $f$ continuous at $(0, 0)$? Are $f_1$ and $f_2$ continuous at $(0, 0)$?

* **40.** Let $f(x, y, z) = \begin{cases} \dfrac{xy^2z}{x^4 + y^4 + z^4}, & \text{if } (x, y, z) \neq (0, 0, 0) \\ 0, & \text{if } (x, y, z) = (0, 0, 0). \end{cases}$

Find $f_1(0, 0, 0)$, $f_2(0, 0, 0)$, and $f_3(0, 0, 0)$. Is $f$ continuous at $(0, 0, 0)$? Are $f_1$, $f_2$, and $f_3$ continuous at $(0, 0, 0)$?

## 4.4 HIGHER-ORDER DERIVATIVES

Partial derivatives of second and higher orders are calculated by taking partial derivatives of already calculated partial derivatives. The order in which the differentiations are performed is indicated in the notations used. If $z = f(x, y)$, we can calculate *four* partial derivatives of second order, namely

two **pure** second partial derivatives with respect to $x$ or $y$,

$$\frac{\partial^2 z}{\partial x^2} = \frac{\partial}{\partial x}\frac{\partial z}{\partial x} = f_{11}(x, y) = f_{xx}(x, y),$$
$$\frac{\partial^2 z}{\partial y^2} = \frac{\partial}{\partial y}\frac{\partial z}{\partial y} = f_{22}(x, y) = f_{yy}(x, y),$$

and two **mixed** second partial derivatives with respect to $x$ and $y$,

$$\frac{\partial^2 z}{\partial x \partial y} = \frac{\partial}{\partial x}\frac{\partial z}{\partial y} = f_{21}(x, y) = f_{yx}(x, y),$$
$$\frac{\partial^2 z}{\partial y \partial x} = \frac{\partial}{\partial y}\frac{\partial z}{\partial x} = f_{12}(x, y) = f_{xy}(x, y).$$

Again we remark that the notations $f_{11}$, $f_{12}$, $f_{21}$, and $f_{22}$ are usually preferable to $f_{xx}$, $f_{xy}$, $f_{yx}$, and $f_{yy}$, although the latter are often used in partial differential equations. Note that $f_{12}$ indicates differentiation of $f$ *first* with respect to its first variable and *then* with respect to its second variable; $f_{21}$ indicates the opposite order of differentiation. The subscript closest to $f$ indicates which differentiation occurs first.

Similarly, if $w = f(x, y, z)$ then

$$\frac{\partial^5 w}{\partial y \partial x \partial y^2 \partial z} = \frac{\partial}{\partial y}\frac{\partial}{\partial x}\frac{\partial}{\partial y}\frac{\partial}{\partial y}\frac{\partial w}{\partial z} = f_{32212}(x, y, z) = f_{zyyxy}(x, y, z).$$

**EXAMPLE 1** Find the four second partial derivatives of $f(x, y) = x^3y^4$.

***SOLUTION***

$$f_1(x, y) = 3x^2y^4, \qquad f_2(x, y) = 4x^3y^3,$$

$$f_{11}(x, y) = \frac{\partial}{\partial x}(3x^2y^4) = 6xy^4, \qquad f_{21}(x, y) = \frac{\partial}{\partial x}(4x^3y^3) = 12x^2y^3,$$

$$f_{12}(x, y) = \frac{\partial}{\partial y}(3x^2y^4) = 12x^2y^3, \qquad f_{22}(x, y) = \frac{\partial}{\partial y}(4x^3y^3) = 12x^3y^2.$$

■

**EXAMPLE 2** Calculate $f_{223}(x, y, z)$, $f_{232}(x, y, z)$, and $f_{322}(x, y, z)$ for the function $f(x, y, z) = e^{x-2y+3z}$.

***SOLUTION***

$$\begin{aligned} f_{223}(x, y, z) &= \frac{\partial}{\partial z}\frac{\partial}{\partial y}\frac{\partial}{\partial y}e^{x-2y+3z} \\ &= \frac{\partial}{\partial z}\frac{\partial}{\partial y}\left(-2e^{x-2y+3z}\right) \\ &= \frac{\partial}{\partial z}\left(4e^{x-2y+3z}\right) = 12\,e^{x-2y+3z}. \end{aligned}$$

$$\begin{aligned} f_{232}(x, y, z) &= \frac{\partial}{\partial y}\frac{\partial}{\partial z}\frac{\partial}{\partial y}e^{x-2y+3z} \\ &= \frac{\partial}{\partial y}\frac{\partial}{\partial z}\left(-2e^{x-2y+3z}\right) \\ &= \frac{\partial}{\partial y}\left(-6e^{x-2y+3z}\right) = 12\,e^{x-2y+3z}. \\ f_{322}(x, y, z) &= \frac{\partial}{\partial y}\frac{\partial}{\partial y}\frac{\partial}{\partial z}e^{x-2y+3z} \\ &= \frac{\partial}{\partial y}\frac{\partial}{\partial y}\left(3e^{x-2y+3z}\right) \\ &= \frac{\partial}{\partial y}\left(-6e^{x-2y+3z}\right) = 12\,e^{x-2y+3z}. \end{aligned}$$

■

In both of the examples above observe that the mixed partial derivatives taken with respect to the same variables but in different orders turned out to be equal. This is not a coincidence. It will always occur for sufficiently smooth functions. In particular the mixed partial derivatives involved are required to be *continuous*. The following theorem presents a more precise statement of this important phenomenon.

**THEOREM 1**

**Equality of mixed partials**

Suppose that two mixed $n$th order partial derivatives of a function $f$ involve the same differentiations but in different orders. If those partials are continuous at a point $P$, and if $f$ and all partials of $f$ of order less than $n$ are continuous in a neighbourhood of $P$, then the two mixed partials are equal at the point $P$.

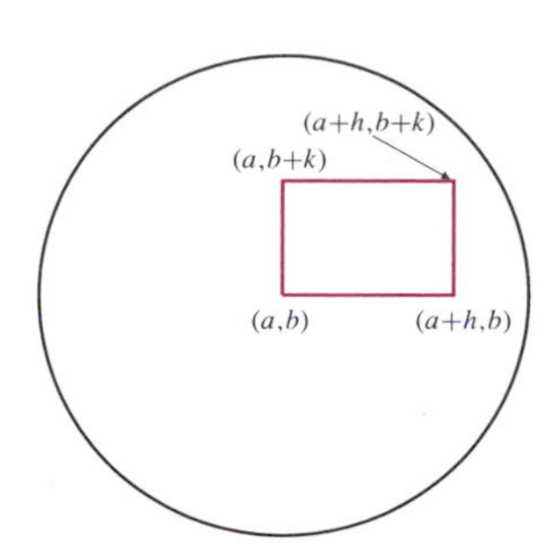


**Figure 4.21** A rectangle contained in the disk where $f$ and certain partials are continuous

***PROOF*** We shall prove only a representative special case, showing the equality of $f_{12}(a,b)$ and $f_{21}(a,b)$ for a function $f$ of two variables, provided $f_{12}$ and $f_{21}$ are defined and $f_1$, $f_2$, and $f$ are continuous throughout a disk of positive radius centred at $(a,b)$, and $f_{12}$ and $f_{21}$ are continuous at $(a,b)$. Let $h$ and $k$ have sufficiently small absolute values that the point $(a+h, b+k)$ lies in this disk. Then so do all points of the rectangle with sides parallel to the coordinate axes and diagonally opposite corners at $(a,b)$ and $(a+h, b+k)$. (See Figure 4.21.)

Let $Q = f(a+h, b+k) - f(a+h, b) - f(a, b+k) + f(a,b)$ and define single-variable functions $u(x)$ and $v(y)$ by

$$u(x) = f(x, b+k) - f(x,b) \qquad \text{and} \qquad v(y) = f(a+h, y) - f(a,y).$$

Evidently $Q = u(a+h) - u(a)$ and also $Q = v(b+k) - v(b)$. By the (single-variable) Mean-Value Theorem, there exists a number $\theta_1$ satisfying $0 < \theta_1 < 1$ (so that $a + \theta_1 h$ lies between $a$ and $a+h$) such that

$$Q = u(a+h) - u(a) = h\,u'(a+\theta_1 h) = h\big[f_1(a+\theta_1 h, b+k) - f_1(a+\theta_1 h, b)\big].$$

Now we apply the Mean-Value Theorem again, this time to $f_1$ considered as a function of its second variable, and obtain another number $\theta_2$ satisfying $0 < \theta_2 < 1$ such that

The Mean-Value Theorem is used four times in this proof, each time to write a difference of the form $g(p+m) - g(p)$ in the form $g'(c)m$, where $c$ is some number between $p$ and $p+m$. It is convenient to write $c$ in the form $p + \theta m$, where $\theta$ is some number between 0 and 1.

$$f_1(a+\theta_1 h, b+k) - f_1(a+\theta_1 h, b) = k\,f_{12}(a+\theta_1 h, b+\theta_2 k).$$

Thus $Q = hk\,f_{12}(a+\theta_1 h, b+\theta_2 k)$. Two similar applications of the Mean-Value Theorem to $Q = v(b+k) - v(b)$ lead to $Q = hk\,f_{21}(a+\theta_3 h, b+\theta_4 k)$, where $\theta_3$ and $\theta_4$ are two numbers each between 0 and 1. Equating these two expressions for $Q$ and cancelling the common factor $hk$, we obtain

$$f_{12}(a+\theta_1 h, b+\theta_2 k) = f_{21}(a+\theta_3 h, b+\theta_4 k).$$

Since $f_{12}$ and $f_{21}$ are continuous at $(a,b)$, we can let $h$ and $k$ approach zero to obtain $f_{12}(a,b) = f_{21}(a,b)$, as required.

Exercise 16 at the end of this section develops an example of a function for which $f_{12}$ and $f_{21}$ exist but are not continuous at $(0,0)$, and for which $f_{12}(0,0) \neq f_{21}(0,0)$.

## The Laplace and Wave Equations

Many important and interesting phenomena are modelled by functions of several variables that satisfy certain *partial differential equations*. In the following examples we encounter two particular partial differential equations that arise frequently in mathematics and the physical sciences. Exercises 17–19 at the end of this section introduce another such equation with important applications.

**EXAMPLE 3** Show that for any real number $k$ the functions

$$z = e^{kx}\cos(ky) \qquad \text{and} \qquad z = e^{kx}\sin(ky)$$

satisfy the partial differential equation

$$\frac{\partial^2 z}{\partial x^2} + \frac{\partial^2 z}{\partial y^2} = 0$$

at every point in the $xy$-plane.

***SOLUTION*** For $z = e^{kx}\cos(ky)$ we have

$$\frac{\partial z}{\partial x} = k\,e^{kx}\cos(ky), \qquad \frac{\partial z}{\partial y} = -k\,e^{kx}\sin(ky),$$
$$\frac{\partial^2 z}{\partial x^2} = k^2\,e^{kx}\cos(ky), \qquad \frac{\partial^2 z}{\partial y^2} = -k^2\,e^{kx}\cos(ky).$$

Thus

$$\frac{\partial^2 z}{\partial x^2} + \frac{\partial^2 z}{\partial y^2} = k^2\,e^{kx}\cos(ky) - k^2\,e^{kx}\cos(ky) = 0.$$

The calculation for $z = e^{kx}\sin(ky)$ is similar. ■

***REMARK*** The partial differential equation in the above example is called the (two-dimensional) **Laplace equation**. A function of two variables having continuous second partial derivatives in a region of the plane is said to be **harmonic** there if it satisfies Laplace's equation. Such functions play a critical role in the theory of differentiable functions of a *complex variable,* and are used to model various physical quantities such as steady-state temperature distributions, fluid flows, and electric and magnetic potential fields. Harmonic functions have many interesting properties. They have derivatives of all orders, and are *analytic,* that is, they are the sums of their (multivariable) Taylor series. Moreover, a harmonic function can achieve maximum and minimum values only on the boundary of its domain. Laplace's equation, and therefore harmonic functions, can be considered in any number of dimensions. (See Exercises 13 and 14 below.)

■ **EXAMPLE 4** If $f$ and $g$ are any twice differentiable functions of one variable, show that

$$w = f(x - ct) + g(x + ct)$$

satisfies the partial differential equation

$$\frac{\partial^2 w}{\partial t^2} = c^2\,\frac{\partial^2 w}{\partial x^2}.$$

***SOLUTION*** Using the Chain Rule for functions of one variable we obtain

$$\frac{\partial w}{\partial t} = -c\,f'(x - ct) + c\,g'(x + ct), \qquad \frac{\partial w}{\partial x} = f'(x - ct) + g'(x + ct),$$
$$\frac{\partial^2 w}{\partial t^2} = c^2\,f''(x - ct) + c^2\,g''(x + ct), \qquad \frac{\partial^2 w}{\partial x^2} = f''(x - ct) + g''(x + ct).$$

Thus $w$ satisfies the given differential equation. ■

***REMARK*** The partial differential equation in the above example is called the (one-dimensional) **wave equation**. If $t$ measures time, then $f(x - ct)$ represents a waveform travelling to the right along the $x$-axis with speed $c$. (See Figure 4.22.) Similarly, $g(x + ct)$ represents a waveform travelling to the left with speed $c$. Unlike the solutions of Laplace's equation that must be infinitely differentiable, solutions of the wave equation need only have enough derivatives to satisfy the differential equation. The functions $f$ and $g$ are otherwise arbitrary.

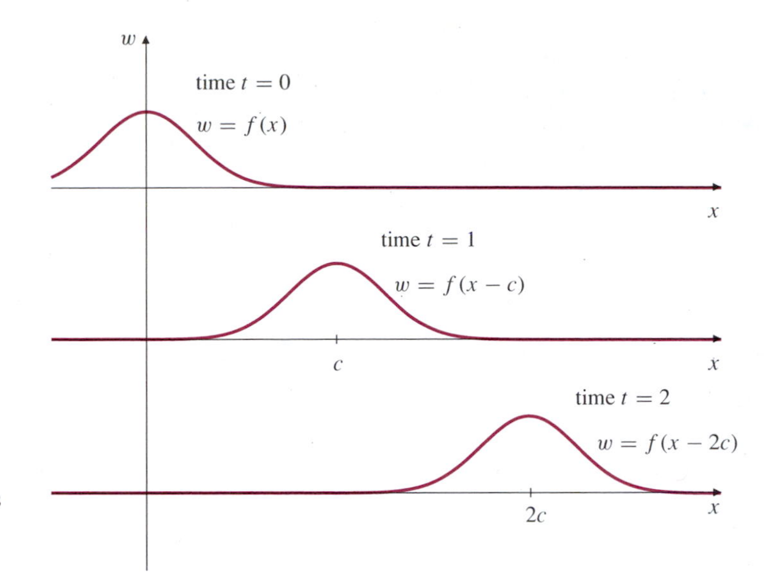


**Figure 4.22** $w = f(x - ct)$ represents a waveform moving to the right with speed $c$

## EXERCISES 4.4

In Exercises 1–6, find all the second partial derivatives of the given function.

**1.** $z = x^2(1 + y^2)$

**2.** $f(x, y) = x^2 + y^2$

**3.** $w = x^3y^3z^3$

**4.** $z = \sqrt{3x^2 + y^2}$

**5.** $z = x\,e^y - y\,e^x$

**6.** $f(x, y) = \ln\big(1 + \sin(xy)\big)$

* **7.** How many mixed partial derivatives of order 3 can a function of three variables have? If they are all continuous, how many different values can they have at one point? Find the mixed partials of order 3 for $f(x, y, z) = x\,e^{xy}\cos(xz)$ that involve two differentiations with respect to $z$ and one with respect to $x$.

Show that the functions in Exercises 8–12 are harmonic in the plane regions indicated.

**8.** $f(x, y) = A(x^2 - y^2) + Bxy$ in the whole plane ($A$ and $B$ are constants.)

**9.** $f(x, y) = 3x^2y - y^3$ in the whole plane (Can you think of another polynomial of degree 3 in $x$ and $y$ that is also harmonic?)

**10.** $f(x, y) = \dfrac{x}{x^2 + y^2}$ everywhere except at the origin

**11.** $f(x, y) = \ln(x^2 + y^2)$ everywhere except at the origin

**12.** $\tan^{-1}(y/x)$ except at points on the $y$-axis

**13.** Show that $w = e^{3x+4y}\sin(5z)$ is harmonic in all of $\mathbb{R}^3$, that is, it satisfies everywhere the 3-dimensional Laplace equation

$$\frac{\partial^2 w}{\partial x^2} + \frac{\partial^2 w}{\partial y^2} + \frac{\partial^2 w}{\partial z^2} = 0.$$

**14.** Assume that $f(x, y)$ is harmonic in the $xy$-plane. Show that each of the functions $z\,f(x, y)$, $x\,f(y, z)$, and $y\,f(z, x)$ is harmonic in the whole of $\mathbb{R}^3$. What condition should the constants $a$, $b$, and $c$ satisfy to ensure that $f(ax + by, cz)$ is harmonic in $\mathbb{R}^3$?

**15.** Suppose the functions $u(x, y)$ and $v(x, y)$ have continuous second partial derivatives and satisfy the **Cauchy–Riemann equations**

$$\frac{\partial u}{\partial x} = \frac{\partial v}{\partial y} \quad \text{and} \quad \frac{\partial v}{\partial x} = -\frac{\partial u}{\partial y}.$$

Show that $u$ and $v$ are both harmonic.

* **16.** Let $F(x, y) = \begin{cases} \dfrac{2xy(x^2 - y^2)}{x^2 + y^2} & \text{if } (x, y) \neq (0, 0) \\ 0 & \text{if } (x, y) = (0, 0) \end{cases}$

Calculate $F_1(x, y)$, $F_2(x, y)$, $F_{12}(x, y)$, and $F_{21}(x, y)$ at points $(x, y) \neq (0, 0)$. Also calculate these derivatives at $(0, 0)$. Observe that $F_{21}(0, 0) = 2$ and $F_{12}(0, 0) = -2$. Does this result contradict Theorem 1? Explain why.

### The heat (diffusion) equation

◆ **17.** Show that the function $u(x, t) = t^{-1/2}\,e^{-x^2/4t}$ satisfies the partial differential equation

$$\frac{\partial u}{\partial t} = \frac{\partial^2 u}{\partial x^2}.$$

This equation is called the (one-dimensional) **heat equation** as it models heat diffusion in an insulated rod (with $u(x, t)$ representing the temperature at position $x$ at time $t$) and other similar phenomena.

◆ **18.** Show that the function $u(x, y, t) = t^{-1}\,e^{-(x^2+y^2)/4t}$ satisfies the two-dimensional heat equation

$$\frac{\partial u}{\partial t} = \frac{\partial^2 u}{\partial x^2} + \frac{\partial^2 u}{\partial y^2}.$$

**19.** By comparing the results of Exercises 17 and 18, guess a solution to the three-dimensional heat equation

$$\frac{\partial u}{\partial t} = \frac{\partial^2 u}{\partial x^2} + \frac{\partial^2 u}{\partial y^2} + \frac{\partial^2 u}{\partial z^2}.$$

Verify your guess.

**Biharmonic functions**

A function $u(x, y)$ with continuous partials of fourth order is **biharmonic** if $\dfrac{\partial^2 u}{\partial x^2} + \dfrac{\partial^2 u}{\partial y^2}$ is a harmonic function.

**20.** Show that $u(x, y)$ is biharmonic if and only if it satisfies the biharmonic equation

$$\frac{\partial^4 u}{\partial x^4} + 2\frac{\partial^4 u}{\partial x^2 \partial y^2} + \frac{\partial^4 u}{\partial y^4} = 0$$

**21.** Verify that $u(x, y) = x^4 - 3x^2y^2$ is biharmonic.

**22.** Show that if $u(x, y)$ is harmonic then $v(x, y) = xu(x, y)$ and $w(x, y) = yu(x, y)$ are biharmonic.

Use the result of Exercise 22 to show that the functions in Exercises 23–25 are biharmonic.

**23.** $x\,e^x \sin y$

**24.** $y \ln(x^2 + y^2)$

**25.** $\dfrac{xy}{x^2 + y^2}$

**26.** Propose a definition of a biharmonic function of three variables, and prove results analogous to those of Exercises 20 and 22 for biharmonic functions $u(x, y, z)$.

## 4.5 THE CHAIN RULE

The Chain Rule for functions of one variable is a formula that gives the derivative of a composition $f(g(x))$ of two functions $f$ and $g$:

$$\frac{d}{dx} f(g(x)) = f'(g(x))g'(x).$$

The situation for several variables is more complicated. If $f$ depends on more than one variable, and any of those variables can be functions of one or more other variables, we cannot expect a simple formula for partial derivatives of the composition to cover all possible cases. We must come to think of the Chain Rule as a *procedure for differentiating compositions* rather than as a formula for their derivatives. In order to motivate a formulation of the Chain Rule for functions of two variables we begin with a concrete example.

**EXAMPLE 1** Suppose you are hiking in a mountainous region for which you have a map. Let $(x, y)$ be the coordinates of your position on the map (that is, the horizontal coordinates of your actual position in the region). Let $z = f(x, y)$ denote the height of land (above sea level, say), at position $(x, y)$. Suppose you are walking along a trail so that your position at time $t$ is given by $x = u(t)$ and $y = v(t)$. (These are parametric equations of the trail on the map.) At time $t$ your altitude above sea level is given by the composite function

$$z = f(u(t), v(t)) = g(t),$$

a function of only one variable. How fast is your altitude changing with respect to time at time $t$?

***SOLUTION*** The answer is the derivative of $g(t)$:

$$\begin{aligned} g'(t) &= \lim_{h\to 0} \frac{g(t+h) - g(t)}{h} = \lim_{h\to 0} \frac{f(u(t+h), v(t+h)) - f(u(t), v(t))}{h} \\ &= \lim_{h\to 0} \frac{f(u(t+h), v(t+h)) - f(u(t), v(t+h))}{h} \\ &\qquad + \lim_{h\to 0} \frac{f(u(t), v(t+h)) - f(u(t), v(t))}{h}. \end{aligned}$$

We added 0 to the numerator of the Newton quotient in a creative way so as to separate the quotient into the sum of two quotients, in the first of which the difference of values of $f$ involves only the first variable of $f$, and in the second of which the difference involves only the second variable of $f$. The single-variable Chain Rule suggests that the sum of the two limits above is

$$g'(t) = f_1\big(u(t), v(t)\big)u'(t) + f_2\big(u(t), v(t)\big)v'(t).$$

■

The above formula is the Chain Rule for $\frac{d}{dt} f\big(u(t), v(t)\big)$. In terms of Leibniz notation we have

**A version of the Chain Rule**

If $z$ is a function of $x$ and $y$ with continuous first partial derivatives, and if $x$ and $y$ are differentiable functions of $t$, then

$$\frac{dz}{dt} = \frac{\partial z}{\partial x}\frac{dx}{dt} + \frac{\partial z}{\partial y}\frac{dy}{dt}.$$

Note that there are two terms in the expression for $dz/dt$ (or $g'(t)$), one arising from each variable of $f$ that depends on $t$.

Now consider a function $f$ of two variables, $x$ and $y$, each of which is in turn a function of two other variables, $s$ and $t$:

$$z = f(x, y), \qquad \text{where} \qquad x = u(s, t) \quad \text{and} \quad y = v(s, t).$$

We can form the composite function

$$z = f\big(u(s, t), v(s, t)\big) = g(s, t).$$

For instance, if $f(x, y) = x^2 + 3y$, where $u(s, t) = st^2$ and $v(s, t) = s - t$, then $g(s, t) = s^2t^4 + 3(s - t)$.

Let us assume that $f$, $u$, and $v$ have first partial derivatives with respect to their respective variables, and that those of $f$ are continuous. Then $g$ has first partial derivatives given by

$$g_1(s, t) = f_1\big(u(s, t), v(s, t)\big)u_1(s, t) + f_2\big(u(s, t), v(s, t)\big)v_1(s, t),$$
$$g_2(s, t) = f_1\big(u(s, t), v(s, t)\big)u_2(s, t) + f_2\big(u(s, t), v(s, t)\big)v_2(s, t).$$

These formulas can be expressed more simply using Leibniz notation:

**Another version of the Chain Rule**

If $z$ is a function of $x$ and $y$ with continuous first partial derivatives, and if $x$ and $y$ depend on $s$ and $t$, then

$$\frac{\partial z}{\partial s} = \frac{\partial z}{\partial x}\frac{\partial x}{\partial s} + \frac{\partial z}{\partial y}\frac{\partial y}{\partial s},$$
$$\frac{\partial z}{\partial t} = \frac{\partial z}{\partial x}\frac{\partial x}{\partial t} + \frac{\partial z}{\partial y}\frac{\partial y}{\partial t}.$$

This can be deduced from the version obtained in Example 1 by allowing $u$ and $v$ there to depend on two variables, but holding one of them fixed while we differentiate with respect to the other. A more formal proof of this simple, but representative case of the Chain Rule will be given in the next section.

The two equations in the box above can be combined into a single matrix equation

$$\begin{pmatrix} \dfrac{\partial z}{\partial s} & \dfrac{\partial z}{\partial t} \end{pmatrix} = \begin{pmatrix} \dfrac{\partial z}{\partial x} & \dfrac{\partial z}{\partial y} \end{pmatrix} \begin{pmatrix} \dfrac{\partial x}{\partial s} & \dfrac{\partial x}{\partial t} \\ \dfrac{\partial y}{\partial s} & \dfrac{\partial y}{\partial t} \end{pmatrix}.$$

We will comment on the significance of this matrix form at the end of the next section.

In general, if $z$ is a function of several "primary" variables, and each of these depends on some "secondary" variables then the partial derivative of $z$ with respect to one of the secondary variables will have several terms, one for the contribution to the derivative arising from each of the primary variables on which $z$ depends.

***REMARK*** Note the significance of the various subscripts denoting partial derivatives in the functional form of the Chain Rule:

$$g_1(s,t) = f_1\big(u(s,t), v(s,t)\big)u_1(s,t) + f_2\big(u(s,t), v(s,t)\big)v_1(s,t).$$

The "1" in $g_1(s,t)$ refers to differentiation with respect to $s$, the first variable on which $g$ depends. By contrast, the "1" in $f_1(u(s,t), v(s,t))$ refers to differentiation with respect to $x$, the first variable on which $f$ depends. (This derivative is then evaluated at $x = u(s,t)$, $y = v(s,t)$.)

**EXAMPLE 2** If $z = \sin(x^2 y)$ where $x = st^2$ and $y = s^2 + \dfrac{1}{t}$, find $\partial z/\partial s$ and $\partial z/\partial t$

(a) by direct substitution and the single-variable form of the Chain Rule, and

(b) by using the (two-variable) Chain Rule.

***SOLUTION*** (a) By direct substitution:

$$\begin{aligned}
z &= \sin\left((st^2)^2\left(s^2 + \frac{1}{t}\right)\right) = \sin(s^4t^4 + s^2t^3), \\
\frac{\partial z}{\partial s} &= (4s^3t^4 + 2st^3)\cos(s^4t^4 + s^2t^3), \\
\frac{\partial z}{\partial t} &= (4s^4t^3 + 3s^2t^2)\cos(s^4t^4 + s^2t^3).
\end{aligned}$$

(b) Using the Chain Rule:

$$\begin{aligned}
\frac{\partial z}{\partial s} &= \frac{\partial z}{\partial x}\frac{\partial x}{\partial s} + \frac{\partial z}{\partial y}\frac{\partial y}{\partial s} \\
&= \big(2xy\cos(x^2y)\big)t^2 + \big(x^2\cos(x^2y)\big)2s \\
&= \left(2st^2\left(s^2 + \frac{1}{t}\right)t^2 + 2s^3t^4\right)\cos(s^4t^4 + s^2t^3) \\
&= (4s^3t^4 + 2st^3)\cos(s^4t^4 + s^2t^3),
\end{aligned}$$

$$\begin{aligned}
\frac{\partial z}{\partial t} &= \frac{\partial z}{\partial x}\frac{\partial x}{\partial t} + \frac{\partial z}{\partial y}\frac{\partial y}{\partial t} \\
&= \big(2xy\cos(x^2y)\big)2st + \big(x^2\cos(x^2y)\big)\left(\frac{-1}{t^2}\right) \\
&= \left(2st^2(s^2 + \frac{1}{t})2st + s^2t^4\left(\frac{-1}{t^2}\right)\right)\cos(s^4t^4 + s^2t^3) \\
&= (4s^4t^3 + 3s^2t^2)\cos(s^4t^4 + s^2t^3).
\end{aligned}$$

Note that we still had to use direct substitution on the derivatives obtained in (b) in order to show that the values were the same as those obtained in (a). ■

**■ EXAMPLE 3** Find $\frac{\partial}{\partial x} f(x^2y, x + 2y)$ and $\frac{\partial}{\partial y} f(x^2y, x + 2y)$ in terms of the partial derivatives of $f$, assuming that these partial derivatives are continuous.

***SOLUTION*** We have

$$\begin{aligned}
\frac{\partial}{\partial x} f(x^2y, x + 2y) &= f_1(x^2y, x + 2y)\frac{\partial}{\partial x}(x^2y) + f_2(x^2y, x + 2y)\frac{\partial}{\partial x}(x + 2y) \\
&= 2xyf_1(x^2y, x + 2y) + f_2(x^2y, x + 2y), \\
\frac{\partial}{\partial y} f(x^2y, x + 2y) &= f_1(x^2y, x + 2y)\frac{\partial}{\partial y}(x^2y) + f_2(x^2y, x + 2y)\frac{\partial}{\partial y}(x + 2y) \\
&= x^2f_1(x^2y, x + 2y) + 2f_2(x^2y, x + 2y).
\end{aligned}$$

■

**■ EXAMPLE 4** Express the partial derivatives of $z = h(s, t) = f\big(g(s, t)\big)$ in terms of the partial derivatives of $f$ and $g$.

***SOLUTION*** The partial derivatives of $h$ can be calculated using the single-variable version of the Chain Rule:

$$\begin{aligned}
h_1(s, t) &= \frac{\partial z}{\partial s} = \frac{dz}{dx}\frac{\partial x}{\partial s} = f'\big(g(s, t)\big)g_1(s, t), \\
h_2(s, t) &= \frac{\partial z}{\partial t} = \frac{dz}{dx}\frac{\partial x}{\partial t} = f'\big(g(s, t)\big)g_2(s, t).
\end{aligned}$$

■

The following example involves a hybrid application of the Chain Rule to a function that depends both directly and indirectly on the variable of differentiation.

**■ EXAMPLE 5** Find $dz/dt$ where $z = f(x, y, t)$, $x = g(t)$, and $y = h(t)$. (Assume that $f$, $g$, and $h$ all have continuous derivatives.)

***SOLUTION*** Since $z$ depends on $t$ through each of the three variables of $f$, there will be three terms in the appropriate Chain Rule:

$$\begin{aligned}
\frac{dz}{dt} &= \frac{\partial z}{\partial x}\frac{dx}{dt} + \frac{\partial z}{\partial y}\frac{dy}{dt} + \frac{\partial z}{\partial t} \\
&= f_1(x, y, t)g'(t) + f_2(x, y, t)h'(t) + f_3(x, y, t).
\end{aligned}$$

■

**REMARK** In the above example we can easily distinguish between the meanings of the symbols $dz/dt$ and $\partial z/\partial t$. If, however, we had been dealing with the situation

$$z = f(x, y, s, t), \qquad \text{where} \quad x = g(s, t) \quad \text{and} \quad y = h(s, t),$$

then the meaning of the symbol $\partial z/\partial t$ would be unclear; it could refer to the simple partial derivative of $f$ with respect to its fourth primary variable (i.e., $f_4(x, y, s, t)$), or it could refer to the derivative of the composite function $f(g(s, t), h(s, t), s, t)$. Three of the four primary variables of $f$ depend on $t$, and therefore contribute to the rate of change of $z$ with respect to $t$. The partial derivative $f_4(x, y, s, t)$ denotes the contribution of only one of these three variables. It is conventional to use $\partial z/\partial t$ to denote the derivative of the composite function with respect to the secondary variable $t$:

$$\begin{aligned}\frac{\partial z}{\partial t} &= \frac{\partial}{\partial t} f(g(s, t), h(s, t), s, t) \\ &= f_1(x, y, s, t)g_2(s, t) + f_2(x, y, s, t)h_2(s, t) + f_4(x, y, s, t).\end{aligned}$$

When necessary, we can denote the contribution coming from the primary variable $t$ by

$$\left(\frac{\partial z}{\partial t}\right)_{x,y,s} = \frac{\partial}{\partial t} f(x, y, s, t) = f_4(x, y, s, t).$$

Here the subscripts denote those primary variables of $f$ whose contributions to the rate of change of $z$ with respect to $t$ are being *ignored*. Of course, in the situation described above, $(\partial z/\partial t)_s$ means the same as $\partial z/\partial t$.

In applications, the variables that contribute to a particular partial derivative will usually be clear from the context. The following example contains such an application. This is an example of a procedure called *differentiation following the motion*.

**EXAMPLE 6** Atmospheric temperature depends on position and time. If we denote position by three spatial coordinates $x$, $y$, and $z$ (measured in km), and time by $t$ (measured in hours), then the temperature $T$ is a function of four variables, $T(x, y, z, t)$.

(a) If a thermometer is attached to a weather balloon that moves through the atmosphere on a path with parametric equations $x = f(t)$, $y = g(t)$, and $z = h(t)$, what is the rate of change at time $T$ of the temperature recorded by the thermometer?

(b) Find the rate of change of the recorded temperature at time $t = 1$ if

$$T(x, y, z, t) = \frac{100}{5 + x^2 + y^2}\left(1 + \sin\frac{\pi t}{12}\right) - 20(1 + z^2)\text{degrees},$$

and if the balloon moves along the curve

$$x = f(t) = t, \qquad y = g(t) = 2t, \qquad z = h(t) = t - (t^4/2).$$

### SOLUTION

(a) Here the rate of change of the thermometer reading must take into account the change in position of the thermometer as well as increasing time. Thus none of the four variables of $T$ can be ignored in the differentiation. The rate is given by

$$\frac{dT}{dt} = \frac{\partial T}{\partial x}\frac{dx}{dt} + \frac{\partial T}{\partial y}\frac{dy}{dt} + \frac{\partial T}{\partial z}\frac{dz}{dt} + \frac{\partial T}{\partial t}.$$

The term $\partial T/\partial t$ refers only to the rate of change of the temperature with respect to time at a fixed position in the atmosphere. The other three terms arise from the motion of the balloon.

(b) The values of the three coordinates and their derivatives at $t = 1$ are $x = 1$, $y = 2$, $z = 1/2$, $dx/dt = 1$, $dy/dt = 2$, and $dz/dt = -1$. Also

$$\begin{aligned}
\frac{\partial T}{\partial x} &= -\frac{200x}{(5 + x^2 + y^2)^2}\left(1 + \sin\frac{\pi t}{12}\right) = -2\left(1 + \sin\frac{\pi}{12}\right),\\
\frac{\partial T}{\partial y} &= -\frac{200y}{(5 + x^2 + y^2)^2}\left(1 + \sin\frac{\pi t}{12}\right) = -4\left(1 + \sin\frac{\pi}{12}\right),\\
\frac{\partial T}{\partial z} &= -40z = -20,\\
\frac{\partial T}{\partial t} &= \frac{100}{5 + x^2 + y^2}\,\frac{\pi}{12}\cos\frac{\pi t}{12} = \frac{10\pi}{12}\cos\frac{\pi}{12}.
\end{aligned}$$

Thus

$$\begin{aligned}
\left.\frac{dT}{dt}\right|_{t=1} &= \left(1 + \sin\frac{\pi}{12}\right)\left(-2(1) - 4(2)\right) - 20(-1) + \frac{10\pi}{12}\cos\frac{\pi}{12}\\
&\approx 9.94;
\end{aligned}$$

the recorded temperature is increasing at a rate of about 9.94 °/h at time $t = 1$. ■

The discussion and examples above show that the Chain Rule for functions of several variables can take different forms depending on the numbers of variables of the various functions being composed. As an aid in determining the correct form of the Chain Rule in a given situation you can construct a chart showing which variables depend on which. Figure 4.23(a) shows such a chart for the temperature function of Example 6. The Chain Rule for $dT/dt$ involves a term for every route from $T$ to $t$ in the chart. The route from $T$ through $x$ to $t$ produces the term $\dfrac{\partial T}{\partial x}\dfrac{dx}{dt}$ and so on.

**EXAMPLE 7** Write the appropriate Chain Rule for $\partial z/\partial x$, where $z = f(u, v, r)$, $u = g(x, y, r)$, $v = h(x, y, r)$, and $r = k(x, y)$.

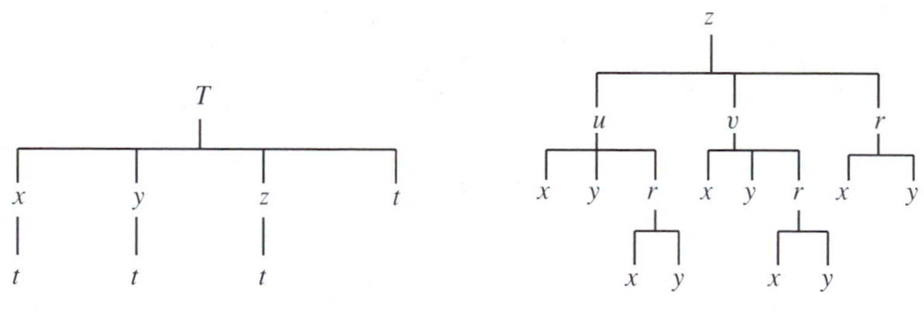


**Figure 4.23**

(a) Chart showing the dependence of $T$ on $t$ in Example 6

(b) Dependence chart for Example 7

**SOLUTION** The appropriate chart is shown in Figure 4.23(b). There are five routes from $z$ to $x$:

$$\frac{\partial z}{\partial x} = \frac{\partial z}{\partial u}\frac{\partial u}{\partial x} + \frac{\partial z}{\partial u}\frac{\partial u}{\partial r}\frac{\partial r}{\partial x} + \frac{\partial z}{\partial v}\frac{\partial v}{\partial x} + \frac{\partial z}{\partial v}\frac{\partial v}{\partial r}\frac{\partial r}{\partial x} + \frac{\partial z}{\partial r}\frac{\partial r}{\partial x}.$$

■

## Homogeneous Functions

A function $f(x_1, \ldots, x_n)$ is said to be **positively homogeneous of degree** $k$ if, for every point $(x_1, x_2, \ldots, x_n)$ in its domain and every real number $t > 0$, we have

$$f(tx_1, tx_2, \ldots, tx_n) = t^k f(x_1, \ldots, x_n).$$

For example

$$f(x, y) = x^2 + xy - y^2 \quad \text{is positively homogeneous of degree 2,}$$
$$f(x, y) = \sqrt{x^2 + y^2} \quad \text{is positively homogeneous of degree 1,}$$
$$f(x, y) = \frac{2xy}{x^2 + y^2} \quad \text{is positively homogeneous of degree 0,}$$
$$f(x, y, z) = \frac{x - y + 5z}{yz - z^2} \quad \text{is positively homogeneous of degree } -1,$$
$$f(x, y) = x^2 + y \quad \text{is not positively homogeneous.}$$

Observe that a positively homogeneous function of degree 0 remains constant along rays from the origin. More generally, along such rays a positively homogeneous function of degree $k$ grows or decays proportionally to the $k$th power of distance from the origin.

**THEOREM 2**

**Euler's Theorem**

If $f(x_1, \ldots, x_n)$ has continuous first partial derivatives and is positively homogeneous of degree $k$, then

$$\sum_{i=1}^{n} x_i f_i(x_1, \ldots, x_n) = kf(x_1, \ldots, x_n).$$

**PROOF** Differentiate the equation $f(tx_1, tx_2, \ldots, tx_n) = t^k f(x_1, \ldots, x_n)$ with respect to $t$ to get

$$x_1 f_1(tx_1, \ldots, tx_n) + x_2 f_2(tx_1, \ldots, tx_n) + \ldots + x_n f_n(tx_1, \ldots, tx_n)$$
$$= k\,t^{k-1} f(x_1, \ldots, x_n).$$

Now substitute $t = 1$ to get the desired result.

Note that Exercises 26–29 in Section 4.3 illustrate this theorem.

## Higher-Order Derivatives

Applications of the Chain Rule to higher-order derivatives can become quite complicated. It is important to keep in mind at each stage which variables are independent of one another.

■ **EXAMPLE 8** Calculate $\dfrac{\partial^2}{\partial x \partial y} f(x^2 - y^2, xy)$ in terms of partial derivatives of the function $f$. Assume that the partials of $f$ are continuous.

**SOLUTION** In this problem symbols for the primary variables on which $f$ depends are not stated explicitly. Let them be $u$ and $v$. The problem therefore asks us to find

$$\frac{\partial^2}{\partial x \partial y} f(u, v), \qquad \text{where} \quad u = x^2 - y^2 \quad \text{and} \quad v = xy.$$

First differentiate with respect to $y$:

$$\frac{\partial}{\partial y} f(u, v) = -2y f_1(u, v) + x f_2(u, v).$$

Now differentiate this result with respect to $x$. Note that the second term on the right is a product to two functions of $x$ so we need to use the Product Rule:

$$\begin{aligned}\frac{\partial^2}{\partial x \partial y} f(u, v) &= -2y\Big(2x f_{11}(u, v) + y f_{12}(u, v)\Big) \\ &\quad + f_2(u, v) + x\Big(2x f_{21}(u, v) + y f_{22}(u, v)\Big) \\ &= f_2(u, v) - 4xy f_{11}(u, v) + 2(x^2 - y^2) f_{12}(u, v) + xy f_{22}(u, v).\end{aligned}$$

In the last step we have used the fact that the mixed partials of $f$ are continuous so we could equate $f_{12}$ and $f_{21}$. ■

Review the above calculation very carefully and make sure you understand what is being done at each step. Note that all the derivatives of $f$ that appear are evaluated at $(u, v) = (x^2 - y^2, xy)$, not at $(x, y)$, because $x$ and $y$ are not themselves the primary variables on which $f$ depends.

**EXAMPLE 9** Show that $f(x^2 - y^2, 2xy)$ is a harmonic function if $f(x, y)$ is harmonic.

**SOLUTION** Let $u = x^2 - y^2$ and $v = 2xy$. If $z = f(u, v)$, then

$$\begin{aligned}\frac{\partial z}{\partial x} &= 2x f_1(u, v) + 2y f_2(u, v), \\ \frac{\partial z}{\partial y} &= -2y f_1(u, v) + 2x f_2(u, v), \\ \frac{\partial^2 z}{\partial x^2} &= 2 f_1(u, v) + 2x\big(2x f_{11}(u, v) + 2y f_{12}(u, v)\big) \\ &\quad + 2y\big(2x f_{21}(u, v) + 2y f_{22}(u, v) \\ &= 2 f_1(u, v) + 4x^2 f_{11}(u, v) + 8xy f_{12}(u, v) + 4y^2 f_{22}(u, v), \\ \frac{\partial^2 z}{\partial y^2} &= -2 f_1(u, v) - 2y\big(-2y f_{11}(u, v) + 2x f_{12}(u, v)\big) \\ &\quad + 2x\big(-2y f_{21}(u, v) + 2x f_{22}(u, v) \\ &= -2 f_1(u, v) + 4y^2 f_{11}(u, v) - 8xy f_{12}(u, v) + 4x^2 f_{22}(u, v).\end{aligned}$$

Therefore

$$\frac{\partial^2 z}{\partial x^2} + \frac{\partial^2 z}{\partial y^2} = 4(x^2 + y^2)\big(f_{11}(u, v) + f_{22}(u, v)\big) = 0$$

because $f$ is given to be harmonic. Thus $z = f(x^2 - y^2, 2xy)$ is a harmonic function of $x$ and $y$. ■

In the following example we show that the two-dimensional Laplace differential equation (see Example 3 in Section 4.4) takes the form

$$\frac{\partial^2 z}{\partial r^2} + \frac{1}{r}\frac{\partial z}{\partial r} + \frac{1}{r^2}\frac{\partial^2 z}{\partial \theta^2} = 0$$

when stated for a function $z$ expressed in terms of polar coordinates $r$ and $\theta$.

**EXAMPLE 10** **Laplace's equation in polar coordinates**

If $z = f(x, y)$ has continuous partial derivatives of second order, and if $x = r\cos\theta$, and $y = r\sin\theta$, show that

$$\frac{\partial^2 z}{\partial r^2} + \frac{1}{r}\frac{\partial z}{\partial r} + \frac{1}{r^2}\frac{\partial^2 z}{\partial \theta^2} = \frac{\partial^2 z}{\partial x^2} + \frac{\partial^2 z}{\partial y^2}.$$

***SOLUTION*** It is possible to do this in two different ways; we can start with either side and use the Chain Rule to show that it is equal to the other side. Here we will calculate the partial derivatives with respect to $r$ and $\theta$ that appear on the left side and express them in terms of partial derivatives with respect to $x$ and $y$. The other approach, involving expressing partial derivatives with respect to $x$ and $y$ in terms of partial derivatives with respect to $r$ and $\theta$, is a little harder. (See Exercise 30 below.) However, we would have to do it that way if we were not given the form of the differential equation in polar coordinates and had to find it.

First note that,

$$\frac{\partial x}{\partial r} = \cos\theta, \qquad \frac{\partial x}{\partial \theta} = -r\sin\theta, \qquad \frac{\partial y}{\partial r} = \sin\theta, \qquad \frac{\partial y}{\partial \theta} = r\cos\theta.$$

Thus

$$\frac{\partial z}{\partial r} = \frac{\partial z}{\partial x}\frac{\partial x}{\partial r} + \frac{\partial z}{\partial y}\frac{\partial y}{\partial r} = \cos\theta\,\frac{\partial z}{\partial x} + \sin\theta\,\frac{\partial z}{\partial y}.$$

**DANGER**
This is a difficult, but important example. Examine each step carefully to make sure you understand what is being done.
**DANGER**

Now differentiate with respect to $r$ again. Remember that $r$ and $\theta$ are independent variables, so that the factors $\cos\theta$ and $\sin\theta$ can be regarded as constants. However, $\partial z/\partial x$ and $\partial z/\partial y$ depend on $x$ and $y$, and therefore on $r$ and $\theta$.

$$\begin{aligned}\frac{\partial^2 z}{\partial r^2} &= \cos\theta\,\frac{\partial}{\partial r}\frac{\partial z}{\partial x} + \sin\theta\,\frac{\partial}{\partial r}\frac{\partial z}{\partial y}\\ &= \cos\theta\left(\cos\theta\,\frac{\partial^2 z}{\partial x^2} + \sin\theta\,\frac{\partial^2 z}{\partial y\partial x}\right) + \sin\theta\left(\cos\theta\,\frac{\partial^2 z}{\partial x\partial y} + \sin\theta\,\frac{\partial^2 z}{\partial y^2}\right)\\ &= \cos^2\theta\,\frac{\partial^2 z}{\partial x^2} + 2\cos\theta\sin\theta\,\frac{\partial^2 z}{\partial x\partial y} + \sin^2\theta\,\frac{\partial^2 z}{\partial y^2}.\end{aligned}$$

We have used the equality of mixed partials in the last line. Similarly,

$$\frac{\partial z}{\partial \theta} = -r\sin\theta\,\frac{\partial z}{\partial x} + r\cos\theta\,\frac{\partial z}{\partial y}.$$

When we differentiate a second time with respect to $\theta$ we can regard $r$ as constant, but each term above is still a product of two functions that depend on $\theta$. Thus

$$\begin{aligned}\frac{\partial^2 z}{\partial \theta^2} &= -r\left(\cos\theta\,\frac{\partial z}{\partial x} + \sin\theta\,\frac{\partial}{\partial \theta}\frac{\partial z}{\partial x}\right) + r\left(-\sin\theta\,\frac{\partial z}{\partial y} + \cos\theta\,\frac{\partial}{\partial \theta}\frac{\partial z}{\partial y}\right)\\ &= -r\,\frac{\partial z}{\partial r} - r\sin\theta\left(-r\sin\theta\,\frac{\partial^2 z}{\partial x^2} + r\cos\theta\,\frac{\partial^2 z}{\partial y\partial x}\right)\\ &\quad + r\cos\theta\left(-r\sin\theta\,\frac{\partial^2 z}{\partial x\partial y} + r\cos\theta\,\frac{\partial^2 z}{\partial y^2}\right)\\ &= -r\,\frac{\partial z}{\partial r} + r^2\left(\sin^2\theta\,\frac{\partial^2 z}{\partial x^2} - 2\sin\theta\cos\theta\,\frac{\partial^2 z}{\partial x\partial y} + \cos^2\theta\,\frac{\partial^2 z}{\partial y^2}\right).\end{aligned}$$

Combining these results, we obtain the desired formula:

$$\frac{\partial^2 z}{\partial r^2} + \frac{1}{r}\frac{\partial z}{\partial r} + \frac{1}{r^2}\frac{\partial^2 z}{\partial \theta^2} = \frac{\partial^2 z}{\partial x^2} + \frac{\partial^2 z}{\partial y^2}.$$

■

## EXERCISES 4.5

In Exercises 1–6, write appropriate versions of the Chain Rule for the indicated derivatives.

**1.** $\partial w/\partial t$ if $w = f(x, y, z)$, where $x = g(s, t)$, $y = h(s, t)$, and $z = k(s, t)$

**2.** $\partial w/\partial t$ if $w = f(x, y, z)$, where $x = g(s)$, $y = h(s, t)$, and $z = k(t)$

**3.** $\partial z/\partial u$ if $z = g(x, y)$, where $y = f(x)$ and $x = h(u, v)$

**4.** $dw/dx$ if $w = f(x, y, z)$, where $y = g(x, z)$ and $z = h(x)$

**5.** $\left(\partial w/\partial x\right)_z$ if $w = f(x, y, z)$ and $y = g(x, z)$

**6.** $dw/dt$ if $w = f(x, y)$, $x = g(r, s)$, $y = h(r, t)$, $r = k(s, t)$, and $s = m(t)$

**7.** If $w = f(x, y, z)$, where $x = g(y, z)$ and $y = h(z)$, state appropriate versions of the Chain Rule for $\dfrac{dw}{dz}$, $\left(\dfrac{\partial w}{\partial z}\right)_x$, and $\left(\dfrac{\partial w}{\partial z}\right)_{x,y}$.

**8.** Use two different methods to calculate $\partial u/\partial t$ if $u = \sqrt{x^2 + y^2}$, $x = e^{st}$, and $y = 1 + s^2 \cos t$.

**9.** Use two different methods to calculate $\partial z/\partial x$ if $z = \tan^{-1}(u/v)$, $u = 2x + y$ and $v = 3x - y$.

**10.** Use two methods to calculate $dz/dt$ given that $z = txy^2$, $x = t + \ln(y + t^2)$ and $y = e^t$.

In Exercises 11–16, find the indicated derivatives, assuming that the function $f(x, y)$ has continuous first partial derivatives.

**11.** $\dfrac{\partial}{\partial x} f(2x, 3y)$

**12.** $\dfrac{\partial}{\partial x} f(2y, 3x)$

**13.** $\dfrac{\partial}{\partial x} f(y^2, x^2)$

**14.** $\dfrac{\partial}{\partial t} f(st^2, s^2 + t)$

**15.** $\dfrac{\partial}{\partial x} f\big(f(x, y), f(x, y)\big)$

**16.** $\dfrac{\partial}{\partial y} f\big(yf(x, t), f(y, t)\big)$

**17.** Suppose that the temperature $T$ in a certain liquid varies with depth $z$ and time $t$ according to the formula $T = e^{-t}z$. Find the rate of change of temperature with respect to time at a point that is moving through the liquid so that at time $t$ its depth is $f(t)$. What is this rate if $f(t) = e^t$? What is happening in this case?

**18.** Suppose the strength $E$ of an electric field in space varies with position $(x, y, z)$ and time $t$ according to the formula $E = f(x, y, z, t)$. Find the rate of change with respect to time of the electric field strength measured by an instrument moving through space along the helix $x = \sin t$, $y = \cos t$, $z = t$.

In Exercises 19–28, assume that $f$ has continuous partial derivatives of all orders.

**19.** If $z = f(x, y)$ where $x = 2s + 3t$ and $y = 3s - 2t$ find (a) $\dfrac{\partial^2 z}{\partial s^2}$, (b) $\dfrac{\partial^2 z}{\partial s \partial t}$, and (c) $\dfrac{\partial^2 z}{\partial t^2}$.

**20.** If $f(x, y)$ is harmonic show that $f(ax + by, bx - ay)$ is also harmonic. (See the Remark following Example 3 in Section 4.4.)

**21.** If $f(x, y)$ is harmonic show that $f(x^3 - 3xy^2, 3x^2y - y^3)$ is also harmonic.

**22.** If $f(x, y)$ is harmonic show that $f\left(\dfrac{x}{x^2 + y^2}, \dfrac{-y}{x^2 + y^2}\right)$ is also harmonic.

**23.** Suppose that $u(x, y)$ and $v(x, y)$ have continuous second partial derivatives and satisfy the Cauchy–Riemann equations

$$\frac{\partial u}{\partial x} = \frac{\partial v}{\partial y} \quad \text{and} \quad \frac{\partial v}{\partial x} = -\frac{\partial u}{\partial y}.$$

Suppose also that $f(u, v)$ is a harmonic function of $u$ and $v$. Show that $f\big(u(x, y), v(x, y)\big)$ is a harmonic function of $x$ and $y$. (*Hint:* $u$ and $v$ are harmonic functions by Exercise 15 in Section 4.4.)

**24.** If $r^2 = x^2 + y^2 + z^2$, verify that $u(x, y, z) = 1/r$ is harmonic throughout $\mathbb{R}^3$ except at the origin.

**25.** If $x = t \sin s$ and $y = t \cos s$ find $\dfrac{\partial^2}{\partial s \partial t} f(x, y)$.

**26.** Find $\dfrac{\partial^3}{\partial x \partial y^2} f(2x + 3y, xy)$ in terms of partial derivatives of $f$.

**27.** Find $\dfrac{\partial^2}{\partial y \partial x} f(y^2, xy, -x^2)$ in terms of partial derivatives of the function $f$.

**28.** Find $\dfrac{\partial^3}{\partial t^2 \partial s} f(s^2 - t, s + t^2)$ in terms of partial derivatives of $f$.

***29.** If $x = e^s \cos t$, $y = e^s \sin t$ and $z = u(x, y) = v(s, t)$ show that

$$\frac{\partial^2 z}{\partial s^2} + \frac{\partial^2 z}{\partial t^2} = (x^2 + y^2)\left(\frac{\partial^2 z}{\partial x^2} + \frac{\partial^2 z}{\partial y^2}\right).$$

***30. (Converting Laplace's equation to polar coordinates)** The transformation to polar coordinates, $x = r\cos\theta$, $y = r \sin\theta$ implies that $r^2 = x^2 + y^2$ and $\tan\theta = y/x$. Use these equations to show that

$$\frac{\partial r}{\partial x} = \cos\theta \qquad \frac{\partial r}{\partial y} = \sin\theta$$
$$\frac{\partial \theta}{\partial x} = -\frac{\sin\theta}{r} \qquad \frac{\partial \theta}{\partial y} = \frac{\cos\theta}{r}.$$

Use these formulas to help you express $\dfrac{\partial^2 u}{\partial x^2} + \dfrac{\partial^2 u}{\partial y^2}$ in terms of partials of $u$ with respect to $r$ and $\theta$, and hence reprove the formula for the Laplace differential equation in polar coordinates given in Example 10.

**31.** If $u(x, y) = r^2 \ln r$, where $r^2 = x^2 + y^2$, verify that $u$ is a biharmonic function by showing that

$$\left(\frac{\partial^2}{\partial x^2} + \frac{\partial^2}{\partial y^2}\right)\left(\frac{\partial^2 u}{\partial x^2} + \frac{\partial^2 u}{\partial y^2}\right) = 0.$$

**32.** If $f(x, y)$ is positively homogeneous of degree $k$ and has continuous partial derivatives of second order, show that

$$x^2 f_{11}(x, y) + 2xy f_{12}(x, y) + y^2 f_{22}(x, y) = k(k-1) f(x, y).$$

* **33.** Generalize the result of Exercise 32 to functions of $n$ variables.

* **34.** Generalize the results of Exercises 32 and 33 to expressions involving $m$th-order partial derivatives of the function $f$.

Exercises 35–36 revisit Exercise 16 of Section 4.4. Let

$$F(x, y) = \begin{cases} \dfrac{2xy(x^2 - y^2)}{x^2 + y^2} & \text{if } (x, y) \neq (0, 0) \\ 0 & \text{if } (x, y) = (0, 0). \end{cases}$$

**35.** (a) Show that $F(x, y) = -F(y, x)$ for all $(x, y)$.

(b) Show that $F_1(x, y) = -F_2(y, x)$ and $F_{12}(x, y) = -F_{21}(y, x)$ for $(x, y) \neq (0, 0)$.

(c) Show that $F_1(0, y) = -2y$ for all $y$, and hence that $F_{12}(0, 0) = 2$.

(d) Deduce that $F_2(x, 0) = 2x$ and $F_{21}(0, 0) = 2$.

**36.** (a) Use Exercise 35(b) to find $F_{12}(x, x)$ for $x \neq 0$.

(b) Is $F_{12}(x, y)$ continuous at $(0, 0)$? Why?

❖ **37.** Use the change of variables $\xi = x + ct$, $\eta = x$ to transform the partial differential equation

$$\frac{\partial u}{\partial t} = c\frac{\partial u}{\partial x}, \qquad (c = \text{constant}),$$

into the simpler equation $\partial v/\partial \eta = 0$, where $v(\xi, \eta) = v(x + ct, x) = u(x, t)$. This equation says that $v(\xi, \eta)$ does not depend on $\eta$, so $v = f(\xi)$ for some arbitrary differentiable function $f$. What is the corresponding "general solution" $u(x, t)$ of the original partial differential equation?

❖ **38.** Having considered Exercise 37, guess a "general solution" $w(r, s)$ of the second-order partial differential equation

$$\frac{\partial^2}{\partial r \partial s} w(r, s) = 0.$$

Your answer should involve two arbitrary functions.

❖ **39.** Use the change of variables $r = x + ct$, $s = x - ct$, $w(r, s) = u(x, t)$ to transform the one-dimensional wave equation

$$\frac{\partial^2 u}{\partial t^2} = c^2 \frac{\partial^2 u}{\partial x^2}$$

to a simpler form. Now use the result of Exercise 38 to find the *general solution* of this wave equation in the form given in Example 4 in Section 4.4.

❖ **40.** Show that the initial-value problem for the one-dimensional wave equation:

$$\begin{cases} u_{tt}(x, t) = c^2 u_{xx}(x, t), \\ u(x, 0) = p(x), \\ u_t(x, 0) = q(x). \end{cases}$$

has the solution

$$u(x, t) = \frac{1}{2}\Big[p(x - ct) + p(x + ct)\Big] + \frac{1}{2c}\int_{x-ct}^{x+ct} q(s)\, ds.$$

(Note that we have used subscripts "$x$" and "$t$" instead of "1" and "2" to denote the partial derivatives here. This is common usage in dealing with partial differential equations.)

***Remark.*** The initial-value problem in Exercise 39 gives the small lateral displacement $u(x, t)$ at position $x$ at time $t$ of a vibrating string held under tension along the $x$-axis. The function $p(x)$ gives the *initial* displacement at position $x$, that is, the displacement at time $t = 0$. Similarly, $q(x)$ gives the initial velocity at position $x$. Observe that the position at time $t$ depends only on values of these initial data at points no further than $ct$ units away. This is consistent with the previous observation that the solutions of the wave equation represent waves travelling with speed $c$.

## 4.6 LINEAR APPROXIMATIONS, DIFFERENTIABILITY, AND DIFFERENTIALS

The tangent line to the graph $y = f(x)$ at $x = a$ provides a convenient approximation for values of $f(x)$ for $x$ near $a$:

$$f(x) \approx L(x) = f(a) + f'(a)(x - a).$$

Here $L(x)$ is the **linearization** of $f$ at $a$; its graph is the tangent line to $y = f(x)$ there. The mere existence of $f'(a)$ is sufficient to guarantee that the error in the approximation $f(x) \approx L(x)$ is small compared to the distance $h = x - a$ between

$a$ and $x$, that is to say,

$$\begin{aligned}\lim_{h\to 0}\frac{f(a+h)-L(a+h)}{h} &= \lim_{h\to 0}\frac{f(a+h)-f(a)-f'(a)h}{h}\\ &= \lim_{h\to 0}\frac{f(a+h)-f(a)}{h}-f'(a)\\ &= f'(a)-f'(a)=0.\end{aligned}$$

Similarly, we can use the tangent plane

$$z = f(a,b)+f_1(a,b)(x-a)+f_2(a,b)(y-b)$$

to $z = f(x,y)$ at $(a,b)$ to define a linear approximation to $f(x,y)$ near $(a,b)$:

The **linearization** (linear approximation) of $f(x,y)$ at $(a,b)$ is given by

$$f(x,y)\approx L(x,y) = f(a,b)+f_1(a,b)(x-a)+f_2(a,b)(y-b).$$

**EXAMPLE 1** Find an approximate value for $f(x,y)=\sqrt{2x^2+e^{2y}}$ at $(2.2, -0.2)$.

***SOLUTION*** It is convenient to use the linearization at $(2,0)$, where the values of $f$ and its partials are easily evaluated:

$$\begin{aligned} & & f(2,0)&=3,\\ f_1(x,y)&=\frac{2x}{\sqrt{2x^2+e^{2y}}}, & f_1(2,0)&=\frac{4}{3},\\ f_2(x,y)&=\frac{e^{2y}}{\sqrt{2x^2+e^{2y}}}, & f_2(2,0)&=\frac{1}{3}.\end{aligned}$$

Thus $L(x,y)=3+\frac{4}{3}(x-2)+\frac{1}{3}(y-0)$, and

$$f(2.2,-0.2)\approx L(2.2,-0.2)=3+\frac{4}{3}(2.2-2)+\frac{1}{3}(-0.2-0)=3.2\,.$$

Unlike the single-variable case, the mere existence of the partial derivatives $f_1(a,b)$ and $f_2(a,b)$ does not even imply that $f$ is continuous at $(a,b)$, let alone that the error in the linearization is small compared to the distance $\sqrt{(x-a)^2+(y-b)^2}$ between $(a,b)$ and $(x,y)$. We adopt this latter condition as our definition of what it means for a function to be *differentiable* at a point.

**DEFINITION 8**

We say that the function $f(x,y)$ is **differentiable** at the point $(a,b)$ if

$$\lim_{(h,k)\to(0,0)}\frac{f(a+h,b+k)-f(a,b)-hf_1(a,b)-kf_2(a,b)}{\sqrt{h^2+k^2}}=0.$$

This definition and the following theorems can be generalized to functions of any number of variables in the obvious way. For the sake of simplicity, we state them for the two-variable case only.

The function $f(x, y)$ is differentiable at the point $(a, b)$ if and only if the surface $z = f(x, y)$ has a *nonvertical tangent plane* at $(a, b)$. This implies that $f_1(a, b)$ and $f_2(a, b)$ must exist, and that $f$ must be continuous at $(a, b)$. (Recall, however, that the existence of the partial derivatives does *not* even imply that $f$ is continuous let alone differentiable. Compare this with the single-variable situation.) In particular, the function is *continuous* wherever it is differentiable. We will prove a two-variable version of the Mean-Value Theorem and use it to show that functions with *continuous* first partial derivatives are differentiable.

**THEOREM 3** **A mean-value theorem**

If $f_1(x, y)$ and $f_2(x, y)$ are continuous in a neighbourhood of the point $(a, b)$, and if the absolute values of $h$ and $k$ are sufficiently small, then there exist numbers $\theta_1$ and $\theta_2$, each between 0 and 1, such that

$$f(a+h, b+k) - f(a, b) = hf_1(a+\theta_1 h, b+k) + kf_2(a, b+\theta_2 k).$$

***PROOF*** The proof of this theorem is very similar to that of Theorem 1 in Section 4.4, so we give only a sketch here. The reader can fill in the details. Write

$$f(a+h, b+k) - f(a, b) = \big(f(a+h, b+k) - f(a, b+k)\big) + \big(f(a, b+k) - f(a, b)\big),$$

and then apply the single-variable Mean-Value Theorem separately to $f(x, b+k)$ on the interval between $a$ and $a+h$, and to $f(a, y)$ on the interval between $b$ and $b+k$ to get the desired result.

**THEOREM 4**

If $f_1$ and $f_2$ are continuous in a neighbourhood of the point $(a, b)$, then $f$ is differentiable at $(a, b)$.

***PROOF*** Using Theorem 3 and the facts that

$$\left|\frac{h}{\sqrt{h^2+k^2}}\right| \le 1 \qquad \text{and} \qquad \left|\frac{k}{\sqrt{h^2+k^2}}\right| \le 1,$$

we estimate

$$\begin{aligned}
&\left|\frac{f(a+h, b+k) - f(a, b) - hf_1(a, b) - kf_2(a, b)}{\sqrt{h^2+k^2}}\right| \\
&\quad = \left|\frac{h}{\sqrt{h^2+k^2}}\big(f_1(a+\theta_1 h, b+k) - f_1(a, b)\big)\right. \\
&\qquad\quad \left. + \frac{k}{\sqrt{h^2+k^2}}\big(f_2(a, b+\theta_2 k) - f_2(a, b)\big)\right| \\
&\quad \le \big|f_1(a+\theta_1 h, b+k) - f_1(a, b)\big| + \big|f_2(a, b+\theta_2 k) - f_2(a, b)\big|.
\end{aligned}$$

Since $f_1$ and $f_2$ are continuous at $(a, b)$ each of these latter terms approaches 0 as $h$ and $k$ approach 0. This is what we needed to prove.

We illustrate differentiability with an example where we can calculate directly the error in the tangent plane approximation.

**EXAMPLE 2** Calculate $f(x+h, y+k) - f(x, y) - f_1(x, y)h - f_2(x, y)k$ if $f(x, y) = x^3 + xy^2$.

***SOLUTION*** Since $f_1(x, y) = 3x^2 + y^2$ and $f_2(x, y) = 2xy$, we have

$$\begin{aligned} f(x+h, y+k) &- f(x, y) - f_1(x, y)h - f_2(x, y)k \\ &= (x+h)^3 + (x+h)(y+k)^2 - x^3 - xy^2 - (3x^2 + y^2)h - 2xyk \\ &= 3xh^2 + h^3 + 2yhk + hk^2 + xk^2. \end{aligned}$$

Observe that the result above is a polynomial in $h$ and $k$ with no term of degree less than two in these variables. Therefore this difference approaches zero like the *square* of the distance $\sqrt{h^2 + k^2}$ from $(x, y)$ to $(x + h, y + k)$ as $(h, k) \to (0, 0)$, so the condition for differentiability is certainly satisfied:

$$\lim_{(h,k)\to(0,0)} \frac{3xh^2 + h^3 + 2yhk + hk^2 + xk^2}{\sqrt{h^2 + k^2}} = 0.$$

This quadratic behaviour is the case for any function $f$ with continuous *second* partial derivatives. (See Exercise 19 at the end of this section.) ■

## Proof of the Chain Rule

We are now able to give a formal statement and proof of a simple but representative case of the Chain Rule for multivariate functions.

**THEOREM 5**

**A Chain Rule**

Let $z = f(x, y)$, where $x = u(s, t)$ and $y = v(s, t)$. Suppose that

(i) $u(a, b) = p$ and $v(a, b) = q$,

(ii) the first partial derivatives of $u$ and $v$ exist at the point $(a, b)$, and

(iii) $f$ is differentiable at the point $(p, q)$.

Then $z = w(s, t) = f(u(s, t), v(s, t))$ has first partial derivatives with respect to $s$ and $t$ at $(a, b)$, and

$$\begin{aligned} w_1(a, b) &= f_1(p, q)u_1(a, b) + f_2(p, q)v_1(a, b), \\ w_2(a, b) &= f_1(p, q)u_2(a, b) + f_2(p, q)v_2(a, b), \end{aligned}$$

that is,

$$\frac{\partial z}{\partial s} = \frac{\partial z}{\partial x}\frac{\partial x}{\partial s} + \frac{\partial z}{\partial y}\frac{\partial y}{\partial s} \qquad \text{and} \qquad \frac{\partial z}{\partial t} = \frac{\partial z}{\partial x}\frac{\partial x}{\partial t} + \frac{\partial z}{\partial y}\frac{\partial y}{\partial t}.$$

***PROOF*** Define a function $E$ of two variables as follows: $E(0, 0) = 0$, and if $(h, k) \neq (0, 0)$, then

$$E(h, k) = \frac{f(p + h, q + k) - f(p, q) - hf_1(p, q) - kf_2(p, q)}{\sqrt{h^2 + k^2}}.$$

Observe that $E(h, k)$ is continuous at $(0, 0)$, because $f$ is differentiable at $(p, q)$. Now

$$f(p + h, q + k) - f(p, q) = hf_1(p, q) + kf_2(p, q) + \sqrt{h^2 + k^2}\, E(h, k).$$

In this formula put $h = u(a + \sigma, b) - u(a, b)$ and $k = v(a + \sigma, b) - v(a, b)$ and divide by $\sigma$ to obtain

$$\begin{aligned} \frac{w(a + \sigma, b) - w(a, b)}{\sigma} &= \frac{f(u(a + \sigma, b), v(a + \sigma, b)) - f(u(a, b), v(a, b))}{\sigma} \\ &= \frac{f(p + h, q + k) - f(p, q)}{\sigma} \\ &= f_1(p, q)\frac{h}{\sigma} + f_2(p, q)\frac{k}{\sigma} + \sqrt{\left(\frac{h}{\sigma}\right)^2 + \left(\frac{k}{\sigma}\right)^2}\, E(h, k). \end{aligned}$$

We want to let $\sigma$ approach 0 in this formula. Note that

$$\lim_{\sigma \to 0} \frac{h}{\sigma} = \lim_{\sigma \to 0} \frac{u(a+\sigma, b) - u(a, b)}{\sigma} = u_1(a, b),$$

and, similarly, $\lim_{\sigma \to 0}(k/\sigma) = v_1(a, b)$. Since $(h, k) \to (0, 0)$ if $\sigma \to 0$, we have

$$w_1(a, b) = f_1(p, q)u_1(a, b) + f_2(p, q)v_1(a, b).$$

The proof for $w_2$ is similar.

## Differentials

If the first partial derivatives of a function $z = f(x_1, \ldots, x_n)$ exist at a point we may construct a **differential** $dz$ or $df$ of the function at that point in a manner similar to that used for functions of one variable:

$$\begin{aligned} dz = df &= \frac{\partial z}{\partial x_1} dx_1 + \frac{\partial z}{\partial x_2} dx_2 + \cdots + \frac{\partial z}{\partial x_n} dx_n \\ &= f_1(x_1, \ldots, x_n)\, dx_1 + \cdots + f_n(x_1, \ldots, x_n)\, dx_n. \end{aligned}$$

Here the differential $dz$ is considered to be a function of the $2n$ independent variables $x_1, x_2, \ldots, x_n, dx_1, dx_2, \ldots, dx_n$.

For a differentiable function $f$, the differential $df$ is an approximation to the change $\Delta f$ in value of the function given by

$$\Delta f = f(x_1 + dx_1, \ldots, x_n + dx_n) - f(x_1, \ldots, x_n).$$

The error in this approximation is small compared to the distance between the two points in the domain of $f$, that is,

$$\frac{\Delta f - df}{\sqrt{(dx_1)^2 + \cdots + (dx_n)^2}} \to 0 \quad \text{if all } dx_i \to 0, \qquad (1 \le i \le n).$$

In this sense, differentials are just another way of looking at linearization.

**EXAMPLE 3** Find the percentage change in the period

$$T = 2\pi \sqrt{\frac{L}{g}}$$

of a simple pendulum if the length $L$ of the pendulum increases by 2% and the acceleration of gravity $g$ decreases by 0.6%.

***SOLUTION*** We calculate the differential of $T$:

$$\begin{aligned} dT &= \frac{\partial T}{\partial L} dL + \frac{\partial T}{\partial g} dg \\ &= \frac{2\pi}{2\sqrt{Lg}} dL - \frac{2\pi\sqrt{L}}{2g^{3/2}} dg. \end{aligned}$$

We are given that $dL = \frac{2}{100} L$ and $dg = -\frac{6}{1{,}000} g$. Thus

$$dT = \frac{1}{100} 2\pi \sqrt{\frac{L}{g}} - \left(-\frac{6}{1{,}000}\right) \frac{2\pi}{2} \sqrt{\frac{L}{g}} = \frac{13}{1{,}000} T.$$

Therefore the period $T$ of the pendulum increases by 1.3%. ■

## Functions from *n*-space to *m*-space

This is an optional topic. A vector $\mathbf{f} = (f_1, f_2, \ldots, f_m)$ of $m$ functions, each depending on $n$ variables $(x_2, x_2, \ldots, x_n)$ defines a *transformation* (that is, a function) from $\mathbb{R}^n$ to $\mathbb{R}^m$; specifically, if $\mathbf{x} = (x_2, x_2, \ldots, x_n)$ is a point in $\mathbb{R}^n$, and

$$\begin{aligned} y_1 &= f_1(x_1, x_2, \ldots, x_n) \\ y_2 &= f_2(x_1, x_2, \ldots, x_n) \\ &\vdots \\ y_m &= f_m(x_1, x_2, \ldots, x_n), \end{aligned}$$

then $\mathbf{y} = (y_1, y_2, \ldots, y_m)$ is the point in $\mathbb{R}^m$ that corresponds to $\mathbf{x}$ under the transformation $\mathbf{f}$. We can write these equations more compactly as

$$\mathbf{y} = \mathbf{f}(\mathbf{x}).$$

Information about the rate of change of $\mathbf{y}$ with respect to $\mathbf{x}$ is contained in the various partial derivatives $\partial y_i/\partial x_j$ $(1 \le i \le m,\ 1 \le j \le n)$, and is conveniently organized into an $m \times n$ matrix, $D\mathbf{f}(\mathbf{x})$, called the **Jacobian matrix** of the transformation $f$:

$$D\mathbf{f}(\mathbf{x}) = \begin{pmatrix} \dfrac{\partial y_1}{\partial x_1} & \dfrac{\partial y_1}{\partial x_2} & \cdots & \dfrac{\partial y_1}{\partial x_n} \\ \dfrac{\partial y_2}{\partial x_1} & \dfrac{\partial y_2}{\partial x_2} & \cdots & \dfrac{\partial y_2}{\partial x_n} \\ \vdots & \vdots & & \vdots \\ \dfrac{\partial y_m}{\partial x_1} & \dfrac{\partial y_m}{\partial x_2} & \cdots & \dfrac{\partial y_m}{\partial x_n} \end{pmatrix}$$

The linear transformation (see Section 3.6) represented by the Jacobian matrix is called **the derivative** of the transformation $\mathbf{f}$.

It is not our purpose to enter into a study of such *vector-valued functions of a vector variable* at this point, but we can observe here that the composition of two such transformations corresponds to forming the matrix product of their Jacobian matrices.

To see this, let $\mathbf{y} = \mathbf{f}(\mathbf{x})$ be a transformation from $\mathbb{R}^n$ to $\mathbb{R}^m$ as described above, and let $\mathbf{z} = \mathbf{g}(\mathbf{y})$ be another such transformation from $\mathbb{R}^m$ to $\mathbb{R}^k$ given by

$$\begin{aligned} z_1 &= g_1(y_1, y_2, \ldots, y_m) \\ z_2 &= g_2(y_1, y_2, \ldots, y_m) \\ &\vdots \\ z_k &= g_k(y_1, y_2, \ldots, y_m), \end{aligned}$$

which has the $k \times m$ Jacobian matrix

$$D\mathbf{g}(\mathbf{y}) = \begin{pmatrix} \dfrac{\partial z_1}{\partial y_1} & \dfrac{\partial z_1}{\partial y_2} & \cdots & \dfrac{\partial z_1}{\partial y_m} \\ \dfrac{\partial z_2}{\partial y_1} & \dfrac{\partial z_2}{\partial y_2} & \cdots & \dfrac{\partial z_2}{\partial y_m} \\ \vdots & \vdots & & \vdots \\ \dfrac{\partial z_k}{\partial y_1} & \dfrac{\partial z_k}{\partial y_2} & \cdots & \dfrac{\partial z_k}{\partial y_m} \end{pmatrix}$$

Then the composition $\mathbf{z} = \mathbf{g} \circ \mathbf{f}(\mathbf{x}) = \mathbf{g}\big(\mathbf{f}(\mathbf{x})\big)$ given by

$$\begin{aligned}
z_1 &= g_1\big(f_1(x_1, \ldots, x_n), \ldots, f_m(x_1, \ldots, x_n)\big) \\
z_2 &= g_2\big(f_1(x_1, \ldots, x_n), \ldots, f_m(x_1, \ldots, x_n)\big) \\
&\ \vdots \\
z_k &= g_k\big(f_1(x_1, \ldots, x_n), \ldots, f_m(x_1, \ldots, x_n)\big)
\end{aligned}$$

has, according to the Chain Rule, the $k \times n$ Jacobian matrix

$$\begin{pmatrix} \frac{\partial z_1}{\partial x_1} & \frac{\partial z_1}{\partial x_2} & \cdots & \frac{\partial z_1}{\partial x_n} \\ \frac{\partial z_2}{\partial x_1} & \frac{\partial z_2}{\partial x_2} & \cdots & \frac{\partial z_2}{\partial x_n} \\ \vdots & \vdots & & \vdots \\ \frac{\partial z_k}{\partial x_1} & \frac{\partial z_k}{\partial x_2} & \cdots & \frac{\partial z_k}{\partial x_n} \end{pmatrix} = \begin{pmatrix} \frac{\partial z_1}{\partial y_1} & \frac{\partial z_1}{\partial y_2} & \cdots & \frac{\partial z_1}{\partial y_m} \\ \frac{\partial z_2}{\partial y_1} & \frac{\partial z_2}{\partial y_2} & \cdots & \frac{\partial z_2}{\partial y_m} \\ \vdots & \vdots & & \vdots \\ \frac{\partial z_k}{\partial y_1} & \frac{\partial z_k}{\partial y_2} & \cdots & \frac{\partial z_k}{\partial y_m} \end{pmatrix} \begin{pmatrix} \frac{\partial y_1}{\partial x_1} & \frac{\partial y_1}{\partial x_2} & \cdots & \frac{\partial y_1}{\partial x_n} \\ \frac{\partial y_2}{\partial x_1} & \frac{\partial y_2}{\partial x_2} & \cdots & \frac{\partial y_2}{\partial x_n} \\ \vdots & \vdots & & \vdots \\ \frac{\partial y_m}{\partial x_1} & \frac{\partial y_m}{\partial x_2} & \cdots & \frac{\partial y_m}{\partial x_n} \end{pmatrix}.$$

This is, in fact, the Chain Rule for compositions of transformations:

$$D(\mathbf{g} \circ \mathbf{f})(\mathbf{x}) = D\mathbf{g}\big(\mathbf{f}(\mathbf{x})\big)D\mathbf{f}(\mathbf{x}).$$

In this context we can regard the function $f(x, y)$, say, as a transformation from $\mathbb{R}^2$ to $\mathbb{R}$. Its derivative is the linear transformation with matrix

$$Df(x, y) = \big(f_1(x, y), f_2(x, y)\big).$$

The transformation $\mathbf{y} = \mathbf{f}(\mathbf{x})$ also defines a vector $d\mathbf{y}$ of differentials of the variables $y_i$ in terms of the vector $d\mathbf{x}$ of differentials of the variables $x_j$. Writing $d\mathbf{y}$ and $d\mathbf{x}$ as column vectors we have

$$d\mathbf{y} = \begin{pmatrix} dy_1 \\ dy_2 \\ \vdots \\ dy_m \end{pmatrix} = \begin{pmatrix} \frac{\partial y_1}{\partial x_1} & \frac{\partial y_1}{\partial x_2} & \cdots & \frac{\partial y_1}{\partial x_n} \\ \frac{\partial y_2}{\partial x_1} & \frac{\partial y_2}{\partial x_2} & \cdots & \frac{\partial y_2}{\partial x_n} \\ \vdots & \vdots & & \vdots \\ \frac{\partial y_m}{\partial x_1} & \frac{\partial y_m}{\partial x_2} & \cdots & \frac{\partial y_m}{\partial x_n} \end{pmatrix} \begin{pmatrix} dx_1 \\ dx_2 \\ \vdots \\ dx_n \end{pmatrix} = D\mathbf{f}(\mathbf{x})d\mathbf{x}.$$

**EXAMPLE 4** Find the Jacobian matrix $D\mathbf{f}(1, 0)$ for the transformation from $\mathbb{R}^2$ to $\mathbb{R}^3$ given by

$$\mathbf{f}(x, y) = \big(xe^y + \cos(\pi y), x^2, x - e^y\big)$$

and use it to find an approximate value for $\mathbf{f}(1.02, 0.01)$.

***SOLUTION*** $D\mathbf{f}(x, y)$ is the $3 \times 2$ matrix whose $j$th row consists of the partial derivatives of the $j$th component of $\mathbf{f}$ with respect to $x$ and $y$. Thus

$$D\mathbf{f}(1, 0) = \begin{pmatrix} e^y & xe^y - \pi \sin(\pi y) \\ 2x & 0 \\ 1 & -e^y \end{pmatrix}\Bigg|_{(1,0)} = \begin{pmatrix} 1 & 1 \\ 2 & 0 \\ 1 & -1 \end{pmatrix}$$

Since $f(1, 0) = (2, 1, 0)$ and $d\mathbf{x} = \begin{pmatrix} 0.02 \\ 0.01 \end{pmatrix}$, we have

$$d\mathbf{f} = D\mathbf{f}(1, 0)\, d\mathbf{x} = \begin{pmatrix} 1 & 1 \\ 2 & 0 \\ 1 & -1 \end{pmatrix} \begin{pmatrix} 0.02 \\ 0.01 \end{pmatrix} = \begin{pmatrix} 0.03 \\ 0.04 \\ 0.01 \end{pmatrix}.$$

Therefore, $\mathbf{f}(1.02, 0.01) \approx (2.03, 1.04, 0.01)$. ■

For transformations between spaces of the same dimension (say from $\mathbb{R}^n$ to $\mathbb{R}^n$) the Jacobian matrices are square, and have determinants. These Jacobian determinants will play a role in our consideration of implicit functions and inverse functions in Section 4.8, and in changes of variables in multiple integrals in Chapter 6.

## EXERCISES 4.6

In Exercises 1–6, use suitable linearizations to find approximate values for the given functions at the points indicated.

**1.** $f(x, y) = x^2 y^3$ at $(3.1, 0.9)$

**2.** $f(x, y) = \tan^{-1}\left(\dfrac{y}{x}\right)$ at $(3.01, 2.99)$

**3.** $f(x, y) = \sin(\pi xy + \ln y)$ at $(0.01, 1.05)$

**4.** $f(x, y) = \dfrac{24}{x^2 + xy + y^2}$ at $(2.1, 1.8)$

**5.** $f(x, y, z) = \sqrt{x + 2y + 3z}$ at $(1.9, 1.8, 1.1)$

**6.** $f(x, y) = x\, e^{y + x^2}$ at $(2.05, -3.92)$

**7.** The edges of a rectangular box are each measured to within an accuracy of 1% of their values. What is the approximate maximum percentage error in

(a) the calculated volume of the box,

(b) the calculated area of one of the faces of the box, and

(c) the calculated length of a diagonal of the box?

**8.** The radius and height of a right-circular conical tank are measured to be 25 ft and 21 ft, respectively. Each measurement is accurate to within 0.5 in. By about how much can the calculated volume of the tank be in error?

**9.** By approximately how much can the calculated area of the conical surface of the tank in Exercise 8 be in error?

**10.** Two sides and the contained angle of a triangular plot of land are measured to be 224 m, 158 m, and 64°, respectively. The length measurements were accurate to within 0.4 m and the angle measurement to within 2°. What is the approximate maximum percentage error if the area of the plot is calculated from these measurements?

**11.** The angle of elevation of the top of a tower is measured at two points $A$ and $B$ on the ground in the same direction from the base of the tower. The angles are 50° at $A$ and 35° at $B$, each measured to within 1°. The distance $AB$ is measured to be 100 m with error at most 0.1%. What is the calculated height of the building, and by about how much can it be in error? To which of the three measurements is the calculated height most sensitive?

**12.** By approximately what percentage will the value of $w = \dfrac{x^2 y^3}{z^4}$ increase or decrease if $x$ increases by 1%, $y$ increases by 2%, and $z$ and increases by 3%?

**13.** Find the Jacobian matrix for the transformation $\mathbf{f}(r, \theta) = (x, y)$, where

$$x = r \cos\theta \quad \text{and} \quad y = r \sin\theta.$$

(Although $(r, \theta)$ can be regarded as *polar coordinates* in the $xy$-plane, they are Cartesian coordinates in their own $r\theta$-plane.)

**14.** Find the Jacobian matrix for the transformation $\mathbf{f}(\rho, \phi, \theta) = (x, y, z)$, where

$$x = \rho \sin\phi \cos\theta,$$
$$y = \rho \sin\phi \sin\theta,$$
$$z = \rho \cos\phi.$$

Here $(\rho, \phi, \theta)$ are *spherical coordinates* in $xyz$-space. They will be formally introduced in Section 6.5.

**15.** Find the Jacobian matrix $D\mathbf{f}(x, y, z)$ for the transformation of $\mathbb{R}^3$ to $\mathbb{R}^2$ given by

$$\mathbf{f}(x, y, z) = (x^2 + yz,\, y^2 - x \ln z).$$

Use $D\mathbf{f}(2, 2, 1)$ to help you find an approximate value for $\mathbf{f}(1.98, 2.01, 1.03)$.

**16.** Find the Jacobian matrix $D\mathbf{g}(1, 3, 3)$ for the transformation of $\mathbb{R}^3$ to $\mathbb{R}^3$ given by

$$\mathbf{g}(r, s, t) = (r^2 s, r^2 t, s^2 - t^2)$$

and use the result to find an approximate value for $\mathbf{g}(0.99, 3.02, 2.97)$.

**17.** Prove that if $f(x, y)$ is differentiable at $(a, b)$, then $f(x, y)$ is continuous at $(a, b)$.

* **18.** Prove the following version of the Mean-Value Theorem: if $f(x, y)$ has first partial derivatives continuous near every point of the straight line segment joining the points $(a, b)$ and $(a + h, b + k)$ then there exists a number $\theta$ satisfying $0 < \theta < 1$ such that

$$\begin{aligned} f(a+h, b+k) = & f(a, b) + hf_1(a+\theta h, b+\theta k) \\ & + kf_2(a+\theta h, b+\theta k). \end{aligned}$$

(*Hint:* apply the single-variable Mean-Value Theorem to $g(t) = f(a + th, b + tk)$.) Why could we not have used this result in place of Theorem 3 to prove Theorem 4 and hence the version of the Chain Rule given in this section?

* **19.** Generalize Exercise 18 as follows: show that if $f(x, y)$ has continuous partial derivatives of second order near the point $(a, b)$, then there exists a number $\theta$ satisfying $0 < \theta < 1$ such that for $h$ and $k$ sufficiently small in absolute value,

$$\begin{aligned} f(a+h, b+k) = & f(a, b) + hf_1(a, b) + kf_2(a, b) \\ & + h^2 f_{11}(a+\theta h, b+\theta k) \\ & + 2hkf_{12}(a+\theta h, b+\theta k) \\ & + k^2 f_{22}(a+\theta h, b+\theta k). \end{aligned}$$

Hence show that there is a constant $K$ such that for all sufficiently small $h$ and $k$,

$$\begin{aligned} & \left| f(a+h, b+k) - f(a, b) - hf_1(a, b) - kf_2(a, b) \right| \\ & \quad \leq K(h^2 + k^2). \end{aligned}$$

## 4.7 GRADIENTS AND DIRECTIONAL DERIVATIVES

A first partial derivative of a function of several variables gives the rate of change of that function with respect to distance measured in the direction of one of the coordinate axes. In this section we will develop a method for finding the rate of change of such a function with respect to distance measured in *any direction* in the domain of the function.

To begin, it is useful to combine the first partial derivatives of a function into a single *vector function* called a **gradient**. For simplicity, we will develop and interpret the gradient for functions of two variables. Extension to functions of three or more variables is straightforward and will be discussed later in the section.

**DEFINITION 9**

At any point $(x, y)$ where the first partial derivatives of the function $f(x, y)$ exist, we define the **gradient vector** $\nabla f(x, y) = \mathbf{grad}\, f(x, y)$ by

$$\nabla f(x, y) = \mathbf{grad}\, f(x, y) = f_1(x, y)\mathbf{i} + f_2(x, y)\mathbf{j}.$$

Recall that $\mathbf{i}$ and $\mathbf{j}$ denote the unit basis vectors from the origin to the points (1,0) and (0,1) respectively. The symbol $\nabla$, called *del* or *nabla*, is a *vector differential operator*:

$$\nabla = \mathbf{i}\frac{\partial}{\partial x} + \mathbf{j}\frac{\partial}{\partial y}.$$

We can *apply* this operator to a function $f(x, y)$ by writing the operator to the left of the function. The result is the gradient of the function

$$\nabla f(x, y) = \left(\mathbf{i}\frac{\partial}{\partial x} + \mathbf{j}\frac{\partial}{\partial y}\right) f(x, y) = f_1(x, y)\mathbf{i} + f_2(x, y)\mathbf{j}.$$

We will make extensive use of the del operator in Chapter 9.

**EXAMPLE 1** If $f(x, y) = x^2 + y^2$ then $\nabla f(x, y) = 2x\mathbf{i} + 2y\mathbf{j}$. In particular, $\nabla f(1, 2) = 2\mathbf{i} + 4\mathbf{j}$. Observe that this vector is perpendicular to the tangent line $x + 2y = 5$ to the circle $x^2 + y^2 = 5$ at $(1, 2)$. This circle is the level curve of $f$ that passes through the point $(1, 2)$. (See Figure 4.24(a).) As the following theorem shows, this perpendicularity is not a coincidence.

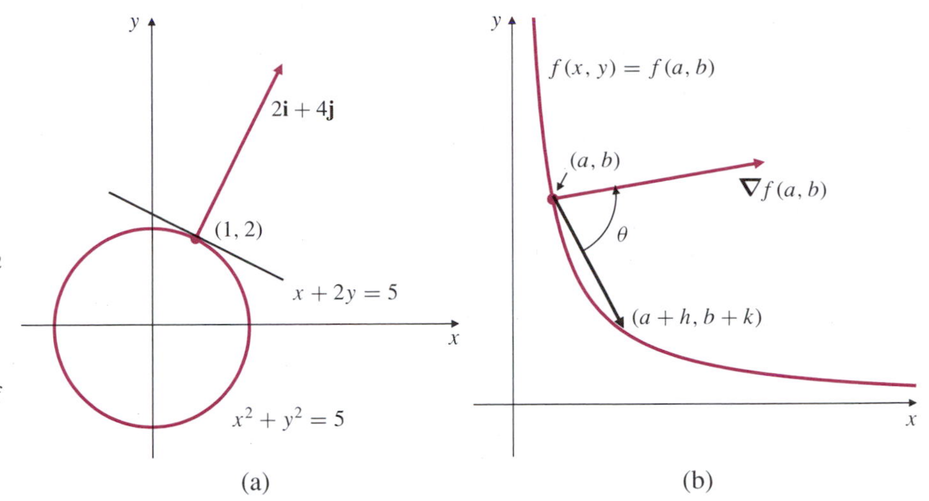


**Figure 4.24**

(a) The gradient of $f(x, y) = x^2 + y^2$ at $(1, 2)$ is normal to the level curve of $f$ through $(1, 2)$

(b) The angle between $\nabla f(a, b)$ and a chord vector to the level curve of $f$ through $(a, b)$ approaches $90^\circ$ as the length of the chord approaches zero

**THEOREM 6**

If $f(x, y)$ is differentiable at the point $(a, b)$ and $\nabla f(a, b) \neq \mathbf{0}$, then $\nabla f(a, b)$ is a normal vector to the level curve of $f$ that passes through $(a, b)$.

***PROOF*** The angle $\theta$ between the vector $\nabla f(a, b)$ and the vector $h\mathbf{i} + k\mathbf{j}$ from the point $(a, b)$ to the point $(a + h, b + k)$ (see Figure 4.24(b)) satisfies

$$\cos\theta = \frac{\nabla f(a, b) \bullet (h\mathbf{i} + k\mathbf{j})}{|\nabla f(a, b)|\sqrt{h^2 + k^2}} = \frac{1}{|\nabla f(a, b)|} \frac{hf_1(a, b) + kf_2(a, b)}{\sqrt{h^2 + k^2}}.$$

If $(a + h, b + k)$ lies on the level curve of $f$ that passes through $(a, b)$, then $f(a + h, b + k) = f(a, b)$, and so, since $f$ is differentiable at $(a, b)$, we have

$$\lim_{h,k\to 0} \frac{hf_1(a, b) + kf_2(a, b)}{\sqrt{h^2 + k^2}}$$
$$= -\lim_{h,k\to 0} \frac{f(a + h, b + k) - f(a, b) - hf_1(a, b) - kf_2(a, b)}{\sqrt{h^2 + k^2}} = 0.$$

Hence $\cos\theta \to 0$ and $\theta \to \pi/2$ as $h, k \to 0$. This says that $\nabla f(a, b)$ is perpendicular to the tangent to the level curve $f(x, y) = f(a, b)$.

## Directional Derivatives

The first partial derivatives $f_1(a, b)$ and $f_2(a, b)$ give the rates of change of $f(x, y)$ at $(a, b)$ measured in the directions of the positive $x$- and $y$-axes, respectively. If we want to know how fast $f(x, y)$ changes value as we move through the domain of $f$ at $(a, b)$ in some other direction we require a more general **directional derivative**. We can specify the direction by means of a nonzero vector. It is most convenient to use a *unit vector*.

**DEFINITION 10**

Let $\mathbf{u} = u\mathbf{i} + v\mathbf{j}$ be a unit vector, so that $u^2 + v^2 = 1$. The **directional derivative** of $f(x, y)$ at $(a, b)$ in the direction of $\mathbf{u}$ is the rate of change of $f(x, y)$ with respect to distance measured at $(a, b)$ along a ray in the direction of $\mathbf{u}$ in the $xy$-plane. This directional derivative is given by

$$D_{\mathbf{u}} f(a, b) = \lim_{h \to 0+} \frac{f(a + hu, b + hv) - f(a, b)}{h}.$$

It is also given by

$$D_{\mathbf{u}} f(a, b) = \frac{d}{dt} f(a + tu, b + tv) \bigg|_{t=0}$$

if the derivative on the right side exists.

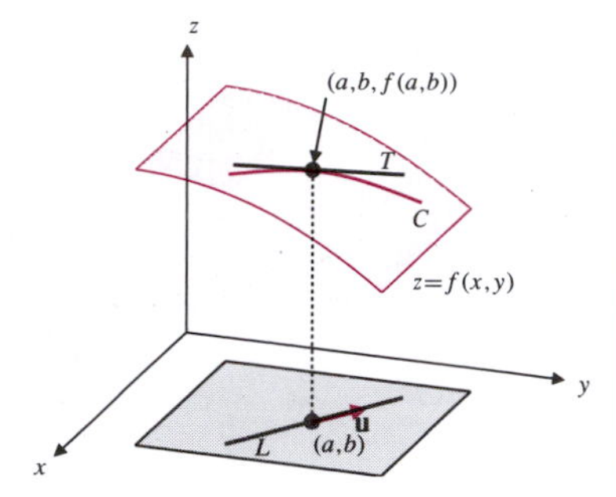


**Figure 4.25** Unit vector **u** determines a line $L$ through $(a, b)$ in the domain of $f$. The vertical plane containing $L$ intersects the graph of $f$ in a curve $C$ whose tangent $T$ at $(a, b, f(a, b))$ has slope $D_{\mathbf{u}} f(a, b)$

Observe that directional derivatives in directions parallel to the coordinate axes are given directly by first partials: $D_{\mathbf{i}} f(a, b) = f_1(a, b)$, $D_{\mathbf{j}} f(a, b) = f_2(a, b)$, $D_{-\mathbf{i}} f(a, b) = -f_1(a, b)$, and $D_{-\mathbf{j}} f(a, b) = -f_2(a, b)$. The following theorem shows how the gradient can be used to calculate any directional derivative.

**THEOREM 7**

**Using the gradient to find directional derivatives**

If $f$ is differentiable at $(a, b)$ and $\mathbf{u} = u\mathbf{i} + v\mathbf{j}$ is a unit vector, then the directional derivative of $f$ at $(a, b)$ in the direction of $\mathbf{u}$ is given by

$$D_{\mathbf{u}} f(a, b) = \mathbf{u} \bullet \nabla f(a, b).$$

***PROOF*** By the Chain Rule:

$$\begin{aligned} D_{\mathbf{u}} f(a, b) &= \frac{d}{dt} f(a + tu, b + tv) \bigg|_{t=0} \\ &= u f_1(a, b) + v f_2(a, b) = \mathbf{u} \bullet \nabla f(a, b). \end{aligned}$$

We already know that having partial derivatives at a point does not imply that a function is continuous there, let alone that it is differentiable. The same can be said about directional derivatives. It is possible for a function to have a directional derivative in every direction at a given point and still not be continuous at that point. See Exercise 41 at the end of this section for an example of such a function.

Given any nonzero vector $\mathbf{v}$, we can always obtain a unit vector in the same direction by dividing $\mathbf{v}$ by its length. The directional derivative of $f$ at $(a, b)$ in the direction of $\mathbf{v}$ is therefore given by

$$D_{\mathbf{v}/|\mathbf{v}|} f(a, b) = \frac{\mathbf{v}}{|\mathbf{v}|} \bullet \nabla f(a, b).$$

**EXAMPLE 2** Find the rate of change of $f(x, y) = y^4 + 2xy^3 + x^2y^2$ at $(0, 1)$ measured in each of the following directions:

(a) $\mathbf{i} + 2\mathbf{j}$, (b) $\mathbf{j} - 2\mathbf{i}$, (c) $3\mathbf{i}$, and (d) $\mathbf{i} + \mathbf{j}$.

**SOLUTION** We calculate

$$\nabla f(x, y) = (2y^3 + 2xy^2)\mathbf{i} + (4y^3 + 6xy^2 + 2x^2y)\mathbf{j},$$
$$\nabla f(0, 1) = 2\mathbf{i} + 4\mathbf{j}.$$

(a) The directional derivative of $f$ at $(0, 1)$ in the direction of $\mathbf{i} + 2\mathbf{j}$ is

$$\frac{\mathbf{i} + 2\mathbf{j}}{|\mathbf{i} + 2\mathbf{j}|} \bullet (2\mathbf{i} + 4\mathbf{j}) = \frac{2 + 8}{\sqrt{5}} = 2\sqrt{5}.$$

Observe that $\mathbf{i} + 2\mathbf{j}$ points in the same direction as $\nabla f(0, 1)$ so the directional derivative is positive and equal to the length of $\nabla f(0, 1)$.

(b) The directional derivative of $f$ at $(0, 1)$ in the direction of $\mathbf{j} - 2\mathbf{i}$ is

$$\frac{-2\mathbf{i} + \mathbf{j}}{|-2\mathbf{i} + \mathbf{j}|} \bullet (2\mathbf{i} + 4\mathbf{j}) = \frac{-4 + 4}{\sqrt{5}} = 0.$$

Since $\mathbf{j} - 2\mathbf{i}$ is perpendicular to $\nabla f(0, 1)$, it is tangent to the level curve of $f$ through $(0, 1)$, so the directional derivative in that direction is zero.

(c) The directional derivative of $f$ at $(0, 1)$ in the direction of $3\mathbf{i}$ is

$$\mathbf{i} \bullet (2\mathbf{i} + 4\mathbf{j}) = 2.$$

As noted previously, the directional derivative of $f$ in the direction of the positive $x$-axis is just $f_1(0, 1)$.

(d) The directional derivative of $f$ at $(0, 1)$ in the direction of $\mathbf{i} + \mathbf{j}$ is

$$\frac{\mathbf{i} + \mathbf{j}}{|\mathbf{i} + \mathbf{j}|} \bullet (2\mathbf{i} + 4\mathbf{j}) = \frac{2 + 4}{\sqrt{2}} = 3\sqrt{2}.$$

If we move along the surface $z = f(x, y)$ through the point $(0, 1, 1)$ in a direction making horizontal angles of 45 degrees with the positive directions of the $x$- and $y$-axes, we would be rising at a rate of $3\sqrt{2}$ vertical units per horizontal unit moved. ■

**REMARK** A direction in the plane can be specified by a polar angle. The direction making angle $\phi$ with the positive direction of the $x$-axis corresponds to the unit vector

$$\mathbf{u}_\phi = \cos\phi\mathbf{i} + \sin\phi\mathbf{j},$$

so the directional derivative of $f$ at $(x, y)$ in that direction is

$$D_\phi f(x, y) = D_{\mathbf{u}_\phi} f(x, y) = \mathbf{u}_\phi \bullet \nabla f(x, y) = f_1(x, y)\cos\phi + f_2(x, y)\sin\phi.$$

Note the use of the symbol $D_\phi f(x, y)$ to denote a derivative of $f$ with respect to *distance* measured in the direction $\phi$.

As observed in the previous example, Theorem 7 provides a useful interpretation for the gradient vector. For any unit vector $\mathbf{u}$ we have

$$D_{\mathbf{u}} f(a, b) = \mathbf{u} \bullet \nabla f(a, b) = |\nabla f(a, b)| \cos\theta,$$

where $\theta$ is the angle between the vectors $\mathbf{u}$ and $\nabla f(a, b)$. Since $\cos\theta$ only takes on values between $-1$ and $1$, $D_{\mathbf{u}} f(a, b)$ only takes on values between $-|\nabla f(a, b)|$ and $|\nabla f(a, b)|$. Moreover, $D_{\mathbf{u}} f(a, b) = -|\nabla f(a, b)|$ if and only if $\mathbf{u}$ points in the opposite direction to $\nabla f(a, b)$ (so that $\cos\theta = -1$), and $D_{\mathbf{u}} f(a, b) = |\nabla f(a, b)|$ if and only if $\mathbf{u}$ points in the same direction as $\nabla f(a, b)$ (so that $\cos\theta = 1$). The directional derivative is zero in the direction $\theta = \pi/2$; this is the direction of the (tangent line to the) level curve of $f$ through $(a, b)$.

We summarize these properties of the gradient as follows:

**Geometric properties of the gradient vector**

(i) At $(a, b)$, $f(x, y)$ increases most rapidly in the direction of the gradient vector $\nabla f(a, b)$. The maximum rate of increase is $|\nabla f(a, b)|$.

(ii) At $(a, b)$, $f(x, y)$ decreases most rapidly in the direction of $-\nabla f(a, b)$. The maximum rate of decrease is $|\nabla f(a, b)|$.

(iii) The rate of change of $f(x, y)$ at $(a, b)$ is zero in directions tangent to the level curve of $f$ that passes through $(a, b)$.

Look again at the topographic map in Figure 4.7 in Section 4.1. The streams on the map flow in the direction of steepest descent, that is, in the direction of $-\nabla f$, where $f$ measures the elevation of land. The streams therefore cross the contours (the level curves of $f$) at right angles.

**EXAMPLE 3** The temperature at position $(x, y)$ in a region of the $xy$-plane is $T\,^\circ$C where

$$T(x, y) = x^2 e^{-y}.$$

In what direction at the point (2, 1) does the temperature increase most rapidly? What is the rate of increase of $f$ in that direction?

***SOLUTION*** We have

$$\nabla T(x, y) = 2x\,e^{-y}\mathbf{i} - x^2 e^{-y}\mathbf{j},$$

$$\nabla T(2, 1) = \frac{4}{e}\mathbf{i} - \frac{4}{e}\mathbf{j} = \frac{4}{e}(\mathbf{i} - \mathbf{j}).$$

At (2, 1), $T(x, y)$ increases most rapidly in the direction of the vector $\mathbf{i} - \mathbf{j}$. The rate of increase in this direction is $|\nabla T(2, 1)| = 4\sqrt{2}/e\,^\circ$C/unit distance. ■

**EXAMPLE 4** A hiker is standing beside a stream on the side of a mountain, examining her map of the region. The height of land (in kilometres) at any point $(x, y)$ is given by

$$h(x, y) = \frac{20}{3 + x^2 + 2y^2},$$

where $x$ and $y$ (also in kilometres) denote the coordinates of the point on the hiker's map. The hiker is at the point (3, 2).

(a) What is the direction of flow of the stream at (3, 2) on the hiker's map? How fast is the stream descending at her location?

(b) Find the equation of the path of the stream on the hiker's map.

(c) At what angle to the path of the stream (on the map) should the hiker set out if she wishes to climb at a 15° inclination to the horizontal?

(d) Make a sketch of the hiker's map, showing some curves of constant elevation, and showing the stream.

## SOLUTION

(a) We begin by calculating the gradient of $h$ and its length at $(3, 2)$:

$$\nabla h(x, y) = -\frac{20}{(3 + x^2 + 2y^2)^2}(2x\mathbf{i} + 4y\mathbf{j}),$$
$$\nabla h(3, 2) = -\frac{1}{20}(6\mathbf{i} + 8\mathbf{j}) = -\frac{1}{10}(3\mathbf{i} + 4\mathbf{j}),$$
$$|\nabla h(3, 2)| = 0.5.$$

The stream is flowing in the direction whose horizontal projection at $(3, 2)$ is $-\nabla h(3, 2)$, that is, in the horizontal direction of the vector $3\mathbf{i} + 4\mathbf{j}$. The stream is descending at a rate of 0.5 vertical units per horizontal unit travelled.

(b) Coordinates on the map are the coordinates $(x, y)$ in the domain of the height function $h$. We can find an equation of the path of the stream on a map of the region by setting up a differential equation for a change of position along the path. If the vector $d\mathbf{r} = dx\,\mathbf{i} + dy\,\mathbf{j}$ is tangent to the path of the stream at point $(x, y)$ on the map then $d\mathbf{r}$ is parallel to $\nabla h(x, y)$. Hence the components of these two vectors are proportional:

$$\frac{dx}{2x} = \frac{dy}{4y} \qquad \text{or} \qquad \frac{dy}{y} = \frac{2dx}{x}.$$

Integrating both sides of this equation, we get $\ln y = 2\ln x + \ln C$, or $y = Cx^2$. Since the path of the stream passes through $(3, 2)$, we have $C = 2/9$ and the equation is $9y = 2x^2$.

(c) Suppose the hiker moves away from $(3, 2)$ in the direction of the unit vector $\mathbf{u}$. She will be ascending at an inclination of $15^\circ$ if the directional derivative of $h$ in the direction of $\mathbf{u}$ is $\tan 15^\circ \approx 0.268$. If $\theta$ is the angle between $\mathbf{u}$ and the upstream direction then

$$0.5\cos\theta = |\nabla h(3, 2)|\cos\theta = D_{\mathbf{u}}h(3, 2) \approx 0.268.$$

Hence $\cos\theta \approx 0.536$ and $\theta \approx 57.6^\circ$. She should set out in a direction making a horizontal angle of about $58^\circ$ with the upstream direction.

(d) A suitable sketch of the map is given in Figure 4.26.

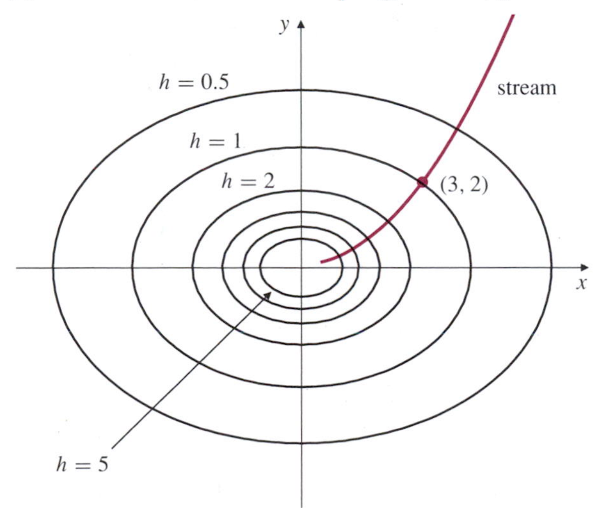


**Figure 4.26** The hiker's map. Unlike most mountains, this one has elliptical cross-sections

**EXAMPLE 5** Find the second directional derivative of $f(x, y)$ in the direction making angle $\phi$ with the positive $x$-axis.

***SOLUTION*** As observed earlier, the first directional derivative is

$$D_\phi f(x, y) = (\cos\phi\mathbf{i} + \sin\phi\mathbf{j}) \bullet \nabla f(x, y) = f_1(x, y)\cos\phi + f_2(x, y)\sin\phi.$$

The second directional derivative is therefore

$$\begin{aligned} D^2_\phi f(x, y) =& D_\phi\left(D_\phi f(x, y)\right) \\ =& (\cos\phi\mathbf{i} + \sin\phi\mathbf{j}) \bullet \nabla\Big(f_1(x, y)\cos\phi + f_2(x, y)\sin\phi\Big) \\ =& \Big(f_{11}(x, y)\cos\phi + f_{21}(x, y)\sin\phi\Big)\cos\phi \\ &+ \Big(f_{12}(x, y)\cos\phi + f_{22}(x, y)\sin\phi\Big)\sin\phi \\ =& f_{11}(x, y)\cos^2\phi + 2f_{12}(x, y)\cos\phi\sin\phi + f_{22}(x, y)\sin^2\phi. \end{aligned}$$

Note that if $\phi = 0$ or $\phi = \pi$ (so the directional derivative is in a direction parallel to the $x$-axis) then $D^2_\phi f(x, y) = f_{11}(x, y)$. Similarly, $D^2_\phi f(x, y) = f_{22}(x, y)$ if $\phi = \pi/2$ or $3\pi/2$. ■

## Rates Perceived by a Moving Observer

Suppose that an observer is moving around in the $xy$-plane measuring the value of a function $f(x, y)$ defined in the plane as he passes through each point $(x, y)$. (For instance, $f(x, y)$ might be the temperature at $(x, y)$.) If the observer is moving with velocity $\mathbf{v}$ at the instant when he passes through the point $(a, b)$, how fast would he observe $f(x, y)$ to be changing at that moment?

At the moment in question the observer is moving in the direction of the unit vector $\mathbf{v}/|\mathbf{v}|$. The rate of change of $f(x, y)$ at $(a, b)$ in that direction is

$$D_{\mathbf{v}/|\mathbf{v}|} f(a, b) = \frac{\mathbf{v}}{|\mathbf{v}|} \bullet \nabla f(a, b)$$

measured in units of $f$ per unit distance in the $xy$-plane. To convert this rate to units of $f$ per unit time, we must multiply by the speed of the observer, $|\mathbf{v}|$ units of distance per unit time. Thus the time rate of change of $f(x, y)$ as measured by the observer passing through $(a, b)$ is

$$|\mathbf{v}| \times \frac{\mathbf{v}}{|\mathbf{v}|} \bullet \nabla f(a, b) = \mathbf{v} \bullet \nabla f(a, b).$$

It is natural to extend our use of the symbol $D_\mathbf{v} f(a, b)$ to represent this rate even though $\mathbf{v}$ is not (necessarily) a unit vector. Thus

> The rate of change of $f(x, y)$ at $(a, b)$ as measured by an observer moving through $(a, b)$ with velocity $\mathbf{v}$ is
>
> $$D_\mathbf{v} f(a, b) = \mathbf{v} \bullet \nabla f(a, b)$$
>
> units of $f$ per unit time.

If the hiker in Example 4 moves away from (3, 2) with horizontal velocity $\mathbf{v} = -\mathbf{i} - \mathbf{j}$ km/h, then she will be rising at a rate

$$\mathbf{v} \bullet \nabla h(3, 2) = (-\mathbf{i} - \mathbf{j}) \bullet \left( -\frac{1}{10}(3\mathbf{i} + 4\mathbf{j}) \right) = \frac{7}{10} \text{ km/h}.$$

As defined here, $D_{\mathbf{v}} f$ is the spatial component of the derivative of $f$ following the motion. See Example 6 in Section 4.5. The rate of change of the reading on the moving thermometer in that example can be expressed as

$$\frac{dT}{dt} = D_{\mathbf{v}} T(x, y, z, t) + \frac{\partial T}{\partial t}$$

where $\mathbf{v}$ is the velocity of the moving thermometer and $D_{\mathbf{v}} T = \mathbf{v} \bullet \nabla T$. The gradient is being taken with respect to the *three spatial variables* only. (See below for the gradient in 3-space.)

## The Gradient in Three and More Dimensions

By analogy with the two-dimensional case, a function $f(x_1, x_2, \ldots, x_n)$ of $n$ variables possessing first partial derivatives has gradient given by

$$\nabla f(x_1, x_2, \ldots, x_n) = \frac{\partial f}{\partial x_1}\mathbf{e}_1 + \frac{\partial f}{\partial x_2}\mathbf{e}_2 + \cdots + \frac{\partial f}{\partial x_n}\mathbf{e}_n,$$

where $\mathbf{e}_j$ is the unit vector from the origin to the unit point on the $j$th coordinate axis. In particular, for a function of three variables,

$$\nabla f(x, y, z) = \frac{\partial f}{\partial x}\mathbf{i} + \frac{\partial f}{\partial y}\mathbf{j} + \frac{\partial f}{\partial z}\mathbf{k}.$$

The level surface of $f(x, y, z)$ passing through $(a, b, c)$ has a tangent plane there if $f$ is differentiable at $(a, b, c)$ and $\nabla f(a, b, c) \neq \mathbf{0}$.

For functions of any number of variables the vector $\nabla f(P_0)$ is normal to the "level surface" of $f$ passing through the point $P_0$ (that is, the (hyper)surface with equation $f(x_1, \ldots, x_n) = f(P_0)$), and, if $f$ is differentiable at $P_0$, the rate of change of $f$ at $P_0$ in the direction of the unit vector $\mathbf{u}$ is given by $\mathbf{u} \bullet \nabla f(P_0)$. Equations of tangent planes to surfaces in 3-space can be found easily with the aid of gradients.

**EXAMPLE 6** Let $f(x, y, z) = x^2 + y^2 + z^2$.

(a) Find $\nabla f(x, y, z)$ and $\nabla f(1, -1, 2)$.

(b) Find an equation of the tangent plane to the sphere $x^2 + y^2 + z^2 = 6$ at the point $(1, -1, 2)$.

(c) What is the maximum rate of increase of $f$ at $(1, -1, 2)$?

(d) What is the rate of change with respect to distance of $f$ at $(1, -1, 2)$ measured in the direction from that point towards the point $(3, 1, 1)$?

***SOLUTION***

(a) $\nabla f(x, y, z) = 2x\mathbf{i} + 2y\mathbf{j} + 2z\mathbf{k}$, so $\nabla f(1, -1, 2) = 2\mathbf{i} - 2\mathbf{j} + 4\mathbf{k}$.

(b) The required tangent plane has $\nabla f(1, -1, 2)$ as normal. (See Figure 4.27(a).) Therefore its equation is given by $2(x - 1) - 2(y + 1) + 4(z - 2) = 0$ or, more simply, $x - y + 2z = 6$.

(c) The maximum rate of increase of $f$ at $(1, -1, 2)$ is $|\nabla f(1, -1, 2)| = 2\sqrt{6}$, and it occurs in the direction of the vector $\mathbf{i} - \mathbf{j} + 2\mathbf{k}$.

(d) The direction from $(1, -1, 2)$ towards $(3, 1, 1)$ is specified by $2\mathbf{i} + 2\mathbf{j} - \mathbf{k}$. The rate of change of $f$ with respect to distance in this direction is

$$\frac{2\mathbf{i} + 2\mathbf{j} - \mathbf{k}}{\sqrt{4+4+1}} \bullet (2\mathbf{i} - 2\mathbf{j} + 4\mathbf{k}) = \frac{4-4-4}{3} = -\frac{4}{3},$$

that is, $f$ decreases at rate $4/3$ of a unit per horizontal unit moved. ■

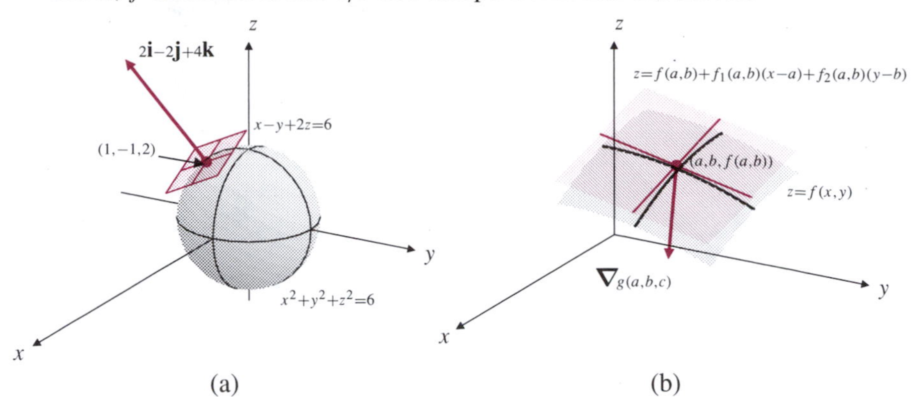


**Figure 4.27**

(a) The tangent plane to $x^2 + y^2 + z^2 = 6$ at $(1, -1, 2)$

(b) The gradient of $f(x, y) - z$ at $(a, b, f(a, b))$ is normal to the tangent plane to $z = f(x, y)$ at that point

■ **EXAMPLE 7** The graph of a *function* $f(x, y)$ of two variables is the graph of the *equation* $z = f(x, y)$ in 3-space. This surface is the level surface $g(x, y, z) = 0$ of the 3-variable function

$$g(x, y, z) = f(x, y) - z.$$

If $f$ is differentiable at $(a, b)$ and $c = f(a, b)$ then $g$ is differentiable at $(a, b, c)$ and

$$\nabla g(a, b, c) = f_1(a, b)\mathbf{i} + f_2(a, b)\mathbf{j} - \mathbf{k}$$

is normal to $g(x, y, z) = 0$ at $(a, b, c)$. (Note that $\nabla g(a, b, c) \neq \mathbf{0}$, since its $z$ component is $-1$.) It follows that the graph of $f$ has nonvertical tangent plane at $(a, b)$ given by

$$f_1(a, b)(x - a) + f_2(a, b)(y - b) - (z - c) = 0,$$

or

$$z = f(a, b) + f_1(a, b)(x - a) + f_2(a, b)(y - b).$$

(See Figure 4.27(b).) This result was obtained by a different argument in Section 4.3. ■

**DANGER**

Make sure you understand the difference between the graph of a function and a level curve or level surface of that function. (See the discussion following this example.) Here the surface $z = f(x, y)$ is the *graph* of the function $f$, but it is also a *level surface* of a *different* function $g$.

**DANGER**

Students sometimes confuse graphs of functions with level curves or surfaces of those functions. In the above example we are talking about a *level surface* of the function $g(x, y, z)$ that happens to coincide with the *graph* of a different function, $f(x, y)$. Do not confuse that surface with the graph of $g$, which is a three-dimensional *hypersurface* in 4-space having equation $w = g(x, y, z)$. Similarly, do not confuse the tangent *plane* to the graph of $f(x, y)$, (that is, the plane obtained in the above example), with the tangent *line* to the level curve of $f(x, y)$ passing through $(a, b)$ and lying in the $xy$-plane. This line has an equation involving only $x$ and $y$, namely $f_1(a, b)(x - a) + f_2(a, b)(y - b) = 0$.

**EXAMPLE 8** Find a vector tangent to the curve of intersection of the two surfaces

$$z = x^2 - y^2 \qquad \text{and} \qquad xyz + 30 = 0$$

at the point $(-3, 2, 5)$.

***SOLUTION*** The coordinates of the given point satisfy the equations of both surfaces so the point lies on the curve of intersection of the two surfaces. A vector tangent to this curve at that point will be perpendicular to the normals to both surfaces, that is, to the vectors

$$\mathbf{n}_1 = \nabla(x^2 - y^2 - z)\Big|_{(-3,2,5)} = 2x\mathbf{i} - 2y\mathbf{j} - \mathbf{k}\Big|_{(-3,2,5)} = -6\mathbf{i} - 4\mathbf{j} - \mathbf{k},$$

$$\mathbf{n}_2 = \nabla(xyz + 30)\Big|_{(-3,2,5)} = (yz\mathbf{i} + xz\mathbf{j} + xy\mathbf{k})\Big|_{(-3,2,5)} = 10\mathbf{i} - 15\mathbf{j} - 6\mathbf{k}.$$

For the tangent vector $\mathbf{T}$ we can therefore use the cross product of these normals

$$\mathbf{T} = \mathbf{n}_1 \times \mathbf{n}_2 = \begin{vmatrix} \mathbf{i} & \mathbf{j} & \mathbf{k} \\ -6 & -4 & -1 \\ 10 & -15 & -6 \end{vmatrix} = 9\mathbf{i} - 46\mathbf{j} + 130\mathbf{k}.$$

## EXERCISES 4.7

In Exercises 1–8, find:

(a) the gradient of the given function at the point indicated,

(b) an equation of the plane tangent to the graph of the given function at the point whose $x$ and $y$ coordinates are given, and

(c) an equation of the straight line tangent, at the given point, to the level curve of the given function passing through that point.

**1.** $f(x, y) = x^2 - y^2$ at $(2, -1)$

**2.** $f(x, y) = \dfrac{x - y}{x + y}$ at $(1, 1)$

**3.** $f(x, y) = \cos(x/y)$ at $(\pi, 4)$

**4.** $f(x, y) = e^{xy}$ at $(2, 0)$

**5.** $f(x, y) = \dfrac{x}{x^2 + y^2}$ at $(1, 2)$

**6.** $f(x, y) = \dfrac{2xy}{x^2 + y^2}$ at $(0, 2)$

**7.** $f(x, y) = \ln(x^2 + y^2)$ at $(1, -2)$

**8.** $f(x, y) = \sqrt{1 + xy^2}$ at $(2, -2)$

In Exercises 9–11, find an equation of the tangent plane to the level surface of the given function that passes through the given point.

**9.** $f(x, y, z) = x^2y + y^2z + z^2x$ at $(1, -1, 1)$

**10.** $f(x, y, z) = \cos(x + 2y + 3z)$ at $(\frac{\pi}{2}, \pi, \pi)$

**11.** $f(x, y, z) = y\,e^{-x^2} \sin z$ at $(0, 1, \pi/3)$

In Exercises 12–17, find the rate of change of the given function at the given point in the specified direction.

**12.** $f(x, y) = 3x - 4y$ at $(0, 2)$ in the direction of the vector $-2\mathbf{i}$

**13.** $f(x, y) = x^2y$ at $(-1, -1)$ in the direction of the vector $\mathbf{i} + 2\mathbf{j}$

**14.** $f(x, y) = \dfrac{x}{1 + y}$ at $(0, 0)$ in the direction of the vector $\mathbf{i} - \mathbf{j}$

**15.** $f(x, y) = x^2 + y^2$ at $(1, -2)$ in the direction making a (positive) angle of $60^\circ$ with the positive $x$-axis

**16.** $f(x, y, z) = (y^2 + \sin z)e^{-x}$ at $(0, 2, \pi)$ in the direction towards the point $(1, 1, 0)$

**17.** $f(x, y, z) = \dfrac{1}{x} + \dfrac{1}{y} + \dfrac{1}{z}$ at $(2, -3, 4)$ in the direction of the vector $\mathbf{i} + \mathbf{j} + \mathbf{k}$

**18.** Let $f(x, y) = \ln|\mathbf{r}|$ where $\mathbf{r} = x\mathbf{i} + y\mathbf{j}$. Show that $\nabla f = \dfrac{\mathbf{r}}{|\mathbf{r}|^2}$.

**19.** Let $f(x, y, z) = |\mathbf{r}|^{-n}$ where $\mathbf{r} = x\mathbf{i} + y\mathbf{j} + z\mathbf{k}$. Show that $\nabla f = \dfrac{-n\mathbf{r}}{|\mathbf{r}|^{n+2}}$.

**20.** Show that, in terms of polar coordinates $(r, \theta)$ (where $x = r\cos\theta$, and $y = r\sin\theta$), the gradient of a function $f(r, \theta)$ is given by

$$\nabla f = \frac{\partial f}{\partial r}\hat{\mathbf{r}} + \frac{1}{r}\frac{\partial f}{\partial \theta}\hat{\boldsymbol{\theta}},$$

where $\hat{\mathbf{r}}$ is a unit vector in the direction of the position vector $\mathbf{r} = x\,\mathbf{i} + y\,\mathbf{j}$, and $\hat{\boldsymbol{\theta}}$ is a unit vector at right angles to $\hat{\mathbf{r}}$ in the direction of increasing $\theta$.

**21.** In what directions at the point $(2, 0)$ does the function $f(x, y) = xy$ have rate of change $-1$? Are there directions in which the rate is $-3$? How about $-2$?

**22.** In what directions at the point $(a, b, c)$ does the function $f(x, y, z) = x^2 + y^2 - z^2$ increase at half of its maximal rate at that point?

**23.** Find $\nabla f(a, b)$ for the differentiable function $f(x, y)$ given the directional derivatives

$$D_{(\mathbf{i}+\mathbf{j})/\sqrt{2}} f(a, b) = 3\sqrt{2} \text{ and } D_{(3\mathbf{i}-4\mathbf{j})/5} f(a, b) = 5.$$

**24.** If $f(x, y)$ is differentiable at $(a, b)$, what condition should angles $\phi_1$ and $\phi_2$ satisfy in order that the gradient $\nabla f(a, b)$ can be determined from the values of the directional derivatives $D_{\phi_1} f(a, b)$ and $D_{\phi_2} f(a, b)$?

**25.** The temperature $T(x, y)$ at points of the $xy$-plane is given by $T(x, y) = x^2 - 2y^2$.

(a) Draw a contour diagram for $T$ showing some isotherms (curves of constant temperature).

(b) In what direction should an ant at position $(2, -1)$ move if it wishes to cool off as quickly as possible?

(c) If the ant moves in that direction at speed $k$ (units distance per unit time), at what rate does it experience the decrease of temperature?

(d) At what rate would the ant experience the decrease of temperature if it moved from $(2, -1)$ at speed $k$ in the direction of the vector $-\mathbf{i} - 2\mathbf{j}$?

(e) Along what curve through $(2, -1)$ should the ant move in order to continue to experience maximum rate of cooling?

**26.** Find an equation of the curve in the $xy$-plane that passes through the point $(1, 1)$ and intersects all level curves of the function $f(x, y) = x^4 + y^2$ at right angles.

**27.** Find an equation of the curve in the $xy$-plane that passes through the point $(2, -1)$ and that intersects every curve with equation of the form $x^2 y^3 = K$ at right angles.

**28.** Find the second directional derivative of $e^{-x^2-y^2}$ at the point $(a, b) \neq (0, 0)$ in the direction directly away from the origin.

**29.** Find the second directional derivative of $f(x, y, z) = xyz$ at $(2, 3, 1)$ in the direction of the vector $\mathbf{i} - \mathbf{j} - \mathbf{k}$.

**30.** Find a vector tangent to the curve of intersection of the two cylinders $x^2 + y^2 = 2$ and $y^2 + z^2 = 2$ at the point $(1, -1, 1)$

**31.** Repeat Exercise 30 for the surfaces $x + y + z = 6$ and $x^2 + y^2 + z^2 = 14$ and the point $(1, 2, 3)$

**32.** The temperature in 3-space is given by

$$T(x, y, z) = x^2 - y^2 + z^2 + xz^2.$$

At time $t = 0$ a fly passes through the point $(1, 1, 2)$, flying along the curve of intersection of the surfaces $z = 3x^2 - y^2$ and $2x^2 + 2y^2 - z^2 = 0$. If the fly's speed is 7, what rate of temperature change does it experience at $t = 0$?

**33.** State and prove a version of Theorem 6 for a function of three variables.

**34.** What is the level surface of $f(x, y, z) = \cos(x + 2y + 3z)$ that passes through $(\pi, \pi, \pi)$? What is the tangent plane to that level surface at that point? (Compare this exercise with Exercise 10 above.)

**35.** If $\nabla f(x, y) = 0$ throughout the disk $x^2 + y^2 < r^2$ prove that $f(x, y)$ is constant throughout the disk.

**36.** Theorem 6 implies that the level curve of $f(x, y)$ passing through $(a, b)$ is smooth (has a tangent line) at $(a, b)$ provided $f$ is differentiable at $(a, b)$ and satisfies $\nabla f(a, b) \neq \mathbf{0}$. Show that the level curve need not be smooth at $(a, b)$ if $\nabla f(a, b) = \mathbf{0}$. (*Hint:* consider $f(x, y) = y^3 - x^2$ at $(0, 0)$.)

**37.** If $\mathbf{v}$ is a nonzero vector, express $D_{\mathbf{v}}(D_{\mathbf{v}} f)$ in terms of the components of $\mathbf{v}$ and the second partials of $f$. What is the interpretation of this quantity for a moving observer?

* **38.** An observer moves so that his position, velocity and acceleration at time $t$ are given by $\mathbf{r}(t) = x(t)\,\mathbf{i} + y(t)\,\mathbf{j} + z(t)\,\mathbf{k}$, $\mathbf{v}(t) = d\mathbf{r}/dt$ and $\mathbf{a}(t) = d\mathbf{v}/dt$. If the temperature in the vicinity of the observer depends only on position, $T = T(x, y, z)$, express the second time derivative of temperature as measured by the observer in terms of $D_{\mathbf{v}}$ and $D_{\mathbf{a}}$.

* **39.** Repeat Exercise 38 but with $T$ depending explicitly on time as well as position: $T = T(x, y, z, t)$.

**40.** Let $f(x, y) = \begin{cases} \dfrac{\sin(xy)}{\sqrt{x^2 + y^2}} & \text{if } (x, y) \neq (0, 0) \\ 0 & \text{if } (x, y) = (0, 0). \end{cases}$

(a) Calculate $\nabla f(0, 0)$.

(b) Use the definition of directional derivative to calculate $D_{\mathbf{u}} f(0, 0)$, where $\mathbf{u} = (\mathbf{i} + \mathbf{j})/\sqrt{2}$.

(b) Is $f(x, y)$ differentiable at $(0, 0)$? Why?

**41.** Let $f(x, y) = \begin{cases} 2x^2 y/(x^4 + y^2) & \text{if } (x, y) \neq (0, 0) \\ 0 & \text{if } (x, y) = (0, 0). \end{cases}$
Use the definition of directional derivative as a limit (Definition 10) to show that $D_{\mathbf{u}} f(0, 0)$ exists for every unit vector $\mathbf{u} = u\mathbf{i} + v\mathbf{j}$ in the plane. Specifically, show that $D_{\mathbf{u}} f(0, 0) = 0$ if $v = 0$ and $D_{\mathbf{u}} f(0, 0) = 2u^2/v$ if $v \neq 0$. However, as was shown in Example 4 in Section 4.2, $f(x, y)$ has no limit as $(x, y) \to (0, 0)$, and so is not continuous there. Even if a function has directional derivatives in all directions at a point it may not be continuous at that point.

# 4.8 IMPLICIT FUNCTIONS

When we study the calculus of functions of one variable we encounter examples of functions that are defined implicitly as solutions of equations in two variables. Suppose, for example, that $F(x, y) = 0$ is such an equation. Suppose that the point $(a, b)$ satisfies the equation, and that $F$ has continuous first partial derivatives (and so is differentiable) at all points near $(a, b)$. Can the equation be solved for $y$ as a function of $x$ near $(a, b)$? That is, does there exist a function $y(x)$ defined in some interval $I = (a - h, a + h)$ (where $h > 0$) satisfying $y(a) = b$ and such that

$$F\left(x, y(x)\right) = 0$$

holds for all $x$ in the interval $I$? If there is such a function $y(x)$ we can try to find its derivative at $x = a$ by differentiating the equation $F(x, y) = 0$ implicitly with respect to $x$, and evaluating the result at $(a, b)$:

$$F_1(x, y) + F_2(x, y)\,\frac{dy}{dx} = 0,$$

so that

$$\left.\frac{dy}{dx}\right|_{x=a} = -\frac{F_1(a, b)}{F_2(a, b)} \qquad \text{provided} \qquad F_2(a, b) \neq 0.$$

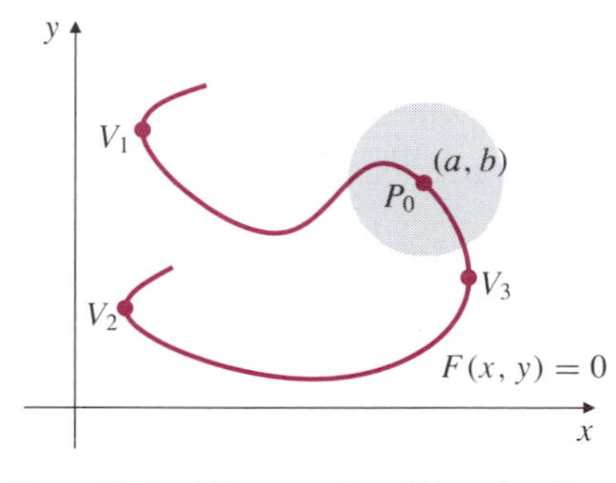


**Figure 4.28** The equation $F(x, y) = 0$ can be solved for $y$ as a function of $x$ near $P_0$, or near any other point except the three points where the curve has a vertical tangent

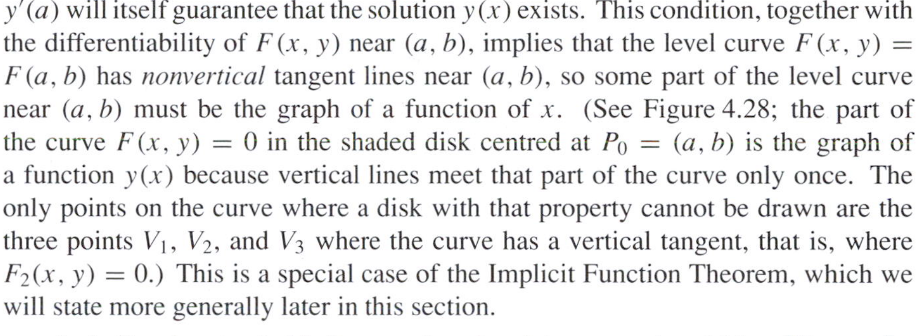

Observe, however, that the condition $F_2(a, b) \neq 0$ required for the calculation of $y'(a)$ will itself guarantee that the solution $y(x)$ exists. This condition, together with the differentiability of $F(x, y)$ near $(a, b)$, implies that the level curve $F(x, y) = F(a, b)$ has *nonvertical* tangent lines near $(a, b)$, so some part of the level curve near $(a, b)$ must be the graph of a function of $x$. (See Figure 4.28; the part of the curve $F(x, y) = 0$ in the shaded disk centred at $P_0 = (a, b)$ is the graph of a function $y(x)$ because vertical lines meet that part of the curve only once. The only points on the curve where a disk with that property cannot be drawn are the three points $V_1$, $V_2$, and $V_3$ where the curve has a vertical tangent, that is, where $F_2(x, y) = 0$.) This is a special case of the Implicit Function Theorem, which we will state more generally later in this section.

A similar situation holds for equations involving several variables. We can, for example, ask whether the equation

$$F(x, y, z) = 0$$

defines $z$ as a function of $x$ and $y$ (say $z = z(x, y)$) near some point $P_0 = (x_0, y_0, z_0)$ satisfying the equation. If so, and if $F$ has continuous first partials near $P_0$, then the partial derivatives of $z$ can be found at $(x_0, y_0)$ by implicit differentiation of the equation $F(x, y, z) = 0$ with respect to $x$ and $y$:

$$F_1(x, y, z) + F_3(x, y, z)\,\frac{\partial z}{\partial x} = 0 \quad \text{and} \quad F_2(x, y, z) + F_3(x, y, z)\,\frac{\partial z}{\partial y} = 0,$$

so that

$$\left.\frac{\partial z}{\partial x}\right|_{(x_0, y_0)} = -\frac{F_1(x_0, y_0, z_0)}{F_3(x_0, y_0, z_0)} \quad \text{and} \quad \left.\frac{\partial z}{\partial y}\right|_{(x_0, y_0)} = -\frac{F_2(x_0, y_0, z_0)}{F_3(x_0, y_0, z_0)},$$

provided $F_3(x_0, y_0, z_0) \neq 0$. Since $F_3$ is the $z$ component of the gradient of $F$, this condition implies that the level surface of $F$ through $P_0$ does not have a horizontal normal vector, so it is not vertical (that is, it is not parallel to the $z$-axis). Therefore, part of the surface near $P_0$ must indeed be the graph of a function $z = z(x, y)$. Similarly, $F(x, y, z) = 0$ can be solved for $x$ as a function of $y$ and $z$ near points where $F_1 \neq 0$, and for $y = y(x, z)$ near points where $F_2 \neq 0$.

**EXAMPLE 1** Near what points on the sphere $x^2 + y^2 + z^2 = 1$ can the equation of the sphere be solved for $z$ as a function of $x$ and $y$? Find $\partial z/\partial x$ and $\partial z/\partial y$ at such points.

***SOLUTION*** The sphere is the level surface $F(x, y, z) = 0$ of the function

$$F(x, y, z) = x^2 + y^2 + z^2 - 1.$$

The above equation can be solved for $z = z(x, y)$ near $P_0 = (x_0, y_0, z_0)$ provided that $P_0$ is not on the *equator* of the sphere, that is, the circle $x^2 + y^2 = 1, \quad z = 0$. The equator consists of those points that satisfy $F_3(x, y, z) = 0$. If $P_0$ is not on the equator, then it is on either the upper or the lower hemisphere. The upper hemisphere has equation $z = z(x, y) = \sqrt{1 - x^2 - y^2}$ and the lower hemisphere has equation $z = z(x, y) = -\sqrt{1 - x^2 - y^2}$.

If $z \neq 0$, we can calculate the partial derivatives of the solution $z = z(x, y)$ by implicitly differentiating the equation of the sphere: $x^2 + y^2 + z^2 = 1$:

$$2x + 2z\frac{\partial z}{\partial x} = 0, \qquad \text{so} \qquad \frac{\partial z}{\partial x} = -\frac{x}{z},$$

$$2y + 2z\frac{\partial z}{\partial y} = 0, \qquad \text{so} \qquad \frac{\partial z}{\partial y} = -\frac{y}{z}.$$

## Systems of Equations

Experience with linear equations shows us that systems of such equations can generally be solved for as many variables as there are equations in the system. We would expect, therefore, that a pair of equations in several variables might determine two of those variables as functions of the remaining ones. For instance we might expect the equations

$$\begin{cases} F(x, y, z, w) = 0 \\ G(x, y, z, w) = 0 \end{cases}$$

to possess, nearby some point that satisfies them, solutions of one or more of the forms

$$\begin{cases} x = x(z, w) \\ y = y(z, w), \end{cases} \qquad \begin{cases} x = x(y, w) \\ z = z(y, w), \end{cases} \qquad \begin{cases} x = x(y, z) \\ w = w(y, z), \end{cases}$$

$$\begin{cases} y = y(x, w) \\ z = z(x, w), \end{cases} \qquad \begin{cases} y = y(x, z) \\ w = w(x, z), \end{cases} \qquad \begin{cases} z = z(x, y) \\ w = w(x, y). \end{cases}$$

Where such solutions exist, we should be able to differentiate the given system of equations implicitly to find partial derivatives of the solutions.

If you are given a single equation $F(x, y, z) = 0$ and asked to find $\partial x/\partial z$, you would understand that $x$ is intended to be a function of the remaining variables $y$ and $z$, so there would be no chance of misinterpreting which variable is to be held constant in calculating the partial derivative. Suppose, however, that you are asked to calculate $\partial x/\partial z$ given the system $F(x, y, z, w) = 0$, $G(x, y, z, w) = 0$. The question implies that $x$ is one of the dependent variables and $z$ is one of the independent variables, but does not imply which of $y$ and $w$ is the other dependent variable and which is the other independent variable. In short, which of the situations

$$\begin{cases} x = x(z, w) \\ y = y(z, w) \end{cases} \qquad \text{and} \qquad \begin{cases} x = x(y, z) \\ w = w(y, z) \end{cases}$$

are we dealing with? As it stands, the question is ambiguous. To avoid this ambiguity we can specify *in the notation for the partial derivative* which variable is to be regarded as the other independent variable and therefore *held fixed* during the differentiation. Thus

$$\left(\frac{\partial x}{\partial z}\right)_w \quad \text{implies the interpretation} \quad \begin{cases} x = x(z, w) \\ y = y(z, w), \end{cases}$$

$$\left(\frac{\partial x}{\partial z}\right)_y \quad \text{implies the interpretation} \quad \begin{cases} x = x(y, z) \\ w = w(y, z). \end{cases}$$

**EXAMPLE 2** Given the equations $F(x, y, z, w) = 0$ and $G(x, y, z, w) = 0$, where $F$ and $G$ have continuous first partial derivatives, calculate $(\partial x/\partial z)_w$.

***SOLUTION*** We differentiate the two equations with respect to $z$, regarding $x$ and $y$ as functions of $z$ and $w$, and holding $w$ fixed:

$$F_1\frac{\partial x}{\partial z} + F_2\frac{\partial y}{\partial z} + F_3 = 0$$
$$G_1\frac{\partial x}{\partial z} + G_2\frac{\partial y}{\partial z} + G_3 = 0$$

(Note that the terms $F_4(\partial w/\partial z)$ and $G_4(\partial w/\partial z)$ are not present because $w$ and $z$ are independent variables, and $w$ is being held fixed during the differentiation.) The pair of equations above is linear in $\partial x/\partial z$ and $\partial y/\partial z$. Eliminating $\partial y/\partial z$ (or using Cramer's Rule, Theorem 5 of Section 3.6) we obtain

$$\left(\frac{\partial x}{\partial z}\right)_w = -\frac{F_3G_2 - F_2G_3}{F_1G_2 - F_2G_1}.$$

In the light of the examples considered above, you should not be too surprised to learn that the nonvanishing of the denominator $F_1G_2 - F_2G_1$ at some point $P_0 = (x_0, y_0, z_0, w_0)$ satisfying the system $F = 0, G = 0$ is sufficient to guarantee that the system does indeed have a solution of the form $x = x(z, w)$, $y = y(z, w)$ near $P_0$. We will not, however, attempt to prove this fact here. ■

**EXAMPLE 3** Let $x$, $y$, $u$, and $v$ be related by the equations

$$\begin{cases} u = x^2 + xy - y^2 \\ v = 2xy + y^2 \end{cases}$$

Find (a) $(\partial x/\partial u)_v$, and (b) $(\partial x/\partial u)_y$ at the point where $x = 2$ and $y = -1$.

***SOLUTION***

(a) To calculate $(\partial x/\partial u)_v$ we regard $x$ and $y$ as functions of $u$ and $v$, and differentiate the given equations with respect to $u$, holding $v$ constant:

$$1 = \frac{\partial u}{\partial u} = (2x + y)\frac{\partial x}{\partial u} + (x - 2y)\frac{\partial y}{\partial u}$$
$$0 = \frac{\partial v}{\partial u} = 2y\frac{\partial x}{\partial u} + (2x + 2y)\frac{\partial y}{\partial u}$$

At $x = 2$, $y = -1$ we have

$$1 = 3\frac{\partial x}{\partial u} + 4\frac{\partial y}{\partial u}$$
$$0 = -2\frac{\partial x}{\partial u} + 2\frac{\partial y}{\partial u}.$$

Eliminating $\partial y/\partial u$ leads to the result $(\partial x/\partial u)_v = 1/7$.

(b) To calculate $(\partial x/\partial u)_y$ we regard $x$ and $v$ as functions of $y$ and $u$, and differentiate the given equations with respect to $u$, holding $y$ constant:

$$1 = \frac{\partial u}{\partial u} = (2x + y)\frac{\partial x}{\partial u}$$
$$\frac{\partial v}{\partial u} = 2y\frac{\partial x}{\partial u}.$$

At $x = 2$, $y = -1$ the first equation immediately gives $(\partial x/\partial u)_y = 1/3$. ■

It often happens that the independent variables in a problem are either clear from the context or can be chosen at the outset. In either case, ambiguity is not likely to occur and we can safely omit subscripts showing which variables are being held constant. The following example is taken from the thermodynamics of an ideal gas.

### ■ EXAMPLE 4 Reversible changes in an ideal gas

The state of a closed system containing $n$ moles of an ideal gas is characterized by three state variables, pressure, volume, and temperature ($P$, $V$, and $T$, respectively), which satisfy the equation of state for an ideal gas:

$$PV = nRT,$$

where $R$ is a universal constant. This equation can be regarded as defining one of the three variables as a function of the other two. The internal energy, $E$, and the entropy, $S$, of the system are thermodynamic quantities that depend on the state variables and hence may be expressed as functions of any two of $P$, $V$, and $T$. Let us choose $T$ and $V$ for the independent variables and so write $E = E(V, T)$ and $S = S(V, T)$. For reversible processes in the system, the first and second laws of thermodynamics imply that infinitesimal changes in these quantities satisfy the differential equation

$$T\,dS = dE + P\,dV.$$

Deduce that for such processes, $E$ is independent of $V$ and so depends only on the temperature $T$.

**SOLUTION** We calculate the differentials $dS$ and $dE$ and substitute them into the differential equation to obtain

$$T\left(\frac{\partial S}{\partial V}\,dV + \frac{\partial S}{\partial T}\,dT\right) = \frac{\partial E}{\partial V}\,dV + \frac{\partial E}{\partial T}\,dT + P\,dV.$$

Divide by $T$, substitute $nR/V$ for $P/T$ (from the equation of state), and collect coefficients of $dV$ and $dT$ on opposite sides of the equation to get

$$\left(\frac{\partial S}{\partial V} - \frac{1}{T}\frac{\partial E}{\partial V} - \frac{nR}{V}\right)dV = \left(\frac{1}{T}\frac{\partial E}{\partial T} - \frac{\partial S}{\partial T}\right)dT.$$

Since $dV$ and $dT$ are independent variables, both coefficients must vanish. Hence

$$\frac{\partial S}{\partial V} = \frac{1}{T}\frac{\partial E}{\partial V} + \frac{nR}{V}$$
$$\frac{\partial S}{\partial T} = \frac{1}{T}\frac{\partial E}{\partial T}.$$

Now differentiate the first of these equations with respect to $T$ and the second with respect to $V$. Using equality of mixed partials for both $S$ and $E$, we obtain the desired result:

$$\begin{aligned}
\frac{\partial}{\partial T}\left(\frac{1}{T}\frac{\partial E}{\partial V} + \frac{nR}{V}\right) &= \frac{\partial^2 S}{\partial T \partial V} = \frac{\partial^2 S}{\partial V \partial T} = \frac{\partial}{\partial V}\left(\frac{1}{T}\frac{\partial E}{\partial T}\right) \\
\frac{-1}{T^2}\frac{\partial E}{\partial V} + \frac{1}{T}\frac{\partial^2 E}{\partial T \partial V} &= \frac{1}{T}\frac{\partial^2 E}{\partial V \partial T} \\
\frac{-1}{T^2}\frac{\partial E}{\partial V} &= 0.
\end{aligned}$$

It follows that $\partial E/\partial V = 0$, and so $E$ is independent of $V$. ■

## Jacobian Determinants

Partial derivatives obtained by implicit differentiation of systems of equations are fractions, the numerators and denominators of which are conveniently expressed in terms of certain determinants called Jacobians.

**DEFINITION 11**

The **Jacobian determinant** (or simply **the Jacobian**) of the two functions, $u = u(x, y)$ and $v = v(x, y)$, with respect to two variables, $x$ and $y$, is the determinant

$$\frac{\partial(u, v)}{\partial(x, y)} = \begin{vmatrix} \dfrac{\partial u}{\partial x} & \dfrac{\partial u}{\partial y} \\ \dfrac{\partial v}{\partial x} & \dfrac{\partial v}{\partial y} \end{vmatrix}.$$

Similarly, the Jacobian of two functions, $F(x, y, \ldots)$ and $G(x, y, \ldots)$, with respect to the variables, $x$ and $y$, is the determinant

$$\frac{\partial(F, G)}{\partial(x, y)} = \begin{vmatrix} \dfrac{\partial F}{\partial x} & \dfrac{\partial F}{\partial y} \\ \dfrac{\partial G}{\partial x} & \dfrac{\partial G}{\partial y} \end{vmatrix} = \begin{vmatrix} F_1 & F_2 \\ G_1 & G_2 \end{vmatrix}.$$

The definition above can be extended in the obvious way to give the Jacobian of $n$ functions (or variables) with respect to $n$ variables. For example the Jacobian of three functions, $F$, $G$, and $H$, with respect to three variables, $x$, $y$, and $z$, is the determinant

$$\frac{\partial(F, G, H)}{\partial(x, y, z)} = \begin{vmatrix} F_1 & F_2 & F_3 \\ G_1 & G_2 & G_3 \\ H_1 & H_2 & H_3 \end{vmatrix}.$$

Jacobians are the determinants of the square Jacobian matrices corresponding to transformations of $\mathbb{R}^n$ to $\mathbb{R}^n$ as discussed briefly in Section 4.6.

■ **EXAMPLE 5** In terms of Jacobians, the value of $(\partial x/\partial z)_w$, obtained from the system of equations

$$F(x, y, z, w) = 0, \qquad G(x, y, z, w) = 0$$

in Example 2, can be expressed in the form

$$\left(\frac{\partial x}{\partial z}\right)_w = -\frac{\dfrac{\partial(F,G)}{\partial(z,y)}}{\dfrac{\partial(F,G)}{\partial(x,y)}}.$$

Observe the pattern here. The denominator is the Jacobian of $F$ and $G$ with respect to the two *dependent* variables, $x$ and $y$. The numerator is the same Jacobian except that the dependent variable $x$ is replaced by the independent variable $z$. ■

The pattern observed above is general. Given $n$ equations in $n+m$ variables,

$$\begin{cases} F_{(1)}(x_1, x_2, \ldots, x_m, y_1, y_2, \ldots, y_n) = 0 \\ F_{(2)}(x_1, x_2, \ldots, x_m, y_1, y_2, \ldots, y_n) = 0 \\ \quad\vdots \\ F_{(n)}(x_1, x_2, \ldots, x_m, y_1, y_2, \ldots, y_n) = 0, \end{cases}$$

we can calculate, for instance,

$$\left(\frac{\partial y_i}{\partial x_j}\right)_{x_1,\ldots,x_{j-1},x_{j+1},\ldots,x_m} = -\frac{\dfrac{\partial(F_{(1)}, F_{(2)}, \ldots, F_{(n)})}{\partial(y_1, \ldots, x_j, \ldots, y_n)}}{\dfrac{\partial(F_{(1)}, F_{(2)}, \ldots, F_{(n)})}{\partial(y_1, \ldots, y_i, \ldots, y_n)}}.$$

This is a consequence of Cramer's Rule (Theorem 5 of Section 3.6) applied to the $n$ linear equations in the $n$ unknowns $\partial y_1/\partial x_j, \ldots, \partial y_n/\partial x_j$ obtained by differentiating each of the equations in the given system with respect to $x_j$.

**EXAMPLE 6** If the equations $x = u^2 + v^2$, and $y = uv$ are solved for $u$ and $v$ in terms of $x$ and $y$, find, where possible,

$$\frac{\partial u}{\partial x}, \quad \frac{\partial u}{\partial y}, \quad \frac{\partial v}{\partial x}, \quad \text{and} \quad \frac{\partial v}{\partial y}.$$

Hence show that $\dfrac{\partial(u,v)}{\partial(x,y)} = 1 \Big/ \dfrac{\partial(x,y)}{\partial(u,v)}$, provided the denominator is not zero.

***SOLUTION*** The given equations can be rewritten in the form

$$F(u,v,x,y) = u^2 + v^2 - x = 0$$
$$G(u,v,x,y) = uv - y = 0.$$

Let

$$J = \frac{\partial(F,G)}{\partial(u,v)} = \begin{vmatrix} 2u & 2v \\ v & u \end{vmatrix} = 2(u^2 - v^2) = \frac{\partial(x,y)}{\partial(u,v)}.$$

If $u^2 \neq v^2$, then $J \neq 0$ and we can calculate the required partial derivatives:

$$\frac{\partial u}{\partial x} = -\frac{1}{J}\frac{\partial(F,G)}{\partial(x,v)} = -\frac{1}{J}\begin{vmatrix} -1 & 2v \\ 0 & u \end{vmatrix} = \frac{u}{2(u^2 - v^2)}$$

$$\frac{\partial u}{\partial y} = -\frac{1}{J}\frac{\partial(F,G)}{\partial(y,v)} = -\frac{1}{J}\begin{vmatrix} 0 & 2v \\ -1 & u \end{vmatrix} = \frac{-2v}{2(u^2 - v^2)}$$

$$\frac{\partial v}{\partial x} = -\frac{1}{J}\frac{\partial(F,G)}{\partial(u,x)} = -\frac{1}{J}\begin{vmatrix} 2u & -1 \\ v & 0 \end{vmatrix} = \frac{-v}{2(u^2 - v^2)}$$

$$\frac{\partial v}{\partial y} = -\frac{1}{J}\frac{\partial(F,G)}{\partial(u,y)} = -\frac{1}{J}\begin{vmatrix} 2u & 0 \\ v & -1 \end{vmatrix} = \frac{2u}{2(u^2 - v^2)}.$$

Thus

$$\frac{\partial(u,v)}{\partial(x,y)} = \frac{1}{J^2}\begin{vmatrix} u & -2v \\ -v & 2u \end{vmatrix} = \frac{1}{J} = \frac{1}{\dfrac{\partial(x,y)}{\partial(u,v)}}.$$ ■

**REMARK** Note in the above example that $\partial u/\partial x \neq 1/(\partial x/\partial u)$. This should be contrasted with the single-variable situation where, if $y = f(x)$ and $dy/dx \neq 0$ then $x = f^{-1}(y)$ and $dx/dy = 1/(dy/dx)$. This is another reason for distinguishing between "$\partial$" and "$d$". It is the Jacobian rather than any single partial derivative that takes the place of the ordinary derivative in such situations.

**REMARK** Let us look briefly at the general case of invertible transformations from $\mathbb{R}^n$ to $\mathbb{R}^n$. Suppose that $\mathbf{y} = \mathbf{f}(\mathbf{x})$ and $\mathbf{z} = \mathbf{g}(\mathbf{y})$ are both functions from $\mathbb{R}^n$ to $\mathbb{R}^n$ whose components have continuous first partial derivatives. As shown in Section 4.6, the Chain Rule implies that

$$\begin{pmatrix} \dfrac{\partial z_1}{\partial x_1} & \cdots & \dfrac{\partial z_1}{\partial x_n} \\ \vdots & \ddots & \vdots \\ \dfrac{\partial z_n}{\partial x_1} & \cdots & \dfrac{\partial z_n}{\partial x_n} \end{pmatrix} = \begin{pmatrix} \dfrac{\partial z_1}{\partial y_1} & \cdots & \dfrac{\partial z_1}{\partial y_n} \\ \vdots & \ddots & \vdots \\ \dfrac{\partial z_n}{\partial y_1} & \cdots & \dfrac{\partial z_n}{\partial y_n} \end{pmatrix} \begin{pmatrix} \dfrac{\partial y_1}{\partial x_1} & \cdots & \dfrac{\partial y_1}{\partial x_n} \\ \vdots & \ddots & \vdots \\ \dfrac{\partial y_n}{\partial x_1} & \cdots & \dfrac{\partial y_n}{\partial x_n} \end{pmatrix}.$$

This is just the Chain Rule for the composition $\mathbf{z} = \mathbf{g}\big(\mathbf{f}(\mathbf{x})\big)$. It follows from Theorem 2(b) of Section 3.6 that the determinants of these matrices satisfy a similar equation:

$$\frac{\partial(z_1 \cdots z_n)}{\partial(x_1 \cdots x_n)} = \frac{\partial(z_1 \cdots z_n)}{\partial(y_1 \cdots y_n)} \frac{\partial(y_1 \cdots y_n)}{\partial(x_1 \cdots x_n)}.$$

If $\mathbf{f}$ is one-to-one and $\mathbf{g}$ is the inverse of $\mathbf{f}$, then $\mathbf{z} = \mathbf{g}\big(\mathbf{f}(\mathbf{x})\big) = \mathbf{x}$, and $\partial(z_1 \cdots z_n)/\partial(x_1 \cdots x_n) = 1$, the determinant of the identity matrix. Thus

$$\frac{\partial(x_1 \cdots x_n)}{\partial(y_1 \cdots y_n)} = \frac{1}{\dfrac{\partial(y_1 \cdots y_n)}{\partial(x_1 \cdots x_n)}}.$$

In fact, the nonvanishing of either of these determinants is sufficient to guarantee that $\mathbf{f}$ is one-to-one and so has an inverse. This is a special case of the Implicit Function Theorem stated below.

We will encounter Jacobians again when we study transformations of coordinates in multiple integrals in Chapter 6.

## The Implicit Function Theorem

The Implicit Function Theorem guarantees that systems of equations can be solved for certain variables as functions of other variables under certain circumstances. Before stating it we consider a simple illustrative example.

**EXAMPLE 7** Consider the system of linear equations

$$F(x, y, s, t) = a_1 x + b_1 y + c_1 s + d_1 t + e_1 = 0$$
$$G(x, y, s, t) = a_2 x + b_2 y + c_2 s + d_2 t + e_2 = 0.$$

This system can be written in matrix form:

$$\mathcal{A}\begin{pmatrix} x \\ y \end{pmatrix} + \mathcal{C}\begin{pmatrix} s \\ t \end{pmatrix} + \mathcal{E} = \begin{pmatrix} 0 \\ 0 \end{pmatrix},$$

where

$$\mathcal{A} = \begin{pmatrix} a_1 & b_1 \\ a_2 & b_2 \end{pmatrix}, \qquad \mathcal{C} = \begin{pmatrix} c_1 & d_1 \\ c_2 & d_2 \end{pmatrix}, \quad \text{and} \quad \mathcal{E} = \begin{pmatrix} e_1 \\ e_2 \end{pmatrix}.$$

The equations can be solved for $x$ and $y$ as functions of $s$ and $t$ provided $\det(\mathcal{A}) \neq 0$, for this implies the existence of the inverse matrix $\mathcal{A}^{-1}$ (Theorem 3 of Section 3.6), so

$$\begin{pmatrix} x \\ y \end{pmatrix} = -\mathcal{A}^{-1}\left(\mathcal{C}\begin{pmatrix} s \\ t \end{pmatrix} + \mathcal{E}\right).$$

Observe that $\det(\mathcal{A}) = \partial(F, G)/\partial(x, y)$, so the nonvanishing of this Jacobian guarantees that the equations can be solved for $x$ and $y$. ■

We conclude this section with a statement, without proof, of a general form of the Implicit Function Theorem.

**THEOREM 8**

### The Implicit Function Theorem

Consider a system of $n$ equations in $n + m$ variables,

$$\begin{cases} F_{(1)}(x_1, x_2, \ldots, x_m, y_1, y_2, \ldots, y_n) = 0 \\ F_{(2)}(x_1, x_2, \ldots, x_m, y_1, y_2, \ldots, y_n) = 0 \\ \quad \vdots \\ F_{(n)}(x_1, x_2, \ldots, x_m, y_1, y_2, \ldots, y_n) = 0, \end{cases}$$

and a point $P_0 = (a_1, a_2, \ldots, a_m, b_1, b_2, \ldots, b_n)$ that satisfies the system. Suppose each of the functions $F_{(i)}$ has continuous first partial derivatives with respect to each of the variables $x_j$ and $y_k$, $(i = 1, \ldots, n,\ j = 1, \ldots, m,\ k = 1, \ldots, n)$, near $P_0$. Finally, suppose that

$$\left.\frac{\partial(F_{(1)}, F_{(2)}, \ldots, F_{(n)})}{\partial(y_1, y_2, \ldots, y_n)}\right|_{P_0} \neq 0.$$

Then the system can be solved for $y_1, y_2, \ldots, y_n$ as functions of $x_1, x_2, \ldots, x_m$ near $P_0$. That is, there exist functions

$$\phi_1(x_1, \ldots, x_m), \ldots, \phi_n(x_1, \ldots, x_m)$$

such that

$$\phi_j(a_1, \ldots, a_m) = b_j, \qquad (j = 1, \ldots, n),$$

and such that the equations

$$F_{(1)}\Big(x_1, \ldots, x_m, \phi_1(x_1, \ldots, x_m), \ldots, \phi_n(x_1, \ldots, x_m)\Big) = 0,$$
$$F_{(2)}\Big(x_1, \ldots, x_m, \phi_1(x_1, \ldots, x_m), \ldots, \phi_n(x_1, \ldots, x_m)\Big) = 0,$$
$$\vdots$$
$$F_{(n)}\Big(x_1, \ldots, x_m, \phi_1(x_1, \ldots, x_m), \ldots, \phi_n(x_1, \ldots, x_m)\Big) = 0,$$

hold for all $(x_1, \ldots, x_m)$ sufficiently near $(a_1, \ldots, a_m)$.

**EXAMPLE 8** Show that the system

$$\begin{cases} xy^2 + xzu + yv^2 = 3 \\ x^3yz + 2xv - u^2v^2 = 2 \end{cases}$$

can be solved for $(u, v)$ as a (vector) function of $(x, y, z)$ near the point $P_0$ where $(x, y, z, u, v) = (1, 1, 1, 1, 1)$, and find the value of $\partial v/\partial y$ for the solution at $(x, y, z) = (1, 1, 1)$.

***SOLUTION*** Let $\begin{cases} F(x, y, z, u, v) = xy^2 + xzu + yv^2 - 3 \\ G(x, y, z, u, v) = x^3yz + 2xv - u^2v^2 - 2 \end{cases}$. Then

$$\left.\frac{\partial(F, G)}{\partial(u, v)}\right|_{P_0} = \left|\begin{vmatrix} xz & 2yv \\ -2uv^2 & 2x - 2u^2v \end{vmatrix}\right|_{P_0} = \begin{vmatrix} 1 & 2 \\ -2 & 0 \end{vmatrix} = 4.$$

Since this Jacobian is not zero, the Implicit Function Theorem assures us that the given equations can be solved for $u$ and $v$ as functions of $x$, $y$, and $z$, that is, for $(u, v) = \mathbf{f}(x, y, z)$. Since

$$\left.\frac{\partial(F, G)}{\partial(u, y)}\right|_{P_0} = \left|\begin{vmatrix} xz & 2xy + v^2 \\ -2uv^2 & x^3z \end{vmatrix}\right|_{P_0} = \begin{vmatrix} 1 & 3 \\ -2 & 1 \end{vmatrix} = 7,$$

we have

$$\left(\frac{\partial v}{\partial y}\right)_{x,z} = -\left.\frac{\dfrac{\partial(F, G)}{\partial(u, y)}}{\dfrac{\partial(F, G)}{\partial(u, v)}}\right|_{P_0} = -\frac{7}{4},$$

by the formula preceding Example 6. ■

## EXERCISES 4.8

In Exercises 1–12, calculate the indicated derivative from the given equation(s). What condition on the variables will guarantee the existence of a solution that has the indicated derivative? Assume that any general functions $F$, $G$, and $H$ have continuous first partial derivatives.

**1.** $\dfrac{dx}{dy}$ if $xy^3 + x^4y = 2$

**2.** $\dfrac{\partial x}{\partial y}$ if $xy^3 = y - z$

**3.** $\dfrac{\partial z}{\partial y}$ if $z^2 + xy^3 = \dfrac{xz}{y}$

**4.** $\dfrac{\partial y}{\partial z}$ if $e^{yz} - x^2z\ln y = \pi$

**5.** $\dfrac{\partial x}{\partial w}$ if $x^2y^2 + y^2z^2 + z^2t^2 + t^2w^2 - xw = 0$

**6.** $\dfrac{dy}{dx}$ if $F(x, y, x^2 - y^2) = 0$

**7.** $\dfrac{\partial u}{\partial x}$ if $G(x, y, z, u, v) = 0$

**8.** $\dfrac{\partial z}{\partial x}$ if $F(x^2 - z^2, y^2 + xz) = 0$

**9.** $\dfrac{\partial w}{\partial t}$ if $H(u^2w, v^2t, wt) = 0$

**10.** $\left(\dfrac{\partial y}{\partial x}\right)_u$ if $xyuv = 1$ and $x + y + u + v = 0$

**11.** $\left(\dfrac{\partial x}{\partial y}\right)_z$ if $x^2 + y^2 + z^2 + w^2 = 1$, and $x + 2y + 3z + 4w = 2$

**12.** $\dfrac{du}{dx}$ if $x^2y + y^2u - u^3 = 0$ and $x^2 + yu = 1$

**13.** If $x = u^3 + v^3$ and $y = uv - v^2$ are solved for $u$ and $v$ in terms of $x$ and $y$ evaluate

$$\frac{\partial u}{\partial x}, \quad \frac{\partial u}{\partial y}, \quad \frac{\partial v}{\partial x}, \quad \frac{\partial v}{\partial y}, \quad \text{and} \quad \frac{\partial(u, v)}{\partial(x, y)}$$

at the point where $u = 1$ and $v = 1$.

**14.** Near what points $(r, s)$ can the transformation

$$x = r^2 + 2s, \qquad y = s^2 - 2r$$

be solved for $r$ and $s$ as functions of $x$ and $y$? Calculate the values of the first partial derivatives of the solution at the origin.

**15.** Evaluate the Jacobian $\partial(x, y)/\partial(r, \theta)$ for the transformation to polar coordinates: $x = r\cos\theta$, $y = r\sin\theta$. Near what points $(r, \theta)$ is the transformation one-to-one and therefore invertible to give $r$ and $\theta$ as functions of $x$ and $y$?

**16.** Evaluate the Jacobian $\partial(x, y, z)/\partial(\rho, \phi, \theta)$ where

$$x = \rho\sin\phi\cos\theta, \quad y = \rho\sin\phi\sin\theta, \quad \text{and } z = \rho\cos\phi.$$

This is the transformation from Cartesian to spherical polar coordinates in 3-space that we will consider in Section 6.5. Near what points is the transformation one-to-one and hence invertible to give $\rho$, $\phi$, and $\theta$ as functions of $x$, $y$, and $z$?

**17.** Show that the equations

$$\begin{cases} xy^2 + zu + v^2 = 3 \\ x^3z + 2y - uv = 2 \\ xu + yv - xyz = 1 \end{cases}$$

can be solved for $x$, $y$, and $z$ as functions of $u$ and $v$ near the point $P_0$ where $(x, y, z, u, v) = (1, 1, 1, 1, 1)$, and find $(\partial y/\partial u)_v$ at $(u, v) = (1, 1)$.

**18.** Show that the equations

$$\begin{cases} xe^y + uz - \cos v = 2 \\ u\cos y + x^2v - yz^2 = 1 \end{cases}$$

can be solved for $u$ and $v$ as functions of $x$, $y$, and $z$ near the point $P_0$ where $(x, y, z) = (2, 0, 1)$ and $(u, v) = (1, 0)$, and find $(\partial u/\partial z)_{x,y}$ at $(x, y, z) = (2, 0, 1)$.

**19.** Find $dx/dy$ from the system

$$F(x, y, z, w) = 0, \quad G(x, y, z, w) = 0, \quad H(x, y, z, w) = 0.$$

**20.** Given the system

$$F(x, y, z, u, v) = 0$$
$$G(x, y, z, u, v) = 0$$
$$H(x, y, z, u, v) = 0,$$

how many possible interpretations are there for $\partial x/\partial y$? Evaluate them.

**21.** Given the system

$$F(x_1, x_2, \ldots, x_8) = 0$$
$$G(x_1, x_2, \ldots, x_8) = 0$$
$$H(x_1, x_2, \ldots, x_8) = 0,$$

how many possible interpretations are there for the partial $\partial x_1/\partial x_2$? Evaluate $\left(\partial x_1/\partial x_2\right)_{x_4, x_6, x_7, x_8}$.

**22.** If $F(x, y, z) = 0$ determines $z$ as a function of $x$ and $y$, calculate $\partial^2 z/\partial x^2$, $\partial^2 z/\partial x\partial y$ and $\partial^2 z/\partial y^2$ in terms of the partial derivatives of $F$.

**23.** If $x = u + v$, $y = uv$ and $z = u^2 + v^2$ define $z$ as a function of $x$ and $y$, find $\partial z/\partial x$, $\partial z/\partial y$ and $\partial^2 z/\partial x\partial y$.

**24.** A certain gas satisfies the law

$$pV = T - \frac{4p}{T^2},$$

where $p$ = pressure, $V$ = volume, and $T$ = temperature.

(a) Calculate $\partial T/\partial p$ and $\partial T/\partial V$ at the point where $p = V = 1$ and $T = 2$.

(b) If measurements of $p$ and $V$ yield the values $p = 1 \pm 0.001$ and $V = 1 \pm 0.002$, find the approximate maximum error in the calculated value $T = 2$.

**25.** If $F(x, y, z) = 0$ show that

$$\left(\frac{\partial x}{\partial y}\right)_z \left(\frac{\partial y}{\partial z}\right)_x \left(\frac{\partial z}{\partial x}\right)_y = -1.$$

Derive analogous results for $F(x, y, z, u) = 0$ and for $F(x, y, z, u, v) = 0$. What is the general case?

* **26.** If the equations $F(x, y, u, v) = 0$ and $G(x, y, u, v) = 0$ are solved for $x$ and $y$ as functions of $u$ and $v$, show that

$$\frac{\partial(x, y)}{\partial(u, v)} = \frac{\partial(F, G)}{\partial(u, v)} \bigg/ \frac{\partial(F, G)}{\partial(x, y)}.$$

* **27.** If the equations $x = f(u, v)$, $y = g(u, v)$ can be solved for $u$ and $v$ in terms of $x$ and $y$, show that

$$\frac{\partial(u, v)}{\partial(x, y)} = 1 \bigg/ \frac{\partial(x, y)}{\partial(u, v)}.$$

(*Hint:* use the result of Exercise 26.)

* **28.** If $x = f(u, v)$, $y = g(u, v)$, $u = h(r, s)$, and $v = k(r, s)$, then $x$ and $y$ can be expressed as functions of $r$ and $s$. Verify by direct calculation that

$$\frac{\partial(x, y)}{\partial(r, s)} = \frac{\partial(x, y)}{\partial(u, v)} \frac{\partial(u, v)}{\partial(r, s)}.$$

This is a special case of the Chain Rule for Jacobians.

* **29.** Two functions, $f(x, y)$ and $g(x, y)$, are said to be functionally dependent if one is a function of the other; that is, if there exists a single-variable function $k(t)$ such that $f(x, y) = k\big(g(x, y)\big)$ for all $x$ and $y$. Show that in this case $\partial(f, g)/\partial(x, y)$ vanishes identically. Assume that all necessary derivatives exist.

* **30.** Prove the converse of the previous exercise as follows: Let $u = f(x, y)$ and $v = g(x, y)$, and suppose that $\partial(u, v)/\partial(x, y) = \partial(f, g)/\partial(x, y)$ is identically zero for all $x$ and $y$. Show that $(\partial u/\partial x)_v$ is identically zero. Hence $u$, considered as a function of $x$ and $v$, is independent of $x$; that is, $u = k(v)$ for some function $k$ of one variable. Why does this imply that $f$ and $g$ are functionally dependent?

## 4.9 TAYLOR SERIES AND APPROXIMATIONS

As is the case for functions of one variable, power series representations and their partial sums (Taylor polynomials) can provide an efficient method for determining the behavior of a smooth function of several variables near a point in its domain. In this section we will look briefly at the extension of Taylor series to such functions. As usual, we will develop the machinery for functions of two variables, but the extension to more variables should be clear.

As a starting point, recall Taylor's Formula for a function $F(x)$ with continuous partial derivatives of order up to $n+1$ on the interval $[a, a+h]$:

$$F(a+h) = F(a)+F'(a)h+\frac{F''(a)}{2!}h^2+\cdots+\frac{F^{(n)}(a)}{n!}h^n+\frac{F^{(n+1)}(X)}{(n+1)!}h^{n+1},$$

where $X$ is some number between $a$ and $a+h$. (The last term in the formula is the *Lagrange* form of the remainder.) In the special case where $a=0$ and $h=1$, this formula becomes

$$F(1) = F(0)+F'(0)+\frac{F''(0)}{2!}+\cdots+\frac{F^{(n)}(0)}{n!}+\frac{F^{(n+1)}(\theta)}{(n+1)!}$$

for some $\theta$ between 0 and 1.

Now suppose that $f(x, y)$ has continuous partial derivatives up to order $n+1$ at all points nearby the line segment joining the points $(a, b)$ and $(a+h, b+k)$ in its domain. We can find the Taylor Formula for $f(a+h, b+k)$ in powers of $h$ and $k$ by applying the one-variable formula above to the function

$$F(t) = f(a+th, b+tk), \qquad 0 \le t \le 1.$$

Clearly $F(0) = f(a, b)$ and $F(1) = f(a+h, b+k)$. Let us calculate some derivatives of $F$:

$$\begin{aligned}
F'(t) =& hf_1(a+th, b+tk) + kf_2(a+th, b+tk),\\
F''(t) =& h^2 f_{11}(a+th, b+tk) + 2hkf_{12}(a+th, b+tk)\\
&+ k^2 f_{22}(a+th, b+tk),\\
F'''(t) =& \left(h^3 f_{111} + 3h^2kf_{112} + 3hk^2 f_{122} + k^3 f_{222}\right)\Big|_{(a+th,b+tk)}.
\end{aligned}$$

The pattern of binomial coefficients is pretty obvious here, but the notation, involving subscripts to denote partial derivatives of $f$, becomes more and more unwieldy as the order of the derivatives increases. The notation can be simplified greatly by using $D_1 f$ and $D_2 f$ to denote the first partials of $f$ with respect to its first and second variables. Since $h$ and $k$ are constant, and mixed partials commute $(D_1 D_2 f = D_2 D_1 f)$, we have

$$(h^2 D_1^2 f + 2hk D_1 D_2 f + k^2 D_2^2 f = (hD_1 + kD_2)^2 f,$$

and so on. Therefore

$$\begin{aligned}
F'(t) &= \big(hD_1 + kD_2\big) f(a+th, b+tk),\\
F''(t) &= \big(hD_1 + kD_2\big)^2 f(a+th, b+tk),\\
F'''(t) &= \big(hD_1 + kD_2\big)^3 f(a+th, b+tk),\\
&\ \ \vdots\\
F^{(m)}(t) &= \big(hD_1 + kD_2\big)^m f(a+th, b+tk).
\end{aligned}$$

In particular, $F^{(m)}(0) = \big(hD_1 + kD_2\big)^m f(a, b)$. Hence the Taylor Formula for $f(a + h, b + k)$ is

$$\begin{aligned} f(a+h, b+k) &= \sum_{m=0}^{n} \frac{1}{m!}\big(hD_1 + kD_2\big)^m f(a, b) + R_n(h, k) \\ &= \sum_{m=0}^{n}\sum_{j=0}^{m} C_{mj}\, D_1^j D_2^{m-j} f(a, b)\, h^j\, k^{m-j} + R_n(h, k), \end{aligned}$$

where, using the binomial expansion, we have

$$C_{mj} = \frac{1}{m!}\binom{m}{j} = \frac{1}{m!}\,\frac{m!}{j!(m-j)!} = \frac{1}{j!(m-j)!},$$

and where the remainder term is given by

$$R_n(h, k) = \frac{1}{(n+1)!}\big(hD_1 + kD_2\big)^{n+1} f(a + \theta h, b + \theta k)$$

for some $\theta$ between 0 and 1. If $f$ has partial derivatives of all orders and

$$\lim_{n\to\infty} R_n(h, k) = 0,$$

then $f(a + h, b + k)$ can be expressed as a Taylor series in powers of $h$ and $k$:

$$f(a+h, b+k) = \sum_{m=0}^{\infty}\sum_{j=0}^{m} \frac{1}{j!(m-j)!}\, D_1^j D_2^{m-j}\, f(a, b)\, h^j\, k^{m-j}.$$

As for functions of one variable, the Taylor polynomial of degree $n$,

$$P_n(x, y) = \sum_{m=0}^{n}\sum_{j=0}^{m} \frac{1}{j!(m-j)!}\, D_1^j D_2^{m-j}\, f(a, b)\, (x-a)^j (y-b)^{m-j},$$

provides the "best" $n$th-degree polynomial approximation to $f(x, y)$ near $(a, b)$. For $n = 1$ this approximation reduces to the tangent plane approximation

$$f(x, y) \approx f(a, b) + f_1(a, b)(x - a) + f_2(a, b)(y - b).$$

**EXAMPLE 1** Find a second-degree polynomial approximation to the function $f(x, y) = \sqrt{x^2 + y^3}$ near the point (1,2) and use it to estimate $\sqrt{(1.02)^2 + (1.97)^3}$.

***SOLUTION*** For the second-degree approximation we need the values of the partial derivatives of $f$ up to second order at (1,2). We have

$$\begin{aligned} f(x, y) &= \sqrt{x^2 + y^3} & f(1, 2) &= 3 \\ f_1(x, y) &= \frac{x}{\sqrt{x^2 + y^3}} & f_1(1, 2) &= \frac{1}{3} \\ f_2(x, y) &= \frac{3y^2}{2\sqrt{x^2 + y^3}} & f_2(1, 2) &= 2 \\ f_{11}(x, y) &= \frac{y^3}{(x^2 + y^3)^{3/2}} & f_{11}(1, 2) &= \frac{8}{27} \\ f_{12}(x, y) &= \frac{-3xy^2}{2(x^2 + y^3)^{3/2}} & f_{12}(1, 2) &= -\frac{2}{9} \\ f_{22}(x, y) &= \frac{12x^2y + 3y^4}{4(x^2 + y^3)^{3/2}} & f_{22}(1, 2) &= \frac{2}{3}. \end{aligned}$$

Thus

$$f(1+h,2+k)=3+\frac{1}{3}h+2k+\frac{1}{2!}\Big(\frac{8}{27}h^2+2(-\frac{2}{9})hk+\frac{2}{3}k^2\Big)$$

or, setting $x=1+h$ and $y=2+k$,

$$f(x,y)=3+\frac{1}{3}(x-1)+2(y-2)+\frac{4}{27}(x-1)^2-\frac{2}{9}(x-1)(y-2)+\frac{1}{3}(y-2)^2.$$

This is the required second-degree Taylor polynomial for $f$ near (1,2). Therefore

$$\begin{aligned}\sqrt{(1.02)^2+(1.97)^3}&=f(1+0.02,2-0.03)\\&\approx 3+\frac{1}{3}(0.02)+2(-0.03)+\frac{4}{27}(0.02)^2\\&\quad-\frac{2}{9}(0.02)(-0.03)+\frac{1}{3}(-0.03)^2\\&\approx 2.9471593\,.\end{aligned}$$

(For comparison purposes: the true value is 2.9471636.... The approximation is accurate to six significant figures.) ■

As observed for functions of one variable, it is not usually necessary to calculate derivatives in order to determine the coefficients in a Taylor series or Taylor polynomial. It is often much easier to perform algebraic manipulations on known series. For instance, the above example could have been done by writing $f$ in the form

$$\begin{aligned}f(1+h,2+k)&=\sqrt{(1+h)^2+(2+k)^3}\\&=\sqrt{9+2h+h^2+12k+6k^2+k^3}\\&=3\sqrt{1+\frac{2h+h^2+12k+6k^2+k^3}{9}}\end{aligned}$$

and then applying the binomial expansion

$$\sqrt{1+t}=1+\frac{1}{2}t+\frac{1}{2!}\left(\frac{1}{2}\right)\left(-\frac{1}{2}\right)t^2+\cdots$$

with $t=\dfrac{2h+h^2+12k+6k^2+k^3}{9}$ to obtain the terms up to second degree in $h$ and $k$.

**EXAMPLE 2** Find the Taylor polynomial of degree 3 in powers of $x$ and $y$ for the function $f(x,y)=e^{x-2y}$.

***SOLUTION*** The required Taylor polynomial will be the Taylor polynomial of degree 3 for $e^t$ evaluated at $t=x-2y$:

$$\begin{aligned}P_3(x,y)&=1+(x-2y)+\frac{1}{2!}(x-2y)^2+\frac{1}{3!}(x-2y)^3\\&=1+x-2y+\frac{1}{2}x^2-2xy+2y^2+\frac{1}{6}x^3-x^2y+2xy^2-\frac{4}{3}y^3.\end{aligned}$$ ■

## Approximating Implicit Functions

In the previous section we saw how to determine whether an equation in several variables could be solved for one of those variables as a function of the others. Even when such a solution is known to exist, it is not usually possible to find an exact formula for it. However, if the equation involves only smooth functions, then the solution will have a Taylor series. We can determine at least the first several coefficients in that series, and thus obtain a useful approximation to the solution. The following example shows the technique.

**EXAMPLE 3** Show that the equation

$$\sin(x+y) = xy + 2x$$

has a solution of the form $y = f(x)$ near $x = 0$ satisfying $f(0) = 0$, and find the terms up to fourth degree for the Taylor series for $f(x)$ in powers of $x$.

***SOLUTION*** The given equation can be written in the form $F(x, y) = 0$, where

$$F(x, y) = \sin(x+y) - xy - 2x.$$

Since $F(0, 0) = 0$ and $F_2(0, 0) = \cos(0) - 0 = 1$, the equation has a solution $y = f(x)$ near $x = 0$ satisfying $f(0) = 0$ by the implicit function theorem. It is not possible to calculate $f(x)$ exactly, but it will have a Maclaurin series of the form

$$y = f(x) = a_1x + a_2x^2 + a_3x^3 + a_4x^4 + \cdots.$$

(There is no constant term because $f(0) = 0$.) We can substitute this series into the given equation and keep track of terms up to degree 4 in order to calculate the coefficients $a_1$, $a_2$, $a_3$ and $a_4$. For the left side we use the Maclaurin series for sin to obtain

$$\begin{aligned}
\sin(x+y) &= \sin\Big((1+a_1)x + a_2x^2 + a_3x^3 + a_4x^4 + \cdots\Big) \\
&= (1+a_1)x + a_2x^2 + a_3x^3 + a_4x^4 + \cdots \\
&\quad - \frac{1}{3!}\Big((1+a_1)x + a_2x^2 + \cdots\Big)^3 + \cdots \\
&= (1+a_1)x + a_2x^2 + \Big(a_3 - \frac{1}{6}(1+a_1)^3\Big)x^3 \\
&\quad + \Big(a_4 - \frac{3}{6}(1+a_1)^2a_2\Big)x^4 + \cdots.
\end{aligned}$$

The right side is

$$xy + 2x = 2x + a_1x^2 + a_2x^3 + a_3x^4 + \cdots.$$

Equating coefficients of like powers of $x$, we obtain

$$\begin{aligned}
1 + a_1 &= 2 & a_1 &= 1 \\
a_2 &= a_1 & a_2 &= 1 \\
a_3 - \frac{1}{6}(1+a_1)^3 &= a_2 & a_3 &= \frac{7}{3} \\
a_4 - \frac{1}{2}(1+a_1)^2a_2 &= a_3 & a_4 &= \frac{13}{3}.
\end{aligned}$$

Thus

$$y = f(x) = x + x^2 + \frac{7}{3}x^3 + \frac{13}{3}x^4 + \cdots.$$

(We could have obtained more terms in the series by keeping track of higher powers of $x$ in the substitution process.) ■

**REMARK** From the series for $f(x)$ obtained above we can determine the values of the first four derivatives of $f$ at $x = 0$. Remember that

$$a_k = \frac{f^{(k)}(0)}{k!}.$$

We have, therefore,

$$f'(0) = a_1 = 1 \qquad f''(0) = 2!a_2 = 2$$
$$f'''(0) = 3!a_3 = 14 \qquad f^{(4)}(0) = 4!a_4 = 104.$$

We could have done the example by first calculating these derivatives by implicit differentiation of the given equation and then determined the series coefficients from them. This would have been much the more difficult way to do it. (Try it and see.)

## EXERCISES 4.9

In Exercises 1–6, find the Taylor series for the given function about the indicated point.

**1.** $f(x, y) = \dfrac{1}{2 + xy^2}$; $(0, 0)$

**2.** $f(x, y) = \ln(1 + x + y + xy)$; $(0, 0)$

**3.** $f(x, y) = \tan^{-1}(x + xy)$; $(0, -1)$

**4.** $f(x, y) = x^2 + xy + y^3$; $(1, -1)$

**5.** $f(x, y) = e^{x^2+y^2}$; $(0, 0)$

**6.** $f(x, y) = \sin(2x + 3y)$; $(0, 0)$

In Exercises 7–12, find Taylor polynomials of the indicated degree for the given functions near the given point.

**7.** $f(x, y) = \dfrac{1}{2 + x - 2y}$, degree 3, near (2,1)

**8.** $f(x, y) = \ln(x^2 + y^2)$, degree 3, near (1,0)

**9.** $f(x, y) = \displaystyle\int_0^{x+y^2} e^{-t^2}\, dt$, degree 3, near (0,0)

**10.** $f(x, y) = \cos(x + \sin y)$, degree 4, near (0,0)

**11.** $f(x, y) = \dfrac{\sin x}{y}$, degree 2, near $(\frac{\pi}{2}, 1)$

**12.** $f(x, y) = \dfrac{1 + x}{1 + x^2 + y^4}$, degree 2, near (0,0)

In Exercises 13–16, show that the given equation has a solution of the form $y = f(x)$ for $x$ near the indicated point $x = a$, and taking on the indicated value at that point. Find the first three nonzero terms of the Taylor series for $f(x)$ in powers of $x - a$.

* **13.** $x \sin y = y + \sin x$, near $x = 0$, with $f(0) = 0$

* **14.** $e^{x+y-1} = 2y + 1$, near $x = 1$, with $f(1) = 0$

* **15.** $\ln(x^2 + y^2) = y - 1$, near $x = 0$, with $f(0) = 1$

* **16.** $\sqrt{1 + xy} = 1 + x + \ln(1 + y)$, near $x = 0$, with $f(0) = 0$

* **17.** Show that the equation $x + 2y + z + e^{2z} = 1$ has a solution of the form $z = f(x, y)$ near $x = 0$, $y = 0$, where $f(0, 0) = 0$. Find the Taylor polynomial of degree 2 for $f(x, y)$ in powers of $x$ and $y$.

* **18.** Use series methods to find the value of the partial derivative $f_{112}(0, 0)$ given that $f(x, y) = \arctan(x + y)$.

* **19.** Use series methods to evaluate

$$\left.\frac{\partial^{4n}}{\partial x^{2n}\partial y^{2n}} \frac{1}{1 + x^2 + y^2}\right|_{(0,0)}.$$

# CHAPTER REVIEW

## Key Ideas

- **What do the following sentences or phrases mean?**
  - $\mathcal{S}$ is the graph of $f(x, y)$.
  - $\mathcal{C}$ is a level curve of $f(x, y)$.
  - $\lim_{(x,y)\to(a,b)} f(x, y) = L$.
  - $f(x, y)$ is continuous at $(a, b)$.
  - the partial derivative $(\partial/\partial x) f(x, y)$
  - the tangent plane to $z = f(x, y)$ at $(a, b)$
  - pure second partials
  - mixed second partials
  - $f(x, y)$ is a harmonic function.
  - $L(x, y)$ is the linearization of $f(x, y)$ at $(a, b)$.
  - the differential of $z = f(x, y)$

◇ $f(x, y)$ is differentiable at $(a, b)$.

◇ the gradient of $f(x, y)$ at $(a, b)$

◇ the directional derivative of $f(x, y)$ at $(a, b)$ in direction $\mathbf{v}$

◇ the Jacobian determinant $\partial(x, y)/\partial(u, v)$

- **Under what conditions are two mixed partial derivatives equal?**
- **State the Chain Rule for $z = f(x, y)$, where $x = g(u, v)$, and $y = h(u, v)$.**
- **Describe the process of calculating partial derivatives of implicitly defined functions.**
- **What is the Taylor series of $f(x, y)$ about $(a, b)$?**

## Review Exercises

**1.** Sketch some level curves of the function $x + \dfrac{4y^2}{x}$.

**2.** Sketch some isotherms (curves of constant temperature) for the temperature function

$$T = \frac{140 + 30x^2 - 60x + 120y^2}{8 + x^2 - 2x + 4y^2} \quad (°\mathrm{C}).$$

What is the coolest location?

**3.** Sketch some level curves of the polynomial function $f(x, y) = x^3 - 3xy^2$. Why do you think the graph of this function is called a *monkey saddle*?

**4.** Let $f(x, y) = \begin{cases} \dfrac{x^3}{x^2 + y^2} & \text{if } (x, y) \neq (0, 0) \\ 0 & \text{if } (x, y) = (0, 0). \end{cases}$
Calculate each of the following partial derivatives or explain why it does not exist: $f_1(0, 0)$, $f_2(0, 0)$, $f_{21}(0, 0)$, $f_{12}(0, 0)$.

**5.** Let $f(x, y) = \dfrac{x^3 - y^3}{x^2 - y^2}$. Where is $f(x, y)$ continuous? To what additional set of points does $f(x, y)$ have a continuous extension? In particular, can $f$ be extended to be continuous at the origin? Can $f$ be defined at the origin in such a way that its first partial derivatives exist there?

**6.** The surface $\mathcal{S}$ is the graph of the function $z = f(x, y)$, where $f(x, y) = e^{x^2-2x-4y^2+5}$.

(a) Find an equation of the tangent plane to $\mathcal{S}$ at the point $(1, -1, 1)$.

(b) Sketch a representative sample of the level curves of the function $f(x, y)$.

**7.** Consider the surface $\mathcal{S}$ with equation $x^2 + y^2 + 4z^2 = 16$.

(a) Find an equation for the tangent plane to $\mathcal{S}$ at the point $(a, b, c)$ on $\mathcal{S}$.

(b) For which points $(a, b, c)$ on $\mathcal{S}$ does the tangent plane to $\mathcal{S}$ at $(a, b, c)$ pass through the point $(0, 0, 4)$? Describe this set of points geometrically.

(c) For which points $(a, b, c)$ on $\mathcal{S}$ is the tangent plane to $\mathcal{S}$ at $(a, b, c)$ parallel to the plane $x + y + 2\sqrt{2}z = 97$?

**8.** Two variable resistors $R_1$ and $R_2$ are connected in parallel so that their combined resistance $R$ is given by

$$\frac{1}{R} = \frac{1}{R_1} + \frac{1}{R_2}.$$

If $R_1 = 100$ ohms $\pm 5\%$ and $R_2 = 25$ ohms $\pm 2\%$, by approximately what percentage can the calculated value of their combined resistance $R = 20$ ohms be in error?

**9.** You have measured two sides of a triangular field and the angle between them. The side measurements are 150 m and 200 m, each accurate to within $\pm 1$ m. The angle measurement is $30°$, accurate to within $\pm 2°$. What area do you calculate for the field and what is your estimate of the maximum percentage error in this area?

**10.** Suppose that $T(x, y, z) = x^3y + y^3z + z^3x$ gives the temperature at the point $(x, y, z)$ in 3-space.

(a) Calculate the directional derivative of $T$ at $(2, -1, 0)$ in the direction towards the point $(1, 1, 2)$.

(b) A fly is moving through space with constant speed 5. At time $t = 0$ the fly crosses the surface $2x^2 + 3y^2 + z^2 = 11$ at right angles at the point $(2, -1, 0)$, moving in the direction of increasing temperature. Find $dT/dt$ at $t = 0$ as experienced by the fly.

**11.** Consider the function $f(x, y, z) = x^2y + yz + z^2$.

(a) Find the directional derivative of $f$ at $(1, -1, 1)$ in the direction of the vector $\mathbf{i} + \mathbf{k}$.

(b) An ant is crawling on the plane $x + y + z = 1$ through $(1, -1, 1)$. Suppose it crawls so as to keep $f$ constant. In what direction is it going as it passes through $(1, -1, 1)$?

(c) Another ant crawls on the plane $x + y + z = 1$, moving in the direction of the greatest rate of increase of $f$. Find its direction as it goes through $(1, -1, 1)$.

**12.** Let $f(x, y, z) = (x^2 + z^2)\sin\dfrac{\pi xy}{2} + yz^2$. Let $P_0$ be the point $(1, 1, -1)$.

(a) Find the gradient of $f$ at $P_0$.

(b) Find the linearization $L(x, y, z)$ of $f$ at $P_0$.

(c) Find an equation for the tangent plane at $P_0$ to the level surface of $f$ through $P_0$.

(d) If a bird flies through $P_0$ with speed 5, heading directly towards the point $(2, -1, 1)$, what is the rate of change of $f$ as seen by the bird as it passes through $P_0$?

(e) In what direction from $P_0$ should the bird fly at speed 5 to experience the greatest rate of increase of $f$?

**13.** Verify that for any constant $k$ the function $u(x, y) = k\big(\ln\cos(x/k) - \ln\cos(y/k)\big)$ satisfies the *minimal surface equation*

$$(1 + u_x^2)u_{yy} - uu_xu_yu_{xy} + (1 + u_y^2)u_{xx} = 0.$$

**14.** The equations $F(x, y, z) = 0$ and $G(x, y, z) = 0$ can define any two of the variables $x$, $y$, and $z$ as functions of the remaining variable. Show that

$$\frac{dx}{dy}\frac{dy}{dz}\frac{dz}{dx} = 1.$$

**15.** The equations $\begin{cases} x = u^3 - uv \\ y = 3uv + 2v^2 \end{cases}$ define $u$ and $v$ as functions of $x$ and $y$ near the point $P$ where $(u, v, x, y) = (-1, 2, 1, 2)$.

(a) Find $\dfrac{\partial u}{\partial x}$ and $\dfrac{\partial u}{\partial y}$ at $P$.

(b) Find the approximate value of $u$ when $x = 1.02$ and $y = 1.97$.

**16.** The equations $\begin{cases} u = x^2 + y^2 \\ v = x^2 - 2xy^2 \end{cases}$ define $x$ and $y$ implicitly as functions of $u$ and $v$ for values of $(x, y)$ near $(1, 2)$ and values of $(u, v)$ near $(5, -7)$.

(a) Find $\dfrac{\partial x}{\partial u}$ and $\dfrac{\partial y}{\partial u}$ at $(u, v) = (5, -7)$.

(b) If $z = \ln(y^2 - x^2)$, find $\dfrac{\partial z}{\partial u}$ at $(u, v) = (5, -7)$.

## Express Yourself

Answer these questions in words, using complete sentences and, if necessary, paragraphs. Use mathematical symbols only where necessary.

1. What are some advantages and disadvantages of using graphs versus level curves to convey pictorial information about the behaviour of a function of two variables? What about a function of three variables?
2. List and discuss some properties of derivatives of functions of one variable that are not shared by partial derivatives of functions of several variables.
3. Describe in geometric terms what it means to say that a function of two variables is differentiable at a point.
4. Describe the geometric properties of the gradient of a function of three variables.
5. What is the difference between a first partial derivative and a directional derivative. If a function has directional derivatives in all directions at a point, must it have all partial derivatives at that point? Is the converse true?

## Challenging Problems

**1.** (a) If the graph of a function $f(x, y)$ that is differentiable at $(a, b)$ contains part of a straight line through $(a, b)$, show that the line lies in the tangent plane to $z = f(x, y)$ at $(a, b)$.

(b) If $g(t)$ is a differentiable function of $t$, describe the surface $z = yg(x/y)$ and show that all its tangent planes pass through the origin.

**2.** A particle moves in 3-space in such a way that its direction of motion at any point is perpendicular to the level surface of

$$f(x, y, z) = 4 - x^2 - 2y^2 + 3z^2$$

through that point. If the path of the particle passes through the point $(1, 1, 8)$, show that it also passes through $(2, 4, 1)$. Does it pass through $(3, 7, 0)$?

**3.** **(The Laplace operator in spherical coordinates)** If $u(x, y, z)$ has continuous second partial derivatives and

$$v(\rho, \phi, \theta) = u(\rho \sin\phi \cos\theta, \rho \sin\phi \sin\theta, \rho \cos\phi),$$

show that

$$\frac{\partial^2 v}{\partial \rho^2} + \frac{2}{\rho}\frac{\partial v}{\partial \rho} + \frac{\cot\phi}{\rho^2}\frac{\partial v}{\partial \phi} + \frac{1}{\rho^2}\frac{\partial^2 v}{\partial \phi^2} + \frac{1}{\rho^2 \sin^2\phi}\frac{\partial^2 v}{\partial \theta^2}$$
$$= \frac{\partial^2 u}{\partial x^2} + \frac{\partial^2 u}{\partial y^2} + \frac{\partial^2 u}{\partial z^2}.$$

You can do this by hand, but it is a lot easier using computer algebra.

**4.** **(Spherically expanding waves)** If $f$ is a twice differentiable function of one variable and $\rho = \sqrt{x^2 + y^2 + z^2}$, show that $u(x, y, z, t) = \dfrac{f(\rho - ct)}{\rho}$ satisfies the three-dimensional wave equation

$$\frac{\partial^2 u}{\partial t^2} = c^2 \left( \frac{\partial^2 u}{\partial x^2} + \frac{\partial^2 u}{\partial y^2} + \frac{\partial^2 u}{\partial z^2} \right).$$

What is the geometric significance of this solution as a function of increasing time $t$? (*Hint:* you may want to use the result of Exercise 3. In this case $v(\rho, \phi, \theta)$ is independent of $\phi$ and $\theta$.)

CHAPTER 5

# Applications of Partial Derivatives

**INTRODUCTION** In this chapter we will discuss some of the ways partial derivatives contribute to the understanding and solution of problems in applied mathematics. Many such problems can be put in the context of determining maximum or minimum values for functions of several variables, and the first four sections of this chapter deal with that subject. The remaining sections discuss some miscellaneous problems involving the differentiation of functions with respect to parameters, and also Newton's method for approximating solutions of systems of nonlinear equations. Much of the material in this chapter may be considered *optional*. Only Sections 5.1–5.3 contain *core material*, and even parts of those sections can be omitted (for example, the discussion of linear programming in Section 5.2).

## 5.1 EXTREME VALUES

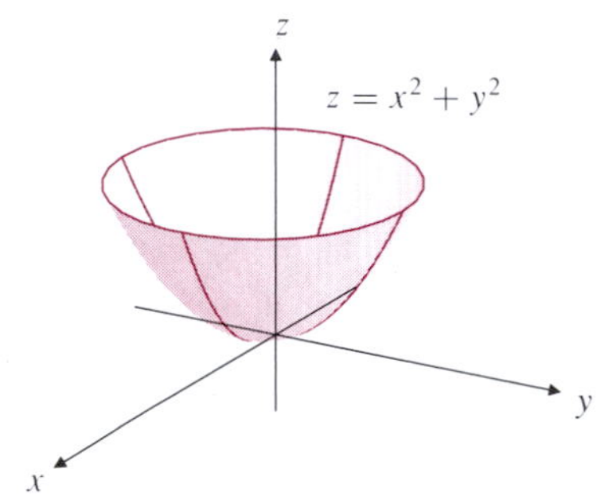


**Figure 5.1** $x^2 + y^2$ has minimum value 0 at the origin

The function $f(x, y) = x^2 + y^2$, part of whose graph is shown in Figure 5.1, has a minimum value of 0; this value occurs at the origin $(0, 0)$. Similarly, the function $g(x, y) = 1 - x^2 - y^2$, part of whose graph appears in Figure 5.2, has a maximum value of 1 at $(0, 0)$. What techniques could be used to discover these facts if they were not evident from a diagram? Finding maximum and minimum values of functions of several variables is, like its single-variable counterpart, the crux of many applications of advanced calculus to problems that arise in other disciplines. Unfortunately, this problem is often much more complicated than in the single-variable case. Our discussion will begin by developing the techniques for functions of two variables. Some of the techniques extend to functions of more variables in obvious ways. The extension of those that do not will be discussed later in the section.

**Figure 5.2** $1 - x^2 - y^2$ has maximum value 1 at the origin

Let us begin by reviewing what we know about the single-variable case. Recall that a function $f(x)$ has a *local maximum value* (or a *local minimum value*) at a point $a$ in its domain if $f(x) \le f(a)$ (or $f(x) \ge f(a)$) for all $x$ in the domain of $f$ that are *sufficiently close* to $a$. If the appropriate inequality holds *for all* $x$ in the domain of $f$, then we say that $f$ has an *absolute maximum* (or *absolute minimum*) value at $a$. Moreover, such local or absolute extreme values can occur only at points of one of the following three types:

(a) critical points — where $f'(x) = 0$, or

(b) singular points — where $f'(x)$ does not exist, or

(c) endpoints of the domain of $f$.

A similar situation exists for functions of several variables. For example, we say that a function of two variables has a **local maximum** or **relative maximum** value at the point $(a, b)$ in its domain if $f(x, y) \le f(a, b)$ for all points $(x, y)$ in the domain of $f$ that are *sufficiently close* to the point $(a, b)$. If the inequality holds *for all* $(x, y)$ in the domain of $f$, then we say that $f$ has a **global maximum** or **absolute maximum**) value at $(a, b)$. Similar definitions obtain for local (relative) and absolute (global) minimum values. In practice, the words *absolute* or *global* are usually omitted, and we refer simply to *the maximum* or *the minimum* value of $f$.

The following theorem shows that there are three possibilities for points where extreme values can occur, analogous to those for the single-variable case.

**THEOREM 1**

**Necessary conditions for extreme values**

A function $f(x, y)$ can have a local or absolute extreme value at a point $(a, b)$ in its domain only if $(a, b)$ is

(a) a **critical point** of $f$, that is, a point satisfying $\nabla f(a, b) = \mathbf{0}$, or

(b) a **singular point** of $f$, that is, a point where $\nabla f(a, b)$ does not exist, or

(c) a **boundary point** of the domain of $f$.

***PROOF*** Suppose that $(a, b)$ belongs to the domain of $f$. If $(a, b)$ is not on the boundary of the domain of $f$, then it must belong to the interior of that domain, and if $(a, b)$ is not a singular point of $f$ then $\nabla f(a, b)$ exists. Finally, if $(a, b)$ is not a critical point of $f$, then $\nabla f(a, b) \neq \mathbf{0}$ and so $f$ has a positive directional derivative in the direction of $\nabla f(a, b)$ and a negative directional derivative in the direction of $-\nabla f(a, b)$. That is, $f$ is increasing as we move from $(a, b)$ in one direction and decreasing as we move in the opposite direction. Hence $f$ cannot have either a maximum or a minimum value at $(a, b)$. Therefore, any point where an extreme value occurs must be either a critical point or a singular point of $f$, or a boundary point of the domain of $f$.

The definitions of boundary and interior can easily be extended to sets in $\mathbb{R}^3$ (or even $\mathbb{R}^n$); the word *disk* in the definition of a boundary point should be replaced with *ball*. Theorem 1 remains valid with unchanged proof for functions of any number of variables. Of course, Theorem 1 does not guarantee that a given function will have any extreme values. It only tells us where to look to find any that may exist. Theorem 2, below, provides conditions that guarantee the existence of absolute maximum and minimum values for a continuous function. It is analogous to the Max-Min Theorem for functions of one variable. The proof is beyond the scope of this book; an interested student should consult an elementary text on mathematical analysis. A set in $\mathbb{R}^n$ is **bounded** if it is contained inside some *ball* $x_1^2 + x_2^2 + \cdots + x_n^2 \leq R^2$ of finite radius $R$. A set on the real line is bounded if it is contained in an interval of finite length.

**THEOREM 2**

**Sufficient conditions for extreme values**

If $f$ is a *continuous* function of $n$ variables whose domain is a *closed* and *bounded* set in $\mathbb{R}^n$, then the range of $f$ is a bounded set of real numbers, and there are points in its domain where $f$ takes on absolute maximum and minimum values.

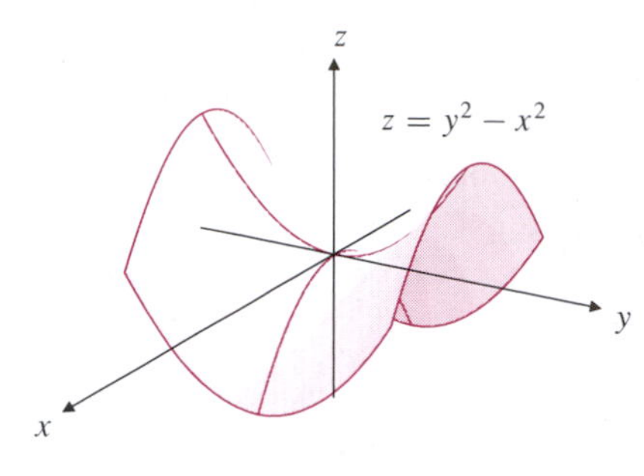


**Figure 5.3** $y^2 - x^2$ has a saddle point at $(0, 0)$

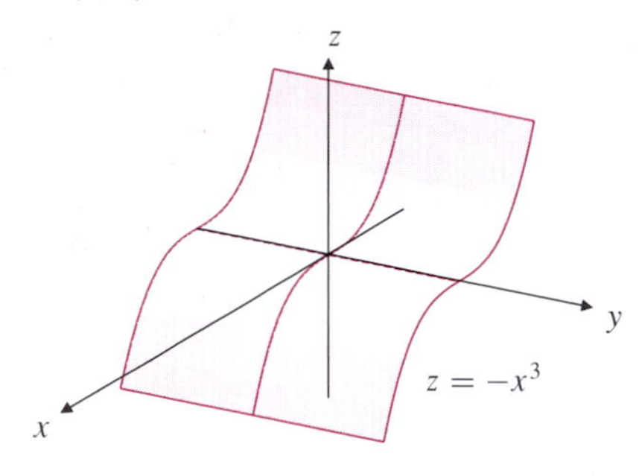


**Figure 5.4** A line of saddle points

**EXAMPLE 1** The function $f(x, y) = x^2 + y^2$ (see Figure 5.1) has a critical point at $(0, 0)$ since $\nabla f = 2x\mathbf{i} + 2y\mathbf{j}$ and both components of $\nabla f$ vanish at (0,0). Since

$$f(x, y) > 0 = f(0, 0) \quad \text{if} \quad (x, y) \neq (0, 0),$$

$f$ must have (absolute) minimum value 0 at that point. If the domain of $f$ is not restricted $f$ has no maximum value. Similarly, $g(x, y) = 1 - x^2 - y^2$ has (absolute) maximum value 1 at its critical point (0,0). (See Figure 5.2.) ■

**EXAMPLE 2** The function $h(x, y) = y^2 - x^2$ also has a critical point at (0,0) but has neither a local maximum nor a local minimum value at that point. Observe that $h(0, 0) = 0$ but $h(x, 0) < 0$ and $h(0, y) > 0$ for all nonzero values of $x$ and $y$. (See Figure 5.3.) The graph of $h$ is a hyperbolic paraboloid. In view of its shape we call the critical point (0,0) a **saddle point** of $h$. ■

In general we will somewhat loosely call any *interior critical* point of the domain of a function $f$ of several variables a **saddle point** if $f$ does not have a local maximum or minimum value there. Even for functions of two variables, the graph will not always look like a saddle near a saddle point. For instance, the function $f(x, y) = -x^3$ has a whole line of *saddle points* along the $y$-axis (see Figure 5.4), though its graph does not resemble a saddle anywhere. These points resemble inflection points of a function of one variable. Saddle points are higher-dimensional analogues of such horizontal inflection points.

**EXAMPLE 3** The function $f(x, y) = \sqrt{x^2 + y^2}$ has no critical points, but does have a singular point at (0,0) where it has a local (and absolute) minimum value, zero. The graph of $f$ is a circular cone. (See Figure 5.5(a).) ■

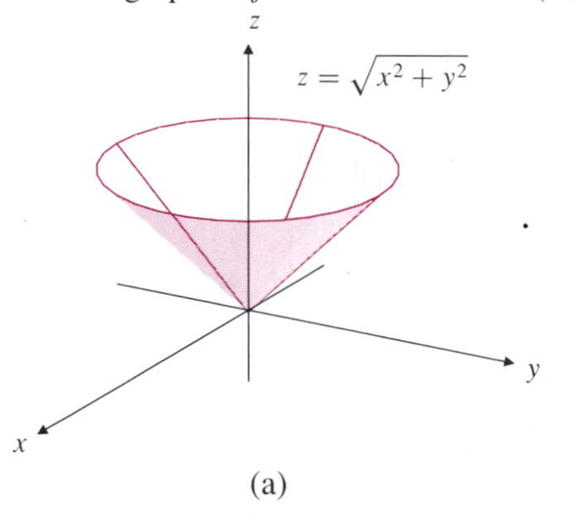


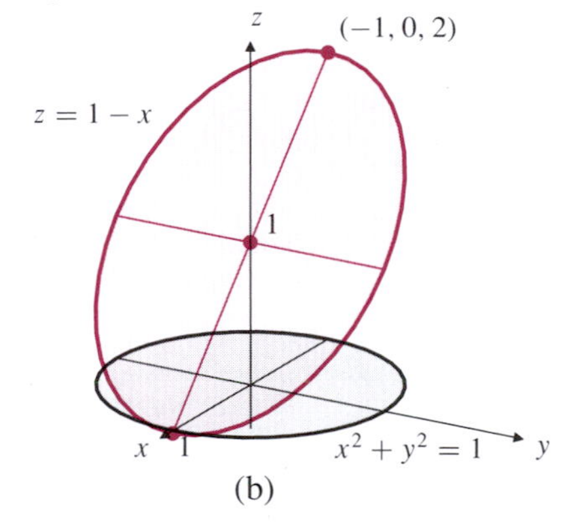


**Figure 5.5**

(a) $\sqrt{x^2 + y^2}$ has a minimum value at the singular point $(0, 0)$

(b) When restricted to the disk $x^2 + y^2 \leq 1$, the function $1 - x$ has maximum and minimum values at boundary points

**EXAMPLE 4** The function $f(x, y) = 1 - x$ is defined everywhere in the $xy$-plane and has no critical or singular points. ($\nabla f(x, y) = -\mathbf{i}$ at every point $(x, y)$.) Therefore $f$ has no extreme values. However, if we restrict the domain of $f$ to the points in the disk $x^2 + y^2 \leq 1$ (a closed bounded set in the $xy$-plane) then $f$ does have absolute maximum and minimum values, as it must by Theorem 2. The maximum value is 2 at the boundary point $(-1, 0)$ and the minimum value is 0 at (1,0). (See Figure 5.5(b).) ■

## Classifying Critical Points

The above examples were very simple ones; it was immediately obvious in each case whether the function had a local maximum or local minimum or a saddle point at the critical or singular point. For more complicated functions, it may be harder to classify the interior critical points. In theory such a classification can always be made by considering the difference

$$\Delta f = f(a+h, b+k) - f(a,b)$$

for small values of $h$ and $k$, where $(a, b)$ is the critical point in question. If the difference is always nonnegative (or nonpositive) for small $h$ and $k$, then $f$ must have a local minimum (or maximum) at $(a, b)$; if the difference is negative for some points $(h, k)$ arbitrarily near (0,0), and positive for others, then $f$ must have a saddle point at $(a, b)$.

**EXAMPLE 5** Find and classify the critical points of $f(x, y) = 2x^3 - 6xy + 3y^2$.

***SOLUTION*** The critical points must satisfy the system of equations:

$$0 = f_1(x, y) = 6x^2 - 6y \iff x^2 = y$$
$$0 = f_2(x, y) = -6x + 6y \iff x = y.$$

Together, these equations imply that $x^2 = x$ so that $x = 0$ or $x = 1$. Therefore the critical points are (0,0) and (1,1).

Consider (0,0). Here $\Delta f$ is given by

$$\Delta f = f(h, k) - f(0, 0) = 2h^3 - 6hk + 3k^2.$$

Since $f(h, 0) - f(0, 0) = 2h^3$ is positive for small positive $h$ and negative for small negative $h$, $f$ cannot have a maximum or minimum value at (0,0). Therefore (0,0) is a saddle point.

Now consider (1,1). Here $\Delta f$ is given by

$$\begin{aligned}\Delta f &= f(1+h, 1+k) - f(1, 1)\\ &= 2(1+h)^3 - 6(1+h)(1+k) + 3(1+k)^2 - (-1)\\ &= 2 + 6h + 6h^2 + 2h^3 - 6 - 6h - 6k - 6hk + 3 + 6k + 3k^2 + 1\\ &= 6h^2 - 6hk + 3k^2 + 2h^3\\ &= 3(h-k)^2 + h^2(3 + 2h).\end{aligned}$$

Both terms in the latter expression are nonnegative if $|h| < 3/2$, and they are not both zero unless $h = k = 0$. Hence $\Delta f > 0$ for small $h$ and $k$, and $f$ has a local minimum value $-1$ at (1,1). ■

The method used to classify critical points in the above example takes on a "brute force" aspect if the function involved is more complicated. However, it provides the basis for developing a *second derivative test* similar to that for functions of one variable. The two-variable version is the subject of the following theorem.

**THEOREM 3**

**A second derivative test**

Suppose that $(a, b)$ is a critical point of $f(x, y)$ interior to the domain of $f$. Suppose also that the second partial derivatives of $f$ are continuous near $(a, b)$ and have at that point the values

$$A = f_{11}(a, b), \quad B = f_{12}(a, b) = f_{21}(a, b), \quad \text{and} \quad C = f_{22}(a, b).$$

(a) If $B^2 < AC$ and $A > 0$, then $f$ has a local minimum value at $(a, b)$.

(b) If $B^2 < AC$ and $A < 0$, then $f$ has a local maximum value at $(a, b)$.

(c) If $B^2 > AC$, then $f$ has a saddle point at $(a, b)$.

(d) If $B^2 = AC$, this test provides no information; $f$ may have a local maximum or a local minimum value or a saddle point at $(a, b)$.

***PROOF*** Since $f_1(a, b) = f_2(a, b) = 0$, and the second partial derivatives of $f$ are continuous near $(a, b)$, the case $n = 1$ of Taylor's Formula (see Section 4.9) gives

$$f(a + h, b + k) - f(a, b) = \frac{1}{2}\left(\bar{A}h^2 + 2\bar{B}hk + \bar{C}k^2\right) = \frac{1}{2}\bar{Q}(h, k),$$

where $\bar{A}$, $\bar{B}$, and $\bar{C}$ are, respectively, the values of the partial derivatives $f_{11}$, $f_{12}$, and $f_{22}$ at some point $(a + \theta h, b + \theta k)$ on the line segment from $(a, b)$ to $(a + h, b + k)$ (i.e., $\theta$ is between 0 and 1), and where $\bar{Q}(h, k) = \bar{A}h^2 + 2\bar{B}hk + \bar{C}k^2$. In order to prove that $f$ has a local minimum (or a local maximum) at $(a, b)$, we need to show that $\bar{Q}(h, k) \geq 0$ (or $\bar{Q}(h, k) \leq 0$) at *all* points $(h, k)$ sufficiently close to $(0, 0)$. If $\bar{Q}(h, k) > 0$ at some points $(h, k)$ arbitrarily close to $(0, 0)$ and $\bar{Q}(h, k) < 0$ at other such points, then $f$ will have a saddle point at $(a, b)$.

Let $Q(h, k) = Ah^2 + 2Bhk + Ck^2$. Since the second partial derivatives of $f$ are continuous at $(a, b)$, any of the inequalities $A > 0$, $A < 0$, $B^2 < AC$, or $B^2 > AC$ satisfied by the coefficients $A$, $B$, and $C$ of $Q$ will imply the same inequalities for the coefficients $\bar{A}$, $\bar{B}$, and $\bar{C}$ of $\bar{Q}$, provided $(h, k)$ is close enough to $(0, 0)$ so that $(a + \theta h, b + \theta k)$ is sufficiently close to $(a, b)$. Hence it suffices to consider the sign of $Q(h, k)$ under the conditions stated in the theorem.

If $A \neq 0$, then completing the square in the quadratic function $Q$, we obtain

$$Q(h, k) = A\left[\left(h + \frac{B}{A}k\right)^2 + \frac{AC - B^2}{A^2}k^2\right].$$

If $B^2 < AC$, then the expression in the square brackets above is a sum of positive expressions, so, for all $(h, k) \neq (0, 0)$, $Q(h, k)$ will have the same sign as $A$, and $f$ will have a local minimum or maximum at $(a, b)$ according to whether $A > 0$ or $A < 0$.

If $B^2 > AC$, then the expression in the square brackets is a difference of positive quantities. It is positive at points $(h, 0)$ and it is negative at points $(-Bk/A, k)$. Thus $f$ must have a saddle point at $(a, b)$.

If $A = 0$ but $B^2 \neq AC$, then $B \neq 0$ and $Q(h, k) = k(2Bh + Ck)$. This has both positive and negative values in the four angular regions lying between the lines $k = 0$ and $2Bh + Ck = 0$. In this case $f$ must have a saddle point.

Consideration of the functions $f(x, y) = x^4 + y^4$, $g(x, y) = -x^4 - y^4$ and $h(x, y) = x^4 - y^4$ will readily convince you that if $B^2 = AC$, then a function can have a minimum, a maximum or a saddle point.

**EXAMPLE 6** Reconsider Example 5 and use the second derivative test to classify the two critical points (0,0) and (1,1) of $f(x, y) = 2x^3 - 6xy + 3y^2$.

***SOLUTION*** We have

$$f_{11}(x, y) = 12x, \quad f_{12}(x, y) = -6, \quad \text{and} \quad f_{22}(x, y) = 6.$$

At (0,0) we therefore have

$$A = 0, \quad B = -6, \quad C = 6, \quad \text{and} \quad B^2 - AC = 36 > 0$$

so (0,0) is a saddle point. At (1,1) we have

$$A = 12 > 0, \quad B = -6, \quad C = 6, \quad \text{and} \quad B^2 - AC = -36 < 0$$

so $f$ must have a local minimum at (1,1).

**EXAMPLE 7** Find and classify the critical points of

$$f(x, y) = xy\,e^{-(x^2+y^2)/2}.$$

Does $f$ have absolute maximum and minimum values? Why?

***SOLUTION*** We begin by calculating the first- and second-order partial derivatives of $f$:

$$\begin{aligned}
f_1(x, y) &= y(1 - x^2)\,e^{-(x^2+y^2)/2},\\
f_2(x, y) &= x(1 - y^2)\,e^{-(x^2+y^2)/2},\\
f_{11}(x, y) &= xy(x^2 - 3)\,e^{-(x^2+y^2)/2},\\
f_{12}(x, y) &= (1 - x^2)(1 - y^2)\,e^{-(x^2+y^2)/2},\\
f_{22}(x, y) &= xy(y^2 - 3)\,e^{-(x^2+y^2)/2}.
\end{aligned}$$

At any critical point $f_1 = 0$ and $f_2 = 0$ so the critical points are the solutions of the system of equations

$$\begin{aligned}
y(1 - x^2) &= 0\\
x(1 - y^2) &= 0.
\end{aligned}$$

The first of these equations says that either $y = 0$ or $x = \pm 1$. The second equation says that either $x = 0$ or $y = \pm 1$. There are five points satisfying both conditions: $(0, 0)$, $(1, 1)$, $(1, -1)$, $(-1, 1)$ and $(-1, -1)$. We classify them using the second derivative test.

At $(0, 0)$ we have $A = C = 0$, $B = 1$, so that $B^2 - AC = 1 > 0$. Thus $f$ has a saddle point at $(0, 0)$.

At $(1, 1)$ and $(-1, -1)$ we have $A = C = -2/e < 0$, $B = 0$. It follows that $B^2 - AC = -4/e^2 < 0$. Thus $f$ has local maximum values at these points. The value of $f$ is $1/e$ at each point.

At $(1, -1)$ and $(-1, 1)$ we have $A = C = 2/e > 0$, $B = 0$. If follows that $B^2 - AC = -4/e^2 < 0$. Thus $f$ has local minimum values at these points. The value of $f$ at each of them is $-1/e$.

Indeed, $f$ has absolute maximum and minimum values—namely, the values obtained above as local extrema. To see why, observe that $f(x, y)$ approaches 0 as the point $(x, y)$ recedes to infinity in any direction, because the negative exponential dominates the power factor $xy$ for large $x^2 + y^2$. Pick a number between 0 and the local maximum value $1/e$ found above, say the number $1/(2e)$. For some $R$, we must have $|f(x, y)| \le 1/(2e)$ whenever $x^2 + y^2 \ge R^2$. On the closed disk $x^2 + y^2 \le R^2$ $f$ must have absolute maximum and minimum values by Theorem 2. These cannot occur on the boundary circle $x^2 + y^2 = R^2$ because $|f|$ is smaller there ($\le 1/(2e)$) than it is at the critical points considered above. Since $f$ has no singular points, the absolute maximum and minimum values for the disk, and therefore for the whole plane, must occur at those critical points. ■

**EXAMPLE 8** Find the dimensions of a rectangular box of given volume $V$ having the least possible surface area.

***SOLUTION*** If the dimensions of the box are $x$, $y$, and $z$, then we want to minimize

$$S = 2xy + 2yz + 2xz$$

subject to the restriction that $xyz = V$, the required volume. We can use this restriction to reduce the number of variables on which $S$ depends, for instance, by substituting

$$z = \frac{V}{xy}.$$

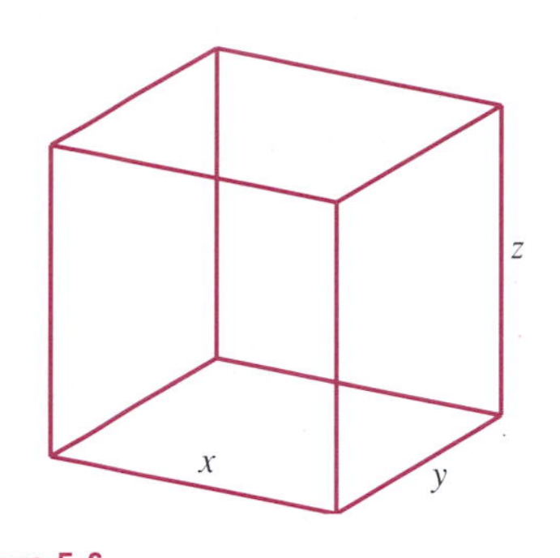


Figure 5.6

Then $S$ becomes a function of the two variables $x$ and $y$:

$$S = S(x, y) = 2xy + \frac{2V}{x} + \frac{2V}{y}.$$

A real box has positive dimensions, so the domain of $S$ should consist of only those points $(x, y)$ that satisfy $x > 0$ and $y > 0$. If either $x$ or $y$ approach 0 or $\infty$, then $S \to \infty$, so the minimum value of $S$ must occur at a critical point. ($S$ has no singular points.) For critical points we solve the equations

$$\begin{aligned} 0 &= \frac{\partial S}{\partial x} = 2y - \frac{2V}{x^2} &\iff&\quad x^2y = V, \\ 0 &= \frac{\partial S}{\partial y} = 2x - \frac{2V}{y^2} &\iff&\quad xy^2 = V. \end{aligned}$$

Thus $x^2y - xy^2 = 0$, or $xy(x - y) = 0$. Since $x > 0$ and $y > 0$, this implies that $x = y$. Therefore $x^3 = V$, and so $x = y = V^{1/3}$. Therefore, $z = V/(xy) = V^{1/3}$. Since there is only one critical point it must minimize $S$. (Why?) The box having minimal surface area is a cube of edge length $V^{1/3}$. ■

***REMARK*** The preceding problem is a *constrained* extreme value problem in three variables; the equation $xyz = V$ is a *constraint* limiting the freedom of $x$, $y$, and $z$. We used the constraint to eliminate one variable, $z$, and so to reduce the problem to a *free* (that is, *unconstrained*) problem in two variables. This is the same technique that was used in *Single-Variable Calculus* to reduce constrained problems in two variables to free problems in one variable. In Section 5.3 we will develop another method for solving constrained extreme value problems.

## Quadratic Forms

It is not immediately apparent how Theorem 3 can be extended to functions of more than two variables. In order to make such an extension we shall rephrase that theorem in terms of *quadratic forms*.

If $\mathbf{u} = u\mathbf{i} + v\mathbf{j}$ is a unit vector, then the *second directional derivative* of $f(x, y)$ at $(a, b)$ in the direction of $\mathbf{u}$ is given by

$$\begin{aligned} D^2_{\mathbf{u}} f(a, b) &= \mathbf{u} \bullet \nabla\Big(\mathbf{u} \bullet \nabla f\Big)(a, b) \\ &= Au^2 + 2Buv + Cv^2, \end{aligned}$$

where $A$, $B$, and $C$ are constants involving $\mathbf{u}$ and the second partial derivatives of $f$ (see Example 5 of Section 4.7, where $\mathbf{u} = \cos\phi\mathbf{i} + \sin\phi\mathbf{j}$.) Being a homogeneous polynomial of degree two, the expression

$$Q(u, v) = Au^2 + 2Buv + Cv^2$$

is called a **quadratic form** in the variables $u$ and $v$. Such a quadratic form is said to be **positive definite** (or **negative definite**) if $Q(u, v) > 0$ (or $Q(u, v) < 0$) for *all* nonzero vectors $\mathbf{u}$. Otherwise it is said to be **indefinite**.

Theorem 3 says that $f$ has a local minimum or a local maximum at the critical point $(a, b)$ if the second directional derivative $Q(u, v) = D^2_{\mathbf{u}} f(a, b)$ is a positive definite or negative definite in the components of $\mathbf{u}$. It says also that $f$ has a saddle point at $(a, b)$ if $Q(u, v)$ is positive for some vectors $\mathbf{u}$ and negative for others. This is not the same as saying $Q$ is indefinite. For instance, $Q$ might satisfy $Q(u, v) \ge 0$ (or $Q(u, v) \le 0$) for all nonzero vectors $\mathbf{u}$, with equality holding for some such vectors, or it can even happen that $Q(u, v)$ vanishes for all vectors $\mathbf{u}$. Such things can happen only in the indeterminate case $B^2 = AC$ of Theorem 3.

An extension of Theorem 3 to functions of any number of variables can be phrased in terms of quadratic forms. For a function $f(x, y, z)$ with continuous second partial derivatives, the second directional derivative at $(a, b, c)$ in the direction of the unit vector $\mathbf{u} = u\mathbf{i} + v\mathbf{j} + w\mathbf{k}$ is the quadratic form

$$Q(u, v, w) = D^2_{\mathbf{u}} f(a, b, c) = Au^2 + Bv^2 + Cw^2 + 2Duv + 2Euw + 2Fvw,$$

where

$$\begin{aligned} A &= f_{11}(a, b, c), & B &= f_{22}(a, b, c), & C &= f_{33}(a, b, c), \\ D &= f_{12}(a, b, c), & E &= f_{13}(a, b, c), & F &= f_{23}(a, b, c). \end{aligned}$$

Again, $f$ will have a local minimum, a local maximum, or a saddle point at the critical point $(a, b, c)$ if $Q(u, v, w)$ is positive definite, negative definite, or has positive values for some vectors $\mathbf{u}$ and negative values for others. This can frequently be determined by a process of successive completion of squares, as the example below shows. If $Q$ does not satisfy one of these conditions, then no *second derivative test* can determine the nature of the critical point. One can still resort to "brute force" methods, as in Example 5.

**EXAMPLE 9** Find and classify the critical points of the function $f(x, y, z) = x^2 y + y^2 z + z^2 - 2x$.

***SOLUTION*** The equations that determine the critical points are

$$\begin{aligned} 0 &= f_1(x, y, z) = 2xy - 2, \\ 0 &= f_2(x, y, z) = x^2 + 2yz, \\ 0 &= f_3(x, y, z) = y^2 + 2z. \end{aligned}$$

The third equation implies $z = -y^2/2$ and the second then implies $y^3 = x^2$. From the first equation we get $y^{5/2} = 1$. Thus $y = 1$ and $z = -\frac{1}{2}$. Since $xy = 1$, we must have $x = 1$. The only critical point is $P = (1, 1, -\frac{1}{2})$. Evaluating the second partial derivatives of $f$ at this point we get

$$A = 2, \quad B = -1, \quad C = 2, \quad D = 2, \quad E = 0, \text{ and } \quad F = 2.$$

Hence the second directional derivative of $f$ at P in the direction of the unit vector $\mathbf{u}$ is

$$\begin{aligned} Q(u, v, w) &= 2u^2 - v^2 + 2w^2 + 4uv + 4vw \\ &= 2(u^2 + 2uv + v^2) + 2(w^2 + 2vw + v^2) - 5v^2 \\ &= 2(u + v)^2 + 2(w + v)^2 - 5v^2. \end{aligned}$$

Observe that $Q(1, 0, 0) > 0$ and $Q(0, 1, 0) < 0$, so $P$ must be a saddle point of the function $f$. ■

**REMARK** There is an orderly procedure for determining whether a quadratic form in any number of variables is positive definite. We shall describe it for the three-variable case but the extension to higher dimensions is similar. The quadratic form

$$Q(u, v, w) = Au^2 + Bv^2 + Cw^2 + 2Duv + 2Euw + 2Fvw$$

can be written as

$$Q(u, v, w) = \mathbf{u}^T \mathcal{Q}\mathbf{u}$$

where

$$\mathbf{u} = \begin{pmatrix} u \\ v \\ w \end{pmatrix} \quad \text{and} \quad \mathcal{Q} = \begin{pmatrix} A & D & E \\ D & B & F \\ E & F & C \end{pmatrix}.$$

$Q$ is positive definite if

$$A > 0, \qquad \begin{vmatrix} A & D \\ D & B \end{vmatrix} > 0, \quad \text{and} \quad \begin{vmatrix} A & D & E \\ D & B & F \\ E & F & C \end{vmatrix} > 0.$$

The reader is encouraged to prove this (see Exercise 28 below) by using the square-completion technique of Example 9. We can deduce an equivalent test for negative definiteness. $Q(u, v, w)$ is negative definite if $-Q(u, v, w)$ is positive definite. Thus $Q$ is negative definite if

$$A < 0, \qquad \begin{vmatrix} A & D \\ D & B \end{vmatrix} > 0, \quad \text{and} \quad \begin{vmatrix} A & D & E \\ D & B & F \\ E & F & C \end{vmatrix} < 0.$$

If $\det(\mathcal{Q}) \neq 0$ but neither $Q$ nor $-Q$ is positive definite, then $Q$ will be positive for some values of $(u, v, w)$ and negative for others.

## EXERCISES 5.1

In Exercises 1–15, find and classify the critical points of the given functions.

**1.** $f(x, y) = x^2 + 2y^2 - 4x + 4y$

**2.** $f(x, y) = xy - x + y$

**3.** $f(x, y) = x^3 + y^3 - 3xy$

**4.** $f(x, y) = x^4 + y^4 - 4xy$

**5.** $f(x, y) = \dfrac{x}{y} + \dfrac{8}{x} - y$

**6.** $f(x, y) = \cos(x + y)$

**7.** $f(x, y) = x \sin y$

**8.** $f(x, y) = \cos x + \cos y$

**9.** $f(x, y) = x^2 y e^{-(x^2+y^2)}$

**10.** $f(x, y) = \dfrac{xy}{2 + x^4 + y^4}$

**11.** $f(x, y) = x e^{-x^3+y^3}$

**12.** $f(x, y) = \dfrac{1}{1 - x + y + x^2 + y^2}$

**13.** $f(x, y) = \left(1 + \dfrac{1}{x}\right)\left(1 + \dfrac{1}{y}\right)\left(\dfrac{1}{x} + \dfrac{1}{y}\right)$

* **14.** $f(x, y, z) = xyz - x^2 - y^2 - z^2$

* **15.** $f(x, y, z) = xy + x^2 z - x^2 - y - z^2$

* **16.** Show that $f(x, y, z) = 4xyz - x^4 - y^4 - z^4$ has a local maximum value at the point (1,1,1).

**17.** Find the maximum and minimum values of $f(x, y) = xy e^{-x^2-y^4}$.

**18.** Find the maximum and minimum values of $f(x, y) = x/(1 + x^2 + y^2)$.

* **19.** Find the maximum and minimum values of $f(x, y, z) = xyz e^{-x^2-y^2-z^2}$. How do you know that such extreme values exist?

**20.** Find the minimum value of $f(x, y) = x + 8y + \dfrac{1}{xy}$ in the first quadrant $x > 0$, $y > 0$. How do you know that a minimum exists?

**21.** Find the dimensions of the rectangular box with no top having given volume $V$ and the least possible total surface area of its five faces.

**22.** The material used to make the bottom of a rectangular box is twice as expensive per unit area as the material used to make the top or side walls. Find the dimensions of the box of given volume $V$ for which the cost of materials is minimum.

**23.** Find the volume of the largest rectangular box (with faces parallel to the coordinate planes) that can be inscribed inside the ellipsoid

$$\frac{x^2}{a^2} + \frac{y^2}{b^2} + \frac{z^2}{c^2} = 1.$$

**24.** Find the three positive numbers $a$, $b$, and $c$, whose sum is 30 and for which the expression $ab^2c^3$ is maximum.

**25.** Find the critical points of the function $z = g(x, y)$ that satisfies the equation $e^{2zx-x^2} - 3e^{2zy+y^2} = 2$.

$*$ **26.** Classify the critical points of the function $g$ in the previous exercise.

$*$ **27.** Let $f(x, y) = (y - x^2)(y - 3x^2)$. Show that the origin is a critical point of $f$ and that the restriction of $f$ to every straight line through the origin has a local minimum value at the origin. (That is, show that $f(x, kx)$ has a local minimum value at $x = 0$ for every $k$, and that $f(0, y)$ has a local minimum value at $y = 0$.) Does $f(x, y)$ have a local minimum value at the origin? What happens to $f$ on the curve $y = 2x^2$? What does the second derivative test say about this situation?

$*$ **28.** Prove the determinant tests for the positive and negative definiteness of the three-variable quadratic forms given in the remark at the end of this section.

$*$ **29.** State conditions on the coefficients $A, B, \ldots, J$ that will guarantee that the quadratic form

$$\begin{aligned} Q(w, x, y, z) = &Aw^2 + Bx^2 + Cy^2 \\ &+ Dz^2 + 2Ewx + 2Fwy \\ &+ 2Gwz + 2Hxy + 2Ixz + 2Jyz \end{aligned}$$

is positive definite. (Generalize the assertion made for the three-variable forms in the text.)

## 5.2 EXTREME VALUES OF FUNCTIONS DEFINED ON RESTRICTED DOMAINS

Much of the previous section was concerned with techniques for determining whether a critical point of a function provides a local maximum or minimum value, or is a saddle point. In this section we address the problem of determining absolute maximum and minimum values for functions that have them — usually functions whose domains are restricted to subsets of $\mathbb{R}^2$ (or $\mathbb{R}^n$) having nonempty interiors. In Example 7 of Section 5.1 we had to *prove* that the given function had absolute extreme values. If, however, we are dealing with a continuous function on a domain that is closed and bounded, then we can rely on Theorem 2 to guarantee the existence of such extreme values, but we will always have to check boundary points as well as any interior critical or singular points to find them. The following examples illustrate the technique.

**EXAMPLE 1** Find the maximum and minimum values of $f(x, y) = 2xy$ on the closed disk $x^2 + y^2 \le 4$.

***SOLUTION*** Since $f$ is continuous and the disk is closed, $f$ must have absolute maximum and minimum values at some points of the disk. The first partial derivatives of $f$ are

$$f_1(x, y) = 2y \qquad \text{and} \qquad f_2(x, y) = 2x,$$

so there are no singular points and the only critical point is (0,0), where $f$ has the value 0.

We must still consider values of $f$ on the boundary circle $x^2 + y^2 = 4$. We can express $f$ as a function of a single variable on this circle, by using a convenient parametrization of the circle, say

$$x = 2\cos t, \qquad y = 2\sin t, \qquad (-\pi \le t \le \pi).$$

We have

$$f\Big(2\cos t, 2\sin t\Big) = 8\cos t \sin t = g(t).$$

We must find any extreme values of $g(t)$. We can do this in either of two ways. If we rewrite $g(t) = 4\sin 2t$, it is clear that $g(t)$ has maximum value 4 (at $t = \frac{\pi}{4}$ and $-\frac{3\pi}{4}$), and minimum value $-4$ (at $t = -\frac{\pi}{4}$ and $\frac{3\pi}{4}$). Alternatively, we can differentiate $g$ to find its critical points:

$$\begin{aligned} 0 = g'(t) = -8\sin^2 t + 8\cos^2 t \quad &\iff \quad \tan^2 t = 1 \\ &\iff \quad t = \pm\frac{\pi}{4} \text{ or } \pm\frac{3\pi}{4}, \end{aligned}$$

which again yield the maximum value 4 and the minimum value $-4$. (It is not necessary to check the endpoints $t = -\pi$ and $t = \pi$; since $g$ is everywhere differentiable and is periodic with period $\pi$, any absolute maximum or minimum will occur at a critical point.)

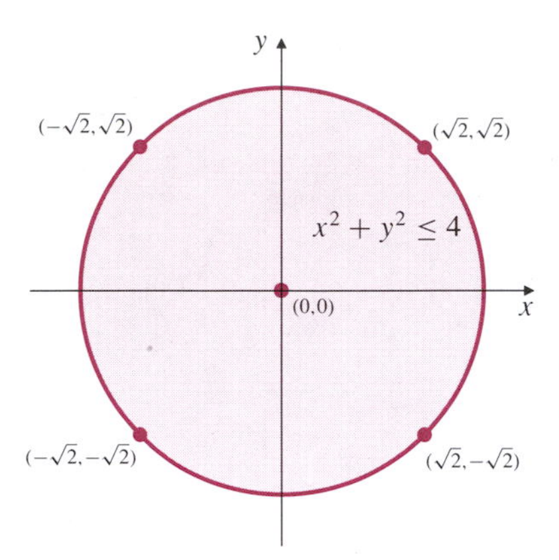


**Figure 5.7** Points that are candidates for extreme values in Example 1

In any event, $f$ has maximum value 4 at the boundary points $(\sqrt{2}, \sqrt{2})$ and $(-\sqrt{2}, -\sqrt{2})$, and minimum value $-4$ at the boundary points $(\sqrt{2}, -\sqrt{2})$ and $(-\sqrt{2}, \sqrt{2})$. It is easily shown by the second derivative test (or otherwise) that the interior critical point (0,0) is a saddle point. ■

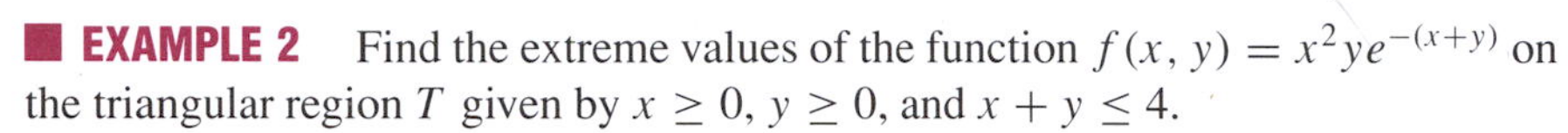

**EXAMPLE 2** Find the extreme values of the function $f(x, y) = x^2 y e^{-(x+y)}$ on the triangular region $T$ given by $x \ge 0$, $y \ge 0$, and $x + y \le 4$.

***SOLUTION*** First we look for critical points:

$$\begin{aligned} 0 = f_1(x, y) = xy(2 - x)e^{-(x+y)} \quad &\iff \quad x = 0, \ y = 0, \text{ or } x = 2, \\ 0 = f_2(x, y) = x^2(1 - y)e^{-(x+y)} \quad &\iff \quad x = 0 \text{ or } y = 1. \end{aligned}$$

The critical points are $(0, y)$ for any $y$ and (2,1). Only (2,1) is an interior point of $T$. (See Figure 5.8.) $f(2, 1) = 4/e^3 \approx 0.199$. The boundary of $T$ consists of three straight line segments. On two of these, the coordinate axes, $f$ is identically zero. The third segment is given by

$$y = 4 - x, \qquad 0 \le x \le 4,$$

so the values of $f$ on this segment can be expressed as a function of $x$ alone:

$$g(x) = f(x, 4 - x) = x^2(4 - x)e^{-4}, \qquad 0 \le x \le 4.$$

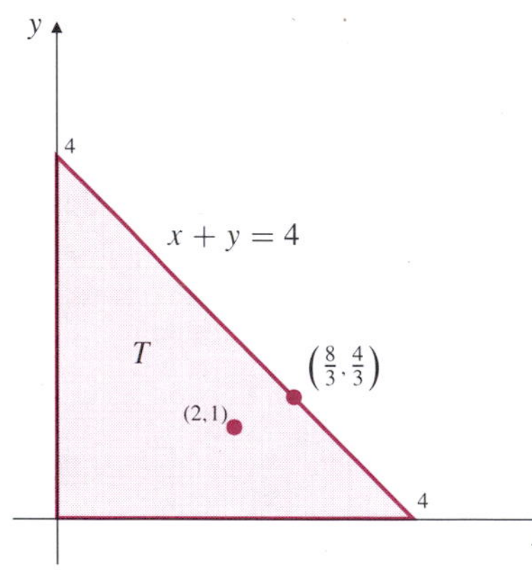


**Figure 5.8** Points of interest in Example 2

Note that $g(0) = g(4) = 0$ and $g(x) > 0$ if $0 < x < 4$. The critical points of $g$ are given by $0 = g'(x) = (8x - 3x^2)e^{-4}$, so they are $x = 0$ and $x = 8/3$. We have

$$g\Big(\frac{8}{3}\Big) = f\Big(\frac{8}{3}, \frac{4}{3}\Big) = \frac{256}{27}e^{-4} \approx 0.174 < f(2, 1).$$

We conclude that the maximum value of $f$ over the region $T$ is $4/e^3$, and that it occurs at the interior critical point (2,1). The minimum value of $f$ is zero and occurs at all points of the two perpendicular boundary segments. Note that $f$ has neither a local maximum nor a local minimum at the boundary point $(8/3, 4/3)$, although $g$ has a local maximum there. Of course that point is not a saddle point of $f$ either. It is not a critical point of $f$. ■

**EXAMPLE 3** Among all triangles with vertices on the unit circle $x^2 + y^2 = 1$, find those that have largest area.

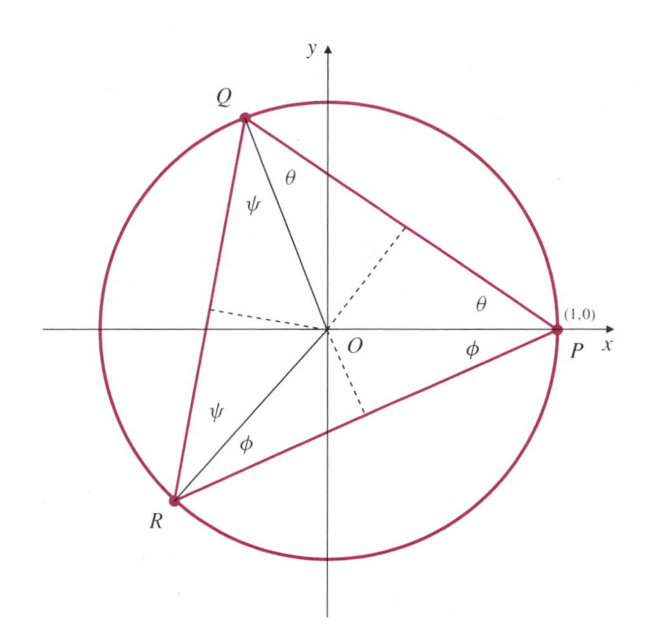


**Figure 5.9** Where should $Q$ and $R$ be to ensure that triangle $PQR$ has maximum area?

***SOLUTION*** Intuition tells us that the equilateral triangles must have the largest area. However, proving this can be quite difficult unless a good choice of variables in which to set up the problem analytically is made. Let one vertex of the triangle be the point $P$ with coordinates (1,0) and let the other two vertices, $Q$ and $R$, be as shown in Figure 5.9. There is no harm in assuming that $Q$ lies on the upper semicircle and $R$ on the lower, and that the origin $O$ is inside triangle $PQR$. Let $PQ$ and $PR$ make angles $\theta$ and $\phi$, respectively, with the negative direction of the $x$-axis. Clearly $0 \le \theta \le \pi/2$ and $0 \le \phi \le \pi/2$. The lines from $O$ to $Q$ and $R$ make equal angles $\psi$ with the line $QR$, where $2\theta + 2\phi + 2\psi = \pi$. Dropping perpendiculars from $O$ to the three sides of the triangle $PQR$, we can write the area $A$ of the triangle as the sum of the areas of six small, right-angled triangles:

$$\begin{aligned} A &= 2 \times \frac{1}{2} \sin\theta \cos\theta + 2 \times \frac{1}{2} \sin\phi \cos\phi + 2 \times \frac{1}{2} \sin\psi \cos\psi \\ &= \frac{1}{2}\big(\sin 2\theta + \sin 2\phi + \sin 2\psi\big). \end{aligned}$$

Since $2\psi = \pi - 2(\theta + \phi)$, we can express $A$ as a function of the two variables $\theta$ and $\phi$:

$$A = A(\theta, \phi) = \frac{1}{2}\big(\sin 2\theta + \sin 2\phi + \sin 2(\theta + \phi)\big).$$

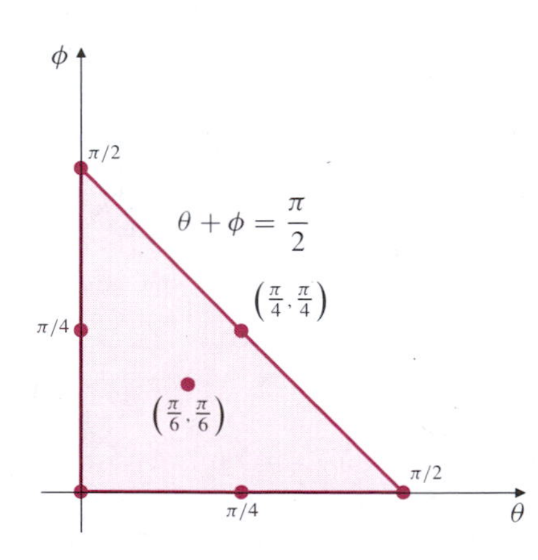


**Figure 5.10** The domain of $A(\theta, \phi)$

The domain of $A$ is the triangle $\theta \ge 0$, $\phi \ge 0$, $\theta + \phi \le \pi/2$. $A = 0$ at the vertices of the triangle and is positive elsewhere. (See Figure 5.10.) We show that the maximum value of $A(\theta, \phi)$ on any edge of the triangle is 1 and occurs at the midpoint of that edge. On the edge $\theta = 0$ we have

$$A(0, \phi) = \frac{1}{2}\big(\sin 2\phi + \sin 2\phi\big) = \sin 2\phi \le 1 = A(0, \pi/4).$$

Similarly, on $\phi = 0$, $A(\theta, 0) \le 1 = A(\pi/4, 0)$. On the edge $\theta + \phi = \pi/2$ we have

$$\begin{aligned} A\Big(\theta, \frac{\pi}{2} - \theta\Big) &= \frac{1}{2}\big(\sin 2\theta + \sin(\pi - 2\theta)\big) \\ &= \sin 2\theta \le 1 = A\Big(\frac{\pi}{4}, \frac{\pi}{4}\Big). \end{aligned}$$

We must now check for any interior critical points of $A(\theta, \phi)$. (There are no singular points.) For critical points we have

$$0 = \frac{\partial A}{\partial \theta} = \cos 2\theta + \cos(2\theta + 2\phi),$$
$$0 = \frac{\partial A}{\partial \phi} = \cos 2\phi + \cos(2\theta + 2\phi),$$

so the critical points satisfy $\cos 2\theta = \cos 2\phi$, and hence $\theta = \phi$. We now substitute this equation into either of the above equations to determine $\theta$:

$$\begin{aligned}
\cos 2\theta + \cos 4\theta &= 0 \\
2\cos^2 2\theta + \cos 2\theta - 1 &= 0 \\
(2\cos 2\theta - 1)(\cos 2\theta + 1) &= 0 \\
\cos 2\theta = \frac{1}{2} \quad \text{or} \quad \cos 2\theta &= -1.
\end{aligned}$$

The only solution leading to an interior point of the domain of $A$ is $\theta = \phi = \pi/6$. Note that

$$A\left(\frac{\pi}{6}, \frac{\pi}{6}\right) = \frac{1}{2}\left(\frac{\sqrt{3}}{2} + \frac{\sqrt{3}}{2} + \frac{\sqrt{3}}{2}\right) = \frac{3\sqrt{3}}{4} > 1,$$

this interior critical point maximizes the area of the inscribed triangle. Finally, observe that for $\theta = \phi = \pi/6$, we have $\psi = \pi/6$ also, so the largest triangle is indeed equilateral. ■

***REMARK*** Since the area $A$ of the inscribed triangle must have a maximum value ($A$ is continuous and its domain is closed and bounded), a strictly geometric argument can be used to show that the largest triangle is equilateral. If an inscribed triangle has two unequal sides, its area can be made larger by moving the common vertex of these two sides along the circle to increase its perpendicular distance from the opposite side of the triangle.

## Linear Programming

Linear programming is a branch of linear algebra that develops systematic techniques for finding maximum or minimum values of a *linear function* subject to several *linear inequality constraints*. Such problems arise frequently in management science and operations research. Because of their linear nature they do not usually involve calculus in their solution; linear programming is frequently presented in courses on *finite mathematics*. We will not attempt any formal study of linear programming here, but will make a few observations for comparison with the more general nonlinear extreme value problems considered above that involve calculus in their solution.

The inequality $ax+by \le c$ is an example of a linear inequality in two variables. The *solution set* of this inequality consists of a half-plane lying on one side of the straight line $ax + by = c$. The solution set of a system of several two-variable linear inequalities is an intersection of such half-planes, so it is a *convex* region of the plane bounded by a *polygonal line*. If it is a bounded set then it is a convex polygon together with its interior. (A set is called **convex** if it contains the entire line segment between any two of its points. On the real line the convex sets are intervals.)

Let us examine a simple concrete example that involves only two variables and a few constraints.

**EXAMPLE 4** Find the maximum value of $F(x, y) = 2x + 7y$ subject to the constraints

$$x + 2y \leq 6, \quad 2x + y \leq 6, \quad x \geq 0, \quad \text{and} \quad y \geq 0.$$

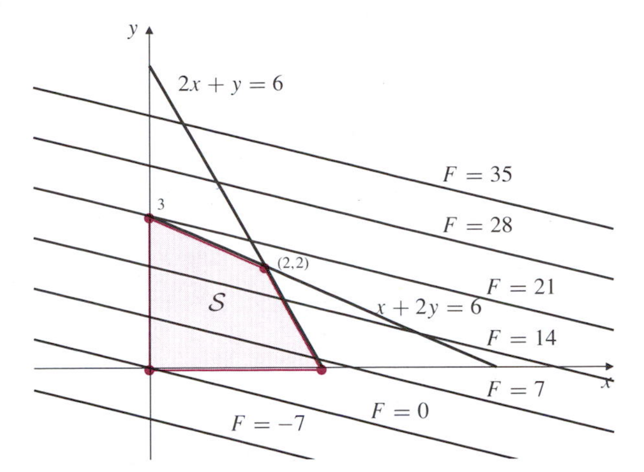


**Figure 5.11** The shaded region is the solution set for the constraint inequalities in Example 4

***SOLUTION*** The solution set $\mathcal{S}$ of the system of four constraint equations is shown in Figure 5.11. It is the quadrilateral region with vertices (0,0), (3,0), (2,2), and (0,3). Several level curves of the linear function $F$ are also shown in the figure. They are parallel straight lines with slope $-\frac{2}{7}$. We want the line that gives $F$ the greatest value and that still intersects $\mathcal{S}$. Evidently this is the line $F = 21$ that passes through the vertex (0,3) of $\mathcal{S}$. The maximum value of $F$ subject to the constraints is 21. ■

As this simple example illustrates, a *linear* function with domain restricted by *linear inequalities* does not achieve maximum or minimum values at points in the interior of its domain (if that domain has an interior). Any such extreme value occurs at a boundary point of the domain, or a set of such boundary points. Where an extreme value occurs at a set of boundary points that set will *always* contain at least one vertex. This phenomenon holds in general for extreme value problems for linear functions in any number of variables with domains restricted by any number of linear inequalities. For problems involving three variables the domain will be a convex region of $\mathbb{R}^3$ bounded by planes. For a problem involving $n$ variables the domain will be a convex region in $\mathbb{R}^n$ bounded by $(n-1)$-dimensional hyperplanes. Such *polyhedral* regions still have vertices (where $n$ hyperplanes intersect), and maximum or minimum values of linear functions subject to the constraints will still occur at subsets of the boundary containing such vertices. These problems can therefore be solved by evaluating the linear function to be extremized (it is called the **objective function**) at all the vertices and selecting the greatest or least value.

In practice, linear programming problems can involve hundreds or even thousands of variables and even more constraints. Such problems need to be solved with computers, but even then it is extremely inefficient, if not impossible, to calculate all the vertices of the constraint solution set and the values of the objective function at them. Much of the study of linear programming therefore centres on devising techniques for getting to (or at least near) the optimizing vertex in as few steps as possible. Usually this involves criteria whereby large numbers of vertices can be rejected on geometric grounds. We will not delve into such techniques here, but will content ourselves with one more example to illustrate, in a very simple case, how the underlying geometry of a problem can be used to reduce the number of vertices that must be considered.

**EXAMPLE 5** A tailor has 230 m of a certain fabric and has orders for up to 20 suits, up to 30 jackets, and up to 40 pairs of slacks to be made from the fabric. Each suit requires 6 m, each jacket 3 m, and each pair of slacks 2 m of the fabric. If the tailor's profit is \$20 per suit, \$14 per jacket, and \$12 per pair of slacks, how many of each should he make to realize the maximum profit from his supply of the fabric?

***SOLUTION*** Suppose he makes $x$ suits, $y$ jackets, and $z$ pairs of slacks. Then his profit will be

$$P = 20x + 14y + 12z.$$

The constraints posed in the problem are

$$\begin{aligned} x &\geq 0, & x &\leq 20, \\ y &\geq 0, & y &\leq 30, \\ z &\geq 0, & z &\leq 40, \end{aligned}$$

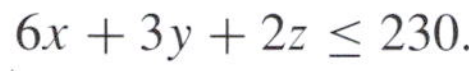

$$6x + 3y + 2z \leq 230.$$

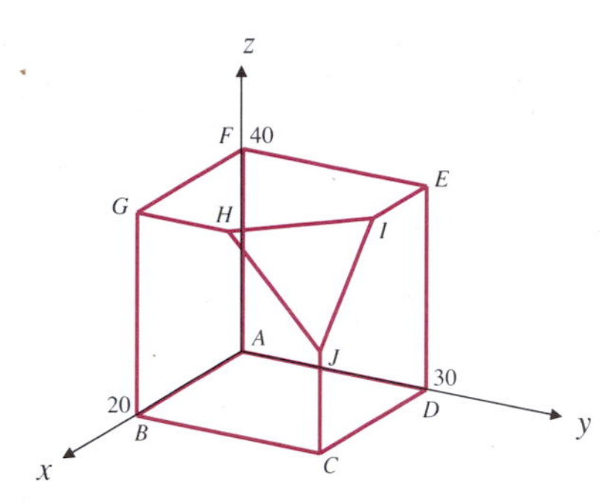


**Figure 5.12** The convex set of points satisfying the constraints in Example 5

The last inequality is due to the limited supply of fabric. The solution set is shown in Figure 5.12. It has ten vertices, $A, B, \ldots, J$. Since $P$ increases in the direction of the vector $\nabla P = 20\mathbf{i} + 14\mathbf{j} + 12\mathbf{k}$ which points into the first octant, its maximum value cannot occur at any of the vertices $A, B, \ldots, G$. (Think about why.) Thus we need look only at the vertices $H$, $I$, and $J$.

$$\begin{aligned} H &= (20, 10, 40), & P &= 1020 \text{ at } H. \\ I &= (10, 30, 40), & P &= 1100 \text{ at } I. \\ J &= (20, 30, 10), & P &= 940 \text{ at } J. \end{aligned}$$

Thus the tailor should make 10 suits, 30 jackets and 40 pairs of slacks to realize the maximum profit, \$1,100, from the fabric. ■

## EXERCISES 5.2

1. Find the maximum and minimum values of $f(x, y) = x - x^2 + y^2$ on the rectangle $0 \leq x \leq 2$, $0 \leq y \leq 1$.

2. Find the maximum and minimum values of $f(x, y) = xy - 2x$ on the rectangle $-1 \leq x \leq 1, 0 \leq y \leq 1$.

3. Find the maximum and minimum values of $f(x, y) = xy - y^2$ on the disk $x^2 + y^2 \leq 1$.

4. Find the maximum and minimum values of $f(x, y) = x + 2y$ on the disk $x^2 + y^2 \leq 1$.

5. Find the maximum value of $f(x, y) = xy - x^3y^2$ over the square $0 \leq x \leq 1, 0 \leq y \leq 1$.

6. Find the maximum and minimum values of $f(x, y) = xy(1 - x - y)$ over the triangle with vertices (0,0), (1,0), and (0,1).

7. Find the maximum and minimum values of $f(x, y) = \sin x \cos y$ on the closed triangular region bounded by the coordinate axes and the line $x + y = 2\pi$.

8. Find the maximum value of
$$f(x, y) = \sin x \sin y \sin(x + y)$$
over the triangle bounded by the coordinate axes and the line $x + y = \pi$.

9. The temperature at all points in the disk $x^2 + y^2 \leq 1$ is given by
$$T = (x + y)\, e^{-x^2-y^2}.$$
Find the maximum and minimum temperatures at points of the disk.

10. Find the maximum and minimum values of
$$f(x, y) = \frac{x - y}{1 + x^2 + y^2}$$
on the upper half-plane $y \geq 0$.

**11.** Find the maximum and minimum values of $xy^2 + yz^2$ over the ball $x^2 + y^2 + z^2 \leq 1$.

**12.** Find the maximum and minimum values of $xz + yz$ over the ball $x^2 + y^2 + z^2 \leq 1$.

**13.** Consider the function $f(x, y) = xy\, e^{-xy}$ with domain the first quadrant: $x \geq 0,\ y \geq 0$. Show that $\lim_{x\to\infty} f(x, kx) = 0$. Does $f$ have a limit as $(x, y)$ recedes arbitrarily far from the origin in the first quadrant? Does $f$ have a maximum value in the first quadrant?

**14.** Repeat Exercise 13 for the function $f(x, y) = xy^2 e^{-xy}$.

**15.** In a certain community there are two breweries in competition, so that sales of each negatively affect the profits of the other. If brewery A produces $x$ litres of beer per month and brewery B produces $y$ litres per month, then brewery A's monthly profit $\$P$ and brewery B's monthly profit $\$Q$ are assumed to be

$$P = 2x - \frac{2x^2 + y^2}{10^6},$$
$$Q = 2y - \frac{4y^2 + x^2}{2 \times 10^6}.$$

Find the sum of the profits of the two breweries if each brewery independently sets its own production level to maximize its own profit, and assumes its competitor does likewise. Find the sum of the profits if the two breweries cooperate to determine their respective productions to maximize that sum.

**16.** Equal angle bends are made at equal distances from the two ends of a 100 m long straight length of fence so the resulting three segment fence can be placed along an existing wall to make an enclosure of trapezoidal shape. What is the largest possible area for such an enclosure?

**17.** Maximize $Q(x, y) = 2x + 3y$ subject to the constraints $x \geq 0$, $y \geq 0$, $y \leq 5$, $x + 2y \leq 12$, and $4x + y \leq 12$.

**18.** Minimize $F(x, y, z) = 2x + 3y + 4z$ subject to $x \geq 0$, $y \geq 0$, $z \geq 0$, $x + y \geq 2$, $y + z \geq 2$, and $x + z \geq 2$.

**19.** A textile manufacturer produces two grades of wool-cotton-polyester fabric. The deluxe grade has composition (by weight) 20% wool, 50% cotton, and 30% polyester, and it sells for \$3 per kilogram. The standard grade has composition 10% wool, 40% cotton, and 50% polyester, and sells for \$2 per kilogram. If he has in stock 2000 kg of wool and 6000 kg each of cotton and polyester, how many kilograms of fabric of each grade should he manufacture to maximize his revenue?

**20.** A 10 hectare parcel of land is zoned for building densities of 6 detached houses per hectare, 8 duplex units per hectare, or 12 apartments per hectare. The developer who owns the land can make a profit of \$40,000 per house, \$20,000 per duplex unit, and \$16,000 per apartment that he builds. Municipal bylaws require him to build at least as many apartments as houses or duplex units. How many of each type of dwelling should he build to maximize his profit?

## 5.3 LAGRANGE MULTIPLIERS

A constrained extreme-value problem is one in which the variables of the function to be maximized or minimized are not completely independent of one another, but must satisfy one or more constraint equations or inequalities. For instance, the problems

$$\textbf{maximize} \quad f(x, y) \quad \textbf{subject to} \quad g(x, y) = C,$$

and

$$\textbf{minimize} \quad f(x, y, z, w) \quad \textbf{subject to} \quad g(x, y, z, w) = C_1,$$
$$\textbf{and} \quad h(x, y, z, w) = C_2$$

have, respectively, one and two constraint equations, while the problem

$$\textbf{maximize} \quad f(x, y, z) \quad \textbf{subject to} \quad g(x, y, z) \leq C$$

has a single constraint inequality.

Generally, inequality constraints can be regarded as restricting the domain of the function to be extremized to a smaller set that still has interior points. Section 5.2 was devoted to such problems. In each of the first three examples of that section we looked for *free* (that is, *unconstrained*) extreme values in the interior of the domain, and then examined the boundary of the domain, which was specified by one or more *constraint equations*. In Example 1 we parametrized the boundary and expressed the function to be extremized as a function of the parameter, thus reducing the boundary case to a free problem in one variable instead of a constrained problem in two variables. In Example 2 the boundary consisted of three line segments, on two of which the function was obviously zero. We solved the equation for the third boundary segment for $y$ in terms of $x$, again in order to express the values of $f(x, y)$ on that segment as a function of one free variable. A similar approach was used in Example 3 to deal with the triangular boundary of the domain of the area function $A(\theta, \phi)$.

The reduction of extremization problems with equation constraints to free problems with fewer independent variables is only feasible when the constraint equations can be solved either explicitly for some variables in terms of others, or parametrically for all variables in terms of some parameters. It is often very difficult or impossible to solve the constraint equations, so we need another technique.

## The Method of Lagrange Multipliers

A technique for finding extreme values of $f(x, y)$ subject to the equality constraint $g(x, y) = 0$ can be based on the following theorem.

**THEOREM 4**

Suppose that $f$ and $g$ have continuous first partial derivatives near the point $P_0 = (x_0, y_0)$ on the curve $\mathcal{C}$ with equation $g(x, y) = 0$. Suppose also that, when restricted to points on $\mathcal{C}$, the function $f(x, y)$ has a local maximum or minimum value at $P_0$. Finally, suppose that

(i) $P_0$ is not an endpoint of $\mathcal{C}$, and

(ii) $\nabla g(P_0) \neq \mathbf{0}$.

Then there exists a number $\lambda_0$ such that $(x_0, y_0, \lambda_0)$ is a critical point of the **Lagrangian function**

$$L(x, y, \lambda) = f(x, y) + \lambda g(x, y).$$

***PROOF*** Together, (i) and (ii) imply that $\mathcal{C}$ is smooth enough to have a tangent line at $P_0$, and that $\nabla g(P_0)$ is normal to that tangent line. If $\nabla f(P_0)$ is not parallel to $\nabla g(P_0)$, then $\nabla f(P_0)$ has a nonzero vector projection $\mathbf{v}$ along the tangent line to $\mathcal{C}$ at $P_0$. (See Figure 5.13.) Therefore $f$ has a positive directional derivative at $P_0$ in the direction of $\mathbf{v}$, and a negative directional derivative in the opposite direction. Thus $f(x, y)$ increases or decreases as we move away from $P_0$ along $\mathcal{C}$ in the direction of $\mathbf{v}$ or $-\mathbf{v}$, and $f$ cannot have a maximum or minimum value at $P_0$. Since we are assuming that $f$ *does* have an extreme value at $P_0$, it must be that $\nabla f(P_0)$ is parallel to $\nabla g(P_0)$. Since $\nabla g(P_0) \neq \mathbf{0}$, there must exist a real number $\lambda_0$ such that $\nabla f(P_0) = -\lambda_0 \nabla g(P_0)$, or

$$\nabla(f + \lambda_0 g)(P_0) = \mathbf{0}.$$

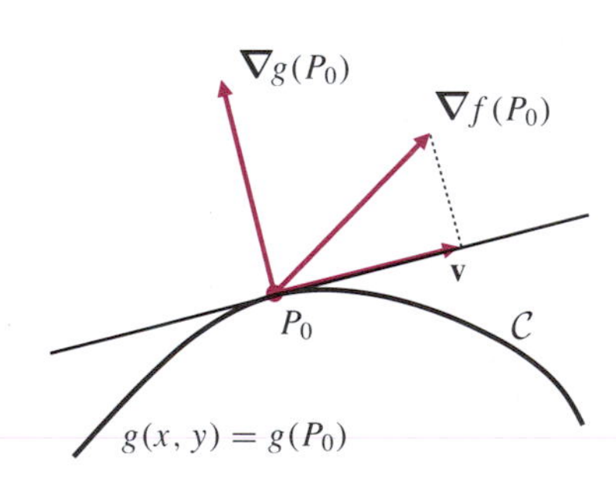


**Figure 5.13** If $\nabla f(P_0)$ is not a multiple of $\nabla g(P_0)$, then $\nabla f(P_0)$ has a nonzero projection $\mathbf{v}$ tangent to the level curve of $g$ through $P_0$

The two components of the above vector equation assert that $\partial L/\partial x = 0$ and $\partial L/\partial y = 0$ at $(x_0, y_0, \lambda_0)$. The third equation that must be satisfied by a critical point of $L$ is $\partial L/\partial \lambda = g(x, y) = 0$. This is satisfied at $(x_0, y_0, \lambda_0)$ because $P_0$ lies on $\mathcal{C}$. Thus $(x_0, y_0, \lambda_0)$ is a critical point of $L(x, y, \lambda)$.

Theorem 4 suggests that to find candidates for points on the curve $g(x, y) = 0$ at which $f(x, y)$ is maximum or minimum, we should look for critical points of the Lagrangian function

$$L(x, y, \lambda) = f(x, y) + \lambda g(x, y).$$

At any critical point of $L$ we must have

$$\left.\begin{aligned} 0 &= \frac{\partial L}{\partial x} = f_1(x, y) + \lambda g_1(x, y), \\ 0 &= \frac{\partial L}{\partial y} = f_2(x, y) + \lambda g_2(x, y), \end{aligned}\right\} \text{ i.e., } \nabla f \text{ is parallel to } \nabla g,$$

$$\text{and} \quad 0 = \frac{\partial L}{\partial \lambda} = g(x, y) \qquad \text{the constraint equation.}$$

Note, however, that it is *assumed* that the constrained problem *has a solution*. Theorem 4 does not guarantee that a solution exists; it only provides a means for finding a solution already known to exist. It is usually necessary to satisfy yourself that the problem you are trying to solve has a solution before using this method to find the solution.

Let us put the method to a concrete test:

**EXAMPLE 1** Find the shortest distance from the origin to the curve $x^2 y = 16$.

***SOLUTION*** The graph of $x^2 y = 16$ is shown in Figure 5.14. There appear to be two points on the curve that are closest to the origin, and no points that are farthest from the origin. (The curve is unbounded.) To find the closest points it is sufficient to minimize the square of the distance from the point $(x, y)$ on the curve to the origin, that is, to solve the problem

$$\textbf{minimize} \quad f(x, y) = x^2 + y^2 \quad \textbf{subject to} \quad g(x, y) = x^2 y - 16 = 0.$$

Let $L(x, y, \lambda) = x^2 + y^2 + \lambda(x^2 y - 16)$. For critical points of $L$ we want

$$0 = \frac{\partial L}{\partial x} = 2x + 2\lambda xy = 2x(1 + \lambda y) \qquad \text{(A)}$$

$$0 = \frac{\partial L}{\partial y} = 2y + \lambda x^2 \qquad \text{(B)}$$

$$0 = \frac{\partial L}{\partial \lambda} = x^2 y - 16. \qquad \text{(C)}$$

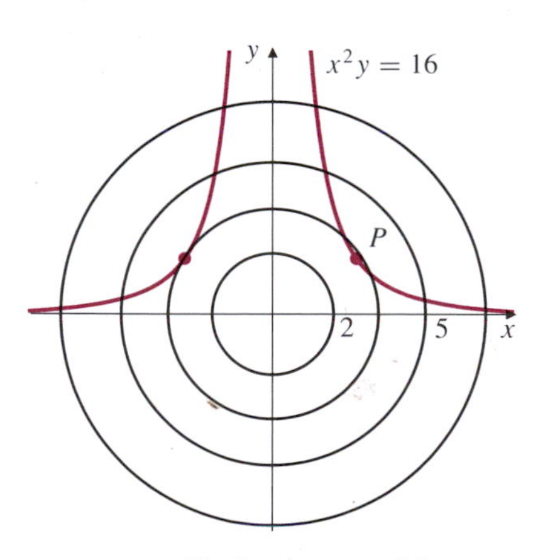


**Figure 5.14** The level curve of the function representing the square of distance from the origin is tangent to the curve $x^2 y = 16$ at the two points on that curve that are closest to the origin

Equation (A) requires that either $x = 0$ or $\lambda y = -1$. However, $x = 0$ is inconsistent with equation (C). Therefore $\lambda y = -1$. From equation (B) we now have

$$0 = 2y^2 + \lambda y x^2 = 2y^2 - x^2.$$

Thus $x = \pm\sqrt{2}y$, and (C) now gives $2y^3 = 16$, so $y = 2$. There are, therefore, two candidates for points on $x^2 y = 16$ closest to the origin, $(\pm 2\sqrt{2}, 2)$. Both of these points are at distance $\sqrt{8 + 4} = 2\sqrt{3}$ units from the origin, so this must be the minimum distance from the origin to the curve. Some level curves of $x^2 + y^2$ are shown, along with the constraint curve $x^2 y = 16$ in Figure 5.14. Observe how the constraint curve is tangent to the level curve passing through the minimizing points $(\pm 2\sqrt{2}, 2)$, reflecting the fact that the two curves have parallel normals there. ■

**REMARK** In the above example we could, of course, have solved the constraint equation for $y = 16/x^2$, substituted into $f$, and thus reduced the problem to one of finding the (unconstrained) minimum value of

$$F(x) = f\left(x, \frac{16}{x^2}\right) = x^2 + \frac{256}{x^4}.$$

The reader is invited to verify that this gives the same result.

The number $\lambda$ that occurs in the Lagrangian function is called a **Lagrange multiplier**. The technique for solving an extreme-value problem with equation constraints by looking for critical points of an unconstrained problem in more variables (the original variables plus a Lagrange multiplier corresponding to each constraint equation) is called **the method of Lagrange multipliers.** It can be expected to give results as long as the function to be maximized or minimized (called the **objective function**) and the constraint equations have *smooth* graphs in a neighbourhood of the points where the extreme values occur, and these points are not on *edges* of those graphs. See Example 3 and Exercise 26 at the end of this section.

**EXAMPLE 2** Find the points on the curve $17x^2 + 12xy + 8y^2 = 100$ that are closest to and farthest away from the origin.

**SOLUTION** The quadratic form on the left side of the equation above is positive definite as can be seen by completing a square. Hence the curve is bounded and must have points closest to and farthest from the origin. (In fact, the curve is an ellipse with centre at the origin and oblique principal axes. The problem asks us to find the ends of the major and minor axes.)

Again we want to extremize $x^2 + y^2$ subject to an equation constraint. The Lagrangian in this case is

$$L(x, y, \lambda) = x^2 + y^2 + \lambda(17x^2 + 12xy + 8y^2 - 100),$$

and its critical points are given by

$$0 = \frac{\partial L}{\partial x} = 2x + \lambda(34x + 12y) \qquad \text{(A)}$$

$$0 = \frac{\partial L}{\partial y} = 2y + \lambda(12x + 16y) \qquad \text{(B)}$$

$$0 = \frac{\partial L}{\partial \lambda} = 17x^2 + 12xy + 8y^2 - 100. \qquad \text{(C)}$$

Solving each of equations (A) and (B) for $\lambda$ and equating the two expressions for $\lambda$ obtained, we get

$$\frac{-2x}{34x + 12y} = \frac{-2y}{12x + 16y} \qquad \text{or} \qquad 12x^2 + 16xy = 34xy + 12y^2.$$

This equation simplifies to

$$2x^2 - 3xy - 2y^2 = 0. \qquad \text{(D)}$$

We multiply equation (D) by 4 and add the result to equation (C) to get $25x^2 = 100$, so that $x = \pm 2$. Finally, we substitute each of these values of $x$ into (D) and obtain (for each) two values of $y$ from the resulting quadratics:

$$\text{For } x = 2: \quad y^2 + 3y - 4 = 0, \quad (y - 1)(y + 4) = 0. \qquad \text{For } x = -2: \quad y^2 - 3y - 4 = 0, \quad (y + 1)(y - 4) = 0.$$

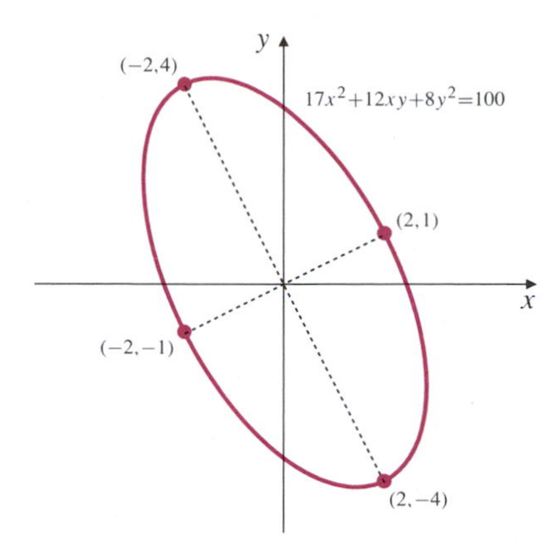


**Figure 5.15** The points on the ellipse that are closest to and farthest from the origin

We therefore obtain four candidate points: $(2, 1)$, $(-2, -1)$, $(2, -4)$, and $(-2, 4)$. The first two points are closest to the origin (they are the ends of the minor axis of the ellipse); the second pair are farthest from the origin (the ends of the major axis). (See Figure 5.15.) ■

Considering the geometric underpinnings of the method of Lagrange multipliers, we would not expect the method to work if the level curves of the functions involved are not smooth, or if the maximum or minimum occurs at an endpoint of the constraint curve. One of the pitfalls of the method is that the level curves of functions may not be smooth, even though the functions themselves are differentiable. Problems can occur where a gradient vanishes, as the following example shows.

■ **EXAMPLE 3** Find the minimum value of $f(x, y) = y$ subject to the constraint equation $g(x, y) = y^3 - x^2 = 0$.

***SOLUTION*** The semicubical parabola $y^3 = x^2$ has a cusp at the origin. (See Figure 5.16.) Clearly, $f(x, y) = y$ has minimum value 0 at that point. Suppose, however, that we try to solve the problem using the method of Lagrange multipliers. The Lagrangian here is

$$L(x, y, \lambda) = y + \lambda(y^3 - x^2),$$

which has critical points given by

$$\begin{aligned} -2\lambda x &= 0, \\ 1 + 3\lambda y^2 &= 0, \\ y^3 - x^2 &= 0. \end{aligned}$$

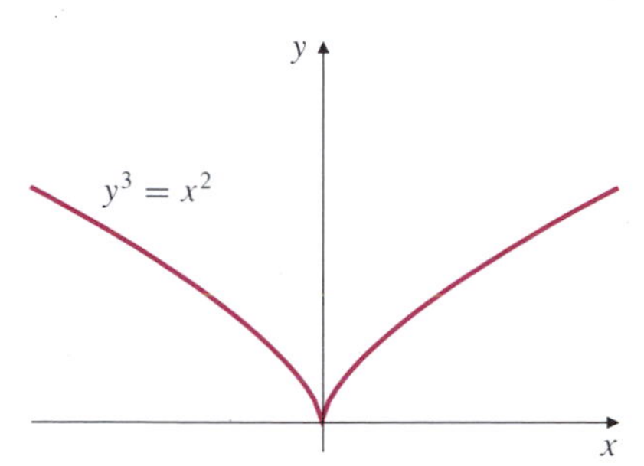


**Figure 5.16** The minimum of $y$ occurs at a point on the curve where the curve has no tangent line

Observe that $y = 0$ cannot satisfy the second equation, and, in fact, the three equations have *no solution* $(x, y, \lambda)$. ■

***REMARK*** The method of Lagrange multipliers breaks down in the above example because $\nabla g = \mathbf{0}$ at the solution point, and therefore the curve $g(x, y) = 0$ need not be smooth there. (In this case, it isn't smooth!) The geometric condition that $\nabla f$ should be parallel to $\nabla g$ at the solution point is meaningless in this case. When applying the method of Lagrange multipliers, be aware that an extreme value may occur at

(i) a critical point of the Lagrangian,

(ii) a point where $\nabla g = \mathbf{0}$,

(iii) a point where $\nabla f$ or $\nabla g$ does not exist, or

(iv) an "endpoint" of the constraint set.

This situation is similar to that for extreme values of a function $f$ of one variable, which can occur at a critical point of $f$, a singular point of $f$, or an endpoint of the domain of $f$.

## Problems With More Than One Constraint

Next consider a three-dimensional problem requiring us to find a maximum or minimum value of a function of three variables subject to two equation constraints:

$$\textbf{extremize} \quad f(x, y, z) \quad \textbf{subject to} \quad g(x, y, z) = 0 \text{ and } h(x, y, z) = 0.$$

Again we assume that the problem has a solution, say at the point $P_0 = (x_0, y_0, z_0)$, that the functions $f$, $g$ and $h$ have continuous first partial derivatives near $P_0$. Also, we assume that $\mathbf{T} = \nabla g(P_0) \times \nabla h(P_0) \neq \mathbf{0}$. These conditions imply that the surfaces $g(x, y, z) = 0$ and $h(x, y, z) = 0$ are smooth near $P_0$ and are not tangent to each other there, so they must intersect in a curve $\mathcal{C}$ that is smooth near $P_0$. The curve $\mathcal{C}$ has tangent vector $\mathbf{T}$ at $P_0$. The same geometric argument used in the proof of Theorem 4 again shows that $\nabla f(P_0)$ must be perpendicular to $\mathbf{T}$. (If not, then it would have a nonzero vector projection along $\mathbf{T}$, and $f$ would have a nonzero directional derivative in the directions $\pm\mathbf{T}$ and would therefore increase and decrease as we moved away from $P_0$ along $\mathcal{C}$ in opposite directions.)

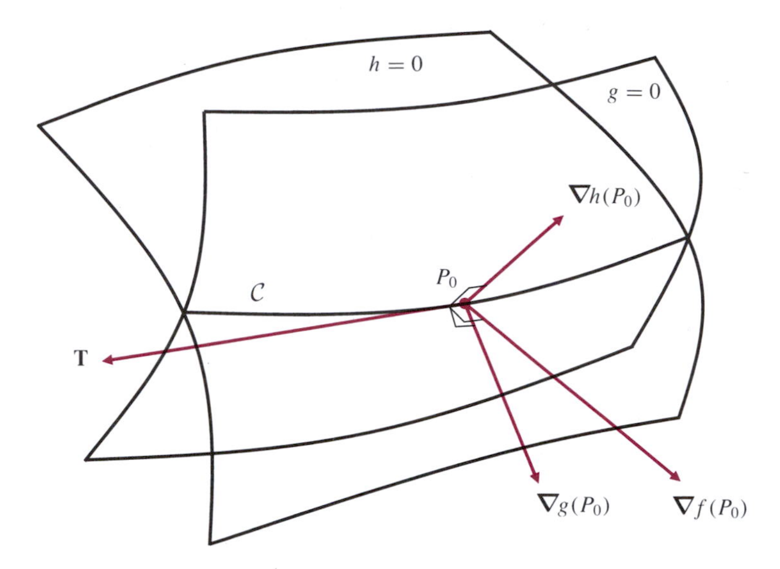


**Figure 5.17** At $P_0$, $\nabla f$, $\nabla g$, and $\nabla h$ are all perpendicular to $\mathbf{T}$. Thus $\nabla f$ is in the plane spanned by $\nabla g$ and $\nabla h$

Since $\nabla g(P_0)$ and $\nabla h(P_0)$ are nonzero and both are perpendicular to $\mathbf{T}$, $\nabla f(P_0)$ must lie in the plane spanned by these two vectors and hence must be a linear combination of them:

$$\nabla f(x, y, z) = -\lambda_0 \nabla g(x, y, z) - \mu_0 \nabla h(x, y, z)$$

for some constants $\lambda_0$ and $\mu_0$. It follows that $(x_0, y_0, z_0, \lambda_0, \mu_0)$ is a critical point of the Lagrangian function

$$L(x, y, z, \lambda, \mu) = f(x, y, z) + \lambda g(x, y, z) + \mu h(x, y, z).$$

We look for triples $(x, y, z)$ that extremize $f(x, y, z)$ subject to the two constraints $g(x, y, z) = 0$ and $h(x, y, z) = 0$ among the points $(x, y, z, \lambda, \mu)$ that are critical points of the above Lagrangian function, and we therefore solve the system of equations

$$\begin{aligned}
f_1(x, y, z) + \lambda g_1(x, y, z) + \mu h_1(x, y, z) &= 0, \\
f_2(x, y, z) + \lambda g_2(x, y, z) + \mu h_2(x, y, z) &= 0, \\
f_3(x, y, z) + \lambda g_3(x, y, z) + \mu h_3(x, y, z) &= 0, \\
g(x, y, z) &= 0, \\
h(x, y, z) &= 0.
\end{aligned}$$

Solving such a system can be very difficult. It should be noted that in using the method of Lagrange multipliers instead of solving the constraint equations, we have traded the problem of having to solve two equations for two variables as *functions* of a third one for a problem of having to solve five equations for *numerical* values of five unknowns.

**EXAMPLE 4** If $f(x, y, z) = xy + 2z$, find the maximum and minimum values of $f$ on the circle that is the intersection of the plane $x + y + z = 0$ and the sphere $x^2 + y^2 + z^2 = 24$.

***SOLUTION*** The function $f$ is continuous and the circle is a closed bounded set in 3-space. Therefore maximum and minimum values must exist. We look for critical points of the Lagrangian

$$L = xy + 2z + \lambda(x + y + z) + \mu(x^2 + y^2 + z^2 - 24).$$

Setting the first partial derivatives of $L$ equal to zero, we obtain

$$y + \lambda + 2\mu x = 0, \qquad \text{(A)}$$
$$x + \lambda + 2\mu y = 0, \qquad \text{(B)}$$
$$2 + \lambda + 2\mu z = 0, \qquad \text{(C)}$$
$$x + y + z = 0, \qquad \text{(D)}$$
$$x^2 + y^2 + z^2 - 24 = 0. \qquad \text{(E)}$$

When none of the equations factors, try to combine two or more of them to produce an equation that does factor.

Subtracting (A) from (B) we get $(x - y)(1 - 2\mu) = 0$. Therefore either $\mu = \frac{1}{2}$ or $x = y$. We analyze both possibilities.

If $\mu = \frac{1}{2}$ we obtain from (B) and (C)

$$x + \lambda + y = 0 \qquad \text{and} \qquad 2 + \lambda + z = 0.$$

Thus $x + y = 2 + z$. Combining this with (D) we get $z = -1$ and $x + y = 1$. Now by (E), $x^2 + y^2 = 24 - z^2 = 23$. Since $x^2 + y^2 + 2xy = (x + y)^2 = 1$, we have $2xy = 1 - 23 = -22$ and $xy = -11$. Now $(x - y)^2 = x^2 + y^2 - 2xy = 23 + 22 = 45$, so $x - y = \pm 3\sqrt{5}$. Combining this with $x + y = 1$ we obtain two critical points arising from $\mu = \frac{1}{2}$, namely $\big((1 + 3\sqrt{5})/2, (1 - 3\sqrt{5})/2, -1\big)$ and $\big((1 - 3\sqrt{5})/2, (1 + 3\sqrt{5})/2, -1\big)$. At both of these points we find that $f(x, y, z) = xy + 2z = -11 - 2 = -13$.

If $x = y$, then (D) implies that $z = -2x$, and (E) then gives $6x^2 = 24$, so $x = \pm 2$. Therefore points $(2, 2, -4)$ and $(-2, -2, 4)$ must be considered. We have $f(2, 2, -4) = 4 - 8 = -4$ and $f(-2, -2, 4) = 4 + 8 = 12$.

We conclude that the maximum value of $f$ on the circle is 12 and the minimum value is $-13$. ■

The method of Lagrange multipliers can be applied to find extreme values of a function of $n$ variables,

$$f(x_1, x_2, \ldots, x_n),$$

subject to $m \le n - 1$ constraints,

$$g_{(1)}(x_1, x_2, \ldots, x_n) = 0,$$
$$g_{(2)}(x_1, x_2, \ldots, x_n) = 0,$$
$$\vdots$$
$$g_{(m)}(x_1, x_2, \ldots, x_n) = 0.$$

Assuming that the problem has a solution at the point $P_0$, that $f$ and all of the functions $g_{(j)}$ have continuous first partial derivatives at all points near $P_0$, and that the intersection of the constraint (hyper)surfaces is smooth near $P_0$, then we should look for $P_0$ among the critical points of the $(n+m)$-variable Lagrangian function

$$\begin{aligned} L(x_1,\ldots,x_n,\lambda_1,\ldots,\lambda_m) = & f(x_1,\ldots,x_n) + \lambda_1 g_{(1)}(x_1,\ldots,x_n) \\ & + \lambda_2 g_{(2)}(x_1,\ldots,x_n) + \cdots + \lambda_m g_{(m)}(x_1,\ldots,x_n). \end{aligned}$$

We will not attempt to prove this general assertion. (A proof can be based on the Implicit Function Theorem.)

## EXERCISES 5.3

1. Use the method of Lagrange multipliers to maximize $x^3y^5$ subject to the constraint $x+y=8$.
2. Find the shortest distance from the point $(3, 0)$ to the parabola $y = x^2$,
   (a) by reducing to an unconstrained problem in one variable, and
   (b) by using the method of Lagrange multipliers.
3. Find the distance from the origin to the plane $x+2y+2z=3$,
   (a) using a geometric argument (no calculus),
   (b) by reducing the problem to an unconstrained problem in two variables, and
   (c) using the method of Lagrange multipliers.
4. Find the maximum and minimum values of the function $f(x,y,z) = x+y-z$ over the sphere $x^2+y^2+z^2=1$.
5. Use the Lagrange multiplier method to find the greatest and least distances from the point $(2, 1, -2)$ to the sphere with equation $x^2+y^2+z^2=1$. (Of course, the answer could be obtained more easily using a simple geometric argument.)
6. Find the shortest distance from the origin to the surface $xyz^2 = 2$.
7. Find $a$, $b$, and $c$ so that the volume $V = 4\pi abc/3$ of an ellipsoid $\dfrac{x^2}{a^2}+\dfrac{y^2}{b^2}+\dfrac{z^2}{c^2}=1$ passing through the point $(1, 2, 1)$ is as small as possible.
8. Find the ends of the major and minor axes of the ellipse $3x^2+2xy+3y^2=16$.
9. Find the maximum and minimum values of $f(x,y,z)=xyz$ on the sphere $x^2+y^2+z^2=12$.
10. Find the maximum and minimum values of $x+2y-3z$ over the ellipsoid $x^2+4y^2+9z^2 \le 108$.
11. Find the maximum and minimum values of the function $f(x,y,z)=x$ over the curve of intersection of the plane $z=x+y$ and the ellipsoid $x^2+2y^2+2z^2=8$.
12. Find the maximum and minimum values of $f(x,y,z)=x^2+y^2+z^2$ on the ellipse formed by the intersection of the cone $z^2=x^2+y^2$ and the plane $x-2z=3$.
13. Find the maximum and minimum values of $f(x,y,z)=4-z$ on the ellipse formed by the intersection of the cylinder $x^2+y^2=8$ and the plane $x+y+z=1$.
14. Find the maximum and minimum values of $f(x,y,z)=x+y^2z$ subject to the constraints $y^2+z^2=2$ and $z=x$.
15. * Use the method of Lagrange multipliers to find the shortest distance between the straight lines $x=y=z$ and $x=-y$, $z=2$. (There are, of course, much easier ways to get the answer. This is an object lesson in the folly of shooting sparrows with cannons.)
16. Find the maximum and minimum values of the $n$-variable function $x_1+x_2+\cdots+x_n$ subject to the constraint $x_1^2+x_2^2+\cdots+x_n^2=1$.
17. Repeat Exercise 16 for the function $x_1+2x_2+3x_3+\cdots+nx_n$ with the same constraint.
18. Find the most economical shape of a rectangular box with no top.
19. Find the maximum volume of a rectangular box with faces parallel to the coordinate planes if one corner is at the origin and the diagonally opposite corner lies on the plane $4x+2y+z=2$.
20. Find the maximum volume of a rectangular box with faces parallel to the coordinate planes if one corner is at the origin and the diagonally opposite corner is on the first octant part of the surface $xy+2yz+3xz=18$.
21. A rectangular box having no top and having a prescribed volume $V$ m$^3$ is to be constructed using two different materials. The material used for the bottom and front of the box is five times as costly (per square metre) as the material used for the back and the other two sides. What should be the dimensions of the box to minimize the cost of materials?
22. * Find the maximum and minimum values of $xy+z^2$ on the ball $x^2+y^2+z^2 \le 1$. Use Lagrange multipliers to treat the boundary case.
23. * Repeat Exercise 22 but handle the boundary case by parametrizing the sphere $x^2+y^2+z^2=1$ using
$$x=\sin\phi\cos\theta, \quad y=\sin\phi\sin\theta, \text{ and } \quad z=\cos\phi,$$
where $0 \le \phi \le \pi$ and $0 \le \theta \le 2\pi$.

*** 24.** If $\alpha$, $\beta$, and $\gamma$ are the angles of a triangle, show that

$$\sin\frac{\alpha}{2}\sin\frac{\beta}{2}\sin\frac{\gamma}{2} \le \frac{1}{8}.$$

For what triangles does equality occur?

*** 25.** Suppose that $f$ and $g$ have continuous first partial derivatives throughout the $xy$-plane, and suppose that $g_2(a, b) \neq 0$. This implies that the equation $g(x, y) = g(a, b)$ defines $y$ implicitly as a function of $x$ nearby the point $(a, b)$. Use the Chain Rule to show that if $f(x, y)$ has a local extreme value at $(a, b)$ subject to the constraint $g(x, y) = g(a, b)$, then for some number $\lambda$ the point $(a, b, \lambda)$ is a critical point of the function

$$L(x, y, \lambda) = f(x, y) + \lambda g(x, y).$$

This constitutes a more formal justification of the method of Lagrange multipliers in this case.

**26.** What is the shortest distance from the point $(0, -1)$ to the curve $y = \sqrt{1 - x^2}$? Can this problem be solved by the Lagrange multiplier method? Why?

**27.** Example 3 showed that the method of Lagrange multipliers might fail to find a point that extremizes $f(x, y)$ subject to the constraint $g(x, y) = 0$ if $\nabla g = \mathbf{0}$ at the extremizing point. Can the method also fail if $\nabla f = \mathbf{0}$ at the extremizing point? Why?

# 5.4 THE METHOD OF LEAST SQUARES

Important optimization problems arise in the statistical analysis of experimental data. Frequently experiments are designed to measure the values of one or more quantities supposed constant, or to demonstrate a supposed functional relationship among variable quantities. Experimental error is usually present in the measurements, and experiments need to be repeated several times in order to arrive at *mean* or *average* values of the quantities being measured.

Consider a very simple example. An experiment to measure a certain physical constant $c$ is repeated $n$ times, yielding the values $c_1, c_2, \ldots, c_n$. If none of the measurements is suspected of being faulty, intuition tells us that we should use the mean value $\bar{c} = (c_1 + c_2 + \cdots + c_n)/n$ as the value of $c$ determined by the experiments. Let us see how this intuition can be justified.

Various methods for determining $c$ from the data values are possible. We could, for instance, choose $c$ to minimize the sum $T$ of its distances from the data points:

$$T = |c - c_1| + |c - c_2| + \cdots + |c - c_n|.$$

This is unsatisfactory for a number of reasons. Since absolute values have singular points it is difficult to determine the minimizing value of $c$. More importantly, $c$ may not be determined uniquely. If $n = 2$ any point in the interval between $c_1$ and $c_2$ will give the same minimum value to $T$. (See Exercise 24 at the end of this section for a generalization of this phenomenon.)

A more promising approach is to minimize the sum $S$ of *squares* of the distances from $c$ to the data points:

$$S = (c - c_1)^2 + (c - c_2)^2 + \cdots + (c - c_n)^2 = \sum_{i=1}^{n}(c - c_i)^2.$$

This function of $c$ is smooth and its (unconstrained) minimum value will occur at a critical point $\bar{c}$ given by

$$0 = \left.\frac{dS}{dc}\right|_{c=\bar{c}} = \sum_{i=1}^{n} 2(\bar{c} - c_i) = 2n\bar{c} - 2\sum_{i=1}^{n} c_i.$$

Thus $\bar{c}$ is the *mean* of the data values:

$$\bar{c} = \frac{1}{n}\sum_{i=1}^{n} c_i = \frac{c_1 + c_2 + \cdots + c_n}{n}.$$

The technique used to obtain $\bar{c}$ above is an example of what is called **the method of least squares.** It has the following geometric interpretation. If the data values $c_1, c_2, \ldots, c_n$ are regarded as components of a vector $\mathbf{c}$ in $\mathbb{R}^n$, and $\mathbf{w}$ is the vector with components $1, 1, \ldots, 1$, then the vector projection of $\mathbf{c}$ in the direction of $\mathbf{w}$,

$$\mathbf{c}_\mathbf{w} = \frac{\mathbf{c} \bullet \mathbf{w}}{|\mathbf{w}|^2}\mathbf{w} = \frac{c_1 + c_2 + \cdots + c_n}{n}\mathbf{w},$$

has all its components equal to the average of the data values. Thus determining $c$ from the data by the method of least squares corresponds to finding the vector projection of the data vector onto the one-dimensional subspace of $\mathbb{R}^n$ spanned by $\mathbf{w}$. Had there been no error in the measurements $c_i$, then $\mathbf{c}$ would have been equal to $c\mathbf{w}$.

## Linear Regression

In scientific investigations it is often believed that the response of a system is a certain kind of function of one or more input variables. An investigator can set up an experiment to measure the response of the system for various values of those variables in order to determine the parameters of the function.

For example, suppose that the response $y$ of a system is suspected to depend on the input $x$ according to the linear relationship

$$y = ax + b,$$

where the values of $a$ and $b$ are unknown. An experiment set up to measure values of $y$ corresponding to several values of $x$ yields $n$ data points, $(x_i, y_i)$, $i = 1, 2, \ldots, n$. If the supposed linear relationship is valid, these data points should lie *approximately* along a straight line, but not exactly on one because of experimental error. Suppose the points are as shown in Figure 5.18. The linear relationship seems reasonable in this case. We want to find values of $a$ and $b$ so that the straight line $y = ax + b$ "best" fits the data.

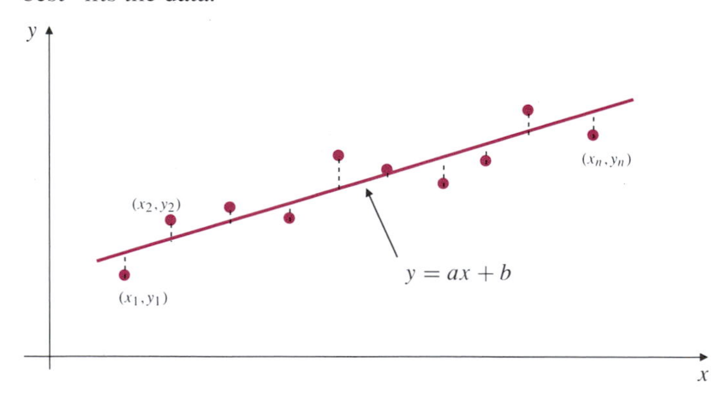


**Figure 5.18** Fitting a straight line through experimental data

In this situation the method of least squares requires that $a$ and $b$ be chosen to minimize the sum $S$ of the squares of the vertical displacements of the data points from the line:

$$S = \sum_{i=1}^{n}(y_i - ax_i - b)^2.$$

This is an unconstrained minimum problem in two variables, $a$ and $b$. The minimum will occur at a critical point of $S$ that satisfies

$$0 = \frac{\partial S}{\partial a} = -2\sum_{i=1}^{n} x_i(y_i - ax_i - b),$$

$$0 = \frac{\partial S}{\partial b} = -2\sum_{i=1}^{n}(y_i - ax_i - b).$$

These equations can be rewritten

$$\begin{aligned}\left(\sum_{i=1}^{n} x_i^2\right)a &+ \left(\sum_{i=1}^{n} x_i\right)b &= \sum_{i=1}^{n} x_i y_i,\\ \left(\sum_{i=1}^{n} x_i\right)a &+ nb &= \sum_{i=1}^{n} y_i.\end{aligned}$$

Solving this pair of linear equations we obtain the desired parameters:

$$a = \frac{n\left(\sum_{i=1}^{n} x_i y_i\right) - \left(\sum_{i=1}^{n} x_i\right)\left(\sum_{i=1}^{n} y_i\right)}{n\left(\sum_{i=1}^{n} x_i^2\right) - \left(\sum_{i=1}^{n} x_i\right)^2} = \frac{\overline{xy} - \bar{x}\bar{y}}{\overline{x^2} - (\bar{x})^2},$$

$$b = \frac{\left(\sum_{i=1}^{n} x_i^2\right)\left(\sum_{i=1}^{n} y_i\right) - \left(\sum_{i=1}^{n} x_i\right)\left(\sum_{i=1}^{n} x_i y_i\right)}{n\left(\sum_{i=1}^{n} x_i^2\right) - \left(\sum_{i=1}^{n} x_i\right)^2} = \frac{\overline{x^2}\bar{y} - \bar{x}\overline{xy}}{\overline{x^2} - (\bar{x})^2}.$$

In these formulas we have used a bar to indicate the mean value of a quantity; thus $\overline{xy} = (1/n)\sum_{i=0}^{n} x_i y_i$, and so on.

This procedure for fitting the "best" straight line through data points by the method of least squares is called **linear regression**, and the line $y = ax + b$ obtained in this way is called the **empirical regression line** corresponding to the data. Some scientific calculators with statistical features provide for linear regression by accumulating the sums of $x_i$, $y_i$, $x_i^2$, and $x_i y_i$ in various registers, and keeping track of the number $n$ of data points entered in another register. At any time it has available the information necessary to calculate $a$ and $b$, and the value of $y$ corresponding to any given $x$.

**EXAMPLE 1** Find the empirical regression line for the data $(x, y) = (0, 2.10)$, $(1,1.92)$, $(2,1.84)$, and $(3,1.71)$, $(4,1.64)$. What is the predicted value of $y$ at $x = 5$?

***SOLUTION*** We have

$$\bar{x} = \frac{0+1+2+3+4}{5} = 2,$$

$$\bar{y} = \frac{2.10 + 1.92 + 1.84 + 1.71 + 1.64}{5} = 1.842,$$

$$\overline{xy} = \frac{(0)(2.10) + (1)(1.92) + (2)(1.84) + (3)(1.71) + (4)(1.64)}{5} = 3.458,$$

$$\overline{x^2} = \frac{0^2 + 1^2 + 2^2 + 3^2 + 4^2}{5} = 6.$$

Therefore

$$a = \frac{3.458 - (2)(1.842)}{6 - 2^2} = -0.113,$$
$$b = \frac{(6)(1.842) - (2)(3.458)}{6 - 2^2} = 2.068,$$

and the empirical regression line is

$$y = 2.068 - 0.113x.$$

The predicted value of $y$ at $x = 5$ is $2.068 - 0.113 \times 5 = 1.503$. ■

Linear regression can also be interpreted in terms of vector projection. The data points define two vectors $\mathbf{x}$ and $\mathbf{y}$ in $\mathbb{R}^n$ with components $x_1, x_2, \ldots, x_n$ and $y_1, y_2, \ldots, y_n$ respectively. Let $\mathbf{w}$ be the vector with components $1, 1, \ldots, 1$. Finding the coefficients $a$ and $b$ for the regression line corresponds to finding the orthogonal projection of $\mathbf{y}$ onto the two-dimensional subspace (plane) in $\mathbb{R}^n$ spanned by $\mathbf{x}$ and $\mathbf{w}$. (See Figure 5.19.) This projection is $\mathbf{p} = a\mathbf{x} + b\mathbf{w}$. In fact, the two equations obtained above by setting the partial derivatives of $S$ equal to zero are just the two conditions

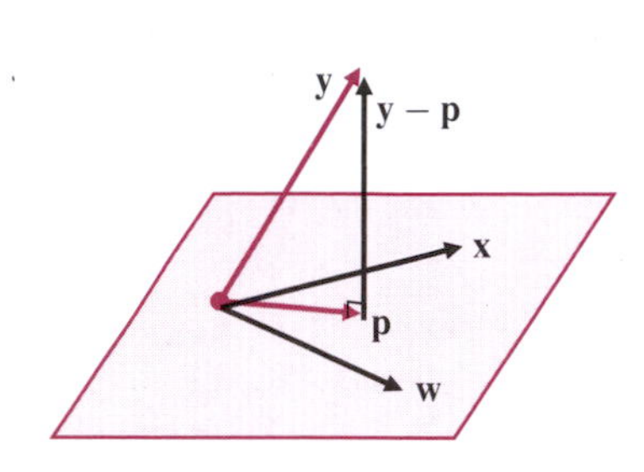


**Figure 5.19** $\mathbf{p} = a\mathbf{x} + b\mathbf{w}$ is the projection of $\mathbf{y}$ onto the plane spanned by $\mathbf{x}$ and $\mathbf{w}$

$$(\mathbf{y} - \mathbf{p}) \bullet \mathbf{x} = 0,$$
$$(\mathbf{y} - \mathbf{p}) \bullet \mathbf{w} = 0,$$

stating that $\mathbf{y}$ minus its projection onto the subspace is perpendicular to the subspace. The angle between $\mathbf{y}$ and this $\mathbf{p}$ provides a measure of how well the empirical regression line fits the data; the smaller the angle the better the fit.

Linear regression can be used to find specific functional relationships of types other than linear if suitable transformations are applied to the data.

■ **EXAMPLE 2** Find the values of constants $K$ and $s$ for which the curve

$$y = Kx^s$$

best fits the experimental data points $(x_i, y_i)$, $i = 1, 2, \ldots, n$. (Assume all data values are positive.)

***SOLUTION*** Observe that the required functional form corresponds to a linear relationship between $\ln y$ and $\ln x$:

$$\ln y = \ln K + s \ln x.$$

If we determine the parameters $a$ and $b$ of the empirical regression line $\eta = a\xi + b$ corresponding to the transformed data $(\xi_i, \eta_i) = (\ln x_i, \ln y_i)$, then $s = a$ and $K = e^b$ are the required values. ■

***REMARK*** It should be stressed that the constants $K$ and $s$ obtained by the method used in the solution above are not the same as those that would be obtained by direct application of the least squares method to the untransformed problem, that is, by minimizing $\sum_{i=1}^{n}(y_i - Kx_i^s)^2$. This latter problem cannot readily be solved. (Try it!)

Generally, the method of least squares is applied to fit an equation in which the response is expressed as a sum of constants times functions of one or more input variables. The constants are determined as critical points of the sum of squared deviations of the actual response values from the values predicted by the equation.

## Applications of the Least Squares Method to Integrals

The method of least squares can be used to find approximations to reasonably well-behaved (say, piecewise continuous) functions as sums of constants times specified functions. The idea is to choose the constants to minimize the *integral* of the square of the difference.

For example, suppose we want to approximate the continuous function $f(x)$ over the interval [0,1] by a linear function $g(x) = px + q$. The method of least squares would require that $p$ and $q$ be chosen to minimize the integral

$$I(p, q) = \int_0^1 \Big(f(x) - px - q\Big)^2 \, dx.$$

Assuming that we can "differentiate through the integral" (we will investigate this issue in Section 5.5) the critical point of $I(p, q)$ can be found from

$$0 = \frac{\partial I}{\partial p} = -2 \int_0^1 x\Big(f(x) - px - q\Big) \, dx,$$
$$0 = \frac{\partial I}{\partial q} = -2 \int_0^1 \Big(f(x) - px - q\Big) \, dx.$$

Thus

$$\frac{p}{3} + \frac{q}{2} = \int_0^1 xf(x) \, dx,$$
$$\frac{p}{2} + q = \int_0^1 f(x) \, dx,$$

and solving this linear system for $p$ and $q$ we get

$$p = \int_0^1 (12x - 6) f(x) \, dx,$$
$$q = \int_0^1 (4 - 6x) f(x) \, dx.$$

The following example concerns the approximation of a function by a **trigonometric polynomial.** Such approximations form the basis for the study of **Fourier series**, which are of fundamental importance in the solution of boundary-value problems for the Laplace, heat, and wave equations, and other partial differential equations that arise in applied mathematics.

**EXAMPLE 3** Use a least squares integral to approximate $f(x)$ by the sum

$$\sum_{k=1}^{n} b_k \sin kx$$

on the interval $0 \le x \le \pi$.

***SOLUTION*** We want to choose the constants to minimize

$$I = \int_0^\pi \Big(f(x) - \sum_{k=1}^{n} b_k \sin kx\Big)^2 \, dx.$$

For each $1 \leq j \leq n$ we have

$$0 = \frac{\partial I}{\partial b_j} = -2 \int_0^{\pi} \left( f(x) - \sum_{k=1}^{n} b_k \sin kx \right) \sin jx \, dx.$$

Thus

$$\sum_{k=1}^{n} b_k \int_0^{\pi} \sin kx \sin jx \, dx = \int_0^{\pi} f(x) \sin jx \, dx.$$

However, if $j \neq k$, then $\sin kx \sin jx$ is an even function, so that

$$\begin{aligned} \int_0^{\pi} \sin kx \sin jx \, dx &= \frac{1}{2} \int_{-\pi}^{\pi} \sin kx \sin jx \, dx \\ &= \frac{1}{4} \int_{-\pi}^{\pi} \big( \cos(k-j)x - \cos(k+j)x \big) \, dx = 0. \end{aligned}$$

If $j = k$, then we have

$$\int_0^{\pi} \sin^2 jx \, dx = \frac{1}{2} \int_0^{\pi} (1 - \cos 2jx) \, dx = \frac{\pi}{2},$$

so that

$$b_j = \frac{2}{\pi} \int_0^{\pi} f(x) \sin jx \, dx.$$

■

***REMARK*** If $f$ is continuous on $0 \leq x \leq \pi$, it can be shown that

$$\lim_{n \to \infty} \int_0^{\pi} \left( f(x) - \sum_{k=1}^{n} b_k \sin kx \right)^2 dx = 0,$$

with the coefficients $b_k$ given as above. It follows that

$$f(x) = \sum_{k=1}^{\infty} b_k \sin kx, \qquad 0 < x < \pi.$$

This is known as a **Fourier sine series** representation of $f(x)$ on the interval $(0, \pi)$. The function $f$ can also be approximated using a sum of cosines (see Exercise 21 below), and such approximations lead to an exact representation of $f$ as a **Fourier cosine series**:

$$f(x) = \frac{a_0}{2} + \sum_{k=1}^{\infty} a_k \cos kx, \qquad 0 < x < \pi,$$

where

$$a_j = \frac{2}{\pi} \int_0^{\pi} f(x) \cos jx \, dx.$$

***REMARK*** Representing a function as the sum of a Fourier series is analogous to representing a vector as a sum of basis vectors. If we think of continuous functions on the interval $[0, \pi]$ as "vectors" with addition and scalar multiplication defined pointwise:

$$(f + g)(x) = f(x) + g(x), \qquad (cf)(x) = cf(x),$$

and with the "dot product" defined as

$$f \bullet g = \int_0^{\pi} f(x)g(x)\,dx,$$

then the functions $e_k(x) = \sqrt{2/\pi}\,\sin kx$ form a "basis." As shown in the example above, $e_j \bullet e_j = 1$, and if $k \neq j$ then $e_k \bullet e_j = 0$. Thus these "basis vectors" are "mutually perpendicular unit vectors." The Fourier coefficients $b_j$ of a function $f$ are the components of $f$ with respect to that basis.

## EXERCISES 5.4

**1.** A generator is to be installed in a factory to supply power to $n$ machines located at positions $(x_i, y_i)$, $i = 1, 2, \ldots, n$. Where should the generator be located to minimize the sum of the squares of its distances from the machines?

**2.** The relationship $y = ax^2$ is known to hold between certain variables. Given the experimental data $(x_i, y_i)$, $i = 1, 2, \ldots, n$, determine a value for $a$ by the method of least squares.

**3.** Repeat Exercise 2 but with the relationship $y = ae^x$.

**4.** Use the method of least squares to find the plane $z = ax + by + c$ that best fits the data $(x_i, y_i, z_i)$, $i = 1, 2, \ldots, n$.

**5.** Repeat Exercise 4 using a vector projection argument instead of the method of least squares.

In Exercises 6–11, show how to adapt linear regression to determine the two parameters $p$ and $q$ so that the given relationship fits the experimental data $(x_i, y_i)$, $i = 1, 2, \ldots, n$. In which of these situations are the values of $p$ and $q$ obtained identical to those obtained by direct application of the method of least squares with no change of variable?

**6.** $y = p + qx^2$ **7.** $y = pe^{qx}$

**8.** $y = \ln(p + qx)$ **9.** $y = px + qx^2$

**10.** $y = \sqrt{px + q}$ **11.** $y = pe^x + qe^{-x}$

**12.** Find the parabola of the form $y = p + qx^2$ that best fits the data $(x, y) = (1, 0.11)$, $(2, 1.62)$, $(3, 4.07)$, $(4, 7.55)$, $(6, 17.63)$, and $(7, 24.20)$. No value of $y$ was measured at $x = 5$. What value would you predict at this point?

**13.** Use the method of least squares to find constants $a$, $b$ and $c$ so that the relationship $y = ax^2 + bx + c$ best describes the experimental data $(x_i, y_i)$, $i = 1, 2, \ldots, n$, $(n \geq 3)$. How is this situation interpreted in terms of vector projection?

**14.** How can the result of Exercise 13 be used to fit a curve of the form $y = pe^x + q + re^{-x}$ through the same data points?

**15.** Find the value of the constant $a$ for which the function $f(x) = ax^2$ best approximates the function $g(x) = x^3$ on the interval $[0, 1]$, in the sense that the integral

$$I = \int_0^1 \big(f(x) - g(x)\big)^2\,dx$$

is minimized. What is the minimum value of $I$?

**16.** Find $a$ to minimize $I = \int_0^{\pi} \big(ax(\pi - x) - \sin x\big)^2\,dx$. What is the minimum value of the integral?

**17.** Repeat Exercise 15 with the function $f(x) = ax^2 + b$ and the same $g$. Find $a$ and $b$.

* **18.** Find $a$, $b$, and $c$ to minimize $\int_0^1 (x^3 - ax^2 - bx - c)^2\,dx$. What is the minimum value of the integral?

* **19.** Find $a$ and $b$ to minimize $\int_0^{\pi} (\sin x - ax^2 - bx)^2\,dx$.

**20.** Find $a$, $b$, and $c$ to minimize the integral

$$J = \int_{-1}^{1} \big(x - a\sin\pi x - b\sin 2\pi x - c\sin 3\pi x\big)^2\,dx.$$

* **21.** Find constants $a_j$, $j = 0, 1, \ldots, n$, to minimize

$$\int_0^{\pi} \left(f(x) - \frac{a_0}{2} - \sum_{k=1}^{n} a_k \cos kx\right)^2 dx.$$

**22.** Find the Fourier sine series for the function $f(x) = x$ on $0 < x < \pi$. To what function would you expect the series to converge in the interval $-\pi < x < 0$?

**23.** Repeat Exercise 22 , but obtaining instead a Fourier cosine series.

**24.** Suppose $x_1, x_2, \ldots, x_n$ satisfy $x_i \leq x_j$ whenever $i < j$. Find $x$ that minimizes $\sum_{i=1}^{n} |x - x_i|$. Treat the cases $n$ odd and $n$ even separately. For what values of $n$ is $x$ unique? (*Hint:* use no calculus in this problem.)

## 5.5 PARAMETRIC PROBLEMS

In this section we will briefly examine three unrelated situations in which one wants to differentiate a function with respect to a parameter rather than one of the basic variables of the function. Such situations arise frequently in mathematics and its applications.

### Differentiating Integrals with Parameters

The Fundamental Theorem of Calculus shows how to differentiate a definite integral with respect to the upper limit of integration:

$$\frac{d}{dx}\int_a^x f(t)\,dt = f(x).$$

We are going to look at a different problem about differentiating integrals. If the integrand of a definite integral also depends on variables other than the variable of integration, then the integral will be a function of those other variables. How are we to find the derivative of such a function? For instance, consider the function $F(x)$ defined by

$$F(x) = \int_a^b f(x,t)\,dt.$$

We would like to be able to calculate $F'(x)$ by taking the derivative inside the integral:

$$F'(x) = \frac{d}{dx}\int_a^b f(x,t)\,dt = \int_a^b \frac{\partial}{\partial x} f(x,t)\,dt.$$

Observe that we use "$d/dx$" outside the integral and "$\partial/\partial x$" inside because the integral is a function of $x$ only, but the integrand $f$ is a function of both $x$ and $t$. If the integrand depends on more than one parameter, then partial derivatives would be needed inside and outside the integral:

$$\frac{\partial}{\partial x}\int_a^b f(x,y,t)\,dt = \int_a^b \frac{\partial}{\partial x} f(x,y,t)\,dt.$$

The operation of taking a derivative with respect to a parameter inside the integral, or *differentiating through the integral* as it is usually called, seems plausible. We differentiate sums term by term, and integrals are the limits of sums. However, both the differentiation and integration operations involve the taking of limits, (limits of Newton quotients for derivatives, limits of Riemann sums for integrals). Differentiating through the integral requires changing the order in which the two limits are taken, and therefore requires justification.

We have already seen another example of change of order of limits. When we assert that two mixed partial derivatives with respect to the same variables are equal,

$$\frac{\partial^2 f}{\partial x \partial y} = \frac{\partial^2 f}{\partial y \partial x},$$

we are, in fact, saying that limits corresponding to differentiation with respect to $x$ and $y$ can be taken in either order with the same result. This is not true in general; we proved it under the assumption that both of the mixed partials were *continuous*. (See Theorem 1 and Exercise 16 of Section 4.4.) In general, some assumptions are required to justify the interchange of limits. The following theorem gives one set of conditions that justify the interchange of limits involved in differentiating through the integral.

**THEOREM 5** **Differentiating through an integral**

Suppose that for every $x$ satisfying $c < x < d$ the following conditions hold:

(i) the integrals

$$\int_a^b f(x,t)\,dt \qquad \text{and} \qquad \int_a^b f_1(x,t)\,dt$$

both exist (either as proper or convergent improper integrals).

(ii) $f_{11}(x,t)$ exists and satisfies

$$|f_{11}(x,t)| \le g(t), \qquad a < t < b,$$

where

$$\int_a^b g(t)\,dt = K < \infty.$$

Then for each $x$ satisfying $c < x < d$ we have

$$\frac{d}{dx}\int_a^b f(x,t)\,dt = \int_a^b \frac{\partial}{\partial x} f(x,t)\,dt.$$

***PROOF*** Let

$$F(x) = \int_a^b f(x,t)\,dt.$$

If $c < x < d$, $h \ne 0$, and $|h|$ is sufficiently small that $c < x+h < d$, then, by Taylor's Formula,

$$f(x+h,t) = f(x,t) + hf_1(x,t) + \frac{h^2}{2} f_{11}(x+\theta h, t)$$

for some $\theta$ between 0 and 1. Therefore

$$\begin{aligned}
&\left|\frac{F(x+h)-F(x)}{h} - \int_a^b f_1(x,t)\,dt\right| \\
&\qquad = \left|\int_a^b \frac{f(x+h,t)-f(x,t)}{h}\,dt - \int_a^b f_1(x,t)\,dt\right| \\
&\qquad \le \int_a^b \left|\frac{f(x+h,t)-f(x,t)}{h} - f_1(x,t)\right| dt \\
&\qquad = \int_a^b \left|\frac{h}{2} f_{11}(x+\theta h, t)\right| dt \\
&\qquad \le \frac{h}{2}\int_a^b g(t)\,dt = \frac{Kh}{2} \to 0 \text{ as } h \to 0.
\end{aligned}$$

Therefore

$$F'(x) = \lim_{h\to 0}\frac{F(x+h)-F(x)}{h} = \int_a^b f_1(x,t)\,dt,$$

which is the desired result.

***REMARK*** It can be shown that the conclusion of Theorem 5 also holds under the sole assumption that $f_1(x,t)$ is continuous on the *closed, bounded* rectangle $c \le x \le d,\ a \le t \le b$. We cannot prove this here—the proof depends on a subtle property called *uniform continuity* possessed by continuous functions on closed bounded sets in $\mathbb{R}^n$. In any event, Theorem 5 is more useful for our purposes because it allows for improper integrals.

**EXAMPLE 1** Evaluate $\displaystyle\int_0^\infty t^n\, e^{-t}\, dt$.

***SOLUTION*** Starting with the convergent improper integral

$$\int_0^\infty e^{-s}\, ds = \lim_{R\to\infty} \left.\frac{e^{-s}}{-1}\right|_0^R = \lim_{R\to\infty}(1 - e^{-R}) = 1$$

we introduce a parameter by substituting $s = xt,\ ds = x\,dt$ (where $x > 0$) and get

$$\int_0^\infty e^{-xt}\, dt = \frac{1}{x}.$$

Now differentiate $n$ times (each resulting integral converges):

$$\begin{aligned}
\int_0^\infty -t\, e^{-xt}\, dt &= -\frac{1}{x^2},\\
\int_0^\infty (-t)^2\, e^{-xt}\, dt &= (-1)^2\frac{2}{x^3},\\
&\ \ \vdots\\
\int_0^\infty (-t)^n\, e^{-xt}\, dt &= (-1)^n\frac{n!}{x^{n+1}}.
\end{aligned}$$

Putting $x = 1$ we get

$$\int_0^\infty t^n\, e^{-t}\, dt = n!.$$

Note that this result could be obtained by integration by parts ($n$ times) or a reduction formula. This method is a little easier. ■

***REMARK*** The reader should check that the function $f(x,t) = t^k\, e^{-xt}$ satisfies the conditions of Theorem 5 for $x > 0$ and $k \ge 0$. We will normally not make a point of this.

**EXAMPLE 2** Evaluate $\displaystyle F(x,y) = \int_0^\infty \frac{e^{-xt} - e^{-yt}}{t}\, dt$ for $x > 0,\ y > 0$.

***SOLUTION*** We have

$$\frac{\partial F}{\partial x} = -\int_0^\infty e^{-xt}\, dt = -\frac{1}{x} \quad\text{and}\quad \frac{\partial F}{\partial y} = \int_0^\infty e^{-yt}\, dt = \frac{1}{y}.$$

It follows that

$$F(x,y) = -\ln x + C_1(y) \quad\text{and}\quad F(x,y) = \ln y + C_2(x).$$

Comparing these two formulas for $F$, we are forced to conclude that $C_1(y) = \ln y + C$ for some constant $C$. Therefore

$$F(x,y) = \ln y - \ln x + C = \ln\frac{y}{x} + C.$$

Since $F(1,1) = 0$, we must have $C = 0$ and $F(x,y) = \ln(y/x)$. ■

***REMARK*** We can combine Theorem 5 and the Fundamental Theorem of Calculus to differentiate an integral with respect to a parameter that appears in the limits of integration as well as in the integrand. If

$$F(x, b, a) = \int_a^b f(x, t)\, dt,$$

then, by the Chain Rule,

$$\frac{d}{dx} F\big(x, b(x), a(x)\big) = \frac{\partial F}{\partial x} + \frac{\partial F}{\partial b}\frac{db}{dx} + \frac{\partial F}{\partial a}\frac{da}{dx}.$$

Accordingly, we have

$$\begin{aligned}&\frac{d}{dx}\int_{a(x)}^{b(x)} f(x, t)\, dt \\ &= \int_{a(x)}^{b(x)} \frac{\partial}{\partial x} f(x, t)\, dt + f\big(x, b(x)\big)b'(x) - f\big(x, a(x)\big)a'(x).\end{aligned}$$

We require that $a(x)$ and $b(x)$ be differentiable at $x$, and for the application of Theorem 5, that $a \le a(x) \le b$ and $a \le b(x) \le b$ for all $x$ satisfying $c < x < d$.

**EXAMPLE 3** Solve the *integral equation*

$$f(x) = a - \int_b^x (x - t) f(t)\, dt.$$

***SOLUTION*** Assume, for the moment, that the equation has a sufficiently well behaved solution to allow for differentiation through the integral. Differentiating twice, we get

$$\begin{aligned}f'(x) &= -(x - x) f(x) - \int_b^x f(t)\, dt = -\int_b^x f(t)\, dt, \\ f''(x) &= -f(x).\end{aligned}$$

The latter equation is the differential equation of simple harmonic motion. Observe that the given equation for $f$ and that for $f'$ imply the initial conditions

$$f(b) = a \qquad \text{and} \qquad f'(b) = 0.$$

Accordingly, we write the general solution of $f''(x) = -f(x)$ in the form

$$f(x) = A\cos(x - b) + B\sin(x - b).$$

The initial conditions then imply $A = a$ and $B = 0$, so the required solution is $f(x) = a\cos(x - b)$. Finally, we note that this function is indeed smooth enough to allow the differentiations through the integral, and so is the solution of the given integral equation. (If you wish, verify it in the integral equation.) ■

## Envelopes

An equation $f(x, y, c) = 0$ that involves a parameter $c$ as well as the variables $x$ and $y$ represents a family of curves in the $xy$-plane. Consider, for instance, the family

$$f(x, y, c) = \frac{x}{c} + cy - 2 = 0.$$

This family consists of straight lines with intercepts $(2c, 2/c)$ on the coordinate axes. Several of these lines are sketched in Figure 5.20. It appears that there is a curve to which all these lines are tangent. This curve is called the envelope of the family of lines.

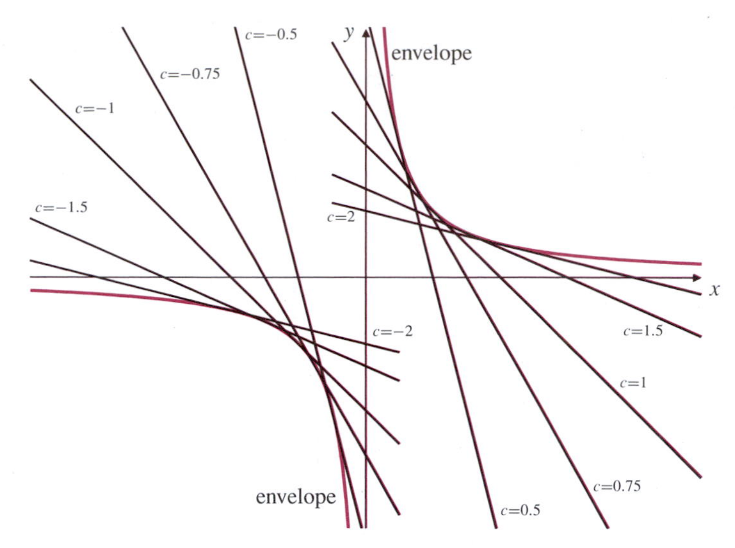


**Figure 5.20** A family of straight lines and their envelope

In general, a curve $\mathcal{C}$ is called the **envelope** of the family of curves with equations $f(x, y, c) = 0$ if, for each value of $c$, the curve $f(x, y, c) = 0$ is tangent to $\mathcal{C}$ at some point depending on $c$.

For the family of lines in Figure 5.20 it appears that the envelope may be the rectangular hyperbola $xy = 1$. We will verify this after developing a method for determining the equation of the envelope of a family of curves. We assume that the function $f(x, y, c)$ has continuous first partials, and that the envelope is a smooth curve.

DANGER
This is a subtle argument. Take your time and try to understand each step in the development.
DANGER

For each $c$, the curve $f(x, y, c) = 0$ is tangent to the envelope at a point $(x, y)$ that depends on $c$. Let us express this dependence in the explicit form $x = g(c)$, $y = h(c)$; these equations are parametric equations of the envelope. Since $(x, y)$ lies on the curve $f(x, y, c) = 0$, we have

$$f\big(g(c), h(c), c\big) = 0.$$

Differentiating this equation with respect to $c$, we obtain

$$f_1 g'(c) + f_2 h'(c) + f_3 = 0, \qquad (*)$$

where the partials of $f$ are evaluated at $\big(g(c), h(c), c\big)$.

The slope of the curve $f(x, y, c)$ at $\big(g(c), h(c), c\big)$ can be obtained by differentiating its equation implicitly with respect to $x$:

$$f_1 + f_2 \frac{dy}{dx} = 0.$$

On the other hand, the slope of the envelope $x = g(c)$, $y = h(c)$ at that point is $dy/dx = h'(c)/g'(c)$. Since the curve and the envelope are tangent at $f\big(g(c), h(c), c\big)$, these slopes must be equal. Therefore

$$f_1 g'(c) + f_2 h'(c) = 0.$$

Combining this with equation $(*)$ we get $f_3(x, y, c) = 0$ at all points of the envelope.

The equation of the envelope can be found by eliminating $c$ between the two equations

$$f(x, y, c) = 0 \qquad \text{and} \qquad \frac{\partial}{\partial c} f(x, y, c) = 0.$$

**EXAMPLE 4** Find the envelope of the family of straight lines

$$f(x, y, c) = \frac{x}{c} + cy - 2 = 0.$$

***SOLUTION*** We eliminate $c$ between the equations

$$f(x, y, c) = \frac{x}{c} + cy - 2 = 0 \qquad \text{and} \qquad f_3(x, y, c) = -\frac{x}{c^2} + y = 0.$$

These equations can be solved for $x = c$ and $y = 1/c$, and hence imply that the envelope is $xy = 1$, as we conjectured earlier.

**EXAMPLE 5** Find the envelope of the family of circles

$$(x - c)^2 + y^2 = c.$$

***SOLUTION*** Here $f(x, y, c) = (x - c)^2 + y^2 - c$. The equation of the envelope is obtained by eliminating $c$ from the pair of equations

$$f(x, y, c) = (x - c)^2 + y^2 - c = 0,$$
$$\frac{\partial}{\partial c} f(x, y, c) = -2(x - c) - 1 = 0.$$

From the second equation, $x = c - \frac{1}{2}$, and then from the first, $y^2 = c - \frac{1}{4}$. Hence the envelope is the parabola

$$x = y^2 - \frac{1}{4}.$$

This envelope and some of the circles in the family are sketched in Figure 5.21.

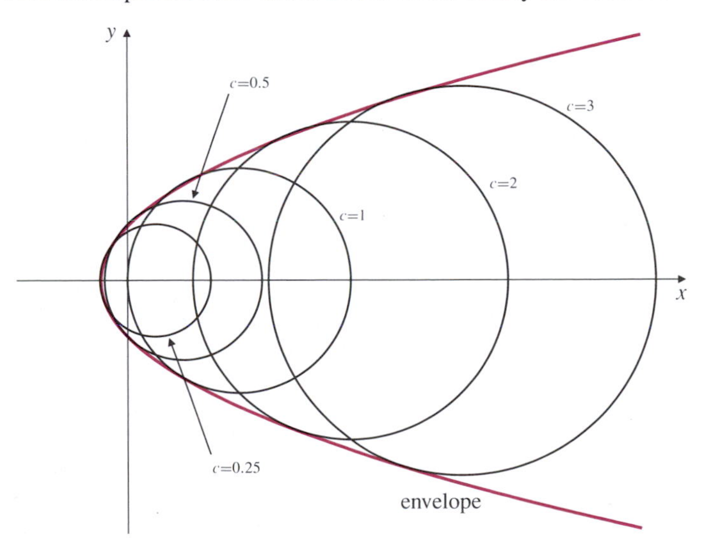


**Figure 5.21** Circles $(x - c)^2 + y^2 = c$ and their envelope

A similar technique can be used to find the envelope of a family of surfaces. This will be a surface tangent to each member of the family.

**EXAMPLE 6** **The Mach cone**

Suppose that sound travels at speed $c$ in still air, and that a supersonic aircraft is travelling at speed $v > c$ along the $x$-axis, so that its position at time $t$ is $(vt, 0, 0)$. Find the envelope at time $t$ of the sound waves created by the aircraft at previous times.

**SOLUTION** The sound created by the aircraft at time $\tau < t$ spreads out as a spherical wave front at speed $c$. The centre of this wave front is $(v\tau, 0, 0)$, the position of the aircraft at time $\tau$. At time $t$ the radius of this wave front is $c(t-\tau)$, so its equation is

$$f(x, y, z, \tau) = (x - v\tau)^2 + y^2 + z^2 - c^2(t-\tau)^2 = 0. \qquad (*)$$

At time $t$ the envelope of all these wave fronts created at earlier times $\tau$ is obtained by eliminating the parameter $\tau$ from the above equation and the equation

$$\frac{\partial}{\partial \tau} f(x, y, z, \tau) = -2v(x - v\tau) + 2c^2(t-\tau) = 0.$$

Solving this latter equation for $\tau$, we get

$$\tau = \frac{vx - c^2 t}{v^2 - c^2}.$$

Thus

$$\begin{aligned} x - v\tau &= x - \frac{v^2 x - vc^2 t}{v^2 - c^2} = \frac{c^2}{v^2 - c^2}(vt - x) \\ t - \tau &= t - \frac{vx - c^2 t}{v^2 - c^2} = \frac{v}{v^2 - c^2}(vt - x). \end{aligned}$$

We substitute these two expressions into equation $(*)$ to eliminate $\tau$:

$$\begin{aligned} \frac{c^4}{(v^2 - c^2)^2}(vt - x)^2 + y^2 + z^2 - \frac{c^2 v^2}{(v^2 - c^2)^2}(vt - x)^2 &= 0 \\ y^2 + z^2 = \frac{c^2}{(v^2 - c^2)^2}(v^2 - c^2)(vt - x)^2 &= \frac{c^2}{v^2 - c^2}(vt - x)^2. \end{aligned}$$

The envelope is the cone

$$x = vt - \frac{\sqrt{v^2 - c^2}}{c}\sqrt{y^2 + z^2},$$

which extends backwards in the $x$ direction from its vertex at $(vt, 0, 0)$, the position of the aircraft at time $t$. This is called the **Mach cone**. The sound of the aircraft can not be heard at any point until the cone reaches that point. ■

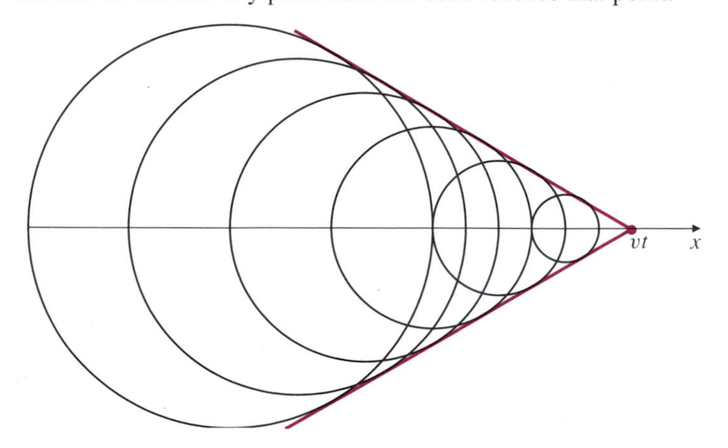


**Figure 5.22** The Mach cone

## Equations with Perturbations

In applied mathematics one frequently encounters intractable equations for which at least approximate solutions are desired. Sometimes such equations result from adding an extra term to what would otherwise be a simple and easily solved equation. This extra term is called a **perturbation** of the simpler equation. Often the perturbation has coefficient smaller than the other terms in the equation, that is, it is a **small perturbation**. When this is the case you can find approximate solutions to the perturbed equation by replacing the small coefficient by a parameter and calculating Maclaurin polynomials in that parameter. One example should serve to clarify the method.

**EXAMPLE 7** Find an approximate solution of the equation

$$y + \frac{1}{50}\ln(1+y) = x^2.$$

***SOLUTION*** Without the logarithm term the equation would clearly have the solution $y = x^2$. Let us replace the coefficient $1/50$ with the parameter $\epsilon$ and look for a solution $y = y(x, \epsilon)$ to the equation

$$y + \epsilon \ln(1+y) = x^2$$

in the form

$$y = y(x, \epsilon) = y(x, 0) + \epsilon y_\epsilon(x, 0) + \frac{\epsilon^2}{2!} y_{\epsilon\epsilon}(x, 0) + \cdots,$$

where the subscripts $\epsilon$ denote derivatives with respect to $\epsilon$. We shall calculate the terms up to second order in $\epsilon$. Evidently $y(x, 0) = x^2$. Differentiating the equation twice with respect to $\epsilon$ and evaluating the results at $\epsilon = 0$, we obtain

$$\frac{\partial y}{\partial \epsilon} + \ln(1+y) + \frac{\epsilon}{1+y}\frac{\partial y}{\partial \epsilon} = 0,$$

$$\frac{\partial^2 y}{\partial \epsilon^2} + \frac{2}{1+y}\frac{\partial y}{\partial \epsilon} + \epsilon\frac{\partial}{\partial \epsilon}\left(\frac{1}{1+y}\frac{\partial y}{\partial \epsilon}\right) = 0,$$

$$y_\epsilon(x, 0) = -\ln(1+x^2),$$

$$y_{\epsilon\epsilon}(x, 0) = \frac{2}{1+x^2}\ln(1+x^2).$$

Hence

$$y(x, \epsilon) = x^2 - \epsilon \ln(1+x^2) + \frac{\epsilon^2}{1+x^2}\ln(1+x^2) + \cdots,$$

and the given equation has the approximate solution

$$y \approx x^2 - \frac{\ln(1+x^2)}{50} + \frac{\ln(1+x^2)}{2500(1+x^2)}.$$

Similar perturbation techniques can be used for systems of equations and for differential equations.

## EXERCISES 5.5

**1.** Let $F(x) = \int_0^1 t^x\,dt = \dfrac{1}{x+1}$ for $x > -1$. By repeated differentiation of $F$ evaluate the integral

$$\int_0^1 t^x (\ln t)^n\,dt.$$

**2.** By replacing $t$ with $xt$ in the well-known integral

$$\int_{-\infty}^{\infty} e^{-t^2}\,dt = \sqrt{\pi},$$

and differentiating with respect to $x$, evaluate

$$\int_{-\infty}^{\infty} t^2 e^{-t^2}\,dt \quad \text{and} \quad \int_{-\infty}^{\infty} t^4 e^{-t^2}\,dt.$$

**3.** Evaluate $\displaystyle\int_{-\infty}^{\infty} \frac{e^{-xt^2} - e^{-yt^2}}{t^2}\,dt$ for $x > 0,\ y > 0$.

**4.** Evaluate $\displaystyle\int_0^1 \frac{t^x - t^y}{\ln t}\,dt$ for $x > -1,\ y > -1$.

**5.** Given that $\displaystyle\int_0^{\infty} e^{-xt}\sin t\,dt = \frac{1}{1+x^2}$ for $x > 0$ (which can be shown by integration by parts) evaluate

$$\int_0^{\infty} te^{-xt}\sin t\,dt \quad \text{and} \quad \int_0^{\infty} t^2e^{-xt}\sin t\,dt.$$

* **6.** Referring to Exercise 5, for $x > 0$ evaluate

$$F(x) = \int_0^{\infty} e^{-xt}\frac{\sin t}{t}\,dt.$$

Show that $\lim_{x\to\infty} F(x) = 0$ and hence evaluate the integral

$$\int_0^{\infty} \frac{\sin t}{t}\,dt = \lim_{x\to 0} F(x).$$

**7.** Evaluate $\displaystyle\int_0^{\infty} \frac{dt}{x^2+t^2}$ and use the result to help you evaluate

$$\int_0^{\infty} \frac{dt}{(x^2+t^2)^2}, \quad \text{and} \quad \int_0^{\infty} \frac{dt}{(x^2+t^2)^3}.$$

* **8.** Evaluate $\displaystyle\int_0^{x} \frac{dt}{x^2+t^2}$ and use the result to help you evaluate

$$\int_0^{x} \frac{dt}{(x^2+t^2)^2}, \quad \text{and} \quad \int_0^{x} \frac{dt}{(x^2+t^2)^3}.$$

**9.** Find $f^{(n+1)}(a)$ if $f(x) = 1 + \displaystyle\int_a^x (x-t)^n f(t)\,dt$.

Solve the integral equations in Exercises 10–12.

**10.** $f(x) = Cx + D + \displaystyle\int_0^x (x-t)f(t)\,dt$

**11.** $f(x) = x + \displaystyle\int_0^x (x-2t)f(t)\,dt$

**12.** $f(x) = 1 + \displaystyle\int_0^1 (x+t)f(t)\,dt$

Find the envelopes of the families of curves in Exercises 13–18.

**13.** $y = 2cx - c^2$

**14.** $y - (x-c)\cos c = \sin c$

**15.** $x\cos c + y\sin c = 1$

**16.** $\dfrac{x}{\cos c} + \dfrac{y}{\sin c} = 1$

**17.** $y = c + (x-c)^2$

**18.** $(x-c)^2 + (y-c)^2 = 1$

**19.** Does every one-parameter family of curves in the plane have an envelope? Try to find the envelope of $y = x^2 + c$.

**20.** For what values of $k$ does the family of curves $x^2 + (y-c)^2 = kc^2$ have an envelope?

**21.** Try to find the envelope of the family $y^3 = (x+c)^2$. Are the curves of the family tangent to the envelope? What have you actually found in this case? Compare with Example 3 of Section 5.3.

* **22.** Show that if a two-parameter family of surfaces $f(x, y, z, \lambda, \mu) = 0$ has an envelope, then the equation of that envelope can be obtained by eliminating $\lambda$ and $\mu$ from the three equations

$$f(x, y, z, \lambda, \mu) = 0,$$
$$\frac{\partial}{\partial\lambda} f(x, y, z, \lambda, \mu) = 0,$$
$$\frac{\partial}{\partial\mu} f(x, y, z, \lambda, \mu) = 0.$$

**23.** Find the envelope of the two-parameter family of planes

$$x\sin\lambda\cos\mu + y\sin\lambda\sin\mu + z\cos\lambda = 1.$$

**24.** Find the envelope of the two-parameter family of spheres

$$(x-\lambda)^2 + (y-\mu)^2 + z^2 = \frac{\lambda^2+\mu^2}{2}.$$

In Exercises 25–27, find the terms up to second power in $\epsilon$ in the solution $y$ of the given equation.

**25.** $y + \epsilon \sin \pi y = x$

**26.** $y^2 + \epsilon e^{-y^2} = 1 + x^2$

**27.** $2y + \dfrac{\epsilon x}{1 + y^2} = 1$

**28.** Use perturbation methods to evaluate $y$ with error less than $10^{-8}$ given that $y + (y^5/100) = 1/2$.

* **29.** Use perturbation methods to find approximate values for $x$ and $y$ from the system

$$x + 2y + \frac{1}{100}e^{-x} = 3, \quad x - y + \frac{1}{100}e^{-y} = 0.$$

Calculate all terms up to second order in $\epsilon = 1/100$.

## 5.6 NEWTON'S METHOD

A frequently encountered problem in applied mathematics is to determine, to some desired degree of accuracy, a root (that is, a solution $r$) of an equation of the form

$$f(r) = 0.$$

Such a root is called a **zero** of the function $f$. In *Single-Variable Calculus* we encountered Newton's method, a simple but powerful method for determining roots of functions that are sufficiently smooth. The method involves *guessing* an approximate value $x_0$ for a root $r$ of the function $f$, and then calculating successive approximations $x_1, x_2, \ldots$, using the formula

$$x_{n+1} = x_n - \frac{f(x_n)}{f'(x_n)}, \qquad n = 0, 1, 2, \cdots.$$

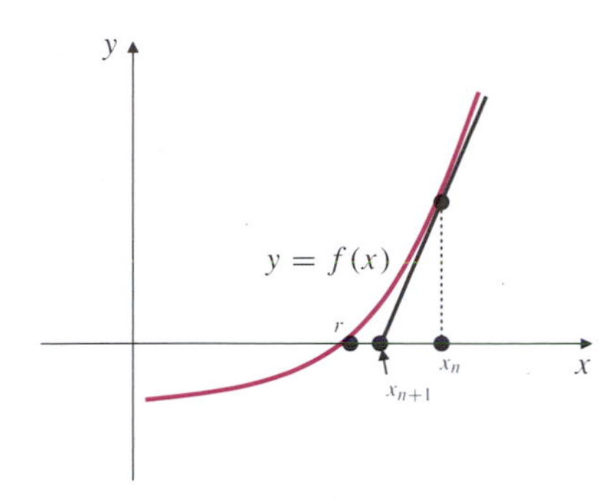

**Figure 5.23** $x_{n+1}$ is the $x$-intercept of the tangent at $x_n$

If the initial guess $x_0$ is not too far from $r$, and if $|f'(x)|$ is *not too small* and $|f''(x)|$ is *not too large* near $r$ then the successive approximations $x_1, x_2, \ldots$ will converge very rapidly to $r$. Recall that each new approximation $x_{n+1}$ is obtained as the $x$-intercept of the tangent line drawn to the graph of $f$ at the previous approximation, $x_n$. The tangent line to the graph $y = f(x)$ at $x = x_n$ has equation

$$y - f(x_n) = f'(x_n)(x - x_n).$$

(See Figure 5.23.) The $x$-intercept, $x_{n+1}$, of this line is determined by setting $y = 0$, $x = x_{n+1}$ in this equation, so is given by the formula in the shaded box above.

Newton's method can be extended to finding solutions of systems of $m$ equations in $m$ variables. We will show here how to adapt the method to find approximations to a solution $(x, y)$ of the pair of equations

$$\begin{cases} f(x, y) = 0 \\ g(x, y) = 0 \end{cases}$$

starting from an initial guess $(x_0, y_0)$. Under auspicious circumstances, we will observe the same rapid convergence of approximations to the root that typifies the single-variable case.

The idea is as follows. The two surfaces $z = f(x, y)$ and $z = g(x, y)$ intersect in a curve which itself intersects the $xy$-plane at the point whose coordinates are the desired solution. If $(x_0, y_0)$ is near that point then the tangent planes to the two surfaces at $(x_0, y_0)$ will intersect in a straight line. This line meets the $xy$-plane at a point $(x_1, y_1)$ that should be even closer to the solution point than was $(x_0, y_0)$. We can easily determine $(x_1, y_1)$. The tangent planes to $z = f(x, y)$ and $z = g(x, y)$ at $(x_0, y_0)$ have equations

$$z = f(x_0, y_0) + f_1(x_0, y_0)(x - x_0) + f_2(x_0, y_0)(y - y_0),$$
$$z = g(x_0, y_0) + g_1(x_0, y_0)(x - x_0) + g_2(x_0, y_0)(y - y_0).$$

The line of intersection of these two planes meets the $xy$-plane at the point $(x_1, y_1)$ satisfying

$$f_1(x_0, y_0)(x_1 - x_0) + f_2(x_0, y_0)(y_1 - y_0) + f(x_0, y_0) = 0,$$
$$g_1(x_0, y_0)(x_1 - x_0) + g_2(x_0, y_0)(y_1 - y_0) + g(x_0, y_0) = 0.$$

Solving these two equations for $x_1$ and $y_1$, we obtain

$$x_1 = x_0 - \left.\frac{fg_2 - f_2 g}{f_1 g_2 - f_2 g_1}\right|_{(x_0, y_0)} = x_0 - \left.\frac{\begin{vmatrix} f & f_2 \\ g & g_2 \end{vmatrix}}{\begin{vmatrix} f_1 & f_2 \\ g_1 & g_2 \end{vmatrix}}\right|_{(x_0, y_0)},$$

$$y_1 = y_0 - \left.\frac{f_1 g - f g_1}{f_1 g_2 - f_2 g_1}\right|_{(x_0, y_0)} = y_0 - \left.\frac{\begin{vmatrix} f_1 & f \\ g_1 & g \end{vmatrix}}{\begin{vmatrix} f_1 & f_2 \\ g_1 & g_2 \end{vmatrix}}\right|_{(x_0, y_0)}.$$

Observe that the denominator in each of these expressions is the Jacobian determinant $\partial(f, g)/\partial(x, y)\big|_{(x_0, y_0)}$. This is another instance where the Jacobian is the appropriate multivariable analogue of the derivative of a function of one variable.

Continuing in this way we generate successive approximations $(x_n, y_n)$ according to the formulas

$$x_{n+1} = x_n - \left.\frac{\begin{vmatrix} f & f_2 \\ g & g_2 \end{vmatrix}}{\begin{vmatrix} f_1 & f_2 \\ g_1 & g_2 \end{vmatrix}}\right|_{(x_n, y_n)},$$

$$y_{n+1} = y_n - \left.\frac{\begin{vmatrix} f_1 & f \\ g_1 & g \end{vmatrix}}{\begin{vmatrix} f_1 & f_2 \\ g_1 & g_2 \end{vmatrix}}\right|_{(x_n, y_n)}.$$

We stop when the desired accuracy has been achieved.

**EXAMPLE 1** Find the four roots of the system

$$x^2 - xy + 2y^2 - 10 = 0, \qquad x^3y^2 - 2 = 0$$

with sufficient accuracy to ensure that the left sides of the equations vanish to the sixth decimal place.

***SOLUTION*** A sketch of the graphs of the two equations (see Figure 5.24) in the $xy$-plane indicates that there are roots near the points $(1, 2.5)$, $(3.5, 0.5)$, $(3, -0.5)$, and $(1, -2)$. Application of Newton's method requires successive computations of the quantities

$$f(x, y) = x^2 - xy + 2y^2 - 10, \quad f_1(x, y) = 2x - y, \quad f_2(x, y) = -x + 4y,$$
$$g(x, y) = x^3y^2 - 2, \quad g_1(x, y) = 3x^2y^2, \quad g_2(x, y) = 2x^3y.$$

Using a calculator or computer we can calculate successive values of $(x_n, y_n)$ starting from each of the four points $(x_0, y_0)$ mentioned above:

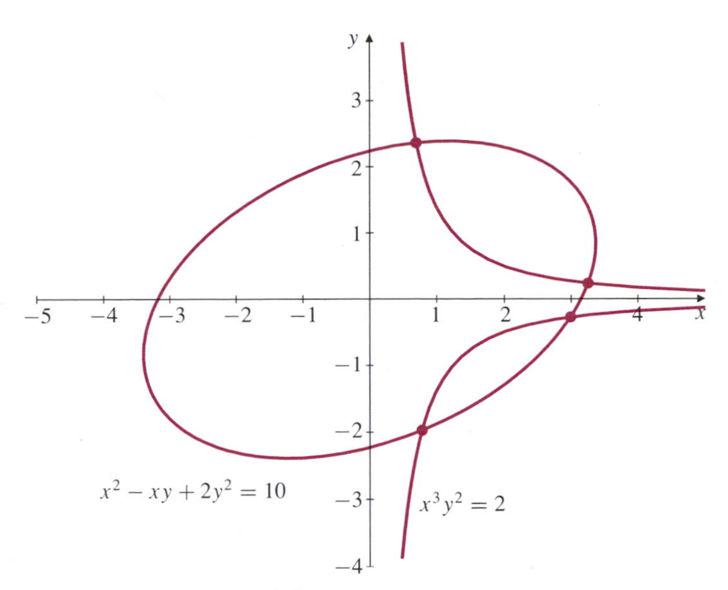


**Figure 5.24** The two graphs intersect at four points

The values in these tables were calculated sequentially in a spreadsheet by the method suggested on page 262. They were rounded for inclusion in the table but the unrounded values were used in subsequent calculations. If you actually use the (rounded) values of $x_n$ and $y_n$ given in the table to calculate $f(x_n, y_n)$ and $g(x_n, y_n)$ your results may vary slightly. For example, the value of $f(x_n, y_n)$ calculated from the values of $x_n$ and $y_n$ given for $n = 5$ in Table 1 is 0.000001.

**Table 1.** Root near (1, 2.5)

| $n$ | $x_n$ | $y_n$ | $f(x_n, y_n)$ | $g(x_n, y_n)$ |
|---|---|---|---|---|
| 0 | 1.000000 | 2.500000 | 1.000000 | 4.250000 |
| 1 | 0.805839 | 2.378102 | 0.043749 | 0.959420 |
| 2 | 0.721568 | 2.365659 | 0.006363 | 0.102502 |
| 3 | 0.710233 | 2.363735 | 0.000114 | 0.001708 |
| 4 | 0.710038 | 2.363701 | 0.000000 | 0.000001 |
| 5 | 0.710038 | 2.363701 | 0.000000 | 0.000000 |

**Table 2.** Root near (3.5, 0.5)

| $n$ | $x_n$ | $y_n$ | $f(x_n, y_n)$ | $g(x_n, y_n)$ |
|---|---|---|---|---|
| 0 | 3.500000 | 0.500000 | 1.000000 | 8.718750 |
| 1 | 3.308687 | 0.337643 | 0.058259 | 2.129355 |
| 2 | 3.273924 | 0.255909 | 0.011728 | 0.298145 |
| 3 | 3.266437 | 0.240187 | 0.000433 | 0.010584 |
| 4 | 3.266148 | 0.239587 | 0.000001 | 0.000015 |
| 5 | 3.266148 | 0.239586 | 0.000000 | 0.000000 |

**Table 3.** Root near (3, −0.5)

| $n$ | $x_n$ | $y_n$ | $f(x_n, y_n)$ | $g(x_n, y_n)$ |
|---|---|---|---|---|
| 0 | 3.000000 | −0.500000 | 1.000000 | 4.750000 |
| 1 | 2.977072 | −0.329806 | 0.062360 | 0.870022 |
| 2 | 3.004434 | −0.275270 | 0.005205 | 0.054976 |
| 3 | 3.006167 | −0.271350 | 0.000027 | 0.000318 |
| 4 | 3.006178 | −0.271327 | 0.000000 | 0.000000 |

**Table 4.** Root near (1, −2)

| $n$ | $x_n$ | $y_n$ | $f(x_n, y_n)$ | $g(x_n, y_n)$ |
|---|---|---|---|---|
| 0 | 1.000000 | −2.000000 | 1.000000 | 2.000000 |
| 1 | 0.847826 | −1.956522 | 0.033554 | 0.332865 |
| 2 | 0.803190 | −1.971447 | 0.001772 | 0.013843 |
| 3 | 0.801180 | −1.972071 | 0.000004 | 0.000028 |
| 4 | 0.801176 | −1.972072 | 0.000000 | 0.000000 |

The desired approximations to the four roots of the system are the $x_n$ and $y_n$ values in the last lines of the above tables. Note the rapidity of convergence in each case. However, many function evaluations are needed for each iteration of the method. For large systems Newton's method is computationally too inefficient to be practical. Other methods requiring more iterations but many fewer calculations per iteration are used in practice. ■

## EXPLORE! Implementing Newton's Method Using a Spreadsheet

A computer spreadsheet is an ideal environment in which to calculate Newton's method approximations. For a pair of equations in two unknowns such as the system in Example 1, you can proceed as follows:

(i) In the first nine cells of the first row (A1–I1) put the labels `n`, `x`, `y`, `f`, `g`, `f1`, `f2`, `g1`, and `g2`.

(ii) In cells A2–A9 put the numbers 0, 1, 2, . . . , 7.

(iii) In cells B2 and C2 put the starting values $x_0$ and $y_0$.

(iv) In cells D2–I2 put formulas for calculating $f(x, y)$, $g(x, y)$, . . . , $g_2(x, y)$ in terms of values of $x$ and $y$ assumed to be in B2 and C2.

(v) In cells B3 and C3 store the Newton's method formulas for calculating $x_1$ and $y_1$ in terms of the values $x_0$ and $y_0$, using values calculated in the second row. For instance, cell B3 should contain the formula

```
+B2-(D2*I2-G2*E2)/(F2*I2-G2*H2).
```

(vi) Replicate the formulas in cells D2–I2 to cells D3–I3.

(vii) Replicate the formulas in cells B3–I3 to the cells B4–I9.

You can now inspect the successive approximations $x_n$ and $y_n$ in columns B and C. To use different starting values, just replace the numbers in cells B2 and C2. To solve a different system of (two) equations, replace the contents of cells D2–I2. You may wish to save this spreadsheet for reuse with the exercises below, or other systems you may want to solve later.

***REMARK*** While a detailed analysis of the convergence of Newton's method approximations is beyond the scope of this book, a few observations can be made. At each step in the approximation process we must divide by $J$, the Jacobian determinant of $f$ and $g$ with respect to $x$ and $y$ evaluated at the most recently obtained approximation. Assuming that the functions and partial derivatives involved in the formulas are continuous, the larger the value of $J$ at the actual solution, the more likely are the approximations to converge to the solution, and to do so rapidly. If $J$ vanishes (or is very small) at the solution, the successive approximations may not converge, even if the initial guess is quite close to the solution. Even if the first partials of $f$ and $g$ are large at the solution, their Jacobian may be small if their gradients are nearly parallel there. Thus we cannot expect convergence to be rapid when the curves $f(x, y) = 0$ and $g(x, y) = 0$ intersect at a very small angle.

Newton's method can be applied to systems of $m$ equations in $m$ variables; the formulas are the obvious generalizations of those for two functions given above.

## EXERCISES 5.6

Find the solutions of the systems in Exercises 1–6, so that the left-hand sides of the equations vanish up to six decimal places. These can be done with the aid of a scientific calculator, but that approach will be very time consuming. It is much easier to program the Newton's method formulas on a computer to generate the required approximations. In each case try to determine reasonable *initial guesses* by sketching graphs of the equations.

**1.** $y - e^x = 0, \quad x - \sin y = 0$

**2.** $x^2 + y^2 - 1 = 0, \quad y - e^x = 0$ (two solutions)

**3.** $x^4 + y^2 - 16 = 0, \quad xy - 1 = 0$ (four solutions)

**4.** $x(1 + y^2) - 1 = 0, \quad y(1 + x^2) - 2 = 0$ (one solution)

**5.** $y - \sin x = 0, \quad x^2 + (y + 1)^2 - 2 = 0$ (two solutions)

**6.** $\sin x + \sin y - 1 = 0, \quad y^2 - x^3 = 0$ (two solutions)

* **7.** Write formulas for obtaining successive Newton's method approximations to a solution of the system

$$f(x, y, z) = 0, \quad g(x, y, z) = 0, \quad h(x, y, z) = 0,$$

starting from an initial guess $(x_0, y_0, z_0)$.

**8.** Use the formulas from Exercise 7 to find the first octant intersection point of the surfaces $y^2 + z^2 = 3$, $x^2 + z^2 = 2$, and $x^2 - z = 0$.

**9.** The equations $y - x^2 = 0$ and $y - x^3 = 0$ evidently have the solutions $x = y = 0$ and $x = y = 1$. Try to obtain these solutions using the two-variable form of Newton's method with starting values (a) $x_0 = y_0 = 0.1$, and (b) $x_0 = y_0 = 0.9$. How many iterations are required to obtain six decimal place accuracy for the appropriate solution in each case? How do you account for the difference in the behaviour of Newton's method for these equations near $(0, 0)$ and $(1, 1)$?

# CHAPTER REVIEW

## Key Ideas

- **What is meant by the following phrases?**
  - ◇ a critical point of $f(x, y)$
  - ◇ a singular point of $f(x, y)$
  - ◇ an absolute maximum value of $f(x, y)$
  - ◇ a local minimum value of $f(x, y)$
  - ◇ a saddle point of $f(x, y)$
  - ◇ a quadratic form
  - ◇ a constraint
  - ◇ linear programming
  - ◇ an envelope of a family of curves
- **State the second derivative test for a critical point of $f(x, y)$.**
- **Describe the method of Lagrange multipliers.**
- **Describe the method of least squares.**
- **Describe Newton's method for two equations.**

## Review Exercises

In Exercises 1–4, find and classify all the critical points of the given functions.

**1.** $xye^{-x+y}$

**2.** $x^2y - 2xy^2 + 2xy$

**3.** $\dfrac{1}{x} + \dfrac{4}{y} + \dfrac{9}{4 - x - y}$

**4.** $x^2y(2 - x - y)$

**5.** Let $f(x, y, z) = x^2 + y^2 + z^2 + 1/(x^2 + y^2 + z^2)$. Does $f$ have a minimum value? If so, what is it and where is it assumed?

**6.** Show that $x^2 + y^2 + z^2 - xy - xz - yz$ has a local minimum value at $(0, 0, 0)$. Is the minimum value assumed anywhere else?

**7.** Find the absolute maximum and minimum values of $f(x, y) = xye^{-x^2-4y^2}$. Justify your answer.

**8.** Let $f(x, y) = (4x^2 - y^2)e^{-x^2+y^2}$.

(a) Find the maximum and minimum values of $f(x, y)$ on the $xy$-plane.

(b) Find the maximum and minimum values of $f(x, y)$ on the wedge-shaped region $0 \le y \le 3x$.

**9.** A wire of length $L$ cm is cut into at most three pieces and each piece is bent into a square. What is the (a) minimum and (b) maximum of the sum of the areas of the squares?

**10.** A delivery service will accept parcels in the shape of rectangular boxes the sum of whose girth and height is at most 120 inches. (The girth is the perimeter of a horizontal cross-section.) What is the largest possible volume of such a box?

**11.** Find the area of the smallest ellipse $(x/a)^2 + (y/b)^2 = 1$ that contains the rectangle $-1 \le x \le 1$, $-2 \le y \le 2$.

**12.** Find the volume of the smallest ellipsoid

$$\frac{x^2}{a^2} + \frac{y^2}{b^2} + \frac{z^2}{c^2} = 1$$

that contains the rectangular box $-1 \le x \le 1$, $-2 \le y \le 2$, $-3 \le z \le 3$.

**13.** Find the volume of the smallest region of the form

$$0 \le z \le a\left(1 - \frac{x^2}{b^2} - \frac{y^2}{c^2}\right)$$

that contains the box $-1 \le x \le 1$, $-2 \le y \le 2$, $0 \le z \le 2$.

**14.** A window has the shape of a rectangle surmounted by an isosceles triangle. What are the dimensions $x$, $y$, and $z$ of the window (see the Figure 5.25) if its perimeter is $L$ and its area is maximum?

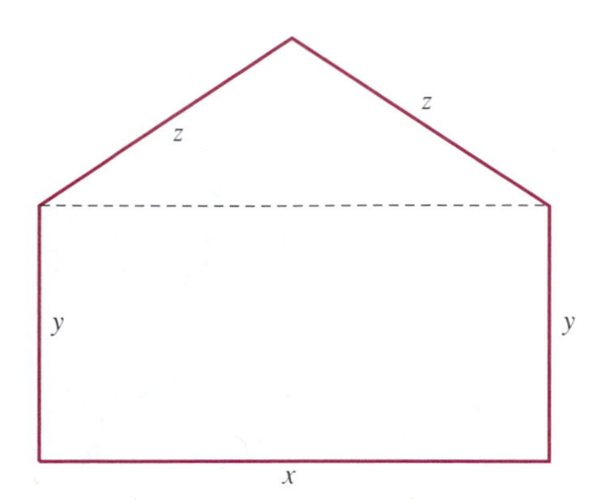


Figure 5.25

**15.** A widget manufacturer determines that if she manufactures $x$ thousands of widgets per month, and sells the widgets for $y$ dollars each, then her monthly profit (in thousands of dollars) will be $P = xy - \frac{1}{27}x^2y^3 - x$. If her factory is capable of producing at most 3,000 widgets per month, and government regulations prevent her from charging more than \$2 per widget, how many should she manufacture, and how much should she charge for each, to maximize her monthly profit?

**16.** Find the envelope of the curves $y = (x - c)^3 + 3c$.

**17.** Find an approximate solution $y(x, \epsilon)$ of the equation $y + \epsilon x e^y = -2x$ having terms up to second degree in $\epsilon$.

**18.** Find the two solutions of the system $xe^y + ye^x = 1$, $x^2 + y^2 = 1$.

## Express Yourself

Answer these questions in words, using complete sentences and, if necessary, paragraphs. Use mathematical symbols only where necessary.

**1.** If $\mathcal{S}$ is a set of points in 3-space, how can you decide whether a particular point is a boundary point of $\mathcal{S}$ or not? Must every nonempty set have boundary points? Can a set consist entirely of boundary points?

**2.** Describe various methods by which you can decide whether a function has a local maximum value or minimum value, or neither, at a critical point.

**3.** Discuss some circumstances under which a function of two variables must possess an absolute maximum value and/or an absolute minimum value.

**4.** What methods are available to enable you to find a maximum or minimum value of a function subject to one or more equality constraints? Discuss the advantages and disadvantages of these methods.

**5.** Why does calculus not play a major role in the solution of linear programming problems?

## Challenging Problems

**1.** **(Fourier series)**
Show that the constants $a_k$, $(k = 0, 1, 2, \ldots, n)$, and $b_k$, $(k = 1, 2, \ldots, n)$, that minimize the integral

$$I_n = \int_{-\pi}^{\pi} \left[ f(x) - \frac{a_0}{2} - \sum_{k=0}^{n} \Big( a_k \cos kx + b_k \sin kx \Big) \right]^2 dx$$

are given by

$$a_k = \frac{1}{\pi} \int_{\pi}^{\pi} f(x) \cos kx \, dx, \qquad b_k = \frac{1}{\pi} \int_{\pi}^{\pi} f(x) \sin kx \, dx.$$

Note that these numbers, called the **Fourier coefficients** of $f$ on $[-\pi, \pi]$, do not depend on $n$. If they can be calculated for all positive integers $k$, then the series

$$\frac{a_0}{2} + \sum_{k=0}^{\infty} \Big( a_k \cos kx + b_k \sin kx \Big)$$

is called the **full range Fourier series** of $f$.

**2.** This is a continuation of Exercise 1. Find the (full range) Fourier coefficients $a_k$ and $b_k$ of

$$f(x) = \begin{cases} 0 & \text{if } -\pi \le x < 0 \\ x & \text{if } 0 \le x \le \pi. \end{cases}$$

What is the minimum value of $I_n$ in this case? How does it behave as $n \to \infty$?

* **3.** Evaluate $\displaystyle\int_0^x \frac{\ln(tx + 1)}{1 + t^2} \, dt$.

* **4.** **(Steiner's problem)** The problem of finding a point in the plane (or a higher-dimensional space) that minimizes the sum of its distances from $n$ given points is very difficult. The case $n = 3$ is known as Steiner's problem. If $P_1P_2P_3$ is a triangle whose largest angle is less than $120°$, there is a point $Q$ inside the triangle so that the lines $QP_1$, $QP_2$, and $QP_3$ make equal $120°$ angles with one another. Show that the sum of the distances from the vertices of the triangle to a point $P$ is minimum when $P = Q$. *Hint:* first show that if $P = (x, y)$ and $P_i = (x_i, y_i)$, then

$$\frac{d|PP_i|}{dx} = \cos\theta_i \quad \text{and} \quad \frac{d|PP_i|}{dy} = \sin\theta_i,$$

where $\theta_i$ is the angle between $\overrightarrow{P_iP}$ and the positive direction of the $x$-axis. Hence show that the minimal point $P$ satisfies two trigonometric equations involving $\theta_1$, $\theta_2$, and $\theta_3$. Then try to show that any two of those angles differ by $\pm 2\pi/3$. Where should $P$ be taken if the triangle has an angle of $120°$ or greater?

CHAPTER 6

# Multiple Integration

**INTRODUCTION** In this chapter we extend the concept of definite integral to functions of several variables. Defined as limits of Riemann sums, like the one-dimensional definite integral, such multiple integrals can be evaluated using successive single definite integrals. They are used to represent and calculate quantities specified in terms of densities in regions of the plane or spaces of higher dimension. In the simplest instance, the volume of a three-dimensional region is given by a *double integral* of its height over the two-dimensional plane region that is its base.

## 6.1 DOUBLE INTEGRALS

The definition of the definite integral, $\int_a^b f(x)\,dx$, is motivated by the *standard area problem*, namely the problem of finding the area of the plane region bounded by the curve $y = f(x)$, the $x$-axis, and the lines $x = a$ and $x = b$. Similarly, we can motivate the double integral of a function of two variables over a domain $D$ in the plane by means of the *standard volume problem* of finding the volume of the three-dimensional region $S$ bounded by the surface $z = f(x, y)$, the $xy$-plane, and the cylinder parallel to the $z$-axis passing through the boundary of $D$. (See Figure 6.1. $D$ is called the **domain of integration**.) We will call such a three-dimensional region $S$ a "solid," although we are not implying that it is filled with any particular substance. We will define the double integral of $f(x, y)$ over the domain $D$,

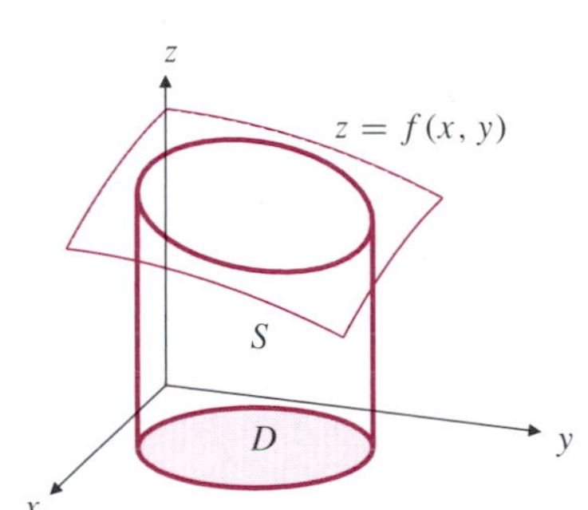


**Figure 6.1** A *solid* region $S$ lying above domain $D$ in the $xy$-plane, and below the surface $z = f(x, y)$.

$$\iint_D f(x, y)\,dA,$$

in such a way that its value will give the volume of the solid $S$ whenever $D$ is a "reasonable" domain and $f$ is a "reasonable" function with positive values.

Let us start with the case where $D$ is a closed rectangle with sides parallel to the coordinate axes in the $xy$-plane, and $f$ is a bounded function on $D$. If $D$ consists of the points $(x, y)$ such that $a \le x \le b$ and $c \le y \le d$, we can form a **partition** $P$ of $D$ into small rectangles by partitioning each of the intervals $[a, b]$ and $[c, d]$, say by points

$$a = x_0 < x_1 < x_2 < \cdots < x_{m-1} < x_m = b,$$
$$c = y_0 < y_1 < y_2 < \cdots < y_{n-1} < y_n = d.$$

The partition $P$ of $D$ then consists of the $mn$ rectangles $R_{ij}$ $(1 \le i \le m,\ 1 \le j \le n)$, consisting of points $(x, y)$ for which $x_{i-1} \le x \le x_i$ and $y_{j-1} \le y \le y_j$. (See Figure 6.2.)

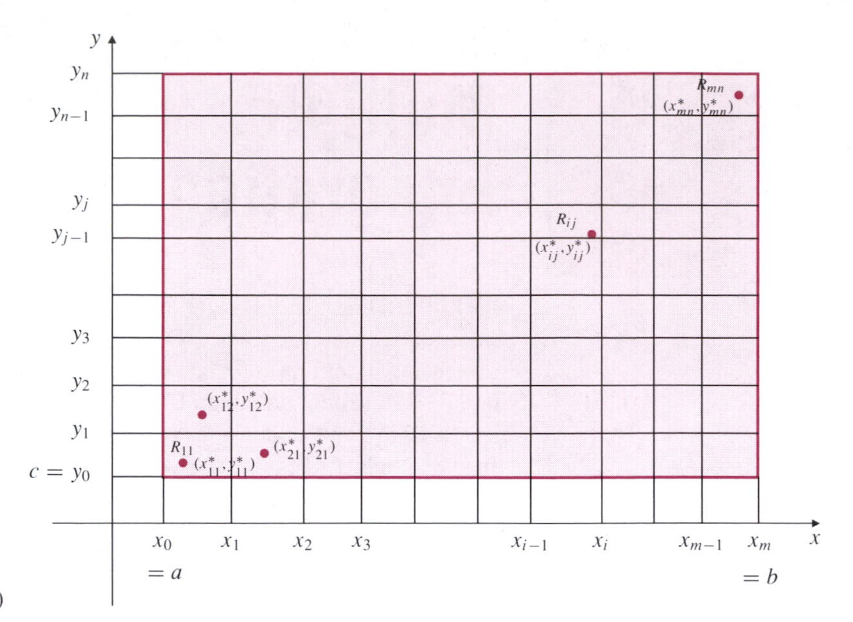


**Figure 6.2** A partition of $D$ (the large shaded rectangle) into smaller rectangles $R_{ij}$ $(1 \le i \le m,\ 1 \le j \le n)$

The rectangle $R_{ij}$ has area

$$\Delta A_{ij} = \Delta x_i \Delta y_j = (x_i - x_{i-1})(y_j - y_{j-1})$$

and *diameter* (that is, diagonal length)

$$\text{diam}(R_{ij}) = \sqrt{(\Delta x_i)^2 + (\Delta y_j)^2} = \sqrt{(x_i - x_{i-1})^2 + (y_j - y_{j-1})^2}.$$

The **norm** of the partition $P$ is the largest of these subrectangle diameters:

$$\|P\| = \max_{\substack{1 \le i \le m \\ 1 \le j \le n}} \text{diam}(R_{ij}).$$

Now we pick an arbitrary point $(x^*_{ij}, y^*_{ij})$ in each of the rectangles $R_{ij}$ and form the **Riemann sum**

$$R(f, P) = \sum_{i=1}^{m} \sum_{j=1}^{n} f(x^*_{ij}, y^*_{ij})\, \Delta A_{ij},$$

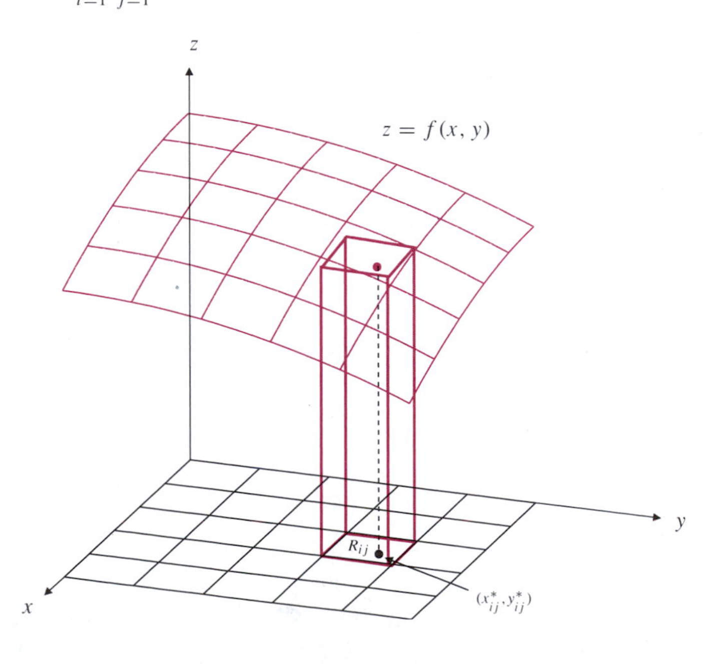


**Figure 6.3** A rectangular box above rectangle $R_{ij}$. The Riemann sum is a sum of volumes of such boxes

which is the sum of $mn$ terms, one for each rectangle in the partition. (Here the *double summation* indicates the sum as $i$ goes from 1 to $m$ of terms, each of which is itself a sum as $j$ goes from 1 to $n$.) The term corresponding to rectangle $R_{ij}$ is, if $f(x^*_{ij}, y^*_{ij}) \geq 0$, the volume of the rectangular box whose base is $R_{ij}$, and whose height is the value of $f$ at $(x^*_{ij}, y^*_{ij})$. (See Figure 6.3.) Therefore, for positive functions $f$, the Riemann sum $R(f, P)$ approximates the volume above $D$ and under the graph of $f$. The double integral of $f$ over $D$ is defined to be the limit of such Riemann sums, provided the limit exists as $\|P\| \to 0$ independently of how the points $(x^*_{ij}, y^*_{ij})$ are chosen. We make this precise in the following definition.

**DEFINITION 1**

**The double integral over a rectangle**

We say that $f$ is **integrable** over the rectangle $D$, and has **double integral**

$$I = \iint_D f(x, y)\, dA,$$

if for every positive number $\epsilon$ there exists a number $\delta$ depending on $\epsilon$, such that

$$|R(f, P) - I| < \epsilon$$

holds for every partition $P$ of $D$ satisfying $\|P\| < \delta$ and for all choices of the points $(x^*_{ij}, y^*_{ij})$ in the subrectangles of $P$.

The "$dA$" that appears in the expression for the double integral is an *area element*. It represents the limit of the $\Delta A = \Delta x\, \Delta y$ in the Riemann sum, and can also be written $dx\, dy$ or $dy\, dx$, the order being unimportant. When we evaluate double integrals by *iteration* in the next section, $dA$ will be replaced with a product of differentials $dx$ and $dy$, and the order will be important.

As is true for functions of one variable, functions that are continuous on $D$ are integrable on $D$. Of course, many bounded but discontinuous functions are also integrable, but an exact description of the class of integrable functions is beyond the scope of this text.

**EXAMPLE 1** Let $D$ be the square $0 \leq x \leq 1, 0 \leq y \leq 1$. Use a Riemann sum corresponding to the partition of $D$ into four smaller squares with points selected at the centre of each to find an approximate value for

$$\iint_D (x^2 + y)\, dA.$$

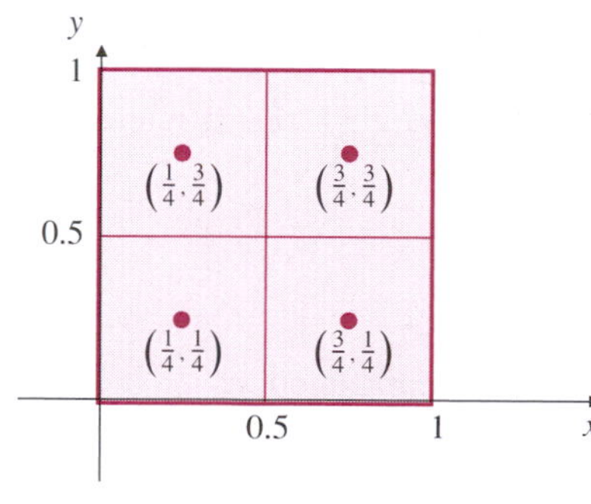


**Figure 6.4** The partitioned square of Example 1

***SOLUTION*** The required partition $P$ is formed by the lines $x = 1/2$ and $y = 1/2$, which divide $D$ into four squares, each of area $\Delta A = 1/4$. The centres of these squares are the points $\left(\frac{1}{4}, \frac{1}{4}\right)$, $\left(\frac{1}{4}, \frac{3}{4}\right)$, $\left(\frac{3}{4}, \frac{1}{4}\right)$, and $\left(\frac{3}{4}, \frac{3}{4}\right)$. (See Figure 6.4.) Therefore, the required approximation is

$$\begin{aligned}\iint_D (x^2 + y)\, dA \approx R(x^2 + y, P) &= \left(\frac{1}{16} + \frac{1}{4}\right)\frac{1}{4} + \left(\frac{1}{16} + \frac{3}{4}\right)\frac{1}{4} \\ &\quad + \left(\frac{9}{16} + \frac{1}{4}\right)\frac{1}{4} + \left(\frac{9}{16} + \frac{3}{4}\right)\frac{1}{4} = \frac{13}{16} = 0.8125.\end{aligned}$$

■

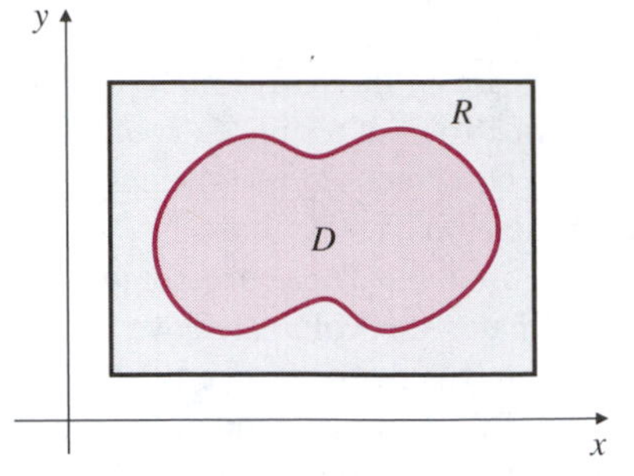


Figure 6.5

## Double Integrals over More General Domains

It is often necessary to use double integrals of bounded functions $f(x, y)$ over domains that are not rectangles. If the domain $D$ is *bounded*, we can choose a rectangle $R$ with sides parallel to the coordinate axes such that $D$ is contained inside $R$. (See Figure 6.5.) If $f(x, y)$ is defined on $D$ we can extend its domain to be $R$ by defining $f(x, y) = 0$ for points in $R$ that are outside of $D$. The integral of $f$ over $D$ can then be defined to be the integral of the extended function over the rectangle $R$.

**DEFINITION 2**

If $f(x, y)$ is defined and bounded on domain $D$, let $\hat{f}$ be the extension of $f$ that is zero everywhere outside $D$:

$$\hat{f}(x, y) = \begin{cases} f(x, y) & \text{if } (x, y) \text{ belongs to } D \\ 0 & \text{if } (x, y) \text{ does not belong to } D. \end{cases}$$

If $D$ is a *bounded* domain, then it is contained in some rectangle $R$ with sides parallel to the coordinate axes. We say that $f$ is **integrable** over $D$ and define the **double integral** of $f$ over $D$ to be

$$\iint_D f(x, y)\, dA = \iint_R \hat{f}(x, y)\, dA$$

provided that $\hat{f}$ is integrable over $R$.

This definition makes sense because the values of $\hat{f}$ in the part of $R$ outside of $D$ are all zero, so do not contribute anything to the value of the integral. However, even if $f$ is continuous on $D$, $\hat{f}$ will not be continuous on $R$ unless $f(x, y) \to 0$ as $(x, y)$ approaches the boundary of $D$. Nevertheless, if $f$ and $D$ are "well-behaved," the integral will exist. We cannot delve too deeply into what constitutes *well-behaved*, but assert, without proof, the following theorem that will assure us that most of the double integrals we encounter do, in fact, exist.

**THEOREM 1**

If $f$ is a continuous function defined on a *closed, bounded* domain $D$ whose boundary consists of finitely many curves of finite length, then $f$ is integrable on $D$.

According to Theorem 2 of Section 5.1, a continuous function is bounded if its domain is closed and bounded. Generally, however, it is not necessary to restrict our domains to be closed. If $D$ is a bounded domain and int($D$) is its interior (an open set), and if $f$ is integrable on $D$, then

$$\iint_D f(x, y)\, dA = \iint_{\text{int}(D)} f(x, y)\, dA.$$

We will discuss *improper double integrals* of unbounded functions or over unbounded domains in Section 6.3.

## Properties of the Double Integral

Some properties of double integrals are analogous to properties of the one-dimensional definite integral and require little comment: if $f$ and $g$ are integrable over $D$, and if $L$ and $M$ are constants, then

(a) **Zero area:** $\iint_D f(x, y)\, dA = 0$ if $D$ has zero area.

(b) **Volume of a cylinder:** $\iint_D 1\, dA =$ area of $D$.

(c) **Integrals representing volumes:**
If $f(x, y) \ge 0$ on $D$, then $\iint_D f(x, y)\, dA = V \ge 0$, where $V$ is the volume of the solid lying vertically above $D$ and below the surface $z = f(x, y)$.

(d) If $f(x, y) \le 0$ on $D$, then $\iint_D f(x, y)\, dA = -V \le 0$, where $V$ is the volume of the solid lying vertically below $D$ and above the surface $z = f(x, y)$.

(e) **Linear dependence on the integrand:**
$$\iint_D \Big(Lf(x, y) + Mg(x, y)\Big)\, dA = L \iint_D f(x, y)\, dA + M \iint_D g(x, y)\, dA.$$

(f) **Inequalities are preserved:**
If $f(x, y) \le g(x, y)$ on $D$, then $\iint_D f(x, y)\, dA \le \iint_D g(x, y)\, dA$.

(g) **The triangle inequality:** $\left|\iint_D f(x, y)\, dA\right| \le \iint_D |f(x, y)|\, dA$.

(h) **Additivity of domains:** If $D_1, D_2, \ldots, D_k$ are nonoverlapping domains on each of which $f$ is integrable, then $f$ is integrable over the union $D = D_1 \cup D_2 \cup \cdots \cup D_k$ and

$$\iint_D f(x, y)\, dA = \sum_{j=1}^{k} \iint_{D_j} f(x, y)\, dA.$$

Nonoverlapping domains can share boundary points, but have no interior points in common.

## Double Integrals by Inspection

As yet we have not said anything about how to *evaluate* a double integral. The main technique for doing this, called *iteration*, will be developed in the next section, but it is worth pointing out that double integrals can sometimes be evaluated using symmetry arguments or by interpreting them as volumes that we already know.

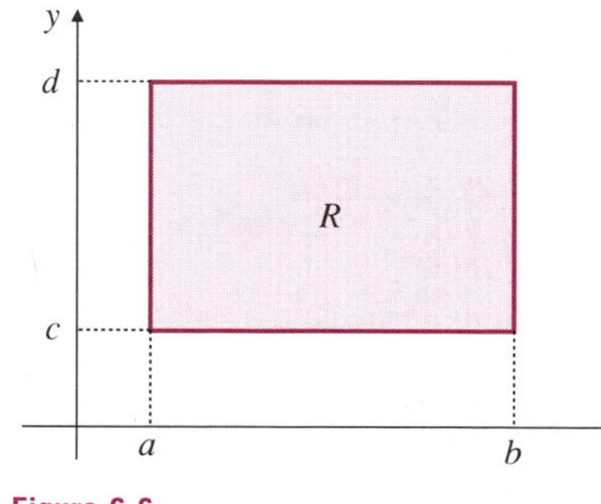

Figure 6.6

**EXAMPLE 2** If $R$ is the rectangle $a \le x \le b$, $c \le y \le d$, then

$$\iint_R 3\, dA = 3 \times \text{area of } R = 3(b - a)(d - c).$$

Here the integrand is $f(x, y) = 3$ and the integral is equal to the volume of the solid box of height 3 whose base is the rectangle $R$. (See Figure 6.6.) ■

**EXAMPLE 3** Evaluate $I = \iint_{x^2+y^2\le 1} (\sin x + y^3 + 3)\, dA$.

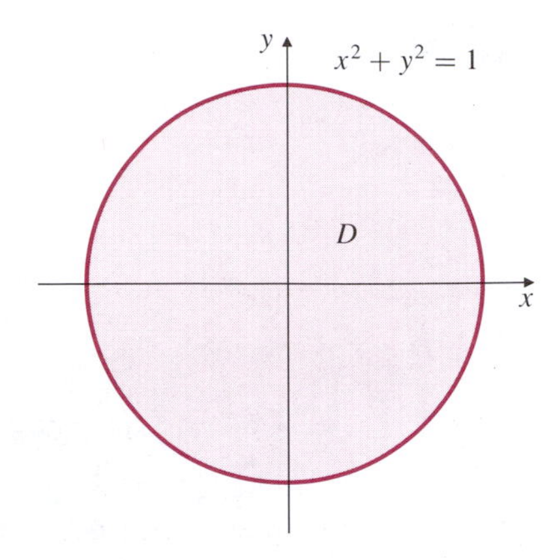

Figure 6.7

***SOLUTION*** The integral can be expressed as the sum of three integrals by property (e) of double integrals:

$$\begin{aligned} I &= \iint_{x^2+y^2\le 1} \sin x\, dA + \iint_{x^2+y^2\le 1} y^3\, dA + \iint_{x^2+y^2\le 1} 3\, dA \\ &= I_1 + I_2 + I_3. \end{aligned}$$

The domain of integration (Figure 6.7) is a circular disk of radius 1 centred at the origin. Since $f(x, y) = \sin x$ is an *odd* function of $x$, its graph bounds as much volume below the $xy$-plane in the region $x < 0$ as it does above the $xy$-plane in the region $x > 0$. These two contributions to the double integral cancel, and so $I_1 = 0$. Note that symmetry of *both* the domain *and* the integrand is necessary for this argument.

Similarly, $I_2 = 0$ because $y^3$ is an odd function and $D$ is symmetric about the $x$-axis.

Finally,

$$I_3 = \iint_D 3\, dA = 3 \times \text{area of } D = 3\pi.$$

Thus $I = 0 + 0 + 3\pi = 3\pi$. ■

**■ EXAMPLE 4** If $D$ is the disk of Example 3, the integral

$$\iint_D \sqrt{1 - x^2 - y^2}\, dA$$

represents the volume of a hemisphere of radius 1, and so has the value $2\pi/3$. ■

When evaluating double integrals, always be alert for situations such as those in the above examples. You can save much time by not trying to calculate an integral whose value should be obvious without calculation.

## EXERCISES 6.1

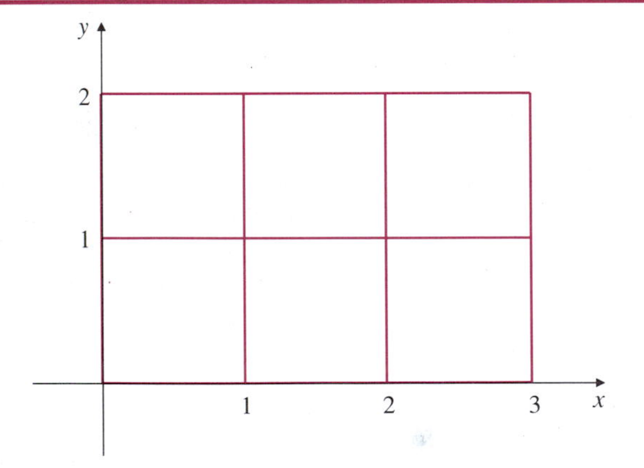

Figure 6.8

Exercises 1–6 refer to the double integral

$$I = \iint_D (5 - x - y)\, dA,$$

where $D$ is the rectangle $0 \le x \le 3$, $0 \le y \le 2$. $P$ is the partition of $D$ into six squares of side 1 as shown in Figure 6.8. In Exercises 1–5, calculate the Riemann sums for $I$ corresponding to the given choices of points $(x^*_{ij}, y^*_{ij})$.

**1.** $(x^*_{ij}, y^*_{ij})$ is the upper left corner of each square.

**2.** $(x^*_{ij}, y^*_{ij})$ is the upper right corner of each square.

**3.** $(x^*_{ij}, y^*_{ij})$ is the lower left corner of each square.

**4.** $(x^*_{ij}, y^*_{ij})$ is the lower right corner of each square.

**5.** $(x^*_{ij}, y^*_{ij})$ is the centre of each square.

**6.** Evaluate $I$ by interpreting it as a volume.

In Exercises 7–10, $D$ is the disk $x^2 + y^2 \le 25$ and $P$ is the partition of the square $-5 \le x \le 5$, $-5 \le y \le 5$ into one hundred $1 \times 1$ squares, as shown in Figure 6.9. Approximate the double integral

$$J = \iint_D f(x, y)\, dA,$$

where $f(x, y) = 1$ by calculating the Riemann sums $R(f, P)$ corresponding to the indicated choice of points in the small squares. *Hint:* using symmetry will make the job easier.

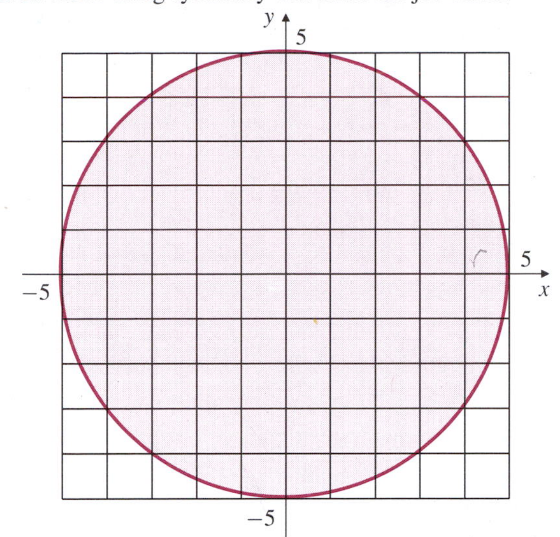


Figure 6.9

**7.** $(x_{ij}^*, y_{ij}^*)$ is the corner of each square closest to the origin.

**8.** $(x_{ij}^*, y_{ij}^*)$ is the corner of each square farthest from the origin.

**9.** $(x_{ij}^*, y_{ij}^*)$ is the centre of each square.

**10.** Evaluate $J$.

**11.** Repeat Exercise 5 using the integrand $e^x$ instead of $5 - x - y$.

**12.** Repeat Exercise 9 using $f(x, y) = x^2 + y^2$ instead of $f(x, y) = 1$.

In Exercises 13–22, evaluate the given double integral by inspection.

**13.** $\iint_R dA$, where $R$ is the rectangle $-1 \le x \le 3$, $-4 \le y \le 1$

**14.** $\iint_D (x + 3)\, dA$, where $D$ is the half-disk $0 \le y \le \sqrt{4 - x^2}$

**15.** $\iint_T (x + y)\, dA$, where $T$ is the parallelogram having the points (2,2), (1, −1), (−2, −2), and (−1, 1) as vertices

**16.** $\iint_{|x|+|y|\le 1} \left(x^3 \cos(y^2) + 3 \sin y - \pi\right) dA$

**17.** $\iint_{x^2+y^2\le 1} (4x^2y^3 - x + 5)\, dA$

**18.** $\iint_{x^2+y^2\le a^2} \sqrt{a^2 - x^2 - y^2}\, dA$

**19.** $\iint_{x^2+y^2\le a^2} (a - \sqrt{x^2 + y^2})\, dA$

**20.** $\iint_S (x + y)\, dA$, where $S$ is the square $0 \le x \le a, 0 \le y \le a$

**21.** $\iint_T (1 - x - y)\, dA$, where $T$ is the triangle with vertices (0,0), (1,0), and (0,1)

**22.** $\iint_R \sqrt{b^2 - y^2}\, dA$, where $R$ is the rectangle $0 \le x \le a, 0 \le y \le b$

## 6.2 ITERATION OF DOUBLE INTEGRALS IN CARTESIAN COORDINATES

The existence of the double integral $\iint_D f(x, y)\, dA$ depends on $f$ and the domain $D$. As we shall see, evaluation of double integrals is easiest when the domain of integration is of *simple* type.

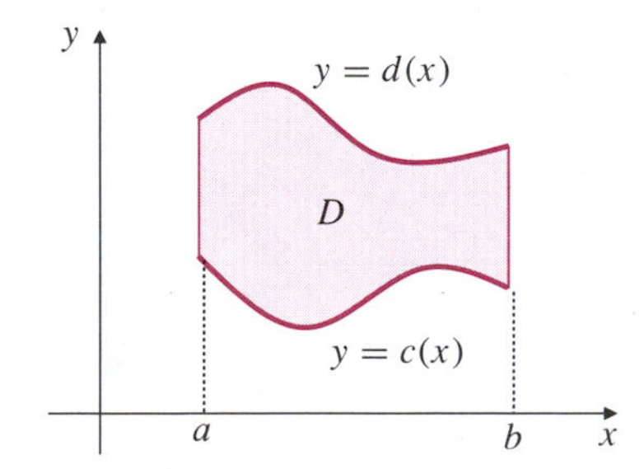


Figure 6.10 A $y$-simple domain

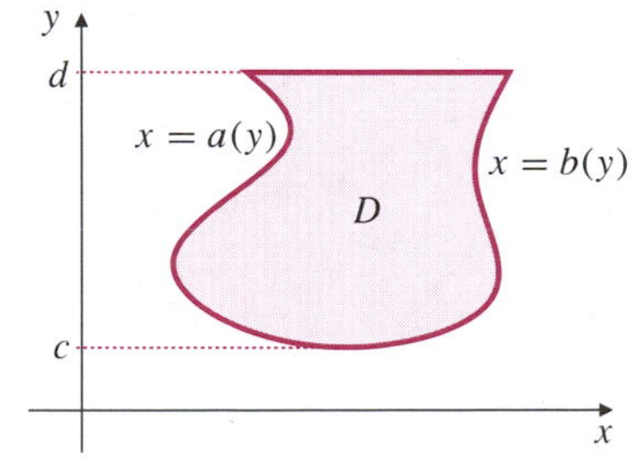


Figure 6.11 An $x$-simple domain

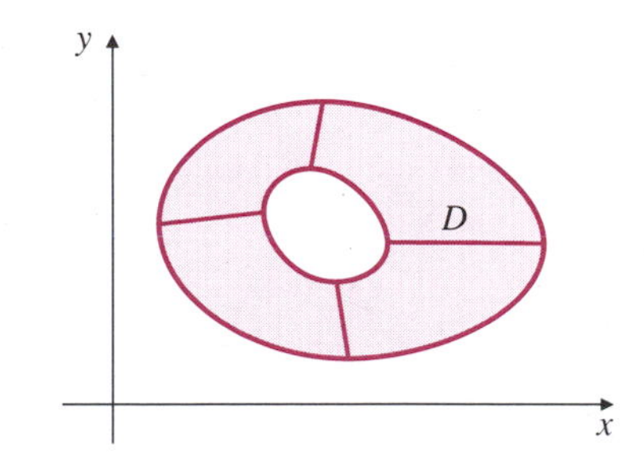


Figure 6.12 A regular domain

We say that the domain $D$ in the $xy$-plane is $y$-**simple** if it is bounded by two vertical lines $x = a$ and $x = b$, and two continuous graphs $y = c(x)$ and $y = d(x)$ between these lines. (See Figure 6.10.) Lines parallel to the $y$-axis intersect a $y$-simple domain in an *interval* (possibly a single point) if at all. Similarly, $D$ is $x$-**simple** if it is bounded by horizontal lines $y = c$ and $y = d$, and two continuous graphs $x = a(y)$ and $x = b(y)$ between these lines. (See Figure 6.11.) Many of the domains over which we will take integrals are $y$-simple, $x$-simple, or both. For example, rectangles, triangles, and disks are both $x$-simple and $y$-simple. Those domains that are neither one or the other will usually be unions of finitely many nonoverlapping subdomains that are both $x$-simple and $y$-simple. We will call such domains **regular**. The shaded region in Figure 6.12 is divided into four subregions, each of which is both $x$-simple and $y$-simple.

It can be shown that a bounded, continuous function $f(x, y)$ is integrable over a bounded $x$-simple or $y$-simple domain, and therefore over any regular domain.

Unlike the examples in the previous section, most double integrals cannot be evaluated by inspection. We need a technique for evaluating double integrals similar to the technique for evaluating single definite integrals in terms of antiderivatives. Since the double integral represents a volume we can evaluate it for simple domains by a slicing technique.

Suppose, for instance, that $D$ is $y$-simple, and is bounded by $x = a$, $x = b$, $y = c(x)$, and $y = d(x)$, as shown in Figure 6.13(a). Then $\iint_D f(x, y)\, dA$ represents the volume of the solid region inside the vertical cylinder through the boundary of $D$ and between the $xy$-plane and the surface $z = f(x, y)$. Consider the cross-section of this solid in the vertical plane perpendicular to the $x$-axis at position $x$. Note that $x$ is constant in that plane. If we use the projections of the $y$- and $z$-axes onto the plane as coordinate axes there, the cross-section is a plane region bounded by vertical lines $y = c(x)$ and $y = d(x)$, by the horizontal line $z = 0$, and by the curve $z = f(x, y)$. The area of the cross-section is therefore given by

$$A(x) = \int_{c(x)}^{d(x)} f(x, y)\, dy.$$

The double integral, $\iint_D f(x, y)\, dA$ is obtained by summing the volumes of "thin" slices of area $A(x)$ and thickness $dx$ between $x = a$ and $x = b$ and is therefore given by

$$\iint_D f(x, y)\, dA = \int_a^b A(x)\, dx = \int_a^b \left( \int_{c(x)}^{d(x)} f(x, y)\, dy \right) dx.$$

Notationally, it is common to omit the large parentheses and write

$$\iint_D f(x, y)\, dA = \int_a^b \int_{c(x)}^{d(x)} f(x, y)\, dy\, dx,$$

or

$$\iint_D f(x, y)\, dA = \int_a^b dx \int_{c(x)}^{d(x)} f(x, y)\, dy.$$

The latter form shows more clearly which variable corresponds to which limits of integration.

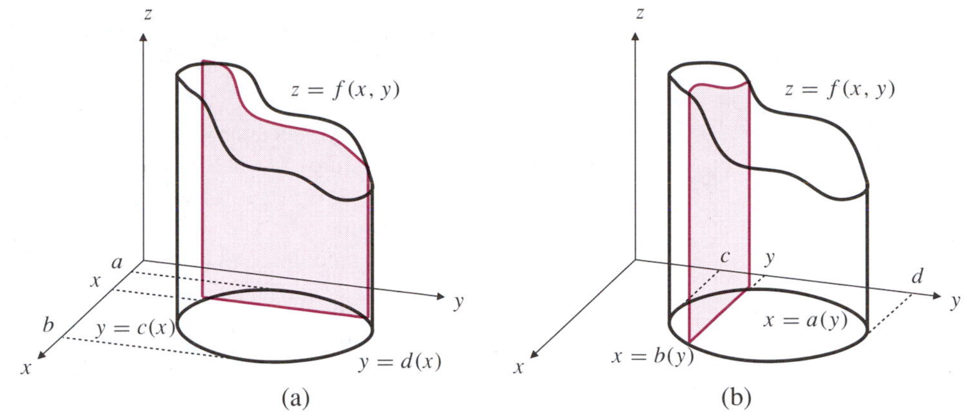

**Figure 6.13**

(a) Integrals over $y$-simple domains should be sliced perpendicular to the $x$-axis

(b) Integrals over $x$-simple domains should be sliced perpendicular to the $y$-axis

The expressions on the right-hand sides of the above formulas are called **iterated** integrals. **Iteration** is the process of reducing the problem of evaluating a double (or multiple) integral to one of evaluating two (or more) successive single definite integrals. In the above iteration, the integral

$$\int_{c(x)}^{d(x)} f(x, y)\, dy$$

is called the **inner** integral since it must be evaluated first. It is evaluated using standard techniques, treating $x$ as a constant. The result of this evaluation is a function of $x$ alone (note that both the integrand and the limits of the inner integral can depend on $x$), and is the integrand of the **outer** integral in which $x$ is the variable of integration.

For double integrals over $x$-simple domains we can slice perpendicular to the $y$ axis and obtain an iterated integral with the outer integral in the $y$ direction. (See Figure 6.13(b).) We summarize the above discussion in the following theorem.

**THEOREM 2**

**Iteration of Double Integrals**

If $f(x, y)$ is continuous on the bounded $y$-simple domain $D$ given by $a \leq x \leq b$ and $c(x) \leq y \leq d(x)$, then

$$\iint_D f(x, y)\, dA = \int_a^b dx \int_{c(x)}^{d(x)} f(x, y)\, dy.$$

Similarly, if $f$ is continuous on the $x$-simple domain $D$ given by $c \leq y \leq d$ and $a(y) \leq x \leq b(y)$, then

$$\iint_D f(x, y)\, dA = \int_c^d dy \int_{a(y)}^{b(y)} f(x, y)\, dx.$$

***REMARK*** The symbol $dA$ in the double integral is replaced in the iterated integrals by the $dx$ and the $dy$. Accordingly, $dA$ is frequently written $dx\, dy$ or $dy\, dx$ even in the double integral. The three expressions

$$\iint_D f(x, y)\, dx\, dy, \qquad \iint_D f(x, y)\, dy\, dx, \qquad \text{and} \qquad \iint_D f(x, y)\, dA$$

all stand for the double integral of $f$ over $D$. Only when the double integral is iterated does the order of $dx$ and $dy$ become important. Later in this chapter we will iterate double integrals in polar coordinates and $dA$ will take the form $r\,dr\,d\theta$.

It is not always necessary to make a three-dimensional sketch of the solid volume represented by a double integral. In order to iterate the integral properly (in one direction or the other) it is usually sufficient to make a sketch of *the domain D* over which the integral is taken. The direction of iteration can be shown by a line along which the inner integral is taken. The following examples illustrate this.

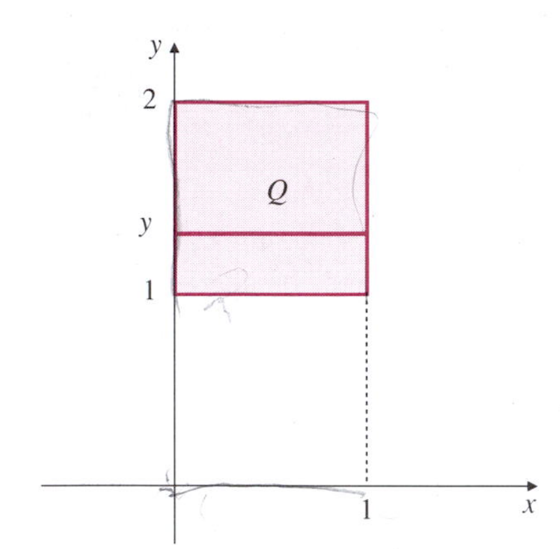

**Figure 6.14** The horizontal line through $Q$ indicates iteration with the inner integral in the $x$ direction

**EXAMPLE 1** Find the volume of the solid lying above the square $Q$ defined by $0 \le x \le 1$, $1 \le y \le 2$, and below the plane $z = 4 - x - y$.

***SOLUTION*** The square $Q$ is both $x$-simple and $y$-simple, so the double integral giving the volume can be iterated in either direction. We will do it both ways just for practice. The horizontal line at height $y$ in Figure 6.14 suggests that we first integrate with respect to $x$ along this line (from 0 to 1) and then integrate the result with respect to $y$ from 1 to 2. Iterating the double integral in this direction, we calculate

$$\begin{aligned}\text{Volume of } Q &= \iint_Q (4 - x - y)\,dA \\ &= \int_1^2 dy \int_0^1 (4 - x - y)\,dx \\ &= \int_1^2 dy \left(4x - \frac{x^2}{2} - xy\right)\Bigg|_0^1 \\ &= \int_1^2 \left(\frac{7}{2} - y\right) dy \\ &= \left(\frac{7y}{2} - \frac{y^2}{2}\right)\Bigg|_1^2 = 2 \text{ cubic units.}\end{aligned}$$

Using the opposite iteration, as suggested by Figure 6.15, we calculate

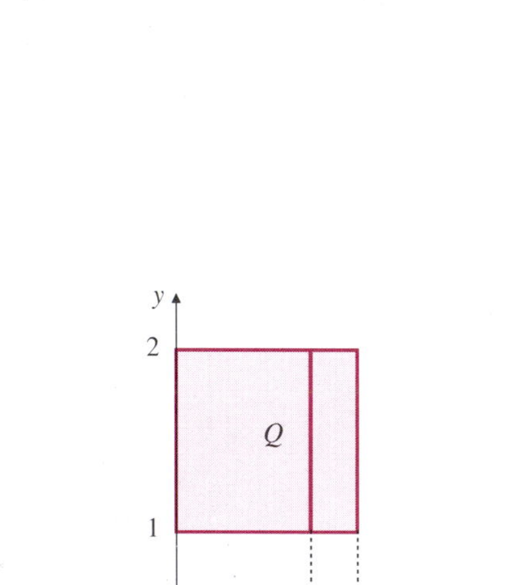

**Figure 6.15** The vertical line through $Q$ indicates iteration with the inner integral in the $y$ direction

$$\begin{aligned}\text{Volume of } Q &= \iint_Q (4 - x - y)\,dA \\ &= \int_0^1 dx \int_1^2 (4 - x - y)\,dy \\ &= \int_0^1 dx \left(4y - xy - \frac{y^2}{2}\right)\Bigg|_1^2 \\ &= \int_0^1 \left(\frac{5}{2} - x\right) dx \\ &= \left(\frac{5x}{2} - \frac{x^2}{2}\right)\Bigg|_0^1 = 2 \text{ cubic units.}\end{aligned}$$

It is comforting to get the same answer both ways! Note that because $Q$ is a rectangle with sides parallel to the coordinate axes, the limits of the inner integrals do not depend on the variables of the outer integrals in either iteration. This cannot be expected to happen with more general domains. ■

**EXAMPLE 2** Evaluate $\iint_T xy\,dA$ over the triangle $T$ with vertices $(0, 0)$, $(1, 0)$, and $(1, 1)$.

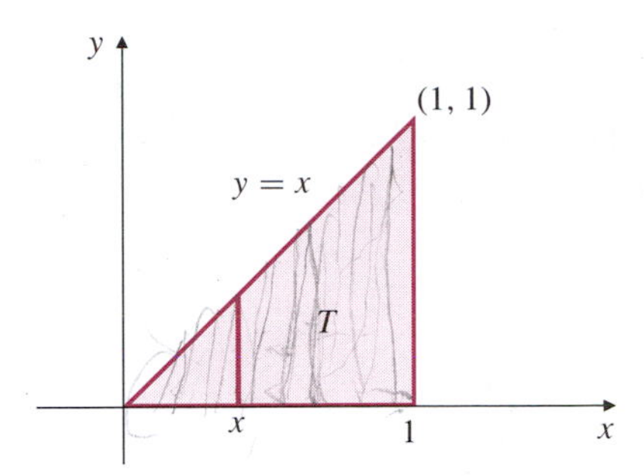


Figure 6.16 The triangular domain $T$ with vertical line indicating iteration with inner integral in the $y$ direction

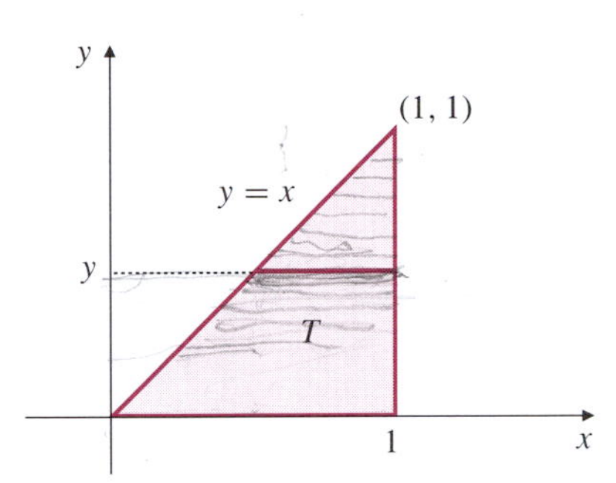


Figure 6.17 The triangular domain $T$ with horizontal line indicating iteration with inner integral in the $x$ direction

**SOLUTION** The triangle $T$ is shown in Figure 6.16. It is both $x$-simple and $y$-simple. Using the iteration corresponding to slicing in the direction shown in the figure, we obtain:

$$\begin{aligned}\iint_T xy\,dA &= \int_0^1 dx \int_0^x xy\,dy \\ &= \int_0^1 dx \left(\frac{xy^2}{2}\right)\bigg|_{y=0}^{y=x} \\ &= \int_0^1 \frac{x^3}{2}\,dx = \frac{x^4}{8}\bigg|_0^1 = \frac{1}{8}.\end{aligned}$$

Iteration in the other direction (Figure 6.17) leads to the same value:

$$\begin{aligned}\iint_T xy\,dA &= \int_0^1 dy \int_y^1 xy\,dx \\ &= \int_0^1 dy \left(\frac{yx^2}{2}\right)\bigg|_{x=y}^{x=1} \\ &= \int_0^1 \frac{y}{2}(1-y^2)\,dy \\ &= \left(\frac{y^2}{4} - \frac{y^4}{8}\right)\bigg|_0^1 = \frac{1}{8}.\end{aligned}$$

■

In both of the examples above the double integral could be evaluated easily using either possible iteration. (We did them both ways just to illustrate that fact.) It often occurs, however, that a double integral is easily evaluated if iterated in one direction, and very difficult, or impossible, if iterated in the other direction. Sometimes you will even encounter iterated integrals whose evaluation requires that they be expressed as double integrals and then reiterated in the opposite direction.

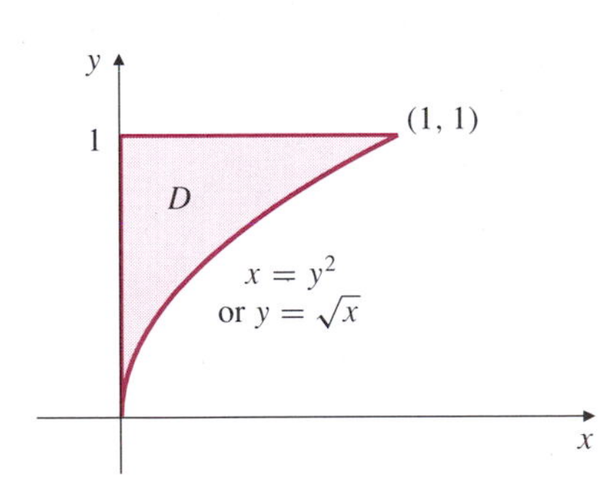


Figure 6.18 The region corresponding to the iterated integral in Example 3

**EXAMPLE 3** Evaluate the iterated integral $I = \displaystyle\int_0^1 dx \int_{\sqrt{x}}^1 e^{y^3}\,dy.$

**SOLUTION** We cannot antidifferentiate $e^{y^3}$ to evaluate the inner integral in this iteration, so we express $I$ as a double integral and identify the region over which it is taken:

$$I = \iint_D e^{y^3}\,dA$$

where $D$ is the region shown in Figure 6.18. Reiterating with the $x$ integration on the inside we get

$$\begin{aligned}I &= \int_0^1 dy \int_0^{y^2} e^{y^3}\,dx \\ &= \int_0^1 e^{y^3}\,dy \int_0^{y^2} dx \\ &= \int_0^1 y^2 e^{y^3}\,dy = \frac{e^{y^3}}{3}\bigg|_0^1 = \frac{e-1}{3}.\end{aligned}$$

■

The following is an example of the calculation of the volume of a somewhat awkward solid. Even though it is not always necessary to sketch solids to find their volumes, you are encouraged to sketch them whenever possible. When we encounter triple integrals over three-dimensional regions later in this chapter it will usually be necessary to sketch the regions. Get as much practice as you can.

**EXAMPLE 4** Sketch and find the volume of the solid bounded by the planes $y = 0$, $z = 0$, and $z = a - x + y$, and the parabolic cylinder $y = a - (x^2/a)$, where $a$ is a positive constant.

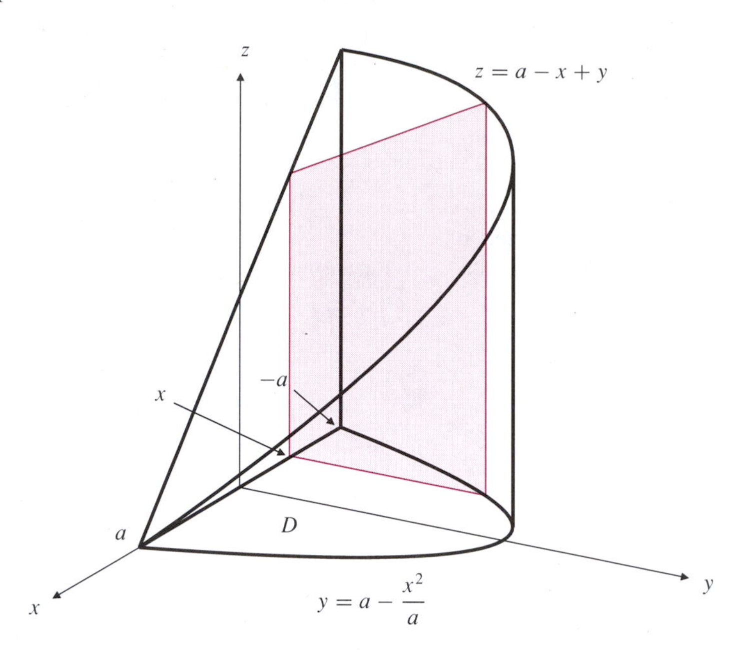


**Figure 6.19** The solid in Example 4, sliced perpendicular to the $x$-axis

***SOLUTION*** The solid is shown in Figure 6.19. Its base is the parabolic segment $D$ in the $xy$-plane bounded by $y = 0$ and $y = a - (x^2/a)$, so the volume of the solid is given by

$$V = \iint_D (a - x + y)\, dA = \iint_D (a + y)\, dA.$$

(Note how we used symmetry to drop the $x$ term from the integrand. This term is an odd function of $x$ and $D$ is symmetric about the $y$-axis.) Iterating the double integral in the direction suggested by the slice shown in the figure, we obtain

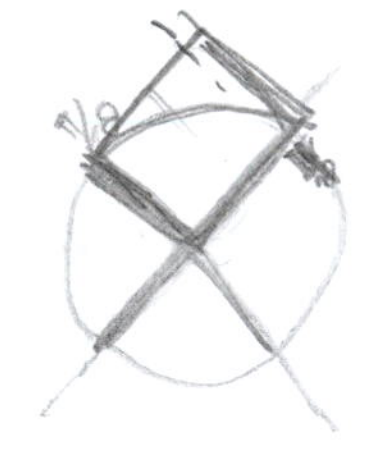

$$\begin{aligned}
V &= \int_{-a}^{a} dx \int_0^{a-(x^2/a)} (a + y)\, dy \\
&= \int_{-a}^{a} \left(ay + \frac{y^2}{2}\right)\bigg|_0^{a-(x^2/a)} dx \\
&= \int_{-a}^{a} \left[a^2 - x^2 + \frac{1}{2}\left(a^2 - 2x^2 + \frac{x^4}{a^2}\right)\right] dx \\
&= 2\int_0^{a} \left[\frac{3}{2}a^2 - 2x^2 + \frac{x^4}{2a^2}\right] dx \\
&= \left(3a^2 x - \frac{4x^3}{3} + \frac{x^5}{5a^2}\right)\bigg|_0^{a} \\
&= 3a^3 - \frac{4}{3}a^3 + \frac{1}{5}a^3 = \frac{28}{15}a^3 \text{ cubic units.}
\end{aligned}$$

■

## EXPLORE! Evaluating Double Integrals Using Computer Algebra

Iterated double integrals can be evaluated with the help of a computer algebra system such as Derive, Maple, or Mathematica. When the definite integrals involved can be evaluated symbolically, these programs will return a symbolic answer. For example, Maple can be given the task of evaluating the iterated integral in Example 4 by issuing the command

```
int(int(a+y,y=0..a-(x^2/a)),x=-a..a);
```

and it will respond with the value $\frac{28}{15}a^3$.

Even when the program cannot evaluate the integrals in symbolic form, you can ask for an approximate value. If charged to evaluate

$$\int_0^1 \int_0^1 e^{((x+y)/2)^4}\, dx\, dy,$$

Maple will respond by returning the integral unevaluated, but the command

```
evalf(Int(Int(exp(((x+y)/2)^4),x=0..1),y=0..1),5);
```

will return the value of the integral to five significant figures, namely 1.1556. The calculation may take a considerable time, depending on your processor.

(a) Evaluate $\iint_T y\sin(x+y)\, dA$ over the triangle $0 \le x \le \pi/2$, $0 \le y \le 1-x$.

(b) Evaluate $\int_0^1 \int_0^x \sqrt{1+e^{xy}}\, dy\, dx$ to five significant figures.

You may wish to use a computer algebra system to check your answers to the exercises below and for other multiple integrals that occur later.

## EXERCISES 6.2

In Exercises 1–4, calculate the given iterated integrals.

**1.** $\int_0^1 dx \int_0^x (xy + y^2)\, dy$ **2.** $\int_0^1 \int_0^y (xy + y^2)\, dx\, dy$

**3.** $\int_0^\pi \int_{-x}^x \cos y\, dy\, dx$ **4.** $\int_0^2 dy \int_0^y y^2 e^{xy}\, dx$

In Exercises 5–14, evaluate the double integrals by iteration.

**5.** $\iint_R (x^2 + y^2)\, dA$, where $R$ is the rectangle $0 \le x \le a$, $0 \le y \le b$

**6.** $\iint_R x^2y^2\, dA$, where $R$ is the rectangle of Exercise 5

**7.** $\iint_S (\sin x + \cos y)\, dA$, where $S$ is the square $0 \le x \le \pi/2$, $0 \le y \le \pi/2$

**8.** $\iint_T (x - 3y)\, dA$, where $T$ is the triangle with vertices $(0,0)$, $(a,0)$, and $(0,b)$

**9.** $\iint_R xy^2\, dA$, where $R$ is the finite region in the first quadrant bounded by the curves $y = x^2$ and $x = y^2$

**10.** $\iint_D x\cos y\, dA$, where $D$ is the finite region in the first quadrant bounded by the coordinate axes and the curve $y = 1 - x^2$

**11.** $\iint_D \ln x\, dA$, where $D$ is the finite region in the first quadrant bounded by the line $2x + 2y = 5$ and the hyperbola $xy = 1$

**12.** $\iint_T \sqrt{a^2 - y^2}\, dA$, where $T$ is the triangle with vertices $(0,0)$, $(a,0)$, and $(a,a)$.

13. $\displaystyle\iint_R \frac{x}{y} e^y \, dA$, where $R$ is the region $0 \le x \le 1$, $x^2 \le y \le x$

14. $\displaystyle\iint_T \frac{xy}{1+x^4} \, dA$, where $T$ is the triangle with vertices $(0,0)$, $(0,1)$, and $(1,1)$

In Exercises 15–18, sketch the domain of integration and evaluate the given iterated integrals.

15. $\displaystyle\int_0^1 dy \int_y^1 e^{-x^2} \, dx$

16. $\displaystyle\int_0^{\pi/2} dy \int_y^{\pi/2} \frac{\sin x}{x} \, dx$

17. $\displaystyle\int_0^1 dx \int_x^1 \frac{y^\lambda}{x^2+y^2} \, dy \quad (\lambda > 0)$

18. $\displaystyle\int_0^1 dx \int_x^{x^{1/3}} \sqrt{1-y^4} \, dy$

In Exercises 19–28, find the volumes of the indicated solids.

19. Under $z = 1 - x^2$ and above the region $0 \le x \le 1$, $0 \le y \le x$

20. Under $z = 1 - x^2$ and above the region $0 \le y \le 1$, $0 \le x \le y$

21. Under $z = 1 - x^2 - y^2$ and above the region $x \ge 0$, $y \ge 0$, $x + y \le 1$

22. Under $z = 1 - y^2$ and above $z = x^2$

23. Under the surface $z = 1/(x+y)$ and above the region in the $xy$-plane bounded by $x = 1$, $x = 2$, $y = 0$, and $y = x$

24. Under the surface $z = x^2 \sin(y^4)$ and above the triangle in the $xy$-plane with vertices $(0,0)$, $(0, \pi^{1/4})$, and $(\pi^{1/4}, \pi^{1/4})$

25. Above the $xy$-plane and under the surface $z = 1 - x^2 - 2y^2$

26. Above the triangle with vertices $(0,0)$, $(a,0)$, and $(0,b)$ and under the plane $z = 2 - (x/a) - (y/b)$

27. Inside the two cylinders $x^2 + y^2 = a^2$ and $y^2 + z^2 = a^2$

28. Inside the cylinder $x^2 + 2y^2 = 8$, above the plane $z = y - 4$ and below the plane $z = 8 - x$

* **29.** Suppose that $f(x,t)$ and $f_1(x,t)$ are continuous on the rectangle $a \le x \le b$ and $c \le t \le d$. Let

$$g(x) = \int_c^d f(x,t)\,dt \quad \text{and} \quad G(x) = \int_c^d f_1(x,t)\,dt.$$

Show that $g'(x) = G(x)$ for $a < x < b$. (*Hint:* evaluate $\int_a^x G(u)\,du$ by reversing the order of iteration. Then differentiate the result. This is a different version of Theorem 5 of Section 5.5.)

* **30.** Let $F'(x) = f(x)$ and $G'(x) = g(x)$ on the interval $a \le x \le b$. Let $T$ be the triangle with vertices $(a,a)$, $(b,a)$, and $(b,b)$. By iterating $\iint_T f(x)g(y)\,dA$ in both directions, show that

$$\int_a^b f(x)G(x)\,dx$$
$$= F(b)G(b) - F(a)G(a) - \int_a^b g(y)F(y)\,dy.$$

(This is an alternative derivation of the formula for integration by parts.)

## 6.3 IMPROPER INTEGRALS AND A MEAN-VALUE THEOREM

In order to simplify matters, the definition of the double integral given in Section 6.1 required that the domain $D$ be bounded, and that the integrand $f$ be bounded on $D$. As in the single-variable case, **improper double integrals** can arise if either the domain of integration is unbounded or the integrand is unbounded near any part of the boundary of the domain.

### Improper Integrals of Positive Functions

If $f(x,y) \ge 0$ on the domain $D$ then such an improper integral must either exist (that is, converge to a finite value) or be infinite (diverge to infinity). Convergence or divergence of improper double integrals of such *positive* functions can be determined by iterating them and determining the convergence or divergence of any single improper integrals that result.

**EXAMPLE 1** Evaluate $I = \displaystyle\iint_R e^{-x^2}\,dA$. Here $R$ is the region where $x \ge 0$ and $-x \le y \le x$. (See Figure 6.20.)

***SOLUTION*** We iterate with the outer integral in the $x$ direction:

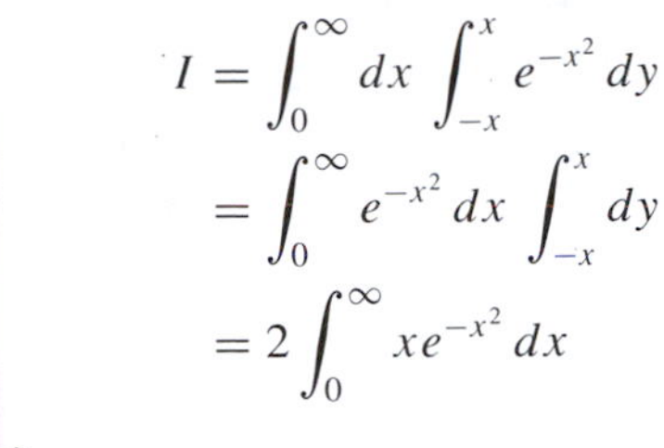

$$\begin{aligned} I &= \int_0^\infty dx \int_{-x}^{x} e^{-x^2}\, dy \\ &= \int_0^\infty e^{-x^2}\, dx \int_{-x}^{x} dy \\ &= 2\int_0^\infty x e^{-x^2}\, dx \end{aligned}$$

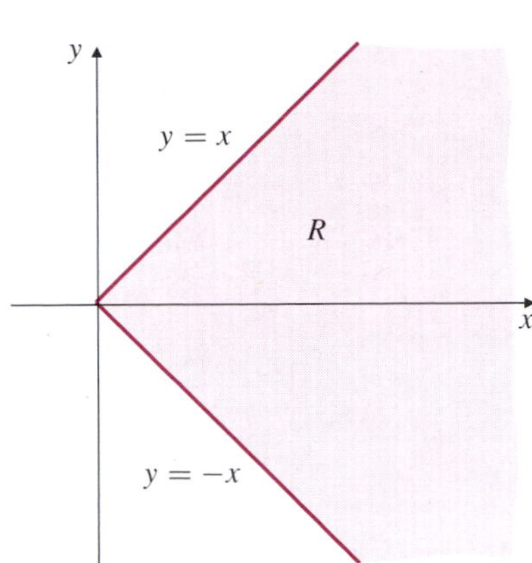


**Figure 6.20** An unbounded sector of the plane

This is an improper integral that can be expressed as a limit:

$$\begin{aligned} I &= 2\lim_{r\to\infty}\int_0^r x e^{-x^2}\, dx \\ &= 2\lim_{r\to\infty}\left(-\frac{1}{2}e^{-x^2}\right)\bigg|_0^r \\ &= \lim_{r\to\infty}(1 - e^{-r^2}) = 1. \end{aligned}$$

The given integral converges; its value is 1. ■

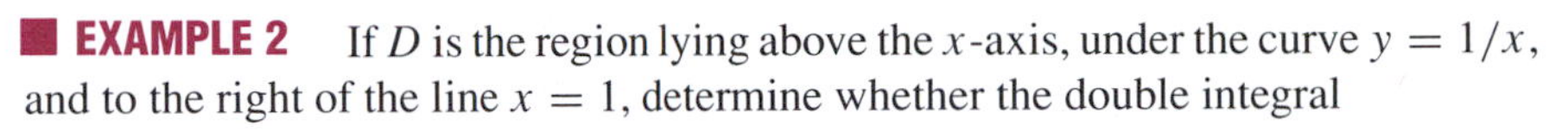

■ **EXAMPLE 2** If $D$ is the region lying above the $x$-axis, under the curve $y = 1/x$, and to the right of the line $x = 1$, determine whether the double integral

$$\iint_D \frac{dA}{x+y}$$

converges or not.

***SOLUTION*** The region $D$ is sketched in Figure 6.21. We have

$$\begin{aligned} \iint_D \frac{dA}{x+y} &= \int_1^\infty dx \int_0^{1/x} \frac{dy}{x+y} \\ &= \int_1^\infty \ln(x+y)\bigg|_{y=0}^{y=1/x} dx \\ &= \int_1^\infty \left(\ln\left(x+\frac{1}{x}\right) - \ln x\right) dx \\ &= \int_1^\infty \ln\left(\frac{x+\frac{1}{x}}{x}\right) dx = \int_1^\infty \ln\left(1+\frac{1}{x^2}\right) dx. \end{aligned}$$

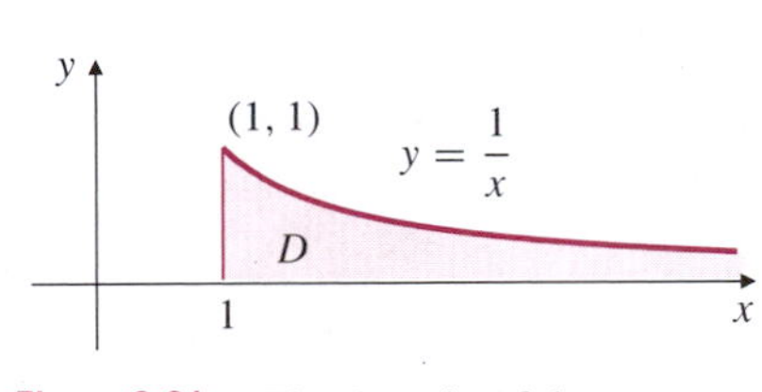


**Figure 6.21** The domain of the integrand in Example 2

It happens that this integral can be evaluated exactly (see Exercise 30 below), but we are only asked to determine whether it converges or not, and that is more easily accomplished by estimating it. Since $0 < \ln(1+u) < u$ if $u > 0$ we have

$$0 < \iint_D \frac{dA}{x+y} < \int_1^\infty \frac{1}{x^2}\, dx = 1.$$

Therefore the given integral converges, and its value lies between 0 and 1. ■

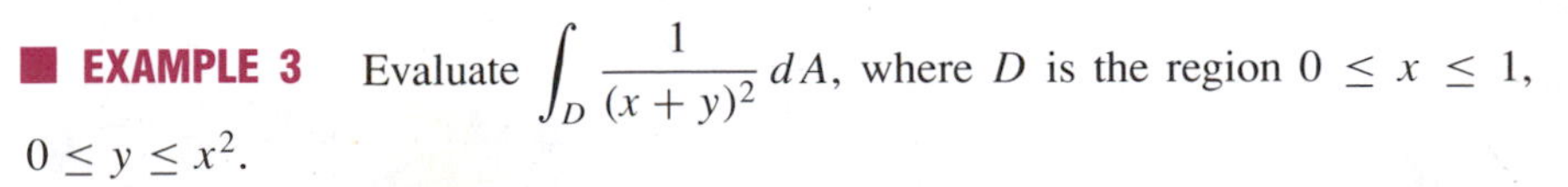

■ **EXAMPLE 3** Evaluate $\displaystyle\int_D \frac{1}{(x+y)^2}\, dA$, where $D$ is the region $0 \le x \le 1$, $0 \le y \le x^2$.

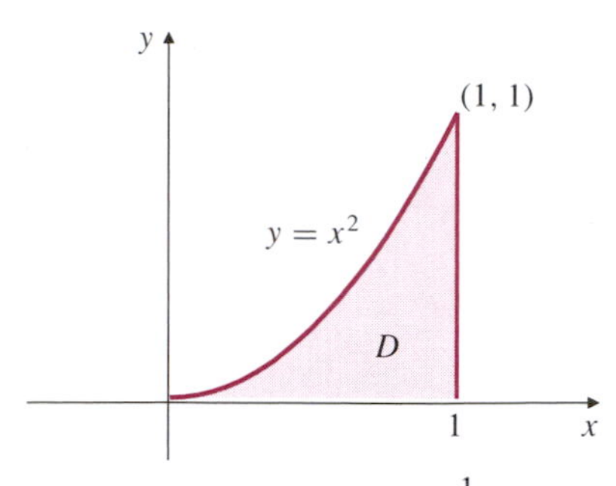


Figure 6.22 The function $\dfrac{1}{(x+y)^2}$ is unbounded on $D$

***SOLUTION*** The integral is improper because the integrand is unbounded as $(x, y)$ approaches $(0, 0)$, a boundary point of $D$. Nevertheless, iteration leads to a proper integral:

$$\begin{aligned}\int_D \frac{1}{(x+y)^2}\,dA &= \lim_{c\to 0+}\int_c^1 dx\int_0^{x^2}\frac{1}{(x+y)^2}\,dy\\ &= \lim_{c\to 0+}\int_c^1 dx\left(-\frac{1}{x+y}\right)\Bigg|_{y=0}^{y=x^2}\\ &= \lim_{c\to 0+}\int_c^1 dx\left(\frac{1}{x}-\frac{1}{x^2+x}\right)dx\\ &= \int_0^1\frac{1}{x+1}\,dx = \ln(x+1)\Big|_0^1 = \ln 2.\end{aligned}$$

■

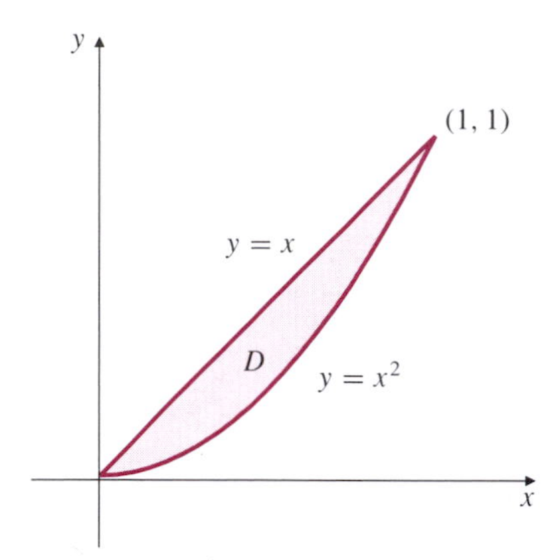


Figure 6.23 $\dfrac{1}{xy}$ is unbounded on $D$

**■ EXAMPLE 4** Determine the convergence or divergence of $I = \displaystyle\iint_D \frac{dA}{xy}$, where $D$ is the bounded region in the first quadrant lying between the line $y = x$ and the parabola $y = x^2$.

***SOLUTION*** The domain $D$ is shown in Figure 6.23. Again the integral is improper because the integrand $1/(xy)$ is unbounded as $(x, y)$ approaches the boundary point $(0, 0)$. We have

$$\begin{aligned}I = \iint_D \frac{dA}{xy} &= \int_0^1\frac{dx}{x}\int_{x^2}^{x}\frac{dy}{y}\\ &= \int_0^1\frac{1}{x}(\ln x - \ln x^2)\,dx = -\int_0^1\frac{\ln x}{x}\,dx.\end{aligned}$$

If we substitute $\ln x = t$ (so that $x = e^t$ and $dx = e^t\,dt$) in this integral, we obtain

$$I = -\int_{-\infty}^{0}\frac{t}{e^t}\,e^t\,dt = -\int_{-\infty}^{0}t\,dt,$$

which diverges to infinity. ■

***REMARK*** In each of the examples above, the integrand was nonnegative on the domain of integration. Nonpositive integrands could have been handled similarly, but we cannot deal here with the convergence of general improper double integrals with integrands $f(x, y)$ that take both positive and negative values on the domain $D$ of the integral. We remark, however, that such an integral cannot converge unless

$$\iint_E f(x, y)\,dA$$

is finite for every bounded, regular subdomain $E$ of $D$. We cannot, in general, determine the convergence of the given integral by looking at the convergence of iterations. The double integral may diverge even if its iterations converge. (See Exercise 23 at the end of this section.) In fact, opposite iterations may even give different values.) This happens because of cancellation of infinite volumes of opposite sign. (Similar behaviour in one dimension is exemplified by the integral $\int_{-1}^{1} dx/x$, which does not exist, although it represents the difference of "equal" but infinite areas.) It can be shown that an improper double integral of $f(x, y)$ over $D$ converges if the integral of $|f(x, y)|$ over $D$ converges:

$$\int_D |f(x, y)|\,dA \text{ converges} \quad\Rightarrow \int_D f(x, y)\,dA \text{ converges.}$$

In this case any iterations will converge to the same value. Such double integrals are called **absolutely convergent** by analogy with absolutely convergent infinite series.

## A Mean-Value Theorem for Double Integrals

Let $D$ be a set in the $xy$-plane that is closed and bounded and has positive area $A = \iint_D dA$. Suppose that $f(x, y)$ is continuous on $D$. Then there exist points $(x_1, y_1)$ and $(x_2, y_2)$ in $D$ where $f$ assumes minimum and maximum values (see Theorem 2 of Section 5.1); that is

$$f(x_1, y_1) \le f(x, y) \le f(x_2, y_2)$$

for all points $(x, y)$ in $D$. If we integrate this inequality over $D$ we obtain

$$\begin{aligned} f(x_1, y_1)A &= \iint_D f(x_1, y_1)\, dA \\ &\le \iint_D f(x, y)\, dA \le \iint_D f(x_2, y_2)\, dA = f(x_2, y_2)A. \end{aligned}$$

Therefore, dividing by $A$, we find that the *number*

$$\bar{f} = \frac{1}{A} \iint_D f(x, y)\, dA$$

lies between the minimum and maximum values of $f$ on $D$:

$$f(x_1, y_1) \le \bar{f} \le f(x_2, y_2).$$

A set $D$ in the plane is said to be **connected** if any two points in it can be joined by a continuous parametric curve $x = x(t)$, $y = y(t)$, $(0 \le t \le 1)$, lying in $D$. Suppose this curve joins $(x_1, y_1)$ (where $t = 0$) and $(x_2, y_2)$ (where $t = 1$). Let $g(t)$ be defined by

$$g(t) = f(x(t), y(t)), \quad 0 \le t \le 1.$$

Then $g$ is continuous and takes the values $f(x_1, y_1)$ at $t = 0$ and $f(x_2, y_2)$ at $t = 1$. By the Intermediate-Value Theorem there exists a number $t_0$ between 0 and 1 such that $\bar{f} = g(t_0) = f(x_0, y_0)$ where $x_0 = x(t_0)$ and $y_0 = y(t_0)$. Thus we have found a point $(x_0, y_0)$ in $D$ such that

$$\frac{1}{\text{area of } D} \iint_D f(x, y)\, dA = f(x_0, y_0).$$

We have therefore proved the following version of the Mean-Value Theorem.

**THEOREM 3**

**A Mean-Value Theorem for Double Integrals**

If the function $f(x, y)$ is continuous on a closed, bounded, connected set $D$ in the $xy$-plane then there exists a point $(x_0, y_0)$ in $D$ such that

$$\iint_D f(x, y)\, dA = f(x_0, y_0) \times (\text{area of } D).$$

By analogy with the definition of average value for one-variable functions, we make the following definition.

**DEFINITION 3**

The **average value** or **mean value** of an integrable function $f(x, y)$ over the set $D$ is the number

$$\bar{f} = \frac{1}{\text{area of } D} \iint_D f(x, y)\, dA.$$

If $f(x, y) \geq 0$ on $D$, then the cylinder with base $D$ and constant height $\bar{f}$ has volume equal to that of the solid region lying above $D$ and below the surface $z = f(x, y)$. It is often very useful to interpret a double integral in terms of the average value of the function which is its integrand.

**EXAMPLE 5** The average value of $x$ over a domain $D$ having area $A$ is

$$\bar{x} = \frac{1}{A} \iint_D x\, dA.$$

Of course, $\bar{x}$ is just the $x$-coordinate of the centroid of the region $D$. ■

**EXAMPLE 6** A large number of points $(x, y)$ are chosen at random in the triangle $T$ with vertices $(0, 0)$, $(1, 0)$, and $(1, 1)$. What is the approximate average value of $x^2 + y^2$ for these points?

***SOLUTION*** The approximate average value of $x^2 + y^2$ for the randomly chosen points will be the average value of that function over the triangle, namely

$$\begin{aligned} \frac{1}{1/2} \iint_T (x^2 + y^2)\, dA &= 2 \int_0^1 dx \int_0^x (x^2 + y^2)\, dy \\ &= 2 \int_0^1 \left( x^2 y + \frac{1}{3} y^3 \right) \Bigg|_0^x dx = \frac{8}{3} \int_0^1 x^3\, dx = \frac{2}{3}. \end{aligned}$$ ■

**EXAMPLE 7** Let $(a, b)$ be an interior point of a domain $D$ on which $f(x, y)$ is continuous. For sufficiently small positive $r$, the closed circular disk $D_r$ with centre at $(a, b)$ and radius $r$ is contained in $D$. Show that

$$\lim_{r \to 0} \frac{1}{\pi r^2} \iint_{D_r} f(x, y)\, dA = f(a, b).$$

***SOLUTION*** If $D_r$ is contained in $D$, then by Theorem 3

$$\frac{1}{\pi r^2} \iint_{D_r} f(x, y)\, dA = f(x_0, y_0)$$

for some point $(x_0, y_0)$ in $D_r$. As $r \to 0$, the point $(x_0, y_0)$ approaches $(a, b)$. Since $f$ is continuous at $(a, b)$, we have $f(x_0, y_0) \to f(a, b)$. Thus

$$\lim_{r \to 0} \frac{1}{\pi r^2} \iint_{D_r} f(x, y)\, dA = f(a, b).$$ ■

## EXERCISES 6.3

In Exercises 1–14, determine whether the given integral converges or not. Try to evaluate those that do exist.

**1.** $\iint_Q e^{-x-y}\, dA$, where $Q$ is the first quadrant of the $xy$-plane

**2.** $\iint_Q \frac{dA}{(1+x^2)(1+y^2)}$, where $Q$ is the first quadrant of the $xy$-plane.

**3.** $\iint_S \frac{y}{1+x^2}\, dA$, where $S$ is the strip $0 < y < 1$ in the $xy$-plane

**4.** $\iint_T \frac{1}{x\sqrt{y}}\, dA$ over the triangle $T$ with vertices $(0, 0)$, $(1, 1)$, and $(1, 2)$

**5.** $\iint_Q \frac{x^2+y^2}{(1+x^2)(1+y^2)}\, dA$, where $Q$ is the first quadrant of the $xy$-plane

**6.** $\iint_H \frac{1}{1+x+y}\, dA$, where $H$ is the half-strip $0 \le x < \infty$, $0 \le y \le 1$

**7.** $\iint_{\mathbb{R}^2} e^{-(|x|+|y|)}\, dA$

**8.** $\iint_{\mathbb{R}^2} e^{-|x+y|}\, dA$

**9.** $\iint_T \frac{1}{x^3} e^{-y/x}\, dA$, where $T$ is the region satisfying $x \ge 1$ and $0 \le y \le x$

**10.** $\iint_T \frac{dA}{x^2+y^2}$, where $T$ is the region in Exercise 9

**11.** $\iint_R \frac{dA}{x^2+y^2}$, where $R$ satisfies $1 \le x < \infty$, and $1 \le y < \infty$

**12.** $\iint_R \frac{dA}{x^4+y^2}$, where $R$ satisfies $1 \le x < \infty$, and $0 \le y \le x^2$

* **13.** $\iint_Q e^{-xy}\, dA$, where $Q$ is the first quadrant of the $xy$-plane

**14.** $\iint_R \frac{1}{x} \sin\frac{1}{x}\, dA$, where $R$ is the region $2/\pi \le x < \infty$, $0 \le y \le 1/x$

**15.** Evaluate

$$I = \iint_S \frac{dA}{x+y},$$

where $S$ is the square $0 \le x \le 1$, $0 \le y \le 1$,

(a) by direct iteration of the double integral,

(b) by using the symmetry of the integrand and the domain to write

$$I = 2\iint_T \frac{dA}{x+y},$$

where $T$ is the triangle with vertices $(0, 0)$, $(1, 0)$, and $(1, 1)$.

**16.** Find the volume of the solid lying above the square $S$ of Exercise 15, and under the surface $z = 2xy/(x^2+y^2)$.

In Exercises 17–22, $a$ and $b$ are given real numbers. $D_k$ is the region $0 \le x \le 1$, $0 \le y \le x^k$, and $R_k$ is the region $1 \le x < \infty$, $0 \le y \le x^k$. Find all real values of $k$ for which the given integral converges.

**17.** $\iint_{D_k} \frac{dA}{x^a}$

**18.** $\iint_{D_k} y^b\, dA$

**19.** $\iint_{R_k} x^a\, dA$

**20.** $\iint_{R_k} \frac{dA}{y^b}$

**21.** $\iint_{D_k} x^a y^b\, dA$

**22.** $\iint_{R_k} x^a y^b\, dA$

* **23.** Evaluate both iterations of the improper integral

$$\iint_S \frac{x-y}{(x+y)^3}\, dA,$$

where $S$ is the square $0 < x < 1$, $0 < y < 1$. Show that the above improper double integral does not exist by considering

$$\iint_T \frac{x-y}{(x+y)^3}\, dA,$$

where $T$ is that part of the square $S$ lying under the line $x = y$.

In Exercises 24–26, find the average value of the given function over the given region.

**24.** $x^2$ over the rectangle $a \le x \le b$, $c \le y \le d$

**25.** $x^2 + y^2$ over the triangle $0 \le x \le a$, $0 \le y \le a - x$

**26.** $1/x$ over the region $0 \le x \le 1$, $x^2 \le y \le \sqrt{x}$

* **27.** Find the average distance from points in the quarter-disk $x^2 + y^2 \le a^2$, $x \ge 0$, $y \ge 0$, to the line $x + y = 0$.

**28.** Does $f(x, y) = x$ have an average value over the region $0 \le x < \infty$, $0 \le y \le \frac{1}{1+x^2}$? If so what is it?

**29.** Does $f(x, y) = xy$ have an average value over the region $0 \le x < \infty$, $0 \le y \le \frac{1}{1+x^2}$? If so what is it?

* **30.** Find the exact value of the integral in Example 2. *Hint:* integrate by parts in $\int_1^\infty \ln\big(1 + (1/x^2)\big)\, dx$.

* **31.** Let $(a, b)$ be an interior point of a domain $D$ on which the function $f(x, y)$ is continuous. For small enough $h^2 + k^2$ the rectangle $R_{hk}$ with vertices $(a, b)$, $(a+h, b)$, $(a, b+k)$, and $(a+h, b+k)$ is contained in $D$. Show that

$$\lim_{(h,k)\to(0,0)} \frac{1}{hk} \iint_{R_{hk}} f(x, y) = f(a, b).$$

(*Hint:* see Example 7.)

* **32.** Suppose that $f_{12}(x,y)$ and $f_{21}(x,y)$ are continuous in a neighbourhood of the point $(a,b)$. Without assuming the equality of these mixed partial derivatives, show that

$$\iint_R f_{12}(x,y)\,dA = \iint_R f_{21}(x,y)\,dA,$$

where $R$ is the rectangle with vertices $(a,b)$, $(a+h,b)$, $(a,b+k)$, and $(a+h,b+k)$ and $h^2+k^2$ is sufficiently small. Now use the result of Exercise 31 to show that $f_{12}(a,b) = f_{21}(a,b)$. (This reproves Theorem 1 of Section 4.4. However, in that theorem we only assumed continuity of the mixed partials *at* $(a,b)$. Here we assume the continuity *at all points sufficiently near* $(a,b)$.)

## 6.4 DOUBLE INTEGRALS IN POLAR COORDINATES

For many applications of double integrals either the domain of integration, or the integrand function, or both may be more easily expressed in terms of polar coordinates than in terms of Cartesian coordinates. Recall that a point $P$ with Cartesian coordinates $(x,y)$ can also be located by its polar coordinates $[r,\theta]$, where $r$ is the distance from $P$ to the origin $O$, and $\theta$ is the angle $OP$ makes with the positive direction of the $x$-axis. (Positive angles $\theta$ are measured counterclockwise.) The polar and Cartesian coordinates of $P$ are related by the transformations

$$x = r\cos\theta, \qquad r^2 = x^2 + y^2,$$
$$y = r\sin\theta, \qquad \tan\theta = y/x.$$

Consider the problem of finding the volume $V$ of the solid region lying above the $xy$-plane and beneath the paraboloid $z = 1 - x^2 - y^2$. Since the paraboloid intersects the $xy$-plane in the circle $x^2 + y^2 = 1$, the volume is given in Cartesian coordinates by

$$V = \iint_{x^2+y^2\le 1} (1 - x^2 - y^2)\,dA = \int_{-1}^{1} dx \int_{-\sqrt{1-x^2}}^{\sqrt{1-x^2}} (1 - x^2 - y^2)\,dy.$$

Evaluating this iterated integral would require considerable effort. However, we can express the same volume in terms of polar coordinates as

$$V = \iint_{r\le 1} (1 - r^2)\,dA.$$

In order to iterate this integral, we have to know the form that the *area element* $dA$ takes in polar coordinates.

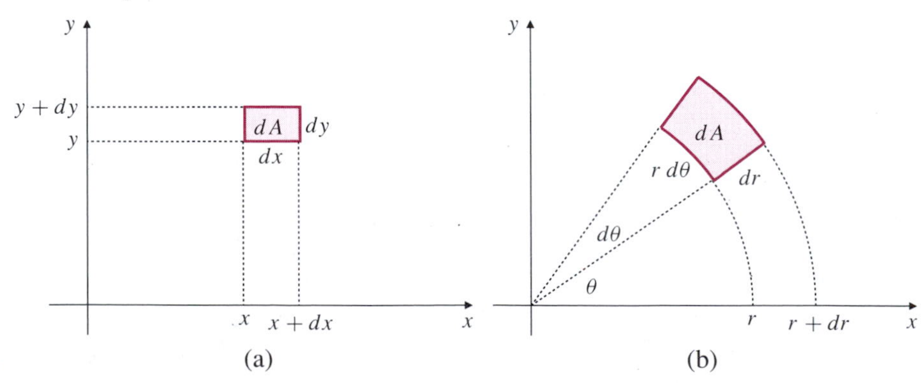


**Figure 6.24**

(a) $dA = dx\,dy$ in Cartesian coordinates

(b) $dA = r\,dr\,d\theta$ in polar coordinates

In the Cartesian formula for $V$, the area element $dA = dx\,dy$ represents the area of the "infinitesimal" region bounded by the coordinate lines at $x$, $x + dx$, $y$, and $y + dy$. (See Figure 6.24(a).) In the polar formula, the area element $dA$ should represent the area of the "infinitesimal" region bounded by the coordinate circles with radii $r$ and $r + dr$, and coordinate rays from the origin at angles $\theta$ and $\theta + d\theta$. (See Figure 6.24(b).) Observe that $dA$ is approximately the area of a rectangle with dimensions $dr$ and $r\,d\theta$. The error in this approximation becomes negligible *compared to the size of* $dA$ as $dr$ and $d\theta$ approach zero. Thus, in transforming a double integral between Cartesian and polar coordinates the area element transforms according to the formula

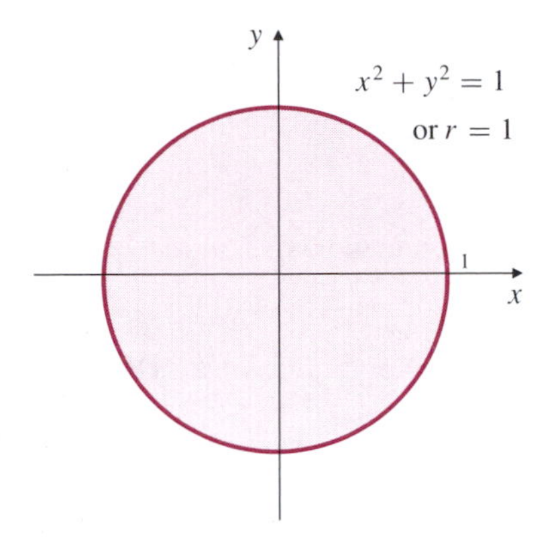

**Figure 6.25** The domain in the $xy$-plane

$$dx\,dy = dA = r\,dr\,d\theta.$$

In order to iterate the polar form of the double integral for $V$ considered above, we can regard the domain of integration as a set in a plane having *Cartesian coordinates* $r$ and $\theta$. In the $xy$ Cartesian plane the domain is a disk $r \le 1$ (see Figure 6.25), but in the $r\theta$ Cartesian plane (with perpendicular $r$- and $\theta$-axes) the domain is the rectangle $R$ specified by $0 \le r \le 1$ and $0 \le \theta \le 2\pi$. (See Figure 6.26.) The area element in the $r\theta$-plane is $dA^* = dr\,d\theta$, so area is not preserved under the transformation to polar coordinates ($dA = r\,dA^*$). Thus the polar integral for $V$ is really a Cartesian integral in the $r\theta$-plane, with integrand modified by the inclusion of an extra factor $r$ to compensate for the change of area. It can be evaluated by standard iteration methods:

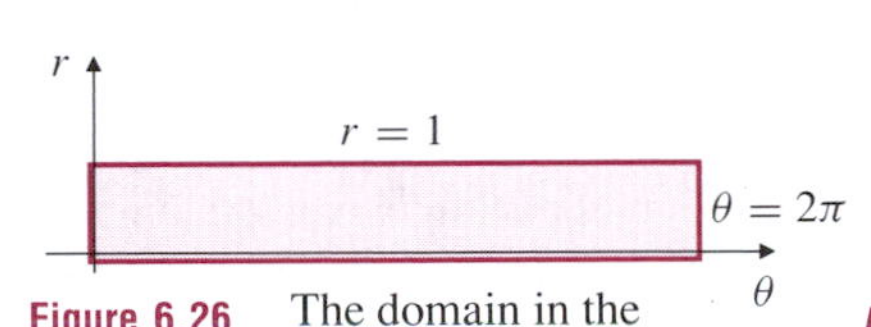

**Figure 6.26** The domain in the $r\theta$-plane

$$\begin{aligned} V = \iint_R (1 - r^2) r\,dA^* &= \int_0^{2\pi} d\theta \int_0^1 (1 - r^2) r\,dr \\ &= \int_0^{2\pi} \left(\frac{r^2}{2} - \frac{r^4}{4}\right)\bigg|_0^1 d\theta = \frac{\pi}{2} \text{ units}^3. \end{aligned}$$

***REMARK*** It is not necessary to sketch the region $R$ in the $r\theta$-plane. We are used to thinking of polar coordinates in terms of distances and angles in the $xy$-plane, and can easily understand from looking at the disk in Figure 6.25 that the iteration of the integral in polar coordinates corresponds to $0 \le \theta \le 2\pi$ and $0 \le r \le 1$. That is, we should be able to write the iteration

$$V = \int_0^{2\pi} d\theta \int_0^1 (1 - r^2) r\,dr$$

directly from consideration of the domain of integration in the $xy$-plane.

**EXAMPLE 1** If $R$ is that part of the annulus $0 < a^2 \le x^2 + y^2 \le b^2$ lying in the first quadrant and below the line $y = x$, evaluate

$$I = \iint_R \frac{y^2}{x^2}\,dA.$$

***SOLUTION*** Figure 6.27 shows the region $R$. It is specified in polar coordinates by $0 \le \theta \le \pi/4$ and $a \le r \le b$. Since

$$\frac{y^2}{x^2} = \frac{r^2 \sin^2\theta}{r^2 \cos^2\theta} = \tan^2\theta,$$

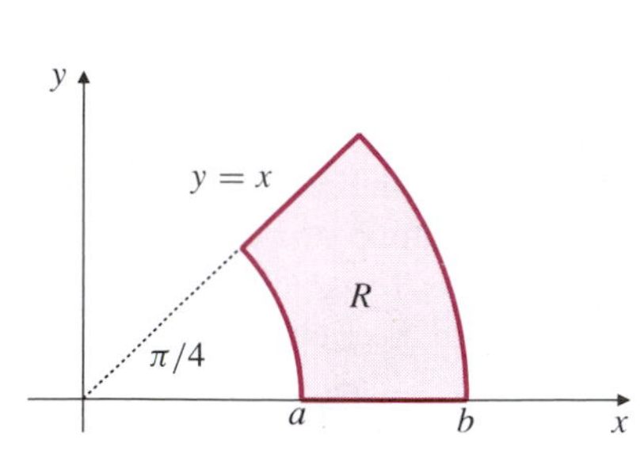


Figure 6.27

we have

$$
\begin{aligned}
I &= \int_0^{\pi/4} \tan^2\theta \, d\theta \int_a^b r \, dr \\
&= \frac{1}{2}(b^2 - a^2) \int_0^{\pi/4} \left(\sec^2\theta - 1\right) d\theta \\
&= \frac{1}{2}(b^2 - a^2)(\tan\theta - \theta)\Big|_0^{\pi/4} \\
&= \frac{1}{2}(b^2 - a^2)\left(1 - \frac{\pi}{4}\right) = \frac{4-\pi}{8}(b^2 - a^2).
\end{aligned}
$$

■

■ **EXAMPLE 2** **Area of a polar region**

Derive the formula for the area of the polar region $R$ bounded by the curve $r = f(\theta)$ and the rays $\theta = \alpha$ and $\theta = \beta$. (See Figure 6.28.)

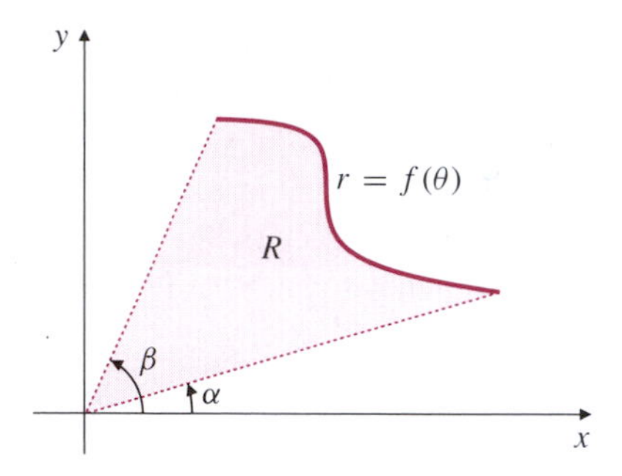


Figure 6.28 A standard area problem for polar coordinates

***SOLUTION*** The area $A$ of $R$ is numerically equal to the volume of a cylinder of height 1 above the region $R$:

$$
\begin{aligned}
A = \iint_R dx\,dy &= \iint_R r\,dr\,d\theta \\
&= \int_\alpha^\beta d\theta \int_0^{f(\theta)} r\,dr = \frac{1}{2}\int_\alpha^\beta \left(f(\theta)\right)^2 d\theta.
\end{aligned}
$$

Observe that the inner integral in the iteration involves integrating $r$ along the ray specified by $\theta$ from 0 to $f(\theta)$. The final result is the formula obtained for the area in *Single-Variable Calculus.* ■

There is no firm rule as to when one should or should not convert a double integral from Cartesian to polar coordinates. In Example 1 above, the conversion was strongly suggested by the shape of the domain, but was also indicated by the fact that the integrand, $y^2/x^2$, becomes a function of $\theta$ alone when converted to polar coordinates. It is usually wise to switch to polar coordinates if the switch simplifies the iteration (that is, if the *domain* is "simpler" when expressed in terms of polar coordinates), even if the form of the integrand is made more complicated.

■ **EXAMPLE 3** Find the volume of the solid lying in the first octant, inside the cylinder $x^2 + y^2 = a^2$, and under the plane $z = y$.

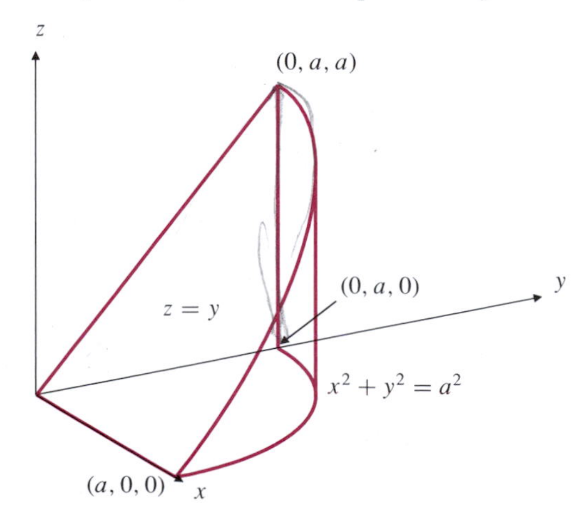


Figure 6.29 This volume is easily calculated using iteration in polar coordinates

**SOLUTION** The solid is shown in Figure 6.29. The base is a quarter disk, which is expressed in polar coordinates by the inequalities $0 \le \theta \le \pi/2$ and $0 \le r \le a$. The height is given by $z = y = r\sin\theta$. The solid has volume

$$V = \int_0^{\pi/2} d\theta \int_0^a (r\sin\theta) r\, dr = \int_0^{\pi/2} \sin\theta\, d\theta \int_0^a r^2\, dr = \frac{1}{3}a^3 \text{ units}^3. \quad \blacksquare$$

The following example establishes the value of a definite integral which plays a very important role in probability theory and statistics. It is interesting that this single-variable integral cannot be evaluated by the techniques of single-variable calculus.

### ■ EXAMPLE 4 A very important integral

Show that

$$\int_{-\infty}^{\infty} e^{-x^2}\, dx = \sqrt{\pi}.$$

**SOLUTION** The improper integral converges, and its value does not depend on what symbol we use for the variable of integration. Therefore we can express the square of the integral as a product of two identical integrals, but with their variables of integration named differently. We then interpret this product as an improper double integral and reiterate it in polar coordinates:

$$\begin{aligned}\left(\int_{-\infty}^{\infty} e^{-x^2}\, dx\right)^2 &= \int_{-\infty}^{\infty} e^{-x^2}\, dx \int_{-\infty}^{\infty} e^{-y^2}\, dy \\ &= \iint_{\mathbb{R}^2} e^{-(x^2+y^2)}\, dA \\ &= \int_0^{2\pi} d\theta \int_0^{\infty} e^{-r^2} r\, dr \\ &= 2\pi \lim_{R\to\infty} \left(-\frac{1}{2}e^{-r^2}\right)\Bigg|_0^R = \pi.\end{aligned}$$

The $r$ integral, a convergent improper integral, was evaluated with the aid of the substitution $u = r^2$. ■

As our final example of iteration in polar coordinates let us try something a little more demanding.

### ■ EXAMPLE 5

Find the volume of the solid region lying inside both the sphere $x^2 + y^2 + z^2 = 4a^2$ and the cylinder $x^2 + y^2 = 2ay$.

**SOLUTION** The sphere is centred at the origin and has radius $2a$. The equation of the cylinder becomes

$$x^2 + (y-a)^2 = a^2$$

if we complete the square in the $y$ terms. Thus it is a vertical circular cylinder of radius $a$ having its axis along the vertical line through $(0, a, 0)$. The $z$-axis lies on the cylinder. One-quarter of the required volume lies in the first octant. This part is shown in Figure 6.30.

If we use polar coordinates in the $xy$-plane, then the sphere has equation $r^2 + z^2 = 4a^2$ and the cylinder has equation $r^2 = 2ar\sin\theta$ or, more simply, $r = 2a\sin\theta$. The first octant portion of the volume lies above the region specified by the inequalities $0 \le \theta \le \pi/2$ and $0 \le r \le 2a\sin\theta$. Therefore the total volume is

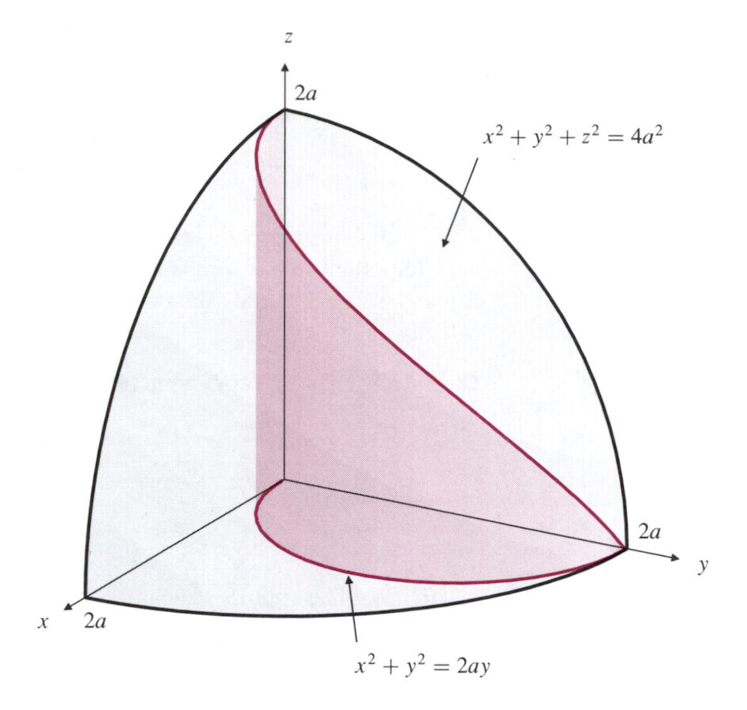


**Figure 6.30** The first octant part of the intersection of the cylinder $x^2 + y^2 = 2ay$ and the sphere $x^2 + y^2 + z^2 = 4a^2$

$$\begin{aligned}
V &= 4\int_0^{\pi/2} d\theta \int_0^{2a\sin\theta} \sqrt{4a^2 - r^2}\, r\, dr \qquad \text{(Let } u = 4a^2 - r^2.\text{)} \\
&= 2\int_0^{\pi/2} d\theta \int_{4a^2\cos^2\theta}^{4a^2} \sqrt{u}\, du \\
&= \frac{4}{3}\int_0^{\pi/2} (8a^3 - 8a^3\cos^3\theta)\, d\theta \qquad \text{(Let } v = \sin\theta.\text{)} \\
&= \frac{16}{3}\pi a^3 - \frac{32}{3}a^3 \int_0^1 (1 - v^2)\, dv \\
&= \frac{16}{3}\pi a^3 - \frac{64}{9}a^3 = \frac{16}{9}(3\pi - 4)a^3 \text{ cubic units.}
\end{aligned}$$

## Change of Variables in Double Integrals

The transformation of a double integral to polar coordinates is just a special case of a general change of variables formula for double integrals. Suppose that $x$ and $y$ are expressed as functions of two other variables $u$ and $v$ by the equations

$$\begin{aligned}
x &= x(u, v) \\
y &= y(u, v).
\end{aligned}$$

We regard these equations as defining a **transformation** (or mapping) from points $(u, v)$ in a $uv$-Cartesian plane to points $(x, y)$ in the $xy$-plane. We say that the transformation is *one-to-one* from the set $S$ in the $uv$-plane *onto* the set $D$ in the $xy$-plane provided:

(i) every point in $S$ gets mapped to a point in $D$,

(ii) every point in $D$ is the image of a point in $S$, and

(iii) different points in $S$ get mapped to different points in $D$.

If the transformation is one-to-one, the defining equations can be solved for $u$ and $v$ as functions of $x$ and $y$, and the resulting **inverse transformation**:

$$u = u(x, y)$$
$$v = v(x, y)$$

is one-to-one from $D$ onto $S$.

Let us assume that the functions $x(u, v)$ and $y(u, v)$ have continuous first partial derivatives, and that the Jacobian determinant

$$\frac{\partial(x, y)}{\partial(u, v)} \neq 0 \qquad \text{at} \quad (u, v).$$

As noted in Section 4.8, the Implicit Function Theorem implies that the transformation is one-to-one near $(u, v)$ and the inverse transformation also has continuous first partial derivatives and nonzero Jacobian satisfying

$$\frac{\partial(u, v)}{\partial(x, y)} = \frac{1}{\dfrac{\partial(x, y)}{\partial(u, v)}} \qquad \text{on } D.$$

**EXAMPLE 6** The transformation $x = r\cos\theta$, $y = r\sin\theta$ to polar coordinates has Jacobian

$$\frac{\partial(x, y)}{\partial(r, \theta)} = \begin{vmatrix} \cos\theta & -r\sin\theta \\ \sin\theta & r\cos\theta \end{vmatrix} = r.$$

Near any point except the origin (where $r = 0$), the transformation is one-to-one. (In fact it is one-to-one from any set in the $r\theta$-plane that does not contain the axis $r = 0$ and lies in, say, the strip $0 \leq \theta < 2\pi$.)

A one-to-one transformation can be used to transform the double integral

$$\iint_D f(x, y)\, dA$$

to a double integral over the corresponding set $S$ in the $uv$-plane. Under the transformation, the integrand $f(x, y)$ becomes $g(u, v) = f\big(x(u, v), y(u, v)\big)$. We must discover how to express the area element $dA = dx\, dy$ in terms of the area element $du\, dv$ in the $uv$-plane.

For any fixed value of $u$, (say $u = c$), the equations

$$x = x(u, v) \quad \text{and} \quad y = y(u, v)$$

define a parametric curve (with $v$ as parameter) in the $xy$-plane. This curve is called a $u$-curve corresponding to the value $u = c$. Similarly, for fixed $v$ the equations define a parametric curve (with parameter $u$) called a $v$-curve. Consider the differential area element bounded by the $u$-curves corresponding to nearby values $u$ and $u + du$ and the $v$-curves corresponding to nearby values $v$ and $v + dv$. Since these curves are smooth, for small values of $du$ and $dv$ the area element is approximately a parallelogram, and its area is approximately

$$dA = |\overrightarrow{PQ} \times \overrightarrow{PR}|,$$

where $P$, $Q$, and $R$ are the points shown in Figure 6.31. The error in this approximation becomes negligible compared to $dA$ as $du$ and $dv$ approach zero.

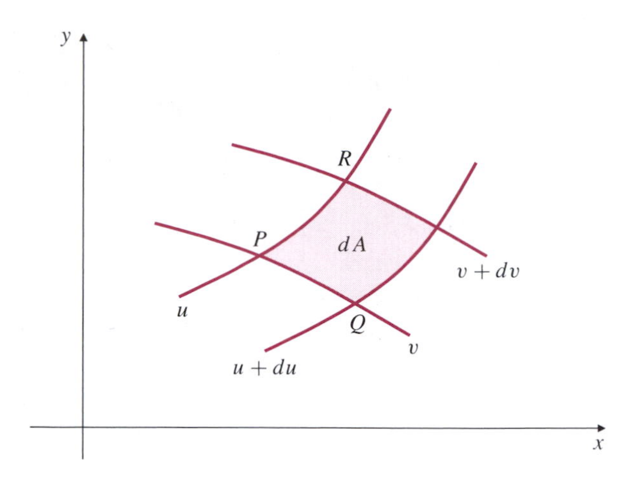


**Figure 6.31** The image in the $xy$ plane of the area element $du\,dv$ in the $uv$-plane

Now $\overrightarrow{PQ} = dx\,\mathbf{i} + dy\,\mathbf{j}$ where

$$dx = \frac{\partial x}{\partial u}du + \frac{\partial x}{\partial v}dv \quad \text{and} \quad dy = \frac{\partial y}{\partial u}du + \frac{\partial y}{\partial v}dv.$$

However, $dv = 0$ along the $v$-curve $PQ$ so

$$\overrightarrow{PQ} = \frac{\partial x}{\partial u}du\,\mathbf{i} + \frac{\partial y}{\partial u}du\,\mathbf{j}.$$

Similarly,

$$\overrightarrow{PR} = \frac{\partial x}{\partial v}dv\,\mathbf{i} + \frac{\partial y}{\partial v}dv\,\mathbf{j}.$$

Hence

$$dA = \left| \begin{vmatrix} \mathbf{i} & \mathbf{j} & \mathbf{k} \\ \frac{\partial x}{\partial u}du & \frac{\partial y}{\partial u}du & 0 \\ \frac{\partial x}{\partial v}dv & \frac{\partial y}{\partial v}dv & 0 \end{vmatrix} \right| = \left| \frac{\partial(x, y)}{\partial(u, v)} \right| du\,dv;$$

that is, the absolute value of the Jacobian $\partial(x, y)/\partial(u, v)$ is the ratio between corresponding area elements in the $xy$-plane and the $uv$-plane:

$$dA = dx\,dy = \left| \frac{\partial(x, y)}{\partial(u, v)} \right| du\,dv.$$

The following theorem summarizes the change of variables procedure for a double integral.

**THEOREM 4**

**Change of variables formula for double integrals**

Let $x = x(u, v)$, $y = y(u, v)$ be a one-to-one transformation from a domain $S$ in the $uv$-plane onto a domain $D$ in the $xy$-plane. Suppose that the functions $x$ and $y$, and their first partial derivatives with respect to $u$ and $v$ are continuous in $S$. If $f(x, y)$ is integrable on $D$, and if $g(u, v) = f(x(u, v), y(u, v))$, then $g$ is integrable on $S$ and

$$\iint_D f(x, y)\,dx\,dy = \iint_S g(u, v) \left| \frac{\partial(x, y)}{\partial(u, v)} \right| du\,dv.$$

***REMARK*** It is not necessary that $S$ or $D$ be closed, or that the transformation be one-to-one on the boundary of $S$. The transformation to polar coordinates maps the rectangle $0 < r < 1$, $0 \le \theta < 2\pi$ one-to-one onto the punctured disk $0 < x^2 + y^2 < 1$ and, as in the first example in this section, we can transform an integral over the closed disk $x^2 + y^2 \le 1$ to one over the closed rectangle $0 \le r \le 1$, $0 \le \theta \le 2\pi$.

**EXAMPLE 7** Use an appropriate change of variables to find the area of the elliptic disk $E$ given by

$$\frac{x^2}{a^2} + \frac{y^2}{b^2} \le 1.$$

***SOLUTION*** Under the transformation $x = au$, $y = bv$, the elliptic disk $E$ is the one-to-one image of the circular disk $D$ given by $u^2 + v^2 \le 1$. Assuming $a > 0$ and $b > 0$, we have

$$dx\,dy = \left|\frac{\partial(x, y)}{\partial(u, v)}\right| du\,dv = \left|\begin{vmatrix} a & 0 \\ 0 & b \end{vmatrix}\right| du\,dv = ab\,du\,dv.$$

Therefore, the area of $E$ is given by

$$\iint_E 1\,dx\,dy = \iint_D ab\,du\,dv = ab \times (\text{area of } D) = \pi ab \text{ square units.}$$

It is often tempting to try to use the change of variable formula to transform the domain of a double integral into a rectangle so that iteration will be easy. As the following example shows, this usually involves defining the inverse transformation ($u$ and $v$ in terms of $x$ and $y$). Remember that inverse transformations have reciprocal Jacobians.

**EXAMPLE 8** Find the area of the finite plane region bounded by the four parabolas $y = x^2$, $y = 2x^2$, $x = y^2$, and $x = 3y^2$.

***SOLUTION*** The region, call it $D$, is sketched in Figure 6.32. Let

$$u = \frac{y}{x^2} \quad \text{and} \quad v = \frac{x}{y^2}.$$

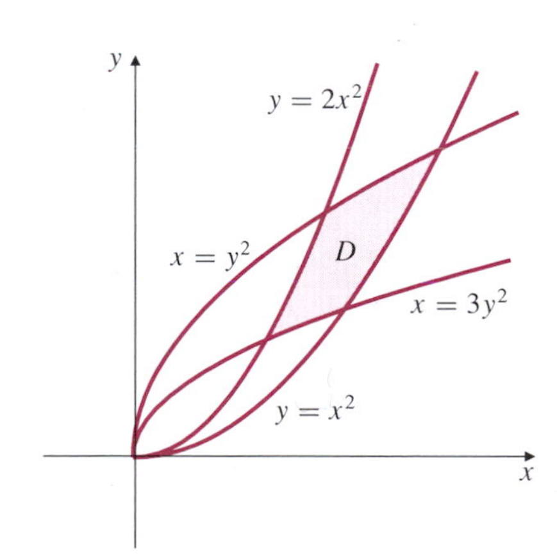

Figure 6.32

Then the region $D$ corresponds to the rectangle $R$ in the $uv$-plane given by $1 \le u \le 2$ and $1 \le v \le 3$. Since

$$\frac{\partial(u, v)}{\partial(x, y)} = \begin{vmatrix} -2y/x^3 & 1/x^2 \\ 1/y^2 & -2x/y^3 \end{vmatrix} = \frac{3}{x^2y^2} = 3u^2v^2,$$

it follows that

$$\left|\frac{\partial(x, y)}{\partial(u, v)}\right| = \frac{1}{3u^2v^2}$$

and so the area of $D$ is given by

$$\iint_D dx\,dy = \iint_R \frac{1}{3u^2v^2}\,du\,dv$$
$$= \frac{1}{3}\int_1^2 \frac{du}{u^2}\int_1^3 \frac{dv}{v^2} = \frac{1}{3} \times \frac{1}{2} \times \frac{2}{3} = \frac{1}{9} \text{ square units.}$$

The following example shows what can happen if the transformation of a double integral is not one-to-one.

**EXAMPLE 9** Let $D$ be the square $0 \le x \le 1$, $0 \le y \le 1$ in the $xy$-plane, and let $S$ be the square $0 \le u \le 1$, $0 \le v \le 1$ in the $uv$-plane. Show that the transformation

$$x = 4u - 4u^2, \qquad y = v$$

maps $S$ onto $D$, and use it to transform the integral $I = \iint_D dx\, dy$. Compare the value of $I$ with that of the transformed integral.

***SOLUTION*** Since $x = 4u - 4u^2 = 1 - (1 - 2u)^2$, the minimum value of $x$ on the interval $0 \le u \le 1$ is 0 (at $u = 0$ and $u = 1$), and the maximum value is 1 (at $u = \frac{1}{2}$). Therefore, $x = 4u - 4u^2$ maps the interval $0 \le u \le 1$ onto the interval $0 \le x \le 1$. Since $y = v$ clearly maps $0 \le v \le 1$ onto $0 \le y \le 1$, the given transformation maps $S$ onto $D$. Since

$$dx\, dy = \left|\frac{\partial(x, y)}{\partial(u, v)}\right| du\, dv = \left|\begin{vmatrix} 4 - 8u & 0 \\ 0 & 1 \end{vmatrix}\right| du\, dv = |4 - 8u|\, du\, dv,$$

transforming $I$ leads to the integral

$$J = \iint_S |4 - 8u|\, du\, dv = 4\int_0^1 dv \int_0^1 |1 - 2u|\, du = 8\int_0^{1/2} (1 - 2u)\, du = 2.$$

However, $I = \iint_D dx\, dy =$ area of $D = 1$. The reason that $J \ne I$ is that the transformation is not one-to-one from $S$ onto $D$; it actually maps $S$ onto $D$ twice. The rectangle $R$ defined by $0 \le u \le \frac{1}{2}$ and $0 \le v \le 1$ is mapped one-to-one onto $D$ by the transformation, so the appropriate transformed integral is $\iint_R |4 - 8u|\, du\, dv$, which is equal to $I$. ■

## EXERCISES 6.4

In Exercises 1–6, evaluate the given double integral over the disk $D$ given by $x^2 + y^2 \le a^2$, where $a > 0$.

**1.** $\displaystyle\iint_D (x^2 + y^2)\, dA$

**2.** $\displaystyle\iint_D \sqrt{x^2 + y^2}\, dA$

**3.** $\displaystyle\iint_D \frac{1}{\sqrt{x^2 + y^2}}\, dA$

**4.** $\displaystyle\iint_D |x|\, dA$

**5.** $\displaystyle\iint_D x^2\, dA$

**6.** $\displaystyle\iint_D x^2y^2\, dA$

In Exercises 7–10, evaluate the given double integral over the quarter-disk $Q$ given by $x \ge 0$, $y \ge 0$, and $x^2 + y^2 \le a^2$, where $a > 0$.

**7.** $\displaystyle\iint_Q y\, dA$

**8.** $\displaystyle\iint_Q (x + y)\, dA$

**9.** $\displaystyle\iint_Q e^{x^2+y^2}\, dA$

**10.** $\displaystyle\iint_Q \frac{2xy}{x^2 + y^2}\, dA$

**11.** Evaluate $\displaystyle\iint_S (x + y)\, dA$, where $S$ is the region in the first quadrant lying inside the disk $x^2 + y^2 \le a^2$ and under the line $y = \sqrt{3}x$.

**12.** Find $\displaystyle\iint_S x\, dA$, where $S$ is the disk segment $x^2 + y^2 \le 2$, $x \ge 1$.

**13.** Evaluate $\displaystyle\iint_T (x^2 + y^2)\, dA$, where $T$ is the triangle with vertices $(0, 0)$, $(1, 0)$, and $(1, 1)$.

**14.** Evaluate $\displaystyle\iint_{x^2+y^2\le 1} \ln(x^2 + y^2)\, dA$.

**15.** Find the average distance from the origin to points in the disk $x^2 + y^2 \le a^2$.

**16.** Find the average value of $e^{-(x^2+y^2)}$ over the annular region $0 < a \le \sqrt{x^2 + y^2} \le b$.

**17.** For what values of $k$, and to what value, does the integral $\displaystyle\iint_{x^2+y^2\le 1} \frac{dA}{(x^2 + y^2)^k}$ converge?

**18.** For what values of $k$, and to what value, does the integral $\displaystyle\iint_{\mathbb{R}^2} \frac{dA}{(1 + x^2 + y^2)^k}$ converge?

**19.** Evaluate $\displaystyle\iint_D xy\, dA$, where $D$ is the plane region satisfying $x \ge 0$, $0 \le y \le x$, and $x^2 + y^2 \le a^2$.

**20.** Evaluate $\iint_C y\, dA$, where $C$ is the upper half of the cardioid disk $r \le 1 + \cos\theta$.

**21.** Find the volume lying between the paraboloids $z = x^2 + y^2$ and $3z = 4 - x^2 - y^2$.

**22.** Find the volume lying inside both the sphere $x^2 + y^2 + z^2 = a^2$ and the cylinder $x^2 + y^2 = ax$.

**23.** Find the volume lying inside both the sphere $x^2 + y^2 + z^2 = 2a^2$ and the cylinder $x^2 + y^2 = a^2$.

* **24.** Find the volume lying outside the cone $z^2 = x^2 + y^2$ and inside the sphere $x^2 + (y - a)^2 + z^2 = a^2$.

**25.** Find the volume lying inside the cylinder $x^2 + (y - a)^2 = a^2$ and between the upper and lower halves of the cone $z^2 = x^2 + y^2$.

**26.** Find the volume of the region lying above the $xy$-plane, inside the cylinder $x^2 + y^2 = 4$ and below the plane $z = x + y + 4$.

* **27.** Find the volume of the region lying inside all three of the circular cylinders $x^2 + y^2 = a^2$, $x^2 + z^2 = a^2$, and $y^2 + z^2 = a^2$. *Hint:* make a good sketch of the first octant part of the region and use symmetry whenever possible.

**28.** Find the volume of the region lying inside the circular cylinder $x^2 + y^2 = 2y$ and inside the parabolic cylinder $z^2 = y$.

* **29.** Many points are chosen at random in the disk $x^2 + y^2 \le 1$. Find the approximate average value of the distance from these points to the nearest side of the smallest square that contains the disk.

* **30.** Find the average value of $x$ over the segment of the disk $x^2 + y^2 \le 4$ lying to the right of $x = 1$. What is the centroid of the segment?

**31.** Find the volume enclosed by the ellipsoid

$$\frac{x^2}{a^2} + \frac{y^2}{b^2} + \frac{z^2}{c^2} = 1.$$

**32.** Find the volume of the region in the first octant below the paraboloid

$$z = 1 - \frac{x^2}{a^2} - \frac{y^2}{b^2}.$$

*Hint:* use the change of variables $x = au$, $y = bv$.

* **33.** Evaluate $\iint_{|x|+|y|\le a} e^{x+y}\, dA$.

**34.** Find $\iint_P (x^2 + y^2)\, dA$, where $P$ is the parallelogram bounded by the lines $x + y = 1$, $x + y = 2$, $3x + 4y = 5$, and $3x + 4y = 6$.

**35.** Find the area of the region in the first quadrant bounded by the curves $xy = 1$, $xy = 4$, $y = x$, and $y = 2x$.

**36.** Evaluate $\iint_R (x^2 + y^2)\, dA$, where $R$ is the region in the first quadrant bounded by $y = 0$, $y = x$, $xy = 1$, and $x^2 - y^2 = 1$.

* **37.** Let $T$ be the triangle with vertices $(0, 0)$, $(1, 0)$, and $(0, 1)$. Evaluate the integral $\iint_T e^{(y-x)/(y+x)}\, dA$,

(a) by transforming to polar coordinates, and

(b) using the change of variables $u = y - x$, $v = y + x$.

**38.** Use the method of Example 7 to find the area of the region inside the ellipse $4x^2 + 9y^2 = 36$ and above the line $2x + 3y = 6$.

* **39.** **(The error function)** The error function, $\mathrm{Erf}(x)$, is defined for $x \ge 0$ by

$$\mathrm{Erf}(x) = \frac{2}{\sqrt{\pi}} \int_0^x e^{-t^2}\, dt.$$

Show that $\displaystyle \Big(\mathrm{Erf}(x)\Big)^2 = \frac{4}{\pi} \int_0^{\pi/4} \Big(1 - e^{-x^2/\cos^2\theta}\Big)\, d\theta.$

Hence deduce that $\mathrm{Erf}(x) \ge \sqrt{1 - e^{-x^2}}$.

* **40.** **(The gamma and beta functions)** The gamma function $\Gamma(x)$ and the beta function $B(x, y)$ are defined by

$$\Gamma(x) = \int_0^\infty t^{x-1} e^{-t}\, dt, \qquad (x > 0),$$

$$B(x, y) = \int_0^1 t^{x-1} (1 - t)^{y-1}\, dt, \qquad (x > 0,\ y > 0).$$

Using integration by parts it is easily shown that

$$\Gamma(x + 1) = x\Gamma(x), \qquad \text{and}$$
$$\Gamma(n + 1) = n!, \qquad (n = 0, 1, 2, \ldots).$$

Deduce the following further properties of these functions:

(a) $\displaystyle \Gamma(x) = 2 \int_0^\infty s^{2x-1} e^{-s^2}\, ds, \qquad (x > 0),$

(b) $\displaystyle \Gamma\left(\frac{1}{2}\right) = \sqrt{\pi}, \qquad \Gamma\left(\frac{3}{2}\right) = \frac{1}{2}\sqrt{\pi},$

(c) If $x > 0$ and $y > 0$, then

$$B(x, y) = 2 \int_0^{\pi/2} \cos^{2x-1}\theta \sin^{2y-1}\theta\, d\theta,$$

(d) $\displaystyle B(x, y) = \frac{\Gamma(x)\Gamma(y)}{\Gamma(x + y)}.$

## 6.5 TRIPLE INTEGRALS

Now that we have seen how to extend definite integration to two-dimensional domains the extension to three (or more) dimensions is straightforward. For a bounded function $f(x, y, z)$ defined on a rectangular box $B$ ($x_0 \leq x \leq x_1$, $y_0 \leq y \leq y_1$, $z_0 \leq z \leq z_1$), the **triple integral** of $f$ over $B$,

$$\iiint_B f(x, y, z)\, dV \qquad \text{or} \qquad \iiint_B f(x, y, z)\, dx\, dy\, dz,$$

can be defined as a suitable limit of Riemann sums corresponding to partitions of $B$ into subboxes by planes parallel to each of the coordinate planes. We omit the details. Triple integrals over more general domains are defined by extending the function to be zero outside the domain and integrating over a rectangular box containing the domain.

All the properties of double integrals mentioned in Section 6.1 have analogues for triple integrals. In particular, a continuous function is integrable over a closed, bounded domain. If $f(x, y, z) = 1$ on the domain $D$ then the triple integral gives the volume of $D$:

$$\text{Volume of } D = \iiint_D dV.$$

The triple integral of a positive function $f(x, y, z)$ can be interpreted as the "hypervolume" (that is, the four-dimensional volume) of a region in 4-space having the set $D$ as its three-dimensional "base" and having its top on the hypersurface $w = f(x, y, z)$. This is not a particularly useful interpretation; many more useful ones arise in applications. For instance, if $\delta(x, y, z)$ represents the density (mass per unit volume) at position $(x, y, z)$ in a substance occupying the domain $D$ in 3-space then the mass $m$ of the solid is the "sum" of mass elements $dm = \delta(x, y, z)\, dV$ occupying volume elements $dV$:

$$\text{mass} = \iiint_D \delta(x, y, z)\, dV.$$

Some triple integrals can be evaluated by inspection, using symmetry and known volumes.

**EXAMPLE 1** Evaluate

$$\iiint_{x^2+y^2+z^2 \leq a^2} (2 + x - \sin z)\, dV.$$

***SOLUTION*** The domain of integration is the ball of radius $a$ centred at the origin. The integral of 2 over this ball is twice the ball's volume, that is $8\pi a^3/3$. The integrals of $x$ and $\sin z$ over the ball are both zero since both functions are odd in one of the variables and the domain is symmetric about each coordinate plane. (For instance, for every volume element $dV$ in the half of the ball where $x > 0$, there is a corresponding element in the other half where $x$ has the same size but the opposite sign. The contributions from these two elements cancel one another.) Thus

$$\iiint_{x^2+y^2+z^2 \leq a^2} (2 + x - \sin z)\, dV = \frac{8}{3}\pi a^3 + 0 + 0 = \frac{8}{3}\pi a^3.$$

Most triple integrals are evaluated by an iteration procedure similar to that used for double integrals. We slice the domain $D$ with a plane parallel to one of the coordinate planes, double integrate the function with respect to two variables over that slice, and then integrate the result with respect to the remaining variable. Some examples should serve to make the procedure clear.

**EXAMPLE 2** Let $B$ be the rectangular box $0 \le x \le a$, $0 \le y \le b$, $0 \le z \le c$. Evaluate

$$I = \iiint_B (xy^2 + z^3)\, dV.$$

***SOLUTION*** As indicated in Figure 6.33(a), we will slice with planes perpendicular to the $z$-axis, so the $z$ integral will be outermost in the iteration. The slices are rectangles, so the double integrals over them can be immediately iterated also. We do it with the $y$ integral outer and the $x$ integral inner, as suggested by the line shown in the slice.

$$\begin{aligned}
I &= \int_0^c dz \int_0^b dy \int_0^a (xy^2 + z^3)\, dx \\
&= \int_0^c dz \int_0^b dy \left( \frac{x^2y^2}{2} + xz^3 \right)\Bigg|_{x=0}^{x=a} \\
&= \int_0^c dz \int_0^b \left( \frac{a^2y^2}{2} + az^3 \right) dy \\
&= \int_0^c dz \left( \frac{a^2y^3}{6} + ayz^3 \right)\Bigg|_{y=0}^{y=b} \\
&= \int_0^c \left( \frac{a^2b^3}{6} + abz^3 \right) dz \\
&= \left( \frac{a^2b^3z}{6} + \frac{abz^4}{4} \right)\Bigg|_{z=0}^{z=c} = \frac{a^2b^3c}{6} + \frac{abc^4}{4}.
\end{aligned}$$

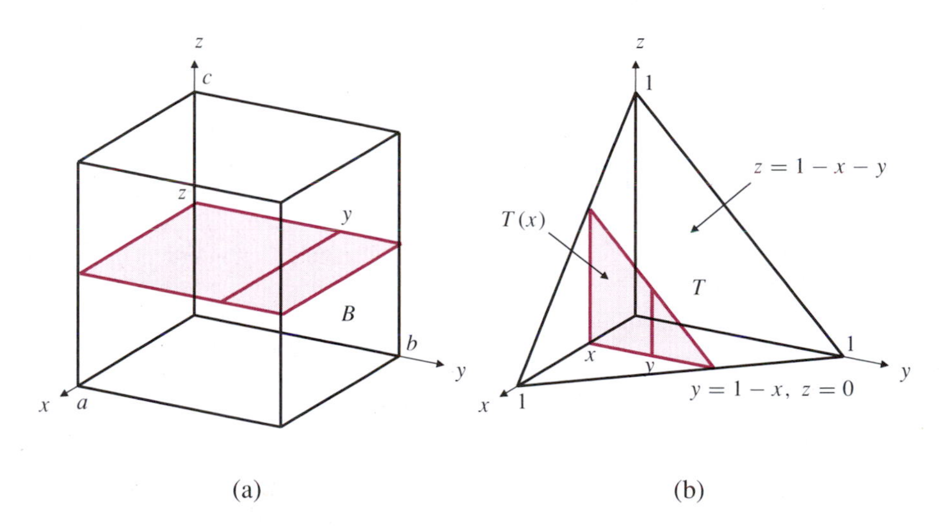

Figure 6.33

(a) The iteration in Example 2

(b) The iteration in Example 3

**EXAMPLE 3** If $T$ is the tetrahedron with vertices $(0, 0, 0)$, $(1, 0, 0)$, $(0, 1, 0)$, and $(0, 0, 1)$, evaluate

$$I = \iiint_T y\, dV.$$

***SOLUTION*** The tetrahedron is shown in Figure 6.33(b). The plane slice in the plane normal to the $x$-axis at position $x$ is the triangle $T(x)$ shown in that figure. $x$ is constant and $y$ and $z$ are variables in the slice. The double integral of $y$ over $T(x)$

is a function of $x$. We evaluate it by integrating first in the $z$ direction and then in the $y$ direction as suggested by the vertical line shown in the slice:

$$\begin{aligned}\iint_{T(x)} y\,dA &= \int_0^{1-x} dy \int_0^{1-x-y} y\,dz \\ &= \int_0^{1-x} y(1-x-y)\,dy \\ &= \left((1-x)\frac{y^2}{2} - \frac{y^3}{3}\right)\Bigg|_0^{1-x} = \frac{1}{6}(1-x)^3.\end{aligned}$$

The value of the triple integral $I$ is the integral of this expression with respect to the remaining variable $x$, to sum the contributions from all such slices between $x = 0$ and $x = 1$:

$$I = \int_0^1 \frac{1}{6}(1-x)^3\,dx = -\frac{1}{24}(1-x)^4\Bigg|_0^1 = \frac{1}{24}.$$ ■

In the above solution we carried out the iteration in two steps in order to show the procedure clearly. In practice, triple integrals are iterated in one step, with no explicit mention made of the double integral over the slice. Thus, using the iteration suggested by Figure 6.33(b) we would immediately write

$$I = \int_0^1 dx \int_0^{1-x} dy \int_0^{1-x-y} y\,dz.$$

The evaluation proceeds as above, starting with the right (that is, inner) integral, followed by the middle integral and then the left (outer) integral. The triple integral represents the "sum" of elements $y\,dV$ over the three-dimensional region $T$. The above iteration corresponds to "summing" (that is, integrating) first along a vertical line (the $z$ integral), then summing these one-dimensional sums in the $y$ direction to get the double sum of all elements in the plane slice, and finally summing these double sums in the $x$ direction to add up the contributions from all the slices. The iteration can be carried out in other directions; there are six possible iterations corresponding to different orders of doing the $x$, $y$, and $z$ integrals. The other five are

$$I = \int_0^1 dx \int_0^{1-x} dz \int_0^{1-x-z} y\,dy,$$

$$I = \int_0^1 dy \int_0^{1-y} dx \int_0^{1-x-y} y\,dz,$$

$$I = \int_0^1 dy \int_0^{1-y} dz \int_0^{1-y-z} y\,dx,$$

$$I = \int_0^1 dz \int_0^{1-z} dx \int_0^{1-x-z} y\,dy,$$

$$I = \int_0^1 dz \int_0^{1-z} dy \int_0^{1-y-z} y\,dx.$$

You should verify these by drawing diagrams analogous to Figure 6.33(b). Of course, all six iterations give the same result.

It is sometimes difficult to visualize the region of 3-space over which a given triple integral is taken. In such situations try to determine the *projection* of that region on one or other of the coordinate planes. For instance, if a region $R$ is bounded by two surfaces with given equations, combining these equations to eliminate one variable will yield the equation of a cylinder (not necessarily circular) with axis parallel to the axis of the eliminated variable. This cylinder will then determine the projection of $R$ onto the coordinate plane perpendicular to that axis. The following example illustrates the use of this technique to find a volume bounded by two surfaces. The volume is expressed as a triple integral with unit integrand.

**EXAMPLE 4** Find the volume of the region $R$ lying below the plane $z = 3 - 2y$ and above the paraboloid $z = x^2 + y^2$.

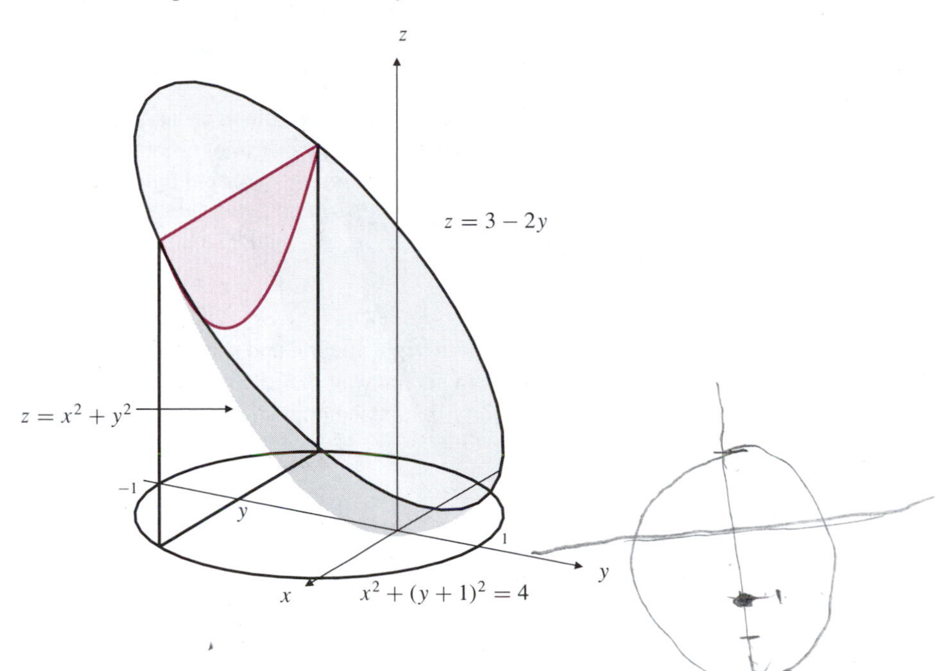


**Figure 6.34** The volume above a paraboloid and under a slanting plane

***SOLUTION*** The region $R$ is shown in Figure 6.34. It is symmetric about the $yz$-plane, and so its volume $V$ may be found by doubling that part lying in the half-space $x > 0$. The two surfaces bounding $R$ intersect on the vertical cylinder $x^2 + y^2 = 3 - 2y$, or $x^2 + (y+1)^2 = 4$, which lies between $y = -3$ and $y = 1$. Thus

$$\begin{aligned}
V = \iiint_R dV &= 2\int_{-3}^{1} dy \int_0^{\sqrt{3-2y-y^2}} dx \int_{x^2+y^2}^{3-2y} dz \\
&= 2\int_{-3}^{1} dy \int_0^{\sqrt{3-2y-y^2}} (3 - 2y - x^2 - y^2)\, dx \\
&= 2\int_{-3}^{1} \left((3 - 2y - y^2)x - \frac{x^3}{3}\right)\Bigg|_0^{\sqrt{3-2y-y^2}} dy \\
&= \frac{4}{3}\int_{-3}^{1} \left(3 - 2y - y^2\right)^{3/2} dy \\
&= \frac{4}{3}\int_{-3}^{1} \left(4 - (y+1)^2\right)^{3/2} dy.
\end{aligned}$$

In the last integral we make the change of variable $y + 1 = 2\sin\theta$ and obtain

$$\begin{aligned} V &= \frac{64}{3}\int_{-\pi/2}^{\pi/2} \cos^4\theta\, d\theta \\ &= \frac{128}{3}\int_0^{\pi/2} \left(\frac{1+\cos 2\theta}{2}\right)^2 d\theta \\ &= \frac{32}{3}\int_0^{\pi/2} \left(1 + 2\cos 2\theta + \frac{1+\cos 4\theta}{2}\right) d\theta = 8\pi \text{ cubic units.} \end{aligned}$$

As was the case for double integrals, it is sometimes necessary to reiterate a given iterated integral so that the integrations are performed in a different order. This task is most easily accomplished if we can translate the given iteration into a sketch of the region of integration. The ability to deduce the shape of the region from the limits in the iterated integral is a skill that one acquires with practice. You should first determine the projection of the region on a coordinate plane — the plane of the two variables in the outer integrals of the given iteration.

It is also possible to reiterate an iterated integral in a different order by manipulating the limits of integration algebraically. We will illustrate both approaches (graphical and algebraic) in the following examples.

**EXAMPLE 5** Express the iterated integral

$$I = \int_0^1 dy \int_y^1 dz \int_0^z f(x, y, z)\, dx$$

as a triple integral and sketch the region over which it is taken. Reiterate the integral in such a way that the integrations are performed in the order: first $y$, then $z$, then $x$ (that is, the opposite order to the given iteration).

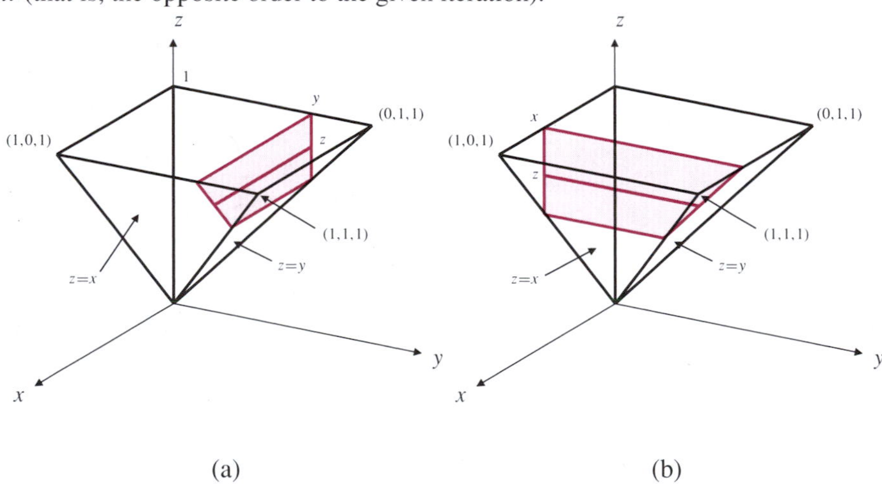


**Figure 6.35**

(a) The solid region for the triple integral in Example 5 sliced corresponding to the given iteration

(b) The same solid sliced to conform to the desired iteration

***SOLUTION*** We express $I$ as an uniterated triple integral:

$$I = \iiint_R f(x, y, z)\, dV.$$

The outer integral in the given iteration shows that the region $R$ lies between the planes $y = 0$ and $y = 1$. For each such value of $y$, $z$ must lie between $y$ and 1. Therefore $R$ lies below the plane $z = 1$ and above the plane $z = y$, and the projection of $R$ onto the $yz$-plane is the triangle with vertices $(0, 0, 0)$, $(0, 0, 1)$, and $(0, 1, 1)$. Through any point $(0, y, z)$ in this triangle, a line parallel to the $x$-axis intersects $R$ between $x = 0$ and $x = z$. Thus the solid is bounded by the five planes $x = 0$, $y = 0$, $z = 1$, $y = z$, and $z = x$. It is sketched in Figure 6.35(a), with slice and line corresponding to the given iteration.

The required iteration corresponds to the slice and line shown in Figure 6.35(b). Therefore it is

$$I = \int_0^1 dx \int_x^1 dz \int_0^z f(x, y, z)\, dy.$$

■

■ **EXAMPLE 6** Use algebra to write an iteration of the integral

$$I = \int_0^1 dx \int_x^1 dy \int_x^y f(x, y, z)\, dz$$

with the order of integrations reversed.

***SOLUTION*** From the given iteration we can write three sets of inequalities satisfied by the outer variable $x$, the middle variable $y$, and the inner variable $z$. We write these in order as follows:

$$\begin{array}{ll} 0 \le x \le 1 & \text{inequalities for } x \\ x \le y \le 1 & \text{inequalities for } y \\ x \le z \le y & \text{inequalities for } z. \end{array}$$

Note that the limits for each variable can be constant or can depend only on variables whose inequalities are on lines above the line for that variable. (In this case, the limits for $x$ must both be constant, those for $y$ can depend on $x$, and those for $z$ can depend on both $x$ and $y$.) This is a requirement for iterated integrals – outer integrals cannot depend on the variables of integration of the inner integrals.

We want to construct an equivalent set of inequalities with those for $z$ on the top line, then those for $y$, then those for $x$ on the bottom line. The limits for $z$ must be constants. From the inequalities above we determine that $0 \le x \le z$ and $z \le y \le 1$. Thus $z$ must satisfy $0 \le z \le 1$. The inequalities for $y$ can depend on $z$. Since $z \le y$ and $y \le 1$, we have $z \le y \le 1$. Finally, the limits for $x$ can depend on both $y$ and $z$. We have $0 \le x$, $x \le y$, and $x \le z$. Since we have already determined that $z \le y$, we must have $0 \le x \le z$. Thus the revised inequalities are

$$\begin{array}{ll} 0 \le z \le 1 & \text{inequalities for } z \\ z \le y \le 1 & \text{inequalities for } y \\ 0 \le x \le z & \text{inequalities for } x \end{array}$$

and the required iteration is

$$I = \int_0^1 dz \int_z^1 dy \int_0^z f(x, y, z)\, dx.$$

■

## EXPLORE! Computer Evaluation of Triple Integrals

Computer algebra systems can be used to evaluate iterated triple integrals just as they could be used to evaluate double integrals. (See Section 6.2.) Try to evaluate the following integrals using such software. If symbolic evaluation is not possible, find a numerical approximation.

(a) $\displaystyle\int_0^1 dx \int_1^{1+x} dy \int_2^{2+xy} (x^2+y^2+z^2)\,dz$

(b) $\displaystyle\int_0^{\pi} dz \int_0^{z^2} dy \int_0^{z^2} \sin z\,dx$

(c) $\displaystyle\int_0^1 dz \int_0^z dy \int_0^y \cos(xyz)\,dx$

## EXERCISES 6.5

In Exercises 1–12, evaluate the triple integrals over the indicated region $R$. Be alert for simplifications and auspicious orders of iteration.

1. $\displaystyle\iiint_R (1+2x-3y)\,dV$, over the box $-a \le x \le a$, $-b \le y \le b$, $-c \le z \le c$

2. $\displaystyle\iiint_B xyz\,dV$ over the box $0 \le x \le 1$, $-2 \le y \le 0$, $1 \le z \le 4$

3. $\displaystyle\iiint_D (3+2xy)\,dV$ over the solid hemispherical dome $D$ given by $x^2+y^2+z^2 \le 4$ and $z \ge 0$

4. $\displaystyle\iiint_R x\,dV$, over the tetrahedron bounded by the coordinate planes and the plane $\dfrac{x}{a}+\dfrac{y}{b}+\dfrac{z}{c}=1$

5. $\displaystyle\iiint_R (x^2+y^2)\,dV$ over the cube $0 \le x, y, z \le 1$

6. $\displaystyle\iiint_R (x^2+y^2+z^2)\,dV$ over the cube of Exercise 5

7. $\displaystyle\iiint_R (xy+z^2)\,dV$, over the set $0 \le z \le 1-|x|-|y|$

8. $\displaystyle\iiint_R yz^2e^{-xyz}\,dV$, over the cube $0 \le x, y, z \le 1$

9. $\displaystyle\iiint_R \sin(\pi y^3)\,dV$, over the pyramid with vertices $(0,0,0)$, $(0,1,0)$, $(1,1,0)$, $(1,1,1)$, and $(0,1,1)$

10. $\displaystyle\iiint_R y\,dV$, over that part of the cube $0 \le x, y, z \le 1$ lying above the plane $y+z=1$ and below the plane $x+y+z=2$

11. $\displaystyle\iiint_R \frac{1}{(x+y+z)^3}\,dV$, over the region bounded by the six planes $z=1$, $z=2$, $y=0$, $y=z$, $x=0$, and $x=y+z$

12. $\displaystyle\iiint_R \cos x \cos y \cos z\,dV$, over the tetrahedron defined by $x \ge 0$, $y \ge 0$, $z \ge 0$, and $x+y+z \le \pi$

13. Evaluate $\displaystyle\iiint_{\mathbb{R}^3} e^{-x^2-2y^2-3z^2}\,dV$. *Hint:* use the result of Example 4 of Section 6.4.

14. Find the volume of the region lying inside the cylinder $x^2+4y^2=4$, above the $xy$-plane and below the plane $z=2+x$.

15. Find $\displaystyle\iiint_T x\,dV$ where $T$ is the tetrahedron bounded by the planes $x=1$, $y=1$, $z=1$ and $x+y+z=2$.

16. Sketch the region $R$ in the first octant of 3-space that has finite volume and is bounded by the surfaces $x=0$, $z=0$, $x+y=1$, and $z=y^2$. Write six different iterations of the triple integral of $f(x,y,z)$ over $R$.

In Exercises 17–20, express the given iterated integral as a triple integral and sketch the region over which it is taken. Reiterate the integral so that the outermost integral is with respect to $x$ and the innermost is with respect to $z$.

* 17. $\displaystyle\int_0^1 dz \int_0^{1-z} dy \int_0^1 f(x,y,z)\,dx$

* 18. $\displaystyle\int_0^1 dz \int_z^1 dy \int_0^y f(x,y,z)\,dx$

* 19. $\displaystyle\int_0^1 dz \int_z^1 dx \int_0^{x-z} f(x,y,z)\,dy$

* 20. $\displaystyle\int_0^1 dy \int_0^{\sqrt{1-y^2}} dz \int_{y^2+z^2}^1 f(x,y,z)\,dx$

21. Repeat Exercise 17 using the method of Example 6.

22. Repeat Exercise 18 using the method of Example 6.

23. Repeat Exercise 19 using the method of Example 6.

24. Repeat Exercise 20 using the method of Example 6.

**25.** Rework Example 5 using the method of Example 6.

**26.** Rework Example 6 using the method of Example 5.

In Exercises 27–28, evaluate the given iterated integral by reiterating it in a different order. (You will need to make a good sketch of the region.)

* **27.** $\int_0^1 dz \int_z^1 dx \int_0^x e^{x^3}\, dy$

* **28.** $\int_0^1 dx \int_0^{1-x} dy \int_y^1 \frac{\sin(\pi z)}{z(2-z)}\, dz$

**29.** Define the average value of an integrable function $f(x, y, z)$ over a region $R$ of 3-space. Find the average value of $x^2 + y^2 + z^2$ over the cube $0 \le x \le 1, 0 \le y \le 1, 0 \le z \le 1$.

* **30.** State a Mean-Value Theorem for triple integrals analogous to Theorem 3 of Section 6.3. Use it to prove that if $f(x, y, z)$ is continuous near the point $(a, b, c)$ and if $B_\epsilon(a, b, c)$ is the ball of radius $\epsilon$ centred at $(a, b, c)$ then

$$\lim_{\epsilon \to 0} \frac{3}{4\pi\epsilon^3} \iiint_{B_\epsilon(a,b,c)} f(x, y, z)\, dV = f(a, b, c).$$

## 6.6 CHANGE OF VARIABLES IN TRIPLE INTEGRALS

The change of variables formula for a double integral extends to triple (or higher-order) integrals. Consider the transformation:

$$\begin{aligned} x &= x(u, v, w), \\ y &= y(u, v, w), \\ z &= z(u, v, w), \end{aligned}$$

where $x$, $y$, and $z$ have continuous first partial derivatives with respect to $u$, $v$, and $w$. Near any point where the Jacobian $\partial(x, y, z)/\partial(u, v, w)$ is nonzero, the transformation scales volume elements according to the formula

$$dV = dx\, dy\, dz = \left| \frac{\partial(x, y, z)}{\partial(u, v, w)} \right| du\, dv\, dw.$$

Thus, if the transformation is one-to-one from a domain $S$ in $uvw$-space onto a domain $D$ in $xyz$-space, and if

$$g(u, v, w) = f\big(x(u, v, w), y(u, v, w), z(u, v, w)\big),$$

then

$$\iiint_D f(x, y, z)\, dx\, dy\, dz = \iiint_S g(u, v, w) \left| \frac{\partial(x, y, z)}{\partial(u, v, w)} \right| du\, dv\, dw.$$

The proof is similar to that of the two-dimensional case given in Section 6.4. (See Exercise 39 at the end of this section.)

**EXAMPLE 1** Under the change of variables $x = au$, $y = bv$, $z = cw$, the solid ellipsoid $E$ given by

$$\frac{x^2}{a^2} + \frac{y^2}{b^2} + \frac{z^2}{c^2} \le 1$$

becomes the ball $B$ given by $u^2 + v^2 + w^2 \le 1$. Since the Jacobian of this transformation is

$$\frac{\partial(x, y, z)}{\partial(u, v, w)} = \begin{vmatrix} a & 0 & 0 \\ 0 & b & 0 \\ 0 & 0 & c \end{vmatrix} = abc$$

the volume of the ellipsoid is given by

$$\begin{aligned}\text{Volume of } E &= \iiint_E dx\,dy\,dz \\ &= \iiint_B abc\,du\,dv\,dw = abc \times (\text{Volume of } B) \\ &= \frac{4}{3}\pi abc \text{ cubic units.}\end{aligned}$$

## Cylindrical Coordinates

Among the most useful alternatives to Cartesian coordinates in 3-space are two coordinate systems that generalize plane polar coordinates. The simpler of these systems is called the system of **cylindrical coordinates**. It uses ordinary plane polar coordinates in the $xy$-plane while retaining the Cartesian $z$ coordinate for measuring vertical distances. Thus each point in 3-space has cylindrical coordinates $[r, \theta, z]$ related to its Cartesian coordinates $(x, y, z)$ by the transformation

$$x = r\cos\theta, \qquad y = r\sin\theta, \qquad z = z.$$

Figure 6.36 shows how a point $P$ is located by its cylindrical coordinates $[r, \theta, z]$ as well as by its Cartesian coordinates $(x, y, z)$. Note that the distance from the origin to $P$ is

$$d = \sqrt{r^2 + z^2} = \sqrt{x^2 + y^2 + z^2}.$$

**EXAMPLE 2** The point with Cartesian coordinates $(1, 1, 1)$ has cylindrical coordinates $[\sqrt{2}, \pi/4, 1]$. The point with Cartesian coordinates $(0, 2, -3)$ has cylindrical coordinates $[2, \pi/2, -3]$. The point with cylindrical coordinates $[4, -\pi/3, 5]$ has Cartesian coordinates $(2, -2\sqrt{3}, 5)$.

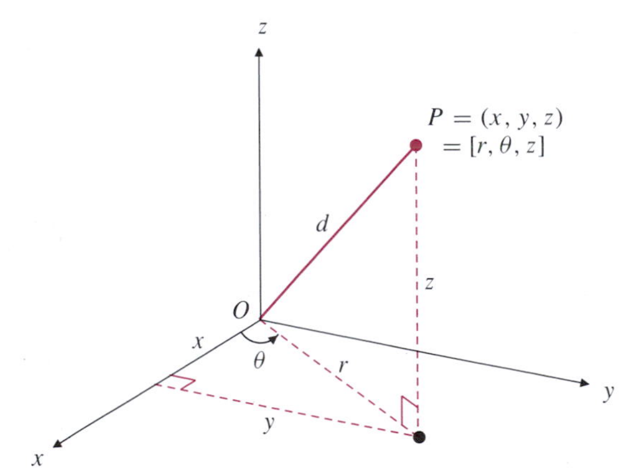


Figure 6.36 The cylindrical coordinates of a point

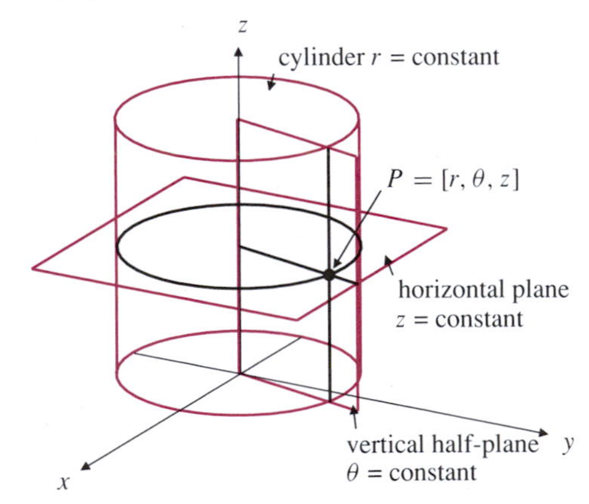


Figure 6.37 The coordinate surfaces for cylindrical coordinates

The coordinate surfaces in cylindrical coordinates are the $r$-surfaces (vertical circular cylinders centred on the $z$-axis), the $\theta$-surfaces (vertical planes containing the $z$-axis), and the $z$-surfaces (planes perpendicular to the $z$-axis). (See Figure 6.37.) Cylindrical coordinates lend themselves to representing domains that are bounded by such surfaces, and in general to problems with axial symmetry (around the $z$-axis).

The **volume element in cylindrical coordinates** is

$$dV = r\,dr\,d\theta\,dz,$$

which is easily seen by examining the infinitesimal "box" bounded by the coordinate surfaces corresponding to values $r, r+dr, \theta, \theta+d\theta, z$, and $z+dz$ (see Figure 6.38), or by calculating the Jacobian

$$\frac{\partial(x, y, z)}{\partial(r, \theta, z)} = \begin{vmatrix} \cos\theta & -r\sin\theta & 0 \\ \sin\theta & r\cos\theta & 0 \\ 0 & 0 & 1 \end{vmatrix} = r.$$

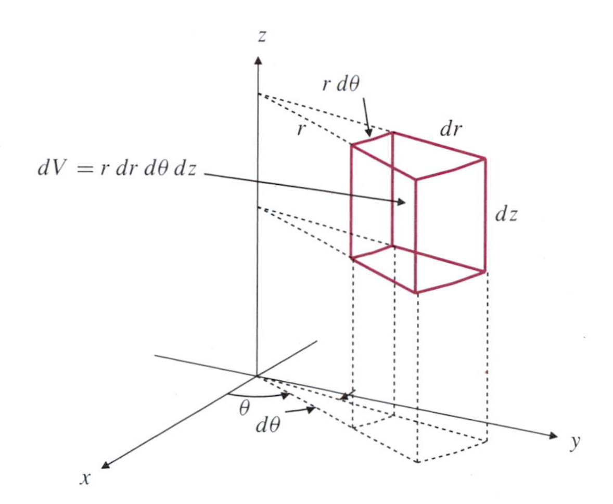


**Figure 6.38** The volume element in cylindrical coordinates

**EXAMPLE 3** Evaluate

$$\iiint_D (x^2 + y^2)\,dV$$

over the first octant region bounded by the cylinders $x^2 + y^2 = 1$ and $x^2 + y^2 = 4$ and the planes $z = 0$, $z = 1$, $x = 0$, and $x = y$.

***SOLUTION*** In terms of cylindrical coordinates the region is bounded by $r = 1$, $r = 2$, $\theta = \pi/4$, $\theta = \pi/2$, $z = 0$, and $z = 1$. (See Figure 6.39. It is a rectangular coordinate box in $r\theta z$-space.) Since the integrand is $x^2 + y^2 = r^2$ the integral is

$$\begin{aligned}\iiint_D (x^2 + y^2)\,dV &= \int_0^1 dz \int_{\pi/4}^{\pi/2} d\theta \int_1^2 r^2\,r\,dr \\ &= (1 - 0)\left(\frac{\pi}{2} - \frac{\pi}{4}\right)\left(\frac{2^4}{4} - \frac{1^4}{4}\right) = \frac{15}{16}\pi.\end{aligned}$$

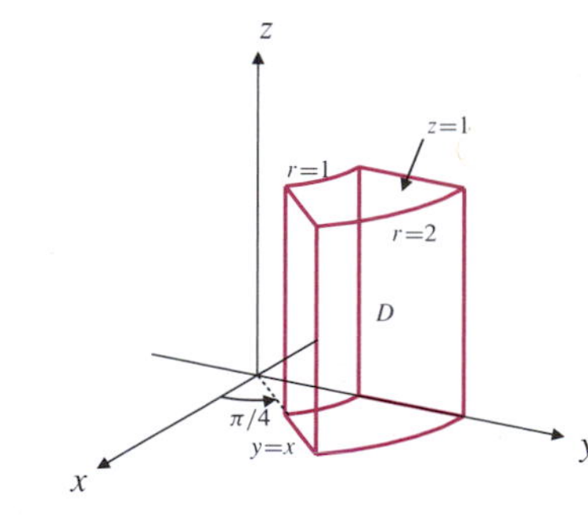


**Figure 6.39**

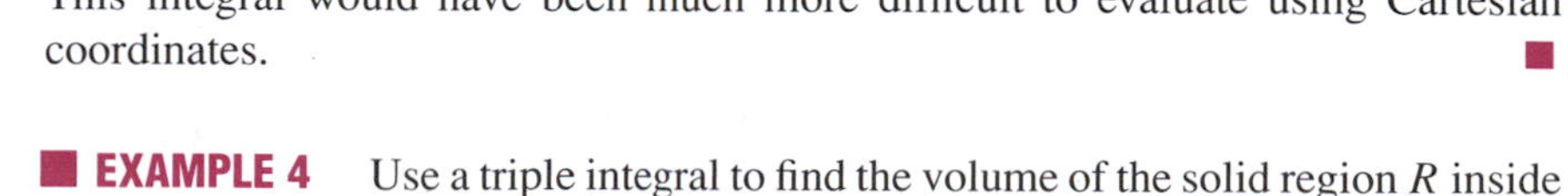

This integral would have been much more difficult to evaluate using Cartesian coordinates. ■

**EXAMPLE 4** Use a triple integral to find the volume of the solid region $R$ inside the sphere $x^2 + y^2 + z^2 = 6$ and above the paraboloid $z = x^2 + y^2$.

***SOLUTION*** One-quarter of the required volume lies in the first octant. It is shown in Figure 6.40. The two surfaces intersect on the vertical cylinder

$$6 - x^2 - y^2 = z^2 = (x^2 + y^2)^2,$$

or, in terms of cylindrical coordinates,

$$\begin{aligned} &r^4 + r^2 - 6 = 0 \\ &(r^2 + 3)(r^2 - 2) = 0. \end{aligned}$$

The only relevant solution to this equation is $r = \sqrt{2}$. Thus the required volume lies above the disk of radius $\sqrt{2}$ centred at the origin in the $xy$-plane. The total volume $V$ is

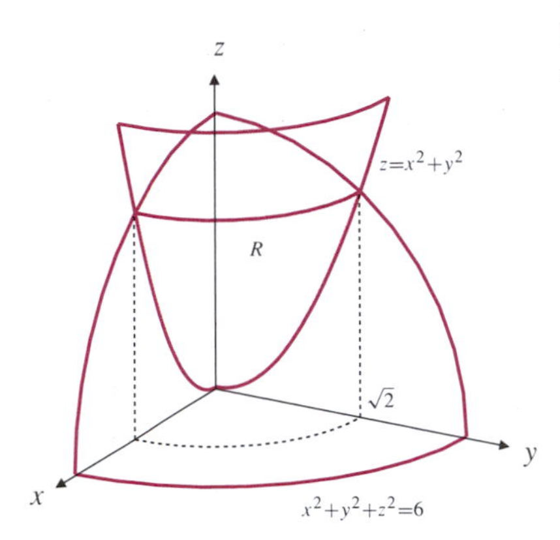


Figure 6.40

$$\begin{aligned} V = \iiint_R dV &= \int_0^{2\pi} d\theta \int_0^{\sqrt{2}} r\,dr \int_{r^2}^{\sqrt{6-r^2}} dz \\ &= 2\pi \int_0^{\sqrt{2}} \left( r\sqrt{6 - r^2} - r^3 \right) dr \\ &= 2\pi \left[ -\frac{1}{3}(6 - r^2)^{3/2} - \frac{r^4}{4} \right]\Bigg|_0^{\sqrt{2}} \\ &= 2\pi \left[ \frac{6\sqrt{6}}{3} - \frac{8}{3} - 1 \right] = \frac{2\pi}{3}(6\sqrt{6} - 11) \text{ cubic units.} \end{aligned}$$

## Spherical Coordinates

In the system of **spherical coordinates** a point $P$ in 3-space is represented by the ordered triple $[\rho, \phi, \theta]$, where $\rho$ is the distance from $P$ to the origin $O$, $\phi$ is the angle the radial line $OP$ makes with the positive direction of the $z$-axis, and $\theta$ is the angle between the plane containing $P$ and the $z$-axis and the $xz$-plane. (See Figure 6.41.) It is conventional to consider spherical coordinates restricted in such a way that $\rho \ge 0$, $0 \le \phi \le \pi$, and $0 \le \theta < 2\pi$ (or $-\pi < \theta \le \pi$). Every point not on the $z$-axis then has exactly one spherical coordinate representation, and the transformation from Cartesian coordinates $(x, y, z)$ to spherical coordinates $[\rho, \phi, \theta]$ is one-to-one off the $z$-axis. Using the right-angled triangles in the figure, we can see that this transformation is given by:

$$\begin{aligned} x &= \rho \sin\phi \cos\theta \\ y &= \rho \sin\phi \sin\theta \\ z &= \rho \cos\phi. \end{aligned}$$

Observe that

$$\rho^2 = x^2 + y^2 + z^2,$$

and that the $r$ coordinate in cylindrical coordinates is related to $\rho$ and $\phi$ by

$$r = \sqrt{x^2 + y^2} = \rho \sin\phi.$$

Thus, also

$$\tan\phi = \frac{r}{z} = \frac{\sqrt{x^2+y^2}}{z} \qquad \text{and} \qquad \tan\theta = \frac{y}{x}.$$

If $\phi = 0$ or $\phi = \pi$, then $r = 0$, so the $\theta$ coordinate is irrelevant at points on the $z$-axis.

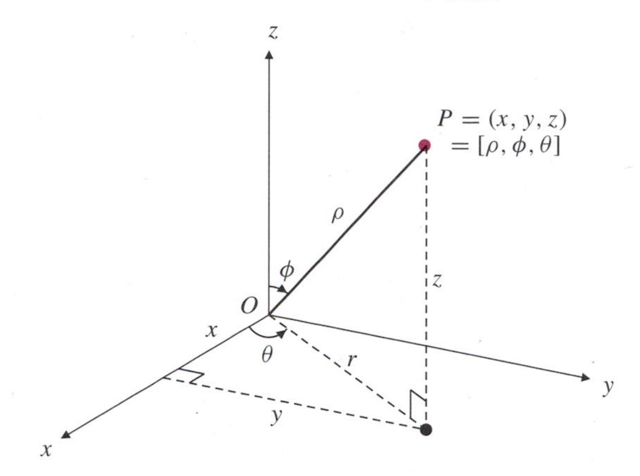


**Figure 6.41** The spherical coordinates of a point

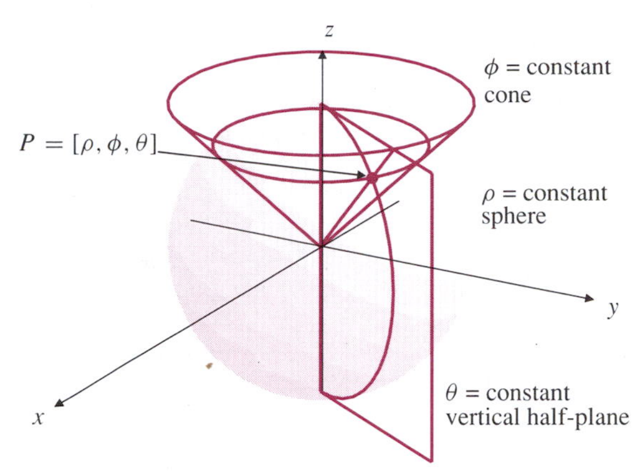


**Figure 6.42** The coordinate surfaces for spherical coordinates

Some coordinate surfaces for spherical coordinates are shown in Figure 6.42. The $\rho$-surfaces ($\rho$ = constant) are spheres centred at the origin; the $\phi$-surfaces ($\phi$ = constant) are circular cones with the $z$-axis as axis; the $\theta$-surfaces ($\theta$ = constant) are vertical half-planes with edge along the $z$-axis. If we take a coordinate system with origin at the centre of the earth, $z$-axis through the north pole, and $x$-axis through the intersection of the Greenwich meridian and the equator, then the intersection of the surface of the earth with the $\phi$-surfaces are the *parallels of latitude*, and the intersection with the $\theta$-surfaces are the *meridians of longitude*. Since latitude is measured from 90° at the north pole to −90° at the south pole, while $\phi$ is measured from 0 at the north pole to $\pi$ (= 180°) at the south pole, the coordinate $\phi$ is frequently referred to as the **colatitude** coordinate. $\theta$ is the **longitude** coordinate. Observe that $\theta$ has the same significance in spherical coordinates as it does in cylindrical coordinates.

**EXAMPLE 5** Find:

(a) the Cartesian coordinates of the point $P$ with spherical coordinates $[2, \pi/3, \pi/2]$, and

(b) the spherical coordinates of the point $Q$ with Cartesian coordinates $(1, 1, \sqrt{2})$.

***SOLUTION***

(a) If $\rho = 2$, $\phi = \pi/3$, and $\theta = \pi/2$, then

$$\begin{aligned} x &= 2\sin(\pi/3)\cos(\pi/2) = 0 \\ y &= 2\sin(\pi/3)\sin(\pi/2) = \sqrt{3} \\ z &= 2\cos(\pi/3) = 1. \end{aligned}$$

The Cartesian coordinates of $P$ are $(0, \sqrt{3}, 1)$.

(b) Given that

$$\rho \sin\phi \cos\theta = x = 1$$
$$\rho \sin\phi \sin\theta = y = 1$$
$$\rho \cos\phi = z = \sqrt{2}$$

we calculate that $\rho^2 = 1 + 1 + 2 = 4$, so $\rho = 2$. Also $r^2 = 1 + 1 = 2$, so $r = \sqrt{2}$. Thus $\tan\phi = r/z = 1$, so $\phi = \pi/4$. Also, $\tan\theta = y/x = 1$, so $\theta = \pi/4$ or $5\pi/4$. Since $x > 0$, we must have $\theta = \pi/4$. The spherical coordinates of $Q$ are $[2, \pi/4, \pi/4]$. ■

**REMARK** You may wonder why we write spherical coordinates in the order $\rho, \phi, \theta$ rather than $\rho, \theta, \phi$. The reason, which will not become apparent until Chapter 9, is so that the triad of unit vectors at any point $P$ pointing in the directions of increasing $\rho$, increasing $\phi$, and increasing $\theta$ form a *right-hand basis* rather than a left-hand one.

The **volume element in spherical coordinates** is

$$dV = \rho^2 \sin\phi \, d\rho \, d\phi \, d\theta.$$

To see this, observe that the infinitesimal coordinate box bounded by the coordinate surfaces corresponding to values $\rho, \rho+d\rho, \phi, \phi+d\phi, \theta$, and $\theta+d\theta$ has dimensions $d\rho$, $\rho\, d\phi$, and $\rho \sin\phi \, d\theta$. (See Figure 6.43.) Alternatively, the Jacobian of the transformation can be calculated:

$$\begin{aligned}\frac{\partial(x, y, z)}{\partial(\rho, \phi, \theta)} &= \begin{vmatrix} \sin\phi\cos\theta & \rho\cos\phi\cos\theta & -\rho\sin\phi\sin\theta \\ \sin\phi\sin\theta & \rho\cos\phi\sin\theta & \rho\sin\phi\cos\theta \\ \cos\phi & -\rho\sin\phi & 0 \end{vmatrix} \\ &= \cos\phi \begin{vmatrix} \rho\cos\phi\cos\theta & -\rho\sin\phi\sin\theta \\ \rho\cos\phi\sin\theta & \rho\sin\phi\cos\theta \end{vmatrix} \\ &\quad + \rho\sin\phi \begin{vmatrix} \sin\phi\cos\theta & -\rho\sin\phi\sin\theta \\ \sin\phi\sin\theta & \rho\sin\phi\cos\theta \end{vmatrix} \\ &= \cos\phi(\rho^2\sin\phi\cos\phi) + \rho\sin\phi(\rho\sin^2\phi) \\ &= \rho^2\sin\phi.\end{aligned}$$

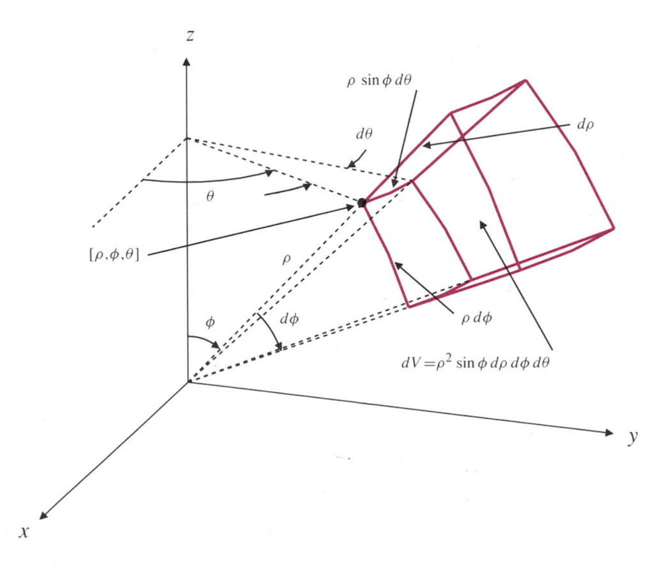


**Figure 6.43** The volume element in spherical coordinates

Spherical coordinates are suited to problems involving spherical symmetry, and in particular, to regions bounded by spheres centred at the origin, circular cones with axes along the $z$-axis, and vertical planes containing the $z$-axis.

**EXAMPLE 6** A solid half-ball $H$ of radius $a$ has density depending on the distance $\rho$ from the centre of the base disk. The density is given by $k(2a - \rho)$, where $k$ is a constant. Find the mass of the half-ball.

***SOLUTION*** Choosing coordinates with origin at the centre of the base and, so that the half-ball lies above the $xy$-plane, we calculate the mass $m$ as follows:

$$\begin{aligned} m &= \iiint_H k(2a - \rho)\,dV = \iiint_H k(2a - \rho)\,\rho^2 \sin\phi\,d\rho\,d\phi\,d\theta \\ &= k\int_0^{2\pi} d\theta \int_0^{\pi/2} \sin\phi\,d\phi \int_0^a (2a - \rho)\,\rho^2\,d\rho \\ &= 2k\pi \times 1 \times \left(\frac{2a}{3}\rho^3 - \frac{1}{4}\rho^4\right)\Big|_0^a = \frac{5}{6}\pi k a^4 \text{ units.} \end{aligned}$$

***REMARK*** In the above example both the integrand and the region of integration exhibited spherical symmetry, so that the choice of spherical coordinates to carry out the integration was most appropriate. The mass could have been evaluated in cylindrical coordinates. The iteration in that system is

$$m = \int_0^{2\pi} d\theta \int_0^a r\,dr \int_0^{\sqrt{a^2-r^2}} k(2a - \sqrt{r^2 + z^2})\,dz$$

and is much harder to evaluate. It is even more difficult in Cartesian coordinates:

$$m = 4\int_0^a dx \int_0^{\sqrt{a^2-x^2}} dy \int_0^{\sqrt{a^2-x^2-y^2}} k(2a - \sqrt{x^2 + y^2 + z^2})\,dz.$$

The choice of coordinate system can greatly affect the difficulty of computation of a multiple integral.

Many problems will have elements of spherical and axial symmetry. In such cases it may not be clear whether it would be better to use spherical or cylindrical coordinates. In such doubtful cases the integrand is usually the best guide. Use cylindrical or spherical coordinates according to whether the integrand involves $x^2 + y^2$ or $x^2 + y^2 + z^2$.

**EXAMPLE 7** The moment of inertia about the $z$-axis of a solid of density $\delta$ occupying the region $R$ is given by the integral

$$I = \iiint_R (x^2 + y^2)\delta\,dV.$$

(See Section 6.7.) Calculate that moment of inertia for a solid of unit density occupying the region inside the sphere $x^2 + y^2 + z^2 = 4a^2$ and outside the cylinder $x^2 + y^2 = a^2$.

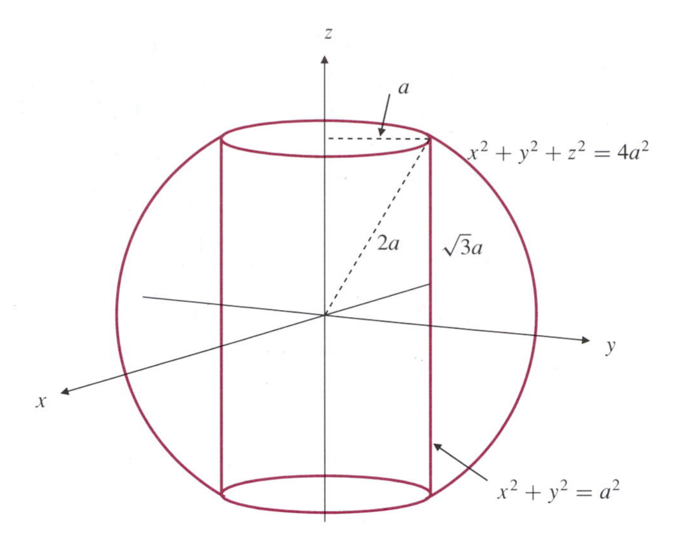


**Figure 6.44** A solid ball with a cylindrical hole through it

**SOLUTION** See Figure 6.44. In terms of spherical coordinates the required moment of inertia is

$$I = 2\int_0^{2\pi} d\theta \int_{\pi/6}^{\pi/2} \sin\phi\, d\phi \int_{a/\sin\phi}^{2a} \rho^2 \sin^2\phi\, \rho^2\, d\rho.$$

In terms of cylindrical coordinates it is

$$I = 2\int_0^{2\pi} d\theta \int_a^{2a} r\, dr \int_0^{\sqrt{4a^2-r^2}} r^2\, dz.$$

The latter formula looks somewhat easier to evaluate. We continue with it. Evaluating the $\theta$ and $z$ integrals, we get

$$I = 4\pi \int_a^{2a} r^3\sqrt{4a^2 - r^2}\, dr.$$

Making the substitution $u = 4a^2 - r^2$, $du = -2r\, dr$, we obtain

$$I = 2\pi \int_0^{3a^2} (4a^2 - u)\sqrt{u}\, du = 2\pi\left(4a^2\frac{u^{3/2}}{3/2} - \frac{u^{5/2}}{5/2}\right)\Bigg|_0^{3a^2} = \frac{44}{5}\sqrt{3}\pi a^5. \ \blacksquare$$

## EXERCISES 6.6

1. Convert the spherical coordinates $[4, \pi/3, 2\pi/3]$ to Cartesian coordinates and to cylindrical coordinates.

2. Convert the Cartesian coordinates $(2, -2, 1)$ to cylindrical coordinates and to spherical coordinates.

3. Convert the cylindrical coordinates $[2, \pi/6, -2]$ to Cartesian coordinates and to spherical coordinates.

4. A point $P$ has spherical coordinates $[1, \phi, \theta]$ and cylindrical coordinates $[r, \pi/4, r]$. Find the Cartesian coordinates of the point.

Describe the sets of points in 3-space that satisfy the equations in Exercises 5–14. Here 4, $\theta$, $\rho$, and $\phi$ denote the appropriate cylindrical or spherical coordinates.

5. $\theta = \pi/2$

6. $\phi = 2\pi/3$

7. $\phi = \pi/2$

8. $\rho = 4$

**9.** $r = 4$

**10.** $\rho = z$

**11.** $\rho = r$

**12.** $\rho = 2x$

**13.** $\rho = 2\cos\phi$

**14.** $r = 2\cos\theta$

In Exercises 15–23, find the volumes of the indicated regions.

**15.** Inside the cone $z = \sqrt{x^2 + y^2}$ and inside the sphere $x^2 + y^2 + z^2 = a^2$

**16.** Above the surface $z = (x^2 + y^2)^{1/4}$ and inside the sphere $x^2 + y^2 + z^2 = 2$

**17.** Between the paraboloids $z = 10 - x^2 - y^2$ and $z = 2(x^2 + y^2 - 1)$

**18.** Inside the paraboloid $z = x^2 + y^2$ and inside the sphere $x^2 + y^2 + z^2 = 12$

**19.** Above the $xy$-plane, inside the cone $z = 2a - \sqrt{x^2 + y^2}$ and inside the cylinder $x^2 + y^2 = 2ay$

**20.** Above the $xy$-plane, under the paraboloid $z = 1 - x^2 - y^2$ and in the wedge $-x \le y \le \sqrt{3}x$

**21.** In the first octant, between the planes $y = 0$ and $y = x$ and inside the ellipsoid $\dfrac{x^2}{a^2} + \dfrac{y^2}{b^2} + \dfrac{z^2}{c^2} = 1$. *Hint:* use the change of variables suggested in Example 1.

* **22.** Bounded by the hyperboloid $\dfrac{x^2}{a^2} + \dfrac{y^2}{b^2} - \dfrac{z^2}{c^2} = 1$ and the planes $z = -c$ and $z = c$

**23.** Above the $xy$-plane and below the paraboloid $z = 1 - \dfrac{x^2}{a^2} - \dfrac{y^2}{b^2}$

* **24.** Find the volume of the region lying above the cone $z = 2\sqrt{x^2 + y^2}$ and beneath the plane $z = 1 - x$. *Hint:* this is rather difficult to do with integrals but can be done by a geometric argument. The region is itself a cone with elliptical base and central axis oblique to the base. The volume of *any* cone is one-third the base area times the height measured perpendicular to the base.

**25.** Evaluate $\displaystyle\iiint_R z(x^2 + y^2)\,dV$, where $R$ is the solid cylinder $0 \le x^2 + y^2 \le a^2$, $0 \le z \le h$.

**26.** Evaluate $\displaystyle\iiint_R (x^2 + y^2 + z^2)\,dV$, where $R$ is the same cylinder as in Exercise 25.

**27.** Find $\displaystyle\iiint_B (x^2 + y^2)\,dV$, where $B$ is the ball given by $x^2 + y^2 + z^2 \le a^2$.

**28.** Find $\displaystyle\iiint_B (x^2 + y^2 + z^2)\,dV$, where $B$ is the ball of Exercise 27.

**29.** Find $\displaystyle\iiint_R (x^2 + y^2 + z^2)\,dV$, where $R$ is the region which lies above the cone $z = c\sqrt{x^2 + y^2}$ and inside the sphere $x^2 + y^2 + z^2 = a^2$.

**30.** Evaluate $\displaystyle\iiint_R (x^2 + y^2)\,dV$ over the region $R$ of Exercise 29.

**31.** Find $\displaystyle\iiint_R z\,dV$ over the region $R$ satisfying $x^2 + y^2 \le z \le \sqrt{2 - x^2 - y^2}$.

**32.** Find $\displaystyle\iiint_R x\,dV$ and $\displaystyle\iiint_R z\,dV$ over that part of the hemisphere $0 \le z \le \sqrt{a^2 - x^2 - y^2}$ that lies in the first octant.

* **33.** Find $\displaystyle\iiint_R x\,dV$ and $\displaystyle\iiint_R z\,dV$ over that part of the cone

$$0 \le z \le h\left(1 - \frac{\sqrt{x^2 + y^2}}{a}\right)$$

that lies in the first octant.

* **34.** For what real values of $\lambda$ and $\mu$ does the integral

$$\iiint_{\mathbb{R}^3} \frac{dV}{(x^2 + y^2)^\lambda (1 + x^2 + y^2 + z^2)^\mu}$$

converge?

* **35.** For what real values of $\lambda$ and $\mu$ does the integral

$$\iiint_{\mathbb{R}^3} \frac{dV}{(x^2 + y^2 + z^2)^\lambda (1 + x^2 + y^2 + z^2)^\mu}$$

converge?

* **36.** Find the volume of the region inside the ellipsoid $\dfrac{x^2}{a^2} + \dfrac{y^2}{b^2} + \dfrac{z^2}{c^2} = 1$ and above the plane $z = b - y$.

* **37.** Show that for cylindrical coordinates the Laplace operator

$$\Delta u = \frac{\partial^2 u}{\partial x^2} + \frac{\partial^2 u}{\partial y^2} + \frac{\partial^2 u}{\partial z^2}$$

is given by

$$\frac{\partial^2 u}{\partial r^2} + \frac{1}{r}\frac{\partial u}{\partial r} + \frac{1}{r^2}\frac{\partial^2 u}{\partial \theta^2} + \frac{\partial^2 u}{\partial z^2}.$$

* **38.** Show that in spherical coordinates the Laplace operator is given by

$$\frac{\partial^2 u}{\partial \rho^2} + \frac{2}{\rho}\frac{\partial u}{\partial \rho} + \frac{\cot\phi}{\rho^2}\frac{\partial u}{\partial \phi} + \frac{1}{\rho^2}\frac{\partial^2 u}{\partial \phi^2} + \frac{1}{\rho^2 \sin^2\phi}\frac{\partial^2 u}{\partial \theta^2}.$$

* **39.** If $x$, $y$, and $z$ are functions of $u$, $v$, and $w$ with continuous first partial derivatives and nonvanishing Jacobian at $(u, v, w)$ show that they map an infinitesimal volume element in $uvw$-space bounded by the coordinate planes $u$, $u + du$, $v$, $v + dv$, $w$, and $w + dw$ into an infinitesimal "parallelepiped" in $xyz$-space having volume

$$dx\,dy\,dz = \left|\frac{\partial(x, y, z)}{\partial(u, v, w)}\right| du\,dv\,dw.$$

*Hint:* adapt the two-dimensional argument given in Section 6.4. What three vectors from the point $P = (x(u, v, w), y(u, v, w), z(u, v, w))$ span the parallelepiped?

## 6.7 APPLICATIONS OF MULTIPLE INTEGRALS

When we express the volume $V$ of a region $R$ in 3-space as an integral,

$$V = \iiint_R dV,$$

we are regarding $V$ as a "sum" of infinitely many *infinitesimal elements of volume,* that is, as the limit of the sum of volumes of smaller and smaller, nonoverlapping subregions into which we subdivide $R$. This idea of representing sums of infinitesimal elements of quantities by integrals has many applications.

For example, if a rigid body of constant density $\delta$ g/cm$^3$ occupies a volume $V$ cm$^3$ then its mass is $m = \delta V$ grams. If the density is not constant but varies continuously over the region $R$ of 3-space occupied by the rigid body, say $\delta = \delta(x, y, z)$, we can still regard the density as being constant on an infinitesimal element of $R$ having volume $dV$. The mass of this element is therefore $dm = \delta(x, y, z)\,dV$ and the mass of the whole body is calculated by integrating these mass elements over $R$:

$$m = \iiint_R \delta(x, y, z)\,dV.$$

Similar formulas apply when the rigid body is one- or two-dimensional and its density is given in units of mass per unit length or per unit area. In such cases single or double integrals are needed to sum the individual elements of mass. All this works because mass is "additive," that is, the mass of a composite object is the sum of the masses of the parts that comprise the object. The surface areas, gravitational forces, moments, and energies we consider in this section all have this additivity property.

### The Surface Area of a Graph

We can use a double integral over a domain $D$ in the $xy$-plane to add up surface area elements and thereby calculate the total area of the surface $z = f(x, y)$ defined for $(x, y)$ in $D$. We assume that $f$ has continuous first partial derivatives in $D$, so that its graph, the surface $S$ with equation $z = f(x, y)$, has a nonvertical tangent plane at $P = \big(x, y, f(x, y)\big)$ if $(x, y)$ belongs to $D$. The vector

$$\mathbf{n} = -f_1(x, y)\mathbf{i} - f_2(x, y)\mathbf{j} + \mathbf{k}$$

is an upward normal to $\mathcal{S}$ at $P$. An area element $dA$ at position $(x, y)$ in the $xy$-plane has a *vertical projection* onto $\mathcal{S}$, in the form of a parallelogram whose area $dS$ is $\sec\gamma$ times the area $dA$, where $\gamma$ is the angle between $\mathbf{n}$ and $\mathbf{k}$. (See Figure 6.45.) Since

$$\cos\gamma = \frac{\mathbf{n} \bullet \mathbf{k}}{|\mathbf{n}||\mathbf{k}|} = \frac{1}{\sqrt{1 + \big(f_1(x, y)\big)^2 + \big(f_2(x, y)\big)^2}},$$

we have

$$dS = \sqrt{1 + \left(\frac{\partial z}{\partial x}\right)^2 + \left(\frac{\partial z}{\partial y}\right)^2}\, dA.$$

Therefore the area of $\mathcal{S}$ is

$$S = \iint_D \sqrt{1 + \left(\frac{\partial z}{\partial x}\right)^2 + \left(\frac{\partial z}{\partial y}\right)^2}\, dA.$$

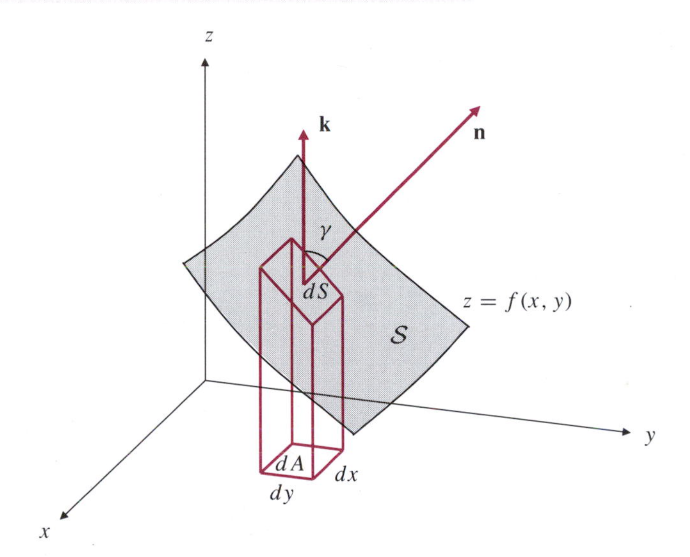


**Figure 6.45** The surface area element $dS$ on the surface $z = f(x, y)$ is $\sec\gamma$ times as large as its projection $dA$ onto the $xy$-plane

**EXAMPLE 1** Find the area of that part of the hyperbolic paraboloid $z = x^2 - y^2$ that lies inside the cylinder $x^2 + y^2 = a^2$.

***SOLUTION*** Since $\partial z/\partial x = 2x$ and $\partial z/\partial y = -2y$, the surface area element is

$$dS = \sqrt{1 + 4x^2 + 4y^2}\, dA = \sqrt{1 + 4r^2}\, r\, dr\, d\theta.$$

The required surface area is the integral of $dS$ over the disk $r \le a$:

$$\begin{aligned} S &= \int_0^{2\pi} d\theta \int_0^a \sqrt{1 + 4r^2}\, r\, dr \qquad \text{(Let } u = 1 + 4r^2.\text{)} \\ &= (2\pi)\frac{1}{8}\int_1^{1+4a^2} \sqrt{u}\, du \\ &= \frac{\pi}{4}\left(\frac{2}{3}\right) u^{3/2}\Big|_1^{1+4a^2} = \frac{\pi}{6}\big((1 + 4a^2)^{3/2} - 1\big) \text{ square units.} \end{aligned}$$

■

## The Gravitational Attraction of a Disk

Newton's universal law of gravitation asserts that two point masses $m_1$ and $m_2$, separated by a distance $s$, attract one another with a force

$$F = \frac{km_1m_2}{s^2},$$

$k$ being a universal constant. The force on each mass is directed towards the other, along the line joining the two masses. Suppose that a flat disk $D$ of radius $a$, occupying the region $x^2 + y^2 = a^2$ of the $xy$-plane, has constant *areal density* $\sigma$ (units of mass per unit area). Let us calculate the total force of attraction that this disk exerts upon a mass $m$ located at the point $(0, 0, b)$ on the positive $z$-axis. The total force is a vector quantity. Although the various mass elements on the disk are in different directions from the mass $m$, symmetry indicates that the net force will be in the direction towards the centre of the disk, that is, towards the origin. Thus the total force will be $-F\mathbf{k}$, where $F$ is the magnitude of the force.

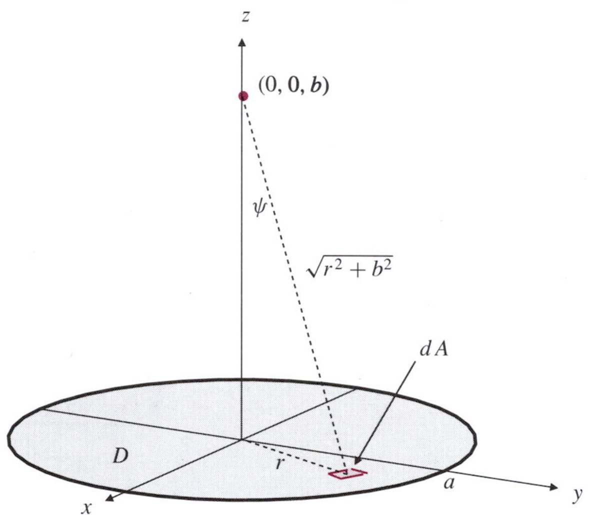


**Figure 6.46** Each mass element $\sigma\, dA$ attracts $m$ along a different line

We will calculate $F$ by integrating the vertical component $dF$ of the force of attraction on $m$ due to the mass $\sigma\, dA$ in an area element $dA$ on the disk. If the area element is at the point with polar coordinates $(r, \theta)$, and if the line from this point to $(0, 0, b)$ makes angle $\psi$ with the $z$-axis as shown in Figure 6.46, then the vertical component of the force of attraction of the mass element $\sigma\, dA$ on $m$ is

$$dF = \frac{km\sigma\, dA}{r^2 + b^2} \cos\psi = km\sigma b \frac{dA}{(r^2 + b^2)^{3/2}}.$$

Accordingly, the total vertical force of attraction of the disk on $m$ is

$$\begin{aligned}
F &= km\sigma b \iint_D \frac{dA}{(r^2 + b^2)^{3/2}} \\
&= km\sigma b \int_0^{2\pi} d\theta \int_0^a \frac{r\, dr}{(r^2 + b^2)^{3/2}} \qquad \text{(Let } u = r^2 + b^2.\text{)} \\
&= \pi km\sigma b \int_{b^2}^{a^2+b^2} u^{-3/2}\, du \\
&= 2\pi km\sigma \left(1 - \frac{b}{\sqrt{a^2 + b^2}}\right).
\end{aligned}$$

**REMARK** If we let $a$ approach infinity in the above formula we obtain the formula $F = 2\pi km\sigma$ for the force of attraction of a plane of areal density $\sigma$ on a mass $m$ located at distance $b$ from the plane. Observe that $F$ does not depend on $b$. Try to reason on physical grounds why this should be so.

**REMARK** The force of attraction on a point mass due to suitably symmetric solid objects (such as spheres, cylinders, and cones) having constant density $\delta$ (units of mass per unit volume) can be found by integrating elements of force contributed by thin, disk-shaped slices of the solid. See Exercises 14–17 at the end of this section.

## Moments and Centres of Mass

Moments, centres of mass, and centroids are introduced in Sections 8.4 and 8.5 of *Single-Variable Calculus*. You might wish to review the introductions to those sections before continuing with the material below.

The centre of mass of a rigid body is that point (fixed in the body) at which the body can be supported so that in the presence of a constant gravitational field it will not experience any unbalanced torques that will cause it to rotate. The torques experienced by a mass element $dm$ in the body can be expressed in terms of the **moments** of $dm$ about the three coordinate planes. If the body occupies a region $R$ in 3-space and has continuous volume density $\delta(x, y, z)$, then the mass element $dm = \delta(x, y, z)\, dV$ that occupies the volume element $dV$ is said to have **moments** $(x - x_0)\, dm$, $(y - y_0)\, dm$, and $(z - z_0)\, dm$ about the planes $x = x_0$, $y = y_0$, and $z = z_0$, respectively. Thus the total moments of the body about these three planes are

$$M_{x=x_0} = \iiint_R (x - x_0)\delta(x, y, z)\, dV = M_{x=0} - x_0 m$$

$$M_{y=y_0} = \iiint_R (y - y_0)\delta(x, y, z)\, dV = M_{y=0} - y_0 m$$

$$M_{z=z_0} = \iiint_R (z - z_0)\delta(x, y, z)\, dV = M_{z=0} - z_0 m$$

where $m = \iiint_R \delta\, dV$ is the mass of the body and $M_{x=0}$, $M_{y=0}$, and $M_{z=0}$ are the moments about the coordinate planes $x = 0$, $y = 0$, and $z = 0$, respectively. The **centre of mass** $\bar{P} = (\bar{x}, \bar{y}, \bar{z})$ of the body is that point for which $M_{x=\bar{x}}$, $M_{y=\bar{y}}$, and $M_{z=\bar{z}}$ are all equal to zero. Thus

**Centre of mass**

The centre of mass of a solid occupying region $R$ of 3-space and having continuous density $\delta(x, y, z)$ (units of mass per unit volume) is the point $(\bar{x}, \bar{y}, \bar{z})$ with coordinates given by

$$\bar{x} = \frac{M_{x=0}}{m} = \frac{\displaystyle\iiint_R x\delta\, dV}{\displaystyle\iiint_R \delta\, dV}, \qquad \bar{y} = \frac{M_{y=0}}{m} = \frac{\displaystyle\iiint_R y\delta\, dV}{\displaystyle\iiint_R \delta\, dV},$$

$$\bar{z} = \frac{M_{z=0}}{m} = \frac{\displaystyle\iiint_R z\delta\, dV}{\displaystyle\iiint_R \delta\, dV}.$$

These formulas can be combined into a single vector formula for the position vector $\bar{\mathbf{r}} = \bar{x}\mathbf{i} + \bar{y}\mathbf{j} + \bar{z}\mathbf{k}$ of the centre of mass in terms of the position vector $\mathbf{r} = x\mathbf{i} + y\mathbf{j} + z\mathbf{k}$ of an arbitrary point in $R$,

$$\bar{\mathbf{r}} = \frac{M_{x=0}\mathbf{i} + M_{y=0}\mathbf{j} + M_{z=0}\mathbf{k}}{m} = \frac{\iiint_R \delta\,\mathbf{r}\,dV}{\iiint_R \delta\,dV},$$

where the integral of the vector function $\delta\,\mathbf{r}$ is understood to mean the vector whose components are the integrals of the components of $\delta\,\mathbf{r}$.

***REMARK*** Similar expressions hold for distributions of mass over regions in the plane, or over intervals on a line. We use the appropriate areal or line densities and double or single definite integrals.

***REMARK*** If the density is constant it cancels out of the expressions for the centre of mass. In this case the centre of mass is a *geometric* property of the region $R$, and is called the **centroid** or **centre of gravity** of that region.

**EXAMPLE 2** Find the centroid of the tetrahedron $T$ bounded by the coordinate planes and the plane $\dfrac{x}{a} + \dfrac{y}{b} + \dfrac{z}{c} = 1$.

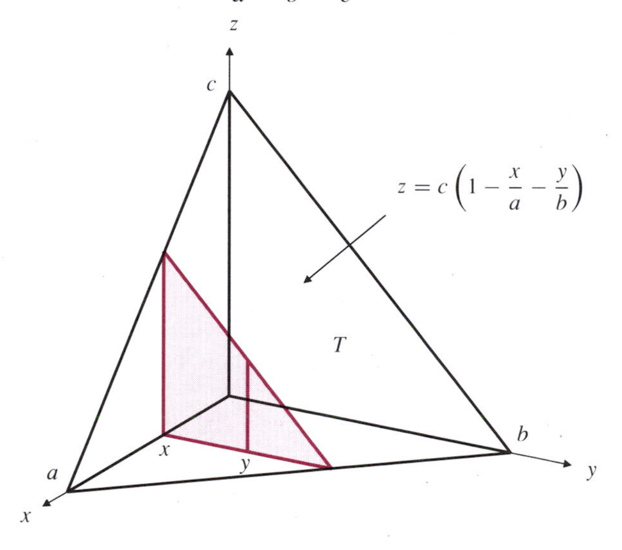


Figure 6.47 Iteration diagram for triple integrals over a tetrahedron

***SOLUTION*** The density is assumed to be constant, and so we may take it to be unity. The mass of $T$ is thus equal to its volume: $m = V = abc/6$. The moment of $T$ about the $yz$-plane is (see Figure 6.47):

$$\begin{aligned}
M_{x=0} &= \iiint_T x\,dV \\
&= \int_0^a x\,dx \int_0^{b(1-\frac{x}{a})} dy \int_0^{c(1-\frac{x}{a}-\frac{y}{b})} dz \\
&= c\int_0^a x\,dx \int_0^{b(1-\frac{x}{a})} \left(1 - \frac{x}{a} - \frac{y}{b}\right) dy
\end{aligned}$$

$$\begin{aligned}
&= c\int_0^a x\left[\left(1-\frac{x}{a}\right)y - \frac{y^2}{2b}\right]\Bigg|_0^{b(1-\frac{x}{a})} dx \\
&= \frac{bc}{2}\int_0^a x\left(1-\frac{x}{a}\right)^2 dx \\
&= \frac{bc}{2}\left[\frac{x^2}{2} - \frac{2}{3}\frac{x^3}{a} + \frac{x^4}{4a^2}\right]\Bigg|_0^a = \frac{a^2bc}{24}.
\end{aligned}$$

Thus $\bar{x} = M_{x=0}/m = a/4$. By symmetry, the centroid of $T$ is $\left(\frac{a}{4}, \frac{b}{4}, \frac{c}{4}\right)$. ■

■ **EXAMPLE 3** Find the centre of mass of a solid occupying the region $S$ that satisfies $x \ge 0$, $y \ge 0$, $z \ge 0$, and $x^2 + y^2 + z^2 \le a^2$, if the density at distance $\rho$ from the origin is $k\rho$.

***SOLUTION*** The mass of the solid is distributed symmetrically in the first octant part of the ball $\rho \le a$ so that the centre of mass, $(\bar{x}, \bar{y}, \bar{z})$, must satisfy $\bar{x} = \bar{y} = \bar{z}$. The mass of the solid is

$$m = \iiint_S k\rho\, dV = \int_0^{\pi/2} d\theta \int_0^{\pi/2} \sin\phi\, d\phi \int_0^a (k\rho)\rho^2\, d\rho = \frac{\pi k a^4}{8}.$$

The moment about the $xy$-plane is

$$\begin{aligned}
M_{z=0} &= \iiint_S zk\rho\, dV = \iiint_S (k\rho)\rho\cos\phi\, \rho^2 \sin\phi\, d\rho\, d\phi\, d\theta \\
&= \frac{k}{2}\int_0^{\pi/2} d\theta \int_0^{\pi/2} \sin(2\phi)\, d\phi \int_0^a \rho^4\, d\rho = \frac{k\pi a^5}{20}.
\end{aligned}$$

Hence $\bar{x} = \dfrac{k\pi a^5}{20}\Big/\dfrac{k\pi a^4}{8} = \dfrac{2a}{5}$ and the centre of mass is $\left(\dfrac{2a}{5}, \dfrac{2a}{5}, \dfrac{2a}{5}\right)$. ■

## Moment of Inertia

The **kinetic energy** of a particle of mass $m$ moving with speed $v$ is

$$\text{KE} = \frac{1}{2}mv^2.$$

The mass of the particle measures its *inertia*, which is twice the energy it has when its speed is one unit.

If the particle is moving in a circle of radius $D$, its motion can be described in terms of its **angular speed**, $\Omega$, measured in radians per unit time. In one revolution the particle travels a distance $2\pi D$ in time $2\pi/\Omega$. Thus its (translational) speed $v$ is related to its angular speed by

$$v = \Omega D.$$

Suppose that a rigid body is rotating with angular speed $\Omega$ about an axis $L$. If (at some instant) the body occupies a region $R$ and has density $\delta = \delta(x, y, z)$ then each mass element $dm = \delta\, dV$ in the body has kinetic energy

$$d\text{KE} = \frac{1}{2}v^2\, dm = \frac{1}{2}\delta\Omega^2 D^2\, dV,$$

where $D = D(x, y, z)$ is the perpendicular distance from the volume element $dV$ to the axis of rotation $L$. The total kinetic energy of the rotating body is therefore

$$\text{KE} = \frac{1}{2}\Omega^2 \iiint_R D^2\delta\, dV = \frac{1}{2} I\Omega^2,$$

where

$$I = \iiint_R D^2\delta\, dV.$$

$I$ is called the **moment of inertia** of the rotating body about the axis $L$. The moment of inertia plays the same role in the expression for kinetic energy of rotation (in terms of angular speed) that the mass does in the expression for kinetic energy of translation (in terms of linear speed). The moment of inertia is twice the kinetic energy of the body when it is rotating with unit angular speed.

If the entire mass of the rotating body were concentrated at a distance $D_0$ from the axis of rotation, then its kinetic energy would be $\frac{1}{2}mD_0^2\Omega^2$. The **radius of gyration** $\bar{D}$ is the value of $D_0$ for which this energy is equal to the actual kinetic energy $\frac{1}{2}I\Omega^2$ of the rotating body. Thus $m\bar{D}^2 = I$ and the radius of gyration is

$$\bar{D} = \sqrt{I/m} = \left(\frac{\displaystyle\iiint_R D^2\delta\, dV}{\displaystyle\iiint_R \delta\, dV}\right)^{1/2}.$$

**EXAMPLE 4 The acceleration of a rolling ball**

(a) Find the moment of inertia and radius of gyration of a solid ball of radius $a$ and constant density $\delta$ about a diameter of that ball.

(b) With what linear acceleration will the ball roll (without slipping) down a plane inclined at angle $\alpha$ to the horizontal?

***SOLUTION***

(a) We take the $z$-axis as the diameter and integrate in cylindrical coordinates over the ball $B$ of radius $a$ centred at the origin. Since the density $\delta$ is constant we have

$$\begin{aligned} I &= \delta \iiint_B r^2\, dV \\ &= \delta \int_0^{2\pi} d\theta \int_0^a r^3\, dr \int_{-\sqrt{a^2-r^2}}^{\sqrt{a^2-r^2}} dz \\ &= 4\pi\delta \int_0^a r^3\sqrt{a^2-r^2}\, dr \qquad \text{(Let } u = a^2 - r^2.\text{)} \\ &= 2\pi\delta \int_0^{a^2} (a^2-u)\sqrt{u}\, du \\ &= 2\pi\delta \left(\frac{2}{3}a^2u^{3/2} - \frac{2}{5}u^{5/2}\right)\Bigg|_0^{a^2} = \frac{8}{15}\pi\delta a^5. \end{aligned}$$

Since the mass of the ball is $m = \frac{4}{3}\pi\delta a^3$, the radius of gyration is

$$\bar{D} = \sqrt{\frac{I}{m}} = \sqrt{\frac{2}{5}}\, a.$$

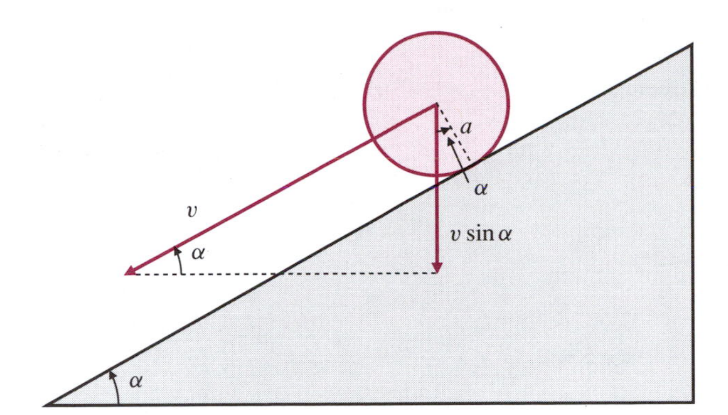


**Figure 6.48** The actual velocity and the vertical velocity of a ball rolling down an incline

(b) We can determine the acceleration of the ball by using conservation of total (kinetic plus potential) energy. When the ball is rolling down the plane with speed $v$ its centre is moving with speed $v$ and losing height at a rate $v \sin\alpha$. (See Figure 6.48.) Since the ball is not slipping, it is rotating about a horizontal axis through its centre with angular speed $\Omega = v/a$. Hence its kinetic energy (due to translation and rotation) is

$$\begin{aligned}\mathrm{KE} &= \frac{1}{2}mv^2 + \frac{1}{2}I\Omega^2 \\ &= \frac{1}{2}mv^2 + \frac{1}{2}\frac{2}{5}ma^2\frac{v^2}{a^2} = \frac{7}{10}mv^2.\end{aligned}$$

When the centre of the ball is at height $h$ (above some reference height) the ball has (gravitational) potential energy

$$\mathrm{PE} = mgh.$$

(This is the work that must be done against a constant gravitational force $F = mg$ to raise it to height $h$.) Since total energy is conserved,

$$\frac{7}{10}mv^2 + mgh = \text{constant}.$$

Differentiating with respect to time $t$, we obtain

$$0 = \frac{7}{10}m\,2v\frac{dv}{dt} + mg\frac{dh}{dt} = \frac{7}{5}mv\frac{dv}{dt} - mgv\sin\alpha.$$

Thus the ball rolls down the incline with acceleration $\dfrac{dv}{dt} = \dfrac{5}{7}g\sin\alpha$. ■

# EXERCISES 6.7

### Surface area problems

Use double integrals to calculate the areas of the surfaces in Exercises 1–9.

**1.** The part of the plane $z = 2x + 2y$ inside the cylinder $x^2 + y^2 = 1$

**2.** The part of the plane $5z = 3x - 4y$ inside the elliptic cylinder $x^2 + 4y^2 = 4$

**3.** The hemisphere $z = \sqrt{a^2 - x^2 - y^2}$

**4.** The half-ellipsoidal surface $z = 2\sqrt{1 - x^2 - y^2}$

**5.** The conical surface $3z^2 = x^2 + y^2$, $0 \le z \le 2$

**6.** The paraboloid $z = 1 - x^2 - y^2$ in the first octant

**7.** The part of the surface $z = y^2$ above the triangle with vertices $(0, 0)$, $(0, 1)$, and $(1, 1)$

**8.** The part of the surface $z = \sqrt{x}$ above the region $0 \le x \le 1$, $0 \le y \le \sqrt{x}$

**9.** The part of the cylindrical surface $x^2 + z^2 = 4$ that lies above the region $0 \le x \le 2$, $0 \le y \le x$

**10.** Show that the parts of the surfaces $z = 2xy$ and $z = x^2 + y^2$ that lie in the same vertical cylinder have the same area.

**11.** Show that the area $S$ of the part of the paraboloid $z = \frac{1}{2}(x^2 + y^2)$ lying above the square $-1 \le x \le 1$, $-1 \le y \le 1$ is given by

$$S = \frac{8}{3}\int_0^{\pi/4} (1 + \sec^2\theta)^{3/2}\, d\theta - \frac{2\pi}{3},$$

and use numerical methods to evaluate the area to three decimal places.

* **12.** The canopy shown in Figure 6.49 is the part of the hemisphere of radius $\sqrt{2}$ centred at the origin that lies above the square $-1 \le x \le 1$, $-1 \le y \le 1$. Find its area. *Hint:* it is possible to get an exact solution by first finding the area of the part of the sphere $x^2 + y^2 + z^2 = 2$ that lies above the plane $z = 1$. If you do the problem directly by integrating the surface area element over the square, you may encounter an integral that you can't evaluate exactly and will have to use numerical methods.

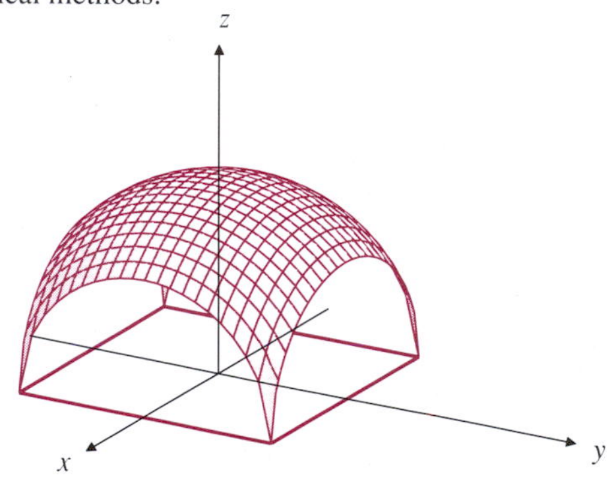


Figure 6.49

### Mass and gravitational attraction

**13.** Find the mass of a spherical planet of radius $a$ whose density at distance $R$ from the centre is $\delta = A/(B + R^2)$.

In Exercises 14–17, find the gravitational attraction that the given object exerts on a mass $m$ located at $(0, 0, b)$. Assume the object has constant density $\delta$. In each case you can obtain the answer by integrating the contributions made by disks of thickness $dz$, making use of the formula for the attraction exerted by the disk obtained in the text.

**14.** The ball $x^2 + y^2 + z^2 \le a^2$, where $a < b$.

**15.** The cylinder $x^2 + y^2 \le a^2$, $0 \le z \le h$, where $h < b$.

**16.** The cone $0 \le z \le b - (\sqrt{x^2 + y^2})/a$.

**17.** The half-ball $0 \le z \le \sqrt{a^2 - x^2 - y^2}$, where $a < b$.

### Centres of mass and centroids

**18.** Find the centre of mass of an object occupying the cube $0 \le x, y, z \le a$ with density given by $\delta = x^2 + y^2 + z^2$.

Find the centroids of the regions in Exercises 19–24.

**19.** The prism $x \ge 0$, $y \ge 0$, $x + y \le 1$, $0 \le z \le 1$

**20.** The unbounded region $0 \le z \le e^{-(x^2+y^2)}$

**21.** The first octant part of the ball $x^2 + y^2 + z^2 \le a^2$

**22.** The region $\sqrt{x^2 + y^2} \le a^2 z \le a^2$, $y \ge 0$

**23.** The region inside the cylinder $x^2 + y^2 = a^2$, above the $xy$-plane and under the plane $z = a + y$

**24.** The region inside the cube $0 \le x, y, z \le 1$ and under the plane $x + y + z = 2$

### Moments of inertia

**25.** Explain *in physical terms* why the acceleration of the ball rolling down the incline in Example 4 does not approach $g$ (the acceleration due to gravity) as the angle of incline, $\alpha$, approaches $90°$.

Find the moments of inertia and radii of gyration of the solid objects in Exercises 26–34. Assume constant density in all cases.

**26.** A circular cylinder of base radius $a$ and height $h$ about the axis of the cylinder

**27.** A circular cylinder of base radius $a$ and height $h$ about a diameter of the base of the cylinder

**28.** A right circular cone of base radius $a$ and height $h$ about the axis of the cone

**29.** A right circular cone of base radius $a$ and height $h$ about a diameter of the base of the cone

**30.** A cube of edge length $a$ about an edge of the cube

**31.** A cube of edge length $a$ about a diagonal of a face of the cube

**32.** A cube of edge length $a$ about a diagonal of the cube

**33.** The rectangular box $-a \le x \le a$, $-b \le y \le b$, $-c \le z \le c$ about the $z$-axis

**34.** The region between the two concentric cylinders $x^2 + y^2 = a^2$ and $x^2 + y^2 = b^2$ (where $0 < a < b$) and between $z = 0$ and $z = c$ about the $z$-axis

**35.** A ball of radius $a$ has constant density $\delta$. A cylindrical hole of radius $b < a$ is drilled through the centre of the ball. Find the mass of the remaining part of the ball, and its moment of inertia about the axis of the hole.

**36.** With what acceleration will a solid cylinder of base radius $a$, height $h$ and constant density $\delta$ roll (without slipping) down a plane inclined at angle $\alpha$ to the horizontal?

**37.** Repeat Exercise 36 for the ball with the cylindrical hole in Exercise 35. Assume that the axis of the hole remains horizontal while the ball rolls.

* **38.** A rigid pendulum of mass $m$ swings about point $A$ on a horizontal axis. Its moment of inertia about that axis is $I$. The centre of mass $C$ of the pendulum is at distance $a$ from $A$. When the pendulum hangs at rest, $C$ is directly under $A$. (Why?) Suppose the pendulum is swinging. Let $\theta = \theta(t)$ measure the angular displacement of the line $AC$ from the vertical at time $t$. ($\theta = 0$ when the pendulum is in its rest position.) Use a conservation of energy argument similar to that in Example 4 to show that

$$\frac{1}{2} I \left(\frac{d\theta}{dt}\right)^2 - mga\cos\theta = \text{constant},$$

and hence, differentiating with respect to $t$, that

$$\frac{d^2\theta}{dt^2} + \frac{mga}{I}\sin\theta = 0.$$

This is a nonlinear differential equation, and it is not easily solved. However, for small oscillations ($|\theta|$ small) we can use the approximation $\sin\theta \approx \theta$. In this case the differential equation is that of simple harmonic motion. What is the period?

* **39.** Let $L_0$ be a straight line passing through the centre of mass of a rigid body $B$ of mass $m$. Let $L_k$ be a straight line parallel to and $k$ units distant from $L_0$. If $I_0$ and $I_k$ are the moments of inertia of $B$ about $L_0$ and $L_k$, respectively, show that

$$I_k = I_0 + k^2 m.$$

Hence a body always has smallest moment of inertia about an axis through its centre of mass. (*Hint:* assume that the $z$-axis coincides with $L_0$, and that $L_k$ passes through the point $(k, 0, 0)$.)

* **40.** Re-establish the expression for the total kinetic energy of the rolling ball in Example 4 by regarding the ball at any instant as rotating about a horizontal line through its point of contact with the inclined plane. Use the result of Exercise 39.

* **41.** **(Products of inertia)** A rigid body with density $\delta$ is placed with its centre of mass at the origin and occupies a region $R$ of 3-space. Suppose the six second moments $P_{xx}$, $P_{yy}$, $P_{zz}$, $P_{xy}$, $P_{xz}$, and $P_{yz}$ are all known, where

$$P_{xx} = \iiint_R x^2 \delta \, dV, \quad P_{xy} = \iiint_R xy\delta \, dV, \quad \ldots.$$

(There exist tables giving these six moments for bodies of many standard shapes. They are called products of inertia.) Show how to express the moment of inertia of the body about any axis through the origin in term of these six second moments. (If this result is combined with that of Exercise 39, the moment of inertia about *any* axis can be found.)

# CHAPTER REVIEW

## Key Ideas

- **What do the following sentences or phrases mean?**
  - ◇ a Riemann sum for $f(x, y)$ on $a \le x \le b, c \le y \le d$
  - ◇ $f(x, y)$ is integrable on $a \le x \le b, c \le y \le d$
  - ◇ the double integral of $f(x, y)$ over $a \le x \le b, c \le y \le d$
  - ◇ iteration of a double integral
  - ◇ the average value of $f(x, y)$ over region $R$
  - ◇ the area element in polar coordinates
  - ◇ a triple integral
  - ◇ the volume element in cylindrical coordinates
  - ◇ the volume element in spherical coordinates
  - ◇ the surface area of the graph of $z = f(x, y)$
  - ◇ the moment of inertia of a solid about an axis
- **Describe how to change variables in a double integral.**
- **How do you calculate the centroid of a solid region?**
- **How do you calculate the moment of inertia of a solid about an axis?**

## Review Exercises

**1.** Evaluate $\iint_R (x + y)\, dA$ over the first-quadrant region lying under $x = y^2$ and above $y = x^2$.

**2.** Evaluate $\iint_P (x^2 + y^2)\, dA$ where $P$ is the parallelogram with vertices $(0, 0)$, $(2, 0)$, $(3, 1)$, and $(1, 1)$.

**3.** Find $\iint_S (y/x)\, dA$ where $S$ is the part of the disk $x^2 + y^2 \le 4$ in the first quadrant and under the line $y = x$.

4. Consider the iterated integral

$$I = \int_0^{\sqrt{3}} dy \int_{y/\sqrt{3}}^{\sqrt{4-y^2}} e^{-x^2-y^2}\, dx.$$

(a) Write $I$ as a double integral $\iint_R e^{-x^2-y^2}\, dA$, and sketch the region $R$ over which the double integral is taken.

(b) Write $I$ as an iterated integral with the order of integrations reversed from that of the given iteration.

(c) Write $I$ as an iterated integral in polar coordinates.

(d) Evaluate $I$.

5. Find the constant $k > 0$ such that the volume of the region lying inside the sphere $x^2+y^2+z^2 = a^2$ and above the cone $z = k\sqrt{x^2+y^2}$ is one-quarter of the volume contained by the whole sphere.

6. Reiterate the integral

$$I = \int_0^2 dy \int_0^y f(x,y)\, dx + \int_2^6 dy \int_0^{\sqrt{6-y}} f(x,y)\, dx$$

with the $y$ integral on the inside.

7. Let $J = \int_0^1 dz \int_0^z dy \int_0^y f(x,y,z)\, dx$. Express $J$ as an iterated integral where the integrations are to be performed in the order $z$ first, then $y$, and then $x$.

8. An object in the shape of a right-circular cone has height 10 m and base radius 5 m. Its density is proportional to the square of the distance from the base, and equals 3,000 kg/m$^3$ at the vertex.

(a) Find the mass of the object.

(b) Express the moment of inertia of the object about its central axis as an iterated integral.

9. Find the average value of $f(t) = \int_t^a e^{-x^2}\, dx$ over the interval $0 \le t \le a$.

10. Find the average value of the function $f(x,y) = \lfloor x+y \rfloor$ over the quarter disk $x \ge 0$, $y \ge 0$, $x^2+y^2 \le 4$. (Recall that $\lfloor x \rfloor$ denotes the greatest integer less than or equal to $x$.)

11. Let $D$ be the smaller of the two solid regions bounded by the surfaces

$$z = \frac{x^2+y^2}{a} \quad \text{and} \quad x^2+y^2+z^2 = 6a^2,$$

where $a$ is a positive constant. Find $\iiint_D (x^2+y^2)\, dV$.

12. Find the moment of inertia about the $z$-axis of a solid $V$ of density 1 if $V$ is specified by the inequalities $0 \le z \le \sqrt{x^2+y^2}$ and $x^2+y^2 \le 2ay$, where $a > 0$.

13. The rectangular solid $0 \le x \le 1$, $0 \le y \le 2$, $0 \le z \le 1$ is cut into two pieces by the plane $2x+y+z = 2$. Let $D$ be the piece that includes the origin. Find the volume of $D$ and $\bar{z}$, the $z$-coordinate of the centroid of $D$.

14. A solid $S$ consists of those points $(x,y,z)$ that lie in the first octant and satisfy $x+y+2z \le 2$ and $y+z \le 1$. Find the volume of $S$ and the $x$-coordinate of its centroid.

15. Find $\iiint_S z\, dV$, where $S$ is the portion of the first octant that is above the plane $x+y-z = 1$ and below the plane $z = 1$.

16. Find the area of that part of the plane $z = 2x$ that lies inside the paraboloid $z = x^2+y^2$.

17. Find the area of that part of the paraboloid $z = x^2+y^2$ that lies below the plane $z = 2x$. Express the answer as a single integral and evaluate it to three decimal places.

* 18. Find the volume of the smaller of the two regions into which the plane $x+y+z = 1$ divides the interior of the ellipsoid $x^2+4y^2+9z^2 = 36$. *Hint:* first change variables so that the ellipsoid becomes a ball. Then replace the plane by a plane with a simpler equation passing the same distance from the origin.

## Express Yourself

Answer these questions in words, using complete sentences and, if necessary, paragraphs. Use mathematical symbols only where necessary.

1. What is the most important factor to consider when deciding what coordinate system to use for iterating a multiple integral? Why? What other factors should also be considered?

2. Some quantities (like area) that are associated with regions in the plane can be expressed as double integrals, and others (like centroids) can not. (Centroids can be expressed as *quotients* of double integrals.) What property of such quantities determines whether or not they can be expressed as double integrals?

3. Iteration of a double integral in Cartesian coordinates corresponds to the technique of finding volumes by slicing introduced in *Single-Variable Calculus*. What technique for evaluating double integrals corresponds to finding volumes by cylindrical shells? Explain.

## Challenging Problems

1. The plane $(x/a)+(y/b)+(z/c) = 1$ (where $a > 0$, $b > 0$, and $c > 0$) divides the solid ellipsoid

$$\frac{x^2}{a^2}+\frac{y^2}{b^2}+\frac{z^2}{c^2} \le 1$$

into two unequal pieces. Find the volume of the smaller piece.

2. Find the area of the part of the plane $(x/a)+(y/b)+(z/c) = 1$ (where $a > 0$, $b > 0$, and $c > 0$) that lies inside the ellipsoid

$$\frac{x^2}{a^2}+\frac{y^2}{b^2}+\frac{z^2}{c^2} \le 1.$$

**3.** (a) Expand $1/(1-xy)$ as a geometric series, and hence show that

$$\int_0^1 \int_0^1 \frac{1}{1-xy}\,dx\,dy = \sum_{n=1}^{\infty} \frac{1}{n^2}.$$

(b) Similarly, express the following integrals as sums of series:

(i) $\displaystyle\int_0^1 \int_0^1 \frac{1}{1+xy}\,dx\,dy,$

(ii) $\displaystyle\int_0^1 \int_0^1 \int_0^1 \frac{1}{1-xyz}\,dx\,dy\,dz,$

(iii) $\displaystyle\int_0^1 \int_0^1 \int_0^1 \frac{1}{1+xyz}\,dx\,dy\,dz.$

**4.** (a) Calculate $G'(y)$ if $G(y) = \displaystyle\int_0^{\infty} \frac{\tan^{-1}(xy)}{x}\,dx.$

(b) Evaluate $\displaystyle\int_0^{\infty} \frac{\tan^{-1}(\pi x) - \tan^{-1} x}{x}\,dx.$ *Hint:* this integral is $G(\pi) - G(1)$.

* **5.** Let $P$ be the parallelepiped bounded by the three pairs of parallel planes $\mathbf{a} \bullet \mathbf{r} = 0$, $\mathbf{a} \bullet \mathbf{r} = d_1 > 0$, $\mathbf{b} \bullet \mathbf{r} = 0$, $\mathbf{b} \bullet \mathbf{r} = d_2 > 0$, $\mathbf{c} \bullet \mathbf{r} = 0$, and $\mathbf{c} \bullet \mathbf{r} = d_3 > 0$, where $\mathbf{a}$, $\mathbf{b}$, and $\mathbf{c}$ are constant vectors, and $\mathbf{r} = x\mathbf{i} + y\mathbf{j} + z\mathbf{k}$. Show that

$$\iiint_P (\mathbf{a} \bullet \mathbf{r})(\mathbf{b} \bullet \mathbf{r})(\mathbf{c} \bullet \mathbf{r})\,dx\,dy\,dz = \frac{(d_1 d_2 d_3)^2}{8|\mathbf{a} \bullet (\mathbf{b} \times \mathbf{c})|}.$$

*Hint:* Make the change of variables $u = \mathbf{a} \bullet \mathbf{r}$, $v = \mathbf{b} \bullet \mathbf{r}$, $w = \mathbf{c} \bullet \mathbf{r}$.

**6.** A hole whose cross-section is a square of side 2 is punched through the middle of a ball of radius 2. Find the volume of the remaining part of the ball.

* **7.** Find the volume bounded by the surface with equation $x^{2/3} + y^{2/3} + z^{2/3} = a^{2/3}$.

* **8.** Find the volume bounded by the surface $|x|^{1/3} + |y|^{1/3} + |z|^{1/3} = |a|^{1/3}$.

CHAPTER 7

# Curves in 3-Space

**INTRODUCTION** This chapter is concerned with functions of a single real variable that have *vector* values. Such functions can be thought of as parametric representations of curves, and we will examine them from both a *kinematic* point of view (involving position, velocity and acceleration of a moving particle) and a *geometric* point of view (involving tangents, normals, curvature, and torsion). Finally, we will work through a simple derivation of Kepler's laws of planetary motion.

Most of the material of this chapter does not rely on the content of Chapters 1 and 2 or 4–6. It reviews, expands upon, and extends to three dimensions the vectorial treatment of plane curves in Section 9.5 of *Single-Variable Calculus*, and makes extensive use of Sections 3.1–3.3. The reason it is placed here is that it sets the stage for the multivariate vector calculus of Chapters 8 and 9.

## 7.1 VECTOR FUNCTIONS OF ONE VARIABLE

In this section we will examine several aspects of differential and integral calculus as applied to **vector-valued functions** of a single real variable. Such functions can be used to represent curves parametrically. It is natural to interpret a vector-valued function of the real variable $t$ as giving the position, at time $t$, of a point or "particle" moving around in space. Derivatives of this *position vector* are then other vector-valued functions giving the velocity and acceleration of the particle. To motivate the study of vector functions we will consider such a vectorial description of motion in 3-space similar to the description of motion in the plane given in Section 9.5 of *Single-Variable Calculus.*

If a particle moves around in 3-space, its motion can be described by giving the three coordinates of its position as functions of time $t$:

$$x = x(t), \qquad y = y(t), \qquad \text{and} \qquad z = z(t).$$

It is more convenient, however, to replace these three equations by a single vector equation,

$$\mathbf{r} = \mathbf{r}(t),$$

giving the position vector of the moving particle as a function of $t$. (Recall that the position vector of a point is the vector from the origin to that point.) In terms of the standard basis vectors **i**, **j**, and **k**, the position of the particle at time $t$ is

$$\text{position:} \quad \mathbf{r} = \mathbf{r}(t) = x(t)\,\mathbf{i} + y(t)\,\mathbf{j} + z(t)\,\mathbf{k}.$$

As $t$ increases, the particle moves along a *path*, a curve $C$ in 3-space. We assume that $C$ is a *continuous curve*; the particle cannot instantaneously jump from one point to a distant point. This is equivalent to requiring that the component functions $x(t)$, $y(t)$, and $z(t)$ are continuous functions of $t$, and we therefore say that $\mathbf{r}(t)$ is a continuous vector function of $t$.

In the time interval from $t$ to $t + \Delta t$ the particle moves from position $\mathbf{r}(t)$ to position $\mathbf{r}(t + \Delta t)$. Therefore its **average velocity** is

$$\frac{\mathbf{r}(t + \Delta t) - \mathbf{r}(t)}{\Delta t},$$

which is a vector parallel to the secant vector from $\mathbf{r}(t)$ to $\mathbf{r}(t + \Delta t)$. If the average velocity has a limit as $\Delta t \to 0$, then we say that $\mathbf{r}$ is **differentiable** at $t$, and we call the limit the (instantaneous) **velocity** of the particle at time $t$. We denote the velocity vector by $\mathbf{v}(t)$.

$$\text{velocity:} \quad \mathbf{v}(t) = \lim_{\Delta t \to 0} \frac{\mathbf{r}(t + \Delta t) - \mathbf{r}(t)}{\Delta t} = \frac{d}{dt}\mathbf{r}(t).$$

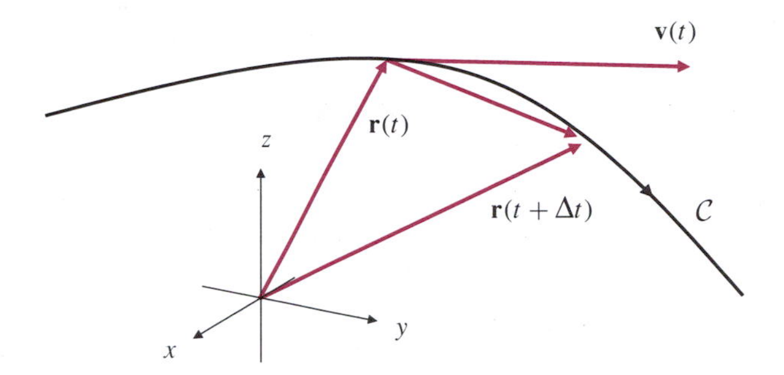


**Figure 7.1** The velocity $\mathbf{v}(t)$ is the derivative of the position $\mathbf{r}(t)$, and is tangent to the path of motion at the point with position vector $\mathbf{r}(t)$

This velocity vector has direction tangent to the path $\mathcal{C}$ at the point $\mathbf{r}(t)$, (see Figure 7.1), and points in the direction of motion. The length of the velocity vector, $v(t) = |\mathbf{v}(t)|$, is called the **speed** of the particle:

$$\text{speed:} \quad v(t) = |\mathbf{v}(t)|.$$

Wherever the velocity vector exists, is continuous, and does not vanish, the path $\mathcal{C}$ is a **smooth** curve, that is, it has a continuously turning tangent line. The path may not be smooth at points where the velocity is zero even if the components of the velocity vector are smooth functions of $t$.

The rules for addition and scalar multiplication of vectors imply that

$$\begin{aligned}
\mathbf{v} &= \frac{d\mathbf{r}}{dt} \\
&= \lim_{\Delta t \to 0} \left( \frac{x(t + \Delta t) - x(t)}{\Delta t}\mathbf{i} + \frac{y(t + \Delta t) - y(t)}{\Delta t}\mathbf{j} + \frac{z(t + \Delta t) - z(t)}{\Delta t}\mathbf{k} \right) \\
&= \frac{dx}{dt}\mathbf{i} + \frac{dy}{dt}\mathbf{j} + \frac{dz}{dt}\mathbf{k}.
\end{aligned}$$

Thus the vector function $\mathbf{r}$ is differentiable at $t$ if and only if its three scalar components, $x$, $y$, and $z$, are differentiable at $t$. In general, vector functions can be differentiated (or integrated) by differentiating (or integrating) their component functions, provided that the basis vectors with respect to which the components are taken are fixed in space and not changing with time. As an example of the termwise integration of a vector function, the position vector of the centre of mass of a mass distribution with density $\delta(x, y, z)$ which occupies a region $R$ in 3-space is, as noted in Section 6.7, given by the vector integral

$$\bar{\mathbf{r}} = \frac{1}{m} \iiint_R \delta \mathbf{r} \, dV,$$

where $m = \iiint_R \delta\, dV$ is the total mass of the distribution and $\mathbf{r}$ is the vector function $\mathbf{r} = x\mathbf{i} + y\mathbf{j} + z\mathbf{k}$.

Continuing our analysis of the moving particle, we define the **acceleration** of the particle to be the time derivative of the velocity:

$$\text{acceleration:} \quad \mathbf{a}(t) = \frac{d\mathbf{v}}{dt} = \frac{d^2\mathbf{r}}{dt^2}.$$

Newton's Second Law of Motion asserts that this acceleration is proportional to, and *in the same direction as*, the force $\mathbf{F}$ causing the motion: if the particle has mass $m$ then the law is expressed by the *vector equation* $\mathbf{F} = m\mathbf{a}$.

**EXAMPLE 1** Describe the curve $\mathbf{r} = t\mathbf{i} + t^2\mathbf{j} + t^3\mathbf{k}$. Find the velocity and acceleration vectors for this curve at $(1, 1, 1)$.

***SOLUTION*** Since the scalar parametric equations for the curve are

$$x = t, \qquad y = t^2, \qquad \text{and} \qquad z = t^3,$$

which satisfy $y = x^2$ and $z = x^3$, the curve is the curve of intersection of the two cylinders $y = x^2$ and $z = x^3$. At any time $t$ the velocity and acceleration vectors are given by

$$\mathbf{v} = \frac{d\mathbf{r}}{dt} = \mathbf{i} + 2t\mathbf{j} + 3t^2\mathbf{k},$$
$$\mathbf{a} = \frac{d\mathbf{v}}{dt} = 2\mathbf{j} + 6t\mathbf{k}.$$

The point $(1, 1, 1)$ on the curve corresponds to $t = 1$, so the velocity and acceleration at that point are $\mathbf{v} = \mathbf{i} + 2\mathbf{j} + 3\mathbf{k}$ and $\mathbf{a} = 2\mathbf{j} + 6\mathbf{k}$. ■

**EXAMPLE 2** Find the velocity, speed, and acceleration, and describe the motion of a particle whose position at time $t$ is

$$\mathbf{r} = 3\cos\omega t\,\mathbf{i} + 4\cos\omega t\,\mathbf{j} + 5\sin\omega t\,\mathbf{k}.$$

***SOLUTION*** The velocity, speed, and acceleration are readily calculated:

$$\mathbf{v} = \frac{d\mathbf{r}}{dt} = -3\omega\sin\omega t\,\mathbf{i} - 4\omega\sin\omega t\,\mathbf{j} + 5\omega\cos\omega t\,\mathbf{k}$$
$$v = |\mathbf{v}| = 5\omega$$
$$\mathbf{a} = \frac{d\mathbf{v}}{dt} = -3\omega^2\cos\omega t\,\mathbf{i} - 4\omega^2\cos\omega t\,\mathbf{j} - 5\omega^2\sin\omega t\,\mathbf{k} = -\omega^2\mathbf{r}.$$

Observe that $|\mathbf{r}| = 5$. Therefore the path of the particle lies on the sphere with equation $x^2 + y^2 + z^2 = 25$. Since $x = 3\cos\omega t$ and $y = 4\cos\omega t$, the path also lies on the vertical plane $4x = 3y$. Hence the particle moves around a circle of radius 5 centred at the origin and lying in the plane $4x = 3y$. Observe also that $\mathbf{r}$ is periodic with period $2\pi/\omega$. Therefore the particle makes one revolution around the circle in time $2\pi/\omega$. The acceleration is always in the direction of $-\mathbf{r}$, that is, towards the origin. The term **centripetal acceleration** is used to describe such a "centre-seeking" acceleration. ■

**EXAMPLE 3** **The projectile problem**

Describe the path followed by a particle experiencing a constant downward acceleration, $-g\mathbf{k}$, caused by gravity. Assume that at time $t = 0$ the position of the particle is $\mathbf{r}_0$ and its velocity is $\mathbf{v}_0$.

***SOLUTION*** If the position of the particle at time $t$ is $\mathbf{r}(t)$, then its acceleration is $d^2\mathbf{r}/dt^2$. If the $z$-axis is vertically upward, then the position of the particle can be found by solving the *initial-value problem*

$$\frac{d^2\mathbf{r}}{dt^2} = -g\mathbf{k}, \qquad \left.\frac{d\mathbf{r}}{dt}\right|_{t=0} = \mathbf{v}_0, \qquad \mathbf{r}(0) = \mathbf{r}_0.$$

We integrate the differential equation twice. Each integration introduces a *vector* constant of integration that we can determine from the given data by evaluating at $t = 0$:

$$\frac{d\mathbf{r}}{dt} = -gt\mathbf{k} + \mathbf{v}_0$$
$$\mathbf{r} = -\frac{gt^2}{2}\mathbf{k} + \mathbf{v}_0 t + \mathbf{r}_0.$$

The latter equation represents a parabola in the vertical plane passing through the point with position vector $\mathbf{r}_0$ and parallel to the vector $\mathbf{v}_0$. (See Figure 7.2.) The parabola has scalar parametric equations

$$x = u_0 t + x_0$$
$$y = v_0 t + y_0$$
$$z = -\frac{gt^2}{2} + w_0 t + z_0,$$

where $r_0 = x_0\mathbf{i} + y_0\mathbf{j} + z_0\mathbf{k}$ and $\mathbf{v}_0 = u_0\mathbf{i} + v_0\mathbf{j} + w_0\mathbf{k}$. ■

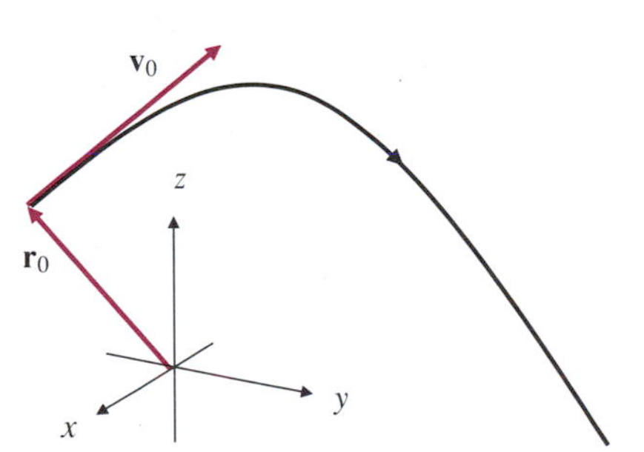


Figure 7.2 The path of a projectile fired from position $\mathbf{r}_0$ with velocity $\mathbf{v}_0$

**EXAMPLE 4** An object moves to the right along the curve $y = x^2$ with constant speed $v = 5$. Find the velocity and acceleration of the object when it is at the point $(1, 1)$.

***SOLUTION*** The position of the object at time $t$ is

$$\mathbf{r} = x\mathbf{i} + x^2\mathbf{j},$$

where $x$, the $x$-coordinate of the object's position, is a function of $t$. The object's velocity, speed, and acceleration at time $t$ are given by

$$\begin{aligned}
\mathbf{v} &= \frac{d\mathbf{r}}{dt} = \frac{dx}{dt}\mathbf{i} + 2x\frac{dx}{dt}\mathbf{j} \\
v = |\mathbf{v}| &= \sqrt{\left(\frac{dx}{dt}\right)^2 + \left(2x\,\frac{dx}{dt}\right)^2} = \frac{dx}{dt}\sqrt{1+4x^2} \\
\mathbf{a} &= \frac{d\mathbf{v}}{dt} = \frac{d^2x}{dt^2}\mathbf{i} + \left(2\left(\frac{dx}{dt}\right)^2 + 2\frac{d^2x}{dt^2}\right)\mathbf{j}.
\end{aligned}$$

(In the speed calculation we used $\sqrt{(dx/dt)^2} = dx/dt$ because the object is moving to the right.) We are given that the speed is constant; $v = 5$. Therefore

$$\frac{dx}{dt} = \frac{5}{\sqrt{1+4x^2}}.$$

When $x = 1$, we have $dx/dt = 5/\sqrt{1+4} = \sqrt{5}$, so the velocity of the object at that point is

$$\mathbf{v} = \sqrt{5}\mathbf{i} + 2\sqrt{5}\mathbf{j}.$$

Now we can calculate

$$\begin{aligned}
\frac{d^2x}{dt^2} &= \frac{d}{dt}\frac{5}{\sqrt{1+4x^2}} = \left(\frac{d}{dx}\frac{5}{\sqrt{1+4x^2}}\right)\frac{dx}{dt} \\
&= -\frac{5}{2(1+4x^2)^{3/2}}(8x)\frac{5}{\sqrt{1+4x^2}} = -\frac{100x}{(1+4x^2)^2}.
\end{aligned}$$

At $x = 1$, we have $d^2x/dt^2 = -4$. Thus the acceleration at that point is

$$\mathbf{a} = -4\mathbf{i} + (10-8)\mathbf{j} = -4\mathbf{i} + 2\mathbf{j}.$$

■

## Differentiating Combinations of Vectors

Vectors and scalars can be combined in a variety of ways to form other vectors or scalars. Vectors can be added and multiplied by scalars, and can be factors in dot and cross products. Appropriate differentiation rules apply to all such combinations of vector and scalar functions; we summarize them in the following theorem. The proof only involves standard differentiation rules applied to the components of the vectors involved; it is left to the reader.

**THEOREM 1**

**Differentiation rules for vector functions**

Let $\mathbf{u}(t)$ and $\mathbf{v}(t)$ be differentiable vector-valued functions, and let $\lambda(t)$ and $f(x, y, z)$ be differentiable scalar-valued functions. Then $\mathbf{u}(t) + \mathbf{v}(t)$, $\lambda(t)\mathbf{u}(t)$, $\mathbf{u}(t) \bullet \mathbf{v}(t)$, $\mathbf{u}(t) \times \mathbf{v}(t)$, $\mathbf{u}\big(\lambda(t)\big)$, and $f\big(\mathbf{u}(t)\big)$ are differentiable, and

(a) $$\frac{d}{dt}\Big(\mathbf{u}(t)+\mathbf{v}(t)\Big)=\mathbf{u}'(t)+\mathbf{v}'(t)$$

(b) $$\frac{d}{dt}\Big(\lambda(t)\mathbf{u}(t)\Big)=\lambda'(t)\mathbf{u}(t)+\lambda(t)\mathbf{u}'(t)$$

(c) $$\frac{d}{dt}\Big(\mathbf{u}(t)\bullet\mathbf{v}(t)\Big)=\mathbf{u}'(t)\bullet\mathbf{v}(t)+\mathbf{u}(t)\bullet\mathbf{v}'(t)$$

(d) $$\frac{d}{dt}\Big(\mathbf{u}(t)\times\mathbf{v}(t)\Big)=\mathbf{u}'(t)\times\mathbf{v}(t)+\mathbf{u}(t)\times\mathbf{v}'(t)$$

(e) $$\frac{d}{dt}\Big(\mathbf{u}\big(\lambda(t)\big)\Big)=\lambda'(t)\mathbf{u}'\big(\lambda(t)\big)$$

(f) $$\frac{d}{dt}\Big(f\big(\mathbf{u}(t)\big)\Big)=\nabla f\big(\mathbf{u}(t)\big)\bullet\mathbf{u}'(t).$$

**REMARK** Formulas (b), (c), and (d) are versions of the Product Rule. Formulas (e) and (f) are versions of the Chain Rule. All have the obvious form. Note that the order of the factors is the same in the terms on both sides of the cross product formula (d). It is essential that the order be preserved because, unlike the dot product or the product of a vector with a scalar, the cross product is *not commutative*.

**REMARK** The formula for the derivative of a cross product is a special case of that for the derivative of a $3 \times 3$ determinant. Since every term in the expansion of a determinant of any order is a product involving one element from each row (or column), the general Product Rule implies that the derivative of an $n \times n$ determinant whose elements are functions will be the sum of $n$ such $n \times n$ determinants, each with the elements of one of the rows (or columns) differentiated. For the $3 \times 3$ case we have

$$\frac{d}{dt}\begin{vmatrix} a_{11}(t) & a_{12}(t) & a_{13}(t) \\ a_{21}(t) & a_{22}(t) & a_{23}(t) \\ a_{31}(t) & a_{32}(t) & a_{33}(t) \end{vmatrix} = \begin{vmatrix} a'_{11}(t) & a'_{12}(t) & a'_{13}(t) \\ a_{21}(t) & a_{22}(t) & a_{23}(t) \\ a_{31}(t) & a_{32}(t) & a_{33}(t) \end{vmatrix}$$
$$+\begin{vmatrix} a_{11}(t) & a_{12}(t) & a_{13}(t) \\ a'_{21}(t) & a'_{22}(t) & a'_{23}(t) \\ a_{31}(t) & a_{32}(t) & a_{33}(t) \end{vmatrix} + \begin{vmatrix} a_{11}(t) & a_{12}(t) & a_{13}(t) \\ a_{21}(t) & a_{22}(t) & a_{23}(t) \\ a'_{31}(t) & a'_{32}(t) & a'_{33}(t) \end{vmatrix}.$$

**EXAMPLE 5** Show that the speed of a moving particle remains constant over an interval of time if and only if the acceleration is perpendicular to the velocity throughout that interval.

**SOLUTION** Since $\big(v(t)\big)^2 = \mathbf{v}(t)\bullet\mathbf{v}(t)$ we have

$$\begin{aligned} 2v(t)\frac{dv}{dt} &= \frac{d}{dt}\big(v(t)\big)^2 = \frac{d}{dt}\mathbf{v}(t)\bullet\mathbf{v}(t) \\ &= \mathbf{a}(t)\bullet\mathbf{v}(t)+\mathbf{v}(t)\bullet\mathbf{a}(t) = 2\mathbf{v}(t)\bullet\mathbf{a}(t). \end{aligned}$$

If we assume that $v(t) \neq 0$, it follows that $dv/dt = 0$ if and only if $\mathbf{v}\bullet\mathbf{a} = 0$. The speed is constant if and only if the velocity is perpendicular to the acceleration. ■

**EXAMPLE 6** If $\mathbf{u}$ is three times differentiable, calculate and simplify the triple product derivative

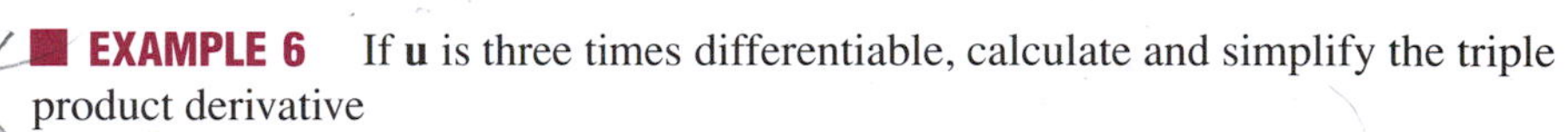

$$\frac{d}{dt}\left(\mathbf{u}\bullet\left(\frac{d\mathbf{u}}{dt}\times\frac{d^2\mathbf{u}}{dt^2}\right)\right).$$

**SOLUTION** Using various versions of the Product Rule, we calculate

$$\begin{aligned}
&\frac{d}{dt}\left(\mathbf{u}\bullet\left(\frac{d\mathbf{u}}{dt}\times\frac{d^2\mathbf{u}}{dt^2}\right)\right)\\
&\quad=\frac{d\mathbf{u}}{dt}\bullet\left(\frac{d\mathbf{u}}{dt}\times\frac{d^2\mathbf{u}}{dt^2}\right)+\mathbf{u}\bullet\left(\frac{d^2\mathbf{u}}{dt^2}\times\frac{d^2\mathbf{u}}{dt^2}\right)+\mathbf{u}\bullet\left(\frac{d\mathbf{u}}{dt}\times\frac{d^3\mathbf{u}}{dt^3}\right)\\
&\quad=0+0+\mathbf{u}\bullet\left(\frac{d\mathbf{u}}{dt}\times\frac{d^3\mathbf{u}}{dt^3}\right)=\mathbf{u}\bullet\left(\frac{d\mathbf{u}}{dt}\times\frac{d^3\mathbf{u}}{dt^3}\right).
\end{aligned}$$

The first term vanishes because $d\mathbf{u}/dt$ is perpendicular to its cross product with another vector; the second term vanishes because of the cross product of identical vectors. ■

## EXERCISES 7.1

In Exercises 1–14, find the velocity, speed and acceleration at time $t$ of the particle whose position is $\mathbf{r}(t)$. Describe the path of the particle.

**1.** $\mathbf{r}=\mathbf{i}+t\mathbf{j}$

**2.** $\mathbf{r}=t^2\mathbf{i}+\mathbf{k}$

**3.** $\mathbf{r}=t^2\mathbf{j}+t\mathbf{k}$

**4.** $\mathbf{r}=\mathbf{i}+t\mathbf{j}+t\mathbf{k}$

**5.** $\mathbf{r}=t^2\mathbf{i}-t^2\mathbf{j}+\mathbf{k}$

**6.** $\mathbf{r}=t\mathbf{i}+t^2\mathbf{j}+t^2\mathbf{k}$

**7.** $\mathbf{r}=a\cos t\,\mathbf{i}+a\sin t\,\mathbf{j}+ct\mathbf{k}$

**8.** $\mathbf{r}=a\cos\omega t\,\mathbf{i}+b\mathbf{j}+a\sin\omega t\,\mathbf{k}$

**9.** $\mathbf{r}=3\cos t\,\mathbf{i}+4\cos t\,\mathbf{j}+5\sin t\,\mathbf{k}$

**10.** $\mathbf{r}=3\cos t\,\mathbf{i}+4\sin t\,\mathbf{j}+t\mathbf{k}$

**11.** $\mathbf{r}=ae^t\mathbf{i}+be^t\mathbf{j}+ce^t\mathbf{k}$

**12.** $\mathbf{r}=at\cos\omega t\,\mathbf{i}+at\sin\omega t\,\mathbf{j}+b\ln t\,\mathbf{k}$

**13.** $\mathbf{r}=e^{-t}\cos(e^t)\mathbf{i}+e^{-t}\sin(e^t)\mathbf{j}-e^t\mathbf{k}$

**14.** $\mathbf{r}=a\cos t\sin t\,\mathbf{i}+a\sin^2 t\,\mathbf{j}+a\cos t\,\mathbf{k}$

**15.** A particle moves around the circle $x^2+y^2=25$ at constant speed, making one revolution in 2 seconds. Find its acceleration when it is at $(3,4)$.

**16.** A particle moves to the right along the curve $y=3/x$. If its speed is 10 when it passes through the point $\left(2,\frac{3}{2}\right)$, what is its velocity at that time?

**17.** A point $P$ moves along the curve of intersection of the cylinder $z=x^2$ and the plane $x+y=2$ in the direction of increasing $y$ with constant speed $v=3$. Find the velocity of $P$ when it is at $(1,1,1)$.

**18.** An object moves along the curve $y=x^2$, $z=x^3$, with constant vertical speed $dz/dt=3$. Find the velocity and acceleration of the object when it is at the point $(2,4,8)$.

**19.** Show that if the dot product of the velocity and acceleration of a moving particle is positive (or negative) then the speed of the particle is increasing (or decreasing).

**20.** Verify the formula for the derivative of a dot product given in Theorem 1(c).

**21.** Verify the formula for the derivative of a $3\times 3$ determinant in the second remark following Theorem 1. Use this formula to verify the formula for the derivative of the cross product in Theorem 1.

**22.** If the position and velocity vectors of a moving particle are always perpendicular show that the path of the particle lies on a sphere.

**23.** Generalize the previous problem to the case where the velocity of the particle is always perpendicular to the line joining the particle to a fixed point $P_0$.

**24.** What can be said about the motion of a particle at a time when its position and velocity satisfy $\mathbf{r}\bullet\mathbf{v}>0$? What can be said when $\mathbf{r}\bullet\mathbf{v}<0$?

In Exercises 25–30, assume that the vector functions encountered have continuous derivatives of all required orders.

**25.** Show that $\dfrac{d}{dt}\left(\dfrac{d\mathbf{u}}{dt}\times\dfrac{d^2\mathbf{u}}{dt^2}\right)=\dfrac{d\mathbf{u}}{dt}\times\dfrac{d^3\mathbf{u}}{dt^3}$.

**26.** Write the Product Rule for $\dfrac{d}{dt}\Big(\mathbf{u}\bullet(\mathbf{v}\times\mathbf{w})\Big)$.

**27.** Write the Product Rule for $\dfrac{d}{dt}\Big(\mathbf{u}\times(\mathbf{v}\times\mathbf{w})\Big)$.

**28.** Expand and simplify: $\dfrac{d}{dt}\left(\mathbf{u}\times\left(\dfrac{d\mathbf{u}}{dt}\times\dfrac{d^2\mathbf{u}}{dt^2}\right)\right)$.

**29.** Expand and simplify: $\dfrac{d}{dt}\Big((\mathbf{u}+\mathbf{u}'')\bullet(\mathbf{u}\times\mathbf{u}')\Big)$.

**30.** Expand and simplify: $\dfrac{d}{dt}\Big((\mathbf{u}\times\mathbf{u}')\bullet(\mathbf{u}'\times\mathbf{u}'')\Big)$.

**31.** If at all times $t$ the position and velocity vectors of a moving particle satisfy $\mathbf{v}(t)=2\mathbf{r}(t)$, and if $\mathbf{r}(0)=\mathbf{r}_0$, find $\mathbf{r}(t)$ and the acceleration $\mathbf{a}(t)$. What is the path of motion?

◈ **32.** Verify that $\mathbf{r}=\mathbf{r}_0\cos(\omega t)+(\mathbf{v}_0/\omega)\sin(\omega t)$ satisfies the initial-value problem

$$\frac{d^2\mathbf{r}}{dt^2}=-\omega^2\mathbf{r},\qquad \mathbf{r}'(0)=\mathbf{v}_0,\qquad \mathbf{r}(0)=\mathbf{r}_0.$$

(It is the unique solution.) Describe the path $\mathbf{r}(t)$. What is the path if $\mathbf{r}_0$ is perpendicular to $\mathbf{v}_0$?

❖**33.** **(Free fall with air resistance)** A projectile falling under gravity and slowed by air resistance proportional to its speed has position satisfying

$$\frac{d^2\mathbf{r}}{dt^2} = -g\mathbf{k} - c\frac{d\mathbf{r}}{dt},$$

where $c$ is a positive constant. If $\mathbf{r} = \mathbf{r}_0$ and $d\mathbf{r}/dt = \mathbf{v}_0$ at time $t = 0$ find $\mathbf{r}(t)$. (*Hint:* let $\mathbf{w} = e^{ct}(d\mathbf{r}/dt)$.) Show that the solution approaches that of the projectile problem given in this section as $c \to 0$.

## 7.2 SOME APPLICATIONS OF VECTOR DIFFERENTIATION

Many interesting problems in mechanics involve the differentiation of vector functions. This section is devoted to a brief discussion of a few of these.

### Motion Involving Varying Mass

The **momentum p** of a moving object is the product of its (scalar) mass $m$ and its (vector) velocity $\mathbf{v}$; $\mathbf{p} = m\mathbf{v}$. Newton's second law of motion states that the rate of change of *momentum* is equal to the external force acting on the object:

$$\mathbf{F} = \frac{d\mathbf{p}}{dt} = \frac{d}{dt}\left(m\mathbf{v}\right).$$

It is only when the mass of the object remains constant that this law reduces to the more familiar $\mathbf{F} = m\mathbf{a}$. When mass is changing you must deal with momentum rather than acceleration.

**EXAMPLE 1** **The changing velocity of a rocket**

A rocket accelerates by burning its onboard fuel. If the exhaust gases are ejected with constant velocity $\mathbf{v}_e$ *relative to the rocket*, and if the rocket ejects $p\%$ of its initial mass while its engines are firing, by what amount will the velocity of the rocket change? Assume the rocket is in deep space so that gravitational and other external forces acting on it can be neglected.

***SOLUTION*** Since the rocket is not acted on by any external forces (i.e., $\mathbf{F} = \mathbf{0}$), Newton's law implies that the total momentum of the rocket and its exhaust gases will remain constant. At time $t$ the rocket has mass $m(t)$ and velocity $\mathbf{v}(t)$. At time $t + \Delta t$ the rocket's mass is $m + \Delta m$ (where $\Delta m < 0$), its velocity is $\mathbf{v} + \Delta\mathbf{v}$, and the mass $-\Delta m$ of exhaust gases has escaped with velocity $\mathbf{v} + \mathbf{v}_e$ (relative to a coordinate system fixed in space). Equating total momenta at $t$ and $t + \Delta t$ we obtain

$$(m + \Delta m)(\mathbf{v} + \Delta\mathbf{v}) - \Delta m(\mathbf{v} + \mathbf{v}_e) = m\mathbf{v}.$$

Simplifying this equation and dividing by $\Delta t$ gives

$$(m + \Delta m)\frac{\Delta\mathbf{v}}{\Delta t} = \frac{\Delta m}{\Delta t}\mathbf{v}_e,$$

and, on taking the limit as $\Delta t \to 0$,

$$m\frac{d\mathbf{v}}{dt} = \frac{dm}{dt}\mathbf{v}_e.$$

Suppose that the engine fires from $t = 0$ to $t = T$. Then by the Fundamental Theorem of Calculus, the velocity of the rocket will change by

$$\begin{aligned}\mathbf{v}(T) - \mathbf{v}(0) &= \int_0^T \frac{d\mathbf{v}}{dt}\,dt = \left(\int_0^T \frac{1}{m}\frac{dm}{dt}\,dt\right)\mathbf{v}_e \\ &= \Big(\ln m(T) - \ln m(0)\Big)\mathbf{v}_e = -\ln\left(\frac{m(0)}{m(T)}\right)\mathbf{v}_e.\end{aligned}$$

Since $m(0) > m(T)$, we have $\ln\big(m(0)/m(T)\big) > 0$ and, as was to be expected, the change in velocity of the rocket is in the opposite direction to the exhaust velocity $\mathbf{v}_e$. If $p\%$ of the mass of the rocket is ejected during the burn then the velocity of the rocket will change by the amount $-\mathbf{v}_e \ln(100/(100-p))$. ■

***REMARK*** It is interesting that this model places no restriction on how great a velocity the rocket can achieve, provided that a sufficiently large percentage of its initial mass is fuel. See Exercise 1 at the end of the section.

## Circular Motion

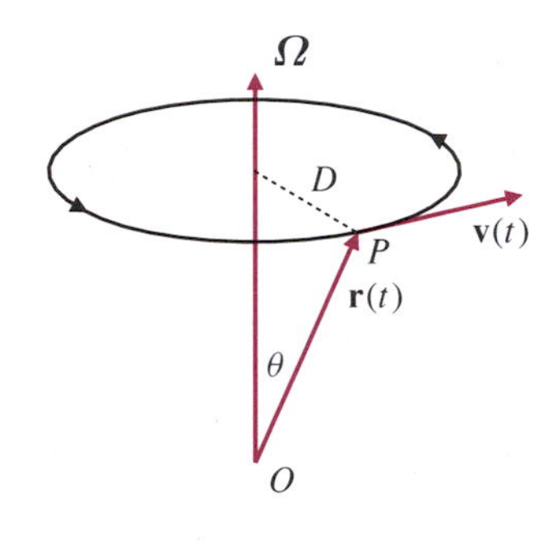


**Figure 7.3** Rotation with angular velocity $\boldsymbol{\Omega}$: $\mathbf{v} = \boldsymbol{\Omega}\times\mathbf{r}$

The angular speed $\Omega$ of a rotating body is its rate of rotation measured in radians per unit time. For instance, a lighthouse lamp rotating at a rate of three revolutions per minute has an angular speed of $\Omega = 6\pi$ radians per minute. It is useful to represent the rate of rotation of a rigid body about an axis in terms of an **angular velocity** vector rather than just the scalar angular speed. The angular velocity vector, $\boldsymbol{\Omega}$, has magnitude equal to the angular speed, $\Omega$, and direction along the axis of rotation such that if the extended right thumb points in the direction of $\boldsymbol{\Omega}$ then the fingers surround the axis in the direction of rotation.

If the origin of the coordinate system is on the axis of rotation, and $\mathbf{r} = \mathbf{r}(t)$ is the position vector at time $t$ of a point $P$ in the rotating body, then $P$ moves around a circle of radius $D = |\mathbf{r}(t)|\sin\theta$, where $\theta$ is the (constant) angle between $\boldsymbol{\Omega}$ and $\mathbf{r}(t)$. (See Figure 7.3.) Thus $P$ travels a distance $2\pi D$ in time $2\pi/\Omega$, and its linear speed is

$$\frac{\text{distance}}{\text{time}} = \frac{2\pi D}{2\pi/\Omega} = \Omega D = |\boldsymbol{\Omega}||\mathbf{r}(t)|\sin\theta = |\boldsymbol{\Omega}\times\mathbf{r}(t)|.$$

Since the direction of $\boldsymbol{\Omega}$ was defined so that $\boldsymbol{\Omega}\times\mathbf{r}(t)$ would point in the direction of motion of $P$, the linear velocity of $P$ at time $t$ is given by

$$\frac{d\mathbf{r}}{dt} = \mathbf{v}(t) = \boldsymbol{\Omega}\times\mathbf{r}(t).$$

■ **EXAMPLE 2** The position vector $\mathbf{r}(t)$ of a moving particle $P$ satisfies the initial-value problem

$$\begin{cases} \dfrac{d\mathbf{r}}{dt} = 2\mathbf{i}\times\mathbf{r}(t) \\ \mathbf{r}(0) = \mathbf{i} + 3\mathbf{j}. \end{cases}$$

Find $\mathbf{r}(t)$ and describe the motion of $P$.

**SOLUTION** There are two ways to solve this problem. We will do it both ways.

METHOD I: By the discussion above, the given differential equation is consistent with rotation about the $x$-axis with angular velocity $2\mathbf{i}$, so that the angular speed is 2 and the motion is counterclockwise as seen from far out on the positive $x$-axis. Therefore the particle $P$ moves on a circle in a plane $x$ = constant and centred on the $x$-axis. Since $P$ is at $(1, 3, 0)$ at time $t = 0$, the plane of motion is $x = 1$ and the radius of the circle is 3. Therefore the circle has a parametric equation of the form

$$\mathbf{r} = \mathbf{i} + 3\cos(\lambda t)\mathbf{j} + 3\sin(\lambda t)\mathbf{k}.$$

$P$ travels once around this circle ($2\pi$ radians) in time $t = 2\pi/\lambda$, so the angular speed is $\lambda$. Therefore $\lambda = 2$ and the motion of the particle is given by

$$\mathbf{r} = \mathbf{i} + 3\cos(2t)\mathbf{j} + 3\sin(2t)\mathbf{k}.$$

METHOD II: Break the given vector differential equation into components:

$$\frac{dx}{dt}\mathbf{i} + \frac{dy}{dt}\mathbf{j} + \frac{dz}{dt}\mathbf{k} = 2\mathbf{i}\times(x\mathbf{i} + y\mathbf{j} + z\mathbf{k}) = -2z\mathbf{j} + 2y\mathbf{k}$$
$$\frac{dx}{dt} = 0, \qquad \frac{dy}{dt} = -2z, \qquad \frac{dz}{dt} = 2y.$$

The first equation implies that $x$ = constant. Since $x(0) = 1$, we have $x(t) = 1$ for all $t$. Differentiate the second equation with respect to $t$ and substitute the third equation. This leads to the equation of simple harmonic motion for $y$,

$$\frac{d^2y}{dt^2} = -2\frac{dz}{dt} = -4y,$$

for which a general solution is

$$y = A\cos(2t) + B\sin(2t).$$

Thus $z = -\frac{1}{2}(dy/dt) = A\sin(2t) - B\cos(2t)$. Since $y(0) = 3$ and $z(0) = 0$, we have $A = 3$ and $B = 0$. Thus the particle $P$ travels counterclockwise around the circular path

$$\mathbf{r} = \mathbf{i} + 3\cos(2t)\mathbf{j} + 3\sin(2t)\mathbf{k}$$

in the plane $x = 1$ with angular speed 2. ■

**REMARK** Newton's second law states that $\mathbf{F} = (d/dt)m\mathbf{v} = d\mathbf{p}/dt$, where $\mathbf{p} = m\mathbf{v}$ is the (linear) momentum of a particle of mass $m$ moving under the influence of a force $\mathbf{F}$. This law may be reformulated in a manner appropriate for describing rotational motion as follows. If $\mathbf{r}(t)$ is the position of the particle at time $t$, then, since $\mathbf{v}\times\mathbf{v} = \mathbf{0}$,

$$\frac{d}{dt}(\mathbf{r}\times\mathbf{p}) = \frac{d}{dt}\Big(\mathbf{r}\times(m\mathbf{v})\Big) = \mathbf{v}\times(m\mathbf{v}) + \mathbf{r}\times\frac{d}{dt}(m\mathbf{v}) = \mathbf{r}\times\mathbf{F}.$$

The quantities $\mathbf{H} = \mathbf{r}\times(m\mathbf{v})$ and $\mathbf{T} = \mathbf{r}\times\mathbf{F}$ are respectively the **angular momentum** of the particle about the origin and the **torque** of $\mathbf{F}$ about the origin. We have shown that

$$\mathbf{T} = \frac{d\mathbf{H}}{dt};$$

the torque of the external forces is equal to the rate of change of the angular momentum of the particle. This is the analogue for rotational motion of $\mathbf{F} = d\mathbf{p}/dt$.

## Rotating Frames and the Coriolis Effect

This is an optional topic. The procedure of differentiating a vector function by differentiating its components is valid only if the basis vectors themselves do not depend on the variable of differentiation. In some situations in mechanics this is not the case. For instance, in modelling large scale weather phenomena the analysis is affected by the fact that a coordinate system fixed with respect to the earth is, in fact, rotating (along with the earth) relative to directions fixed in space.

In order to understand the effect that the rotation of the coordinate system has on representations of velocity and acceleration, let us consider two Cartesian coordinate frames (that is, systems of axes), sharing a common origin, one "fixed" in space and the other rotating with angular velocity $\boldsymbol{\Omega}$ about an axis through that origin. Let $\hat{\mathbf{I}}$, $\hat{\mathbf{J}}$, and $\hat{\mathbf{K}}$ be the standard basis vectors of the fixed coordinate frame, and let $\mathbf{i}$, $\mathbf{j}$, and $\mathbf{k}$ be the basis vectors of the rotating frame. The latter three are constant vectors in their own frame but are functions of time in the fixed frame. We showed above that the velocity of a point with position $\mathbf{r}$ rotating about an axis with angular velocity $\boldsymbol{\Omega}$ is $\boldsymbol{\Omega}\times\mathbf{r}$. Therefore

$$\frac{d\,\mathbf{i}}{dt} = \boldsymbol{\Omega}\times\mathbf{i}, \qquad \frac{d\,\mathbf{j}}{dt} = \boldsymbol{\Omega}\times\mathbf{j}, \qquad \text{and} \qquad \frac{d\,\mathbf{k}}{dt} = \boldsymbol{\Omega}\times\mathbf{k}.$$

Any vector function can be expressed in terms of either basis. If $\mathbf{r}(t)$ is the position vector of a moving particle it can be expressed in the fixed frame as

$$\mathbf{r} = \hat{x}(t)\hat{\mathbf{I}} + \hat{y}(t)\hat{\mathbf{J}} + \hat{z}(t)\hat{\mathbf{K}},$$

or in the rotating frame as

$$\mathbf{r} = x(t)\mathbf{i} + y(t)\mathbf{j} + z(t)\mathbf{k}.$$

The velocity of the moving particle can be expressed in the fixed frame, $\mathcal{F}$, as

$$\mathbf{v}_{\mathcal{F}} = \frac{d\hat{x}}{dt}\hat{\mathbf{I}} + \frac{d\hat{y}}{dt}\hat{\mathbf{J}} + \frac{d\hat{z}}{dt}\hat{\mathbf{K}}.$$

Similarly, the velocity of the particle, measured with respect to the rotating frame, $\mathcal{R}$, is

$$\mathbf{v}_{\mathcal{R}} = \frac{dx}{dt}\mathbf{i} + \frac{dy}{dt}\mathbf{j} + \frac{dz}{dt}\mathbf{k}.$$

These vectors are, however, not equal. (For instance, a point at rest with respect to the rotating frame would be seen as moving with respect to the fixed frame.) If we want to express the fixed-frame velocity of the particle, $\mathbf{v}_{\mathcal{F}}$, in terms of the rotating basis, we must also account for the rate of change of the basis vectors in the rotating frame; using three applications of the Product Rule we have

$$\begin{aligned}
\mathbf{v}_{\mathcal{F}} &= \frac{dx}{dt}\mathbf{i} + x\frac{d\mathbf{i}}{dt} + \frac{dy}{dt}\mathbf{j} + y\frac{d\mathbf{j}}{dt} + \frac{dz}{dt}\mathbf{k} + z\frac{d\mathbf{k}}{dt} \\
&= \mathbf{v}_{\mathcal{R}} + x\boldsymbol{\Omega}\times\mathbf{i} + y\boldsymbol{\Omega}\times\mathbf{j} + z\boldsymbol{\Omega}\times\mathbf{k} \\
&= \mathbf{v}_{\mathcal{R}} + \boldsymbol{\Omega}\times\mathbf{r}.
\end{aligned}$$

This formula shows that differentiation with respect to time has different effects in the two coordinate systems. If we denote by $(d/dt)_{\mathcal{F}}$ and $(d/dt)_{\mathcal{R}}$ the operators of time differentiation in the two frames then the formula above can be written

$$\left(\frac{d}{dt}\right)_{\mathcal{F}}\mathbf{r}(t) = \left(\left(\frac{d}{dt}\right)_{\mathcal{R}} + \boldsymbol{\Omega}\times\right)\mathbf{r}(t),$$

or, in terms of operators,

$$\left(\frac{d}{dt}\right)_{\mathcal{F}} = \left(\left(\frac{d}{dt}\right)_{\mathcal{R}} + \boldsymbol{\Omega}\times\right).$$

These operators can be applied to *any* vector function, and in particular to the function $\mathbf{v}_{\mathcal{F}} = \mathbf{v}_{\mathcal{R}} + \boldsymbol{\Omega}\times\mathbf{r}$. Since $(d/dt)_{\mathcal{F}}\mathbf{v}_{\mathcal{F}}$ is the acceleration $\mathbf{a}_{\mathcal{F}}$ of the particle in the fixed frame, and $(d/dt)_{\mathcal{R}}\mathbf{v}_{\mathcal{R}}$ is the acceleration $\mathbf{a}_{\mathcal{R}}$ in the rotating frame, we calculate, assuming that the angular velocity $\boldsymbol{\Omega}$ of the rotating frame is constant,

$$\begin{aligned}\mathbf{a}_{\mathcal{F}} &= \left(\frac{d}{dt}\right)_{\mathcal{F}}(\mathbf{v}_{\mathcal{R}} + \boldsymbol{\Omega}\times\mathbf{r})\\ &= \left(\left(\frac{d}{dt}\right)_{\mathcal{R}} + \boldsymbol{\Omega}\times\right)(\mathbf{v}_{\mathcal{R}} + \boldsymbol{\Omega}\times\mathbf{r})\\ &= \mathbf{a}_{\mathcal{R}} + 2\boldsymbol{\Omega}\times\mathbf{v}_{\mathcal{R}} + \boldsymbol{\Omega}\times(\boldsymbol{\Omega}\times\mathbf{r}).\end{aligned}$$

The term $2\boldsymbol{\Omega}\times\mathbf{v}_{\mathcal{R}}$ is called the **Coriolis acceleration** and the term $\boldsymbol{\Omega}\times(\boldsymbol{\Omega}\times\mathbf{r})$ is called the **centripetal acceleration**.

In order to interpret these accelerations, suppose that the origin is at the centre of the earth, the fixed frame is fixed with respect to the stars, and the rotating frame is attached to the rotating earth. Suppose that the particle has mass $m$ and is acted upon by an external force $\mathbf{F}$. If we view the moving particle from a vantage point fixed in space (so that we see the fixed frame as unmoving) then, in accordance with Newton's second law, we observe that $\mathbf{F} = m\mathbf{a}_{\mathcal{F}}$. If, instead, we are sitting on the earth, and regarding the rotating frame as stationary, we would still want to regard $\mathbf{a}_{\mathcal{R}}$ as the force per unit mass acting on the particle. Thus we would write

$$\mathbf{a}_{\mathcal{R}} = \frac{\mathbf{F}}{m} - 2\boldsymbol{\Omega}\times\mathbf{v}_{\mathcal{R}} - \boldsymbol{\Omega}\times(\boldsymbol{\Omega}\times\mathbf{r}),$$

and would interpret the $\boldsymbol{\Omega}$ terms as other *forces* per unit mass acting on the particle. The term $-\boldsymbol{\Omega}\times(\boldsymbol{\Omega}\times\mathbf{r})$ is a vector pointing directly away from the axis of rotation of the earth and we would interpret it as a **centrifugal force**. (If the earth were spinning fast enough, we would expect objects on the surface to be flung into space.) The term $-2\boldsymbol{\Omega}\times\mathbf{v}_{\mathcal{R}}$ is at right angles to both the velocity and the polar axis of the earth. We would call this a **Coriolis force**. The centrifugal and Coriolis forces are not "real" forces acting on the particle. They are fictitious forces which compensate for the fact that we are measuring acceleration with respect to a frame which we are regarding as fixed although it is really rotating and hence accelerating.

### EXAMPLE 3 Winds around the eye of a storm

The circulation of winds around a storm centre is an example of the Coriolis effect. The eye of a storm is an area of low pressure sucking air towards it. The direction of rotation of the earth is such that the angular velocity $\boldsymbol{\Omega}$ points north and is parallel to the earth's axis of rotation. At any point $P$ on the surface of the earth we can express $\boldsymbol{\Omega}$ as a sum of tangential (to the earth's surface) and normal components (see Figure 7.4(a)),

$$\boldsymbol{\Omega}(P) = \boldsymbol{\Omega}_T(P) + \boldsymbol{\Omega}_N(P).$$

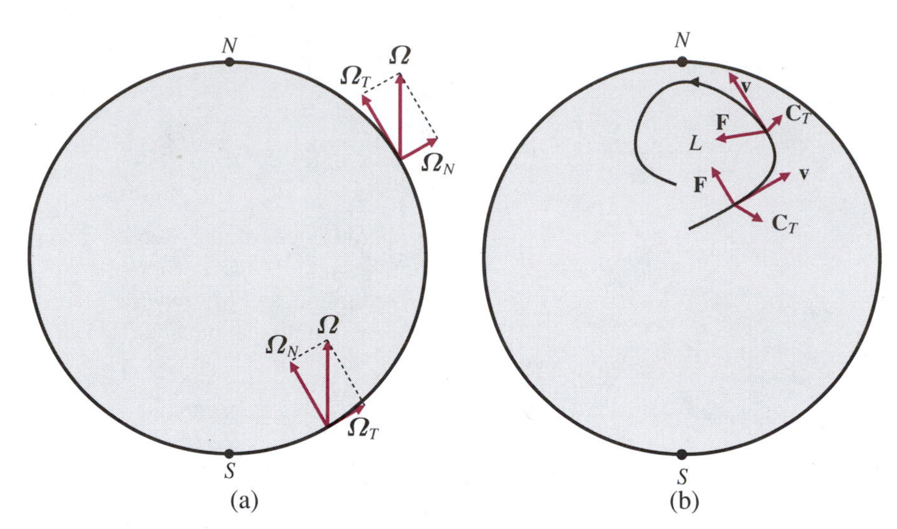


**Figure 7.4**

(a) Tangential and normal components of the angular velocity of the earth in the northern and southern hemispheres

(b) In the northern hemisphere the tangential Coriolis force deflects winds to the right of the path towards the low pressure area $L$ so the winds move counterclockwise around the low

If $P$ is in the northern hemisphere $\boldsymbol{\Omega}_N(P)$ points upwards (away from the centre of the earth). At such a point the Coriolis "force" $\mathbf{C} = -2\boldsymbol{\Omega}(P)\times\mathbf{v}$ on a particle of air moving with horizontal velocity $\mathbf{v}$ would itself have horizontal and normal components

$$\mathbf{C} = -2\boldsymbol{\Omega}_T\times\mathbf{v} - 2\boldsymbol{\Omega}_N\times\mathbf{v} = \mathbf{C}_N + \mathbf{C}_T.$$

The tangential component of the Coriolis force, $\mathbf{C}_T = -2\boldsymbol{\Omega}_N\times\mathbf{v}$ is $90°$ to the right of $\mathbf{v}$(that is, clockwise from $\mathbf{v}$). Therefore, particles of air which are being sucked towards the eye of the storm experience Coriolis deflection to the right and so actually spiral into the eye in a counterclockwise direction. The opposite is true in the southern hemisphere where the normal component of $\boldsymbol{\Omega}_N$ is downward (into the earth). The suction force $\mathbf{F}$, the velocity $\mathbf{v}$, and the component of the Coriolis force tangential to the earth's surface, $\mathbf{C}_T$, are shown at two positions on the path of an air particle spiralling around a low pressure area in the northern hemisphere in Figure 7.4(b). ■

***REMARK*** Strong winds spiralling inwards around low pressure areas are called **cyclones**. Strong winds spiralling outwards around high pressure areas are called **anticyclones**. These latter spiral counterclockwise in the southern hemisphere and clockwise in the northern hemisphere.

## EXERCISES 7.2

**1.** What fraction of its total initial mass would the rocket considered in Example 1 have to burn as fuel in order to accelerate in a straight line from rest to the speed of its own exhaust gases? to twice that speed?

* **2.** When run at maximum power output the motor in a self-propelled tank car can accelerate the full car (mass $M$ kg) along a horizontal track at $a$ m/sec$^2$. The tank is full at time zero but the contents pour out of a hole in the bottom at rate $k$ kg/sec thereafter. If the car is at rest at time zero and full forward power is turned on at that time, how fast will it be moving at any time $t$ before the tank is empty?

❖ **3.** Solve the initial-value problem

$$\frac{d\mathbf{r}}{dt} = \mathbf{k}\times\mathbf{r}, \qquad \mathbf{r}(0) = \mathbf{i} + \mathbf{k}.$$

Describe the curve $\mathbf{r} = \mathbf{r}(t)$.

❖ **4.** An object moves so that its position vector $\mathbf{r}(t)$ satisfies

$$\frac{d\mathbf{r}}{dt} = \mathbf{a}\times\big(\mathbf{r}(t) - \mathbf{b}\big)$$

and $\mathbf{r}(0) = \mathbf{r}_0$. Here $\mathbf{a}$, $\mathbf{b}$, and $\mathbf{r}_0$ are given constant vectors with $\mathbf{a} \neq \mathbf{0}$. Describe the path along which the object moves.

**The Coriolis effect**

* **5.** A satellite is in a low, circular orbit around the earth, passing over the north and south poles. It makes one revolution every two hours. From the point of view of an observer standing on the earth at the equator, what is the approximate

value of the Coriolis force acting on the satellite as it crosses over his head heading south? (Assume the radius of the satellite's orbit is approximately the radius of the earth.) In what direction would the satellite seem to the observer to be moving as it crosses over his head?

* **6.** Describe the tangential and normal components of the Coriolis force on a particle moving with horizontal velocity **v** at (a) the north pole, (b) the south pole, and (c) the equator. In general, what is the effect of the normal component of the Coriolis force near the eye of a storm?

## 7.3 CURVES AND PARAMETRIZATIONS

In this section we will consider curves as geometric objects rather than as paths of moving particles. Everyone has an intuitive idea of what a curve is, but it is difficult to give a formal definition of a curve as a geometric object (that is, as a certain kind of set of points) without involving the concept of parametric representation. We will avoid this difficulty by continuing to regard a curve in 3-space as the set of points whose positions are given by the position vector function

$$\mathbf{r} = \mathbf{r}(t) = x(t)\mathbf{i} + y(t)\mathbf{j} + z(t)\mathbf{k}, \qquad a \le t \le b.$$

However, the parameter $t$ need no longer represent time, or any other specific physical quantity.

Curves can be very pathological. For instance, there exist continuous curves that pass through every point in a cube. It is difficult to think of such a curve as a one-dimensional object. In order to avoid such strange objects we will assume hereafter that the defining function $\mathbf{r}(t)$ has a *continuous* first derivative, $d\mathbf{r}/dt$, which we will continue to call "velocity" and denote by $\mathbf{v}(t)$ by analogy with the physical case where $t$ is time. (We also continue to call $v(t) = |\mathbf{v}(t)|$ the "speed.") As we will see later, this implies that the curve has an *arc length* between any two points corresponding to parameter values $t_1$ and $t_2$; if $t_1 < t_2$ this arc length is

$$\int_{t_1}^{t_2} v(t)\,dt = \int_{t_1}^{t_2} |\mathbf{v}(t)|\,dt = \int_{t_1}^{t_2} \left|\frac{d\mathbf{r}}{dt}\right| dt.$$

Frequently we will want $\mathbf{r}(t)$ to have continuous derivatives of higher order. Whenever needed, we will assume that the "acceleration", $\mathbf{a}(t) = d^2\mathbf{r}/dt^2$, and even the third derivative, $d^3\mathbf{r}/dt^3$, are continuous. Of course, most of the curves we encounter in practice have parametrizations with continuous derivatives of all orders.

It must be noted, however, that no assumptions on the continuity of derivatives of the function $\mathbf{r}(t)$ are sufficient to guarantee that the curve $\mathbf{r} = \mathbf{r}(t)$ is a "smooth" curve.

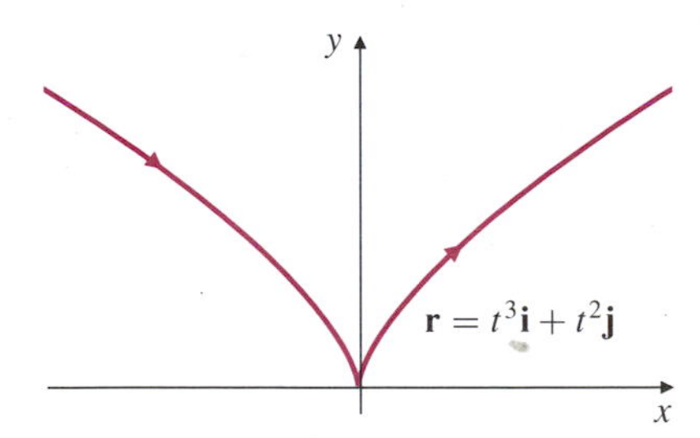


**Figure 7.5** The components of $\mathbf{r}(t)$ are the smooth functions $t^3$ and $t^2$, but the curve with parametrization $\mathbf{r}(t)$ is not smooth at the origin

**EXAMPLE 1** Consider the curve $C$ defined by

$$\mathbf{r} = t^3\,\mathbf{i} + t^2\,\mathbf{j}, \qquad -\infty < t < \infty.$$

Although $t^2$ and $t^3$ have derivatives of all orders at every value of $t$, the curve $C$ is not smooth at the origin. It has a cusp there. (See Figure 7.5.) ■

We will show in the next section that if, besides being continuous, the velocity vector $\mathbf{v}(t)$ is *never the zero vector*, then the curve $\mathbf{r} = \mathbf{r}(t)$ is **smooth** in the sense that it has a continuously turning tangent line. In this context we call a curve, $\mathbf{r} = \mathbf{r}(t)$, $(a \le t \le b)$, **piecewise smooth** if $\mathbf{v}(t)$ is defined, continuous, and nonzero for all but finitely many values of $t$ in $[a, b]$. The curve in Example 1 is piecewise smooth.

Although we have said that a curve is a set of points given by a parametric equation $\mathbf{r} = \mathbf{r}(t)$, there is no *unique* way of representing a given curve parametrically. Just as two cars can travel the same highway at different speeds and stopping and starting at different places, so also different parametrizations can define the same curve. Typically, a given curve can have infinitely many different parametrizations.

**EXAMPLE 2** Show that each of the vector functions

$$\begin{aligned} \mathbf{r}_1(t) &= \sin t\,\mathbf{i} + \cos t\,\mathbf{j}, && (-\pi/2 \le t \le \pi/2), \\ \mathbf{r}_2(t) &= (t-1)\,\mathbf{i} + \sqrt{2t - t^2}\,\mathbf{j}, && (0 \le t \le 2), \quad \text{and} \\ \mathbf{r}_3(t) &= t\sqrt{2-t^2}\,\mathbf{i} + (1-t^2)\,\mathbf{j}, && (-1 \le t \le 1) \end{aligned}$$

all represent the same curve. Describe the curve.

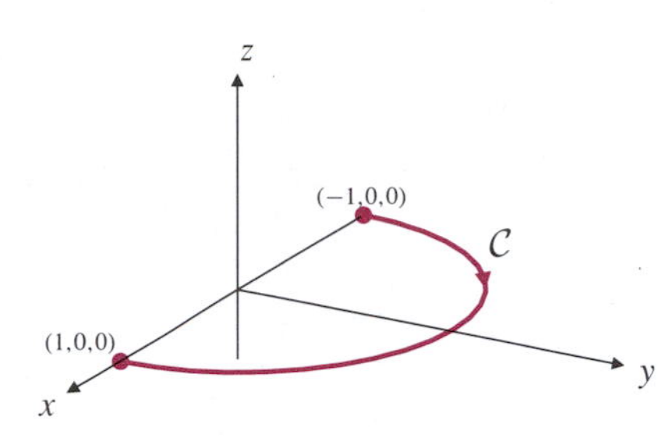


Figure 7.6 Three parametrizations of the semicircle $C$ are given in Example 2

**SOLUTION** The function $\mathbf{r}_1(t)$ starts at the point $(-1, 0, 0)$ with position vector $\mathbf{r}_1(-\pi/2) = -\mathbf{i}$, and ends at the point $(1, 0, 0)$ with position vector $\mathbf{i}$. It lies in the half of the $xy$-plane where $y \ge 0$ (because $\cos t \ge 0$ for $(-\pi/2 \le t \le \pi/2)$). Finally, all points on the curve are at distance 1 from the origin:

$$|\mathbf{r}_1(t)| = (\sin t)^2 + (\cos t)^2 = 1.$$

Therefore, $\mathbf{r}_1(t)$ represents the semicircle $y = \sqrt{1 - x^2}$ in the $xy$-plane.

The other two functions have the same properties: both graphs lie in $y \ge 0$,

$$\begin{aligned} \mathbf{r}_2(0) &= -\mathbf{i}, & \mathbf{r}_2(2) &= \mathbf{i}, & |\mathbf{r}_2(t)| &= (t-1)^2 + 2t - t^2 = 1, \\ \mathbf{r}_3(-1) &= -\mathbf{i}, & \mathbf{r}_3(1) &= \mathbf{i}, & |\mathbf{r}_3(t)| &= t^2(2-t^2) + (1-t^2)^2 = 1. \end{aligned}$$

Thus both functions represent the same semicircle. Of course, all three parametrizations trace out the curve with different velocities. ■

The curve $\mathbf{r} = \mathbf{r}(t)$, $(a \le t \le b)$ is called a **closed curve** if $\mathbf{r}(a) = \mathbf{r}(b)$ that is, if the curve begins and ends at the same point. The curve $\mathcal{C}$ is **non-self-intersecting** if there exists some parametrization $\mathbf{r} = \mathbf{r}(t)$, $(a \le t \le b)$, of $\mathcal{C}$ that is one-to-one except that the endpoints could be the same;

$$\mathbf{r}(t_1) = \mathbf{r}(t_2) \quad a \le t_1 < t_2 \le b \quad \Longrightarrow t_1 = a \quad \text{and} \quad t_2 = b.$$

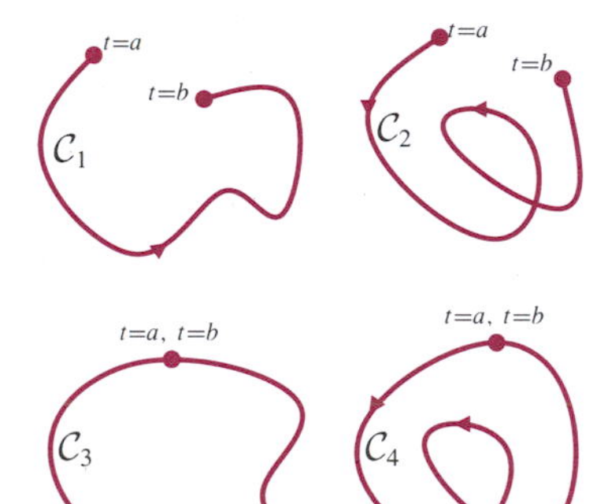


Figure 7.7 Curves $\mathcal{C}_1$ and $\mathcal{C}_3$ are non-self-intersecting
Curves $\mathcal{C}_2$ and $\mathcal{C}_4$ intersect themselves
Curves $\mathcal{C}_1$ and $\mathcal{C}_2$ are not closed
Curves $\mathcal{C}_3$ and $\mathcal{C}_4$ are closed
Curve $\mathcal{C}_3$ is a simple closed curve

Such a curve can be closed, but otherwise does not intersect itself; it is then called a **simple closed curve.** Circles and ellipses are examples of simple closed curves. Every parametrization of a particular curve determines one of two possible **orientations** corresponding to the direction along the curve in which the parameter is increasing. All three parametrizations of the semicircle in the Example 2 orient the semicircle clockwise as viewed from a point above the $xy$-plane. This orientation is shown by the arrowhead on the curve in Figure 7.6. The same semicircle could be given the opposite orientation by, for example, the parametrization

$$\mathbf{r}(t) = \cos t\,\mathbf{i} + \sin t\,\mathbf{j}, \qquad 0 \le t \le \pi.$$

## Parametrizing the Curve of Intersection of Two Surfaces

Frequently a curve is specified as the intersection of two surfaces with given Cartesian equations. We may want to represent the curve by parametric equations. There is no unique way to do do this, but if one of the given surfaces is a cylinder parallel to a coordinate axis (so its equation is independent of one of the variables) we can begin by parametrizing that surface. The following examples clarify the method.

**EXAMPLE 3** Parametrize the curve of intersection of the plane $3x+2y+z=4$ and the elliptic cylinder $x^2+4y^2=4$.

***SOLUTION*** We begin with the equation $x^2+4y^2=4$, which is independent of $z$. It can be parametrized in many ways; one convenient way is

$$x = 2\cos t, \qquad y = \sin t, \qquad (0 \le t \le 2\pi).$$

The equation of the plane can then be solved for $z$, so that $z$ can be expressed in terms of $t$:

$$z = 4 - 3x - 2y = 4 - 6\cos t - 2\sin t.$$

Thus the given surfaces intersect in the curve

$$\mathbf{r} = 2\cos t\,\mathbf{i} + \sin t\,\mathbf{j} + (4 - 6\cos t - 2\sin t)\mathbf{k} \qquad (0 \le t \le 2\pi).$$

■

**EXAMPLE 4** Find a parametric representation of the curve of intersection of the two surfaces

$$x^2 + y + z = 2 \qquad \text{and} \qquad xy + z = 1.$$

***SOLUTION*** Here, neither given equation is independent of a variable, but we can obtain a third equation representing a surface containing the curve of intersection of the two given surfaces by subtracting the two given equations to eliminate $z$:

$$x^2 + y - xy = 1.$$

This equation is readily parametrized. If, for example, we let $x = t$, then

$$t^2 + y(1-t) = 1, \qquad \text{so} \quad y = \frac{1-t^2}{1-t} = 1 + t.$$

Either of the given equations can then be used to express $z$ in terms of $t$:

$$z = 1 - xy = 1 - t(1+t) = 1 - t - t^2.$$

The required parametrization is

$$\mathbf{r} = t\mathbf{i} + (1+t)\mathbf{j} + (1 - t - t^2)\mathbf{k}.$$

Of course, this answer is not unique. Many other parametrizations can be found for the curve. ■

## Arc Length

We now consider how to define and calculate the length of a curve. Let $\mathcal{C}$ be a bounded, continuous curve specified by

$$\mathbf{r} = \mathbf{r}(t), \qquad a \le t \le b.$$

Subdivide the closed interval $[a, b]$ into $n$ subintervals by points

$$a = t_0 < t_1 < t_2 < \cdots < t_{n-1} < t_n = b.$$

The points $\mathbf{r}_i = \mathbf{r}(t_i)$, $(0 \le i \le n)$, subdivide $\mathcal{C}$ into $n$ arcs. If we use the chord length $|\mathbf{r}_i - \mathbf{r}_{i-1}|$ as an approximation to the arc length between $\mathbf{r}_{i-1}$ and $\mathbf{r}_i$, then the sum

$$s_n = \sum_{i=1}^{n} |\mathbf{r}_i - \mathbf{r}_{i-1}|$$

approximates the length of $\mathcal{C}$ by the length of a polygonal line. (See Figure 7.8.) Evidently any such approximation is less than or equal to the actual length of $\mathcal{C}$. We say that $\mathcal{C}$ is **rectifiable** if there exists a constant $K$ such that $s_n \le K$ for every $n$ and every choice of the points $t_i$. In this case, the completeness axiom of the real number system assures us that there will be a smallest such number $K$. We call this smallest $K$ the **length** of $\mathcal{C}$ and denote it by $s$.

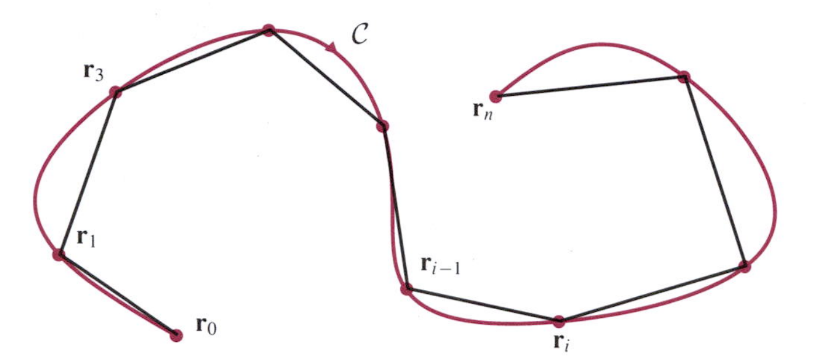


**Figure 7.8** A polygonal approximation to a curve $\mathcal{C}$. The length of the polygonal line cannot exceed the length of the curve

Let $\Delta t_i = t_i - t_{i-1}$ and $\Delta \mathbf{r}_i = \mathbf{r}_i - \mathbf{r}_{i-1}$. Then $s_n$ can be written in the form

$$s_n = \sum_{i=1}^{n} \left| \frac{\Delta \mathbf{r}_i}{\Delta t_i} \right| \Delta t_i .$$

If $\mathbf{r}(t)$ has a continuous derivative $\mathbf{v}(t)$ that does not vanish on $[a, b]$, it can be shown that

$$s = \lim_{\max \Delta t_i \to 0} s_n = \int_a^b \left| \frac{d\mathbf{r}}{dt} \right| dt = \int_a^b |\mathbf{v}(t)|\, dt = \int_a^b v(t)\, dt.$$

In kinematic terms, this formula states that the distance travelled by a moving particle is the integral of the speed.

***REMARK*** Although the above formula is expressed in terms of the parameter $t$, the arc length, as defined above, is a strictly geometric property of the curve $\mathcal{C}$. It is independent of the particular parametrization used to represent $\mathcal{C}$. See Exercise 29 at the end of this section.

If $s(t)$ denotes the arc length of that part of $\mathcal{C}$ corresponding to parameter values in $[a, t]$ then

$$\frac{ds}{dt} = \frac{d}{dt} \int_a^t v(\tau)\, d\tau = v(t)$$

so that the **arc length element** for $\mathcal{C}$ is given by

$$ds = v(t)\, dt = \left| \frac{d}{dt} \mathbf{r}(t) \right| dt.$$

The length of $\mathcal{C}$ is the integral of these arc length elements: we write

$$\int_{\mathcal{C}} ds = \text{length of } \mathcal{C} = \int_a^b v(t)\, dt.$$

Several familiar formulas for arc length follow from the above formula by using specific parametrizations of curves. For instance, the arc length element $ds$ for the Cartesian plane curve $y = f(x)$ on $[a, b]$ is obtained by using $x$ as parameter: here $\mathbf{r} = x\mathbf{i} + f(x)\mathbf{j}$ so $\mathbf{v} = \mathbf{i} + f'(x)\mathbf{j}$ and

$$ds = \sqrt{1 + \big(f'(x)\big)^2}\, dx.$$

Similarly, the arc length element $ds$ for a plane polar curve $r = g(\theta)$ can be calculated from the parametrization

$$\mathbf{r}(\theta) = g(\theta)\cos\theta\mathbf{i} + g(\theta)\sin\theta\mathbf{j}.$$

It is

$$ds = \sqrt{\big(g(\theta)\big)^2 + \big(g'(\theta)\big)^2}\, d\theta.$$

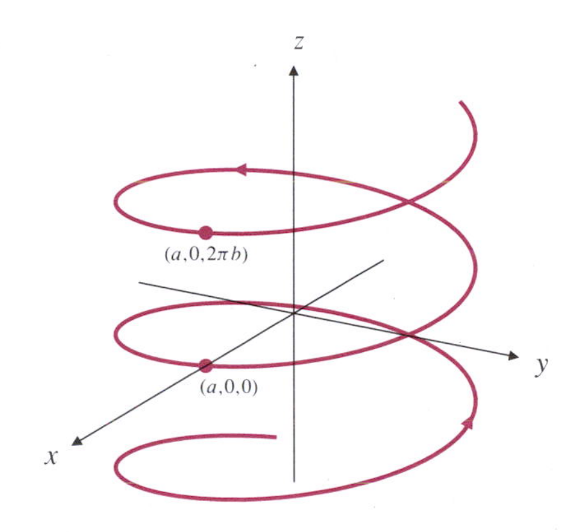


**Figure 7.9** The helix
$x = a\cos t$
$y = a\sin t$
$z = bt$

**EXAMPLE 5** Find the length $s$ of that part of the **circular helix**

$$\mathbf{r} = a\,\cos t\,\mathbf{i} + a\,\sin t\,\mathbf{j} + bt\,\mathbf{k}$$

between the points $(a, 0, 0)$ and $(a, 0, 2\pi b)$.

***SOLUTION*** This curve spirals around the $z$-axis, rising as it turns. (See Figure 7.9.) It lies on the surface of the circular cylinder $x^2 + y^2 = a^2$. We have

$$\mathbf{v} = \frac{d\mathbf{r}}{dt} = -a\,\sin t\,\mathbf{i} + a\cos t\,\mathbf{j} + b\mathbf{k}$$
$$v = \sqrt{a^2 + b^2},$$

so that in terms of the parameter $t$ the helix is traced out at constant speed. The required length $s$ corresponds to parameter interval $[0, 2\pi]$. Thus

$$s = \int_0^{2\pi} v(t)\, dt = \int_0^{2\pi} \sqrt{a^2 + b^2}\, dt = 2\pi\sqrt{a^2 + b^2}.$$ ■

A curve specified parametrically in cylindrical coordinates by

$$r = r(t), \qquad \theta = \theta(t), \qquad \text{and} \qquad z = z(t)$$

has arc length element given by

$$ds = \sqrt{\left(\frac{dr}{dt}\right)^2 + \big(r(t)\big)^2\left(\frac{d\theta}{dt}\right)^2 + \left(\frac{dz}{dt}\right)^2}\, dt,$$

as can be calculated from the Cartesian parametrization

$$\mathbf{r} = r(t)\cos\theta(t)\mathbf{i} + r(t)\sin\theta(t)\mathbf{j} + z(t)\mathbf{k}.$$

(See Exercise 20 below.) Similarly, a curve specified parametrically in spherical coordinates by

$$\rho = \rho(t), \qquad \phi = \phi(t), \qquad \theta = \theta(t)$$

can be expressed in Cartesian form by

$$\mathbf{r} = \rho(t)\sin\phi(t)\cos\theta(t)\mathbf{i} + \rho(t)\sin\phi(t)\sin\theta(t)\mathbf{j} + \rho(t)\cos\phi(t)\mathbf{k},$$

and its arc length element calculated to be

$$ds = \sqrt{\left(\frac{d\rho}{dt}\right)^2 + \big(\rho(t)\big)^2\left(\frac{d\phi}{dt}\right)^2 + \big(\rho(t)\sin\phi(t)\big)^2\left(\frac{d\theta}{dt}\right)^2}\,dt.$$

(See Exercise 21 below.)

**EXAMPLE 6** Describe the curve given in cylindrical coordinates by

$$r = \sqrt{2}\cos t, \qquad \theta = t/\sqrt{2}, \qquad z = \sin t, \qquad (0 \le t \le \pi/2),$$

and find its length.

***SOLUTION*** The curve lies on the surface of the ellipsoid $r^2 + 2z^2 = 2$, that is, $x^2 + y^2 + 2z^2 = 2$, winding eastward as it rises from the point $[\sqrt{2}, 0, 0]$ on the "equator" to the "north pole" $[0, \pi/(2\sqrt{2}), 1]$. (See Figure 7.10.) Its length is

$$\begin{aligned} s &= \int_0^{\pi/2} \sqrt{\left(\frac{dr}{dt}\right)^2 + \big(r(t)\big)^2\left(\frac{d\theta}{dt}\right)^2 + \left(\frac{dz}{dt}\right)^2}\,dt \\ &= \int_0^{\pi/2} \sqrt{2\sin^2 t + (2\cos^2 t)\frac{1}{2} + \cos^2 t}\,dt = \frac{\pi}{\sqrt{2}} \text{ units.} \end{aligned}$$

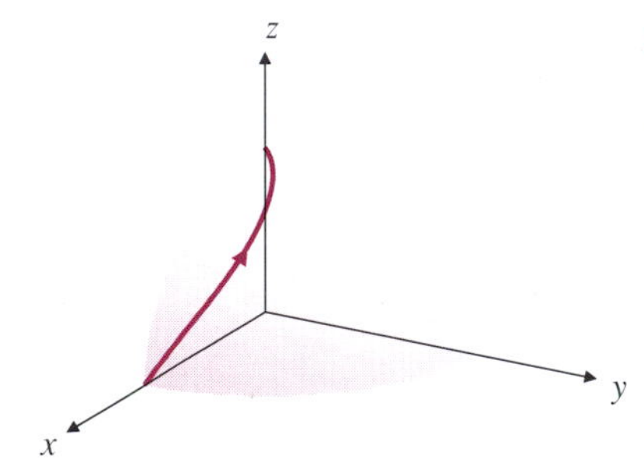


**Figure 7.10** The curve of Example 6

## The Arc-Length Parametrization

The selection of a particular parameter in terms of which to specify a given curve will usually depend on the problem in which the curve arises; there is no one "right way" to parametrize a curve. However, there is one parameter that is "natural" in that it arises from the geometry (shape and size) of the curve itself, and not from any particular coordinate system in which the equation of the curve is to be expressed. This parameter is the *arc length* measured from some particular point (the *initial point*) on the curve. The position vector of an arbitrary point $P$ on the curve can be specified as a function of the arc length $s$ along the curve from the initial point $P_0$ to $P$,

$$\mathbf{r} = \mathbf{r}(s).$$

This equation is called an **arc-length parametrization** or **intrinsic parametrization** of the curve. Since $ds = v(t)\,dt$ for any parametrization $\mathbf{r} = \mathbf{r}(t)$, for the arc-length parametrization we have $ds = v(s)\,ds$. Thus $v(s) = 1$, identically; *a curve parametrized in terms of arc length is traced at unit speed.* Although it is seldom easy (and usually not possible), to find $\mathbf{r}(s)$ explicitly when the curve is given in terms of some other parameter, smooth curves always have such parametrizations (see Exercise 30 at the end of this section), and they will prove useful when we develop the fundamentals of the *differential geometry* for 3-space curves in the next section.

Suppose that a curve is specified in terms of an arbitrary parameter $t$. If the arc length over a parameter interval $[t_0, t]$,

$$s = s(t) = \int_{t_0}^{t} \left| \frac{d}{d\tau}\mathbf{r}(\tau) \right| d\tau,$$

can be evaluated explicitly, and if the equation $s = s(t)$ can be explicitly solved for $t$ as a function of $s$ $(t = t(s))$ then the curve can be reparametrized in terms of arc length by substituting for $t$ in the original parametrization:

$$\mathbf{r} = \mathbf{r}(t(s)).$$

**EXAMPLE 7** Parametrize the circular helix

$$\mathbf{r} = a\cos t\,\mathbf{i} + a\sin t\,\mathbf{j} + bt\,\mathbf{k}$$

in terms of the arc length measured from the point $(a, 0, 0)$ in the direction of increasing $t$.

***SOLUTION*** The initial point corresponds to $t = 0$. As shown in Example 5, we have $ds/dt = \sqrt{a^2 + b^2}$ so

$$s = s(t) = \int_0^t \sqrt{a^2 + b^2}\,d\tau = \sqrt{a^2 + b^2}\,t.$$

Therefore $t = s/\sqrt{a^2 + b^2}$ and the arc-length parametrization is

$$\mathbf{r}(s) = a\cos\left(\frac{s}{\sqrt{a^2 + b^2}}\right)\mathbf{i} + a\sin\left(\frac{s}{\sqrt{a^2 + b^2}}\right)\mathbf{j} + \frac{bs}{\sqrt{a^2 + b^2}}\mathbf{k}.$$

## EXERCISES 7.3

In Exercises 1–4, find the required parametrization of the first quadrant part of the circular arc $x^2 + y^2 = a^2$.

1. In terms of the $y$-coordinate, oriented counterclockwise
2. In terms of the $x$-coordinate, oriented clockwise
3. In terms of the angle between the tangent line and the positive $x$-axis, oriented counterclockwise
4. In terms of arc length measured from $(0, a)$, oriented clockwise
5. The cylinders $z = x^2$ and $z = 4y^2$ intersect in two curves, one of which passes through the point $(2, -1, 4)$. Find a parametrization of that curve using $t = y$ as parameter.
6. The plane $x + y + z = 1$ intersects the cylinder $z = x^2$ in a parabola. Parametrize the parabola using $t = x$ as parameter.

In Exercises 7–10, parametrize the curve of intersection of the given surfaces. *Note*: the answers are not unique.

7. $x^2 + y^2 = 9$ and $z = x + y$
8. $z = \sqrt{1 - x^2 - y^2}$ and $x + y = 1$
9. $z = x^2 + y^2$ and $2x - 4y - z - 1 = 0$

**10.** $yz + x = 1$ and $xz - x = 1$

**11.** The plane $z = 1 + x$ intersects the cone $z^2 = x^2 + y^2$ in a parabola. Try to parametrize the parabola using as parameter: (a) $t = x$, (b) $t = y$, and (c) $t = z$. Which of these choices for $t$ leads to a single parametrization that represents the whole parabola? What is that parametrization? What happens with the other two choices?

* **12.** The plane $x + y + z = 1$ intersects the sphere $x^2 + y^2 + z^2 = 1$ in a circle $\mathcal{C}$. Find the centre $\mathbf{r}_0$ and radius $r$ of $\mathcal{C}$. Also find two perpendicular unit vectors $\hat{\mathbf{v}}_1$ and $\hat{\mathbf{v}}_2$ parallel to the plane of $\mathcal{C}$. (*Hint:* to be specific, show that $\hat{\mathbf{v}}_1 = (\mathbf{i} - \mathbf{j})/\sqrt{2}$ is one such vector, and then find a second that is perpendicular to $\hat{\mathbf{v}}_1$.) Use your results to construct a parametrization of $\mathcal{C}$.

**13.** Find the length of the curve $\mathbf{r} = t^2\mathbf{i} + t^2\mathbf{j} + t^3\mathbf{k}$ from $t = 0$ to $t = 1$.

**14.** For what values of the parameter $\lambda$ is the length $s(T)$ of the curve $\mathbf{r} = t\mathbf{i} + \lambda t^2\mathbf{j} + t^3\mathbf{k}$, $(0 \le t \le T)$ given by $s(T) = T + T^3$?

**15.** Express the length of the curve $\mathbf{r} = at^2\,\mathbf{i} + bt\,\mathbf{j} + c\ln t\,\mathbf{k}$, $(1 \le t \le T)$ as a definite integral. Evaluate the integral if $b^2 = 4ac$.

**16.** Describe the parametric curve $\mathcal{C}$ given by

$$x = a\cos t\sin t, \qquad y = a\sin^2 t, \qquad z = bt.$$

What is the length of $\mathcal{C}$ between $t = 0$ and $t = T > 0$?

**17.** Find the length of the conical helix $\mathbf{r} = t\cos t\,\mathbf{i} + t\sin t\,\mathbf{j} + t\,\mathbf{k}$, $(0 \le t \le 2\pi)$. Why is the curve called a conical helix?

**18.** Describe the intersection of the sphere $x^2 + y^2 + z^2 = 1$ and the elliptic cylinder $x^2 + 2z^2 = 1$. Find the total length of this intersection curve.

**19.** Let $\mathcal{C}$ be the curve $x = e^t\cos t$, $y = e^t\sin t$, $z = t$ between $t = 0$ and $t = 2\pi$. Find the length of $\mathcal{C}$.

**20.** Verify the formula for the arc length element in cylindrical coordinates,

$$ds = \sqrt{\left(\frac{dr}{dt}\right)^2 + \left(r(t)\right)^2\left(\frac{d\theta}{dt}\right)^2 + \left(\frac{dz}{dt}\right)^2}\,dt,$$

given in this section.

* **21.** Verify the formula for the arc length element in spherical coordinates,

$$ds = \sqrt{\left(\frac{d\rho}{dt}\right)^2 + \left(\rho(t)\right)^2\left(\frac{d\phi}{dt}\right)^2 + \left(\rho(t)\sin\phi(t)\right)^2\left(\frac{d\theta}{dt}\right)^2}\,dt,$$

given in this section.

**22.** Find the length of the conical spiral given in cylindrical coordinates by $r = t$, $\theta = t$, and $z = t$, $(0 \le t \le 2\pi)$.

**23.** Find the length of the curve given in spherical coordinates by $\rho = t$, $\phi = t$, and $\theta = t$, $(0 \le t \le \pi/2)$.

* **24.** A cable of length $L$ and circular cross-section of radius $a$ is wound around a cylindrical spool of radius $b$ with no overlapping and so that adjacent windings touch one another. What length of the spool is covered by the cable?

In Exercises 25–28, reparametrize the given curve in the same orientation in terms of arc length measured from the point where $t = 0$.

**25.** $\mathbf{r} = At\mathbf{i} + Bt\mathbf{j} + Ct\mathbf{k}$, $\qquad (A^2 + B^2 + C^2 > 0)$

**26.** $\mathbf{r} = e^t\mathbf{i} + \sqrt{2}t\mathbf{j} - e^{-t}\mathbf{k}$

* **27.** $\mathbf{r} = a\cos^3 t\,\mathbf{i} + a\sin^3 t\,\mathbf{j} + b\cos 2t\,\mathbf{k}$, $\qquad (0 \le t \le \frac{\pi}{2})$

* **28.** $\mathbf{r} = 3t\cos t\,\mathbf{i} + 3t\sin t\,\mathbf{j} + 2\sqrt{2}t^{3/2}\mathbf{k}$

* **29.** Let $\mathbf{r} = \mathbf{r}_1(t)$, $(a \le t \le b)$, and $\mathbf{r} = \mathbf{r}_2(u)$, $(c \le u \le d)$, be two parametrizations of the same curve $\mathcal{C}$, each one-to-one on its domain, and each giving $\mathcal{C}$ the same orientation (so that $\mathbf{r}_1(a) = \mathbf{r}_2(c)$ and $\mathbf{r}_1(b) = \mathbf{r}_2(d)$). Then for each $t$ in $[a, b]$ there is a unique $u = u(t)$ such that $\mathbf{r}_2(u(t)) = \mathbf{r}_1(t)$. Show that

$$\int_a^b \left|\frac{d}{dt}\mathbf{r}_1(t)\right| dt = \int_c^d \left|\frac{d}{du}\mathbf{r}_2(u)\right| du,$$

and thus that the length of $\mathcal{C}$ is independent of parametrization.

* **30.** If the curve $\mathbf{r} = \mathbf{r}(t)$ has continuous, nonvanishing velocity $\mathbf{v}(t)$ on the interval $[a, b]$, and if $t_0$ is some point in $[a, b]$, show that the function

$$s = g(t) = \int_{t_0}^{t} |\mathbf{v}(u)|\,du$$

is an increasing function on $[a, b]$, and so has an inverse:

$$t = g^{-1}(s) \iff s = g(t).$$

Hence show that the curve can be parametrized in terms of arc length measured from $\mathbf{r}(t_0)$.

# 7.4 CURVATURE, TORSION, AND THE FRENET FRAME

In this section and the next we develop the fundamentals of differential geometry of curves in 3-space. We will introduce several new scalar and vector functions associated with a curve $\mathcal{C}$. The most important of these are the curvature and torsion of the curve, and a right-handed triad of mutually perpendicular unit vectors forming a basis at any point on the curve, and called the Frenet frame. The curvature and torsion measure the rate at which a curve is turning (away from its tangent line) and twisting (out of the plane in which it is turning) at any point.

## The Unit Tangent Vector

The velocity vector $\mathbf{v}(t) = d\mathbf{r}/dt$ is tangent to the parametric curve $\mathbf{r} = \mathbf{r}(t)$ at the point $\mathbf{r}(t)$, and points in the direction of the orientation of the curve there. Since we are assuming that $\mathbf{v}(t) \neq \mathbf{0}$ we can find a **unit tangent vector**, $\hat{\mathbf{T}}(t)$, at $\mathbf{r}(t)$ by dividing $\mathbf{v}(t)$ by its length:

$$\hat{\mathbf{T}}(t) = \frac{\mathbf{v}(t)}{v(t)} = \frac{d\mathbf{r}}{dt} \bigg/ \left|\frac{d\mathbf{r}}{dt}\right|.$$

Recall that a curve parametrized in terms of arc length, $\mathbf{r} = \mathbf{r}(s)$, is traced at unit speed; $v(s) = 1$. In terms of the arc-length parametrization, the unit tangent vector is

$$\hat{\mathbf{T}}(s) = \frac{d\mathbf{r}}{ds}.$$

**EXAMPLE 1** Find the unit tangent vector, $\hat{\mathbf{T}}$, for the circular helix of Example 5 of Section 7.3, both in terms of $t$ and in terms of the arc-length parameter $s$.

***SOLUTION*** In terms of $t$ we have

$$\begin{aligned}
\mathbf{r} &= a\cos t\,\mathbf{i} + a\sin t\,\mathbf{j} + bt\mathbf{k}\\
\mathbf{v}(t) &= -a\sin t\,\mathbf{i} + a\cos t\,\mathbf{j} + b\mathbf{k}\\
v(t) &= \sqrt{a^2\sin^2 t + a^2\cos^2 t + b^2} = \sqrt{a^2+b^2}\\
\hat{\mathbf{T}}(t) &= -\frac{a}{\sqrt{a^2+b^2}}\sin t\,\mathbf{i} + \frac{a}{\sqrt{a^2+b^2}}\cos t\,\mathbf{j} + \frac{b}{\sqrt{a^2+b^2}}\mathbf{k}.
\end{aligned}$$

In terms of the arc-length parameter (see Example 7 of Section 7.3)

$$\begin{aligned}
\mathbf{r}(s) &= a\cos\left(\frac{s}{\sqrt{a^2+b^2}}\right)\mathbf{i} + a\sin\left(\frac{s}{\sqrt{a^2+b^2}}\right)\mathbf{j} + \frac{bs}{\sqrt{a^2+b^2}}\mathbf{k}\\
\hat{\mathbf{T}}(s) = \frac{d\mathbf{r}}{ds} &= -\frac{a}{\sqrt{a^2+b^2}}\sin\left(\frac{s}{\sqrt{a^2+b^2}}\right)\mathbf{i} + \frac{a}{\sqrt{a^2+b^2}}\cos\left(\frac{s}{\sqrt{a^2+b^2}}\right)\mathbf{j}\\
&\quad + \frac{b}{\sqrt{a^2+b^2}}\mathbf{k}.
\end{aligned}$$

■

***REMARK*** If the curve $\mathbf{r} = \mathbf{r}(t)$ has a continuous, nonvanishing velocity $\mathbf{v}(t)$ then the unit tangent vector $\hat{\mathbf{T}}(t)$ is a continuous function of $t$. The angle $\theta(t)$ between $\hat{\mathbf{T}}(t)$ and any fixed unit vector $\hat{\mathbf{u}}$ is also continuous in $t$:

$$\theta(t) = \cos^{-1}(\hat{\mathbf{T}}(t) \bullet \hat{\mathbf{u}}).$$

Thus, as asserted previously, the curve is *smooth* in the sense that it has a continuously turning tangent line. The rate of this turning is quantified by the curvature, which we introduce now.

## Curvature and the Unit Normal

In the rest of this section we will deal abstractly with a curve $\mathcal{C}$ parametrized in terms of arc length measured from some point on it:

$$\mathbf{r} = \mathbf{r}(s).$$

In the next section we return to curves with arbitrary parametrizations and apply the principles developed in this section to specific problems. Throughout we assume that the parametric equations of curves have continuous derivatives up to third order on the intervals where they are defined.

Having unit length, the tangent vector $\hat{\mathbf{T}}(s) = d\mathbf{r}/ds$ satisfies $\hat{\mathbf{T}}(s) \bullet \hat{\mathbf{T}}(s) = 1$. Differentiating this equation with respect to $s$ we get

$$2\hat{\mathbf{T}}(s) \bullet \frac{d\hat{\mathbf{T}}}{ds} = 0,$$

so that $d\hat{\mathbf{T}}/ds$ is perpendicular to $\hat{\mathbf{T}}(s)$.

**DEFINITION 1**

**Curvature and radius of curvature**

The **curvature** of $\mathcal{C}$ at the point $\mathbf{r}(s)$ is the length of $d\hat{\mathbf{T}}/ds$ there. It is denoted by the Greek letter $\kappa$ (*kappa*):

$$\kappa(s) = \left|\frac{d\hat{\mathbf{T}}}{ds}\right|.$$

The **radius of curvature**, denoted $\rho$ (the Greek letter *rho*) is the reciprocal of the curvature:

$$\rho(s) = \frac{1}{\kappa(s)}.$$

As we will see below, the curvature of $\mathcal{C}$ at $\mathbf{r}(s)$ measures the rate of turning of the tangent line to the curve there. The radius of curvature is the radius of the circle through $\mathbf{r}(s)$ that most closely approximates the curve $\mathcal{C}$ near that point.

According to its definition, $\kappa(s) \geq 0$ everywhere on $\mathcal{C}$. If $\kappa(s) \neq 0$ we can divide $d\hat{\mathbf{T}}/ds$ by its length, $\kappa(s)$, and obtain a unit vector $\hat{\mathbf{N}}(s)$ in the same direction. This unit vector is called the **unit principal normal** to $\mathcal{C}$ at $\mathbf{r}(s)$, or, more commonly, just the **unit normal**.

$$\hat{\mathbf{N}}(s) = \frac{1}{\kappa(s)} \frac{d\hat{\mathbf{T}}}{ds}$$

Note that $\hat{\mathbf{N}}(s)$ is perpendicular to $\mathcal{C}$ at $\mathbf{r}(s)$, and points in the direction that $\hat{\mathbf{T}}$ and therefore $\mathcal{C}$ is turning. The principal normal is not defined at points where the curvature $\kappa(s)$ is zero. For instance, a straight line has no principal normal. Figure 7.11(a) shows $\hat{\mathbf{T}}$ and $\hat{\mathbf{N}}$ at a point on a typical curve.

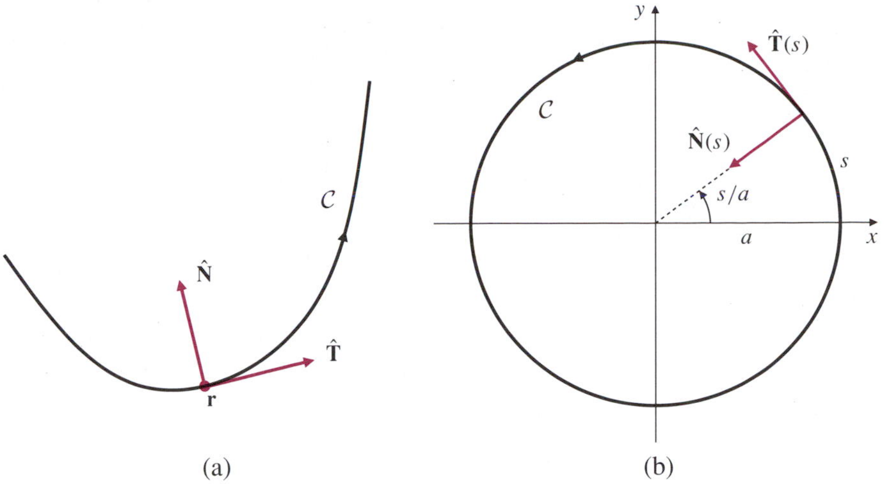

**Figure 7.11**

(a) The unit tangent and principal normal vectors for a curve

(b) The unit tangent and principal normal vectors for a circle

**EXAMPLE 2** Let $a > 0$. Show that the curve $\mathcal{C}$ given by

$$\mathbf{r} = a\cos\left(\frac{s}{a}\right)\mathbf{i} + a\sin\left(\frac{s}{a}\right)\mathbf{j}$$

is a circle in the $xy$-plane having radius $a$ and centre at the origin, and that it is parametrized in terms of arc length. Find the curvature, the radius of curvature, and the unit tangent and principal normal vectors at any point on $\mathcal{C}$.

***SOLUTION*** Since

$$|\mathbf{r}(s)| = a\sqrt{\left(\cos\left(\frac{s}{a}\right)\right)^2 + \left(\sin\left(\frac{s}{a}\right)\right)^2} = a,$$

$\mathcal{C}$ is indeed a circle of radius $a$ centred at the origin in the $xy$-plane. Since the speed

$$\left|\frac{d\mathbf{r}}{ds}\right| = \left|-\sin\left(\frac{s}{a}\right)\mathbf{i} + \cos\left(\frac{s}{a}\right)\mathbf{j}\right| = 1,$$

the parameter $s$ must represent arc length, and hence the unit tangent vector is

$$\hat{\mathbf{T}}(s) = -\sin\left(\frac{s}{a}\right)\mathbf{i} + \cos\left(\frac{s}{a}\right)\mathbf{j}.$$

Therefore

$$\frac{d\hat{\mathbf{T}}}{ds} = -\frac{1}{a}\cos\left(\frac{s}{a}\right)\mathbf{i} - \frac{1}{a}\sin\left(\frac{s}{a}\right)\mathbf{j}$$

and the curvature and radius of curvature at $\mathbf{r}(s)$ are

$$\kappa(s) = \left|\frac{d\hat{\mathbf{T}}}{ds}\right| = \frac{1}{a}, \qquad \rho(s) = \frac{1}{\kappa(s)} = a.$$

Finally, the unit principal normal is

$$\hat{\mathbf{N}}(s) = -\cos\left(\frac{s}{a}\right)\mathbf{i} - \sin\left(\frac{s}{a}\right)\mathbf{j} = -\frac{1}{a}\mathbf{r}(s).$$

Note that the curvature and radius of curvature are constant; the latter is in fact the radius of the circle. The circle and its unit tangent and normal vectors at a typical point are sketched in Figure 7.11(b). Note that $\hat{\mathbf{N}}$ points towards the centre of the circle. ■

***REMARK*** Another observation can be made about the above example. The position vector $\mathbf{r}(s)$ makes angle $\theta = s/a$ with the positive $x$-axis, and therefore $\hat{\mathbf{T}}(s)$ makes the same angle with the positive $y$-axis. Therefore the rate of rotation of $\hat{\mathbf{T}}$ with respect to $s$ is

$$\frac{d\theta}{ds} = \frac{1}{a} = \kappa.$$

That is, $\kappa$ is the rate at which $\hat{\mathbf{T}}$ is turning. This observation extends to a general smooth curve.

**THEOREM 2**

**Curvature is the rate of turning of the unit tangent**

Let $\kappa > 0$ on an interval containing $s$, and let $\Delta\theta$ be the angle between $\hat{\mathbf{T}}(s + \Delta s)$ and $\hat{\mathbf{T}}(s)$, the unit tangent vectors at neighbouring points on the curve. Then

$$\kappa(s) = \lim_{\Delta s \to 0} \frac{\Delta\theta}{\Delta s}.$$

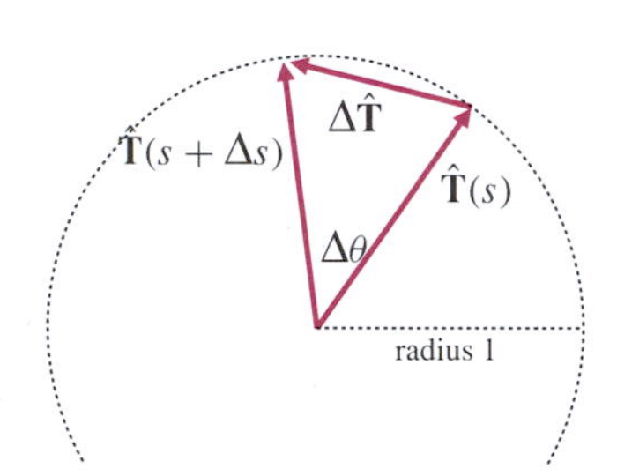


Figure 7.12 $\Delta\hat{\mathbf{T}} \approx \Delta\theta$ for small $\Delta s$

***PROOF*** Let $\Delta\hat{\mathbf{T}} = \hat{\mathbf{T}}(s + \Delta s) - \hat{\mathbf{T}}(s)$. Because both $\hat{\mathbf{T}}(s)$ and $\hat{\mathbf{T}}(s + \Delta s)$ are unit vectors, $|\Delta\hat{\mathbf{T}}/\Delta\theta|$ is the ratio of the length of a chord to the length of the corresponding arc on a circle of radius 1. (See Figure 7.12.) Thus

$$\lim_{\Delta s \to 0} \left| \frac{\Delta\hat{\mathbf{T}}}{\Delta\theta} \right| = 1, \qquad \text{and}$$

$$\kappa(s) = \lim_{\Delta s \to 0} \left| \frac{\Delta\hat{\mathbf{T}}}{\Delta s} \right| = \lim_{\Delta s \to 0} \left| \frac{\Delta\hat{\mathbf{T}}}{\Delta\theta} \right| \left| \frac{\Delta\theta}{\Delta s} \right| = \lim_{\Delta s \to 0} \left| \frac{\Delta\theta}{\Delta s} \right|.$$

The unit tangent $\hat{\mathbf{T}}$ and unit normal $\hat{\mathbf{N}}$ at a point $\mathbf{r}(s)$ on a curve $\mathcal{C}$ are regarded as having their tails at that point. They are perpendicular, and $\hat{\mathbf{N}}$ points in the direction towards which $\hat{\mathbf{T}}(s)$ turns as $s$ increases. The plane passing through $\mathbf{r}(s)$ and containing the vectors $\hat{\mathbf{T}}(s)$ and $\hat{\mathbf{N}}(s)$ is called the **osculating plane** of $\mathcal{C}$ at $\mathbf{r}(s)$ (from the Latin *osculum*, meaning *kiss*). For a *plane curve*, such as the circle in Example 2, the osculating plane is just the plane containing the curve. For more general three-dimensional curves the osculating plane varies from point to point; at any point it is that plane which comes closest to containing the part of the curve near that point. The osculating plane is not properly defined at a point where $\kappa(s) = 0$, though if such points are isolated, it can sometimes be defined as a limit of osculating planes for neighbouring points.

Still assuming that $\kappa(s) \neq 0$, let

$$\mathbf{r}_c(s) = \mathbf{r}(s) + \rho(s)\hat{\mathbf{N}}(s).$$

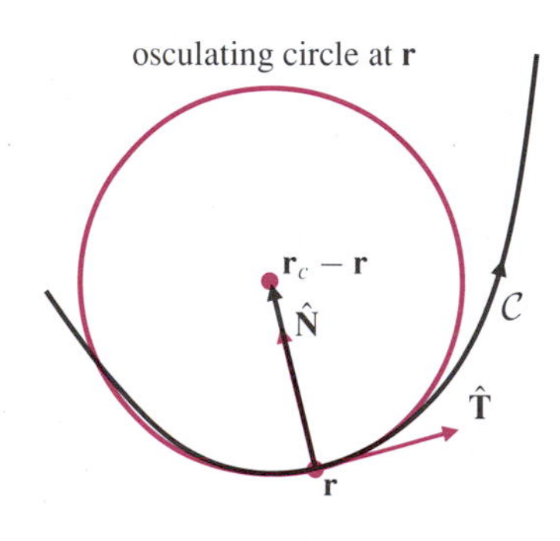


Figure 7.13 An osculating circle

For each $s$ the point with position vector $\mathbf{r}_c(s)$ lies in the osculating plane of $\mathcal{C}$ at $\mathbf{r}(s)$, on the concave side of $\mathcal{C}$ and at distance $\rho(s)$ from $\mathbf{r}(s)$. It is called the **centre of curvature** of $\mathcal{C}$ for the point $\mathbf{r}(s)$. The circle in the osculating plane having centre at the centre of curvature and radius equal to the radius of curvature $\rho(s)$ is called the **osculating circle** for $\mathcal{C}$ at $\mathbf{r}(s)$. Among all circles that pass through the point $\mathbf{r}(s)$, the osculating circle is the one that best describes the behaviour of $\mathcal{C}$ near that point. Of course, the osculating circle of a circle at any point is the same circle. A typical example of an osculating circle is shown in Figure 7.13.

## Torsion and Binormal, The Frenet–Serret Formulas

At any point $\mathbf{r}(s)$ on the curve $\mathcal{C}$ where $\hat{\mathbf{T}}$ and $\hat{\mathbf{N}}$ are defined, a third unit vector, the **unit binormal** $\hat{\mathbf{B}}$, is defined by the formula

$$\hat{\mathbf{B}} = \hat{\mathbf{T}} \times \hat{\mathbf{N}}.$$

Note that $\hat{\mathbf{B}}(s)$ is normal to the osculating plane of $\mathcal{C}$ at $\mathbf{r}(s)$; if $\mathcal{C}$ is a plane curve, then $\hat{\mathbf{B}}$ is a constant vector, independent of $s$. At each point on $\mathcal{C}$, the three vectors $\{\hat{\mathbf{T}}, \hat{\mathbf{N}}, \hat{\mathbf{B}}\}$ constitute a right-handed basis of mutually perpendicular unit vectors like the standard basis $\{\mathbf{i}, \mathbf{j}, \mathbf{k}\}$. (See Figure 7.14.) This basis is called the **Frenet frame** at the point. Note that

$$\hat{\mathbf{B}} \times \hat{\mathbf{T}} = \hat{\mathbf{N}} \text{ and } \hat{\mathbf{N}} \times \hat{\mathbf{B}} = \hat{\mathbf{T}}.$$

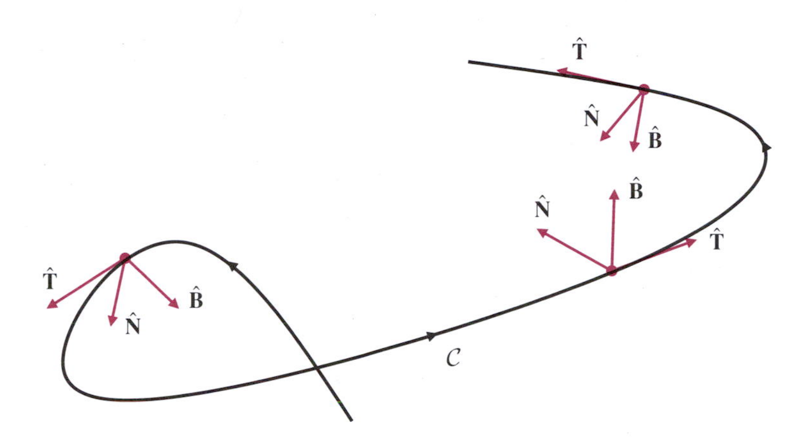


**Figure 7.14** The Frenet frame $\{\hat{\mathbf{T}}, \hat{\mathbf{N}}, \hat{\mathbf{B}}\}$ at some points on $\mathcal{C}$

Since $1 = \hat{\mathbf{B}}(s) \bullet \hat{\mathbf{B}}(s)$, therefore $\hat{\mathbf{B}}(s) \bullet (d\hat{\mathbf{B}}/ds) = 0$, and $d\hat{\mathbf{B}}/ds$ is perpendicular to $\hat{\mathbf{B}}(s)$. Also, differentiating $\hat{\mathbf{B}} = \hat{\mathbf{T}} \times \hat{\mathbf{N}}$ we obtain

$$\frac{d\hat{\mathbf{B}}}{ds} = \frac{d\hat{\mathbf{T}}}{ds} \times \hat{\mathbf{N}} + \hat{\mathbf{T}} \times \frac{d\hat{\mathbf{N}}}{ds} = \kappa \hat{\mathbf{N}} \times \hat{\mathbf{N}} + \hat{\mathbf{T}} \times \frac{d\hat{\mathbf{N}}}{ds} = \hat{\mathbf{T}} \times \frac{d\hat{\mathbf{N}}}{ds}.$$

Therefore $d\hat{\mathbf{B}}/ds$ is also perpendicular to $\hat{\mathbf{T}}$. Being perpendicular to both $\hat{\mathbf{T}}$ and $\hat{\mathbf{B}}$, $d\hat{\mathbf{B}}/ds$ must be parallel to $\hat{\mathbf{N}}$. This fact is the basis for our definition of torsion.

**DEFINITION 2**

**Torsion**

There exists a function $\tau(s)$ such that

$$\frac{d\hat{\mathbf{B}}}{ds} = -\tau(s)\hat{\mathbf{N}}(s).$$

The number $\tau(s)$ is called the **torsion** of $\mathcal{C}$ at $\mathbf{r}(s)$.

The torsion measures the degree of twisting that the curve exhibits near a point, that is, the extent to which the curve fails to be planar. It may be positive or negative, depending on the right-handedness or left-handedness of the twisting. We will present an example later in this section.

Theorem 2 has an analogue for torsion, for which the proof is similar. It states that the absolute value of the torsion, $|\tau(s)|$, at point $\mathbf{r}(s)$ on the curve $\mathcal{C}$ is the rate of turning of the unit binormal:

$$\lim_{\Delta s \to 0} \frac{\Delta\psi}{\Delta s} = |\tau(s)|,$$

where $\Delta\psi$ is the angle between $\hat{\mathbf{B}}(s + \Delta s)$ and $\hat{\mathbf{B}}(s)$.

### EXAMPLE 3 The circular helix

As observed in Example 7 of Section 7.3, the parametric equation

$$\mathbf{r}(s) = a\cos(cs)\mathbf{i} + a\sin(cs)\mathbf{j} + bcs\mathbf{k}, \qquad \text{where } c = \frac{1}{\sqrt{a^2 + b^2}},$$

represents a circular helix wound on the cylinder $x^2 + y^2 = a^2$, and parametrized in terms of arc length. Assume $a > 0$. Find the curvature and torsion functions $\kappa(s)$ and $\tau(s)$ for this helix, and also the unit vectors comprising the Frenet frame at any point $\mathbf{r}(s)$ on the helix.

***SOLUTION*** In Example 1 we calculated the unit tangent vector to be

$$\hat{\mathbf{T}}(s) = -ac\sin(cs)\mathbf{i} + ac\cos(cs)\mathbf{j} + bc\mathbf{k}.$$

Differentiating again leads to

$$\frac{d\hat{\mathbf{T}}}{ds} = -ac^2\cos(cs)\mathbf{i} - ac^2\sin(cs)\mathbf{j},$$

so that the curvature of the helix is

$$\kappa(s) = \left|\frac{d\hat{\mathbf{T}}}{ds}\right| = ac^2 = \frac{a}{a^2 + b^2},$$

and the unit normal vector is

$$\hat{\mathbf{N}}(s) = \frac{1}{\kappa(s)}\frac{d\hat{\mathbf{T}}}{ds} = -\cos(cs)\mathbf{i} - \sin(cs)\mathbf{j}.$$

Now we have

$$\begin{aligned}\hat{\mathbf{B}}(s) = \hat{\mathbf{T}}(s) \times \hat{\mathbf{N}}(s) &= \begin{vmatrix} \mathbf{i} & \mathbf{j} & \mathbf{k} \\ -ac\sin(cs) & ac\cos(cs) & bc \\ -\cos(cs) & -\sin(cs) & 0 \end{vmatrix} \\ &= bc\sin(cs)\mathbf{i} - bc\cos(cs)\mathbf{j} + ac\mathbf{k}.\end{aligned}$$

Differentiating this formula leads to

$$\frac{d\hat{\mathbf{B}}}{ds} = bc^2\cos(cs)\mathbf{i} + bc^2\sin(cs)\mathbf{j} = -bc^2\hat{\mathbf{N}}(s).$$

Therefore, the torsion is given by

$$\tau(s) = -(-bc^2) = \frac{b}{a^2 + b^2}.$$

■

***REMARK*** Observe that the curvature $\kappa(s)$ and the torsion $\tau(s)$ are both constant (i.e. independent of $s$) for a circular helix. In the above example, $\tau > 0$ (assuming that $b > 0$). This corresponds to the fact that the helix is *right-handed.* (See Figure 7.9 in the previous section.) If you grasp the helix with your right hand so your fingers surround it in the direction of increasing $s$ (counterclockwise, looking down from the positive $z$-axis), then your thumb also points in the axial direction corresponding to increasing $s$ (the upward direction). Had we started with a left-handed helix, such as

$$\mathbf{r} = a\sin t\,\mathbf{i} + a\cos t\,\mathbf{j} + bt\mathbf{k}, \qquad (a, b > 0),$$

we would have obtained $\tau = -b/(a^2 + b^2)$.

Making use of the formulas $d\hat{\mathbf{T}}/ds = \kappa\hat{\mathbf{N}}$ and $d\hat{\mathbf{B}}/ds = -\tau\hat{\mathbf{N}}$ we can calculate $d\hat{\mathbf{N}}/ds$ as well:

$$\begin{aligned}\frac{d\hat{\mathbf{N}}}{ds} = \frac{d}{ds}(\hat{\mathbf{B}}\times\hat{\mathbf{T}}) &= \frac{d\hat{\mathbf{B}}}{ds}\times\hat{\mathbf{T}} + \hat{\mathbf{B}}\times\frac{d\hat{\mathbf{T}}}{ds} \\ &= -\tau\hat{\mathbf{N}}\times\hat{\mathbf{T}} + \kappa\hat{\mathbf{B}}\times\hat{\mathbf{N}} = -\kappa\hat{\mathbf{T}} + \tau\hat{\mathbf{B}}.\end{aligned}$$

Together, the three formulas

$$\frac{d\hat{\mathbf{T}}}{ds} = \kappa\hat{\mathbf{N}}$$
$$\frac{d\hat{\mathbf{N}}}{ds} = -\kappa\hat{\mathbf{T}} + \tau\hat{\mathbf{B}}$$
$$\frac{d\hat{\mathbf{B}}}{ds} = -\tau\hat{\mathbf{N}}$$

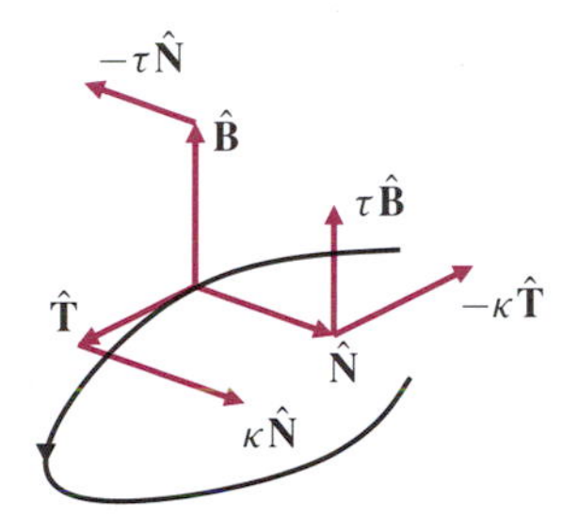

**Figure 7.15** $\hat{\mathbf{T}}$, $\hat{\mathbf{N}}$, and $\hat{\mathbf{B}}$, and their directions of change

are known as **The Frenet–Serret formulas**. They are of fundamental importance in the theory of curves in 3-space. The Frenet–Serret formulas can be written in matrix form as follows:

$$\frac{d}{ds}\begin{pmatrix}\hat{\mathbf{T}}\\ \hat{\mathbf{N}}\\ \hat{\mathbf{B}}\end{pmatrix} = \begin{pmatrix}0 & \kappa & 0\\ -\kappa & 0 & \tau\\ 0 & -\tau & 0\end{pmatrix}\begin{pmatrix}\hat{\mathbf{T}}\\ \hat{\mathbf{N}}\\ \hat{\mathbf{B}}\end{pmatrix}.$$

Using the Frenet–Serret formulas we can show that the shape of a curve with non-vanishing curvature is completely determined by the curvature and torsion functions $\kappa(s)$ and $\tau(s)$.

**THEOREM 3**

**The Fundamental Theorem of Space Curves**

Let $C_1$ and $C_2$ be two curves, both of which have the same non-vanishing curvature function $\kappa(s)$ and the same torsion function $\tau(s)$. Then the curves are congruent. That is, one can be moved rigidly (translated and rotated) so as to coincide exactly with the other.

***PROOF*** We require $\kappa \neq 0$ because $\hat{\mathbf{N}}$ and $\hat{\mathbf{B}}$ are not defined where $\kappa = 0$. Move $C_2$ rigidly so that its initial point coincides with the initial point of $C_1$ and so that the Frenet frames of both curves coincide at that point. Let $\hat{\mathbf{T}}_1$, $\hat{\mathbf{T}}_2$, $\hat{\mathbf{N}}_1$, $\hat{\mathbf{N}}_2$, $\hat{\mathbf{B}}_1$, and $\hat{\mathbf{B}}_2$ be the unit tangents, normals and binormals for the two curves. Let

$$f(s) = \hat{\mathbf{T}}_1(s)\bullet\hat{\mathbf{T}}_2(s) + \hat{\mathbf{N}}_1(s)\bullet\hat{\mathbf{N}}_2(s) + \hat{\mathbf{B}}_1(s)\bullet\hat{\mathbf{B}}_2(s).$$

We calculate the derivative of $f(s)$ using the Product Rule and the Frenet–Serret formulas:

$$\begin{aligned} f'(s) &= \hat{\mathbf{T}}_1' \bullet \hat{\mathbf{T}}_2 + \hat{\mathbf{T}}_1 \bullet \hat{\mathbf{T}}_2' + \hat{\mathbf{N}}_1' \bullet \hat{\mathbf{N}}_2 + \hat{\mathbf{N}}_1 \bullet \hat{\mathbf{N}}_2' + \hat{\mathbf{B}}_1' \bullet \hat{\mathbf{B}}_2 + \hat{\mathbf{B}}_1 \bullet \hat{\mathbf{B}}_2' \\ &= \kappa \hat{\mathbf{N}}_1 \bullet \hat{\mathbf{T}}_2 + \kappa \hat{\mathbf{T}}_1 \bullet \hat{\mathbf{N}}_2 - \kappa \hat{\mathbf{T}}_1 \bullet \hat{\mathbf{N}}_2 + \tau \hat{\mathbf{B}}_1 \bullet \hat{\mathbf{N}}_2 - \kappa \hat{\mathbf{N}}_1 \bullet \hat{\mathbf{T}}_2 \\ &\qquad + \tau \hat{\mathbf{N}}_1 \bullet \hat{\mathbf{B}}_2 - \tau \hat{\mathbf{N}}_1 \bullet \hat{\mathbf{B}}_2 - \tau \hat{\mathbf{B}}_1 \bullet \hat{\mathbf{N}}_2 \\ &= 0. \end{aligned}$$

Therefore $f(s)$ is constant. Since the frames coincide at $s = 0$, the constant must be 3:

$$\hat{\mathbf{T}}_1(s) \bullet \hat{\mathbf{T}}_2(s) + \hat{\mathbf{N}}_1(s) \bullet \hat{\mathbf{N}}_2(s) + \hat{\mathbf{B}}_1(s) \bullet \hat{\mathbf{B}}_2(s) = 3.$$

However, each dot product cannot exceed 1 since the factors are unit vectors. Therefore each dot product must be equal to 1. In particular, $\hat{\mathbf{T}}_1(s) \bullet \hat{\mathbf{T}}_2(s) = 1$ for all $s$, and hence

$$\frac{d\mathbf{r}_1}{ds} = \hat{\mathbf{T}}_1(s) = \hat{\mathbf{T}}_2(s) = \frac{d\mathbf{r}_2}{ds}.$$

Integrating with respect to $s$ and using the fact that both curves start from the same point when $s = 0$, we obtain

$$\mathbf{r}_1(s) = \mathbf{r}_2(s)$$

for all $s$, which is what we wanted to show.

◆

***REMARK*** It is a consequence of the above theorem that any curve having nonzero constant curvature and constant torsion must, in fact, be a circle (if the torsion is zero) or a circular helix (if the torsion is nonzero). See Exercises 7 and 8 below.

## EXERCISES 7.4

Find the unit tangent vector $\hat{\mathbf{T}}(t)$ for the curves in Exercises 1–4.

**1.** $\mathbf{r} = t\mathbf{i} - 2t^2\mathbf{j} + 3t^3\mathbf{k}$

**2.** $\mathbf{r} = a \sin \omega t\, \mathbf{i} + a \cos \omega t\, \mathbf{k}$

**3.** $\mathbf{r} = \cos t \sin t\, \mathbf{i} + \sin^2 t\, \mathbf{j} + \cos t\, \mathbf{k}$

**4.** $\mathbf{r} = a \cos t\, \mathbf{i} + b \sin t\, \mathbf{j} + t\mathbf{k}$

**5.** Show that if $\kappa(s) = 0$ for all $s$, then the curve $\mathbf{r} = \mathbf{r}(s)$ is a straight line.

**6.** Show that if $\tau(s) = 0$ for all $s$, then the curve $\mathbf{r} = \mathbf{r}(s)$ is a plane curve. (*Hint:* show that $\mathbf{r}(s)$ lies in the plane through $\mathbf{r}(0)$ with normal $\hat{\mathbf{B}}(0)$.)

**7.** Show that if $\kappa(s) = C$ is a positive constant, and $\tau(s) = 0$, for all $s$, then the curve $\mathbf{r} = \mathbf{r}(s)$ is a circle. (*Hint:* find a circle having the given constant curvature. Then use Theorem 3.)

**8.** Show that if the curvature $\kappa(s)$ and the torsion $\tau(s)$ are both nonzero constants, then the curve $\mathbf{r} = \mathbf{r}(s)$ is a circular helix. (*Hint:* find a helix having the given curvature and torsion.)

# 7.5 CURVATURE AND TORSION FOR GENERAL PARAMETRIZATIONS

The formulas developed above for curvature and torsion as well as for the unit normal and binormal vectors are not very useful if the curve we want to analyze is not expressed in terms of the arc length parameter. We will now consider how to find these quantities in terms of a general parametrization $\mathbf{r} = \mathbf{r}(t)$. We will express them all in terms of the velocity, $\mathbf{v}(t)$, the speed, $v(t) = |\mathbf{v}(t)|$, and the acceleration, $\mathbf{a}(t)$. First observe that

$$\begin{aligned}\mathbf{v} &= \frac{d\mathbf{r}}{dt} = \frac{d\mathbf{r}}{ds}\frac{ds}{dt} = v\hat{\mathbf{T}}\\ \mathbf{a} &= \frac{d\mathbf{v}}{dt} = \frac{dv}{dt}\hat{\mathbf{T}} + v\frac{d\hat{\mathbf{T}}}{dt}\\ &= \frac{dv}{dt}\hat{\mathbf{T}} + v\frac{d\hat{\mathbf{T}}}{ds}\frac{ds}{dt} = \frac{dv}{dt}\hat{\mathbf{T}} + v^2\kappa\hat{\mathbf{N}}\\ \mathbf{v}\times\mathbf{a} &= v\frac{dv}{dt}\hat{\mathbf{T}}\times\hat{\mathbf{T}} + v^3\kappa\hat{\mathbf{T}}\times\hat{\mathbf{N}} = v^3\kappa\hat{\mathbf{B}}\end{aligned}$$

Note that $\hat{\mathbf{B}}$ is in the direction of $\mathbf{v}\times\mathbf{a}$. From these formulas we obtain useful formulas for $\hat{\mathbf{T}}$, $\hat{\mathbf{B}}$, and $\kappa$:

$$\hat{\mathbf{T}} = \frac{\mathbf{v}}{v}, \qquad \hat{\mathbf{B}} = \frac{\mathbf{v}\times\mathbf{a}}{|\mathbf{v}\times\mathbf{a}|}, \qquad \kappa = \frac{|\mathbf{v}\times\mathbf{a}|}{v^3}.$$

We could calculate $\hat{\mathbf{N}}$ from the formula for $\mathbf{a}$ but it is easier to get it directly from $\hat{\mathbf{T}}$ and $\hat{\mathbf{B}}$:

$$\hat{\mathbf{N}} = \hat{\mathbf{B}}\times\hat{\mathbf{T}}.$$

The torsion remains to be calculated. Observe that

$$\frac{d\mathbf{a}}{dt} = \frac{d}{dt}\left(\frac{dv}{dt}\hat{\mathbf{T}} + v^2\kappa\hat{\mathbf{N}}\right).$$

This differentiation will produce several terms. The only one that involves $\hat{\mathbf{B}}$ is the one that comes from evaluating $v^2\kappa(d\hat{\mathbf{N}}/dt) = v^3\kappa(d\hat{\mathbf{N}}/ds) = v^3\kappa(\tau\hat{\mathbf{B}} - \kappa\hat{\mathbf{T}})$. Therefore

$$\frac{d\mathbf{a}}{dt} = \lambda\hat{\mathbf{T}} + \mu\hat{\mathbf{N}} + v^3\kappa\tau\hat{\mathbf{B}},$$

for certain scalars $\lambda$ and $\mu$. Since $\mathbf{v}\times\mathbf{a} = v^3\kappa\hat{\mathbf{B}}$, it follows that

$$(\mathbf{v}\times\mathbf{a}) \bullet \frac{d\mathbf{a}}{dt} = (v^3\kappa)^2\tau = |\mathbf{v}\times\mathbf{a}|^2\tau.$$

Hence

$$\tau = \frac{(\mathbf{v}\times\mathbf{a}) \bullet (d\mathbf{a}/dt)}{|\mathbf{v}\times\mathbf{a}|^2}.$$

**EXAMPLE 1** Find the curvature, the torsion, and the Frenet frame at a general point on the curve

$$\mathbf{r} = (t + \cos t)\mathbf{i} + (t - \cos t)\mathbf{j} + \sqrt{2}\sin t\mathbf{k}.$$

Describe this curve.

***SOLUTION*** We calculate the various quantities using the recipe given above. First the preliminaries:

$$\begin{aligned}
\mathbf{v} &= (1 - \sin t)\mathbf{i} + (1 + \sin t)\mathbf{j} + \sqrt{2}\cos t\mathbf{k} \\
\mathbf{a} &= -\cos t\mathbf{i} + \cos t\mathbf{j} - \sqrt{2}\sin t\mathbf{k} \\
\frac{d\mathbf{a}}{dt} &= \sin t\mathbf{i} - \sin t\mathbf{j} - \sqrt{2}\cos t\mathbf{k} \\
\mathbf{v}\times\mathbf{a} &= \begin{vmatrix} \mathbf{i} & \mathbf{j} & \mathbf{k} \\ 1 - \sin t & 1 + \sin t & \sqrt{2}\cos t \\ -\cos t & \cos t & -\sqrt{2}\sin t \end{vmatrix} \\
&= -\sqrt{2}(1 + \sin t)\mathbf{i} - \sqrt{2}(1 - \sin t)\mathbf{j} + 2\cos t\mathbf{k} \\
(\mathbf{v}\times\mathbf{a}) \bullet \frac{d\mathbf{a}}{dt} &= -\sqrt{2}\sin t(1 + \sin t) + \sqrt{2}\sin t(1 - \sin t) - 2\sqrt{2}\cos^2 t \\
&= -2\sqrt{2} \\
v = |\mathbf{v}| &= \sqrt{2 + 2\sin^2 t + 2\cos^2 t} = 2 \\
|\mathbf{v}\times\mathbf{a}| &= \sqrt{2(2 + 2\sin^2 t) + 4\cos^2 t} = \sqrt{8} = 2\sqrt{2}.
\end{aligned}$$

Thus we have

$$\begin{aligned}
\kappa &= \frac{|\mathbf{v}\times\mathbf{a}|}{v^3} = \frac{2\sqrt{2}}{8} = \frac{1}{2\sqrt{2}} \\
\tau &= \frac{(\mathbf{v}\times\mathbf{a}) \bullet (d\mathbf{a}/dt)}{|\mathbf{v}\times\mathbf{a}|^2} = \frac{-2\sqrt{2}}{(2\sqrt{2})^2} = -\frac{1}{2\sqrt{2}} \\
\hat{\mathbf{T}} &= \frac{\mathbf{v}}{v} = \frac{1 - \sin t}{2}\mathbf{i} + \frac{1 + \sin t}{2}\mathbf{j} + \frac{1}{\sqrt{2}}\cos t\mathbf{k} \\
\hat{\mathbf{B}} &= \frac{\mathbf{v}\times\mathbf{a}}{|\mathbf{v}\times\mathbf{a}|} = -\frac{1 + \sin t}{2}\mathbf{i} - \frac{1 - \sin t}{2}\mathbf{j} + \frac{1}{\sqrt{2}}\cos t\mathbf{k} \\
\hat{\mathbf{N}} &= \hat{\mathbf{B}}\times\hat{\mathbf{T}} = -\frac{1}{\sqrt{2}}\cos t\mathbf{i} + \frac{1}{\sqrt{2}}\cos t\mathbf{j} - \sin t\mathbf{k}.
\end{aligned}$$

Since the curvature and torsion are both constant (they are therefore constant when expressed in terms of any parametrization) the curve must be a circular helix by Theorem 3. It is left-handed, since $\tau < 0$. By Example 3 in Section 7.4, it is congruent to the helix

$$\mathbf{r} = a\cos t\mathbf{i} + a\sin t\mathbf{j} + bt\mathbf{k},$$

provided that $a/(a^2 + b^2) = 1/(2\sqrt{2}) = -b/(a^2 + b^2)$. Solving these equations gives $a = \sqrt{2}$ and $b = -\sqrt{2}$, so the helix is wound on a cylinder of radius $\sqrt{2}$. The axis of this cylinder is the line $x = y$, $z = 0$, as can be seen by inspecting the components of $\mathbf{r}(t)$. ■

**EXAMPLE 2** **Curvature of the graph of a function of one variable**

Find the curvature of the plane curve with equation $y = f(x)$ at an arbitrary point $(x, f(x))$ on the curve.

***SOLUTION*** The graph can be parametrized $\mathbf{r} = x\mathbf{i} + f(x)\mathbf{j}$. Thus

$$\begin{aligned} \mathbf{v} &= \mathbf{i} + f'(x)\mathbf{j}, \\ \mathbf{a} &= f''(x)\mathbf{j} \\ \mathbf{v}\times\mathbf{a} &= f''(x)\mathbf{k}. \end{aligned}$$

Therefore the curvature is

$$\kappa(x) = \frac{|f''(x)|}{\left(1 + (f'(x))^2\right)^{3/2}}.$$

## Tangential and Normal Acceleration

In the formula obtained above for the acceleration in terms of the unit tangent and normal,

$$\mathbf{a} = \frac{dv}{dt}\hat{\mathbf{T}} + v^2\kappa\hat{\mathbf{N}},$$

the term $(dv/dt)\hat{\mathbf{T}}$ is called the **tangential** acceleration and the term $v^2\kappa\hat{\mathbf{N}}$ is called the **normal** or **centripetal** acceleration. This latter component is directed towards the centre of curvature and its magnitude is $v^2\kappa = v^2/\rho$. Highway and railway designers attempt to bank curves in such a way that the resultant of the corresponding "centrifugal force," $-m(v^2/\rho)\hat{\mathbf{N}}$, and the weight, $-mg\mathbf{k}$, of the vehicle will be normal to the surface at a desired speed. (See Exercise 18 below.)

**EXAMPLE 3** A bead of mass $m$ slides without friction down a wire bent in the shape of the curve $y = x^2$, under the influence of the gravitational force $-mg\mathbf{j}$. The speed of the bead is $v$ as it passes through the point $(1, 1)$. Find, at that instant, the magnitude of the normal acceleration of the bead and the rate of change of its speed.

***SOLUTION*** By Example 2, the curvature of $y = x^2$ at $(1, 1)$ is

$$\kappa = \left.\frac{2}{(1 + 4x^2)^{3/2}}\right|_{x=1} = \frac{2}{5\sqrt{5}}.$$

Thus the magnitude of the normal acceleration of the bead at that point is $v^2\kappa = 2v^2/(5\sqrt{5})$.

The rate of change of the speed, $dv/dt$, is the tangential component of the acceleration, and is due entirely to the tangential component of the gravitational force since there is no friction:

$$\frac{dv}{dt} = g\cos\theta = g(-\mathbf{j}) \bullet \hat{\mathbf{T}},$$

where $\theta$ is the angle between $\hat{\mathbf{T}}$ and $-\mathbf{j}$. (See Figure 7.16.) Since the slope of $y = x^2$ at $(1, 1)$ is 2, we have $\hat{\mathbf{T}} = -(\mathbf{i} + 2\mathbf{j})/\sqrt{5}$, and therefore $dv/dt = 2g/\sqrt{5}$.

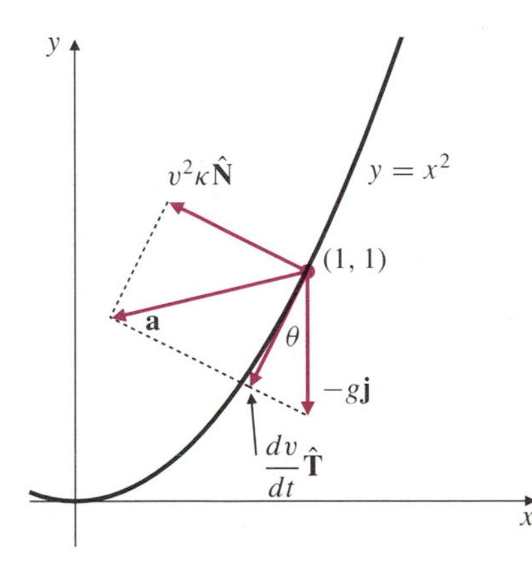


**Figure 7.16** The acceleration of a sliding bead

***REMARK*** The definition of centripetal acceleration given above is consistent with the one that arose in the discussion of rotating frames in Section 7.2. If $\mathbf{r}(t)$ is the position of a moving particle at time $t$ we can regard the motion at any instant as being a rotation about the centre of curvature, so that the angular velocity must be $\boldsymbol{\Omega} = \Omega\hat{\mathbf{B}}$. The linear velocity is $\mathbf{v} = \boldsymbol{\Omega}\times(\mathbf{r} - \mathbf{r}_c) = v\hat{\mathbf{T}}$, so the speed is $v = \Omega\rho$, and $\boldsymbol{\Omega} = (v/\rho)\hat{\mathbf{B}}$. As developed in Section 7.2, the centripetal acceleration is

$$\boldsymbol{\Omega}\times(\boldsymbol{\Omega}\times(\mathbf{r} - \mathbf{r}_c)) = \boldsymbol{\Omega}\times\mathbf{v} = \frac{v^2}{\rho}\hat{\mathbf{B}}\times\hat{\mathbf{T}} = \frac{v^2}{\rho}\hat{\mathbf{N}}.$$

## Evolutes

The centre of curvature $\mathbf{r}_c(t)$ of a given curve can itself trace out another curve as $t$ varies. This curve is called the **evolute** of the given curve $\mathbf{r}(t)$.

**EXAMPLE 4** Find the evolute of the exponential spiral

$$\mathbf{r} = ae^{-t}\cos t\,\mathbf{i} + ae^{-t}\sin t\,\mathbf{j}.$$

***SOLUTION*** The curve is a plane curve so $\tau = 0$. We will take a shortcut to the curvature and the unit normal without calculating $\mathbf{v}\times\mathbf{a}$. First we calculate

$$\begin{aligned}
\mathbf{v} &= ae^{-t}\Big(-(\cos t + \sin t)\mathbf{i} - (\sin t - \cos t)\mathbf{j}\Big)\\
\frac{ds}{dt} &= v = \sqrt{2}ae^{-t}\\
\hat{\mathbf{T}}(t) &= \frac{1}{\sqrt{2}}\Big(-(\cos t + \sin t)\mathbf{i} - (\sin t - \cos t)\mathbf{j}\Big)\\
\frac{d\hat{\mathbf{T}}}{ds} &= \frac{1}{(ds/dt)}\frac{d\hat{\mathbf{T}}}{dt} = \frac{1}{2ae^{-t}}\Big((\sin t - \cos t)\mathbf{i} - (\cos t + \sin t)\mathbf{j}\Big)\\
\kappa(t) &= \left|\frac{d\hat{\mathbf{T}}}{ds}\right| = \frac{1}{\sqrt{2}ae^{-t}}.
\end{aligned}$$

It follows that the radius of curvature is $\rho(t) = \sqrt{2}ae^{-t}$. Since $d\hat{\mathbf{T}}/ds = \kappa\hat{\mathbf{N}}$, therefore $\hat{\mathbf{N}} = \rho(d\hat{\mathbf{T}}/ds)$. The centre of curvature is

$$\begin{aligned}
\mathbf{r}_c(t) &= \mathbf{r}(t) + \rho(t)\hat{\mathbf{N}}(t)\\
&= \mathbf{r}(t) + \rho^2\frac{d\hat{\mathbf{T}}}{ds}\\
&= ae^{-t}\Big(\cos t\,\mathbf{i} + \sin t\,\mathbf{j}\Big)\\
&\quad + 2a^2e^{-2t}\frac{1}{2ae^{-t}}\Big((\sin t - \cos t)\mathbf{i} - (\cos t + \sin t)\mathbf{j}\Big)\\
&= ae^{-t}\Big(\sin t\,\mathbf{i} - \cos t\,\mathbf{j}\Big)\\
&= ae^{-t}\Big(\cos(t - \frac{\pi}{2})\mathbf{i} + \sin(t - \frac{\pi}{2})\mathbf{j}\Big).
\end{aligned}$$

Thus, interestingly, the evolute of the exponential spiral is the same exponential spiral rotated 90° clockwise in the plane. (See Figure 7.17(a).) ■

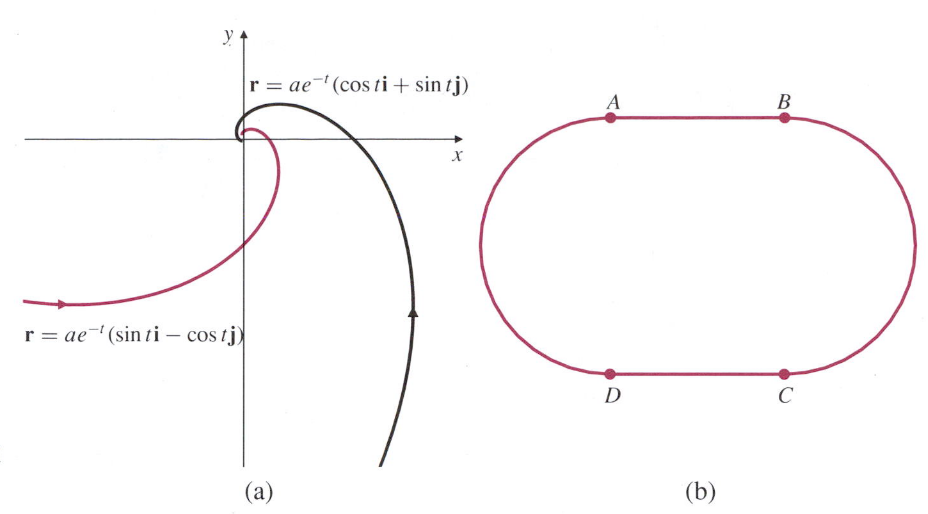


**Figure 7.17**

(a) The evolute of an exponential spiral is another exponential spiral

(b) The shape of a model train track

## An Application to Track Design

Model trains frequently come with two kinds of track sections, straight and curved. The curved sections are arcs of a circle of radius $R$, and the track is intended to be laid out in the shape shown in Figure 7.17(b); $AB$ and $CD$ are straight, and $BC$ and $DA$ are semicircles. The track looks smooth, but is it smooth enough?

The track is held together by friction and occasionally it can come apart as the train is racing around. It is especially likely to come apart at the points $A$, $B$, $C$, and $D$. To see why, assume that the train is travelling at constant speed $v$. Then the tangential acceleration, $(dv/dt)\hat{\mathbf{T}}$, is zero and the total acceleration is just the centripetal acceleration, $\mathbf{a} = (v^2/\rho)\hat{\mathbf{N}}$. Therefore, $|\mathbf{a}| = 0$ along the straight sections, and $|\mathbf{a}| = v^2\kappa = v^2/R$ on the semicircular sections. The acceleration is *discontinuous* at the points $A$, $B$, $C$, and $D$, and the reactive force exerted by the train on the track is also discontinuous at these points. There is a "shock" or "jolt" as the train enters or leaves a curved part of the track. In order to avoid such stress points, tracks should be designed so that the curvature varies continuously from point to point.

**EXAMPLE 5** Existing track along the negative $x$-axis, and along the ray $y = x - 1$, $x \geq 2$, is to be joined smoothly by track along the transition curve $y = f(x)$, $0 \leq x \leq 2$, where $f(x)$ is a polynomial of degree as small as possible. Find $f(x)$ so that a train moving along the track will not experience discontinuous acceleration at the joins.

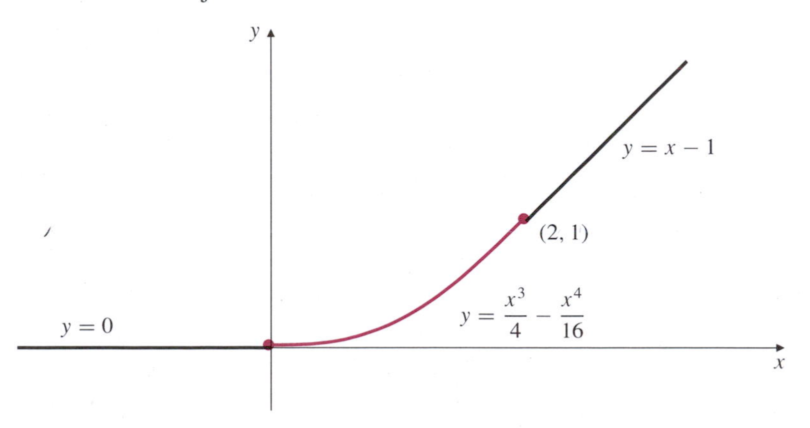


**Figure 7.18** Joining two straight tracks with a curved track

**SOLUTION** The situation is shown in Figure 7.18. The polynomial $f(x)$ must be chosen so that the track is continuous, has continuous slope, and has continuous curvature at $x = 0$ and $x = 1$. Since the curvature of $y = f(x)$ is

$$\kappa = f''(x)\Big(1 + (f'(x))^2\Big)^{-3/2},$$

we need only arrange that $f$, $f'$, and $f''$ take the same values at $x = 0$ and $x = 2$ that the straight sections do there:

$$\begin{aligned} f(0) &= 0, & f'(0) &= 0, & f''(0) &= 0, \\ f(2) &= 1, & f'(2) &= 1, & f''(2) &= 0. \end{aligned}$$

These six independent conditions suggest we should try a polynomial of degree five involving six arbitrary coefficients:

$$\begin{aligned} f(x) &= A + Bx + Cx^2 + Dx^3 + Ex^4 + Fx^5 \\ f'(x) &= B + 2Cx + 3Dx^2 + 4Ex^3 + 5Fx^4 \\ f''(x) &= 2C + 6Dx + 12Ex^2 + 20Fx^3. \end{aligned}$$

The three conditions at $x = 0$ imply that $A = B = C = 0$. Those at $x = 2$ imply that

$$\begin{aligned} 8D + 16E + \phantom{1}32F &= f(2) = 1 \\ 12D + 32E + \phantom{1}80F &= f'(2) = 1 \\ 12D + 48E + 160F &= f''(2) = 0. \end{aligned}$$

This system has solution $D = 1/4$, $E = -1/16$, and $F = 0$, so we should use $f(x) = (x^3/4) - (x^4/16)$. ■

**REMARK** Road and railroad builders do not usually use polynomial graphs as transition curves. Other kinds of curves called **clothoids** and **lemniscates** are usually used. (See Exercise 7 in the Review Exercises at the end of this chapter.)

## EXERCISES 7.5

Find the radius of curvature of the curves in Exercises 1–4 at the points indicated.

1. $y = x^2$ at $x = 0$ and at $x = \sqrt{2}$
2. $y = \cos x$ at $x = 0$ and at $x = \pi/2$
3. $\mathbf{r} = 2t\mathbf{i} + (1/t)\mathbf{j} - 2t\mathbf{k}$ at $(2, 1, -2)$
4. $\mathbf{r} = t^3\mathbf{i} + t^2\mathbf{j} + t\mathbf{k}$ at the point where $t = 1$

Find the Frenet frames $\{\hat{\mathbf{T}}, \hat{\mathbf{N}}, \hat{\mathbf{B}}\}$ for the curves in Exercises 5–8 at the points indicated.

5. $\mathbf{r} = t\mathbf{i} + t^2\mathbf{j} + 2\mathbf{k}$ at $(1, 1, 2)$
6. $\mathbf{r} = t\mathbf{i} - t^2\mathbf{j}$ at $(1, -1, 0)$
7. $\mathbf{r} = t^2\mathbf{i} + t^3\mathbf{j} + t\mathbf{k}$ at $(0, 0, 0)$
8. $\mathbf{r} = t\mathbf{i} + t^2\mathbf{j} + t\mathbf{k}$ at $(1, 1, 1)$

In Exercises 9–12, find the unit tangent, normal, and binormal vectors, and the curvature and torsion at a general point on the given curve.

9. $\mathbf{r} = t\mathbf{i} + \dfrac{t^2}{2}\mathbf{j} + \dfrac{t^3}{3}\mathbf{k}$
10. $\mathbf{r} = t\mathbf{i} + 2t\mathbf{j} + \sin t\,\mathbf{k}$
11. $\mathbf{r} = t(\cos t\,\mathbf{i} + \sin t\,\mathbf{j} + \mathbf{k})$
12. $\mathbf{r} = e^t(\cos t\,\mathbf{i} + \sin t\,\mathbf{j} + \mathbf{k})$
13. Find the curvature and torsion of the parametric curve

$$\begin{cases} x = 2 + \sqrt{2}\cos t \\ y = 1 - \sin t \\ z = 3 + \sin t \end{cases}$$

at an arbitrary point $t$. What is the curve?

**14.** A particle moves along the plane curve $y = \sin x$ in the direction of increasing $x$ with constant horizontal speed $dx/dt = k$. Find the tangential and normal components of the acceleration of the particle when it is at position $x$.

**15.** Find the unit tangent, normal and binormal, and the curvature and torsion for the curve

$$\mathbf{r} = \sin t \cos t\,\mathbf{i} + \sin^2 t\,\mathbf{j} + \cos t\,\mathbf{k}$$

at the points (i) $t = 0$ and (ii) $t = \pi/4$.

**16.** A particle moves on an elliptical path in the $xy$-plane so that its position at time $t$ is $\mathbf{r} = a\cos t\mathbf{i} + b\sin t\mathbf{j}$. Find the tangential and normal components of its acceleration at time $t$. At what points is the tangential acceleration zero?

**17.** Find the maximum and minimum values for the curvature of the ellipse $x = a\cos t$, $y = b\sin t$, where $a > b > 0$.

**18.** A level, curved road lies along the curve $y = x^2$ in the horizontal $xy$-plane. Find the normal acceleration at position $(x, x^2)$ of a vehicle travelling along the road with constant speed $v_0$. Find, as a function of $x$, the angle at which the road should be banked (i.e. the angle between the vertical and the normal to the surface of the road) so that the result of the centrifugal and gravitational ($-mg\mathbf{k}$) forces acting on the vehicle is always normal to the surface of the road.

**19.** Find the curvature of the plane curve $y = e^x$ at $x$. Find the equation of the evolute of this curve.

**20.** Show that the curvature of the plane polar graph $r = f(\theta)$ at a general point $\theta$ is

$$\kappa(\theta) = \frac{|2\big(f'(\theta)\big)^2 + \big(f(\theta)\big)^2 - f(\theta)f''(\theta)|}{\big[\big(f'(\theta)\big)^2 + \big(f(\theta)\big)^2\big]}.$$

**21.** Find the curvature of the cardioid $r = a(1 - \cos\theta)$.

**22.** Find the curve $\mathbf{r} = \mathbf{r}(t)$ for which $\kappa(t) = 1$ and $\tau(t) = 1$ for all $t$, and $\mathbf{r}(0) = \hat{\mathbf{T}}(0) = \mathbf{i}$, $\hat{\mathbf{N}}(0) = \mathbf{j}$, and $\hat{\mathbf{B}}(0) = \mathbf{k}$.

**23.** Suppose the curve $\mathbf{r} = \mathbf{r}(t)$ satisfies

$$\frac{d\mathbf{r}}{dt} = \mathbf{c}\times\mathbf{r}(t),$$

where $\mathbf{c}$ is a constant vector. Use curvature and torsion to show that the curve is the circle in which the plane through $\mathbf{r}(0)$ normal to $\mathbf{c}$ intersects a sphere with radius $|\mathbf{r}(0)|$ centred at the origin.

**24.** Find the evolute of the circular helix $\mathbf{r} = a\cos t\,\mathbf{i} + a\sin t\,\mathbf{j} + bt\mathbf{k}$.

**25.** Find the evolute of the parabola $y = x^2$.

**26.** Find the evolute of the ellipse $x = 2\cos t$, $y = \sin t$.

**27.** Find the polynomial $f(x)$ of lowest degree so that track along $y = f(x)$ from $x = -1$ to $x = 1$ joins with existing straight tracks $y = -1$, $x \le -1$ and $y = 1$, $x \ge 1$ sufficiently smoothly that a train moving at constant speed will not experience discontinuous acceleration at the joins.

* **28.** Help out model train manufacturers. Design a track segment $y = f(x)$, $-1 \le x \le 0$, to provide a jolt-free link between a straight track section $y = 1$, $x \le -1$, and a semicircular arc section $x^2 + y^2 = 1$, $x \ge 0$.

* **29.** If the position $\mathbf{r}$, velocity $\mathbf{v}$, and acceleration $\mathbf{a}$ of a moving particle satisfy $\mathbf{a}(t) = \lambda(t)\mathbf{r}(t) + \mu(t)\mathbf{v}(t)$, where $\lambda(t)$ and $\mu(t)$ are scalar functions of time $t$, and if $\mathbf{v}\times\mathbf{a} \ne \mathbf{0}$, show that the path of the particle lies in a plane.

## 7.6 KEPLER'S LAWS OF PLANETARY MOTION

The German mathematician and astronomer, Johannes Kepler (1571–1630), was a student and colleague of Danish astronomer Tycho Brahe (1546–1601). Over a lifetime of observing the positions of planets without the aid of a telescope, Brahe compiled a vast amount of data, which Kepler analyzed. Although Polish astronomer Nicolaus Copernicus (1473–1543) had postulated that the earth and other planets moved around the sun, the religious and philosophical climate in Europe at the end of the sixteenth century still favoured explaining the motion of heavenly bodies in terms of circular orbits around the earth. It was known that planets such as Mars could not move on circular orbits centred at the earth, but models were proposed in which they moved on other circles (epicycles) whose centres moved on circles centred at the earth.

Brahe's observations of Mars were sufficiently detailed that Kepler realized that no simple model based on circles could be made to conform very closely with the actual orbit. He was, however, able to fit a more general ellipse with one focus at the sun, and, based on this success and on Brahe's data on other planets, he formulated the following three laws of planetary motion.

**Kepler's Laws**

1. The planets move on elliptical orbits with the sun at one focus.
2. The radial line from the sun to a planet sweeps out equal areas in equal times.
3. The squares of the periods of revolution of the planets around the sun are proportional to the cubes of the major axes of their orbits.

Kepler's statement of the third law actually says that the squares of the periods of revolution of the planets are proportional to the cubes of their mean distances from the sun. The mean distance of points on an ellipse from a focus of the ellipse is equal to the semi-major axis. (See Exercise 17 at the end of this section.) Therefore the two statements are equivalent.

The choice of ellipses was reasonable once it became clear that circles would not work. The properties of the conic sections were well understood, having been developed by the Greek mathematician Apollonius of Perga around 200 BC. Nevertheless, based, as it was, on observations rather than theory, Kepler's formulation of his laws without any causal explanation was a truly remarkable feat. The theoretical underpinnings came later when Newton, with the aid of his newly created calculus, showed that Kepler's laws implied an inverse square gravitational force. (See Review Exercises 14–16 at the end of this chapter.) Newton believed that his universal gravitational law also implied Kepler's laws, but his writings fail to provide a proof that is convincing by todays standards.[1]

Later in this section we will derive Kepler's laws from the gravitational law by an elegant method that exploits vector differentiation to the fullest. First, however, we need to attend to some preliminaries.

## Ellipses in Polar Coordinates

The polar coordinates $[r, \theta]$ of a point in the plane whose distance from the origin is $\varepsilon$ times its distance from the line $x = p$ (see Figure 7.19) satisfy $r = \varepsilon(p - r\cos\theta)$, or, solving for $r$,

$$r = \frac{\ell}{1 + \varepsilon\cos\theta},$$

where $\ell = \varepsilon p$. For $0 \le \varepsilon < 1$ this equation represents an ellipse having **eccentricity** $\varepsilon$. (It is a circle if $\varepsilon = 0$.) To see this, let us transform the equation to Cartesian coordinates:

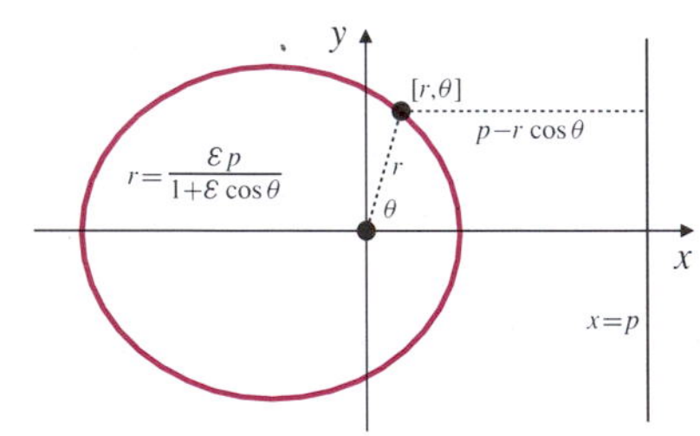

**Figure 7.19** An ellipse with focus at the origin, directrix $x = p$, and eccentricity $\varepsilon$

$$x^2 + y^2 = r^2 = \varepsilon^2(p - r\cos\theta)^2 = \varepsilon^2(p - x)^2 = \varepsilon^2(p^2 - 2px + x^2).$$

With some algebraic manipulation this equation can be juggled into the form

$$\frac{\left(x + \dfrac{\varepsilon\ell}{1-\varepsilon^2}\right)^2}{\left(\dfrac{\ell}{1-\varepsilon^2}\right)^2} + \frac{y^2}{\left(\dfrac{\ell}{\sqrt{1-\varepsilon^2}}\right)^2} = 1,$$

[1] There are interesting articles debating the historical significance of Newton's work by Robert Weinstock, Curtis Wilson, and others in *The College Mathematics Journal,* volume 25 #3, 1994.

which can be recognized as an ellipse with **centre** at the point $C = (-c, 0)$, where $c = \varepsilon\ell/(1-\varepsilon^2)$, and semi-axes $a$ and $b$ given by

$$a = \frac{\ell}{1-\varepsilon^2}, \qquad \textbf{(semi-major axis)},$$
$$b = \frac{\ell}{\sqrt{1-\varepsilon^2}}, \qquad \textbf{(semi-minor axis)}.$$

(See Section 9.1 of *Single-Variable Calculus* for a discussion of ellipses and other conic sections in Cartesian coordinates.)

The Cartesian equation of the ellipse shows that the curve is symmetric about the lines $x = -c$ and $y = 0$, and so has a second focus at $F = (-2c, 0)$ and a second directrix with equation $x = -2c - p$. (See Figure 7.20.) The ends of the major axis are $A = (a - c, 0)$ and $A' = (-a - c, 0)$, and the ends of the minor axis are $B = (-c, b)$ and $B' = (-c, -b)$.

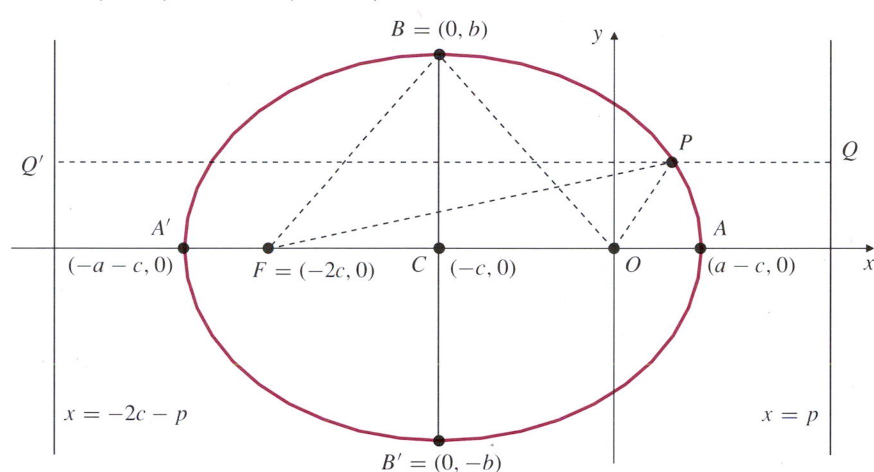


**Figure 7.20** The sum of the distances from any point $P$ on the ellipse to the two foci $O$ and $F$ is constant, $\varepsilon$ times the distance between the directrices

If $P$ is any point on the ellipse then the distance $OP$ is $\varepsilon$ times the distance $PQ$ from $P$ to the right directrix. Similarly, the distance $FQ$ is $\varepsilon$ times the distance $Q'P$ from $P$ to the left directrix. Thus the sum of the focal radii $OP + FP$ is the constant $\varepsilon Q'Q = \varepsilon(2c + 2p)$, regardless of where $P$ is on the ellipse. Taking for $P$ the two points $A$ and $B$ we get for this sum,

$$2a = (a - c) + (a + c) = OA + FA = OB + FB = 2\sqrt{b^2 + c^2}.$$

It follows that

$$a^2 = b^2 + c^2, \qquad c = \sqrt{a^2 - b^2} = \frac{\ell\varepsilon}{1-\varepsilon^2} = \varepsilon a.$$

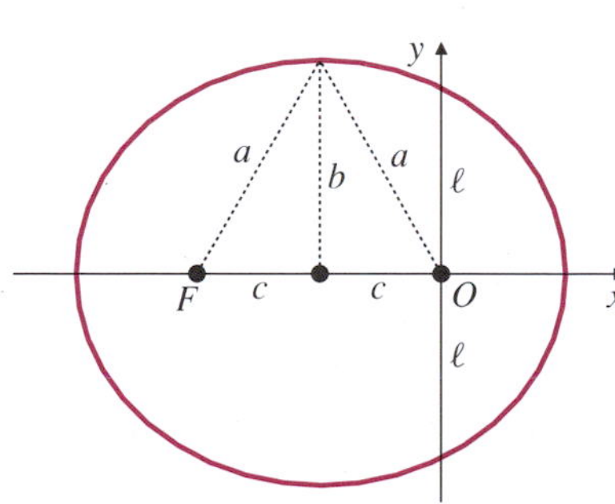


**Figure 7.21** Some parameters of an ellipse

The number $\ell$ is called the **semi-latus rectum** of the ellipse; the latus rectum is the width measured along the line through a focus, perpendicular to the major axis. (See Figure 7.21.)

**REMARK** The polar equation $r = \ell/(1 + \varepsilon\cos\theta)$ represents a *bounded* curve only if $\varepsilon < 1$: in this case we have $\ell/(1+\varepsilon) \le r \le \ell/(1-\varepsilon)$ for all directions $\theta$. If $\varepsilon = 1$ the equation represents a parabola, and if $\varepsilon > 1$ a hyperbola. It is possible for objects to travel on parabolic or hyperbolic orbits, but they will approach the sun only once, rather than continue to loop around it. Some comets have hyperbolic orbits.

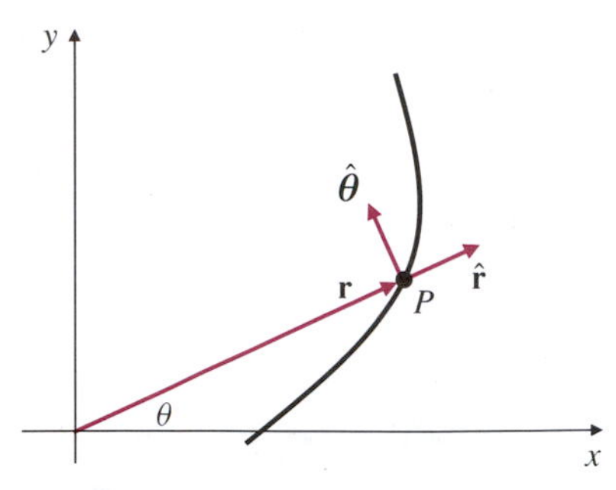


**Figure 7.22** Basis vectors in the direction of increasing $r$ and $\theta$

## Polar Components of Velocity and Acceleration

Let $\mathbf{r}(t)$ be the position vector at time $t$ of a particle $P$ moving in the $xy$-plane. We construct two unit vectors at $P$, the vector $\hat{\mathbf{r}}$ points in the direction of the position vector $\mathbf{r}$, and the vector $\hat{\boldsymbol{\theta}}$ is rotated 90° counterclockwise from $\hat{\mathbf{r}}$. (See Figure 7.22.) If $P$ has polar coordinates $[r, \theta]$ then $\hat{\mathbf{r}}$ points in the direction of increasing $r$ at $P$, and $\hat{\boldsymbol{\theta}}$ points in the direction of increasing $\theta$. Evidently

$$\hat{\mathbf{r}} = \cos\theta\,\mathbf{i} + \sin\theta\,\mathbf{j}$$
$$\hat{\boldsymbol{\theta}} = -\sin\theta\,\mathbf{i} + \cos\theta\,\mathbf{j}.$$

Note that $\hat{\mathbf{r}}$ and $\hat{\boldsymbol{\theta}}$ do not depend on $r$ but only on $\theta$:

$$\frac{d\hat{\mathbf{r}}}{d\theta} = \hat{\boldsymbol{\theta}} \qquad \text{and} \qquad \frac{d\hat{\boldsymbol{\theta}}}{d\theta} = -\hat{\mathbf{r}}.$$

The pair $\{\hat{\mathbf{r}}, \hat{\boldsymbol{\theta}}\}$ form a reference frame (a basis) at $P$ so that vectors in the plane can be expressed in terms of these two unit vectors. The $\hat{\mathbf{r}}$ component of a vector is called the **radial component**, and the $\hat{\boldsymbol{\theta}}$ component is called the **transverse component**. The frame varies from point to point, so we must remember that $\hat{\mathbf{r}}$ and $\hat{\boldsymbol{\theta}}$ are both functions of $t$. In terms of this moving frame, the position $\mathbf{r}(t)$ of $P$ can be expressed very simply:

$$\mathbf{r} = r\hat{\mathbf{r}},$$

where $r = r(t) = |\mathbf{r}(t)|$ is the distance from $P$ to the origin at time $t$.

We are going to differentiate this equation with respect to $t$ in order to express the velocity and acceleration of $P$ in terms of the moving frame. Along the path of motion, $\mathbf{r}$ can be regarded as a function of either $\theta$ or $t$; $\theta$ is itself a function of $t$. So as to avoid confusion, let us adopt a notation that is used extensively in mechanics and that resembles the notation originally used by Newton in his calculus.

A dot over a quantity denotes the time derivative of that quantity. Two dots denotes the second derivative with respect to time. Thus:

$$\dot{u} = du/dt \qquad \text{and} \qquad \ddot{u} = d^2u/dt^2.$$

First let us record the time derivatives of the vectors $\hat{\mathbf{r}}$ and $\hat{\boldsymbol{\theta}}$. By the Chain Rule, we have

$$\dot{\hat{\mathbf{r}}} = \frac{d\hat{\mathbf{r}}}{d\theta}\frac{d\theta}{dt} = \dot{\theta}\hat{\boldsymbol{\theta}}$$
$$\dot{\hat{\boldsymbol{\theta}}} = \frac{d\hat{\boldsymbol{\theta}}}{d\theta}\frac{d\theta}{dt} = -\dot{\theta}\hat{\mathbf{r}}.$$

Now the velocity of $P$ is

$$\mathbf{v} = \dot{\mathbf{r}} = \frac{d}{dt}(r\hat{\mathbf{r}}) = \dot{r}\hat{\mathbf{r}} + r\dot{\theta}\hat{\boldsymbol{\theta}}.$$

**Polar components of velocity**

The **radial component of velocity** is $\dot{r}$.
The **transverse component of velocity** is $r\dot{\theta}$.

Since $\hat{\mathbf{r}}$ and $\hat{\boldsymbol{\theta}}$ are perpendicular unit vectors, the speed of $P$ is given by

$$v = |\mathbf{v}| = \sqrt{\dot{r}^2 + r^2\dot{\theta}^2}.$$

Similarly, the acceleration of $P$ can be expressed in terms of radial and transverse components:

$$\begin{aligned}\mathbf{a} = \dot{\mathbf{v}} = \ddot{\mathbf{r}} &= \frac{d}{dt}(\dot{r}\hat{\mathbf{r}} + r\dot{\theta}\hat{\boldsymbol{\theta}})\\ &= \ddot{r}\hat{\mathbf{r}} + \dot{r}\dot{\theta}\hat{\boldsymbol{\theta}} + \dot{r}\dot{\theta}\hat{\boldsymbol{\theta}} + r\ddot{\theta}\hat{\boldsymbol{\theta}} - r\dot{\theta}^2\hat{\mathbf{r}}\\ &= (\ddot{r} - r\dot{\theta}^2)\hat{\mathbf{r}} + (r\ddot{\theta} + 2\dot{r}\dot{\theta})\hat{\boldsymbol{\theta}}.\end{aligned}$$

**Polar components of acceleration**

The **radial component of acceleration** is $\ddot{r} - r\dot{\theta}^2$.
The **transverse component of acceleration** is $r\ddot{\theta} + 2\dot{r}\dot{\theta}$.

## Central Forces and Kepler's Second Law

Polar coordinates are most appropriate for analyzing motion due to a **central force** that is always directed towards (or away from) a single point, the origin: $\mathbf{F} = \lambda\mathbf{r}$, where the scalar $\lambda$ depends on the position $\mathbf{r}$ of the object. If the velocity and acceleration of the object are $\mathbf{v} = \dot{\mathbf{r}}$ and $\mathbf{a} = \dot{\mathbf{v}}$, then Newton's second law of motion ($\mathbf{F} = m\mathbf{a}$) says that $\mathbf{a}$ is parallel to $\mathbf{r}$. Therefore

$$\frac{d}{dt}(\mathbf{r}\times\mathbf{v}) = \dot{\mathbf{r}}\times\mathbf{v} + \mathbf{r}\times\dot{\mathbf{v}} = \mathbf{v}\times\mathbf{v} + \mathbf{r}\times\mathbf{a} = \mathbf{0} + \mathbf{0} = \mathbf{0},$$

and $\mathbf{r}\times\mathbf{v} = \mathbf{h}$, a constant vector representing the object's angular momentum per unit mass about the origin. This says that $\mathbf{r}$ is always perpendicular to $\mathbf{h}$, so motion due to a central force always takes place in a *plane* through the origin.

If we choose the $z$-axis to be in the direction of $\mathbf{h}$ and let $|\mathbf{h}| = h$, then $\mathbf{h} = h\mathbf{k}$, and the path of the object is in the $xy$-plane. In this case the position and velocity of the object satisfy

$$\mathbf{r} = r\hat{\mathbf{r}} \qquad \text{and} \qquad \mathbf{v} = \dot{r}\hat{\mathbf{r}} + r\dot{\theta}\hat{\boldsymbol{\theta}}.$$

Since $\hat{\mathbf{r}}\times\hat{\boldsymbol{\theta}} = \mathbf{k}$, we have

$$h\mathbf{k} = \mathbf{r}\times\mathbf{v} = r\dot{r}\hat{\mathbf{r}}\times\hat{\mathbf{r}} + r^2\dot{\theta}\hat{\mathbf{r}}\times\hat{\boldsymbol{\theta}} = r^2\dot{\theta}\mathbf{k}.$$

Hence, for any motion under a central force,

$$r^2\dot{\theta} = h, \qquad \text{(a constant for the path of motion).}$$

This formula is equivalent to Kepler's second law; if $A(t)$ is the area in the plane of motion bounded by the orbit and radial lines $\theta = \theta_0$ and $\theta = \theta(t)$ then

$$A(t) = \frac{1}{2}\int_{\theta_0}^{\theta(t)} r^2\,d\theta,$$

so that

$$\frac{dA}{dt} = \frac{dA}{d\theta}\frac{d\theta}{dt} = \frac{1}{2}r^2\dot{\theta} = \frac{h}{2}.$$

Thus area is being swept out at the constant rate $h/2$, and equal areas are swept out in equal times. Note that this law does not depend on the magnitude or direction of the force on the moving object other than the fact that it is *central*. You can also derive the equation $r^2\dot{\theta} = h$ (constant) directly from the fact that the transverse accleration is zero:

$$(d/dt)(r^2\dot{\theta}) = 2r\dot{r}\dot{\theta} + r^2\ddot{\theta} = r(2\dot{r}\dot{\theta} + r\ddot{\theta}) = 0.$$

**EXAMPLE 1** An object moves along the polar curve $r = \theta^{-2}$ under the influence of a force attracting it towards the origin. If the speed of the object is $v_0$ at the instant when $\theta = 1$, find the magnitude of the acceleration of the object at any point on its path as a function of its distance $r$ from the origin.

***SOLUTION*** Since the force is central, we know that the transverse acceleration is zero and that $r^2\dot{\theta} = h$ is constant. Differentiating the equation of the path with respect to time twice, and expressing the results in terms of $r$, we obtain

$$\begin{aligned}\dot{r} &= -2\theta^{-3}\dot{\theta} = -2r^{3/2}\dot{\theta} = -2r^{-1/2}h\\ \ddot{r} &= r^{-3/2}\dot{r}h = -2r^{-2}h^2.\end{aligned}$$

Hence the radial component of acceleration is

$$a_r = \ddot{r} - r(\dot{\theta})^2 = -2r^{-2}h^2 - rh^2r^{-4} = -h^2\left(\frac{2}{r^2} + \frac{1}{r^3}\right).$$

At $\theta = 1$ we have $r = 1$ and the square of the speed is

$$v_0^2 = (\dot{r})^2 + (r\dot{\theta})^2 = 4h^2 + h^2 = 5h^2.$$

Thus $h^2 = v_0^2/5$, and the magnitude of the acceleration of the object is

$$|a_r| = \frac{v_0^2}{5}\left(\frac{2}{r^2} + \frac{1}{r^3}\right).$$

## Derivation of Kepler's First and Third Laws

The planets and the sun move around their common centre of mass. Since the sun is vastly more massive than the planets that centre of mass is quite close to the sun. For example, the joint centre of mass of the sun and the earth lies inside the sun. For the following derivation we will take the sun and a planet as *point masses* and consider the sun to be fixed at the origin. We will specify the directions of the coordinate axes later, when the need arises.

According to Newton's law of gravitation, the force that the sun exerts on a planet of mass $m$ whose position vector is $\mathbf{r}$ is

$$\mathbf{F} = -\frac{km}{r^2}\hat{\mathbf{r}} = -\frac{km}{r^3}\mathbf{r},$$

where $k$ is a positive constant depending on mass of the sun, and $\hat{\mathbf{r}} = \mathbf{r}/r$.

As observed above, the fact that the force on the planet is always directed towards the origin implies that $\mathbf{r}\times\mathbf{v}$ is constant. We choose the direction of the $z$-axis so that $\mathbf{r}\times\mathbf{v} = h\mathbf{k}$, so the motion will be in the $xy$-plane and $r^2\dot{\theta} = h$. We have not yet specified the directions of the $x$- and $y$-axes, but will do so shortly. Using polar coordinates in the $xy$-plane, we calculate

$$\frac{d\mathbf{v}}{d\theta} = \frac{\dot{\mathbf{v}}}{\dot{\theta}} = \frac{-\dfrac{k}{r^2}\hat{\mathbf{r}}}{\dfrac{h}{r^2}} = -\frac{k}{h}\hat{\mathbf{r}}.$$

Since $d\hat{\boldsymbol{\theta}}/d\theta = -\hat{\mathbf{r}}$, we can integrate the differential equation above to find $\mathbf{v}$:

$$\mathbf{v} = -\frac{k}{h}\int \hat{\mathbf{r}}\,d\theta = \frac{k}{h}\hat{\boldsymbol{\theta}} + \mathbf{C},$$

where $\mathbf{C}$ is a vector constant of integration. Therefore, we have shown that

$$|\mathbf{v} - \mathbf{C}| = \frac{k}{h}.$$

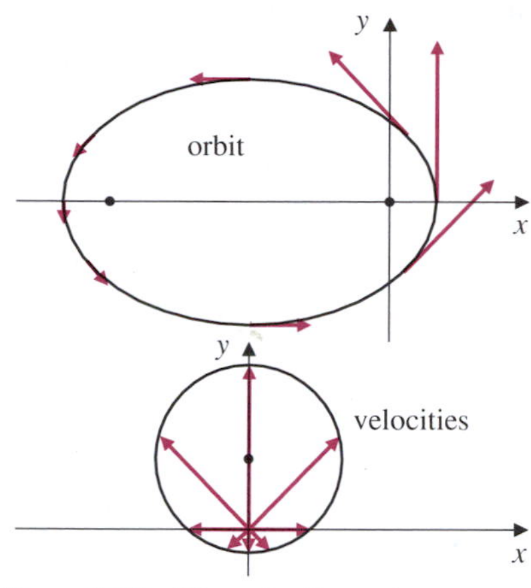


**Figure 7.23** The velocity vectors define a circle

This result, known as **Hamilton's Theorem**, says that as a planet moves around its orbit, its velocity vector (when positioned with its tail at the origin) traces out a *circle*. It is perhaps surprising that there is a circle associated with the orbit of a planet after all. Only it is not the *position* vector that moves on a circle, but the *velocity* vector. (See Figure 7.23.)

Recall that so far we have specified only the position of the origin and the direction of the $z$-axis. Therefore the $xy$-plane is determined, but not the directions of the $x$-axis or the $y$-axis. Let us choose these axes in the $xy$-plane so that $\mathbf{C}$ is in the direction of the $y$-axis; say $\mathbf{C} = (\varepsilon k/h)\mathbf{j}$, where $\varepsilon$ is a positive constant. We therefore have

$$\mathbf{v} = \frac{k}{h}(\hat{\boldsymbol{\theta}} + \varepsilon\mathbf{j}).$$

The position of the $x$-axis is now determined by the fact that the three vectors $\mathbf{i}$, $\mathbf{j}$, and $\mathbf{k}$ are mutually perpendicular and form a right-handed basis. We calculate $\mathbf{r}\times\mathbf{v}$ again. Remember that $\mathbf{r} = r\cos\theta\mathbf{i} + r\sin\theta\mathbf{j}$, and also $\mathbf{r} = r\hat{\mathbf{r}}$:

$$\begin{aligned} h\mathbf{k} = \mathbf{r}\times\mathbf{v} &= \frac{k}{h}(r\hat{\mathbf{r}}\times\hat{\boldsymbol{\theta}} + r\varepsilon\cos\theta\mathbf{i}\times\mathbf{j} + r\varepsilon\sin\theta\mathbf{j}\times\mathbf{j}) \\ &= \frac{k}{h}r(1+\varepsilon\cos\theta)\mathbf{k}. \end{aligned}$$

Thus $h = \frac{kr}{h}(1+\varepsilon\cos\theta)$, or, solving for $r$,

$$r = \frac{h^2/k}{1+\varepsilon\cos\theta}.$$

This is the polar equation of the orbit. If $\varepsilon < 1$ it is an ellipse with one focus at the origin (the sun) and with parameters given by

| | |
|---|---|
| **Semi-latus rectum:** | $\ell = \dfrac{h^2}{k}$ |
| **Semi-major axis:** | $a = \dfrac{h^2}{k(1-\varepsilon^2)} = \dfrac{\ell}{1-\varepsilon^2}$ |
| **Semi-minor axis:** | $b = \dfrac{h^2}{k\sqrt{1-\varepsilon^2}} = \dfrac{\ell}{\sqrt{1-\varepsilon^2}}$ |
| **Semi-focal separation:** | $c = \sqrt{a^2-b^2} = \dfrac{\varepsilon\ell}{1-\varepsilon^2}.$ |

We have deduced Kepler's first law! The choices we made for the coordinate axes result in **perihelion** (the point on the orbit which is closest to the sun) being on the positive $x$-axis ($\theta = 0$).

**EXAMPLE 2** A planet's orbit has eccentricity $\varepsilon$ (where $0 < \varepsilon < 1$) and its speed at perihelion is $v_P$. Find its speed $v_A$ at **aphelion** (the point on its orbit farthest from the sun).

***SOLUTION*** At perihelion and aphelion the planet's radial velocity $\dot{r}$ is zero (since $r$ is minimum or maximum), so the velocity is entirely transverse. Thus $v_P = r_P\dot{\theta}_P$ and $v_A = r_A\dot{\theta}_A$. Since $r^2\dot{\theta} = h$ has the same value at all points of the orbit, we have

$$r_P v_P = r_P^2\dot{\theta}_P = h = r_A^2\dot{\theta}_A = r_A v_A.$$

The planet's orbit has equation

$$r = \frac{\ell}{1 + \varepsilon\cos\theta},$$

so perihelion corresponds to $\theta = 0$ and aphelion to $\theta = \pi$:

$$r_P = \frac{\ell}{1+\varepsilon} \qquad \text{and} \qquad r_A = \frac{\ell}{1-\varepsilon}.$$

Therefore, $v_A = \dfrac{r_P}{r_A}\, v_P = \dfrac{1-\varepsilon}{1+\varepsilon}\, v_P.$ ■

We can obtain Kepler's third law from the other two as follows. Since the radial line from the sun to a planet sweeps out area at a constant rate $h/2$, the total area $A$ enclosed by the orbit is $A = (h/2)T$ where $T$ is the period of revolution. The area of an ellipse with semi-axes $a$ and $b$ is $A = \pi ab$. Since $b^2 = \ell a = h^2 a/k$, we have

$$T^2 = \frac{4}{h^2}A^2 = \frac{4}{h^2}\pi^2 a^2 b^2 = \frac{4\pi^2}{k}\, a^3.$$

Note how the final expression for $T^2$ does not depend on $h$, which is a constant for the orbit of any one planet, but varies from planet to planet. The constant $4\pi^2/k$ does not depend on the particular planet. ($k$ depends on the mass of the sun and a universal gravitational constant.) Thus

$$T^2 = \frac{4\pi^2}{k}a^3$$

says that the square of the period of a planet is proportional to the length, $2a$, of the major axis of the orbit, the proportionality extending over all the planets. This is Kepler's third law. Modern astronomical data show that $T^2/a^3$ varies by only about three-tenths of one percent over the nine known planets.

## Conservation of Energy

Solving the second-order differential equation of motion $\mathbf{F} = m\ddot{\mathbf{r}}$ to find the orbit of a planet requires two integrations. In the above derivation we exploited properties of the cross product to make these integrations easy. More traditional derivations of Kepler's laws usually begin with separating the radial and transverse components in the equation of motion:

$$\ddot{r} - r\dot{\theta}^2 = -\frac{k}{r^2}, \qquad r\ddot{\theta} + 2\dot{r}\dot{\theta} = 0.$$

As observed earlier, the second equation above implies that $r^2\dot{\theta} = h =$ constant, which is Kepler's second law. This can be used to eliminate $\theta$ from the first equation to give

$$\ddot{r} - \frac{h^2}{r^3} = -\frac{k}{r^2}.$$

Therefore

$$\frac{d}{dt}\left(\frac{\dot{r}^2}{2} + \frac{h^2}{2r^2}\right) = \dot{r}\left(\ddot{r} - \frac{h^2}{r^3}\right) = -\frac{k}{r^2}\dot{r}.$$

If we integrate this equation we obtain

$$\frac{1}{2}\left(\dot{r}^2 + \frac{h^2}{r^2}\right) - \frac{k}{r} = E.$$

This is a **conservation of energy** law. The first term on the left is $v^2/2$, the kinetic energy (per unit mass) of the planet. The term $-k/r$ is the potential energy per unit mass. It is difficult to integrate this equation and to find $r$ as a function of $t$. In any event, we really want $r$ as a function of $\theta$ so that we can recognize that we have an ellipse. A way to obtain this is suggested in Exercise 18 below.

***REMARK*** The procedure used above to demonstrate Kepler's laws in fact shows that if any object moves under the influence of a force that attracts it towards the origin (or repels it away from the origin) and has magnitude proportional to the reciprocal of the square of distance from the origin, then the object must move in a plane orbit whose shape is a conic section. If the total energy $E$ defined above is negative, then the orbit is *bounded* and must therefore be an ellipse. If $E = 0$, the orbit is a parabola. If $E > 0$, then the orbit is a hyperbola. Hyperbolic orbits are typical for repulsive forces, but may also occur for attractions if the object has high enough velocity (exceeding the *escape velocity*). See Exercise 22 below for an example.

## EXERCISES 7.6

1. Fill in the details of the calculation suggested in the text to transform the polar equation of an ellipse, $r = \ell/(1 + \varepsilon\cos\theta)$, where $0 < \varepsilon < 1$, to Cartesian coordinates in a form showing the centre and semi-axes explicitly.

2. A particle moves on the circle with polar equation $r = k$. $(k > 0)$. What are the radial and transverse components of its velocity and acceleration? Show that the transverse component of the acceleration is equal to the rate of change of the speed of the particle.

3. Find the radial and transverse components of velocity and acceleration of a particle moving at unit speed along the exponential spiral $r = e^{\theta}$. Express your answers in terms of the angle $\theta$.

4. If a particle moves along the polar curve $r = \theta$ under the influence of a central force attracting it to the origin, find the magnitude of the acceleration as a function of $r$ and the speed of the particle.

* 5. A particle moves along the curve $2y = x^2 - 1$ in such a way that its acceleration is always directed towards the origin. Show that its speed is proportional to $1/\sqrt{r}$, and that the magnitude of its acceleration is proportional to $1/r^2$, where $r$ is its distance from the origin.

6. The mean distance from the earth to the sun is

approximately 150 million km. Halley's Comet approaches perihelion (comes closest to the sun) in its elliptical orbit approximately every 76 years. Estimate the major axis of the orbit of Halley's Comet.

**7.** The mean distance from the moon to the earth is about 385,000 km, and its period of revolution around the earth is about 27 days (the siderial month). At approximately what distance from the centre of the earth, and in what plane, should a communications satellite be inserted into circular orbit if it must remain directly above the the same position on the earth at all times?

**8.** An asteroid is in a circular orbit around the sun. If its period of revolution is $T$ find the radius of its orbit.

* **9.** If the asteroid Exercise 8 is instantaneously stopped in its orbit it will fall towards the sun. How long will it take to get there? *Hint:* you can do this question easily if instead you regard the asteroid as *almost* stopped, so that it goes into a highly eccentric elliptical orbit whose major axis is a bit greater than the radius of the original circular orbit.

**10.** Find the eccentricity of an asteroid's orbit if the asteroid's speed at perihelion is twice its speed at aphelion.

**11.** Show that the orbital speed of a planet is constant if and only if the orbit is circular. *Hint:* use the conservation of energy identity.

**12.** A planet's distance from the sun at perihelion is 80% of its distance at aphelion. Find the ratio of its speeds at perihelion and aphelion, and the eccentricity of its orbit.

* **13.** As a result of a collision, an asteroid originally in a circular orbit about the sun suddenly has its velocity cut in half, so that it falls into an elliptical orbit with maximum distance from the sun equal to the radius of the original circular orbit. Find the eccentricity of its new orbit.

**14.** If the speeds of a planet at perihelion and aphelion are $v_P$ and $v_A$, what is its speed when it is at the ends of the minor axis of its orbit?

**15.** What fraction of its "year" (i.e. the period of its orbit) does a planet spend traversing the half of its orbit that is closest to the sun? Give your answer in terms of the eccentricity $\varepsilon$ of the planet's orbit.

* **16.** Suppose that a planet is travelling at speed $v_0$ at an instant when it is at distance $r_0$ from the sun. Show that the period of the planet's orbit is

$$T = \frac{2\pi}{\sqrt{k}}\left(\frac{2}{r_0} - \frac{v_0^2}{k}\right)^{-3/2}.$$

*Hint:* the quantity $\dfrac{k}{r} - \dfrac{1}{2}v^2$ is constant at all points of the orbit, as shown in the discussion of conservation of energy. Find the value of this expression at perihelion in terms of the semi-major axis, $a$.

* **17.** The sum of the distances from a point $P$ on an ellipse $\mathcal{E}$ to the foci of $\mathcal{E}$ is the constant $2a$, the length of the major axis of the ellipse. Use this fact in a *geometric* argument to show that the mean distance from points $P$ to one focus of $\mathcal{E}$ is $a$. That is, show that

$$\frac{1}{c(\mathcal{E})}\int_{\mathcal{E}} r\,ds = a,$$

where $c(\mathcal{E})$ is the circumference of $\mathcal{E}$ and $r$ is the distance from a point on $\mathcal{E}$ to one focus.

* **18.** The result of eliminating $\theta$ between the equations for the radial and transverse components of acceleration for a planet is

$$\ddot{r} - \frac{h^2}{r^3} = -\frac{k}{r^2}.$$

Show that the change of dependent and independent variables:

$$r(t) = \frac{1}{u(\theta)}, \qquad \theta = \theta(t),$$

transforms this equation to the simpler equation

$$\frac{d^2u}{d\theta^2} + u = \frac{k}{h^2}.$$

Show that the solution of this equation is

$$u = \frac{k}{h^2}\big(1 + \varepsilon\cos(\theta - \theta_0)\big),$$

where $\varepsilon$ and $\theta_0$ are constants. Hence show that the orbit is elliptical if $|\varepsilon| < 1$.

* **19.** Use the technique of Exercise 18 to find the trajectory of an object of unit mass attracted to the origin by a force of magnitude $f(r) = k/r^3$. Are there any orbits that do not approach infinity or the origin as $t \to \infty$?

**20.** Use the conservation of energy formula to show that if $E < 0$, the orbit must be bounded; i.e. it cannot get arbitrarily far away from the origin.

* **21.** If $\varepsilon > 1$ then the equation

$$r = \frac{\ell}{1 + \varepsilon\cos\theta}$$

represents a hyperbola rather than an ellipse. Sketch the hyperbola, find its centre and the directions of its asymptotes, and determine its semi-transverse axis, its semi-conjugate axis, and semi-focal separation in terms of $\ell$ and $\varepsilon$. (See Section 9.1 of *Single-Variable Calculus* for definitions of terms.)

$*$ **22.** A meteor travels from infinity on a hyperbolic orbit passing near the sun. At a very large distance from the sun it has speed $v_\infty$. The asymptotes of its orbit pass at perpendicular distance $D$ from the sun. (See Figure 7.24.) Show that the angle $\delta$ through which the meteor's path is deflected by the gravitational attraction of the sun is given by

$$\cot\left(\frac{\delta}{2}\right) = \frac{Dv_\infty^2}{k}.$$

(*Hint:* you will need the result of Exercise 21.) The same analysis, and result, holds for electrostatic attraction or repulsion; $f(r) = \pm k/r^2$ in that case also. The constant $k$ depends on the charges of two particles and $r$ is the distance between them.

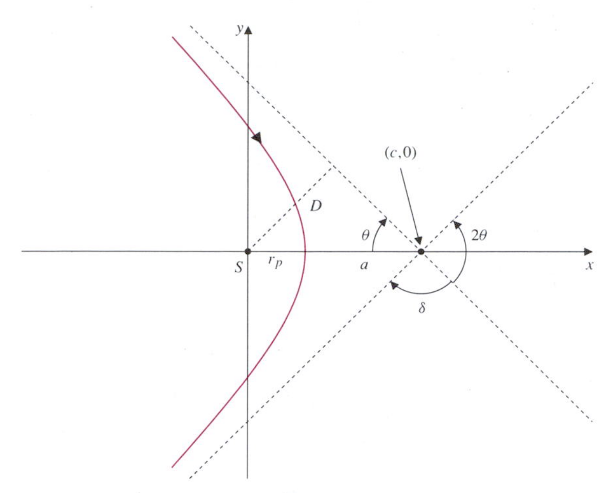


**Figure 7.24** Path of a meteor

# CHAPTER REVIEW

## Key Ideas

- **What is a vector function of a real variable, and why does it represent a curve?**
- **State the Product Rule for the derivative of $\mathbf{u}(t) \bullet \big(\mathbf{v}(t) \times \mathbf{w}(t)\big)$**
- **What do the following phrases mean?**
  - ◇ angular velocity
  - ◇ angular momentum
  - ◇ centripetal acceleration
  - ◇ Coriolis acceleration
  - ◇ arc-length parametrization
  - ◇ central force
- **Find the following quantities associated with a parametric curve $\mathcal{C}$ with parametrization $\mathbf{r} = \mathbf{r}(t)$, $(a \le t \le b)$.**
  - ◇ the velocity $\mathbf{v}(t)$
  - ◇ the speed $v(t)$
  - ◇ the arc length
  - ◇ the acceleration $\mathbf{a}(t)$
  - ◇ the unit tangent $\hat{\mathbf{T}}(t)$
  - ◇ the unit normal $\hat{\mathbf{N}}(t)$
  - ◇ the curvature $\kappa(t)$
  - ◇ the radius of curvature $\rho(t)$
  - ◇ the osculating plane
  - ◇ the osculating circle
  - ◇ the unit binormal $\hat{\mathbf{B}}(t)$
  - ◇ the torsion $\tau(t)$
  - ◇ the tangential acceleration
  - ◇ the normal acceleration
  - ◇ the evolute
- **State the Frenet–Serret formulas.**
- **State Kepler's laws of planetary motion.**
- **What are the radial and transverse components of velocity and acceleration?**

## Review Exercises

**1.** If $\mathbf{r}(t)$, $\mathbf{v}(t)$, and $\mathbf{a}(t)$ represent the position, velocity, and acceleration at time $t$ of a particle moving in 3-space, and if, at every time $t$, the $\mathbf{a}$ is perpendicular to both $\mathbf{r}$ and $\mathbf{v}$, show that the vector $\mathbf{r}(t) - t\mathbf{v}(t)$ has constant length.

**2.** Describe the parametric curve

$$\mathbf{r} = t\cos t\,\mathbf{i} + t\sin t\,\mathbf{j} + (2\pi - t)\mathbf{k},$$

$(0 \le t \le 2\pi)$ and find its length.

**3.** A particle moves along the curve of intersection of the surfaces $y = x^2$ and $z = 2x^3/3$ with constant speed $v = 6$. It is moving in the direction of increasing $x$. Find its velocity and acceleration when it is at the point $(1, 1, 2/3)$.

**4.** A particle moves along the curve $y = x^2$ in the $xy$-plane so that at time $t$ its speed is $v = t$. Find its acceleration at time $t = 3$ if it is at the point $(\sqrt{2}, 2)$ at that time.

**5.** Find the curvature and torsion at a general point of the curve $\mathbf{r} = e^t\mathbf{i} + \sqrt{2}t\mathbf{j} + e^{-t}\mathbf{k}$.

**6.** A particle moves on the curve of Exercise 5 so that it is at position $\mathbf{r}(t)$ at time $t$. Find its normal acceleration and tangential acceleration at any time $t$. What is its minimum speed?

**7. (A clothoid curve)** The plane curve $\mathcal{C}$ in Figure 7.25 has parametric equations

$$x(s) = \int_0^s \cos\frac{kt^2}{2}\,dt \quad \text{and} \quad y(s) = \int_0^s \sin\frac{kt^2}{2}\,dt.$$

Verify that $s$ is, in fact, arc length along $\mathcal{C}$ measured from $(0, 0)$, and that the curvature of $\mathcal{C}$ is given by $\kappa(s) = ks$. Because the curvature changes linearly with distance along the curve, such curves, called *clothoids*, are useful for joining track sections of different curvatures.

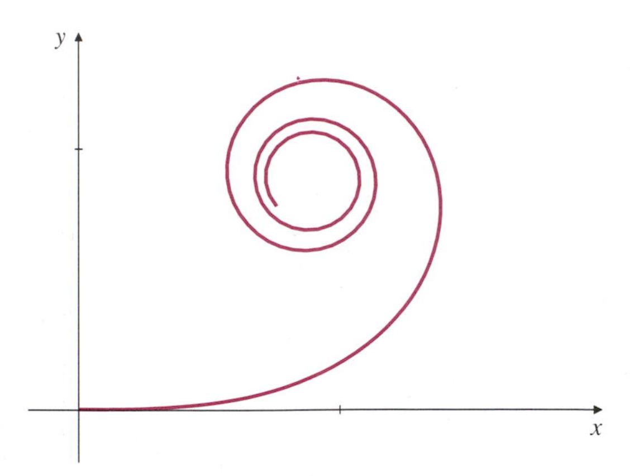


Figure 7.25 A clothoid curve

8. A particle moves along the polar curve $r = e^{-\theta}$ with constant angular speed $\dot{\theta} = k$. Express its velocity and acceleration in terms of radial and transverse components depending only on the distance $r$ from the origin.

**Some properties of cycloids**

Exercises 9–12 all deal with the cycloid

$$\mathbf{r} = a(t - \sin t)\mathbf{i} + a(1 - \cos t)\mathbf{j}.$$

Recall that this curve is the path of a point on the circumference of a circle of radius $a$ rolling along the $x$-axis. (See Section 9.2 in *Single-Variable Calculus*.)

9. Find the arc length $s = s(T)$ of the part of the cycloid from $t = 0$ to $t = T \le 2\pi$.

10. Find the arc-length parametrization $\mathbf{r} = \mathbf{r}(s)$ of the arch $0 \le t \le 2\pi$ of the cycloid, with $s$ measured from the point $(0, 0)$.

11. Find the *evolute* of the cycloid, that is, find parametric equations of the centre of curvature $\mathbf{r} = \mathbf{r}_c(t)$ of the cycloid. Show that the evolute is the same cycloid translated $\pi a$ units to the right and $2a$ units downward.

12. A string of length $4a$ has one end fixed at the origin and is wound along the arch of the cycloid to the right of the origin. Since that arch has total length $8a$, the free end of the string lies at the highest point $A$ of the arch. Find the path followed by the free end $Q$ of the string as it is unwound from the cycloid and is held taught during the unwinding. (See Figure 7.26.) If the string leaves the cycloid at $P$, then

$$(\text{arc } OP) + PQ = 4a.$$

The path of $Q$ is called the *involute* of the cycloid. Show that, like the evolute, the involute is also a translate of the original cycloid. In fact, the cycloid is the evolute of its involute.

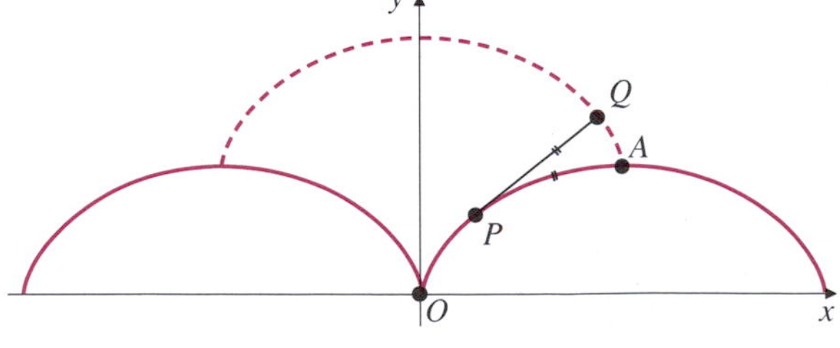


Figure 7.26

13. Let $P$ be a point in 3-space with spherical coordinates $(\rho, \phi, \theta)$. Suppose that $P$ is not on the $z$-axis. Find a triad of mutually perpendicular unit vectors, $\{\hat{\boldsymbol{\rho}}, \hat{\boldsymbol{\phi}}, \hat{\boldsymbol{\theta}}\}$, at $P$ in the directions of increasing $\rho$, $\phi$, and $\theta$ respectively. Is the triad right- or left-handed?

**Kepler's laws imply Newton's law of gravitation**

In Exercises 14–16, it is *assumed* that a planet of mass $m$ moves in an elliptical orbit $r = \ell/(1 + \varepsilon \cos\theta)$, with focus at the origin (the sun), under the influence of a force $\mathbf{F} = \mathbf{F}(\mathbf{r})$ that depends only on the position of the planet.

14. Assuming Kepler's second law, show that $\mathbf{r} \times \mathbf{v} = \mathbf{h}$ is constant, and hence that $r^2\dot{\theta} = h$ is constant.

15. Use Newton's second law of motion ($\mathbf{F} = m\ddot{\mathbf{r}}$) to show that $\mathbf{r} \times \mathbf{F}(\mathbf{r}) = \mathbf{0}$. Therefore $\mathbf{F}(\mathbf{r})$ is parallel to $\mathbf{r}$: $\mathbf{F}(\mathbf{r}) = -f(\mathbf{r})\,\hat{\mathbf{r}}$, for some scalar-valued function $f(\mathbf{r})$, and transverse component of $\mathbf{F}(\mathbf{r})$ is zero.

16. By direct calculation of the radial acceleration of the planet, show that $f(\mathbf{r}) = mh^2/(\ell r^2)$, where $r = |\mathbf{r}|$. Thus $\mathbf{F}$ is an attraction to the origin, proportional to the mass of the planet, and inversely proportional to the square of its distance from the sun.

## Express Yourself

Answer these questions in words, using complete sentences and, if necessary, paragraphs. Use mathematical symbols only where necessary.

1. Why is it better to know the parametric equations of the path of a moving object (parametrized in terms of time), than it is to know a single equation satisfied by the coordinates of all points on the path?

2. What is the difference between the radial and centripetal (normal) components of the acceleration of a moving particle? Between the transverse and tangential components? How are these components related if the particle is moving at constant speed on a circle centred at the origin?

3. Why is it extremely unlikely that a meteor passing near the sun is on a parabolic path?

## Challenging Problems

1. Let $P$ be a point on the surface of the earth at 45° north latitude. Use a coordinate system with origin at $P$ and basis vectors $\mathbf{i}$ and $\mathbf{j}$ pointing east and north, respectively, so that $\mathbf{k}$ points vertically upward.

   (a) Express the angular velocity $\boldsymbol{\Omega}$ of the earth in terms of the basis vectors at $P$. What is the magnitude $\Omega$ of $\boldsymbol{\Omega}$ in rad/s?

   (b) Find the Coriolis acceleration $\mathbf{a}_C = 2\boldsymbol{\Omega} \times \mathbf{v}$ of an object falling vertically with speed $v$ above $P$.

(c) If the object in (b) drops from rest from a height of 100 metres above $P$, approximately where will it strike the ground? Ignore air resistance, but not the Coriolis acceleration. Since the Coriolis acceleration is much smaller than the gravitational acceleration in magnitude, you can use the vertical velocity as a good approximation to the actual velocity of the object at any time during its fall.

* **2. (The spin of a baseball)** When a ball is thrown with spin about an axis that is not parallel to its velocity, it experiences a lateral acceleration due to differences in friction along its sides. This spin acceleration is given by

$$\mathbf{a}_s = k\mathbf{S} \times \mathbf{v},$$

where $\mathbf{v}$ is the velocity of the ball, $\mathbf{S}$ is the angular velocity of its spin, and $k$ is a positive constant depending on the surface of the ball. Suppose that a ball for which $k = 0.001$ is thrown horizontally along the $x$-axis with an initial speed of 70 ft/s and a spin of 1,000 radians/s about a vertical axis. Its velocity $\mathbf{v}$ must satisfy

$$\begin{cases} \dfrac{d\mathbf{v}}{dt} = (0.001)(1,000\mathbf{k}) \times \mathbf{v} - 32\mathbf{k} = \mathbf{k} \times \mathbf{v} - 32\mathbf{k} \\ \mathbf{v}(0) = 70\mathbf{i}, \end{cases}$$

since the acceleration of gravity is 32 ft/s$^2$.

(a) Show that the components of $\mathbf{v} = v_1\mathbf{i} + v_2\mathbf{j} + v_3\mathbf{k}$ satisfy

$$\begin{cases} \dfrac{dv_1}{dt} = -v_2 \\ v_1(0) = 70 \end{cases} \quad \begin{cases} \dfrac{dv_2}{dt} = v_1 \\ v_2(0) = 0 \end{cases} \quad \begin{cases} \dfrac{dv_3}{dt} = -32 \\ v_3(0) = 0. \end{cases}$$

(b) Solve these equations, and find the position of the ball $t$ seconds after it is thrown. Assume that it is thrown from the origin at time $t = 0$.

(c) At $t = 1/5$ s, how far, and in what direction, has the ball deviated from the parabolic path it would have followed if it had been thrown without spin?

* **3. (Charged particles moving in magnetic fields)** Magnetic fields exert forces on moving charged particles. If a particle of mass $m$ and charge $q$ is moving with velocity $\mathbf{v}$ in a magnetic field $\mathbf{B}$, then it experiences a force $\mathbf{F} = q\mathbf{v} \times \mathbf{B}$, and hence its velocity is governed by the equation

$$m\frac{d\mathbf{v}}{dt} = q\mathbf{v} \times \mathbf{B}.$$

For this exercise, suppose that the magnetic field is constant and vertical, say $\mathbf{B} = B\mathbf{k}$ (as, for example, in a cathode-ray tube). If the moving particle has initial velocity $\mathbf{v}_0$, then its velocity at time $t$ is determined by

$$\begin{cases} \dfrac{d\mathbf{v}}{dt} = \omega\mathbf{v} \times \mathbf{k}, \qquad \text{where } \omega = \dfrac{qB}{m} \\ \mathbf{v}(0) = \mathbf{v}_0. \end{cases}$$

(a) Show that $\mathbf{v} \bullet \mathbf{k} = \mathbf{v}_0 \bullet \mathbf{k}$ and $|\mathbf{v}| = |\mathbf{v}_0|$ for all $t$.

(b) Let $\mathbf{w}(t) = \mathbf{v}(t) - (\mathbf{v}_0 \bullet \mathbf{k})\mathbf{k}$, so that $\mathbf{w}$ is perpendicular to $\mathbf{k}$ for all $t$. Show that $\mathbf{w}$ satisfies

$$\begin{cases} \dfrac{d^2\mathbf{w}}{dt^2} = -\omega^2\mathbf{w} \\ \mathbf{w}(0) = \mathbf{v}_0 - (\mathbf{v}_0 \bullet \mathbf{k})\mathbf{k} \\ \mathbf{w}'(0) = \omega\mathbf{v}_0 \times \mathbf{k}. \end{cases}$$

(c) Solve the initial-value problem in (b) for $\mathbf{w}(t)$, and hence find $\mathbf{v}(t)$.

(d) Find the position vector $\mathbf{r}(t)$ of the particle at time $t$ if it is at the origin at time $t = 0$. Verify that the path of the particle is, in general, a circular helix. Under what circumstances is the path a straight line? a circle?

* **4. (The tautochrone)** The parametric equations

$$x = a(\theta - \sin\theta) \quad \text{and} \quad y = a(\cos\theta - 1)$$

(for $0 \le \theta \le 2\pi$) describe an arch of the cycloid followed by a point on a circle of radius $a$ rolling along the underside of the $x$-axis. Suppose the curve is made of wire along which a bead can slide without friction. (See Figure 7.27.) If the bead slides from rest under gravity, starting at a point having parameter value $\theta_0$, show that the time it takes for the bead to fall to the lowest point on the arch (corresponding to $\theta = \pi$) is a *constant*, independent of the starting position $\theta_0$. Thus two such beads released simultaneously from different positions along the wire will always collide at the lowest point. For this reason, the cycloid is sometimes called the *tautochrone*, from the Greek for "constant time." *Hint:* when the bead has fallen from height $y(\theta_0)$ to height $y(\theta)$ its speed is $v = \sqrt{2g\big(y(\theta_0) - y(\theta)\big)}$. (Why?) The time for the bead to fall to the bottom is

$$T = \int_{\theta=\theta_0}^{\theta=\pi} \frac{1}{v}\,ds,$$

where $ds$ is the arc length element along the cycloid.

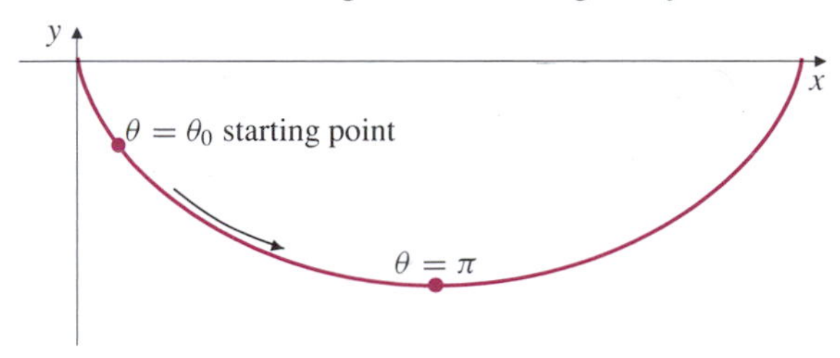


Figure 7.27

* **5.** **(The Drop of Doom)** An amusement park ride at the West Edmonton Mall in Alberta, Canada, gives thrill seekers a taste of free-fall. It consists of a car moving along a track consisting of straight vertical and horizontal sections joined by a smooth curve. The car drops from the top and falls vertically under gravity for $10 - 2\sqrt{2} \approx 7.2$ m before entering the curved section at $B$. It falls another $2\sqrt{2} \approx 2.8$ m as it whips around the curve and into the horizontal section $DE$ at ground level, where brakes are applied to stop it. (Thus, the total vertical drop from $A$ to $D$ or $E$ is 10 m, a figure, like the others in this problem, chosen for mathematical convenience rather than engineering precision.) For purposes of this problem it is helpful to take the coordinate axes at a 45° angle to the vertical, so that the two straight sections of the track lie along the graph $y = |x|$. The curved section then goes from $(-2, 2)$ to $(2, 2)$, and can be taken to be symmetric about the $y$-axis. With this coordinate system, the gravitational acceleration is in the direction of $\mathbf{i} - \mathbf{j}$. (See Figure 7.28.)

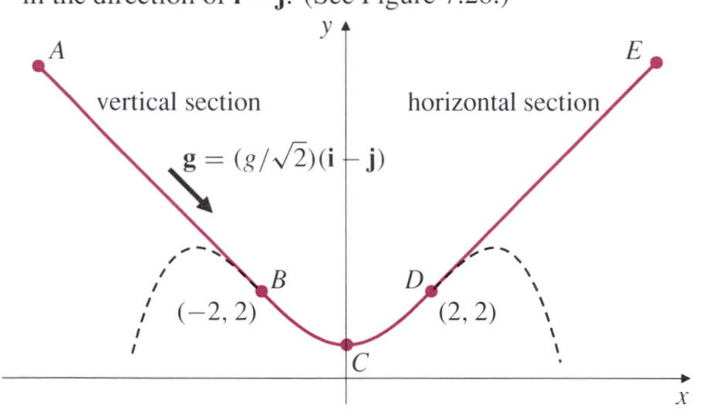


Figure 7.28

(a) Find a fourth-degree polynomial whose graph can be used to link the two straight sections of track without producing discontinuous accelerations for the falling car. (Why is fourth degree adequate?)

(b) Ignoring friction and air resistance, how fast is the car moving when it enters the curve at $B$? at the midpoint $C$ of the curve? and when it leaves the curve at $D$?

(c) Find the magnitude of the normal acceleration and of the total acceleration of the car as it passes through $C$.

* **6.** **(A chase problem)** A fox and a hare are running in the $xy$-plane. Both are running at the same speed $v$. The hare is running up the $y$-axis; at time $t = 0$ it is at the origin. The fox is always running straight towards the hare. At time $t = 0$ the fox is at the point $(a, 0)$, where $a > 0$. Let the fox's position at time $t$ be $\big(x(t), y(t)\big)$.

(a) Verify that the tangent to the fox's path at time $t$ has slope

$$\frac{dy}{dx} = \frac{y(t) - vt}{x(t)}.$$

(b) Show that the equation of the path of the fox satisfies the equation

$$x\frac{d^2y}{dx^2} = \sqrt{1 + \left(\frac{dy}{dx}\right)^2}.$$

*Hint:* differentiate the equation in (a) with respect to $t$. On the left side note that $(d/dt) = (dx/dt)(d/dx)$.

(c) Solve the equation in (b) by substituting $u(x) = dy/dx$ and separating variables. Note that $y = 0$ and $u = 0$ when $x = a$.

**7.** Suppose the earth is a perfect sphere of radius $a$. You set out from the point on the equator whose spherical coordinates are $(\rho, \phi, \theta) = (a, \pi/2, 0)$ and travel on the surface of the earth at constant speed $v$, always moving towards the northeast (45° east of north).

(a) Will you ever get to the north pole? If so, how long will it take to get there?

(b) Find the functions $\phi(t)$ and $\theta(t)$ that are the angular spherical coordinates of your position at time $t > 0$.

(c) How many times does your path cross the meridian $\theta = 0$?

CHAPTER 8

# Vector Fields

**INTRODUCTION** This chapter and the next are concerned mainly with vector-valued functions of a vector variable, typically functions whose domains and ranges lie in the plane or in 3-space. Such functions are frequently called *vector fields*. Applications of vector fields often involve integrals taken, not along axes or over regions in the plane or 3-space, but rather over curves and surfaces. We will introduce such line and surface integrals in this chapter. The next chapter will be devoted to developing analogues of the Fundamental Theorem of Calculus for integrals of vector fields.

## 8.1 VECTOR AND SCALAR FIELDS

In Chapters 4–6 we extended the concepts of single-variable calculus to scalar-valued functions of a vector variable (that is, functions of several real variables), and in Chapter 7 we considered vector-valued functions of a single (scalar) variable, that is, curves. It would appear that the time has come to talk about vector-valued functions of a vector variable. In general, this would mean studying functions from $\mathbb{R}^m$ to $\mathbb{R}^n$. However, we do not intend to be this general; we will restrict our attention to cases where $m$ and $n$ are 2 or 3, where the most interesting applications arise.

A function whose domain and range are subsets of Euclidean 3-space, $\mathbb{R}^3$, is called a **vector field.** Thus a vector field, $\mathbf{F}$, associates a vector $\mathbf{F}(x, y, z)$ with each point $(x, y, z)$ in its domain. The three components of $\mathbf{F}$ are scalar-valued (real-valued) functions $F_1(x, y, z)$, $F_2(x, y, z)$, and $F_3(x, y, z)$, and $\mathbf{F}(x, y, z)$ can be expressed in terms of the standard basis in $\mathbb{R}^3$ as

$$\mathbf{F}(x, y, z) = F_1(x, y, z)\mathbf{i} + F_2(x, y, z)\mathbf{j} + F_3(x, y, z)\mathbf{k}.$$

(Note that the subscripts here represent *components* of a vector, *not* partial derivatives.) If $F_3(x, y, z) = 0$ and $F_1$ and $F_2$ are independent of $z$ then $\mathbf{F}$ reduces to

$$\mathbf{F}(x, y) = F_1(x, y)\mathbf{i} + F_2(x, y)\mathbf{j}$$

and so is called a **plane vector field** or a vector field in the $xy$-plane. We will frequently make use of position vectors in the arguments of vector fields. The position vector of $(x, y, z)$ is $\mathbf{r} = x\mathbf{i} + y\mathbf{j} + z\mathbf{k}$, and we can write $\mathbf{F}(\mathbf{r})$ as a shorthand for $\mathbf{F}(x, y, z)$. In the context of discussion of vector fields, a scalar-valued function of a vector variable (that is, a function of several variables in the context of chapters 4–6) is frequently called a **scalar field.** Thus the components of a vector field are scalar fields.

Many of the results we prove about vector fields require that the field be smooth in some sense. We will call a vector field **smooth** wherever its component scalar fields have continuous partial derivatives of all orders. (For most purposes, however, second order would be sufficient.)

Vector fields arise in many situations in applied mathematics. Let us list some:

(a) The gravitational field $\mathbf{F}(x, y, z)$ due to some object is the force of attraction that the object exerts on a unit mass located at position $(x, y, z)$.

(b) The electrostatic force field $\mathbf{E}(x, y, z)$ due to an electrically charged object is the electrical force that the object exerts on a unit charge at position $(x, y, z)$. (The force may be either an attraction or a repulsion.)

(c) The velocity field $\mathbf{v}(x, y, z)$ in a moving fluid (or solid) is the velocity of motion of the particle at position $(x, y, z)$. If the motion is not "steady state" then the velocity field will also depend on time: $\mathbf{v} = \mathbf{v}(x, y, z, t)$.

(d) The gradient $\nabla f(x, y, z)$, of any scalar field $f$ gives the direction and magnitude of the greatest rate of increase of $f$ at $(x, y, z)$. In particular, a *temperature gradient*, $\nabla T(x, y, z)$, is a vector field giving the direction and magnitude of the greatest rate of increase of temperature $T$ at the point $(x, y, z)$ in a heat conducting medium. *Pressure gradients* provide similar information about the variation of pressure in a fluid such as an air mass or an ocean.

(e) The unit radial and unit transverse vectors $\hat{\mathbf{r}}$ and $\hat{\boldsymbol{\theta}}$ are examples of vector fields in the $xy$-plane. Both are defined at all points of the plane except the origin.

### EXAMPLE 1 The gravitational field of a point mass

The gravitational force field due to a point mass $m$ located at point $P_0$ having position $\mathbf{r}_0$ is

$$\begin{aligned} \mathbf{F}(x, y, z) = \mathbf{F}(\mathbf{r}) &= \frac{-km}{|\mathbf{r} - \mathbf{r}_0|^3}(\mathbf{r} - \mathbf{r}_0) \\ &= -km \frac{(x - x_0)\mathbf{i} + (y - y_0)\mathbf{j} + (z - z_0)\mathbf{k}}{\left((x - x_0)^2 + (y - y_0)^2 + (z - z_0)^2\right)^{3/2}}. \end{aligned}$$

Observe that $\mathbf{F}$ points towards the point $\mathbf{r}_0$ and has magnitude

$$|\mathbf{F}| = km/|\mathbf{r} - \mathbf{r}_0|^2.$$

Some of the vectors in a plane section of the field are shown graphically in Figure 8.1. Each represents the value of the field at the position of its tail. Observe how the lengths of the vectors indicate that the strength of the force increases the closer you get to $P_0$. ■

***REMARK*** The electrostatic field $\mathbf{F}$ due to a point charge $q$ at $P_0$ is given by the same formula as the gravitational field above, except with $-m$ replaced by $q$. The reason for the opposite sign is that like charges repel each other whereas masses attract each other.

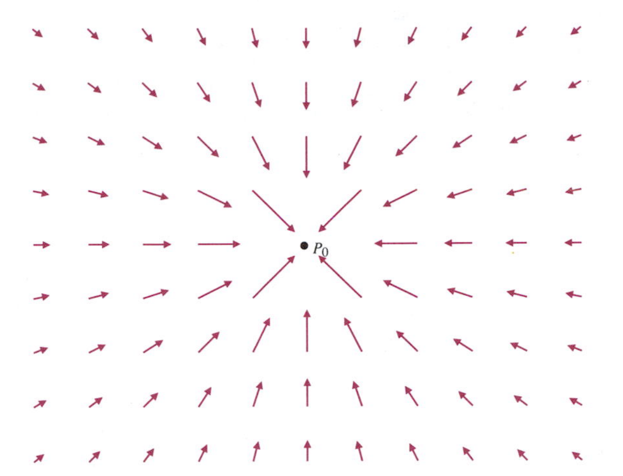

**Figure 8.1** The gravitational field of a point mass located at $P_0$

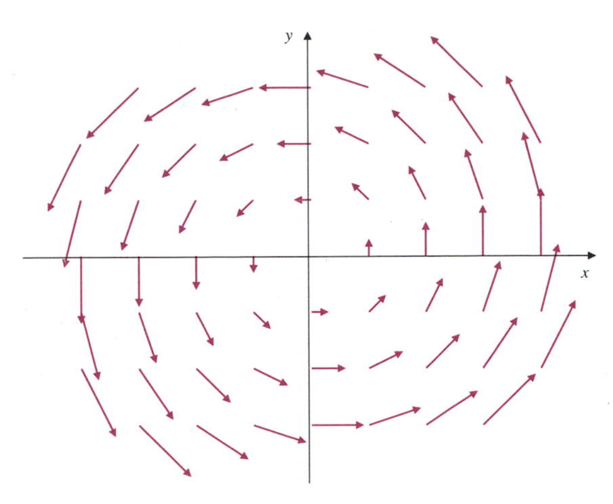

**Figure 8.2** The velocity field of a rigid body rotating about the $z$-axis

**EXAMPLE 2** The velocity field of a solid rotating about the $z$-axis with angular velocity $\boldsymbol{\Omega} = \Omega\mathbf{k}$ is

$$\mathbf{v}(x, y, z) = \mathbf{v}(\mathbf{r}) = \boldsymbol{\Omega}\times\mathbf{r} = -\Omega y\mathbf{i} + \Omega x\mathbf{j}.$$

Being the same in all planes normal to the $z$-axis, $\mathbf{v}$ can be regarded as a plane vector field. Some vectors of the field are shown in Figure 8.2. ■

## Field Lines (Integral Curves)

The graphical representation of vector fields such as those shown in Figures 8.1 and 8.2, and the wind velocity field over a hill shown in Figure 8.3 suggest a pattern of motion through space or in the plane. Whether or not the field is a velocity field, we can interpret it as such and ask what path will be followed by a particle, initially at some point, whose velocity is given by the field. The path will be a curve to which the field is tangent at every point. Such curves are called **field lines** or **integral curves** for the given vector field. In the specific case where the vector field gives the velocity in a fluid flow, the field lines are called **streamlines** of the flow; some of these are shown for the air flow in Figure 8.3. For a force field the field lines are called **lines of force.**

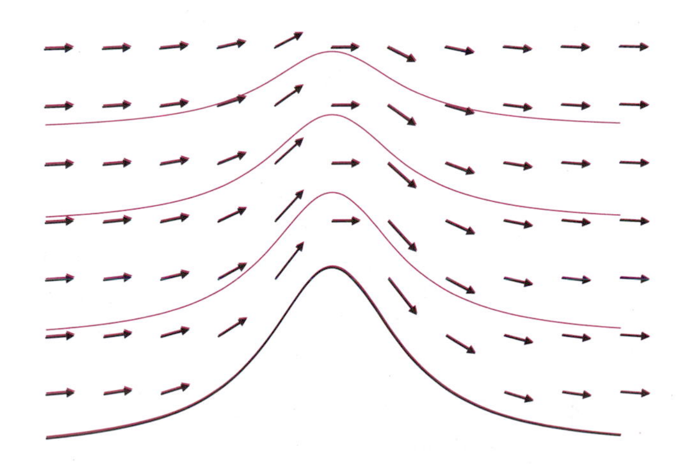
**Figure 8.3** The velocity field of wind blowing over a hill

The field lines of $\mathbf{F}$ do not depend on the magnitude of $\mathbf{F}$ at any point, but only on the direction of the field. If the field line through some point has parametric equation $\mathbf{r} = \mathbf{r}(t)$ then its tangent vector $d\mathbf{r}/dt$ must be parallel to $\mathbf{F}(\mathbf{r}(t))$ for all $t$. Thus

$$\frac{d\mathbf{r}}{dt} = \lambda(t)\mathbf{F}\big(\mathbf{r}(t)\big).$$

For *some* vector fields this differential equation can be integrated to find the field lines. If we break the equation into components,

$$\frac{dx}{dt} = \lambda(t)F_1(x, y, z), \qquad \frac{dy}{dt} = \lambda(t)F_2(x, y, z), \qquad \frac{dz}{dt} = \lambda(t)F_3(x, y, z),$$

we can obtain equivalent differential expressions for $\lambda(t)\,dt$ and hence write the differential equation for the field lines in the form

$$\frac{dx}{F_1(x, y, z)} = \frac{dy}{F_2(x, y, z)} = \frac{dz}{F_3(x, y, z)}.$$

If multiplication of these differential equations by some function puts them in the form

$$P(x)\,dx = Q(y)\,dy = R(z)\,dz,$$

then we can integrate all three expressions to find the field lines.

**EXAMPLE 3** Find the field lines of the gravitational force field of Example 1:

$$\mathbf{F}(x, y, z) = -km\,\frac{(x - x_0)\mathbf{i} + (y - y_0)\mathbf{j} + (z - z_0)\mathbf{k}}{\Big((x - x_0)^2 + (y - y_0)^2 + (z - z_0)^2\Big)^{3/2}}.$$

***SOLUTION*** The vector in the numerator of the fraction gives the direction of $\mathbf{F}$. Therefore, the field lines satisfy the system

$$\frac{dx}{x - x_0} = \frac{dy}{y - y_0} = \frac{dz}{z - z_0}.$$

Integrating all three expressions leads to

$$\ln|x - x_0| + \ln C_1 = \ln|y - y_0| + \ln C_2 = \ln|z - z_0| + \ln C_3,$$

or, on taking exponentials,

$$C_1(x - x_0) = C_2(y - y_0) = C_3(z - z_0).$$

This represents two families of planes all passing through $P_0 = (x_0, y_0, z_0)$. The field lines are the intersections of planes from each of the families and so are straight lines through the point $P_0$. (This is a *two-parameter* family of lines — any one of the constants $C_i$ that is nonzero can be divided out of the equations above.) The nature of the field lines should also be apparent from the plot of the vector field in Figure 8.1. ■

**EXAMPLE 4** Find the field lines of the velocity field $\mathbf{v} = \Omega(-y\mathbf{i} + x\mathbf{j})$ of Example 2.

**SOLUTION** The field lines satisfy the differential equation

$$\frac{dx}{-y} = \frac{dy}{x}.$$

We can separate variables in this equation to get $x\,dx = -y\,dy$. Integration then gives $x^2/2 = -y^2/2 + C/2$, or $x^2 + y^2 = C$. Thus the field lines are circles centred at the origin in the $xy$-plane, as is also apparent from the vector field plot in Figure 8.2. If we regard $\mathbf{v}$ as a vector field in 3-space, we find that the field lines are horizontal circles centred on the $z$-axis:

$$\begin{cases} x^2 + y^2 = C_1 \\ z = C_2. \end{cases}$$

Our ability to find field lines depends on our ability to solve differential equations, and, in 3-space, systems of differential equations.

**EXAMPLE 5** Find the field lines of $\mathbf{F} = xz\mathbf{i} + 2x^2z\mathbf{j} + x^2\mathbf{k}$.

**SOLUTION** The field lines satisfy $\dfrac{dx}{xz} = \dfrac{dy}{2x^2z} = \dfrac{dz}{x^2}$, or, equivalently

$$dy = 2x\,dx \qquad \text{and} \qquad dy = 2z\,dz.$$

The level curves are the curves of intersection of the two families $y = x^2 + C_1$ and $y = z^2 + C_2$ of parabolic cylinders.

## Vector Fields in Polar Coordinates

A vector field in the plane can be expressed in terms of polar coordinates in the form

$$\mathbf{F} = \mathbf{F}(r, \theta) = F_r(r, \theta)\hat{\mathbf{r}} + F_\theta(r, \theta)\hat{\boldsymbol{\theta}},$$

where $\hat{\mathbf{r}}$ and $\hat{\boldsymbol{\theta}}$, defined everywhere except at the origin by

$$\hat{\mathbf{r}} = \cos\theta\mathbf{i} + \sin\theta\mathbf{j}$$
$$\hat{\boldsymbol{\theta}} = -\sin\theta\mathbf{i} + \cos\theta\mathbf{j},$$

are unit vectors in the direction of increasing $r$ and $\theta$ at $[r, \theta]$. Note that $\hat{\boldsymbol{\theta}}$ is just $\hat{\mathbf{r}}$ rotated 90° counterclockwise. Also note that we are using $F_r$ and $F_\theta$ to denote the components of $\mathbf{F}$ with respect to the basis $\{\hat{\mathbf{r}}, \hat{\boldsymbol{\theta}}\}$; the subscripts to not indicate partial derivatives. Here $F_r(r, \theta)$ is called the *radial* component of $\mathbf{F}$, and $F_\theta(r, \theta)$ is called the *transverse* component.

A curve with polar equation $r = r(\theta)$ can be expressed in vector parametric form

$$\mathbf{r} = r\hat{\mathbf{r}},$$

as we did in Section 7.6. This curve is a field line of $\mathbf{F}$ if its differential tangent vector

$$d\mathbf{r} = dr\,\hat{\mathbf{r}} + r\,\frac{d\hat{\mathbf{r}}}{d\theta}\,d\theta = dr\,\hat{\mathbf{r}} + r\,d\theta\,\hat{\boldsymbol{\theta}}$$

is parallel to the field vector $\mathbf{F}(r, \theta)$ at any point except the origin, that is, if $r = f(\theta)$ satisfies the differential equation

$$\frac{dr}{F_r(r, \theta)} = \frac{r\,d\theta}{F_\theta(r, \theta)}.$$

In specific cases we can find the field lines by solving this equation.

**EXAMPLE 6** Sketch the vector field $\mathbf{F}(r, \theta) = \hat{\mathbf{r}} + \hat{\boldsymbol{\theta}}$, and find its field lines.

***SOLUTION*** At each point $[r, \theta]$, the field vector bisects the angle between $\hat{\mathbf{r}}$ and $\hat{\boldsymbol{\theta}}$, and so makes a counterclockwise angle of 45° with $\hat{\mathbf{r}}$. All of the vectors in the field have the same length, $\sqrt{2}$. Some of the field vectors are shown in Figure 8.4(a). They suggest that the field lines will spiral outward from the origin.

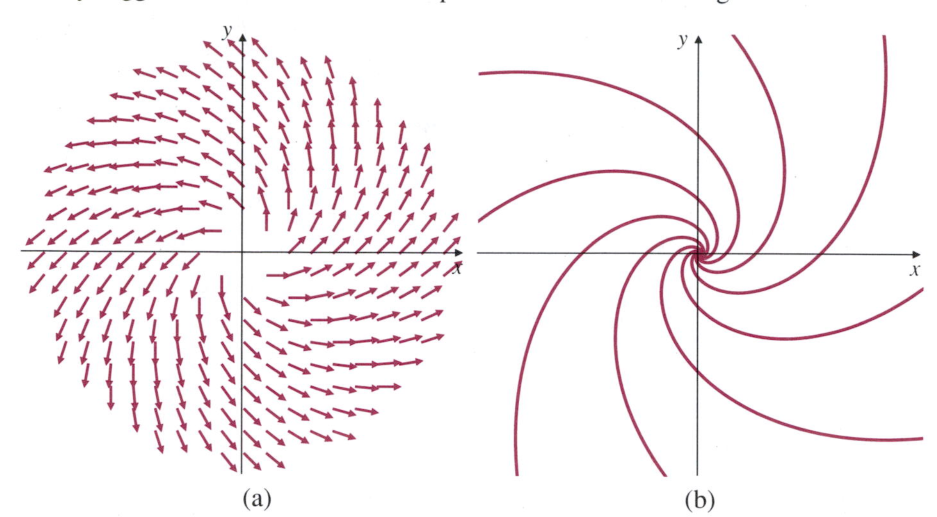


**Figure 8.4**

(a) The vector field $\mathbf{F} = \hat{\mathbf{r}} + \hat{\boldsymbol{\theta}}$

(b) Field lines of $\mathbf{F} = \hat{\mathbf{r}} + \hat{\boldsymbol{\theta}}$

Since $F_r(r, \theta) = F_\theta(r, \theta) = 1$ for this field, the field lines satisfy $dr = r\, d\theta$, or, dividing by $d\theta$, $dr/d\theta = r$. This is the differential equation of exponential growth, and has solution $r = Ke^{\theta}$, or, equivalently, $r = e^{\theta+\alpha}$, where $\alpha = \ln K$ is a constant. Several such curves are shown in Figure 8.4(b). ■

## EXERCISES 8.1

In Exercises 1–8, sketch the given plane vector field and determine its field lines.

1. $\mathbf{F}(x, y) = x\mathbf{i} + x\mathbf{j}$
2. $\mathbf{F}(x, y) = x\mathbf{i} + y\mathbf{j}$
3. $\mathbf{F}(x, y) = y\mathbf{i} + x\mathbf{j}$
4. $\mathbf{F}(x, y) = \mathbf{i} + \sin x\,\mathbf{j}$
5. $\mathbf{F}(x, y) = e^x\mathbf{i} + e^{-x}\mathbf{j}$
6. $\mathbf{F}(x, y) = \nabla(x^2 - y)$
7. $\mathbf{F}(x, y) = \nabla \ln(x^2 + y^2)$
8. $\mathbf{F}(x, y) = \cos y\,\mathbf{i} - \cos x\,\mathbf{j}$

In Exercises 9–16, describe the streamlines of the given velocity fields.

9. $\mathbf{v}(x, y, z) = y\mathbf{i} - y\mathbf{j} - y\mathbf{k}$
10. $\mathbf{v}(x, y, z) = x\mathbf{i} + y\mathbf{j} - x\mathbf{k}$
11. $\mathbf{v}(x, y, z) = y\mathbf{i} - x\mathbf{j} + \mathbf{k}$
12. $\mathbf{v}(x, y, z) = \dfrac{x\mathbf{i} + y\mathbf{j}}{(1 + z^2)(x^2 + y^2)}$
13. $\mathbf{v}(x, y, z) = xz\mathbf{i} + yz\mathbf{j} + x\mathbf{k}$
14. $\mathbf{v}(x, y, z) = e^{xyz}(x\mathbf{i} + y^2\mathbf{j} + z\mathbf{k})$
15. $\mathbf{v}(x, y) = x^2\mathbf{i} - y\mathbf{j}$

*16. $\mathbf{v}(x, y) = x\mathbf{i} + (x + y)\mathbf{j}$ *Hint:* let $y = xv(x)$.

In Exercises 17–20, determine the field lines of the given polar vector fields.

17. $\mathbf{F} = \hat{\mathbf{r}} + r\hat{\boldsymbol{\theta}}$
18. $\mathbf{F} = \hat{\mathbf{r}} + \theta\hat{\boldsymbol{\theta}}$
19. $\mathbf{F} = 2\hat{\mathbf{r}} + \theta\hat{\boldsymbol{\theta}}$
20. $\mathbf{F} = r\hat{\mathbf{r}} - \hat{\boldsymbol{\theta}}$

## 8.2 CONSERVATIVE FIELDS

Since the gradient of a scalar field is a vector field it is natural to ask whether every vector field is the gradient of a scalar field. Given a vector field $\mathbf{F}(x, y, z)$, does there exist a scalar field $\phi(x, y, z)$ such that

$$\mathbf{F}(x, y, z) = \nabla\phi(x, y, z) = \frac{\partial\phi}{\partial x}\mathbf{i} + \frac{\partial\phi}{\partial y}\mathbf{j} + \frac{\partial\phi}{\partial z}\mathbf{k}\,?$$

The answer in general is "no." Only special vector fields can be written in this way.

**DEFINITION 1**

> If $\mathbf{F}(x, y, z) = \nabla\phi(x, y, z)$ in a domain $D$, then we say that $\mathbf{F}$ is a **conservative** vector field in $D$, and we call the function $\phi$ a **(scalar) potential** for $\mathbf{F}$ on $D$. Similar definitions hold in the plane or in $n$-space.

Like antiderivatives, potentials are not determined uniquely; arbitrary constants can be added to them. Note that $\mathbf{F}$ is **conservative in a domain** $D$ if and only if $\mathbf{F} = \nabla\phi$ at *every* point of $D$; the potential $\phi$ cannot have any singular points in $D$.

The equation $F_1(x, y, z)\,dx + F_2(x, y, z)\,dy + F_3(x, y, z)\,dz = 0$ is called an **exact** differential equation if the left side is the differential of a scalar function $\phi(x, y, z)$:

$$d\phi = F_1(x, y, z)\,dx + F_2(x, y, z)\,dy + F_3(x, y, z)\,dz.$$

In this case the differential equation has solutions given by $\phi(x, y, z) = C$ (constant). Observe that the differential equation is exact if and only if the vector field $\mathbf{F} = F_1\mathbf{i} + F_2\mathbf{j} + F_3\mathbf{k}$ is conservative, and that $\phi$ is the potential of $\mathbf{F}$. Exact differential equations in two variables are discussed in Section 10.3.

Being scalar fields rather than vector fields, potentials for conservative vector fields are easier to manipulate algebraically than are the vector fields themselves. For instance, a sum of potential functions is the potential function for the sum of the corresponding vector fields. A vector field can always be computed from its potential function by taking the gradient.

**EXAMPLE 1** **The gravitational field of a point mass is conservative**

Show that the gravitational field $\mathbf{F}(\mathbf{r}) = -km(\mathbf{r} - \mathbf{r}_0)/|\mathbf{r} - \mathbf{r}_0|^3$ of Example 1 in Section 8.1 is conservative wherever it is defined (that is, everywhere in $\mathbb{R}^3$ except at the origin), by showing that

$$\phi(x, y, z) = \frac{km}{|\mathbf{r} - \mathbf{r}_0|} = \frac{km}{\sqrt{(x - x_0)^2 + (y - y_0)^2 + (z - z_0)^2}}$$

is a potential function for $\mathbf{F}$.

***SOLUTION*** Observe that

$$\frac{\partial\phi}{\partial x} = \frac{-km(x - x_0)}{\Big((x - x_0)^2 + (y - y_0)^2 + (z - z_0)^2\Big)^{3/2}} = \frac{-km(x - x_0)}{|\mathbf{r} - \mathbf{r}_0|^3} = F_1(\mathbf{r}),$$

and similar formulas hold for the other partial derivatives of $\phi$. It follows that $\nabla\phi(x, y, z) = \mathbf{F}(x, y, z)$ for $(x, y, z) \neq (0, 0, 0)$, and $\mathbf{F}$ is conservative except at the origin. ■

***REMARK*** It is not necessary to write the expression $km/|\mathbf{r} - \mathbf{r}_0|$ in terms of the components of $\mathbf{r} - \mathbf{r}_0$ as we did in Example 1 in order to calculate its partial derivatives. Here is a useful formula for the derivative of the length of a vector function $\mathbf{F}$ with respect to a variable $x$:

$$\frac{\partial}{\partial x}|\mathbf{F}| = \frac{\mathbf{F} \bullet \left(\dfrac{\partial}{\partial x}\mathbf{F}\right)}{|\mathbf{F}|}.$$

To see why this is true, express $|\mathbf{F}| = \sqrt{\mathbf{F} \bullet \mathbf{F}}$ and calculate its derivative using the Chain Rule and the Product Rule:

$$\frac{\partial}{\partial x}|\mathbf{F}| = \frac{\partial}{\partial x}\sqrt{\mathbf{F} \bullet \mathbf{F}} = \frac{1}{2\sqrt{\mathbf{F} \bullet \mathbf{F}}}\, 2\mathbf{F} \bullet \left(\frac{\partial}{\partial x}\mathbf{F}\right) = \frac{\mathbf{F} \bullet \left(\frac{\partial}{\partial x}\mathbf{F}\right)}{|\mathbf{F}|}.$$

Compare this with the derivative of an absolute value of a function of one variable:

$$\frac{d}{dx}|f(x)| = \operatorname{sgn}(f(x))\, f'(x) = \frac{f(x)}{|f(x)|}\, f'(x).$$

In the context of Example 1 we have

$$\frac{\partial}{\partial x}\frac{km}{|\mathbf{r} - \mathbf{r}_0|} = \frac{-km}{|\mathbf{r} - \mathbf{r}_0|^2}\frac{\partial}{\partial x}|\mathbf{r} - \mathbf{r}_0| = \frac{-km(x - x_0)}{|\mathbf{r} - \mathbf{r}_0|^3},$$

with similar expressions for the other partials of $km/|\mathbf{r} - \mathbf{r}_0|$.

**EXAMPLE 2** Show that the velocity field $\mathbf{v} = -\Omega y\mathbf{i} + \Omega x\mathbf{j}$ of rigid body rotation about the $z$ axis (see Example 2 of Section 8.1) is not conservative.

***SOLUTION*** There are two ways to show that no potential for $\mathbf{v}$ can exist. One way is to try to find a potential $\phi(x, y)$ for the vector field. We require

$$\frac{\partial \phi}{\partial x} = -\Omega y \qquad \text{and} \qquad \frac{\partial \phi}{\partial y} = \Omega x.$$

The first of these equations implies that $\phi(x, y) = -\Omega xy + C_1(y)$. (We have integrated with respect to $x$; the constant can still depend on $y$.) Similarly, the second equation implies that $\phi(x, y) = \Omega xy + C_2(x)$. Therefore, we must have $-\Omega xy + C_1(y) = \Omega xy + C_2(x)$, or $2\Omega xy = C_1(y) - C_2(x)$ for all $(x, y)$. This is not possible for any choice of the single-variable functions $C_1(y)$ and $C_2(x)$ unless $\Omega = 0$.

Alternatively, if $\mathbf{v}$ has a potential $\phi$, then we can form the mixed partial derivatives of $\phi$ from the two equations above and get

$$\frac{\partial^2 \phi}{\partial y \partial x} = -\Omega \qquad \text{and} \qquad \frac{\partial^2 \phi}{\partial x \partial y} = \Omega.$$

This is not possible if $\Omega \neq 0$ because the smoothness of $\mathbf{v}$ implies that its potential should be smooth, so the mixed partials should be equal. Thus no such $\phi$ can exist; $\mathbf{v}$ is not conservative. ■

Example 2 suggests a condition that must be satisfied by any conservative plane vector field.

**Necessary condition for a conservative plane vector field**

If $\mathbf{F}(x, y) = F_1(x, y)\mathbf{i} + F_2(x, y)\mathbf{j}$ is a conservative vector field in a domain $D$ of the $xy$-plane, then the condition

$$\frac{\partial}{\partial y}F_1(x, y) = \frac{\partial}{\partial x}F_2(x, y)$$

must be satisfied at all points of $D$.

To see this, observe that

$$F_1\mathbf{i} + F_2\mathbf{j} = \mathbf{F} = \nabla\phi = \frac{\partial\phi}{\partial x}\mathbf{i} + \frac{\partial\phi}{\partial y}\mathbf{j}$$

implies the two scalar equations

$$F_1 = \frac{\partial\phi}{\partial x} \qquad \text{and} \qquad F_2 = \frac{\partial\phi}{\partial y},$$

and since the mixed partial derivatives of $\phi$ should be equal,

$$\frac{\partial F_1}{\partial y} = \frac{\partial^2\phi}{\partial y\partial x} = \frac{\partial^2\phi}{\partial x\partial y} = \frac{\partial F_2}{\partial x}.$$

A similar condition obtains for vector fields in 3-space.

**Necessary conditions for a conservative vector field in 3-space**

If $\mathbf{F}(x, y, z) = F_1(x, y, z)\mathbf{i} + F_2(x, y, z)\mathbf{j} + F_3(x, y, z)\mathbf{k}$ is a conservative vector field in a domain $D$ in 3-space, then we must have, everywhere in $D$,

$$\frac{\partial}{\partial y}F_1 = \frac{\partial}{\partial x}F_2, \qquad \frac{\partial}{\partial z}F_1 = \frac{\partial}{\partial x}F_3, \qquad \frac{\partial}{\partial z}F_2 = \frac{\partial}{\partial y}F_3.$$

## Equipotential Surfaces and Curves

If $\phi(x, y, z)$ is a potential function for the conservative vector field $\mathbf{F}$ then the *level surfaces* $\phi(x, y, z) = C$ of $\phi$ are called **equipotential surfaces** of $\mathbf{F}$. Since $\mathbf{F} = \nabla\phi$ is normal to these surfaces (wherever it does not vanish), the field lines of $\mathbf{F}$ always intersect the equipotential surfaces at right angles. For instance, the equipotential surfaces of the gravitational force field of a point mass are spheres centred at the point; these spheres are normal to the field lines, which are straight lines passing through the point. Similarly, for a conservative plane vector field, the *level curves* of the potential function are called **equipotential curves** of the vector field. They are the *orthogonal trajectories* of the field lines; that is, they intersect the field lines at right angles.

**EXAMPLE 3** Show that the vector field $\mathbf{F}(x, y) = x\mathbf{i} - y\mathbf{j}$ is conservative and find a potential function for it. Describe the field lines and the equipotential curves.

***SOLUTION*** Since $\partial F_1/\partial y = 0 = \partial F_2/\partial x$ everywhere in $\mathbb{R}^2$, we would expect $\mathbf{F}$ to be conservative. Any potential function $\phi$ must satisfy

$$\frac{\partial\phi}{\partial x} = F_1 = x \qquad \text{and} \qquad \frac{\partial\phi}{\partial y} = F_2 = -y.$$

The first of these equations gives

$$\phi(x, y) = \int x\,dx = \frac{1}{2}x^2 + C_1(y).$$

Observe that, since the integral is taken with respect to $x$, the "constant" of integration is allowed to depend on the other variable. Now we use the second equation to get

$$-y = \frac{\partial\phi}{\partial y} = C_1{}'(y) \quad\Rightarrow\quad C_1(y) = -\frac{1}{2}y^2 + C_2.$$

Thus $\mathbf{F}$ is conservative and, for any constant $C_2$,

$$\phi(x, y) = \frac{x^2 - y^2}{2} + C_2$$

is a potential function for $\mathbf{F}$. The field lines of $\mathbf{F}$ satisfy

$$\frac{dx}{x} = -\frac{dy}{y} \quad\Rightarrow\quad \ln|x| = -\ln|y| + \ln C_3 \quad\Rightarrow\quad xy = C_3.$$

The field lines of $\mathbf{F}$ are thus rectangular hyperbolas with the coordinate axes as asymptotes. The equipotential curves constitute another family of rectangular hyperbolas, $x^2 - y^2 = C_4$, with the lines $x = \pm y$ as asymptotes. Curves of the two families intersect at right angles. (See Figure 8.5.) Note, however, that $\mathbf{F}$ does not specify a direction at the origin and the orthogonality breaks down there; in fact neither family has a unique curve through that point. ■

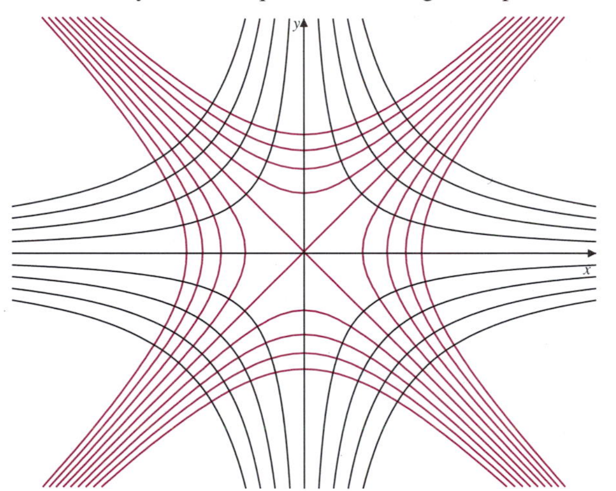


**Figure 8.5** The field lines (black) and equipotential curves (colour) for the field $\mathbf{F} = x\mathbf{i} - y\mathbf{j}$

***REMARK*** In the above example we constructed the potential $\phi$ by first integrating $\partial\phi/\partial x = F_1$. We could equally well have started by integrating $\partial\phi/\partial y = F_2$, in which case the constant of integration would have depended on $x$. In the end the same $\phi$ would have emerged.

**EXAMPLE 4** Decide whether the vector field

$$\mathbf{F} = \left(xy - \sin z\right)\mathbf{i} + \left(\frac{1}{2}x^2 - \frac{e^y}{z}\right)\mathbf{j} + \left(\frac{e^y}{z^2} - x\cos z\right)\mathbf{k}$$

is conservative in $D = \{(x, y, z) : z \neq 0\}$, and find a potential if it is.

***SOLUTION*** Note that $\mathbf{F}$ is not defined when $z = 0$. However, since

$$\frac{\partial F_1}{\partial y} = x = \frac{\partial F_2}{\partial x}, \quad \frac{\partial F_1}{\partial z} = -\cos z = \frac{\partial F_3}{\partial x}, \quad\text{and}\quad \frac{\partial F_2}{\partial z} = \frac{e^y}{z^2} = \frac{\partial F_3}{\partial y},$$

$\mathbf{F}$ may still be conservative in domains not intersecting the $xy$-plane $z = 0$. If so, its potential $\phi$ should satisfy

$$\frac{\partial \phi}{\partial x} = xy - \sin z, \qquad \frac{\partial \phi}{\partial y} = \frac{1}{2}x^2 - \frac{e^y}{z}, \quad \text{and} \quad \frac{\partial \phi}{\partial z} = \frac{e^y}{z^2} - x\cos z. \qquad (*)$$

From the first equation of $(*)$,

$$\phi(x, y, z) = \int (xy - \sin z)\,dx = \frac{1}{2}x^2 y - x \sin z + C_1(y, z).$$

(Again note that the constant of integration can be a function of any parameters of the integrand; it is constant only with respect to the variable of integration.) Using the second equation of $(*)$, we obtain

$$\frac{1}{2}x^2 - \frac{e^y}{z} = \frac{\partial \phi}{\partial y} = \frac{1}{2}x^2 + \frac{\partial C_1(y, z)}{\partial y}.$$

Thus

$$C_1(y, z) = -\int \frac{e^y}{z}\,dy = -\frac{e^y}{z} + C_2(z)$$

and

$$\phi(x, y, z) = \frac{1}{2}x^2 y - x \sin z - \frac{e^y}{z} + C_2(z).$$

Finally, using the third equation of $(*)$,

$$\frac{e^y}{z^2} - x\cos z = \frac{\partial \phi}{\partial z} = -x\cos z + \frac{e^y}{z^2} + C_2{}'(z).$$

Thus $C_2{}'(z) = 0$ and $C_2(z) = C$ (a constant). Indeed, $\mathbf{F}$ is conservative and, for any constant $C$,

$$\phi(x, y, z) = \frac{1}{2}x^2 y - x \sin z - \frac{e^y}{z} + C$$

is a potential function for $\mathbf{F}$ in the given domain $D$. $C$ may have different values in the two regions $z > 0$ and $z < 0$ whose union constitutes $D$. ■

**REMARK** If, in the above solution, the differential equation for $C_1(y, z)$ had involved $x$, or if that for $C_2(z)$ had involved either $x$ or $y$, we would not have been able to find $\phi$. This did not happen because of the three conditions on the partials of $F_1$, $F_2$, and $F_3$ verified at the outset.

**REMARK** The existence of a potential for a vector field depends on the *topology* of the domain of the field (that is, whether the domain has *holes* in it, and what kind of holes) as well as on the structure of the components of the field itself. (Even if the necessary conditions given above are satisfied, a vector field may not be conservative in a domain that has *holes*.) We will be probing further into the nature of conservative vector fields in Section 8.4 and in the next chapter, and will eventually show that the above *necessary conditions* are also *sufficient* to guarantee that $\mathbf{F}$ is conservative if the domain of $\mathbf{F}$ satisfies certain conditions. At this point, however, we give an example in which a plane vector field fails to be conservative on a domain where the necessary condition is, nevertheless, satisfied.

**EXAMPLE 5** For $(x, y) \neq (0, 0)$ define a vector field $\mathbf{F}(x, y)$ and a scalar field $\theta(x, y)$ as follows:

$$\mathbf{F}(x, y) = \left(\frac{-y}{x^2 + y^2}\right)\mathbf{i} + \left(\frac{x}{x^2 + y^2}\right)\mathbf{j}$$

$$\theta(x, y) = \text{the polar angle } \theta \text{ of } (x, y) \text{ such that } 0 \leq \theta < 2\pi.$$

Thus $x = r\cos\theta(x, y)$ and $y = r\sin\theta(x, y)$, where $r^2 = x^2 + y^2$. Verify the following:

(a) $\dfrac{\partial}{\partial y}F_1(x, y) = \dfrac{\partial}{\partial x}F_2(x, y)$ for $(x, y) \neq (0, 0)$.

(b) $\nabla\theta(x, y) = \mathbf{F}(x, y)$, for all $(x, y) \neq (0, 0)$ such that $0 < \theta < 2\pi$.

(c) $\mathbf{F}$ is not conservative on the whole $xy$-plane excluding the origin.

***SOLUTION***

(a) We have $F_1 = \dfrac{-y}{x^2 + y^2}$ and $F_2 = \dfrac{x}{x^2 + y^2}$. Thus

$$\begin{aligned}\frac{\partial}{\partial y}F_1(x, y) &= \frac{\partial}{\partial y}\left(-\frac{y}{x^2 + y^2}\right) = \frac{y^2 - x^2}{(x^2 + y^2)^2} = \frac{\partial}{\partial x}\left(\frac{x}{x^2 + y^2}\right)\\ &= \frac{\partial}{\partial x}F_2(x, y)\end{aligned}$$

for all $(x, y) \neq (0, 0)$.

(b) Differentiate the equations $x = r\cos\theta$ and $y = r\sin\theta$ implicitly with respect to $x$ to obtain

$$\begin{aligned}1 &= \frac{\partial x}{\partial x} = \frac{\partial r}{\partial x}\cos\theta - r\sin\theta\frac{\partial\theta}{\partial x},\\ 0 &= \frac{\partial y}{\partial x} = \frac{\partial r}{\partial x}\sin\theta + r\cos\theta\frac{\partial\theta}{\partial x}.\end{aligned}$$

Eliminating $\partial r/\partial x$ from this pair of equations and solving for $\partial\theta/\partial x$ leads to

$$\frac{\partial\theta}{\partial x} = -\frac{r\sin\theta}{r^2} = -\frac{y}{x^2 + y^2} = F_1.$$

Similarly, differentiation with respect to $y$ produces

$$\frac{\partial\theta}{\partial y} = \frac{x}{x^2 + y^2} = F_2.$$

These formulas hold only if $0 < \theta < 2\pi$; $\theta$ is not even continuous on the positive $x$-axis; if $x > 0$, then

$$\lim_{y \to 0+}\theta(x, y) = 0 \qquad \text{but} \qquad \lim_{y \to 0-}\theta(x, y) = 2\pi.$$

Thus $\nabla\theta = \mathbf{F}$ holds everywhere in the plane except at points $(x, 0)$ where $x \geq 0$.

(c) Suppose that $\mathbf{F}$ is conservative on the whole plane excluding the origin. Then $\mathbf{F} = \nabla\phi$ there, for some scalar function $\phi(x, y)$. Then $\nabla(\theta - \phi) = 0$ for $0 < \theta < 2\pi$, and $\theta - \phi = C$ (constant), or $\theta = \phi + C$. The left side of this equation is discontinuous along the positive $x$-axis but the right side is not. Therefore the two sides cannot be equal. This contradiction shows that $\mathbf{F}$ cannot be conservative on the whole plane excluding the origin. ■

**REMARK** Observe that the origin $(0, 0)$ is a *hole* in the domain of $\mathbf{F}$ in the above example. While $\mathbf{F}$ satisfies the necessary condition for being conservative everywhere except at this hole, you must remove from the domain of $\mathbf{F}$ a half-line (ray), or, more generally, a curve from the origin to infinity in order to get a potential function for $\mathbf{F}$. $\mathbf{F}$ is *not* conservative on any domain containing a curve which surrounds the origin. Exercises 22–24 of Section 8.4 will shed further light on this situation.

## Sources, Sinks, and Dipoles

Imagine that 3-space is filled with an incompressible fluid emitted by a point source at the origin at a volume rate $dV/dt = 4\pi m$. (We say that the origin is a **source** of strength $m$.) By symmetry, the fluid flows outward on radial lines from the origin with equal speed at equal distances from the origin in all directions, and the fluid emitted at the origin at some instant $t = 0$ will at later time $t$ be spread over a spherical surface of radius $r = r(t)$. All the fluid inside that sphere was emitted in the time interval $[0, t]$, so we have

$$\frac{4}{3}\pi r^3 = 4\pi m t.$$

Differentiating this equation with respect to $t$ we obtain $r^2(dr/dt) = m$, and the outward speed of the fluid at distance $r$ from the origin is $v(r) = m/r^2$. The velocity field of the moving fluid is therefore

$$\mathbf{v}(\mathbf{r}) = v(r)\frac{\mathbf{r}}{|\mathbf{r}|} = \frac{m}{r^3}\mathbf{r}.$$

This velocity field is conservative (except at the origin) and has potential

$$\phi(\mathbf{r}) = -\frac{m}{r}.$$

A **sink** is a negative source. A sink of strength $m$ at the origin (which annihilates or sucks up fluid at a rate $dV/dt = 4\pi m$) has velocity field and potential given by

$$\mathbf{v}(\mathbf{r}) = -\frac{m}{r^3}\mathbf{r} \qquad \text{and} \qquad \phi(\mathbf{r}) = \frac{m}{r}.$$

The potentials or velocity fields of sources or sinks located at other points are obtained by translation of these formulas; for instance, the velocity field of a source of strength $m$ at the point with position vector $\mathbf{r}_0$ is

$$\mathbf{v}(\mathbf{r}) = -\nabla\left(\frac{m}{|\mathbf{r}-\mathbf{r}_0|}\right) = \frac{m}{|\mathbf{r}-\mathbf{r}_0|^3}(\mathbf{r}-\mathbf{r}_0).$$

This should be compared with the gravitational force field due to a mass $m$ at the origin. The two are the same except for sign and a constant related to units of measurement. For this reason we regard a point mass as a sink for its own gravitational field. Similarly, the electrostatic field due to a point charge $q$ at $\mathbf{r}_0$ is the field of a source (or sink if $q < 0$) of strength proportional to $q$; if units of measurement are suitably chosen we have

$$\mathbf{E}(\mathbf{r}) = -\nabla\left(\frac{q}{|\mathbf{r}-\mathbf{r}_0|}\right) = \frac{q}{|\mathbf{r}-\mathbf{r}_0|^3}(\mathbf{r}-\mathbf{r}_0).$$

In general, the field lines of a vector field converge at a source or sink of that field.

A **dipole** is a system consisting of a source and a sink of equal strength $m$ separated by a short distance $\ell$. The product $\mu = m\ell$ is called the **dipole moment** and the line containing the source and sink is called the **axis** of the dipole. Real physical dipoles, such as magnets, are frequently modelled by *ideal* dipoles that are the limits of such real dipoles as $m \to \infty$ and $\ell \to 0$ in such a way that the dipole moment $\mu$ remains constant.

**EXAMPLE 6** Calculate the velocity field, $\mathbf{v}(x, y, z)$, associated with a dipole of moment $\mu$ located at the origin and having axis along the $z$-axis.

***SOLUTION*** We start with a source of strength $m$ at position $(0, 0, \ell/2)$ and a sink of strength $m$ at $(0, 0, -\ell/2)$. The potential of this system is

$$\phi(\mathbf{r}) = -m\left(\frac{1}{\left|\mathbf{r} - \frac{1}{2}\ell\mathbf{k}\right|} - \frac{1}{\left|\mathbf{r} + \frac{1}{2}\ell\mathbf{k}\right|}\right).$$

The potential of the ideal dipole is the limit of the potential of this system as $m \to \infty$ and $\ell \to 0$ in such a way that $m\ell = \mu$:

$$\begin{aligned}
\phi(\mathbf{r}) &= \lim_{\substack{\ell\to 0\\ ml=\mu}} -m\left(\frac{\left|\mathbf{r} + \frac{1}{2}\ell\mathbf{k}\right| - \left|\mathbf{r} - \frac{1}{2}\ell\mathbf{k}\right|}{\left|\mathbf{r} + \frac{1}{2}\ell\mathbf{k}\right|\left|\mathbf{r} - \frac{1}{2}\ell\mathbf{k}\right|}\right)\\
&= -\frac{\mu}{|\mathbf{r}|^2}\lim_{\ell\to 0}\frac{\left|\mathbf{r} + \frac{1}{2}\ell\mathbf{k}\right| - \left|\mathbf{r} - \frac{1}{2}\ell\mathbf{k}\right|}{\ell}
\end{aligned}$$

(now use l'Hôpital's Rule and the rule for differentiating lengths of vectors)

$$\begin{aligned}
&= -\frac{\mu}{|\mathbf{r}|^2}\lim_{\ell\to 0}\frac{\dfrac{\left(\mathbf{r} + \frac{1}{2}\ell\mathbf{k}\right)\bullet\frac{1}{2}\mathbf{k}}{\left|\mathbf{r} + \frac{1}{2}\ell\mathbf{k}\right|} + \dfrac{\left(\mathbf{r} - \frac{1}{2}\ell\mathbf{k}\right)\bullet\left(-\frac{1}{2}\mathbf{k}\right)}{\left|\mathbf{r} - \frac{1}{2}\ell\mathbf{k}\right|}}{1}\\
&= -\frac{\mu}{|\mathbf{r}|^2}\lim_{\ell\to 0}\left(\frac{\frac{1}{2}z + \frac{1}{4}\ell}{\left|\mathbf{r} + \frac{1}{2}\ell\mathbf{k}\right|} + \frac{\frac{1}{2}z - \frac{1}{4}\ell}{\left|\mathbf{r} - \frac{1}{2}\ell\mathbf{k}\right|}\right)\\
&= -\frac{\mu z}{|\mathbf{r}|^3}.
\end{aligned}$$

The required velocity field is the gradient of this potential. We have

$$\begin{aligned}
\frac{\partial\phi}{\partial x} &= \frac{3\mu z}{|\mathbf{r}|^4}\frac{\mathbf{r}\bullet\mathbf{i}}{|\mathbf{r}|} = \frac{3\mu xz}{|\mathbf{r}|^5}\\
\frac{\partial\phi}{\partial y} &= \frac{3\mu yz}{|\mathbf{r}|^5}, \qquad \text{similarly}\\
\frac{\partial\phi}{\partial z} &= -\frac{\mu}{|\mathbf{r}|^3} + \frac{3\mu z^2}{|\mathbf{r}|^5} = \frac{\mu(2z^2 - x^2 - y^2)}{|\mathbf{r}|^5}\\
\mathbf{v}(\mathbf{r}) = \nabla\phi(\mathbf{r}) &= \frac{\mu}{|\mathbf{r}|^5}\big(3xz\mathbf{i} + 3yz\mathbf{j} + (2z^2 - x^2 - y^2)\mathbf{k}\big).
\end{aligned}$$

Some streamlines for a plane cross-section containing the $z$-axis are shown in Figure 8.6.

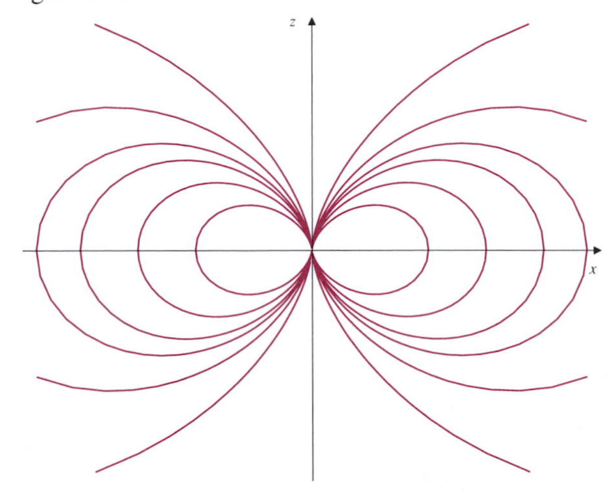


**Figure 8.6** Streamlines of a dipole

## EXERCISES 8.2

In Exercises 1–6, determine whether the given vector field is conservative and find a potential if it is conservative.

**1.** $\mathbf{F}(x, y, z) = x\mathbf{i} - 2y\mathbf{j} + 3z\mathbf{k}$

**2.** $\mathbf{F}(x, y, z) = y\mathbf{i} + x\mathbf{j} + z^2\mathbf{k}$

**3.** $\mathbf{F}(x, y) = \dfrac{x\mathbf{i} - y\mathbf{j}}{x^2 + y^2}$ **4.** $\mathbf{F}(x, y) = \dfrac{x\mathbf{i} + y\mathbf{j}}{x^2 + y^2}$

**5.** $\mathbf{F}(x, y, z) = (2xy - z^2)\mathbf{i} + (2yz + x^2)\mathbf{j} - (2zx - y^2)\mathbf{k}$

**6.** $\mathbf{F}(x, y, z) = e^{x^2+y^2+z^2}(xz\mathbf{i} + yz\mathbf{j} + xy\mathbf{k})$

**7.** Find the three-dimensional vector field with potential $\phi(\mathbf{r}) = \frac{1}{|\mathbf{r}-\mathbf{r}_0|^2}$.

**8.** Calculate $\nabla \ln |\mathbf{r}|$, where $\mathbf{r} = x\mathbf{i} + y\mathbf{j} + z\mathbf{k}$.

**9.** Show that the vector field

$$\mathbf{F}(x, y, z) = \frac{2x}{z}\mathbf{i} + \frac{2y}{z}\mathbf{j} - \frac{x^2 + y^2}{z^2}\mathbf{k}$$

is conservative, and find its potential. Describe the equipotential surfaces. Find the field lines of $\mathbf{F}$.

**10.** Repeat Exercise 9 for the field

$$\mathbf{F}(x, y, z) = \frac{2x}{z}\mathbf{i} + \frac{2y}{z}\mathbf{j} + \left(1 - \frac{x^2 + y^2}{z^2}\right)\mathbf{k}.$$

**11.** Find the velocity field due to two sources of strength $m$, one located at $(0, 0, \ell)$ and the other at $(0, 0, -\ell)$. Where is the velocity zero? Find the velocity at any point $(x, y, 0)$ in the $xy$-plane. Where in the $xy$-plane is the speed greatest?

* **12.** Find the velocity field for a system consisting of a source of strength 2 at the origin and a sink of strength 1 at $(0, 0, 1)$. Show that the velocity is vertical at all points of a certain sphere. Sketch the streamlines of the flow.

Exercises 13–18 provide an analysis of two-dimensional sources and dipoles similar to that developed for three dimensions in the text.

**13.** In 3-space filled with an incompressible fluid we say that the $z$-axis is a **line source** of strength $m$ if every interval $\Delta z$ along that axis emits fluid at volume rate $dV/dt = 2\pi m \Delta z$. The fluid then spreads out symmetrically in all directions perpendicular to the $z$-axis. Show that the velocity field of the flow is

$$\mathbf{v} = \frac{m}{x^2 + y^2}(x\mathbf{i} + y\mathbf{j}).$$

**14.** The flow in Exercise 13 is two-dimensional because $\mathbf{v}$ depends only on $x$ and $y$ and has no component in the $z$ direction. Regarded as a *plane* vector field, it is the field of a two-dimensional point source of strength $m$ located at the origin. (That is, fluid is emitted at the origin at the *areal rate* $dA/dt = 2\pi m$.) Show that the vector field is conservative and find a potential function $\phi(x, y)$ for it.

* **15.** Find the potential, $\phi$, and the field, $\mathbf{F} = \nabla\phi$, for a two-dimensional dipole at the origin, with axis in the $y$ direction and dipole moment $\mu$. Such a dipole is the limit of a system consisting of a source of strength $m$ at $(0, \ell/2)$ and a sink of strength $m$ at $(0, -\ell/2)$, as $\ell \to 0$ and $m \to \infty$ such that $m\ell = \mu$.

* **16.** Show that the equipotential curves of the two-dimensional dipole in Exercise 15 are circles tangent to the $x$-axis at the origin.

* **17.** Show that the streamlines (field lines) of the two-dimensional dipole in Exercises 15 and 16 are circles tangent to the $y$-axis at the origin. *Hint:* it is possible to do this geometrically. If you choose to do it by setting up a differential equation you may find the change of dependent variable

$$y = vx, \qquad \frac{dy}{dx} = v + x\frac{dv}{dx}$$

useful for integrating the equation.

* **18.** Show that the velocity field of a line source of strength $2m$ can be found by integrating the (three-dimensional) velocity field of a point source of strength $m\,dz$ at $(0, 0, z)$ over the whole $z$-axis. Why does the integral correspond to a line source of strength $2m$ rather than strength $m$? Can the potential of the line source be obtained by integrating the potentials of the point sources?

**19.** Show that the gradient of a function expressed in terms of polar coordinates in the plane is

$$\nabla\phi(r, \theta) = \frac{\partial\phi}{\partial r}\hat{\mathbf{r}} + \frac{1}{r}\frac{\partial\phi}{\partial\theta}\hat{\boldsymbol{\theta}}.$$

(This is a repeat of Exercise 20 in Section 4.7.)

**20.** Use the result of Exercise 19 to show that a necessary condition for the vector field

$$\mathbf{F}(r, \theta) = F_r(r, \theta)\hat{\mathbf{r}} + F_\theta(r, \theta)\hat{\boldsymbol{\theta}}$$

(expressed in terms of polar coordinates) to be conservative is that

$$\frac{\partial F_r}{\partial\theta} - r\frac{\partial F_\theta}{\partial r} = F_\theta.$$

**21.** Show that $\mathbf{F} = r\sin 2\theta\,\hat{\mathbf{r}} + r\cos 2\theta\,\hat{\boldsymbol{\theta}}$ is conservative, and find a potential for it.

**22.** For what values of the constants $\alpha$ and $\beta$ is the vector field

$$\mathbf{F} = r^2\cos\theta\,\hat{\mathbf{r}} + \alpha r^\beta\sin\theta\,\hat{\boldsymbol{\theta}}$$

conservative? Find a potential for $\mathbf{F}$ if $\alpha$ and $\beta$ have these values.

# 8.3 LINE INTEGRALS

If a wire stretched out along the $x$-axis from $x = a$ to $x = b$ has constant *line density* $\delta$ (units of mass per unit length), then the total mass of the wire will be $m = \delta(b - a)$. If the density of the wire is not constant, but varies continuously from point to point (for example, because the thickness of the wire is not uniform), then we must find the total mass of the wire by "summing" (that is, integrating) differential elements of mass $dm = \delta(x)\, dx$:

$$m = \int_a^b \delta(x)\, dx.$$

In general, the definite integral, $\int_a^b f(x)\, dx$, represents the *total amount* of a quantity distributed along the $x$-axis between $a$ and $b$ in terms of the *line density*, $f(x)$, of that quantity at point $x$. The amount of the quantity in an *infinitesimal* interval of length $dx$ at $x$ is $f(x)\, dx$, and the integral adds up these infinitesimal contributions (or *elements*) to give the total amount of the quantity. Similarly, the integrals $\iint_D f(x, y)\, dA$ and $\iiint_R f(x, y, z)\, dV$ represent the total amounts of quantities distributed over regions $D$ in the plane and $R$ in 3-space in terms of the *areal* or *volume* densities of these quantities.

It may happen that a quantity is distributed with specified line density along a *curve* in the plane or in 3-space, or with specified areal density over a *surface* in 3-space. In such cases we require *line integrals* or *surface integrals* to add up the contributing elements and calculate the total quantity. We examine line integrals in these next two sections and surface integrals in Sections 8.5 and 8.6.

Let $\mathcal{C}$ be a bounded, continuous parametric curve in $\mathbb{R}^3$. Recall (from Section 7.1) that $\mathcal{C}$ is a *smooth curve* if it has a parametrization of the form

$$\mathbf{r} = \mathbf{r}(t) = x(t)\mathbf{i} + y(t)\mathbf{j} + z(t)\mathbf{k}, \qquad a \le t \le b,$$

with "velocity" vector $\mathbf{v} = d\mathbf{r}/dt$ continuous and nonzero. We will call $\mathcal{C}$ a **smooth arc** if it is a smooth curve with *finite* parameter interval $[a, b]$.

In Section 7.3 we saw how to calculate the length of $\mathcal{C}$ by subdividing it into short arcs using points corresponding to parameter values

$$a = t_0 < t_1 < t_2 < \cdots < t_{n-1} < t_n = b,$$

adding up the lengths $|\Delta\mathbf{r}_i| = |\mathbf{r}_i - \mathbf{r}_{i-1}|$ of line segments joining these points, and taking the limit as the maximum distance between adjacent points approached zero. The length was denoted

$$\int_{\mathcal{C}} ds$$

and is a special example of a line integral along $\mathcal{C}$ having integrand 1.

The line integral of a general function $f(x, y, z)$ can be defined similarly. We choose a point $(x_i^*, y_i^*, z_i^*)$ on the $i$th subarc and form the Riemann sum

$$S_n = \sum_{i=1}^{n} f(x_i^*, y_i^*, z_i^*)\, |\Delta\mathbf{r}_i|.$$

If this sum has a limit as max $|\Delta \mathbf{r}_i| \to 0$, independent of the particular choices of the points $(x_i^*, y_i^*, z_i^*)$, then we call this limit the line integral of $f$ along $\mathcal{C}$, and denote it by

$$\int_{\mathcal{C}} f(x, y, z)\, ds.$$

If $\mathcal{C}$ is a smooth arc and if $f$ is continuous on $\mathcal{C}$, then the limit will certainly exist; its value is given by a definite integral of a continuous function, as shown in the next paragraph. It will also exist (for continuous $f$) if $\mathcal{C}$ is **piecewise smooth**, consisting of finitely many smooth arcs linked end to end; in this case the line integral of $f$ along $\mathcal{C}$ is the sum of the line integrals of $f$ along each of the smooth arcs. Improper line integrals can also be considered, where $f$ has discontinuities or where the length of an arc or curve is not finite.

## Evaluating Line Integrals

The length of $\mathcal{C}$ was evaluated by expressing the arc length element $ds = |d\mathbf{r}/dt|\, dt$ in terms of a parametrization $\mathbf{r} = \mathbf{r}(t)$, $(a \le t \le b)$ of the curve, and integrating this from $t = a$ to $t = b$:

$$\text{length of } \mathcal{C} = \int_{\mathcal{C}} ds = \int_a^b \left| \frac{d\mathbf{r}}{dt} \right| dt.$$

More general line integrals are evaluated similarly:

$$\int_{\mathcal{C}} f(x, y, z)\, ds = \int_a^b f\big(\mathbf{r}(t)\big) \left| \frac{d\mathbf{r}}{dt} \right| dt.$$

Of course, all of the above discussion applies equally well to line integrals of functions $f(x, y)$ along curves $\mathcal{C}$ in the $xy$-plane.

***REMARK*** It should be noted that the value of the line integral of a function $f$ along a curve $\mathcal{C}$ depends on $f$ and $\mathcal{C}$, but not on the particular way $\mathcal{C}$ is parametrized. If $\mathbf{r} = \mathbf{r}^*(u)$, $\alpha \le u \le \beta$, is another parametrization of the same smooth curve $\mathcal{C}$, then any point $\mathbf{r}(t)$ on $\mathcal{C}$ can be expressed in terms of the new parametrization as $\mathbf{r}^*(u)$, where $u$ depends on $t$: $u = u(t)$. If $\mathbf{r}^*(u)$ traces $\mathcal{C}$ in the same direction as $\mathbf{r}(t)$ then $u(a) = \alpha$, $u(b) = \beta$, and $du/dt \ge 0$; if $\mathbf{r}^*(u)$ traces $\mathcal{C}$ in the opposite direction then $u(a) = \beta$, $u(b) = \alpha$, and $du/dt \le 0$. In either event,

$$\int_a^b f\big(\mathbf{r}(t)\big) \left| \frac{d\mathbf{r}}{dt} \right| dt = \int_a^b f\big(\mathbf{r}^*(u(t))\big) \left| \frac{d\mathbf{r}^*}{du} \frac{du}{dt} \right| dt = \int_\alpha^\beta f\big(\mathbf{r}^*(u)\big) \left| \frac{d\mathbf{r}^*}{du} \right| du.$$

Thus the line integral is *independent of parametrization* of the curve $\mathcal{C}$. The following example illustrates this fact.

**EXAMPLE 1** A circle of radius $a > 0$ has centre at the origin in the $xy$-plane. Let $\mathcal{C}$ be the half of this circle lying in the half-plane $y \ge 0$. Use two different parametrizations of $\mathcal{C}$ to find the moment of $\mathcal{C}$ about $y = 0$.

***SOLUTION*** We are asked to calculate $\displaystyle\int_{\mathcal{C}} y\, ds$.

$\mathcal{C}$ can be parametrized $\mathbf{r} = a \cos t\mathbf{i} + a \sin t\mathbf{j}$, $(0 \le t \le \pi)$. Therefore

$$\frac{d\mathbf{r}}{dt} = -a \sin t\mathbf{i} + a \cos t\mathbf{j} \qquad \text{and} \qquad \left| \frac{d\mathbf{r}}{dt} \right| = a,$$

and the moment of $\mathcal{C}$ about $y = 0$ is

$$\int_{\mathcal{C}} y\,ds = \int_0^{\pi} a \sin t \, a\,dt = -a^2 \cos t \Big|_0^{\pi} = 2a^2.$$

$\mathcal{C}$ can also be parametrized $\mathbf{r} = x\mathbf{i} + \sqrt{a^2 - x^2}\mathbf{j}$, $(-a \le x \le a)$, for which we have

$$\frac{d\mathbf{r}}{dx} = \mathbf{i} - \frac{x}{\sqrt{a^2 - x^2}}\mathbf{j},$$

$$\left|\frac{d\mathbf{r}}{dx}\right| = \sqrt{1 + \frac{x^2}{a^2 - x^2}} = \frac{a}{\sqrt{a^2 - x^2}}.$$

Thus the moment of $\mathcal{C}$ about $y = 0$ is

$$\int_{\mathcal{C}} y\,ds = \int_{-a}^{a} \sqrt{a^2 - x^2}\,\frac{a}{\sqrt{a^2 - x^2}}\,dx = a \int_{-a}^{a} dx = 2a^2.$$

It is comforting to get the same answer using different parametrizations. Unlike the line integrals of vector fields considered in the next section, the line integrals of scalar fields considered here do not depend on the direction (orientation) of $\mathcal{C}$. The two parametrizations of the semicircle were in opposite directions but still gave the same result. ■

Line integrals frequently lead to definite integrals that are very difficult or impossible to evaluate without using numerical techniques. Only very simple curves and ones that have been contrived to lead to simple expressions for $ds$ are amenable to exact calculation of line integrals.

**EXAMPLE 2** Find the centroid of the circular helix $\mathcal{C}$ given by

$$\mathbf{r} = a \cos t\,\mathbf{i} + a \sin t\,\mathbf{j} + bt\,\mathbf{k}, \qquad 0 \le t \le 2\pi.$$

***SOLUTION*** As we observed in Example 5 of Section 7.3, $ds = \sqrt{a^2 + b^2}\,dt$ for this helix. On the helix we have $z = bt$, so its moment about $z = 0$ is

$$M_{z=0} = \int_{\mathcal{C}} z\,ds = b\sqrt{a^2 + b^2} \int_0^{2\pi} t\,dt = 2\pi^2 b\sqrt{a^2 + b^2}.$$

Since the helix has length $L = 2\pi\sqrt{a^2 + b^2}$, the $z$-component of its centroid is $M_{z=0}/L = \pi b$. The moment of the helix about $x = 0$ is

$$M_{x=0} = \int_{\mathcal{C}} x\,ds = a\sqrt{a^2 + b^2} \int_0^{2\pi} \cos t\,dt = 0,$$

$$M_{y=0} = \int_{\mathcal{C}} y\,ds = a\sqrt{a^2 + b^2} \int_0^{2\pi} \sin t\,dt = 0.$$

Thus the centroid is $(0, 0, \pi b)$. ■

Sometimes a curve, along which a line integral is to be taken, is specified as the intersection of two surfaces with given equations. It is normally necessary to parametrize the curve in order to evaluate a line integral. Recall from Section 7.3 that if one of the surfaces is a cylinder parallel to one of the coordinate axes it is usually easiest to begin by parametrizing that cylinder. (Otherwise, combine the equations to eliminate one variable and thus obtain such a cylinder on which the curve lies.)

**EXAMPLE 3** Find the mass of a wire lying along the first octant part $\mathcal{C}$ of the curve of intersection of the elliptic paraboloid $z = 2 - x^2 - 2y^2$ and the parabolic cylinder $z = x^2$ between $(0, 1, 0)$ and $(1, 0, 1)$ if the density of the wire at position $(x, y, z)$ is $\delta(x, y, z) = xy$.

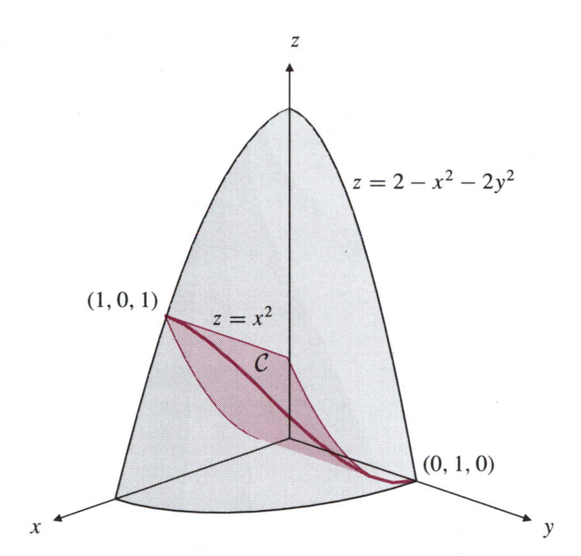


**Figure 8.7** The curve of intersection of $z = x^2$ and $z = 2 - x^2 - 2y^2$

***SOLUTION*** We need a convenient parametrization of $\mathcal{C}$. Since the curve $\mathcal{C}$ lies on the cylinder $z = x^2$ and $x$ goes from 0 to 1, we can let $x = t$ and $z = t^2$. Thus $2y^2 = 2 - x^2 - z = 2 - 2t^2$, so $y^2 = 1 - t^2$. Since $\mathcal{C}$ lies in the first octant, it can be parametrized by

$$x = t, \qquad y = \sqrt{1 - t^2}, \qquad z = t^2, \qquad (0 \le t \le 1).$$

Then $dx/dt = 1$, $dy/dt = -t/\sqrt{1 - t^2}$, and $dz/dt = 2t$ so

$$ds = \sqrt{1 + \frac{t^2}{1 - t^2} + 4t^2}\, dt = \frac{\sqrt{1 + 4t^2 - 4t^4}}{\sqrt{1 - t^2}}\, dt.$$

Hence the mass of the wire is

$$\begin{aligned}
m = \int_{\mathcal{C}} xy\, ds &= \int_0^1 t\sqrt{1 - t^2}\frac{\sqrt{1 + 4t^2 - 4t^4}}{\sqrt{1 - t^2}}\, dt \\
&= \int_0^1 t\sqrt{1 + 4t^2 - 4t^4}\, dt \qquad \text{Let } u = t^2. \\
&= \frac{1}{2}\int_0^1 \sqrt{1 + 4u - 4u^2}\, du \\
&= \frac{1}{2}\int_0^1 \sqrt{2 - (2u - 1)^2} \qquad \text{Let } v = 2u - 1. \\
&= \frac{1}{4}\int_{-1}^1 \sqrt{2 - v^2}\, dv = \frac{1}{2}\int_0^1 \sqrt{2 - v^2}\, dv \\
&= \frac{1}{2}\left(\frac{\pi}{4} + \frac{1}{2}\right) = \frac{\pi + 2}{8}.
\end{aligned}$$

(The final integral above was evaluated by interpreting it as the area of part of a circle. You are invited to supply the details. It can also be done by the substitution $v = \sqrt{2}\sin w$.) ■

## EXERCISES 8.3

1. Show that the curve $\mathcal{C}$ given by
$$\mathbf{r} = a\cos t\sin t\,\mathbf{i} + a\sin^2 t\,\mathbf{j} + a\cos t\,\mathbf{k}, \quad (0 \le t \le \tfrac{\pi}{2}),$$
lies on a sphere centred at the origin. Find $\int_{\mathcal{C}} z\,ds$.

2. Let $\mathcal{C}$ be the conical helix with parametric equations $x = t\cos t$, $y = t\sin t$, $z = t$, $(0 \le t \le 2\pi)$. Find $\int_{\mathcal{C}} z\,ds$.

3. Find the mass of a wire along the curve
$$\mathbf{r} = 3t\mathbf{i} + 3t^2\mathbf{j} + 2t^3\mathbf{k}, \quad (0 \le t \le 1),$$
if the density at $\mathbf{r}(t)$ is $1 + t$ gm/unit length.

4. Show that the curve $\mathcal{C}$ in Example 3 also has parametrization $x = \cos t$, $y = \sin t$, $z = \cos^2 t$, $(0 \le t \le \pi/2)$, and recalculate the mass of the wire in that example using this parametrization.

5. Find the moment of inertia about the $z$-axis (that is, the value of $\int_{\mathcal{C}} (x^2 + y^2)\,ds$), for a wire of constant density $\delta$ lying along the curve $\mathcal{C}$: $\mathbf{r} = e^t\cos t\,\mathbf{i} + e^t\sin t\,\mathbf{j} + t\mathbf{k}$, from $t = 0$ to $t = 2\pi$.

6. Evaluate $\int_{\mathcal{C}} e^z\,ds$, where $\mathcal{C}$ is the curve in Exercise 5.

7. Find $\int_{\mathcal{C}} x^2\,ds$ along the line of intersection of the two planes $x - y + z = 0$, and $x + y + 2z = 0$, from the origin to the point $(3, 1, -2)$.

8. Find $\int_{\mathcal{C}} \sqrt{1 + 4x^2z^2}\,ds$, where $\mathcal{C}$ is the curve of intersection of the surfaces $x^2 + z^2 = 1$ and $y = x^2$.

9. Find the mass and centre of mass of a wire bent in the shape of the circular helix $x = \cos t$, $y = \sin t$, $z = t$, $(0 \le t \le 2\pi)$ if the wire has line density given by $\delta(x, y, z) = z$.

10. Repeat Exercise 9 for the part of the wire corresponding to $0 \le t \le \pi$.

11. Find the moment of inertia about the $y$-axis, that is,
$$\int_{\mathcal{C}} (x^2 + z^2)\,ds,$$
of the curve $x = e^t$, $y = \sqrt{2}\,t$, $z = e^{-t}$, $(0 \le t \le 1)$.

12. Find the centroid of the curve in Exercise 11.

* 13. Find $\int_{\mathcal{C}} x\,ds$ along the first octant part of the curve of intersection of the cylinder $x^2 + y^2 = a^2$ and the plane $z = x$.

* 14. Find $\int_{\mathcal{C}} z\,ds$ along the part of the curve $x^2 + y^2 + z^2 = 1$, $x + y = 1$, where $z \ge 0$.

* 15. Find $\displaystyle\int_{\mathcal{C}} \frac{ds}{(2y^2 + 1)^{3/2}}$, where $\mathcal{C}$ is the parabola $z^2 = x^2 + y^2$, $x + z = 1$.

16. Express as a definite integral, but do not try to evaluate, the value of $\int_{\mathcal{C}} xyz\,ds$, where $\mathcal{C}$ is the curve $y = x^2$, $z = y^2$ from $(0, 0, 0)$ to $(2, 4, 16)$.

* 17. The function
$$E(k, \phi) = \int_0^{\phi} \sqrt{1 - k^2\sin^2 t}\,dt$$
is called the **elliptic integral function of the second kind**. The **complete elliptic integral** of the second kind is the function $E(k) = E(k, \pi/2)$. In terms of these functions, express the length of one complete revolution of the elliptic helix
$$x = a\cos t, \qquad y = b\sin t, \qquad z = ct,$$
where $0 < a < b$. What is the length of that part of the helix lying between $t = 0$ and $t = T$ where $0 < T < \pi/2$?

* 18. Evaluate $\displaystyle\int_{\mathcal{L}} \frac{ds}{x^2 + y^2}$, where $\mathcal{L}$ is the entire straight line with equation $Ax + By = C$, $(C \ne 0)$. *Hint:* use the symmetry of the integrand to replace the line with a line having a simpler equation but giving the same value to the integral.

## 8.4 LINE INTEGRALS OF VECTOR FIELDS

In elementary physics the **work** done by a constant force of magnitude $F$ in moving an object a distance $d$ is defined to be the product of $F$ and $d$: $W = Fd$. There is, however, a catch to this; it is understood that the force is exerted in the direction of motion of the object. If the object moves in a direction different from that of the force (because of some other forces acting on it) then the work done by the particular force is the product of the distance moved and the component of the force in the direction of motion. For instance, the work done by gravity in causing a 10 kg crate to slide 5 m down a ramp inclined at 45° to the horizontal is $W = 50g/\sqrt{2}$ N·m, (where $g = 9.8$ m/s$^2$), since the scalar projection of the $10g$ N gravitational force on the crate in the direction of the ramp is $10g/\sqrt{2}$ N.

The work done by a *variable* force $\mathbf{F}(x, y, z) = \mathbf{F}(\mathbf{r})$, that depends continuously on position, in moving an object along a smooth curve $\mathcal{C}$ is the integral of *work elements* $dW$. The element $dW$ corresponding to arc length element $ds$ at position $\mathbf{r}$ on $\mathcal{C}$ is $ds$ times the tangential component of the force $\mathbf{F}(\mathbf{r})$ along $\mathcal{C}$ in the direction of motion:

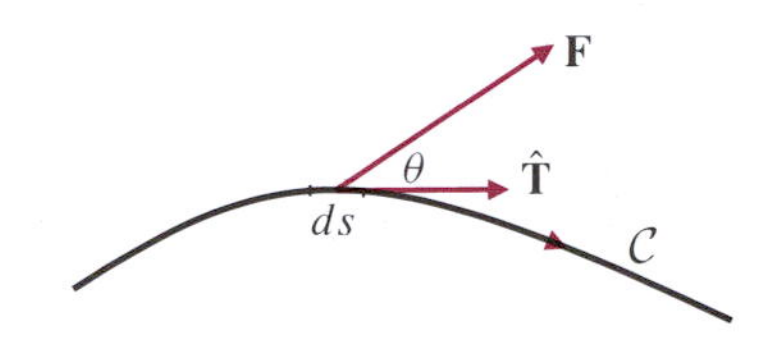


**Figure 8.8** $dW = \mathbf{F}\cos\theta\, ds = \mathbf{F} \bullet \hat{\mathbf{T}}\, ds$

$$dW = \mathbf{F}(\mathbf{r}) \bullet \hat{\mathbf{T}}\, ds.$$

Thus the total work done by $\mathbf{F}$ in moving the object along $\mathcal{C}$ is

$$W = \int_{\mathcal{C}} \mathbf{F} \bullet \hat{\mathbf{T}}\, ds = \int_{\mathcal{C}} \mathbf{F} \bullet d\mathbf{r},$$

where we use the vector differential $d\mathbf{r}$ as a convenient shorthand for $\hat{\mathbf{T}}\, ds$. Since

$$\hat{\mathbf{T}} = \frac{d\mathbf{r}}{ds} = \frac{dx}{ds}\mathbf{i} + \frac{dy}{ds}\mathbf{j} + \frac{dz}{ds}\mathbf{k}$$

we have

$$d\mathbf{r} = \hat{\mathbf{T}}\, ds = dx\,\mathbf{i} + dy\,\mathbf{j} + dz\,\mathbf{k},$$

and the work $W$ can be written in terms of the components of $\mathbf{F}$ as

$$W = \int_{\mathcal{C}} F_1\, dx + F_2\, dy + F_3\, dz.$$

In general, if $\mathbf{F} = F_1\mathbf{i} + F_2\mathbf{j} + F_3\mathbf{k}$ is a continuous vector field, and $\mathcal{C}$ is a smooth curve, then the **line integral of the tangential component of F along an oriented curve** $\mathcal{C}$ is

$$\int_{\mathcal{C}} \mathbf{F} \bullet d\mathbf{r} = \int_{\mathcal{C}} \mathbf{F} \bullet \hat{\mathbf{T}}\, ds = \int_{\mathcal{C}} F_1(x, y, z)\, dx + F_2(x, y, z)\, dy + F_3(x, y, z)\, dz.$$

Such a line integral is sometimes called, somewhat improperly, the line integral of $\mathbf{F}$ along $\mathcal{C}$. (It is not the line integral of $\mathbf{F}$, which should have a vector value, but rather the line integral of the *tangential component* of $\mathbf{F}$, which has a scalar value.) Unlike the line integral considered in the previous section, this line integral depends on the direction of the orientation of $\mathcal{C}$; reversing the direction of $\mathcal{C}$ causes this line integral to change sign.

If $\mathcal{C}$ is a closed curve, the line integral of the tangential component of $\mathbf{F}$ around $\mathcal{C}$ is also called the *circulation* of $\mathbf{F}$ around $\mathcal{C}$. The fact that the curve is closed is often indicated by a small circle drawn on the integral sign;

$$\oint_{\mathcal{C}} \mathbf{F} \bullet d\mathbf{r}$$

denotes the circulation of $\mathbf{F}$ around the closed curve $\mathcal{C}$.

Like the line integrals studied in the previous section, a line integral of a continuous vector field is converted into an ordinary definite integral by using a parametrization of the path of integration. For a smooth arc

$$\mathbf{r} = \mathbf{r}(t) = x(t)\mathbf{i} + y(t)\mathbf{j} + z(t)\mathbf{k}, \qquad (a \le t \le b),$$

we have

$$\begin{aligned}\int_{\mathcal{C}} \mathbf{F} \bullet d\mathbf{r} &= \int_a^b \mathbf{F} \bullet \frac{d\mathbf{r}}{dt}\, dt \\ &= \int_a^b \bigg[ F_1\big(x(t), y(t), z(t)\big)\frac{dx}{dt} + F_2\big(x(t), y(t), z(t)\big)\frac{dy}{dt} \\ &\qquad + F_3\big(x(t), y(t), z(t)\big)\frac{dz}{dt}\bigg]\, dt.\end{aligned}$$

Although this type of line integral changes sign if the orientation of $\mathcal{C}$ is reversed, it is otherwise independent of the particular parametrization used for $\mathcal{C}$. Again, a line integral over a piecewise smooth path is the sum of the line integrals over the individual smooth arcs constituting that path.

**EXAMPLE 1** Let $\mathbf{F}(x, y) = y^2\mathbf{i} + 2xy\mathbf{j}$. Evaluate the line integral

$$\int_{\mathcal{C}} \mathbf{F} \bullet d\mathbf{r}$$

from (0, 0) to (1, 1) along

(a) the straight line $y = x$,

(b) the curve $y = x^2$, and

(c) the piecewise smooth path consisting of the straight line segments from (0, 0) to (0, 1) and from (0, 1) to (1, 1).

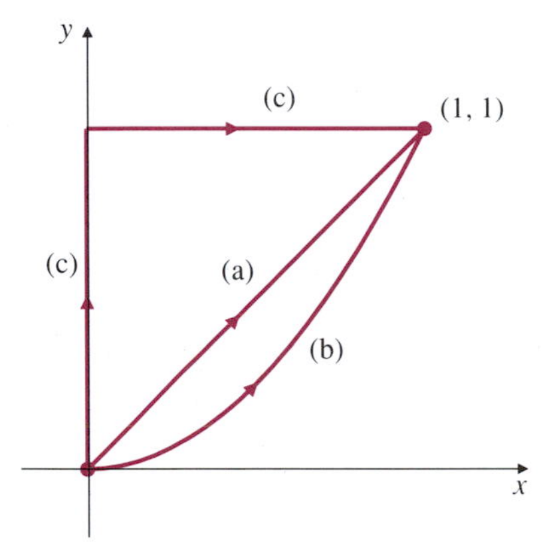


**Figure 8.9** Three paths from (0, 0) to (1, 1)

***SOLUTION*** The three paths are shown in Figure 8.9. The straight path (a) can be parametrized $\mathbf{r} = t\mathbf{i} + t\mathbf{j}$, $0 \le t \le 1$. Thus $d\mathbf{r} = dt\mathbf{i} + dt\mathbf{j}$ and

$$\mathbf{F} \bullet d\mathbf{r} = (t^2\mathbf{i} + 2t^2\mathbf{j}) \bullet (\mathbf{i} + \mathbf{j})dt = 3t^2\, dt.$$

Therefore,

$$\int_{\mathcal{C}} \mathbf{F} \bullet d\mathbf{r} = \int_0^1 3t^2\, dt = t^3\bigg|_0^1 = 1.$$

The parabolic path (b) can be parametrized $\mathbf{r} = t\mathbf{i} + t^2\mathbf{j}$, $0 \le t \le 1$, so that $d\mathbf{r} = dt\mathbf{i} + 2t\, dt\mathbf{j}$. Thus

$$\mathbf{F} \bullet d\mathbf{r} = (t^4\mathbf{i} + 2t^3\mathbf{j}) \bullet (\mathbf{i} + 2t\mathbf{j})\, dt = 5t^4\, dt,$$

and

$$\int_{\mathcal{C}} \mathbf{F} \bullet d\mathbf{r} = \int_0^1 5t^4\, dt = t^5\bigg|_0^1 = 1.$$

The third path (c) is made up of two segments, and we parametrize each separately. Let us use $y$ as parameter on the vertical segment (where $x = 0$ and $dx = 0$), and $x$ as parameter on the horizontal segment (where $y = 1$ and $dy = 0$):

$$\begin{aligned}\int_{\mathcal{C}} \mathbf{F} \bullet d\mathbf{r} &= \int_{\mathcal{C}} y^2\, dx + 2xy\, dy \\ &= \int_0^1 (1)\, dx + \int_0^1 (0)\, dy = 1.\end{aligned}$$

In view of these results, we might ask whether $\int_{\mathcal{C}} \mathbf{F} \bullet d\mathbf{r}$ is the same along *every* path from (0, 0) to (1, 1). ■

**EXAMPLE 2** Let $\mathbf{F} = y\mathbf{i} - x\mathbf{j}$. Find $\int_{\mathcal{C}} \mathbf{F} \bullet d\mathbf{r}$ from $(1, 0)$ to $(0, -1)$ along

(a) the straight line segment joining these points, and

(b) three-quarters of the circle of unit radius centred at the origin and traversed counterclockwise.

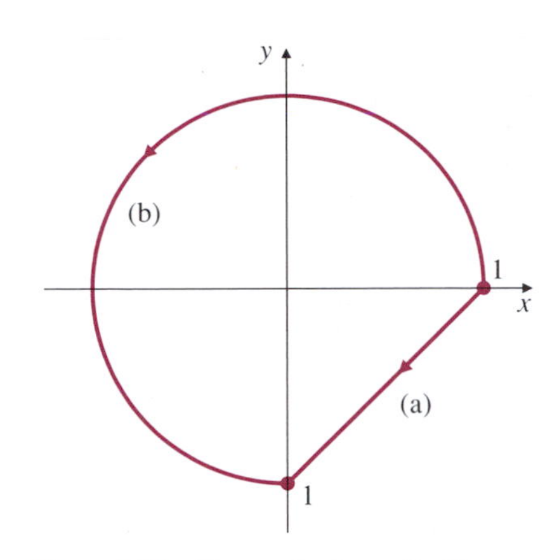


Figure 8.10 Two paths from $(1, 0)$ to $(0, -1)$

***SOLUTION*** Both paths are shown in Figure 8.10. The straight path (a) can be parametrized

$$\mathbf{r} = (1 - t)\mathbf{i} - t\mathbf{j}, \qquad 0 \le t \le 1.$$

Thus $d\mathbf{r} = -dt\mathbf{i} - dt\mathbf{j}$, and

$$\int_{\mathcal{C}} \mathbf{F} \bullet d\mathbf{r} = \int_0^1 \Big((-t)(-dt) - (1 - t)(-dt)\Big) = \int_0^1 dt = 1.$$

The circular path (b) can be parametrized

$$\mathbf{r} = \cos t\,\mathbf{i} + \sin t\,\mathbf{j}, \qquad 0 \le t \le \frac{3\pi}{2},$$

so that $d\mathbf{r} = -\sin t\,dt\mathbf{i} + \cos t\,dt\mathbf{j}$. Therefore

$$\mathbf{F} \bullet d\mathbf{r} = -\sin^2 t\,dt - \cos^2 t\,dt = -dt,$$

and we have

$$\int_{\mathcal{C}} \mathbf{F} \bullet d\mathbf{r} = -\int_0^{3\pi/2} dt = -\frac{3\pi}{2}.$$

In this case the line integral depends on the path to from $(1, 0)$ to $(0, -1)$ along which the integral is taken. ■

Some readers may have noticed that in Example 1 above the vector field $\mathbf{F}$ is conservative, while in Example 2 it is not. Theorem 1 below confirms the link between *independence of path* for a line integral of the tangential component of a vector field and the existence of a scalar potential function for that field. This and subsequent theorems require specific assumptions on the nature of the domain of the vector field $\mathbf{F}$, so we need to formulate some definitions.

## Connected and Simply Connected Domains

Recall that a set $S$ in the plane (or in 3-space) is open if every point in $S$ is the centre of a disk (or a ball) having positive radius and contained in $S$. If $S$ is open and $B$ is a set (possibly empty) of boundary points of $S$, then the set $D = S \cup B$ is called a **domain**. A domain cannot contain isolated points. It may be closed, but it must have interior points near any of its boundary points. (See the early part of Section 4.1 for a discussion of open and closed sets, and interior and boundary points.)

**DEFINITION 2**

> A domain $D$ in the plane (or in 3-space) is said to be **connected**, if for every pair of points $P$ and $Q$ in $D$, there exists a piecewise smooth curve in $D$ from $P$ to $Q$.

For instance, the set of points $(x, y)$ in the plane satisfying $x > 0$, $y > 0$, and $x^2 + y^2 \le 4$ is a connected domain, but the set of points satisfying $|x| > 1$ is not connected. (There is no path from $(-2, 0)$ to $(2, 0)$ lying entirely in $|x| > 1$.) The set of points $(x, y, z)$ in 3-space satisfying $0 < z < 1/(x^2 + y^2)$ is a connected domain but the set satisfying $z \ne 0$ is not.

**DEFINITION 3**

A connected domain $D$ is said to be **simply connected** if every *closed curve* in $D$ can be continuously shrunk to a point in $D$ without any part ever passing out of $D$.

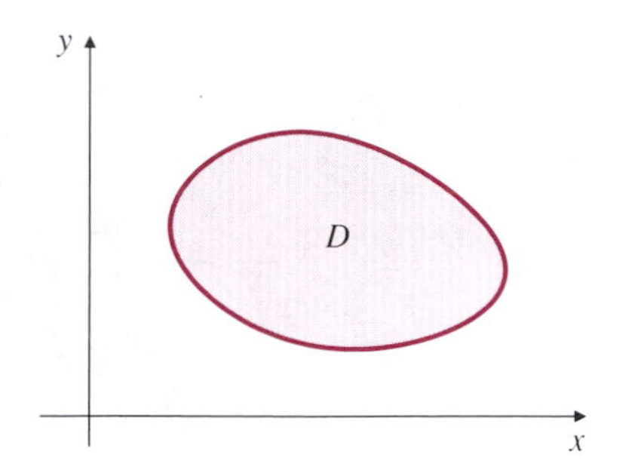

**Figure 8.11** A simply connected domain

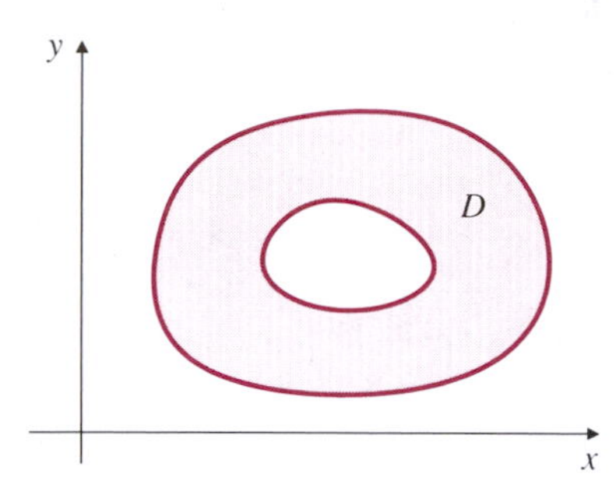

**Figure 8.12** A connected domain that is not simply connected

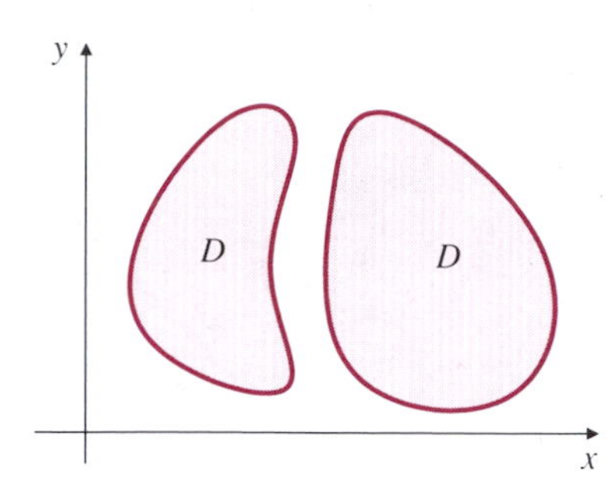

**Figure 8.13** A domain that is not connected

Figure 8.11 shows a simply connected domain in the plane. Figure 8.12 shows a connected but not simply connected domain. (A closed curve surrounding the hole cannot be shrunk to a point without passing out of $D$.) The domain in Figure 8.13 is not even connected. It has two *components*, and if $P$ and $Q$ are points in different components they cannot be joined by a curve that lies in $D$.

In the plane, a simply connected domain $D$ can have no holes, not even holes consisting of a single point. The interior of every non-self-intersecting closed curve in such a domain $D$ lies in $D$. For instance, the domain of the function $1/(x^2 + y^2)$ is not simply connected because the origin does not belong to it. (The origin is a "hole" in that domain.) In 3-space simple connectedness is a little more tricky to characterize. The set of points $(x, y, z)$ satisfying $1 < x^2 + y^2 + z^2 < 4$ (the region between two concentric spheres) is simply connected because any closed curve in that set can be shrunk to a point without having to leave that set. However, the interior of a doughnut (a torus) is not simply connected; a circle in the doughnut surrounding the doughnut hole cannot be shrunk to a point without part of it passing outside the doughnut. The set of all points lying off the $z$-axis is not simply connected. In general, each of the following conditions characterizes simply connected domains $D$:

(i) Any closed curve in $D$ is the boundary of a "surface" lying wholly in $D$.

(ii) If $C_1$ and $C_2$ are two curves in $D$ having the same endpoints, then $C_1$ can be continuously deformed into $C_2$, remaining in $D$ throughout the deformation process.

## Independence of Path

**THEOREM 1**

**Independence of path**

Let $D$ be an open, connected domain, and let $\mathbf{F}$ be a smooth vector field defined on $D$. Then the following three statements are *equivalent* in the sense that if any one of them is true, then so are the other two:

(a) $\mathbf{F}$ is conservative in $D$.

(b) $\oint_C \mathbf{F} \bullet d\mathbf{r} = 0$ for every piecewise smooth, closed curve $C$ in $D$.

(c) Given any two points $P_0$ and $P_1$ in $D$, $\int_{\mathcal{C}} \mathbf{F} \bullet d\mathbf{r}$ has the same value for all piecewise smooth curves in $D$ starting at $P_0$ and ending at $P_1$.

**PROOF** We will show that (a) implies (b), that (b) implies (c), and that (c) implies (a). It then follows that any one implies the other two.

Suppose (a) is true. Then $\mathbf{F} = \nabla\phi$ for some scalar potential function $\phi$ defined in $D$. Therefore,

$$\begin{aligned}\mathbf{F} \bullet d\mathbf{r} &= \left(\frac{\partial\phi}{\partial x}\mathbf{i} + \frac{\partial\phi}{\partial y}\mathbf{j} + \frac{\partial\phi}{\partial z}\mathbf{k}\right) \bullet \left(dx\,\mathbf{i} + dy\,\mathbf{j} + dz\,\mathbf{k}\right) \\ &= \frac{\partial\phi}{\partial x}\,dx + \frac{\partial\phi}{\partial y}\,dy + \frac{\partial\phi}{\partial y}\,dz = d\phi.\end{aligned}$$

If $\mathcal{C}$ is any piecewise smooth, closed curve, parametrized, say, by $\mathbf{r} = \mathbf{r}(t)$, $(a \le t \le b)$, then $\mathbf{r}(a) = \mathbf{r}(b)$, and

$$\int_{\mathcal{C}} \mathbf{F} \bullet d\mathbf{r} = \int_a^b \frac{d\phi\big(\mathbf{r}(t)\big)}{dt}\,dt = \phi\big(\mathbf{r}(b)\big) - \phi\big(\mathbf{r}(a)\big) = 0.$$

Thus (a) implies (b).

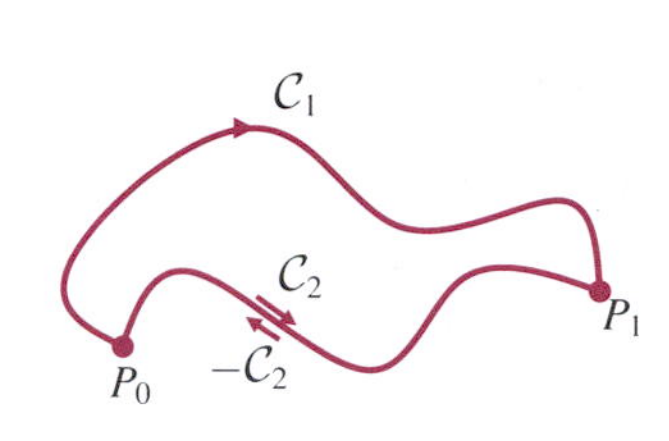

**Figure 8.14** $\mathcal{C}_1 - \mathcal{C}_2 = \mathcal{C}_1 + (-\mathcal{C}_2)$ is a closed curve

Now suppose (b) is true. Let $P_0$ and $P_1$ be two points in $D$, and let $\mathcal{C}_1$ and $\mathcal{C}_2$ be two piecewise smooth curves in $D$ from $P_0$ to $P_1$. Let $\mathcal{C} = \mathcal{C}_1 - \mathcal{C}_2$ denote the closed curve going from $P_0$ to $P_1$ along $\mathcal{C}_1$ and then back to $P_0$ along $\mathcal{C}_2$ in the opposite direction. (See Figure 8.14.) Since we are assuming that (b) is true, we have

$$0 = \oint_{\mathcal{C}} \mathbf{F} \bullet d\mathbf{r} = \int_{\mathcal{C}_1} \mathbf{F} \bullet d\mathbf{r} - \int_{\mathcal{C}_2} \mathbf{F} \bullet d\mathbf{r}.$$

Therefore,

$$\int_{\mathcal{C}_1} \mathbf{F} \bullet d\mathbf{r} = \int_{\mathcal{C}_2} \mathbf{F} \bullet d\mathbf{r},$$

and we have proved that (b) implies (c).

Finally, suppose that (c) is true. Let $P_0 = (x_0, y_0, z_0)$ be a fixed point in the domain $D$, and let $P = (x, y, z)$ be an arbitrary point in that domain. Define a function $\phi$ by

$$\phi(x, y, z) = \int_{\mathcal{C}} \mathbf{F} \bullet d\mathbf{r},$$

where $\mathcal{C}$ is some piecewise smooth curve in $D$ from $P_0$ to $P$. (Under the hypotheses of the theorem such a curve exists, and, since we are assuming (c), the integral has the same value for all such curves. Therefore, $\phi$ is well defined in $D$.) We will show that $\nabla\phi = \mathbf{F}$ and thus establish that $\mathbf{F}$ is conservative and has potential $\phi$.

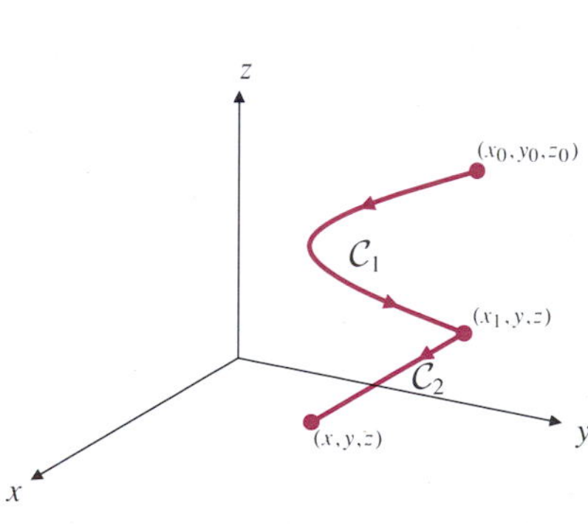

**Figure 8.15** A special path from $P_0$ to $P_1$

It is sufficient to show that $\partial\phi/\partial x = F_1(x, y, z)$; the other two components are treated similarly. Since $D$ is open there is a ball of positive radius centred at $P$ and contained in $D$. Pick a point $(x_1, y, z)$ in this ball having $x_1 < x$. Note that the line from this point to $P$ is parallel to the $x$-axis. Since we are free to choose the curve $\mathcal{C}$ in the integral defining $\phi$, let us choose it to consist of two segments, $\mathcal{C}_1$ which is piecewise smooth and goes from $(x_0, y_0, z_0)$ to $(x_1, y, z)$, and $\mathcal{C}_2$, a straight line segment from $(x_1, y, z)$ to $(x, y, z)$. (See Figure 8.15.) Then

$$\phi(x, y, z) = \int_{\mathcal{C}_1} \mathbf{F} \bullet d\mathbf{r} + \int_{\mathcal{C}_2} \mathbf{F} \bullet d\mathbf{r}.$$

The first integral does not depend on $x$, so its derivative with respect to $x$ is zero. The straight line path for the second integral is parametrized by $\mathbf{r} = t\mathbf{i} + y\mathbf{j} + z\mathbf{k}$, where $x_1 \le t \le x$ so $d\mathbf{r} = dt\,\mathbf{i}$ and

$$\frac{\partial \phi}{\partial x} = \frac{\partial}{\partial x}\int_{C_2} \mathbf{F} \bullet d\mathbf{r} = \frac{\partial}{\partial x}\int_{x_1}^{x} F_1(t, y, z)\,dt = F_1(x, y, z),$$

which is what we wanted. Thus $\mathbf{F} = \nabla\phi$ is conservative, and (c) implies (a).

***REMARK*** It is very easy to evaluate the line integral of the tangential component of a *conservative* vector field along a curve $\mathcal{C}$, when you know a potential for $\mathbf{F}$. If $\mathbf{F} = \nabla\phi$, and $\mathcal{C}$ goes from $P_0$ to $P_1$, then

$$\int_{\mathcal{C}} \mathbf{F} \bullet d\mathbf{r} = \int_{\mathcal{C}} d\phi = \phi(P_1) - \phi(P_0).$$

As noted above, the value of the integral depends only on the endpoints of $\mathcal{C}$.

***REMARK*** In the next chapter we will add another item to the list of three conditions shown to be equivalent in Theorem 1, provided that the domain $D$ is *simply connected*. For such a domain each of the above three conditions in the theorem is equivalent to

$$\frac{\partial F_1}{\partial y} = \frac{\partial F_2}{\partial x}, \qquad \frac{\partial F_1}{\partial z} = \frac{\partial F_3}{\partial x} \qquad \text{and} \qquad \frac{\partial F_2}{\partial z} = \frac{\partial F_3}{\partial y}.$$

We already know that these equations are satisfied on a domain where $\mathbf{F}$ is conservative. In Chapter 9 we will show that if these three equations hold on a simply connected domain, then $\mathbf{F}$ is conservative on that domain.

**EXAMPLE 3** For what values of the constants $A$ and $B$ is the vector field

$$\mathbf{F} = Ax\sin(\pi y)\mathbf{i} + \big(x^2\cos(\pi y) + Bye^{-z}\big)\mathbf{j} + y^2e^{-z}\mathbf{k}$$

conservative? For this choice of $A$ and $B$, evaluate $\int_{\mathcal{C}} \mathbf{F} \bullet d\mathbf{r}$, where $\mathcal{C}$ is

(a) the curve $\mathbf{r} = \cos t\mathbf{i} + \sin 2t\mathbf{j} + \sin^2 t\mathbf{k}$, $(0 \le t \le 2\pi)$, and

(b) the curve of intersection of the paraboloid $z = x^2 + 4y^2$ and the plane $z = 3x - 2y$ from $(0, 0, 0)$ to $(1, 1/2, 2)$.

***SOLUTION*** $\mathbf{F}$ cannot be conservative unless

$$\frac{\partial F_1}{\partial y} = \frac{\partial F_2}{\partial x}, \qquad \frac{\partial F_1}{\partial z} = \frac{\partial F_3}{\partial x}, \qquad \text{and} \qquad \frac{\partial F_2}{\partial z} = \frac{\partial F_3}{\partial y},$$

that is, unless

$$A\pi x\cos(\pi y) = 2x\cos(\pi y), \qquad 0 = 0, \qquad \text{and} \qquad -Bye^{-z} = 2ye^{-z}.$$

Thus we require that $A = 2/\pi$ and $B = -2$. In this case, it is easily checked that

$$\mathbf{F} = \nabla\phi, \quad \text{where} \quad \phi = \left(\frac{x^2\sin(\pi y)}{\pi} - y^2e^{-z}\right).$$

For the curve (a) we have $\mathbf{r}(0) = \mathbf{i} = \mathbf{r}(2\pi)$, so this curve is a closed curve, and

$$\int_C \mathbf{F} \bullet d\mathbf{r} = \oint_C \nabla\phi \bullet d\mathbf{r} = 0.$$

Since the curve (b) starts at $(0, 0, 0)$ and ends at $(1, 1/2, 2)$, we have

$$\int_C \mathbf{F} \bullet d\mathbf{r} = \left(\frac{x^2 \sin(\pi y)}{\pi} - y^2 e^{-z}\right)\Bigg|_{(0,0,0)}^{(1,1/2,2)} = \frac{1}{\pi} - \frac{1}{4e^2}.$$

The following example shows how to exploit the fact that

$$\int_C \mathbf{F} \bullet d\mathbf{r}$$

is easily evaluated for conservative $\mathbf{F}$ even if the $\mathbf{F}$ we want to integrate isn't quite conservative.

**EXAMPLE 4** Evaluate $I = \oint_C (e^x \sin y + 3y)dx + (e^x \cos y + 2x - 2y)dy$ counterclockwise around the ellipse $4x^2 + y^2 = 4$.

***SOLUTION*** $I = \oint_C \mathbf{F} \bullet d\mathbf{r}$, where $\mathbf{F}$ is the vector field

$$\mathbf{F} = \big(e^x \sin y + 3y\big)\mathbf{i} + \big(e^x \cos y + 2x - 2y\big)\mathbf{j}.$$

This vector field is not conservative, but it would be if the $3y$ term in $F_1$ were $2y$ instead; specifically, if

$$\phi(x, y) = e^x \sin y + 2xy - y^2,$$

then $\mathbf{F} = \nabla\phi + y\mathbf{i}$, the sum of a conservative part and a non-conservative part. Therefore, we have

$$I = \oint_C \nabla\phi \bullet d\mathbf{r} + \oint_C y\, dx.$$

The first integral is zero since $\nabla\phi$ is conservative and $C$ is closed. For the second integral we parametrize $C$ by $x = \cos t$, $y = 2\sin t$, $(0 \le t \le 2\pi)$, and obtain

$$I = \oint_C y\, dx = -2\int_0^{2\pi} \sin^2 t\, dt = -2\int_0^{2\pi} \frac{1 - \cos(2t)}{2}\, dt = -2\pi.$$

## EXERCISES 8.4

In Exercises 1–6, evaluate the line integral of the tangential component of the given vector field along the given curve.

1. $\mathbf{F}(x, y) = xy\mathbf{i} - x^2\mathbf{j}$ along $y = x^2$ from $(0, 0)$ to $(1, 1)$
2. $\mathbf{F}(x, y) = \cos x\, \mathbf{i} - y\mathbf{j}$ along $y = \sin x$ from $(0, 0)$ to $(\pi, 0)$
3. $\mathbf{F}(x, y, z) = y\mathbf{i} + z\mathbf{j} - x\mathbf{k}$ along the straight line from $(0,0,0)$ to $(1,1,1)$
4. $\mathbf{F}(x, y, z) = z\mathbf{i} - y\mathbf{j} + 2x\mathbf{k}$ along the curve $x = t$, $y = t^2$, $z = t^3$ from $(0, 0, 0)$ to $(1, 1, 1)$
5. $\mathbf{F}(x, y, z) = yz\mathbf{i} + xz\mathbf{j} + xy\mathbf{k}$ from $(-1, 0, 0)$ to $(1, 0, 0)$ along either direction of the curve of intersection of the cylinder $x^2 + y^2 = 1$ and the plane $z = y$

6. $\mathbf{F}(x, y, z) = (x - z)\mathbf{i} + (y - z)\mathbf{j} - (x + y)\mathbf{k}$ along the polygonal path from $(0, 0, 0)$ to $(1, 0, 0)$ to $(1, 1, 0)$ to $(1, 1, 1)$

7. Find the work done by the force field

$$\mathbf{F} = (x + y)\mathbf{i} + (x - z)\mathbf{j} + (z - y)\mathbf{k}$$

in moving an object from $(1, 0, -1)$ to $(0, -2, 3)$ along any smooth curve.

8. Evaluate $\oint_C x^2y^2\,dx + x^3y\,dy$ counterclockwise around the square with vertices $(0, 0)$, $(1, 0)$, $(1, 1)$ and $(0, 1)$.

9. Evaluate

$$\int_C e^{x+y}\sin(y+z)\,dx + e^{x+y}\Big(\sin(y+z) + \cos(y+z)\Big)\,dy + e^{x+y}\cos(y+z)\,dz$$

along the straight line segment from (0,0,0) to $(1, \frac{\pi}{4}, \frac{\pi}{4})$.

10. The field $\mathbf{F} = (axy + z)\mathbf{i} + x^2\mathbf{j} + (bx + 2z)\mathbf{k}$ is conservative. Find $a$ and $b$, and find a potential for $\mathbf{F}$. Also, evaluate $\int_C \mathbf{F} \bullet d\mathbf{r}$, where $C$ is the curve from $(1, 1, 0)$ to $(0, 0, 3)$ that lies on the intersection of the surfaces $2x + y + z = 3$ and $9x^2 + 9y^2 + 2z^2 = 18$ in the octant $x \geq 0$, $y \geq 0$, $z \geq 0$.

11. Determine the values of $A$ and $B$ for which the vector field

$$\mathbf{F} = Ax\ln z\,\mathbf{i} + By^2z\,\mathbf{j} + \left(\frac{x^2}{z} + y^3\right)\mathbf{k}$$

is conservative. If $C$ is the straight line from $(1, 1, 1)$ to $(2, 1, 2)$, find

$$\int_C 2x\ln z\,dx + 2y^2z\,dy + y^3\,dz.$$

12. Find the work done by the force field

$$\mathbf{F} = (y^2\cos x + z^3)\mathbf{i} + (2y\sin x - 4)\mathbf{j} + (3xz^2 + 2)\mathbf{k}$$

in moving a particle along the curve $x = \sin^{-1}t$, $y = 1 - 2t$, $z = 3t - 1$, $(0 \leq t \leq 1)$.

13. If $C$ is the intersection of $z = \ln(1 + x)$ and $y = x$ from $(0, 0, 0)$ to $(1, 1, \ln 2)$, evaluate

$$\int_C (2x\sin(\pi y) - e^z)\,dx + (\pi x^2\cos(\pi y) - 3e^z)\,dy - xe^z\,dz.$$

14. Is each of the following sets a domain? a connected domain? a simply connected domain?

(a) the set of points $(x, y)$ in the plane such that $x > 0$ and $y \geq 0$

(b) the set of points $(x, y)$ in the plane such that $x = 0$ and $y \geq 0$

(c) the set of points $(x, y)$ in the plane such that $x \neq 0$ and $y > 0$

(d) the set of points $(x, y, z)$ in 3-space such that $x^2 > 1$

(e) the set of points $(x, y, z)$ in 3-space such that $x^2 + y^2 > 1$

(f) the set of points $(x, y, z)$ in 3-space such that $x^2 + y^2 + z^2 > 1$

In Exercises 15–19, evaluate the closed line integrals

$$\text{(a)}\quad \oint_C x\,dy, \qquad\qquad \text{(b)}\quad \oint_C y\,dx$$

around the given curves, all oriented counterclockwise.

15. The circle $x^2 + y^2 = a^2$

16. The ellipse $\dfrac{x^2}{a^2} + \dfrac{y^2}{b^2} = 1$

17. The boundary of the half disk $x^2 + y^2 \leq a^2$, $y \geq 0$

18. The boundary of the square with vertices $(0, 0)$, $(1, 0)$, $(1, 1)$, and $(0, 1)$

19. The triangle with vertices $(0, 0)$, $(a, 0)$ and $(0, b)$

20. On the basis of your results for Exercises 15–19, guess the values of the closed line integrals

$$\text{(a)}\quad \oint_C x\,dy, \qquad\qquad \text{(b)}\quad \oint_C y\,dx$$

for any non-self-intersecting closed curve in the $xy$-plane. Prove your guess in the case that $C$ bounds a region of the plane which is both $x$-simple and $y$-simple. (See Section 6.1.)

21. If $f$ and $g$ are scalar fields with continuous first partial derivatives in a connected domain $D$, show that

$$\int_C f\nabla g \bullet d\mathbf{r} + \int_C g\nabla f \bullet d\mathbf{r} = f(Q)g(Q) - f(P)g(P)$$

for any piecewise smooth curve in $D$ from $P$ to $Q$.

22. Evaluate

$$\frac{1}{2\pi}\oint_C \frac{-y\,dx + x\,dy}{x^2 + y^2}$$

(a) counterclockwise around the circle $x^2 + y^2 = a^2$,

(b) clockwise around the square with vertices $(-1, -1)$, $(-1, 1)$, $(1, 1)$, and $(1, -1)$,

(c) counterclockwise around the boundary of the region $1 \leq x^2 + y^2 \leq 2$, $y \geq 0$.

23. Review Example 5 in Section 8.2, in which it was shown that

$$\frac{\partial}{\partial y}\left(\frac{-y}{x^2 + y^2}\right) = \frac{\partial}{\partial x}\left(\frac{x}{x^2 + y^2}\right),$$

for all $(x, y) \neq (0, 0)$. Why does this result, together with that of Exercise 22, not contradict the final assertion in the remark following Theorem 1?

* **24.** Let $\mathcal{C}$ be a piecewise smooth curve in the $xy$-plane which does not pass through the origin. Let $\theta = \theta(x, y)$ be the polar angle coordinate of the point $P = (x, y)$ on $\mathcal{C}$, not restricted to an interval of length $2\pi$, but varying continuously as $P$ moves from one end of $\mathcal{C}$ to the other. As in Example 5 of Section 8.2, it happens that

$$\nabla\theta = -\frac{y}{x^2 + y^2}\mathbf{i} + \frac{x}{x^2 + y^2}\mathbf{j}.$$

If, in addition, $\mathcal{C}$ is a closed curve, show that

$$w(\mathcal{C}) = \frac{1}{2\pi}\oint_{\mathcal{C}} \frac{x\,dy - y\,dx}{x^2 + y^2}$$

has an integer value. $w$ is called the **winding number** of $\mathcal{C}$ about the origin.

# 8.5 SURFACES AND SURFACE INTEGRALS

This section and the next are devoted to integrals of functions defined over surfaces in 3-space. Before we can begin, it is necessary to make more precise just what is meant by the term "surface." Until now we have been treating surfaces in an intuitive way, either as the graphs of functions $f(x, y)$, or as the graphs of equations $f(x, y, z) = 0$.

A smooth curve is considered to be a *one-dimensional* object because points on it can be located by giving *one coordinate* (for instance, the distance from an endpoint). Therefore the curve can be defined as the range of a vector-valued function of a single real variable. Similarly, a surface is a *two-dimensional* object; points on it can be located by using *two coordinates*, and the surface can be defined as the range of a vector-valued function of two real variables. We will call certain such functions parametric surfaces.

## Parametric Surfaces

**DEFINITION 4**

A **parametric surface** in 3-space is a continuous function **r** defined on some rectangle $R$ given by $a \leq u \leq b$, $c \leq v \leq d$ in the $uv$-plane, and having values in 3-space:

$$\mathbf{r}(u, v) = x(u, v)\mathbf{i} + y(u, v)\mathbf{j} + z(u, v)\mathbf{k}, \qquad (u, v) \text{ in } R.$$

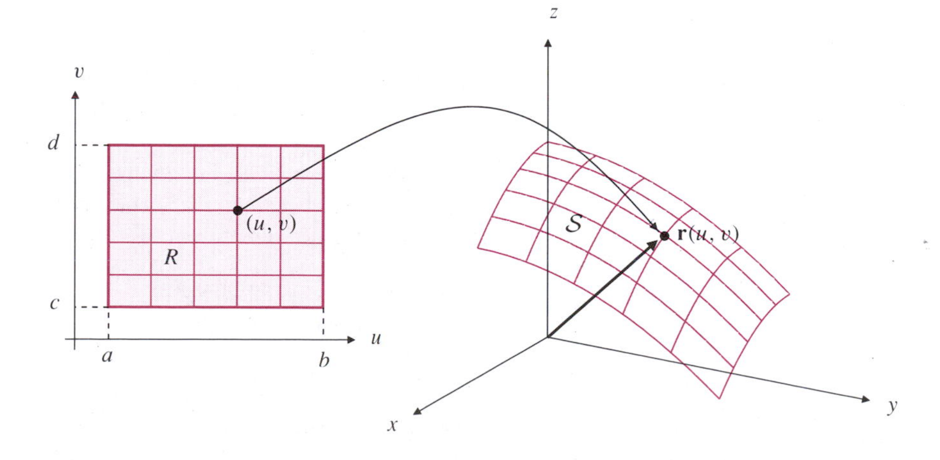


**Figure 8.16** A parametric surface $\mathcal{S}$ defined on parameter region $R$. The *contour curves* on $\mathcal{S}$ correspond to the rulings of $R$

Actually, we think of the *range* of the function $\mathbf{r}(u, v)$ as being the parametric surface. It is a set $\mathcal{S}$ of points $(x, y, z)$ in 3-space whose position vectors are the vectors $\mathbf{r}(u, v)$ for $(u, v)$ in $R$. (See Figure 8.16.) If $\mathbf{r}$ is one-to-one, then the surface does not intersect itself. In this case $\mathbf{r}$ maps the boundary of the rectangle $R$ (the four edges) onto a curve in 3-space, which we call the **boundary of the parametric surface**. The requirement that $R$ be a rectangle is made only to simplify the discussion. Any connected, closed, bounded set in the $uv$-plane, having well-defined area, and consisting of an open set together with its boundary points would do as well. Thus, we will from time to time consider parametric surfaces over closed disks, triangles, or other such domains in the $uv$-plane. Being the range of a continuous function defined on a closed, bounded set, a parametric surface is always bounded in 3-space.

**EXAMPLE 1** The graph of $z = f(x, y)$, where $f$ has the rectangle $R$ as its domain, can be represented as the parametric surface

$$\mathbf{r} = \mathbf{r}(u, v) = u\mathbf{i} + v\mathbf{j} + f(u, v)\mathbf{k}$$

for $(u, v)$ in $R$. Its scalar parametric equations are

$$x = u, \qquad y = v, \qquad z = f(u, v), \qquad (u, v) \text{ in } R.$$

For such graphs it is sometimes convenient to identify the $uv$-plane with the $xy$-plane, and write the equation of the surface in the form

$$\mathbf{r} = x\mathbf{i} + y\mathbf{j} + f(x, y)\mathbf{k}, \qquad (x, y) \text{ in } R.$$

**EXAMPLE 2** Describe the surface

$$\mathbf{r} = a\cos u \sin v\,\mathbf{i} + a\sin u \sin v\,\mathbf{j} + a\cos v\,\mathbf{k}, \qquad (0 \le u \le 2\pi, \quad 0 \le v \le \pi/2),$$

where $a > 0$. What is its boundary?

***SOLUTION*** Observe that if $x = a\cos u \sin v$, $y = a\sin u \sin v$, and $z = a\cos v$, then $x^2 + y^2 + z^2 = a^2$. Thus the given parametric surface lies on the sphere of radius $a$ centred at the origin. (Observe that $u$ and $v$ are the spherical coordinates $\theta$ and $\phi$ on the sphere.) The restrictions on $u$ and $v$ allow $(x, y)$ to be any point in the disk $x^2 + y^2 \le a^2$, but force $z \ge 0$. Thus the surface is the *upper half* of the sphere. The given parametrization is one-to-one on the open rectangle $0 < u < 2\pi$, $0 < v < \pi/2$, but not on the closed rectangle, since the edges $u = 0$ and $u = 2\pi$ get mapped onto the same points, and the entire edge $v = 0$ collapses to a single point. The boundary of the surface is still the circle $x^2 + y^2 = a^2$, $z = 0$, and corresponds to the edge $v = \pi/2$ of the rectangle.

***REMARK*** Surface parametrizations that are one-to-one only in the interior of the parameter domain $R$ are still reasonable representations of the surface. However, as in Example 2, the boundary of the surface may be obtained from only part of the boundary of $R$, or there may be no boundary at all, in which case the surface is called a **closed surface**. For example, if the domain of $\mathbf{r}$ in Example 2 is extended to allow $0 \le v \le \pi$, then the surface becomes the entire sphere of radius $a$ centred at the origin. The sphere is a closed surface, having no boundary curves.

***REMARK*** Like parametrizations of curves, parametrizations of surfaces are not unique. The hemisphere in Example 2 can also be parametrized

$$\mathbf{r}(u, v) = u\mathbf{i} + v\mathbf{j} + \sqrt{a^2 - u^2 - v^2}\,\mathbf{k} \qquad \text{for} \quad u^2 + v^2 \le a^2.$$

Here the domain of $\mathbf{r}$ is a closed disk of radius $a$.

**EXPLORE!** Tubes Around Curves

If $\mathbf{r} = \mathbf{F}(t)$, $a \le t \le b$, is a parametric curve $\mathcal{C}$ in 3-space having unit normal $\hat{\mathbf{N}}(t)$ and binormal $\hat{\mathbf{B}}(t)$, then the parametric surface

$$\mathbf{r} = \mathbf{F}(u) + s\cos v\,\hat{\mathbf{N}}(u) + s\sin v\,\hat{\mathbf{B}}(u), \qquad a \le u \le b, \qquad 0 \le v \le 2\pi,$$

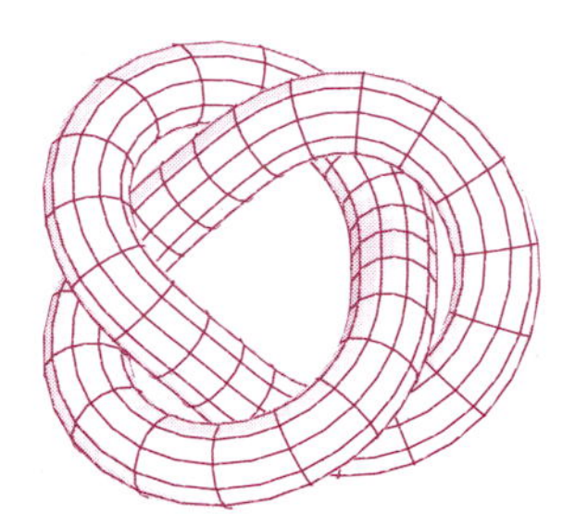

Figure 8.17 A trefoil knot

is a tube-shaped surface of radius $s$ centred along the curve $\mathcal{C}$. (Why?) Use a three-dimensional graphing program with parametric surface capabilities to plot tubes of radius $s = 0.25$ about the following curves:

(a) the circle $\mathbf{r} = \cos t\,\mathbf{i} + \sin t\,\mathbf{j}$, $0 \le t \le 2\pi$,

(b) the helix $\mathbf{r} = \cos t\,\mathbf{i} + \sin t\,\mathbf{j} + (t/2\pi)\mathbf{k}$, $0 \le t \le 2\pi$, and

(c) the trefoil knot $\mathbf{r} = (1 + 0.3\cos(3t))(\cos(2t)\mathbf{i} + \sin(2t)\mathbf{j}) + 0.35\sin(3t)\mathbf{k}$, $0 \le t \le 2\pi$. You may want to use a computer algebra program to calculate $\hat{\mathbf{N}}$ and $\hat{\mathbf{B}}$ for part (c).

## Composite Surfaces

If two parametric surfaces are joined together along part or all of their boundary curves, the result is called a **composite surface**, or, thinking geometrically, just a **surface**. For example, a sphere can be obtained by joining two hemispheres along their boundary circles. In general, composite surfaces can be obtained by joining a finite number of parametric surfaces pairwise along edges. The surface of a cube consists of the six square faces joined in pairs along the edges of the cube. This surface is closed since there are no unjoined edges to comprise the boundary. If the top square face is removed, the remaining five form the surface of a cubical box with no top. The top edges of the four side faces now constitute the boundary of this composite surface. (See Figure 8.18.)

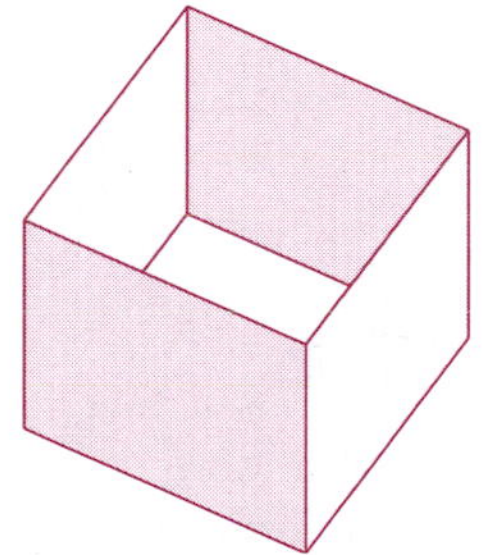

Figure 8.18 A composite surface obtained by joining five smooth parametric surfaces (squares) in pairs along edges. The four unpaired edges at the tops of the side faces make up the boundary of the composite surface

## Surface Integrals

In order to define integrals of functions defined on a surface as limits of Riemann sums, we need to refer to the *areas* of regions on the surface. It is more difficult to define the area of a curved surface than it is to define the length of a curve. However, you will likely have a good idea of what area means for a region lying in a plane, and we examined briefly the problem of finding the area of the graph of a function $f(x, y)$ in Section 6.7. We will avoid difficulties by assuming that all the surfaces we will encounter are "smooth enough" that they can be subdivided into small pieces each of which is approximately planar. We can then approximate the surface area of each piece by a plane area and add up the approximations to get a Riemann sum approximation to the area of the whole surface. We will make more precise definitions of "smooth surface" and "surface area" later in this section. For the moment, we assume the reader has an intuitive feel for what they mean.

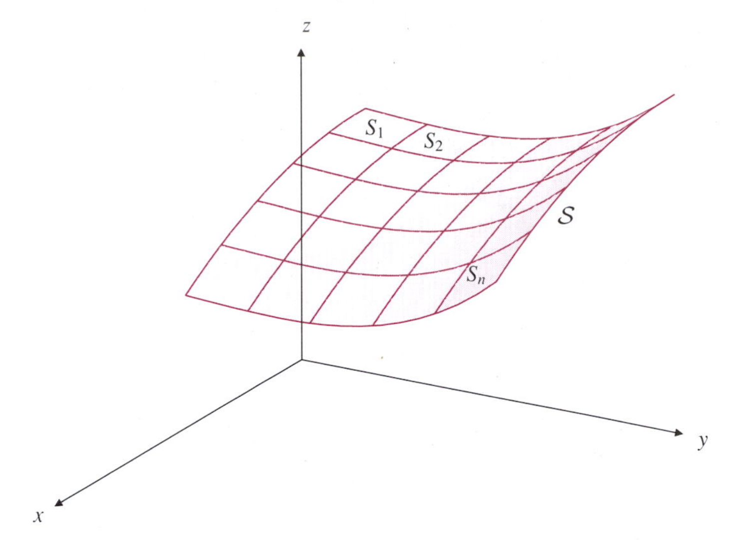


**Figure 8.19** A partition of a parametric surface into many nonoverlapping pieces

Let $\mathcal{S}$ be a smooth surface of finite area in $\mathbb{R}^3$, and let $f(x, y, z)$ be a bounded function defined at all points of $\mathcal{S}$. If we subdivide $\mathcal{S}$ into small, nonoverlapping pieces, say $S_1, S_2, \ldots, S_n$, where $S_i$ has area $\Delta S_i$ (see Figure 8.19), we can form a **Riemann sum** $R_n$ for $f$ on $\mathcal{S}$ by choosing arbitrary points $(x_i, y_i, z_i)$ in $S_i$ and letting

$$R_n = \sum_{i=1}^{n} f(x_i, y_i, z_i)\,\Delta S_i.$$

If such Riemann sums have a unique limit as the diameters of all the pieces $S_i$ approach zero, independently of how the points $(x_i, y_i, z_i)$ are chosen, then we say that $f$ is **integrable** on $\mathcal{S}$, and denote the limit by

$$\iint_{\mathcal{S}} f(x, y, z)\,dS.$$

## Smooth Surfaces, Normals, and Area Elements

A surface is smooth if it has a unique tangent plane at any nonboundary point $P$. A nonzero vector **n** normal to that tangent plane at $P$ is said to be normal to the surface at $P$. The following somewhat technical definition makes this precise.

**DEFINITION 5**

A set $\mathcal{S}$ in 3-space is a **smooth surface** if any point $P$ in $\mathcal{S}$ has a neighbourhood $N$ (an open ball of positive radius centred at $P$) which is the domain of a smooth function $g(x, y, z)$ satisfying:

(i) $N \cap S = \{Q \in N : g(Q) = 0\}$, and

(ii) $\nabla g(Q) \neq \mathbf{0}$, if $Q$ is in $N \cap S$.

For example, the cone $x^2 + y^2 = z^2$, with the origin removed, is a smooth surface. Note that $\nabla(x^2 + y^2 - z^2) = \mathbf{0}$ at the origin, and the cone is not smooth there, since it does not have a unique tangent plane.

A parametric surface cannot satisfy the condition of the smoothness definition at its boundary points, but will be called **smooth** if that condition is satisfied at all non-boundary points.

We can find the normal to a smooth parametric surface defined on parameter domain $R$ as follows. If $(u_0, v_0)$ is a point in the interior of $R$, then $\mathbf{r} = \mathbf{r}(u, v_0)$ and $\mathbf{r} = \mathbf{r}(u_0, v)$ are two curves on $\mathcal{S}$, intersecting at $\mathbf{r}_0 = \mathbf{r}(u_0, v_0)$, and having, at that point, tangent vectors

$$\left.\frac{\partial \mathbf{r}}{\partial u}\right|_{(u_0,v_0)} \quad \text{and} \quad \left.\frac{\partial \mathbf{r}}{\partial v}\right|_{(u_0,v_0)}$$

respectively. Assuming these two tangent vectors are not parallel, their cross product $\mathbf{n}$, which is not zero, is *normal* to $\mathcal{S}$ at $\mathbf{r}_0$. Furthermore, the *area element* on $\mathcal{S}$ bounded by the four curves $\mathbf{r} = \mathbf{r}(u_0, v)$, $\mathbf{r} = \mathbf{r}(u_0 + du, v)$, $\mathbf{r} = \mathbf{r}(u, v_0)$, and $\mathbf{r} = \mathbf{r}(u, v_0 + dv)$, (see Figure 8.20), is an infinitesimal parallelogram spanned by the vectors $(\partial\mathbf{r}/\partial u)\, du$, and $(\partial\mathbf{r}/\partial v)\, dv$ (at $(u_0, v_0)$), and hence has area

$$dS = \left|\frac{\partial \mathbf{r}}{\partial u} \times \frac{\partial \mathbf{r}}{\partial v}\right| du\, dv.$$

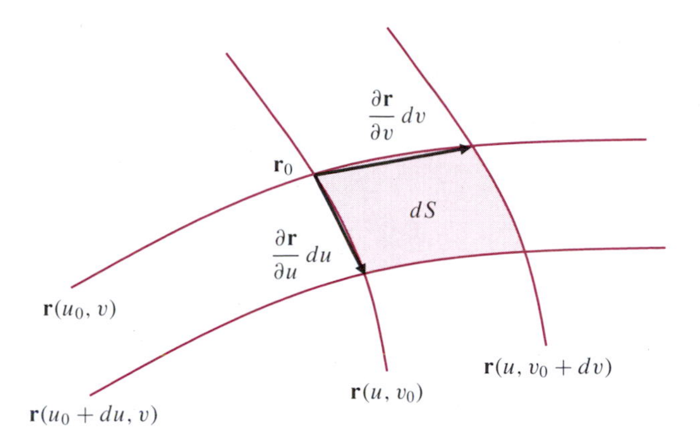


**Figure 8.20** An area element $dS$ on a parametric surface

Let us express the normal vector $\mathbf{n}$ and the area element $dS$ in terms of the components of $\mathbf{r}$. Since

$$\frac{\partial \mathbf{r}}{\partial u} = \frac{\partial x}{\partial u}\mathbf{i} + \frac{\partial y}{\partial u}\mathbf{j} + \frac{\partial z}{\partial u}\mathbf{k} \quad \text{and} \quad \frac{\partial \mathbf{r}}{\partial v} = \frac{\partial x}{\partial v}\mathbf{i} + \frac{\partial y}{\partial v}\mathbf{j} + \frac{\partial z}{\partial v}\mathbf{k},$$

the **normal vector** to $\mathcal{S}$ at $\mathbf{r}(u, v)$ is

$$\begin{aligned}\mathbf{n} = \frac{\partial \mathbf{r}}{\partial u} \times \frac{\partial \mathbf{r}}{\partial v} &= \begin{vmatrix} \mathbf{i} & \mathbf{j} & \mathbf{k} \\ \frac{\partial x}{\partial u} & \frac{\partial y}{\partial u} & \frac{\partial z}{\partial u} \\ \frac{\partial x}{\partial v} & \frac{\partial y}{\partial v} & \frac{\partial z}{\partial v} \end{vmatrix} \\ &= \frac{\partial(y, z)}{\partial(u, v)}\mathbf{i} + \frac{\partial(z, x)}{\partial(u, v)}\mathbf{j} + \frac{\partial(x, y)}{\partial(u, v)}\mathbf{k}.\end{aligned}$$

Also, the **area element** at a point $\mathbf{r}(u, v)$ on the surface is given by

$$\begin{aligned}dS &= \left|\frac{\partial \mathbf{r}}{\partial u} \times \frac{\partial \mathbf{r}}{\partial v}\right| du\, dv \\ &= \sqrt{\left(\frac{\partial(y, z)}{\partial(u, v)}\right)^2 + \left(\frac{\partial(z, x)}{\partial(u, v)}\right)^2 + \left(\frac{\partial(x, y)}{\partial(u, v)}\right)^2}\, du\, dv.\end{aligned}$$

The area of the surface itself is the "sum" of these area elements.

$$\text{Area of } \mathcal{S} = \iint_{\mathcal{S}} dS.$$

**EXAMPLE 3** The graph $z = g(x, y)$ of a function $g$ with continuous first partial derivatives in a domain $D$ of the $xy$-plane can be regarded as a parametric surface $\mathcal{S}$ with parametrization

$$x = u, \qquad y = v, \qquad z = g(u, v), \qquad (u, v) \text{ in } D.$$

In this case

$$\frac{\partial(y,z)}{\partial(u,v)} = -g_1(u,v), \qquad \frac{\partial(z,x)}{\partial(u,v)} = -g_2(u,v), \qquad \text{and} \qquad \frac{\partial(x,y)}{\partial(u,v)} = 1,$$

and, since the parameter region coincides with the domain $D$ of $g$, the surface integral of $f(x, y, z)$ over $\mathcal{S}$ can be expressed as a double integral over $D$:

$$\iint_{\mathcal{S}} f(x,y,z)\, dS = \iint_D f\big(x,y,g(x,y)\big)\sqrt{1 + \big(g_1(x,y)\big)^2 + \big(g_2(x,y)\big)^2}\, dx\, dy.$$

As observed in Section 6.7, this formula can also be justified geometrically. The vector $\mathbf{n} = -g_1(x, y)\mathbf{i} - g_2(x, y)\mathbf{j} + \mathbf{k}$ is normal to $\mathcal{S}$ and makes angle $\gamma$ with the positive $z$-axis, where

$$\cos\gamma = \frac{\mathbf{n} \bullet \mathbf{k}}{|\mathbf{n}|} = \frac{1}{\sqrt{1 + \big(g_1(x,y)\big)^2 + \big(g_2(x,y)\big)^2}}.$$

The surface area element $dS$ must have area $1/\cos\gamma$ times the area $dx\,dy$ of its perpendicular projection onto the $xy$-plane. (See Figure 8.21.) ■

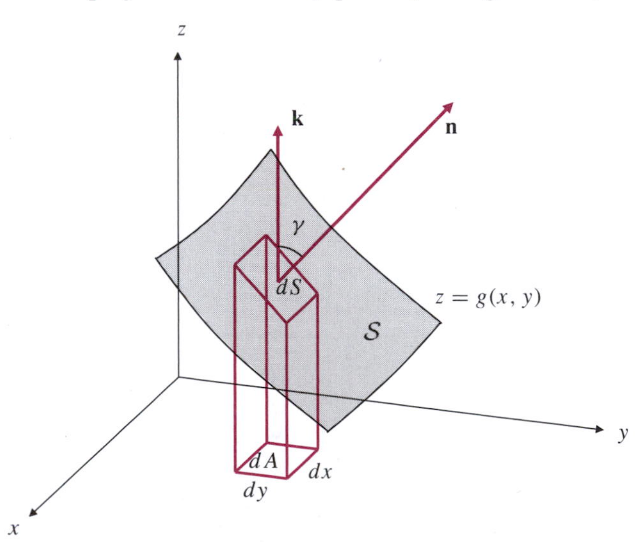


Figure 8.21 The surface area element $dS$ and its projection onto the $xy$-plane

## Evaluating Surface Integrals

We illustrate the use of the formulas given above for $dS$ in calculating surface integrals.

**EXAMPLE 4** Evaluate $\iint_{\mathcal{S}} z\,dS$ over the conical surface $z = \sqrt{x^2 + y^2}$ between $z = 0$ and $z = 1$.

**SOLUTION** Since $z^2 = x^2 + y^2$ on the surface $\mathcal{S}$, we have $\partial z/\partial x = x/z$ and $\partial z/\partial y = y/z$. Therefore

$$dS = \sqrt{1 + \frac{x^2}{z^2} + \frac{y^2}{z^2}}\,dx\,dy = \sqrt{\frac{z^2 + z^2}{z^2}}\,dx\,dy = \sqrt{2}\,dx\,dy.$$

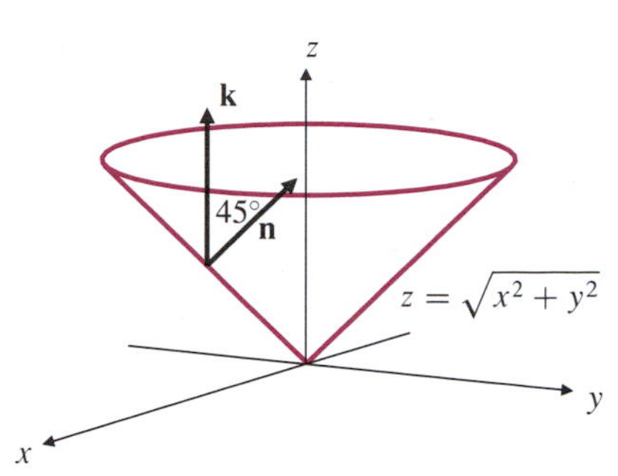


**Figure 8.22** $dS = \sqrt{2}\,dx\,dy$ on this cone

(Note that we could have anticipated this result since the normal to the cone always makes an angle of $\gamma = 45^\circ$ with the positive $z$-axis; see Figure 8.22. Therefore $dS = dx\,dy/\cos 45^\circ = \sqrt{2}\,dx\,dy$.) Since $z = \sqrt{x^2 + y^2} = r$ on the conical surface, it is easiest to carry out the integration in polar coordinates:

$$\iint_{\mathcal{S}} z\,dS = \sqrt{2}\iint_{x^2+y^2\le 1} z\,dx\,dy$$

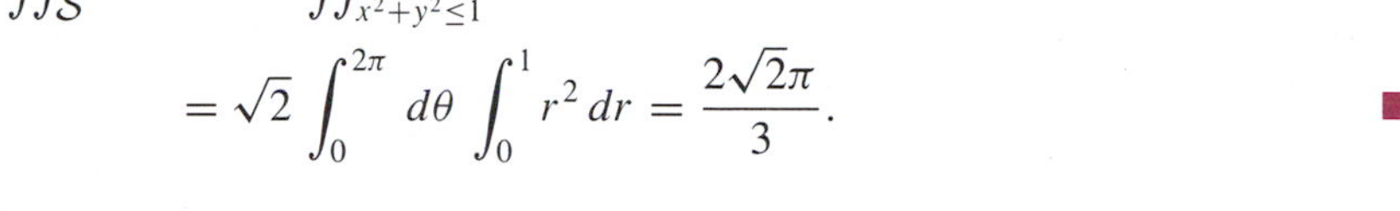

$$= \sqrt{2}\int_0^{2\pi} d\theta \int_0^1 r^2\,dr = \frac{2\sqrt{2}\pi}{3}.$$

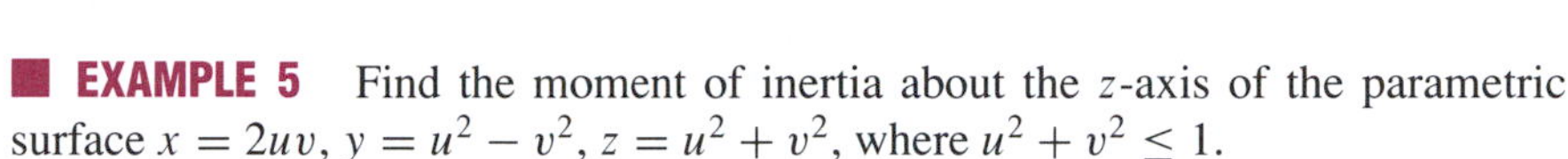

**EXAMPLE 5** Find the moment of inertia about the $z$-axis of the parametric surface $x = 2uv$, $y = u^2 - v^2$, $z = u^2 + v^2$, where $u^2 + v^2 \le 1$.

**SOLUTION** We are asked to find $\iint_{\mathcal{S}} (x^2 + y^2)\,dS$. We have

$$\frac{\partial(x, y)}{\partial(u, v)} = \begin{vmatrix} 2v & 2u \\ 2u & -2v \end{vmatrix} = -4(u^2 + v^2),$$

$$\frac{\partial(x, z)}{\partial(u, v)} = \begin{vmatrix} 2v & 2u \\ 2u & 2v \end{vmatrix} = 4(v^2 - u^2),$$

$$\frac{\partial(y, z)}{\partial(u, v)} = \begin{vmatrix} 2u & -2v \\ 2u & 2v \end{vmatrix} = 8uv.$$

Therefore, the surface area element on $\mathcal{S}$ is given by

$$\begin{aligned} dS &= 4\sqrt{(u^2 + v^2)^2 + (v^2 - u^2)^2 + 4u^2v^2}\,du\,dv \\ &= 4\sqrt{2(u^4 + v^4 + 2u^2v^2)}\,du\,dv = 4\sqrt{2}(u^2 + v^2)\,du\,dv. \end{aligned}$$

Now $x^2 + y^2 = 4u^2v^2 + (u^2 - v^2)^2 = (u^2 + v^2)^2$. Thus

$$\begin{aligned} \iint_{\mathcal{S}} (x^2 + y^2)\,dS &= \iint_{u^2+v^2\le 1} (u^2 + v^2)^2\, 4\sqrt{2}(u^2 + v^2)\,du\,dv \\ &= 4\sqrt{2}\int_0^{2\pi} d\theta \int_0^1 r^6\, r\,dr \qquad \text{(using polar coordinates)} \\ &= \sqrt{2}\pi. \end{aligned}$$

This is the required moment of inertia.

Even though most surfaces we encounter can be easily parametrized, it is usually possible to obtain the surface area element $dS$ geometrically rather than relying on the parametric formula. As we have seen above, if a surface has a one-to-one projection onto a region in the $xy$-plane, then the area element $dS$ on the surface can be expressed as

$$dS = \left|\frac{1}{\cos\gamma}\right| dx\,dy = \frac{|\mathbf{n}|}{|\mathbf{n} \bullet \mathbf{k}|}\,dx\,dy,$$

where $\gamma$ is the angle between the normal vector $\mathbf{n}$ to $\mathcal{S}$ and the positive $z$-axis. This formula is useful no matter how we obtain $\mathbf{n}$.

Consider a surface $\mathcal{S}$ with equation of the form $F(x, y, z) = 0$. As we discovered in Section 4.7, if $F$ has continuous first partial derivatives that do not all vanish at a point $(x, y, z)$ on $\mathcal{S}$, then the nonzero vector

$$\mathbf{n} = \nabla F(x, y, z)$$

is normal to $\mathcal{S}$ at that point. Since $\mathbf{n} \bullet \mathbf{k} = F_3(x, y, z)$, if $\mathcal{S}$ has a one-to-one projection onto the domain $D$ in the $xy$-plane, then

$$dS = \left|\frac{\nabla F(x, y, z)}{F_3(x, y, z)}\right| dx\,dy,$$

and the surface integral of $f(x, y, z)$ over $\mathcal{S}$ can be expressed as a double integral over the domain $D$:

$$\iint_{\mathcal{S}} f(x, y, z)\,dS = \iint_D f\big(x, y, g(x, y)\big)\left|\frac{\nabla F(x, y, z)}{F_3(x, y, z)}\right| dx\,dy.$$

Of course, there are analogous formulas for area elements of surfaces (and integrals over surfaces) with one-to-one projections onto the $xz$-plane or the $yz$-plane. ($F_3$ is replaced by $F_2$ and $F_1$, respectively.)

**EXAMPLE 6** Find the moment about $z = 0$, that is, $\iint_{\mathcal{S}} z\,dS$, where $\mathcal{S}$ is the hyperbolic bowl $z^2 = 1 + x^2 + y^2$ between the planes $z = 1$ and $z = \sqrt{5}$.

***SOLUTION*** $\mathcal{S}$ is given by $F(x, y, z) = 0$, where $F(x, y, z) = x^2 + y^2 - z^2 + 1$. It lies above the disk $x^2 + y^2 \le 4$ in the $xy$-plane. We have $\nabla F = 2x\mathbf{i} + 2y\mathbf{j} - 2z\mathbf{k}$, and $F_3 = -2z$. Hence, on $\mathcal{S}$, we have

$$z\,dS = z\,\frac{\sqrt{4x^2 + 4y^2 + 4z^2}}{2z}\,dx\,dy = \sqrt{1 + 2(x^2 + y^2)}\,dx\,dy,$$

and the required moment is

$$\begin{aligned}\iint_{\mathcal{S}} z\,dS &= \iint_{x^2+y^2\le 4} \sqrt{1 + 2(x^2 + y^2)}\,dx\,dy\\ &= \int_0^{2\pi} d\theta \int_0^2 \sqrt{1 + 2r^2}\,r\,dr = \frac{\pi}{3}(1 + 2r^2)^{3/2}\Big|_0^2 = \frac{26\pi}{3}.\end{aligned}$$

The next example illustrates a technique that can often reduce the effort needed to integrate over a cylindrical surface.

**EXAMPLE 7** Find the area of that part of the cylinder $x^2 + y^2 = 2ay$ that lies inside the sphere $x^2 + y^2 + z^2 = 4a^2$.

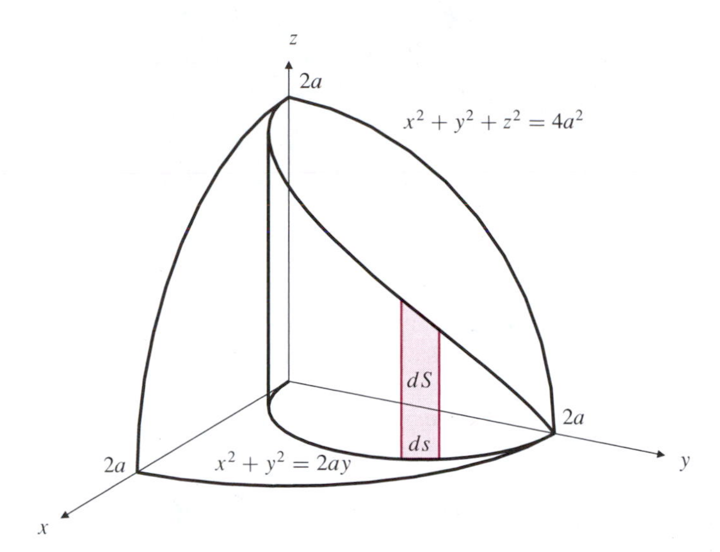


**Figure 8.23** An area element on a cylinder. The $z$-coordinate has already been integrated

***SOLUTION*** One quarter of the required area lies in the first octant. (See Figure 8.23.) Since the cylinder is generated by vertical lines, we can express an area element $dS$ on it in terms of the length element $ds$ along the curve $\mathcal{C}$ in the $xy$-plane having equation $x^2 + y^2 = 2ay$:

$$dS = z\,ds = \sqrt{4a^2 - x^2 - y^2}\,ds.$$

In expressing $dS$ this way, we have already integrated $dz$, so only a single integral is needed to sum these area elements. Again it is convenient to use polar coordinates in the $xy$-plane. In terms of polar coordinates the curve $\mathcal{C}$ has equation $r = 2a\sin\theta$. Thus $dr/d\theta = 2a\cos\theta$ and $ds = \sqrt{r^2 + (dr/d\theta)^2}\,d\theta = 2a\,d\theta$. Therefore the total surface area of that part of the cylinder that lies inside the sphere is given by

$$\begin{aligned} A &= 4\int_0^{\pi/2} \sqrt{4a^2 - r^2}\,2a\,d\theta \\ &= 8a\int_0^{\pi/2} \sqrt{4a^2 - 4a^2\sin^2\theta}\,d\theta \\ &= 16a^2\int_0^{\pi/2} \cos\theta\,d\theta = 16a^2 \text{ square units.} \end{aligned}$$

■

***REMARK*** The area calculated in Example 7 can also be calculated by projecting the cylindrical surface in Figure 8.23 into the $yz$-plane. (This is the only coordinate plane you can use. Why?) See Exercise 6 below.

In spherical coordinates, $\phi$ and $\theta$ can be used as parameters on the spherical surface $R = a$. The area element on that surface can therefore be expressed in terms of these coördinates.

Area element on the sphere $\rho = a$: $\quad dS = a^2 \sin\phi\,d\phi\,d\theta.$

(See Figure 6.43 in Section 6.6 and Exercise 2 at the end of this section.)

**■ EXAMPLE 8** Find $\displaystyle\iint_S z^2\,dS$ over the hemisphere $z = \sqrt{a^2 - x^2 - y^2}$.

**SOLUTION** Since $z = a\cos\phi$ and the hemisphere corresponds to $0 \le \theta \le 2\pi$, and $0 \le \phi \le \frac{\pi}{2}$, we have

$$\begin{aligned}\iint_S z^2\,dS &= \int_0^{2\pi} d\theta \int_0^{\pi/2} a^2\cos^2\phi\, a^2\sin\phi\,d\phi \\ &= 2\pi a^4\Big(-\frac{1}{3}\cos^3\phi\Big)\Big|_0^{\pi/2} = \frac{2\pi a^4}{3}.\end{aligned}$$

Finally, if a composite surface $\mathcal{S}$ is composed of *smooth parametric surfaces* joined pairwise along their edges, then we call $\mathcal{S}$ a **piecewise smooth surface**. The surface integral of a function $f$ over a piecewise smooth surface $\mathcal{S}$ is the sum of the surface integrals of $f$ over the individual smooth surfaces comprising $\mathcal{S}$. We will encounter an example of this in the next section.

## The Attraction of a Spherical Shell

In Section 6.7 we calculated the gravitational attraction of a disk in the $xy$-plane on a mass $m$ located at position $(0, 0, b)$ on the $z$-axis. Here we undertake a similar calculation of the attractive force exerted on $m$ by a spherical shell of radius $a$ and areal density $\sigma$ (units of mass per unit area) centred at the origin. This calculation would be more difficult if we try to do it by integrating the vertical component of the force on $m$ as we did in Section 6.7. It is greatly simplified if, instead, we use an integral to find the total *gravitational potential* $\Phi(0, 0, z)$ due to the sphere at position $(0, 0, z)$, and then calculate the force on $m$ as $\mathbf{F} = m\nabla\Phi(0, 0, b)$.

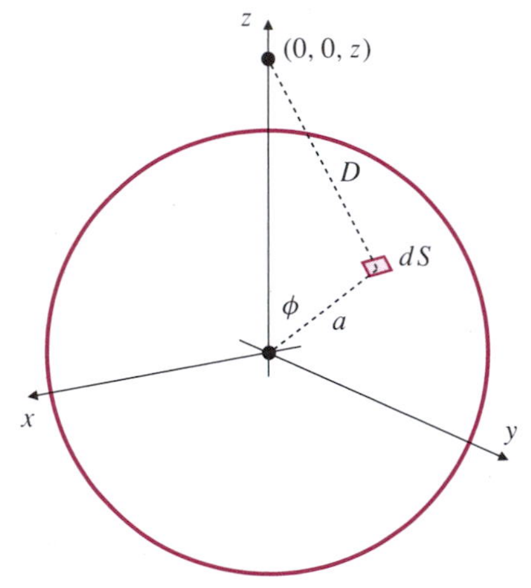


**Figure 8.24** The attraction of a sphere

The distance from the point with spherical coordinates $[a, \phi, \theta]$ to the point $(0, 0, z)$ on the positive $z$-axis is (see Figure 8.24)

$$D = \sqrt{a^2 + z^2 - 2az\cos\phi}.$$

The area element $dS = a^2\sin\phi\,d\phi\,d\theta$ at $[a, \phi, \theta]$ has mass $dm = \sigma\,dS$, and its gravitational potential at $(0, 0, z)$ is (see Example 1 in Section 8.2)

$$d\Phi(0, 0, z) = \frac{k\,dm}{D} = \frac{k\sigma a^2\sin\phi\,d\phi\,d\theta}{\sqrt{a^2 + z^2 - 2az\cos\phi}}.$$

For the total potential at $(0, 0, z)$ due to the sphere, we integrate $d\Phi$ over the surface of the sphere. Making the change of variables $u = a^2 + z^2 - 2az\cos\phi$, $du = 2az\sin\phi\,d\phi$, we obtain

$$\begin{aligned}\Phi(0, 0, z) &= k\sigma a^2\int_0^{2\pi} d\theta \int_0^{\pi} \frac{\sin\phi\,d\phi}{\sqrt{a^2 + z^2 - 2az\cos\phi}} \\ &= 2\pi k\sigma a^2\int_{(z-a)^2}^{(z+a)^2} \frac{1}{\sqrt{u}}\,\frac{du}{2az} \\ &= \frac{2\pi k\sigma a}{z}\sqrt{u}\,\Big|_{(z-a)^2}^{(z+a)^2} \\ &= \frac{2\pi k\sigma a}{z}\Big(z + a - |z - a|\Big) = \begin{cases} 4\pi k\sigma a^2/z & \text{if } z > a \\ 4\pi k\sigma a & \text{if } z < a. \end{cases}\end{aligned}$$

The potential is constant inside the sphere, and decreases proportionally to $1/z$ outside. The force on a mass $m$ located at $(0, 0, b)$ is, therefore,

$$\mathbf{F} = m\nabla\Phi(0, 0, b) = \begin{cases} -(4\pi km\sigma a^2/b^2)\mathbf{k} & \text{if } b > a \\ \mathbf{0} & \text{if } b < a. \end{cases}$$

We are led to the somewhat surprising result that if the mass $m$ is anywhere inside the sphere, the net force of attraction of the sphere on it is zero. This is to be expected at the centre of the sphere, but away from the centre it appears that the larger forces due to parts of the sphere close to $m$ are exactly cancelled by smaller forces due to parts further away; these further parts have larger area and therefore larger total mass. If $m$ is outside the sphere, the sphere attracts it with a force of magnitude

$$F = \frac{kmM}{b^2},$$

where $M = 4\pi\sigma a^2$ is the total mass of the sphere. This is the same force that would be exerted by a point mass with the same mass as the sphere and located at the centre of the sphere.

**REMARK** A solid ball of constant density, or density depending only on the distance from the centre (for instance, a planet), can be regarded as being made up of mass elements that are concentric spheres of constant density. Therefore, the attraction of such a ball on a mass $m$ located outside the ball will also be the same as if the whole mass of the ball were concentrated at its centre. However, the attraction on a mass $m$ located somewhere inside the ball will be that produced by only the part of the ball that is closer to the centre than $m$ is. The maximum force of attraction will occur when $m$ is right at the surface of the ball. If the density is constant, the magnitude of the force increases linearly with the distance from the centre (why?) up to the surface and then decreases with the square of the distance as $m$ recedes from the ball. (See Figure 8.25.)

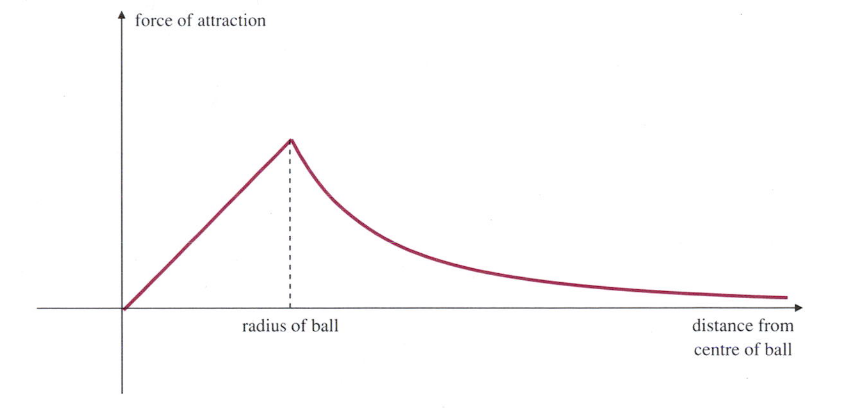


**Figure 8.25** The force of attraction of a homogeneous solid ball on a particle located at varying distances from the centre of the ball

**REMARK** All of the above discussion also holds for the electrostatic attraction or repulsion of a point charge by a uniform charge density over a spherical shell, which is also governed by an inverse square law. In particular there is no net electrostatic force on a charge located inside the shell.

## EXERCISES 8.5

1. Verify that on the curve with polar equation $r = g(\theta)$ the arc length element is given by

$$ds = \sqrt{(g(\theta))^2 + (g'(\theta))^2}\, d\theta.$$

What is the area element on the vertical cylinder given in terms of cylindrical coordinates by $r = g(\theta)$?

2. Verify that on the spherical surface $x^2 + y^2 + z^2 = a^2$ the area element is given in terms of spherical coordinates by $dS = a^2 \sin\phi \, d\phi \, d\theta$.

3. Find the area of the part of the plane $Ax + By + Cz = D$

lying inside the elliptic cylinder

$$\frac{x^2}{a^2} + \frac{y^2}{b^2} = 1.$$

**4.** Find the area of the part of the sphere $x^2 + y^2 + z^2 = 4a^2$ that lies inside the cylinder $x^2 + y^2 = 2ay$.

**5.** State formulas for the surface area element $dS$ for the surface with equation $F(x, y, z) = 0$ valid for the case where the surface has a one-to-one projection on (a) the $xz$-plane, and (b) the $yz$-plane.

**6.** Repeat the area calculation of Example 7 by projecting the part of the surface shown in Figure 8.23 onto the $yz$-plane, and using the formula in Exercise 5(b).

**7.** Find $\iint_{\mathcal{S}} x\, dS$ over the part of the parabolic cylinder $z = x^2/2$ that lies inside the first octant part of the cylinder $x^2 + y^2 = 1$.

**8.** Find the area of the part of the cone $z^2 = x^2 + y^2$ that lies inside the cylinder $x^2 + y^2 = 2ay$.

**9.** Find the area of the part of the cylinder $x^2 + y^2 = 2ay$ that lies outside the cone $z^2 = x^2 + y^2$.

**10.** Find the area of the part of the cylinder $x^2 + z^2 = a^2$ that lies inside the cylinder $y^2 + z^2 = a^2$.

**11.** A circular cylinder of radius $a$ is circumscribed about a sphere of radius $a$ so that the cylinder is tangent to the sphere along the equator. Two planes, each perpendicular to the axis of the cylinder, intersect the sphere and the cylinder in circles. Show that the area of that part of the sphere between the two planes is equal to the area of the part of the cylinder between the two planes. Thus the area of the part of a sphere between two parallel planes that intersect it depends only on the radius of the sphere and the distance between the planes, and not on the particular position of the planes.

**12.** Let $0 < a < b$. In terms of the elliptic integral functions defined in Exercise 17 of Section 8.3, find the area of that part of each of the cylinders $x^2 + z^2 = a^2$ and $y^2 + z^2 = b^2$ that lies inside the other cylinder.

**13.** Find $\iint_{\mathcal{S}} y\, dS$, where $\mathcal{S}$ is the part of the plane $z = 1 + y$ that lies inside the cone $z = \sqrt{2(x^2 + y^2)}$.

**14.** Find $\iint_{\mathcal{S}} y\, dS$, where $\mathcal{S}$ is the part of the cone $z = \sqrt{2(x^2 + y^2)}$ that lies below the plane $z = 1 + y$.

**15.** Find $\iint_{\mathcal{S}} xz\, dS$, where $\mathcal{S}$ is the part of the surface $z = x^2$ that lies in the first octant of 3-space and inside the paraboloid $z = 1 - 3x^2 - y^2$.

**16.** Find the mass of the part of the surface $z = \sqrt{2xy}$ that lies above the region $0 \le x \le 5$, $0 \le y \le 2$, if the areal density of the surface is $\sigma(x, y, z) = kz$.

**17.** Find the total charge on the surface

$$\mathbf{r} = e^u \cos v\mathbf{i} + e^u \sin v\mathbf{j} + u\mathbf{k},$$

$0 \le u \le 1$, $0 \le v \le \pi$, if the charge density on the surface is $\delta = \sqrt{1 + e^{2u}}$.

Exercises 18–19 concern **spheroids**, which are ellipsoids with two of their three semi-axes equal, say $a = b$:

$$\frac{x^2}{a^2} + \frac{y^2}{a^2} + \frac{z^2}{c^2} = 1.$$

* **18.** Find the surface area of a **prolate spheroid** where $0 < a < c$. A prolate spheroid has its two shorter semi-axes equal, like a "pro football."

* **19.** Find the surface area of an **oblate spheroid** where $0 < c < a$. An oblate spheroid has its two longer semi-axes equal, like the earth.

**20.** Describe the parametric surface

$$x = au\cos v, \qquad y = au\sin v, \qquad z = bv,$$

$(0 \le u \le 1,\ 0 \le v \le 2\pi)$, and find its area.

* **21.** Evaluate $\iint_{\mathcal{P}} \frac{dS}{(x^2 + y^2 + z^2)^{3/2}}$, where $\mathcal{P}$ is the plane with equation $Ax + By + Cz = D$, $(D \ne 0)$.

**22.** A spherical shell of radius $a$ is centred at the origin. Find the centroid of that part of the sphere that lies in the first octant.

**23.** Find the centre of mass of a right-circular conical shell of base radius $a$, height $h$, and constant areal density $\sigma$.

* **24.** Find the gravitational attraction of a hemispherical shell of radius $a$ and constant areal density $\sigma$ on a mass $m$ located at the centre of the base of the hemisphere.

* **25.** Find the gravitational attraction of a circular cylindrical shell of radius $a$, height $h$, and constant areal density $\sigma$ on a mass $m$ located on the axis of the cylinder $b$ units above the base.

In Exercises 26–28, find the moment of inertia and radius of gyration of the given object about the given axis. Assume constant areal density $\sigma$ in each case.

**26.** A cylindrical shell of radius $a$ and height $h$ about the axis of the cylinder

**27.** A spherical shell of radius $a$ about a diameter

**28.** A right-circular conical shell of base radius $a$ and height $h$ about the axis of the cone

**29.** With what acceleration will the spherical shell of Exercise 27 roll down a plane inclined at angle $\alpha$ to the horizontal? (Compare your result with that of Example 4(b) of Section 6.7.)

# 8.6 SURFACE INTEGRALS OF VECTOR FIELDS

Surface integrals of normal components of vector fields play a very important role in vector calculus, similar to the role played by line integrals of tangential components of vector fields. Before we consider such surface integrals we need to define the *orientation* of a surface.

## Oriented Surfaces

A smooth surface $\mathcal{S}$ in 3-space is said to be **orientable** if there exists a unit vector field $\hat{\mathbf{N}}(P)$ defined on $\mathcal{S}$, which varies continuously as $P$ ranges over $\mathcal{S}$ and which is everywhere normal to $\mathcal{S}$. Any such vector field $\hat{\mathbf{N}}(P)$ determines an **orientation** of $\mathcal{S}$. The surface must have two sides since $\hat{\mathbf{N}}(P)$ can have only one value at each point $P$. The side out of which $\hat{\mathbf{N}}$ points is called the **positive side**; the other side is the **negative side**. An **oriented surface** is a smooth surface together with a particular choice of orienting unit normal vector field $\hat{\mathbf{N}}(P)$.

For example, if we define $\hat{\mathbf{N}}$ on the smooth surface $z = f(x, y)$ by

$$\hat{\mathbf{N}} = \frac{-f_1(x, y)\mathbf{i} - f_2(x, y)\mathbf{j} + \mathbf{k}}{\sqrt{1 + (f_1(x, y))^2 + (f_2(x, y))^2}},$$

then the top of the surface is the positive side. (See Figure 8.26.)

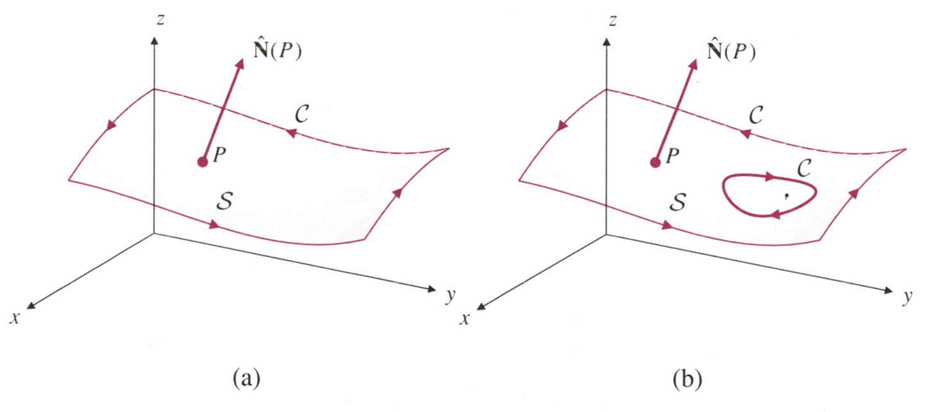


**Figure 8.26** The boundary curves of an oriented surface are themselves oriented with the surface on the left

A smooth or piecewise smooth surface may be **closed** (that is, it may have no boundary), or it may have one or more boundary curves. (The unit normal vector field $\hat{\mathbf{N}}(P)$ need not be defined at points of the boundary curves.)

An oriented surface $\mathcal{S}$ **induces an orientation** on any of its boundary curves $\mathcal{C}$; if we stand on the positive side of the surface $\mathcal{S}$ and walk around $\mathcal{C}$ in the direction of its orientation, then $\mathcal{S}$ will be on our left side. (See Figure 8.26(a) and (b).)

A *piecewise smooth* surface is **orientable** if, whenever two smooth component surfaces join along a common boundary curve $\mathcal{C}$, they induce *opposite* orientations along $\mathcal{C}$. This forces the normals $\hat{\mathbf{N}}$ to be on the same side of adjacent components. For instance, the surface of a cube is a piecewise smooth, closed surface, consisting of six smooth surfaces (the square faces) joined along edges. (See Figure 8.27.) If all of the faces are oriented so that their normals $\hat{\mathbf{N}}$ point out of the cube (or if they all point into the cube), then surface of the cube itself is oriented.

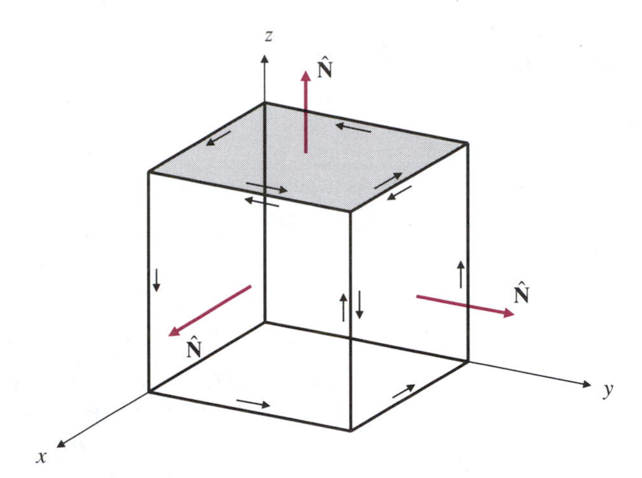


**Figure 8.27** The surface of the cube is orientable; adjacent faces induce opposite orientations on their common edge

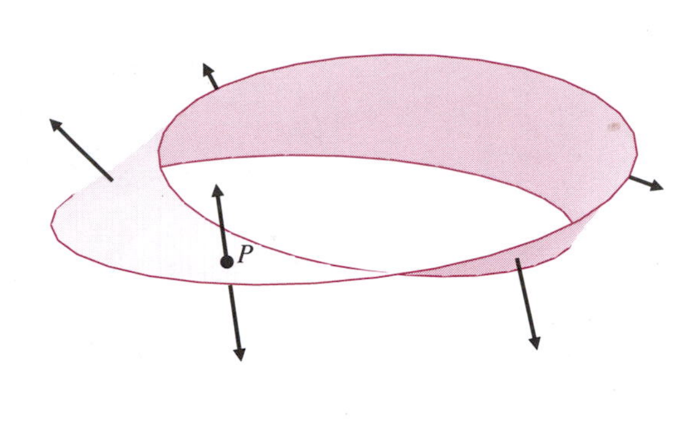


**Figure 8.28** The Möbius band is not orientable; it has only one "side"

Not every surface can be oriented, even if it appears smooth. An orientable surface must have two sides. For example, a Möbius band, consisting of a strip of paper with ends joined together to form a loop, but with one end given a half twist before the ends are joined, has only one side (make one and see), so it cannot be oriented. (See Figure 8.28.) If a nonzero vector is moved around the band, starting at point $P$, so that it is always normal to the surface, then it can return to its starting position pointing in the opposite direction.

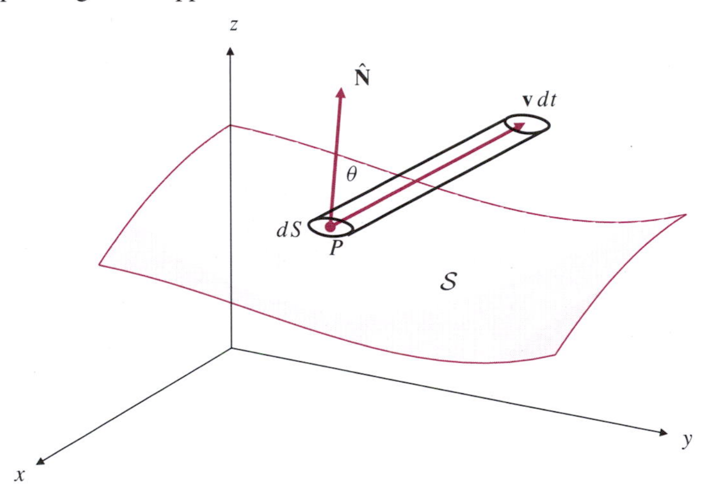


**Figure 8.29** The fluid crossing $dS$ in time $dt$ fills the tube

## The Flux of a Vector Field Across a Surface

Suppose 3-space is filled with an incompressible fluid that is flowing around with velocity field $\mathbf{v}$. Let $\mathcal{S}$ be an imaginary, smooth, oriented surface in 3-space. (We called $\mathcal{S}$ *imaginary* because it does not provide a barrier to the motion of the fluid. It is fixed in space, not moving with the fluid, and the fluid can freely move through it.) Let us calculate the rate at which fluid flows across $\mathcal{S}$. Let $dS$ be a small area element at point $P$ on the surface. The fluid crossing that element between time $t$ and time $t + dt$ occupies a cylinder of base area $dS$ and height $|\mathbf{v}(P)|\,dt\,\cos\theta$, where $\theta$ is the angle between $\mathbf{v}(P)$ and the normal $\hat{\mathbf{N}}(P)$. (See Figure 8.29.) This cylinder has volume $\mathbf{v}(P) \bullet \hat{\mathbf{N}}(P)\,dS\,dt$. The rate at which fluid is crossing $dS$ is

$\mathbf{v}(P) \bullet \hat{\mathbf{N}}(P)\, dS$, and the total rate at which it is crossing $\mathcal{S}$ is given by the surface integral

$$\iint_{\mathcal{S}} \mathbf{v} \bullet \hat{\mathbf{N}}\, dS \qquad \text{or} \qquad \iint_{\mathcal{S}} \mathbf{v} \bullet d\mathbf{S},$$

where we use $d\mathbf{S}$ to represent the vector surface area element $\hat{\mathbf{N}}\, dS$.

**DEFINITION 6**

**Flux of a vector field across an oriented surface**

Given any continuous vector field $\mathbf{F}$, the integral of the normal component of $\mathbf{F}$ over the oriented surface $\mathcal{S}$,

$$\iint_{\mathcal{S}} \mathbf{F} \bullet \hat{\mathbf{N}}\, dS \qquad \text{or} \qquad \iint_{\mathcal{S}} \mathbf{F} \bullet d\mathbf{S}$$

is called the **flux** of $\mathbf{F}$ across $\mathcal{S}$.

When the surface is closed, the flux integral can be denoted by

$$\oiint_{\mathcal{S}} \mathbf{F} \bullet \hat{\mathbf{N}}\, dS \qquad \text{or} \qquad \oiint_{\mathcal{S}} \mathbf{F} \bullet d\mathbf{S}.$$

In this case we refer to the flux of $\mathbf{F}$ *out of* $\mathcal{S}$ if $\hat{\mathbf{N}}$ is the unit *exterior* normal, and the flux *into* $\mathcal{S}$ if $\hat{\mathbf{N}}$ is the unit *interior* normal.

**EXAMPLE 1** Find the flux of the vector field $\mathbf{F} = m\mathbf{r}/|\mathbf{r}|^3$ out of a sphere $\mathcal{S}$ of radius $a$ centred at the origin. (Here $\mathbf{r} = x\mathbf{i} + y\mathbf{j} + z\mathbf{k}$.)

***SOLUTION*** Since $\mathbf{F}$ is the field associated with a source of strength $m$ at the origin (which produces $4\pi m$ units of fluid per unit time at the origin), the answer must be $4\pi m$. Let us calculate it anyway. We use spherical coordinates. At any point $\mathbf{r}$ on the sphere, with spherical coordinates $[a, \phi, \theta]$, the unit outward normal is $\hat{\mathbf{r}} = \mathbf{r}/|\mathbf{r}|$. Since the vector field is $\mathbf{F} = m\hat{\mathbf{r}}/a^2$ on the sphere, and since an area element is $dS = a^2 \sin\phi\, d\phi\, d\theta$, the flux of $\mathbf{F}$ out of the sphere is

$$\oiint_{\mathcal{S}} \left(\frac{m}{a^2}\hat{\mathbf{r}}\right) \bullet \hat{\mathbf{r}}\, a^2 \sin\phi\, d\phi\, d\theta = m \int_0^{2\pi} d\theta \int_0^{\pi} \sin\phi\, d\phi = 4\pi m.$$

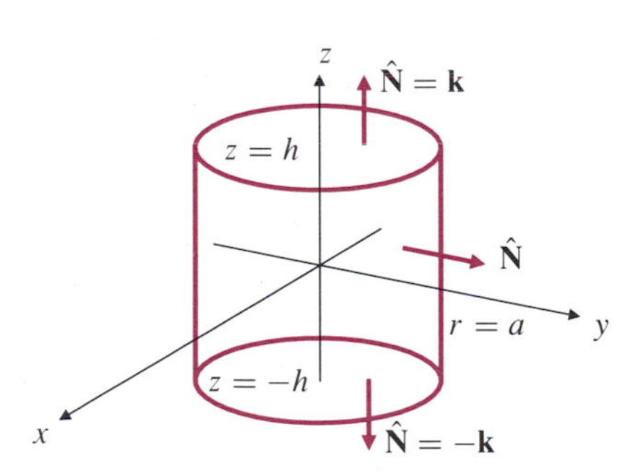


**Figure 8.30** The three components of the surface of a solid cylinder with their outward normals

**EXAMPLE 2** Calculate the total flux of $\mathbf{F} = x\mathbf{i} + y\mathbf{j} + z\mathbf{k}$ outward through the surface of the solid cylinder $x^2 + y^2 \le a^2$, $-h \le z \le h$.

***SOLUTION*** The cylinder is shown in Figure 8.30. Its surface consists of top and bottom disks and the cylindrical side wall. We calculate the flux of $\mathbf{F}$ out of each. Naturally, we use cylindrical coordinates.

On the top disk we have $z = h$, $\hat{\mathbf{N}} = \mathbf{k}$, and $dS = r\, dr\, d\theta$. Therefore, $\mathbf{F} \bullet \hat{\mathbf{N}}\, dS = hr\, dr\, d\theta$ and

$$\iint_{\text{top}} \mathbf{F} \bullet \hat{\mathbf{N}}\, dS = h \int_0^{2\pi} d\theta \int_0^{a} r\, dr = \pi a^2 h.$$

On the bottom disk we have $z = -h$, $\hat{\mathbf{N}} = -\mathbf{k}$, and $dS = r\, dr\, d\theta$. Therefore, $\mathbf{F} \bullet \hat{\mathbf{N}}\, dS = hr\, dr\, d\theta$ and

$$\iint_{\text{bottom}} \mathbf{F} \bullet \hat{\mathbf{N}}\, dS = \iint_{\text{top}} \mathbf{F} \bullet \hat{\mathbf{N}}\, dS = \pi a^2 h.$$

On the cylindrical wall $\mathbf{F} = a\cos\theta\,\mathbf{i} + a\sin\theta\,\mathbf{j} + z\mathbf{k}$, $\hat{\mathbf{N}} = \cos\theta\,\mathbf{i} + \sin\theta\,\mathbf{j}$, and $dS = a\,d\theta\,dz$. Thus $\mathbf{F} \bullet \hat{\mathbf{N}}\,dS = a^2\,d\theta\,dz$ and

$$\iint_{\text{cylwall}} \mathbf{F} \bullet \hat{\mathbf{N}}\,dS = a^2 \int_0^{2\pi} d\theta \int_{-h}^{h} dz = 4\pi a^2 h.$$

The total flux of $\mathbf{F}$ out of the surface $\mathcal{S}$ of the cylinder is the sum of these three contributions:

$$\oiint_{\mathcal{S}} \mathbf{F} \bullet \hat{\mathbf{N}}\,dS = 6\pi a^2 h.$$

■

Let $\mathcal{S}$ be a smooth, oriented surface with a one-to-one projection onto a domain $D$ in the $xy$-plane, and with equation of the form $G(x, y, z) = 0$. In Section 8.5 we showed that the surface area element on $\mathcal{S}$ could be written in the form

$$dS = \left|\frac{\nabla G}{G_3}\right| dx\,dy,$$

and hence surface integrals over $\mathcal{S}$ could be reduced to double integrals over the domain $D$. Flux integrals can be treated likewise. Depending on the orientation of $\mathcal{S}$, the unit normal $\hat{\mathbf{N}}$ can be written as

$$\hat{\mathbf{N}} = \pm\frac{\nabla G}{|\nabla G|}.$$

Thus the vector area element $d\mathbf{S}$ can be written

$$d\mathbf{S} = \hat{\mathbf{N}}\,dS = \pm\frac{\nabla G(x, y, z)}{G_3(x, y, z)}\,dx\,dy.$$

The sign must be chosen to give $\mathcal{S}$ the desired orientation. If $G_3 > 0$ and we want the positive side of $\mathcal{S}$ to face upward, we should use the "+" sign. Of course, similar formulas apply for surfaces with one-to-one projections onto the other coordinate planes.

■ **EXAMPLE 3** Find the flux of $z\mathbf{i} + x^2\mathbf{k}$ upward through that part of the surface $z = x^2 + y^2$ lying above the square $R$ defined by $-1 \le x \le 1$ and $-1 \le y \le 1$.

***SOLUTION*** For $F(x, y, z) = z - x^2 - y^2$ we have $\nabla F = -2x\mathbf{i} - 2y\mathbf{j} + \mathbf{k}$ and $F_3 = 1$. Thus

$$d\mathbf{S} = (-2x\mathbf{i} - 2y\mathbf{j} + \mathbf{k})\,dx\,dy,$$

and the required flux is

$$\begin{aligned}\iint_{\mathcal{S}} (z\mathbf{i} + x^2\mathbf{k}) \bullet d\mathbf{S} &= \iint_R \left(-2x(x^2 + y^2) + x^2\right) dx\,dy \\ &= \int_{-1}^{1} dx \int_{-1}^{1} (x^2 - 2x^3 - 2xy^2)\,dx\,dy \\ &= \int_{-1}^{1} 2x^2\,dx = \frac{4}{3}.\end{aligned}$$

(Two of the three terms in the double integral had zero integrals because of symmetry.) ■

For a surface $\mathcal{S}$ with equation $z = f(x, y)$ we have

$$\hat{\mathbf{N}} = \pm \frac{-\dfrac{\partial f}{\partial x}\mathbf{i} - \dfrac{\partial f}{\partial y}\mathbf{j} + \mathbf{k}}{\sqrt{1 + \left(\dfrac{\partial f}{\partial x}\right)^2 + \left(\dfrac{\partial f}{\partial y}\right)^2}}, \qquad \text{and}$$

$$dS = \sqrt{1 + \left(\frac{\partial f}{\partial x}\right)^2 + \left(\frac{\partial f}{\partial y}\right)^2}\, dx\, dy,$$

so that the vector area element on $\mathcal{S}$ is given by

$$d\mathbf{S} = \hat{\mathbf{N}}\, dS = \pm\left(-\frac{\partial f}{\partial x}\mathbf{i} - \frac{\partial f}{\partial y}\mathbf{j} + \mathbf{k}\right) dx\, dy.$$

Again, the $+$ sign corresponds to an upward normal.

For a general parametric surface $\mathbf{r} = \mathbf{r}(u, v)$, the unit normal $\hat{\mathbf{N}}$ and area element $dS$ were calculated in Section 8.5:

$$\hat{\mathbf{N}} = \pm \frac{\dfrac{\partial \mathbf{r}}{\partial u} \times \dfrac{\partial \mathbf{r}}{\partial v}}{\left|\dfrac{\partial \mathbf{r}}{\partial u} \times \dfrac{\partial \mathbf{r}}{\partial v}\right|}, \qquad dS = \left|\frac{\partial \mathbf{r}}{\partial u} \times \frac{\partial \mathbf{r}}{\partial v}\right| du\, dv.$$

Thus the vector area element is

$$d\mathbf{S} = \hat{\mathbf{N}}\, dS = \pm\left(\frac{\partial \mathbf{r}}{\partial u} \times \frac{\partial \mathbf{r}}{\partial v}\right) du\, dv.$$

**EXAMPLE 4** Find the flux of $\mathbf{F} = y\mathbf{i} - x\mathbf{j} + 4\mathbf{k}$ upward through $\mathcal{S}$, where $\mathcal{S}$ is the part of the surface $z = 1 - x^2 - y^2$ lying in the first octant of 3-space.

***SOLUTION*** The vector area element corresponding to the upward normal on $\mathcal{S}$ is

$$d\mathbf{S} = \left(-\frac{\partial z}{\partial x}\mathbf{i} - \frac{\partial z}{\partial y}\mathbf{j} + \mathbf{k}\right) dx\, dy = (2x\mathbf{i} + 2y\mathbf{j} + \mathbf{k})\, dx\, dy.$$

The projection of $\mathcal{S}$ onto the $xy$-plane is the quarter-circular disk $Q$ given by $x^2 + y^2 \le 1$, $x \ge 0$, and $y \ge 0$. Thus the flux of $\mathbf{F}$ upward through $\mathcal{S}$ is

$$\begin{aligned}\iint_{\mathcal{S}} \mathbf{F} \bullet d\mathbf{S} &= \iint_{Q} (2xy - 2xy + 4)\, dx\, dy \\ &= 4 \times (\text{area of } Q) = \pi.\end{aligned}$$

■

**EXAMPLE 5** Find the flux of

$$\mathbf{F} = \frac{2x\mathbf{i} + 2y\mathbf{j}}{x^2 + y^2} + \mathbf{k}$$

downward through the surface $\mathcal{S}$ defined parametrically by

$$\mathbf{r} = u\cos v\mathbf{i} + u\sin v\mathbf{j} + u^2\mathbf{k}, \qquad (0 \le u \le 1,\ 0 \le v \le 2\pi).$$

**SOLUTION** First we calculate $d\mathbf{S}$.

$$\frac{\partial \mathbf{r}}{\partial u} = \cos v\mathbf{i} + \sin v\mathbf{j} + 2u\mathbf{k}$$
$$\frac{\partial \mathbf{r}}{\partial v} = -u\sin v\mathbf{i} + u\cos v\mathbf{j}$$
$$\frac{\partial \mathbf{r}}{\partial u} \times \frac{\partial \mathbf{r}}{\partial v} = -2u^2\cos v\mathbf{i} - 2u^2\sin v\mathbf{j} + u\mathbf{k}$$

Since $u \ge 0$ on $\mathcal{S}$, the latter expression is an upward normal. We want a downward normal, so we use

$$d\mathbf{S} = (2u^2\cos v\mathbf{i} + 2u^2\sin v\mathbf{j} - u\mathbf{k})\,du\,dv.$$

On $\mathcal{S}$ we have

$$\mathbf{F} = \frac{2x\mathbf{i} + 2y\mathbf{j}}{x^2 + y^2} + \mathbf{k} = \frac{2u\cos v\mathbf{i} + 2u\sin v\mathbf{j}}{u^2} + \mathbf{k},$$

so the downward flux of $\mathbf{F}$ through $\mathcal{S}$ is

$$\iint_{\mathcal{S}} \mathbf{F} \bullet d\mathbf{S} = \int_0^{2\pi} dv \int_0^1 (4u - u)\,du = 3\pi.$$ ■

## EXERCISES 8.6

1. Find the flux of $\mathbf{F} = x\mathbf{i} + z\mathbf{j}$ out of the tetrahedron bounded by the coordinate planes and the plane $x + 2y + 3z = 6$.

2. Find the flux of $\mathbf{F} = x\mathbf{i} + y\mathbf{j} + z\mathbf{k}$ outward across the sphere $x^2 + y^2 + z^2 = a^2$.

3. Find the flux of the vector field of Exercise 2 out of the surface of the rectangular box $0 \le x \le a$, $0 \le y \le b$, $0 \le z \le c$.

4. Find the flux of the vector field $\mathbf{F} = y\mathbf{i} + z\mathbf{k}$ out across the boundary of the solid cone $0 \le z \le 1 - \sqrt{x^2 + y^2}$.

5. Find the flux of $\mathbf{F} = x\mathbf{i} + y\mathbf{j} + z\mathbf{k}$ upward through the part of the surface $z = a - x^2 - y^2$ lying above the plane $z = b$, where $b < a$.

6. Find the flux of $\mathbf{F} = x\mathbf{i} + x\mathbf{j} + \mathbf{k}$ upward through the part of the surface $z = x^2 - y^2$ lying inside the cylinder $x^2 + y^2 = a^2$.

7. Find the flux of $\mathbf{F} = y^3\mathbf{i} + z^2\mathbf{j} + x\mathbf{k}$ downward through the part of the surface $z = 4 - x^2 - y^2$ that lies above the plane $z = 2x + 1$.

8. Find the flux of $\mathbf{F} = z^2\mathbf{k}$ upward through the part of the sphere $x^2 + y^2 + z^2 = a^2$ in the first octant of 3-space.

9. Find the flux of $\mathbf{F} = x\mathbf{i} + y\mathbf{j}$ upward through the part of the surface $z = 2 - x^2 - 2y^2$ that lies above the $xy$-plane.

10. Find the flux of $\mathbf{F} = 2x\mathbf{i} + y\mathbf{j} + z\mathbf{k}$ upward through the surface $\mathbf{r} = u^2v\mathbf{i} + uv^2\mathbf{j} + v^3\mathbf{k}$, $(0 \le u \le 1, 0 \le v \le 1)$.

11. Find the flux of $\mathbf{F} = x\mathbf{i} + y\mathbf{j} + z^2\mathbf{k}$ upward through the surface $u\cos v\,\mathbf{i} + u\sin v\,\mathbf{j} + u\,\mathbf{k}$, $(0 \le u \le 2, 0 \le v \le \pi)$.

12. Find the flux of $\mathbf{F} = yz\mathbf{i} - xz\mathbf{j} + (x^2 + y^2)\mathbf{k}$ upward through the surface $\mathbf{r} = e^u\cos v\,\mathbf{i} + e^u\sin v\,\mathbf{j} + u\,\mathbf{k}$, where $0 \le u \le 1$ and $0 \le v \le \pi$.

13. Find the flux of $\mathbf{F} = m\mathbf{r}/|\mathbf{r}|^3$ out of the surface of the cube $-a \le x, y, z \le a$.

* 14. Find the flux of the vector field of Exercise 13 out of the box $1 \le x, y, z \le 2$. *Note:* This problem can be solved very easily using the Divergence Theorem of Section 9.3; the required flux is, in fact, zero. However, the object here is to do it by direct calculation of the surface integrals involved, and as such it is quite difficult. By symmetry, it is sufficient to evaluate the net flux out of the cube through any one of the three pairs of opposite faces; that is, you must calculate the flux through only two faces, say $z = 1$ and $z = 2$. Be prepared to work very hard to evaluate these integrals! When they are done you may find the identities

$$2\arctan a = \arctan\left(\frac{2a}{1 - a^2}\right), \qquad \text{and}$$
$$\arctan a + \arctan\left(\frac{1}{a}\right) = \frac{\pi}{2}$$

useful for showing that the net flux is zero.

15. Define the flux of a *plane* vector field across a piecewise smooth *curve*. Find the flux of $\mathbf{F} = x\mathbf{i} + y\mathbf{j}$ outward across

 (a) the circle $x^2 + y^2 = a^2$.

 (b) the boundary of the square $-1 \le x, y \le 1$.

16. Find the flux of

$$\mathbf{F} = -\frac{x\mathbf{i} + y\mathbf{j}}{x^2 + y^2}$$

*inward* across each of the two curves in the previous exercise.

17. If $\mathcal{S}$ is a smooth, oriented surface in 3-space and $\hat{\mathbf{N}}$ is the unit vector field determining the orientation of $\mathcal{S}$, show that the flux of $\hat{\mathbf{N}}$ across $\mathcal{S}$ is the area of $\mathcal{S}$.

* 18. The Divergence Theorem presented in Section 9.3 implies that the flux of a constant vector field across any oriented, piecewise smooth, closed surface is zero. Prove this now for (a) a rectangular box and (b) a sphere.

# CHAPTER REVIEW

## Key Ideas

- **What do the following phrases mean?**
  - ◇ a vector field
  - ◇ a scalar field
  - ◇ a field line
  - ◇ a conservative field
  - ◇ a scalar potential
  - ◇ an equipotential
  - ◇ a source
  - ◇ a dipole
  - ◇ a connected domain
  - ◇ a simply connected domain
  - ◇ a parametric surface
  - ◇ an orientable surface
  - ◇ the line integral of $f$ along curve $\mathcal{C}$
  - ◇ the line integral of the tangential component of $\mathbf{F}$ along $\mathcal{C}$
  - ◇ the flux of a vector field through a surface
- **How are the field lines of a conservative field related to its equipotential curves or surfaces?**
- **How is a line integral of a scalar field calculated?**
- **How is a line integral of the tangential component of a vector field calculated?**
- **When is a line integral between two points independent of the path joining those points?**
- **How is a surface integral of a scalar field calculated?**
- **How is the flux of a vector field through a surface calculated?**

## Review Exercises

1. Find $\displaystyle\int_{\mathcal{C}} \frac{1}{y}\, ds$, where $\mathcal{C}$ is the curve

$$x = t, \quad y = 2e^t, \quad z = e^{2t}, \quad (-1 \le t \le 1).$$

2. Let $\mathcal{C}$ be the part of the curve of intersection of the surfaces $z = x + y^2$ and $y = 2x$ from the origin to the point $(2, 4, 18)$. Evaluate $\displaystyle\int_{\mathcal{C}} 2y\,dx + x\,dy + 2\,dz$.

3. Find $\displaystyle\iint_{\mathcal{S}} x\,dS$ where $\mathcal{S}$ is that part of the cone $z = \sqrt{x^2 + y^2}$ in the region $0 \le x \le 1 - y^2$.

4. Find $\displaystyle\iint_{\mathcal{S}} xyz\,dS$ over the part of the plane $x + y + z = 1$ lying in the first octant.

5. Find the flux of $x^2y\mathbf{i} - 10xy^2\mathbf{j}$ upward through the surface $z = xy, 0 \le x \le 1, 0 \le y \le 1$.

6. Find the flux of $x\mathbf{i} + y\mathbf{j} + z\mathbf{k}$ downward through the part of the plane $x + 2y + 3z = 6$ lying in the first octant.

7. A bead of mass $m$ slides down a wire in the shape of the curve

$$x = a\sin t, \quad y = a\cos t, \quad z = bt,$$

where $0 \le t \le 6\pi$.

 (a) What is the work done by the gravitational force $\mathbf{F} = -mg\mathbf{k}$ on the bead during its descent?

 (b) What is the work done against a resistance of constant magnitude $R$ which directly opposes the motion of the bead during its descent?

8. For what values of the constants $a$, $b$, and $c$ can you determine the value of the integral $I$ of the tangential component of

$$\mathbf{F} = (axy + 3yz)\mathbf{i} + (x^2 + 3xz + by^2z)\mathbf{j} + (bxy + cy^3)\mathbf{k}$$

along a curve from $(0, 1, -1)$ to $(2, 1, 1)$ without knowing exactly which curve? What is the value of the integral?

9. Let $\mathbf{F} = (x^2/y)\mathbf{i} + y\mathbf{j} + \mathbf{k}$.

 (a) Find the field line of $\mathbf{F}$ that passes through $(1, 1, 0)$ and show that it also passes through $(e, e, 1)$.

 (b) Find $\displaystyle\int_{\mathcal{C}} \mathbf{F} \bullet d\mathbf{r}$ where $\mathcal{C}$ is the part of the field line in (a) from $(1, 1, 0)$ to $(e, e, 1)$.

10. Consider the vector fields

$$\mathbf{F} = (1 + x)e^{x+y}\mathbf{i} + (xe^{x+y} + 2y)\mathbf{j} - 2z\mathbf{k},$$
$$\mathbf{G} = (1 + x)e^{x+y}\mathbf{i} + (xe^{x+y} + 2z)\mathbf{j} - 2y\mathbf{k}.$$

 (a) Show that $\mathbf{F}$ is conservative by finding a potential for it.

 (b) Evaluate $\displaystyle\int_{\mathcal{C}} \mathbf{G} \bullet d\mathbf{r}$, where $\mathcal{C}$ is given by

$$\mathbf{r} = (1 - t)e^t\mathbf{i} + t\mathbf{j} + 2t\mathbf{k}, \quad (0 \le t \le 1),$$

 by taking advantage of the similarity between $\mathbf{F}$ and $\mathbf{G}$.

**11.** Find a plane vector field $\mathbf{F}(x, y)$ that satisfies all of the following conditions:

(i) The field lines of $\mathbf{F}$ are the curves $xy = C$.

(ii) $|\mathbf{F}(x, y)| = 1$ if $(x, y) \neq (0, 0)$.

(iii) $\mathbf{F}(1, 1) = (\mathbf{i} - \mathbf{j})/\sqrt{2}$.

(iv) $\mathbf{F}$ is continuous except at $(0, 0)$.

**12.** Let $\mathcal{S}$ be the part of the surface of the cylinder $y^2 + z^2 = 16$ that lies in the first octant and between the planes $x = 0$ and $x = 5$. Find the flux of $3z^2x\mathbf{i} - x\mathbf{j} - y\mathbf{k}$ away from the $x$-axis through $\mathcal{S}$.

## Express Yourself

Answer these questions in words, using complete sentences and, if necessary, paragraphs. Use mathematical symbols only where necessary.

**1.** Discuss the physical significance of the field lines and the equipotential surfaces of a temperature gradient field.

**2.** What do the vector fields representing the gravitational attraction of a point mass, the electrostatic attraction of a point charge, and the velocity field of a point sink have in common? Mention as many characteristics as you can.

**3.** Describe some reasons why it is useful to know when a vector field is conservative, and to know a potential function in that case.

**4.** Describe the relationship between the orientation of a composite surface and the orientations of the boundary curves of its component smooth surfaces.

## Challenging Problems

* **1.** Find the centroid of the surface

$$\mathbf{r} = (2 + \cos v)(\cos u\mathbf{i} + \sin u\mathbf{j}) + \sin v\mathbf{k},$$

where $0 \leq u \leq 2\pi$ and $0 \leq v \leq \pi$. Describe this surface.

* **2.** A smooth surface $\mathcal{S}$ is given parametrically by

$$\begin{aligned}\mathbf{r} = &(\cos 2u)(2 + v\cos u)\mathbf{i} \\ &+ (\sin 2u)(2 + v\cos u)\mathbf{j} + v\sin u\mathbf{k},\end{aligned}$$

where $0 \leq u \leq 2\pi$ and $-1 \leq v \leq 1$. Show that for *every* smooth vector field $\mathbf{F}$ on $\mathcal{S}$,

$$\iint_{\mathcal{S}} \mathbf{F} \bullet \hat{\mathbf{N}}\, dS = 0,$$

where $\hat{\mathbf{N}} = \hat{\mathbf{N}}(u, v)$ is a unit normal vector field on $\mathcal{S}$ that depends continuously on $(u, v)$. How do you explain this? *Hint:* try to describe what the surface $\mathcal{S}$ looks like.

* **3.** Recalculate the gravitational force exerted by a sphere of radius $a$ and areal density $\sigma$ centred at the origin on a point mass located at $(0, 0, b)$ by directly integrating the vertical component of the force due to an area element $dS$, rather than by integrating the potential as we did in the last part of Section 8.5. You will have to be quite creative in dealing with the resulting integral.

CHAPTER 9

# Vector Calculus

**INTRODUCTION** In this chapter we develop two- and three-dimensional analogues of the one-dimensional Fundamental Theorem of Calculus. These analogues — Green's Theorem, Gauss's Divergence Theorem, and Stokes's Theorem — are of great importance both theoretically and in applications. They are phrased in terms of certain differential operators, divergence and curl, which are related to the gradient operator encountered in Section 4.7. The operators are introduced and their properties are derived in Sections 9.1 and 9.2. The rest of the chapter deals with the generalizations of the Fundamental Theorem of Calculus and their applications.

## 9.1 GRADIENT, DIVERGENCE, AND CURL

First-order information about the rate of change of a 3-dimensional scalar field, $f(x, y, z)$, is contained in the three first partial derivatives $\partial f/\partial x$, $\partial f/\partial y$, and $\partial f/\partial z$. The gradient,

$$\mathbf{grad}\, f(x, y, z) = \nabla f(x, y, z) = \frac{\partial f}{\partial x}\mathbf{i} + \frac{\partial f}{\partial y}\mathbf{j} + \frac{\partial f}{\partial z}\mathbf{k},$$

collects this information into a single vector-valued "derivative" of $f$. We would like to develop similar ways of conveying information about the rate of change of vector fields.

First-order information about the rate of change of the vector field

$$\mathbf{F}(x, y, z) = F_1(x, y, z)\mathbf{i} + F_2(x, y, z)\mathbf{j} + F_3(x, y, z)\mathbf{k}$$

is contained in nine first partial derivatives, three for each of the three components of $\mathbf{F}$:

$$\begin{array}{ccc} \dfrac{\partial F_1}{\partial x} & \dfrac{\partial F_1}{\partial y} & \dfrac{\partial F_1}{\partial z} \\[2ex] \dfrac{\partial F_2}{\partial x} & \dfrac{\partial F_2}{\partial y} & \dfrac{\partial F_2}{\partial z} \\[2ex] \dfrac{\partial F_3}{\partial x} & \dfrac{\partial F_3}{\partial y} & \dfrac{\partial F_3}{\partial z}. \end{array}$$

(Again we stress that $F_1$, $F_2$, and $F_3$ denote the components of $\mathbf{F}$, not partial derivatives.) Two special combinations of these derivatives organize this information in particularly useful ways, as the gradient does for scalar fields. These are the **divergence** of $\mathbf{F}$ (**div** $\mathbf{F}$), and the **curl** of $\mathbf{F}$ (**curl** $\mathbf{F}$), defined as follows:

**Divergence and curl**

$$\mathbf{div}\,\mathbf{F} = \nabla\bullet\mathbf{F} = \frac{\partial F_1}{\partial x} + \frac{\partial F_2}{\partial y} + \frac{\partial F_3}{\partial z},$$

$$\begin{aligned}\mathbf{curl}\,\mathbf{F} &= \nabla\times\mathbf{F}\\ &= \left(\frac{\partial F_3}{\partial y} - \frac{\partial F_2}{\partial z}\right)\mathbf{i} + \left(\frac{\partial F_1}{\partial z} - \frac{\partial F_3}{\partial x}\right)\mathbf{j} + \left(\frac{\partial F_2}{\partial x} - \frac{\partial F_1}{\partial y}\right)\mathbf{k}\\ &= \begin{vmatrix} \mathbf{i} & \mathbf{j} & \mathbf{k} \\ \frac{\partial}{\partial x} & \frac{\partial}{\partial y} & \frac{\partial}{\partial z} \\ F_1 & F_2 & F_3 \end{vmatrix}.\end{aligned}$$

Note that the divergence of a vector field is a scalar field, while the curl is another vector field. Also observe the notation $\nabla\bullet\mathbf{F}$ and $\nabla\times\mathbf{F}$, which we will sometimes use instead of $\mathbf{div}\,\mathbf{F}$ and $\mathbf{curl}\,\mathbf{F}$. This makes use of the *vector differential operator*,

$$\nabla = \mathbf{i}\frac{\partial}{\partial x} + \mathbf{j}\frac{\partial}{\partial y} + \mathbf{k}\frac{\partial}{\partial z}$$

frequently called *del* or *nabla*. Just as the gradient of the scalar field $f$ can be regarded as *formal scalar multiplication* of $\nabla$ by $f$, so also can the divergence and curl of $\mathbf{F}$ be regarded as *formal dot* and *cross* products of $\nabla$ with $\mathbf{F}$. When using $\nabla$ the order of "factors" is important; the quantities on which $\nabla$ acts must appear to the right of $\nabla$. For instance, $\nabla\bullet\mathbf{F}$ and $\mathbf{F}\bullet\nabla$ do not mean the same thing; the former is a scalar field, the latter is a scalar differential operator.

**EXAMPLE 1** Find the divergence and curl of the vector field

$$\mathbf{F} = xy\mathbf{i} + (y^2 - z^2)\mathbf{j} + yz\mathbf{k}.$$

***SOLUTION*** We have

$$\mathbf{div}\,\mathbf{F} = \nabla\bullet\mathbf{F} = \frac{\partial}{\partial x}(xy) + \frac{\partial}{\partial y}(y^2 - z^2) + \frac{\partial}{\partial z}(yz) = y + 2y + y = 4y,$$

$$\begin{aligned}\mathbf{curl}\,\mathbf{F} = \nabla\times\mathbf{F} &= \begin{vmatrix} \mathbf{i} & \mathbf{j} & \mathbf{k} \\ \frac{\partial}{\partial x} & \frac{\partial}{\partial y} & \frac{\partial}{\partial z} \\ xy & y^2 - z^2 & yz \end{vmatrix}\\ &= \left[\frac{\partial}{\partial y}(yz) - \frac{\partial}{\partial z}(y^2 - z^2)\right]\mathbf{i} + \left[\frac{\partial}{\partial z}(xy) - \frac{\partial}{\partial x}(yz)\right]\mathbf{j}\\ &\quad + \left[\frac{\partial}{\partial x}(y^2 - z^2) - \frac{\partial}{\partial y}(xy)\right]\mathbf{k} = 3z\mathbf{i} - x\mathbf{k}.\end{aligned}$$

The divergence and curl of a two-dimensional vector field are also defined: if $\mathbf{F}(x, y) = F_1(x, y)\mathbf{i} + F_2(x, y)\mathbf{j}$, then

$$\mathbf{div}\,\mathbf{F} = \frac{\partial F_1}{\partial x} + \frac{\partial F_2}{\partial y},$$

$$\mathbf{curl}\,\mathbf{F} = \left(\frac{\partial F_2}{\partial x} - \frac{\partial F_1}{\partial y}\right)\mathbf{k}.$$

Note that the curl of a two-dimensional vector field is still a 3-vector and is perpendicular to the plane of the field. Although **div** and **grad** are defined in all dimensions, **curl** is defined only in three dimensions and in the plane (provided we allow values in three dimensions).

**EXAMPLE 2** Find the divergence and curl of $\mathbf{F} = xe^y\mathbf{i} - ye^x\mathbf{j}$.

***SOLUTION*** We have

$$\begin{aligned}
\mathbf{div}\,\mathbf{F} &= \nabla \bullet \mathbf{F} = \frac{\partial}{\partial x}(xe^y) + \frac{\partial}{\partial y}(-ye^x) = e^y - e^x, \\
\mathbf{curl}\,\mathbf{F} &= \nabla \times \mathbf{F} = \left(\frac{\partial}{\partial x}(-ye^x) - \frac{\partial}{\partial y}(xe^y)\right)\mathbf{k} \\
&= -(ye^x + xe^y)\mathbf{k}.
\end{aligned}$$

## Interpretation of the Divergence

The value of the divergence of a vector field $\mathbf{F}$ at point $P$ is, loosely speaking, a measure of the rate at which the field "diverges" or "spreads away" from $P$. This spreading away can be measured by the flux out of a small closed surface surrounding $P$. For instance, $\mathbf{div}\,\mathbf{F}(P)$ is the limit of the *flux per unit volume* out of smaller and smaller spheres centred at $P$.

**THEOREM 1**

**The divergence as flux density**

If $\hat{\mathbf{N}}$ is the unit outward normal on the sphere $\mathcal{S}_\epsilon$ of radius $\epsilon$ centred at point $P$, and if $\mathbf{F}$ is a smooth three-dimensional vector field, then

$$\mathbf{div}\,\mathbf{F}(P) = \lim_{\epsilon \to 0^+} \frac{3}{4\pi\epsilon^3} \oiint_{\mathcal{S}_\epsilon} \mathbf{F} \bullet \hat{\mathbf{N}}\, dS.$$

***PROOF*** Without loss of generality we assume that $P$ is at the origin. We want to expand $\mathbf{F}$ in a Taylor series about the origin (a Maclaurin series). As shown in Section 4.9 for a function of two variables, the Maclaurin series for a scalar-valued function of three variables takes the form

$$f(x, y, z) = f(0, 0, 0) + \left.\frac{\partial f}{\partial x}\right|_{(0,0,0)} x + \left.\frac{\partial f}{\partial y}\right|_{(0,0,0)} y + \left.\frac{\partial f}{\partial z}\right|_{(0,0,0)} z + \cdots,$$

where "$\cdots$" represents terms of second and higher degree in $x$, $y$, and $z$. If we apply this formula to the components of $\mathbf{F}$, we obtain

$$\mathbf{F}(x, y, z) = \mathbf{F}_0 + \mathbf{F}_1 x + \mathbf{F}_2 y + \mathbf{F}_3 z + \cdots$$

where

$$\begin{aligned}
\mathbf{F}_0 &= \mathbf{F}(0, 0, 0) \\
\mathbf{F}_1 &= \left.\frac{\partial \mathbf{F}}{\partial x}\right|_{(0,0,0)} = \left.\left(\frac{\partial F_1}{\partial x}\mathbf{i} + \frac{\partial F_2}{\partial x}\mathbf{j} + \frac{\partial F_3}{\partial x}\mathbf{k}\right)\right|_{(0,0,0)} \\
\mathbf{F}_2 &= \left.\frac{\partial \mathbf{F}}{\partial y}\right|_{(0,0,0)} = \left.\left(\frac{\partial F_1}{\partial y}\mathbf{i} + \frac{\partial F_2}{\partial y}\mathbf{j} + \frac{\partial F_3}{\partial y}\mathbf{k}\right)\right|_{(0,0,0)} \\
\mathbf{F}_3 &= \left.\frac{\partial \mathbf{F}}{\partial z}\right|_{(0,0,0)} = \left.\left(\frac{\partial F_1}{\partial z}\mathbf{i} + \frac{\partial F_2}{\partial z}\mathbf{j} + \frac{\partial F_3}{\partial z}\mathbf{k}\right)\right|_{(0,0,0)}
\end{aligned}$$

and again the "$\cdots$" represents the second- and higher-degree terms in $x$, $y$, and $z$. The unit normal on $\mathcal{S}_\epsilon$ is $\hat{\mathbf{N}} = (x\mathbf{i} + y\mathbf{j} + z\mathbf{k})/\epsilon$, so we have

$$\begin{aligned}\mathbf{F} \bullet \hat{\mathbf{N}} = \frac{1}{\epsilon}\Big(&\mathbf{F}_0 \bullet \mathbf{i}\, x + \mathbf{F}_0 \bullet \mathbf{j}\, y + \mathbf{F}_0 \bullet \mathbf{k}\, z \\ &+ \mathbf{F}_1 \bullet \mathbf{i}\, x^2 + \mathbf{F}_1 \bullet \mathbf{j}\, xy + \mathbf{F}_1 \bullet \mathbf{k}\, xz \\ &+ \mathbf{F}_2 \bullet \mathbf{i}\, xy + \mathbf{F}_2 \bullet \mathbf{j}\, y^2 + \mathbf{F}_2 \bullet \mathbf{k}\, yz \\ &+ \mathbf{F}_3 \bullet \mathbf{i}\, xz + \mathbf{F}_3 \bullet \mathbf{j}\, yz + \mathbf{F}_3 \bullet \mathbf{k}\, z^2 + \cdots\Big).\end{aligned}$$

We integrate each term within the parentheses over $\mathcal{S}_\epsilon$. By symmetry,

$$\begin{aligned}&\oiint_{\mathcal{S}_\epsilon} x\, dS = \oiint_{\mathcal{S}_\epsilon} y\, dS = \oiint_{\mathcal{S}_\epsilon} z\, dS = 0, \\ &\oiint_{\mathcal{S}_\epsilon} xy\, dS = \oiint_{\mathcal{S}_\epsilon} xz\, dS = \oiint_{\mathcal{S}_\epsilon} yz\, dS = 0.\end{aligned}$$

Also, by symmetry,

$$\begin{aligned}\oiint_{\mathcal{S}_\epsilon} x^2\, dS &= \oiint_{\mathcal{S}_\epsilon} y^2\, dS = \oiint_{\mathcal{S}_\epsilon} z^2\, dS \\ &= \frac{1}{3}\oiint_{\mathcal{S}_\epsilon} (x^2 + y^2 + z^2)\, dS = \frac{1}{3}(\epsilon^2)(4\pi\epsilon^2) = \frac{4}{3}\pi\epsilon^4,\end{aligned}$$

and the higher-degree terms have surface integrals involving $\epsilon^5$ and higher powers. Thus

$$\begin{aligned}\frac{3}{4\pi\epsilon^3}\oiint_{\mathcal{S}_\epsilon} \mathbf{F} \bullet \hat{\mathbf{N}}\, dS &= \mathbf{F}_1 \bullet \mathbf{i} + \mathbf{F}_2 \bullet \mathbf{j} + \mathbf{F}_3 \bullet \mathbf{k} + \epsilon\,(\cdots) \\ &= \boldsymbol{\nabla} \bullet \mathbf{F}(0, 0, 0) + \epsilon\,(\cdots) \\ &\to \boldsymbol{\nabla} \bullet \mathbf{F}(0, 0, 0)\end{aligned}$$

as $\epsilon \to 0^+$. This is what we wanted to show.

**REMARK** The spheres $\mathcal{S}_\epsilon$ in the above theorem can be replaced by other contracting families of piecewise smooth surfaces. For instance, if $B$ is the surface of a rectangular box with dimensions $\Delta x$, $\Delta y$, and $\Delta z$ containing $P$ then

$$\mathbf{div}\,\mathbf{F}(P) = \lim_{\Delta x, \Delta y, \Delta z \to 0} \frac{1}{\Delta x \Delta y \Delta z} \oiint_B \mathbf{F} \bullet \hat{\mathbf{N}}\, dS.$$

See Exercise 12 at the end of this section.

**REMARK** In two dimensions, the value $\mathbf{div}\,\mathbf{F}(P)$ represents the limiting *flux per unit area* outward across small, non-self-intersecting closed curves which enclose $P$. See Exercise 13 at the end of this section.

Let us return again to the interpretation of a vector field as a velocity field of a moving incompressible fluid. If the total flux of the velocity field outward across the boundary surface of a domain is positive (or negative), then the fluid must be produced (or annihilated) within that domain.

The vector field $\mathbf{F} = x\mathbf{i} + y\mathbf{j} + z\mathbf{k}$ of Example 2 in Section 8.6 has constant divergence, $\nabla \bullet \mathbf{F} = 3$. In that example we showed that the flux of $\mathbf{F}$ out of a certain cylinder of base radius $a$ and height $2h$ is $6\pi a^2 h$, which is three times the volume of the cylinder. Exercises 2 and 3 of Section 8.6 confirm similar results for the flux of $\mathbf{F}$ out of other domains. This leads to another interpretation for the divergence; $\mathbf{div}\,\mathbf{F}(P)$ is the *source strength per unit volume* of $\mathbf{F}$ at $P$. With this interpretation, we would expect, even for a vector field $\mathbf{F}$ with nonconstant divergence, that the total flux of $\mathbf{F}$ out the surface $\mathcal{S}$ of a domain $D$ would be equal to the total source strength of $\mathbf{F}$ within $D$, that is,

$$\oiint_{\mathcal{S}} \mathbf{F} \bullet \hat{\mathbf{N}}\, dS = \iiint_D \nabla \bullet \mathbf{F}\, dV.$$

This is the **Divergence Theorem**, which we will prove in Section 9.3.

**EXAMPLE 3** Verify that the vector field $\mathbf{F} = m\mathbf{r}/|\mathbf{r}|^3$, due to a source of strength $m$ at $(0, 0, 0)$, has zero divergence at all points in $\mathbb{R}^3$ except the origin. What would you expect to be the total flux of $\mathbf{F}$ outward across the boundary surface of a domain $D$ if the origin lies outside $D$? if the origin is inside $D$?

***SOLUTION*** Since

$$\mathbf{F}(x, y, z) = \frac{m}{r^3}\Big(x\mathbf{i} + y\mathbf{j} + z\mathbf{k}\Big), \qquad \text{where} \quad r^2 = x^2 + y^2 + z^2,$$

and since $\partial r/\partial x = x/r$, we have

$$\frac{\partial F_1}{\partial x} = m\,\frac{\partial}{\partial x}\left(\frac{x}{r^3}\right) = m\,\frac{r^3 - 3xr^2\left(\frac{x}{r}\right)}{r^6} = m\,\frac{r^2 - 3x^2}{r^5}.$$

Similarly,

$$\frac{\partial F_2}{\partial y} = m\,\frac{r^2 - 3y^2}{r^5} \qquad \text{and} \qquad \frac{\partial F_3}{\partial z} = m\,\frac{r^2 - 3z^2}{r^5}.$$

Adding these up, we get $\nabla \bullet \mathbf{F}(x, y, z) = 0$ if $r > 0$.

If the origin lies outside the domain $D$ then the source density of $\mathbf{F}$ in $D$ is zero, so we would expect the total flux of $\mathbf{F}$ out of $D$ to be zero. If the origin lies inside $D$ then $D$ contains a source of strength $m$ (producing $4\pi m$ cubic units of fluid per unit time) and so we would expect the flux out of $D$ to be $4\pi m$. See Example 1 and Exercises 9 and 10 of Section 8.6 for specific examples. ■

## Distributions and Delta Functions

If $\ell(x)$ represents the line density (mass per unit length) of mass distributed on the $x$-axis, then the total mass so distributed is

$$m = \int_{-\infty}^{\infty} \ell(x)\, dx.$$

Now suppose that the only mass on the axis is a "point mass" $m = 1$ located at the origin. Then at all other points $x \neq 0$, the density is $\ell(x) = 0$, but we must still have

$$\int_{-\infty}^{\infty} \ell(x)\, dx = m = 1,$$

so $\ell(0)$ must be infinite. This is an ideal situation, a mathematical model. No real *function* $\ell(x)$ can have such properties; if a function is zero everywhere except at a single point then any integral of that function will be zero. (Why?) (Also, no real mass can occupy just a single point.) Nevertheless, it is very useful to model real, isolated masses as point masses, and to model their densities using **generalized functions** (also called **distributions**).

We can think of the density of a point mass 1 at $x = 0$ as the limit of large densities concentrated on small intervals. For instance, if

$$d_n(x) = \begin{cases} n/2 & \text{if } |x| \le 1/n \\ 0 & \text{if } |x| > 1/n \end{cases}$$

(see Figure 9.1), then for any smooth function $f(x)$ defined on $\mathbb{R}$ we have

$$\int_{-\infty}^{\infty} d_n(x)\, f(x)\, dx = \frac{n}{2} \int_{-1/n}^{1/n} f(x)\, dx.$$

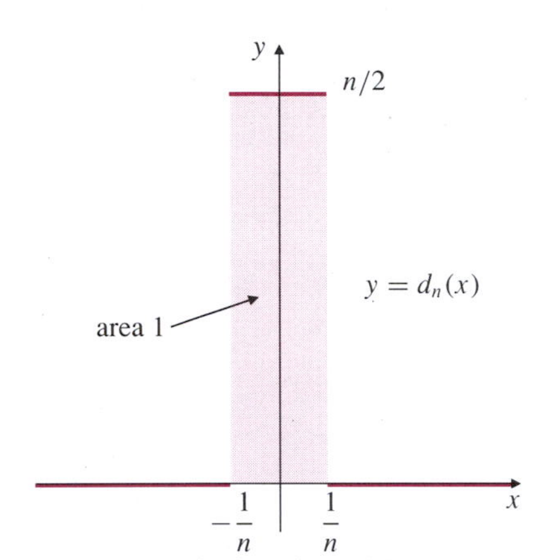


**Figure 9.1** The functions $d_n(x)$ converge to $\delta(x)$ as $n \to \infty$

Replace $f(x)$ in the integral on the right with its Maclaurin series:

$$f(x) = f(0) + \frac{f'(0)}{1!}x + \frac{f''(0)}{2!}x^2 + \cdots.$$

Since

$$\int_{-1/n}^{1/n} x^k\, dx = \begin{cases} 2/((k+1)n^{k+1}) & \text{if } k \text{ is even} \\ 0 & \text{if } k \text{ is odd,} \end{cases}$$

we can take the limit as $n \to \infty$ and obtain

$$\lim_{n\to\infty} \int_{-\infty}^{\infty} d_n(x)\, f(x)\, dx = f(0).$$

**DEFINITION 1**

The **Dirac distribution** $\delta(x)$ (also called the **Dirac delta function**, although it is really not a function) is the limit of the sequence $d_n(x)$ as $n \to \infty$. It is defined by the requirement that

$$\int_{-\infty}^{\infty} \delta(x) f(x)\, dx = f(0)$$

for every smooth function $f(x)$.

A formal change of variables shows that the delta function also satisfies

$$\int_{-\infty}^{\infty} \delta(x - t) f(t)\, dt = f(x).$$

**EXAMPLE 4** In view of the fact that $\mathbf{F}(\mathbf{r}) = m\mathbf{r}/|\mathbf{r}|^3$ satisfies $\mathbf{div}\,\mathbf{F}(x, y, z) = 0$ for $(x, y, z) \ne (0, 0, 0)$ but produces a flux of $4\pi m$ out of any sphere centred at the origin, we can regard $\mathbf{div}\,\mathbf{F}(x, y, z)$ as a distribution

$$\mathbf{div}\,\mathbf{F}(x, y, z) = 4\pi m \delta(x)\delta(y)\delta(z).$$

In particular, integrating this distribution against $f(x, y, z) = 1$ over $\mathbb{R}^3$, we have

$$\iiint_{\mathbb{R}^3} \mathbf{div}\,\mathbf{F}(x, y, z)\,dV = 4\pi m \int_{-\infty}^{\infty} \delta(x)\,dx \int_{-\infty}^{\infty} \delta(y)\,dy \int_{-\infty}^{\infty} \delta(z)\,dz$$
$$= 4\pi m.$$

The integral can equally well be taken over *any domain* in $\mathbb{R}^3$ that contains the origin in its interior, and the result will be the same. If the origin is outside the domain the result will be zero. We will reexamine this situation after establishing the Divergence Theorem in Section 9.3. ■

A formal study of distributions is beyond the scope of this book and is usually undertaken in more advanced textbooks on differential equations and engineering mathematics.

## Interpretation of the Curl

Roughly speaking, $\mathbf{curl}\,\mathbf{F}(P)$ measures the extent to which the vector field $\mathbf{F}$ "swirls" around $P$.

**EXAMPLE 5** Consider the velocity field,

$$\mathbf{v} = -\Omega y\mathbf{i} + \Omega x\mathbf{j},$$

of a solid rotating with angular speed $\Omega$ about the $z$-axis, that is, with angular velocity $\boldsymbol{\Omega} = \Omega\mathbf{k}$. Calculate the circulation of this field around a circle $\mathcal{C}_\epsilon$ in the $xy$-plane centred at any point $(x_0, y_0)$, having radius $\epsilon$, and oriented counterclockwise. What is the relationship between this circulation and the curl of $\mathbf{v}$?

***SOLUTION*** The indicated circle has parametrization

$$\mathbf{r} = (x_0 + \epsilon\cos t)\mathbf{i} + (y_0 + \epsilon\sin t)\mathbf{j}, \qquad (0 \le t \le 2\pi),$$

and the circulation of $\mathbf{v}$ around it is given by

$$\oint_{\mathcal{C}_\epsilon} \mathbf{v} \bullet d\mathbf{r} = \int_0^{2\pi} \Big(-\Omega(y_0 + \epsilon\sin t)(-\epsilon\sin t) + \Omega(x_0 + \epsilon\cos t)(\epsilon\cos t)\Big)\,dt$$
$$= \int_0^{2\pi} \Big(\Omega\epsilon(y_0\sin t + x_0\cos t) + \Omega\epsilon^2\Big)\,dt$$
$$= 2\Omega\pi\epsilon^2.$$

Since

$$\mathbf{curl}\,\mathbf{v} = \nabla\times\mathbf{v} = \Big(\frac{\partial}{\partial x}(\Omega x) - \frac{\partial}{\partial y}(-\Omega y)\Big)\mathbf{k} = 2\Omega\mathbf{k} = 2\boldsymbol{\Omega},$$

the circulation is the product of $(\mathbf{curl}\,\mathbf{v}) \bullet \mathbf{k}$ and the area bounded by $\mathcal{C}_\epsilon$. Note that this circulation is constant for circles of any fixed radius; it does not depend on the position of the centre. ■

The calculations in the example above suggest that the curl of a vector field is a measure of the *circulation per unit area* in planes normal to the curl. A more precise version of this conjecture is stated in Theorem 2 below. We will not prove this theorem now, because a proof at this stage would be quite complicated. (However, see Exercise 14 at the end of this section for a special case.) A simple proof can be based on Stokes's Theorem. (See Exercise 22 in Section 9.4.)

**THEOREM 2**

### The curl as circulation density

If $\mathbf{F}$ is a smooth vector field and $\mathcal{C}_\epsilon$ is a circle of radius $\epsilon$ centred at point $P$ and bounding a disk $\mathcal{S}_\epsilon$ with unit normal $\hat{\mathbf{N}}$ (and orientation inherited from $\mathcal{C}_\epsilon$ — see Figure 9.2), then

$$\lim_{\epsilon \to 0^+} \frac{1}{\pi\epsilon^2} \oint_{\mathcal{C}_\epsilon} \mathbf{F} \bullet d\mathbf{r} = \hat{\mathbf{N}} \bullet \mathbf{curl}\,\mathbf{F}(P).$$

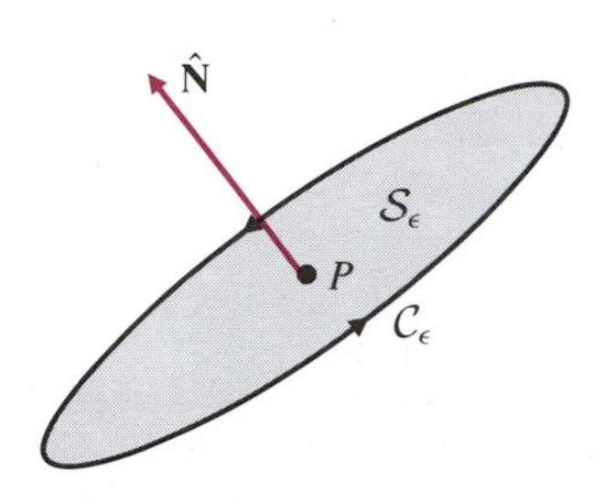


Figure 9.2

Example 5 also suggests the following definition for the *local* angular velocity of a moving fluid.

The **local angular velocity** at point $P$ in a fluid moving with velocity field $\mathbf{v}(P)$ is given by

$$\boldsymbol{\Omega}(P) = \tfrac{1}{2}\mathbf{curl}\,\mathbf{v}(P).$$

Theorem 2 states that the local angular velocity $\boldsymbol{\Omega}(P)$ is that vector whose component in the direction of any unit vector $\hat{\mathbf{N}}$ is one-half of the limiting circulation per unit area around the (oriented) boundary circles of small disks centred at $P$ and having normal $\hat{\mathbf{N}}$.

Not all vector fields with nonzero curl *appear* to circulate. The velocity field for the rigid body rotation considered in Example 5 appears to circulate around the axis of rotation, but the circulation around a circle in a plane perpendicular to that axis turned out to be independent of the position of the circle; it depended only on its area. The circle need not even surround the axis of rotation. The following example investigates a fluid velocity field whose streamlines are *straight lines*, but which still has nonzero, constant curl, and therefore constant local angular velocity.

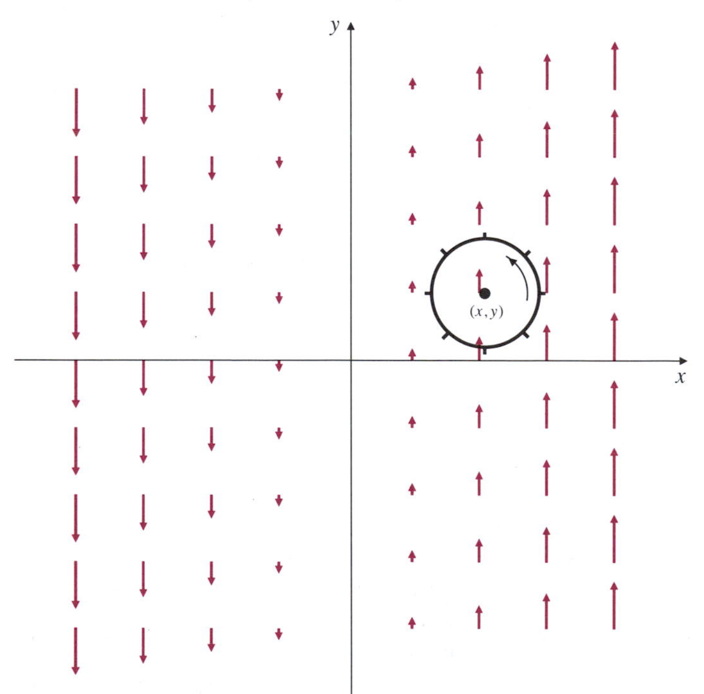


Figure 9.3 The paddle wheel is not only carried along but is set rotating by flow

**EXAMPLE 6** Consider the velocity field $\mathbf{v} = x\mathbf{j}$ of a fluid moving in the $xy$-plane. Evidently, particles of fluid are moving along lines parallel to the $y$-axis. However, $\mathbf{curl}\,\mathbf{v}(x, y) = \mathbf{k}$, and $\boldsymbol{\Omega}(x, y) = \frac{1}{2}\mathbf{k}$. A small "paddle wheel" of radius $\epsilon$ placed with its centre at position $(x, y)$ in the fluid (see Figure 9.3) will be carried along with the fluid at velocity $x\mathbf{i}$, but will also be set rotating with angular velocity $\boldsymbol{\Omega}(x, y)$ which is independent of its position. This angular velocity is due to the fact that the velocity of the fluid along the right side of the wheel exceeds that along the left side.

## EXERCISES 9.1

In Exercises 1–11, calculate $\mathbf{div}\,\mathbf{F}$ and $\mathbf{curl}\,\mathbf{F}$ for the given vector fields.

**1.** $\mathbf{F} = x\mathbf{i} + y\mathbf{j}$

**2.** $\mathbf{F} = y\mathbf{i} + x\mathbf{j}$

**3.** $\mathbf{F} = y\mathbf{i} + z\mathbf{j} + x\mathbf{k}$

**4.** $\mathbf{F} = yz\mathbf{i} + xz\mathbf{j} + xy\mathbf{k}$

**5.** $\mathbf{F} = x\mathbf{i} + x\mathbf{k}$

**6.** $\mathbf{F} = xy^2\mathbf{i} - yz^2\mathbf{j} + zx^2\mathbf{k}$

**7.** $\mathbf{F} = f(x)\mathbf{i} + g(y)\mathbf{j} + h(z)\mathbf{k}$

**8.** $\mathbf{F} = f(z)\mathbf{i} - f(z)\mathbf{j}$

**9.** $\mathbf{F}(r, \theta) = r\mathbf{i} + \sin\theta\mathbf{j}$, where $(r, \theta)$ are polar coordinates in the plane

**10.** $\mathbf{F} = \hat{\mathbf{r}} = \cos\theta\mathbf{i} + \sin\theta\mathbf{j}$

**11.** $\mathbf{F} = \hat{\boldsymbol{\theta}} = -\sin\theta\mathbf{i} + \cos\theta\mathbf{j}$.

* **12.** Let $\mathbf{F}$ be a smooth, 3-dimensional vector field. If $B_{a,b,c}$ is the surface of the box $-a \le x \le a$, $-b \le y \le b$, $-c \le z \le c$, with outward normal $\hat{\mathbf{N}}$, show that

$$\lim_{a,b,c\to 0+} \frac{1}{8abc} \oiint_{B_{a,b,c}} \mathbf{F} \bullet \hat{\mathbf{N}}\, dS = \nabla \bullet \mathbf{F}(0, 0, 0).$$

* **13.** Let $\mathbf{F}$ be a smooth 2-dimensional vector field. If $\mathcal{C}_\epsilon$ is the circle of radius $\epsilon$ centred at the origin, and $\hat{\mathbf{N}}$ is the unit outward normal to $\mathcal{C}_\epsilon$, show that

$$\lim_{\epsilon\to 0+} \frac{1}{\pi\epsilon^2} \oint_{\mathcal{C}_\epsilon} \mathbf{F} \bullet \hat{\mathbf{N}}\, ds = \mathbf{div}\,\mathbf{F}(0, 0).$$

* **14.** Prove Theorem 2 in the special case that $\mathcal{C}_\epsilon$ is the circle in the $xy$-plane with parametrization

$$x = \epsilon\cos\theta, \quad y = \epsilon\sin\theta, \quad (0 \le \theta \le 2\pi).$$

In this case $\hat{\mathbf{N}} = \mathbf{k}$. *Hint:* expand $\mathbf{F}(x, y, z)$ in a vector Taylor series about the origin as in the proof of Theorem 1, and calculate the circulation of individual terms around $\mathcal{C}_\epsilon$.

# 9.2 SOME IDENTITIES INVOLVING GRAD, DIV, AND CURL

There are numerous identities involving the functions

$$\mathbf{grad}\, f(x, y, z) = \nabla f(x, y, z) = \frac{\partial f}{\partial x}\mathbf{i} + \frac{\partial f}{\partial y}\mathbf{j} + \frac{\partial f}{\partial z}\mathbf{k},$$

$$\mathbf{div}\,\mathbf{F}(x, y, z) = \nabla \bullet \mathbf{F}(x, y, z) = \frac{\partial F_1}{\partial x} + \frac{\partial F_2}{\partial y} + \frac{\partial F_3}{\partial z},$$

$$\mathbf{curl}\,\mathbf{F}(x, y, z) = \nabla \times \mathbf{F}(x, y, z) = \begin{vmatrix} \mathbf{i} & \mathbf{j} & \mathbf{k} \\ \dfrac{\partial}{\partial x} & \dfrac{\partial}{\partial y} & \dfrac{\partial}{\partial z} \\ F_1 & F_2 & F_3 \end{vmatrix},$$

and the **Laplacian operator**, $\nabla^2 = \nabla \bullet \nabla$, defined for a scalar field $\phi$ by

$$\nabla^2\phi = \nabla \bullet \nabla\phi = \mathbf{div}\,\mathbf{grad}\,\phi = \frac{\partial^2\phi}{\partial x^2} + \frac{\partial^2\phi}{\partial y^2} + \frac{\partial^2\phi}{\partial z^2},$$

and for a vector field $\mathbf{F} = F_1\mathbf{i} + F_2\mathbf{j} + F_3\mathbf{k}$ by

$$\nabla^2\mathbf{F} = (\nabla^2 F_1)\mathbf{i} + (\nabla^2 F_2)\mathbf{j} + (\nabla^2 F_3)\mathbf{k}.$$

(The Laplacian operator, $\nabla^2 = (\partial^2/\partial x^2) + (\partial^2/\partial y^2) + (\partial^2/\partial z^2)$, is denoted by $\Delta$ in some books.) Recall that a function $\phi$ is called **harmonic** in a domain $D$ if $\nabla^2\phi = 0$ throughout $D$. (See Section 4.4.)

We collect the most important identities together in the following theorem. Most of them are forms of the Product Rule. We will prove a few of the identities to illustrate the techniques involved (mostly brute-force calculation) and leave the rest as exercises. Note that two of the identities involve quantities like $(\mathbf{G} \bullet \nabla)\mathbf{F}$; this represents the vector obtained by applying the scalar differential operator $\mathbf{G} \bullet \nabla$ to the vector field $\mathbf{F}$:

$$(\mathbf{G} \bullet \nabla)\mathbf{F} = G_1 \frac{\partial \mathbf{F}}{\partial x} + G_2 \frac{\partial \mathbf{F}}{\partial y} + G_3 \frac{\partial \mathbf{F}}{\partial z}.$$

**THEOREM 3**

**Vector differential identities**

Let $\phi$ and $\psi$ be scalar fields and $\mathbf{F}$ and $\mathbf{G}$ be vector fields, all assumed to be sufficiently smooth that all the partial derivatives in the identities are continuous. Then the following identities hold:

(a) $\nabla(\phi\psi) = \phi\nabla\psi + \psi\nabla\phi$

(b) $\nabla \bullet (\phi\mathbf{F}) = (\nabla\phi) \bullet \mathbf{F} + \phi(\nabla \bullet \mathbf{F})$

(c) $\nabla \times (\phi\mathbf{F}) = (\nabla\phi) \times \mathbf{F} + \phi(\nabla \times \mathbf{F})$

(d) $\nabla \bullet (\mathbf{F} \times \mathbf{G}) = (\nabla \times \mathbf{F}) \bullet \mathbf{G} - \mathbf{F} \bullet (\nabla \times \mathbf{G})$

(e) $\nabla \times (\mathbf{F} \times \mathbf{G}) = (\nabla \bullet \mathbf{G})\mathbf{F} + (\mathbf{G} \bullet \nabla)\mathbf{F} - (\nabla \bullet \mathbf{F})\mathbf{G} - (\mathbf{F} \bullet \nabla)\mathbf{G}$

(f) $\nabla(\mathbf{F} \bullet \mathbf{G}) = \mathbf{F} \times (\nabla \times \mathbf{G}) + \mathbf{G} \times (\nabla \times \mathbf{F}) + (\mathbf{F} \bullet \nabla)\mathbf{G} + (\mathbf{G} \bullet \nabla)\mathbf{F}$

(g) $\nabla \bullet (\nabla \times \mathbf{F}) = 0$ (**div curl** $= 0$)

(h) $\nabla \times (\nabla\phi) = \mathbf{0}$ (**curl grad** $= \mathbf{0}$)

(i) $\nabla \times (\nabla \times \mathbf{F}) = \nabla(\nabla \bullet \mathbf{F}) - \nabla^2\mathbf{F}$
(**curl curl** $=$ **grad div** $-$ Laplacian)

Identities (a)–(f) are versions of the Product Rule, and are first-order identities involving only one application of $\nabla$. Identities (g)–(i) are second-order identities. Identities (g) and (h) are equivalent to the equality of mixed partial derivatives and are especially important for understanding of **div** and **curl**.

***PROOF*** We will prove only identities (c), (e), and (g). The remaining proofs are similar to these.

(c) The first component (**i** component) of $\nabla \times (\phi\mathbf{F})$ is

$$\frac{\partial}{\partial y}(\phi F_3) - \frac{\partial}{\partial z}(\phi F_2) = \frac{\partial \phi}{\partial y} F_3 - \frac{\partial \phi}{\partial z} F_2 + \phi \frac{\partial F_3}{\partial y} - \phi \frac{\partial F_2}{\partial z}.$$

The first two terms on the right constitute the first component of $(\nabla\phi) \times \mathbf{F}$ and the last two terms constitute the first component of $\phi(\nabla \times \mathbf{F})$. Therefore the first components of both sides of identity (c) are equal. The equality of the other components follows similarly.

(e) Again, it is sufficient to show that the first components of the vectors on both sides of the identity are equal. To calculate the first component of $\nabla \times (\mathbf{F} \times \mathbf{G})$ we need the second and third components of $\mathbf{F} \times \mathbf{G}$, which are

$$(\mathbf{F} \times \mathbf{G})_2 = F_3G_1 - F_1G_3 \quad \text{and} \quad (\mathbf{F} \times \mathbf{G})_3 = F_1G_2 - F_2G_1.$$

The first component of $\nabla\times(\mathbf{F}\times\mathbf{G})$ is therefore

$$\begin{aligned}\frac{\partial}{\partial y}&(F_1G_2 - F_2G_1) - \frac{\partial}{\partial z}(F_3G_1 - F_1G_3)\\ &= \frac{\partial F_1}{\partial y}G_2 + F_1\frac{\partial G_2}{\partial y} - \frac{\partial F_2}{\partial y}G_1 - F_2\frac{\partial G_1}{\partial y} - \frac{\partial F_3}{\partial z}G_1\\ &\quad - F_3\frac{\partial G_1}{\partial z} + \frac{\partial F_1}{\partial z}G_3 + F_1\frac{\partial G_3}{\partial z}.\end{aligned}$$

The first components of the four terms on the right side of identity (e) are

$$\begin{aligned}((\nabla\bullet\mathbf{G})\mathbf{F})_1 &= F_1\frac{\partial G_1}{\partial x} + F_1\frac{\partial G_2}{\partial y} + F_1\frac{\partial G_3}{\partial z}\\ ((\mathbf{G}\bullet\nabla)\mathbf{F})_1 &= \frac{\partial F_1}{\partial x}G_1 + \frac{\partial F_1}{\partial y}G_2 + \frac{\partial F_1}{\partial z}G_3\\ -((\nabla\bullet\mathbf{F})\mathbf{G})_1 &= -\frac{\partial F_1}{\partial x}G_1 - \frac{\partial F_2}{\partial y}G_1 - \frac{\partial F_3}{\partial z}G_1\\ -((\mathbf{F}\bullet\nabla)\mathbf{G})_1 &= -F_1\frac{\partial G_1}{\partial x} - F_2\frac{\partial G_1}{\partial y} - F_3\frac{\partial G_1}{\partial z}.\end{aligned}$$

When we add up all the terms in these four expressions some cancel out and we are left with the same terms as in the first component of $\nabla\times(\mathbf{F}\times\mathbf{G})$.

(g) This is a straightforward calculation involving the equality of mixed partial derivatives:

$$\begin{aligned}\nabla\bullet(\nabla\times\mathbf{F}) &= \frac{\partial}{\partial x}\left(\frac{\partial F_3}{\partial y} - \frac{\partial F_2}{\partial z}\right) + \frac{\partial}{\partial y}\left(\frac{\partial F_1}{\partial z} - \frac{\partial F_3}{\partial x}\right)\\ &\quad + \frac{\partial}{\partial z}\left(\frac{\partial F_2}{\partial x} - \frac{\partial F_1}{\partial y}\right)\\ &= \frac{\partial^2 F_3}{\partial x\partial y} - \frac{\partial^2 F_2}{\partial x\partial z} + \frac{\partial^2 F_1}{\partial y\partial z} - \frac{\partial^2 F_3}{\partial y\partial x} + \frac{\partial^2 F_2}{\partial z\partial x} - \frac{\partial^2 F_1}{\partial z\partial y}\\ &= 0.\end{aligned}$$

**REMARK** Two *triple product* identities for vectors were previously presented in Exercises 18 and 23 of Section 3.3:

$$\begin{aligned}&\mathbf{a}\bullet(\mathbf{b}\times\mathbf{c}) = \mathbf{b}\bullet(\mathbf{c}\times\mathbf{a}) = \mathbf{c}\bullet(\mathbf{a}\times\mathbf{b}),\\ &\mathbf{a}\times(\mathbf{b}\times\mathbf{c}) = (\mathbf{a}\bullet\mathbf{c})\mathbf{b} - (\mathbf{a}\bullet\mathbf{b})\mathbf{c}.\end{aligned}$$

While these are useful identities, they *cannot* be used to give simpler proofs of the identities in Theorem 3 by replacing one or other of the vectors with $\nabla$. (Why?)

## EXPLORE! Vector Calculus Using Computer Algebra

Computer algebra programs are capable of doing vector algebra and vector calculus, though with some systems the commands that must be used to achieve a desired effect can only be discovered by patient reading of on-line and off-line documentation. For example, the following comands will cause Maple V to verify identity (i) of Theorem 3.

```
with(linalg);
f := vector([f1(x,y,z), f2(x,y,z), f3(x,y,z)]);
v := [x, y, z];
evalm(curl(curl(f,v),v) - grad(diverge(f,v),v)
    + map(laplacian,f,v));
```

Maple will respond with the result `[0 0 0]`, indicating its agreement that

$$\nabla\times(\nabla\times\mathbf{F}) - \nabla(\nabla\bullet\mathbf{F}) + \nabla^2\mathbf{F} = \mathbf{0}.$$

The `evalm` function causes the operators of addition and scalar multiplication of vectors to be applied to the components. Maple requires that the variables with respect to which the operators curl, grad, div (which Maple calls `diverge`), and laplacian are taken must be shown explicitly. These variables are stored in the vector `v`. Because Maple's `laplacian` function is defined for scalar rather than vector arguments, it must be `map`ped onto the components of `f`.

Use Maple, (or some other computer algebra system) to verify the other identities of Theorem 3. Sometimes Maple will generate vectors whose components have not been properly simplified. The command `map(simplify,g);` will simplify the components of a vector `g`. Expressions like $(\mathbf{G}\bullet\nabla)\mathbf{F}$ appearing in identities (e) and (f) can be constructed in Maple using

```
g[1]*map(diff,f,x)+g[2]*map(diff,f,y)
    +g[3]*map(diff,f,z),
```

provided `f` and `g` have both been declared as vectors with three components that are functions of $x$, $y$, and $z$.

## Scalar and Vector Potentials

Two special terms are used to describe vector fields for which either the divergence or the curl is identically zero.

**DEFINITION 2**

**Solenoidal and irrotational vector fields**

A vector field $\mathbf{F}$ is called **solenoidal** in a domain $D$ if $\mathbf{div}\,\mathbf{F} = 0$ in $D$.
A vector field $\mathbf{F}$ is called **irrotational** in a domain $D$ if $\mathbf{curl}\,\mathbf{F} = 0$ in $D$.

Part (h) of Theorem 3 says that $\mathbf{F} = \mathbf{grad}\,\phi \implies \mathbf{curl}\,\mathbf{F} = 0$. Thus

**Every conservative vector field is irrotational.**

Part (g) of Theorem 3 says that $\mathbf{F} = \mathbf{curl}\,\mathbf{G} \implies \mathbf{div}\,\mathbf{F} = 0$. Thus

**The curl of any vector field is solenoidal.**

The *converses* of these assertions hold if the domain of $\mathbf{F}$ satisfies certain conditions.

**THEOREM 4**

If $\mathbf{F}$ is a smooth, irrotational vector field on a simply connected domain $D$ then $\mathbf{F} = \nabla\phi$ for some scalar potential function defined on $D$, and so $\mathbf{F}$ is conservative.

**THEOREM 5**

If $\mathbf{F}$ is a smooth, solenoidal vector field on a domain $D$ with the property that every closed surface in $D$ bounds a domain contained in $D$, then $\mathbf{F} = \mathbf{curl}\,\mathbf{G}$ for some vector field $\mathbf{G}$ defined on $D$. Such a vector field $\mathbf{G}$ is called a **vector potential** of the vector field $\mathbf{F}$.

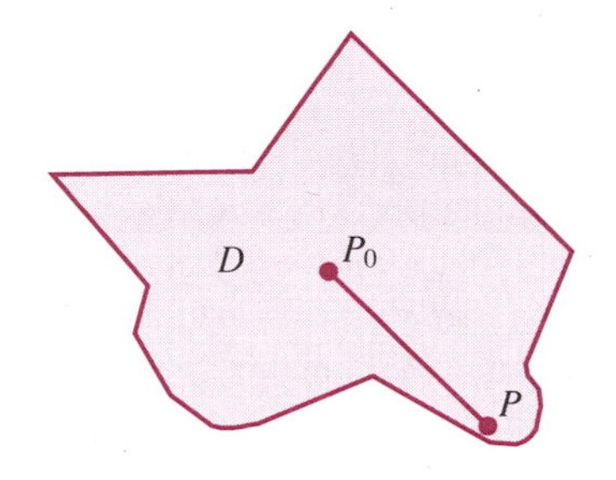


**Figure 9.4** The line segment from $P_0$ to any point in $D$ lies in $D$

We cannot prove these result in their full generality at this point. However, both theorems have simple proofs in the special case where the domain $D$ is **star-like**. A star-like domain is one for which there exists a point $P_0$ such that the line segment from $P_0$ to any point $P$ in $D$ lies wholly in $D$. (See Figure 9.4.) Both proofs are *constructive* in that they tell you how to find a potential.

***Proof of Theorem 4 for star-like domains.*** Without loss of generality, we can assume that $P_0$ is the origin. If $P = (x, y, z)$ is any point in $D$, then the straight line segment

$$\mathbf{r}(t) = tx\mathbf{i} + ty\mathbf{j} + tz\mathbf{k}, \qquad 0 \le t \le 1$$

from $P_0$ to $P$ lies in $D$. Define the function $\phi$ on $D$ by

$$\begin{aligned}\phi(x, y, z) &= \int_0^1 \mathbf{F}\big(\mathbf{r}(t)\big) \bullet \frac{d\mathbf{r}}{dt}\,dt \\ &= \int_0^1 \Big(xF_1(\xi, \eta, \zeta) + yF_2(\xi, \eta, \zeta) + zF_3(\xi, \eta, \zeta)\Big)\,dt,\end{aligned}$$

where $\xi = tx$, $\eta = ty$, and $\zeta = tz$. We calculate $\partial\phi/\partial x$, making use of the fact that $\mathbf{curl}\,\mathbf{F} = \mathbf{0}$ to replace $(\partial/\partial\xi)F_2(\xi, \eta, \zeta)$ with $(\partial/\partial\eta)F_1(\xi, \eta, \zeta)$ and $(\partial/\partial\xi)F_3(\xi, \eta, \zeta)$ with $(\partial/\partial\zeta)F_1(\xi, \eta, \zeta)$:

$$\begin{aligned}\frac{\partial\phi}{\partial x} &= \int_0^1 \left(F_1(\xi, \eta, \zeta) + tx\frac{\partial F_1}{\partial\xi} + ty\frac{\partial F_2}{\partial\xi} + tz\frac{\partial F_3}{\partial\xi}\right) dt \\ &= \int_0^1 \left(F_1(\xi, \eta, \zeta) + tx\frac{\partial F_1}{\partial\xi} + ty\frac{\partial F_1}{\partial\eta} + tz\frac{\partial F_1}{\partial\zeta}\right) dt \\ &= \int_0^1 \frac{d}{dt}\Big(t\,F_1(\xi, \eta, \zeta)\Big)\,dt \\ &= \Big(t\,F_1(tx, ty, tz)\Big)\Big|_0^1 = F_1(x, y, z).\end{aligned}$$

Similarly, $\partial\phi/\partial y = F_2$ and $\partial\phi/\partial z = F_3$. Thus $\nabla\phi = \mathbf{F}$.

The details of the proof of Theorem 5 for star-like domains are similar to those of Theorem 4, and we relegate the proof to Exercise 18 at the end of this section.

Note that vector potentials, when they exist, are *very* non-unique. Since $\mathbf{curl}\ \mathbf{grad}\,\phi$ is identically zero (Theorem 3(h)), an arbitrary conservative field can be added to $\mathbf{G}$ without changing the value of $\mathbf{curl}\ \mathbf{G}$. The following example illustrates just how much freedom you have in making simplifying assumptions when trying to find a vector potential.

**EXAMPLE 1** Show that the vector field $\mathbf{F} = (x^2+yz)\mathbf{i} - 2y(x+z)\mathbf{j} + (xy+z^2)\mathbf{k}$ is solenoidal in $\mathbb{R}^3$ and find a vector potential for it.

***SOLUTION*** Since $\mathbf{div}\,\mathbf{F} = 2x - 2(x+z) + 2z = 0$ in $\mathbb{R}^3$, $\mathbf{F}$ is solenoidal. A vector potential $\mathbf{G}$ for $\mathbf{F}$ must satisfy $\mathbf{curl}\,\mathbf{G} = \mathbf{F}$, that is,

$$\frac{\partial G_3}{\partial y} - \frac{\partial G_2}{\partial z} = x^2 + yz,$$
$$\frac{\partial G_1}{\partial z} - \frac{\partial G_3}{\partial x} = -2xy - 2yz,$$
$$\frac{\partial G_2}{\partial x} - \frac{\partial G_1}{\partial y} = xy + z^2.$$

The three components of $\mathbf{G}$ have nine independent first-partial derivatives, so there are nine "degrees of freedom" involved in their determination. The three equations above use up three of these nine degrees of freedom. That leaves six. Let us try to find a solution $\mathbf{G}$ with $G_2 = 0$ identically. This means that all three first partials of $G_2$ are zero so we have used up three degrees of freedom in making this assumption. We have three left. The first equation now implies that

$$G_3 = \int (x^2 + yz)\,dy = x^2 y + \frac{1}{2}y^2 z + M(x, z).$$

(Since we were integrating with respect to $y$, the constant of integration can still depend on $x$ and $z$.) We make a second simplifying assumption, that $M(x, z) = 0$. This uses up two more degrees of freedom, leaving one. From the second equation we have

$$\frac{\partial G_1}{\partial z} = \frac{\partial G_3}{\partial x} - 2xy - 2yz = 2xy - 2xy - 2yz = -2yz,$$

so

$$G_1 = -2\int yz\,dz = -yz^2 + N(x, y).$$

We can not assume that $N(x, y) = 0$ identically, because that would require two degrees of freedom and we have only one. However, the third equation implies

$$xy + z^2 = -\frac{\partial G_1}{\partial y} = z^2 - \frac{\partial N}{\partial y}.$$

Thus, $(\partial/\partial y)N(x, y) = -xy$; observe that the terms involving $z$ have cancelled out. This happened because $\mathbf{div}\,\mathbf{F} = 0$. Had $\mathbf{F}$ not been solenoidal, we could not have determined $N$ as a function of $x$ and $y$ only from the above equation. As it is, however, we have

$$N(x, y) = -\int xy\,dy = -\frac{1}{2}xy^2 + P(x).$$

We can use our last degree of freedom to choose $P(x)$ to be identically zero, and hence obtain

$$\mathbf{G} = -\left(yz^2 + \frac{xy^2}{2}\right)\mathbf{i} + \left(x^2 y + \frac{y^2 z}{2}\right)\mathbf{k}$$

as the required vector potential for $\mathbf{F}$. You can check that $\mathbf{curl}\,\mathbf{G} = \mathbf{F}$. Of course, other choices of simplifying assumptions would have led to very different functions $\mathbf{G}$, which would have been equally correct. ∎

## EXERCISES 9.2

1. Prove part (a) of Theorem 3.
2. Prove part (b) of Theorem 3.
3. Prove part (d) of Theorem 3.
4. Prove part (f) of Theorem 3.
5. Prove part (h) of Theorem 3.
6. Prove part (i) of Theorem 3.

* **7.** Given that the field lines of the vector field $\mathbf{F}(x, y, z)$ are parallel straight lines, can you conclude anything about **div F**? about **curl F**?

**8.** Let $\mathbf{r} = x\mathbf{i} + y\mathbf{j} + z\mathbf{k}$ and let $\mathbf{c}$ be a constant vector. Show that $\nabla \bullet (\mathbf{c} \times \mathbf{r}) = 0$, $\nabla \times (\mathbf{c} \times \mathbf{r}) = 2\mathbf{c}$, and $\nabla(\mathbf{c} \bullet \mathbf{r}) = \mathbf{c}$.

**9.** Let $\mathbf{r} = x\mathbf{i} + y\mathbf{j} + z\mathbf{k}$ and let $r = |\mathbf{r}|$. If $f$ is a differentiable function of one variable show that

$$\nabla \bullet (f(r)\mathbf{r}) = rf'(r) + 3f(r).$$

Find $f(r)$ if $f(r)\mathbf{r}$ is solenoidal for $r \neq 0$.

**10.** If the smooth vector field $\mathbf{F}$ is both irrotational and solenoidal on $\mathbb{R}^3$, show that the three components of $\mathbf{F}$ and the scalar potential for $\mathbf{F}$ are all harmonic functions in $\mathbb{R}^3$.

**11.** If $\mathbf{r} = x\mathbf{i} + y\mathbf{j} + z\mathbf{k}$ and $\mathbf{F}$ is smooth, show that

$$\nabla \times (\mathbf{F} \times \mathbf{r}) = \mathbf{F} - (\nabla \bullet \mathbf{F})\mathbf{r} + \nabla(\mathbf{F} \bullet \mathbf{r}) - \mathbf{r} \times (\nabla \times \mathbf{F}).$$

In particular, if $\nabla \bullet \mathbf{F} = 0$ and $\nabla \times \mathbf{F} = \mathbf{0}$, then

$$\nabla \times (\mathbf{F} \times \mathbf{r}) = \mathbf{F} + \nabla(\mathbf{F} \bullet \mathbf{r}).$$

**12.** If $\phi$ and $\psi$ are harmonic functions, show that $\phi \nabla \psi - \psi \nabla \phi$ is solenoidal.

**13.** If $\phi$ and $\psi$ are smooth scalar fields show that

$$\nabla \times (\phi \nabla \psi) = -\nabla \times (\psi \nabla \phi) = \nabla \phi \times \nabla \psi.$$

**14.** Verify the identity

$$\nabla \bullet \Big(f(\nabla g \times \nabla h)\Big) = \nabla f \bullet (\nabla g \times \nabla h),$$

for smooth scalar fields $f$, $g$, and $h$.

**15.** If the vector fields $\mathbf{F}$ and $\mathbf{G}$ are smooth and conservative, show that $\mathbf{F} \times \mathbf{G}$ is solenoidal. Find a vector potential for $\mathbf{F} \times \mathbf{G}$.

**16.** Find a vector potential for $\mathbf{F} = -y\mathbf{i} + x\mathbf{j}$.

**17.** Show that $\mathbf{F} = xe^{2z}\mathbf{i} + ye^{2z}\mathbf{j} - e^{2z}\mathbf{k}$ is a solenoidal vector field, and find a vector potential for it.

* **18.** Suppose **div F** $= 0$ in a domain $D$ any point $P$ of which can by joined to the origin by a straight line segment in $D$. Let $\mathbf{r} = tx\mathbf{i} + ty\mathbf{j} + tz\mathbf{k}$, $(0 \leq t \leq 1)$, be a parametrization of the line segment from the origin to $(x, y, z)$ in $D$. If

$$\mathbf{G}(x, y, z) = \int_0^1 t\mathbf{F}(\mathbf{r}(t)) \times \frac{d\mathbf{r}}{dt}\, dt,$$

show that **curl G** $= \mathbf{F}$ throughout $D$. *Hint:* it is enough to check the first components of **curl G** and **F**. Proceed in a manner similar to the proof of Theorem 4.

## 9.3 THE DIVERGENCE THEOREM

The Fundamental Theorem of Calculus,

$$\int_a^b \frac{d}{dx} f(x)\, dx = f(b) - f(a),$$

expresses the integral, taken over the interval $[a, b]$, of the derivative of a single-variable function, $f$, as a *sum* of values of that function at the *oriented boundary* of the interval $[a, b]$, that is, at the two endpoints $a$ and $b$, the former providing a *negative* contribution and the latter a *positive* one. The line integral of a conservative vector field over a curve $\mathcal{C}$ from $A$ to $B$,

$$\int_{\mathcal{C}} \nabla \phi \bullet d\mathbf{r} = \phi(B) - \phi(A),$$

has a similar interpretation; $\nabla \phi$ is a derivative, and the curve $\mathcal{C}$, although lying in a two- or three-dimensional space, is intrinsically a one-dimensional object, and the points $A$ and $B$ constitute its boundary.

In this section and the next we will extend the Fundamental Theorem of Calculus to double and triple integrals of certain "derivatives" of vector fields over two- and three-dimensional structures, and will discover, in each case, that the integral is equal to a "sum" (actually an integral of one lower dimension) over the boundary of the structure. The principal generalizations are three in number: Green's Theorem applies to domains in the plane and their bounding curves, Stokes's Theorem applies to surfaces in 3-space and their bounding curves, and Gauss's Theorem (the Divergence Theorem) applies to domains in 3-space and their bounding surfaces. We start with the Divergence Theorem.

In the Divergence Theorem, the integral of the *derivative* $\mathbf{div}\,\mathbf{F} = \boldsymbol{\nabla} \bullet \mathbf{F}$ over a domain in 3-space is expressed as the flux of $\mathbf{F}$ out of the surface of that domain. The theorem holds, and we therefore state it, for a general class of domains in $\mathbb{R}^3$ that are bounded by piecewise smooth closed surfaces. However, we will restrict our proof to domains of a special type. Extending the concept of an $x$-simple plane domain defined in Section 6.2, we say the three-dimensional domain $D$ is **$x$-simple** if it is bounded by a piecewise smooth surface $\mathcal{S}$ and if every straight line parallel to the $x$-axis and passing through an interior point of $D$ meets $\mathcal{S}$ at exactly two points. Similar definitions hold for $y$-simple and $z$-simple, and we call the domain $D$ **regular** if it is a union of finitely many, nonoverlapping subdomains, each of which is $x$-simple, $y$-simple, and $z$-simple.

**THEOREM 6**

**The Divergence Theorem (Gauss's Theorem)**

Let $D$ be a regular, 3-dimensional domain whose boundary $\mathcal{S}$ is an oriented, closed surface with unit normal field $\hat{\mathbf{N}}$ pointing out of $D$. If $\mathbf{F}$ is a smooth vector field defined on $D$ then

$$\iiint_D \mathbf{div}\,\mathbf{F}\,dV = \oiint_{\mathcal{S}} \mathbf{F} \bullet \hat{\mathbf{N}}\,dS.$$

***PROOF*** Since the domain $D$ is a union of finitely many non-overlapping domains that are $x$-simple, $y$-simple, and $z$-simple, it is sufficient to prove the theorem for a subdomain of $D$ with this property. To see this, suppose, for instance, that $D$ and $\mathcal{S}$ are each divided into two parts, $D_1$ and $D_2$, and $\mathcal{S}_1$ and $\mathcal{S}_2$, by a surface $\mathcal{S}^*$ slicing through $D$. (See Figure 9.5.) $\mathcal{S}^*$ is part of the boundary of both $D_1$ and $D_2$, but the exterior normals, $\hat{\mathbf{N}}_1$ and $\hat{\mathbf{N}}_2$, of the two subdomains point in opposite directions on either side of $\mathcal{S}^*$. If the formula in the theorem holds for both subdomains,

$$\iiint_{D_1} \mathbf{div}\,\mathbf{F}\,dV = \oiint_{\mathcal{S}_1 \cup \mathcal{S}^*} \mathbf{F} \bullet \hat{\mathbf{N}}_1\,dS$$
$$\iiint_{D_2} \mathbf{div}\,\mathbf{F}\,dV = \oiint_{\mathcal{S}_2 \cup \mathcal{S}^*} \mathbf{F} \bullet \hat{\mathbf{N}}_2\,dS,$$

then, adding these equations, we get

$$\iiint_D \mathbf{div}\,\mathbf{F}\,dV = \oiint_{\mathcal{S}_1 \cup \mathcal{S}_2} \mathbf{F} \bullet \hat{\mathbf{N}}\,dS = \oiint_{\mathcal{S}} \mathbf{F} \bullet \hat{\mathbf{N}}\,dS;$$

the contributions from $\mathcal{S}^*$ cancel out because on that surface $\hat{\mathbf{N}}_2 = -\hat{\mathbf{N}}_1$.

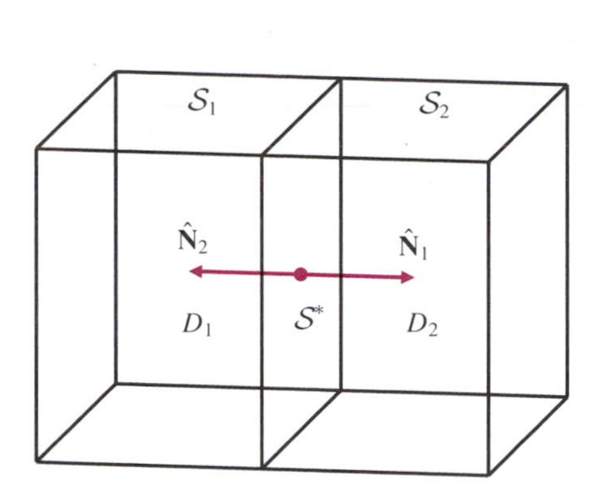


**Figure 9.5** A union of abutting domains

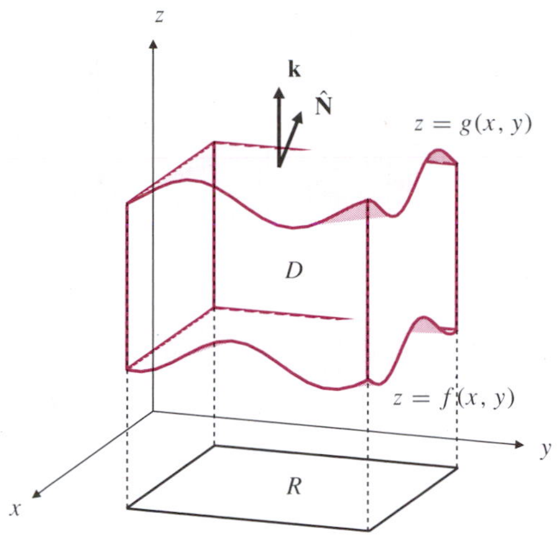

**Figure 9.6** A $z$-simple domain

For the rest of this proof we assume, therefore, that $D$ is $x$-, $y$-, and $z$-simple. Since $D$ is $z$-simple, it lies between the graphs of two functions defined on a region $R$ in the $xy$-plane; if $(x, y, z)$ is in $D$ then $(x, y)$ is in $R$ and $f(x, y) \le z \le g(x, y)$. (See Figure 9.6.) We have

$$\begin{aligned}\iiint_D \frac{\partial F_3}{\partial z}\, dV &= \iint_R dx\, dy \int_{f(x,y)}^{g(x,y)} \frac{\partial F_3}{\partial z}\, dz \\ &= \iint_R \Big(F_3\big(x, y, g(x, y)\big) - F_3\big(x, y, f(x, y)\big)\Big)\, dx\, dy.\end{aligned}$$

Now

$$\oiint_{\mathcal{S}} \mathbf{F} \bullet \hat{\mathbf{N}}\, dS = \oiint_{\mathcal{S}} \Big(F_1\, \mathbf{i} \bullet \hat{\mathbf{N}} + F_2\, \mathbf{j} \bullet \hat{\mathbf{N}} + F_3\, \mathbf{k} \bullet \hat{\mathbf{N}}\Big)\, dS.$$

Only the last term involves $F_3$, and it can be split into three integrals, over the top surface $z = g(x, y)$, the bottom surface $z = f(x, y)$, and vertical side wall lying above the boundary of $R$:

$$\oiint_{\mathcal{S}} F_3(x, y, z)\, \mathbf{k} \bullet \hat{\mathbf{N}}\, dS = \left(\iint_{\text{top}} + \iint_{\text{bottom}} + \iint_{\text{side}}\right) F_3(x, y, z)\, \mathbf{k} \bullet \hat{\mathbf{N}}\, dS.$$

On the side wall, $\mathbf{k} \bullet \hat{\mathbf{N}} = 0$, so that integral is zero. On the top surface, $z = g(x, y)$, the vector area element is

$$\hat{\mathbf{N}}\, d\mathbf{S} = \left(-\frac{\partial g}{\partial x}\mathbf{i} - \frac{\partial g}{\partial y}\mathbf{j} + \mathbf{k}\right) dx\, dy.$$

Accordingly,

$$\iint_{\text{top}} F_3(x, y, z)\, \mathbf{k} \bullet \hat{\mathbf{N}}\, dS = \iint_R F_3\big(x, y, g(x, y)\big)\, dx\, dy.$$

Similarly, we have

$$\iint_{\text{bottom}} F_3(x, y, z)\, \mathbf{k} \bullet \hat{\mathbf{N}}\, dS = -\iint_R F_3\big(x, y, f(x, y)\big)\, dx\, dy;$$

the negative sign occurs because $\hat{\mathbf{N}}$ points down rather than up on the bottom. Thus we have shown that

$$\iiint_D \frac{\partial F_3}{\partial z}\,dV = \oiint_S F_3\,\mathbf{k} \bullet \hat{\mathbf{N}}\,dS.$$

Similarly, because $D$ is also $x$-simple and $y$-simple,

$$\iiint_D \frac{\partial F_1}{\partial x}\,dV = \oiint_S F_1\,\mathbf{i} \bullet \hat{\mathbf{N}}\,dS$$
$$\iiint_D \frac{\partial F_2}{\partial y}\,dV = \oiint_S F_2\,\mathbf{j} \bullet \hat{\mathbf{N}}\,dS.$$

Adding these three results we get

$$\iiint_D \operatorname{\mathbf{div}} \mathbf{F}\,dV = \oiint_S \mathbf{F} \bullet \hat{\mathbf{N}}\,dS.$$

The Divergence Theorem can be used in both directions to simplify explicit calculations of surface integrals or volumes. We give examples of each.

**EXAMPLE 1** Let $\mathbf{F} = bxy^2\mathbf{i} + bx^2y\mathbf{j} + (x^2 + y^2)z^2\mathbf{k}$, and let $\mathcal{S}$ be the closed surface bounding the solid cylinder $R$ defined by $x^2 + y^2 \le a^2$ and $0 \le z \le b$. Find $\oiint_{\mathcal{S}} \mathbf{F} \bullet d\mathbf{S}$.

***SOLUTION*** By the Divergence Theorem,

$$\begin{aligned}\oiint_{\mathcal{S}} \mathbf{F} \bullet d\mathbf{S} &= \iiint_R \operatorname{\mathbf{div}} \mathbf{F}\,dV = \iiint_R (x^2 + y^2)(b + 2z)\,dV \\ &= \int_0^b (b + 2z)\,dz \int_0^{2\pi} d\theta \int_0^a r^2\,r\,dr \\ &= (b^2 + b^2)2\pi(a^4/4) = \pi a^4 b^2.\end{aligned}$$

**EXAMPLE 2** Evaluate $\oiint_{\mathcal{S}} (x^2+y^2)\,dS$ where $\mathcal{S}$ is the sphere $x^2+y^2+z^2 = a^2$. Use the Divergence Theorem.

***SOLUTION*** On $\mathcal{S}$ we have

$$\hat{\mathbf{N}} = \frac{\mathbf{r}}{a} = \frac{x\mathbf{i} + y\mathbf{j} + z\mathbf{k}}{a}.$$

We would like to choose $\mathbf{F}$ so that $\mathbf{F} \bullet \hat{\mathbf{N}} = x^2 + y^2$. Observe that $\mathbf{F} = a(x\mathbf{i} + y\mathbf{j})$ will do. If $B$ is the ball bounded by $\mathcal{S}$, then

$$\begin{aligned}\oiint_{\mathcal{S}} (x^2 + y^2)\,dS &= \oiint_{\mathcal{S}} \mathbf{F} \bullet \hat{\mathbf{N}}\,dS = \iiint_B \operatorname{\mathbf{div}} \mathbf{F}\,dV \\ &= \iiint_B 2a\,dV = (2a)\frac{4}{3}\pi a^3 = \frac{8}{3}\pi a^4.\end{aligned}$$

**EXAMPLE 3** By using the Divergence Theorem with $\mathbf{F} = x\mathbf{i}+y\mathbf{j}+z\mathbf{k}$, calculate the volume of a cone having base area $A$ and height $h$. The base can be any smoothly bounded plane region.

***SOLUTION*** Let the vertex of the cone be at the origin and the base in the plane $z = h$ as shown in Figure 9.7. The solid cone $C$ has surface consisting of two parts, the conical wall $\mathcal{S}$, and the base region $D$ that has area $A$. Since $\mathbf{F}(x, y, z)$ points directly away from the origin at any point $(x, y, z) \neq (0, 0, 0)$, we have $\mathbf{F} \bullet \hat{\mathbf{N}} = 0$ on $\mathcal{S}$. On $D$, we have $\hat{\mathbf{N}} = \mathbf{k}$ and $z = h$, so $\mathbf{F} \bullet \hat{\mathbf{N}} = z = h$ on the base of the cone. Since $\mathbf{div}\,\mathbf{F}(x, y, z) = 1 + 1 + 1 = 3$, we have, by the Divergence Theorem,

$$3V = \iiint_C \mathbf{div}\,\mathbf{F}\,dV = \iint_{\mathcal{S}} \mathbf{F} \bullet \hat{\mathbf{N}}\,dS + \iint_D \mathbf{F} \bullet \hat{\mathbf{N}}\,dS$$
$$= 0 + h \iint_D dS = Ah.$$

Thus $V = \frac{1}{3}Ah$, the well-known formula for the volume of a cone. ■

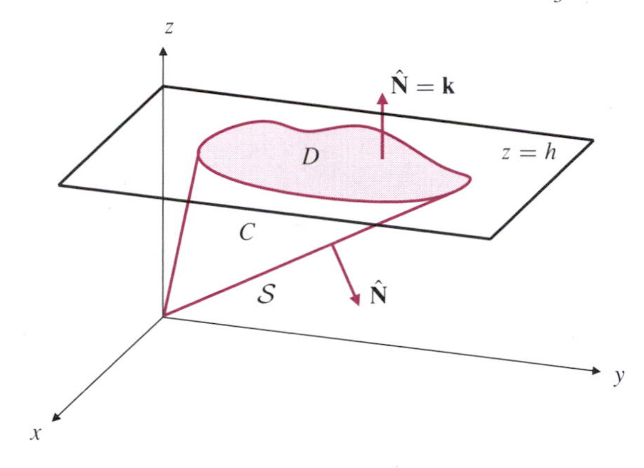


**Figure 9.7** A cone with an arbitrarily shaped base

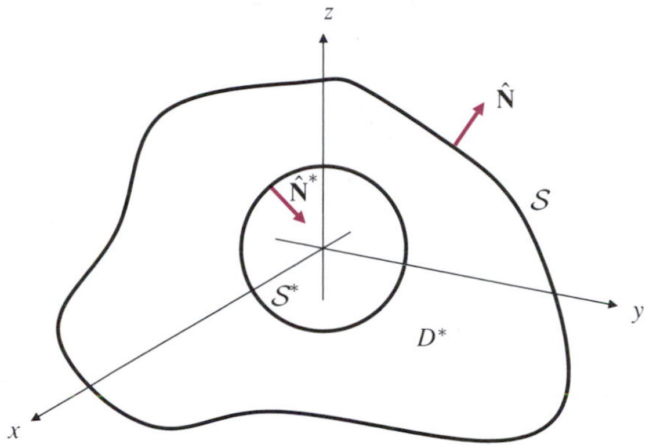

**Figure 9.8** A solid domain with a spherical cavity

**EXAMPLE 4** Let $\mathcal{S}$ be the surface of an arbitrary regular domain $D$ in 3-space that contains the origin in its interior. Find

$$\oiint_{\mathcal{S}} \mathbf{F} \bullet \hat{\mathbf{N}}\,dS,$$

where $\mathbf{F}(\mathbf{r}) = m\mathbf{r}/|\mathbf{r}|^3$ and $\hat{\mathbf{N}}$ is the unit outward normal on $\mathcal{S}$.

***SOLUTION*** Since $\mathbf{F}$, and therefore $\mathbf{div}\,\mathbf{F}$, is undefined at the origin, we cannot apply the Divergence Theorem directly. To overcome this problem we use a little trick. Let $\mathcal{S}^*$ be a small sphere centred at the origin bounding a ball contained wholly in $D$. (See Figure 9.8.) Let $\hat{\mathbf{N}}^*$ be the unit normal on $\mathcal{S}^*$ pointing *into* the sphere, and let $D^*$ be that part of $D$ which lies outside $\mathcal{S}^*$. As shown in Example 3 of Section 9.1, $\mathbf{div}\,\mathbf{F} = 0$ on $D^*$. Also,

$$\oiint_{\mathcal{S}^*} \mathbf{F} \bullet \hat{\mathbf{N}}^*\,dS = -4\pi m,$$

is the flux of $\mathbf{F}$ *inward* through the sphere $\mathcal{S}^*$. (See Example 1 of Section 8.6.) Therefore

$$0 = \iiint_{D^*} \mathbf{div}\,\mathbf{F}\,dV = \oiint_{\mathcal{S}} \mathbf{F} \bullet \hat{\mathbf{N}}\,dS + \oiint_{\mathcal{S}^*} \mathbf{F} \bullet \hat{\mathbf{N}}^*\,dS$$
$$= \oiint_{\mathcal{S}} \mathbf{F} \bullet \hat{\mathbf{N}}\,dS - 4\pi m,$$

and so $\displaystyle\oiint_{\mathcal{S}} \mathbf{F} \bullet \hat{\mathbf{N}}\, dS = 4\pi m.$ ■

**EXAMPLE 5** Find the flux of $\mathbf{F} = x\mathbf{i} + y^2\mathbf{j} + z\mathbf{k}$ upwards through the first-octant part $\mathcal{S}$ of the cylindrical surface $x^2 + y^2 = a^2$, $0 \le y \le b$.

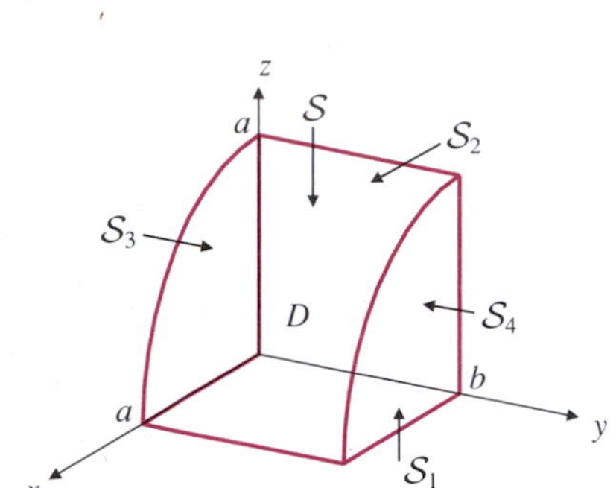


**Figure 9.9** The boundary of domain $D$ has five faces, one curved and four planar

***SOLUTION*** $\mathcal{S}$ is one of five surfaces that form the boundary of the solid region $D$ shown in Figure 9.9. The other four surfaces are planar: $\mathcal{S}_1$ lies in the plane $z = 0$, $\mathcal{S}_2$ lies in the plane $x = 0$, $\mathcal{S}_3$ lies in the plane $y = 0$, and $\mathcal{S}_4$ lies in the plane $y = b$. Orient all these surfaces with normal $\hat{\mathbf{N}}$ pointing out of $D$. On $\mathcal{S}_1$ we have $\hat{\mathbf{N}} = -\mathbf{k}$, so $\mathbf{F} \bullet \hat{\mathbf{N}} = -z = 0$ on $\mathcal{S}_1$. Similarly, $\mathbf{F} \bullet \hat{\mathbf{N}} = 0$ on $\mathcal{S}_2$ and $\mathcal{S}_3$. On $\mathcal{S}_4$, $y = b$ and $\hat{\mathbf{N}} = \mathbf{j}$, so $\mathbf{F} \bullet \hat{\mathbf{N}} = y^2 = b^2$ there. If $\mathcal{S}_{\text{tot}}$ denotes the whole boundary of $D$, then

$$\begin{aligned}\oiint_{\mathcal{S}_{\text{tot}}} \mathbf{F} \bullet \hat{\mathbf{N}}\, dS &= \oiint_{\mathcal{S}} \mathbf{F} \bullet \hat{\mathbf{N}}\, dS + 0 + 0 + 0 + \oiint_{\mathcal{S}_4} \mathbf{F} \bullet \hat{\mathbf{N}}\, dS \\ &= \oiint_{\mathcal{S}} \mathbf{F} \bullet \hat{\mathbf{N}}\, dS + \frac{\pi a^2 b^2}{4}.\end{aligned}$$

On the other hand, by the Divergence Theorem,

$$\oiint_{\mathcal{S}_{\text{tot}}} \mathbf{F} \bullet \hat{\mathbf{N}}\, dS = \iiint_D \mathbf{div}\,\mathbf{F}\, dV = \iiint_D (2 + 2y)\, dV = 2V + 2V\bar{y},$$

where $V = \pi a^2 b/4$ is the volume of $D$, and $\bar{y} = b/2$ is the $y$-coordinate of the centroid of $D$. Combining these results, we find that the flux of $\mathbf{F}$ upward through $\mathcal{S}$ is

$$\oiint_{\mathcal{S}} \mathbf{F} \bullet \hat{\mathbf{N}}\, dS = \frac{2\pi a^2 b}{4}\left(1 + \frac{b}{2}\right) - \frac{\pi a^2 b^2}{4} = \frac{\pi a^2 b}{2}.$$ ■

Among the examples above, Example 4 is the most significant, and the one that best represents the way that the Divergence Theorem is used in practice. It is predominantly a theoretical tool, rather than a tool for calculation. We will look at some applications in the Section 9.5.

## Variants of the Divergence Theorem

Other versions of the Fundamental Theorem of Calculus can be derived from the Divergence Theorem. Two are given in the following theorem:

**THEOREM 7**

If $D$ satisfies the conditions of of the Divergence Theorem and has surface $\mathcal{S}$, and if $\mathbf{F}$ is a smooth vector field, and $\phi$ is a smooth scalar field, then:

(a) $$\iiint_D \mathbf{curl}\,\mathbf{F}\, dV = -\oiint_{\mathcal{S}} \mathbf{F} \times \hat{\mathbf{N}}\, dS,$$

(b) $$\iiint_D \mathbf{grad}\,\phi\, dV = \oiint_{\mathcal{S}} \phi \hat{\mathbf{N}}\, dS.$$

***PROOF*** Observe that both of these formulas are equations of *vectors*. They are derived by applying the Divergence Theorem to $\mathbf{F} \times \mathbf{c}$ and $\phi\mathbf{c}$, respectively, where $\mathbf{c}$ is an arbitrary constant vector. We give the details for formula (a) and leave (b) as an exercise.

Using Theorem 3(d), we calculate

$$\nabla \bullet (\mathbf{F} \times \mathbf{c}) = (\nabla \times \mathbf{F}) \bullet \mathbf{c} - \mathbf{F} \bullet (\nabla \times \mathbf{c}) = (\nabla \times \mathbf{F}) \bullet \mathbf{c}.$$

Also, by the scalar triple product identity (see Exercise 18 of Section 3.3),

$$(\mathbf{F}\times\mathbf{c}) \bullet \hat{\mathbf{N}} = (\hat{\mathbf{N}}\times\mathbf{F}) \bullet \mathbf{c} = -(\mathbf{F}\times\hat{\mathbf{N}}) \bullet \mathbf{c}.$$

Therefore

$$\begin{aligned}\left(\iiint_D \mathbf{curl}\,\mathbf{F}\,dV + \oiint_S \mathbf{F}\times\hat{\mathbf{N}}\,dS\right) \bullet \mathbf{c} \\ &= \iiint_D (\nabla\times\mathbf{F}) \bullet \mathbf{c}\,dV - \oiint_S (\mathbf{F}\times\mathbf{c}) \bullet \hat{\mathbf{N}}\,dS \\ &= \iiint_D \mathbf{div}\,(\mathbf{F}\times\mathbf{c})\,dV - \oiint_S (\mathbf{F}\times\mathbf{c}) \bullet \hat{\mathbf{N}}\,dS = 0.\end{aligned}$$

Since $\mathbf{c}$ is arbitrary, the vector in the large parentheses must be the zero vector. (If $\mathbf{c} \bullet \mathbf{a} = 0$ for every vector $\mathbf{c}$ then $\mathbf{a} = \mathbf{0}$.) This establishes formula (a).

The following example illustrates how the Divergence Theorem is used to establish a two-dimensional version of itself. We will need this two-dimensional version in the next section.

### EXAMPLE 6 A two-dimensional Divergence Theorem

Let $R$ be a domain in the $xy$-plane with piecewise smooth boundary curve $\mathcal{C}$. If

$$\mathbf{F} = F_1(x, y)\mathbf{i} + F_2(x, y)\mathbf{j}$$

is a smooth vector field show that

$$\iint_R \mathbf{div}\,\mathbf{F}\,dA = \oint_{\mathcal{C}} \mathbf{F} \bullet \hat{\mathbf{N}}\,ds,$$

where $\hat{\mathbf{N}}$ is the unit *outward* normal on $\mathcal{C}$.

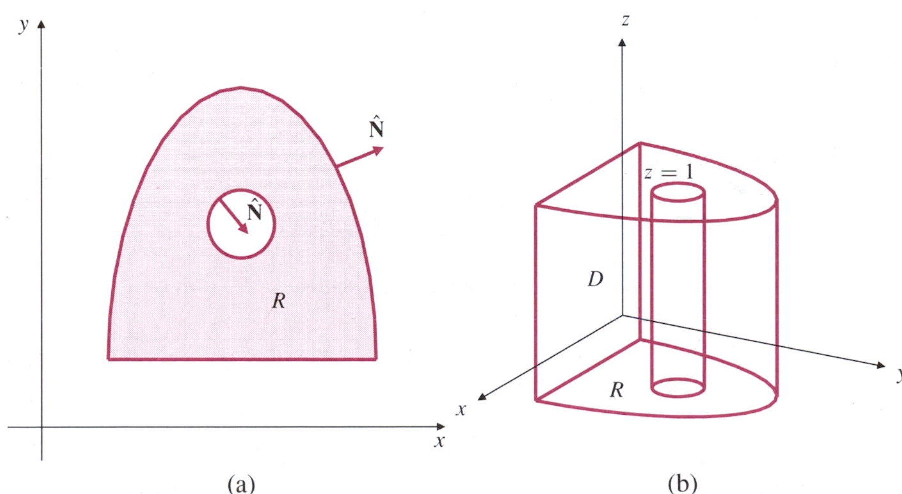


Figure 9.10

***SOLUTION*** A typical domain $R$ and its boundary normals are shown in part (a) of Figure 9.10. Let $D$ be the vertical cylinder whose base is the region $R$ and whose top is in the plane $z = 1$. (See Figure 9.10(b).) We apply the 3-dimensional Divergence Theorem to $\mathbf{F}$ (considered as a 3-dimensional vector field) on $D$. We have

$$\iiint_D \operatorname{div} \mathbf{F}\, dV = \int_0^1 dz \iint_R \left( \frac{\partial F_1(x, y)}{\partial x} + \frac{\partial F_2(x, y)}{\partial y} \right) dA$$
$$= \iint_R \operatorname{div} \mathbf{F}\, dA.$$

On the top and bottom surfaces of the cylinder $D$, the normals $\hat{\mathbf{N}}$ are $\mathbf{k}$ and $-\mathbf{k}$ respectively. Since the $z$-component $F_3$ is identically zero, $\mathbf{F} \bullet \hat{\mathbf{N}}$ vanishes on these two surfaces. On the vertical cylindrical wall, the outward normal is given by $\hat{\mathbf{N}}(x, y, z) = \hat{\mathbf{N}}(x, y)$, the normal to $\mathcal{C}$, and the area element is $dS = dz\, ds$. Since $\mathbf{F}$ is independent of $z$ we have, denoting the whole surface of $D$ by $\mathcal{S}$,

$$\oiint_{\mathcal{S}} \mathbf{F} \bullet \hat{\mathbf{N}}\, dS = \iint_{\text{cylwall}} \mathbf{F} \bullet \hat{\mathbf{N}}\, dS = \int_0^1 dz \oint_{\mathcal{C}} \mathbf{F} \bullet \hat{\mathbf{N}}\, ds = \oint_{\mathcal{C}} \mathbf{F} \bullet \hat{\mathbf{N}}\, ds.$$

Therefore,

$$\iint_R \operatorname{div} \mathbf{F}\, dA = \iiint_D \operatorname{div} \mathbf{F}\, dV = \oiint_{\mathcal{S}} \mathbf{F} \bullet \hat{\mathbf{N}}\, dS = \oint_{\mathcal{C}} \mathbf{F} \bullet \hat{\mathbf{N}}\, ds$$

as claimed. ■

## EXERCISES 9.3

In Exercises 1–4, use the Divergence Theorem to calculate the flux of the given vector field out of the sphere $\mathcal{S}$ with equation $x^2 + y^2 + z^2 = a^2$, where $a > 0$.

**1.** $\mathbf{F} = x\mathbf{i} - 2y\mathbf{j} + 4z\mathbf{k}$

**2.** $\mathbf{F} = ye^z\mathbf{i} + x^2e^z\mathbf{j} + xy\mathbf{k}$

**3.** $\mathbf{F} = (x^2 + y^2)\mathbf{i} + (y^2 - z^2)\mathbf{j} + z\mathbf{k}$

**4.** $\mathbf{F} = x^3\mathbf{i} + 3yz^2\mathbf{j} + (3y^2z + x^2)\mathbf{k}$

In Exercises 5–8, evaluate the flux of $\mathbf{F} = x^2\mathbf{i} + y^2\mathbf{j} + z^2\mathbf{k}$ outward across the boundary of the given solid region.

**5.** The ball $(x - 2)^2 + y^2 + (z - 3)^2 \le 9$

**6.** The solid ellipsoid $x^2 + y^2 + 4(z - 1)^2 \le 4$

**7.** The tetrahedron $x + y + z \le 3$, $x \ge 0$, $y \ge 0$, $z \ge 0$

**8.** The cylinder $x^2 + y^2 \le 2y$, $0 \le z \le 4$

**9.** Let $A$ be the area of a region $D$ forming part of the surface of a sphere of radius $R$ centred at the origin, and let $V$ be the volume of the solid cone $C$ consisting of all points on line segments joining the centre of the sphere to points in $D$. Show that

$$V = \frac{1}{3}AR$$

by applying the Divergence Theorem to $\mathbf{F} = x\mathbf{i} + y\mathbf{j} + z\mathbf{k}$.

**10.** Let $\phi(x, y, z) = xy + z^2$. Find the flux of $\nabla\phi$ upwards through the triangular planar surface $\mathcal{S}$ with vertices at $(a, 0, 0)$, $(0, b, 0)$, and $(0, 0, c)$.

**11.** A conical domain with vertex $(0, 0, b)$ and axis along the $z$-axis has as base a disk of radius $a$ in the $xy$-plane. Find the flux of

$$\mathbf{F} = (x + y^2)\mathbf{i} + (3x^2y + y^3 - x^3)\mathbf{j} + (z + 1)\mathbf{k}$$

upward through the conical part of the surface of the domain.

**12.** Find the flux of $\mathbf{F} = (y + xz)\mathbf{i} + (y + yz)\mathbf{j} - (2x + z^2)\mathbf{k}$ upward through the first octant part of the sphere $x^2 + y^2 + z^2 = a^2$.

**13.** Let $D$ be the region $x^2 + y^2 + z^2 \le 4a^2$, $x^2 + y^2 \ge a^2$. The surface $\mathcal{S}$ of $D$ consists of a cylindrical part, $\mathcal{S}_1$, and a spherical part, $\mathcal{S}_2$. Evaluate the flux of

$$\mathbf{F} = (x + yz)\mathbf{i} + (y - xz)\mathbf{j} + (z - e^x \sin y)\mathbf{k}$$

out of $D$ through (a) the whole surface $\mathcal{S}$, (b) the surface $\mathcal{S}_1$, and (c) the surface $\mathcal{S}_2$.

**14.** Evaluate $\iint_{\mathcal{S}} (3xz^2\mathbf{i} - x\mathbf{j} - y\mathbf{k}) \bullet \hat{\mathbf{N}}\, dS$, where $\mathcal{S}$ is that part of the cylinder $y^2 + z^2 = 1$ which lies in the first octant and between the planes $x = 0$ and $x = 1$.

**15.** A solid region $R$ has volume $V$ and centroid at the point $(\bar{x}, \bar{y}, \bar{z})$. Find the flux of

$$\mathbf{F} = (x^2 - x - 2y)\mathbf{i} + (2y^2 + 3y - z)\mathbf{j} - (z^2 - 4z + xy)\mathbf{k}$$

out of $R$ through its surface.

**16.** The plane $x + y + z = 0$ divides the cube $-1 \le x \le 1$, $-1 \le y \le 1$, $-1 \le z \le 1$ into two parts. Let the lower part (with one vertex at $(-1, -1, -1)$) be $D$. Sketch $D$. Note that it has seven faces, one of which is hexagonal. Find the flux of $\mathbf{F} = x\mathbf{i} + y\mathbf{j} + z\mathbf{k}$ out of $D$ through each of its faces.

**17.** Let $\mathbf{F} = (x^2 + y + 2 + z^2)\mathbf{i} + (e^{x^2} + y^2)\mathbf{j} + (3 + x)\mathbf{k}$. Let $a > 0$ and let $\mathcal{S}$ be the part of the spherical surface $x^2 + y^2 + z^2 = 2az + 3a^2$ that is above the $xy$-plane. Find the flux of $\mathbf{F}$ outward across $\mathcal{S}$.

**18.** A pile of wet sand having total volume $5\pi$ covers the disk $x^2 + y^2 \le 1$, $z = 0$. The momentum of water vapour is given by $\mathbf{F} = \mathbf{grad}\,\phi + \mu\mathbf{curl}\,\mathbf{G}$, where $\phi = x^2 - y^2 + z^2$ is the water concentration, $\mathbf{G} = \frac{1}{3}(-y^3\mathbf{i} + x^3\mathbf{j} + z^3\mathbf{k})$, and $\mu$ is a constant. Find the flux of $\mathbf{F}$ upward through the top surface of the sand pile.

In Exercises 19–29, $D$ is a three-dimensional domain satisfying the conditions of the Divergence Theorem, and $\mathcal{S}$ is its surface. $\hat{\mathbf{N}}$ is the unit outward (from $D$) normal field on $\mathcal{S}$. The functions $\phi$ and $\psi$ are smooth scalar fields on $D$. Also, $\partial\phi/\partial n$ denotes the first directional derivative of $\phi$ in the direction of $\hat{\mathbf{N}}$ at any point on $\mathcal{S}$:

$$\frac{\partial\phi}{\partial n} = \nabla\phi \bullet \hat{\mathbf{N}}.$$

**19.** Show that $\displaystyle\oiint_{\mathcal{S}} \mathbf{curl}\,\mathbf{F} \bullet \hat{\mathbf{N}}\,dS = 0$, where $\mathbf{F}$ is an arbitrary smooth vector field.

**20.** Show that the volume $V$ of $D$ is given by

$$V = \frac{1}{3}\oiint_{\mathcal{S}} (x\mathbf{i} + y\mathbf{j} + z\mathbf{k}) \bullet \hat{\mathbf{N}}\,dS.$$

**21.** If $D$ has volume $V$ show that

$$\bar{\mathbf{r}} = \frac{1}{2V}\oiint_{\mathcal{S}} (x^2 + y^2 + z^2)\hat{\mathbf{N}}\,dS$$

is the position vector of the centre of gravity of $D$.

**22.** Show that $\displaystyle\oiint_{\mathcal{S}} \nabla\phi \times \hat{\mathbf{N}}\,dS = 0$.

**23.** If $\mathbf{F}$ is a smooth vector field on $D$, show that

$$\iiint_D \phi\,\mathbf{div}\,\mathbf{F}\,dV + \iiint_D \nabla\phi \bullet \mathbf{F}\,dV = \oiint_{\mathcal{S}} \phi\mathbf{F} \bullet \hat{\mathbf{N}}\,dS.$$

(*Hint:* use Theorem 3(b) from Section 9.2.)

**Properties of the Laplacian operator**

**24.** If $\nabla^2\phi = 0$ in $D$ and $\phi(x, y, z) = 0$ on $\mathcal{S}$, show that $\phi(x, y, z) = 0$ in $D$. *Hint:* Let $\mathbf{F} = \nabla\phi$ in Exercise 23.

**25.** **(Uniqueness for the Dirichlet problem)** The Dirichlet problem for the Laplacian operator is the boundary-value problem

$$\begin{cases} \nabla^2 u(x, y, z) = f(x, y, z) & \text{on } D \\ u(x, y, z) = g(x, y, z) & \text{on } \mathcal{S}, \end{cases}$$

where $f$ and $g$ are given functions defined on $D$ and $\mathcal{S}$, respectively. Show that this problem can have at most one solution $u(x, y, z)$. *Hint:* suppose there are two solutions, $u$ and $v$, and apply Exercise 24 to their difference $\phi = u - v$.

**26.** **(The Neumann problem)** If $\nabla^2\phi = 0$ in $D$ and $\partial\phi/\partial n = 0$ on $\mathcal{S}$, show that $\nabla\phi(x, y, z) = 0$ on $D$. The Neumann problem for the Laplacian operator is the boundary-value problem

$$\begin{cases} \nabla^2 u(x, y, z) = f(x, y, z) & \text{on } D \\ \dfrac{\partial}{\partial n} u(x, y, z) = g(x, y, z) & \text{on } \mathcal{S} \end{cases}$$

where $f$ and $g$ are given functions defined on $D$ and $\mathcal{S}$ respectively. Show that if $D$ is connected, then any two solutions of the Neumann problem must differ by a constant on $D$.

**27.** Verify that $\displaystyle\iiint_D \nabla^2\phi\,dV = \oiint_{\mathcal{S}} \frac{\partial\phi}{\partial n}\,dS$.

**28.** Verify that

$$\iiint_D \left(\phi\nabla^2\psi - \psi\nabla^2\phi\right) dV = \oiint_{\mathcal{S}} \left(\phi\frac{\partial\psi}{\partial n} - \psi\frac{\partial\phi}{\partial n}\right) dS.$$

**29.** By applying the Divergence Theorem to $\mathbf{F} = \phi\mathbf{c}$, where $\mathbf{c}$ is an arbitrary constant vector, show that

$$\iiint_D \nabla\phi\,dV = \oiint_{\mathcal{S}} \phi\hat{\mathbf{N}}\,dS.$$

* **30.** Let $P_0$ be a fixed point and for each $\epsilon > 0$ let $D_\epsilon$ be a domain with boundary $\mathcal{S}_\epsilon$ satisfying the conditions of the Divergence Theorem. Suppose that the maximum distance from $P_0$ to points $P$ in $D_\epsilon$ approaches zero as $\epsilon \to 0^+$. If $D_\epsilon$ has volume $\text{vol}(D_\epsilon)$, show that

$$\lim_{\epsilon\to 0^+} \frac{1}{\text{vol}(D_\epsilon)} \oiint_{\mathcal{S}_\epsilon} \mathbf{F} \bullet \hat{\mathbf{N}}\,dS = \mathbf{div}\,\mathbf{F}(P_0).$$

This generalizes Theorem 1 of Section 9.1.

# 9.4 GREEN'S THEOREM AND STOKES'S THEOREM

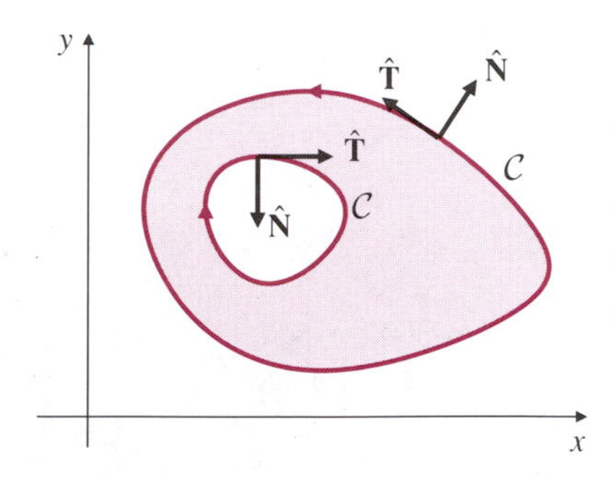


**Figure 9.11** A plane domain with positively oriented boundary

## Green's Theorem in the Plane

Green's Theorem is a two-dimensional version of the Fundamental Theorem of Calculus, similar to Example 6 of the previous section, but rephrased so that the boundary line integral involves the tangential rather than the normal component of $\mathbf{F}$. This requires that we specify an orientation for the boundary curves $\mathcal{C}$ of the region $R$. We require this orientation to be **positive**, in the sense that if we move around $\mathcal{C}$ in the direction of its orientation, $R$ will be on our left. This corresponds to regarding $R$ as an oriented surface with normal $\mathbf{k}$. The region $R$ need not be simply connected; it can have holes, and the boundaries of the holes constitute part of the boundary of the region. (See Figure 9.11.) The "outer" boundary of $R$ is oriented counterclockwise, while the boundaries of holes are oriented clockwise.

**THEOREM 8**

**Green's Theorem**

Let $R$ be a closed region in the $xy$-plane whose boundary, $\mathcal{C}$, consists of one or more piecewise smooth, non-self-intersecting, closed curves that are positively oriented. If $\mathbf{F} = F_1(x, y)\mathbf{i} + F_2(x, y)\mathbf{j}$ is a smooth vector field on $R$, then

$$\oint_{\mathcal{C}} F_1(x, y)\,dx + F_2(x, y)\,dy = \iint_R \left(\frac{\partial F_2}{\partial x} - \frac{\partial F_1}{\partial y}\right) dA.$$

***PROOF*** Because of the orientation of $\mathcal{C}$, the unit exterior normal $\hat{\mathbf{N}}$ on $\mathcal{C}$ (pointing out of $R$) is $\hat{\mathbf{N}} = \hat{\mathbf{T}} \times \mathbf{k}$, where $\hat{\mathbf{T}}$ is the unit tangent. If $\mathcal{C}$ is parametrized in terms of arc length, then

$$\hat{\mathbf{T}} = \frac{dx}{ds}\mathbf{i} + \frac{dy}{ds}\mathbf{j} \qquad \text{and} \qquad \hat{\mathbf{N}} = \frac{dy}{ds}\mathbf{i} - \frac{dx}{ds}\mathbf{j}.$$

Let $\mathbf{G} = F_2(x, y)\mathbf{i} - F_1(x, y)\mathbf{j}$. Then $\mathbf{div}\,\mathbf{G} = (\partial F_2/\partial x) - (\partial F_1/\partial y)$, and $\mathbf{G} \bullet \hat{\mathbf{N}} = \mathbf{F} \bullet \hat{\mathbf{T}}$. Applying the two-dimensional Divergence Theorem from Example 6 of Section 9.3, we obtain

$$\begin{aligned}\oint_{\mathcal{C}} F_1(x, y)\,dx + F_2(x, y)\,dy &= \oint_{\mathcal{C}} \mathbf{F} \bullet d\mathbf{r} = \oint_{\mathcal{C}} \mathbf{F} \bullet \hat{\mathbf{T}}\,ds \\ &= \oint_{\mathcal{C}} \mathbf{G} \bullet \hat{\mathbf{N}}\,ds \\ &= \iint_R \mathbf{div}\,\mathbf{G}\,dA \\ &= \iint_R \left(\frac{\partial F_2}{\partial x} - \frac{\partial F_1}{\partial y}\right) dA,\end{aligned}$$

as required.

***REMARK*** If we regard the region $R$ as a surface in 3-space, its normal is $\hat{\mathbf{N}} = \mathbf{k}$. The formula in Green's Theorem can therefore be written

$$\oint_{\mathcal{C}} \mathbf{F} \bullet d\mathbf{r} = \iint_R \mathbf{curl}\,\mathbf{F} \bullet \hat{\mathbf{N}}\,dS.$$

Stokes's Theorem given below generalizes this to nonplanar surfaces.

***REMARK*** The expression $F_1(x, y)\,dx + F_2(x, y)\,dy$ is called a **differential form**. This differential form is said to be **exact** if it is equal to $d\phi$ for some function $\phi(x, y)$. Evidently the differential form is exact if the corresponding vector field $\mathbf{F} = F_1\mathbf{i} + F_2\mathbf{j}$ is conservative, in which case $\phi$ is the scalar potential of $\mathbf{F}$. The line integral of an exact differential form around a closed curve is zero.

**EXAMPLE 1** **Area bounded by a simple closed curve**

For any of the three vector fields

$$\mathbf{F} = x\mathbf{j}, \qquad \mathbf{F} = -y\mathbf{i}, \qquad \text{and} \qquad \mathbf{F} = \frac{1}{2}(-y\mathbf{i} + x\mathbf{j}),$$

we have $(\partial F_2/\partial x) - (\partial F_1/\partial y) = 1$. If $C$ is a positively oriented, piecewise smooth, simple closed curve bounding a region $R$ in the plane, then by Green's Theorem,

$$\oint_C x\,dy = -\oint_C y\,dx = \frac{1}{2}\oint_C x\,dy - y\,dx = \iint_R 1\,dA = \text{area of } R.$$

**EXAMPLE 2** Evaluate $I = \oint_C (x - y^3)\,dx + (y^3 + x^3)\,dy$, where $C$ is the positively oriented boundary of the quarter-disk $Q$: $0 \le x^2 + y^2 \le a^2$, $x \ge 0$, $y \ge 0$.

***SOLUTION*** We use Green's Theorem to calculate $I$:

$$\begin{aligned} I &= \iint_Q \left(\frac{\partial}{\partial x}(y^3 + x^3) - \frac{\partial}{\partial y}(x - y^3)\right) dA \\ &= 3\iint_Q (x^2 + y^2)\,dA = 3\int_0^{\pi/2} d\theta \int_0^a r^3\,dr = \frac{3}{8}\pi a^4. \end{aligned}$$

**EXAMPLE 3** Let $C$ be a positively oriented, simple closed curve in the $xy$-plane, bounding a region $R$ and not passing through the origin. Show that

$$\oint_C \frac{-y\,dx + x\,dy}{x^2 + y^2} = \begin{cases} 0 & \text{if the origin is outside } R \\ 2\pi & \text{if the origin is inside } R. \end{cases}$$

***SOLUTION*** First, if $(x, y) \ne (0, 0)$, then, by direct calculation,

$$\frac{\partial}{\partial x}\left(\frac{x}{x^2 + y^2}\right) - \frac{\partial}{\partial y}\left(\frac{-y}{x^2 + y^2}\right) = 0.$$

If the origin is not in $R$, then Green's Theorem implies that

$$\oint_C \frac{-y\,dx + x\,dy}{x^2 + y^2} = \iint_R \left[\frac{\partial}{\partial x}\left(\frac{x}{x^2 + y^2}\right) - \frac{\partial}{\partial y}\left(\frac{-y}{x^2 + y^2}\right)\right] dx\,dy = 0.$$

Now suppose the origin is in $R$. Since it is assumed that the origin is not on $C$, it must be an interior point of $R$. The interior of $R$ is open, so there exists $\epsilon > 0$ such that the circle $C_\epsilon$ of radius $\epsilon$ centred at the origin is in the interior of $R$. Let $C_\epsilon$ be oriented negatively (clockwise). By direct calculation (see Exercise 22(a) of Section 8.4) it is easily shown that

$$\oint_{C_\epsilon} \frac{-y\,dx + x\,dy}{x^2 + y^2} = -2\pi.$$

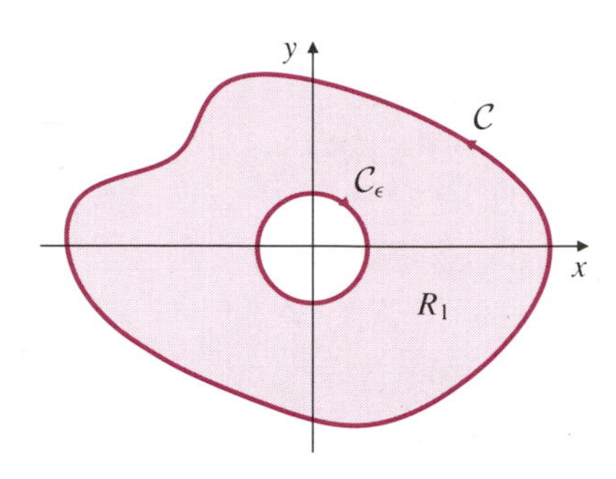

Figure 9.12

Together $\mathcal{C}$ and $\mathcal{C}_\epsilon$ form the positively oriented boundary of a region $R_1$ that excludes the origin. (See Figure 9.12.) So, by Green's Theorem,

$$\oint_{\mathcal{C}} \frac{-y\,dx + x\,dy}{x^2+y^2} + \oint_{\mathcal{C}_\epsilon} \frac{-y\,dx + x\,dy}{x^2+y^2} = 0.$$

The desired result now follows:

$$\oint_{\mathcal{C}} \frac{-y\,dx + x\,dy}{x^2+y^2} = -\oint_{\mathcal{C}_\epsilon} \frac{-y\,dx + x\,dy}{x^2+y^2} = -(-2\pi) = 2\pi. \qquad \blacksquare$$

## Stokes's Theorem

Stokes's Theorem is a generalization of Green's Theorem that applies to surfaces (not necessarily planar) in 3-space.

**THEOREM 9**

**Stokes's Theorem**

Let $\mathcal{S}$ be a piecewise smooth, oriented surface in 3-space, having unit normal field $\hat{\mathbf{N}}$, and having boundary $\mathcal{C}$ consisting of one or more piecewise smooth, closed curves with orientation inherited from $\mathcal{S}$. If $\mathbf{F}$ is a smooth vector field defined on an open set containing $\mathcal{S}$, then

$$\oint_{\mathcal{C}} \mathbf{F} \bullet d\mathbf{r} = \iint_{\mathcal{S}} \mathbf{curl}\,\mathbf{F} \bullet \hat{\mathbf{N}}\,dS.$$

***PROOF*** An argument similar to that given at the beginning of the proof of the Divergence Theorem shows that if $\mathcal{S}$ is decomposed into finitely many nonoverlapping subsurfaces then it is sufficient to prove that the formula above holds for each of them. (If subsurfaces $\mathcal{S}_1$ and $\mathcal{S}_2$ meet along the curve $\mathcal{C}^*$, then $\mathcal{C}^*$ inherits opposite orientations as part of the boundaries of $\mathcal{S}_1$ and $\mathcal{S}_2$, so the line integrals along $\mathcal{C}^*$ cancel out. See Figure 9.13(a).) We can subdivide $\mathcal{S}$ into enough smooth subsurfaces that each one has a one-to-one normal projection onto a coordinate plane. We will establish the formula for one such subsurface, which we will now call $\mathcal{S}$.

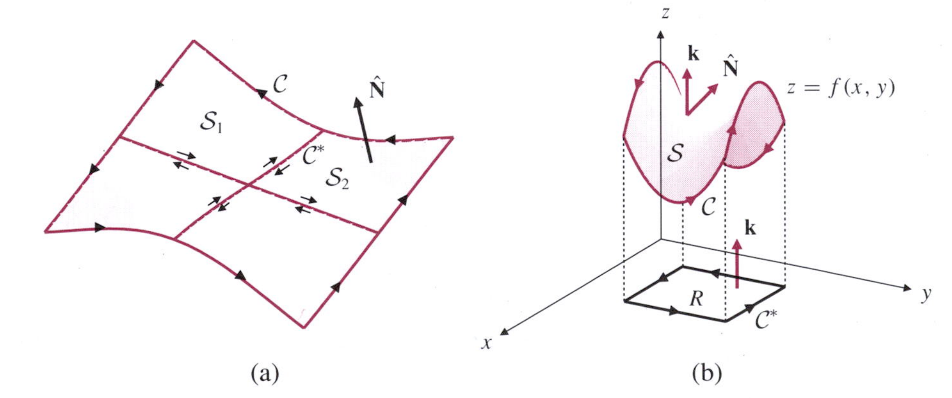

Figure 9.13

(a) Stokes's Theorem holds for a composite surface comprised of non-overlapping subsurfaces for which it is true

(b) A surface with a one-to-one projection on the $xy$-plane

Without loss of generality, assume that $\mathcal{S}$ has a one-to-one normal projection onto the $xy$-plane, and that its normal field $\hat{\mathbf{N}}$ points upward. Therefore, on $\mathcal{S}$, $z$ is a smooth function of $x$ and $y$ defined for $(x, y)$ in a region $R$ of the $xy$-plane. The boundaries $\mathcal{C}$ of $\mathcal{S}$, and $\mathcal{C}^*$ of R, are both oriented counterclockwise as seen from a point high on the $z$-axis. (See Figure 9.13(b).) The normal field on $\mathcal{S}$ is

$$\hat{\mathbf{N}} = \frac{-\dfrac{\partial z}{\partial x}\mathbf{i} - \dfrac{\partial z}{\partial y}\mathbf{j} + \mathbf{k}}{\sqrt{1 + \left(\dfrac{\partial z}{\partial x}\right)^2 + \left(\dfrac{\partial z}{\partial y}\right)^2}},$$

and the surface area element on $\mathcal{S}$ is expressed in terms of the area element $dA = dx\,dy$ in the $xy$-plane as

$$dS = \sqrt{1 + \left(\frac{\partial z}{\partial x}\right)^2 + \left(\frac{\partial z}{\partial y}\right)^2}\,dA.$$

Therefore,

$$\iint_{\mathcal{S}} \mathbf{curl}\,\mathbf{F} \bullet \hat{\mathbf{N}}\,dS = \iint_R \left[\left(\frac{\partial F_3}{\partial y} - \frac{\partial F_2}{\partial z}\right)\left(-\frac{\partial z}{\partial x}\right) + \left(\frac{\partial F_1}{\partial z} - \frac{\partial F_3}{\partial x}\right)\left(-\frac{\partial z}{\partial y}\right) + \left(\frac{\partial F_2}{\partial x} - \frac{\partial F_1}{\partial y}\right)\right] dA.$$

Since $z$ is a function of $x$ and $y$ on $\mathcal{C}$, we have $dz = \dfrac{\partial z}{\partial x}\,dx + \dfrac{\partial z}{\partial y}\,dy$. Thus

$$\begin{aligned}
\oint_{\mathcal{C}} \mathbf{F} \bullet d\mathbf{r} &= \oint_{\mathcal{C}^*} \left[F_1(x, y, z)\,dx + F_2(x, y, z)\,dy + F_3(x, y, z)\left(\frac{\partial z}{\partial x}\,dx + \frac{\partial z}{\partial y}\,dy\right)\right] \\
&= \oint_{\mathcal{C}^*} \left(\left[F_1(x, y, z) + F_3(x, y, z)\frac{\partial z}{\partial x}\right] dx + \left[F_2(x, y, z) + F_3(x, y, z)\frac{\partial z}{\partial y}\right] dy\right).
\end{aligned}$$

We now apply Green's Theorem in the $xy$-plane to obtain

$$\begin{aligned}
\oint_{\mathcal{C}} \mathbf{F} \bullet d\mathbf{r} &= \iint_R \left(\frac{\partial}{\partial x}\left[F_2(x, y, z) + F_3(x, y, z)\frac{\partial z}{\partial y}\right] - \frac{\partial}{\partial y}\left[F_1(x, y, z) + F_3(x, y, z)\frac{\partial z}{\partial x}\right]\right) dA \\
&= \iint_R \left(\frac{\partial F_2}{\partial x} + \frac{\partial F_2}{\partial z}\frac{\partial z}{\partial x} + \frac{\partial F_3}{\partial x}\frac{\partial z}{\partial y} + \frac{\partial F_3}{\partial z}\frac{\partial z}{\partial x}\frac{\partial z}{\partial y} + F_3\frac{\partial^2 z}{\partial x \partial y}\right. \\
&\qquad \left. - \frac{\partial F_1}{\partial y} - \frac{\partial F_1}{\partial z}\frac{\partial z}{\partial y} - \frac{\partial F_3}{\partial y}\frac{\partial z}{\partial x} - \frac{\partial F_3}{\partial z}\frac{\partial z}{\partial y}\frac{\partial z}{\partial x} - F_3\frac{\partial^2 z}{\partial y \partial x}\right) dA.
\end{aligned}$$

Observe that four terms in the final integrand cancel out, and the remaining terms are equal to the terms in the expression for $\iint_{\mathcal{S}} \mathbf{curl}\,\mathbf{F} \bullet \hat{\mathbf{N}}\,dS$ calculated above. Therefore the proof is complete.

***REMARK*** If $\mathbf{curl\,F} = \mathbf{0}$ on a domain $D$ with the property that every piecewise smooth, non-self-intersecting, closed curve in $D$ is the boundary of a piecewise smooth surface in $D$, then Stokes's Theorem assures us that $\oint_{\mathcal{C}} \mathbf{F} \bullet d\mathbf{r} = 0$ for *every* such curve $\mathcal{C}$, and therefore $\mathbf{F}$ must be conservative. A simply connected domain $D$ does have the property specified above. We will not attempt a formal proof of this topological fact here, but it should seem plausible if you recall the definition of simple connectedness. A closed curve $\mathcal{C}$ in a simply connected domain $D$ must be able to shrink to a point in $D$ without ever passing out of $D$. In so shrinking, it traces out a surface in $D$. This is why Theorem 4 of Section 9.2 is valid for simply connected domains.

**EXAMPLE 4** Evaluate $\oint_{\mathcal{C}} \mathbf{F} \bullet d\mathbf{r}$, where $\mathbf{F} = -y^3\mathbf{i} + x^3\mathbf{j} - z^3\mathbf{k}$, and $\mathcal{C}$ is the curve of intersection of the cylinder $x^2 + y^2 = 1$ and the plane $2x + 2y + z = 3$ oriented so as to have a counterclockwise projection onto the $xy$-plane.

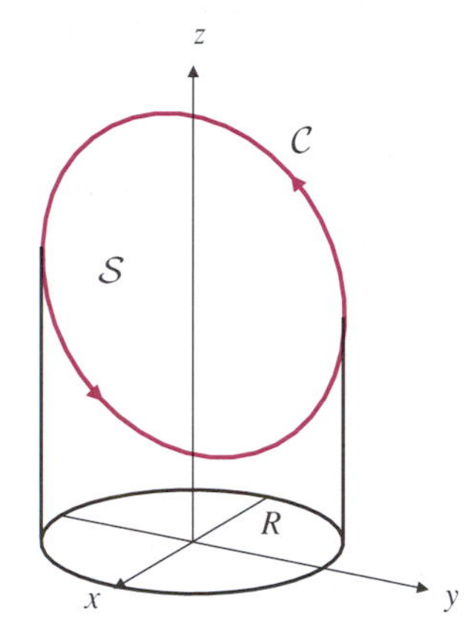


Figure 9.14

***SOLUTION*** $\mathcal{C}$ is the oriented boundary of an elliptic disk $\mathcal{S}$ that lies in the plane $2x + 2y + z = 3$ and has the circular disk $R$: $x^2 + y^2 \le 1$ as projection onto the $xy$-plane. (See Figure 9.14.) On $\mathcal{S}$ we have

$$\hat{\mathbf{N}}\, dS = (2\mathbf{i} + 2\mathbf{j} + \mathbf{k})\, dx\, dy.$$

Also,

$$\mathbf{curl\,F} = \begin{vmatrix} \mathbf{i} & \mathbf{j} & \mathbf{k} \\ \dfrac{\partial}{\partial x} & \dfrac{\partial}{\partial y} & \dfrac{\partial}{\partial z} \\ -y^3 & x^3 & -z^3 \end{vmatrix} = 3(x^2 + y^2)\mathbf{k}.$$

Thus, by Stokes's Theorem,

$$\begin{aligned} \oint_{\mathcal{C}} \mathbf{F} \bullet d\mathbf{r} &= \iint_{\mathcal{S}} \mathbf{curl\,F} \bullet \hat{\mathbf{N}}\, dS \\ &= \iint_{R} 3(x^2 + y^2)\, dx\, dy = 2\pi \int_0^1 3r^2\, r\, dr = \frac{3\pi}{2}. \end{aligned}$$

As with the Divergence Theorem, the principal importance of Stokes's Theorem is as a theoretical tool. However, it can also simplify the calculation of circulation integrals such as the one in the previous example. It is not difficult to imagine integrals whose evaluation would be impossibly difficult without the use of Stokes's Theorem or the Divergence Theorem. In the following example we use Stokes's Theorem twice, but the result could be obtained just as easily by using the Divergence Theorem.

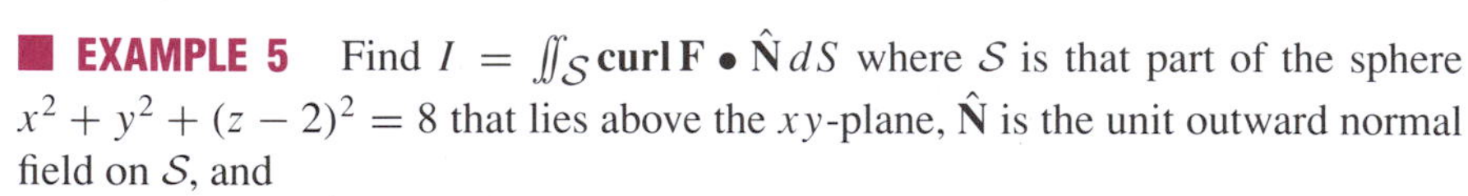

**EXAMPLE 5** Find $I = \iint_{\mathcal{S}} \mathbf{curl\,F} \bullet \hat{\mathbf{N}}\, dS$ where $\mathcal{S}$ is that part of the sphere $x^2 + y^2 + (z-2)^2 = 8$ that lies above the $xy$-plane, $\hat{\mathbf{N}}$ is the unit outward normal field on $\mathcal{S}$, and

$$\mathbf{F} = y^2 \cos xz\, \mathbf{i} + x^3 e^{yz}\mathbf{j} - e^{-xyz}\mathbf{k}.$$

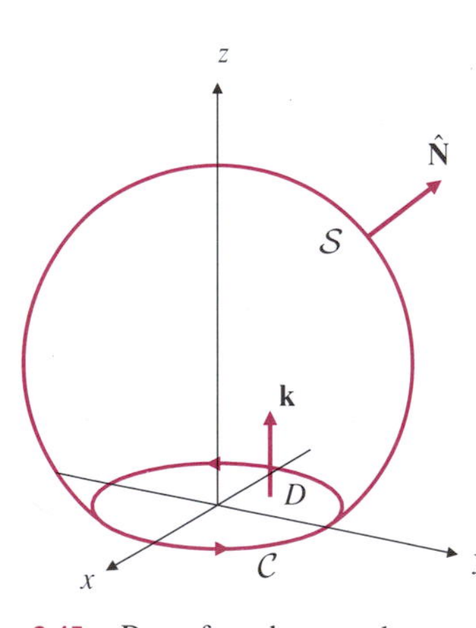


**Figure 9.15** Part of a sphere, and a disk with the same boundary

***SOLUTION*** The boundary, $\mathcal{C}$, of $\mathcal{S}$ is the circle $x^2 + y^2 = 4$ in the $xy$-plane, oriented counterclockwise as seen from the positive $z$-axis. (See Figure 9.15.) This curve is also the oriented boundary of the plane disk $D$: $x^2 + y^2 \le 4$, $z = 0$, with normal field $\hat{\mathbf{N}} = \mathbf{k}$. Thus, two applications of Stokes's Theorem give

$$I = \iint_{\mathcal{S}} \mathbf{curl\,F} \bullet \hat{\mathbf{N}}\, dS = \oint_{\mathcal{C}} \mathbf{F} \bullet d\mathbf{r} = \iint_{D} \mathbf{curl\,F} \bullet \mathbf{k}\, dA.$$

On $D$ we have

$$\begin{aligned}\mathbf{curl}\,\mathbf{F}\bullet\mathbf{k} &= \left(\frac{\partial}{\partial x}\left(x^3e^{yz}\right) - \frac{\partial}{\partial y}\left(y^2\cos xz\right)\right)\Bigg|_{z=0}\\ &= 3x^2 - 2y.\end{aligned}$$

By symmetry, $\iint_D y\,dA = 0$, so

$$I = 3\iint_D x^2\,dA = 3\int_0^{2\pi}\cos^2\theta\,d\theta\int_0^2 r^3\,dr = 12\pi.$$

***REMARK*** A surface $\mathcal{S}$ satisfying the conditions of Stokes's Theorem may no longer do so if a single point is removed from it. An isolated boundary point of a surface is not an orientable curve, and Stokes's Theorem may therefore break down for such a surface. Consider, for example, the vector field

$$\mathbf{F} = \frac{\hat{\boldsymbol{\theta}}}{r} = -\frac{y}{x^2+y^2}\mathbf{i} + \frac{x}{x^2+y^2}\mathbf{j},$$

which is defined on the *punctured disk* $D$ satisfying $0 < x^2 + y^2 \le a^2$. (See Example 3 above.) If $D$ is oriented with upward normal $\mathbf{k}$, then its boundary consists of the oriented, smooth, closed curve, $\mathcal{C}$, given by $x = a\cos\theta$, $y = a\sin\theta$, $(0 \le \theta \le 2\pi)$, and the isolated point $(0, 0)$. We have

$$\begin{aligned}\oint_{\mathcal{C}}\mathbf{F}\bullet d\mathbf{r} &= \int_0^{2\pi}\left(\frac{-\sin\theta}{a}\mathbf{i} + \frac{\cos\theta}{a}\mathbf{j}\right)\bullet(-a\sin\theta\mathbf{i} + a\cos\theta\mathbf{j})\,d\theta\\ &= \int_0^{2\pi}(\sin^2\theta + \cos^2\theta)\,d\theta = 2\pi.\end{aligned}$$

However,

$$\mathbf{curl}\,\mathbf{F} = \frac{\partial}{\partial x}\left(\frac{x}{x^2+y^2}\right) - \frac{\partial}{\partial y}\left(-\frac{y}{x^2+y^2}\right) = 0$$

identically on $D$. Thus

$$\iint_D \mathbf{curl}\,\mathbf{F}\bullet\hat{\mathbf{N}}\,dS = 0,$$

and the conclusion of Stokes's Theorem fails in this case.

## EXERCISES 9.4

1. Evaluate $\oint_{\mathcal{C}}(\sin x + 3y^2)\,dx + (2x - e^{-y^2})\,dy$, where $\mathcal{C}$ is the boundary of the half-disk $x^2 + y^2 \le a^2$, $y \ge 0$, oriented counterclockwise.

2. Evaluate $\oint_{\mathcal{C}}(x^2 - xy)\,dx + (xy - y^2)\,dy$ clockwise around the triangle with vertices $(0, 0)$, $(1, 1)$ and $(2, 0)$.

3. Evaluate $\oint_{\mathcal{C}}\left(x\sin(y^2) - y^2\right)dx + \left(x^2y\cos(y^2) + 3x\right)dy$, where $\mathcal{C}$ is the counterclockwise boundary of the trapezoid with vertices $(0, -2)$, $(1, -1)$, $(1, 1)$, and $(0, 2)$.

4. Evaluate $\oint_{\mathcal{C}} x^2y\,dx - xy^2\,dy$, where $\mathcal{C}$ is the clockwise boundary of the region $0 \le y \le \sqrt{9 - x^2}$.

5. Use a line integral to find the plane area enclosed by the curve $\mathbf{r} = a\cos^3 t\,\mathbf{i} + b\sin^3 t\,\mathbf{j}$, $0 \le t \le 2\pi$.

6. Evaluate $\oint_{\mathcal{C}}\mathbf{F}\bullet d\mathbf{r}$, where $\mathbf{F} = ye^x\mathbf{i} + (x + e^x)\mathbf{j} + z^2\mathbf{k}$ and

$\mathcal{C}$ is the curve

$$\mathbf{r} = (1+\cos t)\mathbf{i} + (1+\sin t)\mathbf{j} + (1-\sin t-\cos t)\mathbf{k},$$

where $0 \le t \le 2\pi$.

**7.** Sketch the plane curve $\mathcal{C}$: $\mathbf{r} = \sin t\,\mathbf{i} + \sin 2t\,\mathbf{j}$, $(0 \le t \le 2\pi)$. Evaluate $\oint_{\mathcal{C}} \mathbf{F} \bullet d\mathbf{r}$, where $\mathbf{F} = ye^{x^2}\mathbf{i} + x^3e^y\mathbf{j}$.

**8.** If $\mathcal{C}$ is the positively oriented boundary of a plane region R having area $A$ and centroid $(\bar{x}, \bar{y})$, interpret geometrically the line integral $\oint_{\mathcal{C}} \mathbf{F} \bullet d\mathbf{r}$, where:
(a) $\mathbf{F} = x^2\mathbf{j}$, (b) $\mathbf{F} = xy\mathbf{i}$, and (c) $\mathbf{F} = y^2\mathbf{i} + 3xy\mathbf{j}$.

**9.** Evaluate $\oint_{\mathcal{C}} xy\,dx + yz\,dy + zx\,dz$ around the triangle with vertices $(1, 0, 0)$, $(0, 1, 0)$ and $(0, 0, 1)$, oriented clockwise as seen from the point $(1, 1, 1)$.

**10.** Evaluate $\oint_{\mathcal{C}} y\,dx - x\,dy + z^2\,dz$ around the curve $\mathcal{C}$ of intersection of the cylinders $z = y^2$ and $x^2 + y^2 = 4$, oriented counterclockwise as seen from a point high on the $z$-axis.

**11.** Evaluate $\iint_{\mathcal{S}} \mathbf{curl}\,\mathbf{F} \bullet \hat{\mathbf{N}}\,dS$, where $\mathcal{S}$ is the hemisphere $x^2 + y^2 + z^2 = a^2$, $z \ge 0$ with outward normal, and $\mathbf{F} = 3y\mathbf{i} - 2xz\mathbf{j} + (x^2 - y^2)\mathbf{k}$.

**12.** Evaluate $\iint_{\mathcal{S}} \mathbf{curl}\,\mathbf{F} \bullet \hat{\mathbf{N}}\,dS$, where $\mathcal{S}$ is the surface $x^2 + y^2 + 2(z-1)^2 = 6$, $z \ge 0$, $\hat{\mathbf{N}}$ is the unit outward (away from the origin) normal on $\mathcal{S}$, and

$$\mathbf{F} = (xz - y^3\cos z)\mathbf{i} + x^3e^z\mathbf{j} + xyz\,e^{x^2+y^2+z^2}\mathbf{k}.$$

**13.** Use Stokes's Theorem to show that

$$\oint_{\mathcal{C}} y\,dx + z\,dy + x\,dz = \sqrt{3}\pi a^2,$$

where $\mathcal{C}$ is the suitably oriented intersection of the surfaces $x^2 + y^2 + z^2 = a^2$ and $x + y + z = 0$.

**14.** Evaluate $\oint_{\mathcal{C}} \mathbf{F} \bullet d\mathbf{r}$ around the curve

$$\mathbf{r} = \cos t\,\mathbf{i} + \sin t\,\mathbf{j} + \sin 2t\,\mathbf{k}, \quad (0 \le t \le 2\pi),$$

where

$$\mathbf{F} = (e^x - y^3)\mathbf{i} + (e^y + x^3)\mathbf{j} + e^z\mathbf{k}.$$

*Hint:* show that $\mathcal{C}$ lies on the surface $z = 2xy$.

**15.** Find the circulation of $\mathbf{F} = -y\mathbf{i} + x^2\mathbf{j} + z\mathbf{k}$ around the oriented boundary of the part of the paraboloid $z = 9 - x^2 - y^2$ lying above the $xy$-plane and having normal field pointing upward.

**16.** Evaluate $\oint_{\mathcal{C}} \mathbf{F} \bullet d\mathbf{r}$, where

$$\mathbf{F} = ye^x\mathbf{i} + (x^2 + e^x)\mathbf{j} + z^2e^z\mathbf{k},$$

and $\mathcal{C}$ is the curve

$$\mathbf{r}(t) = (1+\cos t)\mathbf{i} + (1+\sin t)\mathbf{j} + (1-\cos t-\sin t)\mathbf{k}$$

for $0 \le t \le 2\pi$. *Hint:* use Stokes's Theorem, observing that $\mathcal{C}$ lies in a certain plane, and has a circle as its projection onto the $xy$-plane. The integral can also be evaluated by using the techniques of Section 8.4.

**17.** Let $\mathcal{C}_1$ be the straight line joining $(-1, 0, 0)$ to $(1, 0, 0)$ and let $\mathcal{C}_2$ be the semicircle $x^2 + y^2 = 1$, $z = 0$, $y \ge 0$. Let $\mathcal{S}$ be a smooth surface joining $\mathcal{C}_1$ to $\mathcal{C}_2$ having upward normal, and let

$$\mathbf{F} = (\alpha x^2 - z)\mathbf{i} + (xy + y^3 + z)\mathbf{j} + \beta y^2(z+1)\mathbf{k}.$$

Find the values of $\alpha$ and $\beta$ for which $I = \iint_{\mathcal{S}} \mathbf{F} \bullet d\mathbf{S}$ is independent of the choice of $\mathcal{S}$, and find the value of $I$ for these values of $\alpha$ and $\beta$.

**18.** Let $\mathcal{C}$ be the curve $(x-1)^2 + 4y^2 = 16$, $2x + y + z = 3$, oriented counterclockwise when viewed from high on the $z$-axis. Let

$$\mathbf{F} = (z^2 + y^2 + \sin x^2)\mathbf{i} + (2xy + z)\mathbf{j}) + (xz + 2yz)\mathbf{k}.$$

Evaluate $\oint_{\mathcal{C}} \mathbf{F} \bullet d\mathbf{r}$.

**19.** If $\mathcal{C}$ is the oriented boundary of surface $\mathcal{S}$ and $\phi$ and $\psi$ are arbitrary smooth scalar fields, show that

$$\oint_{\mathcal{C}} \phi\nabla\psi \bullet d\mathbf{r} = -\oint_{\mathcal{C}} \psi\nabla\phi \bullet d\mathbf{r} = \iint_{\mathcal{S}} (\nabla\phi \times \nabla\psi) \bullet \hat{\mathbf{N}}\,dS.$$

Is $\nabla\phi \times \nabla\psi$ solenoidal? Find a vector potential for it.

**20.** Let $\mathcal{C}$ be a closed, non-self-intersecting, piecewise smooth plane curve in $\mathbb{R}^3$, which lies in a plane with unit normal $\hat{\mathbf{N}} = a\mathbf{i} + b\mathbf{j} + c\mathbf{k}$, and which has orientation inherited from that of the plane. Show that the plane area enclosed by $\mathcal{C}$ is

$$\frac{1}{2}\oint_{\mathcal{C}} (bz - cy)\,dx + (cx - az)\,dy + (ay - bx)\,dz.$$

**21.** We deduced Green's Theorem from the two-dimensional version of the Divergence Theorem. Reverse the argument and use Green's Theorem to show that

$$\iint_R \operatorname{div} \mathbf{F}\, dA = \oint_C \mathbf{F} \bullet \hat{\mathbf{N}}\, ds,$$

where $C$ is the oriented boundary of the plane domain $R$.

**22.** Use Stokes's Theorem to prove Theorem 2 of Section 9.1.

# 9.5 SOME PHYSICAL APPLICATIONS OF VECTOR CALCULUS

In this section we will show how the theory developed in this chapter can be used to model concrete applied mathematical problems. We will look at two areas of application, fluid dynamics and electromagnetism, and will develop a few of the fundamental vector equations underlying these disciplines. Our purpose is to illustrate the techniques of vector calculus in applied contexts, rather than to provide any complete or even coherent introductions to the disciplines themselves.

## Fluid Dynamics

Suppose that a region of 3-space is filled with a fluid (liquid or gas) in motion. Two approaches can be taken to describe the motion. We could attempt to determine the position, $\mathbf{r} = \mathbf{r}(a, b, c, t)$ at any time $t$, of a "particle" of fluid which was located at the point $(a, b, c)$ at time $t = 0$. This is the Lagrange approach. Alternatively, we could attempt to determine the velocity, $\mathbf{v}(x, y, z, t)$, the density, $\delta(x, y, z, t)$, and other physical variables such as the pressure, $p(x, y, z, t)$, at any time $t$ at any point $(x, y, z)$ in the region occupied by the fluid. This is the Euler approach.

We will examine the latter method, and describe how the Divergence Theorem can be used to translate some fundamental physical laws into equivalent mathematical equations. We assume throughout that the velocity, density, and pressure vary smoothly in all their variables, and that the fluid is an *ideal fluid*, that is, nonviscous (it doesn't stick to itself), homogeneous, and isotropic (it has the same properties at all points and in all directions). Such properties are not always shared by real fluids, and so we are dealing with a simplified mathematical model that does not always correspond exactly to the behaviour of real fluids.

Consider an imaginary closed surface $\mathcal{S}$ in the fluid, bounding a domain $D$. We call $\mathcal{S}$ "imaginary" because it is not a barrier that impedes the flow of the fluid in any way—it is just a means to concentrate our attention on a particular part of the fluid It is fixed in space and does not move with the fluid. Let us assume that the fluid is being neither created nor destroyed anywhere (in particular, there are no sources or sinks), and so the law of **conservation of mass** tells us that the rate of change of the mass of fluid in $D$ equals the rate at which fluid enters $D$ across $\mathcal{S}$.

The mass of fluid in volume element $dV$ at position $(x, y, z)$ at time $t$ is $\delta(x, y, z, t)\, dV$, so the mass in $D$ at time $t$ is $\iiint_D \delta\, dV$. This mass changes at rate

$$\frac{\partial}{\partial t} \iiint_D \delta\, dV = \iiint_D \frac{\partial \delta}{\partial t}\, dV.$$

As we noted in Section 8.6, the volume of fluid passing *out* of $D$ through area element $dS$ at position $(x, y, z)$ in the interval from time $t$ to $t + dt$ is given by $\mathbf{v}(x, y, z, t) \bullet \hat{\mathbf{N}}\, dS\, dt$, where $\hat{\mathbf{N}}$ is the unit normal at $(x, y, z)$ on $\mathcal{S}$ pointing out of $D$. Hence the mass crossing $dS$ outwards in that time interval is $\delta\mathbf{v} \bullet \hat{\mathbf{N}}\, dS\, dt$, and the *rate* at which mass is flowing out of $D$ across $\mathcal{S}$ at time $t$ is

$$\oint\!\!\!\oint_{\mathcal{S}} \delta\mathbf{v} \bullet \hat{\mathbf{N}}\, dS.$$

The rate at which mass is flowing *into* $D$ is the negative of the above rate. Since mass is conserved, we must have

$$\iiint_D \frac{\partial\delta}{\partial t}\, dV = -\oint\!\!\!\oint_{\mathcal{S}} \delta\mathbf{v} \bullet \hat{\mathbf{N}}\, dS = -\iiint_D \mathbf{div}\,(\delta\mathbf{v})\, dV,$$

where we have used the Divergence Theorem to replace the surface integral with a volume integral. Thus

$$\iiint_D \left(\frac{\partial\delta}{\partial t} + \mathbf{div}\,(\delta\mathbf{v})\right) dV = 0.$$

This equation must hold for *any* domain $D$ in the fluid.

If a continuous function $f$ satisfies $\iiint_D f(P)\, dV = 0$ for every domain $D$, then $f(P) = 0$ at all points $P$, for if there were a point $P_0$ such that $f(P_0) \neq 0$, (say $f(P_0) > 0$), then, by continuity, $f$ would be positive at all points in some sufficiently small ball $B$ centred at $P_0$, and $\iiint_B f(P)\, dV$ would be greater than 0. Applying this principle, we must have

$$\frac{\partial\delta}{\partial t} + \mathbf{div}\,(\delta\mathbf{v}) = 0$$

throughout the fluid. This is called the **equation of continuity** for the fluid. It is equivalent to conservation of mass. Observe that if the fluid is **incompressible** then $\delta$ is a constant, independent of both time and spatial position. In this case $\partial\delta/\partial t = 0$ and $\mathbf{div}\,(\delta\mathbf{v}) = \delta\, \mathbf{div}\, \mathbf{v}$. Therefore the equation of continuity for an incompressible fluid is simply

$$\mathbf{div}\, \mathbf{v} = 0.$$

The motion of the fluid is governed by Newton's second law, which asserts that the rate of change of momentum of any part of the fluid is equal to the sum of the forces applied to that part. Again let us consider the part of the fluid in a domain $D$. At any time $t$ its momentum is $\iiint_D \delta\mathbf{v}\, dV$, and is changing at rate

$$\iiint_D \frac{\partial}{\partial t}(\delta\mathbf{v})\, dV.$$

This change is due partly to momentum crossing $\mathcal{S}$ into or out of $D$ (the momentum of the fluid crossing $\mathcal{S}$), partly to the pressure exerted on the fluid in $D$ by the fluid outside, and partly to any external *body forces* (such as gravity or electromagnetic forces) acting on the fluid. Let us examine each of these causes in turn.

Momentum is transferred across $\mathcal{S}$ into $D$ at rate

$$-\oint\!\!\!\oint_{\mathcal{S}} \mathbf{v}(\delta\mathbf{v} \bullet \hat{\mathbf{N}})\, dS.$$

The pressure on the fluid in $D$ is exerted across $\mathcal{S}$ in the direction of the inward normal $-\hat{\mathbf{N}}$. Thus this part of the force on the fluid in $D$ is

$$-\oiint_{\mathcal{S}} p\hat{\mathbf{N}}\, dS.$$

The body forces are best expressed in terms of the *force density* (force per unit mass), $\mathbf{F}$. The total body force on the fluid in $D$ is therefore

$$\iiint_D \delta\mathbf{F}\, dV.$$

Newton's second law now implies that

$$\iiint_D \frac{\partial}{\partial t}(\delta\mathbf{v})\, dV = -\oiint_{\mathcal{S}} \mathbf{v}(\delta\mathbf{v}\bullet\hat{\mathbf{N}})\, dS - \oiint_{\mathcal{S}} p\hat{\mathbf{N}}\, dS + \iiint_D \delta\mathbf{F}\, dV.$$

Again we would like to convert the surface integrals to triple integrals over $D$. If we use the results of Exercise 29 of Section 9.3 and Exercise 2 at the end of this section, we get

$$\oiint_{\mathcal{S}} p\hat{\mathbf{N}}\, dS = \iiint_D \nabla p\, dV,$$
$$\oiint_{\mathcal{S}} \mathbf{v}(\delta\mathbf{v}\bullet\hat{\mathbf{N}})\, dS = \iiint_D \Big(\delta(\mathbf{v}\bullet\nabla)\mathbf{v} + \mathbf{v}\,\mathbf{div}\,(\delta\mathbf{v})\Big)\, dV.$$

Accordingly, we have

$$\iiint_D \left(\delta\frac{\partial\mathbf{v}}{\partial t} + \mathbf{v}\frac{\partial\delta}{\partial t} + \mathbf{v}\,\mathbf{div}\,(\delta\mathbf{v}) + \delta(\mathbf{v}\bullet\nabla)\mathbf{v} + \nabla p - \delta\mathbf{F}\right) dV = 0.$$

The second and third terms in the integrand cancel out by virtue of the continuity equation. Since $D$ is arbitrary, we must therefore have

$$\delta\frac{\partial\mathbf{v}}{\partial t} + \delta(\mathbf{v}\bullet\nabla)\mathbf{v} = -\nabla p + \delta\mathbf{F}.$$

This is the **equation of motion** of the fluid. Observe that it is not a *linear* partial differential equation; the second term on the left is not linear in $\mathbf{v}$.

## Electrostatics

In 3-space there are defined two vector fields which determine the electric and magnetic forces that would be experienced by a unit charge at a particular point if it is moving with unit speed. (These vector fields are determined by electric charges and currents present in the space.) A charge $q_0$ at position $\mathbf{r} = x\mathbf{i} + y\mathbf{j} + z\mathbf{k}$ moving with velocity $\mathbf{v}_0$ experiences an electric force $q_0\mathbf{E}(\mathbf{r})$, where $\mathbf{E}$ is the **electric field**, and a magnetic force $\mu q_0\mathbf{v}_0\times\mathbf{H}(\mathbf{r})$, where $\mathbf{H}$ is the **magnetic field** and $\mu$ is a physical constant. We will look briefly at each of these fields, but will restrict ourselves to considering *static* situations, electric fields produced by static charge distributions, and magnetic fields produced by constant electric currents. Let us begin with the former.

Experimental evidence shows that the value of the electric field at any point $\mathbf{r}$ is the vector sum of the fields caused by any elements of charge located in 3-space. A "point charge" $q$ at position $\mathbf{s} = \xi\mathbf{i} + \eta\mathbf{j} + \zeta\mathbf{k}$ generates the electric field

$$\mathbf{E}(\mathbf{r}) = \frac{kq}{4\pi}\frac{\mathbf{r}-\mathbf{s}}{|\mathbf{r}-\mathbf{s}|^3}, \qquad \text{(Coulomb's Law)},$$

where $k$ is a physical constant. This is just the field due to a point source of strength $kq/4\pi$ at $\mathbf{s}$. Except at $\mathbf{r} = \mathbf{s}$ the field is conservative, with potential

$$\phi(\mathbf{r}) = -\frac{kq}{4\pi}\frac{1}{|\mathbf{r}-\mathbf{s}|},$$

so for $\mathbf{r} \neq \mathbf{s}$ we have $\mathbf{curl}\,\mathbf{E} = \mathbf{0}$. Also $\mathbf{div}\,\mathbf{E} = 0$, except at $\mathbf{r} = \mathbf{s}$ where it is infinite; in terms of the Dirac distribution, $\mathbf{div}\,\mathbf{E} = kq\delta(x-\xi)\delta(y-\eta)\delta(z-\zeta)$. (See Section 9.1.) The flux of $\mathbf{E}$ outward across the surface $\mathcal{S}$ of any region $R$ containing $q$ is

$$\oiint_{\mathcal{S}} \mathbf{E} \bullet \hat{\mathbf{N}}\, dS = kq,$$

by analogy with Example 4 of Section 9.3.

Given a *charge distribution* of density $\rho(\xi, \eta, \zeta)$ in 3-space (so that the charge in volume element $dV = d\xi\, d\eta\, d\zeta$ at $\mathbf{s}$ is $dq = \rho\, dV$), the flux of $\mathbf{E}$ out of $\mathcal{S}$ due to the charge in $R$ is

$$\oiint_{\mathcal{S}} \mathbf{E} \bullet \hat{\mathbf{N}}\, dS = k \iiint_R dq = k \iiint_R \rho\, dV.$$

If we apply the Divergence Theorem to the surface integral, we obtain

$$\iiint_R (\mathbf{div}\,\mathbf{E} - k\rho)\, dV = 0,$$

and since R is an arbitrary region,

$$\mathbf{div}\,\mathbf{E} = k\rho.$$

This is one of the fundamental equations of electromagnetic theory.

The potential due to a charge distribution of density $\rho(\mathbf{s})$ in the region R is

$$\begin{aligned}\phi(\mathbf{r}) &= -\frac{k}{4\pi}\iiint_R \frac{dq}{|\mathbf{r}-\mathbf{s}|}\, dV \\ &= -\frac{k}{4\pi}\iiint_R \frac{\rho(\xi,\eta,\zeta)\, d\xi\, d\eta\, d\zeta}{\sqrt{(x-\xi)^2+(y-\eta)^2+(z-\zeta)^2}}.\end{aligned}$$

If $\rho$ is continuous and vanishes outside a bounded region, the triple integral is convergent everywhere, so $\mathbf{E} = \nabla\phi$ is conservative throughout 3-space. Thus, at all points,

$$\mathbf{curl}\,\mathbf{E} = \mathbf{0}.$$

Since $\mathbf{div}\,\mathbf{E} = \mathbf{div}\,\nabla\phi = \nabla^2\phi$, the potential $\phi$ satisfies **Poisson's equation**

$$\nabla^2\phi = k\rho.$$

In particular, $\phi$ is a harmonic function in regions of space where no charge is distributed.

## Magnetostatics

Magnetic fields are produced by moving charges, that is, by currents. Suppose that a constant electric current, $I$, is flowing in a filament along the curve $\mathcal{F}$. It has been determined experimentally that the magnetic field produced at position $\mathbf{r} = x\mathbf{i} + y\mathbf{j} + z\mathbf{k}$ by the elements of current $dI = I\,ds$ along the filament add vectorially, and that the element at position $\mathbf{s} = \xi\mathbf{i} + \eta\mathbf{j} + \zeta\mathbf{k}$ produces the field

$$d\mathbf{H}(\mathbf{r}) = \frac{I}{4\pi}\frac{d\mathbf{s}\times(\mathbf{r}-\mathbf{s})}{|\mathbf{r}-\mathbf{s}|^3}, \qquad \text{(the Biot–Savart Law)},$$

where $d\mathbf{s} = \hat{\mathbf{T}}\,ds$, $\hat{\mathbf{T}}$ being the unit tangent to $\mathcal{F}$ in the direction of the current. Under the reasonable assumption that charge is not created or destroyed anywhere, the filament $\mathcal{F}$ must form a closed circuit, and the total magnetic field at $\mathbf{r}$ due to the current flowing in the circuit is

$$\mathbf{H} = \frac{I}{4\pi}\oint_{\mathcal{F}}\frac{d\mathbf{s}\times(\mathbf{r}-\mathbf{s})}{|\mathbf{r}-\mathbf{s}|^3}.$$

Let $\mathbf{A}$ be a vector field be defined by

$$\mathbf{A}(\mathbf{r}) = \frac{I}{4\pi}\oint_{\mathcal{F}}\frac{d\mathbf{s}}{|\mathbf{r}-\mathbf{s}|},$$

for all $\mathbf{r}$ not on the filament $\mathcal{F}$. If we make use of the fact that

$$\nabla\left(\frac{1}{|\mathbf{r}-\mathbf{s}|}\right) = -\frac{\mathbf{r}-\mathbf{s}}{|\mathbf{r}-\mathbf{s}|^3},$$

and the vector identity $\nabla\times(\phi\mathbf{F}) = (\nabla\phi)\times\mathbf{F} + \phi(\nabla\times\mathbf{F})$, (with $\mathbf{F}$ the vector $d\mathbf{s}$, which does not depend on $\mathbf{r}$), we can calculate the curl of $\mathbf{A}$:

$$\nabla\times\mathbf{A} = \frac{I}{4\pi}\oint_{\mathcal{F}}\nabla\left(\frac{1}{|\mathbf{r}-\mathbf{s}|}\right)\times d\mathbf{s} = \frac{I}{4\pi}\oint_{\mathcal{F}}-\frac{\mathbf{r}-\mathbf{s}}{|\mathbf{r}-\mathbf{s}|^3}\times d\mathbf{s} = \mathbf{H}(\mathbf{r}).$$

Thus $\mathbf{A}$ is a vector potential for $\mathbf{H}$, and $\mathbf{div}\,\mathbf{H} = 0$ at points off the filament. We can also verify by calculation that $\mathbf{curl}\,\mathbf{H} = \mathbf{0}$ off the filament. (See Exercises 7–9 at the end of this section.)

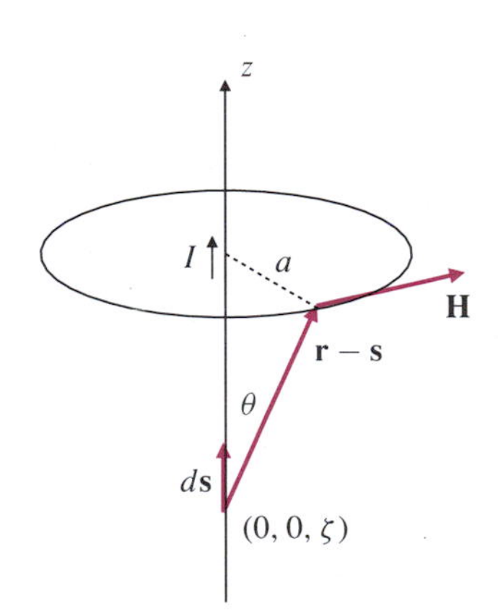


Figure 9.16

Imagine a circuit consisting of a straight filament along the $z$-axis with return at infinite distance. The field $\mathbf{H}$ at a finite point will then just be due to the current along the $z$-axis, where the current $I$ is flowing in the direction of $\mathbf{k}$, say. The currents in all elements $d\mathbf{s}$ produce at $\mathbf{r}$ fields in the same direction, normal to the plane containing $\mathbf{r}$ and the $z$-axis. (See Figure 9.16.) Therefore the field strength $H = |\mathbf{H}|$ at a distance $a$ from the $z$-axis is obtained by integrating the elements

$$dH = \frac{I}{4\pi}\frac{\sin\theta\,d\zeta}{a^2+(\zeta-z)^2} = \frac{I}{4\pi}\frac{a\,d\zeta}{\left(a^2+(\zeta-z)^2\right)^{3/2}}.$$

We have

$$\begin{aligned} H &= \frac{Ia}{4\pi}\int_{-\infty}^{\infty}\frac{d\zeta}{\left(a^2+(\zeta-z)^2\right)^{3/2}} \qquad \text{(Let } \zeta - z = a\tan\phi.\text{)} \\ &= \frac{I}{4\pi a}\int_{-\pi/2}^{\pi/2}\cos\phi\,d\phi = \frac{I}{2\pi a}. \end{aligned}$$

The field lines of **H** are evidently horizontal circles centred on the $z$-axis. If $\mathcal{C}_a$ is such a circle, having radius $a$, then the circulation of **H** around $\mathcal{C}_a$ is

$$\oint_{\mathcal{C}_a} \mathbf{H} \bullet d\mathbf{r} = \frac{I}{2\pi a} 2\pi a = I.$$

Observe that the circulation calculated above is independent of $a$. In fact, if $\mathcal{C}$ is any closed curve that encircles the $z$-axis once counterclockwise (as seen from above), then $\mathcal{C}$ and $-\mathcal{C}_a$ comprise the oriented boundary of a washer-like surface $\mathcal{S}$ with a hole in it through which the filament passes. Since $\mathbf{curl\,H} = \mathbf{0}$ on $\mathcal{S}$, Stokes's Theorem guarantees that

$$\oint_{\mathcal{C}} \mathbf{H} \bullet d\mathbf{r} = \oint_{\mathcal{C}_a} \mathbf{H} \bullet d\mathbf{r} = I.$$

Furthermore, when $\mathcal{C}$ is very small (and therefore very close to the filament) most of the contribution to the circulation of **H** around it comes from that part of the filament which is very close to $\mathcal{C}$. It therefore does not matter whether the filament is straight, or infinitely long. For any closed-loop filament carrying a current, the circulation of the magnetic field around the oriented boundary of a surface through which the filament passes is equal to the current flowing in the loop. This is **Ampère's circuital law**. The surface is oriented with normal on the side out of which the current is flowing.

Now let us replace the filament with a more general current specified by a vector density, **J**. This means that at any point **s** the current is flowing in the direction $\mathbf{J}(\mathbf{s})$, and that the current crossing an area element $dS$ with unit normal $\hat{\mathbf{N}}$ is $\mathbf{J} \bullet \hat{\mathbf{N}}\, dS$. The circulation of **H** around the boundary $\mathcal{C}$ of surface $\mathcal{S}$ is equal to the total current flowing across $\mathcal{S}$, so

$$\oint_{\mathcal{C}} \mathbf{H} \bullet d\mathbf{r} = \iint_{\mathcal{S}} \mathbf{J} \bullet \hat{\mathbf{N}}\, dS.$$

By using Stokes's Theorem, we can replace the line integral with another surface integral, and so obtain

$$\iint_{\mathcal{S}} (\mathbf{curl\,H} - \mathbf{J}) \bullet \hat{\mathbf{N}}\, dS = 0.$$

Since $\mathcal{S}$ is arbitrary, we must have, at all points,

$$\mathbf{curl\,H} = \mathbf{J},$$

which is the pointwise version of Ampère's circuital law. It can be readily checked that

$$\mathbf{A}(\mathbf{r}) = \frac{1}{4\pi} \iiint_R \frac{\mathbf{J}(\mathbf{s})}{|\mathbf{r} - \mathbf{s}|}\, dV,$$

is a vector potential for the magnetic field **H**. Here $R$ is the region of 3-space where **J** is nonzero. If **J** is continuous and vanishes outside a bounded set, then the triple integral converges for all **r**, and **H** is everywhere solenoidal:

$$\mathbf{div\,H} = 0.$$

**REMARK** Even in more general situations we would expect to have $\mathbf{div\,H} = 0$. This corresponds to the fact that the field lines of **H** are always closed curves—there are no *known* magnetic *sources* or *sinks* (that is, magnetic *monopoles*).

## EXERCISES 9.5

1. **(Gauss's Law)** Show that the flux of the electric field $\mathbf{E}$ outward through a closed surface $\mathcal{S}$ in 3-space is $k$ times the total charge enclosed by $\mathcal{S}$.

2. By breaking the vector $\mathbf{F}(\mathbf{G} \bullet \hat{\mathbf{N}})$ into its separate components and applying the Divergence Theorem to each separately, show that

$$\oiint_{\mathcal{S}} \mathbf{F}(\mathbf{G} \bullet \hat{\mathbf{N}})\, dS = \iiint_D \Big(\mathbf{F}\,\mathbf{div}\,\mathbf{G} + (\mathbf{G} \bullet \nabla)\mathbf{F}\Big)\, dV,$$

where $\hat{\mathbf{N}}$ is the unit outward normal on the surface $\mathcal{S}$ of the domain $D$.

3. The electric charge density, $\rho$, in 3-space depends on time as well as position if charge is moving around. The motion is described by the current density, $\mathbf{J}$. Derive the **continuity equation**

$$\frac{\partial \rho}{\partial t} = -\mathbf{div}\,\mathbf{J},$$

from the fact that charge is conserved.

4. If $\mathbf{b}$ is a constant vector, show that

$$\nabla\left(\frac{1}{|\mathbf{r} - \mathbf{b}|}\right) = -\frac{\mathbf{r} - \mathbf{b}}{|\mathbf{r} - \mathbf{b}|^3}.$$

5. If $\mathbf{a}$ and $\mathbf{b}$ are constant vectors, show that for $\mathbf{r} \neq \mathbf{b}$,

$$\mathbf{div}\left(\mathbf{a} \times \frac{\mathbf{r} - \mathbf{b}}{|\mathbf{r} - \mathbf{b}|^3}\right) = 0.$$

*Hint:* use identities (d) and (h) from Theorem 3 of Section 9.2.

6. Use the result of Exercise 5 to give an alternative proof that

$$\mathbf{div} \oint_{\mathcal{F}} \frac{d\mathbf{s} \times (\mathbf{r} - \mathbf{s})}{|\mathbf{r} - \mathbf{s}|^3} = 0.$$

Note that **div** refers to the $\mathbf{r}$ variable.

7. If $\mathbf{a}$ and $\mathbf{b}$ are constant vectors, show that for $\mathbf{r} \neq \mathbf{b}$,

$$\mathbf{curl}\left(\mathbf{a} \times \frac{\mathbf{r} - \mathbf{b}}{|\mathbf{r} - \mathbf{b}|^3}\right) = -(\mathbf{a} \bullet \nabla)\frac{\mathbf{r} - \mathbf{b}}{|\mathbf{r} - \mathbf{b}|^3}.$$

*Hint:* use identity (e) from Theorem 3 of Section 9.2.

8. If $\mathbf{F}$ is any smooth vector field, show that

$$\oint_{\mathcal{F}} (d\mathbf{s} \bullet \nabla)\mathbf{F}(\mathbf{s}) = \mathbf{0}$$

around any closed loop $\mathcal{F}$. *Hint:* the gradients of the components of $\mathbf{F}$ are conservative.

9. Verify that if $\mathbf{r}$ does not lie on $\mathcal{F}$, then

$$\mathbf{curl} \oint_{\mathcal{F}} \frac{d\mathbf{s} \times (\mathbf{r} - \mathbf{s})}{|\mathbf{r} - \mathbf{s}|^3} = \mathbf{0}.$$

Here **curl** is taken with respect to the $\mathbf{r}$ variable.

10. Verify the formula $\mathbf{curl}\,\mathbf{A} = \mathbf{H}$, where $\mathbf{A}$ is the magnetic vector potential defined in terms of the steady state current density $\mathbf{J}$.

11. If $\mathbf{A}$ is the vector potential for the magnetic field produced by a steady current in a closed-loop filament, show that $\mathbf{div}\,\mathbf{A} = 0$ off the filament.

12. If $\mathbf{A}$ is the vector potential for the magnetic field produced by a steady, continuous current density, show that $\mathbf{div}\,\mathbf{A} = 0$ everywhere. Hence show that $\mathbf{A}$ satisfies the vector Poisson equation: $\nabla^2 \mathbf{A} = -\mathbf{J}$.

* 13. **(Archimedes' principle)** A solid occupying region $R$ with surface $\mathcal{S}$ is immersed in a liquid of constant density $\delta$. The pressure at depth $h$ in the liquid is $\delta g h$, so the pressure satisfies $\nabla p = \delta \mathbf{g}$, where $\mathbf{g}$ is the (vector) constant acceleration of gravity. Over each surface element $dS$ on $\mathcal{S}$ the pressure of the fluid exerts a force $-p\hat{\mathbf{N}}\, dS$ on the solid. Show that the resultant "buoyancy force" on the solid is

$$\mathbf{B} = -\iiint_R \delta \mathbf{g}\, dV.$$

Thus the buoyancy force has the same magnitude as, and opposite direction to, the weight of the liquid displaced by the solid. This is Archimedes' principle.

* 14. **(Archimedes' principle (continued))** Extend the result of Exercise 13 to the case where the solid is only partly submerged in the fluid.

* 15. The heat content of a volume element $dV$ within a homogeneous solid is $\delta c T\, dV$, where $\delta$ and $c$ are constants (the density and specific heat of the solid material) and $T = T(x, y, z, t)$ is the temperature at time $t$ at position $(x, y, z)$ in the solid. Heat always flows in the direction of the negative temperature gradient, and at a rate proportional to the size of that gradient. Thus the rate of flow of heat energy across a surface element $dS$ with normal $\hat{\mathbf{N}}$ is $-k\nabla T \bullet \hat{\mathbf{N}}\, dS$, where $k$ is also a constant depending on the material of the solid (the coefficient of thermal conductivity). Use "conservation of heat energy" to show that for any region $R$ with surface $\mathcal{S}$ within the solid

$$\delta c \iiint_R \frac{\partial T}{\partial t}\, dV = k \oiint_{\mathcal{S}} \nabla T \bullet \hat{\mathbf{N}}\, dS,$$

where $\hat{\mathbf{N}}$ is the unit outward normal on $\mathcal{S}$. Hence show that heat flow within the solid is governed by the partial differential equation

$$\frac{\partial T}{\partial t} = \frac{k}{\delta c}\nabla^2 T = \frac{k}{\delta c}\left(\frac{\partial^2 T}{\partial x^2} + \frac{\partial^2 T}{\partial y^2} + \frac{\partial^2 T}{\partial z^2}\right).$$

## 9.6 ORTHOGONAL CURVILINEAR COORDINATES

In this optional section we will derive formulas for the gradient of a scalar field and the divergence and curl of a vector field in terms of coordinate systems more general than the Cartesian coordinate system used in the earlier sections of this chapter. In particular, we will express these quantities in terms of the cylindrical and spherical coordinate systems introduced in Section 6.6.

We denote by $xyz$-space, the usual system of Cartesian coordinates $(x, y, z)$ in $\mathbb{R}^3$. A different system of coordinates $[u, v, w]$ in $xyz$-space can be defined by a continuous transformation of the form

$$x = x(u, v, w), \qquad y = y(u, v, w), \qquad z = z(u, v, w).$$

If the transformation is one-to-one from a region $D$ in $uvw$-space onto a region $R$ in $xyz$-space, then a point in $R$ can be represented by a triple $[u, v, w]$, the (Cartiesian) coordinates of the unique point $Q$ in $uvw$-space which the transformation maps to $P$. In this case we say that the transformation defines a **curvilinear coordinate system** in $R$, and call $[u, v, w]$ the **curvilinear coordinates** of $P$ with respect to that system. Note that $[u, v, w]$ are Cartesian coordinates in their own space ($uvw$-space); they are curvilinear coordinates in $xyz$-space.

Typically, we relax the requirement that the transformation defining a curvilinear coordinate system be one-to-one, that is, that every point $P$ in $R$ should have a unique set of curvilinear coordinates. It is reasonable to require the transformation to be only *locally one-to-one*. Thus there may be more than one point $Q$ that gets mapped to a point $P$ by the transformation, but only one in any suitably small subregion of $D$. For example, in the plane polar coordinate system

$$x = r\cos\theta, \qquad y = r\sin\theta,$$

the transformation is locally one-to-one from $D$, the half of the $r\theta$-plane where $0 < r < \infty$, to the region $R$ consisting of all points in the $xy$-plane except the origin. Although, say, $[1, 0]$ and $[1, 2\pi]$ are polar coordinates of the same point in the $xy$-plane, they are not close together in $D$. Observe, however, that there is still a problem with the origin, which can be represented by $[0, \theta]$ for *any* $\theta$. Since the transformation is not even locally one-to-one at $r = 0$, we regard the origin of the $xy$-plane as a **singular point** for the polar coordinate system in the plane.

**EXAMPLE 1** The **cylindrical coordinate system** $[r, \theta, z]$ in $\mathbb{R}^3$ is defined by the transformation

$$x = r\cos\theta, \qquad y = r\sin\theta, \qquad z = z,$$

where $r \ge 0$. (See Section 6.6.) This transformation maps the half-space $D$ given by $r > 0$ onto all of $xyz$-space excluding the $z$-axis, and is locally one-to-one. We regard $[r, \theta, z]$ as cylindrical polar coordinates in all of $xyz$-space but regard points on the $z$-axis as singular points of the system since the points $[0, \theta, z]$ are identical for any $\theta$. ■

**EXAMPLE 2** The **spherical coordinate system** $[\rho, \phi, \theta]$ is defined by the transformation

$$x = \rho\sin\phi\cos\theta, \qquad y = \rho\sin\phi\sin\theta, \qquad z = \rho\cos\phi,$$

where $\rho \geq 0$ and $0 \leq \phi \leq \pi$. (See Section 6.6.) The transformation maps the region $D$ in $\rho\phi\theta$-space given by $\rho > 0$, $0 < \phi < \pi$ in a locally one-to-one way onto $xyz$-space excluding the $z$-axis. The point with Cartesian coordinates $(0, 0, z)$ can be represented by the spherical coordinates $[0, \phi, \theta]$ for arbitrary $\phi$ and $\theta$ if $z = 0$, by $[z, 0, \theta]$ for arbitrary $\theta$ if $z > 0$, and by $[|z|, \pi, \theta]$ for arbitrary $\theta$ if $z < 0$. Thus all points of the $z$-axis are singular for the spherical coordinate system. ■

## Coordinate Surfaces and Coordinate Curves

Let $[u, v, w]$ be a curvilinear coordinate system in $xyz$-space, and let $P_0$ be a nonsingular point for the system. Thus the transformation

$$x = x(u, v, w), \qquad y = y(u, v, w), \qquad z = z(u, v, w)$$

is locally one-to-one near $P_0$. Let $P_0$ have curvilinear coordinates $[u_0, v_0, w_0]$. The plane with equation $u = u_0$ in $uvw$-space gets mapped by the transformation to a surface in $xyz$-space passing through $P_0$. We call this surface a $u$-surface and still refer to it by the equation $u = u_0$; it has parametric equations

$$x = x(u_0, v, w), \qquad y = y(u_0, v, w), \qquad z = z(u_0, v, w)$$

with parameters $v$ and $w$. Similarly, the $v$-surface $v = v_0$ and the $w$-surface $w = w_0$ pass through $P_0$; they are the images of the planes $v = v_0$ and $w = w_0$ in $uvw$-space.

**Orthogonal curvilinear coordinates**

We say that $[u, v, w]$ is an **orthogonal curvilinear coordinate system** in $xyz$-space if, for every nonsingular point $P_0$ in $xyz$-space, the three **coordinate surfaces**, $u = u_0$, $v = v_0$, and $w = w_0$, intersect at $P_0$ at mutually right angles.

It is tacitly assumed that the coordinate surfaces are smooth at all nonsingular points, so we are really assuming that their normal vectors are mutually perpendicular. Figure 9.17 shows the coordinate surfaces through $P_0$ for a typical orthogonal curvilinear coordinate system.

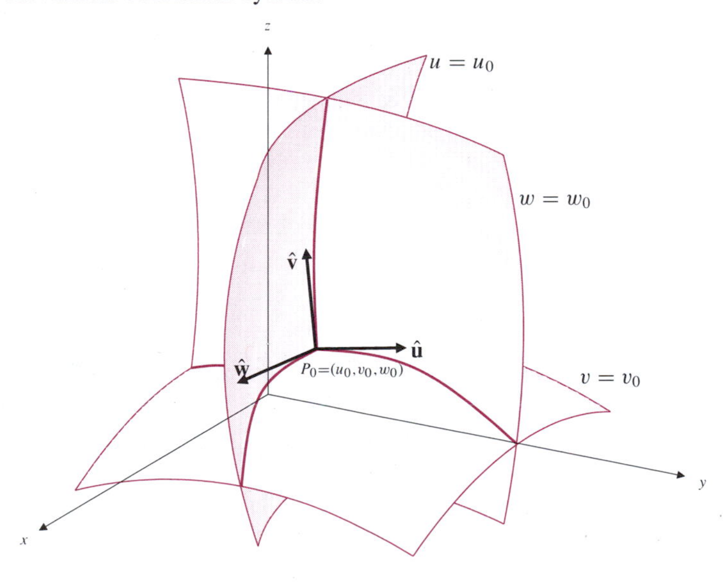


**Figure 9.17** $u$-, $v$-, and $w$-coordinate surfaces

Pairs of coordinate surfaces through a point intersect along a **coordinate curve** through that point. For example, the coordinate surfaces $v = v_0$ and $w = w_0$ intersect along the $u$-**curve** with parametric equations

$$x = x(u, v_0, w_0), \quad y = y(u, v_0, w_0), \quad \text{and} \quad z = z(u, v_0, w_0),$$

where the parameter is $u$. A unit vector $\hat{\mathbf{u}}$ tangent to the $u$-curve through $P_0$ is normal to the coordinate surface $u = u_0$ there. Similar statements hold for unit vectors $\hat{\mathbf{v}}$ and $\hat{\mathbf{w}}$. For an orthogonal curvilinear coordinate system, the three vectors $\hat{\mathbf{u}}$, $\hat{\mathbf{v}}$, and $\hat{\mathbf{w}}$, form a basis mutually perpendicular unit vectors at any nonsingular point $P$. (See Figure 9.17.) We call this basis the **local basis** at $P$.

**EXAMPLE 3** For the cylindrical coordinate system, (Figure 9.18), the coordinate surfaces are:

| | |
|---|---|
| circular cylinders with axis along the z-axis | ($r$-surfaces), |
| vertical half-planes radiating from the $z$-axis | ($\theta$-surfaces), |
| horizontal planes | ($z$-surfaces). |

The coordinate curves are:

| | |
|---|---|
| horizontal straight half-lines radiating from the $z$-axis | ($r$-curves), |
| horizontal circles with centres on the $z$-axis | ($\theta$-curves), |
| vertical straight lines | ($z$-curves). ■ |

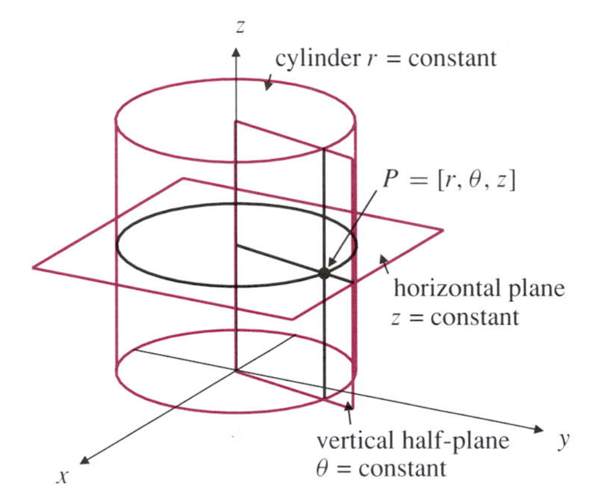


**Figure 9.18** The coordinate surfaces for cylindrical coordinates

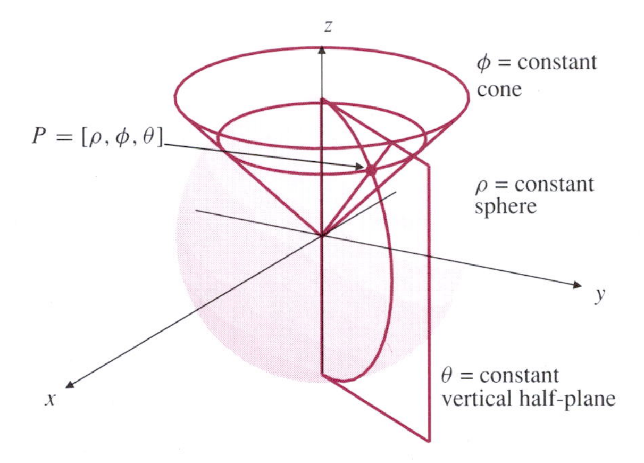


**Figure 9.19** The coordinate surfaces for spherical coordinates

**EXAMPLE 4** For the spherical coordinate system, (see Figure 9.19), the coordinate surfaces are:

| | |
|---|---|
| spheres centred at the origin | ($\rho$-surfaces), |
| vertical circular cones with vertices at the origin | ($\phi$-surfaces), |
| vertical half-planes radiating from the $z$-axis | ($\theta$-surfaces). |

The coordinate curves are:

| | |
|---|---|
| half-lines radiating from the origin | ($\rho$-curves), |
| vertical semi-circles with centres at the origin | ($\phi$-curves), |
| horizontal circles with centres on the $z$-axis | ($\theta$-curves). ■ |

## Scale Factors and Differential Elements

For the rest of this section we assume that $[u, v, w]$ are **orthogonal** curvilinear coordinates in $xyz$-space defined via the transformation

$$x = x(u, v, w), \qquad y = y(u, v, w), \qquad z = z(u, v, w).$$

We also assume that the the coordinate surfaces are smooth at any nonsingular point, and that the local basis vectors $\hat{\mathbf{u}}$, $\hat{\mathbf{v}}$, and $\hat{\mathbf{w}}$ at any such point form a right-handed triad. This is the case for both cylindrical and spherical coordinates. For spherical coordinates, this is the reason we chose the order of the coordinates as $[\rho, \phi, \theta]$ rather than $[\rho, \theta, \phi]$.

The **position vector** of a point $P$ in $xyz$-space can be expressed in terms of the curvilinear coordinates:

$$\mathbf{r} = x(u, v, w)\mathbf{i} + y(u, v, w)\mathbf{j} + z(u, v, w)\mathbf{k}.$$

If we hold $v = v_0$ and $w = w_0$ fixed and let $u$ vary, then $\mathbf{r} = \mathbf{r}(u, v_0, w_0)$ defines a $u$-curve in $xyz$-space. At any point $P$ on this curve, the vector

$$\frac{\partial \mathbf{r}}{\partial u} = \frac{\partial x}{\partial u}\mathbf{i} + \frac{\partial y}{\partial u}\mathbf{j} + \frac{\partial z}{\partial u}\mathbf{k}$$

is tangent to the $u$-curve at $P$. In general, the three vectors

$$\frac{\partial \mathbf{r}}{\partial u}, \qquad \frac{\partial \mathbf{r}}{\partial v}, \qquad \text{and} \qquad \frac{\partial \mathbf{r}}{\partial w}$$

are tangent, respectively, to the the $u$-curve, the $v$-curve, and the $w$-curve through $P$. They are also normal, respectively, to the $u$-surface, the $v$-surface, and the $w$-surface through $P$, so they are mutually perpendicular. (See Figure 9.17.) The lengths of these tangent vectors are called the **scale factors** of the coordinate system.

The **scale factors** of the orthogonal curvilinear coordinate system $[u, v, w]$ are the three functions

$$h_u = \left|\frac{\partial \mathbf{r}}{\partial u}\right|, \qquad h_v = \left|\frac{\partial \mathbf{r}}{\partial v}\right|, \qquad h_w = \left|\frac{\partial \mathbf{r}}{\partial w}\right|.$$

The scale factors are nonzero at a nonsingular point $P$ of the coordinate system, so the local basis at $P$ can be obtained by dividing the tangent vectors to the coordinate curves by their lengths. As noted previously, we denote the local basis vectors by $\hat{\mathbf{u}}$, $\hat{\mathbf{v}}$, and $\hat{\mathbf{w}}$. Thus

$$\frac{\partial \mathbf{r}}{\partial u} = h_u\hat{\mathbf{u}}, \qquad \frac{\partial \mathbf{r}}{\partial v} = h_v\hat{\mathbf{v}}, \qquad \text{and} \qquad \frac{\partial \mathbf{r}}{\partial w} = h_w\hat{\mathbf{w}}.$$

The basis vectors $\hat{\mathbf{u}}$, $\hat{\mathbf{v}}$, and $\hat{\mathbf{w}}$ will form a right-handed triad provided we have chosen a suitable order for the coordinates $u$, $v$, and $w$.

**EXAMPLE 5** For cylindrical coordinates we have $\mathbf{r} = r\cos\theta\mathbf{i} + r\sin\theta\mathbf{j} + z\mathbf{k}$, so

$$\frac{\partial \mathbf{r}}{\partial r} = \cos\theta\,\mathbf{i} + \sin\theta\,\mathbf{j}, \qquad \frac{\partial \mathbf{r}}{\partial \theta} = -r\sin\theta\,\mathbf{i} + r\cos\theta\,\mathbf{j}, \qquad \text{and} \qquad \frac{\partial \mathbf{r}}{\partial z} = \mathbf{k}.$$

Thus the scale factors for the cylindrical coordinate system are given by

$$h_r = \left|\frac{\partial \mathbf{r}}{\partial r}\right| = 1, \qquad h_\theta = \left|\frac{\partial \mathbf{r}}{\partial \theta}\right| = r, \qquad \text{and} \qquad h_z = \left|\frac{\partial \mathbf{r}}{\partial z}\right| = 1,$$

and the local basis consists of the vectors

$$\hat{\mathbf{r}} = \cos\theta\,\mathbf{i} + \sin\theta\,\mathbf{j}, \qquad \hat{\boldsymbol{\theta}} = -\sin\theta\,\mathbf{i} + \cos\theta\,\mathbf{j}, \qquad \hat{\mathbf{z}} = \mathbf{k}.$$

See Figure 9.20. The local basis is right-handed. ■

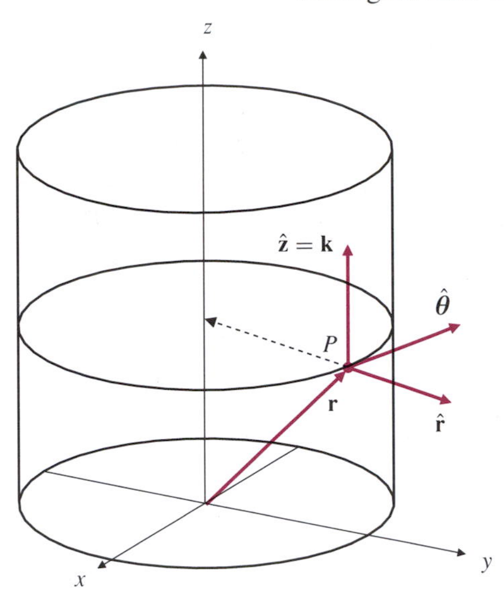


**Figure 9.20** The local basis for cylindrical coordinates

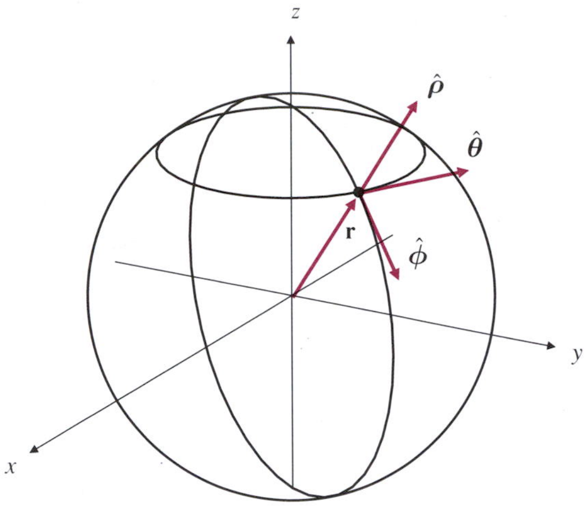

**Figure 9.21** The local basis for spherical coordinates

■ **EXAMPLE 6** For spherical coordinates we have

$$\mathbf{r} = \rho\sin\phi\cos\theta\mathbf{i} + \rho\sin\phi\sin\theta\mathbf{j} + \rho\cos\phi\mathbf{k}.$$

Thus the tangent vectors to the coordinate curves are

$$\frac{\partial\mathbf{r}}{\partial\rho} = \sin\phi\cos\theta\,\mathbf{i} + \sin\phi\sin\theta\,\mathbf{j} + \cos\phi\,\mathbf{k},$$

$$\frac{\partial\mathbf{r}}{\partial\phi} = \rho\cos\phi\cos\theta\,\mathbf{i} + \rho\cos\phi\sin\theta\,\mathbf{j} - \rho\sin\phi\,\mathbf{k},$$

$$\frac{\partial\mathbf{r}}{\partial\theta} = -\rho\sin\phi\sin\theta\,\mathbf{i} + \rho\sin\phi\cos\theta\,\mathbf{j},$$

and the scale factors are given by

$$h_\rho = \left|\frac{\partial\mathbf{r}}{\partial\rho}\right| = 1, \quad h_\phi = \left|\frac{\partial\mathbf{r}}{\partial\phi}\right| = \rho, \quad \text{and} \quad h_\theta = \left|\frac{\partial\mathbf{r}}{\partial\theta}\right| = \rho\sin\phi.$$

The local basis consists of the vectors

$$\hat{\boldsymbol{\rho}} = \sin\phi\cos\theta\,\mathbf{i} + \sin\phi\sin\theta\,\mathbf{j} + \cos\phi\,\mathbf{k}$$

$$\hat{\boldsymbol{\phi}} = \cos\phi\cos\theta\,\mathbf{i} + \cos\phi\sin\theta\,\mathbf{j} - \sin\phi\,\mathbf{k}$$

$$\hat{\boldsymbol{\theta}} = -\sin\theta\,\mathbf{i} + \cos\theta\,\mathbf{j}.$$

See Figure 9.21. The local basis is right-handed. ■

The volume element in an orthogonal curvilinear coordinate system is the volume of an infinitesimal *coordinate box* bounded by pairs of $u$-, $v$-, and $w$-surfaces corresponding to values $u$ and $u+du$, $v$ and $v+dv$, and $w$ and $w+dw$, respectively. See Figure 9.22. Since these coordinate surfaces are assumed smooth, and since they intersect at right angles, the coordinate box is rectangular, and is spanned by the vectors

$$\frac{\partial\mathbf{r}}{\partial u}\,du = h_u\,du\,\hat{\mathbf{u}}, \quad \frac{\partial\mathbf{r}}{\partial v}\,dv = h_v\,dv\,\hat{\mathbf{v}}, \quad \text{and} \quad \frac{\partial\mathbf{r}}{\partial w}\,dw = h_w\,dw\,\hat{\mathbf{w}}.$$

Therefore the volume element is given by

$$dV = h_u h_v h_w \, du \, dv \, dw.$$

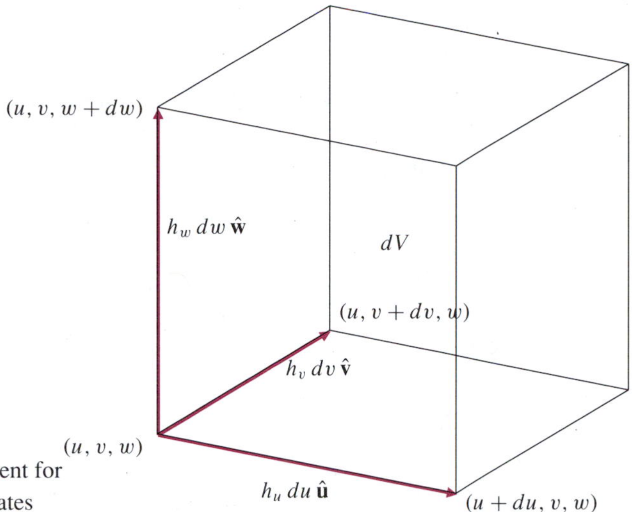


**Figure 9.22** The volume element for orthogonal curvilinear coordinates

Furthermore, the surface area elements on the $u$-, $v$-, and $w$-surfaces are the areas of the appropriate faces of the coordinate box:

**Area elements on coordinate surfaces**

$$dS_u = h_v h_w \, dv \, dw, \qquad dS_v = h_u h_w \, du \, dw, \qquad dS_w = h_u h_v \, du \, dv.$$

The arc length elements along the $u$-, $v$-, and $w$-coordinate curves are the edges of the coordinate box:

**Arc length elements on coordinate curves**

$$ds_u = h_u \, du, \qquad ds_v = h_v \, dv, \qquad ds_w = h_w \, dw.$$

**EXAMPLE 7** For cylindrical coordinates, the volume element, as shown in Section 6.6, is

$$dV = h_r h_\theta h_z \, dr \, d\theta \, dz = r \, dr \, d\theta \, dz.$$

The surface are elements on the cylinder $r$ = constant, the half-plane $\theta$ = constant, and the plane $z$ = constant are, respectively,

$$dS_r = r \, d\theta \, dz, \qquad dS_\theta = dr \, dz, \qquad \text{and} \qquad dS_z = r \, dr \, d\theta.$$

■

**EXAMPLE 8** For spherical coordinates, the volume element, as shown in Section 6.6, is

$$dV = h_\rho h_\phi h_\theta \, d\rho \, d\phi \, d\theta = \rho^2 \sin\phi \, d\rho \, d\phi \, d\theta.$$

The area element on the sphere $\rho$ = constant is

$$dS_\rho = h_\phi h_\theta \, d\phi \, d\theta = \rho^2 \sin\phi \, d\phi \, d\theta.$$

The area element on the cone $\phi$ = constant is

$$dS_\phi = h_\rho h_\theta \, d\rho \, d\theta = \rho \sin\phi \, d\rho \, d\theta.$$

The area element on the half-plane $\theta$ = constant is

$$dS_\theta = h_\rho h_\phi \, d\rho \, d\phi = \rho \, d\rho \, d\phi.$$

■

## Grad, Div, and Curl in Orthogonal Curvilinear Coordinates

The gradient $\nabla f$ of a scalar field $f$ can be expressed in terms of the local basis at any point $P$ with curvilinear coordinates $[u, v, w]$ in the form

$$\nabla f = F_u \hat{\mathbf{u}} + F_v \hat{\mathbf{v}} + F_w \hat{\mathbf{w}}.$$

In order to determine the coefficients $F_u$, $F_v$, and $F_w$ in this formula, we will compare two expressions for the directional derivative of $f$ along an arbitrary curve in $xyz$-space.

If the curve $\mathcal{C}$ has parametrization $\mathbf{r} = \mathbf{r}(s)$ in terms of arc length, then the directional derivative of $f$ along $\mathcal{C}$ is given by

$$\frac{df}{ds} = \frac{\partial f}{\partial u}\frac{du}{ds} + \frac{\partial f}{\partial v}\frac{dv}{ds} + \frac{\partial f}{\partial w}\frac{dw}{ds}.$$

On the other hand, this directional derivative is also given by $df/ds = \nabla f \bullet \hat{\mathbf{T}}$, where $\hat{\mathbf{T}}$ is the unit tangent vector to $\mathcal{C}$. We have

$$\begin{aligned}\hat{\mathbf{T}} = \frac{d\mathbf{r}}{ds} &= \frac{\partial \mathbf{r}}{\partial u}\frac{du}{ds} + \frac{\partial \mathbf{r}}{\partial v}\frac{dv}{ds} + \frac{\partial \mathbf{r}}{\partial w}\frac{dw}{ds}\\ &= h_u \frac{du}{ds}\hat{\mathbf{u}} + h_v \frac{dv}{ds}\hat{\mathbf{v}} + h_w \frac{dw}{ds}\hat{\mathbf{w}}.\end{aligned}$$

Thus

$$\frac{df}{ds} = \nabla f \bullet \hat{\mathbf{T}} = F_u h_u \frac{du}{ds} + F_v h_v \frac{dv}{ds} + F_w h_w \frac{dw}{ds}.$$

Comparing these two expressions for $df/ds$ along $\mathcal{C}$, we see that

$$F_u h_u = \frac{\partial f}{\partial u}, \qquad F_v h_v = \frac{\partial f}{\partial v}, \qquad F_w h_w = \frac{\partial f}{\partial w}.$$

Therefore, we have shown that

**The gradient in orthogonal curvilinear coordinates**

$$\nabla f = \frac{1}{h_u}\frac{\partial f}{\partial u}\hat{\mathbf{u}} + \frac{1}{h_v}\frac{\partial f}{\partial v}\hat{\mathbf{v}} + \frac{1}{h_w}\frac{\partial f}{\partial w}\hat{\mathbf{w}}.$$

**EXAMPLE 9** In terms of cylindrical coordinates, the gradient of the scalar field $f(r, \theta, z)$ is

$$\nabla f(r, \theta, z) = \frac{\partial f}{\partial r}\hat{\mathbf{r}} + \frac{1}{r}\frac{\partial f}{\partial \theta}\hat{\boldsymbol{\theta}} + \frac{\partial f}{\partial z}\mathbf{k}.$$

■

**EXAMPLE 10** In terms of spherical coordinates, the gradient of the scalar field $f(\rho, \phi, \theta)$ is

$$\nabla f(\rho, \phi, \theta) = \frac{\partial f}{\partial \rho}\hat{\boldsymbol{\rho}} + \frac{1}{\rho}\frac{\partial f}{\partial \phi}\hat{\boldsymbol{\phi}} + \frac{1}{\rho \sin\phi}\frac{\partial f}{\partial \theta}\hat{\boldsymbol{\theta}}.$$

Now consider a vector field $\mathbf{F}$ expressed in terms of the curvilinear coordinates:

$$\mathbf{F}(u, v, w) = F_u(u, v, w)\hat{\mathbf{u}} + F_v(u, v, w)\hat{\mathbf{v}} + F_w(u, v, w)\hat{\mathbf{w}}.$$

The flux of $\mathbf{F}$ out of the infinitesimal coordinate box of Figure 9.22 is the sum of the fluxes of $\mathbf{F}$ out of the three pairs of opposite surfaces of the box. The flux out of the $u$-surfaces corresponding to $u$ and $u + du$ is

$$\begin{aligned}
&\mathbf{F}(u + du, v, w) \bullet \hat{\mathbf{u}}\, dS_u - \mathbf{F}(u, v, w) \bullet \hat{\mathbf{u}}\, dS_u \\
&= \big(F_u(u + du, v, w)h_v(u + du, v, w)h_w(u + du, v, w) \\
&\quad - F_u(u, v, w)h_v(u, v, w)h_w(u, v, w)\big)\, dv\, dw \\
&= \frac{\partial}{\partial u}\big(h_v h_w F_u\big)\, du\, dv\, dw.
\end{aligned}$$

Similar expressions hold for the fluxes out of the other pairs of coordinate surfaces.

The divergence at $P$ of $\mathbf{F}$ is the flux *per unit volume* out of the infinitesimal coordinate box at $P$. Thus it is given by

**The divergence in orthogonal curvilinear coordinates**

$$\begin{aligned}
\mathbf{div}\, \mathbf{F}(u, v, w) = &\frac{1}{h_u h_v h_w}\bigg[\frac{\partial}{\partial u}\big(h_v h_w F_u(u, v, w)\big) \\
&+ \frac{\partial}{\partial v}\big(h_u h_w F_v(u, v, w)\big) + \frac{\partial}{\partial w}\big(h_u h_v F_w(u, v, w)\big)\bigg].
\end{aligned}$$

**EXAMPLE 11** For cylindrical coordinates, $h_r = h_z = 1$, and $h_\theta = r$. Thus the divergence of $\mathbf{F} = F_r\hat{\mathbf{r}} + F_\theta\hat{\boldsymbol{\theta}} + F_z\mathbf{k}$ is

$$\begin{aligned}
\mathbf{div}\, \mathbf{F} &= \frac{1}{r}\left[\frac{\partial}{\partial r}\big(rF_r\big) + \frac{\partial}{\partial \theta}F_\theta + \frac{\partial}{\partial z}\big(rF_z\big)\right] \\
&= \frac{\partial F_r}{\partial r} + \frac{1}{r}F_r + \frac{1}{r}\frac{\partial F_\theta}{\partial \theta} + \frac{\partial F_z}{\partial z}.
\end{aligned}$$

**EXAMPLE 12** For spherical coordinates, $h_\rho = 1$, $h_\phi = \rho$, and $h_\theta = \rho\sin\phi$. The divergence of the vector field $\mathbf{F} = F_\rho\hat{\boldsymbol{\rho}} + F_\phi\hat{\boldsymbol{\phi}} + F_\theta\hat{\boldsymbol{\theta}}$ is

$$\begin{aligned}
\mathbf{div}\, \mathbf{F} &= \frac{1}{\rho^2\sin\phi}\left[\frac{\partial}{\partial \rho}(\rho^2\sin\phi\, F_\rho) + \frac{\partial}{\partial \phi}(\rho\sin\phi\, F_\phi) + \frac{\partial}{\partial \theta}(\rho\, F_\theta)\right] \\
&= \frac{1}{\rho^2}\frac{\partial}{\partial \rho}(\rho^2 F_\rho) + \frac{1}{\rho\sin\phi}\frac{\partial}{\partial \phi}(\sin\phi\, F_\phi) + \frac{1}{\rho\sin\phi}\frac{\partial F_\theta}{\partial \theta} \\
&= \frac{\partial F_\rho}{\partial \rho} + \frac{2}{\rho}F_\rho + \frac{1}{\rho}\frac{\partial F_\phi}{\partial \phi} + \frac{\cot\phi}{\rho}F_\phi + \frac{1}{\rho\sin\phi}\frac{\partial F_\theta}{\partial \theta}.
\end{aligned}$$

To calculate the curl of a function expressed in terms of orthogonal curvilinear coordinates we can make use of some previously obtained vector identities. First, observe that the gradient of the scalar field $f(u, v, w) = u$ is $\hat{\mathbf{u}}/h_u$, so that $\hat{\mathbf{u}} = h_u \nabla u$. Similarly, $\hat{\mathbf{v}} = h_v \nabla v$ and $\hat{\mathbf{w}} = h_w \nabla w$. Therefore, the vector field

$$\mathbf{F} = F_u \hat{\mathbf{u}} + F_v \hat{\mathbf{v}} + F_w \hat{\mathbf{w}}$$

can be written in the form

$$\mathbf{F} = F_u h_u \nabla u + F_v h_v \nabla v + F_w h_w \nabla w.$$

Using the identity $\mathbf{curl}\,(f \nabla g) = \nabla f \times \nabla g$ (see Exercise 13 of Section 9.2) we can calculate the curl of each term in the expression above. We have

$$\begin{aligned}
\mathbf{curl}\left(F_u h_u \nabla u\right) &= \nabla(F_u h_u) \times \nabla u \\
&= \left[\frac{1}{h_u}\frac{\partial}{\partial u}(F_u h_u)\hat{\mathbf{u}} + \frac{1}{h_v}\frac{\partial}{\partial v}(F_u h_u)\hat{\mathbf{v}} + \frac{1}{h_w}\frac{\partial}{\partial w}(F_u h_u)\hat{\mathbf{w}}\right] \times \frac{\hat{\mathbf{u}}}{h_u} \\
&= \frac{1}{h_u h_w}\frac{\partial}{\partial w}(F_u h_u)\hat{\mathbf{v}} - \frac{1}{h_u h_v}\frac{\partial}{\partial v}(F_u h_u)\hat{\mathbf{w}} \\
&= \frac{1}{h_u h_v h_w}\left[\frac{\partial}{\partial w}(F_u h_u)(h_v \hat{\mathbf{v}}) - \frac{\partial}{\partial v}(F_u h_u)(h_w \hat{\mathbf{w}})\right].
\end{aligned}$$

We have used the facts that $\hat{\mathbf{u}} \times \hat{\mathbf{u}} = \mathbf{0}$, $\hat{\mathbf{v}} \times \hat{\mathbf{u}} = -\hat{\mathbf{w}}$, and $\hat{\mathbf{w}} \times \hat{\mathbf{u}} = \hat{\mathbf{v}}$ to obtain the result above. This is why we assumed that the curvilinear coordinate system was right-handed.

Corresponding expressions can be calculated for the other two terms in the formula for $\mathbf{curl}\,\mathbf{F}$. Combining the three terms, we conclude that the curl of

$$\mathbf{F} = F_u \hat{\mathbf{u}} + F_v \hat{\mathbf{v}} + F_w \hat{\mathbf{w}}$$

is given by

**The curl in orthogonal curvilinear coordinates**

$$\mathbf{curl}\,\mathbf{F}(u, v, w) = \frac{1}{h_u h_v h_w}\begin{vmatrix} h_u \hat{\mathbf{u}} & h_v \hat{\mathbf{v}} & h_w \hat{\mathbf{w}} \\ \dfrac{\partial}{\partial u} & \dfrac{\partial}{\partial v} & \dfrac{\partial}{\partial w} \\ F_u h_u & F_v h_v & F_w h_w \end{vmatrix}.$$

**EXAMPLE 13** For cylindrical coordinates, the curl of $\mathbf{F} = F_r \hat{\mathbf{r}} + F_\theta \hat{\boldsymbol{\theta}} + F_z \mathbf{k}$ is given by

$$\begin{aligned}
\mathbf{curl}\,\mathbf{F} &= \frac{1}{r}\begin{vmatrix} \hat{\mathbf{r}} & r\hat{\boldsymbol{\theta}} & \mathbf{k} \\ \dfrac{\partial}{\partial r} & \dfrac{\partial}{\partial \theta} & \dfrac{\partial}{\partial z} \\ F_r & rF_\theta & F_z \end{vmatrix} \\
&= \left(\frac{1}{r}\frac{\partial F_z}{\partial \theta} - \frac{\partial F_\theta}{\partial z}\right)\hat{\mathbf{r}} + \left(\frac{\partial F_r}{\partial z} - \frac{\partial F_z}{\partial r}\right)\hat{\boldsymbol{\theta}} + \left(\frac{\partial F_\theta}{\partial r} + \frac{F_\theta}{r} - \frac{1}{r}\frac{\partial F_r}{\partial \theta}\right)\mathbf{k}.
\end{aligned}$$

**EXAMPLE 14** For spherical coordinates, the curl of $\mathbf{F} = F_\rho\hat{\boldsymbol{\rho}} + F_\phi\hat{\boldsymbol{\phi}} + F_\theta\hat{\boldsymbol{\theta}}$ is given by

$$\begin{aligned}
\mathbf{curl}\,\mathbf{F} &= \frac{1}{\rho^2\sin\phi}\begin{vmatrix} \hat{\boldsymbol{\rho}} & \rho\hat{\boldsymbol{\phi}} & \rho\sin\phi\hat{\boldsymbol{\theta}} \\ \dfrac{\partial}{\partial\rho} & \dfrac{\partial}{\partial\phi} & \dfrac{\partial}{\partial\theta} \\ F_\rho & \rho F_\phi & \rho\sin\phi F_\theta \end{vmatrix} \\
&= \frac{1}{\rho\sin\phi}\left[\frac{\partial}{\partial\phi}(\sin\phi F_\theta) - \frac{\partial F_\phi}{\partial\theta}\right]\hat{\boldsymbol{\rho}} \\
&\quad + \frac{1}{\rho\sin\phi}\left[\frac{\partial F_\rho}{\partial\theta} - \sin\phi\frac{\partial}{\partial\rho}(\rho F_\theta)\right]\hat{\boldsymbol{\phi}} \\
&\quad + \frac{1}{\rho}\left[\frac{\partial}{\partial\rho}(\rho F_\phi) - \frac{\partial F_\rho}{\partial\phi}\right]\hat{\boldsymbol{\theta}} \\
&= \frac{1}{\rho\sin\phi}\left[(\cos\phi)F_\theta + (\sin\phi)\frac{\partial F_\theta}{\partial\phi} - \frac{\partial F_\phi}{\partial\theta}\right]\hat{\boldsymbol{\rho}} \\
&\quad + \frac{1}{\rho\sin\phi}\left[\frac{\partial F_\rho}{\partial\theta} - (\sin\phi)F_\theta - (\rho\sin\phi)\frac{\partial F_\theta}{\partial\rho}\right]\hat{\boldsymbol{\phi}} \\
&\quad + \frac{1}{\rho}\left[F_\phi + \rho\frac{\partial F_\phi}{\partial\rho} - \frac{\partial F_\rho}{\partial\phi}\right]\hat{\boldsymbol{\theta}}.
\end{aligned}$$

## EXERCISES 9.6

In Exercises 1–2, calculate the gradients of the given scalar fields expressed in terms of cylindrical or spherical coordinates.

**1.** $f(r, \theta, z) = r\theta z$

**2.** $f(\rho, \phi, \theta) = \rho\phi\theta$

In Exercises 3–8, calculate **div F** and **curl F** for the given vector fields expressed in terms of cylindrical coordinates or spherical coordinates.

**3.** $\mathbf{F}(r, \theta, z) = r\hat{\mathbf{r}}$

**4.** $\mathbf{F}(r, \theta, z) = r\hat{\boldsymbol{\theta}}$

**5.** $\mathbf{F}(\rho, \phi, \theta) = \sin\phi\hat{\boldsymbol{\rho}}$

**6.** $\mathbf{F}(\rho, \phi, \theta) = \rho\hat{\boldsymbol{\phi}}$

**7.** $\mathbf{F}(\rho, \phi, \theta) = \rho\hat{\boldsymbol{\theta}}$

**8.** $\mathbf{F}(\rho, \phi, \theta) = \rho^2\hat{\boldsymbol{\rho}}$

**9.** Let $x = x(u, v)$, $y = y(u, v)$ define orthogonal curvilinear coordinates $(u, v)$ in the $xy$-plane. Find the scale factors, local basis vectors, and area element for the system of coordinates $(u, v)$.

**10.** Continuing the previous exercise, express the gradient of a scalar field $f(u, v)$, and the divergence and curl of a vector field $\mathbf{F}(u, v)$ in terms of the curvilinear coordinates.

**11.** Express the gradient of the scalar field $f(r, \theta)$, and the divergence and curl of a vector field $\mathbf{F}(r, \theta)$ in terms of plane polar coordinates $(r, \theta)$.

**12.** The transformation

$$x = a\cosh u\cos v \qquad y = a\sinh u\sin v$$

defines **elliptical coordinates** in the $xy$-plane. This coordinate system has singular points at $x = \pm a$, $y = 0$.

(a) Show that the $v$-curves, $u$ = constant, are ellipses with foci at the singular points.

(b) Show that the $u$-curves, $v$ = constant, are hyperbolas with foci at the singular points.

(c) Show that the $u$-curve and the $v$-curve through a nonsingular point intersect at right angles.

(d) Find the scale factors $h_u$ and $h_v$, and the area element $dA$ for the elliptical coordinate system.

**13.** Describe the coordinate surfaces and coordinate curves of the system of elliptical cylindrical coordinates in $xyz$-space defined by

$$x = a\cosh u\cos v, \qquad y = a\sinh u\sin v, \qquad z = z.$$

**14.** The Laplacian $\nabla^2 f$ of a scalar field $f$ can be calculated as **div** $\nabla f$. Use this method to calculate the Laplacian of the function $f(r, \theta, z)$ expressed in terms of cylindrical coordinates. (This repeats Exercise 22 of Section 6.5.)

**15.** Calculate the Laplacian $\nabla^2 f = \mathbf{div}\,\nabla f$ for the function $f(\rho, \phi, \theta)$, expressed in terms of spherical coordinates. (This repeats Exercise 38 of Section 6.6, but is now much easier.)

**16.** Calculate the Laplacian $\nabla^2 f = \mathbf{div}\,\nabla f$ for a function $f(u, v, w)$ expressed in terms of arbitrary orthogonal curvilinear coordinates $(u, v, w)$.

# CHAPTER REVIEW

## Key Ideas

- **What do the following phrases and statements mean?**
  - ◇ the divergence of a vector field $\mathbf{F}$
  - ◇ the curl of a vector field $\mathbf{F}$
  - ◇ $\mathbf{F}$ is solenoidal.
  - ◇ $\mathbf{F}$ is irrotational.
  - ◇ a scalar potential
  - ◇ a vector potential
  - ◇ orthogonal curvilinear coordinates
- **State the following theorems:**
  - ◇ the Divergence Theorem
  - ◇ Green's Theorem
  - ◇ Stokes's theorem

## Review Exercises

**1.** If $\mathbf{F} = x^2z\mathbf{i} + (y^2z + 3y)\mathbf{j} + x^2\mathbf{k}$, find the flux of $\mathbf{F}$ across the part of the ellipsoid $x^2 + y^2 + 4z^2 = 16$, where $z \ge 0$, oriented with upward normal.

**2.** Let $\mathcal{S}$ be the part of the cylinder $x^2 + y^2 = 2ax$ between the horizontal planes $z = 0$ and $z = b$, where $b > 0$. Find the flux of $\mathbf{F} = x\mathbf{i} + \cos(z^2)\mathbf{j} + e^z\mathbf{k}$ outward through $\mathcal{S}$.

**3.** Find $\oint_C (3y^2 + 2xe^{y^2})\,dx + (2x^2ye^{y^2})\,dy$ counterclockwise around the boundary of the parallelogram with vertices $(0, 0)$, $(2, 0)$, $(3, 1)$, and $(1, 1)$.

**4.** If $\mathbf{F} = -z\mathbf{i} + x\mathbf{j} + y\mathbf{k}$, what are the possible values of $\oint_C \mathbf{F} \bullet d\mathbf{r}$ around circles of radius $a$ in the plane $2x + y + 2z = 7$?

**5.** Let $\mathbf{F}$ be a smooth vector field in 3-space and suppose that for every $a > 0$, the flux of $\mathbf{F}$ out of the sphere of radius $a$ centred at the origin is $\pi(a^3 + 2a^4)$. Find the divergence of $\mathbf{F}$ at the origin.

**6.** Let $\mathbf{F} = -y\mathbf{i} + x\cos(1 - x^2 - y^2)\mathbf{j} + yz\mathbf{k}$. Find the flux of $\mathbf{curl}\,\mathbf{F}$ upward through a surface whose boundary is the curve $x^2 + y^2 = 1$, $z = 2$.

**7.** Let $\mathbf{F}(\mathbf{r}) = r^\lambda\mathbf{r}$, where $\mathbf{r} = x\mathbf{i}+y\mathbf{j}+z\mathbf{k}$ and $r = |\mathbf{r}|$. For what value(s) of $\lambda$ is $\mathbf{F}$ solenoidal on an open subset of 3-space? Is $\mathbf{F}$ solenoidal on all of 3-space for any value of $\lambda$?

**8.** Given that $\mathbf{F}$ satisfies $\mathbf{curl}\,\mathbf{F} = \mu\mathbf{F}$ on 3-space, where $\mu$ is a nonzero constant, show that $\nabla^2\mathbf{F} + \mu^2\mathbf{F} = \mathbf{0}$.

**9.** Let $P$ be a polyhedron in 3-space having $n$ planar faces, $F_1$, $F_2, \ldots, F_n$. Let $\mathbf{N}_i$ be normal to $F_i$ in the direction outward from $P$ and let $\mathbf{N}_i$ have length equal to the area of face $F_i$. Show that

$$\sum_{i=1}^{n} \mathbf{N}_i = \mathbf{0}.$$

Also, state a version of this result for a plane polygon $P$.

**10.** Around what simple, closed curve $C$ in the $xy$-plane does the vector field

$$\mathbf{F} = (2y^3 - 3y + xy^2)\mathbf{i} + (x - x^3 + x^2y)\mathbf{j}$$

have the greatest circulation?

**11.** Through what closed, oriented surface in $\mathbb{R}^3$ does the vector field

$$\mathbf{F} = (4x + 2x^3z)\mathbf{i} - y(x^2 + z^2)\mathbf{j} - (3x^2z^2 + 4y^2z)\mathbf{k}$$

have the greatest flux?

**12.** Find the maximum value of

$$\oint_C \mathbf{F} \bullet d\mathbf{r}$$

where $\mathbf{F} = xy^2\mathbf{i} + (3z - xy^2)\mathbf{j} + (4y - x^2y)\mathbf{k}$, and $C$ is a simple closed curve in the plane $x + y + z = 1$ oriented counterclockwise as seen from high on the $z$-axis. What curve $C$ gives this maximum?

## Express Yourself

Answer these questions in words, using complete sentences and, if necessary, paragraphs. Use mathematical symbols only where necessary.

**1.** Discuss the appropriateness of the term "divergence" of a vector field. In what sense does a vector field with positive divergence actually diverge?

**2.** Why are the Divergence Theorem, Stokes's Theorem, and Green's Theorem considered to be versions of the Fundamental Theorem of Calculus? What do these three theorems have in common with the Fundamental Theorem and with each other?

**3.** Describe in words (without using symbols) what the Divergence Theorem says? Do the same for Stokes's Theorem.

**4.** A vector field $\mathbf{F}$ is defined throughout 3-space. An intelligent ant living in the $xy$-plane knows the values of $\mathbf{F}$ at every point in that plane, but nowhere else. Can the ant compute the curl of $\mathbf{F}$? Can she compute any components of the curl of $\mathbf{F}$? Give reasons for your answers.

## Challenging Problems

**1. (The expanding universe)** Let $\mathbf{v}$ be the large-scale velocity field of matter in the universe. (*Large-scale* means on the scale of intergalactic distances — *small-scale* motion such as that of planetary systems about their suns, and even stars about galactic centres has been averaged out.) Assume that $\mathbf{v}$ is a smooth vector field. According to present astronomical theory, the distance between any two points is increasing, and the rate of increase is proportional to the distance between the points. The constant of proportionality, $C$, is called *Hubble's constant*. In terms of $\mathbf{v}$, if $\mathbf{r}_1$ and $\mathbf{r}_2$ are two points then

$$\big(\mathbf{v}(\mathbf{r}_2) - \mathbf{v}(\mathbf{r}_1)\big) \bullet (\mathbf{r}_2 - \mathbf{r}_1) = C|\mathbf{r}_2 - \mathbf{r}_1|^2.$$

Show that **div v** is constant, and find the value of the constant in terms of Hubble's constant. *Hint:* find the flux of $\mathbf{v}(\mathbf{r})$ out of a sphere of radius $\epsilon$ centred at $\mathbf{r}_1$ and take the limit as $\epsilon$ approaches zero.

2. **(Solid angle)** Two rays from a point $P$ determine an angle at $P$ whose measure in radians is equal to the length of the arc of the circle of radius 1 with centre at $P$ lying between the two rays. Similarly, an arbitrarily shaped half-cone $K$ with vertex at $P$ determines a **solid angle** at $P$ whose measure in **steradians** (*stereo* + *radians*) is the area of that part of the sphere of radius 1 with centre at $P$ lying within $K$. For example, the first octant of $\mathbb{R}^3$ is a half-cone with vertex at the origin. It determines a solid angle at the origin measuring

$$4\pi \times \frac{1}{8} = \frac{\pi}{2} \text{ steradians},$$

since the area of the unit sphere is $4\pi$.

(a) Find the steradian measure of the solid angle at the vertex of a right-circular half-cone whose generators make angle $\alpha$ with its central axis.

(b) If a smooth, oriented surface intersects the general half-cone $K$, but not at its vertex $P$, let $S$ be the part of the surface lying within $K$. Orient $S$ with normal pointing away from $P$. Show that the steradian measure of the solid angle at $P$ determined by $K$ is the flux of $\mathbf{r}/|\mathbf{r}|^3$ through $S$, where $\mathbf{r}$ is the vector from $P$ to the point $(x, y, z)$.

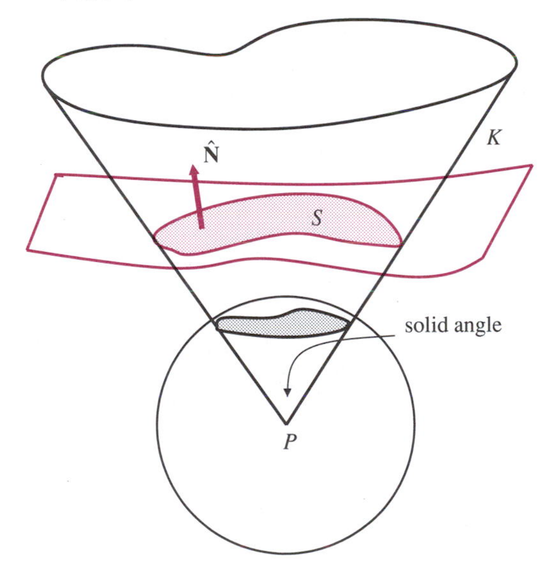


Figure 9.23

### Integrals over moving domains

By the Fundamental Theorem of Calculus, the derivative with respect to time $t$ of an integral of $f(x, t)$ over a "moving interval" $[a(t), b(t)]$ is given by

$$\frac{d}{dt}\int_{a(t)}^{b(t)} f(x,t)\,dx = \int_{a(t)}^{b(t)} \frac{\partial}{\partial t} f(x,t)\,dx + f(b(t),t)\frac{db}{dt} - f(a(t),t)\frac{da}{dt}.$$

The next three problems, suggested to the author by Luigi Quartapelle of the Politecnico di Milano, provide various extensions of this one-dimensional result to higher dimensions. As the calculations are somewhat lengthy, you may want to try to get some help from a computer algebra system.

* 3. **(Rate of change of circulation along a moving curve)**

(a) Let $\mathbf{F}(\mathbf{r}, t)$ be a smooth vector field in $\mathbb{R}^3$ depending on a parameter $t$, and let

$$\mathbf{G}(s,t) = \mathbf{F}(\mathbf{r}(s,t),t) = \mathbf{F}(x(s,t), y(s,t), z(x,t), t),$$

where $\mathbf{r}(s,t) = x(s,t)\mathbf{i} + y(s,t)\mathbf{j} + z(s,t)\mathbf{k}$ has continuous partial derivatives of second order. Show that

$$\frac{\partial}{\partial t}\left(\mathbf{G} \bullet \frac{\partial \mathbf{r}}{\partial s}\right) - \frac{\partial}{\partial s}\left(\mathbf{G} \bullet \frac{\partial \mathbf{r}}{\partial t}\right) = \frac{\partial \mathbf{F}}{\partial t} \bullet \frac{\partial \mathbf{r}}{\partial s} + \left((\nabla \times \mathbf{F}) \times \frac{\partial \mathbf{r}}{\partial t}\right) \bullet \frac{\partial \mathbf{r}}{\partial s}.$$

Here the curl $\nabla \times \mathbf{F}$ is taken with respect to the position vector $\mathbf{r}$.

(b) For fixed $t$ (which you can think of as time), $\mathbf{r} = \mathbf{r}(s,t)$, $(a \le s \le b)$, represents parametrically a curve $C_t$ in $\mathbb{R}^3$. The curve moves as $t$ varies; the velocity of any point on $C_t$ is $\mathbf{v}_C(s,t) = \partial\mathbf{r}/\partial t$. Show that

$$\frac{d}{dt}\int_{C_t} \mathbf{F} \bullet d\mathbf{r} = \int_{C_t} \frac{\partial \mathbf{F}}{\partial t} \bullet d\mathbf{r} + \int_{C_t} \big((\nabla \times \mathbf{F}) \times \mathbf{v}_C\big) \bullet d\mathbf{r} + \mathbf{F}(\mathbf{r}(b,t),t) \bullet \mathbf{v}_C(b,t) - \mathbf{F}(\mathbf{r}(a,t),t) \bullet \mathbf{v}_C(a,t).$$

*Hint:* write

$$\begin{aligned}\frac{d}{dt}\int_{C_t} \mathbf{F} \bullet d\mathbf{r} &= \int_a^b \frac{\partial}{\partial t}\left(\mathbf{G} \bullet \frac{\partial \mathbf{r}}{\partial s}\right) ds \\ &= \int_a^b \left[\frac{\partial}{\partial s}\left(\mathbf{G} \bullet \frac{\partial \mathbf{r}}{\partial t}\right) + \left(\frac{\partial}{\partial t}\left(\mathbf{G} \bullet \frac{\partial \mathbf{r}}{\partial s}\right) - \frac{\partial}{\partial s}\left(\mathbf{G} \bullet \frac{\partial \mathbf{r}}{\partial t}\right)\right)\right] ds.\end{aligned}$$

Now use the result of (a).

* 4. **(Rate of change of flux through a moving surface)** Let $S_t$ be a moving surface in $\mathbb{R}^3$ smoothly parametrized (for each $t$) by

$$\mathbf{r} = \mathbf{r}(u,v,t) = x(u,v,t)\mathbf{i} + y(u,v,t)\mathbf{j} + z(u,v,t)\mathbf{k},$$

where $(u, v)$ belongs to a parameter region $R$ in the $uv$-plane. Let $\mathbf{F}(\mathbf{r}, t) = F_1\mathbf{i} + F_2\mathbf{j} + F_3\mathbf{k}$ be a smooth 3-vector function and let $\mathbf{G}(u,v,t) = \mathbf{F}(\mathbf{r}(u,v,t),t)$.

(a) Show that

$$\begin{aligned}&\frac{\partial}{\partial t}\left(\mathbf{G}\bullet\left[\frac{\partial\mathbf{r}}{\partial u}\times\frac{\partial\mathbf{r}}{\partial v}\right]\right)-\frac{\partial}{\partial u}\left(\mathbf{G}\bullet\left[\frac{\partial\mathbf{r}}{\partial t}\times\frac{\partial\mathbf{r}}{\partial v}\right]\right)\\&\quad-\frac{\partial}{\partial v}\left(\mathbf{G}\bullet\left[\frac{\partial\mathbf{r}}{\partial u}\times\frac{\partial\mathbf{r}}{\partial t}\right]\right)\\&=\frac{\partial\mathbf{F}}{\partial t}\bullet\left[\frac{\partial\mathbf{r}}{\partial u}\times\frac{\partial\mathbf{r}}{\partial v}\right]+(\nabla\bullet\mathbf{F})\frac{\partial\mathbf{r}}{\partial t}\bullet\left[\frac{\partial\mathbf{r}}{\partial u}\times\frac{\partial\mathbf{r}}{\partial v}\right].\end{aligned}$$

(b) If $C_t$ is the boundary of $S_t$ with orientation corresponding to that of $S_t$, use Green's Theorem to show that

$$\begin{aligned}&\iint_R\left[\frac{\partial}{\partial u}\left(\mathbf{G}\bullet\left[\frac{\partial\mathbf{r}}{\partial t}\times\frac{\partial\mathbf{r}}{\partial v}\right]\right)\right.\\&\quad\left.+\frac{\partial}{\partial v}\left(\mathbf{G}\bullet\left[\frac{\partial\mathbf{r}}{\partial u}\times\frac{\partial\mathbf{r}}{\partial t}\right]\right)\right]du\,dv\\&=\oint_{C_t}\left(\mathbf{F}\times\frac{\partial\mathbf{r}}{\partial t}\right)\bullet d\mathbf{r}.\end{aligned}$$

(c) Combine the results of (a) and (b) to show that

$$\begin{aligned}&\frac{d}{dt}\iint_{S_t}\mathbf{F}\bullet\hat{\mathbf{N}}\,dS\\&=\iint_{S_t}\frac{\partial\mathbf{F}}{\partial t}\bullet\hat{\mathbf{N}}\,dS+\iint_{S_t}(\nabla\bullet\mathbf{F})\mathbf{v}_S\bullet\hat{\mathbf{N}}\,dS\\&\quad+\oint_{C_t}(\mathbf{F}\times\mathbf{v}_C)\bullet d\mathbf{r},\end{aligned}$$

where $\mathbf{v}_S=\partial\mathbf{r}/\partial t$ on $S_t$ is the velocity of $S_t$, $\mathbf{v}_C=\partial\mathbf{r}/\partial t$ on $C_t$ is the velocity of $C_t$, and $\hat{\mathbf{N}}$ is the unit normal field on $S_t$ corresponding to its orientation.

* **5. (Rate of change of integrals over moving volumes)** Let $S_t$ be the position at time $t$ of a smooth, closed surface in $\mathbb{R}^3$ that varies smoothly with $t$, and bounds at any time $t$ a region $D_t$. If $\hat{\mathbf{N}}(\mathbf{r},t)$ denotes the unit outward (from $D_t$) normal field on $S_t$, and $\mathbf{v}_S(\mathbf{r},t)$ is the velocity of the point $\mathbf{r}$ on $S_t$ at time $t$, show that

$$\frac{d}{dt}\iiint_{D_t}f\,dV=\iiint_{D_t}\frac{\partial f}{\partial t}\,dV+\oiint_{S_t}f\mathbf{v}_S\bullet\hat{\mathbf{N}}\,dS$$

holds for smooth functions $f(\mathbf{r},t)$.
*Hint:* let $\Delta D_t$ consist of the points through which $S_t$ passes as $t$ increases to $t+\Delta t$. The volume element $dV$ in $\Delta D_t$ can be expressed in terms of the area element $dS$ on $S_t$ by

$$dV=\mathbf{v}\bullet\hat{\mathbf{N}}\,dS\,\Delta t.$$

Show that

$$\begin{aligned}&\frac{1}{\Delta t}\left[\iiint_{D_{t+\Delta t}}f(\mathbf{r},t+\Delta t)\,dV-\iiint_{D_t}f(\mathbf{r},t)\,dV\right]\\&=\iiint_{D_t}\frac{f(\mathbf{r},t+\Delta t)-f(\mathbf{r},t)}{\Delta t}\,dV\\&\quad+\frac{1}{\Delta t}\iiint_{\Delta D_t}f(\mathbf{r},t)\,dV\\&\quad+\iiint_{\Delta D_t}\frac{f(\mathbf{r},t+\Delta t)-f(\mathbf{r},t)}{\Delta t}\,dV,\end{aligned}$$

and show that the last integral $\to 0$ as $\Delta t\to 0$.

CHAPTER 10

# Ordinary Differential Equations

**INTRODUCTION** A **differential equation** (or **DE**) is an equation that involves one or more derivatives of an unknown function. Solving the differential equation means finding a function (or every such function) that satisfies the differential equation.

Many physical laws and relationships between quantities studied in various scientific disciplines are expressed mathematically as differential equations. For example, Newton's second law of motion ($F = ma$) states that the position $x(t)$ at time $t$ of an object of constant mass $m$ subjected to a force $F(t)$ must satisfy the differential equation (equation of motion):

$$m\frac{d^2x}{dt^2} = F(t).$$

Similarly, the biomass $m(t)$ at time $t$ of a bacterial culture growing in a uniformly supporting medium changes at a rate proportional to the biomass:

$$\frac{dm}{dt} = km(t),$$

which is the differential equation of exponential growth (or, if $k < 0$, exponential decay). Because differential equations arise so extensively in the abstract modelling of concrete phenomena, such equations and techniques for solving them are at the heart of applied mathematics. Indeed, most of the existing mathematical literature is either directly involved with differential equations, or is motivated by problems arising in the study of such equations. Because of this, we have introduced various differential equations, terms for their description, and techniques for their solution at several places in the development of calculus throughout this book. This final chapter provides a more unified framework for a brief introduction to the study of ordinary differential equations. This chapter is, of necessity, relatively short. Students of mathematics and its applications usually take one or more full courses on differential equations, and even then hardly scratch the surface of the subject.

## 10.1 CLASSIFYING DIFFERENTIAL EQUATIONS

Differential equations are classified in several ways. The most significant classification is based on the number of variables with respect to which derivatives appear in the equation. An **ordinary differential equation (ODE)** is one that involves derivatives with respect to only one variable. Both of the examples given above are ordinary differential equations. A **partial differential equation (PDE)** is one that involves partial derivatives of the unknown function with respect to more than one variable. For example, the **one-dimensional wave equation**

$$\frac{\partial^2 u}{\partial t^2} = c^2\frac{\partial^2 u}{\partial x^2}$$

models the lateral displacement $u(x, t)$ at position $x$ at time $t$ of a stretched vibrating string. (See Section 4.4.) We will not discuss partial differential equations in this chapter.

Differential equations are also classified with respect to **order**. The order of a differential equation is the order of the highest order derivative present in the equation. The one-dimensional wave equation is a second-order PDE. The following example records the order of two ODEs.

**EXAMPLE 1**

$$\frac{d^2y}{dx^2} + x^3y = \sin x \qquad \text{has order 2,}$$

$$\frac{d^3y}{dx^3} + 4x\left(\frac{dy}{dx}\right)^2 = y\frac{d^2y}{dx^2} + e^y \qquad \text{has order 3.}$$

Like any equation, a differential equation can be written in the form $F = 0$, where $F$ is a function. For an ODE, the function $F$ can depend on the independent variable (usually called $x$ or $t$), the unknown function (usually $y$), and any derivatives of the unknown function up to the order of the equation. For instance, an $n$th-order ODE can be written in the form

$$F(x, y, y', y'', \ldots, y^{(n)}) = 0.$$

An important special class of differential equations are those that are **linear**. An $n$th-order linear ODE has the form

$$a_n(x)y^{(n)}(x) + a_{n-1}(x)y^{(n-1)}(x) + \cdots$$
$$+ a_2(x)y''(x) + a_1(x)y'(x) + a_0(x)y(x) = f(x).$$

Each term in the expression on the left side is the product of a *coefficient* that is a function of $x$, and a second factor that is either $y$ or one of the derivatives of $y$. The term on the right does not depend on $y$; it is called the **nonhomogeneous term**. Observe that no term on the left side involves any power of $y$ or its derivatives other than the first power, and $y$ and its derivatives are never multiplied together.

A linear ODE is said to be **homogeneous** if all of its terms involve the unknown function $y$, that is, if $f(x) = 0$. If $f(x)$ is not identically zero, the equation is **nonhomogeneous**.

**EXAMPLE 2** In Example 1 the first DE,

$$\frac{d^2y}{dx^2} + x^3y = \sin x,$$

is linear. Here the coefficients are $a_2(x) = 1$, $a_1(x) = 0$, $a_0(x) = x^3$, and the nonhomogeneous term is $f(x) = \sin x$. Although it can be written in the form

$$\frac{d^3y}{dx^3} + 4x\left(\frac{dy}{dx}\right)^2 - y\frac{d^2y}{dx^2} - e^y = 0,$$

the second equation is *not linear* (we say it is **nonlinear**) because the second term involves the square of a derivative of $y$, the third term involves the product of $y$ and one of its derivatives, and the fourth term is not $y$ times a function of $x$. The equation

$$(1 + x^2)\frac{d^3y}{dx^3} + \sin x\frac{d^2y}{dx^2} - 4\frac{dy}{dx} + y = 0$$

is a linear equation of order 3. The coefficients are $a_3(x) = 1 + x^2$, $a_2(x) = \sin x$, $a_1(x) = -4$, and $a_0(x) = 1$. Since $f(x) = 0$, this equation is *homogeneous*.

The following theorem states that any *linear combination* of solutions of a linear, homogeneous DE is also a solution. This is an extremely important fact about linear, homogeneous DEs.

**THEOREM 1**

If $y = y_1(x)$ and $y = y_2(x)$ are two solutions of the linear, homogeneous DE

$$a_n y^{(n)} + a_{n-1} y^{(n-1)} + \cdots + a_2 y'' + a_1 y' + a_0 y = 0$$

then so is the linear combination

$$y = Ay_1(x) + By_2(x)$$

for any values of the constants $A$ and $B$.

***PROOF*** We are given that

$$\begin{aligned} &a_n y_1^{(n)} + a_{n-1} y_1^{(n-1)} + \cdots + a_2 y_1'' + a_1 y_1' + a_0 y_1 = 0, \qquad \text{and} \\ &a_n y_2^{(n)} + a_{n-1} y_2^{(n-1)} + \cdots + a_2 y_2'' + a_1 y_2' + a_0 y_2 = 0. \end{aligned}$$

Multiplying the first equation by $A$ and the second by $B$ and adding the two gives

$$\begin{aligned} &a_n (Ay_1^{(n)} + By_2^{(n)}) + a_{n-1}(Ay_1^{(n-1)} + By_2^{(n-1)}) \\ &\qquad + \cdots + a_2(Ay_1'' + By_2'') + a_1(Ay_1' + By_2') + a_0(Ay_1 + By_2) = 0. \end{aligned}$$

Thus $y = Ay_1(x) + By_2(x)$ is also a solution of the equation.

The same kind of proof can be used to verify the following theorem.

**THEOREM 2**

If $y = y_1(x)$ is a solution of the linear, homogeneous equation

$$a_n y^{(n)} + a_{n-1} y^{(n-1)} + \cdots + a_2 y'' + a_1 y' + a_0 y = 0$$

and $y = y_2(x)$ is a solution of the linear, nonhomogeneous equation

$$a_n y^{(n)} + a_{n-1} y^{(n-1)} + \cdots + a_2 y'' + a_1 y' + a_0 y = f(x),$$

then $y = y_1(x) + y_2(x)$ is also a solution of the same linear, nonhomogeneous equation.

We will make extensive use of the two theorems above when we discuss second-order linear equations in Sections 10.6–10.8.

**EXAMPLE 3** Verify that $y = \sin 2x$ and $y = \cos 2x$ satisfy the DE $y'' + 4y = 0$. Find a solution $y(x)$ of that DE that satisfies $y(0) = 2$ and $y'(0) = -4$.

***SOLUTION*** If $y = \sin 2x$, then $y'' = \dfrac{d}{dx}(2\cos 2x) = -4\sin 2x = -4y$. Thus $y'' + 4y = 0$. A similar calculation shows that $y = \cos 2x$ also satisfies the DE. Since the DE is linear and homogeneous, the function

$$y = A\sin 2x + B\cos 2x$$

is a solution for any values of the constants $A$ and $B$. We want $y(0) = 2$, so we need $2 = A\sin 0 + B\cos 0 = B$. Thus $B = 2$. Also,

$$y' = 2A\cos 2x - 2B\sin 2x.$$

We want $y'(0) = -4$, so $-4 = 2A\cos 0 - 2B\sin 0 = 2A$. Thus $A = -2$ and the required solution is $y = -2\sin 2x + 2\cos 2x$. ■

**REMARK** Let $P_n(r)$ be the $n$th-degree polynomial in the variable $r$ given by

$$P_n(r) = a_n(x)r^n + a_{n-1}(x)r^{n-1} + \cdots + a_2(x)r^2 + a_1(x)r + a_0(x),$$

with coefficients depending on the variable $x$. We can write the $n$th-order linear ODE with coefficients $a_k(x)$, $(0 \le k \le n)$, and nonhomogeneous term $f(x)$ in the form

$$P_n(D)y(x) = f(x)$$

where $D$ stands for the *differential operator* $d/dx$. The left side of the equation above denotes the application of the $n$th-order differential operator

$$P_n(D) = a_n(x)D^n + a_{n-1}(x)D^{n-1} + \cdots + a_2(x)D^2 + a_1(x)D + a_0(x)$$

to the function $y(x)$. For example,

$$a_k(x)D^k y(x) = a_k(x)\frac{d^k y}{dx^k}.$$

It is often useful to write linear DEs in terms of differential operators in this way.

**REMARK** Unfortunately, the term *homogeneous* is used in more than one way in the study of differential equations. Certain ODEs that are not necessarily linear are called homogeneous for a different reason than the one applying for linear equations above. We will encounter equations of this type in Section 10.3.

## EXERCISES 10.1

In Exercises 1–10, state the order of the given DE, and whether it is linear or nonlinear. If it is linear, is it homogeneous or nonhomogeneous?

**1.** $\dfrac{dy}{dx} = 5y$

**2.** $\dfrac{d^2y}{dx^2} + x = y$

**3.** $y\dfrac{dy}{dx} = x$

**4.** $y''' + xy' = x \sin x$

**5.** $y'' + x \sin x \, y' = y$

**6.** $y'' + 4y' - 3y = 2y^2$

**7.** $\dfrac{d^3y}{dt^3} + t\dfrac{dy}{dt} + t^2 y = t^3$

**8.** $\cos x \dfrac{dx}{dt} + x \sin t = 0$

**9.** $y^{(4)} + e^x y'' = x^3 y'$

**10.** $x^2 y'' + e^x y' = \dfrac{1}{y}$

**11.** Verify that $y = \cos x$ and $y = \sin x$ are solutions of the DE $y'' + y = 0$. Are any of the following functions solutions? (a) $\sin x - \cos x$, (b) $\sin(x + 3)$, (c) $\sin 2x$? Justify your answers.

**12.** Verify that $y = e^x$ and $y = e^{-x}$ are solutions of the DE $y'' - y = 0$. Are any of the following functions solutions? (a) $\cosh x = \frac{1}{2}(e^x + e^{-x})$, (b) $\cos x$, (c) $x^e$? Justify your answers.

**13.** $y_1 = \cos(kx)$ is a solution of $y'' + k^2 y = 0$. Guess and verify another solution $y_2$ that is not a multiple of $y_1$. Then find a solution that satisfies $y(\pi/k) = 3$ and $y'(\pi/k) = 3$.

**14.** $y_1 = e^{kx}$ is a solution of $y'' - k^2 y = 0$. Guess and verify another solution $y_2$ that is not a multiple of $y_1$. Then find a solution that satisfies $y(1) = 0$ and $y'(1) = 2$.

**15.** Find a solution of $y'' + y = 0$ that satisfies $y(\pi/2) = 2y(0)$ and $y(\pi/4) = 3$. *Hint:* see Exercise 11.

**16.** Find two values of $r$ such that $y = e^{rx}$ is a solution of $y'' - y' - 2y = 0$. Then find a solution of the equation that satisfies $y(0) = 1$, $y'(0) = 2$.

**17.** Verify that $y = x$ is a solution of $y'' + y = x$, and find a solution $y$ of this DE that satisfies $y(\pi) = 1$ and $y'(\pi) = 0$. *Hint:* use Exercise 11 and Theorem 2.

**18.** Verify that $y = -e$ is a solution of $y'' - y = e$, and find a solution $y$ of this DE that satisfies $y(1) = 0$ and $y'(1) = 1$. *Hint:* use Exercise 12 and Theorem 2.

# 10.2 FIRST-ORDER SEPARABLE AND HOMOGENEOUS EQUATIONS

In the next three sections we will develop techniques for solving first-order ODEs, mostly ones of the form

$$\frac{dy}{dx} = f(x, y).$$

Solving differential equations typically involves integration; indeed, the process of solving a DE is called *integrating* the DE. Nevertheless, solving DEs is usually more complicated than just writing down an integral and evaluating it. The only kind of DE that can be solved that way is the simplest kind of first-order, linear DE that can be written in the form

$$\frac{dy}{dx} = f(x).$$

The solution is then just the antiderivative of $f$:

$$y = \int f(x)\, dx.$$

## Separable Equations

The next most simple kind of equation to solve is a so-called **separable equation**. A separable equation is one of the form

$$\frac{dy}{dx} = f(x)g(y).$$

We can rewrite such an equation in the equivalent differential form

$$\frac{dy}{g(y)} = f(x)\, dx,$$

where the variables have been *separated* on opposite sides of the equation. This was possible because the derivative $dy/dx$ was a product of a function of $x$ alone times a function of $y$ alone, rather than a more general function of the two variables $x$ and $y$. We solve the separated equation by integrating both sides,

$$\int \frac{dy}{g(y)} = \int f(x)\, dx,$$

remembering to include a constant of integration on one side after the integrals are evaluated. (Why don't you need to include a constant of integration on both sides?) Assuming both of the integrals can be evaluated, we can obtain a **general solution** to the given DE in the form

$$G(y) = F(x) + C.$$

It may or may not be possible to solve this equation *explicitly* for $y$ as a function of $x$, but the equation will normally define $y(x)$ at least *implicitly.* The graphs of the curves $G(y) = F(x) + C$ for various values of $C$ are called **solution curves** of the given DE. A single **initial condition** of the form $y = b$ when $x = a$ (or, equivalently, $y(a) = b$) will serve to single out a unique solution curve from among those given by the general solution.

**EXAMPLE 1** Solve the equation $\frac{dy}{dx} = \frac{x}{y}$.

***SOLUTION*** We rewrite the equation in the form

$$y\,dy = x\,dx,$$

and integrate both sides to get

$$\frac{1}{2}y^2 = \frac{1}{2}x^2 + C,$$

or $y^2 - x^2 = C_1$ where $C_1 = 2C$ is an arbitrary constant. The solution curves are rectangular hyperbolas with asymptotes $y = x$ and $y = -x$. ■

**EXAMPLE 2** Solve the initial-value problem

$$\begin{cases} \dfrac{dy}{dx} = x^2y^3 \\ y(1) = 3. \end{cases}$$

***SOLUTION*** The differential equation becomes $\frac{dy}{y^3} = x^2\,dx$, so $\int \frac{dy}{y^3} = \int x^2\,dx$ and

$$\frac{-1}{2y^2} = \frac{x^3}{3} + C.$$

Since $y = 3$ when $x = 1$, we have $-\frac{1}{18} = \frac{1}{3} + C$ and $C = -\frac{7}{18}$. Substituting this value into the above solution and solving for $y$, we obtain

$$y(x) = \frac{3}{\sqrt{7 - 6x^3}}.$$

This solution is valid for $x \le \left(\frac{7}{6}\right)^{1/3}$. ■

**EXAMPLE 3** Solve the logistic equation:

$$\frac{dy}{dt} = ky\left(1 - \frac{y}{L}\right).$$

Here $y(t)$ is the size of an animal population at time $t$, $k$ is a positive constant related to the fertility of the animals, and $L$ is the steady-state population size that can be sustained by the available food supply. Assume that $0 < y(t) < L$ for all $t$.

***SOLUTION*** The equation can be separated to give

$$\frac{L\,dy}{y(L-y)} = k\,dt,$$

and solved by integrating both sides. Expanding the left side in partial fractions and integrating, we get

$$\int \left(\frac{1}{y} + \frac{1}{L-y}\right) dy = kt + C.$$

Assuming that $0 < y < L$, we therefore obtain

$$\ln y - \ln(L - y) = kt + C,$$
$$\ln\left(\frac{y}{L - y}\right) = kt + C.$$

We can solve this equation for $y$ by first taking exponentials of both sides:

$$\frac{y}{L - y} = e^{kt+C} = C_1 e^{kt}$$
$$y = (L - y)C_1 e^{kt}$$
$$y = \frac{C_1 L e^{kt}}{1 + C_1 e^{kt}},$$

where $C_1 = e^C$. ■

Many applied problems lead to separable equations. Three extended examples are given in Section 8.9 of *Single-Variable Calculus*. These concern solution concentrations, the rate of a chemical reaction, and determining a family of curves each of which intersects all the curves of a second family at right angles.

## First-Order Homogeneous Equations

A first-order DE of the form

$$\frac{dy}{dx} = f\left(\frac{y}{x}\right)$$

is said to be **homogeneous**. This is a *different* use of the term homogeneous from that in the previous section, which applied only to linear equations. Here homogeneous refers to the fact that $y/x$, and therefore $g(x, y) = f(y/x)$ is *homogeneous of degree 0* in the sense described after Example 7 in Section 4.5. Such a homogeneous equation can be transformed into a separable equation (and therefore solved) by means of a change of dependent variable. If we set

$$v = \frac{y}{x}, \qquad \text{or equivalently} \qquad y = xv(x),$$

then we have

$$\frac{dy}{dx} = v + x\frac{dv}{dx}$$

and the original differential equation transforms into

$$\frac{dv}{dx} = \frac{f(v) - v}{x},$$

which is separable.

**■ EXAMPLE 4** Solve the equation

$$\frac{dy}{dx} = \frac{x^2 + xy}{xy + y^2}.$$

**SOLUTION** The equation is homogeneous. (Divide the numerator and denominator of the right-hand side by $x^2$ to see this.) If $y = vx$ the equation becomes

$$v + x\frac{dv}{dx} = \frac{1+v}{v+v^2} = \frac{1}{v},$$

or

$$x\frac{dv}{dx} = \frac{1-v^2}{v}.$$

Separating variables and integrating, we calculate

$$\begin{aligned}
\int \frac{v\,dv}{1-v^2} &= \int \frac{dx}{x} && \text{let } u = 1 - v^2\\
-\frac{1}{2}\int \frac{du}{u} &= \int \frac{dx}{x}\\
-\ln|u| &= 2\ln|x| + C_1 = \ln C_2 x^2 && (C_1 = \ln C_2)\\
\frac{1}{|u|} &= C_2 x^2\\
|1 - v^2| &= \frac{C_3}{x^2} && (C_3 = 1/C_2)\\
\left|1 - \frac{y^2}{x^2}\right| &= \frac{C_3}{x^2}.
\end{aligned}$$

The solution is best expressed in the form $x^2 - y^2 = C_4$. However, near points where $y \neq 0$, the equation can be solved for $y$ as a function of $x$. ∎

## EXERCISES 10.2

Solve the differential equations in Exercises 1–20.

1. $\dfrac{dy}{dx} = \dfrac{y}{2x}$
2. $\dfrac{dy}{dx} = \dfrac{3y-1}{x}$
3. $\dfrac{dy}{dx} = \dfrac{x^2}{y^2}$
4. $\dfrac{dy}{dx} = x^2y^2$
5. $\dfrac{dy}{dx} = \dfrac{x^2}{y^3}$
6. $\dfrac{dy}{dx} = x^2y^3$
7. $\dfrac{dY}{dt} = tY$
8. $\dfrac{dx}{dt} = e^x \sin t$
9. $\dfrac{dy}{dx} = 1 - y^2$
10. $\dfrac{dy}{dx} = 1 + y^2$
11. $\dfrac{dy}{dt} = 2 + e^y$
12. $\dfrac{dy}{dx} = y^2(1-y)$
13. $\dfrac{dy}{dx} = \sin x \cos^2 y$
14. $x\dfrac{dy}{dx} = y \ln x$
15. $\dfrac{dy}{dx} = \dfrac{x+y}{x-y}$
16. $\dfrac{dy}{dx} = \dfrac{xy}{x^2+2y^2}$
17. $\dfrac{dy}{dx} = \dfrac{x^2+xy+y^2}{x^2}$
18. $\dfrac{dy}{dx} = \dfrac{x^3+3xy^2}{3x^2y+y^3}$
19. $x\dfrac{dy}{dx} = y + x\cos^2\left(\dfrac{y}{x}\right)$
20. $\dfrac{dy}{dx} = \dfrac{y}{x} - e^{-y/x}$
21. Find an equation of the curve in the $xy$-plane which passes through the point (2,3), and has, at every point $(x, y)$ on it, slope equal to $2x/(1+y^2)$.
22. Repeat the previous problem for the point (1,3) and slope $1 + (2y/x)$.
23. Show that the change of variables $\xi = x - x_0$, $\eta = y - y_0$ transforms the equation
$$\frac{dy}{dx} = \frac{ax+by+c}{ex+fy+g}$$
into the homogeneous equation
$$\frac{d\eta}{d\xi} = \frac{a\xi + b\eta}{e\xi + f\eta}$$
provided $(x_0, y_0)$ is the solution of the system
$$\begin{aligned} ax + by + c &= 0\\ ex + fy + g &= 0. \end{aligned}$$
24. Use the technique of the previous exercise to solve the equation $\dfrac{dy}{dx} = \dfrac{x+2y-4}{2x-y-3}$.

# 10.3 EXACT EQUATIONS AND INTEGRATING FACTORS

A first-order differential equation expressed in differential form as

$$M(x, y)\,dx + N(x, y)\,dy = 0,$$

which is equivalent to $\dfrac{dy}{dx} = -\dfrac{M(x, y)}{N(x, y)}$, is said to be **exact** if the left-hand side is the differential of a function $\phi(x, y)$:

$$d\phi(x, y) = M(x, y)\,dx + N(x, y)\,dy.$$

The function $\phi$ is called an **integral function** of the differential equation. The level curves $\phi(x, y) = C$ of $\phi$ are the **solution curves** of the differential equation. For example, the differential equation

$$x\,dx + y\,dy = 0$$

has solution curves given by

$$x^2 + y^2 = C$$

since $d(x^2 + y^2) = 2(x\,dx + y\,dy) = 0$.

**REMARK** The condition that the differential equation $M\,dx + N\,dy = 0$ should be exact is just the condition that the vector field

$$\mathbf{F} = M(x, y)\,\mathbf{i} + N(x, y)\,\mathbf{j}$$

should be *conservative*; the integral function of the differential equation is then the potential function of the vector field. (See Section 8.2.)

A **necessary condition** for the exactness of the DE $M\,dx + N\,dy = 0$ is that

$$\frac{\partial M}{\partial y} = \frac{\partial N}{\partial x};$$

this just says that the mixed partial derivatives $\dfrac{\partial^2 \phi}{\partial x \partial y}$ and $\dfrac{\partial^2 \phi}{\partial y \partial x}$ of the integral function $\phi$ must be equal.

Once you know that an equation is exact, you can often guess the integral function. In any event, $\phi$ can always be found by the same method used to find the potential of a conservative vector field in Section 8.2.

**EXAMPLE 1** Verify that the DE

$$(2x + \sin y - ye^{-x})\,dx + (x\cos y + \cos y + e^{-x})\,dx = 0$$

is exact and find its solution curves.

**SOLUTION** Here $M = 2x + \sin y - ye^{-x}$ and $N = x\cos y + \cos y + e^{-x}$. Since

$$\frac{\partial M}{\partial y} = \cos y - e^{-x} = \frac{\partial N}{\partial x},$$

the DE is exact. We want to find $\phi$ so that

$$\frac{\partial \phi}{\partial x} = M = 2x + \sin y - ye^{-x} \quad \text{and} \quad \frac{\partial \phi}{\partial y} = N = x\cos y + \cos y + e^{-x}.$$

Integrate the first equation with respect to $x$, being careful to allow the constant of integration to depend on $y$:

$$\phi(x, y) = \int (2x + \sin y - ye^{-x})\,dx = x^2 + x\sin y + ye^{-x} + C_1(y).$$

Now substitute this expression into the second equation:

$$x\cos y + \cos y + e^{-x} = \frac{\partial \phi}{\partial y} = x\cos y + e^{-x} + C_1'(y).$$

Thus $C_1'(y) = \cos y$, and $C_1(y) = \sin y + C_2$. (It is because the original DE was exact that the equation for $C_1'(y)$ turned out to be independent of $x$; this had to happen or we could not have found $C_1$ as a function of $y$ only.) Choosing $C_2 = 0$, we find that $\phi(x, y) = x^2 + x\sin y + ye^{-x} + \sin y$ is an integral function for the given DE. The solution curves for the DE are the level curves

$$x^2 + x\sin y + ye^{-x} + \sin y = C.$$

■

## Integrating Factors

Any ordinary differential equation of order 1 and degree 1 can be expressed in differential form: $M\,dx + N\,dy = 0$. However, this latter equation will usually not be exact. It *may* be possible to multiply the equation by an **integrating factor** $\mu(x, y)$ so that the resulting equation

$$\mu(x, y)\,M(x, y)\,dx + \mu(x, y)\,N(x, y)\,dy = 0$$

is exact. In general such integrating factors are difficult to find; they must satisfy the partial differential equation

$$M(x, y)\frac{\partial \mu}{\partial y} - N(x, y)\frac{\partial \mu}{\partial x} = \mu(x, y)\left(\frac{\partial N}{\partial x} - \frac{\partial M}{\partial y}\right),$$

which follows from the necessary condition for exactness stated above. We will not try to solve this equation here.

Sometimes it happens that a differential equation has an integrating factor depending on only one of the two variables. Suppose, for instance, that $\mu(x)$ is an integrating factor for $M\,dx + N\,dy = 0$. Then $\mu(x)$ must satisfy the ordinary differential equation

$$N(x, y)\frac{d\mu}{dx} = \mu(x)\left(\frac{\partial M}{\partial y} - \frac{\partial N}{\partial x}\right),$$

or,

$$\frac{1}{\mu(x)}\frac{d\mu}{dx} = \frac{\dfrac{\partial M}{\partial y} - \dfrac{\partial N}{\partial x}}{N(x, y)}.$$

This equation can be solved (by integration) for $\mu$ as a function of $x$ alone *provided that the right-hand side is independent of* $y$.

**EXAMPLE 2** Show that $(x + y^2)\,dx + xy\,dy = 0$ has an integrating factor depending only on $x$, find it, and solve the equation.

***SOLUTION*** Here $M = x + y^2$ and $N = xy$. Since

$$\frac{\dfrac{\partial M}{\partial y} - \dfrac{\partial N}{\partial x}}{N(x, y)} = \frac{2y - y}{xy} = \frac{1}{x}$$

does not depend on $y$, the equation has an integrating factor depending only on $x$. This factor is given by $d\mu/\mu = dx/x$. Evidently $\mu = x$ is a suitable integrating factor; if we multiply the given differential equation by $x$, we obtain

$$0 = (x^2 + xy^2)\,dx + x^2y\,dy = d\left(\frac{x^3}{3} + \frac{x^2y^2}{2}\right).$$

The solution is therefore $2x^3 + 3x^2y^2 = C$. ■

***REMARK*** Of course, it may be possible to find an integrating factor depending on $y$ instead of $x$. See Exercises 7–9 below. It is also possible to look for integrating factors that depend on specific combinations of $x$ and $y$, for instance, $xy$. See Exercise 10.

## EXERCISES 10.3

Show that the DEs in Exercises 1–4 are exact, and solve them.

1. $(xy^2 + y)\,dx + (x^2y + x)\,dy = 0$
2. $(e^x \sin y + 2x)\,dx + (e^x \cos y + 2y)\,dy = 0$
3. $e^{xy}(1 + xy)\,dx + x^2e^{xy}\,dy = 0$
4. $\left(2x + 1 - \dfrac{y^2}{x^2}\right) dx + \dfrac{2y}{x}\,dy = 0$

Show that the DEs in Exercises 5–6 admit integrating factors that are functions of $x$ alone. Then solve the equations.

5. $(x^2 + 2y)\,dx - x\,dy = 0$
6. $(xe^x + x \ln y + y)\,dx + \left(\dfrac{x^2}{y} + x \ln x + x \sin y\right) dy = 0$
7. What condition must the coefficients $M(x, y)$ and $N(x, y)$ satisfy if the equation $M\,dx + N\,dy = 0$ is to have an integrating factor of the form $\mu(y)$, and what DE must the integrating factor satisfy?
8. Find an integrating factor of the form $\mu(y)$ for the equation
$$2y^2(x + y^2)\,dx + xy(x + 6y^2)\,dy = 0,$$
and hence solve the equation. *Hint:* see Exercise 7.
9. Find an integrating factor of the form $\mu(y)$ for the equation $y\,dx - (2x + y^3e^y)\,dy = 0$, and hence solve the equation. *Hint:* see Exercise 7.
10. What condition must the coefficients $M(x, y)$ and $N(x, y)$ satisfy if the equation $M\,dx + N\,dy = 0$ is to have an integrating factor of the form $\mu(xy)$, and what DE must the integrating factor satisfy?
11. Find an integrating factor of the form $\mu(xy)$ for the equation
$$\left(x \cos x + \frac{y^2}{x}\right) dx - \left(\frac{x \sin x}{y} + y\right) dy = 0,$$
and hence solve the equation. *Hint:* see Exercise 10.

# 10.4 FIRST-ORDER LINEAR EQUATIONS

A first-order **linear** differential equation is one of the type

$$\frac{dy}{dx} + p(x)y = q(x),$$

where $p(x)$ and $q(x)$ are given functions, which we assume to be continuous. The equation is *homogeneous* (in the sense described in Section 10.1) provided that $q(x) = 0$. In that case, the given linear equation is separable:

$$\frac{dy}{y} = -p(x)\,dx,$$

which can be solved by integrating both sides. Nonhomogeneous first-order linear equations can be solved by the following procedure involving the calculation of an integrating factor.

Let $\mu(x)$ be any antiderivative of $p(x)$:

$$\mu(x) = \int p(x)\,dx, \qquad \text{and} \qquad \frac{d\mu}{dx} = p(x).$$

If $y = y(x)$ satisfies the given equation, then we calculate, using the Product Rule,

$$\begin{aligned}\frac{d}{dx}\big(e^{\mu(x)}y(x)\big) &= e^{\mu(x)}\frac{dy}{dx} + e^{\mu(x)}\frac{d\mu}{dx}\,y(x)\\ &= e^{\mu(x)}\left(\frac{dy}{dx} + p(x)y\right) = e^{\mu(x)}\,q(x).\end{aligned}$$

Therefore,

$$e^{\mu(x)}\,y(x) = \int e^{\mu(x)}\,q(x)\,dx$$

or

$$y(x) = e^{-\mu(x)}\int e^{\mu(x)}\,q(x)\,dx.$$

We reuse this method, rather than the final formula, in the following examples. We call $e^{\mu(x)}$ an **integrating factor** of the given differential equation because, if we multiply the equation by $e^{\mu(x)}$, the left side becomes the derivative of $e^{\mu(x)}y(x)$.

**EXAMPLE 1** Solve $\dfrac{dy}{dx} + \dfrac{y}{x} = 1$ for $x > 0$.

***SOLUTION*** Here $p(x) = 1/x$ and $q(x) = 1$. We want $d\mu/dx = 1/x$, so that $\mu(x) = \ln x$ (for $x > 0$). Thus $e^{\mu(x)} = x$ and we calculate

$$\frac{d}{dx}\,(e^{\mu(x)}y) = \frac{d}{dx}\,(xy) = x\,\frac{dy}{dx} + y = x\left(\frac{dy}{dx} + \frac{y}{x}\right) = x.$$

Therefore

$$xy = \int x\,dx = \frac{1}{2}x^2 + C.$$

Finally,

$$y = \frac{1}{x}\left(\frac{1}{2}x^2 + C\right) = \frac{x}{2} + \frac{C}{x}.$$

This function is a solution of the given equation for any value of the constant $C$. ■

**EXAMPLE 2** Solve $\dfrac{dy}{dx} + xy = x^3$.

***SOLUTION*** Here $p(x) = x$ so $\mu(x) = x^2/2$. We calculate

$$\frac{d}{dx}\left(e^{x^2/2}y\right) = e^{x^2/2}\frac{dy}{dx} + e^{x^2/2}xy = x^3e^{x^2/2}.$$

Thus,

$$\begin{aligned} e^{x^2/2}y &= \int x^3 e^{x^2/2}\,dx \qquad \begin{array}{lll} \text{Let} & U = x^2 & dV = x\,e^{x^2/2}\,dx \\ \text{Then} & dU = 2x\,dx & V = e^{x^2/2} \end{array} \\ &= x^2 e^{x^2/2} - 2\int x\,e^{x^2/2}\,dx \\ &= x^2 e^{x^2/2} - 2\,e^{x^2/2} + C, \end{aligned}$$

and, finally,

$$y = x^2 - 2 + Ce^{-x^2/2}.$$

Example 8 of Section 8.9 in *Single-Variable Calculus* illustrates the application of the technique for solving a first-order linear DE to a typical *stream of payments* problem. We conclude this section with an applied example involving current flow in an electric circuit.

**EXAMPLE 3** **An inductance-resistance circuit**

An electric circuit (Figure 10.1) contains a constant DC voltage source of $V$ volts, a switch, a resistor of size $R$ ohms, and an inductor of size $L$ henrys. The circuit has no capacitance. The switch, initially open so that no current is flowing, is closed at time $t = 0$ so that current begins to flow at that time. If the inductance $L$ were zero, the current would suddenly jump from 0 amperes when $t < 0$ to $I = V/R$ amperes when $t > 0$. However, if $L > 0$ the current cannot change instantaneously; it will depend on time $t$. Let the current $t$ seconds after the switch is closed be $I(t)$ amperes. It is known that $I(t)$ satisfies the initial-value problem

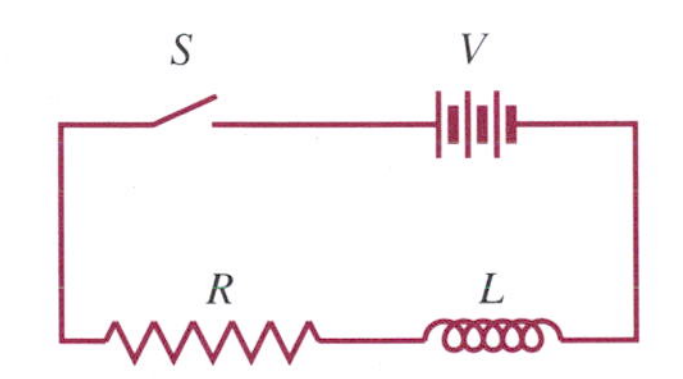


**Figure 10.1** An inductance-resistance circuit

$$\begin{cases} L\dfrac{dI}{dt} + RI = V \\ I(0) = 0. \end{cases}$$

Find $I(t)$. What is $\lim_{t\to\infty} I(t)$? How long does it take after the switch is closed for the current to rise to 90% of its limiting value?

***SOLUTION*** The DE can be written in the form $\dfrac{dI}{dt} + \dfrac{R}{L}I = \dfrac{V}{L}$. It is linear and has integrating factor $e^{\mu(t)}$, where

$$\mu(t) = \int \frac{R}{L}\,dt = \frac{Rt}{L}.$$

Therefore

$$\begin{aligned} \frac{d}{dt}\left(e^{Rt/L}I\right) &= e^{Rt/L}\left(\frac{dI}{dt} + \frac{R}{L}I\right) = e^{Rt/L}\frac{V}{L} \\ e^{Rt/L}I &= \frac{V}{L}\int e^{Rt/L}\,dt = \frac{V}{R}e^{Rt/L} + C \\ I(t) &= \frac{V}{R} + Ce^{-Rt/L}. \end{aligned}$$

Since $I(0) = 0$, we have $0 = (V/R) + C$, so $C = -V/R$. Thus the current flowing at any time $t > 0$ is

$$I(t) = \frac{V}{R}\left(1 - e^{-Rt/L}\right).$$

It is clear from this solution that $\lim_{t\to\infty} I(t) = V/R$; the *steady state* current is the current that would flow if the inductance were zero.

$I(t)$ will be 90% of this limiting value when

$$\frac{V}{R}\left(1 - e^{-Rt/L}\right) = \frac{90}{100}\frac{V}{R}.$$

This equation implies that $e^{-Rt/L} = 1/10$, or $t = (L \ln 10)/R$. The current will grow to 90% of its limiting value in $(L \ln 10)/R$ seconds. ■

## EXERCISES 10.4

Solve the differential equations in Exercises 1–6.

**1.** $\dfrac{dy}{dx} - \dfrac{2y}{x} = x^2$

**2.** $\dfrac{dy}{dx} + \dfrac{2y}{x} = \dfrac{1}{x^2}$

**3.** $\dfrac{dy}{dx} + 2y = 3$

**4.** $\dfrac{dy}{dx} + y = e^x$

**5.** $\dfrac{dy}{dx} + y = x$

**6.** $\dfrac{dy}{dx} + 2e^x y = e^x$

Solve the initial-value problems in Exercises 7–10.

**7.** $\begin{cases} \dfrac{dy}{dt} + 10y = 1 \\ y(1/10) = 2/10 \end{cases}$

**8.** $\begin{cases} \dfrac{dy}{dx} + 3x^2 y = x^2 \\ y(0) = 1 \end{cases}$

**9.** $\begin{cases} y' + (\cos x)y = 2xe^{-\sin x} \\ y(\pi) = 0 \end{cases}$

**10.** $\begin{cases} x^2 y' + y = x^2 e^{1/x} \\ y(1) = 3e \end{cases}$

**11.** An object of mass $m$ falling near the surface of the earth is retarded by air resistance proportional to its velocity so that, according to Newton's second law of motion,

$$m\frac{dv}{dt} = mg - kv,$$

where $v = v(t)$ is the velocity of the object at time $t$, and $g$ is the acceleration of gravity near the surface of the earth. Assuming that the object falls from rest at time $t = 0$, that is, $v(0) = 0$, find the velocity $v(t)$ for any $t > 0$ (up until the object strikes the ground). Show $v(t)$ approaches a limit as $t \to \infty$. Do you need the explicit formula for $v(t)$ to determine this limiting velocity?

**12.** Repeat Exercise 11 assuming that the air resistance is proportional to the square of the velocity so that the equation of motion is

$$m\frac{dv}{dt} = mg - kv^2.$$

(Note that this equation is not linear, but it is separable.)

**13.** A savings account has an initial balance of \$900 and earns interest at a nominal annual rate of 5% paid continuously into the account. The account is also being continuously depleted by service charges at a rate of \$50/year. Write a DE satisfied by the balance $y(t)$ in the account after $t$ years. Find the balance in the account after five years.

# 10.5 EXISTENCE, UNIQUENESS, AND NUMERICAL METHODS

A general first-order differential equation of the form

$$\frac{dy}{dx} = f(x, y)$$

specifies a slope $f(x, y)$ at every point $(x, y)$ in the domain of $f$, and therefore represents a **slope field**. Such a slope field can be represented graphically by drawing short line segments of the indicated slope at many points in the $xy$-plane. Slope fields resemble vector fields, but the segments are usually drawn having the same length and without arrowheads. Figure 10.2 portrays the slope field for the differential equation

$$\frac{dy}{dx} = x - y.$$

Solving a typical initial-value problem

$$\begin{cases} \dfrac{dy}{dx} = f(x, y) \\ y(x_0) = y_0 \end{cases}$$

involves finding a function $y = \phi(x)$ such that

$$\phi'(x) = f\big(x, \phi(x)\big) \qquad \text{and} \qquad \phi(x_0) = y_0.$$

The graph of the equation $y = \phi(x)$ is a curve passing through $(x_0, y_0)$ that is tangent to the slope-field at each point. Such curves are called **solution curves** of the differential equation. Figure 10.2 shows four solution curves for $y' = x - y$ corresponding to the initial conditions $y(0) = C$, where $C = -2, -1, 0$, and 1.

The DE $y' = x - y$ is linear, and can be solved explicitly by the method of Section 10.4. Indeed, the solution satisfying $y(0) = C$ is $y = x - 1 + (C + 1)e^{-x}$. Most differential equations of the form $y' = f(x, y)$ cannot be solved for $y$ as an explicit function of $x$, so we must use numerical approximation methods to find the value of a solution function $\phi(x)$ at particular points.

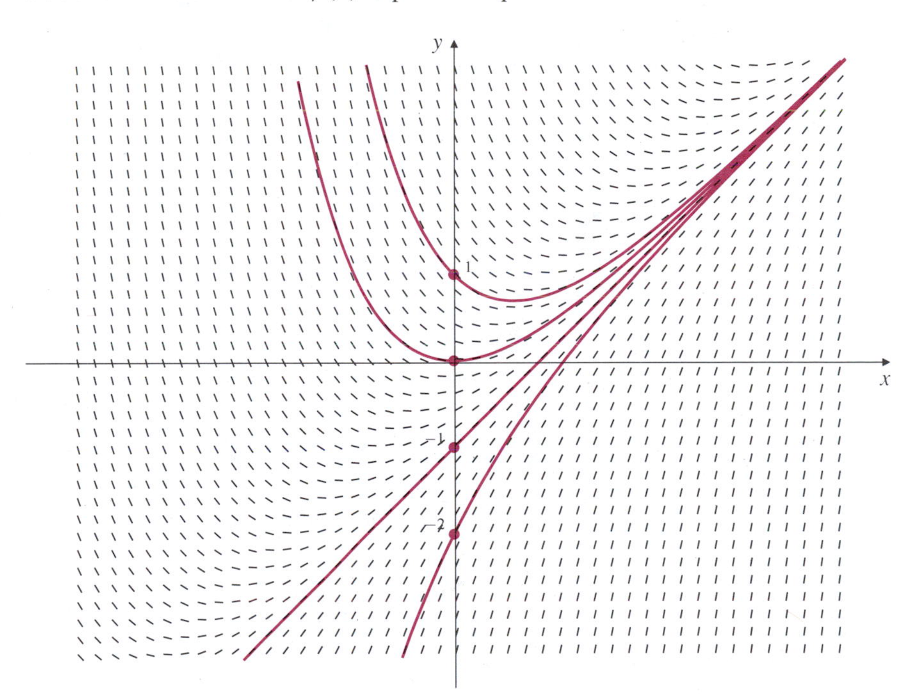


**Figure 10.2** The slope field for the DE $y' = x - y$ and four solution curves for this DE

## Existence and Uniqueness of Solutions

Even if we cannot calculate an explicit solution of an initial-value problem, it is important to know when the problem has a solution, and whether that solution is unique.

**THEOREM 3** **An existence and uniqueness theorem for first-order initial-value problems**

Suppose that $f(x, y)$ and $f_2(x, y) = (\partial/\partial y)f(x, y)$ are continuous on a rectangle $R$ of the form $a \le x \le b$, $c \le y \le d$, containing the point $(x_0, y_0)$ in its interior. Then there exists a number $\delta > 0$ and a *unique* function $\phi(x)$ defined and having a continuous derivative on the interval $(x_0 - \delta, x_0 + \delta)$ such that $\phi(x_0) = y_0$ and $\phi'(x) = f\big(x, \phi(x)\big)$ for $x_0 - \delta < x < x_0 + \delta$. In other words, the initial-value problem

$$\begin{cases} \dfrac{dy}{dx} = f(x, y) \\ y(x_0) = y_0 \end{cases} \qquad (*)$$

has a unique solution on $(x_0 - \delta, x_0 + \delta)$.

We give only an outline of the proof here. Any solution $y = \phi(x)$ of the initial-value problem $(*)$ must also satisfy the **integral equation**

$$\phi(x) = y_0 + \int_{x_0}^{x} f\big(t, \phi(t)\big)\, dt, \qquad (**)$$

and, conversely, any solution of the integral equation $(**)$ must also satisfy the initial-value problem $(*)$. A sequence of approximations $\phi_n(x)$ to a solution of $(**)$ can be constructed as follows:

$$\begin{aligned} \phi_0(x) &= y_0 \\ \phi_{n+1}(x) &= y_0 + \int_{x_0}^{x} f\big(t, \phi_n(t)\big)\, dt \qquad \text{for} \quad n = 0, 1, 2, \ldots \end{aligned}$$

(These are called **Picard iterations.**) The proof of Theorem 3 involves showing that

$$\lim_{n\to\infty} \phi_n(x) = \phi(x)$$

exists on an interval $(x_0 - \delta, x_0 + \delta)$, and that the resulting limit $\phi(x)$ satisfies the integral equation $(**)$. The details can be found in more advanced texts on differential equations and analysis.

***REMARK*** Some initial-value problems can have non-unique solutions. For example, the functions $y_1(x) = x^3$ and $y_2(x) = 0$ both satisfy the initial-value problem

$$\begin{cases} \dfrac{dy}{dx} = 3y^{2/3} \\ y(0) = 0. \end{cases}$$

In this case $f(x, y) = 3y^{2/3}$ is continuous on the whole $xy$-plane. However $\partial f/\partial y = 2y^{-1/3}$ is not continuous on the $x$-axis, and is therefore not continuous on any rectangle containing $(0, 0)$ in its interior. The conditions of Theorem 3 are not satisfied and the initial-value problem has a solution but not a unique one.

**REMARK** The unique solution $y = \phi(x)$ to the initial-value problem (∗) guaranteed by Theorem 3 may not be defined on the whole interval $[a, b]$, because it can "escape" from the rectangle $R$ through the top or bottom edges. Even if $f(x, y)$ and $(\partial/\partial y) f(x, y)$ are continuous on the whole $xy$-plane, the solution may not be defined on the whole real line. For example

$$y = \frac{1}{1-x} \quad \text{satisfies the initial-value problem} \quad \begin{cases} \dfrac{dy}{dx} = y^2 \\ y(0) = 1 \end{cases}$$

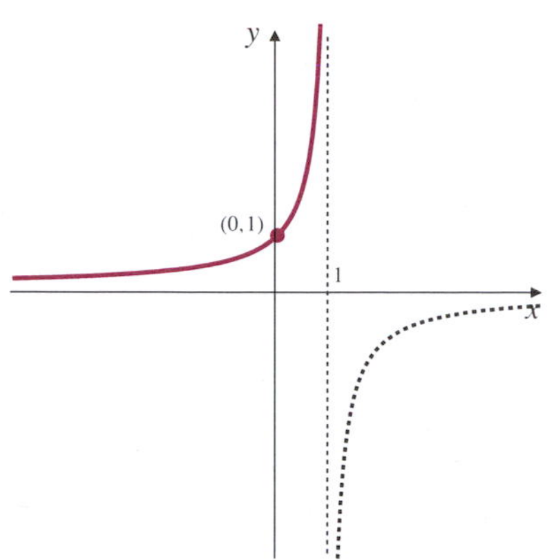


**Figure 10.3** The solution to $y' = y^2$, $y(0) = 1$ is the part of the curve $y = 1/(1-x)$ to the left of the vertical asymptote at $x = 1$

but only for $x < 1$. Starting from (0, 1), we can follow the solution curve as far as we want to the left of $x = 0$, but to the right of $x = 0$ the curve recedes to $\infty$ as $x \to 1-$. It makes no sense to regard the part of the curve to the right of $x = 1$ as part of the solution curve to the initial-value problem.

## Numerical Methods

Suppose that the conditions of Theorem 3 are satisfied, so we know that the initial-value problem

$$\begin{cases} \dfrac{dy}{dx} = f(x, y) \\ y(x_0) = y_0 \end{cases}$$

has a unique solution $y = \phi(x)$ on some interval containing $x_0$. Even if we cannot solve the differential equation and find $\phi(x)$ explicitly, we can still try to find approximate values $y_n$ for $\phi(x_n)$ at a sequence of points

$$x_0, \quad x_1 = x_0 + h, \quad x_2 = x_0 + 2h, \quad x_3 = x_0 + 3h, \quad \ldots$$

starting at $x_0$. Here $h > 0$ (or $h < 0$) is called the **step size** of the approximation scheme. In the remainder of this section we will describe three methods for constructing the approximations $\{y_n\}$, namely

1. The Euler method,
2. The improved Euler method, and
3. The fourth-order Runge–Kutta method.

Each of these methods starts with the given value of $y_0$ and provides a formula for constructing $y_{n+1}$ when you know $y_n$. The three methods are listed above in increasing order of the complexity of their formulas, but the more complicated formulas produce much better approximations for any given step size $h$.

**The Euler method** involves approximating the solution curve $y = \phi(x)$ by a polygonal line (a sequence of straight line segments joined end to end), where each segment has horizontal length $h$, and has slope determined by the value of $f(x, y)$ at the end of the previous segment. Thus, if $x_n = x_0 + nh$, then

$$\begin{aligned} y_1 &= y_0 + f(x_0, y_0)h \\ y_2 &= y_1 + f(x_1, y_1)h \\ y_3 &= y_2 + f(x_2, y_2)h \end{aligned}$$

and, in general,

**Iteration formulas for Euler's method**

$$x_{n+1} = x_n + h, \qquad y_{n+1} = y_n + hf(x_n, y_n).$$

**EXAMPLE 1** Use Euler's method to find approximate values for the solution of the initial-value problem

$$\begin{cases} \dfrac{dy}{dx} = x - y \\ y(0) = 1 \end{cases}$$

on the interval [0, 1] using

(a) five steps of size $h = 0.2$, and

(b) ten steps of size $h = 0.1$.

Calculate the error at each step, given that the problem (which involves a linear equation and so can be solved explicitly) has solution $y = \phi(x) = x - 1 + 2e^{-x}$.

**SOLUTION**

(a) Here we have $f(x, y) = x - y$, $x_0 = 0$, $y_0 = 1$, and $h = 0.2$, so that

$$x_n = \frac{n}{5}, \qquad y_{n+1} = y_n + 0.2(x_n - y_n),$$

and the error is $e_n = \phi(x_n) - y_n$ for $n = 0, 1, 2, 3, 4$, and 5. The results of the calculation, which was done easily using a computer spreadsheet program, are presented in Table 1.

**Table 1.** Euler approximations with $h = 0.2$

| $n$ | $x_n$ | $y_n$ | $f(x_n, y_n)$ | $y_{n+1}$ | $e_n = \phi(x_n) - y_n$ |
|---|---|---|---|---|---|
| 0 | 0.0 | 1.000000 | −1.000000 | 0.800000 | 0.000000 |
| 1 | 0.2 | 0.800000 | −0.600000 | 0.680000 | 0.037462 |
| 2 | 0.4 | 0.680000 | −0.280000 | 0.624000 | 0.060640 |
| 3 | 0.6 | 0.624000 | −0.024000 | 0.619200 | 0.073623 |
| 4 | 0.8 | 0.619200 | 0.180800 | 0.655360 | 0.079458 |
| 5 | 1.0 | 0.655360 | 0.344640 | | 0.080399 |

The exact solution $y = \phi(x)$ and the polygonal line representing the Euler approximation are shown in Figure 10.4. The approximation lies below the solution curve, as is reflected in the positive values in the last column of Table 1, representing the error at each step.

(b) Here we have $h = 0.1$, so that

$$x_n = \frac{n}{10}, \qquad y_{n+1} = y_n + 0.1(x_n - y_n)$$

for $n = 0, 1, \ldots, 10$. Again we present the results in tabular form:

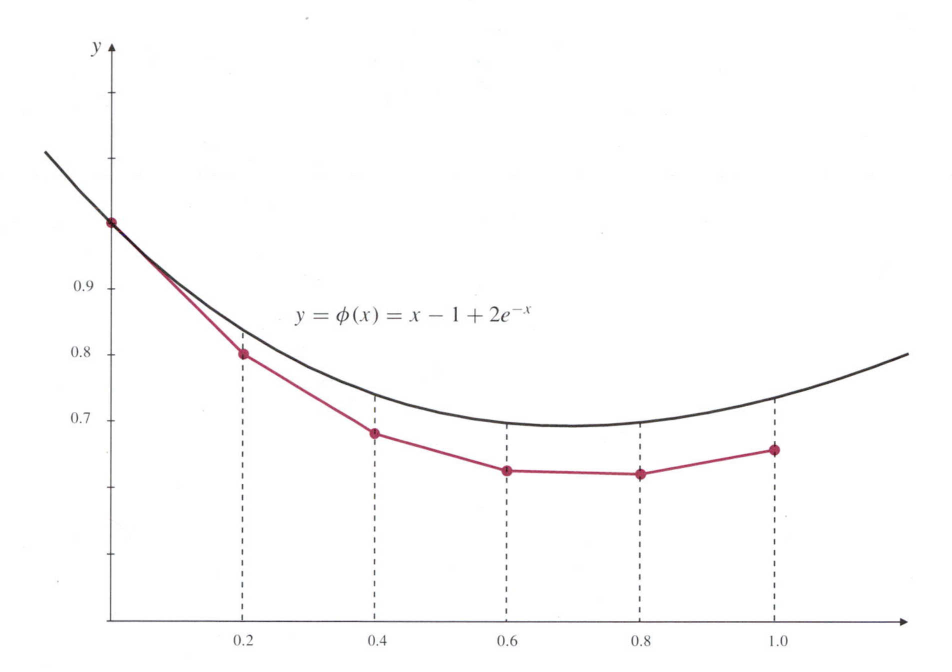


**Figure 10.4** The solution $y = \phi(x)$ to $y' = x - y$, $y(0) = 1$, and an Euler approximation to it on $[0, 1]$ with step size $h = 0.2$

**Table 2.** Euler approximations with $h = 0.1$

| $n$ | $x_n$ | $y_n$ | $f(x_n, y_n)$ | $y_{n+1}$ | $e_n = \phi(x_n) - y_n$ |
|---|---|---|---|---|---|
| 0 | 0.0 | 1.000000 | −1.000000 | 0.900000 | 0.000000 |
| 1 | 0.1 | 0.900000 | −0.800000 | 0.820000 | 0.009675 |
| 2 | 0.2 | 0.820000 | −0.620000 | 0.758000 | 0.017462 |
| 3 | 0.3 | 0.758000 | −0.458000 | 0.712200 | 0.023636 |
| 4 | 0.4 | 0.712200 | −0.312200 | 0.680980 | 0.028440 |
| 5 | 0.5 | 0.680980 | −0.180980 | 0.662882 | 0.032081 |
| 6 | 0.6 | 0.662882 | −0.062882 | 0.656594 | 0.034741 |
| 7 | 0.7 | 0.656594 | 0.043406 | 0.660934 | 0.036577 |
| 8 | 0.8 | 0.660934 | 0.139066 | 0.674841 | 0.037724 |
| 9 | 0.9 | 0.674841 | 0.225159 | 0.697357 | 0.038298 |
| 10 | 1.0 | 0.697357 | 0.302643 | | 0.038402 |

Observe that the error at the end of the first step is about one-quarter of the error at the end of the first step in part (a), but the final error at $x = 1$ is only about half as large as in part (a). This behaviour is characteristic of Euler's Method. ■

If we decrease the step size $h$ it takes more steps ($n = |x - x_0|/h$) to get from the starting point $x_0$ to a particular value $x$ where we want to know the value of the solution. For Euler's method it can be shown that the error at each step decreases on average proportionally to $h^2$, but the errors can accumulate from step to step, so the error at $x$ can be expected to decrease proportionally to $nh^2 = |x - x_0|h$. This is consistent with the results of Example 1. Decreasing $h$ and so increasing $n$ is costly in terms of computing resourses, so we would like to find ways of reducing the error without increasing the step size. This is similar to developing better techniques than the Trapezoid Rule for evaluating definite integrals numerically.

**The improved Euler method** is a step in this direction. The accuracy of the Euler method is hampered by the fact that the slope of each segment in the approximating polygonal line is determined by the value of $f(x, y)$ at one endpoint of the segment. Since $f$ varies along the segment, we would expect to do better by using, say, the average value of $f(x, y)$ at the two ends of the segment, that is, by calculating $y_{n+1}$ from the formula

$$y_{n+1} = y_n + h\frac{f(x_n, y_n) + f(x_{n+1}, y_{n+1})}{2}.$$

Unfortunately, $y_{n+1}$ appears on both sides of this equation, and we can't usually solve the equation for $y_{n+1}$. We can get around this difficulty by replacing $y_{n+1}$ on the right side by its Euler approximation $y_n + hf(x_n, y_n)$. The resulting formula is the basis for the improved Euler method.

**Iteration formulas for the improved Euler method**

$$x_{n+1} = x_n + h$$
$$u_{n+1} = y_n + h\,f(x_n, y_n)$$
$$y_{n+1} = y_n + h\,\frac{f(x_n, y_n) + f(x_{n+1}, u_{n+1})}{2}.$$

**EXAMPLE 2** Use the improved Euler method with $h = 0.2$ to find approximate values for the solution to the initial-value problem of Example 1 on $[0, 1]$. Compare the errors with those obtained by the Euler method.

***SOLUTION*** Table 3 summarizes the calculation of five steps of the improved Euler method for $f(x, y) = x - y$, $x_0 = 0$, and $y_0 = 1$.

**Table 3.** Improved Euler approximations with $h = 0.2$

| $n$ | $x_n$ | $y_n$ | $u_{n+1}$ | $y_{n+1}$ | $e_n = \phi(x_n) - y_n$ |
|---|---|---|---|---|---|
| 0 | 0.0 | 1.000000 | 0.800000 | 0.840000 | 0.000000 |
| 1 | 0.2 | 0.840000 | 0.712000 | 0.744800 | −0.002538 |
| 2 | 0.4 | 0.744800 | 0.675840 | 0.702736 | −0.004160 |
| 3 | 0.6 | 0.702736 | 0.682189 | 0.704244 | −0.005113 |
| 4 | 0.8 | 0.704244 | 0.723395 | 0.741480 | −0.005586 |
| 5 | 1.0 | 0.741480 | 0.793184 | | −0.005721 |

Observe that the errors are considerably less than one-tenth those obtained in Example 1(a). Of course, more calculations are necessary at each step, but the number of evaluations of $f(x, y)$ required is only twice the number required for Example 1(a). As for numerical integration, if $f$ is complicated, it is these function evaluations that constitute most of the computational "cost" of computing numerical solutions. ■

***REMARK*** It can be shown for well-behaved functions $f$ that the error at each step in the improved Euler method is bounded by a multiple of $h^3$, rather than $h^2$ as for the (unimproved) Euler method. Thus the cumulative error at $x$ can be bounded by a constant times $|x - x_0|h^2$. If Example 2 is repeated with ten steps of size $h = 0.1$, the error at $n = 10$ (that is, at $x = 1$) is $-0.001323$, which is about one-fourth the size of the error at $x = 1$ with $h = 0.2$.

**The fourth-order Runge–Kutta method** further improves upon the improved Euler method, but at the expense of requiring more complicated calculations at each step. It requires four evaluations of $f(x, y)$ at each step, but the error at each step is less than a constant times $h^5$, so the cumulative error decreases like $h^4$ as $h$ decreases. Like the improved Euler method, this method involves calculating a certain kind of average slope for each segment in the polygonal approximation to the solution to the initial-value problem. We present the appropriate formulas below, but cannot derive them here.

**Iteration formulas for the Runge–Kutta method**

$$
\begin{aligned}
x_{n+1} &= x_n + h \\
p_n &= f(x_n, y_n) \\
q_n &= f\left(x_n + \frac{h}{2}, y_n + \frac{h}{2}p_n\right) \\
r_n &= f\left(x_n + \frac{h}{2}, y_n + \frac{h}{2}q_n\right) \\
s_n &= f(x_n + h, y_n + hr_n) \\
y_{n+1} &= y_n + h\,\frac{p_n + 2q_n + 2r_n + s_n}{6}.
\end{aligned}
$$

**EXAMPLE 3** Use the fourth-order Runge-Kutta method with $h = 0.2$ to find approximate values for the solution to the initial-value problem of Example 1 on $[0, 1]$. Compare the errors with those obtained by the Euler and improved Euler methods.

***SOLUTION*** Table 4 summarizes the calculation of five steps of the Runge–Kutta method for $f(x, y) = x - y$, $x_0 = 0$, and $y_0 = 1$ according to the formulas above. The table does not show the values of the intermediate quantities $p_n$, $q_n$, $r_n$, and $s_n$, but columns for these quantities were included in the spreadsheet in which the calculations were made.

**Table 4.** Fourth-order Runge–Kutta approximations with $h = 0.2$

| $n$ | $x_n$ | $y_n$ | $e_n = \phi(x_n) - y_n$ |
|---|---|---|---|
| 0 | 0.0 | 1.000000 | 0.0000000 |
| 1 | 0.2 | 0.837467 | −0.0000052 |
| 2 | 0.4 | 0.740649 | −0.0000085 |
| 3 | 0.6 | 0.697634 | −0.0000104 |
| 4 | 0.8 | 0.698669 | −0.0000113 |
| 5 | 1.0 | 0.735770 | −0.0000116 |

The errors here are about 1/500 of the size of the errors obtained with the improved Euler method, and about 1/7,000 of the size of the errors obtained with the Euler method. This great improvement was achieved at the expense of doubling the number of function evaluations required in the improved Euler method and quadrupling the number required in the Euler method. If we use 10 steps of size $h = 0.1$ in the Runge–Kutta method, the error at $x = 1$ is reduced to $-6.66482 \times 10^{-7}$, which is less than 1/16 of its value when $h = 0.2$. ■

Our final example shows what can happen with numerical approximations to a solution that is unbounded.

**EXAMPLE 4** Obtain solutions at $x = 0.4$, $x = 0.8$, and $x = 1.0$ for solutions to the initial-value problem

$$
\begin{cases} y' = y^2 \\ y(0) = 1 \end{cases}
$$

using all three methods described above, and using step sizes $h = 0.2$, $h = 0.1$, and $h = 0.05$ for each method. What do the results suggest about the values of the solution at these points? Compare the results with the actual solution $y = 1/(1-x)$.

**SOLUTION** The various approximations are calculated using the various formulas described above for $f(x, y) = y^2$, $x_0 = 0$, and $y_0 = 1$. The results are presented in Table 5.

**Table 5.** Comparing methods and step sizes for $y' = y^2$, $y(0) = 1$

| | $h = 0.2$ | $h = 0.1$ | $h = 0.05$ |
|---|---|---|---|
| Euler | | | |
| $x = 0.4$ | 1.488000 | 1.557797 | 1.605224 |
| $x = 0.8$ | 2.676449 | 3.239652 | 3.793197 |
| $x = 1.0$ | 4.109124 | 6.128898 | 9.552668 |
| Improved Euler | | | |
| $x = 0.4$ | 1.640092 | 1.658736 | 1.664515 |
| $x = 0.8$ | 4.190396 | 4.677726 | 4.897519 |
| $x = 1.0$ | 11.878846 | 22.290765 | 43.114668 |
| Runge–Kutta | | | |
| $x = 0.4$ | 1.666473 | 1.666653 | 1.666666 |
| $x = 0.8$ | 4.965008 | 4.996663 | 4.999751 |
| $x = 1.0$ | 41.016258 | 81.996399 | 163.983395 |

Little useful information can be read from the Euler results. The improved Euler results suggest that the solution exists at $x = 0.4$ and $x = 0.8$, but likely not at $x = 1$. The Runge–Kutta results confirm this, and suggest that $y(0.4) = 5/3$ and $y(0.8) = 5$, which are the correct values provided by the actual solution $y = 1/(1-x)$. They also suggest very strongly that the solution "blows up" at (or near) $x = 1$. ■

## EXERCISES 10.5

A computer is almost essential for doing most of these exercises. The calculations are easily done with a spreadsheet program in which formulas for calculating the various quantities involved can be replicated down columns to automate the iteration process.

1. Use the Euler method with step sizes (a) $h = 0.2$, (b) $h = 0.1$, and (c) $h = 0.05$ to approximate $y(2)$ given that $y' = x + y$ and $y(1) = 0$.

2. Repeat Exercise 1 using the improved Euler method.

3. Repeat Exercise 1 using the Runge–Kutta method.

4. Use the Euler method with step sizes (a) $h = 0.2$, and (b) $h = 0.1$ to approximate $y(2)$ given that $y' = xe^{-y}$ and $y(0) = 0$.

5. Repeat Exercise 4 using the improved Euler method.

6. Repeat Exercise 4 using the Runge–Kutta method.

7. Use the Euler method with (a) $h = 0.2$, (b) $h = 0.1$, and (c) $h = 0.05$ to approximate $y(1)$ given that $y' = \cos y$ and $y(0) = 0$.

8. Repeat Exercise 7 using the improved Euler method.

9. Repeat Exercise 7 using the Runge–Kutta method.

10. Use the Euler method with (a) $h = 0.2$, (b) $h = 0.1$, and (c) $h = 0.05$ to approximate $y(1)$ given that $y' = \cos(x^2)$ and $y(0) = 0$.

11. Repeat Exercise 10 using the improved Euler method.

12. Repeat Exercise 10 using the Runge–Kutta method.

Solve the integral equations in Exercises 13–14 by rephrasing them as initial-value problems.

13. $y(x) = 2 + \int_1^x \big(y(t)\big)^2\,dt$. *Hint:* find $\dfrac{dy}{dx}$ and $y(1)$.

14. $u(x) = 1 + 3\int_2^x t^2 u(t)\,dt$. *Hint:* find $\dfrac{du}{dx}$ and $u(2)$.

15. The methods of this section can be used to approximate definite integrals numerically. For example,

$$I = \int_a^b f(x)\,dx$$

is given by $I = y(b)$, where

$$y' = f(x), \qquad \text{and} \qquad y(a) = 0.$$

Show that one step of the Runge–Kutta method with $h = b - a$ gives the same result for $I$ as does Simpson's Rule with two subintervals of length $h/2$.

**16.** If $\phi(0) = A \geq 0$ and $\phi'(x) \geq k\phi(x)$ on $[0, X]$, where $k > 0$ and $X > 0$ are constants, show that $\phi(x) \geq Ae^{kx}$ on $[0, X]$. *Hint:* calculate $(d/dx)(\phi(x)/e^{kx})$.

* **17.** Consider the three initial-value problems

$$\begin{array}{lll} \text{(A):} & u' = u^2 & u(0) = 1 \\ \text{(B):} & y' = x + y^2 & y(0) = 1 \\ \text{(C):} & v' = 1 + v^2 & v(0) = 1 \end{array}$$

(a) Show that the solution of (B) remains between the solutions of (A) and (C) on any interval $[0, X]$ where solutions of all three problems exist. *Hint:* we must have $u(x) \geq 1$, $y(x) \geq 1$, and $v(x) \geq 1$ on $[0, X]$. (Why?) Apply the result of Exercise 16 to $\phi = y - u$ and to $\phi = v - y$.

(b) Find explicit solutions for problems (A) and (C). What can you conclude about the solution to problem (B).

(c) Use the Runge–Kutta method with $h = 0.05$, $h = 0.02$, and $h = 0.01$ to approximate the solution to (B) on $[0, 1]$. What can you conclude now?

# 10.6 DIFFERENTIAL EQUATIONS OF SECOND ORDER

The general second-order ordinary differential equation is of the form

$$F\left(\frac{d^2y}{dx^2}, \frac{dy}{dx}, y, x\right) = 0$$

for some function $F$ of four variables. When such an equation can be solved explicitly for $y$ as a function of $x$, the solution typically involves two integrations and therefore two arbitrary constants. A unique solution usually results from prescribing the values of the solution $y$ and its first derivative $y' = dy/dx$ at a particular point. Such a prescription constitutes an **initial-value problem** for the second-order equation.

## Equations Reducible to First Order

A second-order equation of the form

$$F\left(\frac{d^2y}{dx^2}, \frac{dy}{dx}, x\right) = 0$$

that does not involve the unknown function $y$ explicitly (except through its derivatives) can be reduced to a first order equation by a change of dependent variable; if $v = dy/dx$, then the equation can be written

$$F\left(\frac{dv}{dx}, v, x\right) = 0.$$

This first-order equation in $v$ may be amenable to the techniques described in earlier sections. If an explicit solution $v = v(x)$ can be found and integrated, then the function

$$y = \int v(x)\, dx$$

is an explicit solution of the given equation.

**EXAMPLE 1** Solve the initial-value problem

$$\frac{d^2y}{dx^2} = x\left(\frac{dy}{dx}\right)^2, \qquad y(0) = 1, \qquad y'(0) = -2.$$

**SOLUTION** If we let $v = dy/dx$, the given differential equation becomes

$$\frac{dv}{dx} = xv^2,$$

which is a separable first-order equation. Thus

$$\begin{aligned} \frac{dv}{v^2} &= x\,dx \\ -\frac{1}{v} &= \frac{x^2}{2} + \frac{C_1}{2} \\ v &= -\frac{2}{x^2 + C_1}. \end{aligned}$$

The initial condition $y'(0) = -2$ implies that $v(0) = -2$ and so $C_1 = 1$. Therefore

$$y = -2 \int \frac{dx}{x^2 + 1} = -2 \tan^{-1} x + C_2.$$

The initial condition $y(0) = 1$ implies that $C_2 = 1$ so the solution of the given initial-value problem is $y = 1 - 2 \tan^{-1} x$. ■

A second-order equation of the form

$$F\left(\frac{d^2y}{dx^2}, \frac{dy}{dx}, y\right) = 0$$

that does not explicitly involve the independent variable $x$ can be reduced to a first-order equation by a change of both dependent and independent variables. Again let $v = dy/dx$, but regard $v$ as a function of $y$ rather than $x$; $v = v(y)$. Then

$$\frac{d^2y}{dx^2} = \frac{dv}{dx} = \frac{dv}{dy}\frac{dy}{dx} = v\frac{dv}{dy}$$

by the Chain Rule. Hence the given differential equation becomes

$$F\left(v\frac{dv}{dy}, v, y\right) = 0$$

which is a first-order equation for $v$ as a function of $y$. If this equation can be solved for $v = v(y)$ there still remains the problem of solving the separable equation $(dy/dx) = v(y)$ for $y$ as a function of $x$.

**EXAMPLE 2** Solve the equation $y\dfrac{d^2y}{dx^2} = \left(\dfrac{dy}{dx}\right)^2$.

**SOLUTION** The change of variable $dy/dx = v(y)$ leads to the equation

$$yv\frac{dv}{dy} = v^2,$$

which is separable, $dv/v = dy/y$, and has solution $v = C_1 y$. The equation

$$\frac{dy}{dx} = C_1 y$$

is again separable, and leads to

$$\begin{aligned} \frac{dy}{y} &= C_1\,dx \\ \ln|y| &= C_1 x + C_2 \\ y &= \pm e^{C_1 x + C_2} = C_3 e^{C_1 x}. \end{aligned}$$

■

## Second-Order Linear Equations

The most frequently encountered ordinary differential equations arising in applications are second-order linear equations. The general second-order linear equation is of the form

$$a_2(x)\frac{d^2y}{dx^2} + a_1(x)\frac{dy}{dx} + a_0(x)y = f(x).$$

As remarked in Section 10.1, if $f(x) = 0$ identically, then we say that the equation is **homogeneous**. If the coefficients $a_2(x)$, $a_1(x)$, and $a_0(x)$ are continuous on an interval and $a_2(x) \neq 0$ there, then the homogeneous equation

$$a_2(x)\frac{d^2y}{dx^2} + a_1(x)\frac{dy}{dx} + a_0(x)y = 0$$

has a general solution of the form

$$y_h = C_1y_1(x) + C_2y_2(x)$$

where $y_1(x)$ and $y_2(x)$ are two **independent** solutions, that is, two solutions with the property that $C_1y_1(x) + C_2y_2(x) = 0$ for all $x$ in the interval only if $C_1 = C_2 = 0$. (We will not prove this here.)

Whenever one solution, $y_1(x)$, of a homogeneous linear second-order equation is known, another independent solution (and therefore the general solution) can be found by substituting $y = v(x)y_1(x)$ into the differential equation. This leads to a first-order, linear, separable equation for $v'$.

**EXAMPLE 3** Show that $y_1 = e^{-2x}$ is a solution of $y'' + 4y' + 4y = 0$, and find the general solution of this equation.

***SOLUTION*** Since $y_1' = -2e^{-2x}$ and $y_1'' = 4e^{-2x}$, we have

$$y_1'' + 4y_1' + 4y_1 = e^{-2x}(4 - 8 + 4) = 0,$$

so $y_1$ is indeed a solution of the given differential equation. To find the general solution, try $y = y_1v = e^{-2x}v(x)$. We have

$$\begin{aligned} y' &= -2e^{-2x}v + e^{-2x}v' \\ y'' &= 4e^{-2x}v - 4e^{-2x}v' + e^{-2x}v''. \end{aligned}$$

Substituting these expressions into the given DE we obtain

$$\begin{aligned} 0 &= y'' + 4y' + 4y \\ &= e^{-2x}(4v - 4v' + v'' - 8v + 4v' + 4v) = e^{-2x}v''. \end{aligned}$$

Thus $y = y_1v$ is a solution provided $v''(x) = 0$. This equation for $v$ has the general solution $v = C_1 + C_2x$, so the given equation has the general solution

$$y = C_1e^{-2x} + C_2xe^{-2x} = C_1y_1(x) + C_2y_2(x),$$

where $y_2 = xe^{-2x}$ is a second solution of the DE, independent of $y_1$. ■

By Theorem 2 of Section 10.1, the general solution of the second-order, linear, nonhomogeneous equation (with $f(x) \neq 0$) is of the form

$$y = y_p(x) + y_h(x)$$

where $y_p(x)$ is any particular solution of the nonhomogeneous equation and $y_h(x)$ is the general solution (as described above) of the corresponding homogeneous equation. In Section 10.8 we will discuss the solution of nonhomogeneous linear equations. First, however, in Section 10.7 we concentrate on some special classes of homogeneous, linear equations.

## EXERCISES 10.6

**1.** Show that $y = e^x$ is a solution of $y'' - 3y' + 2y = 0$, and find the general solution of this DE.

**2.** Show that $y = e^{-2x}$ is a solution of $y'' - y' - 6y = 0$, and find the general solution of this DE.

**3.** Show that $y = x$ is a solution of $x^2y'' + 2xy' - 2y = 0$ on the interval $(0, \infty)$, and find the general solution on this interval.

**4.** Show that $y = x^2$ is a solution of $x^2y'' - 3xy' + 4y = 0$ on the interval $(0, \infty)$, and find the general solution on this interval.

**5.** Show that $y = x$ is a solution of the differential equation $x^2y'' - (2x + x^2)y' + (2 + x)y = 0$, and find the general solution of this equation.

**6.** Show that $y = x^{-1/2}\cos x$ is a solution of the Bessel equation with $\nu = 1/2$:

$$x^2y'' + xy' + \left(x^2 - \frac{1}{4}\right)y = 0.$$

Find the general solution of this equation.

**First-order systems**

**7.** A system of $n$ first order, linear, differential equations in $n$ unknown functions $y_1, y_2, \cdots, y_n$ is written

$$\begin{aligned} y_1' &= a_{11}(x)y_1 + a_{12}(x)y_2 + \cdots + a_{1n}(x)y_n + f_1(x) \\ y_2' &= a_{21}(x)y_1 + a_{22}(x)y_2 + \cdots + a_{2n}(x)y_n + f_2(x) \\ &\vdots \\ y_n' &= a_{n1}(x)y_1 + a_{n2}(x)y_2 + \cdots + a_{nn}(x)y_n + f_n(x). \end{aligned}$$

Such a system is called an $n \times n$ **first-order linear system** and can be rewritten in vector-matrix form as $\mathbf{y}' = \mathcal{A}(x)\mathbf{y} + \mathbf{f}(x)$, where

$$\mathbf{y}(x) = \begin{pmatrix} y_1(x) \\ \vdots \\ y_n(x) \end{pmatrix}, \quad \mathbf{f}(x) = \begin{pmatrix} f_1(x) \\ \vdots \\ f_n(x) \end{pmatrix},$$

$$\mathcal{A}(x) = \begin{pmatrix} a_{11}(x) & \cdots & a_{1n}(x) \\ \vdots & \ddots & \vdots \\ a_{n1}(x) & \cdots & a_{nn}(x) \end{pmatrix}.$$

Show that the second-order, linear equation $y'' + a_1(x)y' + a_0(x)y = f(x)$ can be transformed into a $2 \times 2$ first-order system with $y_1 = y$ and $y_2 = y'$ having

$$\mathcal{A}(x) = \begin{pmatrix} 0 & 1 \\ -a_0(x) & -a_1(x) \end{pmatrix}, \quad \mathbf{f}(x) = \begin{pmatrix} 0 \\ f(x) \end{pmatrix}.$$

**8.** Generalize the previous exercise to transform an $n$th-order linear equation

$$y^{(n)} + a_{n-1}(x)y^{(n-1)} + a_{n-2}(x)y^{(n-2)} + \cdots + a_0(x)y = f(x)$$

into an $n \times n$ first-order system.

**9.** If $\mathcal{A}$ is an $n \times n$ constant matrix, and if there exists a scalar $\lambda$ and a nonzero constant vector $\mathbf{v}$ for which $\mathcal{A}\mathbf{v} = \lambda\mathbf{v}$ show that $\mathbf{y} = C_1e^{\lambda x}\mathbf{v}$ is a solution of the homogeneous system $\mathbf{y}' = \mathcal{A}\mathbf{y}$.

**10.** Show that the determinant $\begin{vmatrix} 2-\lambda & 1 \\ 2 & 3-\lambda \end{vmatrix}$ is zero for two distinct values of $\lambda$. For each of these values find a nonzero vector $\mathbf{v}$ which satisfies the condition $\begin{pmatrix} 2 & 1 \\ 2 & 3 \end{pmatrix}\mathbf{v} = \lambda\mathbf{v}$. Hence solve the system

$$y_1' = 2y_1 + y_2, \qquad y_2' = 2y_1 + 3y_2.$$

# 10.7 LINEAR DIFFERENTIAL EQUATIONS WITH CONSTANT COEFFICIENTS

In this section we will solve second-order, linear, homogeneous differential equations of the form

$$a\,y'' + b\,y' + cy = 0, \qquad (*)$$

where $a$, $b$, and $c$ are constants and $a \neq 0$, as well as higher-order equations of this type.

Equations of type $(*)$ arise in many applications of mathematics. The equation of simple harmonic motion,

$$y'' + \omega^2 y = 0,$$

is of type $(*)$ (with $a = 1$, $b = 0$ and $c = \omega^2$) and governs the vibration of a mass suspended by an elastic spring. If the motion of the mass is impeded by a resistance proportional to its velocity (supplied, say, by friction with the air) then the equation of motion would look like $(*)$ with a positive constant $b$. The mass will still vibrate if the resistance is not too large ($b^2 < 4ac$) but the amplitude of the vibration will decay exponentially with time, as we will see shortly.

Equation $(*)$ also arises in the study of current flowing in an electric circuit having resistors, capacitors, and inductors connected in series. In this case $a$ measures the inductance, $b$ the resistance, and $c$ the reciprocal of the capacitance of the circuit. In both the mechanical (vibrating mass) and electrical applications the constants $a$, $b$, and $c$ are positive. Also, the independent variable in all these applications is time, so we will regard the function $y$ in $(*)$ as a function of $t$ rather than $x$. Thus a *prime* ($'$) denotes $d/dt$ rather than $d/dx$.

Let us try to find a solution of equation $(*)$ having the form $y = e^{rt}$. Substituting this expression into the equation, we obtain

$$ar^2e^{rt} + bre^{rt} + ce^{rt} = 0.$$

Since $e^{rt}$ does not vanish, $y = e^{rt}$ will be a solution of the differential equation $(*)$ if and only if $r$ satisfies the quadratic **auxiliary equation**

$$ar^2 + br + c = 0, \qquad (**)$$

which has roots given by the **quadratic formula:**

$$r = \frac{-b \pm \sqrt{b^2 - 4ac}}{2a} = -\frac{b}{2a} \pm \frac{\sqrt{D}}{2a}.$$

Here $D = b^2 - 4ac$ is the **discriminant** of the auxiliary equation $(**)$.

There are three cases to consider, depending on whether $D$ is positive, zero, or negative.

**CASE I.** Suppose $D = b^2 - 4ac > 0$. Then the auxiliary equation has two different real roots, $r_1$ and $r_2$, given by

$$r_1 = \frac{-b - \sqrt{D}}{2a}, \qquad r_2 = \frac{-b + \sqrt{D}}{2a}.$$

(Often these roots can be found easily by factoring the left side of the auxiliary equation.) In this case $y = y_1(t) = e^{r_1 t}$ and $y = y_2(t) = e^{r_2 t}$ are independent solutions of the differential equation $(*)$, and so

$$y = C_1 e^{r_1 t} + C_2 e^{r_2 t},$$

is also a solution for any choice of the constants $C_1$ and $C_2$ by Theorem 1 of Section 10.1. Since the differential equation is of second order and this solution involves two arbitrary constants we suspect it is the **general solution**, that is, that every solution of the differential equation can be written in this form. Exercise 18 at the end of this section outlines a way to prove this fact.

**EXAMPLE 1** Find the general solution of $y'' + y' - 2y = 0$.

***SOLUTION*** The auxiliary equation is $r^2 + r - 2 = 0$, or $(r + 2)(r - 1) = 0$. The auxiliary roots are $r_1 = -2$ and $r_2 = 1$. Hence the general solution of the differential equation is

$$y = C_1 e^{-2t} + C_2 e^t.$$

In order to facilitate discussion of the other two cases, $D = 0$, and $D < 0$, let us make a change of dependent variable in the differential equation $(*)$ to transform it into a similar equation with no first-order derivative. This can be achieved by letting

$$y(t) = e^{kt}u(t), \qquad \text{where } k = -\frac{b}{2a}.$$

Differentiating twice, using the Product Rule, we obtain

$$y'(t) = e^{kt}[u'(t) + ku(t)]$$
$$y''(t) = e^{kt}[u''(t) + 2ku'(t) + k^2u(t)].$$

Substituting these expressions into the differential equation $(*)$, we obtain

$$e^{kt}\big(au'' + (2ak + b)u' + (ak^2 + bk + c)u\big) = 0.$$

Since $2ak = -b$, the coefficient of $u'$ in this equation is 0 and that of $u$ is

$$ak^2 + bk + c = \frac{ab^2}{4a^2} - \frac{b^2}{2a} + c = \frac{4ac - b^2}{4a} = \frac{-D}{4a}.$$

Dividing by $a$ and $e^{kt}$, we are left with the **reduced equation**

$$u'' - \left(\frac{D}{4a^2}\right)u = 0.$$

**CASE II.** Suppose $D = b^2 - 4ac = 0$. Then the auxiliary equation has two equal roots $r_1 = r_2 = -b/(2a) = k$. Moreover, the reduced equation is $u'' = 0$, which has general solution $u = A + Bt$, where $A$ and $B$ are any constants. Thus, the general solution to the differential equation $(*)$ for the case $D = 0$ is

$$y = A e^{kt} + Bt\, e^{kt}, \qquad \text{where } k = -\frac{b}{2a}.$$

**EXAMPLE 2** Find the general solution of $y'' + 6y' + 9y = 0$.

***SOLUTION*** The auxiliary equation is $r^2 + 6r + 9 = 0$, or $(r + 3)^2 = 0$, which has identical roots $r = -3$. According to the discussion above, the general solution of the differential equation is

$$y = C_1 e^{-3t} + C_2 t e^{-3t}.$$

**CASE III.** Suppose $D = b^2 - 4ac < 0$. Then the auxiliary equation has complex roots, and the reduced equation for $u$ is

$$u'' + \omega^2 u = 0, \qquad \text{where} \quad \omega = \frac{\sqrt{-D}}{2a}.$$

We recognize this as the equation of simple harmonic motion. It has the general solution

$$u = C_1 \cos(\omega t) + C_2 \sin(\omega t),$$

where $C_1$ and $C_2$ are arbitrary constants. Therefore the given differential equation has the general solution

$$y = C_1 e^{kt} \cos(\omega t) + C_2 e^{kt} \sin(\omega t),$$
$$\text{where} \quad k = -\frac{b}{2a} \text{ and } \omega = \frac{\sqrt{4ac - b^2}}{2a}.$$

***REMARK*** In Case III, where $b^2 - 4ac < 0$, the auxiliary equation $(**)$ has complex conjugate roots given by

$$r = \frac{-b \pm \sqrt{b^2 - 4ac}}{2a} = k \pm i\omega,$$

where $i$ is the imaginary unit ($i^2 = -1$). Thus the values of $k$ and $\omega$ needed to write the general solution of $(*)$ can be read immediately from the solution of the auxiliary equation.

**EXAMPLE 3** Find the general solution of $y'' + 4y' + 13y = 0$.

***SOLUTION*** The auxiliary equation is $r^2 + 4r + 13 = 0$, which has solutions

$$r = \frac{-4 \pm \sqrt{16 - 52}}{2} = \frac{-4 \pm \sqrt{-36}}{2} = -2 \pm 3i.$$

Thus $k = -2$ and $\omega = 3$. The general solution of the given differential equation is

$$y = C_1 e^{-2t} \cos(3t) + C_2 e^{-2t} \sin(3t).$$

***REMARK*** If $a$ and $c$ are positive, but $b = 0$, then equation $(*)$ is the differential equation of simple harmonic motion and has oscillatory solutions of fixed amplitude. If $b > 0$ but $b^2 < 4ac$ (Case III), the solutions still oscillate, but the amplitude diminishes exponentially as $t \to \infty$ because of the factor $e^{kt} = e^{-(b/2a)t}$. A system whose behaviour is modelled by such an equation is said to exhibit **damped harmonic motion**. If $b^2 = 4ac$ (Case II) the system is said to be **critically damped**, and if $b^2 > 4ac$ (Case I) it is **overdamped**. In these latter cases the behaviour is no longer oscillatory. (See Figure 10.5. Imagine a mass suspended by a spring in a jar of molasses.)

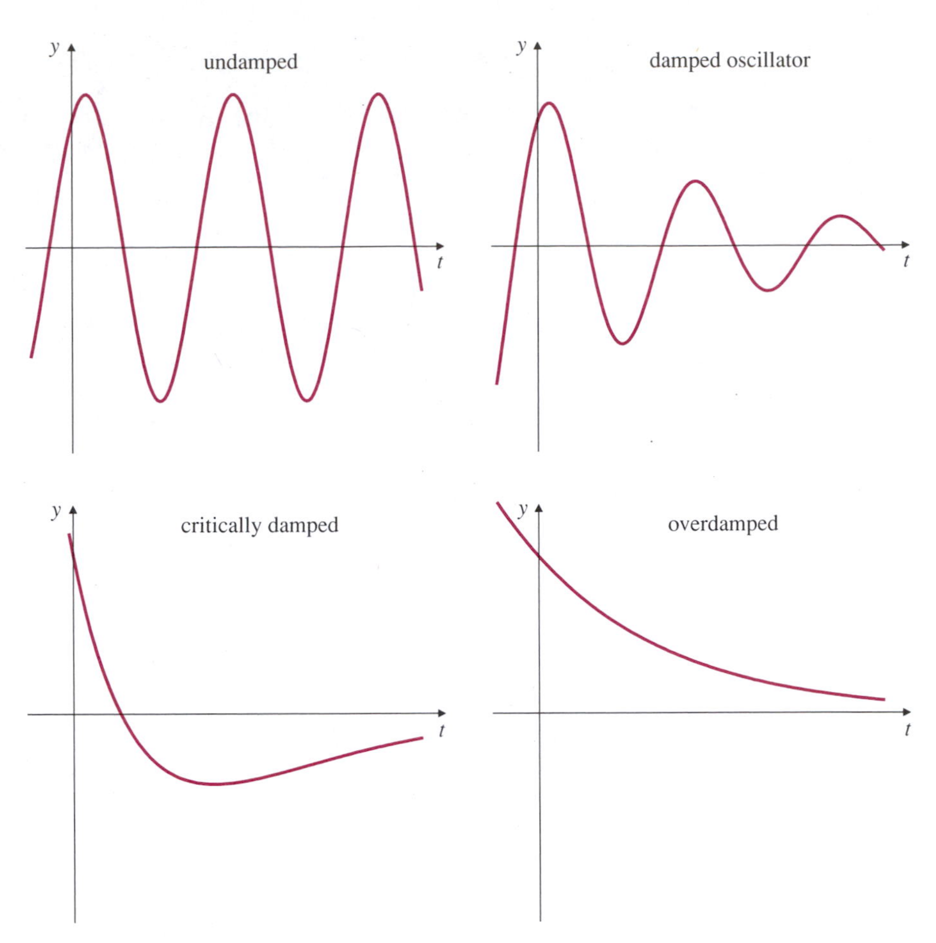


**Figure 10.5**
Undamped oscillator ($b = 0$)
Damped oscillator ($b > 0$, $b^2 < 4ac$)
Critically damped case ($b^2 = 4ac$)
Overdamped case ($b^2 > 4ac$)

Here is a typical initial-value problem for a differential equation of type (∗).

**EXAMPLE 4** Solve the initial-value problem $\begin{cases} y'' + 2y' + 2y = 0 \\ y(0) = 2 \\ y'(0) = -3. \end{cases}$

***SOLUTION*** The auxiliary equation for the DE $y'' + 2y' + 2y = 0$ is $r^2 + 2r + 2 = 0$, and has roots

$$r = \frac{-2 \pm \sqrt{4 - 8}}{2} = -1 \pm i.$$

Thus Case III applies: $k = -1$ and $\omega = 1$. The differential equation has the general solution

$$y = C_1 e^{-t} \cos t + C_2 e^{-t} \sin t.$$

Also

$$\begin{aligned} y' &= e^{-t}\big(-C_1 \cos t - C_2 \sin t - C_1 \sin t + C_2 \cos t\big) \\ &= (C_2 - C_1)\, e^{-t} \cos t - (C_1 + C_2)\, e^{-t} \sin t. \end{aligned}$$

Applying the initial conditions $y(0) = 2$ and $y'(0) = -3$, we obtain $C_1 = 2$ and $C_2 - C_1 = -3$. Hence $C_2 = -1$ and the initial-value problem has the solution

$$y = 2e^{-t} \cos t - e^{-t} \sin t.$$

■

## Constant-Coefficient Equations of Higher Order

The techniques described above for solving second-order linear, homogeneous DEs with constant coefficients can also be applied to equations of higher order. We describe the procedure without offering any proofs. If

$$P_n(r) = a_n r^n + a_{n-1} r^{n-1} + \cdots + a_2 r^2 + a_1 r + a_0$$

is a polynomial of degree $n$ with constant coefficients $a_j$ $(0 \le j \le n)$ and $a_n \ne 0$, then the DE

$$P_n(D)y = 0, \qquad (*)$$

where $D = d/dt$ (or $D = d/dx$) can be solved by substituting $y = e^{rx}$ and obtaining the *auxiliary equation* $P_n(r) = 0$. This polynomial equation has $n$ roots (see Appendix II) some of which may be equal, and some or all of which can be complex. If the coefficients of the polynomial $P_n(r)$ are all real then any complex roots must occur in complex conjugate pairs $a \pm ib$, where $a$ and $b$ are real.

The general solution of $(*)$ can be expressed as a *linear combination* of $n$ independent particular solutions

$$y = C_1 y_1(x) + C_2 y_2(x) + \cdots + C_n y_n(x),$$

where the $C_j$ are arbitrary constants. The independent solutions $y_1, y_2, \ldots, y_n$ are constructed as follows:

1. If $r_1$ is an $k$-fold root of the auxiliary equation (that is, if $(r - r_1)^k$ is a factor of $P_n(r)$), then

$$e^{r_1 t}, \quad te^{r_1 t}, \quad t^2 e^{r_1 t}, \quad \ldots, \quad t^{k-1} e^{r_1 t}$$

   are $k$ independent solutions of $(*)$.

2. If $r = a + ib$ and $r = a - ib$ ($a$ and $b$ real) constitute a $k$-fold pair of complex conjugate roots of the auxiliary equation (that is, if $[(r - a)^2 + b^2]^k$ is a factor of $P_n(r)$), then

$$\begin{matrix} e^{at}\cos bt, & te^{at}\cos bt, & \ldots, & t^{k-1}e^{at}\cos bt, \\ e^{at}\sin bt, & te^{at}\sin bt, & \ldots, & t^{k-1}e^{at}\sin bt \end{matrix}$$

   are $2k$ independent solutions of $(*)$.

**EXAMPLE 5** Solve (a) $y^{(4)} - 16y = 0$, and (b) $y^{(5)} - 2y^{(4)} + y^{(3)} = 0$.

***SOLUTION*** The auxiliary equation for (a) is $r^4 - 16 = 0$, which factors down to $(r - 2)(r + 2)(r^2 + 4) = 0$, and hence has roots $r = 2, -2, 2i$, and $-2i$. Thus the DE (a) has general solution

$$y = C_1 e^{2t} + C_2 e^{-2t} + C_3 \cos(2t) + C_4 \sin(2t)$$

for arbitrary constants $C_1$, $C_2$, $C_3$, and $C_4$.

The auxiliary equation for (b) is $r^5 - 2r^4 + r^3$, which factors to $r^3(r - 1)^2 = 0$, and so has roots $r = 0,\ 0,\ 0,\ 1,\ 1$. The general solution of the $DE$ (b) is

$$y = C_1 + C_2 t + C_3 t^2 + C_4 e^t + C_5 te^t,$$

where $C_1, \ldots, C_5$ are arbitrary constants. ■

**EXAMPLE 6** What is the order and the general solution of the constant-coefficient, linear, homogeneous DE whose auxiliary equation is

$$(r+4)^3(r^2+4r+13)^2 = 0?$$

***SOLUTION*** The auxiliary equation has degree 7 so the DE is of seventh order. Since $r^2+4r+13 = (r+2)^2+9$, which has roots $-2\pm 3i$, the DE must have the general solution

$$\begin{aligned} y = &C_1e^{-4t} + C_2te^{-4t} + C_3t^2e^{-4t} \\ &+ C_4e^{-2t}\cos(3t) + C_5e^{-2t}\sin(3t) + C_6te^{-2t}\cos(3t) + C_7te^{-2t}\sin(3t). \end{aligned}$$

## Euler (Equidimensional) Equations

A homogeneous, linear equation of the form

$$ax^2\frac{d^2y}{dx^2} + bx\frac{dy}{dx} + cy = 0$$

is called an **Euler equation** or an **equidimensional equation**, the latter term being appropriate since all the terms in the equation have the same dimension (that is, they are measured in the same units), provided that the constants $a$, $b$, and $c$ all have the same dimension. The coefficients of an Euler equation are *not constant*, but the technique for solving these equations is similar to that for solving equations with constant coefficients, so we include these equations in this section. As in the case of constant coefficient equations, we assume that the constants $a$, $b$, and $c$ are real numbers, and that $a \neq 0$. Even so, the leading coefficient, $ax^2$, does vanish at $x = 0$ (which is called a **singular point** of the equation), and this can cause solutions to fail to be defined at $x = 0$. We will solve the equation in the interval $x > 0$; the same solution will also hold for $x < 0$ provided we replace $x$ by $|x|$ in the solution.

Let us search for solutions in $x > 0$ given by powers of $x$; if

$$y = x^r, \qquad \frac{dy}{dx} = rx^{r-1}, \qquad \frac{d^2y}{dx^2} = r(r-1)x^{r-2},$$

then the Euler equation becomes

$$\big(ar(r-1) + br + c\big)x^r = 0.$$

This will be satisfied for all $x > 0$ provided that $r$ satisfies the **auxiliary equation**

$$ar^2 + (b-a)r + c = 0.$$

As for constant coefficient equations, there are three possibilities.

If $(b-a)^2 \geq 4ac$ then the auxiliary equation has two real roots

$$r_1 = \frac{a-b+\sqrt{(b-a)^2-4ac}}{2a},$$
$$r_2 = \frac{a-b-\sqrt{(b-a)^2-4ac}}{2a}.$$

In this case, the Euler equation has the general solution

$$y = C_1x^{r_1} + C_2x^{r_2}, \qquad (x > 0).$$

The general solution is usually quoted in the form

$$y = C_1|x|^{r_1} + C_2|x|^{r_2},$$

which is valid in any interval not containing $x = 0$, and may even be valid on intervals containing the origin if, for example, $r_1$ and $r_2$ are nonnegative integers.

**EXAMPLE 7** Solve the initial-value problem

$$2x^2y'' - xy' - 2y = 0, \qquad y(1) = 5, \qquad y'(1) = 0.$$

***SOLUTION*** The auxiliary equation is $2r(r-1)-r-2=0$, that is, $2r^2-3r-2=0$, or $(r-2)(2r+1)=0$, and has roots $r=2$ and $r=-(1/2)$. Thus, the general solution of the differential equation (valid for $x>0$) is

$$y = C_1x^2 + C_2x^{-1/2}.$$

The initial conditions imply that

$$5 = y(1) = C_1 + C_2, \quad \text{and} \quad 0 = y'(1) = 2C_1 - \frac{1}{2}C_2.$$

Therefore $C_1 = 1$ and $C_2 = 4$, and the initial-value problem has solution

$$y = x^2 + \frac{4}{\sqrt{x}}, \qquad (x > 0).$$

If $(b-a)^2 = 4ac$ then the auxiliary equation has one double root, namely the root $r = (a-b)/2a$. It is left to the reader to verify that in this case the transformation $y = x^r v(x)$ leads to the general solution

$$y = C_1x^r + C_2x^r \ln x, \qquad (x > 0),$$

or, more generally,

$$y = C_1|x|^r + C_2|x|^r \ln|x|, \qquad (x \neq 0).$$

If $(b-a)^2 < 4ac$ then the auxiliary equation has complex conjugate roots

$$r = \alpha \pm i\beta, \qquad \text{where} \quad \alpha = \frac{a-b}{2a}, \quad \beta = \frac{\sqrt{4ac-(b-a)^2}}{2a}.$$

The corresponding powers $x^r$ can be expressed in real form in a manner similar to that used for constant coefficient equations; we have

$$\begin{aligned} x^{\alpha \pm i\beta} = e^{(\alpha \pm i\beta)\ln x} &= e^{\alpha \ln x}\big[\cos(\beta \ln x) \pm i \sin(\beta \ln x)\big] \\ &= x^\alpha \cos(\beta \ln x) \pm i x^\alpha \sin(\beta \ln x). \end{aligned}$$

Accordingly, the Euler equation has the general solution

$$y = C_1|x|^\alpha \cos(\beta \ln|x|) + C_2|x|^\alpha \sin(\beta \ln|x|).$$

**EXAMPLE 8** Solve the DE $x^2y'' - 3xy' + 13y = 0$.

***SOLUTION*** The DE has the auxiliary equation $r(r-1) - 3r + 13 = 0$, that is, $r^2 - 4r + 13 = 0$, which has roots $r = 2 \pm 3i$. The DE, therefore, has the general solution

$$y = C_1x^2\cos(3\ln|x|) + C_2x^2\sin(3\ln|x|).$$

## EXERCISES 10.7

In Exercises 1–12, find the general solutions for the given equations.

**1.** $y'' + 7y' + 10y = 0$ **2.** $y'' - 2y' - 3y = 0$

**3.** $y'' + 2y' = 0$ **4.** $4y'' - 4y' - 3y = 0$

**5.** $y'' + 8y' + 16y = 0$ **6.** $y'' - 2y' + y = 0$

**7.** $y'' - 6y' + 10y = 0$ **8.** $9y'' + 6y' + y = 0$

**9.** $y'' + 2y' + 5y = 0$ **10.** $y'' - 4y' + 5y = 0$

**11.** $y'' + 2y' + 3y = 0$ **12.** $y'' + y' + y = 0$

In Exercises 13–16, solve the given initial-value problems.

**13.** $\begin{cases} 2y'' + 5y' - 3y = 0 \\ y(0) = 1 \\ y'(0) = 0. \end{cases}$ **14.** $\begin{cases} y'' + 10y' + 25y = 0 \\ y(1) = 0 \\ y'(1) = 2. \end{cases}$

**15.** $\begin{cases} y'' + 4y' + 5y = 0 \\ y(0) = 2 \\ y'(0) = 2. \end{cases}$ **16.** $\begin{cases} y'' + y' + y = 0 \\ y(2\pi/\sqrt{3}) = 0 \\ y'(2\pi/\sqrt{3}) = 1. \end{cases}$

**17.** If $a > 0$, $b > 0$, and $c > 0$, prove that all solutions of the differential equation $ay'' + by' + cy = 0$ satisfy $\lim_{t\to\infty} y(t) = 0$.

***18.** Prove that the solution given in the discussion of Case I, namely $y = A\,e^{r_1t} + B\,e^{r_2t}$, is the general solution for that case as follows: first let $y = e^{r_1t}u$ and show that $u$ satisfies the equation

$$u'' - (r_2 - r_1)u' = 0.$$

Then let $v = u'$, so that $v$ must satisfy $v' = (r_2 - r_1)v$. The general solution of this equation is $v = C\,e^{(r_2-r_1)t}$. Hence find $u$ and $y$.

Find general solutions of the DEs in Exercises 19–22.

**19.** $y''' - 4y'' + 3y' = 0$

**20.** $y^{(4)} - 2y'' + y = 0$ **21.** $y^{(4)} + 2y'' + y = 0$

**22.** $y^{(4)} + 4y^{(3)} + 6y'' + 4y' + y = 0$

**23.** Show that $y = e^{2t}$ is a solution of

$$y''' - 2y' - 4y = 0$$

(where $'$ denotes $d/dt$) and find the general solution of this DE.

**24.** Write the general solution of the linear, constant-coefficient DE having auxiliary equation $(r^2 - r - 2)^2(r^2 - 4)^2 = 0$.

Find general solutions to the Euler equations in Exercises 25–30.

**25.** $x^2y'' - xy' + y = 0$ **26.** $x^2y'' - xy' - 3y = 0$

**27.** $x^2y'' + xy' - y = 0$ **28.** $x^2y'' - xy' + 5y = 0$

**29.** $x^2y'' + xy' = 0$ **30.** $x^2y'' + xy' + y = 0$

***31.** Solve the DE $x^3y''' + xy' - y = 0$ in the interval $x > 0$.

# 10.8 NONHOMOGENEOUS LINEAR EQUATIONS

We now consider the problem of solving the nonhomogeneous second-order differential equation

$$a_2(x)\frac{d^2y}{dx^2} + a_1(x)\frac{dy}{dx} + a_0(x)y = f(x). \qquad (*)$$

We assume that two independent solutions, $y_1(x)$ and $y_2(x)$, of the corresponding homogeneous equation

$$a_2(x)\frac{d^2y}{dx^2} + a_1(x)\frac{dy}{dx} + a_0(x)y = 0$$

are known. The function $y_h(x) = C_1 y_1(x) + C_2 y_2(x)$, which is the general solution of the homogeneous equation, is called the **complementary function** for the nonhomogeneous equation. Theorem 2 of Section 10.1 suggests that the general solution of the nonhomogeneous equation is of the form

$$y = y_p(x) + y_h(x) = y_p(x) + C_1 y_1(x) + C_2 y_2(x),$$

where $y_p(x)$ is any **particular solution** of the nonhomogeneous equation. All we need to do is find *one solution* of the nonhomogeneous equation and we can write the general solution.

There are two common methods for finding a particular solution $y_p$ of the nonhomogeneous equation $(*)$:

1. The method of undetermined coefficients, and
2. The method of variation of parameters.

The first of these hardly warrants being called a *method*; it just involves making an educated guess about the form of the solution as a sum of terms with unknown coefficients and substituting this guess into the equation to determine the coefficients. This method works well for simple DEs, especially ones with constant coefficients. The nature of the *guess* depends on the nonhomogeneous term $f(x)$, but can also be affected by the solution of the corresponding homogeneous equation. A few examples will illustrate the ideas involved.

**EXAMPLE 1** Find the general solution of $y'' + y' - 2y = 4x$.

***SOLUTION*** Because the nonhomogeneous term $f(x) = 4x$ is a first-degree polynomial, we "guess" that a particular solution can be found which is also such a polynomial. Thus we try

$$y = Ax + B, \qquad y' = A, \qquad y'' = 0.$$

Substituting these expressions into the given DE we obtain

$$\begin{aligned} 0 + A - 2(Ax + B) &= 4x \qquad \text{or} \\ -(2A + 4)x + (A - 2B) &= 0. \end{aligned}$$

This latter equation will be satisfied for all $x$ provided $2A + 4 = 0$ and $A - 2B = 0$. Thus we require $A = -2$ and $B = -1$; a particular solution of the given DE is

$$y_p(x) = -2x - 1.$$

Since the corresponding homogeneous equation $y'' + y' - 2y = 0$ has auxiliary equation $r^2 + r - 2 = 0$ with roots $r = 1$ and $r = -2$, the given DE has the general solution

$$y = y_p(x) + C_1 e^x + C_2 e^{-2x} = -2x - 1 + C_1 e^x + C_2 e^{-2x}.$$

■

**EXAMPLE 2** Find general solutions of the equations (where $'$ denotes $d/dt$)

(a) $y'' + 4y = \sin t$,

(b) $y'' + 4y = \sin(2t)$,

(c) $y'' + 4y = \sin t + \sin(2t)$.

### *SOLUTION*

(a) Let us look for a particular solution of the form

$$\begin{aligned} y &= A\sin t + B\cos t \qquad \text{so that} \\ y' &= A\cos t - B\sin t \\ y'' &= -A\sin t - B\cos t. \end{aligned}$$

Substituting these expressions into the DE $y'' + 4y = \sin t$, we get

$$-A\sin t - B\cos t + 4A\sin t + 4B\cos t = \sin t,$$

which is satisfied for all $x$ if $3A = 1$ and $3B = 0$. Thus $A = 1/3$ and $B = 0$. Since the homogeneous equation $y'' + 4y = 0$ has general solution $y = C_1\cos(2t) + C_2\sin(2t)$, the given nonhomogeneous equation has the general solution

$$y = \frac{1}{3}\sin t + C_1\cos(2t) + C_2\sin(2t).$$

(b) Motivated by our success in part (a), we might be tempted to try for a particular solution of the form $y = A\sin(2t) + B\cos(2t)$, but that won't work, because this function is a solution of the homogeneous equation, so we would get $y'' + 4y = 0$ for any choice of $A$ and $B$. In this case it is useful to try

$$y = At\sin(2t) + Bt\cos(2t).$$

We have

$$\begin{aligned} y' &= A\sin(2t) + 2At\cos(2t) + B\cos(2t) - 2Bt\sin(2t) \\ &= (A - 2Bt)\sin(2t) + (B + 2At)\cos(2t) \\ y'' &= -2B\sin(2t) + 2(A - 2Bt)\cos(2t) + 2A\cos(2t) \\ &\quad - 2(B + 2At)\sin(2t) \\ &= -4(B + At)\sin(2t) + 4(A - Bt)\cos(2t). \end{aligned}$$

Substituting into $y'' + 4y = \sin(2t)$ leads to

$$\begin{aligned} -4(B + At)\sin(2t) + 4(A - Bt)\cos(2t) + 4At\sin(2t) + 4Bt\cos(2t) \\ = \sin(2t). \end{aligned}$$

Observe that the terms involving $t\sin(2t)$ and $t\cos(2t)$ cancel out and we are left with

$$-4B\sin(2t) + 4A\cos(2t) = \sin(2t),$$

which is satisfied for all $x$ if $A = 0$ and $B = -1/4$. Hence, the general solution for part (b) is

$$y = -\frac{1}{4}t\cos(2t) + C_1\cos(2t) + C_2\sin(2t).$$

(c) Since the homogeneous equation is the same for (a), (b), and (c), and the nonhomogeneous term in equation (c) is the sum of the nonhomogeneous terms in equations (a) and (b), the sum of particular solutions of (a) and (b) is a particular solution of (c). (This is because the equation is *linear*.) Thus the general solution of equation (c) is

$$y = \frac{1}{3}\sin t - \frac{1}{4}t\cos(2t) + C_1\cos(2t) + C_2\sin(2t).$$

■

We summarize the appropriate forms to try for particular solutions of constant-coefficient equations as follows:

**Trial solutions for constant-coefficient equations**

Let $A_n(x)$, $B_n(x)$, and $P_n(x)$ denote the $n$th-degree polynomials

$$\begin{aligned} A_n(x) &= a_0 + a_1x + a_2x^2 + \cdots + a_nx^n \\ B_n(x) &= b_0 + b_1x + b_2x^2 + \cdots + b_nx^n \\ P_n(x) &= p_0 + p_1x + p_2x^2 + \cdots + p_nx^n \end{aligned}$$

To find a particular solution $y_p(x)$ of the second-order linear, constant-coefficient, nonhomogeneous DE

$$a_2\frac{d^2y}{dx^2} + a_1\frac{dy}{dx} + a_0y = f(x)$$

use the following forms:

If $f(x) = P_n(x)$ try $y_p = x^m A_n(x)$.

If $f(x) = P_n(x)e^{rx}$ try $y_p = x^m A_n(x)e^{rx}$.

If $f(x) = P_n(x)e^{rx}\cos(kx)$ try $y_p = x^m e^{rx}[A_n(x)\cos(kx) + B_n(x)\sin(kx)]$.

If $f(x) = P_n(x)e^{rx}\sin(kx)$ try $y_p = x^m e^{rx}[A_n(x)\cos(kx) + B_n(x)\sin(kx)]$.

where $m$ is the smallest of the integers 0, 1, and 2, that ensures that no term of $y_p$ is a solution of the corresponding homogeneous equation

$$a_2\frac{d^2y}{dx^2} + a_1\frac{dy}{dx} + a_0y = 0.$$

## EXPLORE! Resonance

For $\lambda > 0$ let $y_\lambda(t)$ be the solution of the initial-value problem

$$\begin{cases} y'' + y = \sin(\lambda t) \\ y(0) = 0 \\ y'(0) = 1. \end{cases}$$

(a) Find $y_\lambda(t)$ for $\lambda \neq 1$.

(b) Find $y_1(t)$ and compare it with $\lim_{\lambda\to 1} y_\lambda(t)$.

(c) For what values of $\lambda > 0$ is $y_\lambda(t)$ a bounded function of $t$?

(d) Graph the function $y = y_\lambda(t)$ on the interval $0 \le t \le 50$ for various values of $\lambda$ including, say, $\lambda = 0.2, 0.5, 0.8, 0.9, 1.0, 1.2, 1.2, 1.5, 2.0$. Use a large $y$-range for the graphs, say $-20 \le y \le 20$. What happens to the maximum value of $|y_\lambda(t)|$ on $0 \le t \le 50$ as $\lambda$ approacues 1?

The phenomenon illustrated here is called **resonance**. If you push a child on a swing, the swing will rise highest if your pushes are timed to have the same frequency as the natural frequency of the swing. Resonance is used in the design of tuning circuits of radios; the circuit is tuned (ususally by a variable capacitor) so that its natural frequency of oscillation is the frequency of the station being tuned in. The circuit then responds much more strongly to the signal received from that station than to others on different frequencies.

## Variation of Parameters

A more formal method for finding a particular solution $y_p(x)$ of the nonhomogeneous equation when we know two independent solutions, $y_1(x)$ and $y_2(x)$, of the homogeneous equation is to replace the constants in the complementary function by functions, that is, search for $y_p$ in the form

$$y_p = u_1(x)y_1(x) + u_2(x)y_2(x).$$

Requiring $y_p$ to satisfy the given nonhomogeneous DE provides one equation that must be satisfied by the two unknown functions $u_1$ and $u_2$. We are free to require them to satisfy a second equation also. To simplify the calculations below, we choose this second equation to be

$$u_1'(x)y_1(x) + u_2'(x)y_2(x) = 0.$$

Now we have

$$\begin{aligned} y_p' &= u_1'y_1 + u_1y_1' + u_2'y_2 + u_2y_2' = u_1y_1' + u_2y_2' \\ y_p'' &= u_1'y_1' + u_1y_1'' + u_2'y_2' + u_2y_2''. \end{aligned}$$

Substituting these expressions into the given DE we obtain

$$\begin{aligned} a_2(u_1'y_1' + u_2'y_2') &+ u_1(a_2y_1'' + a_1y_1' + a_0y_1) + u_2(a_2y_2'' + a_1y_2' + a_0y_2) \\ &= a_2(u_1'y_1' + u_2'y_2') = f(x), \end{aligned}$$

because $y_1$ and $y_2$ satisfy the homogeneous equation. Therefore $u_1'$ and $u_2'$ satisfy the pair of equations

$$\begin{aligned} u_1'(x)y_1(x) + u_2'(x)y_2(x) &= 0 \\ u_1'(x)y_1'(x) + u_2'(x)y_2'(x) &= \frac{f(x)}{a_2(x)}. \end{aligned}$$

We can solve these two equations for the unknown functions $u_1'$ and $u_2'$ by Cramer's Rule (Theorem 5 of Section 3.5), or otherwise, and obtain

$$u_1' = -\frac{y_2(x)}{W(x)}\frac{f(x)}{a_2(x)}, \qquad u_2' = \frac{y_1(x)}{W(x)}\frac{f(x)}{a_2(x)},$$

where $W(x)$, called the **Wronskian** of $y_1$ and $y_2$, is the determinant

$$W(x) = \begin{vmatrix} y_1(x) & y_2(x) \\ y_1'(x) & y_2'(x) \end{vmatrix}.$$

Then $u_1$ and $u_2$ can be found by integration.

**EXAMPLE 3** Find the general solution of $y'' - 3y' + 2y = 4x$.

***SOLUTION*** First we solve the homogeneous equation $y'' - 3y' + 2y = 0$, which has auxiliary equation $r^2 - 3r + 2 = 0$ with roots $r = 1$ and $r = 2$. Therefore two independent solutions of the homogeneous equation are $y_1 = e^x$ and $y_2 = e^{2x}$, and the complementary function is

$$y_h = C_1 e^x + C_2 e^{2x}.$$

A particular solution $y_p(x)$ of the nonhomogeneous equation can be found in the form

$$y_p = u_1(x)e^x + u_2(x)e^{2x}$$

where $u_1$ and $u_2$ satisfy

$$\begin{aligned} u_1' e^x + 2u_2' e^{2x} &= 4x \\ u_1' e^x + u_2' e^{2x} &= 0. \end{aligned}$$

We solve these linear equations for $u_1'$ and $u_2'$ and then integrate to obtain

$$\begin{aligned} u_1' &= -4xe^{-x} & u_2' &= 4xe^{-2x} \\ u_1 &= 4(x+1)e^{-x} & u_2 &= -(2x+1)e^{-2x}. \end{aligned}$$

Hence $y_p = 4x + 4 - (2x + 1) = 2x + 3$ is a particular solution of the nonhomogeneous equation, and the general solution is

$$y = 2x + 3 + C_1 e^x + C_2 e^{2x}.$$

■

***REMARK*** This method for solving the nonhomogeneous equation is called the **method of variation of parameters.** It is completely general and extends to higher order equations in a reasonable way, but it is computationally somewhat difficult. We could have found $y_p$ more easily had we "guessed" that it would be of the form $y_p = Ax + B$ and substituted this into the differential equation to get

$$\begin{aligned} -3A + 2(Ax + B) &= 4x \\ \text{or} \quad 2Ax + (2B - 3A) &= 4x. \end{aligned}$$

The only way this latter equation can be satisfied for all $x$ is to have $2A = 4$ and $2B - 3A = 0$, that is, $A = 2$ and $B = 3$.

## EXERCISES 10.8

Find general solutions for the nonhomogeneous equations in Exercises 1–12 by the method of undetermined coefficients.

**1.** $y'' + y' - 2y = 1$

**2.** $y'' + y' - 2y = x$

**3.** $y'' + y' - 2y = e^{-x}$

**4.** $y'' + y' - 2y = e^x$

**5.** $y'' + 2y' + 5y = x^2$

**6.** $y'' + 4y = x^2$

**7.** $y'' - y' - 6y = e^{-2x}$

**8.** $y'' + 4y' + 4y = e^{-2x}$

**9.** $y'' + 2y' + 2y = e^x \sin x$

**10.** $y'' + 2y' + 2y = e^{-x} \sin x$

**11.** $y'' + y' = 4 + 2x + e^{-x}$

**12.** $y'' + 2y' + y = xe^{-x}$

**13.** Repeat Exercise 3 using the method of variation of parameters.

**14.** Repeat Exercise 4 using the method of variation of parameters.

**15.** Find a particular solution of the form $y = Ax^2$ for the Euler equation $x^2y'' + xy' - y = x^2$, and hence obtain the general solution of this equation on the interval $(0, \infty)$.

**16.** For what values of $r$ can the Euler equation $x^2y'' + xy' - y = x^r$ be solved by the method of the previous exercise. Find a particular solution for each such $r$.

**17.** Try to guess the form of a particular solution for $x^2y'' + xy' - y = x$ and hence obtain the general solution for this equation on the interval $(0, \infty)$.

**18.** Use variation of parameters to solve $x^2y'' + xy' - y = x$.

**19.** Consider the nonhomogeneous, linear equation

$$x^2y'' - (2x + x^2)y' + (2 + x)y = x^3.$$

Use the fact that $y_1(x) = x$ and $y_2(x) = xe^x$ are independent solutions of the corresponding homogeneous equation (see Exercise 5 of Section 10.6) to find the general solution of this nonhomogeneous equation.

**20.** Consider the nonhomogeneous, Bessel equation

$$x^2y'' + xy' + \left(x^2 - \frac{1}{4}\right) y = x^{3/2}.$$

Use the fact that $y_1(x) = x^{-1/2}\cos x$ and $y_2(x) = x^{-1/2}\sin x$ are independent solutions of the corresponding homogeneous equation (see Exercise 6 of Section 10.6) to find the general solution of this nonhomogeneous equation.

## 10.9 SERIES SOLUTIONS

Many of the second-order, linear, differential equations that arise in applications do not have constant coefficients and are not Euler equations. If the coefficient functions of such an equation are sufficiently well behaved we can often find solutions in the form of power series (Taylor series). Such series solutions are frequently used to define new functions, whose properties are deduced partly from the fact that they solve particular differential equations. For example, Bessel functions of order $\nu$ are defined to be certain series solutions of Bessel's differential equation

$$x^2y'' + xy' + (x^2 - \nu^2)y = 0.$$

Series solutions for second-order homogeneous linear differential equations are most easily found near an **ordinary point** of the equation. This is a point $x = a$ such that the equation can be expressed in the form

$$y'' + p(x)y' + q(x)y = 0$$

where the functions $p(x)$ and $q(x)$ are **analytic** at $x = a$. (A function $f$ is analytic at $x = a$ if $f(x)$ can be expressed as the sum of its Taylor series in powers of $x - a$ in an interval of positive radius centred at $x = a$.) Thus we assume

$$p(x) = \sum_{n=0}^{\infty} p_n(x - a)^n, \qquad q(x) = \sum_{n=0}^{\infty} q_n(x - a)^n$$

with both series converging in some interval of the form $a - R < x < a + R$. Frequently $p(x)$ and $q(x)$ are polynomials and so are analytic everywhere. A change of independent variable $\xi = x - a$ will put the point $x = a$ at the origin $\xi = 0$, so we can assume that $a = 0$.

The following example illustrates the technique of series solution around an ordinary point.

**EXAMPLE 1** Find two independent solutions in powers of $x$ for the Hermite equation

$$y'' - 2xy' + \nu y = 0.$$

For what values of $\nu$ does the equation have a polynomial solution?

**SOLUTION** We try

$$y = \sum_{n=0}^{\infty} a_n x^n = a_0 + a_1 x + a_2 x^2 + a_3 x^3 + \cdots$$

$$y' = \sum_{n=1}^{\infty} n a_n x^{n-1}$$

$$y'' = \sum_{n=2}^{\infty} n(n-1) a_n x^{n-2} = \sum_{n=0}^{\infty} (n+2)(n+1) a_{n+2} x^n.$$

(We have replaced $n$ by $n+2$ in order to get $x^n$ in the sum for $y''$.) We substitute these expressions into the differential equation to get

$$\sum_{n=0}^{\infty} (n+2)(n+1) a_{n+2} x^n - 2 \sum_{n=1}^{\infty} n a_n x^n + \nu \sum_{n=0}^{\infty} a_n x^n = 0$$

$$\text{or} \quad 2a_2 + \nu a_0 + \sum_{n=1}^{\infty} \Big[ (n+2)(n+1) a_{n+2} - (2n - \nu) a_n \Big] x^n = 0.$$

This identity holds for all $x$ provided that the coefficient of every power of $x$ vanishes; that is,

$$a_2 = -\frac{\nu a_0}{2}, \qquad a_{n+2} = \frac{(2n-\nu) a_n}{(n+2)(n+1)}, \qquad (n = 1, 2, \cdots).$$

The latter of these formulas is called a **recurrence relation**.

We can choose $a_0$ and $a_1$ to have any values; then the above conditions determine all the remaining coefficients $a_n$, $(n \geq 2)$. We can get one solution by choosing, for instance, $a_0 = 1$ and $a_1 = 0$. Then, by the recurrence relation,

$$a_3 = 0, \quad a_5 = 0, \quad a_7 = 0, \quad \cdots, \qquad \text{and}$$

$$\begin{aligned}
a_2 &= -\frac{\nu}{2} \\
a_4 &= \frac{(4-\nu) a_2}{4 \times 3} = -\frac{\nu(4-\nu)}{2 \times 3 \times 4} = -\frac{\nu(4-\nu)}{4!} \\
a_6 &= \frac{(8-\nu) a_4}{6 \times 5} = -\frac{\nu(4-\nu)(8-\nu)}{6!} \\
&\cdots
\end{aligned}$$

The pattern is obvious here:

$$a_{2n} = -\frac{\nu(4-\nu)(8-\nu)\cdots(4n-4-\nu)}{(2n)!}, \qquad (n = 1, 2, \cdots).$$

One solution to the Hermite equation is

$$y_1 = 1 + \sum_{n=1}^{\infty} -\frac{\nu(4-\nu)(8-\nu)\cdots(4n-4-\nu)}{(2n)!} x^{2n}.$$

We observe that if $\nu = 4n$ for some nonnegative integer $n$ then $y_1$ is an even polynomial of degree $2n$, for $a_{2n+2} = 0$ and all subsequent even coefficients therefore also vanish.

The second solution, $y_2$, can be found in the same way, by choosing $a_0 = 0$ and $a_1 = 1$. It is

$$y_2 = x + \sum_{n=1}^{\infty} \frac{(2-\nu)(6-\nu)\cdots(4n-2-\nu)}{(2n+1)!} x^{2n+1},$$

and is an odd polynomial of degree $2n+1$ if $\nu = 4n+2$.

Both of these series solutions converge for all $x$. The Ratio Test can be applied directly to the recurrence relation. Since consecutive nonzero terms of each series are of the form $a_n x^n$ and $a_{n+2}x^{n+2}$, we calculate

$$\rho = \lim_{n\to\infty} \left| \frac{a_{n+2}x^{n+2}}{x_n x^n} \right| = |x|^2 \lim_{n\to\infty} \left| \frac{a_{n+2}}{a_n} \right| = |x|^2 \lim_{n\to\infty} \left| \frac{2n-\nu}{(n+2)(n+1)} \right| = 0$$

for every $x$, so the series converges by the Ratio Test. ■

If $x = a$ is not an ordinary point of the equation

$$y'' + p(x)y' + q(x)y = 0$$

then it is called a **singular point** of that equation. This means that at least one of the functions $p(x)$ and $q(x)$ is not analytic at $x = a$. If, however, $(x-a)p(x)$ and $(x-a)^2q(x)$ are analytic at $x = a$ then the singular point is said to be a **regular singular point.** For example, the origin $x = 0$ is a regular singular point of any Euler equation, and also of Bessel's equation,

$$x^2y'' + xy' + (x^2 - \nu^2)y = 0,$$

since $p(x) = 1/x$ and $q(x) = (x^2-\nu^2)/x^2$ satisfy $xp(x) = 1$ and $x^2q(x) = x^2-\nu^2$, which are both polynomials and therefore analytic.

The solutions of differential equations are usually not analytic at singular points. However, it is still possible to find at least one series solution about such a point. The method involves searching for a series solution of the form $x^\mu$ times a power series, that is,

$$y = \sum_{n=0}^{\infty} a_n(x-a)^{n+\mu}, \qquad \text{where } a_0 \neq 0.$$

Substitution into the differential equation produces a quadratic **indicial equation**, which determines one or two values of $\mu$ for which such solutions can be found, and a **recurrence relation** enabling the coefficients $a_n$ to be calculated for $n \geq 1$. If the indicial roots are not equal, and do not differ by an integer, two independent solutions can be calculated. If the indicial roots are equal, or differ by an integer, one such solution can be calculated (corresponding to the larger indicial root), and a second solution can be found by using the technique described in Section 10.6. These calculations can be quite difficult. The reader is referred to standard texts on differential equations for more discussion and examples. We will content ourselves here with one final example.

**■ EXAMPLE 2** Find one solution, in powers of $x$, of Bessel's equation of order $\nu = 1$, namely

$$x^2y'' + xy' + (x^2 - 1)y = 0.$$

**SOLUTION** We try

$$y = \sum_{n=0}^{\infty} a_n x^{\mu+n}$$

$$y' = \sum_{n=0}^{\infty} (\mu+n) a_n x^{\mu+n-1}$$

$$y'' = \sum_{n=0}^{\infty} (\mu+n)(\mu+n-1) a_n x^{\mu+n-2}.$$

Substituting these expressions into the Bessel equation, we get

$$\sum_{n=0}^{\infty} \Big[ \big((\mu+n)(\mu+n-1) + (\mu+n) - 1\big) a_n x^n + a_n x^{n+2} \Big] = 0$$

$$\sum_{n=0}^{\infty} \Big[ (\mu+n)^2 - 1 \Big] a_n x^n + \sum_{n=2}^{\infty} a_{n-2} x^n = 0$$

$$(\mu^2 - 1)a_0 + \big((\mu+1)^2 - 1\big) a_1 x + \sum_{n=2}^{\infty} \Big[ \big((\mu+n)^2 - 1\big) a_n + a_{n-2} \Big] x^n = 0.$$

All of the terms must vanish. Since $a_0 \neq 0$ (we may take $a_0 = 1$) we obtain

$$\mu^2 - 1 = 0, \qquad \text{the indicial equation}$$

$$[(\mu+1)^2 - 1] a_1 = 0,$$

$$a_n = -\frac{a_{n-2}}{(\mu+n)^2 - 1}, \quad (n \geq 2). \qquad \text{the recurrence relation}$$

Evidently $\mu = \pm 1$, and therefore $a_1 = 0$. If we take $\mu = 1$, then the recurrence relation is $a_n = -a_{n-2}/(n)(n+2)$. Thus

$$a_3 = 0, \quad a_5 = 0, \quad a_7 = 0, \quad \cdots$$

$$a_2 = \frac{-1}{2 \times 4}, \quad a_4 = \frac{1}{2 \times 4 \times 4 \times 6}, \quad a_6 = \frac{-1}{2 \times 4 \times 4 \times 6 \times 6 \times 8}, \quad \cdots.$$

Again the pattern is obvious:

$$a_{2n} = \frac{(-1)^n}{2^{2n} n!(n+1)!}$$

and one solution of the Bessel equation of order 1 is

$$y = \sum_{n=0}^{\infty} \frac{(-1)^n}{2^{2n} n!(n+1)!} x^{2n+1}.$$

By the Ratio Test, this series converges for all $x$. ■

**REMARK** Observe that if we tried to calculate a second solution using $\mu = -1$ we would get the recurrence relation

$$a_n = -\frac{a_{n-2}}{n(n-2)},$$

and we would be unable to calculate $a_2$. This shows what can happen if the indicial roots differ by an integer.

## EXERCISES 10.9

1. Find the general solution of $y'' = (x-1)^2 y$ in the form of a power series $y = \sum_{n=0}^{\infty} a_n (x-1)^n$.

2. Find the general solution of $y'' = xy$ in the form of a power series $y = \sum_{n=0}^{\infty} a_n x^n$ with $a_0$ and $a_1$ arbitrary.

3. Find the first three nonzero terms in a power series solution in powers of $x$ for the initial-value problem $y'' + (\sin x)y = 0$, $y(0) = 1$, $y'(0) = 0$.

4. Find the solution, in powers of $x$, for the initial-value problem

$$(1-x^2)y'' - xy' + 9y = 0, \quad y(0) = 0, \quad y'(0) = 1.$$

5. Find two power series solutions in powers of $x$ for $3xy'' + 2y' + y = 0$.

6. Find one power series solution for the Bessel equation of order $\nu = 0$, that is, the equation $xy'' + y' + xy = 0$.

# CHAPTER REVIEW

## Key Ideas

- **What do the following phrases mean?**
  - ◇ an ordinary DE
  - ◇ a partial DE
  - ◇ the general solution of a DE
  - ◇ a linear combination of solutions of a DE
  - ◇ the order of a DE
  - ◇ a linear DE
  - ◇ a separable DE
  - ◇ an exact DE
  - ◇ an integrating factor
  - ◇ a constant coefficient DE
  - ◇ an Euler equation
  - ◇ an auxiliary equation
- **Describe how to solve:**
  - ◇ a separable DE
  - ◇ a first-order, linear DE
  - ◇ a homogeneous, first-order DE
  - ◇ a constant coefficient DE
  - ◇ an Euler equation
- **What conditions imply that an initial-value problem for a first-order DE has a unique solution near the initial point?**
- **Describe the following methods for solving first-order DEs numerically:**
  - ◇ the Euler method
  - ◇ the improved Euler method
  - ◇ the fourth-order Runge–Kutta method
- **Describe the following methods for solving a nonhomogeneous, linear DE:**
  - ◇ undetermined coefficients
  - ◇ variation of parameters
- **What are an ordinary point and a regular singular point of a linear, second-order DE? Describe how series can be used to solve such an equation near such a point.**

## Review Exercises

Find general solutions of the differential equations in Exercises 1–16.

1. $\dfrac{dy}{dx} = 2xy$

2. $\dfrac{dy}{dx} = e^{-y}\sin x$

3. $\dfrac{dy}{dx} = x + 2y$

4. $\dfrac{dy}{dx} = \dfrac{x^2+y^2}{2xy}$

5. $\dfrac{dy}{dx} = \dfrac{x+y}{y-x}$

6. $\dfrac{dy}{dx} = -\dfrac{y+e^x}{x+e^y}$

7. $\dfrac{d^2y}{dt^2} = \left(\dfrac{dy}{dt}\right)^2$

8. $2\dfrac{d^2y}{dt^2} + 5\dfrac{dy}{dt} + 2y = 0$

9. $4y'' - 4y' + 5y = 0$

10. $2x^2y'' + y = 0$

11. $t^2\dfrac{d^2y}{dt^2} - t\dfrac{dy}{dt} + 5y = 0$

12. $\dfrac{d^3y}{dt^3} + 8\dfrac{d^2y}{dt^2} + 16\dfrac{dy}{dt} = 0$

13. $\dfrac{d^2y}{dx^2} - 5\dfrac{dy}{dx} + 6y = e^x + e^{3x}$

14. $\dfrac{d^2y}{dx^2} - 5\dfrac{dy}{dx} + 6y = xe^{2x}$

15. $\dfrac{d^2y}{dx^2} + 2\dfrac{dy}{dx} + y = x^2$

16. $x^2\dfrac{d^2y}{dx^2} - 2y = x^3$

Solve the initial-value problems in Exercises 17–26.

17. $\begin{cases} \dfrac{dy}{dx} = \dfrac{x^2}{y^2} \\ y(2) = 1 \end{cases}$

18. $\begin{cases} \dfrac{dy}{dx} = \dfrac{y^2}{x^2} \\ y(2) = 1 \end{cases}$

19. $\begin{cases} \dfrac{dy}{dx} = \dfrac{xy}{x^2+y^2} \\ y(0) = 1 \end{cases}$

20. $\begin{cases} \dfrac{dy}{dx} + (\cos x)y = 2\cos x \\ y(\pi) = 1 \end{cases}$

21. $\begin{cases} y'' + 3y' + 2y = 0 \\ y(0) = 1 \\ y'(0) = 2 \end{cases}$

22. $\begin{cases} y'' + 2y' + (1+\pi^2)y = 0 \\ y(1) = 0 \\ y'(1) = \pi \end{cases}$

23. $\begin{cases} y'' + 10y' + 25y = 0 \\ y(1) = e^{-5} \\ y'(1) = 0 \end{cases}$

24. $\begin{cases} x^2y'' - 3xy' + 4y = 0 \\ y(e) = e^2 \\ y'(e) = 0 \end{cases}$

25. $\begin{cases} \dfrac{d^2y}{dt^2} + 4y = 8e^{2t} \\ y(0) = 1 \\ y'(0) = -2 \end{cases}$

26. $\begin{cases} 2\dfrac{d^2y}{dx^2} + 5\dfrac{dy}{dx} - 3y = 6 + 7e^{x/2} \\ y(0) = 0 \\ y'(0) = 1 \end{cases}$

**27.** For what values of the constants $A$ and $B$ is the equation

$$[(x+A)e^x \sin y+\cos y]\,dx+x[e^x \cos y+B \sin y]\,dy = 0$$

exact? What is the general solution of the equation if $A$ and $B$ have these values?

**28.** Find a value of $n$ for which $x^n$ is an integrating factor for

$$(x^2 + 3y^2)\,dx + xy\,dy = 0$$

and solve the equation.

**29.** Show that $y = x$ is a solution of

$$x^2y'' - x(2 + x\cot x)y' + (2 + x\cot x)y = 0$$

and find the general solution of this equation.

**30.** Use the method of variation of parameters and the result of the previous exercise to find the general solution of the nonhomogeneous equation

$$x^2y'' - x(2 + x\cot x)y' + (2 + x\cot x)y = x^3 \sin x.$$

**31.** Suppose that $f(x, y)$ and $\dfrac{\partial}{\partial y} f(x, y)$ are continuous on the whole $xy$-plane, and that $f(x, y)$ is bounded there, say $|f(x, y)| \le K$. Show that no solution of $y' = f(x, y)$ can have a vertical asymptote. Describe a region in the plane in which the solution to the initial-value problem

$$\begin{cases} y' = f(x, y) \\ y(x_0) = y_0 \end{cases}$$

must remain.

## Express Yourself

Answer these questions in words, using complete sentences and, if necessary, paragraphs. Use mathematical symbols only where necessary.

1. Why do differential equations play a very important role in mathematics?

2. What does it mean to say that a differential equation is linear? List some properties enjoyed by all linear differential equations but not, in general, by nonlinear ones.

3. Describe two different meanings of the word *homogeneous* as it applies to differential equations.

4. Discuss the similarities and the differences between the concepts of solving a differential equation and solving an algebraic equation.

5. What do we mean when we say that two different solutions of a linear differential equation are independent on an interval? How can knowing two independent solutions of a second-order linear, homogeneous equation help you find a solution of a nonhomogeneous equation?

Appendix I

# Complex Numbers

Many of the problems to which mathematics is applied involve the solution of equations. Over the centuries the number system had to be expanded many times to provide solutions for more and more kinds of equations. The natural numbers

$$\mathbb{N} = \{1, 2, 3, 4, \ldots\}$$

are inadequate for the solutions of equations of the form

$$x + n = m, \qquad (m, n \in \mathbb{N}).$$

Zero and negative numbers can be added to create the integers

$$\mathbb{Z} = \{\ldots, -3, -2, -1, 0, 1, 2, 3, \ldots\}$$

in which that equation has the solution $x = m - n$ even if $m < n$. (Historically, this extension of the number system came much later than some of those mentioned below.) Some equations of the form

$$nx = m, \qquad (m, n \in \mathbb{Z}, \quad n \neq 0)$$

cannot be solved in the integers. Another extension is made to to include numbers of the form $m/n$, thus producing the set of rational numbers

$$\mathbb{Q} = \left\{ \frac{m}{n} : m, n \in \mathbb{Z}, \quad n \neq 0 \right\}.$$

Every linear equation

$$ax = b, \qquad (a, b \in \mathbb{Q}, \quad a \neq 0)$$

has a solution $x = b/a$ in $\mathbb{Q}$, but the quadratic equation

$$x^2 = 2$$

has no solution in $\mathbb{Q}$. Another extension enriches the rational numbers to the real numbers $\mathbb{R}$ in which some equations like $x^2 = 2$ have solutions. However, other quadratic equations, for instance,

$$x^2 = -1$$

do not have solutions even in the real numbers, so the extension process is not complete. In order to be able to solve any quadratic equation, we need to extend the real number system to a larger set, which we call **the complex number system**. In this appendix we will define complex numbers and develop some of their basic properties. In Appendix II we will briefly consider functions whose domains and ranges consist of sets of complex numbers. We will also show that every polynomial equation can be solved in the complex number system.

The calculus of functions of a complex variable, called **complex analysis**, is very rich and leads to discoveries which have no counterparts for functions of real variables. We will hardly scratch the surface of the subject in these pages.

## Definition of Complex Numbers

We begin by defining the symbol $i$, called **the imaginary unit**[1], to have the property

$$i^2 = -1.$$

Thus we could also call $i$ the **square root of** $-1$ and denote it $\sqrt{-1}$. Of course, $i$ is not a real number; no real number has a negative square.

**DEFINITION 1**

> A **complex number** is an expression of the form
>
> $$a + bi \qquad \text{or} \qquad a + ib$$
>
> where $a$ and $b$ are *real numbers*, and $i$ is the imaginary unit.

For example $3 + 2i$, $\frac{7}{2} - \frac{2}{3}i$, $i\pi = 0 + i\pi$, and $-3 = -3 + 0i$ are all complex numbers. The last of these examples shows that every real number can be regarded as a complex number. (We will normally use $a + bi$ unless $b$ is a complicated expression, in which case we will write $a + ib$ instead. Either form is acceptable.)

It is often convenient to represent a complex number by a single letter; $w$ and $z$ are frequently used for this purpose. If $a$, $b$, $x$, and $y$ are real numbers, and

$$w = a + bi, \qquad \text{and} \qquad z = x + yi,$$

then we can refer to the complex numbers $w$ and $z$. Note that $w = z$ if and only if $a = x$ and $b = y$. Of special importance are the complex numbers

$$0 = 0 + 0i, \qquad 1 = 1 + 0i, \qquad \text{and} \qquad i = 0 + 1i.$$

**DEFINITION 2**

> If $z = x + yi$ is a complex number (where $x$ and $y$ are real), we call $x$ the **real part** of $z$ and denote it $\mathrm{Re}\,(z)$. We call $y$ the **imaginary part** of $z$ and denote it $\mathrm{Im}\,(z)$:
>
> $$\mathrm{Re}\,(z) = \mathrm{Re}\,(x + yi) = x, \qquad \mathrm{Im}\,(z) = \mathrm{Im}\,(x + yi) = y.$$

Note that both the real and imaginary parts of a complex number are real numbers.

$$\mathrm{Re}\,(3 - 5i) = 3 \qquad \mathrm{Im}\,(3 - 5i) = -5$$
$$\mathrm{Re}\,(2i) = \mathrm{Re}\,(0 + 2i) = 0 \qquad \mathrm{Im}\,(2i) = \mathrm{Im}\,(0 + 2i) = 2$$
$$\mathrm{Re}\,(-7) = \mathrm{Re}\,(-7 + 0i) = -7 \qquad \mathrm{Im}\,(-7) = \mathrm{Im}\,(-7 + 0i) = 0.$$

## Graphical Representation of Complex Numbers

Since complex numbers are constructed from pairs of real numbers (their real and imaginary parts) it is natural to represent complex numbers graphically as points in a Cartesian plane. We use the point with coordinates $(a, b)$ to represent the complex number $w = a + ib$. In particular, the origin $(0, 0)$ represents the complex number 0, the point $(1, 0)$ represents the complex number $1 = 1 + 0i$, and the point $(0, 1)$ represents the point $i = 0 + 1i$. (See Figure I.1.)

[1] In some fields, for example, electrical engineering, the imaginary unit is denoted $j$ instead of $i$. Like "negative", "surd", and "irrational", the term "imaginary" suggests the distrust which greeted the new kinds of numbers when they were first introduced.

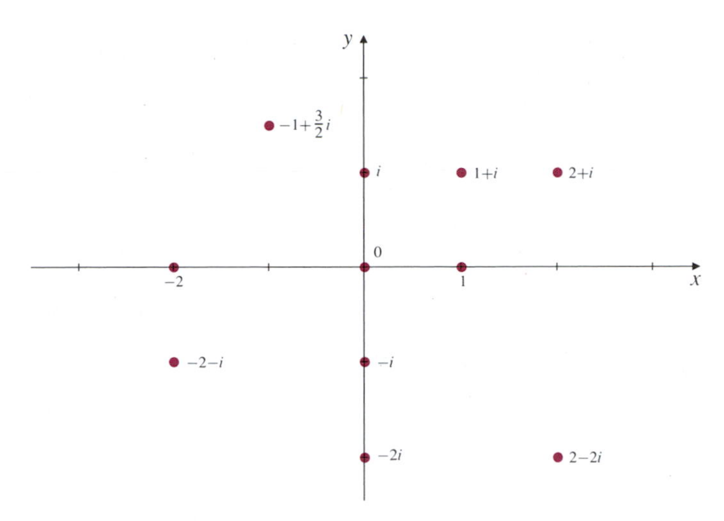


**Figure I.1** An Argand diagram representing the complex plane

Such a representation of complex numbers as points in a plane is called an **Argand diagram**. Since each complex number is represented by a unique point in the plane, the set of all complex numbers is often referred to as **the complex plane**. The sumbol $\mathbb{C}$ is used to represent the set of all complex numbers and, equivalently, the complex plane:

$$\mathbb{C} = \{x + yi \ : \ x,\ y,\ \in\ \mathbb{R}\}.$$

The points on the $x$-axis of the complex plane correspond to real numbers ($x = x + 0i$), so the $x$-axis is called the **real axis**. The points on the $y$-axis correspond to **pure imaginary** numbers ($yi = 0 + yi$), so the $y$-axis is called the **imaginary axis**.

It is often useful to consider the *polar coordinates* of a point in the complex plane.

**DEFINITION 3**

The distance from the origin to the point $(a, b)$ corresponding to the complex number $w = a + bi$ is called the **modulus** of $w$ and is denoted by $|w|$ or $|a + bi|$:

$$|w| = |a + bi| = \sqrt{a^2 + b^2}.$$

If the line from the origin to $(a, b)$ makes angle $\theta$ with the positive direction of the real axis (with positive angles measured counterclockwise), then we call $\theta$ an **argument** of the complex number $w = a + bi$, and denote it by $\arg(w)$ or $\arg(a + bi)$.

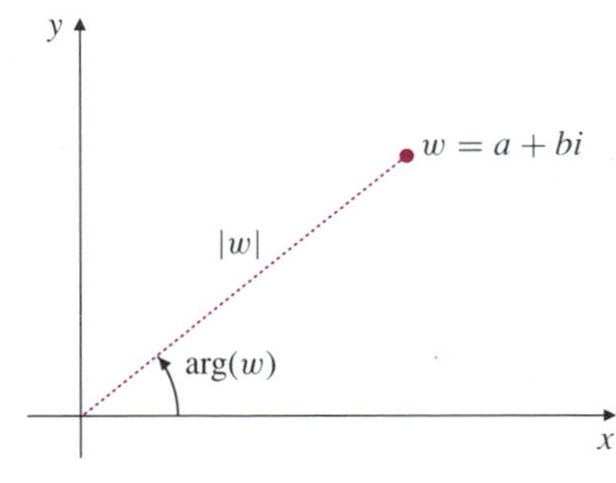


**Figure I.2** The modulus and argument of a complex number

The modulus of a complex number is always real and nonnegative. It is positive unless the complex number is 0. Modulus plays a similar role for complex numbers that absolute value does for real numbers. Indeed, sometimes modulus is called absolute value.

Arguments of complex numbers are not unique. If $w = a + bi \neq 0$ then any two possible values for $\arg(w)$ differ by an integer multiple of $2\pi$. The symbol $\arg(w)$ actually represents not a single number, but a set of numbers. When we write $\arg(w) = \theta$ we are saying that the set $\arg(w)$ contains all numbers of the form $\theta + 2k\pi$, where $k$ is an integer.

If $w = a + bi$ where $a = \text{Re}(w) \neq 0$, then

$$\tan \arg(w) = \tan \arg(a + bi) = \frac{b}{a}.$$

This means that $\tan\theta = b/a$ for every $\theta$ in the set $\arg(w)$.

It is sometimes convenient to restrict $\theta = \arg(w)$ to an interval of length $2\pi$, say, the interval $0 \leq \theta < 2\pi$, or $-\pi < \theta \leq \pi$, so that nonzero complex numbers will have unique arguments. The value of $\arg(w)$ in the interval $0 \leq \theta < 2\pi$ is called the **principal argument** of $w$ and is denoted $\text{Arg}(w)$. Every complex number $w$ except 0 has a unique principal argument $\text{Arg}(w)$.

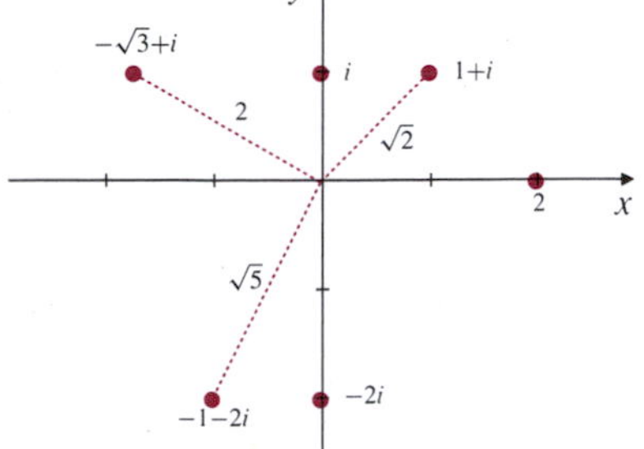


**Figure I.3** Some complex numbers with their moduli

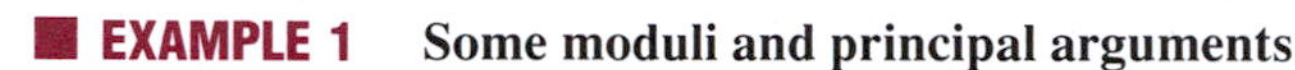

**EXAMPLE 1 Some moduli and principal arguments**

See Figure I.3.

$$|2| = 2 \qquad \text{Arg}(2) = 0$$
$$|1 + i| = \sqrt{2} \qquad \text{Arg}(1 + i) = \pi/4$$
$$|i| = 1 \qquad \text{Arg}(i) = \pi/2$$
$$|-2i| = 2 \qquad \text{Arg}(-2i) = 3\pi/2$$
$$|-\sqrt{3} + i| = 2 \qquad \text{Arg}(-\sqrt{3} + i) = 5\pi/6$$
$$|-1 - 2i| = \sqrt{5} \qquad \text{Arg}(-1 - 2i) = \pi + \tan^{-1}(2).$$

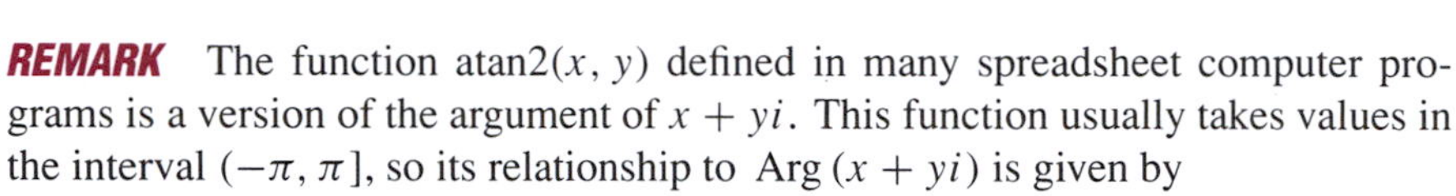

**REMARK** The function atan2$(x, y)$ defined in many spreadsheet computer programs is a version of the argument of $x + yi$. This function usually takes values in the interval $(-\pi, \pi]$, so its relationship to $\text{Arg}(x + yi)$ is given by

$$\text{Arg}(x + yi) = \begin{cases} \text{atan2}(x, y) & \text{if } y \geq 0, \\ \text{atan2}(x, y) + 2\pi & \text{if } y < 0 \end{cases}$$

Given the modulus $r = |w|$ and any value of the argument $\theta = \arg(w)$ of a complex number $w = a + bi$, we have $a = r\cos\theta$ and $b = r\sin\theta$, so $w$ can be expressed in terms of its modulus and argument as

$$w = r\cos\theta + i\,r\sin\theta.$$

The expression on the right side is called the **polar representation** of $w$.

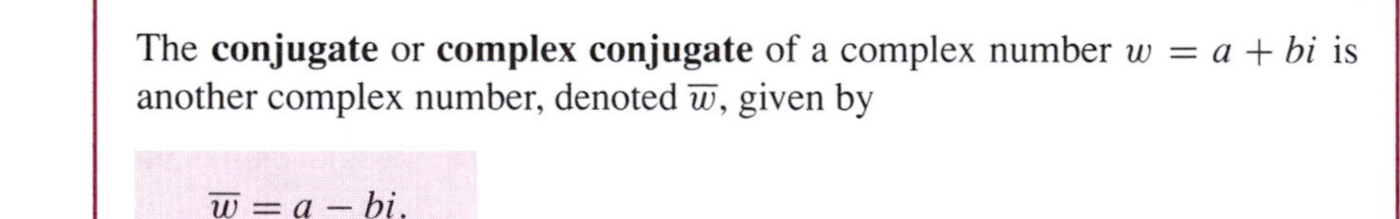

The **conjugate** or **complex conjugate** of a complex number $w = a + bi$ is another complex number, denoted $\overline{w}$, given by

$$\overline{w} = a - bi.$$

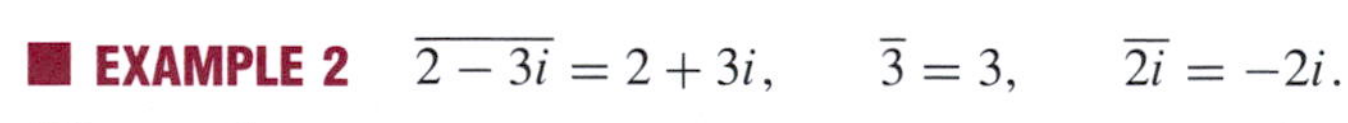

**EXAMPLE 2** $\overline{2 - 3i} = 2 + 3i, \qquad \overline{3} = 3, \qquad \overline{2i} = -2i.$

Observe that

$$\text{Re}(\overline{w}) = \text{Re}(w) \qquad |\overline{w}| = |w|$$
$$\text{Im}(\overline{w}) = -\text{Im}(w) \qquad \arg(\overline{w}) = -\arg(w) \quad \text{or} \quad 2\pi - \arg(w).$$

In an Argand diagram the point $\overline{w}$ is the reflection of the point $w$ in the real axis. (See Figure I.4.)

**Figure I.4** A complex number and its conjugate are mirror images of each other in the real axis

Note that $w$ is real ($\text{Im}(w) = 0$) if and only if $\overline{w} = w$. Also, $w$ is pure imaginary ($\text{Re}(w) = 0$) if and only if $\overline{w} = -w$. (Here $-w = -a - bi$ if $w = a + bi$.)

## Complex Arithmetic

Like real numbers, complex numbers can be added, subtracted, multiplied, and divided. Two complex numbers are added or subtracted as though they are two-dimensional vectors whose components are their real and imaginary parts.

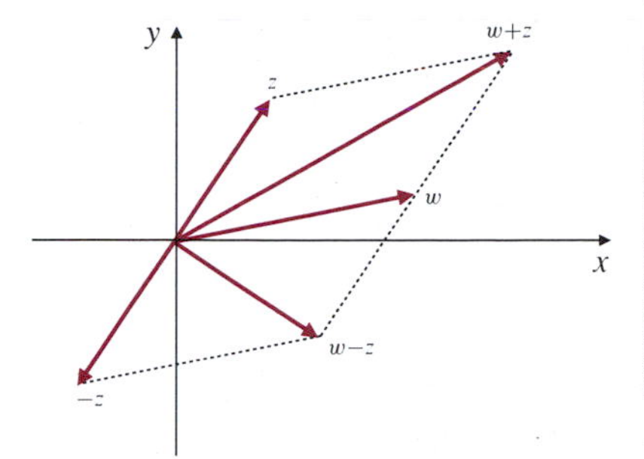


**Figure I.5** Complex numbers are added and subtracted vectorially. Observe the parallelograms

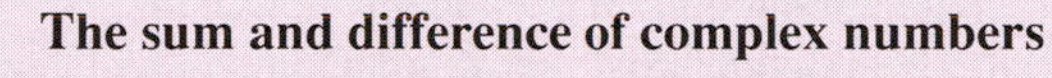

**The sum and difference of complex numbers**

If $w = a + bi$ and $z = x + yi$ where $a$, $b$, $x$, and $y$ are real numbers, then

$$w + z = (a + x) + (b + y)i$$
$$w - z = (a - x) + (b - y)i.$$

In an Argand diagram the points $w + z$ and $w - z$ are the points whose position vectors are, respectively, the sum and difference of the position vectors of the points $w$ and $z$. (See Figure I.5.) In particular, the complex number $a + bi$ is the sum of the real number $a = a + 0i$, and the pure imaginary number $bi = 0 + bi$.

Complex addition obeys the same rules as real addition: if $w_1$, $w_2$, and $w_3$ are three complex numbers, the following are easily verified:

| | |
|---|---|
| $w_1 + w_2 = w_2 + w_1$ | Addition is commutative. |
| $(w_1 + w_2) + w_3 = w_1 + (w_2 + w_3)$ | Addition is associative. |
| $\lvert w_1 \pm w_2\rvert \le \lvert w_1\rvert + \lvert w_2\rvert$ | the triangle inequality |

Note that $|w_1 - w_2|$ is the distance between the two points $w_1$ and $w_2$ in the complex plane. Thus the triangle inequality says that in the triangle with vertices $w_1$, $\mp w_2$ and 0, the length of one side is less than the sum of the other two.

It is also easily verified that the conjugate of a sum (or difference) is the sum (or difference) of the conjugates:

$$\overline{w + z} = \overline{w} + \overline{z}.$$

**EXAMPLE 3**

(a) If $w = 2 + 3i$ and $z = 4 - 5i$, then

$$w + z = (2 + 4) + (3 - 5)i = 6 - 2i$$
$$w - z = (2 - 4) + (3 - (-5))i = -2 + 8i.$$

(b) $3i + (1 - 2i) - (2 + 3i) + 5 = 4 - 2i$. ■

Multiplication of the complex numbers $w = a + bi$ and $z = x + yi$ is carried out by formally multiplying the binomial expressions and replacing $i^2$ by $-1$:

$$\begin{aligned} wz &= (a + bi)(x + yi) = ax + ayi + bxi + byi^2 \\ &= (ax - by) + (ay + bx)i. \end{aligned}$$

**The product of complex numbers**

If $w = a + bi$ and $z = x + yi$ where $a$, $b$, $x$, and $y$ are real numbers, then

$$wz = (ax - by) + (ay + bx)i.$$

**EXAMPLE 4**

(a) $(2+3i)(1-2i) = 2 - 4i + 3i - 6i^2 = 8 - i$.

(b) $i(5-4i) = 5i - 4i^2 = 4 + 5i$.

(c) $(a+bi)(a-bi) = a^2 - abi + abi - b^2i^2 = a^2 + b^2$. ■

Part (c) of the example above shows that the square of the modulus of a complex number is the product of that number with its complex conjugate:

$$w\,\overline{w} = |w|^2.$$

Complex multiplication has many properties in common with real multiplication. In particular, if $w_1$, $w_2$, and $w_3$ are complex numbers, then

| | |
|---|---|
| $w_1w_2 = w_2w_1$ | Multiplication is commutative. |
| $(w_1w_2)w_3 = w_1(w_2w_3)$ | Multiplication is associative. |
| $w_1(w_2 + w_3) = w_1w_2 + w_1w_3$ | Multiplication distributes over addition. |

The conjugate of a product is the product of the conjugates:

$$\overline{wz} = \overline{w}\,\overline{z}.$$

To see this, let $w = a + bi$ and $z = x + yi$. Then

$$\begin{aligned}\overline{wz} &= \overline{(ax - by) + (ay + bx)i} \\ &= (ax - by) - (ay + bx)i \\ &= (a - bi)(x - yi) = \overline{w}\,\overline{z}.\end{aligned}$$

It is particularly easy to determine the product of complex numbers expressed in polar form. If

$$w = r(\cos\theta + i\sin\theta), \qquad \text{and} \qquad z = s(\cos\phi + i\sin\phi),$$

where $r = |w|$, $\theta = \arg(w)$, $s = |z|$, and $\phi = \arg(z)$, then

$$\begin{aligned}wz &= rs(\cos\theta + i\sin\theta)(\cos\phi + i\sin\phi) \\ &= rs\big((\cos\theta\cos\phi - \sin\theta\sin\phi) + i(\sin\theta\cos\phi + \cos\theta\sin\phi)\big) \\ &= rs\big(\cos(\theta+\phi) + i\sin(\theta+\phi)\big).\end{aligned}$$

(See Figure I.6.) Since arguments are only determined up to integer multiples of $2\pi$, we have proved that

**The modulus and argument of a product**

$$|wz| = |w||z|, \qquad \text{and} \qquad \arg(wz) = \arg(w) + \arg(z).$$

The second of these equations says that the set $\arg(wz)$ consists of all numbers $\theta + \phi$ where $\theta$ belongs to the set $\arg(w)$ and $\phi$ to the set $\arg(z)$.

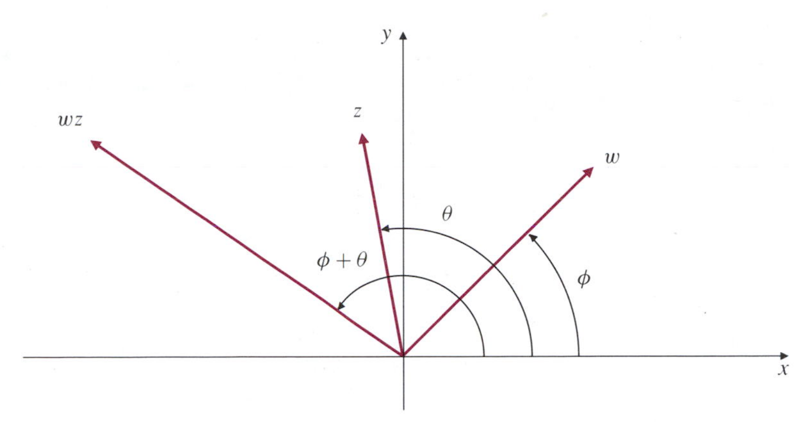


**Figure I.6** The argument of a product is the sum of the arguments of the factors

More generally, if $w_1, w_2, \ldots w_n$ are complex numbers then

$$|w_1 w_2 \cdots w_n| = |w_1||w_2| \cdots |w_n|$$
$$\arg(w_1 w_2 \cdots w_n) = \arg(w_1) + \arg(w_2) + \cdots + \arg(w_n).$$

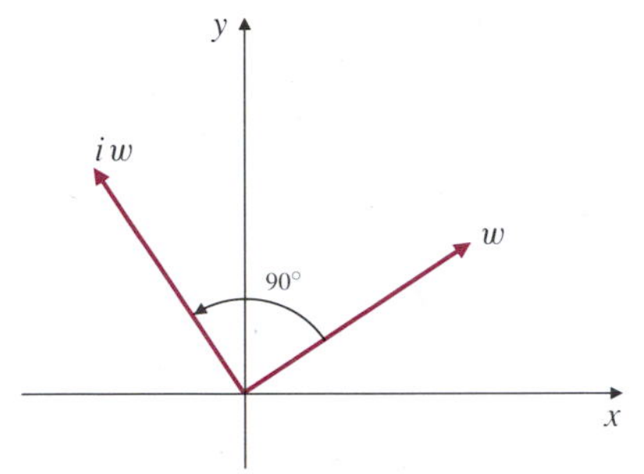


**Figure I.7** Multiplication by $i$ corresponds to counterclockwise rotation by 90°

Multiplication of a complex number by $i$ has a particularly simple geometric interpretation in an Argand diagram. Since $|i| = 1$ and $\arg(i) = \pi/2$, multiplication of $w = a + bi$ by $i$ leaves the modulus of $w$ unchanged but increases its argument by $\pi/2$. Thus multiplication by $i$ rotates the position vector of $w$ counterclockwise by 90° about the origin.

Let $z = \cos\theta + i\sin\theta$. Then $|z| = 1$ and $\arg(z) = \theta$. Since the modulus of a product is the product of the moduli of the factors and the argument of a product is the sum of the arguments of the factors, we have $|z^n| = |z|^n = 1$ and $\arg(z^n) = n\arg(z) = n\theta$. Thus

$$z^n = \cos n\theta + i\sin n\theta,$$

and we have proved

**THEOREM 1**

**De Moivre's Theorem**

$$\left(\cos\theta + i\sin\theta\right)^n = \cos n\theta + i\sin n\theta.$$

**EXAMPLE 5** Express $(1+i)^5$ in the form $a + bi$.

***SOLUTION*** Since $|(1+i)^5| = |1+i|^5 = \left(\sqrt{2}\right)^5 = 4\sqrt{2}$, and $\arg\left((1+i)^5\right) = 5\arg(1+i) = \dfrac{5\pi}{4}$, we have

$$(1+i)^5 = 4\sqrt{2}\left(\cos\frac{5\pi}{4} + i\sin\frac{5\pi}{4}\right) = 4\sqrt{2}\left(-\frac{1}{\sqrt{2}} - \frac{1}{\sqrt{2}}i\right) = -4 - 4i.$$

■

De Moivre's Theorem can be used to generate trigonometric identities for multiples of an angle. For example, for $n = 2$ we have

$$\cos 2\theta + i\sin 2\theta = \left(\cos\theta + i\sin\theta\right)^2 = \cos^2\theta - \sin^2\theta + 2i\cos\theta\sin\theta,$$

Thus $\cos 2\theta = \cos^2\theta - \sin^2\theta$, and $\sin 2\theta = 2\sin\theta\cos\theta$.

The **reciprocal** of the nonzero complex number $w = a + bi$ can be calculated by multiplying the numerator and denominator of the reciprocal expression by the conjugate of $w$:

$$w^{-1} = \frac{1}{w} = \frac{1}{a+bi} = \frac{a-bi}{(a+bi)(a-bi)} = \frac{a-bi}{a^2+b^2} = \frac{\overline{w}}{|w|^2}.$$

Since $|\overline{w}| = |w|$, and $\arg(\overline{w}) = -\arg(w)$, we have

$$\left|\frac{1}{w}\right| = \frac{|\overline{w}|}{|w|^2} = \frac{1}{|w|}, \qquad \text{and} \qquad \arg\left(\frac{1}{w}\right) = -\arg(w).$$

The **quotient** $z/w$ of two complex numbers $z = x + yi$ and $w = a + bi$ is the product of $z$ and $1/w$, so

$$\frac{z}{w} = \frac{z\overline{w}}{|w|^2} = \frac{(x+yi)(a-bi)}{a^2+b^2} = \frac{xa+yb+i(ya-xb)}{a^2+b^2}.$$

We have:

**The modulus and argument of a quotient**

$$\left|\frac{z}{w}\right| = \frac{|z|}{|w|} \qquad \text{and} \qquad \arg\left(\frac{z}{w}\right) = \arg(z) - \arg(w).$$

The set $\arg(z/w)$ consists of all numbers $\theta - \phi$ where $\theta$ belongs to the set $\arg(z)$ and $\phi$ to the set $\arg(w)$.

**EXAMPLE 6** Simplify (a) $\dfrac{2+3i}{4-i}$. (b) $\dfrac{i}{1+i\sqrt{3}}$.

***SOLUTION***

(a) $$\frac{2+3i}{4-i} = \frac{(2+3i)(4+i)}{(4-i)(4+i)} = \frac{8-3+(2+12)i}{4^2+1^2} = \frac{5}{17} + \frac{14}{17}i.$$

(b) $$\frac{i}{1+i\sqrt{3}} = \frac{i(1-i\sqrt{3})}{(1+i\sqrt{3})(1-i\sqrt{3})} = \frac{\sqrt{3}+i}{1^2+3} = \frac{\sqrt{3}}{4} + \frac{1}{4}i.$$

Alternatively, since $|1 + i\sqrt{3}| = 2$ and $\arg(1 + i\sqrt{3}) = \tan^{-1}\sqrt{3} = \dfrac{\pi}{3}$, the quotient in (b) has modulus $\dfrac{1}{2}$ and argument $\dfrac{\pi}{2} - \dfrac{\pi}{3} = \dfrac{\pi}{6}$. Thus

$$\frac{i}{1+i\sqrt{3}} = \frac{1}{2}\left(\cos\frac{\pi}{6} + i\sin\frac{\pi}{6}\right) = \frac{\sqrt{3}}{4} + \frac{1}{4}i.$$

## Roots of Complex Numbers

If $a$ is a positive real number, there are two distinct real numbers whose square is $a$. These are usually denoted

$\sqrt{a}$ (the positive square root of $a$), and

$-\sqrt{a}$ (the negative square root of $a$).

Every nonzero complex number $z = x + yi$ (where $x^2 + y^2 > 0$) also has two square roots; if $w_1$ is a complex number such that $w_1^2 = z$ then $w_2 = -w_1$ also satisfies $w_2^2 = z$. Again we would like to single out one of these roots and call it $\sqrt{z}$.

Let $r = |z|$, so that $r > 0$. Let $\theta = \operatorname{Arg}(z)$. Thus $0 \le \theta < 2\pi$. Since

$$z = r\big(\cos\theta + i\sin\theta\big),$$

the complex number

$$w = \sqrt{r}\left(\cos\frac{\theta}{2} + i\sin\frac{\theta}{2}\right)$$

clearly satisfies $w^2 = z$. We call this $w$ the **principal square root** of $z$, and denote it $\sqrt{z}$. The two solutions of the equation $w^2 = z$ are, thus, $w = \sqrt{z}$ and $w = -\sqrt{z}$.

Observe that the imaginary part of $\sqrt{z}$ is always nonnegative since $\sin\dfrac{\theta}{2} \ge 0$ for $0 \le \theta < 2\pi$; in fact, since $\dfrac{\theta}{2} < \pi$, $\sqrt{z}$ is real only if $z$ is real ($\theta = 0$) in which case $\sqrt{z}$ is positive.

**EXAMPLE 7**

(a) $\sqrt{4} = \sqrt{4(\cos 0 + i\sin 0)} = 2.$

(b) $\sqrt{i} = \sqrt{1\left(\cos\dfrac{\pi}{2} + i\sin\dfrac{\pi}{2}\right)} = \cos\dfrac{\pi}{4} + i\sin\dfrac{\pi}{4} = \dfrac{1}{\sqrt{2}} + \dfrac{1}{\sqrt{2}}i.$

(c) $\sqrt{-4i} = \sqrt{4\left(\cos\dfrac{3\pi}{2} + i\sin\dfrac{3\pi}{2}\right)} = 2\left(\cos\dfrac{3\pi}{4} + i\sin\dfrac{3\pi}{4}\right) = -\sqrt{2}+i\sqrt{2}.$

(d) $\sqrt{-\dfrac{1}{2} + i\dfrac{\sqrt{3}}{2}} = \sqrt{\cos\dfrac{2\pi}{3} + i\sin\dfrac{2\pi}{3}} = \cos\dfrac{\pi}{3} + i\sin\dfrac{\pi}{3} = \dfrac{1}{2} + \dfrac{\sqrt{3}}{2}i.$ ■

Given a nonzero complex number $z$ we can find $n$ distinct complex numbers $w$ that satisfy $w^n = z$. These $n$ numbers are called $n$th roots of $z$. For example, if $z = 1 = \cos 0 + i\sin 0$, then each of the numbers

$$\begin{aligned}
w_1 &= 1 \\
w_2 &= \cos\frac{2\pi}{n} + i\sin\frac{2\pi}{n} \\
w_3 &= \cos\frac{4\pi}{n} + i\sin\frac{4\pi}{n} \\
w_4 &= \cos\frac{6\pi}{n} + i\sin\frac{6\pi}{n} \\
&\vdots \\
w_n &= \cos\frac{2(n-1)\pi}{n} + i\sin\frac{2(n-1)\pi}{n}
\end{aligned}$$

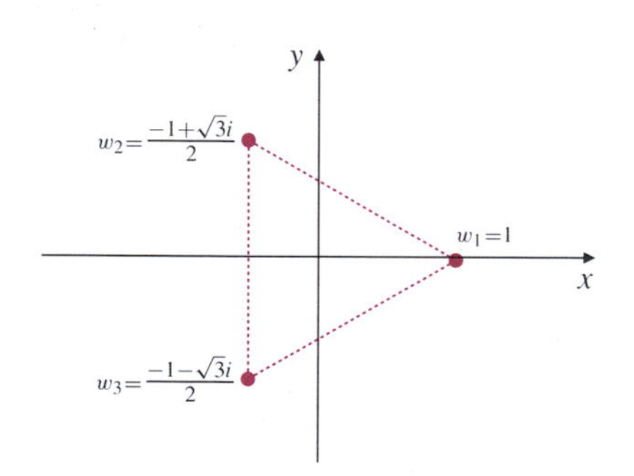


**Figure I.8** The cube roots of unity

satisfies $w^n = 1$ so is an $n$th root of 1. (These numbers are usually called the $n$th roots of unity.) Figure I.8 shows the three cube roots of 1. Observe that they are at the three vertices of an equilateral triangle with centre at the origin and one vertex at 1. In general, the $n$ $n$th roots of unity lie on a circle of radius 1 centred at the origin, at the vertices of a regular $n$-sided polygon with one vertex at 1.

If $z$ is any nonzero complex number, and $\theta$ is the principal argument of $z$ $(0 \le \theta < 2\pi)$ then the number

$$w_1 = |z|^{1/n}\left(\cos\frac{\theta}{n} + i\sin\frac{\theta}{n}\right)$$

is called the **principal** $n$th root of $z$. All the $n$th roots of $z$ are on the circle of radius $|z|^{1/n}$ centred at the origin, and are at the vertices of a regular $n$-sided polygon with one vertex at $w_1$. (See Figure I.9.) The other $n$th roots are

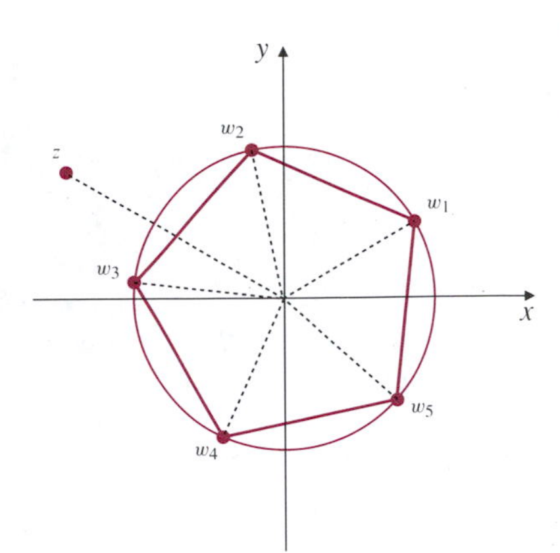


**Figure I.9** The five 5th roots of $z$

$$w_2 = |z|^{1/n}\left(\cos\frac{\theta+2\pi}{n} + i\sin\frac{\theta+2\pi}{n}\right)$$

$$w_3 = |z|^{1/n}\left(\cos\frac{\theta+4\pi}{n} + i\sin\frac{\theta+4\pi}{n}\right)$$

$$\vdots$$

$$w_n = |z|^{1/n}\left(\cos\frac{\theta+2(n-1)\pi}{n} + i\sin\frac{\theta+2(n-1)\pi}{n}\right).$$

We can obtain all $n$ of the $n$th roots of $z$ by multiplying the principal $n$th root by the $n$th roots of unity.

**EXAMPLE 8** Find the 4th roots of $-4$. Sketch them in an Argand diagram.

***SOLUTION*** Since $|-4|^{1/4} = \sqrt{2}$ and $\arg(-4) = \pi$, the principal 4th root of $-4$ is

$$w_1 = \sqrt{2}\left(\cos\frac{\pi}{4} + i\sin\frac{\pi}{4}\right) = 1 + i.$$

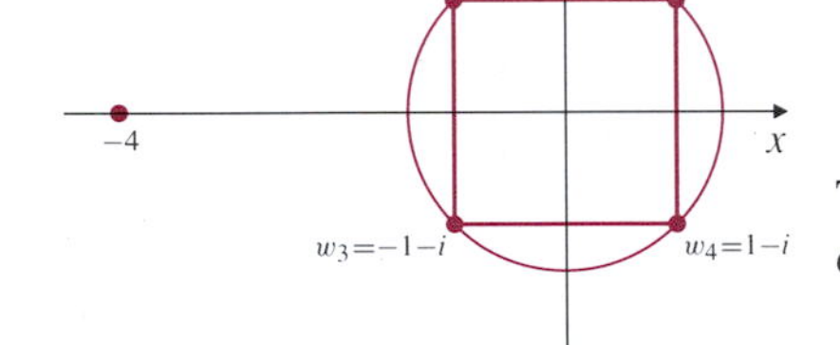


**Figure I.10** The four 4th roots of $-4$

The other three 4th roots are at the vertices of a square with centre at the origin and one vertex at $1 + i$. (See Figure I.10.) Thus the other roots are

$$w_2 = -1 + i, \qquad w_3 = -1 - i, \qquad w_4 = 1 - i.$$

## EXERCISES

In Exercises 1–4, find the real and imaginary parts (Re $(z)$ and Im $(z)$) of the given complex numbers $z$, and sketch the position of each number in the complex plane (i.e. in an Argand diagram).

**1.** $z = -5 + 2i$

**2.** $z = 4 - i$

**3.** $z = -\pi i$

**4.** $z = -6$

In Exercises 5–15, find the modulus $r = |z|$ and the principal argument $\theta = \text{Arg}(z)$ of each given complex number $z$, and express $z$ in terms of $r$ and $\theta$.

**5.** $z = -1 + i$

**6.** $z = -2$

**7.** $z = 3i$

**8.** $z = -5i$

**9.** $z = 1 + 2i$

**10.** $z = -2 + i$

**11.** $z = -3 - 4i$

**12.** $z = 3 - 4i$

**13.** $z = \sqrt{3} - i$

**14.** $z = -\sqrt{3} - 3i$

**15.** $z = 3\cos\frac{4\pi}{5} + 3i\sin\frac{4\pi}{5}$

**16.** If $\text{Arg}(z) = 7\pi/4$ and $\text{Arg}(w) = \pi/2$, find $\text{Arg}(zw)$.

**17.** If $\text{Arg}(z) = \pi/6$ and $\text{Arg}(w) = \pi/4$, find $\text{Arg}(z/w)$.

In Exercises 18–23, express in the form $z = x + yi$ the complex number $z$ whose modulus and argument is given.

**18.** $|z| = 2, \quad \arg(z) = \pi$

**19.** $|z| = 5, \quad \arg(z) = \tan^{-1}\frac{3}{4}$

**20.** $|z| = 1, \quad \arg(z) = \frac{3\pi}{4}$

**21.** $|z| = \pi, \quad \arg(z) = \frac{\pi}{6}$

**22.** $|z| = 0, \quad \arg(z) = 1$

**23.** $|z| = \frac{1}{2}, \quad \arg(z) = -\frac{\pi}{3}$

In Exercises 24–27, find the complex conjugates of the given complex numbers.

**24.** $5+3i$

**25.** $-3-5i$

**26.** $4i$

**27.** $2-i$

Describe, geometrically, (or make a sketch of) the set of points $z$ in the complex plane satisfying the given equations or inequalities in Exercises 28–33.

**28.** $|z|=2$

**29.** $|z|\le 2$

**30.** $|z-2i|\le 3$

**31.** $|z-3+4i|\le 5$

**32.** $\arg z=\dfrac{\pi}{3}$

**33.** $\pi\le \arg(z)\le \dfrac{7\pi}{4}$

Simplify the expressions in Exercises 34–43.

**34.** $(2+5i)+(3-i)$

**35.** $i-(3-2i)+(7-3i)$

**36.** $(4+i)(4-i)$

**37.** $(1+i)(2-3i)$

**38.** $(a+bi)(\overline{2a-bi})$

**39.** $(2+i)^3$

**40.** $\dfrac{2-i}{2+i}$

**41.** $\dfrac{1+3i}{2-i}$

**42.** $\dfrac{1+i}{i(2+3i)}$

**43.** $\dfrac{(1+2i)(2-3i)}{(2-i)(3+2i)}$

**44.** Prove that $\overline{z+w}=\overline{z}+\overline{w}$.

**45.** Prove that $\overline{\left(\dfrac{z}{w}\right)}=\dfrac{\overline{z}}{\overline{w}}$.

**46.** Express each of the complex numbers $z=3+i\sqrt{3}$ and $w=-1+i\sqrt{3}$ in polar form (that is, in terms of its modulus and argument). Use these expressions to calculate $zw$ and $z/w$.

**47.** Repeat the previous exercise for $z=-1+i$ and $w=3i$.

**48.** Use de Moivre's Theorem to find a trigonometric identity for $\cos 3\theta$ in terms of $\cos\theta$, and one for $\sin 3\theta$ in terms of $\sin\theta$.

**49.** Describe the solutions, if any, of the equations (a) $\overline{z}=2/z$, and (b) $\overline{z}=-2/z$.

**50.** For positive real numbers $a$ and $b$ it is always true that $\sqrt{ab}=\sqrt{a}\sqrt{b}$. Does a similar identity hold for $\sqrt{zw}$, where $z$ and $w$ are complex numbers? (*Hint:* consider $z=w=-1$.)

**51.** Find the three cube roots of $-1$.

**52.** Find the three cube roots of $-8i$.

**53.** Find the three cube roots of $-1+i$.

**54.** Find all the fourth roots of 4.

**55.** Find all complex solutions of the equation $z^4+1-i\sqrt{3}=0$.

**56.** Find all solutions of $z^5+a^5=0$, where $a$ is a positive real number.

* **57.** Show that the sum of the $n$ $n$th roots of unity is zero. (*Hint:* show that these roots are all powers of the principal root.)

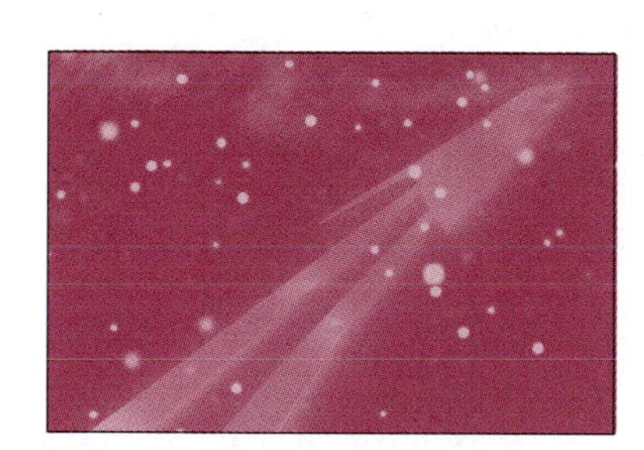

## Appendix II
# Complex Functions

Most of this book is concerned with developing the properties of **real functions**, that is, functions of one or more real variables, having values that are themselves real numbers or vectors with real components. The definition of *function* can be paraphrased to allow for complex-valued functions of a complex variable.

**DEFINITION 1**

A **complex function** $f$ is a rule that assigns a unique complex number $f(z)$ to each number $z$ in some set of complex numbers (called the **domain** of the function).

Typically, we will use $z=x+yi$ to denote a general point in the domain of a complex function, and $w=u+vi$ to denote the value of the function at $z$; if $w=f(z)$ then the real and imaginary parts of $w$ ($u=\text{Re}(w)$ and $v=\text{Im}(w)$) are real-valued functions of $z$, and hence real-valued functions of the two real variables $x$ and $y$:

$$u=u(x,y),\qquad v=v(x,y).$$

For example, the complex function $f(z) = z^2$, whose domain is the whole complex plane $\mathbb{C}$, assigns the value $z^2$ to the complex number $z$. If $w = z^2$ (where $w = u + vi$ and $z = x + yi$), then

$$u + vi = (x + yi)^2 = x^2 - y^2 + 2xyi,$$

so that

$$u = \operatorname{Re}(z^2) = x^2 - y^2, \qquad \text{and} \qquad v = \operatorname{Im}(z^2) = 2xy.$$

It is not convenient to draw the *graph* of a complex function. The graph of $w = f(z)$ would have to be drawn in a four-dimensional (real) space since two dimensions (a $z$-plane) are required for the independent variable and two more dimensions (a $w$-plane) are required for the dependent variable. Instead, we can graphically represent the behaviour of a complex function $w = f(z)$ by drawing the $z$-plane and the $w$-plane separately, and showing the image in the $w$-plane of certain, appropriately chosen sets of points in the $z$-plane. For example, Figure II.1 illustrates the fact that for the function $w = z^2$ the image of the quarter-disk $|z| \le a$, $0 \le \arg(z) \le \frac{\pi}{2}$ is the half-disk $|w| \le a^2$, $0 \le \arg(w) \le \pi$. To see why this is so, observe that if $z = r(\cos\theta + i\sin\theta)$, then $w = r^2(\cos 2\theta + i\sin 2\theta)$. Thus the function maps the circle $|z| = r$ onto the circle $|w| = r^2$, and the radial line $\arg(z) = \theta$ onto the radial line $\arg(w) = 2\theta$.

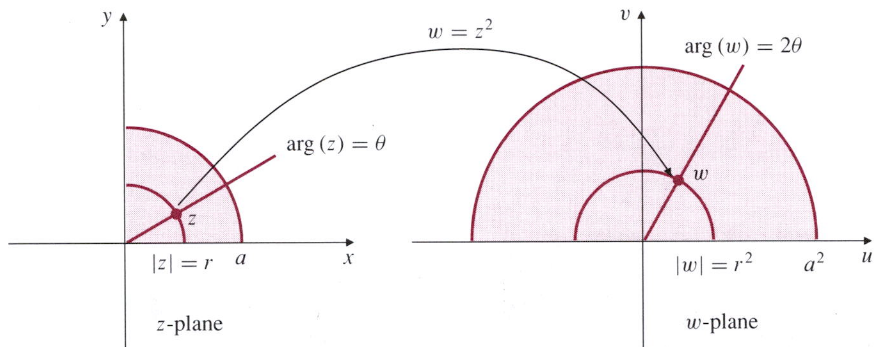


**Figure II.1** The function $w = z^2$ maps a quarter-disk of radius $a$ to a half-disk of radius $a^2$ by squaring the modulus and doubling the argument of each point $z$

## Limits and Continuity

The concepts of limit and continuity carry over from real functions to complex functions in an obvious way providing we use $|z_1 - z_2|$ as the distance between the complex numbers $z_1$ and $z_2$. We say that

$$\lim_{z \to z_0} f(z) = \lambda$$

provided we can ensure that $|f(z) - \lambda|$ is as small as we wish by taking $z$ sufficiently close to $z_0$. Formally,

**DEFINITION 2**

We say that $f(z)$ tends to the **limit** $\lambda$ as $z$ approaches $z_0$, and we write

$$\lim_{z \to z_0} f(z) = \lambda,$$

if for every positive real number $\epsilon$ there exists a positive real number $\delta$, (depending on $\epsilon$), such that

$$0 < |z - z_0| < \delta \qquad \Longrightarrow \qquad |f(z) - \lambda| < \epsilon.$$

**DEFINITION 3**

> The complex function $f(z)$ is **continuous** at $z = z_0$ if $\lim_{z \to z_0} f(z)$ exists and equals $f(z_0)$.

All the laws of limits and continuity apply as for real functions. Polynomials, that is, functions of the form

$$P(z) = a_0 + a_1 z + a_2 z^2 + \cdots + a_n z^n,$$

are continuous at every point of the complex plane. Rational functions, that is, functions of the form

$$R(z) = \frac{P(z)}{Q(z)},$$

where $P(z)$ and $Q(z)$ are polynomials, are continuous everywhere except at points where $Q(z) = 0$. Integer powers $z^n$ are continuous except at the origin if $n < 0$. The situation for fractional powers is more complicated. For example, $\sqrt{z}$ (the principal square root) is continuous except at points $z = x > 0$. The function $f(z) = \bar{z}$ is continuous everywhere, because

$$|\bar{z} - \overline{z_0}| = |\overline{z - z_0}| = |z - z_0|.$$

## The Complex Derivative

The definition of derivative is the same as for real functions:

**DEFINITION 4**

> The complex function $f$ is **differentiable** at $z$, and has **derivative** $f'(z)$ there, provided
>
> $$\lim_{h \to 0} \frac{f(z+h) - f(z)}{h} = f'(z)$$
>
> exists.

Note, however, that in this definition $h$ is a complex number. The limit must exist no matter how $h$ approaches 0 in the complex plane. This fact has profound implications. The existence of a derivative in this sense forces the function $f$ to be much better behaved than is necessary for a differentiable real function. For example, it can be shown that if $f'(z)$ exists for all $z$ in an open region $D$ in $\mathbb{C}$, then $f$ has derivatives of *all* orders throughout $D$. Moreover, such a function is the sum of its Taylor series

$$f(z) = f(z_0) + f'(z_0)(z - z_0) + \frac{f''(z_0)}{2!}(z - z_0)^2 + \cdots$$

about any point $z_0$ in $D$; the series has positive radius of convergence $R$, and converges in the disk $|z - z_0| < R$. For this reason, complex functions that are differentiable on open sets in $\mathbb{C}$ are usually called **analytic functions.** It is beyond the scope of this introductory appendix to prove these assertions. They are proved in courses and texts on complex analysis.

The usual differentiation rules apply:

$$\frac{d}{dz}\big(Af(z)+Bg(z)\big) = Af'(z)+Bg'(z)$$
$$\frac{d}{dz}\big(f(z)g(z)\big) = f'(z)g(z)+f(z)g'(z)$$
$$\frac{d}{dz}\left(\frac{f(z)}{g(z)}\right) = \frac{g(z)f'(z)-f(z)g'(z)}{\big(g(z)\big)^2}$$
$$\frac{d}{dz}f\big(g(z)\big) = f'\big(g(z)\big)g'(z).$$

As one would expect, the derivative of $f(z)=z^n$ is $f'(z)=nz^{n-1}$.

**EXAMPLE 1** Show that the function $f(z)=\overline{z}$ is not differentiable at any point.

***SOLUTION*** We have

$$\begin{aligned} f'(z) &= \lim_{h\to 0}\frac{\overline{z+h}-\overline{z}}{h} \\ &= \lim_{h\to 0}\frac{\overline{z+h-z}}{h} = \lim_{h\to 0}\frac{\overline{h}}{h}. \end{aligned}$$

But $\overline{h}/h = 1$ if $h$ is real, and $\overline{h}/h=-1$ if $h$ is pure imaginary. Since there are real and pure imaginary numbers arbitrarily close to 0, the limit above does not exist, and so $f'(z)$ does not exist. ■

The following theorem links the existence of the derivative of a complex function $f(z)$ with certain properties of its real and imaginary parts $u(x,y)$ and $v(x,y)$.

**THEOREM 1**

**The Cauchy–Riemann equations**

If $f(z)=u(x,y)+iv(x,y)$ is differentiable at $z=x+yi$ then $u$ and $v$ satisfy the Cauchy-Riemann equations

$$\frac{\partial u}{\partial x}=\frac{\partial v}{\partial y}, \qquad \frac{\partial v}{\partial x}=-\frac{\partial u}{\partial y}.$$

Conversely, if $u$ and $v$ are sufficiently smooth, (say, if they have continuous second partial derivatives near $(x,y)$), and if $u$ and $v$ satisfy the Cauchy–Riemann equations at $(x,y)$, then $f$ is differentiable at $z=x+yi$, and

$$f'(z)=\frac{\partial u}{\partial x}+i\frac{\partial v}{\partial x}.$$

***PROOF*** First assume that $f$ is differentiable at $z$. Letting $h=s+ti$, we have

$$\begin{aligned} f'(z) &= \lim_{h\to 0}\frac{f(z+h)-f(z)}{h} \\ &= \lim_{(s,t)\to(0,0)}\left[\frac{u(x+s,y+t)-u(x,y)}{s+it}+i\,\frac{v(x+s,y+t)-v(x,y)}{s+it}\right]. \end{aligned}$$

The limit must be independent of the path along which $h$ approaches 0. Letting $t=0$, so that $h=s$ approaches 0 along the real axis, we obtain

$$f'(z)=\lim_{s\to 0}\left[\frac{u(x+s,y)-u(x,y)}{s}+i\,\frac{v(x+s,y)-v(x,y)}{s}\right]=\frac{\partial u}{\partial x}+i\frac{\partial v}{\partial x}.$$

Similarly, letting $s = 0$, so that $h = ti$ approaches 0 along the imaginary axis, we obtain

$$\begin{aligned} f'(z) &= \lim_{t\to 0}\left[\frac{u(x, y+t) - u(x, y)}{it} + i\,\frac{v(x, y+t) - v(x, y)}{it}\right] \\ &= \lim_{t\to 0}\left[\frac{v(x, y+t) - v(x, y)}{t} - i\,\frac{u(x, y+t) - u(x, y)}{t}\right] \\ &= \frac{\partial v}{\partial y} - i\,\frac{\partial u}{\partial y}. \end{aligned}$$

Equating these two expressions for $f'(z)$ we see that

$$\frac{\partial u}{\partial x} = \frac{\partial v}{\partial y}, \qquad \frac{\partial v}{\partial x} = -\frac{\partial u}{\partial y}.$$

To prove the converse, we use the result of Exercise 18 of Section 4.6. Since $u$ and $v$ are assumed to have continuous second partial derivatives, we must have

$$\begin{aligned} u(x+s, y+t) - u(x, y) &= s\frac{\partial u}{\partial x} + t\frac{\partial u}{\partial y} + O(s^2+t^2) \\ v(x+s, y+t) - v(x, y) &= s\frac{\partial v}{\partial x} + t\frac{\partial v}{\partial y} + O(s^2+t^2), \end{aligned}$$

where we have used the "Big-O" notation; the expression $O(\lambda)$ denotes a term satisfying $|O(\lambda)| \le K|\lambda|$ for some constant $K$. Thus, if $u$ and $v$ satisfy the Cauchy–Riemann equations,

$$\begin{aligned} \frac{f(z+h) - f(z)}{h} &= \frac{s\dfrac{\partial u}{\partial x} + t\dfrac{\partial u}{\partial y} + i\left(s\dfrac{\partial v}{\partial x} + t\dfrac{\partial v}{\partial y}\right) + O(s^2+t^2)}{s+it} \\ &= \frac{(s+it)\dfrac{\partial u}{\partial x} + i(s+it)\dfrac{\partial v}{\partial x}}{s+it} + O(\sqrt{s^2+t^2}) \\ &= \frac{\partial u}{\partial x} + i\,\frac{\partial v}{\partial x} + O(\sqrt{s^2+t^2}). \end{aligned}$$

Thus we may let $h = s + ti$ approach 0 and obtain

$$f'(z) = \frac{\partial u}{\partial x} + i\,\frac{\partial v}{\partial x}.$$

◆

It follows immediately from the Cauchy–Riemann equations that the real and imaginary parts of a differentiable complex function are real harmonic functions:

$$\frac{\partial^2 u}{\partial x^2} + \frac{\partial^2 u}{\partial y^2} = 0, \qquad \frac{\partial^2 v}{\partial x^2} + \frac{\partial^2 v}{\partial y^2} = 0.$$

(See Exercise 15 of Section 4.4.)

## The Exponential Function

Consider the function

$$f(z) = e^x \cos y + i e^x \sin y,$$

where $z = x + yi$. The real and imaginary parts of $f(z)$,

$$u(x, y) = \operatorname{Re}(f(z)) = e^x \cos y, \qquad \text{and} \qquad v(x, y) = \operatorname{Im}(f(z)) = e^x \sin y$$

satisfy the Cauchy–Riemann equations

$$\frac{\partial u}{\partial x} = e^x \cos y = \frac{\partial v}{\partial y}, \qquad \text{and} \qquad \frac{\partial v}{\partial x} = e^x \sin y = -\frac{\partial u}{\partial y}$$

everywhere in the $z$-plane. Therefore $f(z)$ is differentiable (analytic) everywhere, and satisfies

$$f'(z) = \frac{\partial u}{\partial x} + i\frac{\partial v}{\partial x} = e^x \cos y + i e^x \sin y = f(z).$$

Evidently $f(0) = 1$, and $f(z) = e^x$ if $z = x$ is a real number. It is therefore natural to denote the function $f(z)$ as the exponential function $e^z$.

**The complex exponential function**

$$e^z = e^x(\cos y + i \sin y) \qquad \text{for} \quad z = x + yi.$$

In particular, if $z = yi$ is pure imaginary, then

$$e^{yi} = \cos y + i \sin y,$$

a fact that can also be obtained by separating the real and imaginary parts of the Maclaurin series for $e^{yi}$:

$$\begin{aligned} e^{yi} &= 1 + (yi) + \frac{(yi)^2}{2!} + \frac{(yi)^3}{3!} + \frac{(yi)^4}{4!} + \frac{(yi)^5}{5!} + \cdots \\ &= \left(1 - \frac{y^2}{2!} + \frac{y^4}{4!} - \cdots\right) + i\left(y - \frac{y^3}{3!} + \frac{y^5}{5!} - \cdots\right) \\ &= \cos y + i \sin y. \end{aligned}$$

Observe that

$$\begin{aligned} |e^z| &= \sqrt{e^{2x}\left(\cos^2 y + \sin^2 y\right)} = e^x, \\ \arg(e^z) &= \arg(e^{yi}) = \arg(\cos y + i \sin y) = y, \\ \overline{e^z} &= e^x \cos y - i e^x \sin y = e^x \cos(-y) + i e^x \sin(-y) = e^{\bar{z}}. \end{aligned}$$

In summary:

**Properties of the exponential function**

If $z = x + yi$ then $\overline{e^z} = e^{\bar{z}}$. Also,

$$\begin{aligned} \operatorname{Re}(e^z) &= e^x \cos y, \qquad & |e^z| &= e^x, \\ \operatorname{Im}(e^z) &= e^x \sin y, \qquad & \arg(e^z) &= y. \end{aligned}$$

**EXAMPLE 2** Sketch the image in the $w$-plane of the rectangle

$$R : a \le x \le b, \quad c \le y \le d$$

in the $z$-plane under the transformation $w = e^z$.

***SOLUTION*** The vertical lines $x = a$ and $x = b$ get mapped to the concentric circles $|w| = e^a$ and $|w| = e^b$. The horizontal lines $y = c$ and $y = d$ get mapped to the radial lines $\arg(w) = c$ and $\arg(w) = d$. Thus the rectangle $R$ gets mapped to the polar region $P$ shown in Figure II.2. ■

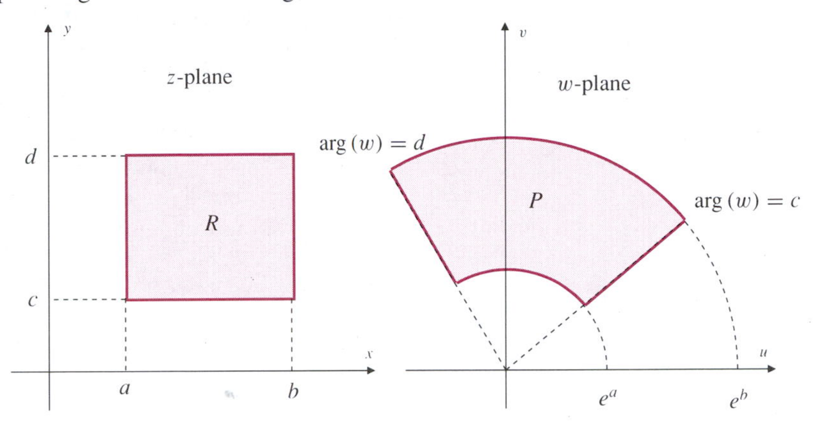


**Figure II.2** Under the exponential function $w = e^z$ vertical lines get mapped to circles centred at the origin and horizontal lines get mapped to half-lines radiating from the origin

Note that if $d - c \ge 2\pi$, then the image of $R$ will be the entire annular region $e^a \le |w| \le e^b$, which may be covered more than once. The exponential function $e^z$ is periodic with period $2\pi i$:

$$e^{z+2\pi i} = e^z \qquad \text{for all} \quad z,$$

and is therefore not one-to-one on the whole complex plane. However, $w = e^z$ is one-to-one from any horizontal strip of the form

$$-\infty < x < \infty, \qquad c < y \le c + 2\pi$$

onto the whole $w$-plane.

## The Fundamental Theorem of Algebra

As observed at the beginning of this appendix, extending the number system to include complex numbers allows larger classes of equations to have solutions. We conclude this appendix by verifying that polynomial equations always have solutions in the complex numbers.

A **complex polynomial** of degree $n$ is a function of the form

$$P_n(z) = a_n z^n + a_{n-1} z^{n-1} + \cdots + a_2 z^2 + a_1 z + a_0,$$

where $a_0, a_1, \ldots, a_n$ are complex numbers and $a_n \ne 0$. The numbers $a_i$ $(0 \le i \le n)$ are called the **coefficients** of the polynomial. If they are all real numbers then $P_n(x)$ is called a real polynomial.

A complex number $z_0$ that satisfies the equation $P(z_0) = 0$ is called a **zero** or **root** of the polynomial. Every polynomial of degree 1 has a zero: if $a_1 \ne 0$, then $a_1 z + a_0$ has zero $z = -a_0/a_1$. This zero is real if $a_1$ and $a_0$ are both real.

Similarly, every complex polynomial of degree 2 has two zeros. If the polynomial is

$$P_2(z) = a_2z^2 + a_1z + a_0$$

(where $a_2 \neq 0$), then the zeros are given by the *quadratic formula*

$$z = z_1 = \frac{-a_1 - \sqrt{a_1^2 - 4a_2a_0}}{2a_2}, \quad \text{and} \quad z = z_2 = \frac{-a_1 + \sqrt{a_1^2 - 4a_2a_0}}{2a_2}.$$

In this case $P_2(z)$ has two linear factors:

$$P_2(z) = a_2(z - z_1)(z - z_2).$$

Even if $a_1^2 - 4a_2a_0 = 0$, so that $z_1 = z_2$, we still regard the polynomial as having two (equal) zeros, one corresponding to each factor. If the coefficients $a_0$, $a_1$, and $a_2$ are all real numbers the zeros will be real provided $a_1^2 \geq 4a_2a_0$. When real coefficients satisfy $a_1^2 < 4a_2a_0$ then the zeros are complex, in fact complex conjugates: $z_2 = \overline{z_1}$.

**EXAMPLE 3** Solve the equation $z^2 + 2iz - (1 + i) = 0$.

***SOLUTION*** The zeros of this equation are

$$\begin{aligned} z &= \frac{-2i \pm \sqrt{-4 + 4(1+i)}}{2} \\ &= -i \pm \sqrt{i} \\ &= -i \pm \frac{1+i}{\sqrt{2}} = \frac{1}{\sqrt{2}}\left(1 + (1 - \sqrt{2})i\right) \quad \text{or} \quad -\frac{1}{\sqrt{2}}\left(1 + (1 + \sqrt{2})i\right). \end{aligned}$$

The Fundamental Theorem of Algebra asserts that every complex polynomial of positive degree has a complex zero.

**THEOREM 2**

**The Fundamental Theorem of Algebra**

If $P(z) = a_nz^n + a_{n-1}z^{n-1} + \cdots + a_1z + a_0$ is a complex polynomial of degree $n \geq 1$, then there exists a complex number $z_1$ such that $P(z_1) = 0$.

***PROOF*** (We will only give an informal sketch of the proof.) We can assume that the coefficient of $z^n$ in $P(z)$ is $a_n = 1$ since we can divide the equation $P(z) = 0$ by $a_n$ without changing its solutions. We can also assume that $a_0 \neq 0$; if $a_0 = 0$, then $z = 0$ is certainly a zero of $P(z)$. Thus we deal with the polynomial

$$P(z) = z^n + Q(z)$$

where $Q(z)$ is a polynomial of degree less than $n$ having a nonzero constant term. If $R$ is sufficiently large, then $|Q(z)|$ will be less than $R^n$ for all numbers $z$ satisfying $|z| = R$. As $z$ moves around the circle $|z| = R$ in the $z$-plane, $w = z^n$ moves around the circle $|w| = R^n$ in the $w$-plane ($n$ times). Since the distance from $z^n$ to $P(z)$ is equal to $|P(z) - z^n| = |Q(z)| < R^n$, it follows that the image of the circle $|z| = R$ under the transformation $w = P(z)$ is a curve that winds around the origin $n$ times. (If you walk around a circle of radius $r$ $n$ times, with your dog on a leash of length less than $r$, and your dog returns to his starting point, then he must also go around the centre of the circle $n$ times.) This situation is illustrated for the particular case

$$P(z) = z^3 + z^2 - iz + 1, \qquad |z| = 2$$

in Figure II.3. The image of $|z| = 2$ is the large curve in the $w$-plane that winds around the origin three times. As $R$ decreases, the curve traced out by $w = P(z)$ for $|z| = R$ changes continuously. For $R$ close to 0, it is a small curve staying close to the constant term $a_0$ of $P(z)$. For small enough $R$ the curve will not enclose the origin. (In Figure II.3 the image of $|z| = 0.3$ is the small curve staying close to the point 1 in the $w$-plane.) Thus for some value of $R$, say $R = R_1$, the curve must pass through the origin. That is, there must be a complex number $z_1$, with $|z_1| = R_1$, such that $P(z_1) = 0$.

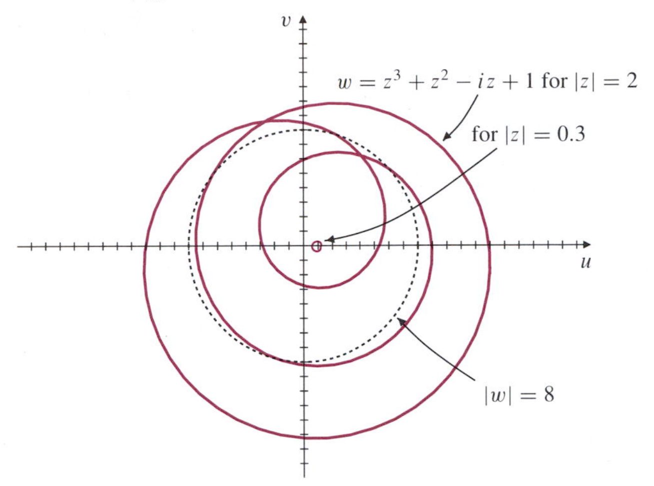


**Figure II.3** The image of the circle $|z| = 2$ winds around the origin in the $w$-plane three times, but the image of $|z| = 0.3$ does not wind around the origin at all

***REMARK*** The above proof suggests that there should be $n$ such solutions of the equation $P(z) = 0$; the curve has to go from winding around the origin $n$ times to winding around the origin 0 times as $R$ decreases towards 0. We can establish this as follows. $P(z_1) = 0$ implies that $z - z_1$ is a factor of $P(z)$:

$$P(z) = (z - z_1)P_{n-1}(z),$$

where $P_{n-1}$ is a polynomial of degree $n - 1$. If $n > 1$, then $P_{n-1}$ must also have a zero, $z_2$, by the Fundamental Theorem. We can continue this argument inductively to obtain $n$ zeros and factor $P(z)$ into a product of the constant $a_n$ and $n$ linear factors:

$$P(z) = a_n(z - z_1)(z - z_2)\cdots(z - z_n).$$

Of course, some of the zeros can be equal.

***REMARK*** If $P$ is a real polynomial, that is, one whose coefficients are all real numbers, then $\overline{P(z)} = P(\overline{z})$. Therefore, if $z_1$ is a non-real zero of $P(z)$, then so is $z_2 = \overline{z_1}$:

$$P(z_2) = P(\overline{z_1}) = \overline{P(z_1)} = \overline{0} = 0.$$

Real polynomials can have complex zeros, but they must always occur in complex conjugate pairs. Every real polynomial of odd degree must have at least one real zero.

**EXAMPLE 4** Show that $z_1 = -i$ is a zero of the polynomial

$$P(z) = z^4 + 5z^3 + 7z^2 + 5z + 6,$$

and find all the other zeros of this polynomial.

**SOLUTION** First observe that $P(z_1) = P(-i) = 1 + 5i - 7 - 5i + 6 = 0$, so $z_1 = -i$ is indeed a zero. Since the coefficients of $P(z)$ are real, $z_2 = i$ must also be a zero. Thus $z + i$ and $z - i$ are factors of $P(z)$, and so is

$$(z + i)(z - i) = z^2 + 1.$$

Dividing $P(z)$ by $z^2 + 1$ we obtain

$$\frac{P(z)}{z^2 + 1} = z^2 + 5z + 6 = (z + 2)(z + 3).$$

Thus the four zeros of $P(z)$ are $z_1 = -i$, $z_2 = i$, $z_3 = -2$, and $z_4 = -3$. ■

## EXERCISES

In Exercises 1–12, the $z$-plane region $D$ consists of the complex numbers $z = x + yi$ that satisfy the given conditions. Describe (or sketch) the image $R$ of $D$ in the $w$-plane under the given function $w = f(z)$.

**1.** $0 \le x \le 1,\ 0 \le y \le 2; \quad w = \bar{z}.$

**2.** $x + y = 1; \quad w = \bar{z}.$

**3.** $1 \le |z| \le 2,\ \frac{\pi}{2} \le \arg z \le \frac{3\pi}{4}; \quad w = z^2.$

**4.** $0 \le |z| \le 2,\ 0 \le \arg(z) \le \frac{\pi}{2}; \quad w = z^3.$

**5.** $0 < |z| \le 2,\ 0 \le \arg(z) \le \frac{\pi}{2}; \quad w = \frac{1}{z}.$

**6.** $\frac{\pi}{4} \le \arg(z) \le \frac{\pi}{3}; \quad w = -iz.$

**7.** $\arg(z) = -\frac{\pi}{3}; \quad w = \sqrt{z}.$

**8.** $x = 1; \quad w = z^2.$

**9.** $y = 1; \quad w = z^2.$

**10.** $x = 1; \quad w = \frac{1}{z}.$

**11.** $-\infty < x < \infty,\ \frac{\pi}{4} \le y \le \frac{\pi}{2}; \quad w = e^z.$

**12.** $0 < x < \frac{\pi}{2},\ 0 < y < \infty; \quad w = e^{iz}.$

In Exercises 13–16, verify that the real and imaginary parts of each function $f(z)$ satisfy the Cauchy–Riemann equations, and thus find $f'(z)$.

**13.** $f(z) = z^2.$

**14.** $f(z) = z^3.$

**15.** $f(z) = \frac{1}{z}.$

**16.** $f(z) = e^{z^2}.$

**17.** Use the fact that $e^{yi} = \cos y + i \sin y$ (for real $y$) to show that

$$\cos y = \frac{e^{yi} + e^{-yi}}{2} \quad \text{and} \quad \sin y = \frac{e^{yi} - e^{-yi}}{2i}.$$

The previous exercise suggests that we define complex functions

$$\cos z = \frac{e^{zi} + e^{-zi}}{2} \quad \text{and} \quad \sin z = \frac{e^{zi} - e^{-zi}}{2i},$$

as well as extend the definitions of the hyperbolic functions to

$$\cosh z = \frac{e^z + e^{-z}}{2} \quad \text{and} \quad \sinh z = \frac{e^z - e^{-z}}{2}.$$

Exercises 18–26 develop properties of these functions and relationships among them.

**18.** Show that $\cos z$ and $\sin z$ are periodic with period $2\pi$, and that $\cosh z$ and $\sinh z$ are periodic with period $2\pi i$.

**19.** Show that $(d/dz)\sin z = \cos z$, and $(d/dz)\cos z = -\sin z$. What are the derivatives of $\sinh z$ and $\cosh z$?

**20.** Verify the identities $\cos z = \cosh(iz)$, and $\sin z = -i \sinh(iz)$. What are the corresponding identities for $\cosh z$ and $\sinh(z)$ in terms of cos and sin?

**21.** Find all complex zeros of $\cos z$ (that is, all solutions of $\cos z = 0$).

**22.** Find all complex zeros of $\sin z$.

**23.** Find all complex zeros of $\cosh z$ and $\sinh z$.

**24.** Show that $\operatorname{Re}(\cosh z) = \cosh x \cos y$ and $\operatorname{Im}(\cosh z) = \sinh x \sin y$.

**25.** Find the real and imaginary parts of $\sinh z$.

**26.** Find the real and imaginary parts of $\cos z$ and $\sin z$.

Find the zeros of the polynomials in Exercises 27–32.

**27.** $P(z) = z^2 + 2iz$

**28.** $P(z) = z^2 - 2z + i$

**29.** $P(z) = z^2 + 2z + 5$

**30.** $P(z) = z^2 - 2iz - 1$

**31.** $P(z) = z^3 - 3iz^2 - 2z$

**32.** $P(z) = z^4 - 2z^2 + 4$

**33.** The polynomial $P(z) = z^4 + 1$ has two pairs of complex conjugate zeros. Find them, and hence express $P(z)$ as a product of two quadratic factors with real coefficients.

In Exercises 34–36, check that the given number $z_1$ is a zero of the given polynomial, and find all the zeros of the polynomial.

**34.** $P(z) = z^4 - 4z^3 + 12z^2 - 16z + 16; \quad z_1 = 1 - \sqrt{3}\,i.$

**35.** $P(z) = z^5 + 3z^4 + 4z^3 + 4z^2 + 3z + 1; \quad z_1 = i.$

**36.** $P(z) = z^5 - 2z^4 - 8z^3 + 8z^2 + 31z - 30;$ $z_1 = -2 + i.$

**37.** Show that the image of the circle $|z| = 2$ under the mapping $w = z^4 + z^3 - 2iz - 3$ winds around the origin in the $w$-plane four times.

# Answers to Odd-Numbered Exercises

## Chapter 1

### Section 1.1 (page 35)

**1.** bounded, positive, increasing, convergent to 2
**3.** bounded, positive, convergent to 4
**5.** bounded below, positive, increasing, divergent to infinity
**7.** bounded below, positive, increasing, divergent to infinity
**9.** bounded, positive, decreasing, convergent to 0

**11.** divergent
**13.** divergent
**15.** $\infty$
**17.** 0
**19.** 1
**21.** $e^{-3}$
**23.** 0
**25.** $1/2$
**27.** 0
**29.** 0
**31.** $\lim_{n\to\infty} a_n = 5$
**33.** If $\{a_n\}$ is (ultimately) decreasing, then either it is bounded below, and therefore convergent, or else it is unbounded below and therefore divergent to negative infinity.

### Section 1.2 (page 42)

**1.** $\dfrac{1}{2}$
**3.** $\dfrac{1}{(2+\pi)^8\big((2+\pi)^2-1\big)}$
**5.** $\dfrac{25}{4,416}$
**7.** $\dfrac{8e^4}{e-2}$
**9.** diverges to $\infty$
**11.** $\dfrac{3}{4}$
**13.** $\dfrac{1}{3}$
**15.** div. to $\infty$
**17.** div. to $\infty$
**19.** diverges
**21.** 14 m
**25.** If $\{a_n\}$ is ultimately negative, then the series $\sum a_n$ must either converge (if its partial sums are bounded below), or diverge to $-\infty$ (if its partial sums are not bounded below).
**27.** false, e.g. $\sum \dfrac{(-1)^n}{2^n}$
**29.** true
**31.** true

### Section 1.3 (page 51)

**1.** converges
**3.** diverges to $\infty$
**5.** converges
**7.** diverges to $\infty$
**9.** converges
**11.** diverges to $\infty$
**13.** diverges to $\infty$
**15.** converges
**17.** diverges to $\infty$
**19.** converges
**21.** diverges to $\infty$
**23.** converges
**25.** converges
**27.** converges
**35.** no info from ratio test, but series diverges to infinity since all terms exceed 1.

### Section 1.4 (page 57)

**1.** conv. conditionally
**3.** conv. conditionally
**5.** diverges
**7.** conv. absolutely
**9.** conv. conditionally
**11.** diverges
**13.** 999
**15.** 13
**17.** converges absolutely if $-1 < x < 1$, conditionally if $x = -1$, diverges elsewhere
**19.** converges absolutely if $0 < x < 2$, conditionally if $x = 2$, diverges elsewhere
**21.** converges absolutely if $-2 < x < 2$, conditionally if $x = -2$, diverges elsewhere
**23.** converges absolutely if $-\frac{7}{2} < x < \frac{1}{2}$, conditionally if $x = -\frac{7}{2}$, diverges elsewhere
**25.** converges absolutely if $x \ge \frac{1}{2}$, diverges if $x < \frac{1}{2}$, $x \ne 0$, undefined at $x = 0$
**27.** converges absolutely if $-1 < x < 0$ or $1 < x < 2$, diverges elsewhere
**29.** AST does not apply directly, but does if we remove all the 0 terms; series converges conditionally
**31.** (a) false, e.g. $a_n = \dfrac{(-1)^n}{n}$,
(b) false, e.g. $a_n = \dfrac{\sin(n\pi/2)}{n}$, (see Exercise 25),
(c) true
**33.** converges absolutely for $-1 < x < 1$, conditionally if $x = -1$, diverges elsewhere

### Section 1.5 (page 63)

**1.** $s_n + \dfrac{1}{9(n+1)^9} \le s \le s_n + \dfrac{1}{9n^9}$; $n = 2$
**3.** $s_n + \dfrac{2}{\sqrt{n+1}} \le s \le s_n + \dfrac{2}{\sqrt{n}}$; $n = 63$
**5.** $s_n + \dfrac{1}{e^{n+1}} \le s \le s_n + \dfrac{1}{e^n}$; $n = 6$
**7.** $0 < s - s_n \le \dfrac{n+2}{2^n(n+1)!(2n+3)}$; $n = 4$

**9.** $0 < s - s_n \le \dfrac{2^n(4n^2+6n+2)}{(2n)!(4n^2+6n)}$; $n = 4$

**11.** (b) $s \le \dfrac{2}{k(1-k)}$, $k = \frac{1}{2}$,

(c) $0 < s - s_n < \dfrac{(1+k)^{n+1}}{2^n k(1-k)}$, $k = \dfrac{n+2-\sqrt{n^2+8}}{2(n-1)}$ for $n \ge 2$

**13.** (a) 10, (b) 5, (c) 0.765

## Review Exercises (page 64)

**1.** conv. to $-1/3$ **3.** conv. to 0

**5.** div. to $\infty$ **7.** $\lim_{n\to\infty} a_n = \sqrt{2}$

**9.** $4\sqrt{2}/(\sqrt{2}-1)$ **11.** 2

**13.** converges **15.** div. to $\infty$

**17.** converges **19.** div. to $\infty$

**21.** converges **23.** conv. abs.

**25.** conv. cond.

**27.** conv. abs. for $x$ in $(-1, 5)$, cond. for $x = -1$, div. elsewhere

**29.** diverges at $x = (\pi/2) + 2n\pi$, conv. cond. at $x = -(\pi/2) + 2n\pi$, conv. abs. elsewhere

**31.** 1.202 **33.** 0.316

## Challenging Problems (page 64)

**5.** (c) 1.645

# Chapter 2

## Section 2.1 (page 76)

**1.** centre 0, radius 1, interval $(-1, 1)$

**3.** centre $-2$, radius 2, interval $[-4, 0)$

**5.** centre $\frac{3}{2}$, radius $\frac{1}{2}$, interval $(1, 2)$

**7.** centre 0, radius $\infty$, interval $(-\infty, \infty)$

**9.** $\dfrac{1}{(1-x)^3} = \displaystyle\sum_{n=0}^{\infty} \frac{(n+1)(n+2)}{2} x^n$, $(-1 < x < 1)$

**11.** $\dfrac{1}{(1-x)^2} = \displaystyle\sum_{n=0}^{\infty} (n+1)x^n$, $(-1 < x < 1)$

**13.** $\dfrac{1}{(2-x)^2} = \displaystyle\sum_{n=0}^{\infty} \frac{n+1}{2^{n+2}} x^n$, $(-2 < x < 2)$

**15.** $\ln(2-x) = \ln 2 - \sum_{n=1}^{\infty} \dfrac{x^n}{2^n n}$, $(-2 \le x < 2)$

**17.** $\dfrac{1}{x^2} = \displaystyle\sum_{n=0}^{\infty} \frac{n+1}{2^{n+2}} (x+2)^n$, $(-4 < x < 0)$

**19.** $\dfrac{x^3}{1-2x^2} = \displaystyle\sum_{n=0}^{\infty} 2^n x^{2n+3}$, $\left(-\dfrac{1}{\sqrt{2}} < x < \dfrac{1}{\sqrt{2}}\right)$

**21.** $\left(-\frac{1}{4}, \frac{1}{4}\right)$; $\dfrac{1}{1+4x}$

**23.** $[-1, 1)$; $\frac{1}{3}$ if $x = 0$, $-\dfrac{1}{x^3}\ln(1-x) - \dfrac{1}{x^2} - \dfrac{1}{2x}$ otherwise

**25.** $(-1, 1)$; $\dfrac{2}{(1-x^2)^2}$ **27.** $3/4$

**29.** $\pi^2(\pi+1)/(\pi-1)^3$ **31.** $\ln(3/2)$

## Section 2.2 (page 84)

**1.** $e^{3x+1} = \sum_{n=0}^{\infty} \dfrac{3^n e}{n!} x^n$, (all $x$)

**3.** $\sin\left(x - \dfrac{\pi}{4}\right) = \dfrac{1}{\sqrt{2}} \displaystyle\sum_{n=0}^{\infty} (-1)^n \left[-\frac{x^{2n}}{(2n)!} + \frac{x^{2n+1}}{(2n+1)!}\right]$, (all $x$)

**5.** $x^2 \sin\left(\dfrac{x}{3}\right) = \sum_{n=0}^{\infty} \dfrac{(-1)^n}{3^{2n+1}(2n+1)!} x^{2n+3}$, (all $x$)

**9.** $\dfrac{1+x^3}{1+x^2} = 1 - x^2 + \displaystyle\sum_{n=2}^{\infty} (-1)^n \left(x^{2n-1} + x^{2n}\right)$, $(-1 < x < 1)$

**11.** $\ln \dfrac{1-x}{1+x} = -2 \displaystyle\sum_{n=1}^{\infty} \frac{x^{2n-1}}{2n-1}$, $(-1 < x < 1)$

**13.** $\cosh x - \cos x = 2 \sum_{n=0}^{\infty} \dfrac{x^{4n+2}}{(4n+2)!}$, (all $x$)

**15.** $e^{-2x} = e^2 \sum_{n=0}^{\infty} \dfrac{(-1)^n 2^n}{n!} (x+1)^n$, (all $x$)

**17.** $\cos x = \sum_{n=0}^{\infty} \dfrac{(-1)^{n+1}}{(2n)!} (x-\pi)^{2n}$, (all $x$)

**19.** $\ln 4 + \sum_{n=1}^{\infty} \dfrac{(-1)^{n-1}}{4^n n} (x-2)^n$, $(-2 < x \le 6)$

**21.** $\sin x - \cos x = \sqrt{2} \sum_{n=0}^{\infty} \dfrac{(-1)^n}{(2n+1)!} \left(x - \dfrac{\pi}{4}\right)^{2n+1}$, (all $x$)

**23.** $\dfrac{1}{x^2} = \dfrac{1}{4} \displaystyle\sum_{n=0}^{\infty} \frac{n+1}{2^n} (x+2)^n$, $(-4 < x < 0)$

**25.** $(x-1) + \sum_{n=2}^{\infty} \dfrac{(-1)^n}{n(n-1)} (x-1)^n$, $(0 \le x \le 2)$

**27.** $1 + \dfrac{x^2}{2} + \dfrac{5x^4}{24}$ **29.** $x + \dfrac{x^2}{2} - \dfrac{x^3}{6}$

**31.** $1 + \dfrac{x}{2} - \dfrac{x^2}{8}$ **33.** $e^{x^2}$ (all $x$)

**35.** $\dfrac{e^x - e^{-x}}{2x} = \dfrac{\sinh x}{x}$ if $x \ne 0$, 1 if $x = 0$

**37.** (a) $1 + x + x^2$, (b) $3 + 3(x-1) + (x-1)^2$

## Section 2.3 (page 88)

**1.** 1.22140 **3.** 3.32011

**5.** 0.99619 **7.** $-0.10533$

**9.** 0.42262 **11.** 1.54306

**13.** $I(x) = \sum_{n=0}^{\infty} \dfrac{(-1)^n}{(2n+1)(2n+1)!} x^{2n+1}$, (all $x$)

**15.** $K(x) = \sum_{n=0}^{\infty} \dfrac{(-1)^n}{(n+1)^2} x^{n+1}$, $(-1 \le x \le 1)$

**17.** $M(x) = \sum_{n=0}^{\infty} \dfrac{(-1)^n}{(2n+1)(4n+1)} x^{4n+1}$, $(-1 \le x \le 1)$

**19.** 0.946 **21.** 2

**23.** $-3/25$ **25.** 0

**27.** $y = \sum_{n=0}^{\infty} (-1)^n \left[ \dfrac{2^n n!}{(2n)!} x^{2n} + \dfrac{1}{2^{n-1} n!} x^{2n+1} \right]$

### Section 2.4 (page 96)

**1.** $\dfrac{1}{2} - \dfrac{x-2}{4} + \dfrac{(x-2)^2}{8} - \dfrac{(x-2)^3}{16} + \dfrac{(x-2)^4}{32}$

**3.** $1 - \dfrac{x^2}{2} + \dfrac{x^4}{24}$

**5.** $\dfrac{1}{\sqrt{2}} \left[ 1 + \left(x - \dfrac{\pi}{4}\right) - \dfrac{1}{2!}\left(x - \dfrac{\pi}{4}\right)^2 - \dfrac{1}{3!}\left(x - \dfrac{\pi}{4}\right)^3 + \dfrac{1}{4!}\left(x - \dfrac{\pi}{4}\right)^4 + \dfrac{1}{5!}\left(x - \dfrac{\pi}{4}\right)^5 - \dfrac{1}{6!}\left(x - \dfrac{\pi}{4}\right)^6 \right]$

**7.** $e^2 \left[ 1 - (x+2) + \dfrac{(x+2)^2}{2!} - \cdots + (-1)^n \dfrac{(x+2)^n}{n!} \right]$

**9.** $1 + \dfrac{x^2}{2} + \dfrac{5x^4}{24}$

**11.** (a) $3+2x+x^2$, (b) $3+2x+x^2$, (c) $3-2(x+2)+(x+2)^2$, (d) $18 + 8(x-3) + (x-3)^2$

**13.** polynomials of degree at most $n$

**17.** $\dfrac{1}{720}(0.2)^7$ **19.** $\dfrac{1}{120}(0.5)^5$

**21.** $\dfrac{4 \sec^2(0.1) \tan^2(0.1) + 2 \sec^4(0.1)}{4!\, 10^4}$

**23.** $\dfrac{24}{120(1.95)^5(20)^5}$

**25.** $2^x = \sum_{n=0}^{\infty} \dfrac{(x \ln 2)^n}{n!}$, all $x$

**27.** $\sin x = \sum_{n=0}^{\infty} (-1)^n \dfrac{x^{2n+1}}{(2n+1)!}$, all $x$

**29.** $\dfrac{1}{1-x} = \sum_{n=0}^{\infty} x^n$, $-1 < x < 1$

**31.** $\dfrac{x}{2+3x} = \sum_{n=1}^{\infty} (-1)^{n-1} 3^{n-1} \left(\dfrac{x}{2}\right)^n$, $-\dfrac{2}{3} < x < \dfrac{2}{3}$

**33.** $\sin x = \dfrac{1}{2} \sum_{n=0}^{\infty} \dfrac{c_n}{n!} \left(x - \dfrac{\pi}{6}\right)^n$, (for all $x$), where $c_n = (-1)^{n/2}$ if $n$ is even, and $c_n = (-1)^{(n-1)/2}\sqrt{3}$ if $n$ is odd

**35.** $\ln x = \sum_{n=1}^{\infty} (-1)^{n-1} \dfrac{(x-1)^n}{n}$, $0 < x \le 2$

**37.** $\dfrac{1}{x} = -\dfrac{1}{2} \sum_{n=0}^{\infty} \left(\dfrac{x+2}{2}\right)^n$, $-4 < x < 0$

### Section 2.5 (page 100)

**1.** $\sqrt{1+x} = 1 + \sum_{n=1}^{\infty} \dfrac{(-1)^{n-1}\, 1 \times 3 \times 5 \times \cdots \times (2n-3)}{2^n n!} x^n$, $|x| < 1$

**3.** $\sqrt{4+x} = 2 + \dfrac{x}{4} + 2 \sum_{n=2}^{\infty} (-1)^{n-1} \dfrac{1 \times 3 \times 5 \times \cdots \times (2n-3)}{2^{3n} n!} x^n$, $(-4 < x \le 4)$

**5.** $\sum_{n=0}^{\infty} (n+1) x^n$, $|x| < 1$

### Section 2.6 (page 106)

**1.** $2\pi/3$ **3.** $\pi$

**5.** $2 \sum_{n=1}^{\infty} (-1)^{n-1} (\sin(nt))/n$

**7.** $\dfrac{1}{4} - \sum_{n=1}^{\infty} \left( \dfrac{2\cos((2n-1)\pi t)}{(2n-1)^2 \pi^2} + \dfrac{(-1)^n \sin(n\pi t)}{n\pi} \right)$

**9.** 1

**11.** $2 \sum_{n=1}^{\infty} \dfrac{(-1)^n}{n\pi} \sin(n\pi t)$

**13.** $\pi^2/8$

### Review Exercises (page 107)

**1.** $\sum_{n=0}^{\infty} x^n / 3^{n+1}$, $|x| < 3$

**3.** $1 + \sum_{n=1}^{\infty} (-1)^{n-1} x^{2n} / (n e^n)$, $-\sqrt{e} < x \le \sqrt{e}$

**5.** $\sum_{n=0}^{\infty} (-1)^n 2^{2n+1} x^{4n+1} / (2n+1)!$, $x \ne 0$

**7.** $x + \sum_{n=1}^{\infty} (-1)^n 2^{2n-1} x^{2n+1} / (2n)!$, all $x$

**9.** $(1/2) + \sum_{n=1}^{\infty} \dfrac{(-1)^n\, 1 \times 4 \times 7 \times \cdots \times (3n-2) x^n}{2 \times 24^n\, n!}$, $-8 < x \le 8$

**11.** $\sum_{n=0}^{\infty} (-1)^n (x-\pi)^n / \pi^{n+1}$, $0 < x < 2\pi$

**13.** $-(1/e) + \sum_{n=2}^{\infty} \dfrac{(n-1)(x+1)^n}{n!\, e}$, all $x$

**15.** $1 + 2x + 3x^2 + \frac{10}{3}x^3$ **17.** $1 - \frac{1}{2}x^2 + \frac{5}{24}x^4$

**19.** $x + \frac{1}{2}x^3 + \frac{5}{24}x^5$ **21.** $\begin{cases} \cos\sqrt{x} & \text{if } x \ge 0 \\ \cosh\sqrt{|x|} & \text{if } x < 0 \end{cases}$

**23.** $\pi^2/(\pi-1)^2$ **25.** $\ln(e/(e-1))$

**27.** $1/14$ **29.** 3, 0.49386

**31.** $\sum_{n=1}^{\infty} \dfrac{2}{n} \sin(nt)$

## Chapter 3

### Section 3.1 (page 114)

**1.** 3 units **3.** $\sqrt{6}$ units

**5.** $|z|$ units; $\sqrt{y^2+z^2}$ units

**7.** $\cos^{-1}(-4/9) \approx 116.39°$

**9.** $\sqrt{3}/2$ sq. units **11.** $\sqrt{n-1}$ units

**13.** the half-space containing the origin and bounded by the plane passing through $(0,-1,0)$ perpendicular to the $y$-axis.

**15.** the vertical plane (parallel to the $z$-axis) passing through $(1,0,0)$ and $(0,1,0)$.

**17.** the sphere of radius 2 centred at $(1,-2,3)$.

**19.** the solid circular cylinder of radius 2 with axis along the $x$-axis.

**21.** the parabolic cylinder generated by translating the parabola $z=y^2$ in the $yz$-plane in the direction of the $x$-axis .

**23.** the plane through the points $(6,0,0)$, $(0,3,0)$ and $(0,0,2)$.

**25.** the straight line through $(1,0,0)$ and $(1,1,1)$.

**27.** the circle in which the sphere of radius 2 centred at the origin intersects the sphere of radius 2 with centre $(2,0,0)$.

**29.** the ellipse in which the plane $z=x$ intersects the circular cylinder of radius 1 and axis along the $z$-axis.

**31.** the part of the solid circular cylinder of radius 1 and axis along the $z$-axis lying above or on the plane $z=y$.

**33.** the part of the solid ball of radius 1 centred at the origin which lies above or on the parabolic cylinder $z=x^2$.

## Section 3.2 (page 119)

**1.** a) $6\mathbf{i}-10\mathbf{k}$, $8\mathbf{j}$, $-3\mathbf{i}+20\mathbf{j}+5\mathbf{k}$
b) $5\sqrt{2}$, $5\sqrt{2}$
c) $\frac{3}{5\sqrt{2}}\mathbf{i}\pm\frac{4}{5\sqrt{2}}\mathbf{j}-\frac{1}{\sqrt{2}}\mathbf{k}$ d) 18
e) $\cos^{-1}(9/25)\approx 68.9^\circ$ f) $18/5\sqrt{2}$
g) $(27/25)\mathbf{i}+(36/25)\mathbf{j}-(9/5)\mathbf{k}$

**3.** a) $3\mathbf{i}-3\mathbf{k}$, $-\mathbf{i}-4\mathbf{j}+3\mathbf{k}$, $-4\mathbf{i}-10\mathbf{j}+9\mathbf{k}$
b) $\sqrt{5}$, $\sqrt{17}$
c) $\frac{1}{\sqrt{5}}\mathbf{i}-\frac{2}{\sqrt{5}}\mathbf{j}$, $\frac{2}{\sqrt{17}}\mathbf{i}+\frac{2}{\sqrt{17}}\mathbf{j}-\frac{3}{\sqrt{17}}\mathbf{k}$ d) $-2$
e) $\cos^{-1}(-2/\sqrt{85})\approx 102.5^\circ$ f) $-2/\sqrt{17}$
g) $-(2/5)\mathbf{i}+(4/5)\mathbf{j}$

**5.** from southwest at $50\sqrt{2}$ km/h.

**7.** head at angle $\theta$ to the east of $AC$, where
$$\theta=\sin^{-1}\frac{3}{2\sqrt{1+4k^2}}.$$
The trip not possible if $k<\frac{1}{4}\sqrt{5}$. If $k>\frac{1}{4}\sqrt{5}$ there is a second possible heading, $\pi-\theta$, but the trip will take longer.

**9.** $t=2$

**11.** $\cos^{-1}(2/\sqrt{6})\approx 35.26^\circ$, $90^\circ$

**13.** $(\mathbf{i}+\mathbf{j}+\mathbf{k})/\sqrt{3}$

**15.** $\lambda=1/2$, midpoint, $\lambda=2/3$, 2/3 of way from $P_1$ to $P_2$, $\lambda=-1$, $P_1$ is midway between this point and $P_2$.

**17.** plane through point with position vector $(b/|\mathbf{a}|^2)\mathbf{a}$ perpendicular to $\mathbf{a}$.

**19.** $\mathbf{x}=2\mathbf{i}-3\mathbf{j}-4\mathbf{k}$

**21.** $(|\mathbf{u}|\mathbf{v}+|\mathbf{v}|\mathbf{u})/\big||\mathbf{u}|\mathbf{v}+|\mathbf{v}|\mathbf{u}\big|$

**27.** $\mathbf{u}=(\mathbf{w}\bullet\mathbf{a}/|\mathbf{a}|^2)\mathbf{a}$, $\mathbf{v}=\mathbf{w}-\mathbf{u}$

**29.** $\mathbf{x}=(\mathbf{a}+K\hat{\mathbf{u}})/(2r)$, $\mathbf{y}=(\mathbf{a}-K\hat{\mathbf{u}})/(2s)$, where $K=\sqrt{|\mathbf{a}|^2-4rst}$ and $\hat{\mathbf{u}}$ is any unit vector

## Section 3.3 (page 127)

**1.** $5\mathbf{i}+13\mathbf{j}+7\mathbf{k}$ **3.** $\sqrt{6}$ sq. units

**5.** $\pm\frac{1}{3}(2\mathbf{i}-2\mathbf{j}+\mathbf{k})$ **15.** 4/3 cubic units

**17.** $k=-6$

**19.** $\lambda=\dfrac{\mathbf{x}\bullet(\mathbf{v}\times\mathbf{w})}{\mathbf{u}\bullet(\mathbf{v}\times\mathbf{w})}$, $\mu=\dfrac{\mathbf{x}\bullet(\mathbf{w}\times\mathbf{u})}{\mathbf{u}\bullet(\mathbf{v}\times\mathbf{w})}$, $\nu=\dfrac{\mathbf{x}\bullet(\mathbf{u}\times\mathbf{v})}{\mathbf{u}\bullet(\mathbf{v}\times\mathbf{w})}$

**21.** $\mathbf{u}\times(\mathbf{v}\times\mathbf{w})=-2\mathbf{i}+7\mathbf{j}-4\mathbf{k}$, $(\mathbf{u}\times\mathbf{v})\times\mathbf{w}=\mathbf{i}+9\mathbf{j}+9\mathbf{k}$; the first is in the plane of $\mathbf{v}$ and $\mathbf{w}$, the second is in the plane of $\mathbf{u}$ and $\mathbf{v}$.

## Section 3.4 (page 135)

**1.** a) $x^2+y^2+z^2=z^2$; b) $x+y+z=x+y+z$; c) $x^2+y^2+z^2=-1$

**3.** $x-y+2z=0$ **5.** $7x+5y-z=12$

**7.** $x-5y-3z=-7$ **9.** $x+6y-5z=17$

**11.** $(\mathbf{r}_1-\mathbf{r}_2)\bullet[(\mathbf{r}_1-\mathbf{r}_3)\times(\mathbf{r}_1-\mathbf{r}_4)]=0$

**13.** planes passing through the line $x=0$, $y+z=1$ (except the plane $y+z=1$ itself)

**15.** $\mathbf{r}=(1+2t)\mathbf{i}+(2-3t)\mathbf{j}+(3-4t)\mathbf{k}$, $(-\infty<t<\infty)$
$x=1+2t, y=2-3t, z=3-4t$, $(-\infty<t<\infty)$
$$\frac{x-1}{2}=\frac{y-2}{-3}=\frac{z-3}{-4}$$

**17.** $\mathbf{r}=t(7\mathbf{i}-6\mathbf{j}-5\mathbf{k})$; $x=7t$, $y=-6t$, $z=-5t$; $x/7=-y/6=-z/5$

**19.** $\mathbf{r}=\mathbf{i}+2\mathbf{j}-\mathbf{k}+t(\mathbf{i}+\mathbf{j}+\mathbf{k})$;
$x=1+t$, $y=2+t$, $z=-1+t$;
$x-1=y-2=z+1$

**21.** $\frac{x-4}{-5}=\frac{y}{3}$, $z=7$

**23.** $\mathbf{r}_i\neq\mathbf{r}_j$, $(i,j=1,\cdots,4,\ i\neq j)$,
$\mathbf{v}=(\mathbf{r}_1-\mathbf{r}_2)\times(\mathbf{r}_3-\mathbf{r}_4)\neq 0$, $(\mathbf{r}_1-\mathbf{r}_3)\bullet\mathbf{v}=0$.

**25.** $7\sqrt{2}/10$ units **27.** $18/\sqrt{69}$ units

**29.** all lines parallel to the $xy$-plane and passing through $(x_0, y_0, z_0)$.

**31.** $(x,y,z)$ satisfies the quadratic if either $A_1x+B_1y+C_1z=D_1$ or $A_2x+B_2y+C_2z=D_2$.

## Section 3.5 (page 139)

**1.** ellipsoid centred at the origin with semiaxes 6, 3 and 2 along the $x$-, $y$- and $z$-axes respectively.

**3.** circle with centre $(1,-2,3)$ and radius $1/\sqrt{2}$.

**5.** elliptic paraboloid with vertex at the origin, axis along the $z$-axis, and cross-section $x^2+2y^2 = 1$ in the plane $z = 1$.

**7.** hyperboloid of two sheets with vertices $(\pm 2, 0, 0)$ and circular cross-sections in planes $x = c$, $(c^2 > 4)$.

**9.** hyperbolic paraboloid — same as $z = x^2 - y^2$ but rotated 45° about the $z$-axis (counterclockwise as seen from above).

**11.** hyperbolic cylinder parallel to the $y$-axis, intersecting the $xz$-plane in the hyperbola $(x^2/4) - z^2 = 1$.

**13.** parabolic cylinder parallel to the $y$-axis.

**15.** circular cone with vertex $(2, 3, 1)$, vertical axis, and semi-vertical angle 45°.

**17.** circle in the plane $x + y + z = 1$ having centre $(1/3, 1/3, 1/3)$ and radius $\sqrt{11/3}$.

**19.** a parabola in the plane $z = 1 + x$ having vertex at $(-1/2, 0, 1/2)$ and axis along the line $z = 1 + x$, $y = 0$.

**21.** $\dfrac{y}{b} - \dfrac{z}{c} = \lambda\left(1 - \dfrac{x}{a}\right)$, $\quad \dfrac{y}{b} + \dfrac{z}{c} = \dfrac{1}{\lambda}\left(1 + \dfrac{x}{a}\right)$;
$\dfrac{y}{b} - \dfrac{z}{c} = \mu\left(1 + \dfrac{x}{a}\right)$, $\quad \dfrac{y}{b} + \dfrac{z}{c} = \dfrac{1}{\mu}\left(1 - \dfrac{x}{a}\right)$

**23.** $\mathbf{a} = \mathbf{i} \pm \mathbf{k}$ (or any multiple)

### Section 3.6 (page 147)

**1.** $\begin{pmatrix} 6 & 7 \\ 5 & -3 \\ 1 & 1 \end{pmatrix}$

**3.** $\begin{pmatrix} aw + by & ax + bz \\ cw + dy & cx + dz \end{pmatrix}$

**5.** $\mathcal{A}\mathcal{A}^T = \begin{pmatrix} 4 & 3 & 2 & 1 \\ 3 & 3 & 2 & 1 \\ 2 & 2 & 2 & 1 \\ 1 & 1 & 1 & 1 \end{pmatrix} \quad \mathcal{A}^2 = \begin{pmatrix} 1 & 2 & 3 & 4 \\ 0 & 1 & 2 & 3 \\ 0 & 0 & 1 & 2 \\ 0 & 0 & 0 & 1 \end{pmatrix}$

**7.** 36 **15.** $\begin{pmatrix} 1 & -1 & 0 \\ 0 & 1 & -1 \\ 0 & 0 & 1 \end{pmatrix}$

**17.** $x = 1,\ y = 2,\ z = 3$

**19.** $x_1 = 1,\ x_2 = 2,\ x_3 = -1,\ x_4 = -2$

### Review Exercises (page 148)

**1.** plane parallel to $y$-axis through $(3, 0, 0)$ and $(0, 0, 1)$

**3.** all points on or above the plane through the origin with normal $\mathbf{i} + \mathbf{j} + \mathbf{k}$

**5.** circular paraboloid with vertex at $(0, 1, 0)$ and axis along the $y$-axis, opening in the direction of increasing $y$

**7.** hyperbolic paraboloid

**9.** points inside the ellipsoid with vertices at $(\pm 2, 0, 0)$, $(0, \pm 2, 0$, and $(0, 0, \pm 1)$

**11.** cone with axis along the $x$-axis, vertex at the origin, and elliptical cross-sections perpendicular to its axis

**13.** oblique circular cone (elliptic cone). Cross-sections in horizontal planes $z = k$ are circles of radius 1 with centres at $(k, 0, k)$

**15.** horizontal line through $(0, 0, 3)$ and $(2, -1, 3)$

**17.** circle of radius 1 centred at $(1, 1, 1)$ in plane normal to $\mathbf{i} + \mathbf{j} + \mathbf{k}$

**19.** $2x - y + 3z = 0$ **21.** $2x + 5y + 3z = 2$

**23.** $7x + 4y - 8z = 6$

**25.** $\mathbf{r} = (2 + 3t)\mathbf{i} + (1 + t)\mathbf{j} - (1 + 2t)\mathbf{k}$

**27.** $x = 3t,\ y = -2t,\ z = 4t$

**29.** $(\mathbf{r}_2 - \mathbf{r}_1) \times (\mathbf{r}_3 - \mathbf{r}_1) = \mathbf{0}$

**31.** $(3/2)\sqrt{34}$ sq. units

**33.** $\mathcal{A}^{-1} = \begin{pmatrix} 1 & 0 & 0 & 0 \\ -2 & 1 & 0 & 0 \\ 1 & -2 & 1 & 0 \\ 0 & 1 & -2 & 1 \end{pmatrix}$

## Chapter 4

### Section 4.1 (page 155)

**1.** all $(x, y)$ with $x \neq y$ **3.** all $(x, y)$ except $(0, 0)$

**5.** all $(x, y)$ satisfying $4x^2 + 9y^2 \geq 36$

**7.** all $(x, y)$ with $xy > -1$

**9.** all $(x, y, z)$ except $(0, 0, 0)$

**11.** $z = f(x, y) = x$ **13.** $z = f(x, y) = y^2$

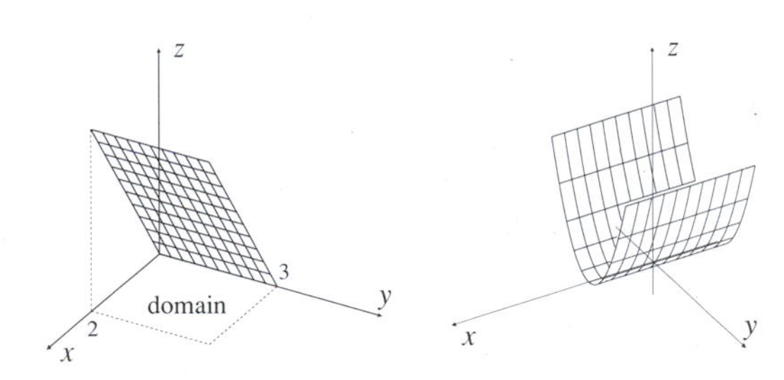


**15.** $f(x, y) = \sqrt{x^2 + y^2}$ **17.** $f(x, y) = |x| + |y|$

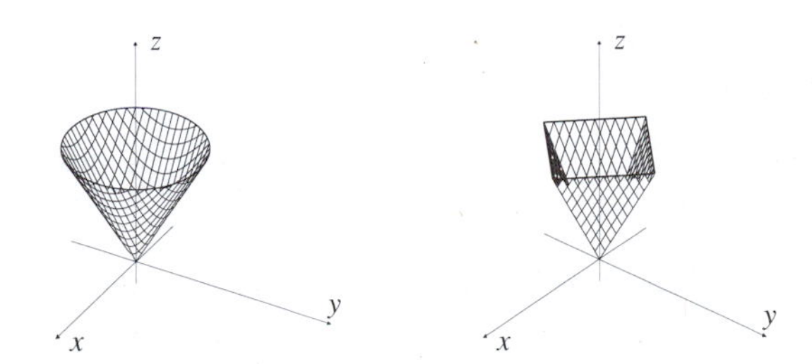

**19.** $f(x, y) = x - y = C$

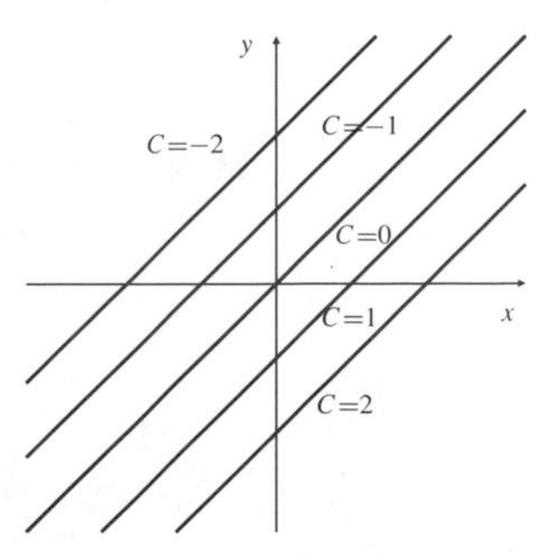


**21.** $f(x, y) = xy = C$

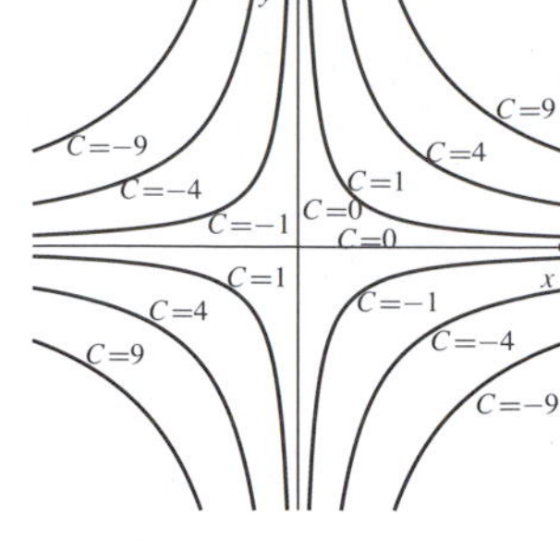


**23.** $f(x, y) = \dfrac{x - y}{x + y} = C$ **25.** $f(x, y) = xe^{-y} = C$

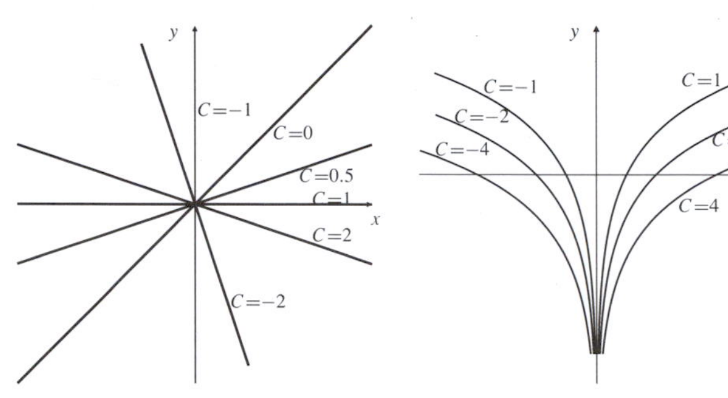


**27.** At $B$, because the contours are closer together there

**29.** a plane containing the $y$-axis, sloping uphill in the $x$ direction

**31.** a right-circular cone with base in the $xy$-plane and vertex at height 5 on the $z$-axis

**33.** No, different curves of the family must not intersect in the region.

**35.** (a) $\sqrt{x^2 + y^2}$, (b) $(x^2 + y^2)^{1/4}$, (c) $x^2 + y^2$, (d) $e^{\sqrt{x^2+y^2}}$

**37.** spheres centred at the origin

**39.** circular cylinders with axis along the $z$-axis

**41.** regular octahedra with vertices on the coordinate axes

## Section 4.2 (page 161)

**1.** bdry (0,0) and $x^2 + y^2 = 1$; interior$= S$; $S$ open

**3.** bdry of $S$ is $S$; interior empty; $S$ is closed

**5.** bdry — the spheres $x^2 + y^2 + z^2 = 1$ and $x^2 + y^2 + z^2 = 4$; interior — points between these spheres; $S$ is closed

**7.** bdry of $S$ is $S$, namely the line $x = y = z$; interior is empty; $S$ closed

**9.** 2

**11.** does not exist

**13.** $-1$

**15.** 0

**17.** does not exist

**19.** 0

**21.** $f(0, 0) = 1$

**23.** all $(x, y)$ such that $x \neq \pm y$; no; yes; yes $f(x, x) = \frac{1}{2x}$ makes $f$ continuous at $(x, x)$ for $x \neq 0$; no, $f$ has no continuous extension to the line $x + y = 0$.

**31.** a surface having no tears in it, meeting vertical lines through points of the region exactly once

## Section 4.3 (page 168)

**1.** $f_1(x, y) = f_1(3, 2) = 1$, $f_2(x, y) = f_2(3, 2) = -1$

**3.** $f_1 = 3x^2y^4z^5$, $f_2 = 4x^3y^3z^5$, $f_3 = 5x^3y^4z^4$
All three vanish at $(0, -1, -1)$.

**5.** $\dfrac{\partial z}{\partial x} = \dfrac{-y}{x^2 + y^2}$, $\dfrac{\partial z}{\partial y} = \dfrac{x}{x^2 + y^2}$,
At $(-1, 1)$: $\dfrac{\partial z}{\partial x} = -\dfrac{1}{2}$, $\dfrac{\partial z}{\partial y} = -\dfrac{1}{2}$

**7.** $f_1 = \sqrt{y}\cos(x\sqrt{y})$, $f_2 = \dfrac{x\cos(x\sqrt{y})}{2\sqrt{y}}$,
At $(\pi/3, 4)$: $f_1 = -1$, $f_2 = -\pi/24$

**9.** $\dfrac{\partial w}{\partial x} = y \ln z\, x^{(y \ln z - 1)}$, $\dfrac{\partial w}{\partial y} = \ln x \ln z\, x^{y \ln z}$,
$\dfrac{\partial w}{\partial z} = \dfrac{y \ln x}{z} x^{y \ln z}$
At $(e, 2, e)$: $\dfrac{\partial w}{\partial x} = \dfrac{\partial w}{\partial z} = 2e$, $\dfrac{\partial w}{\partial y} = e^2$.

**11.** $f_1(0, 0 = 2$, $f_2(0, 0) = -1/3$

**13.** $z = -4x - 2y - 3$; $\frac{x+2}{-4} = \frac{y-1}{-2} = \frac{z-3}{-1}$

**15.** $z = \dfrac{1}{\sqrt{2}}\left(1 - \dfrac{x - \pi}{4} + \dfrac{\pi}{16}(y - 4)\right)$;
$$\frac{x - \pi}{-1/4\sqrt{2}} = \frac{y - 4}{\pi/16\sqrt{2}} = \frac{z - 1/\sqrt{2}}{-1}$$

**17.** $z = \frac{2}{5} + \frac{3x}{25} - \frac{4y}{25}$; $\frac{x-1}{3} = \frac{y-2}{-4} = \frac{z-1/5}{-25}$

**19.** $z = \ln 5 + \frac{2}{5}(x - 1) - \frac{4}{5}(y + 2)$;
$$\frac{x - 1}{2/5} = \frac{y + 2}{-4/5} = \frac{z - \ln 5}{-1}$$

**21.** $z = \dfrac{x + y}{2} - \dfrac{\pi}{4}$; $2(x - 1) = 2(y + 1) = -z - \dfrac{\pi}{4}$

**23.** $(0, 0)$, $(1, 1)$, $(-1, -1)$

**33.** $w = f(a, b, c) + f_1(a, b, c)(x - a) + f_2(a, b, c)(y - b) + f_3(a, b, c)(z - c)$

**35.** $\sqrt{7}/4$ units

**37.** $f_1(0, 0) = 1$, $f_2(0, 0)$ does not exist.

**39.** $f$ is continuous at $(0, 0)$; $f_1$ and $f_2$ are not.

## Section 4.4 (page 173)

**1.** $\dfrac{\partial^2 z}{\partial x^2} = 2(1 + y^2)$, $\dfrac{\partial^2 z}{\partial x \partial y} = 4xy$, $\dfrac{\partial^2 z}{\partial y^2} = 2x^2$

**3.** $\dfrac{\partial^2 w}{\partial x^2} = 6xy^3z^3, \quad \dfrac{\partial^2 w}{\partial y^2} = 6x^3yz^3,$
$\dfrac{\partial^2 w}{\partial z^2} = 6x^3y^3z, \quad \dfrac{\partial^2 w}{\partial x \partial y} = 9x^2y^2z^3,$
$\dfrac{\partial^2 w}{\partial x \partial z} = 9x^2y^3z^2, \quad \dfrac{\partial^2 w}{\partial y \partial z} = 9x^3y^2z^2$

**5.** $\dfrac{\partial^2 z}{\partial x^2} = -y\,e^x, \quad \dfrac{\partial^2 z}{\partial x \partial y} = e^y - e^x, \quad \dfrac{\partial^2 z}{\partial y^2} = x\,e^y$

**7.** $27,\ 10,\ x^2e^{xy}\big(xz\sin xz - (3+xy)\cos xz\big)$

**19.** $u(x, y, z, t) = t^{-3/2}e^{-(x^2+y^2+z^2)/4t}$

## Section 4.5 (page 183)

**1.** $\dfrac{\partial w}{\partial t} = f_1g_2 + f_2h_2 + f_3k_2$

**3.** $\dfrac{\partial z}{\partial u} = g_1h_1 + g_2f'h_1$ **5.** $\dfrac{\partial w}{\partial x}\Big|_z = f_1 + f_2g_1$

**7.** $\dfrac{dw}{dz} = f_1g_1h' + f_1g_2 + f_2h' + f_3,$
$\dfrac{\partial w}{\partial z}\Big|_x = f_2h' + f_3,$
$\dfrac{\partial w}{\partial z}\Big|_{x,y} = f_3$

**9.** $\dfrac{\partial z}{\partial x} = \dfrac{-5y}{13x^2 - 2xy + 2y^2}$

**11.** $2f_1(2x, 3y)$ **13.** $2x\ f_2(y^2, x^2)$

**15.** $f_1(x, y)\Big(f_1\big(f(x, y), f(x, y)\big) + f_2\big(f(x, y), f(x, y)\big)\Big)$

**17.** $dT/dt = e^{-t}\big(f'(t) - f(t)\big)$; $dT/dt = 0$ if $f(t) = e^t$: in this case the decrease in $T$ with time (at fixed depth) is exactly balanced by the increase in $T$ with depth.

**19.** $4f_{11} + 12f_{12} + 9f_{22}, \quad 6f_{11} + 5f_{12} - 6f_{22},$
$9f_{11} - 12f_{12} + 4f_{22}$

**25.** $f_1\cos s - f_2\sin s + f_{11}\,t\cos s\sin s$
$+ f_{12}\,t(\cos^2 s - \sin^2 s) - f_{22}\,t\sin s\cos s$

**27.** $f_2 + 2y^2f_{12} + xyf_{22} - 4xyf_{31} - 2x^2f_{32}$;
all derivatives at $(y^2, xy, -x^2)$

**33.** $\sum_{i,j=1}^n x_i x_j f_{ij}(x_1, \cdots, x_n) = k(k-1)\,f(x_1, \cdots, x_n)$

**37.** $u(x, y) = f(x + ct)$

## Section 4.6 (page 191)

**1.** 6.9 **3.** 0.0814

**5.** 2.967 **7.** (a) 3%, (b) 2%, (c) 1%

**9.** 8.88 ft$^2$

**11.** 169 m, 24 m, most sensitive to angle at $B$

**13.** $\begin{pmatrix} \cos\theta & -r\sin\theta \\ \sin\theta & r\cos\theta \end{pmatrix}$

**15.** $\begin{pmatrix} 2x & z & y \\ -\ln z & 2y & -x/z \end{pmatrix}$, $(5.99, 3.98)$

## Section 4.7 (page 201)

**1.** $4\mathbf{i} + 2\mathbf{j}; \quad z = 4x + 2y - 3; \quad 2x + y = 3$

**3.** $(-\mathbf{i} + (\pi/4)\mathbf{j})/4\sqrt{2}; \quad 4x - \pi y + 16\sqrt{2}z = 16;$
$4x - \pi y = 0$

**5.** $(3\mathbf{i} - 4\mathbf{j})/25; \quad 3x - 4y - 25z + 10 = 0;$
$3x - 4y + 5 = 0$

**7.** $(2\mathbf{i} - 4\mathbf{j})/5;\ 2x - 4y - 5z = 10 - 5\ln 5;\ x - 2y = 5$

**9.** $x + y - 3z = -3$ **11.** $\sqrt{3}y + z = \sqrt{3} + \pi/3$

**13.** $\dfrac{4}{\sqrt{5}}$ **15.** $1 - 2\sqrt{3}$

**17.** $\dfrac{-61}{144\sqrt{3}}$

**21.** in directions making angles $-30°$ or $-150°$ with positive $x$-axis; no; $-\mathbf{j}$.

**23.** $7\mathbf{i} - \mathbf{j}$

**25.** a)

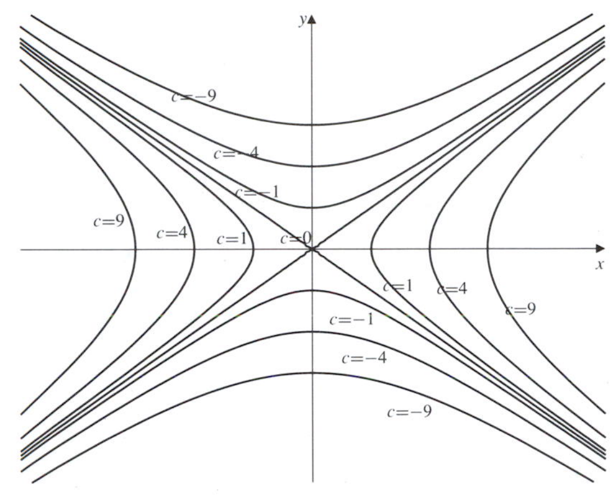


b) in direction $-\mathbf{i} - \mathbf{j}$
c) $4\sqrt{2}k$ deg/unit time
d) $12k/\sqrt{5}$ deg/unit time
e) $x^2y = -4$

**27.** $3x^2 - 2y^2 = 10$ **29.** $-4/3$

**31.** $\mathbf{i} - 2\mathbf{j} + \mathbf{k}$

**37.** $D_{\mathbf{v}}(D_{\mathbf{v}}f) = v_1^2f_{11} + v_2^2f_{22} + v_3^2f_{33} + 2v_1v_2f_{12} + 2v_1v_3f_{13}$
$+ 2v_2v_3f_{23}.$
This is the second derivative of $f$ as measured by an observer moving with velocity $\mathbf{v}$.

**39.** $\dfrac{\partial^2 T}{\partial t^2} + 2D_{\mathbf{v}(t)}\left(\dfrac{\partial T}{\partial t}\right) + D_{\mathbf{a}(t)}T + D_{\mathbf{v}(t)}(D_{\mathbf{v}(t)}T)$

## Section 4.8 (page 211)

**1.** $-\dfrac{x^4 + 3xy^2}{y^3 + 4x^3y}, \quad y \neq 0,\ y^2 \neq -4x^3$

**3.** $\dfrac{3xy^4 + xz}{xy - 2y^2z}, \quad y \neq 0,\ x \neq 2yz$

**5.** $\dfrac{x - 2t^2 w}{2xy^2 - w}$, $w \neq 2xy^2$ **7.** $-\dfrac{\partial G/\partial x}{\partial G/\partial u}$, $\dfrac{\partial G}{\partial u} \neq 0$

**9.** $-\dfrac{v^2 H_2 + w H_3}{u^2 H_1 + t H_3}$, $u^2 H_1 + t H_3 \neq 0$, all derivatives at $(u^2 w, v^2 t, wt)$

**11.** $\dfrac{2w - 4y}{4x - w}$, $4x \neq w$ **13.** $\frac{1}{6}, \frac{1}{2}, \frac{1}{6}, -\frac{1}{2}, -\frac{1}{6}$

**15.** $r$; all points except the origin

**17.** $-3/2$

**19.** $-\dfrac{\partial(F, G, H)}{\partial(y, z, w)} \Big/ \dfrac{\partial(F, G, H)}{\partial(x, z, w)}$

**21.** 15; $-\dfrac{\partial(F, G, H)}{\partial(x_2, x_3, x_5)} \Big/ \dfrac{\partial(F, G, H)}{\partial(x_1, x_3, x_5)}$

**23.** $2(u + v)$, $-2$, $0$

### Section 4.9 (page 217)

**1.** $\displaystyle\sum_{n=0}^{\infty} (-1)^n \frac{x^n y^{2n}}{2^{n+1}}$

**3.** $\displaystyle\sum_{n=0}^{\infty} (-1)^n \frac{x^{2n+1}(y+1)^{2n+1}}{2n+1}$

**5.** $\displaystyle\sum_{n=0}^{\infty} \sum_{k=0}^{n} \frac{1}{k!(n-k)!} x^{2k} y^{2n-2k}$

**7.** $\frac{1}{2} - \frac{1}{4}(x-2) + \frac{1}{2}(y-1) + \frac{1}{8}(x-2)^2 - \frac{1}{2}(x-2)(y-1) + \frac{1}{2}(y-1)^2 - \frac{1}{16}(x-2)^3 + \frac{3}{8}(x-2)^2(y-1) - \frac{3}{4}(x-2)(y-1)^2 + \frac{1}{2}(y-1)^3$

**9.** $x + y^2 - \frac{x^3}{3}$

**11.** $1 - (y-1) + (y-1)^2 - \frac{1}{2}(x - \frac{\pi}{2})^2$

**13.** $-x - x^2 - (5/6)x^3$ **15.** $1 - x^2 - \frac{1}{2}x^4$

**17.** $-\dfrac{x}{3} - \dfrac{2y}{3} - \dfrac{2x^2}{27} - \dfrac{8xy}{27} - \dfrac{8y^2}{27}$

**19.** $\dfrac{[(2n)!]^3}{(n!)^2}$

### Review Exercises (page 218)

**1.**

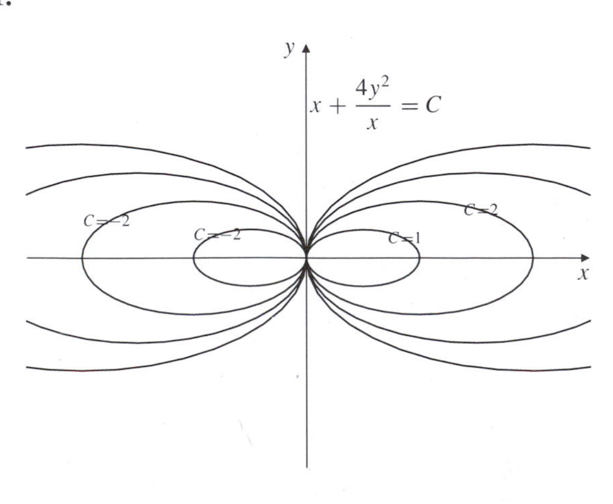


**3.**

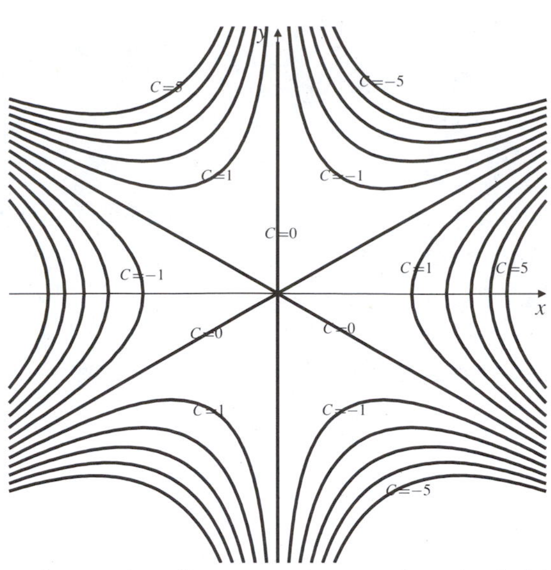


**5.** cont. except on lines $x = \pm y$; can be extended to $x = y$ except at the origin; if $f(0,0) = 0$ then $f_1(0,0) = f_2(0,0) = 1$

**7.** (a) $ax + by + 4cz = 16$, (b) the circle $z = 1$, $x^2 + y^2 = 12$, (c) $\pm(2, 2, \sqrt{2})$

**9.** $7{,}500$ m$^2$, $7.213\%$

**11.** (a) $-1/\sqrt{2}$, (b) dir. of $\pm(\mathbf{i} + 3\mathbf{j} - 4\mathbf{k}$, (c) dir. of $-7\mathbf{i} + 5\mathbf{j} + 2\mathbf{k}$

**15.** (a) $\partial u/\partial x = -5$, $\partial u/\partial y = 1$, (b) $-1.13$

## Chapter 5

### Section 5.1 (page 228)

**1.** $(2, -1)$, loc. (abs) min.

**3.** $(0, 0)$, saddle pt; $(1, 1)$, loc. min.

**5.** $(-4, 2)$, loc. max.

**7.** $(0, n\pi)$, $n = 0, \pm 1, \pm 2, \cdots$, all saddle points

**9.** $(0, a)$, $(a > 0)$, loc min; $(0, a)$, $(a < 0)$, loc max; $(0, 0)$ saddle point; $(\pm 1, 1/\sqrt{2})$, loc. (abs) max; $(\pm 1, -1/\sqrt{2})$, loc. (abs) min.

**11.** $(3^{-1/3}, 0)$, saddle pt.

**13.** $(-1, -1)$, $(1, -1)$, $(-1, 1)$, saddle pts; $(-3, -3)$, loc. min.

**15.** $(1, 1, \frac{1}{2})$, saddle pt.

**17.** $(0, 0)$, saddle pt; $(\frac{1}{\sqrt{2}}, \frac{1}{\sqrt{2}})$, $(-\frac{1}{\sqrt{2}}, -\frac{1}{\sqrt{2}})$, loc. (abs) max; $(\frac{1}{\sqrt{2}}, -\frac{1}{\sqrt{2}})$, $(-\frac{1}{\sqrt{2}}, \frac{1}{\sqrt{2}})$, loc. (abs) min.

**19.** max $e^{-3/2}/2\sqrt{2}$, min $-e^{-3/2}/2\sqrt{2}$; $f$ is continuous everywhere, and $f(x, y, z) \to 0$ as $x^2 + y^2 + z^2 \to \infty$.

**21.** length = width = $(2V)^{1/3}$, height = $(V/4)^{1/3}$

**23.** $8abc/(3\sqrt{3})$ cu. units

**25.** CPs are $(\sqrt{\ln 3}, -\sqrt{\ln 3})$ and $(-\sqrt{\ln 3}, \sqrt{\ln 3})$.

**27.** $f$ does not have a local minimum at $(0,0)$; the second derivative test is inconclusive $(B^2 = AC)$.

**29.** $A > 0$, $\begin{vmatrix} A & E \\ E & B \end{vmatrix} > 0$, $\begin{vmatrix} A & E & F \\ E & B & H \\ F & H & C \end{vmatrix} > 0$,

$\begin{vmatrix} A & E & F & G \\ E & B & H & I \\ F & H & C & J \\ G & I & J & D \end{vmatrix} > 0$

### Section 5.2 (page 234)

**1.** max $5/4$, min $-2$

**3.** max $(\sqrt{2}-1)/2$, min $-(\sqrt{2}+1)/2$.

**5.** max $2/3\sqrt{3}$, min 0

**7.** max 1, min $-1$

**9.** max $1/\sqrt{e}$, min $-1/\sqrt{e}$

**11.** max $4/9$, min $-4/9$

**13.** no limit; yes, max $f = e^{-1}$ (at all points of the curve $xy = 1$)

**15.** \$625,000, \$733,333

**17.** max $37/2$ at $(7/4, 5)$

**19.** 6667 kg deluxe, 6667 kg standard

### Section 5.3 (page 242)

**1.** 84, 375

**3.** 1 unit

**5.** max 4 units, min 2 units

**7.** $a = \pm\sqrt{3}$, $b = \pm 2\sqrt{3}$, $c = \pm\sqrt{3}$

**9.** max 8, min $-8$

**11.** max 2, min $-2$

**13.** max 7, min $-1$

**15.** $\dfrac{2\sqrt{6}}{3}$ units

**17.** $\pm\sqrt{\dfrac{3n(n+1)}{2(2n+1)}}$

**19.** $\frac{1}{6} \times \frac{1}{3} \times \frac{2}{3}$

**21.** width = $\left(\dfrac{2V}{15}\right)^{1/3}$, depth = 3×width, height = $\dfrac{5}{2}$×width

**23.** max 1, min $-\frac{1}{2}$

**27.** method will not fail if $\nabla f = \mathbf{0}$ at extreme point; but we will have $\lambda = 0$.

### Section 5.4 (page 249)

**1.** at $(\bar{x}, \bar{y})$ where $\bar{x} = \left(\sum_{i=1}^n x_i\right)/n$, $\bar{y} = \left(\sum_{i=1}^n y_i\right)/n$

**3.** $a = \left(\sum_{i=1}^n y_i e^{x_i}\right) \Big/ \left(\sum_{i=1}^n e^{2x_i}\right)$

**5.** If $A = \sum x_i^2$, $B = \sum x_i y_i$, $C = \sum x_i$, $D = \sum y_i^2$, $E = \sum y_i$, $F = \sum x_i z_i$, $G = \sum y_i z_i$, and $H = \sum z_i$, then

$\Delta = \begin{vmatrix} A & B & C \\ B & D & E \\ C & E & n \end{vmatrix}$, $a = \dfrac{1}{\Delta}\begin{vmatrix} F & B & C \\ G & D & E \\ H & E & n \end{vmatrix}$,

$b = \dfrac{1}{\Delta}\begin{vmatrix} A & F & C \\ B & G & E \\ C & H & n \end{vmatrix}$, $c = \dfrac{1}{\Delta}\begin{vmatrix} A & B & F \\ B & D & G \\ C & E & H \end{vmatrix}$

**7.** Use linear regression to fit $\eta = a + bx$ to the data $(x_i, \ln y_i)$. Then $p = e^a$, $q = b$. These are not the same values as would be obtained by minimizing the expression $\sum(y_i - pe^{qx_i})^2$.

**9.** Use linear regression to fit $\eta = a + b\xi$ to the data $\left(x_i, \dfrac{y_i}{x_i}\right)$. Then $p = a, q = b$. Not the same as minimizing $\sum(y_i - px_i - qx_i^2)^2$.

**11.** Use linear regression to fit $\eta = a + b\xi$ to the data $\left(e^{-2x_i}, \dfrac{y_i}{e^{x_i}}\right)$. Then $p = a, q = b$. Not the same as minimizing $\sum(y_i - pe^{x_i} - qe^{-x_i})^2$. Other answers possible.

**13.** If $A = \sum x_i^4$, $B = \sum x_i^3$, $C = \sum x_i^2$, $D = \sum x_i$, $H = \sum x_i^2 y_i$, $I = \sum x_i y_i$, and $J = \sum y_i$, then

$\Delta = \begin{vmatrix} A & B & C \\ B & C & D \\ C & D & n \end{vmatrix}$, $a = \dfrac{1}{\Delta}\begin{vmatrix} H & B & C \\ I & C & D \\ J & D & n \end{vmatrix}$,

$b = \dfrac{1}{\Delta}\begin{vmatrix} A & H & C \\ B & I & D \\ C & J & n \end{vmatrix}$, $c = \dfrac{1}{\Delta}\begin{vmatrix} A & B & H \\ B & C & I \\ C & D & J \end{vmatrix}$

**15.** $a = 5/6$, $I = 1/252$

**17.** $a = 15/16$, $b = -1/16$, $I = 1/448$

**19.** $a = \frac{20}{\pi^5}(\pi^2 - 16)$, $b = \frac{12}{\pi^4}(20 - \pi^2)$

**21.** $a_k = \frac{2}{\pi} \int_0^\pi f(x) \cos kx\, dx$, $(k = 0, 1, 2, \cdots)$

**23.** $\pi - \frac{4}{\pi} \sum_{k=0}^\infty \frac{\cos((2k+1)x)}{(2k+1)^2}$; $-x$

### Section 5.5 (page 258)

**1.** $\frac{(-1)^n n!}{(x+1)^{n+1}}$

**3.** $2\sqrt{\pi}(\sqrt{y} - \sqrt{x})$

**5.** $\frac{2x}{(1+x^2)^2}$; $\frac{(6x^2-2)}{(1+x^2)^3}$

**7.** $\frac{\pi}{2x}$, assume $x > 0$; $\frac{\pi}{4x^3}$; $\frac{3\pi}{16x^5}$

**9.** $n!$

**11.** $f(x) = \int_0^x e^{-t^2/2}\, dt$

**13.** $y = x^2$

**15.** $x^2 + y^2 = 1$

**17.** $y = x - \frac{1}{4}$

**19.** no

**21.** no; a line of singular points

**23.** $x^2 + y^2 + z^2 = 1$

**25.** $y = x - \epsilon \sin(\pi x) + \frac{\pi\epsilon^2}{2} \sin(2\pi x) + \cdots$

**27.** $y = \frac{1}{2} - \frac{2}{5}\epsilon x - \frac{16}{125}\epsilon^2 x^2 + \cdots$

**29.** $x \approx 1 - \frac{1}{100e} - \frac{1}{30000e^2}$, $y \approx 1 - \frac{1}{30000e^2}$

### Section 5.6 (page 263)

**1.** (0.797105, 2.219107)

**3.** ($\pm$0.2500305, $\pm$3.9995115), ($\pm$1.9920783, $\pm$0.5019883)

**5.** (0.3727730, 0.3641994), ($-$1.4141606, $-$0.9877577)

**7.** $x = x_0 - \frac{\Delta_1}{\Delta}$, $y = y_0 - \frac{\Delta_2}{\Delta}$, $z = z_0 - \frac{\Delta_3}{\Delta}$, where $\Delta = \left.\frac{\partial(f,g,h)}{\partial(x,y,z)}\right|_{(x_0,y_0,z_0)}$ and $\Delta_i$ is $\Delta$ with the $i$th column replaced with $\begin{matrix} f \\ g \\ h \end{matrix}$

**9.** 18 iterations near (0, 0), 4 iterations near (1, 1); the two curves are tangent at (0,0), but not at (1,1).

### Review Exercises (page 263)

**1.** (0, 0) saddle pt., (1, $-$1) loc. min.

**3.** (2/3, 4/3) loc. min; (2, $-$4) and ($-$1, 2) saddle points

**5.** yes, 2, on the sphere $x^2 + y^2 + z^2 = 1$

**7.** max $1/(4e)$, min $-1/(4e)$

**9.** (a) $L^2/48$ cm$^2$, (b) $L^2/16$ cm$^2$

**11.** $4\pi$ sq. units

**13.** $16\pi$ cu. units

**17.** $y \approx -2x - \epsilon x e^{-2x} + \epsilon^2 x^2 e^{-4x}$

### Challenging Problems (page 264)

**3.** $\frac{1}{2}\ln(1+x^2)\tan^{-1}x$

## Chapter 6

### Section 6.1 (page 270)

**1.** 15

**3.** 21

**5.** 15

**7.** 96

**9.** 80

**11.** 36.6258

**13.** 20

**15.** 0

**17.** $5\pi$

**19.** $\frac{\pi a^3}{3}$

**21.** $\frac{1}{6}$

### Section 6.2 (page 277)

**1.** 5/24

**3.** 4

**5.** $\frac{ab(a^2+b^2)}{3}$

**7.** $\pi$

**9.** $\frac{3}{56}$

**11.** $\frac{33}{8}\ln 2 - \frac{45}{16}$

**13.** $\frac{e-2}{2}$

**15.** $\frac{1}{2}\left(1 - \frac{1}{e}\right)$; region is a triangle with vertices (0,0), (1,0) and (1,1)

**17.** $\frac{\pi}{4\lambda}$; region is a triangle with vertices (0,0), (0,1) and (1,1)

**19.** 1/4 cu. units

**21.** 1/3 cu. units

**23.** ln 2 cu. units

**25.** $\frac{\pi}{2\sqrt{2}}$ cu. units

**27.** $\frac{16a^3}{3}$ cu. units

### Section 6.3 (page 283)

**1.** converges to 1

**3.** converges to $\pi/2$

**5.** diverges to $\infty$

**7.** converges to 4

**9.** converges to $1 - \frac{1}{e}$

**11.** diverges to $\infty$

**13.** diverges to $\infty$

**15.** converges to 2 ln 2

**17.** $k > a - 1$

**19.** $k < -1 - a$

**21.** $k > -\frac{1+a}{1+b}$ (provided $b > -1$)

**23.** $\frac{1}{2}$, $-\frac{1}{2}$ (different answers are possible because the *double integral* does not exist.)

**25.** $\frac{a^2}{3}$

**27.** $\frac{4\sqrt{2}a}{3\pi}$

**29.** yes, $1/(2\pi)$

### Section 6.4 (page 292)

**1.** $\pi a^4/2$

**3.** $2\pi a$

**5.** $\pi a^4/4$

**7.** $a^3/3$

**9.** $\pi(e^{a^2} - 1)/4$

**11.** $\frac{(\sqrt{3}+1)a^3}{6}$

**13.** $\frac{1}{3}$

**15.** $\frac{2a}{3}$

**17.** $k < 1$; $\frac{\pi}{1-k}$

**19.** $\frac{a^4}{16}$

**21.** $\frac{2\pi}{3}$ cu. units

**23.** $\frac{4\pi(2\sqrt{2}-1)a^3}{3}$ cu. units

**25.** $\frac{64a^3}{9}$ cu. units

**27.** $16[1 - (1/\sqrt{2})]a^3$ cu. units

**29.** $1 - \frac{4\sqrt{2}}{3\pi}$ units

**31.** $\frac{4}{3}\pi abc$ cu. units

**33.** $2a\sinh a$

**35.** $\frac{3\ln 2}{2}$ sq. units

**37.** $\frac{1}{4}(e - e^{-1})$

### Section 6.5 (page 300)

**1.** $8abc$

**3.** $16\pi$

**5.** 2/3

**7.** 1/15

**9.** $2/(3\pi)$

**11.** $\frac{3}{16}\ln 2$

**13.** $\pi\sqrt{\frac{\pi}{6}}$

**15.** 1/8

**17.** $\int_0^1 dx \int_0^1 dy \int_0^{1-y} f(x,y,z)\,dz$

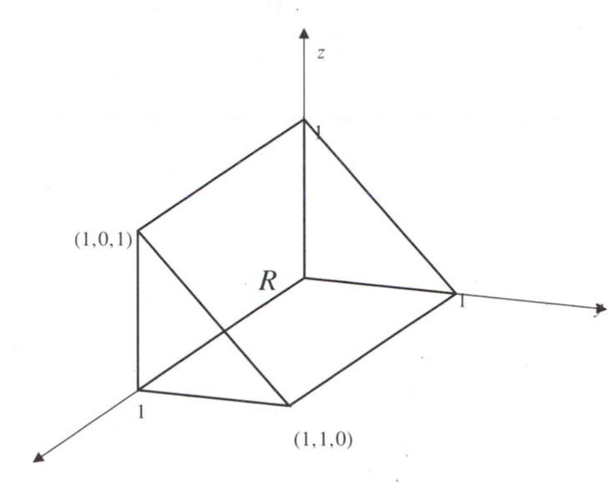


**19.** $\int_0^1 dx \int_0^x dy \int_0^{x-y} f(x,y,z)\,dz$

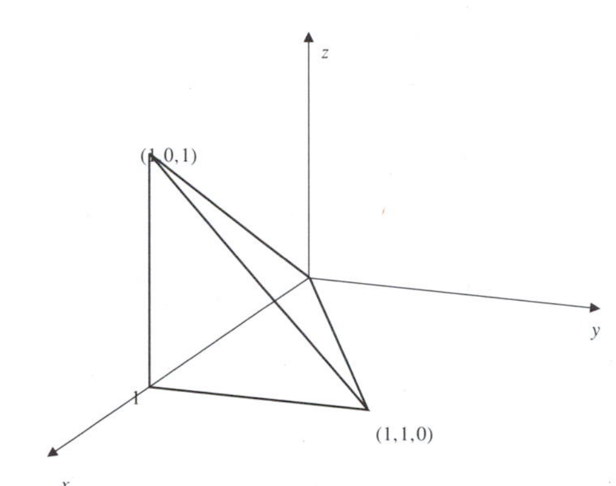


**27.** $(e-1)/3$

**29.** $\overline{f} = \dfrac{1}{\text{vol}(R)} \iiint_R f\,dV$; 1

## Section 6.6 (page 308)

**1.** Cartesian: $(-\sqrt{3}, 3, 2)$; cylindrical: $[2\sqrt{3}, 2\pi/3, 2]$

**3.** Cartesian: $(\sqrt{3}, 1, -2)$; spherical: $[2\sqrt{2}, 3\pi/4, \pi/6]$

**5.** the half-plane $x = 0$, $y > 0$

**7.** the $xy$-plane

**9.** the circular cylinder of radius 4 with axis along the $z$-axis

**11.** the $xy$-plane

**13.** sphere of radius 1 with centre $(0, 0, 1)$

**15.** $\frac{2}{3}\pi a^3\left(1 - \frac{1}{\sqrt{2}}\right)$ cu. units

**17.** $24\pi$ cu. units

**19.** $(2\pi - \dfrac{32}{9})a^3$ cu. units.

**21.** $\dfrac{abc}{3}\tan^{-1}\dfrac{a}{b}$ cu. units

**23.** $\frac{\pi ab}{2}$ cu. units

**25.** $\dfrac{\pi a^4 h^2}{4}$

**27.** $\dfrac{8\pi a^5}{15}$

**29.** $\dfrac{2\pi a^5}{5}\left(1 - \dfrac{c}{\sqrt{c^2+1}}\right)$

**31.** $\dfrac{7\pi}{12}$

**33.** $\frac{ha^3}{12}$; $\frac{\pi a^2 h^2}{48}$

**35.** $\lambda < \frac{3}{2}$, $\mu + \lambda > \frac{3}{2}$

## Section 6.7 (page 318)

**1.** $3\pi$ sq. units

**3.** $2\pi a^2$ sq. units

**5.** $24\pi/\sqrt{3}$ sq. units

**7.** $(5\sqrt{5} - 1)/12$ sq. units

**9.** 4 sq. units

**11.** 5.123

**13.** $4\pi A\left[a - \sqrt{B}\tan^{-1}\left(\frac{a}{\sqrt{B}}\right)\right]$ units

**15.** $2\pi km\delta(h + \sqrt{a^2 + (b-h)^2} - \sqrt{a^2 + b^2})$

**17.** $\dfrac{2\pi km\delta}{3b^2}\left[2b^3 + a^3 - (2b^2 - a^2)\sqrt{a^2 + b^2}\right]$

**19.** $\left(\frac{1}{3}, \frac{1}{3}, \frac{1}{2}\right)$

**21.** $\left(\frac{3a}{8}, \frac{3a}{8}, \frac{3a}{8}\right)$

**23.** $\left(0, \frac{a}{4}, \frac{5a}{8}\right)$

**25.** The model still involves angular acceleration to spin the ball—it doesn't just fall. Part of the gravitational energy goes to producing this spin even in the limiting case.

**27.** $I = \pi\delta a^2 h\left(\frac{h^2}{3} + \frac{a^2}{4}\right)$, $\quad \bar{D} = \left(\frac{h^2}{3} + \frac{a^2}{4}\right)^{1/2}$

**29.** $I = \frac{\pi\delta a^2 h}{3}\left(\frac{2h^2+3a^2}{20}\right)$, $\quad \bar{D} = \left(\frac{2h^2+3a^2}{20}\right)^{1/2}$

**31.** $I = \frac{5a^5\delta}{12}$, $\quad \bar{D} = \sqrt{\frac{5}{12}}\,a$

**33.** $I = \frac{8}{3}\delta abc(a^2 + b^2)$, $\bar{D} = \sqrt{\frac{a^2+b^2}{3}}$

**35.** $m = \frac{4\pi}{3}\delta(a^2 - b^2)^{3/2}$, $I = \frac{1}{5}m(2a^2 + 3b^2)$

**37.** $\dfrac{5a^2 g\sin\alpha}{7a^2 + 3b^2}$

**41.** The moment of inertia about the line $\mathbf{r}(t) = At\mathbf{i} + Bt\mathbf{j} + Ct\mathbf{k}$ is

$$\frac{1}{A^2 + B^2 + C^2}\big((B^2 + C^2)P_{xx} + (A^2 + C^2)P_{yy} + (A^2 + B^2)P_{zz} - 2ABP_{xy} - 2ACP_{xz} - 2BCP_{yz}\big).$$

## Review Exercises (page 319)

**1.** $3/10$

**3.** $\ln 2$

**5.** $k = 1/\sqrt{3}$

**7.** $\displaystyle\int_0^1 dx \int_x^1 dy \int_y^1 f(x,y,z)\,dz$

**9.** $(1 - e^{-a^2})/(2a)$

**11.** $\dfrac{8\pi}{15}(18\sqrt{6} - 41)a^5$

**13.** vol = $7/12$, $\bar{z} = 11/28$

**15.** $17/24$

**17.** $\displaystyle\frac{1}{6}\int_0^{\pi/2}\left[(1 + 16\cos^2\theta)^{3/2} - 1\right]d\theta \approx 7.904$ sq. units

### Challenging Problems (page 320)

1. $\pi abc\left(\frac{2}{3}-\frac{8}{9\sqrt{3}}\right)$ cu. units

3. (b) (i) $\sum_{n=1}^{\infty}(-1)^{n-1}1/n^2$, (ii) $\sum_{n=1}^{\infty}1/n^3$, (iii) $\sum_{n=1}^{\infty}(-1)^{n-1}1/n^3$

7. $4\pi|a|^3/35$ cu. units

## Chapter 7

### Section 7.1 (page 328)

1. $\mathbf{v}=\mathbf{j}$, $v=1$, $\mathbf{a}=\mathbf{0}$, path is the line $x=1$, $z=0$

3. $\mathbf{v}=2t\mathbf{j}+\mathbf{k}$, $v=\sqrt{4t^2+1}$, $\mathbf{a}=2\mathbf{j}$, path is the parabola $y=z^2$, in the $yz$-plane

5. $\mathbf{v}=2t\mathbf{i}-2t\mathbf{j}$, $v=2\sqrt{2}t$, $\mathbf{a}=2\mathbf{i}-2\mathbf{j}$, path is the straight half-line $x+y=0$, $z=1$, $(x\ge 0)$

7. $\mathbf{v}=-a\sin t\mathbf{i}+a\cos t\mathbf{j}+c\mathbf{k}$, $v=\sqrt{a^2+c^2}$, $\mathbf{a}=-a\cos t\mathbf{i}-a\sin t\mathbf{j}$, path is a circular helix

9. $\mathbf{v}=-3\sin t\,\mathbf{i}-4\sin t\,\mathbf{j}+5\cos t\,\mathbf{k}$, $v=5$, $\mathbf{a}=-\mathbf{r}$, path is the circle of intersection of the plane $4x=3y$ with the sphere $x^2+y^2+z^2=25$

11. $\mathbf{a}=\mathbf{v}=\mathbf{r}$, $v=\sqrt{a^2+b^2+c^2}\,e^t$, path is the straight line $\frac{x}{a}=\frac{y}{b}=\frac{z}{c}$

13. $\mathbf{v}=-(e^{-t}\cos e^t+\sin e^t)\mathbf{i}$
$\quad+(-e^{-t}\sin e^t+\cos e^t)\mathbf{j}-e^t\mathbf{k}$
$v=\sqrt{1+e^{-2t}+e^{2t}}$
$\mathbf{a}=[(e^{-t}-e^t)\cos e^t+\sin e^t]\mathbf{i}$
$\quad+[(e^{-t}-e^t)\sin e^t-\cos e^t]\mathbf{j}-e^t\mathbf{k}$
The path is a spiral lying on the surface $z=-1/\sqrt{x^2+y^2}$

15. $\mathbf{a}=-3\pi^2\mathbf{i}-4\pi^2\mathbf{j}$ 17. $\sqrt{3/2}(-\mathbf{i}+\mathbf{j}-2\mathbf{k})$

27. $\frac{d}{dt}\big(\mathbf{u}\times(\mathbf{v}\times\mathbf{w})\big)=\frac{d\mathbf{u}}{dt}\times(\mathbf{v}\times\mathbf{w})$
$\quad+\mathbf{u}\times\left(\frac{d\mathbf{v}}{dt}\times\mathbf{w}\right)+\mathbf{u}\times\left(\mathbf{v}\times\frac{d\mathbf{w}}{dt}\right)$

29. $\mathbf{u}'''\bullet(\mathbf{u}\times\mathbf{u}')$

31. $\mathbf{r}=\mathbf{r}_0e^{2t}$, $\mathbf{a}=4\mathbf{r}_0e^{2t}$; the path is a straight line through the origin in the direction of $\mathbf{r}_0$

### Section 7.2 (page 334)

1. $\frac{e-1}{e}$, $\frac{e^2-1}{e^2}$

3. $\mathbf{r}=\cos t\mathbf{i}+\sin t\mathbf{j}+\mathbf{k}$; the curve is a circle of radius 1 in the plane $z=1$

5. $\frac{\pi^2R}{72}$ towards the ground, where $R$ is the radius of the earth; 4.76° west of south

### Section 7.3 (page 341)

1. $x=\sqrt{a^2-t^2}$, $y=t$, $0\le t\le a$

3. $x=a\sin\theta$, $y=-a\cos\theta$, $\frac{\pi}{2}\le\theta\le\pi$

5. $\mathbf{r}=-2t\mathbf{i}+t\mathbf{j}+4t^2\mathbf{k}$

7. $\mathbf{r}=3\cos t\mathbf{i}+3\sin t\mathbf{j}+3(\cos t+\sin t)\mathbf{k}$

9. $\mathbf{r}=(1+2\cos t)\mathbf{i}-2(1-\sin t)\mathbf{j}$
$\quad+(9+4\cos t-8\sin t)\mathbf{k}$

11. Choice (b) leads to $\mathbf{r}=\frac{t^2-1}{2}\mathbf{i}+t\mathbf{j}+\frac{t^2+1}{2}\mathbf{k}$, which represents the whole parabola. Choices (a) and (c) lead to separate parametrizations for the halves $y\ge 0$ and $y\le 0$ of the parabola. For (a) these are $\mathbf{r}=t\mathbf{i}\pm\sqrt{1+2t}\mathbf{j}+(1+t)\mathbf{k}$, $(t\ge -1/2)$

13. $(17\sqrt{17}-16\sqrt{2})/27$ units

15. $\int_1^T\frac{\sqrt{4a^2t^4+b^2t^2+c^2}}{t}\,dt$ units;
$a(T^2-1)+c\ln T$ units

17. $\pi\sqrt{2+4\pi^2}+\ln(\sqrt{2}\pi+\sqrt{1+2\pi^2})$ units

19. $\sqrt{2e^{4\pi}+1}-\sqrt{3}+\frac{1}{2}\ln\frac{e^{4\pi}+1-\sqrt{2e^{4\pi}+1}}{e^{4\pi}}$
$-\frac{1}{2}\ln(2-\sqrt{3})$ units

23. approx 2.3712 units

25. $\mathbf{r}=\frac{1}{\sqrt{A^2+B^2+C^2}}(As\mathbf{i}+Bs\mathbf{j}+Cs\mathbf{k})$

27. $\mathbf{r}=a\left(1-\frac{s}{K}\right)^{3/2}\mathbf{i}+a\left(\frac{s}{K}\right)^{3/2}\mathbf{j}+b\left(1-\frac{2s}{K}\right)\mathbf{k}$,
$0\le s\le K$, $K=(\sqrt{9a^2+16b^2})/2$

### Section 7.4 (page 350)

1. $\hat{\mathbf{T}}=\frac{1}{\sqrt{1+16t^2+81t^4}}(\mathbf{i}-4t\mathbf{j}+9t^2\mathbf{k})$

3. $\hat{\mathbf{T}}=\frac{1}{\sqrt{1+\sin^2 t}}(\cos 2t\mathbf{i}+\sin 2t\mathbf{j}-\sin t\mathbf{k})$

### Section 7.5 (page 356)

1. 1/2, 27/2 3. $27/(4\sqrt{2})$

5. $\hat{\mathbf{T}}=(\mathbf{i}+2\mathbf{j})/\sqrt{5}$, $\hat{\mathbf{N}}=(-2\mathbf{i}+\mathbf{j})/\sqrt{5}$, $\hat{\mathbf{B}}=\mathbf{k}$

7. $\hat{\mathbf{T}}=\mathbf{k}$, $\hat{\mathbf{N}}=\mathbf{i}$, $\hat{\mathbf{B}}=\mathbf{j}$

9. $\hat{\mathbf{T}}=\frac{1}{\sqrt{1+t^2+t^4}}(\mathbf{i}+t\mathbf{j}+t^2\mathbf{k})$,
$\hat{\mathbf{B}}=\frac{1}{\sqrt{t^4+4t^2+1}}(t^2\mathbf{i}-2t\mathbf{j}+\mathbf{k})$,
$\hat{\mathbf{N}}=\frac{-(t+2t^3)\mathbf{i}+(1-t^4)\mathbf{j}+(t^3+2t)\mathbf{k}}{\sqrt{t^4+4t^2+1}\sqrt{1+t^2+t^4}}$,
$\kappa=\frac{\sqrt{t^4+4t^2+1}}{(t^4+t^2+1)^{3/2}}$, $\tau=\frac{2}{t^4+4t^2+1}$

11. $\hat{\mathbf{T}}=\frac{(\cos t-t\sin t)\mathbf{i}+(\sin t+t\cos t)\mathbf{j}+\mathbf{k}}{\sqrt{2+t^2}}$,
$\hat{\mathbf{N}}=-\Big(\frac{(4+t^2)\sin t+(3t+t^3)\cos t)\mathbf{i}}{\sqrt{2+t^2}\sqrt{8+5t^2+t^4}}$
$\quad+\frac{((4+t^2)\cos t-(3t+t^3)\sin t)\mathbf{j}-t\mathbf{k}}{\sqrt{2+t^2}\sqrt{8+5t^2+t^4}}$,
$\hat{\mathbf{B}}=\frac{-(2\cos t-t\sin t)\mathbf{i}-(2\sin t+t\cos t)\mathbf{j}+(2+t^2)\mathbf{k}}{\sqrt{8+5t^2+t^4}}$,

$$\kappa = \frac{\sqrt{8+5t^2+t^4}}{(2+t^2)^{3/2}}, \qquad \tau = \frac{6+t^2}{8+5t^2+t^4}$$

**13.** $\kappa(t) = 1/\sqrt{2}$, $\tau(t) = 0$, curve is a circle in the plane $y+z=4$, having centre $(2,1,3)$ and radius $\sqrt{2}$

**15.** i) $\hat{\mathbf{T}} = \mathbf{i}$, $\quad \hat{\mathbf{N}} = \dfrac{2\mathbf{j}-\mathbf{k}}{\sqrt{5}}$,

$\hat{\mathbf{B}} = \dfrac{\mathbf{j}+2\mathbf{k}}{\sqrt{5}}$, $\quad \kappa = \sqrt{5}$, $\quad \tau = 0$

ii) $\hat{\mathbf{T}} = \sqrt{\frac{2}{3}}(\mathbf{j} - \frac{1}{\sqrt{2}}\mathbf{k})$, $\quad \hat{\mathbf{B}} = \frac{1}{\sqrt{13}}(-\mathbf{i}+2\mathbf{j}+2\sqrt{2}\mathbf{k})$,

$\hat{\mathbf{N}} = -\frac{1}{\sqrt{39}}(6\mathbf{i}+\mathbf{j}+\sqrt{2}\mathbf{k})$, $\quad \kappa = \frac{2\sqrt{39}}{9}$, $\quad \tau = -\frac{6\sqrt{2}}{13}$

**17.** max $a/b^2$, min $b/a^2$

**19.** $\kappa = \dfrac{e^x}{(1+e^{2x})^{3/2}}$,

$\mathbf{r} = (x - 1 - e^{2x})\mathbf{i} + (2e^x + e^{-x})\mathbf{j}$

**21.** $\frac{3}{2\sqrt{2ar}}$

**25.** $\mathbf{r} = -4x^3\mathbf{i} + (3x^2 + \frac{1}{2})\mathbf{j}$

**27.** $f(x) = \frac{1}{8}(15x - 10x^3 + 3x^5)$

## Section 7.6 (page 365)

**3.** velocity: $1/\sqrt{2}$, $1/\sqrt{2}$;
acceleration: $-e^{-\theta}/2$, $e^{-\theta}/2$.

**7.** 42,777 km, the equatorial plane

**9.** $\frac{T}{4\sqrt{2}}$ **13.** 3/4

**15.** $(1/2) - (\epsilon/\pi)$

**19.** $r = A\sec\omega(\theta - \theta_0)$, $\quad \omega^2 = 1 - (k/h^2)$ if $k < h^2$,
$r = 1/(A + B\theta)$ if $k = h^2$,
$r = Ae^{\omega\theta} + Be^{-\omega\theta}$, $\quad \omega^2 = (k/h^2) - 1$, $\quad$ if $k > h^2$;
there are no bounded orbits that do not approach the origin except in the case $k = h^2$ if $B = 0$ when there are circular orbits. (Now aren't you glad gravitation is an inverse square rather than an inverse cube attraction?)

**21.** centre $\left(\dfrac{\ell\epsilon}{\epsilon^2-1}, 0\right)$;

asymptotes in directions $\theta = \pm\cos^{-1}\left(-\dfrac{1}{\epsilon}\right)$;

semi-transverse axis $a = \dfrac{\ell}{\epsilon^2-1}$;

semi-conjugate axis $b = \dfrac{\ell}{\sqrt{\epsilon^2-1}}$;

semi-focal separation $c = \dfrac{\ell\epsilon}{\epsilon^2-1}$.

## Review Exercises (page 367)

**3.** $\mathbf{v} = 2(\mathbf{i}+2\mathbf{j}+2\mathbf{k})$, $\mathbf{a} = (8/3)(-2\mathbf{i}-\mathbf{j}+2\mathbf{k})$

**5.** $\kappa = \tau = \sqrt{2}/(e^t + e^{-t})^2$

**9.** $4a(1 - \cos(T/2))$ units

**13.** $\hat{\boldsymbol{\rho}} = \sin\phi\cos\theta\mathbf{i} + \sin\phi\sin\theta\mathbf{j} + \cos\phi\mathbf{k}$

$\hat{\boldsymbol{\phi}} = \cos\phi\cos\theta\mathbf{i} + \cos\phi\sin\theta\mathbf{j} - \sin\phi\mathbf{k}$

$\hat{\boldsymbol{\theta}} = -\sin\theta\mathbf{i} + \cos\theta\mathbf{j}$

right-handed

## Challenging Problems (page 368)

**1.** (a) $\boldsymbol{\Omega} = \Omega\dfrac{\mathbf{j}+\mathbf{k}}{\sqrt{2}}$, $\Omega \approx 7.272 \times 10^{-5}$.
(b) $\mathbf{a}_C = -\sqrt{2}\Omega v\mathbf{i}$.
(c) about 15.5 cm west of $P$.

**3.**
(c) $\mathbf{v}(t) = (\mathbf{v}_0 - (\mathbf{v}_0 \bullet \mathbf{k})\mathbf{k})\cos(\omega t) + (\mathbf{v}_0 \times \mathbf{k})\sin(\omega t) + (\mathbf{v}_0 \bullet \mathbf{k})\mathbf{k}$.
(d) Straight line if $\mathbf{v}_0$ is parallel to $\mathbf{k}$, circle if $\mathbf{v}_0$ is perpendicular to $\mathbf{k}$.

**5.** (a) $y = (48 + 24x^2 - x^4)/64$

**7.** (a) Yes, time $\pi a/(v\sqrt{2})$, (b) $\phi = \dfrac{\pi}{2} - \dfrac{vt}{a\sqrt{2}}$,

$\theta = \ln\left[\sec\left(\dfrac{vt}{a\sqrt{2}}\right) + \tan\left(\dfrac{vt}{a\sqrt{2}}\right)\right]$.

(c) infinitely often

# Chapter 8

## Section 8.1 (page 376)

**1.** field lines: $y = x + C$

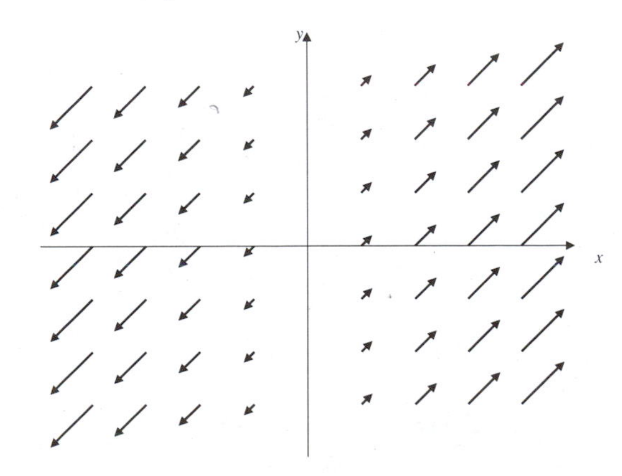

**3.** field lines: $y^2 = x^2 + C$

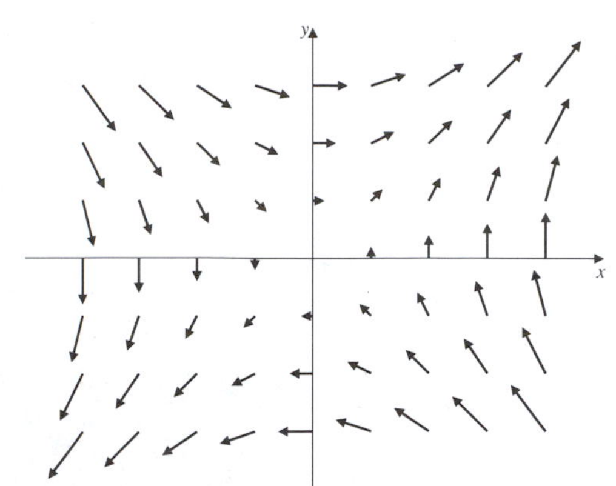


**5.** field lines: $y = -\frac{1}{2}e^{-2x} + C$

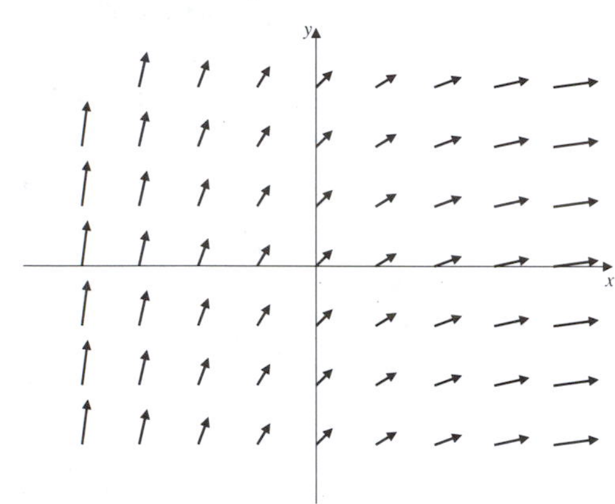


**7.** field lines: $y = Cx$

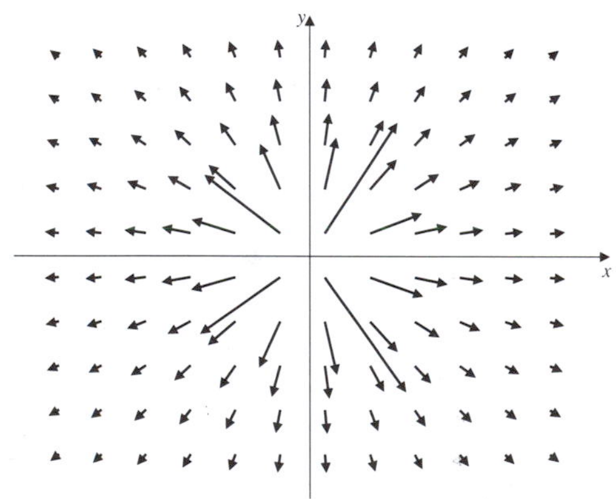


**9.** streamlines are lines parallel to $\mathbf{i} - \mathbf{j} - \mathbf{k}$

**11.** streamlines: $x^2 + y^2 = a^2$, $x = a\sin(z - b)$ (spirals)

**13.** $y = C_1 x$, $2x = z^2 + C_2$

**15.** $y = Ce^{1/x}$ **17.** $r = \theta + C$

**19.** $r = C\theta^2$

## Section 8.2 (page 385)

**1.** conservative; $\dfrac{x^2}{2} - y^2 + \dfrac{3z^2}{2}$

**3.** not conservative

**5.** conservative; $x^2y + y^2z - z^2x$

**7.** $-2\frac{\mathbf{r}-\mathbf{r}_0}{|\mathbf{r}-\mathbf{r}_0|^4}$

**9.** $(x^2 + y^2)/z$; equipotential surfaces are paraboloids $z = C(x^2+y^2)$; field lines are ellipses $x^2+y^2+2z^2 = A$, $y = Bx$ in vertical planes through the origin

**11.** $\mathbf{v} = \dfrac{m(x\mathbf{i} + y\mathbf{j} + (z - \ell)\mathbf{k})}{[x^2 + y^2 + (z - \ell)^2]^{3/2}} + \dfrac{m(x\mathbf{i} + y\mathbf{j} + (z + \ell)\mathbf{k})}{[x^2 + y^2 + (z + \ell)^2]^{3/2}}$, $\mathbf{v} = \mathbf{0}$ only at the origin; $\mathbf{v}(x, y, 0) = \dfrac{2m(x\mathbf{i} + y\mathbf{j})}{(x^2 + y^2 + \ell^2)^{3/2}}$; speed maximum on the circle $x^2 + y^2 = \ell^2/2$, $z = 0$

**15.** $\phi = -\frac{\mu y}{r^2}$, $\mathbf{F} = \frac{\mu(2xy\mathbf{i}+(y^2-x^2)\mathbf{j})}{r^4}$, $(r^2 = x^2 + y^2)$

**21.** $\phi = \frac{1}{2}r^2 \sin 2\theta$

## Section 8.3 (page 390)

**1.** $\frac{a^2}{2}\left(\sqrt{2} + \ln(1 + \sqrt{2})\right)$ **3.** 8 gm

**5.** $\frac{\delta}{6}\left((2e^{4\pi} + 1)^{3/2} - 3^{3/2}\right)$

**7.** $3\sqrt{14}$

**9.** $m = 2\sqrt{2}\pi^2$, $(0, -1/\pi, 4\pi/3)$

**11.** $(e^6 + 3e^4 - 3e^2 - 1)/(3e^3)$

**13.** $\left(\sqrt{2} + \ln(\sqrt{2} + 1)\right)a^2/2$

**15.** $\pi/\sqrt{2}$

**17.** $4\sqrt{b^2 + c^2}\, E\left(\sqrt{\dfrac{b^2 - a^2}{b^2 + c^2}}\right)$; $\sqrt{b^2 + c^2}\, E\left(\sqrt{\dfrac{b^2 - a^2}{b^2 + c^2}}, T\right)$

## Section 8.4 (page 397)

**1.** $-1/4$ **3.** $1/2$

**5.** 0 **7.** $19/2$

**9.** $e^{1+(\pi/4)}$

**11.** $A = 2$, $B = 3$; $4\ln s - \frac{1}{2}$

**13.** $-13/2$ **15.** a) $\pi a^2$, b) $-\pi a^2$

**17.** a) $\dfrac{\pi a^2}{2}$, b) $-\dfrac{\pi a^2}{2}$ **19.** a) $ab/2$, b) $-ab/2$

**23.** The plane with origin removed is not simply connected.

## Section 8.5 (page 409)

**1.** $dS = ds\,dz = \sqrt{(g(\theta))^2 + (g'(\theta))^2}\, d\theta\, dz$

**3.** $\frac{\pi ab\sqrt{A^2+B^2+C^2}}{|C|}$ sq. units $(C \neq 0)$

**5.** (a) $dS = |\nabla F/F_1|\, dx\, dz$, (b) $dS = |\nabla F/F_2|\, dy\, dz$

**7.** $\frac{\pi}{8}$ **9.** $16a^2$ sq. units

**13.** $2\pi$ **15.** $1/96$

**17.** $\pi(3e + e^3 - 4)/3$

**19.** $2\pi a^2 + \dfrac{2\pi ac^2}{\sqrt{a^2 - c^2}} \ln\left(\dfrac{a + \sqrt{a^2 - c^2}}{c}\right)$ sq. units

**21.** $2\pi\sqrt{A^2 + B^2 + C^2}/|D|$

**23.** one-third of the way from the base to the vertex on the axis

**25.** $2\pi k\sigma ma\left(\frac{1}{\sqrt{a^2+(b-h)^2}} - \frac{1}{\sqrt{a^2+b^2}}\right)$

**27.** $I = \frac{8}{3}\pi\sigma a^4;\ \bar{D} = \sqrt{\frac{2}{3}}\,a$

**29.** $\frac{3}{5}\,g \sin\alpha$

### Section 8.6 (page 416)

**1.** 6 **3.** $3abc$

**5.** $\pi(3a^2 - 4ab + b^2)/2$ **7.** $4\pi$

**9.** $2\sqrt{2}\pi$ **11.** $4\pi/3$

**13.** $4\pi m$ **15.** a) $2\pi a^2$, b) 8

### Review Exercises (page 417)

**1.** $(3e/2) - (3/(2e))$ **3.** $8\sqrt{2}/15$

**5.** 1

**7.** (a) $6\pi mgb$, (b) $6\pi R\sqrt{a^2 + b^2}$

**9.** (b) $e^2$ **11.** $(x\mathbf{i} - y\mathbf{j})/\sqrt{x^2 + y^2}$

### Challenging Problems (page 418)

**1.** centroid $(0, 0, 2/\pi)$; upper half of the surface of the torus obtained by rotating the circle $(x-2)^2+z^2 = 1$, $y = 0$, about the $z$-axis

## Chapter 9

### Section 9.1 (page 427)

**1.** $\textbf{div}\,\mathbf{F} = 2,\ \textbf{curl}\,\mathbf{F} = \mathbf{0}$

**3.** $\textbf{div}\,\mathbf{F} = 0,\ \textbf{curl}\,\mathbf{F} = -\mathbf{i} - \mathbf{j} - \mathbf{k}$

**5.** $\textbf{div}\,\mathbf{F} = 1,\ \textbf{curl}\,\mathbf{F} = -\mathbf{j}$

**7.** $\textbf{div}\,\mathbf{F} = f'(x) + g'(y) + h'(z),\quad \textbf{curl}\,\mathbf{F} = \mathbf{0}$

**9.** $\textbf{div}\,\mathbf{F} = \cos\theta\left(1 + \dfrac{1}{r}\cos\theta\right);$

$\textbf{curl}\,\mathbf{F} = -\sin\theta\left(1 + \dfrac{1}{r}\cos\theta\right)\mathbf{k}$

**11.** $\textbf{div}\,\mathbf{F} = 0;\quad \textbf{curl}\,\mathbf{F} = (1/r)\mathbf{k}$

### Section 9.2 (page 433)

**7.** $\textbf{div}\,\mathbf{F}$ can have any value, $\textbf{curl}\,\mathbf{F}$ must be normal to $\mathbf{F}$

**9.** $f(r) = Cr^{-3}$

**15.** If $\mathbf{F} = \nabla\phi$ and $\mathbf{G} = \nabla\psi$ then $\nabla\times(\phi\nabla\psi) = \mathbf{F}\times\mathbf{G}$.

**17.** $\mathbf{G} = ye^{2z}\mathbf{i} + xye^{2z}\mathbf{k}$ is one possible vector potential.

### Section 9.3 (page 440)

**1.** $4\pi a^3$ **3.** $(4/3)\pi a^3$

**5.** $360\pi$ **7.** $81/4$

**11.** $\frac{2}{3}\pi a^2 b + \frac{3}{10}\pi a^4 b + \pi a^2$

**13.** (a) $12\sqrt{3}\pi a^3$, (b) $-4\sqrt{3}\pi a^3$, (c) $16\sqrt{3}\pi a^3$

**15.** $(6 + 2\bar{x} + 4\bar{y} - 2\bar{z})V$ **17.** $9\pi a^2$

### Section 9.4 (page 447)

**1.** $\pi a^2 - 4a^3$ **3.** 9

**5.** $\dfrac{3\pi ab}{8}$ sq. units

**7.** 0

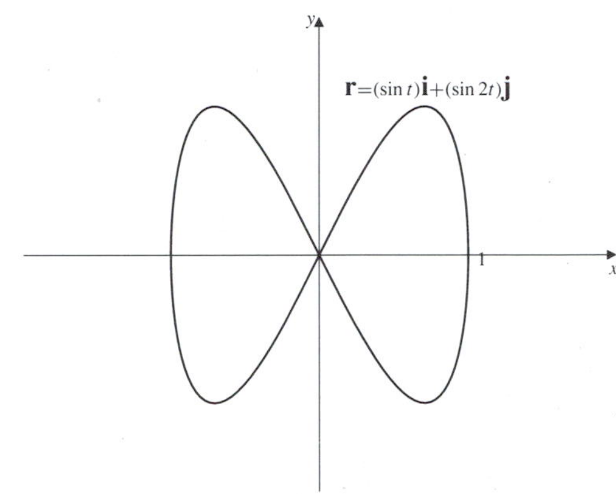


**9.** $1/2$ **11.** $-3\pi a^2$

**15.** $9\pi$

**17.** $\alpha = -\frac{1}{2},\ \beta = -3,\ I = -\frac{3}{8}\pi$

**19.** yes, $\phi\nabla\psi$

### Section 9.6 (page 465)

**1.** $\nabla f = \theta z\hat{\mathbf{r}} + z\hat{\boldsymbol{\theta}} + r\theta\mathbf{k}$

**3.** $\textbf{div}\,\mathbf{F} = 2,\ \textbf{curl}\,\mathbf{F} = \mathbf{0}$

**5.** $\textbf{div}\,\mathbf{F} = \dfrac{2\sin\phi}{\rho},\ \textbf{curl}\,\mathbf{F} = -\dfrac{\cos\phi}{\rho}\hat{\boldsymbol{\theta}}$

**7.** $\textbf{div}\,\mathbf{F} = 0,\ \textbf{curl}\,\mathbf{F} = \cot\phi\hat{\boldsymbol{\rho}} - 2\hat{\boldsymbol{\phi}}$

**9.** scale factors: $h_u = \left|\dfrac{\partial\mathbf{r}}{\partial u}\right|,\ h_v = \left|\dfrac{\partial\mathbf{r}}{\partial v}\right|$

local basis: $\hat{\mathbf{u}} = \dfrac{1}{h_u}\dfrac{\partial\mathbf{r}}{\partial u},\ \hat{\mathbf{v}} = \dfrac{1}{h_v}\dfrac{\partial\mathbf{r}}{\partial v}$

area element: $dA = h_u h_v\, du\, dv$

**11.** $\nabla f(r,\theta) = \dfrac{\partial f}{\partial r}\hat{\mathbf{r}} + \dfrac{1}{r}\dfrac{\partial r}{\partial\theta}\hat{\boldsymbol{\theta}}$

$\nabla\bullet\mathbf{F}(r,\theta) = \dfrac{\partial F_r}{\partial r} + \dfrac{1}{r}F_r + \dfrac{1}{r}\dfrac{\partial F_\theta}{\partial\theta}$

$\nabla\times\mathbf{F}(r,\theta) = \left(\dfrac{\partial F_\theta}{\partial r} + \dfrac{1}{r}F_\theta - \dfrac{1}{r}\dfrac{\partial F_r}{\partial\theta}\right)\mathbf{k}$

**13.** $u$-surfaces: vertical elliptic cylinders with focal axes at $x = \pm a$, $y = 0$
$v$-surfaces: vertical hyperbolic cylinders with focal axes at $x = \pm a$, $y = 0$
$z$-surfaces: horizontal planes
$u$-curves: horizontal hyperbolas with foci $x = \pm a$, $y = 0$
$v$-curves: horizontal ellipses with foci $x = \pm a$, $y = 0$
$z$-curves: vertical straight lines

**15.** $$\nabla f = \frac{\partial^2 f}{\partial \rho^2} + \frac{2}{\rho}\frac{\partial f}{\partial \rho} + \frac{1}{\rho^2}\frac{\partial^2 f}{\partial \phi^2} + \frac{\cot\phi}{\rho^2}\frac{\partial f}{\partial \phi} + \frac{1}{\rho^2 \sin^2\phi}\frac{\partial^2 f}{\partial \theta^2}$$

### Review Exercises (page 466)

**1.** $128\pi$ **3.** $-6$

**5.** $3/4$ **7.** $\lambda = -3$, no

**11.** the ellipsoid $x^2 + 4y^2 + z^2 = 4$ with outward normal

### Challenging Problems (page 466)

**1.** $\mathrm{div}\,\mathbf{v} = 3C$

## Chapter 10

### Section 10.1 (page 472)

**1.** 1, linear, homogeneous **3.** 1, nonlinear

**5.** 2, linear, homogeneous
**7.** 3, linear, nonhomogeneous

**9.** 4, linear, homogeneous
**11.** (a) and (b) are solutions, (c) is not

**13.** $y_2 = \sin(kx)$, $y = -3(\cos(kx) + (3/k)\sin(kx))$
**15.** $y = \sqrt{2}(\cos x + 2\sin x)$
**17.** $y = x + \sin x + (\pi - 1)\cos x$

### Section 10.2 (page 476)

**1.** $y^2 = Cx$ **3.** $x^3 - y^3 = C$

**5.** $\dfrac{y^4}{4} = \dfrac{x^3}{3} + C$ **7.** $Y = Ce^{t^2/2}$

**9.** $y = \dfrac{Ce^{2x} - 1}{Ce^{2x} + 1}$ **11.** $y = -\ln\left(Ce^{-2t} - \frac{1}{2}\right)$

**13.** $y = \tan^{-1}(C - \cos x) + n\pi$
**15.** $2\tan^{-1}(y/x) = \ln(x^2 + y^2) + C$
**17.** $\tan^{-1}(y/x) = \ln|x| + C$
**19.** $y = x\tan^{-1}(\ln|Cx|)$ **21.** $y^3 + 3y - 3x^2 = 24$

### Section 10.3 (page 479)

**1.** $2xy + x^2y^2 = C$ **3.** $xe^{xy} = C$

**5.** $\ln|x| - \frac{y}{x^2} = C$

**7.** $\dfrac{1}{M}\left(\dfrac{\partial N}{\partial x} - \dfrac{\partial M}{\partial y}\right)$ must depend only on $y$

**9.** $\dfrac{1}{M}\left(\dfrac{\partial N}{\partial x} - \dfrac{\partial M}{\partial y}\right)$ must depend only on $y$.
$x - y^2e^y = Cy^2$

**11.** $\dfrac{1}{\mu}\dfrac{d\mu}{dx} = \dfrac{\dfrac{\partial N}{\partial x} - \dfrac{\partial M}{\partial y}}{xM - yN}$ must depend only on $xy$;
$\dfrac{\sin x}{y} - \dfrac{y}{x} = C$

### Section 10.4 (page 482)

**1.** $y = x^3 + Cx^2$ **3.** $y = \frac{3}{2} + Ce^{-2x}$

**5.** $y = x - 1 + Ce^{-x}$ **7.** $y = (1 + e^{1-10t})/10$

**9.** $y = (x^2 - \pi^2)e^{-\sin x}$

**13.** $\dfrac{dy}{dt} = 0.05y - 50$, \$871.60

### Section 10.5 (page 490)

**1.** (a) 1.97664, (b) 2.187485, (c) 2.306595
**3.** (a) 2.436502, (b) 2.436559, (c) 2.436563
**5.** (a) 1.097897, (b) 1.098401
**7.** (a) 0.89441, (b) 0.87996, (c) 0.872831
**9.** (a) 0.865766, (b) 0.865769, (c) 0.865769
**11.** (a) 0.898914, (b) 0.903122, (c) 0.904174
**13.** $y = 2/(3 - 2x)$
**17.** (b) $u = 1/(1-x)$, $v = \tan(x + \frac{\pi}{4})$. $y(x)$ is defined at least on $[0, \pi/4)$ and satisfies $1/(1-x) \le y(x) \le \tan(x + \frac{\pi}{4})$ there.

### Section 10.6 (page 494)

**1.** $y = C_1e^x + C_2e^{2x}$ **3.** $y = C_1x + \dfrac{C_2}{x^2}$

**5.** $y = C_1x + C_2xe^x$

### Section 10.7 (page 502)

**1.** $y = C_1e^{-5t} + C_2e^{-2t}$ **3.** $y = C_1 + C_2e^{-2t}$
**5.** $y = (C_1 + C_2t)e^{-4t}$
**7.** $y = (C_1\cos t + C_2\sin t)e^{3t}$
**9.** $y = (C_1\cos 2t + C_2\sin 2t)e^{-t}$
**11.** $y = (C_1\cos\sqrt{2}t + C_2\sin\sqrt{2}t)e^{-t}$
**13.** $y = \frac{6}{7}e^{t/2} + \frac{1}{7}e^{-3t}$
**15.** $y = e^{-2t}(2\cos t + 6\sin t)$
**19.** $y = C_1 + C_2e^t + C_3e^{3t}$
**21.** $y = C_1\cos t + C_2\sin t + C_3t\cos t + C_4t\sin t$

**23.** $y = C_1e^{2t} + C_2e^{-t}\cos t + C_3e^{-t}\sin t$

**25.** $y = Ax + Bx\ln x$ **27.** $y = Ax + \dfrac{B}{x}$

**29.** $y = A + B\ln x$

## Section 10.8 (page 507)

**1.** $y = -\dfrac{1}{2} + C_1e^x + C_2e^{-2x}$

**3.** $y = -\dfrac{1}{2}e^{-x} + C_1e^x + C_2e^{-2x}$

**5.** $y = -\frac{2}{125} - \frac{4x}{25} + \frac{x^2}{5} + C_1e^{-x}\cos(2x) + C_2e^{-x}\sin(2x)$

**7.** $y = -\dfrac{1}{5}xe^{-2x} + C_1e^{-2x} + C_2e^{3x}$

**9.** $y = \dfrac{1}{8}e^x(\sin x - \cos x) + e^{-x}(C_1\cos x + C_2\sin x)$

**11.** $y = 2x + x^2 - xe^{-x} + C_1 + C_2e^{-x}$

**15.** $y_p = \dfrac{x^2}{3}, \quad y = \dfrac{x^2}{3} + C_1x + \dfrac{C_2}{x}$

**17.** $y = \dfrac{1}{2}x\ln x + C_1x + \dfrac{C_2}{x}$

**19.** $y = -x^2 + C_1x + C_2xe^x$

## Section 10.9 (page 512)

**1.** $$y = a_0\left(1 + \sum_{k=1}^{\infty}\frac{(x-1)^{4k}}{4(k!)(3)(7)\cdots(4k-1)}\right) + a_1\left(x - 1 + \sum_{k=1}^{\infty}\frac{(x-1)^{4k+1}}{4(k!)(5)(9)\cdots(4k+1)}\right)$$

**3.** $y = 1 - \frac{1}{6}x^3 + \frac{1}{120}x^5 + \cdots$

**5.** $$y_1 = 1 + \sum_{k=1}^{\infty}\frac{(-1)^kx^k}{(k!)(2)(5)(8)\cdots(3k-1)},$$
$$y_2 = x^{1/3}\left(1 + \sum_{k=0}^{\infty}\frac{(-1)^kx^k}{(k!)(4)(7)\cdots(3k+1)}\right)$$

## Review Exercises (page 512)

**1.** $y = Ce^{x^2}$ **3.** $y = Ce^{2x} - \dfrac{x}{2} - \dfrac{1}{4}$

**5.** $x^2 + 2xy - y^2 = C$ **7.** $y = C_1 - \ln|t + C_2|$

**9.** $y = e^{x/2}(C_2\cos x + C_2\sin x)$

**11.** $y = C_1t\cos(2\ln|t|) + C_2t\sin(2\ln|t|)$

**13.** $y = \frac{1}{2}e^x + xe^{3x} + C_1e^{2x} + C_2e^{3x}$

**15.** $y = x^2 - 4x + 6 + C_1e^{-x} + C_2xe^{-x}$

**17.** $y = (x^3 - 7)^{1/3}$ **19.** $y = e^{x^2/2y^2}$

**21.** $y = 4e^{-t} - 3e^{-2t}$ **23.** $y = (5t - 4)e^{-5t}$

**25.** $y = e^{2t} - 2\sin(2t)$

**27.** $A = 1, \; B = -1, \quad x(e^x\sin y + \cos y) = C$

**29.** $y = C_1x + C_2x\cos x$

## Appendix I Complex Numbers (page A-10)

**1.** $\mathrm{Re}(z) = -5, \; \mathrm{Im}(z) = 2$

**3.** $\mathrm{Re}(z) = 0, \; \mathrm{Im}(z) = -\pi$

**5.** $|z| = \sqrt{2}, \; \theta = 3\pi/4$ **7.** $|z| = 3, \; \theta = \pi/2$

**9.** $|z| = \sqrt{5}, \; \theta = \tan^{-1}2$

**11.** $|z| = 5, \; \theta = \pi + \tan^{-1}(4/3)$

**13.** $|z| = 2, \; \theta = 11\pi/6$ **15.** $|z| = 3, \; \theta = 4\pi/5$

**17.** $23\pi/12$ **19.** $4 + 3i$

**21.** $\frac{\pi\sqrt{3}}{2} + \frac{\pi}{2}i$ **23.** $\frac{1}{4} - \frac{\sqrt{3}}{4}i$

**25.** $-3 + 5i$ **27.** $2 + i$

**29.** closed disk, radius 2, centre 0

**31.** closed disk, radius 5, centre $3 - 4i$

**33.** closed plane sector lying under $y = 0$ and to the left of $y = -x$

**35.** 4 **37.** $5 - i$

**39.** $2 + 11i$ **41.** $-\frac{1}{5} + \frac{7}{5}i$

**43.** 1

**47.** $zw = -3 - 3i, \; \dfrac{z}{w} = \dfrac{1+i}{3}$

**49.** (a) circle $|z| = \sqrt{2}$, (b) no solutions

**51.** $-1, \; \frac{1}{2} \pm \frac{\sqrt{3}}{2}\mathbf{i}$

**53.** $2^{1/6}(\cos\theta + i\sin\theta)$ where $\theta = \pi/4, 11\pi/12, 19\pi/12$

**55.** $\pm 2^{1/4}\left(\frac{\sqrt{3}}{2} + \frac{1}{2}i\right), \; \pm 2^{1/4}\left(\frac{1}{2} - \frac{\sqrt{3}}{2}i\right)$

## Appendix II Complex Functions (page A-20)

**1.** $0 \le \mathrm{Re}(w) \le 1, \; -2 \le \mathrm{Im}(w) \le 0$

**3.** $1 \le |w| \le 4, \; \pi \le \arg w \le \dfrac{3\pi}{2}$

**5.** $\dfrac{1}{2} \le |w| < \infty, \; -\dfrac{\pi}{2} \le \arg w \le 0$

**7.** $\arg(w) = 5\pi/6$ **9.** parabola $v^2 = 4u + 4$

**11.** $u \ge 0, \; v \ge u$ **13.** $f'(z) = 2z$

**15.** $f'(z) = -1/z^2$

**19.** $\dfrac{d}{dz}\sinh z = \cosh z, \; \dfrac{d}{dz}\cosh z = \sinh z$

**21.** $z = \dfrac{\pi}{2} + k\pi, \; (k \in \mathbb{Z})$

**23.** zeros of $\cosh z$: $z = i\left(\dfrac{\pi}{2} + k\pi\right) \; (k \in \mathbb{Z})$
zeros of $\sinh z$: $z = k\pi i \; (k \in \mathbb{Z})$

**25.** $\Re(\sinh z) = \sinh x \cos y,\ \Im(\sinh z) = \cosh x \sin y$

**27.** $z = 0,\ -2i$

**29.** $z = -1 \pm 2i$

**31.** $z = 0,\ i,\ 2i$

**33.** $z = \dfrac{1 \pm i}{\sqrt{2}},\ z = \dfrac{-1 \pm i}{\sqrt{2}}$

$z^4 + 1 = (z^2 + \sqrt{2}z + 1)(z^2 - \sqrt{2}z + 1)$

**35.** $z = -1,\ -1,\ -1,\ i,\ -i$

# Index

## INTEGRATION RULES

$$\int (Af(x) + Bg(x))\,dx = A\int f(x)\,dx + B\int g(x)\,dx$$

$$\int f'(g(x))\,g'(x)\,dx = f(g(x)) + C$$

$$\int U(x)\,dV(x) = U(x)\,V(x) - \int V(x)\,dU(x)$$

$$\int_a^b f'(x)\,dx = f(b) - f(a)$$

$$\frac{d}{dx}\int_a^x f(t)\,dt = f(x)$$

## ELEMENTARY INTEGRALS

$$\int x^r\,dx = \frac{1}{r+1}x^{r+1} + C \text{ if } r \neq -1$$

$$\int \frac{dx}{x} = \ln|x| + C$$

$$\int e^x\,dx = e^x + C$$

$$\int a^x\,dx = \frac{a^x}{\ln a} + C$$

$$\int \sin x\,dx = -\cos x + C$$

$$\int \cos x\,dx = \sin x + C$$

$$\int \sec^2 x\,dx = \tan x + C$$

$$\int \csc^2 x\,dx = -\cot x + C$$

$$\int \sec x \tan x\,dx = \sec x + C$$

$$\int \csc x \cot x\,dx = -\csc x + C$$

$$\int \tan x\,dx = \ln|\sec x| + C$$

$$\int \cot x\,dx = \ln|\sin x| + C$$

$$\int \sec x\,dx = \ln|\sec x + \tan x| + C$$

$$\int \csc x\,dx = \ln|\csc x - \cot x| + C$$

$$\int \frac{dx}{\sqrt{a^2 - x^2}} = \sin^{-1}\frac{x}{a} + C$$

$$\int \frac{dx}{a^2 + x^2} = \frac{1}{a}\tan^{-1}\frac{x}{a} + C$$

$$\int \frac{dx}{a^2 - x^2} = \frac{1}{2a}\ln\left|\frac{x+a}{x-a}\right| + C$$

$$\int \frac{dx}{x\sqrt{x^2 - a^2}} = \frac{1}{a}\sec^{-1}\left|\frac{x}{a}\right| + C$$

## TRIGONOMETRIC INTEGRALS

$$\int \sin^2 x\,dx = \frac{x}{2} - \frac{1}{4}\sin 2x + C$$

$$\int \cos^2 x\,dx = \frac{x}{2} + \frac{1}{4}\sin 2x + C$$

$$\int \tan^2 x\,dx = \tan x - x + C$$

$$\int \cot^2 x\,dx = -\cot x - x + C$$

$$\int \sec^3 x\,dx = \frac{1}{2}\sec x \tan x + \frac{1}{2}\ln|\sec x + \tan x| + C$$

$$\int \csc^3 x\,dx = -\frac{1}{2}\csc x \cot x + \frac{1}{2}\ln|\csc x - \cot x| + C$$

$$\int \sin ax\,\sin bx\,dx = \frac{\sin(a-b)x}{2(a-b)} - \frac{\sin(a+b)x}{2(a+b)} + C \text{ if } a^2 \neq b^2$$

$$\int \cos ax\,\cos bx\,dx = \frac{\sin(a-b)x}{2(a-b)} + \frac{\sin(a+b)x}{2(a+b)} + C \text{ if } a^2 \neq b^2$$

$$\int \sin ax\,\cos bx\,dx = -\frac{\cos(a-b)x}{2(a-b)} - \frac{\cos(a+b)x}{2(a+b)} + C \text{ if } a^2 \neq b^2$$

$$\int \sin^n x\,dx = -\frac{1}{n}\sin^{n-1} x\,\cos x + \frac{n-1}{n}\int \sin^{n-2} x\,dx$$

$$\int \cos^n x\,dx = \frac{1}{n}\cos^{n-1} x\,\sin x + \frac{n-1}{n}\int \cos^{n-2} x\,dx$$

$$\int \tan^n x\,dx = \frac{1}{n-1}\tan^{n-1} x - \int \tan^{n-2} x\,dx \text{ if } n \neq 1$$

$$\int \cot^n x\,dx = \frac{-1}{n-1}\cot^{n-1} x - \int \cot^{n-2} x\,dx \text{ if } n \neq 1$$

$$\int \sec^n x\,dx = \frac{1}{n-1}\sec^{n-2} x\,\tan x + \frac{n-2}{n-1}\int \sec^{n-2} x\,dx \text{ if } n \neq 1$$

$$\int \csc^n x\,dx = \frac{-1}{n-1}\csc^{n-2} x\,\cot x + \frac{n-2}{n-1}\int \csc^{n-2} x\,dx \text{ if } n \neq 1$$

$$\int \sin^n x\,\cos^m x\,dx = -\frac{\sin^{n-1} x\,\cos^{m+1} x}{n+m} + \frac{n-1}{n+m}\int \sin^{n-2} x\,\cos^m x\,dx \text{ if } n \neq -m$$

$$\int \sin^n x\,\cos^m x\,dx = \frac{\sin^{n+1} x\,\cos^{m-1} x}{n+m} + \frac{m-1}{n+m}\int \sin^n x\,\cos^{m-2} x\,dx \text{ if } m \neq -n$$

$$\int x\,\sin x\,dx = \sin x - x\,\cos x + C$$

$$\int x\,\cos x\,dx = \cos x + x\,\sin x + C$$

$$\int x^n\,\sin x\,dx = -x^n\,\cos x + n\int x^{n-1}\cos x\,dx$$

$$\int x^n\,\cos x\,dx = x^n\,\sin x - n\int x^{n-1}\sin x\,dx$$

## INTEGRALS INVOLVING $\sqrt{x^2 \pm a^2}$

$$\int \sqrt{x^2 \pm a^2}\,dx = \frac{x}{2}\sqrt{x^2 \pm a^2} \pm \frac{a^2}{2}\ln|x + \sqrt{x^2 \pm a^2}| + C$$

$$\int \frac{dx}{\sqrt{x^2 \pm a^2}} = \ln|x + \sqrt{x^2 \pm a^2}| + C$$

$$\int \frac{\sqrt{x^2 + a^2}}{x}\,dx = \sqrt{x^2 + a^2} - a\ln\left(\frac{a + \sqrt{x^2 + a^2}}{x}\right) + C$$

$$\int \frac{\sqrt{x^2 - a^2}}{x}\,dx = \sqrt{x^2 - a^2} - a\sec^{-1}\frac{x}{a} + C$$

$$\int x^2\sqrt{x^2 \pm a^2}\,dx = \frac{x}{8}(2x^2 \pm a^2)\sqrt{x^2 \pm a^2} - \frac{a^4}{8}\ln|x + \sqrt{x^2 \pm a^2}| + C$$

$$\int \frac{x^2}{\sqrt{x^2 \pm a^2}}\,dx = \frac{x}{2}\sqrt{x^2 \pm a^2} \mp \frac{a^2}{2}\ln|x + \sqrt{x^2 \pm a^2}| + C$$

$$\int \frac{\sqrt{x^2 \pm a^2}}{x^2}\,dx = -\frac{\sqrt{x^2 \pm a^2}}{x} + \ln|x + \sqrt{x^2 \pm a^2}| + C$$

$$\int \frac{dx}{x^2\sqrt{x^2 \pm a^2}} = \mp\frac{\sqrt{x^2 \pm a^2}}{a^2x} + C$$

$$\int \frac{dx}{(x^2 \pm a^2)^{3/2}} = \frac{\pm x}{a^2\sqrt{x^2 \pm a^2}} + C$$

$$\int (x^2 \pm a^2)^{3/2}\,dx = \frac{x}{8}(2x^2 \pm 5a^2)\sqrt{x^2 \pm a^2} + \frac{3a^4}{8}\ln|x + \sqrt{x^2 \pm a^2}| + C$$

## INTEGRALS INVOLVING $\sqrt{a^2 - x^2}$

$$\int \sqrt{a^2 - x^2}\,dx = \frac{x}{2}\sqrt{a^2 - x^2} + \frac{a^2}{2}\sin^{-1}\frac{x}{a} + C$$

$$\int \frac{\sqrt{a^2 - x^2}}{x}\,dx = \sqrt{a^2 - x^2} - a\ln\left|\frac{a + \sqrt{a^2 - x^2}}{x}\right| + C$$

$$\int \frac{x^2}{\sqrt{a^2 - x^2}}\,dx = -\frac{x}{2}\sqrt{a^2 - x^2} + \frac{a^2}{2}\sin^{-1}\frac{x}{a} + C$$

$$\int x^2\sqrt{a^2 - x^2}\,dx = \frac{x}{8}(2x^2 - a^2)\sqrt{a^2 - x^2} + \frac{a^4}{8}\sin^{-1}\frac{x}{a} + C$$

$$\int \frac{dx}{x^2\sqrt{a^2 - x^2}} = -\frac{\sqrt{a^2 - x^2}}{a^2x} + C$$

$$\int \frac{\sqrt{a^2 - x^2}}{x^2}\,dx = -\frac{\sqrt{a^2 - x^2}}{x} - \sin^{-1}\frac{x}{a} + C$$

$$\int \frac{dx}{x\sqrt{a^2 - x^2}} = -\frac{1}{a}\ln\left|\frac{a + \sqrt{a^2 - x^2}}{x}\right| + C$$

$$\int \frac{dx}{(a^2 - x^2)^{3/2}} = \frac{x}{a^2\sqrt{a^2 - x^2}} + C$$

$$\int (a^2 - x^2)^{3/2}\,dx = \frac{x}{8}(5a^2 - 2x^2)\sqrt{a^2 - x^2} + \frac{3a^4}{8}\sin^{-1}\frac{x}{a} + C$$

## INTEGRALS OF INVERSE TRIGONOMETRIC FUNCTIONS

$$\int \sin^{-1}x\,dx = x\sin^{-1}x + \sqrt{1 - x^2} + C$$

$$\int \tan^{-1}x\,dx = x\tan^{-1}x - \frac{1}{2}\ln(1 + x^2) + C$$

$$\int \sec^{-1}x\,dx = x\sec^{-1}x - \ln|x + \sqrt{x^2 - 1}| + C$$

$$\int x\sin^{-1}x\,dx = \frac{1}{4}(2x^2 - 1)\sin^{-1}x + \frac{x}{4}\sqrt{1 - x^2} + C$$

$$\int x\tan^{-1}x\,dx = \frac{1}{2}(x^2 + 1)\tan^{-1}x - \frac{x}{2} + C$$

$$\int x\sec^{-1}x\,dx = \frac{x^2}{2}\sec^{-1}x - \frac{1}{2}\sqrt{x^2 - 1} + C$$

$$\int x^n\sin^{-1}x\,dx = \frac{x^{n+1}}{n+1}\sin^{-1}x - \frac{1}{n+1}\int \frac{x^{n+1}}{\sqrt{1 - x^2}}\,dx + C \text{ if } n \neq -1$$

$$\int x^n\tan^{-1}x\,dx = \frac{x^{n+1}}{n+1}\tan^{-1}x - \frac{1}{n+1}\int \frac{x^{n+1}}{1 + x^2}\,dx + C \text{ if } n \neq -1$$

$$\int x^n\sec^{-1}x\,dx = \frac{x^{n+1}}{n+1}\sec^{-1}x - \frac{1}{n+1}\int \frac{x^n}{\sqrt{x^2 - 1}}\,dx + C \text{ if } n \neq -1$$

## EXPONENTIAL AND LOGARITHMIC INTEGRALS

$$\int xe^x\,dx = (x - 1)e^x + C$$

$$\int x^ne^x\,dx = x^ne^x - n\int x^{n-1}e^x\,dx$$

$$\int \ln x\,dx = x\ln x - x + C$$

$$\int x^n\ln x\,dx = \frac{x^{n+1}}{n+1}\ln x - \frac{x^{n+1}}{(n+1)^2} + C$$

$$\int x^n(\ln x)^m\,dx = \frac{x^{n+1}}{n+1}(\ln x)^m - \frac{m}{n+1}\int x^n(\ln x)^{m-1}\,dx$$

$$\int e^{ax}\sin bx\,dx = \frac{e^{ax}}{a^2 + b^2}(a\sin bx - b\cos bx) + C$$

$$\int e^{ax}\cos bx\,dx = \frac{e^{ax}}{a^2 + b^2}(a\cos bx + b\sin bx) + C$$

## INTEGRALS OF HYPERBOLIC FUNCTIONS

$$\int \sinh x\,dx = \cosh x + C$$

$$\int \cosh x\,dx = \sinh x + C$$

$$\int \tanh x\,dx = \ln(\cosh x) + C$$

$$\int \coth x\,dx = \ln|\sinh x| + C$$

$$\int \operatorname{sech} x\,dx = \tan^{-1}|\sinh x| + C$$

$$\int \operatorname{csch} x\,dx = \ln\left|\tanh\frac{x}{2}\right| + C$$

$$\int \sinh^2 x\,dx = \frac{1}{4}\sinh 2x - \frac{x}{2} + C$$

$$\int \cosh^2 x\,dx = \frac{1}{4}\sinh 2x + \frac{x}{2} + C$$

$$\int \tanh^2 x\,dx = x - \tanh x + C$$

$$\int \coth^2 x\,dx = x - \coth x + C$$

$$\int \operatorname{sech}^2 x\,dx = \tanh x + C$$

$$\int \operatorname{csch}^2 x\,dx = -\coth x + C$$

$$\int \operatorname{sech} x\tanh x\,dx = -\operatorname{sech} x + C$$

$$\int \operatorname{csch} x\coth x\,dx = -\operatorname{csch} x + C$$